Selected Titles in This Series

57 **Marc Levine,** Mixed motives, 1998
56 **Leonid I. Korogodski and Yan S. Soibelman,** Algebras of functions on quantum groups: Part I, 1998
55 **J. Scott Carter and Masahico Saito,** Knotted surfaces and their diagrams, 1998
54 **Casper Goffman, Togo Nishiura, and Daniel Waterman,** Homeomorphisms in analysis, 1997
53 **Andreas Kriegl and Peter W. Michor,** The convenient setting of global analysis, 1997
52 **V. A. Kozlov, V. G. Maz'ya, and J. Rossmann,** Elliptic boundary value problems in domains with point singularities, 1997
51 **Jan Malý and William P. Ziemer,** Fine regularity of solutions of elliptic partial differential equations, 1997
50 **Jon Aaronson,** An introduction to infinite ergodic theory, 1997
49 **R. E. Showalter,** Monotone operators in Banach space and nonlinear partial differential equations, 1997
48 **Paul-Jean Cahen and Jean-Luc Chabert,** Integer-valued polynomials, 1997
47 **A. D. Elmendorf, I. Kriz, M. A. Mandell, and J. P. May (with an appendix by M. Cole),** Rings, modules, and algebras in stable homotopy theory, 1997
46 **Stephen Lipscomb,** Symmetric inverse semigroups, 1996
45 **George M. Bergman and Adam O. Hausknecht,** Cogroups and co-rings in categories of associative rings, 1996
44 **J. Amorós, M. Burger, K. Corlette, D. Kotschick, and D. Toledo,** Fundamental groups of compact Kähler manifolds, 1996
43 **James E. Humphreys,** Conjugacy classes in semisimple algebraic groups, 1995
42 **Ralph Freese, Jaroslav Ježek, and J. B. Nation,** Free lattices, 1995
41 **Hal L. Smith,** Monotone dynamical systems: an introduction to the theory of competitive and cooperative systems, 1995
40.3 **Daniel Gorenstein, Richard Lyons, and Ronald Solomon,** The classification of the finite simple groups, number 3, 1998
40.2 **Daniel Gorenstein, Richard Lyons, and Ronald Solomon,** The classification of the finite simple groups, number 2, 1995
40.1 **Daniel Gorenstein, Richard Lyons, and Ronald Solomon,** The classification of the finite simple groups, number 1, 1994
39 **Sigurdur Helgason,** Geometric analysis on symmetric spaces, 1994
38 **Guy David and Stephen Semmes,** Analysis of and on uniformly rectifiable sets, 1993
37 **Leonard Lewin, Editor,** Structural properties of polylogarithms, 1991
36 **John B. Conway,** The theory of subnormal operators, 1991
35 **Shreeram S. Abhyankar,** Algebraic geometry for scientists and engineers, 1990
34 **Victor Isakov,** Inverse source problems, 1990
33 **Vladimir G. Berkovich,** Spectral theory and analytic geometry over non-Archimedean fields, 1990
32 **Howard Jacobowitz,** An introduction to CR structures, 1990
31 **Paul J. Sally, Jr. and David A. Vogan, Jr., Editors,** Representation theory and harmonic analysis on semisimple Lie groups, 1989
30 **Thomas W. Cusick and Mary E. Flahive,** The Markoff and Lagrange spectra, 1989
29 **Alan L. T. Paterson,** Amenability, 1988
28 **Richard Beals, Percy Deift, and Carlos Tomei,** Direct and inverse scattering on the line, 1988
27 **Nathan J. Fine,** Basic hypergeometric series and applications, 1988
26 **Hari Bercovici,** Operator theory and arithmetic in H^∞, 1988

(Continued in the back of this publication)

Mathematical Surveys and Monographs

Volume 57

Mixed Motives

Marc Levine

AMERICAN MATHEMATICAL SOCIETY
Providence, Rhode Island

1991 *Mathematics Subject Classification.* Primary 19E15, 14C25;
Secondary 14C15, 14C17, 14C40, 19D45, 19E08, 19E20.

Research supported in part by the National Science Foundation and the Deutsche Forschungsgemeinschaft.

ABSTRACT. The author constructs and describes a triangulated category of mixed motives over an arbitrary base scheme. The resulting cohomology theory satisfies the Bloch-Ogus axioms; if the base scheme is a smooth scheme of dimension at most one over a field, this cohomology theory agrees with Bloch's higher Chow groups. Most of the classical constructions of cohomology can be made in the motivic setting, including Chern classes from higher K-theory, push-forward for proper maps, Riemann-Roch, duality, as well as an associated motivic homology, Borel-Moore homology and cohomology with compact supports. The motivic category admits a realization functor for each Bloch-Ogus cohomology theory which satisfies certain axioms; as examples the author constructs Betti, etale, and Hodge realizations over smooth base schemes.

This book is a combination of foundational constructions in the theory of motives, together with results relating motivic cohomology with more explicit constructions, such as Bloch's higher Chow groups. It is aimed at research mathematicians interested in algebraic cycles, motives and K-theory, starting at the graduate level. It presupposes a basic background in algebraic geometry and commutative algebra.

Library of Congress Cataloging-in-Publication Data

Levine, Marc, 1952–
Mixed motives / Marc Levine.
p. cm. — (Mathematical surveys and monographs, ISSN 0076-5376 ; v. 57)
Includes bibliographical references and indexes.
ISBN 0-8218-0785-4 (acid-free) | Softcover ISBN 1-4704-8019-0
1. Motives (Mathematics) I. Title. II. Series: Mathematical surveys and monographs ; no. 57.
QA564.L48 1998
516.3′5—dc21 — 98-4734
CIP

Printed in the United States of America.
Reprinted by the American Mathematical Society, 2025

♾ The paper used in this book is acid-free and falls within the guidelines
established to ensure permanence and durability.
Visit the AMS home page at URL: http://www.ams.org/

11 10 9 8 7 6 5 4 3 2 30 29 28 27 26 25

To Ute, Anna, and Rebecca

Contents

Preface

This monograph is a study of triangulated categories of mixed motives over a base scheme S, whose construction is based on the rough ideas I originally outlined in a lecture at the J.A.M.I. conference on K-theory and number theory, held at the Johns Hopkins University in April of 1990. The essential principle is that one can form a categorical framework for *motivic cohomology* by first forming a tensor category from the category of smooth quasi-projective schemes over S, with morphisms generated by algebraic cycles, pull-back maps and external products, imposing the relations of functoriality of cycle pull-back and compatibility of cycle products with the external product, then taking the homotopy category of complexes in this tensor category, and finally localizing to impose the axioms of a Bloch-Ogus cohomology theory, e.g., the homotopy axiom, the Künneth isomorphism, Mayer-Vietoris, and so on.

Remarkably, this quite formal construction turns out to give the same cohomology theory as that given by Bloch's higher Chow groups [**19**], (at least if the base scheme is Spec of a field, or a smooth curve over a field). In particular, this puts the theory of the classical Chow ring of cycles modulo rational equivalence in a categorical context.

Following the identification of the categorical motivic cohomology as the higher Chow groups, we go on to show how the familiar constructions of cohomology: Chern classes, projective push-forward, the Riemann-Roch theorem, Poincaré duality, as well as homology, Borel-Moore homology and compactly supported cohomology, have their counterparts in the motivic category. The category of *Chow motives* of smooth projective varieties, with morphisms being the rational equivalence classes of correspondences, embeds as a full subcategory of our construction.

Our motivic category is specially constructed to give *realization functors* for Bloch-Ogus cohomology theories. As particular examples, we construct realization functors for classical singular cohomology, étale cohomology, and Hodge (Deligne) cohomology. We also have versions over a smooth base scheme, the Hodge realization using Saito's category of algebraic mixed Hodge modules. We put the Betti, étale and Hodge relations together to give the "motivic" realization into the category of mixed realizations, as described by Deligne [**32**], Jannsen [**71**], and Huber [**67**].

The various realizations of an object in the motivic category allow one to relate and unite parallel phenomena in different cohomology theories. A central example is Beilinson's motivic polylogarithm, together with its Hodge and étale realizations (see [**9**] and [**13**]). Beilinson's original construction uses the weight-graded pieces of the rational K-theory of a certain cosimplicial scheme over $\mathbb{P}^1$ minus $\{0, 1, \infty\}$ as a replacement for the motivic object; essentially the same construction gives rise to the motivic polylogarithm as an object in our category of motives over $\mathbb{P}^1$ minus $\{0, 1, \infty\}$, with the advantage that one acquires some integral information.

There have been a number of other constructions of triangulated motivic categories in the past few years, inspired by the conjectural framework for mixed motives set out by Beilinson [**10**] and Deligne [**32**], [**33**]. In addition to the approach via mixed realizations mentioned above, constructions of triangulated categories of motives have been given by Hanamura [**63**] and Voevodsky [**124**]. Deligne has suggested that the category of $\mathbb{Q}$-mixed Tate motives might be accessible via a direct construction of the "motivic Lie algebra"; the motivic Tate category would then be given as the category of representations of this Lie algebra. Along these lines, Bloch and Kriz [**17**] attempt to realize the category of mixed Tate motives as the category of co-representations of an explicit Lie co-algebra, built from Bloch's cycle complex. Kriz and May [**81**] have given a construction of a triangulated category of mixed Tate motives (with $\mathbb{Z}$-coefficients) from co-representations of the "May algebra" given by Bloch's cycle complex. The Bloch-Kriz category has derived category which is equivalent to the $\mathbb{Q}$-version of the triangulated category constructed by Kriz and May, if one assumes the Beilinson-Soulé vanishing conjectures.

We are able to compare our construction with that of Voevodsky, and show that, when the base is a perfect field admitting resolution of singularities, the two categories are equivalent. Although it seems that Hanamura's construction should give an equivalent category, we have not been able to describe an equivalence. Relating our category to the motivic Lie algebra of Bloch and Kriz, or the triangulated category of Kriz and May, is another interesting open problem.

Besides the categorical constructions mentioned above, there have been constructions of motivic cohomology which rely on the axioms for motivic complexes set down by Lichtenbaum [**90**] and Beilinson [**9**], many of which rely on a motivic interpretation of the polylogarithm functions. This began with the Bloch-Wigner dilogarithm function, leading to a construction of weight two motivic cohomology via the Bloch-Suslin complex ([**40**] and [**119**]) and Lichtenbaum's weight two motivic complex [**89**]. Pushing these ideas further has led to the Grassmann cycle complex of Beilinson, MacPherson, and Schechtman [**15**], as well as the motivic complexes of Goncharov ([**50**], [**51**], [**52**]), and the categorical construction of Beilinson, Goncharov, Schechtman, and Varchenko [**14**]. Although we have the polylogarithm as an object in our motivic category, it is at present unclear how these constructions fit in with our category.

While writing this book, the hospitality of the University of Essen allowed me the luxury of a year of undisturbed scholarship in lively mathematical surroundings, for which I am most grateful; I also would like to thank Northeastern University for the leave of absence which made that visit possible. Special and heartfelt thanks are due to Hélène Esnault and Eckart Viehweg for their support and encouragement. The comments of Spencer Bloch, Annette Huber, and Rick Jardine were most helpful and are greatly appreciated. I thank the reviewer for taking the time to go through the manuscript and for suggesting a number of improvements. Last, but not least, I wish to thank the A.M.S., especially Sergei Gelfand, Sarah Donnelly, and Deborah Smith, for their invaluable assistance in bringing this book to press.

Boston Marc Levine
November, 1997

Part I

Motives

Introduction: Part I

The categorical framework for the universal cohomology theory of algebraic varieties is the *category of mixed motives*. This category has yet to be constructed, although many of its desired properties have been described (see [**10**] and [**1**], especially [**70**]). Here is a partial list of the expected properties:

1. For each scheme S, one has the category of mixed motives over S, $\mathcal{MM}_S$; $\mathcal{MM}_S$ is an abelian tensor category with a duality involution. For each map of schemes $f: T \to S$, one has the functors f^*, f_*, $f^!$ and $f_!$, corresponding to the familiar functors for sheaves, and satisfying the standard relations of functoriality, adjointness, and duality.
2. For each S, there is a functor (natural in S)
$$M: (\mathbf{Sm}/S)^{\mathrm{op}} \to \mathcal{MM}_S,$$
where $\mathbf{Sm}/S$ is the category of smooth S-schemes; $M(X)$ is the *motive* of X.
3. There are external products
$$M(X) \otimes M(Y) \to M(X \times_S Y)$$
which are isomorphisms (the *Künneth isomorphism*).
4. There are objects $\mathbb{Z}(q)$, $q = 0, \pm 1, \pm 2, \ldots$ in $\mathcal{MM}_S$, the *Tate objects*, with $\mathbb{Z}(0)$ the unit for the tensor product and $\mathbb{Z}(a) \otimes \mathbb{Z}(b) \cong \mathbb{Z}(a+b)$.
5. Using the Künneth isomorphism to define the product, the groups
$$H^p_\mu(X, \mathbb{Z}(q)) := \mathrm{Ext}^p_{\mathcal{MM}_S}(\mathbb{Z}(0), M(X) \otimes \mathbb{Z}(q))$$
form a bi-graded ring which satisfies the axioms of a Bloch-Ogus cohomology theory: Mayer-Vietoris for Zariski open covers, homotopy property, projective bundle formula, etc.
6. There are Chern classes from algebraic K-theory
$$c^{q,p}: K_{2q-p}(X) \to H^p_\mu(X, \mathbb{Z}(q))$$
which induce an isomorphism
$$K_{2q-p}(X)^{(q)} \cong H^p_\mu(X, \mathbb{Z}(q)) \otimes \mathbb{Q},$$
with $K_*(-)^{(q)}$ the weight q eigenspace of the Adams operations.
7. The cohomology theory $H^p_\mu(X, \mathbb{Z}(q))$ is *universal*: Each Bloch-Ogus cohomology theory $X \mapsto H^*(X, \Gamma(*))$ gives rise to a natural transformation $H^*_\mu(-, \mathbb{Z}(*)) \to H^*(-, \Gamma(*))$.
8. $\mathcal{MM}_S \otimes \mathbb{Q}$ is a Tannakian category, with the $\mathbb{Q}$-Betti or $\mathbb{Q}_l$-étale realization giving a fiber functor.

9. There is a natural *weight filtration* on the objects of $\mathcal{MM}_S \otimes \mathbb{Q}$; morphisms in $\mathcal{MM}_S \otimes \mathbb{Q}$ are strictly compatible with the filtration, and the corresponding graded objects gr^W_* are semi-simple.

Assuming one had the category $\mathcal{MM}_S$, one could hope to realize the motivic cohomology theory $H^*_\mu(-, \mathbb{Z}(*))$ as the cohomology of some natural complexes, the *motivic complexes*. Lichtenbaum (for the étale topology) [**90**] and Beilinson (for the Zariski topology) [**9**] have outlined the desired properties of these complexes. In [**19**], Bloch has given a candidate for the Zariski version, and thereby a candidate, the *higher Chow groups* $\mathrm{CH}^q(X, 2q-p)$, for the motivic cohomology $H^p_\mu(X, \mathbb{Z}(p))$.

Rather than attempting the construction of $\mathcal{MM}_S$, we consider a more modest problem: The construction of a triangulated tensor category which has the expected properties of the bounded derived category of $\mathcal{MM}_S$.

To be more specific, for a reduced scheme S, let $\mathbf{Sm}_S$ denote the category of smooth quasi-projective S-schemes. We construct for each reduced scheme S a triangulated tensor category $\mathcal{DM}(S)$; sending S to $\mathcal{DM}(S_{\mathrm{red}})$ defines a pseudo-functor

$$\mathcal{DM}(-): \mathbf{Sch}^{\mathrm{op}} \to \mathbf{TT},$$

where $\mathbf{TT}$ is the category of triangulated tensor categories. This gives the contravariant functoriality in (1).

The category $\mathcal{DM}(S)$ is generated (as a triangulated category) by objects $\mathbb{Z}_X(q)$, $X \in \mathbf{Sm}_S$, $q \in \mathbb{Z}$, together with the adjunction of summands corresponding to idempotent endomorphisms.

There is an exact duality involution

$$(-)^D: \mathcal{DM}(S)^{\mathrm{pr,op}} \to \mathcal{DM}(S)^{\mathrm{pr}},$$

where $\mathcal{DM}(S)^{\mathrm{pr}}$ is the pseudo-abelian hull of the full triangulated tensor subcategory of $\mathcal{DM}(S)$ generated by objects $\mathbb{Z}_X(q)$, for $X \to S$ projective. This makes $\mathcal{DM}(S)^{\mathrm{pr}}$ into a rigid triangulated tensor category. If $S = \mathrm{Spec}\, k$, with k a perfect field admitting resolution of singularities, then $\mathcal{DM}(S)^{\mathrm{pr}} = \mathcal{DM}(S)$, giving the duality property in (1).

We reinterpret (5) by setting

$$H^p(X, \mathbb{Z}(q)) := \mathrm{Hom}_{\mathcal{DM}(S)}(\mathbb{Z}(0), \mathbb{Z}_X(q)[p]).$$

The properties (2)-(5) expected of motivic cohomology are then realized by properties satisfied by the objects $\mathbb{Z}_X(q)$ in the category $\mathcal{DM}(S)$. This includes:

(i) *Functoriality.* Sending X to $\mathbb{Z}_X(q)$ for fixed q extends to a functor

$$\mathbb{Z}_{(-)}(q): \mathbf{Sm}_S^{\mathrm{op}} \to \mathcal{DM}(S).$$

We set $M(X) := \mathbb{Z}_X(0)$.

(ii) *Homotopy.* The projection $p_1: X \times_S \mathbb{A}^1 \to X$ gives an isomorphism

$$p_1^*: \mathbb{Z}_X(q) \to \mathbb{Z}_{X \times_S \mathbb{A}^1}(q).$$

(iii) *Künneth isomorphism.* There are external products, giving natural isomorphisms

$$\mathbb{Z}_X(a) \otimes \mathbb{Z}_Y(b) \cong \mathbb{Z}_{X \times_S Y}(a+b);$$

$\mathbb{Z}_S$ is the unit for the tensor product structure.

(iv) *Gysin morphism.* Let $i: Z \to X$ be a smooth closed codimension q embedding in $\mathbf{Sm}_S$, with complement $j: U \to X$. Then there is a natural distinguished triangle

$$\mathbb{Z}_Z(-q)[-2q] \xrightarrow{i_*} \mathbb{Z}_X(0) \xrightarrow{j^*} \mathbb{Z}_U(0) \to \mathbb{Z}_Z(-q)[-2q+1].$$

(v) *Mayer-Vietoris.* Write $X \in \mathbf{Sm}_S$ as a union of Zariski open subschemes, $X = U \cup V$. Then there is a natural distinguished triangle

$$\mathbb{Z}_X(0) \xrightarrow{j_U^* \oplus j_V^*} \mathbb{Z}_U(0) \oplus \mathbb{Z}_V(0) \xrightarrow{j_{U,U\cap V}^* - j_{V,U\cap V}^*} \mathbb{Z}_{U\cap V}(0) \longrightarrow \mathbb{Z}_X(0)[1].$$

The functoriality (i), isomorphisms (ii) and (iii), and the distinguished triangles (iv) and (v) then translate into the standard properties of a Bloch-Ogus cohomology theory.

We have Chern classes as in (6); in case the base is a field, or is a smooth curve over a field, the Chern character defines an isomorphism of rational motivic cohomology with weight-graded K-theory, as required by (6).

For a Bloch-Ogus twisted duality theory Γ, defined via cohomology of a complex of $\mathcal{A}$-valued sheaves for a Grothendieck topology $\mathfrak{T}$ on $\mathbf{Sm}_S$, satisfying certain natural axioms, the motivic triangulated category $\mathcal{DM}(S)$ admits a realization functor

$$\Re_\Gamma : \mathcal{DM}(S) \to \mathbf{D}^+(\mathrm{Sh}^{\mathcal{A}}_{\mathfrak{T}}(S)).$$

We have the Betti, étale and Hodge realizations. Thus, the category $\mathcal{DM}(S)$ satisfies a version of the property (7).

We have not investigated the Tannakian property in (8), or the property (9) (see, however, [**62**]).

In Chapter I, we construct the motivic DG tensor category $\mathcal{A}_{\mathrm{mot}}(S)$ and the triangulated motivic category $\mathcal{DM}(S)$, and describe their basic properties.

We examine the motivic cohomology theory:

$$H^p(X, \mathbb{Z}(q)) := \mathrm{Hom}_{\mathcal{DM}(\mathcal{V}}(\mathbb{Z}_S, \mathbb{Z}_X(p)[q])$$

in Chapter II. We define the Chow group of an object Γ of $\mathcal{DM}(S)$, $\mathcal{CH}(\Gamma)$, as well as the cycle class map

$$\mathrm{cl}_\Gamma : \mathcal{CH}(\Gamma) \to \mathrm{Hom}_{\mathcal{DM}(S)}(1, \Gamma),$$

and give a criterion for cl_Γ to be an isomorphism for all Γ in $\mathcal{DM}(S)$. We verify this criterion in case $S = \mathrm{Spec}\, k$, or S a smooth curve over k, where k is a field. This shows in particular that (in these cases) the motivic cohomology $H^p(X, \mathbb{Z}(q))$ agrees with Bloch's higher Chow groups $\mathrm{CH}^q(X, 2q - p)$, which puts the higher Chow groups in a categorical framework. Assuming the above mentioned criterion is satisfied, we derive a number of additional useful consequences for the motivic cohomology, such as the existence of a Gersten resolution for the associated (Zariski) cohomology sheaves.

Chapter III deals with the relationship between motivic cohomology and K-theory. We construct Chern classes with values in motivic cohomology, for both K_0 and higher K-theory, satisfying the standard properties, e.g., Whitney product formula, projective bundle formula, etc. We also construct push-forward maps in motivic cohomology for a projective morphism, and verify the standard properties, including functoriality and the projection formula. Both the Chern classes, and the projective push-forward maps are constructed not just for smooth varieties, but also for *diagrams* of smooth varieties. We prove the Riemann-Roch theorem without denominators, and the usual Riemann-Roch theorem. As an application, we show that the Chern character gives an isomorphism of rational motivic cohomology with weight-graded K-theory, for motives over a field or a smooth curve over a field.

In Chapter IV we examine duality in a tensor category and in a triangulated tensor category, and apply this to the construction of the duality involution on the full subcategory $\mathcal{DM}(S)^{\mathrm{pr}}$ of $\mathcal{DM}(S)$ generated by smooth projective S-schemes in $\mathbf{Sm}_S$. Combined with the operation of cup-product by cycle classes, this gives the action of correspondences as homomorphisms in the category $\mathcal{DM}(S)$, and leads to a fully faithful embedding of the category of graded Chow motives (over a field k) into $\mathcal{DM}(\mathrm{Spec}\, k)$.

We define the homological motive, the Borel-Moore motive and the compactly supported motive. We also relate the motive of X with compact support to a "motive of $\bar{X}$ relative to infinity" if X admits a compactification $\bar{X}$ as a smooth projective S-scheme with a complement a normal crossing scheme.

We then examine extensions of the motivic theory to non-smooth S-schemes. We give a construction of the Borel-Moore motive and the motive with compact support for certain non-smooth S-schemes; as an application we prove a Riemann-Roch theorem for singular varieties. We give a construction of the (cohomological) motive of k-scheme of finite type, for k a perfect field admitting resolution of singularities, using the theory of cubical hyperresolutions.

Chapter V deals with realization of the motivic category. We describe the construction of the realization functor $\Re_\Gamma$ associated to a cohomology theory $\Gamma(*)$; we need to give a somewhat different characterization of the cohomology theory from that of Bloch-Ogus [**20**] or Gillet [**46**], but it seems that this type of cohomology theory is general enough for many applications. We construct the Betti, étale and Hodge realizations of $\mathcal{DM}(\mathcal{V})$ in subsequent sections; we also give the realization to Saito's category of mixed Hodge modules [**110**] (over a smooth base) and to a version of Jannsen's category [**71**] of mixed absolute Hodge complexes.

In Chapter VI we examine various known "motivic" constructions, and reinterpret them in the category $\mathcal{DM}$. We look at Milnor K-theory, prove the motivic Steinberg relation, and give a version of Beilinson's polylogarithm. We also relate the category $\mathcal{DM}(\mathrm{Spec}\, k)$ to Voevodsky's motivic category $DM_{gm}(k)$ [**124**] (k a perfect field admitting resolution of singularities), and show the two categories are equivalent.

There are two appendices. In Appendix A, we give a review of a part of the theory of equi-dimensional cycles due to Suslin-Voevodsky [**117**]. In Appendix B, we collect some foundational notions and results on algebraic K-theory.

We have collected in a second portion of this volume the various categorical constructions necessary for the paper; we refer the reader to the introduction of Part II for an overview.

CHAPTER I

The Motivic Category

This chapter begins with the construction of the motivic DG category $\mathcal{A}_{\mathrm{mot}}(\mathcal{V})$. We construct the triangulated motivic category $\mathcal{DM}(\mathcal{V})$ and describe its basic properties in Section 2; we also define the motives of various types of diagrams of schemes, e.g., simplicial schemes, cosimplicial schemes, n-cubes of schemes, as well as giving a general construction for an arbitrary finite diagram. In Section 3, we define the fundamental motivic cycles functor, and discuss its connection with the morphisms in the homotopy category of complexes $\mathbf{K}^b(\mathcal{A}_{\mathrm{mot}}(\mathcal{V}))$.

The rough idea of the construction of $\mathcal{DM}(S)$ is as follows: Naively, one might attempt to construct $\mathcal{DM}(S)$ by the following process (for simplicity, assume the base S is Spec of a field):

(i) Form the additive category generated by $\mathbf{Sm}_S^{\mathrm{op}} \times \mathbb{Z}$; denote the object (X, n) by $\mathbb{Z}_X(n)$, and the morphism $p^{\mathrm{op}} \times \mathrm{id}_n : \mathbb{Z}_X(n) \to \mathbb{Z}_Y(n)$ corresponding to a morphism $p: Y \to X$ in $\mathbf{Sm}_S$ by $p^* : \mathbb{Z}_X(n) \to \mathbb{Z}_Y(n)$.

(ii) For each algebraic cycle Z of codimension d on X, adjoin a map of degree $2d$

$$[Z]: \mathbb{Z}_S \to \mathbb{Z}_X(d),$$

with the relation of linearity: $[nZ + mW] = n[Z] + m[W]$.

(iii) Impose the relation of functoriality for the cycle maps,

$$p^* \circ [Z] = [p^*(Z)],$$

where $p: Y \to X$ is a map in $\mathbf{Sm}_S$, and Z is a cycle on X for which $p^*(Z)$ is defined.

This constructs an additive category $\mathcal{A}$ which has the objects, morphisms and relations needed to generate $\mathcal{DM}(S)$. The product of schemes over S, $(X, Y) \mapsto X \times_S Y$, extends to give $\mathcal{A}$ the structure of a tensor category with unit $\mathbb{Z}_S$. The construction then continues:

(iv) Form the differential graded category of bounded complexes $\mathbf{C}^b(\mathcal{A})$ and the triangulated homotopy category $\mathbf{K}^b(\mathcal{A})$. The product $\times$ on $\mathcal{A}$ extends to give $\mathbf{K}^b(\mathcal{A})$ the structure of a triangulated tensor category.

(v) Localize the category $\mathbf{K}^b(\mathcal{A})$ to impose the relations of a Bloch-Ogus cohomology theory, e.g.:

  (a) (Homotopy) Invert the map $p_1^* : \mathbb{Z}_X(q) \to \mathbb{Z}_{X \times_S \mathbb{A}^1_S}(q)$.

  (b) (Mayer-Vietoris) Suppose $X = U \cup V$, where $j: U \to X$, $k: V \to X$ are open subschemes. Let $i_U : U \cap V \to U$, $i_V : U \cap V \to V$ be the inclusions; the map

$$j^* \oplus k^* : \mathbb{Z}_X(q) \to \mathbb{Z}_U(q) \oplus \mathbb{Z}_V(q)$$

extends to the map

$$j^* \oplus k^* : \mathbb{Z}_X(q) \to \text{cone}(i_U^* - i_V^*)[-1].$$

Invert this map.

(c) Continue inverting maps until the various axioms of a Bloch-Ogus cohomology theory are satisfied.

(vi) This forms a triangulated tensor category; take the pseudo-abelian hull to give the triangulated tensor category $\mathcal{DM}(S)$.

There are several problems with this naive approach. The first is that the relation (iii) is only given for cycles Z for which the pull-back $p^*(Z)$ is defined.

Classically, this type of problem is solved by imposing an adequate equivalence relation on cycles, giving fully defined pull-backs on the resulting groups of cycle classes. If one does this on the categorical level, one loses the interesting data given by the relations among the relations, and all such higher order relations. To avoid this, we make the operation of cycle pull-back fully defined and functorial by refining the category $\mathbf{Sm}_S$, adjoining to a scheme X the data of a map $f : X' \to X$. For such a pair (X, f), we have the group of cycles $\mathcal{Z}(X)_f$ consisting of those cycles Z for which the pull-back $f^*(Z)$ is defined. We assemble such pairs (X, f) into a category $\mathcal{L}(\mathbf{Sm}_S)$ for which the assignment $(X, f) \mapsto \mathcal{Z}(X)_f$ forms a functor.

Second, one would like a Bloch-Ogus cohomology theory $\Gamma(*)$ on $\mathbf{Sm}_S$ to give rise to a realization functor $\Re_\Gamma$ from $\mathbf{D}^b(\mathcal{A})$ to the appropriate derived category of sheaves on the base S. In attempting to do this, one runs into two related problems:

1. For Z a cycle on X, the cycle class of Z with respect to the Γ-cohomology is represented by a cocycle in the appropriate representing cochain complex, but the choice of representing cocycle is not canonical. Thus, the pull-back of this representing cocycle is not functorial, but only functorial up to homotopy.
2. For most cohomology theories, the cup products are defined by associative products on representing cochain complexes, but these products are usually only commutative up to homotopy; the tensor product we have defined above on $\mathcal{A}$ is, however, strictly commutative.

The problem (1) is solved by replacing strict identities with identities up to homotopy; in categorical terms, one replaces the additive category sketched above with a differential graded category. The problem (2) is more subtle, and is solved by replacing the unit in $\mathcal{A}$ with a "fat unit" $\mathfrak{e}$. This fat unit generates a DG tensor subcategory $\mathbb{E}$, in which the various symmetry isomorphisms are made trivial, up to homotopy and all higher homotopies, in as free a manner as possible. This absorbs the usual cohomology operations, so that the motivic DG category becomes homotopy equivalent to a model which is only commutative up to homotopy and all higher homotopies.

Having made these technical modifications, one can still view the motivic category as being built out of the geometry inherent in the category of smooth quasi-projective S-schemes and the algebraic cycles on such schemes, extended by formally taking complexes, and then superimposing the homological algebra of the localized homotopy category. From this point of view, all properties of the motivic category flow from the mixing of homological algebra with the geometry of schemes and algebraic cycles. In fact, for motives over a field, we actually recover the naive description of the motivic category, once we identify the resulting motivic cohomology with Bloch's higher Chow groups (see the introduction to Chapter IV for further details).

1. The motivic DG category

1.1. The category $\mathcal{L}(\mathcal{V})$

By *scheme*, we will mean a noetherian separated scheme. For a scheme S, an S-scheme W is *essentially of finite type over* S if W is the localization of a scheme of finite type over S. Let $\mathbf{Sch}_S$ denote the category of schemes over S, and $\mathbf{Sm}_S$ the full subcategory of smooth quasi-projective S-schemes. We let $\mathbf{Sm}_S^{\text{ess}}$ denote the full subcategory of $\mathbf{Sch}_S$ of localizations of schemes in $\mathbf{Sm}_S$.

1.1.1. Let S be a reduced scheme, and let $\mathcal{V}$ be a strictly full subcategory of $\mathbf{Sm}_S^{\text{ess}}$. We assume that S is in $\mathcal{V}$ and that $\mathcal{V}$ is closed under the operations of product over S and disjoint union. In particular, the category $\mathcal{V}$ is a symmetric monoidal subcategory of $\mathbf{Sch}_S$.

1.1.2. DEFINITION. Let $\mathcal{L}(\mathcal{V})$ denote the category of equivalence classes of pairs (X, f), where X is an object of $\mathcal{V}$ and $f: X' \to X$ is a map in $\mathbf{Sm}_S^{\text{ess}}$, such that there is a section $s: X \to X'$ to f, with s a smooth morphism; two pairs $(X, f: X' \to X)$, $(X, g: X'' \to X)$ being equivalent if there is an isomorphism, $h: X' \to X''$, with $f = g \circ h$.

For $(X, f: X' \to X)$ and $(Y, g: Y' \to Y)$ in $\mathcal{L}(\mathcal{V})$, $\mathrm{Hom}_{\mathcal{L}(\mathcal{V})}((Y, g), (X, f))$ is the subset of $\mathrm{Hom}_{\mathcal{V}}(Y, X)$ defined by the following condition: A morphism $p: Y \to X$ in $\mathcal{V}$ gives a morphism $p: (Y, g) \to (X, f)$ in $\mathcal{L}(\mathcal{V})$ if there is a flat map $q: Y' \to X'$ over S making the diagram

$$\begin{array}{ccc} Y' & \xrightarrow{q} & X' \\ {\scriptstyle g}\downarrow & & \downarrow{\scriptstyle f} \\ Y & \xrightarrow[p]{} & X \end{array}$$

commute. Composition is induced from the composition of morphisms in $\mathbf{Sch}_S$; this is well-defined since the composition of flat morphisms is flat.

1.1.3. The condition that a morphism $f: X' \to X$ have a smooth section $s: X \to X'$ is the same as saying that we can write X' as a disjoint union $X' = X'_0 \coprod X'_1$ such that the restriction of f to $f_0: X'_0 \to X$ is an isomorphism. Indeed, a section s must be a closed embedding, and a smooth closed embedding is both open and closed. Thus, each object of $\mathcal{L}(\mathcal{V})$ is equivalent to a pair of the form $(X, f \cup \mathrm{id}_X)$, where $f: Z \to X$ is a map in $\mathbf{Sm}_S^{\text{ess}}$. We also note that each morphism $f: X \to Y$ in $\mathcal{V}$ can be lifted to a morphism in $\mathcal{L}(\mathcal{V})$; for example,

$$f: (X, \mathrm{id}_X) \to (Y, f \cup \mathrm{id}_Y)$$

is one such lifting.

1.1.4. If (X, f), (Y, g) are in $\mathcal{L}(\mathcal{V})$, then $(X \times_S Y, f \times g)$ is also in $\mathcal{L}(\mathcal{V})$, as smooth sections $s: X \to X'$ to f, $t: Y \to Y'$ to g determine the smooth section $s \times t$ to $f \times g$. For (X, f), (Y, g) and (Z, h) in $\mathcal{L}(\mathcal{V})$, we let $(X, f) \times (Y, g)$ denote the object

$(X \times_S Y, f \times g)$, and we let

$$((X,f) \times (Y,g)) \times (Z,h) \xrightarrow{a_{(X,f),(Y,g),(Z,h)}} (X,f) \times ((Y,g) \times (Z,h)),$$

$$(X,f) \times (Y,g) \xrightarrow{t_{(X,f),(Y,g)}} (Y,g) \times (X,f)$$

be the isomorphisms induced by the associativity and symmetry isomorphisms in $\mathbf{Sm}_S^{\mathrm{ess}}$.

The proof of the following proposition is elementary:

1.1.5. PROPOSITION. *(i) The category $\mathcal{L}(\mathcal{V})$ with product $\times$, symmetry t, associativity a and unit (S, id_S) is a symmetric monoidal category.*
(ii) The projection $p_1 : \mathcal{L}(\mathcal{V}) \to \mathcal{V}$ defines a faithful symmetric monoidal functor.

1.2. Cycles for the category $\mathcal{L}(\mathcal{V})$

For a smooth S-scheme X, essentially of finite type over S, we have the subgroup $\mathcal{Z}^d(X/S)$ of the group of relative codimension d cycles on X (see Appendix A, Definition 2.2.1(ii)); for a cycle W, we let $\mathrm{supp}(W)$ denote the support of W.

1.2.1. DEFINITION. Let $(X, f : X' \to X)$ be in $\mathcal{L}(\mathcal{V})$. We let $\mathcal{Z}^d(X)_f$ denote the subgroup of $\mathcal{Z}^d(X/S)$ consisting of $W \in \mathcal{Z}^d(X/S)$ such that $f^*(W)$ is defined, i.e.,

$$\mathrm{codim}_{X'}(f^{-1}(\mathrm{supp}(W))) \geq d.$$

The reason for constructing the category $\mathcal{L}(\mathcal{V})$ is that pull-back of cycles is now defined for arbitrary morphisms, without the need of passing to rational equivalence. This is more precisely expressed in

1.2.2. LEMMA. *(i) Suppose $p : (Y,g) \to (X,f)$ is a map in $\mathcal{L}(\mathcal{V})$. Then for each Z in $\mathcal{Z}^d(X)_f$, the cycle-theoretic pull-back $p^*(Z)$ is defined, and is in $\mathcal{Z}^d(Y)_g$.*
(ii) Let $(W,h) \xrightarrow{q} (Y,g) \xrightarrow{p} (X,f)$ be a sequence of maps in $\mathcal{L}(\mathcal{V})$, and let Z be in $\mathcal{Z}^d(X)_f$. Then

$$(p \circ q)^*(Z) = q^*(p^*(Z)).$$

PROOF. It suffices to prove (i) for effective cycles Z. Let $s : Y \to Y'$ be the smooth section to $g : Y' \to Y$. By definition, we have a commutative diagram

$$\begin{array}{ccc} X' & \xleftarrow{q} & Y' \\ {\scriptstyle f}\downarrow & & {\scriptstyle s}\uparrow\downarrow{\scriptstyle g} \\ X & \xleftarrow[p]{} & Y \end{array}$$

with q flat. By assumption, the cycle $f^*(Z)$ is defined. As q is flat and s are smooth, this implies that $(q \circ s)^*(f^*(Z))$ is defined. We have

$$f \circ q \circ s = p \circ g \circ s = p;$$

by (Appendix A, Theorem 2.3.1(iv)), $p^*(Z)$ is defined and is in $\mathcal{Z}^d(Y/S)$. Similarly, the cycle $q^*(f^*(Z))$ is defined; as $f \circ q = p \circ g$, the same argument shows that $g^*(p^*(Z))$ is defined, hence $p^*(Z)$ is in $\mathcal{Z}^d(Y)_g$, completing the proof of (i).

The assertion (ii) follows from (Appendix A, Theorem 2.3.1(v)). □

1.3. The category $\mathcal{L}(\mathcal{V})^*$

We consider a set $\mathcal{S}$ as a category with objects $\mathcal{S}$ and only the identity morphisms.

1.3.1. In the category $\mathcal{L}(\mathcal{V})^{\mathrm{op}} \times \mathbb{Z}$, denote the object $((X, f), n)$ by $X(n)_f$; for a morphism $p: (Y, g) \to (X, f)$ in $\mathcal{L}(\mathcal{V})$, denote the corresponding morphism

$$p^{\mathrm{op}} \times \mathrm{id}_n : X(n)_f \to Y(n)_g$$

by p^*. Giving $\mathbb{Z}$ the structure of a symmetric monoidal category with operation $+$ gives $\mathcal{L}(\mathcal{V})^{\mathrm{op}} \times \mathbb{Z}$ the structure of a symmetric monoidal category with symmetry

$$t_{X(n)_f, Y(m)_g} = t^*_{(Y,g),(X,f)} \times \mathrm{id}_{n+m}.$$

1.3.2. DEFINITION. Form the category $\mathcal{L}(\mathcal{V})^*$ by adjoining morphisms and relations to $\mathcal{L}(\mathcal{V})^{\mathrm{op}} \times \mathbb{Z}$ as follows: For (X, f) and (Y, g) in $\mathcal{V}^*$, with $i: X \to X \coprod Y$ the inclusion, we adjoin the morphism

$$i_* : X(n)_f \to (X \coprod Y)(n)_{f \coprod g}.$$

The relations imposed among the morphisms are:

(a) If $i: X \to X \coprod Y$, $j: X \coprod Y \to X \coprod Y \coprod Z$ are the natural inclusions, then

$$(i \circ j)_* = i_* \circ j_*.$$

(b) Let $p_i: (Y_i, g_i) \to (X_i, f_i)$, $i = 1, 2$, be morphisms in $\mathcal{L}(\mathcal{V})$, and let $i_{Y_1} : Y_1 \to Y_1 \coprod Y_2$ and $i_{X_1} : X_1 \to X_1 \coprod X_2$ be the natural inclusions. Then

$$i_{Y_1 *} \circ p_1^* = (p_1 \coprod p_2)^* \circ i_{X_1 *}.$$

(c) For $i: X \to X \coprod \emptyset$ the canonical isomorphism, we have $i^* \circ i_* = \mathrm{id}$.

1.3.3. One extends the symmetric monoidal structure on $\mathcal{L}(\mathcal{V})^{\mathrm{op}} \times \mathbb{Z}$ to one on $\mathcal{L}(\mathcal{V})^*$ by defining

$$i_{X*} \times \mathrm{id}^* : X(n)_f \times Z(k)_h \to (X(n)_f \coprod Y(m)_g) \times Z(k)_h$$

to be the composition

$$\begin{aligned} X(n)_f \times Z(k)_h &= (X \times_S Z)(n + k)_{f \times h} \\ &\xrightarrow{i_{X \times_S Z *}} (X \times_S Z)(n + k)_{f \times h} \coprod (Y \times_S Z)(m + k)_{g \times h} \\ &\cong (X(n)_f \coprod Y(m)_g) \times Z(k)_h. \end{aligned}$$

The map $\mathrm{id}^* \times i_{X*}$ is defined similarly. One checks that the uniquely defined extension of $\times$ to a product $\times$ on $\mathcal{L}(\mathcal{V})^*$ does indeed define the structure of a symmetric monoidal category on $\mathcal{L}(\mathcal{V})^*$. In particular, the canonical functor $\mathcal{L}(\mathcal{V})^{\mathrm{op}} \times \mathbb{Z} \to \mathcal{L}(\mathcal{V})^*$ is a symmetric monoidal functor.

The notation for the maps p^* and i_* is rather ambiguous, as we have deleted the dependence on the sets of maps and the integer n. This will usually be clear from the context. There are some special cases for which it is useful to have another notation for various morphisms; for instance, let $(X, f: X' \to X)$ be an object of $\mathcal{L}(\mathcal{V})$, and let $g: Z \to X$ be a morphism in $\mathcal{V}$. This gives us the map $f \cup g: X' \coprod Z \to X$ and the object $(X, f \cup g)$ of $\mathcal{L}(\mathcal{V})$. The identity on X gives the $\mathcal{L}(\mathcal{V})$-morphism $\mathrm{id}_X : (X, f) \to (X, f \cup g)$. We denote the corresponding $\mathcal{L}(\mathcal{V})^*$-morphism id_X^* by

$$\rho_{f,g} : X(n)_{f \cup g} \to X(n)_f. \tag{1.3.3.1}$$

1.3.4. REMARK. The identity in the symmetric monoidal category $\mathcal{L}(\mathcal{V})$ is the object (S, id_S). We will systematically identify the schemes $S \times_S X$ and $X \times_S S$ with X via the appropriate projection; this gives us the identities in $\mathcal{L}(\mathcal{V})$:

$$(X, f) \times (S, \mathrm{id}_S) = (X, f) \qquad (S, \mathrm{id}_S) \times (X, f) = (X, f).$$

This makes $\mathcal{L}(\mathcal{V})$ into a symmetric monoidal category with *strict unit* (S, id_S), i.e., the multiplication maps

$$\mu^r : (X, f) \times (S, \mathrm{id}_S) \to (X, f), \; \mu^l : (S, \mathrm{id}_S) \times (X, f) \to (X, f)$$

are the identity maps. Similarly, this makes $\mathcal{L}(\mathcal{V})^*$ into a symmetric monoidal category with strict unit $S(0)_{\mathrm{id}_S}$.

1.4. The construction of the motivic DG tensor category

We now proceed to define a differential graded tensor category $\mathcal{A}_{\mathrm{mot}}(\mathcal{V})$ in a series of steps.

1.4.1. DEFINITION. Let $\mathcal{A}_1(\mathcal{V})$ be the free additive category on $\mathcal{L}(\mathcal{V})^*$, with the following relations; we denote $X(d)_f$ as an object of $\mathcal{A}_1(\mathcal{V})$ by $\mathbb{Z}_X(d)_f$.

(i) Let $\emptyset$ be the empty scheme. The canonical map of $\mathbb{Z}_\emptyset(d)_f$ to 0 is an isomorphism.

(ii) for (X, f) and (Y, g) in $\mathcal{L}(\mathcal{V})$, let $i_X : X \to X \coprod Y$, $i_Y : Y \to X \coprod Y$ be the natural inclusions, and let $\Gamma = \mathbb{Z}_{(X \coprod Y)}(n)_{(f \coprod g)}$. Then

$$i_{X*} \circ i_X^* + i_{Y*} \circ i_Y^* = \mathrm{id}_\Gamma.$$

1.4.2. One checks that the linear extension of the product $\times : \mathcal{L}(\mathcal{V})^* \times \mathcal{L}(\mathcal{V})^* \to \mathcal{L}(\mathcal{V})^*$ descends to the product $\times : \mathcal{A}_1(\mathcal{V}) \otimes_\mathbb{Z} \mathcal{A}_1(\mathcal{V}) \to \mathcal{A}_1(\mathcal{V})$, making $\mathcal{A}_1(\mathcal{V})$ into a tensor category; the associativity and symmetry isomorphisms are given by the corresponding maps in $\mathcal{L}(\mathcal{V})^*$.

1.4.3. Let $(\mathcal{C}, \times, t)$ be a tensor category without unit. We recall from (Part II, Chapter I, §2.4.2 and §2.4.3) the construction of the universal commutative external product on $(\mathcal{C}, \times, t)$, i.e., a tensor category without unit $(\mathcal{C}^{\otimes,c}, \otimes, \tau)$, together with an additive functor $i : \mathcal{C} \to \mathcal{C}^{\otimes,c}$ and a natural transformation $\boxtimes : \otimes \circ (i \otimes_\mathbb{Z} i) \to i \circ \times$ of the functors

$$\otimes \circ (i \otimes_\mathbb{Z} i), i \circ \times : \mathcal{C} \otimes_\mathbb{Z} \mathcal{C} \to \mathcal{C}^{\otimes,c}.$$

The natural transformation $\boxtimes$ is associative and commutative (*cf.* Part II, Chapter I, Definition 2.4.1). The category $\mathcal{C}^{\otimes,c}$ is gotten from the free tensor category on $\mathcal{C}$, $(\mathcal{C}^\otimes, \otimes, \tau)$, by adjoining morphisms $\boxtimes_{X,Y} : X \otimes Y \to X \times Y$ for each pair of objects X and Y, and imposing the relations of

1. (Naturality) For $f : X \to X'$, $g : Y \to Y'$ in $\mathcal{C}$, we have
$$\boxtimes_{X',Y'} \circ (f \otimes g) = (f \times g) \circ \boxtimes_{X,Y},$$
2. (Associativity) For X, Y and Z in $\mathcal{C}$, we have
$$\boxtimes_{X \times Y, Z} \circ (\boxtimes_{X,Y} \otimes \mathrm{id}_Z) = \boxtimes_{X, Y \times Z} \circ (\mathrm{id}_X \otimes \boxtimes_{Y,Z}),$$
3. (Commutativity) For X and Y in $\mathcal{C}$, we have
$$t_{X,Y} \circ \boxtimes_{X,Y} = \boxtimes_{Y,X} \circ \tau_{X,Y}.$$

1.4.4. DEFINITION. Let $(\mathcal{A}_2(\mathcal{V}), \otimes, \tau)$ be the universal commutative external product on $\mathcal{A}_1(\mathcal{V})$: $\mathcal{A}_2(\mathcal{V}) := \mathcal{A}_1(\mathcal{V})^{\otimes,c}$, with external products $\boxtimes_{X,Y} : X \otimes Y \to X \times Y$.

1.4.5. We recall from (Part II, Chapter II, §3.1), the *homotopy one point DG tensor category* $(\mathbb{E}, \otimes, \tau)$. $\mathbb{E}$ has the following properties (see Part II, Chapter II, Proposition 3.1.12)

1. $\mathbb{E}$ is a DG tensor category without unit. There is an object $\mathfrak{e}$ of $\mathbb{E}$ which generates the objects of $\mathbb{E}$, i.e., the objects of $\mathbb{E}$ are finite direct sums of the tensors powers $\mathfrak{e}^{\otimes a}$, $a = 1, 2 \ldots$.
2. We have $\mathrm{Hom}_{\mathbb{E}}(\mathfrak{e}^{\otimes m}, \mathfrak{e}^{\otimes n})^q = 0$ if $n \neq m$, or if $n = m$ and $q > 0$. We have
$$\mathrm{Hom}_{\mathbb{E}}(\mathfrak{e}^{\otimes n}, \mathfrak{e}^{\otimes n})^0 \cong \mathbb{Z}[S_n],$$
the isomorphism sending a permutation $\sigma \in S_n$ to the symmetry isomorphism $\tau_\sigma : \mathfrak{e}^{\otimes n} \to \mathfrak{e}^{\otimes n}$. This gives the Hom-module $\mathrm{Hom}_{\mathbb{E}}(\mathfrak{e}^{\otimes n}, \mathfrak{e}^{\otimes n})^q$ the structure of a module over $\mathbb{Z}[S_n]$ by left or right composition.
3. For $q < 0$, $\mathrm{Hom}_{\mathbb{E}}(\mathfrak{e}^{\otimes n}, \mathfrak{e}^{\otimes n})^q$ is a free $\mathbb{Z}[S_n]$-module by both left and right composition (or is zero).
4. The cohomology of the Hom-complex is given by
$$H^q(\mathrm{Hom}_{\mathbb{E}}(\mathfrak{e}^{\otimes n}, \mathfrak{e}^{\otimes n})^*) = \begin{cases} \mathbb{Z} & \text{with generator } \mathrm{id}_{\mathfrak{e}^{\otimes n}} \text{ for } q = 0, \\ 0 & \text{for } q \neq 0. \end{cases}$$

We consider $\mathcal{A}_2(\mathcal{V})$ as a DG tensor category without unit, where all differentials are zero. Let $\mathcal{A}_2(\mathcal{V})[\mathbb{E}]$ denote the coproduct as DG tensor categories without unit.

1.4.6. DEFINITION. Let $\mathcal{A}_3(\mathcal{V})$ be the DG tensor category formed from $\mathcal{A}_2(\mathcal{V})[\mathbb{E}]$ by adjoining maps as follows: Let (X, f) be in $\mathcal{L}(\mathcal{V})$, and let Z be a non-zero element of $\mathcal{Z}^d(X)_f$. Then we adjoin the map of degree $2d$:

$$[Z] : \mathfrak{e} \to \mathbb{Z}_X(d)_f. \tag{1.4.6.1}$$

For $Z = 0 \in \mathcal{Z}^d(X)_f$, define the map $[Z] : \mathfrak{e} \dot{\to} \mathbb{Z}_X(d)_f$ to be the zero map.

1.4.7. *The cycles functor.* We now adjoin homotopies to the category $\mathcal{A}_3(\mathcal{V})$ which make the various cycle maps behave as cycle maps should. We require the preliminary construction of the *cycles functor* $\mathcal{Z}_1$ on $\mathcal{A}_1(\mathcal{V})$.

For each q, let

$$\mathcal{Z}^q : \mathcal{L}(\mathcal{V})^{\mathrm{op}} \to \mathbf{Ab} \tag{1.4.7.1}$$

be the functor

$$\mathcal{Z}^q(X, f) = \mathcal{Z}^q(X)_f$$
$$\mathcal{Z}^q(p) = p^*,$$

which is well-defined by Lemma 1.2.2. The functors (1.4.7.1) for $q = 0, 1, \ldots$ give rise to the functor

$$\mathcal{Z} : \mathcal{L}(\mathcal{V})^* \to \mathbf{Ab}, \tag{1.4.7.2}$$

defined on objects by $\mathcal{Z}(X(q)_f) = \mathcal{Z}^q(X)_f$. The definition of $\mathcal{Z}$ on morphisms is given by

$$\mathcal{Z}(j^*) = j^*; \mathcal{Z}(i_*) = i_*.$$

It is immediate that $\mathcal{Z}$ respects the relations of Definition 1.3.2, and is thus well-defined. The functor (1.4.7.2) extends to the functor

$$\mathcal{Z}_1 : \mathcal{A}_1(\mathcal{V}) \to \mathbf{Ab}, \tag{1.4.7.3}$$

using the additive structure of $\mathbf{Ab}$.

1.4.8. DEFINITION. Form the DG tensor category without unit $\mathcal{A}_4(\mathcal{V})$ by adjoining the following morphisms to $\mathcal{A}_3(\mathcal{V})$:

(i) Let (Y,g), (X,f) be in $\mathcal{L}(\mathcal{V})$, and let $p\colon \mathbb{Z}_X(q)_f \to \mathbb{Z}_Y(q)_g$ be a map in $\mathcal{A}_1(\mathcal{V})$. Let Z be a non-zero cycle in $\mathcal{Z}^q(X)_f$. From (1.4.7.3), we have the cycle $\mathcal{Z}_1(p)(Z) \in \mathcal{Z}^q(Y)_g$. Then we adjoin the map of degree $2q-1$:

$$h_{X,Y,[Z],p}\colon \mathfrak{e} \to \mathbb{Z}_Y(q)_g$$

with

$$dh_{X,Y,[Z],p} = p \circ [Z] - [\mathcal{Z}_1(p)(Z)].$$

(ii) Let (Y,g), (X,f) be in $\mathcal{L}(\mathcal{V})$, and let $(W,r) = (X,f) \times (Y,g)$. Take cycles $Z \in \mathcal{Z}^q(X)_f$ and $T \in \mathcal{Z}^{q'}(Y)_g$. Let $\Gamma = \mathbb{Z}_X(q)_f$ and $\Delta = \mathbb{Z}_Y(q')_g$, giving the product $\Gamma \times \Delta = \mathbb{Z}_W(q+q')_r$. Write 1 for $\mathbb{Z}_S(0)_{\mathrm{id}_S}$. From (Appendix A, Remark 2.3.3), we have the product cycle $Z \times_{/S} T$ in $\mathcal{Z}^{q+q'}(W)_q$. Then we adjoin the morphisms of degree $2(q+q')-1$,

$$h^l_{X,Y,[Z],[T]}\colon \mathfrak{e} \otimes \mathfrak{e} \to \mathbb{Z}_W(q+q')_r,$$
$$h^r_{X,Y,[Z],[T]}\colon \mathfrak{e} \otimes \mathfrak{e} \to \mathbb{Z}_W(q+q')_r,$$

with

$$dh^l_{X,Y,[Z],[T]} = \boxtimes_{\Gamma,\Delta} \circ ([Z] \otimes [T]) - \boxtimes_{\Gamma\times\Delta,1} \circ ([(Z \times_{/S} T)] \otimes [S]),$$
$$dh^r_{X,Y,[Z],[T]} = \boxtimes_{\Gamma,\Delta} \circ ([Z] \otimes [T]) - \boxtimes_{1,\Gamma\times\Delta} \circ ([S] \otimes [Z \times_{/S} T]).$$

Here

$$[Z] : \mathfrak{e} \to \mathbb{Z}_X(q)_f,\ [T] : \mathfrak{e} \to \mathbb{Z}_Y(q')_g,$$
$$[Z \times_{/S} T]\colon \mathfrak{e} \to \mathbb{Z}_W(q+q')_r,\ [S]\colon \mathfrak{e} \to 1$$

are the cycle maps defined in Definition 1.4.6, and

$$\boxtimes_{\Gamma,\Delta}\colon \Gamma \otimes \Delta \to \Gamma \times \Delta = \mathbb{Z}_W(q+q')_r,$$
$$\boxtimes_{\Gamma\times\Delta,1}\colon (\Gamma \times \Delta) \otimes 1 \to (\Gamma \times \Delta) \times 1 = \Gamma \times \Delta,$$
$$\boxtimes_{1,\Gamma\times\Delta}\colon 1 \otimes (\Gamma \times \Delta) \to 1 \times (\Gamma \times \Delta) = \Gamma \times \Delta$$

are the external products.

(iii) Let (X,f) be in $\mathcal{L}(\mathcal{V})$, let Z and Z' be elements of $\mathcal{Z}^q(X)_f$, and let n, n' be in $\mathbb{Z}$. Adjoin the map of degree $2q-1$:

$$h_{n,n',[Z],[Z']}\colon \mathfrak{e} \to \mathbb{Z}_X(q)_f$$

with

$$dh_{n,n',[Z],[Z']} = [nZ + n'Z'] - n[Z] - n'[Z'].$$

1.4.9. DEFINITION. Let $\mathcal{A}_5(\mathcal{V})$ denote the category gotten from $\mathcal{A}_4(\mathcal{V})$ by successively adjoining morphisms $h\colon \mathfrak{e}^{\otimes k} \to \mathbb{Z}_X(n)_f$ as follows:

Let $\mathcal{A}_5(\mathcal{V})^{(0)} := \mathcal{A}_4(\mathcal{V})$. Suppose we have formed the DG tensor category without unit $\mathcal{A}_5(\mathcal{V})^{(r-1)}$, $r \geq 1$. Let $\mathcal{A}_5(\mathcal{V})^{(r,0)} := \mathcal{A}_5(\mathcal{V})^{(r-1)}$, and suppose we have formed $\mathcal{A}_5(\mathcal{V})^{(r,k-1)}$ for some $k \geq 1$. Form the DG tensor category $\mathcal{A}_5(\mathcal{V})^{(r,k)}$ by adjoining morphisms of degree $2n-r-1$,

$$h_s : \mathfrak{e}^{\otimes k} \to \mathbb{Z}_X(n)_f,$$

to $\mathcal{A}_4(\mathcal{V})^{(r,k-1)}$, with $dh_s = s$, for each non-zero morphism $s: \mathfrak{e}^{\otimes k} \to \mathbb{Z}_X(n)_f$ in $\mathcal{A}_4(\mathcal{V})^{(r,k-1)}$ such that s has degree $2n - r$ and $ds = 0$. Let

$$\mathcal{A}_4(\mathcal{V})^{(r)} := \varinjlim_k \mathcal{A}_4(\mathcal{V})^{(r,k)},$$

$$\mathcal{A}_5(\mathcal{V}) := \varinjlim_r \mathcal{A}_4(\mathcal{V})^{(r)}.$$

1.4.10. DEFINITION. $\mathcal{A}_{\mathrm{mot}}(\mathcal{V})$ is defined to be the full additive subcategory of $\mathcal{A}_5(\mathcal{V})$ generated by tensor products of objects of the form $\mathbb{Z}_X(n)_f$, or $\mathfrak{e}^{\otimes a} \otimes \mathbb{Z}_X(n)_f$.

It follows immediately from the definition of the tensor product in $\mathcal{A}_5(\mathcal{V})$ that $\mathcal{A}_{\mathrm{mot}}(\mathcal{V})$ is a DG tensor subcategory of the DG tensor category without unit $\mathcal{A}_5(\mathcal{V})$.

1.4.11. REMARK. We denote the object $\mathbb{Z}_S(0)_{\mathrm{id}_S}$ of $\mathcal{A}_{\mathrm{mot}}(\mathcal{V})$ by 1. Let $h: \mathfrak{e}^{\otimes a} \to \mathbb{Z}_X(n)_f$ be a morphism in $\mathcal{A}_5(\mathcal{V})$. We let $h^S: \mathfrak{e}^{\otimes a} \otimes 1 \to \mathbb{Z}_X(n)_f$ denote the composition

$$\mathfrak{e}^{\otimes a} \otimes 1 \xrightarrow{h \otimes \mathrm{id}_1} \mathbb{Z}_X(n)_f \otimes 1 \xrightarrow{\boxtimes_{\mathbb{Z}_X(n)_f, 1}} \mathbb{Z}_X(n)_f.$$

It follows directly from (Part II, Chapter I, Proposition 2.5.2), that the map

$$\mathrm{Hom}_{\mathcal{A}_5(\mathcal{V})}(\mathfrak{e}^{\otimes a}, \Gamma) \to \mathrm{Hom}_{\mathcal{A}_{\mathrm{mot}}(\mathcal{V})}(\mathfrak{e}^{\otimes a} \otimes 1, \Gamma)$$
$$f \mapsto f^S$$

is an isomorphism for all Γ in $\mathcal{A}_1(\mathcal{V})$. We sometimes omit the ^{S} in the notation if the context makes the meaning clear.

1.4.12. DEFINITION. For $n = 4, 5$ and $n = \mathrm{mot}$, we let $\mathcal{A}^0_n(\mathcal{V})$ denote the graded tensor category gotten from $\mathcal{A}_n(\mathcal{V})$ by sending to zero all the maps of Definition 1.4.8 and Definition 1.4.9, and the morphisms of degree $p < 0$ in the category $\mathbb{E}$, as well as their differentials. We let

$$H_n: \mathcal{A}_n(\mathcal{V}) \to \mathcal{A}^0_n(\mathcal{V}) \tag{1.4.12.1}$$

denote the canonical DG functor.

We note that the natural map $\mathcal{A}^0_4(\mathcal{V}) \to \mathcal{A}^0_5(\mathcal{V})$ is an isomorphism, and that $\mathcal{A}^0_{\mathrm{mot}}(\mathcal{V})$ is the full tensor subcategory of $\mathcal{A}^0_5(\mathcal{V})$ generated by the objects of $\mathcal{A}_{\mathrm{mot}}(\mathcal{V})$. Furthermore, $\mathcal{A}^0_4(\mathcal{V})$ is isomorphic to the graded tensor category gotten from $\mathcal{A}_3(\mathcal{V})$ by imposing the relations (see Definition 1.4.8 for notation):

(i) Let (Y, g), (X, f) be in $\mathcal{L}(\mathcal{V})$, and let $p: \mathbb{Z}_X(d)_f \to \mathbb{Z}_Y(d)_g$ be a map in $\mathcal{A}_1(\mathcal{V})$. Let Z be a cycle in $\mathcal{Z}^d(X)_f$. Then

$$p \circ [Z] = [\mathcal{Z}_1(p)(Z)].$$

(ii) Let (Y, g), (X, f) be in $\mathcal{L}(\mathcal{V})$, and let $(W, h) = (X, f) \times (Y, g)$. Take Z in $\mathcal{Z}^d(X)_f$ and T in $\mathcal{Z}^e(Y)_g$. Let $\Gamma = \mathbb{Z}_X(d)_f$, $\Delta = \mathbb{Z}_Y(e)_g$, so $\Gamma \times \Delta = \mathbb{Z}_W(d + e)_h$. Then

$$\boxtimes_{\Gamma,\Delta} \circ ([Z] \otimes [T]) = \boxtimes_{\Gamma\times\Delta,1} \circ ([Z \times_S T] \otimes [|S|]),$$
$$\boxtimes_{\Gamma,\Delta} \circ ([Z] \otimes [T]) = \boxtimes_{1,\Gamma\times\Delta} \circ ([|S|] \otimes [Z \times_S T]).$$

(iii) Let (X, f) be in $\mathcal{L}(\mathcal{V})$, let Z and Z' be elements of $\mathcal{Z}^d(X)_f$, and let n, n' be in $\mathbb{Z}$. Then

$$[nZ + n'Z'] = n[Z] + n'[Z'].$$

(iv) Let $\tau_{\mathfrak{e},\mathfrak{e}}:\mathfrak{e}\otimes\mathfrak{e}\to\mathfrak{e}\otimes\mathfrak{e}$ be the symmetry isomorphism. Then

$$\tau_{\mathfrak{e},\mathfrak{e}} = \mathrm{id}_{\mathfrak{e}\otimes\mathfrak{e}}.$$

2. The triangulated motivic category

In this section, we construct the main object of our study. The idea is quite simple: We have all the necessary morphisms and relations among them in the category $\mathcal{A}_{\mathrm{mot}}(\mathcal{V})$. We construct a triangulated tensor category from $\mathcal{A}_{\mathrm{mot}}(\mathcal{V})$ by taking the homotopy category of the category of bounded complexes on $\mathcal{A}_{\mathrm{mot}}(\mathcal{V})$ (see Part II, Chapter II, §1.2 and §2.1). We then localize this category, forcing the various axioms of a Bloch-Ogus cohomology theory, suitably interpreted, to be valid. Finally, we form the pseudo-abelian hull.

2.1. The definition of the triangulated motivic category

We recall from (Part II, Chapter II, Definition 1.2.7) the functor $\mathbf{C}^b(-)$ from DG categories to DG categories, which associates to a DG category $\mathcal{A}$ the category $\mathbf{C}^b(\mathcal{A})$ of bounded complexes in $\mathcal{A}$. We have the functor $\mathbf{K}^b(-) := \mathbf{C}^b(-)/\mathrm{Htp}$, which gives a functor from DG categories to triangulated categories (see Part II, Chapter II, Definition 1.2.7 and Proposition 2.1.6.4). We apply these functors to the categories constructed in Section 1.

We denote the categories $\mathbf{C}^b(\mathcal{A}_{\mathrm{mot}}(\mathcal{V}))$ and $\mathbf{K}^b(\mathcal{A}_{\mathrm{mot}}(\mathcal{V}))$ by $\mathbf{C}^b_{\mathrm{mot}}(\mathcal{V})$ and $\mathbf{K}^b_{\mathrm{mot}}(\mathcal{V})$.

2.1.1. We recall from (Part II, Chapter II, §2.1 and §2.3) the notions of a *triangulated category* $\mathcal{A}$, a *thick subcategory* $\mathcal{B}$ of $\mathcal{A}$, and the triangulated category $\mathcal{A}/\mathcal{B}$ formed by localizing $\mathcal{A}$ with respect to $\mathcal{B}$. We recall as well the notions of *triangulated tensor category* $\mathcal{A}$, a *thick tensor subcategory* $\mathcal{B}$ of $\mathcal{A}$, and the triangulated tensor category $\mathcal{A}/\mathcal{B}$ formed by localizing $\mathcal{A}$ with respect to $\mathcal{B}$.

If $\mathcal{S} = \{h_i: X_i \to Y_i \mid i \in I\}$ is collection of morphisms in a triangulated category $\mathcal{A}$, we let $\mathcal{A}(\mathcal{S})$ be the thick subcategory generated by the objects Z which fit into a distinguished triangle $X \xrightarrow{h} Y \to Z \to X[1]$ with $h \in \mathcal{S}$, and call $\mathcal{A}/\mathcal{A}(\mathcal{S})$ the triangulated category formed by inverting the morphisms in $\mathcal{S}$. Similarly, if $\mathcal{A}$ is a triangulated tensor category, we let $\mathcal{A}(\mathcal{S})^{\otimes}$ be the thick tensor subcategory generated by the objects Z as above. We call $\mathcal{A}/\mathcal{A}(\mathcal{S})^{\otimes}$ the triangulated tensor category formed by inverting the morphisms in $\mathcal{S}$.

2.1.2. Suppose we have a morphism $f: A \to B$ in a DG category $\mathcal{C}$, with $df = 0$. We denote the object $\mathrm{cone}(f)[-1]$ of $\mathbf{C}^b(\mathcal{C})$ by

$$A \xrightarrow{f} B[-1] \quad \text{or} \quad \begin{pmatrix} A \\ f\downarrow \\ B[-1] \end{pmatrix}.$$

2.1.3. Let $(X, f: X' \to X)$ be in $\mathcal{L}(\mathcal{V})$, let $\hat{X}$ be a closed subset of X, and let $j: U \to X$ be the inclusion of the complement $X \setminus \hat{X}$. We write j^*f for the map $p_1: U \times_X X' \to U$. Suppose that the maps

$$j: U \to X$$
$$j^*f: U \times_X X' \to U$$

are in $\mathcal{V}$. Define the object $\mathbb{Z}_{X,\hat{X}}(n)_f$ of $\mathbf{C}^b_{\mathrm{mot}}(\mathcal{V})$ by

$$\mathbb{Z}_{X,\hat{X}}(n)_f := \mathrm{cone}(j^*: \mathbb{Z}_X(n)_f \to \mathbb{Z}_U(n)_{j^*f})[-1]. \tag{2.1.3.1}$$

If (Y, g) is in $\mathcal{L}(\mathcal{V})$, if $\hat{Y}$ is a closed subset of Y, with complement $i: V := Y \setminus \hat{Y} \to Y$, and if the maps i and i^*g are in $\mathcal{V}$, then each map $p: (X, f) \to (Y, g)$ in $\mathcal{L}(\mathcal{V})$, with $p^{-1}(\hat{Y}) \subset \hat{X}$, induces the map

$$p^*: \mathbb{Z}_{Y,\hat{Y}}(n)_g \to \mathbb{Z}_{X,\hat{X}}(n)_f, \tag{2.1.3.2}$$

defined as the map of complexes

$$\begin{pmatrix} \mathbb{Z}_Y(n)_g \\ i^* \downarrow \\ \mathbb{Z}_V(n)_{i^*g}[-1] \end{pmatrix} \xrightarrow[p^*[-1]]{\;p^*\;} \begin{pmatrix} \mathbb{Z}_X(n)_f \\ j^* \downarrow \\ \mathbb{Z}_U(n)_{i^*f}[-1] \end{pmatrix}.$$

If $Z \in \mathcal{Z}^n(X)_f$ is a cycle on X, supported on $\hat{X}$, we have the map (see Definition 1.4.8)

$$\begin{aligned} &h_{Z,j^*}: \mathfrak{e} \to \mathbb{Z}_U(d)_{j^*A}[2n-1], \\ &dh_{Z,j^*} = j^* \circ [Z] - [j^*Z] = j^* \circ [Z]. \end{aligned}$$

The pair $([Z], h_{Z,j^*})$ then defines the *cycle map with support*

$$[Z]_{\hat{X}}: \mathfrak{e} \to \mathbb{Z}_{X,\hat{X}}(n)_f[2n] \tag{2.1.3.3}$$

in the category $\mathbf{C}^b(\mathcal{A}_5(\mathcal{V}))$. These cycle maps with support are functorial in the category $\mathbf{K}^b(\mathcal{A}_5(\mathcal{V}))$.

Let X be a smooth quasi-projective S-scheme, and let $\hat{X}$ be a closed subset of X with irreducible components $\hat{X}_1, \ldots, \hat{X}_s$. We let $|\hat{X}|$ be the cycle on X defined by $|\hat{X}| = \sum_{i=1}^s 1 \cdot X_i$

2.1.4. DEFINITION. Let $\mathcal{V}$ be a strictly full subcategory of $\mathbf{Sm}_S^{\mathrm{ess}}$ satisfying the following conditions:

(i) $\mathcal{V}$ is closed under finite products over S and finite disjoint union; in particular, S and the empty scheme are in $\mathcal{V}$.
(ii) If X is in $\mathcal{V}$, and $j: U \to X$ an open subscheme of X, then U is in $\mathcal{V}$.
(iii) If X is in $\mathcal{V}$ and $E \to X$ is a vector bundle, then E and the projective bundle $\mathbb{P}(E)$ are in $\mathcal{V}$.
(iv) If $i: Z \to X$ is a closed embedding in $\mathcal{V}$, then the blow-up of X along Z is in $\mathcal{V}$.

Form the triangulated tensor category $\mathbf{D}^b_{\mathrm{mot}}(\mathcal{V})$ from $\mathbf{K}^b_{\mathrm{mot}}(\mathcal{V})$ by inverting the following morphisms:

(a) *Homotopy.* Let $p: (X, f) \to (Y, g)$ be a map in $\mathcal{L}(\mathcal{V})$, where $p: X \to Y$ is the inclusion of a closed codimension one subscheme. Let $\hat{Y} \subset Y$ be a closed subset of Y, and let $\hat{X} = p^{-1}(\hat{Y})$ (scheme-theoretic pull-back). Suppose that $\hat{X}$ is in $\mathbf{Sm}_S^{\mathrm{ess}}$, and that we have an isomorphism $q: \hat{X} \times_S \mathbb{A}^1_S \to \hat{Y}$, making the diagram

$$\begin{array}{ccc} \hat{X} & \xrightarrow{p} & \hat{Y} \\ {\scriptstyle p_1}\uparrow & & \uparrow{\scriptstyle q} \\ \hat{X}\times 0 & \subset & \hat{X}\times_S \mathbb{A}^1_S \end{array}$$

commute. Then invert the map

$$p^*: \mathbb{Z}_{Y,\hat{Y}}(n)_g \to \mathbb{Z}_{X,\hat{X}}(n)_f.$$

(b) *Excision.* Let (X, f) be in $\mathcal{L}(\mathcal{V})$, $\hat{X}$ a closed subset of X, $j: U \to X$ an open subscheme containing $\hat{X}$. Invert the map

$$j^*: \mathbb{Z}_{X,\hat{X}}(n)_f \to \mathbb{Z}_{U,\hat{X}}(n)_j^* f.$$

(c) *Künneth isomorphism.* Let X and Y be in $\mathcal{A}_1(\mathcal{V})$. Invert the map

$$\boxtimes_{X,Y}: X \otimes Y \to X \times Y.$$

(d) *Gysin isomorphism.* Let $p: (P, g) \to (X, f)$ be a map in $\mathcal{L}(\mathcal{V})$, and suppose $p: P \to X$ is a smooth morphism of relative dimension d. Suppose we have a section $s: X \to P$ to p with $|s(X)|$ in $\mathcal{Z}^d(P)_g$. Let

$$\alpha: \mathfrak{e} \otimes \mathbb{Z}_X(n-d)_f[-2d] \to \mathbb{Z}_{P\times_S P, s(X)\times_S P}(n)_{g\times g}$$

denote the composition

$$\begin{aligned} \mathfrak{e} \otimes \mathbb{Z}_X(n-d)_f[-2d] &\xrightarrow{[|s(X)|]_{s(X)}\otimes p^*} \mathbb{Z}_{P,s(X)}(d)_g \otimes \mathbb{Z}_P(n-d)_g \\ &\xrightarrow{\boxtimes} \mathbb{Z}_{P\times_S P, s(X)\times_S P}(n)_{g\times g}. \end{aligned}$$

Let ρ be the map (1.3.3.1)

$$\rho_{g\times g,\Delta}: \mathbb{Z}_{P\times_S P, s(X)\times_S P}(n)_{g\times g\cup\Delta} \to \mathbb{Z}_{P\times_S P, s(X)\times_S P}(n)_{g\times g},$$

where $\Delta: P \to P \times_S P$ is the diagonal. Invert the map

$$\begin{aligned} \begin{pmatrix} \alpha & -\rho \\ 0 & \Delta^* \end{pmatrix} : \mathfrak{e} \otimes \mathbb{Z}_X(n-d)_f[-2d] \oplus \mathbb{Z}_{P\times_S P, s(X)\times_S P}(n)_{g\times g\cup\Delta} \\ \to \mathbb{Z}_{P\times_S P, s(X)\times_S P}(n)_{g\times g} \oplus \mathbb{Z}_{P,s(X)}(n)_g. \end{aligned}$$

(e) *Moving lemma.* Let (X, f) be in $\mathcal{L}(\mathcal{V})$, and let $g: Z \to X$ be a morphism in $\mathcal{V}$. Invert the morphism (1.3.3.1)

$$\rho_{f,g}: \mathbb{Z}_X(n)_{f\cup g} \to \mathbb{Z}_X(n)_f.$$

(f) *Unit.* Invert the map

$$[|S|] \otimes \mathrm{id} : \mathfrak{e} \otimes \mathbb{Z}_S(0) \to \mathbb{Z}_S(0) \otimes \mathbb{Z}_S(0).$$

2.1.5. For a pre-additive category $\mathcal{A}$, and a commutative ring R, we let $\mathcal{A} \otimes R$ denote the pre-additive category with the same objects as $\mathcal{A}$, and with

$$\mathrm{Hom}_{\mathcal{A}\otimes R}(X, Y) = \mathrm{Hom}_{\mathcal{A}}(X, Y) \otimes_{\mathbb{Z}} R.$$

If $\mathcal{A}$ is a DG tensor category, then $\mathcal{A} \otimes R$ is in a natural way an R-DG tensor category; if $\mathcal{A}$ is a triangulated (tensor) category, and R is a localization of $\mathbb{Z}$, then $\mathcal{A} \otimes R$ is in a natural way a triangulated (tensor) category.

Let R be a commutative ring, flat over $\mathbb{Z}$. Let $\mathbf{C}^b_{\mathrm{mot}}(\mathcal{V})_R = \mathbf{C}^b(\mathcal{A}_{\mathrm{mot}}(\mathcal{V}) \otimes R)$ and let $\mathbf{K}^b_{\mathrm{mot}}(\mathcal{V})_R$ be the homotopy category of $\mathbf{C}^b_{\mathrm{mot}}(\mathcal{V})_R$. Let $\mathbf{D}^b_{\mathrm{mot}}(\mathcal{V})_R$ be the localization of $\mathbf{K}^b_{\mathrm{mot}}(\mathcal{V})_R$ with respect to the thick tensor subcategory generated by the morphisms in Definition 2.1.4. We note that the natural map $\mathbf{D}^b_{\mathrm{mot}}(\mathcal{V}) \otimes R \to \mathbf{D}^b_{\mathrm{mot}}(\mathcal{V})_R$ is an equivalence of triangulated tensor categories if R is a localization of $\mathbb{Z}$.

2.1.6. DEFINITION. Let R be a commutative ring, flat over $\mathbb{Z}$. Let $\mathcal{DM}(\mathcal{V})_R$ be the pseudo-abelian hull $[\mathbf{D}^b_{\mathrm{mot}}(\mathcal{V})_R]_\#$ of $\mathbf{D}^b_{\mathrm{mot}}(\mathcal{V})_R$ (see Part II, Chapter II, Definition 2.4.1 and Theorem 2.4.7). We call $\mathcal{DM}(\mathcal{V})_R$ the *triangulated motivic category of $\mathcal{V}$ with R coefficients.* We set

$$\mathcal{DM}(S)_R := \mathcal{DM}(\mathbf{Sm}_S)_R.$$

We have the fully faithful embedding $\#: \mathbf{D}^b_{\mathrm{mot}}(\mathcal{V})_R \to \mathcal{DM}(\mathcal{V})_R$. We will often denote the category $\mathcal{DM}(\mathcal{V})_R$ by $\mathcal{DM}$, when the reference to R and $\mathcal{V}$ is understood. We let $R_X(n)_f$ denote the image of $\mathbb{Z}_X(n)_f$ in $\mathcal{DM}(\mathcal{V})_R$ or in $\mathbf{D}^b_{\mathrm{mot}}(\mathcal{V})_R$.

2.2. Properties of motives

We begin with a list of fundamental properties of the objects $\mathbb{Z}_{X,\hat{X}}(q)_f$ in $\mathcal{DM}(\mathcal{V})$; for a commutative ring R, flat over $\mathbb{Z}$, the analogous statements are valid for the category $\mathbf{D}^b_{\mathrm{mot}}(\mathcal{V})_R$ and $\mathcal{DM}(\mathcal{V})_R$ as well. For $\hat{X}$ a closed subset of $X \in \mathcal{V}$, we write $\mathbb{Z}_{X,\hat{X}}(q)$ for $\mathbb{Z}_{X,\hat{X}}(q)_{\mathrm{id}}$ and $\mathbb{Z}_{X,\hat{X}}$ for $\mathbb{Z}_{X,\hat{X}}(0)$.

2.2.1. *Homotopy.* If we take $(Y, g) = (X \times_S \mathbb{A}^1, \mathrm{id} \cup i_0)$, where $i_0: X \to X \times_S \mathbb{A}^1$ is the zero section, we have the map $i_0: (X, \mathrm{id}_X) \to (Y, g)$ in $\mathcal{L}(\mathcal{V})$. The homotopy axiom (Definition 2.1.4(a)), with $\hat{X} = X$, $\hat{Y} = Y$, gives the isomorphism

$$i_0^*: \mathbb{Z}_{X\times_S\mathbb{A}^1}(0)_{\mathrm{id}\cup i_0} \to \mathbb{Z}_X.$$

If we now apply the moving lemma (Definition 2.1.4(e)), we get the isomorphism in $\mathcal{DM}$

$$\rho^{-1}_{\mathrm{id},i_0} \circ i_0^*: \mathbb{Z}_{X\times_S\mathbb{A}^1} \to \mathbb{Z}_X.$$

This then implies that the pull-back by the projection $p: X \times \mathbb{A}^1 \to X$ gives the isomorphism

$$p^*: \mathbb{Z}_X \to \mathbb{Z}_{X\times_S\mathbb{A}^1}.$$

More generally, if $\hat{X}$ is a closed subset of X with complement $j: U \to X$, we have the commutative diagram

$$\begin{array}{ccccccc}
\mathbb{Z}_U[-1] & \longrightarrow & \mathbb{Z}_{X,\hat{X}} & \longrightarrow & \mathbb{Z}_X & \longrightarrow & \mathbb{Z}_U \\
\downarrow{\scriptstyle p^*} & & \downarrow{\scriptstyle p^*} & & \downarrow{\scriptstyle p^*} & & \downarrow{\scriptstyle p^*} \\
\mathbb{Z}_{U\times_S\mathbb{A}^1}[-1] & \longrightarrow & \mathbb{Z}_{X\times_S\mathbb{A}^1,\hat{X}\times_S\mathbb{A}^1} & \longrightarrow & \mathbb{Z}_{X\times_S\mathbb{A}^1} & \longrightarrow & \mathbb{Z}_{U\times_S\mathbb{A}^1}
\end{array}$$

with the rows distinguished triangles. Thus the map

$$p^*: \mathbb{Z}_{X,\hat{X}} \to \mathbb{Z}_{X\times_S\mathbb{A}^1,\hat{X}\times_S\mathbb{A}^1}$$

is also an isomorphism.

2.2.2. *Moving lemma.* Suppose $\hat{X}$ is a closed subset of a scheme $X \in \mathcal{V}$. Since the objects $\mathbb{Z}_{X,\hat{X}}(q)_f$, for varying f, are, by the isomorphism of Definition 2.1.4(e), all canonically isomorphic to $\mathbb{Z}_{X,\hat{X}}(q)$, we will denote all these objects by $\mathbb{Z}_{X,\hat{X}}(q)$, when the explicit use of the auxiliary f is not required. We let

$$\rho_f: \mathbb{Z}_{X,\hat{X}}(q)_f \to \mathbb{Z}_{X,\hat{X}}(q)$$

denote the canonical isomorphism.

2.2.3. *Tate Twist.* Let Γ be an object of $\mathcal{DM}$, and let q be an integer. Denote $\mathbb{Z}_S(q) \otimes \Gamma$ by $\Gamma(q)$. The isomorphism of Definition 2.1.4(c) gives rise to canonical isomorphisms

$$\mu^l_\Gamma: \Gamma(0) = \mathbb{Z}_S \otimes \Gamma \xrightarrow{\boxtimes_{\mathbb{Z}_S,\Gamma}} \Gamma,$$

$$\mu^r_\Gamma: \Gamma \otimes \mathbb{Z}_S \xrightarrow{\boxtimes_{\Gamma,\mathbb{Z}_S}} \Gamma,$$

and a canonical isomorphism

$$\mathbb{Z}_S(a) \otimes \Gamma(b) \to \Gamma(a+b).$$

For $\Gamma = \mathbb{Z}_{X,\hat{X}}(n)$, we have the canonical isomorphism

$$\mathbb{Z}_{X,\hat{X}}(n)(a) \xrightarrow{\boxtimes_{S,X}} \mathbb{Z}_{X,\hat{X}}(a+n).$$

2.2.4. *Unit.* We denote the object $\mathbb{Z}_S$ by 1. We let

$$\nu_a: \mathfrak{e}^{\otimes a} \otimes 1 \to 1 \tag{2.2.4.1}$$

denote the composition

$$\mathfrak{e}^{\otimes a} \otimes 1 \xrightarrow{[S]^{\otimes a}\otimes \mathrm{id}_1} 1^{\otimes a+1} \xrightarrow{\boxtimes_{1,\ldots,1}} 1.$$

By the morphisms inverted in Definition 2.1.4(c),(f), ν_a is an isomorphism.

2.2.5. *Gysin morphism.* Let $\mathbf{K}^b_{\mathrm{mot}}(\mathcal{V})^1$ denote the category formed from the triangulated tensor category $\mathbf{K}^b_{\mathrm{mot}}(\mathcal{V})$ by inverting the morphisms of Definition 2.1.4(e) and (f). Let (X,f), (Y,g) be in $\mathcal{L}(\mathcal{V})$, and let Z be in $\mathcal{Z}^q(X)_f$, supported on a closed subset W, giving the cycle map with support (2.1.3.3). We let $\cup[Z]_W$ denote the composition

$$\begin{aligned}\mathbb{Z}_Y(n)_g \xrightarrow{(\mu^l\circ([S]\otimes\mathrm{id})^{-1}} \mathfrak{e}\otimes\mathbb{Z}_Y(n)_g \xrightarrow{[Z]_W\otimes\mathrm{id}} &\mathbb{Z}_{X,W}(d)_f[2d]\otimes\mathbb{Z}_Y(n)_g\\ \xrightarrow{\boxtimes_{X,Y}} &\mathbb{Z}_{X\times_S Y,W\times_S Y}(n+d)_{f\times g}[2d].\end{aligned}$$

For $p: (P,B) \to (X,f)$, $s: X \to P$, and ρ as in Definition 2.1.4(d), we denote the composition

$$\begin{aligned}\mathbb{Z}_X(-d)_f[-2d] \xrightarrow{p^*} \mathbb{Z}_P(-d)_g[-2d] &\xrightarrow{\cup[|s(X)|]_{s(X)}} \mathbb{Z}_{P\times_S P,s(X)\times_S P}(0)_{g\times g}\\ &\xrightarrow{\rho^{-1}} \mathbb{Z}_{P\times_S P,s(\hat{X})\times_S P}(0)_{g\times g\cup\Delta}[2d] \xrightarrow{\Delta^*} \mathbb{Z}_{P,s(X)}(0)_g\end{aligned} \tag{2.2.5.1}$$

by $\cup[|s(X)|]\circ p^*$. In the category $\mathbf{K}^b_{\mathrm{mot}}(\mathcal{V})^1$, inverting the map of Definition 2.1.4(d) is the same as inverting the map (2.2.5.1).

2.2.6. *Mayer-Vietoris.* Let (X, f) be in $\mathcal{L}(\mathcal{V})$, and let $j_U : U \to X$, $j_V : V \to X$ be open subschemes. Let

$$j_{U,U\cap V} : U\cap V \to U, \quad j_{V,U\cap V} : U\cap V \to U, \quad j_{U\cap V} : U\cap V \to X$$

be the inclusions. It follows from the inversion of the maps in Definition 2.1.4(b) that there is a natural distinguished triangle

$$(2.2.6.1) \quad \mathbb{Z}_X(n)_f \xrightarrow{(j_U^*, j_V^*)} \mathbb{Z}_U(n)_{j_U^* f} \oplus \mathbb{Z}_V(n)_{j_V^* f} \xrightarrow{j_{U,U\cap V}^* - j_{V,U\cap V}^*} \mathbb{Z}_{U\cap V}(n)_{j_{U\cap V}^* f} \longrightarrow \mathbb{Z}_X(n)_f[1]$$

in $\mathcal{DM}$.

2.2.7. *Motivic cohomology.* Let X be a scheme in $\mathcal{V}$, $\hat{X}$ a closed subset. The *motivic cohomology of X with support in $\hat{X}$* is defined as

$$H^p_{\hat{X}}(X, \mathbb{Z}(q)) := \mathrm{Hom}_{\mathcal{DM}}(1, \mathbb{Z}_{X,\hat{X}}(q)[p]).$$

More generally, for an object Γ of $\mathcal{DM}_R$, define the motivic cohomology of Γ by

$$H^p(\Gamma, R(q)) = \mathrm{Hom}_{\mathcal{DM}_R}(1, \Gamma(q)[p]).$$

This is compatible with the above definition because of the Tate twist isomorphism §2.2.3.

2.2.8. *Mod n motivic cohomology.* For Γ in $\mathbf{C}^b_{\mathrm{mot}}(\mathcal{V})$, define $\Gamma \otimes^L \mathbb{Z}/n$ as

$$\Gamma \otimes^L \mathbb{Z}/n := \mathrm{cone}(\Gamma \xrightarrow{\times n} \Gamma),$$

and the mod-n motivic cohomology of Γ as

$$H^p(\Gamma, \mathbb{Z}/n(q)) := H^p(\Gamma \otimes^L \mathbb{Z}/n, \mathbb{Z}(q)).$$

For $\Gamma = \mathbb{Z}_{X,\hat{X}}(0)$, this gives us the mod-$n$ motivic cohomology of X (with support in $\hat{X}$)

$$H^p_{\hat{X}}(X, \mathbb{Z}/n(q)) := H^p(\mathbb{Z}_{X,\hat{X}}(q) \otimes^L \mathbb{Z}/n).$$

The distinguished triangle

$$\Gamma \xrightarrow{\times n} \Gamma \to \Gamma \otimes^L \mathbb{Z}/n \to \Gamma[1]$$

gives rise to the short exact "universal coefficient" sequence

$$0 \to H^p(\Gamma, \mathbb{Z}(q))/n \to H^p(\Gamma, \mathbb{Z}/n(q)) \to {}_nH^{p+1}(\Gamma, \mathbb{Z}(q)) \to 0,$$

where ${}_nH^{p+1}(\Gamma, \mathbb{Z}(q))$ is the n-torsion subgroup of $H^{p+1}(\Gamma, \mathbb{Z}(q))$.

2.2.9. *Motives and motives with support.* Let $\mathbf{P}\mathcal{V}$ denote the category of pairs $(X, \hat{X})$, where $\hat{X}$ a closed subset of X, and X is in $\mathcal{V}$. A morphism $p : (X, \hat{X}) \to (Y, \hat{Y})$ is a morphism $p : X \to Y$ with $p^{-1}(\hat{Y}) \subset \hat{X}$. We define the category $\mathbf{P}\mathcal{L}(\mathcal{V})$ similarly as the category of triples $(X, \hat{X}, f)$ with $(X, \hat{X}) \in \mathbf{P}\mathcal{V}$, and $(X, f) \in \mathcal{L}(\mathcal{V})$. Morphisms $p : (X, \hat{X}, f) \to (Y, \hat{Y}, g)$ are maps $p : X \to Y$ such that $p : (X, f) \to (Y, g)$ is a morphism in $\mathcal{L}(\mathcal{V})$ and $p : (X, \hat{X}) \to (Y, \hat{Y})$ is a morphism in $\mathbf{P}\mathcal{V}$.

The maps p^* of (2.1.3.2) induce maps on the motivic cohomology as follows. If $p:(Y,\hat{Y},g)\to(X,\hat{X},f)$ is a map in $\mathbf{P}\mathcal{L}(\mathcal{V})$, we have the composition

$$\mathbb{Z}_{X,\hat{X}}(a)[b]\xrightarrow{\rho_f^{-1}}\mathbb{Z}_{X,\hat{X}}(a)_f[b]\xrightarrow{p^*}\mathbb{Z}_{Y,\hat{Y}}(a)_g[b]\xrightarrow{\rho_g}\mathbb{Z}_{Y,\hat{Y}}(a)[b]$$

in $\mathcal{DM}$. This defines the functor

$$\mathbb{Z}(a)[b]:\mathbf{P}\mathcal{L}(\mathcal{V})^{\mathrm{op}}\to\mathcal{DM}$$
$$(X,\hat{X},f)\mapsto\mathbb{Z}_{X,\hat{X}}(a)_f[b].$$

If we make another choice of f and g, with the same underlying map $p:Y\to X$ in $\mathcal{V}$, the resulting composition is the same. We can take for example $f=\mathrm{id}_X$, $g=\mathrm{id}_Y\cup p$. Thus the functor $\mathbb{Z}(a)[b]$ descends to the functor

(2.2.9.1)
$$\mathbb{Z}(a)[p]:\mathbf{P}\mathcal{V}^{\mathrm{op}}\to\mathcal{DM}$$
$$(X,\hat{X})\mapsto\mathbb{Z}_{X,\hat{X}}(a)[b].$$

We call the object $\mathbb{Z}_{X,\hat{X}}$ the *motive of X with support in $\hat{X}$*; the object $\mathbb{Z}_X$ is called the *motive* of X.

Composing $\mathbb{Z}(q)[p]$ with the functor $\mathrm{Hom}_{\mathcal{DM}}(1,-)$ gives the motivic cohomology functor $H^p(-,\mathbb{Z}(q)):\mathbf{P}\mathcal{V}^{\mathrm{op}}\to\mathbf{Ab}$.

If $\hat{X}'\subset\hat{X}$ are closed subsets of $X\in\mathcal{V}$, then $\mathrm{id}_X:(X,\hat{X},\mathrm{id}_X)\to(X,\hat{X}',\mathrm{id}_X)$ induces the map $\mathrm{id}_X^*:\mathbb{Z}_{X,\hat{X}'}\to\mathbb{Z}_{X,\hat{X}}$ which we denote by

$$i_{\hat{X}'\subset\hat{X}*}:\mathbb{Z}_{X,\hat{X}'}\to\mathbb{Z}_{X,\hat{X}}.$$

2.2.10. *Mayer-Vietoris and localization for motives with support.* The distinguished triangle of §2.2.6 gives rise to the Mayer-Vietoris distinguished triangle for the union of two closed subsets: If $\hat{X}=\hat{X}_1\cup\hat{X}_2$ are closed subsets of $X\in\mathcal{V}$, let $\hat{X}_{12}$ be the intersection $\hat{X}_1\cap\hat{X}_2$. We have the distinguished triangle

$$\mathbb{Z}_{X,\hat{X}_{12}}\xrightarrow{(i_{\hat{X}_{12}\subset\hat{X}_1*},-i_{\hat{X}_{12}\subset\hat{X}_2*})}\mathbb{Z}_{X,\hat{X}_1}\oplus\mathbb{Z}_{X,\hat{X}_2}$$
$$\xrightarrow{i_{\hat{X}_1\subset\hat{X}*}+i_{\hat{X}_2\subset\hat{X}*}}\mathbb{Z}_{X,\hat{X}}\to\mathbb{Z}_{X,\hat{X}_{12}}[1].$$

We have as well the localization distinguished triangle: If F and $\hat{X}$ are closed subsets of $X\in\mathcal{V}$, if $j:U\to X$ is the complement $X\backslash F$, and if $\hat{U}=\hat{X}\cap U$, then we have the distinguished triangle

(2.2.10.1)
$$\mathbb{Z}_{X,F}\xrightarrow{i_{F\subset F\cup\hat{X}*}}\mathbb{Z}_{X,F\cup\hat{X}}\xrightarrow{j^*}\mathbb{Z}_{U,\hat{U}}\to\mathbb{Z}_{X,F}[1].$$

In particular, taking $\hat{X}=X$, we have the distinguished triangle

(2.2.10.2)
$$\mathbb{Z}_{X,F}\xrightarrow{i_{F\subset X*}}\mathbb{Z}_X\xrightarrow{j^*}\mathbb{Z}_U\to\mathbb{Z}_{X,F}[1].$$

2.2.11. *Products.* The tensor product operation gives rise to external products in cohomology. Indeed, the operation $\otimes$ gives rise to the map

$$\mathrm{Hom}_{\mathcal{DM}}(Z,X[p])\otimes_{\mathbb{Z}}\mathrm{Hom}_{\mathcal{DM}}(W,Y[p'])\to\mathrm{Hom}_{\mathcal{DM}}(Z\otimes W,X\otimes Y[p+p']),$$

for X,Y, Z and W in $\mathcal{DM}$. In particular, we have the map

$$\mathrm{Hom}_{\mathcal{DM}}(1,\mathbb{Z}_X(q)[p])\otimes_{\mathbb{Z}}\mathrm{Hom}_{\mathcal{DM}}(1,\mathbb{Z}_Y(q')[p'])$$
$$\to\mathrm{Hom}_{\mathcal{DM}}(1\otimes 1,\mathbb{Z}_X(q)\otimes\mathbb{Z}_Y(q')[p+p']).$$

Composing with the morphism $\boxtimes_{X,Y}:\mathbb{Z}_X(q)\otimes\mathbb{Z}_Y(q')\to\mathbb{Z}_{X\times_S Y}(q+q')$, and the inverse of the multiplication isomorphism $\mu=\boxtimes_{1,1}:1\otimes 1\to 1$, we get the map

$$\cup_{X,Y}:H^p(X,\mathbb{Z}(q))\otimes_{\mathbb{Z}}H^{p'}(Y,\mathbb{Z}(q'))\to H^{p+p'}(X\times_S Y,\mathbb{Z}(q+q')).$$

If we take $X=Y$, we can compose with the pullback by the diagonal to get products in cohomology

$$\cup_X:H^p(X,\mathbb{Z}(q))\otimes_{\mathbb{Z}}H^{p'}(X,\mathbb{Z}(q'))\to H^{p+p'}(X,\mathbb{Z}(q+q')).$$

More generally, suppose we have closed subsets $\hat{X}$ of X and $\hat{Y}$ of Y, and let $j_U:U:=X\setminus\hat{X}\to X$ and $j_V:V:=Y\setminus\hat{Y}\to Y$ be the complements. Letting $\mathbb{Z}_{X,\hat{X}}(q)\times\mathbb{Z}_{Y,\hat{Y}}(q')$ denote the complex

$$\mathbb{Z}_{X\times_S Y}(q+q')\xrightarrow{((j_U\times\mathrm{id}_Y)^*,(\mathrm{id}_X\times j_V)^*)}\mathbb{Z}_{U\times_S Y}(q+q')\oplus\mathbb{Z}_{X\times_S V}(q+q')$$
$$\xrightarrow{(\mathrm{id}_U\times j_V)^*-(j_U\times\mathrm{id}_V)^*}\mathbb{Z}_{U\times_S V}(q+q'),$$

the external products $\boxtimes$ give the isomorphism

$$\mathbb{Z}_{X,\hat{X}}(q)\otimes\mathbb{Z}_{Y,\hat{Y}}(q')\xrightarrow{\boxtimes}\mathbb{Z}_{X,\hat{X}}(q)\times\mathbb{Z}_{Y,\hat{Y}}(q'). \tag{2.2.11.1}$$

By Mayer-Vietoris (2.2.6), the map

$$\mathbb{Z}_{U\times_S Y\cup X\times_S V}(q+q')\xrightarrow{((j_U\times\mathrm{id}_Y)^*,(\mathrm{id}_X\times j_V)^*)} \tag{2.2.11.2}$$
$$\mathrm{cone}\big(\mathbb{Z}_{U\times_S Y}(q+q')\oplus\mathbb{Z}_{X\times_S V}(q+q')$$
$$\xrightarrow{(\mathrm{id}_U\times j_V)^*-(j_U\times\mathrm{id}_V)^*}\mathbb{Z}_{U\times_S V}(q+q')\big)[-1]$$

is an isomorphism in $\mathcal{DM}(\mathcal{V})$. The map (2.2.11.2), together with the identity map on $\mathbb{Z}_{X\times_S Y}(q+q')$, gives the map

$$\theta^{\hat{X},\hat{Y}}_{X,Y}:\mathbb{Z}_{X\times_S Y,\hat{X}\times\hat{Y}}(q+q')\to\mathbb{Z}_{X,\hat{X}}(q)\times\mathbb{Z}_{Y,\hat{Y}}(q');$$

θ is therefore an isomorphism in $\mathcal{DM}(\mathcal{V})$ as well. Composing θ^{-1} with the external product (2.2.11.1) gives us the isomorphism

$$\boxtimes^{\hat{X},\hat{Y}}_{X,Y}:\mathbb{Z}_{X,\hat{X}}(q)\otimes\mathbb{Z}_{Y,\hat{Y}}(q')\to\mathbb{Z}_{X\times_S Y,\hat{X}\times\hat{Y}}$$

in $\mathcal{DM}(\mathcal{V})$.

As above, this gives us the external cup products

$$\cup^{\hat{X},\hat{Y}}_{X,Y}:H^p_{\hat{X}}(X,\mathbb{Z}(q))\otimes_{\mathbb{Z}}H^{p'}_{\hat{Y}}(Y,\mathbb{Z}(q'))\to H^{p+p'}_{\hat{X}\times_S\hat{Y}}(X\times_S Y,\mathbb{Z}(q+q')), \tag{2.2.11.3}$$

and, for $X=Y$, the cup product

$$\cup^{\hat{X},\hat{Y}}_X:H^p_{\hat{X}}(X,\mathbb{Z}(q))\otimes_{\mathbb{Z}}H^{p'}_{\hat{X}'}(X,\mathbb{Z}(q'))\to H^{p+p'}_{\hat{X}\cap\hat{X}'}(X,\mathbb{Z}(q+q')).$$

2.2.12. *The Lefschetz motive.* Let $i_0:S\to\mathbb{P}^1_S$ and $i_1:S\to\mathbb{P}^1_S$ be the sections with constant value $(1:0)$, $(1:1)$, respectively, and let L be the image in $\mathcal{DM}$ of the object $\mathrm{cone}(i_1^*:\mathbb{Z}_{\mathbb{P}^1_S}(0)_{(i_1,\mathrm{id})}\to\mathbb{Z}_S)[-1]$ of $\mathbf{C}^b_{\mathrm{mot}}(\mathcal{V})$. By the Gysin isomorphism (applied to the section i_0 to the projection $\mathbb{P}^1_S\to S$), we have the isomorphism in $\mathcal{DM}$

$$\mathbb{Z}_{\mathbb{P}^1_S,(1:0)}(0)_{(i_i,\mathrm{id})}\cong\mathbb{Z}_S(-1)[-2]. \tag{2.2.12.1}$$

Letting $j: \mathbb{A}^1_S \to \mathbb{P}^1_S$ be the inclusion of $\mathbb{A}^1_S$ as the open subscheme $\mathbb{P}^1_S \setminus \{i_0(S)\}$, we have the commutative diagram

$$\begin{array}{ccc} \mathbb{Z}_{\mathbb{P}^1_S}(0)_{(i_1,\mathrm{id})} & \xrightarrow{j^*} & \mathbb{Z}_{\mathbb{A}^1_S}(0)_{(i_1,\mathrm{id})} \\ & {\scriptstyle i_1^*}\searrow \quad \swarrow{\scriptstyle i_1^*} & \\ & \mathbb{Z}_S & \end{array}$$

with the right-hand map i_1^* an isomorphism by the homotopy axiom. This, together with (2.2.12.1), gives the isomorphism $L \cong \mathbb{Z}_S(-1)[-2]$ in $\mathcal{DM}$; as the map i_1 is split by the projection $\mathbb{P}^1_S \to S$, we have the isomorphism $\mathbb{Z}_{\mathbb{P}^1_S} \cong \mathbb{Z}_S \oplus L$.

A similar argument, applied to the inclusion $i_n: \mathbb{P}^{n-1}_S \to \mathbb{P}^n_S$ as the hyperplane $X_n = 0$, gives the isomorphism $\mathbb{Z}_{\mathbb{P}^n_S} \cong \mathbb{Z}_S \oplus \mathbb{Z}_{\mathbb{P}^{n-1}_S}(-1)[-2]$. By induction, this gives the isomorphism in $\mathcal{DM}$

$$\mathbb{Z}_{\mathbb{P}^n_S} \cong \oplus_{i=0}^n L^{\otimes i}.$$

2.3. Motivic pull-back

In this section, we examine the functoriality of the categories $\mathcal{DM}(\mathcal{V})$ in the category $\mathcal{V}$.

2.3.1. If $p: T \to S$ is a map of schemes, we let $p^*: \mathbf{Sch}_S \to \mathbf{Sch}_T$ denote the functor $X \mapsto X \times_S T$.

Let $\mathcal{V}$ be a subcategory of $\mathbf{Sch}_S$ and $\mathcal{W}$ a subcategory of $\mathbf{Sch}_T$ which satisfy the conditions of Definition 2.1.4(i)-(iv), so that $\mathcal{DM}(\mathcal{V})$ and $\mathcal{DM}(\mathcal{W})$ are defined. Suppose that p^* restricts to a functor $p^*: \mathcal{V} \to \mathcal{W}$. We proceed to construct an exact tensor functor

(2.3.1.1) $$\mathcal{DM}(p^*): \mathcal{DM}(\mathcal{V}) \to \mathcal{DM}(\mathcal{W}).$$

2.3.2. We first define the functor of DG tensor categories

(2.3.2.1) $$\mathcal{A}_{\mathrm{mot}}(p^*): \mathcal{A}_{\mathrm{mot}}(\mathcal{V}) \to \mathcal{A}_{\mathrm{mot}}(\mathcal{W}).$$

On objects, (2.3.2.1) is given by

(2.3.2.2) $$\mathcal{A}_{\mathrm{mot}}(p^*)(\mathbb{Z}_X(a)_f) = \mathbb{Z}_{p^*(X)}(a)_{p^*(f)}.$$

On morphisms $h^*: \mathbb{Z}_X(a)_f \to \mathbb{Z}_Y(a)_g$ for a map $h: (Y, g) \to (X, f)$ in $\mathcal{L}(\mathcal{V})$, (2.3.2.1) is given by

(2.3.2.3) $$\mathcal{A}_{\mathrm{mot}}(p^*)(h^*) = p^*(h)^*.$$

If $i: X \to X \coprod Y$ is the inclusion, let $\tilde{p}^*(i): p^*(X) \to p^*(Y) \coprod p^*(Z)$ be the map induced by $p^*(i)$ and the canonical isomorphism $p^*(Y \coprod Z) \cong p^*(Y) \coprod p^*(Z)$, and define

(2.3.2.4) $$\mathcal{A}_{\mathrm{mot}}(p^*)(i_*) = \tilde{p}^*(i)_*.$$

Applying Definition 1.4.1 and Definition 1.4.4, the formulas (2.3.2.2), (2.3.2.3) and (2.3.2.4) extend canonically to define the tensor functor

(2.3.2.5) $$\mathcal{A}_2(p^*): \mathcal{A}_2(\mathcal{V}) \to \mathcal{A}_2(\mathcal{W}).$$

Now, let Z be an element of $\mathcal{Z}^q(X)_f$ for some $(X, f) \in \mathcal{L}(\mathcal{V})$. We have the pull-back homomorphism defined in (Appendix A, Lemma 2.2.3),

$$p^*: \mathcal{Z}^q(X/S) \to \mathcal{Z}^q(X \times_S T/T).$$

2.3.3. LEMMA. *Let Z be in $\mathcal{Z}^q(X)_f$. Then $p^*(Z)$ is in $\mathcal{Z}^q(p^*(X))_{p^*(f)}$, hence sending Z to $p^*(Z)$ defines a homomorphism*

$$p^*: \mathcal{Z}^q(X)_f \to \mathcal{Z}^q(p^*(X))_{p^*(f)}$$

PROOF. Write f as $f: X' \to X$, giving the map $p^*(f) = f \times \mathrm{id}_T : X' \times_S T \to X \times_S T$ in $\mathcal{W}$. By (Appendix A, Theorem 2.3.1(iv)), we have

$$p^*(f)^*(p^*(Z)) = p^*(f^*(Z)),$$

and $p^*(f)^*(p^*(Z))$ is in $\mathcal{Z}^q(p^*(X')/T)$. Thus $p^*(Z)$ is in $\mathcal{Z}^q(p^*(X))_{p^*(f)}$, completing the proof. □

2.3.4. By (Appendix A, Theorem 2.3.1), the map p^* on cycles is compatible with pull-back by maps in $\mathcal{L}(\mathcal{V})$, and we have the functoriality

$$(p \circ q)^* = q^* \circ p^* \tag{2.3.4.1}$$

for a sequence of maps $R \xrightarrow{q} T \xrightarrow{p} S$ of reduced noetherian schemes.

Set $\mathcal{A}_3(p^*)([Z]) := [p^*(Z)]$, where $[Z]: \mathfrak{e} \to \mathbb{Z}_X(q)_f[2q]$ is the map associated to $Z \in \mathcal{Z}^q(X)_f$ (see Definition 1.4.6). This defines the extension of (2.3.2.5) to the graded tensor functor

$$\mathcal{A}_3(p^*): \mathcal{A}_3(\mathcal{V}) \to \mathcal{A}_3(\mathcal{W}). \tag{2.3.4.2}$$

For the maps h_* defined in Definition 1.4.8, we define

$$\mathcal{A}_4(p^*)(h_{a,b,\ldots}) = h_{\mathcal{A}_3(p^*)(a),\mathcal{A}_3(p^*)(b),\ldots}. \tag{2.3.4.3}$$

This gives the extension of (2.3.4.2) to the DG tensor functor

$$\mathcal{A}_4(p^*): \mathcal{A}_4(\mathcal{V}) \to \mathcal{A}_4(\mathcal{W}). \tag{2.3.4.4}$$

Similarly, for the maps h_f adjoined in Definition 1.4.9, we inductively define

$$\mathcal{A}_5(p^*)^{(r,k)}(h_f) = h_{\mathcal{A}_5(p^*)^{(r,k-1)}(f)}. \tag{2.3.4.5}$$

This gives the extension of (2.3.4.4) to the DG tensor functor

$$\mathcal{A}_5(p^*): \mathcal{A}_5(\mathcal{V}) \to \mathcal{A}_5(\mathcal{W}); \tag{2.3.4.6}$$

restricting (2.3.4.6) to the full subcategory $\mathcal{A}_{\mathrm{mot}}(\mathcal{V})$ gives the desired DG tensor functor

$$\mathcal{A}_{\mathrm{mot}}(p^*): \mathcal{A}_{\mathrm{mot}}(\mathcal{V}) \to \mathcal{A}_{\mathrm{mot}}(\mathcal{W}). \tag{2.3.4.7}$$

2.3.5. Applying the functor $\mathbf{K}^b$ to (2.3.4.7) gives rise to the exact tensor functor $\mathbf{K}^b_{\mathrm{mot}}(p^*): \mathbf{K}^b_{\mathrm{mot}}(\mathcal{V}) \to \mathbf{K}^b_{\mathrm{mot}}(\mathcal{W})$; passing to the respective localizations gives the exact tensor functor $\mathbf{D}^b_{\mathrm{mot}}(p^*): \mathbf{D}^b_{\mathrm{mot}}(\mathcal{V}) \to \mathbf{D}^b_{\mathrm{mot}}(\mathcal{W})$. Finally, taking the pseudo-abelian hull gives the exact tensor functor

$$\mathcal{DM}(p^*): \mathcal{DM}(\mathcal{V}) \to \mathcal{DM}(\mathcal{W})$$

as desired.

2.3.6. THEOREM. *Suppose we have a sequence of morphisms of reduced noetherian schemes* $R \xrightarrow{q} T \xrightarrow{p} S$ *and subcategories* $\mathcal{V}_R$ *of* $\mathbf{Sm}_R^{\text{ess}}$, $\mathcal{V}_T$ *of* $\mathbf{Sm}_T^{\text{ess}}$, *and* $\mathcal{V}_S$ *of* $\mathbf{Sm}_S^{\text{ess}}$, *such that the functors* q^* *and* p^* *restrict to functors* $p^*: \mathcal{V}_S \to \mathcal{V}_T$, $q^*: \mathcal{V}_T \to \mathcal{V}_R$. *Suppose in addition that the categories* $\mathcal{DM}(\mathcal{V}_S)$, $\mathcal{DM}(\mathcal{V}_T)$ *and* $\mathcal{DM}(\mathcal{V}_R)$ *are defined. Then there is a canonical natural isomorphism*

$$\theta_{p,q}: \mathcal{DM}((p \circ q)^*) \to \mathcal{DM}(q^*) \circ \mathcal{DM}(p^*)$$

satisfying the associativity identity of a pseudo-functor.

PROOF. As is well known, the canonical isomorphism

$$\theta_{p,q}(X): (p \circ q)^*(X) \to q^*(p^*(X)); \quad X \in \mathbf{Sch}_S,$$

makes the operation of pull-back into a pseudo-functor. The same identity thus holds for pull-back in the categories $\mathcal{L}(-)$. This then implies that sending p to the tensor functor $\mathcal{A}_2(p^*)$ (2.3.2.5) defines a pseudo-functor to tensor categories. Using the functoriality (2.3.4.1), we see that (2.3.4.2) defines a pseudo-functor to DG tensor categories; the identities (2.3.4.3) and (2.3.4.5) defining the extension of $\mathcal{A}_3(p^*)$ to $\mathcal{A}_4(p^*)$ and $\mathcal{A}_5(p^*)$ likewise imply that (2.3.4.7) defines a pseudo-functor to DG tensor categories. As the functor $\mathcal{DM}(p^*)$ is gotten from (2.3.4.7) by applying natural operations, sending $\mathcal{V}$ to $\mathcal{DM}(\mathcal{V})$ defines a pseudo-functor to triangulated tensor categories, as desired. □

For a reduced noetherian scheme S, we may take $\mathcal{V}$ equal to the category $\mathbf{Sm}_S$; recall that we have defined $\mathcal{DM}(S) := \mathcal{DM}(\mathbf{Sm}_S)$.

2.3.7. THEOREM. *Sending* S *to* $\mathcal{DM}(S_{\text{red}})$ *and* $p: T \to S$ *to* $\mathcal{DM}(p^*_{\text{red}})$ *defines the pseudo-functor*

$$\mathcal{DM}(-): \mathbf{Sch}^{\text{op}} \to \mathbf{TT},$$

where **Sch** *is the category of noetherian schemes, and* **TT** *is the category of triangulated tensor categories.*

2.4. Motives of cosimplicial schemes

In this, and the subsequent three remaining subsections of this section, we describe how to form objects of $\mathcal{DM}(\mathcal{V})$ associated to various functors to $\mathcal{V}$, e.g., cosimplicial objects of $\mathcal{V}$, simplicial objects of $\mathcal{V}$, etc. We include this material here as a reference to be used throughout the text; as such, we suggest skipping over these subsections on the first reading, referring back to them as needed.

We apply the constructions of (Part II, Chapter III, Lemma 1.1.5 and §1.1.1-§1.1.4) to certain (co)simplicial objects in $\mathcal{V}$; we refer the reader to (Part II, Chapter III, *loc. cit.*) for the notations used in this and the next few subsections.

We recall the category Δ with objects the ordered sets $[n] := \{0 < \ldots < n\}$, and maps the order-preserving maps of sets. For a category $\mathcal{C}$, we have the category c.s.$\mathcal{C}$ of cosimplicial objects of $\mathcal{C}$, i.e., functors $F: \Delta \to \mathcal{C}$, and the category $\mathfrak{s}\mathcal{C}$ of simplicial objects of $\mathcal{C}$, i.e., functors $F: \Delta^{\text{op}} \to \mathcal{C}$.

We have the full subcategory $\Delta^{\leq n}$ of Δ, with objects $[k]$, $0 \leq k \leq n$. The category of functors $\Delta^{\leq n} \to \mathcal{C}$ is the category c.s.$^{\leq n}\mathcal{C}$ of *n-truncated* cosimplicial objects of $\mathcal{C}$; the category s.$^{\leq n}\mathcal{C}$ of n-truncated simplicial objects of $\mathcal{C}$ is defined similarly.

We let $\mathbb{Z}\mathcal{C}$ denote the additive category generated by $\mathcal{C}$, i.e., objects are formal, finite direct sums of objects of $\mathcal{C}$, and $\text{Hom}_{\mathbb{Z}\mathcal{C}}(X, Y) := \mathbb{Z}[\text{Hom}_{\mathcal{C}}(X, Y)]$.

2.4.1. *Very smooth cosimplicial schemes.* Let $X^*: \Delta \to \mathcal{V}$ be a cosimplicial object in $\mathcal{V}$. We call X^* *very smooth* if the maps $X(\sigma_i^m): X^m \to X^{m-1}$ are all flat (where σ_i^m are the codegeneracy maps *cf.* Part II, (III.1.1.1.2)). We now describe a lifting of X^* to a cosimplicial object

$$(X^*, f^*_{X^*}): \Delta \to \mathcal{L}(\mathcal{V}). \tag{2.4.1.1}$$

For each $n \geq 0$, let $X^{\leq n}$ be the disjoint union

$$X^{\leq n} = \coprod_{\substack{g:[m]\to[n] \\ g \text{ injective, order-preserving}}} X^m,$$

and let $f^n_{X^*}: X^{\leq n} \to X^n$ be the map which is $X(g): X^m \to X^n$ on the component indexed by g. This determines the object $(X^n, f^n_{X^*})$ of $\mathcal{L}(\mathcal{V})$.

For a morphism $p: [m] \to [n]$ in Δ, we have the unique factorization of p in Δ as

$$[m] \xrightarrow{p_{\text{surj}}} [m'] \xrightarrow{p_{\text{inj}}} [n],$$

with p_{surj} surjective and p_{inj} injective. Now let $h: [n'] \to [n]$ be a map in Δ, and let $g: [m'] \to [n']$ be an injective map in Δ. We have the factorization of $(g \circ h)$ as

$$[m] \xrightarrow{(g\circ h)_{\text{surj}}} [m_{g,h}] \xrightarrow{(g\circ h)_{\text{inj}}} [n]; \tag{2.4.1.2}$$

as each surjective map in Δ is a composition of the maps σ_i^j, the morphism

$$X((g \circ h)_{\text{surj}}): X^{m'} \to X^{m_{g,h}}$$

is a flat morphism. Let $i_{g,h}: X^{m_{g,h}} \to X^{\leq n}$ be the inclusion as the component indexed by the map $(g \circ h)_{\text{inj}}$. Let $q(h): X^{\leq n'} \to X^{\leq n}$ be the map which on the component $X^{m'}$ indexed by $g: [m'] \to [n']$ is the composition $i_{g,h} \circ X((g \circ f)_{\text{surj}})$; $q(h)$ is thus a flat morphism in $\mathcal{V}$.

The factorization (2.4.1.2) gives us the commutative diagram

$$\begin{array}{ccc} X^{\leq n'} & \xrightarrow{q(h)} & X^{\leq n} \\ {\scriptstyle f^{n'}_{X^*}}\downarrow & & \downarrow{\scriptstyle f^n_{X^*}} \\ X^{n'} & \xrightarrow[X(h)]{} & X^n; \end{array}$$

as $q(h)$ is flat, we have the morphism

$$X(h): (X^{n'}, f^{n'}_{X^*}) \to (X^n, f^n_{X^*})$$

in $\mathcal{L}(\mathcal{V})$. Thus, sending n to $(X^n, f^n_{X^*})$, h to $X(h)$, defines the desired functor (2.4.1.1).

2.4.2. *Motives associated to cosimplicial schemes.* We have the functor

$$\begin{aligned} &\mathbb{Z}(q): \mathcal{L}(\mathcal{V})^{\text{op}} \to \mathcal{A}_{\text{mot}}(\mathcal{V}) \\ &\mathbb{Z}(q)((X, f)) = \mathbb{Z}_X(0)_f, \end{aligned} \tag{2.4.2.1}$$

which extends to the functors

$$\mathbf{C}^b(\mathbb{Z}(q)): \mathbf{C}^b(\mathbb{Z}\mathcal{L}(\mathcal{V})^{\text{op}}) \to \mathbf{C}^b_{\text{mot}}(\mathcal{V})^*$$
$$\mathbf{K}^b(\mathbb{Z}(q)): \mathbf{K}^b(\mathbb{Z}\mathcal{L}(\mathcal{V})^{\text{op}}) \to \mathbf{K}^b_{\text{mot}}(\mathcal{V})^*.$$

We have the inclusion functors

$$j_N : \Delta^{\le N} \to \Delta$$
$$j_{N',N} : \Delta^{\le N} \to \Delta^{\le N'};\quad N < N'.$$

If $X^* : \Delta \to \mathcal{V}$ is a very smooth cosimplicial object, we have the object

$$\mathbb{Z}_N^{\oplus}(j_N^*(X^*, f_{X^*}^*))$$

of $\mathbf{C}^b(\mathcal{L}(\mathbb{Z}\mathcal{V})^{\mathrm{op}})$ (*cf.* (2.4.1.1) and Part II, §1.1.3 and (III.1.1.1.3)); we define $\mathbb{Z}_{X^*}^{\le N}(0)$ in $\mathbf{C}^b_{\mathrm{mot}}(\mathcal{V})^*$ by

$$\mathbb{Z}_{X^*}^{\le N}(0) := \mathbf{C}^b(\mathbb{Z}(0))(\mathbb{Z}_N^{\oplus}(j_N^*(X^*, f_{X^*}^*))). \tag{2.4.2.2}$$

Explicitly, $\mathbb{Z}_{X^*}^{\le N}(0)$ in degree $-m$ is the direct sum

$$[\mathbb{Z}_{X^*}^{\le N}(0)]^{-m} = \bigoplus_{f:[m]\to[N]} \mathbb{Z}_{X^m}(0)_{f^m},$$

where $f:[m] \to [N]$ runs over injective maps in Δ.

Sending X^* to $\mathbb{Z}_{X^*}^{\le N}(0)$ determines the functor

$$\mathbb{Z}^{\le N}(0) : \mathrm{c.s.}\mathcal{V}_{\text{very smooth}} \to \mathbf{D}^b_{\mathrm{mot}}(\mathcal{V})^*$$

from the category of very smooth cosimplicial schemes in $\mathcal{V}$ to $\mathbf{D}^b_{\mathrm{mot}}(\mathcal{V})^*$.

The natural map

$$\chi^{N,n} : \mathbb{Z}_n^{\oplus}(j_n^*(X^*, f_{X^*}^*)) \to \mathbb{Z}_N^{\oplus}(j_N^*(X^*, f_{X^*}^*))$$

induced by the map $f:[n] \to [N]$, $f(i) = i$ (*cf.* Part II, (III.1.1.4.1)) give rise to the natural map in $\mathbf{C}^b_{\mathrm{mot}}(\mathcal{V})^*$

$$\chi^{N,n}(?) : \mathbb{Z}^{\le n}(0)(?) \to \mathbb{Z}^{\le N}(0)(?), \tag{2.4.2.3}$$

and define the natural transformation

$$\chi^{N,n} : \mathbb{Z}^{\le n}(0) \to \mathbb{Z}^{\le N}(0). \tag{2.4.2.4}$$

2.4.3. REMARK. Suppose we have a cosimplicial object $X^* : \Delta \to \mathcal{V}$, not necessarily very smooth. One can modify the construction of §2.4.1 and §2.4.2 to define the motive associated to each truncation $X^{*\le N} : \Delta^{\le N} \to \mathcal{V}$ of X^*: one replaces the map $f_{X^*}^n : X^{\le n} \to X^n$ with the map $f_{X^*}^{n\le N} : X^{\le n \le N} \to X^n$, where

$$X^{\le n\le N} := \coprod_{\substack{g:[k]\to[n]\\ k\le N}} X^k.$$

As we won't be using this construction, we omit the details.

2.5. Motives of simplicial schemes

We describe objects of $\mathcal{DM}(\mathcal{V})$ associated to simplicial objects of $\mathcal{L}(\mathcal{V})$ and $\mathcal{V}$ for later use. We refer the reader to (Part II, Chapter III, §1.1.1-1.1.4) for the various notations.

2.5.1. If we have a functor $(X, f): \Delta^{\leq n \mathrm{op}} \to \mathcal{L}(\mathcal{V})$, we may compose $(X, f)^{\mathrm{op}}$ with the functor (2.4.2.1), forming the functor

$$\mathbb{Z}(q) \circ (X, f): \Delta^{\leq n} \to \mathcal{A}_{\mathrm{mot}}(\mathcal{V}). \tag{2.5.1.1}$$

We let $\mathbb{Z}_X(q)_f \in \mathbf{C}^b_{\mathrm{mot}}(\mathcal{V})$ be the object of the category of complexes $\mathbf{C}^b_{\mathrm{mot}}(\mathcal{V})$ associated to the functor (2.5.1.1), i.e., $\mathbb{Z}_X(q)^s_f := \mathbb{Z}_{X([s])}(q)_{f([s])}$, and d^s is the alternating sum of the maps $X(\delta^s_i)^*$. For $0 \leq m \leq n$, we let $\mathbb{Z}_X(q)^{m\leq *}_f$ be the truncation of $\mathbb{Z}_X(q)_f$ to degree $\geq m$. We sometimes denote $\mathbb{Z}_X(q)^{m\leq *}_f$ as $\mathbb{Z}_X(q)^{m\leq * \leq n}_f$ if we want to refer to n explicitly.

Sending (X, f) to $\mathbb{Z}_X(q)_f$ or $\mathbb{Z}_X(q)^{m\leq *}_f$ defines the functors

$$\begin{aligned} &\mathbb{Z}_{(-)}(q)_{(-)}: \mathrm{s.}^{\leq n}\mathcal{L}(\mathcal{V}) \to \mathbf{C}^b_{\mathrm{mot}}(\mathcal{V}) \\ &\mathbb{Z}_{(-)}(q)^{m\leq *}_{(-)}: \mathrm{s.}^{\leq n}\mathcal{L}(\mathcal{V}) \to \mathbf{C}^b_{\mathrm{mot}}(\mathcal{V}); \end{aligned} \tag{2.5.1.2}$$

we may also consider $\mathbb{Z}_{(-)}(q)_{(-)}$ and $\mathbb{Z}_{(-)}(q)^{m\leq *}_{(-)}$ as functors with values in $\mathbf{K}^b_{\mathrm{mot}}(\mathcal{V})$ or $\mathbf{D}^b_{\mathrm{mot}}(\mathcal{V})$ as the need arises. Clearly the functors (2.5.1.2) factor through the functor

$$\begin{aligned} \mathrm{s.}^{\leq n}\mathcal{L}(\mathcal{V}) = \mathrm{c.s.}^{\leq n}\mathcal{L}(\mathcal{V})^{\mathrm{op}} &\to \mathbf{C}^b(\mathbb{Z}\mathcal{L}(\mathcal{V})^{\mathrm{op}}) \\ X &\mapsto X^*, \end{aligned}$$

where X^* is the complex associated to X.

2.5.2. *Lifting simplicial objects to* $\mathcal{L}(\mathcal{V})$. Let X be a truncated simplicial object in $\mathcal{V}$: $X: \Delta^{\mathrm{op}\leq n} \to \mathcal{V}$. For each $m \leq n$, we let $f_{m,n}$ be the map

$$f_{m,n} = \coprod_{\substack{f:[m]\to[k]\\ 0\leq k\leq n}} X(f) \quad \coprod_{\substack{f:[m]\to[k]\\ 0\leq k\leq n}} X_k \to X_m.$$

As in §2.4.1, for each $g: [k] \to [m]$ in Δ, the map $X(g): X_m \to X_k$ lifts to the map $X(g): (X_m, f_{m,n}) \to (X_k, f_{k,n})$ in $\mathcal{L}(\mathcal{V})$. We let

$$(X, f_X): \Delta^{\leq n \mathrm{op}} \to \mathcal{L}(\mathcal{V}) \tag{2.5.2.1}$$

be the functor lifting X with $(X, f_X)_m = (X_m, f_{m,n})$.

2.5.3. *Motives.* We have the composition

$$\mathbb{Z}(q) \circ (X, f_X): \Delta^{\leq n} \to \mathcal{A}_{\mathrm{mot}}(\mathcal{V}) \tag{2.5.3.1}$$

of (2.5.2.1) with the functor (2.4.2.1); we let $\mathbb{Z}_X(q)^{m\leq *}$ be the truncated complex in $\mathbf{C}^b_{\mathrm{mot}}(\mathcal{V})$ associated to (2.5.3.1), as in §2.5.1.

Sending X to $\mathbb{Z}_X(q)^{m\leq *}$ defines a functor

$$\mathbb{Z}_{(-)}(q)^{m\leq *}: (\mathrm{s.}^{\leq n}\mathcal{V})^{\mathrm{op}} \to \mathbf{D}^b_{\mathrm{mot}}(\mathcal{V}). \tag{2.5.3.2}$$

Indeed, given a map $p: Y \to X$ in $\mathrm{s.}^{\leq n}\mathcal{V}$, let $g_m: X'_m \to X_m$ be the map

$$g_m = \coprod_{\substack{f:[m]\to[k]\\ 0\leq k\leq n}} X(f) \cup p_m \circ Y(f) \quad \coprod_{\substack{f:[m]\to[k]\\ 0\leq k\leq n}} X_k \coprod Y_k \to X_m.$$

This then defines the lifting of X to an object (X, g) of $\mathrm{s.}^{\leq n}\mathcal{L}(\mathcal{V})$, so that the map p lifts to $p: (Y, f_Y) \to (X, g)$ and the identity on X lifts to $i: (X, f_X) \to (X, g)$.

This gives the diagram in $\mathbf{C}^b_{\mathrm{mot}}(\mathcal{V})$

$$\begin{array}{ccc} \mathbb{Z}_X(q)^{m\leq *}_g & \xrightarrow{p^*} & \mathbb{Z}_Y(q)^{m\leq *} \\ {\scriptstyle i^*}\downarrow & & \\ \mathbb{Z}_X(q)^{m\leq *} & & \end{array}$$

with i^* an isomorphism in $\mathbf{D}^b_{\mathrm{mot}}(\mathcal{V})$ by Definition 2.1.4(e). As in §2.2.9, this defines the functor (2.5.3.2). Taking $m = 0$ gives the functor

$$\mathbb{Z}_{(-)}(q): (\mathrm{s.}^{\leq n}\mathcal{V})^{\mathrm{op}} \to \mathbf{D}^b_{\mathrm{mot}}(\mathcal{V})$$
$$X \mapsto \mathbb{Z}_X(q) := \mathbb{Z}_X(q)^{0\leq *}.$$

For $n' \geq n$, and $X: \Delta^{\leq n'\mathrm{op}} \to \mathcal{V}$, we have the canonical map

$$\rho_{n',n}: (X, f_X) \circ j_{n',n} \to (X \circ j_{n',n}, f_{X\circ j_{n',n}});$$

this defines the natural map

$$\rho^*_{n',n}: \mathbb{Z}_X(q)^{m\leq *\leq n'} \to \mathbb{Z}_{X\circ j_{n',n}}(q)^{m\leq *\leq n}. \tag{2.5.3.3}$$

For example, this gives us the map

$$\pi_m := \rho_{n,m}[m]: \mathbb{Z}_X(q)^{m\leq *\leq n}[m] \to \mathbb{Z}_X(q)^{m\leq *\leq m}[m] = \mathbb{Z}_X(q)^m_{f_m}. \tag{2.5.3.4}$$

2.5.4. *Motives of non-degenerate simplicial schemes.* We have the subcategory $\Delta_{\mathrm{n.d.}}$ of Δ, with the same objects, but where we only allow *injective* order-preserving maps. We call a functor $X_*: \Delta^{\mathrm{op}}_{\mathrm{n.d.}} \to \mathcal{C}$ (resp. $X^*: \Delta_{\mathrm{n.d.}} \to \mathcal{C}$) a *non-degenerate simplicial object (resp. non-degenerate cosimplicial object) of* $\mathcal{C}$.

Let (X_*, f_*) be a non-degenerate simplicial object of $\mathcal{L}(\mathcal{V})$, with $(X_*, f_*)_m = (X_m, f_m)$ for $m = 0, 1, \ldots$. This gives us the non-degenerate cosimplicial object $\mathbb{Z}_{X_*}(q)_{f_*}$ of $\mathcal{A}_{\mathrm{mot}}(\mathcal{V})^*$ with $\mathbb{Z}_{X_*}(q)^m_{f_*} = \mathbb{Z}_{X_m}(q)_{f_m}$. For each $N \geq 0$, we may then form the truncated complex $\mathbb{Z}_{X_*}(q)^{*\leq N}_{f_*}$ in $\mathbf{C}^b_{\mathrm{mot}}(\mathcal{V})^*$, which is $\mathbb{Z}_{X_*}(q)^m_{f_m}$ in degree m, and has the usual alternating sum as coboundary.

2.5.5. *Motivic cohomology of simplicial schemes.* For a truncated simplicial object of $\mathcal{V}$, $X: \Delta^{\leq n\mathrm{op}} \to \mathcal{V}$, we have the motive $\mathbb{Z}_X(q)$; we define the *motivic cohomology* of X by

$$H^p(X, \mathbb{Z}(q)) := \mathrm{Hom}_{\mathcal{DM}(\mathcal{V})}(1, \mathbb{Z}_X(q)[p]).$$

Let $(\mathbb{N}, \leq)$ denote the category with set of objects $\mathbb{N}$ and a unique morphism $n \to n'$ for each $n \leq n'$. For a simplicial object $X: \Delta^{\mathrm{op}} \to \mathcal{V}$ of $\mathcal{V}$, the maps (2.5.3.3) give the functor $\mathbb{Z}_X(q): (\mathbb{N}, \leq)^{\mathrm{op}} \to \mathbf{C}^b_{\mathrm{mot}}(\mathcal{V})$ with

$$\mathbb{Z}_X(q)(n) := \mathbb{Z}_{X\circ j_n}(q)$$
$$\mathbb{Z}_X(q)(n \leq n') := \rho^*_{n',n}: \mathbb{Z}_X(q)(n') \to \mathbb{Z}_X(q)(n).$$

We define the *motivic cohomology* of X by

$$\begin{aligned} H^p(X, \mathbb{Z}(q)) &:= \varprojlim_{(\mathbb{N},\leq)^{\mathrm{op}}} [n \mapsto \mathrm{Hom}_{\mathcal{DM}(\mathcal{V})}(1, \mathbb{Z}_X(q)(n)[p])] \\ &= \varprojlim_{(\mathbb{N},\leq)^{\mathrm{op}}} [n \mapsto H^p(X \circ j_n, \mathbb{Z}(q))]. \end{aligned}$$

Similarly, if we have an n-truncated non-degenerate simplicial object of $\mathcal{L}(\mathcal{V})$, $(X, f): \Delta^{\leq n\mathrm{op}}_{\mathrm{n.d.}} \to \mathcal{L}(\mathcal{V})$, define $H^p((X, f), \mathbb{Z}(q)) := \mathrm{Hom}_{\mathcal{DM}(\mathcal{V})}(1, \mathbb{Z}_X(q)_f[p])$, and if we have a (non-truncated) non-degenerate simplicial object $(X,f)\colon \Delta^{\mathrm{op}}_{\mathrm{n.d.}} \to \mathcal{L}(\mathcal{V})$ of $\mathcal{L}(\mathcal{V})$, define

$$H^p((X,f),\mathbb{Z}(q)) := \varprojlim_{(\mathbb{N},\leq)^{\mathrm{op}}} [n \mapsto H^p((X,f)\circ j_n, \mathbb{Z}(q))].$$

2.5.6. *Products.* The category $\mathcal{V}^{\mathrm{op}}$ is a symmetric monoidal category with operation $\times_S$, and has the commutative multiplication (see Part II, Chapter III, §1.2.2) given by the opposite of the diagonal $\Delta_X\colon X \to X\times_S X$. Similarly, the category $\mathcal{L}(\mathcal{V})^{\mathrm{op}}$ is a symmetric monoidal category with product $\times$, and the projection $\mathcal{L}(\mathcal{V})^{\mathrm{op}} \to \mathcal{V}^{\mathrm{op}}$ is a symmetric monoidal functor. By the results of (Part II, Chapter III, §1.2.1) and the external products given in (Part II, (III.1.2.1.4)), we have the natural products in $\mathbf{C}^b_{\mathrm{mot}}(\mathcal{V})$:

$$\text{(2.5.6.1)}\qquad \mathbb{Z}(*)(\cup_{(X,f),(Y,g)})\colon \mathbb{Z}_X(q)_f^{m\leq *\leq n} \otimes \mathbb{Z}_Y(q')_g \to \mathbb{Z}_{X\times_S Y}(q+q')_{f\times g}^{m\leq *\leq n},$$

for (X,f) and (Y,g) in $\mathrm{s.}^{\leq n}\mathcal{L}(\mathcal{V})$. These products are associative and graded-commutative. Taking $(X,f) = (X,f_X)$, and $(Y,g) = (Y,f_Y)$ gives the natural associative, graded-commutative products

$$\text{(2.5.6.2)}\qquad \mathbb{Z}(*)(\cup_{X,Y})\colon \mathbb{Z}_X(q)^{m\leq *\leq n} \otimes \mathbb{Z}_Y(q') \to \mathbb{Z}_{X\times_S Y}(q+q')^{m\leq *\leq n}.$$

Applying the functors $\mathbb{Z}(*)$ to the cup product map of (Part II, (III.1.2.3.2)) produces the associative multiplications

$$\text{(2.5.6.3)}\qquad \mathbb{Z}(*)(m^n(X^{m\leq *\leq n}))\colon \mathbb{Z}_X(q)^{m\leq *\leq n} \otimes \mathbb{Z}_X(q') \to \mathbb{Z}_X(q+q')^{m\leq *\leq n}$$

in $\mathbf{D}^b_{\mathrm{mot}}(\mathcal{V})$; if $m=0$, this multiplication is (graded) commutative.

The products (2.5.6.3) give $H^*(X,\mathbb{Z}(*)) := \oplus_{p,q} H^p(X,\mathbb{Z}(q))$ the structure of a bi-graded ring (without unit), graded-commutative in p. For $m\leq n$, the products (2.5.6.3) make the bi-graded $\mathbb{Z}$-module $\oplus_{p,q} H^p(\mathbb{Z}_X(q)^{m\leq *\leq n})$ a bi-graded module over $H^*(X,\mathbb{Z}(*))$. Additionally, the various maps defined by changing n or m are ring homomorphisms, or module homomorphisms, as appropriate; this follows from the commutativity of the diagrams (Part II, (III.1.2.3.3)-(III.1.2.3.5)). We often write the maps (2.5.6.2) and (2.5.6.3) as $\cup_{X,Y}$ and $\cup_X$, respectively.

Let $j\colon V \to Y$ be an open simplicial subscheme of Y, and let $\hat{Y}_m := Y_m \setminus V_m$. Define the motive with support $\mathbb{Z}_{Y,\hat{Y}}(q)_g$, as in (2.1.3.1), to be the shifted cone of the map j^*:

$$\mathbb{Z}_{Y,\hat{Y}}(q)_g := \mathrm{cone}(j^*\colon \mathbb{Z}_Y(q)_g \to \mathbb{Z}_V(q)_{j^*g})[-1].$$

As the maps (2.5.6.1) are natural in (Y,g), they induce the map

$$\text{(2.5.6.4)}\qquad \cup_{X,Y} := \mathbb{Z}(*)(\cup_{X,Y})\colon \mathbb{Z}_X(q)\otimes \mathbb{Z}_{Y,\hat{Y}}(q') \to \mathbb{Z}_{X\times_S Y, X\times_S \hat{Y}}(q+q').$$

The products for $H^*(X,\mathbb{Z}(*))$ and the external products (2.5.6.4) extend to the case of (non-truncated) simplicial schemes by taking the projective limit.

2.6. Cubes and relative motives

We give a discussion of n-cubes in a category, and the construction of relative motives and relative motivic cohomology

2.6.1. *n-cubes.* Let $<n>$ be opposite of the category of subsets of the finite set $\{1, \dots, n\}$, i.e., an object of $<n>$ is a subset I of $\{1, \dots, n\}$, and there is a morphism $J \to I$ if and only if $J \supset I$. The category $<n>$ is called the *n-cube.* For a category $\mathcal{C}$ we have the category $\mathcal{C}^{<n>}$, the category of *n-cubes in* $\mathcal{C}$, being the category of functors $X\colon <n> \to \mathcal{C}$.

2.6.2. *Lifting n-cubes to $\mathcal{L}(\mathcal{V})$.* Let

$$\begin{aligned} X_*\colon <n> &\to \mathcal{V}, \\ I &\mapsto X_I, \end{aligned}$$

be a functor and let $(X_\emptyset, f_\emptyset\colon X' \to X_\emptyset)$ be a lifting of $X_\emptyset$ to an object of $\mathcal{L}(\mathcal{V})$. For each $I \subset \{1, \dots, n\}$, form the cartesian diagram

$$\begin{array}{ccc} X'_I := X' \times_{X_\emptyset} X_I & \xrightarrow{p_1} & X' \\ {\scriptstyle f_I := p_2}\downarrow & & \downarrow{\scriptstyle f_\emptyset} \\ X_I & \xrightarrow[X_{I\supset\emptyset}]{} & X_\emptyset. \end{array}$$

The maps $X_{J\supset I}$ induce the maps $X'_{J\supset I}\colon X'_J \to X'_I$ defining the n-cube

$$X'_*\colon <n> \to \mathbf{Sch}_S;$$

the maps f_I give the map of n-cubes $f_*\colon X'_* \to X_*$.

Supposing that the X'_I are in $\mathbf{Sm}_S^{\mathrm{ess}}$ for all I, we define the lifting of X_* to a functor

$$(X_*, f_*^X)\colon <n> \to \mathcal{L}(\mathcal{V}) \tag{2.6.2.1}$$

by setting

$$\begin{aligned} X'_I &:= \coprod_{J\supset I} X'_J, \\ f_I^X &:= \cup_{J\supset I} X_{J\supset I} \circ f_J\colon X'_I \to X_I \end{aligned}$$

(compare with (2.4.1.1)).

2.6.3. *Motives of n-cubes.* We apply the functor $\mathbb{Z}(0)\colon \mathcal{L}(\mathcal{V})^{\mathrm{op}} \to \mathcal{A}_{\mathrm{mot}}(\mathcal{V})$ to the functor (2.6.2.1). We then form the complex with $\oplus_{|I|=s}\mathbb{Z}_{X_I}(0)_{f_I^X}$ in degree s, and differential

$$\partial^s\colon \bigoplus_{|I|=s} \mathbb{Z}_{X_I}(0)_{g_I^X} \to \bigoplus_{|I|=s+1} \mathbb{Z}_{X_I}(0)_{f_I^X}$$

given by setting

$$\begin{aligned} \partial^s_{I,i}\colon \mathbb{Z}_{X_I}(0)_{f_I^X} &\to \mathbb{Z}_{X_{I\cup\{i\}}}(0)_{f^X_{I\cup\{i\}}} \\ \partial^s_{I,i} &= \begin{cases} X^*_{I\cup\{i\}\supset I} & \text{for } i \notin I \\ 0 & \text{for } i \in I \end{cases} \\ \partial^s &= \sum_{|I|=s} \sum_{i=1}^n (-1)^i \partial^s_{I,i}. \end{aligned}$$

We denote the resulting object of $\mathbf{C}^b_{\mathrm{mot}}(\mathcal{V})$ by $\mathbb{Z}_{X_*}(0)_{f_\emptyset}$.

If we take $f_\emptyset = \mathrm{id}_{X_\emptyset}$, then we have the lifting (X_*, f_*^X) of X_* and the object $\mathbb{Z}_{X_*}(0) := \mathbb{Z}_{X_*}(0)_{\mathrm{id}_{X_\emptyset}}$ of $\mathbf{C}^b_{\mathrm{mot}}(\mathcal{V})$.

2.6.4. *n-cubes and cones.* The main utility of the n-cube follows from the elementary remark that the category of n-cubes in a category $\mathcal{C}$ is equivalent to the category of maps of $n-1$-cubes in $\mathcal{C}$, the equivalence being given by associating to a map of $n-1$-cubes, $f_*: X_*^- \to X_*^+$, the n-cube $X(f_*)_*$ with

$$X(f_*)_I = \begin{cases} X_I^- & \text{if } n \notin I, \\ X_{I\setminus\{n\}}^+ & \text{if } n \in I, \end{cases}$$

$$X(f_*)_{J\supset I} = \begin{cases} X_{J\supset I}^- & \text{if } n \notin J, \\ X_{J\setminus\{n\}\supset I\setminus\{n\}}^+ & \text{if } n \in I, \end{cases}$$

$$X(f_*)_{I\cup\{n\}\supset I} = f_I \text{ for } I \subset \{1, \dots, n-1\}$$

(this unique determines $X(f_*)_*$). If we have an n-cube X_* in $\mathcal{L}(\mathcal{V})$, which we may then write as $X_* = X(f_*)_*$ for the uniquely determined map f_* of $n-1$-cubes in $\mathcal{L}(\mathcal{V})$, we have the identity

$$\mathbb{Z}_{X_*}(0) = \operatorname{cone}(\mathbb{Z}_{f_*}(0) : \mathbb{Z}_{X_*^-}(0) \to \mathbb{Z}_{X_*^+}(0))[-1]. \tag{2.6.4.1}$$

Thus, each n-cube in $\mathcal{L}(\mathcal{V})$ gives rise to a sequence of linked distinguished triangles in $\mathbf{K}^b_{\mathrm{mot}}(\mathcal{V})$.

As a simple, but useful, application of the above cone sequence, we have

2.6.5. LEMMA. *(i) Let $g: X_* \to Y_*$ be a map of n-cubes in $\mathcal{L}(\mathcal{V})$ such that $g_I^*: \mathbb{Z}_{Y_I} \to \mathbb{Z}_{X_I}$ is an isomorphism in $\mathbf{D}^b_{\mathrm{mot}}(\mathcal{V})$ for all $I \subset \{1, \dots, n\}$. Then $g^*: \mathbb{Z}_{Y_*} \to \mathbb{Z}_{X_*}$ is an isomorphism in $\mathbf{D}^b_{\mathrm{mot}}(\mathcal{V})$.*
(ii) Let $X_*: \langle n \rangle \to \mathcal{L}(\mathcal{V})$ be an n-cube in $\mathcal{L}(\mathcal{V})$. Suppose that

$$X(I \cup \{n\} \supset I)^* : \mathbb{Z}_{X_I} \to \mathbb{Z}_{X_{I\cup\{n\}}}$$

is an isomorphism in $\mathbf{D}^b_{\mathrm{mot}}(\mathcal{V})$ for all $I \subset \{1, \dots, n-1\}$. Then $\mathbb{Z}_{X_*}$ is isomorphic to 0 in $\mathbf{D}^b_{\mathrm{mot}}(\mathcal{V})$.

PROOF. The second assertion follows from the first, using the distinguished triangle coming from the cone sequence (2.6.4.1). The first assertion follows using the same distinguished triangle and induction on n. □

2.6.6. *Relative motives.* Suppose we have a smooth S-scheme X, with subschemes $D_1, \dots, D_n \subset X$. For each index $I = (1 \le i_1 < \dots < i_s \le n)$, let D_I be the subscheme of X, $D_I := D_{i_1} \cap \dots \cap D_{i_s}$.

Suppose we have a lifting $(X, f: X' \to X)$ of X to $\mathcal{L}(\mathcal{V})$ such that each D_I is in $\mathcal{V}$ and the pull-backs $f_I := p_2: X' \times_X D_I \to D_I$ are in $\mathbf{Sm}_S^{\mathrm{ess}}$. We let

$$(X; D_1, \dots, D_n)_* : \langle n \rangle \to \mathcal{V}$$

be the n-cube in $\mathcal{V}$ with $(X; D_1, \dots, D_n)_I = D_I$; for $J \subset I$, we let

$$(X; D_1, \dots, D_n)_{J\subset I} : D_J \to D_I$$

be the inclusion. Applying the construction described in §2.6.2 gives the lifting of the n-cube $(X; D_1, \dots, D_n)_*$ to the n-cube $((X; D_1, \dots, D_n)_*, f_*^X): \langle n \rangle \to \mathcal{L}(\mathcal{V})$, which in turn gives us the object $\mathbb{Z}_{(X;D_1,\dots,D_n)}(0)_f$ of $\mathbf{C}^b_{\mathrm{mot}}(\mathcal{V})$; the identification (2.6.4.1) of $\mathbb{Z}_{X_*}(0)_f$ as a cone gives us the *relativization distinguished triangle* in

$\mathbf{K}^b_{\mathrm{mot}}(\mathcal{V})$,

$$(2.6.6.1)\quad \mathbb{Z}_{(X;D_1,\dots,D_n)}(0)_f \to \mathbb{Z}_{(X;D_1,\dots,D_{n-1})}(0)_f \to \mathbb{Z}_{(D_n;D_{1,n},\dots,D_{n-1,n})}(0)_{f_n} \xrightarrow{i_n} \mathbb{Z}_{(X;D_1,\dots,D_n)}(0)_f[1].$$

We call the object $\mathbb{Z}_{(X;D_1,\dots,D_n)}(0)$ of $\mathcal{DM}(\mathcal{V})$ the *motive of* X, *relative to* $D_1,\dots,D_n$. We define the *relative motivic cohomology* by

$$H^p(X;D_1,\dots,D_n,\mathbb{Z}(q)) := \mathrm{Hom}_{\mathcal{DM}(\mathcal{V})}(1,\mathbb{Z}_{(X;D_1,\dots,D_n)}(q)[p]).$$

Let $j\colon U\to X$ be the inclusion of an open subscheme, with complement W in X. Writing $D_i^U := D_i\cap U$, the collection of maps $j_I\colon D_I^U\to D_I$ gives the map of objects of $\mathbf{C}^b_{\mathrm{mot}}(\mathcal{V})$

$$j^*\colon \mathbb{Z}_{(X;D_1,\dots,D_n)} \to \mathbb{Z}_{(U;D_1^U,\dots,D_n^U)}.$$

We define the *relative motive with support*, $\mathbb{Z}_{(X;D_1,\dots,D_n),W}$, as the cone

$$\mathbb{Z}_{(X;D_1,\dots,D_n),W} := \mathrm{cone}(j^*\colon \mathbb{Z}_{(X;D_1,\dots,D_n)} \to \mathbb{Z}_{(U;D_1^U,\dots,D_n^U)})[-1].$$

This gives us the *localization distinguished triangle*

$$(2.6.6.2)\quad \mathbb{Z}_{(X;D_1,\dots,D_n),W} \to \mathbb{Z}_{(X;D_1,\dots,D_n)} \to \mathbb{Z}_{(U;D_1^U,\dots,D_n^U)} \to \mathbb{Z}_{(X;D_1,\dots,D_n),W}[1].$$

2.6.7. *Functorialities.* Suppose we have $(X;D_1,\dots,D_n)$ and $(Y;E_1,\dots,E_m)$ satisfying the conditions of §2.6.6, and a map $f\colon X\to Y$ such that

$$f(D_i)\subset E_{\alpha(i)};\quad \alpha(i)\in\{1,\dots,m\};\ i=1,\dots,n.$$

Let $\alpha\colon <n> \to <m>$ be the map on the subsets of $\{1,\dots,n\}$ induced by α. Define $f^*\colon \mathbb{Z}_{(Y;E_1,\dots,E_m)} \to \mathbb{Z}_{(X;D_1,\dots,D_n)}$ by the maps $f^*_{|D_I}\colon \mathbb{Z}_{E_{\alpha(I)}} \to \mathbb{Z}_{D_I}$, together with the zero maps on $\mathbb{Z}_{E_J}$ for J not in the image of α.

One easily shows that two different maps α give homotopic maps of complexes, and that $(f\circ g)^* = g^*\circ f^*$.

2.7. Motives of diagrams

We refer the reader to (Part II, Chapter III, Section 3) for the notions related to homotopy limits.

2.7.1. *Adjoining a disjoint base-point.* For a category $\mathcal{C}$ with an initial object $\emptyset$, we let $\mathcal{C}^+$ be the category gotten from $\mathcal{C}$ by adjoining a final object $*$, and making the canonical morphism $\emptyset\to *$ an isomorphism. Heuristically, we have just adjoined a disjoint base-point to each object of $\mathcal{C}$.

Given a functor $F\colon\mathcal{C}\to\mathcal{A}$ such that $\mathcal{A}$ has an initial object $\emptyset_{\mathcal{A}}$ and final object $*_{\mathcal{A}}$ which are isomorphic, and such that $F(\emptyset)=\emptyset_{\mathcal{A}}$, we extend F to $F\colon\mathcal{C}^+\to\mathcal{A}$ by sending $*$ to $*_{\mathcal{A}}$. In particular, each functor $F\colon\mathcal{C}\to\mathcal{A}$ to an additive category $\mathcal{A}$, with $F(\emptyset)=0$, extends canonically to the functor $F\colon\mathcal{C}^+\to\mathcal{A}$.

If $\mathcal{C}$ has a product $\times$, we extend the operation $\times$ to $\mathcal{C}^+$ by taking the *smash product*

$$X\wedge Y = \begin{cases} X\times Y; & \text{for } X\neq * \text{ and } Y\neq * \\ *; & \text{otherwise.}\end{cases}$$

Similarly, a coproduct $\coprod$ on $\mathcal{C}$ extends to the pointed version $X \vee Y = X \coprod Y$ for $X \neq *$ and $Y \neq *$, and $X \vee * := X$, $* \vee Y := Y$.

2.7.2. *The motive of a diagram.* Let I be a finite category and let $X: I \to \mathcal{V}^+$ be a functor. We write $\mathcal{L}(\mathcal{V}^+)$ for $\mathcal{L}(\mathcal{V})^+$. We lift X to a functor $(X, f_X): I \to \mathcal{L}(\mathcal{V}^+)$ by setting

$$X'(i) := \bigvee_{f: j \to i \in I/i} X(j),$$

and letting $f_X(i): X'(i) \to X(i)$ be the union of the maps $X(f): X(j) \to X(i)$.

Given a map $s: i \to i'$, the map $X'(s): X'(i) \to X'(i')$ is defined to be the union of the identity maps $X(j) \to X(j)$, where we send the component $f: j \to i$ to the component $s \circ f: j \to i'$. Thus, (X, f_X) is indeed a lifting.

Composing with the functor $\mathbb{Z}(q): \mathcal{L}(\mathcal{V}^+)^{\mathrm{op}} \to \mathcal{A}_{\mathrm{mot}}(\mathcal{V})$, we have the functor $\mathbb{Z}(q) \circ (X, f_X): I^{\mathrm{op}} \to \mathcal{A}_{\mathrm{mot}}(\mathcal{V})$. We let $\mathbb{Z}_X(q)$ be the non-degenerate homotopy limit (see Part II, Chapter III, §3.2.2 and §3.2.7)

$$\mathbb{Z}_X(q) := \operatorname*{holim}_{I^{\mathrm{op}},\ \mathrm{n.d.}} \mathbb{Z}(q) \circ (X, f_X) \in \mathbf{C}^b_{\mathrm{mot}}(\mathcal{V}).$$

More generally, if $(X, f): I \to \mathcal{L}(\mathcal{V}^+)$ is a functor, we have the object

$$\mathbb{Z}_X(q)_f := \operatorname*{holim}_{I^{\mathrm{op}},\ \mathrm{n.d.}} \mathbb{Z}(q) \circ (X, f)$$

of $\mathbf{C}^b_{\mathrm{mot}}(\mathcal{V})$

2.7.3. *The holim distinguished triangle.* Let $(X, f): I \to \mathcal{L}(\mathcal{V}^+)$ be a functor, and let $i \in I$ be a maximal element (minimal in I^{op}). Recall the category $I^{i/}$ of morphisms $s: i \to j$, $j \neq i$, in I, and the functor $(X^{i/}, f^{i/}): I^{i/} \to \mathcal{L}(\mathcal{V}^+)$,

$$(X^{i/}, f^{i/})(s: i \to j) := (X(j), f(j)).$$

The homotopy limit distinguished triangle (Part II, Chapter III, §3.2.9) gives us the distinguished triangle in $\mathbf{K}^b_{\mathrm{mot}}(\mathcal{V})$

$$\mathbb{Z}_X(q)_f \to \mathbb{Z}_{X(i)}(q)_{f(i)} \oplus \mathbb{Z}_{X|I \setminus \{i\}}(q)_{f|I\setminus\{i\}} \to \mathbb{Z}_{X^{i/}}(q)_{f^{i/}} \to \mathbb{Z}_X(q)_f[1].$$

In particular, for $X: I \to \mathcal{V}^+$ a functor, we have the distinguished triangle in $\mathbf{K}^b_{\mathrm{mot}}(\mathcal{V})$

$$\mathbb{Z}_X(q)_{f_X} \to \mathbb{Z}_{X(i)}(q)_{f_X(i)} \oplus \mathbb{Z}_{X|I\setminus\{i\}}(q)_{f_X|I\setminus\{i\}} \to \mathbb{Z}_{X^{i/}}(q)_{(f_X)^{i/}} \to \mathbb{Z}_X(q)_{f_X}[1].$$

The identity map on each $X(i)$ gives the natural transformation of functors $(X_{|I\setminus\{i\}}, f_{X_{|I\setminus\{i\}}}) \to (X, f_X)_{|I\setminus\{i\}}$, which is an isomorphism in $\mathbf{D}^b_{\mathrm{mot}}(\mathcal{V})$ when evaluated at $j \in I \setminus \{i\}$. By (Part II, Chapter III, Proposition 3.2.10), the map on the holim's $\mathbb{Z}_{X_{|I\setminus\{i\}}}(q)_{f_{X_{|I\setminus\{i\}}}} \to \mathbb{Z}_{X_{|I\setminus\{i\}}}(q)$ is an isomorphism in $\mathbf{D}^b_{\mathrm{mot}}(\mathcal{V})$. Similarly, we have the natural isomorphism

$$\mathbb{Z}_{X^{i/}}(q)_{(f_X)^{i/}} \cong \mathbb{Z}_{X^{i/}}(q),$$

giving us the distinguished triangle in $\mathbf{D}^b_{\mathrm{mot}}(\mathcal{V})$

$$\mathbb{Z}_X(q) \to \mathbb{Z}_{X(i)}(q) \oplus \mathbb{Z}_{X|I\setminus\{i\}}(q) \to \mathbb{Z}_{X^{i/}}(q) \to \mathbb{Z}_X(q)[1]. \tag{2.7.3.1}$$

Using (2.7.3.1) and induction on $\dim I$ and $|\mathcal{N}(I)_{\mathrm{n.d.}}([\dim I])|$, one proves, for example, the *homotopy property*: The map

$$p^*: \mathbb{Z}_X(q) \to \mathbb{Z}_{X \times \mathbb{A}^1}(q)$$

is an isomorphism in $\mathbf{D}^b_{\text{mot}}(\mathcal{V})$, the *Künneth isomorphism*: For Y in $\mathcal{V}$, the external product

$$\boxtimes : \mathbb{Z}_X(q) \otimes \mathbb{Z}_Y(q') \to \mathbb{Z}_{X \times Y}(q+q')$$

is an isomorphism in $\mathbf{D}^b_{\text{mot}}(\mathcal{V})$, the *moving lemma*: The map

$$\mathbb{Z}_X(q)_{f \cup g} \to \mathbb{Z}_X(q)_f$$

is an isomorphism in $\mathbf{D}^b_{\text{mot}}(\mathcal{V})$, etc. Using the moving lemma, one gets contravariant functoriality: If $f : Y \to X$ is a map of functors $X, Y : I \to \mathcal{V}^+$, define $f^* : \mathbb{Z}_X(q) \to \mathbb{Z}_Y(q)$ as the composition

$$\mathbb{Z}_X(q) = \mathbb{Z}_X(q)_{f_X} \cong \mathbb{Z}_X(q)_{f_X \cup f} \xrightarrow{f^*} \mathbb{Z}_Y(q)_{f_Y} = \mathbb{Z}_Y(q).$$

2.7.4. *Products.* The formula for products for the homotopy limit given in (Part II, Chapter III, §3.4.4) give external products

$$\boxtimes_{(X,f),(Y,g)} : \mathbb{Z}_X(q)_f \otimes \mathbb{Z}_Y(q')_g \to \mathbb{Z}_{X \times_S Y}(q+q')_{f \times g}$$

in $\mathbf{C}^b_{\text{mot}}(\mathcal{V})$ for functors $(X,f),(Y,g) : I \to \mathcal{L}(\mathcal{V}^+)$. These products are associative in $\mathbf{C}^b_{\text{mot}}(\mathcal{V})$ and commutative in $\mathbf{K}^b_{\text{mot}}(\mathcal{V})$. Taking $(X,f) = (Y,g) = (X, f_X)$, and pulling back by the diagonal gives the associative, commutative cup product $\cup_X : \mathbb{Z}_X(q) \otimes \mathbb{Z}_X(q') \to \mathbb{Z}_X(q+q')$ in $\mathbf{D}^b_{\text{mot}}(\mathcal{V})$.

2.7.5. *Motivic cohomology.* For a functor $X : I \to \mathcal{V}^+$, define the motivic cohomology of X to be the motivic cohomology of $\mathbb{Z}_X$:

$$H^p(X, \mathbb{Z}(q)) := \operatorname{Hom}_{\mathbf{D}^b_{\text{mot}}(\mathcal{V})}(1, \mathbb{Z}_X(q)[p]).$$

By the moving lemma, one gets the same definition if one chooses another lifting of X to a functor to $\mathcal{L}(\mathcal{V}^+)$. The products defined in (2.7.4) give $H^*(X, \mathbb{Z}(*))$ the structure of a (possibly non-unital) bi-graded ring, graded-commutative with respect to the cohomology degree. If X has values in $\mathcal{V}$, then the structure morphism $p_X : X \to S_I$, where S_I is the constant functor with value the base-scheme S, gives the unit $p_X^* : \mathbb{Z}_S = 1 \to \mathbb{Z}_X$.

2.7.6. REMARK. The constructions of motives of truncated simplicial schemes, truncated cosimplicial schemes and n-cubes of schemes can all be rephrased in terms of the homotopy limit construction of this section, but only up to isomorphism in $\mathbf{D}^b_{\text{mot}}(\mathcal{V})$. The actual representatives in $\mathbf{C}^b_{\text{mot}}(\mathcal{V})$ will in general be different; as this can cause some difficulty in making explicit comparisons and computations, we find it useful to pick and choose among the various methods for constructing isomorphic motives.

3. Structure of the motivic categories

In this section, we prove some basic structural results for the motivic categories $\mathcal{A}_{\text{mot}}(\mathcal{V})$, $\mathbf{K}^b_{\text{mot}}(\mathcal{V})$, $\mathbf{D}^b_{\text{mot}}(\mathcal{V})_R$ and $\mathcal{DM}(\mathcal{V})_R$.

3.1. Structure of the motivic DG category

3.1.1. We begin with a description of $\mathcal{A}_1(\mathcal{V})$, and its image in $\mathcal{A}_{\text{mot}}(\mathcal{V})$.

3.1.2. LEMMA. *In* $\mathcal{A}_1(\mathcal{V})$, $(\mathbb{Z}_{X \amalg Y}(n)_{A \amalg B}, i_{X*}, i_{Y*}, i_X^*, i_Y^*)$ *is the bi-product of* $\mathbb{Z}_X(n)_f$ *and* $\mathbb{Z}_Y(n)_g$.

PROOF. We have the diagram

$$\begin{array}{ccccc} \emptyset & \longrightarrow & \emptyset\coprod Y & = & Y \\ \downarrow & & \downarrow & & \\ X & \longrightarrow & X\coprod Y. & & \end{array}$$

Applying the relations from Definition 1.3.2 and Definition 1.4.1, we find $i_Y^* \circ i_{X*} = 0$. Similarly, $i_X^* \circ i_{Y*} = 0$. Applying the relations of Definition 1.3.2 to the diagram

$$\begin{array}{ccc} X & \longrightarrow & X \\ \downarrow & & \downarrow \\ X & \longrightarrow & X\coprod Y \end{array}$$

shows that $i_X^* \circ i_{X*} = \mathrm{id}_{\mathbb{Z}_X(n)_f}$. Similarly $i_Y^* \circ i_{Y*} = \mathrm{id}_{\mathbb{Z}_Y(n)_g}$. The relation of Definition 1.4.1 completes the proof. □

3.1.3. LEMMA. *Let (X, f) and (Y, g) be in $\mathcal{L}(\mathcal{V})$, with X and Y connected. Then* $\mathrm{Hom}_{\mathcal{A}_1(\mathcal{V})}(\mathbb{Z}_X(n)_f, \mathbb{Z}_Y(n)_g)$ *is the free $\mathbb{Z}$-module on* $\mathrm{Hom}_{\mathcal{L}(\mathcal{V})^{\mathrm{op}}}((X, f), (Y, g))$

PROOF. We have the natural map

$$\Xi\colon \mathrm{Hom}_{\mathcal{L}(\mathcal{V})^{\mathrm{op}}}((X, f), (Y, g)) \to \mathrm{Hom}_{\mathcal{A}_1(\mathcal{V})}(\mathbb{Z}_X(n)_f, \mathbb{Z}_Y(n)_g).$$

It is clear that, for X and Y connected, the image of Ξ generates the $\mathbb{Z}$-module $\mathrm{Hom}_{\mathcal{A}_1(\mathcal{V})}(\mathbb{Z}_X(n)_f, \mathbb{Z}_Y(n)_g)$. Form the additive category $\mathcal{C}$ with objects finite direct sums of objects $\mathbb{Z}_X(n)_f$ for $\mathbb{Z}_X(n)_f$ in $\mathcal{A}_1(\mathcal{V})$ with X non-empty and connected. Morphisms in $\mathcal{C}$ are given by taking $\mathrm{Hom}_{\mathcal{C}}(\mathbb{Z}_X(n)_f, \mathbb{Z}_{(}n)_g)$ to be the free $\mathbb{Z}$-module on $\mathrm{Hom}_{\mathcal{L}(\mathcal{V})^{\mathrm{op}}}((X, f), (Y, g))$, for X and Y non-empty and connected, and in general by taking direct sums. The composition law is induced by that of $\mathcal{L}(\mathcal{V})^{\mathrm{op}}$. Sending maps of the form i_* to the corresponding inclusion on the direct sum in $\mathcal{C}$ defines a functor $F\colon \mathcal{A}_1(\mathcal{V}) \to \mathcal{C}$; one sees directly that $F((X, f), (Y, g)) \circ \Xi$ is the natural inclusion

$$\mathrm{Hom}_{\mathcal{L}(\mathcal{V})^{\mathrm{op}}}((X, f), (Y, g)) \to \mathbb{Z}[\mathrm{Hom}_{\mathcal{L}(\mathcal{V})^{\mathrm{op}}}((X, f), (Y, g))].$$

This shows that $\mathrm{Hom}_{\mathcal{L}(\mathcal{V})^{\mathrm{op}}}((X, f), (Y, g))$ is an independent set (over $\mathbb{Z}$) in the $\mathbb{Z}$-module $\mathrm{Hom}_{\mathcal{A}_1(\mathcal{V})}(\mathbb{Z}_X(n)_f, \mathbb{Z}_Y(n)_g)$, completing the proof. □

We have the canonical "inclusion" functor $\iota_0\colon \mathcal{A}_1(\mathcal{V}) \to \mathcal{A}_{\mathrm{mot}}(\mathcal{V})$; for $k = 1, 2 \ldots$, let $\iota_k\colon \mathcal{A}_1(\mathcal{V}) \to \mathcal{A}_{\mathrm{mot}}(\mathcal{V})$ be the functor

$$X \mapsto \mathfrak{e}^{\otimes k} \otimes X; \quad f \mapsto \mathrm{id}_{\mathfrak{e}^{\otimes k}} \otimes f.$$

3.1.4. LEMMA. *The functors ι_k, $k = 0, 1, \ldots,$ are faithful embeddings.*

PROOF. This follows directly from (Part II, Chapter I, Proposition 2.5.2). □

3.1.5. The results of (Part II, Chapter I, §2.4), give a description of the morphisms in the category $\mathcal{A}_2(\mathcal{V})$. We let $i\colon \mathcal{A}_1(\mathcal{V}) \to \mathcal{A}_2(\mathcal{V}) = \mathcal{A}_1(\mathcal{V})^{\otimes, c}$ denote the canonical functor. From (Part II, Chapter I, Proposition 2.4.5), the functor i is fully faithful. In addition, we have a functor of tensor categories without unit, $\rho\colon \mathcal{A}_2(\mathcal{V}) \to \mathcal{A}_1(\mathcal{V})$, with $\rho \circ i = \mathrm{id}$, and a natural transformation $\boxtimes\colon \mathrm{id}_{\mathcal{A}_2(\mathcal{V})} \to i \circ \rho$.

For $n = 3, 4, 5$, let

$$i_n\colon \mathcal{A}_n(\mathcal{V})^* \to \mathcal{A}_n(\mathcal{V})$$

be the full DG subcategory generated by the objects $\mathfrak{e}^{\otimes k} \otimes X$ and $\mathfrak{e}^{\otimes k}$, for X in $\mathcal{A}_1(\mathcal{V})$ and $k \geq 0$. We let

$$i_{\mathrm{mot}}: \mathcal{A}_{\mathrm{mot}}(\mathcal{V})^* \to \mathcal{A}_{\mathrm{mot}}(\mathcal{V})$$

be the full DG subcategory generated by the objects $\mathfrak{e}^{\otimes k} \otimes X$, for X in $\mathcal{A}_1(\mathcal{V})$, and $k \geq 0$.

It follows from (Part II, Chapter I, Proposition 2.5.3), that, for $n = 3, 4, 5$, the tensor structure on $\mathcal{A}_1(\mathcal{V})$ extends to a graded tensor structure without unit on $\mathcal{A}_n(\mathcal{V})^*$, the functor ρ extends to a graded tensor functor $r_n: \mathcal{A}_n(\mathcal{V}) \to \mathcal{A}_n(\mathcal{V})^*$ with $r_n \circ i_n = \mathrm{id}$, and the natural transformation $\boxtimes$ extends to a natural transformation $\boxtimes_n: \mathrm{id}_{\mathcal{A}_n(\mathcal{V})} \to i_n \circ r_n$. One checks that r_5 and $\boxtimes_5$ restrict to give the functor and natural transformation

$$\begin{aligned} r_{\mathrm{mot}}&: \mathcal{A}_{\mathrm{mot}}(\mathcal{V}) \to \mathcal{A}_{\mathrm{mot}}(\mathcal{V})^* \\ \boxtimes_{\mathrm{mot}}&: \mathrm{id}_{\mathcal{A}_{\mathrm{mot}}(\mathcal{V})} \to i_{\mathrm{mot}} \circ r_{\mathrm{mot}}. \end{aligned} \tag{3.1.5.1}$$

3.1.6. LEMMA. *For $n = 3, 4, 5$, the DG categories $\mathcal{A}_n(\mathcal{V})^*$, with the given tensor structure, are DG tensor categories without unit, and the DG category $\mathcal{A}_{\mathrm{mot}}(\mathcal{V})^*$ is a DG tensor category with unit 1. The functors r_n, (resp., natural transformations $\boxtimes_n$), for $n = 3, 4, 5$ and for $n =$ mot are DG tensor functors (resp. natural transformations of DG tensor functors).*

PROOF. For Γ and Δ in $\mathcal{A}_1(\mathcal{V})$, the symmetries $t_{\Gamma,\Delta}: \Gamma \times \Delta \to \Delta \times \Gamma$ and $\tau_{\Gamma,\Delta}: \Gamma \otimes \Delta \to \Delta \otimes \Gamma$, and the external product $\boxtimes_{\Gamma,\Delta}: \Gamma \otimes \Delta \to \Gamma \times \Delta$ are morphisms in the tensor category $\mathcal{A}_2(\mathcal{V})$, hence, as morphisms in the DG tensor categories $\mathcal{A}_n(\mathcal{V})$, $n = 3, 4, 5$, these maps are morphisms of degree 0, with zero differential. From the explicit expression for graded tensor product structure on $\mathcal{A}_n(\mathcal{V})^*$, $n = 3, 4, 5$, given in the proof of (Part II, Chapter I, Proposition 2.5.3), one sees that this tensor structure respects the differential structure (i.e., that the Leibniz rule is satisfied), and similarly, that the functors r_n respect the differential structure. The analogous result for $n =$ mot follows from the case $n = 5$.

Similarly, it follows from (Part II, Chapter I, Proposition 2.5.3), that $\mathcal{A}_{\mathrm{mot}}(\mathcal{V})^*$ is an tensor category with unit 1; arguing as above, we see that the unit respects the differential structure, completing the proof. □

3.2. The motivic cycles functor

We now show how the operation of taking the group of cycles of various codimension becomes a functor on $\mathcal{A}_{\mathrm{mot}}(\mathcal{V})$.

3.2.1. We start with the cycles functor (1.4.7.3). We define the functor of additive categories

$$\mathcal{Z}_2: \mathcal{A}_2(\mathcal{V}) \to \mathbf{Ab} \tag{3.2.1.1}$$

by $\mathcal{Z}_2 = \mathcal{Z}_1 \circ \rho$.

3.2.2. LEMMA. *The functor (3.2.1.1) extends to a functor of graded additive categories*

$$\mathcal{Z}_3: \mathcal{A}_3(\mathcal{V}) \to \mathbf{GrAb}$$

which satisfies

(i) $\mathcal{Z}_3$ factors as a composition

$$\mathcal{A}_3(\mathcal{V}) \xrightarrow{r_3} \mathcal{A}_3(\mathcal{V})^* \xrightarrow{\mathcal{Z}_3^*} \mathbf{GrAb}. \tag{3.2.2.1}$$

(ii) We have

$$\mathcal{Z}_3^*(\mathfrak{e}^{\otimes k} \otimes \mathbb{Z}_X(n)) = \mathcal{Z}_1(\mathbb{Z}_X(n)_f)[-2n] = \mathcal{Z}^n(X)_f[-2n]; \quad \mathcal{Z}_3^*(\mathfrak{e}^{\otimes k}) = \mathbb{Z},$$

with $\mathcal{Z}_1(\mathbb{Z}_X(n))$ and $\mathbb{Z}$ being concentrated in degree 0.
(iii) For X in $\mathcal{A}_1(\mathcal{V})$, and for $h\colon \mathfrak{e}^{\otimes k} \otimes X \to \mathfrak{e}^{\otimes k} \otimes X$ a morphism of the form $\tau \otimes \mathrm{id}_X$, where $\tau\colon \mathfrak{e}^{\otimes k} \to \mathfrak{e}^{\otimes k}$ is a symmetry isomorphism in the category $\mathbb{E}$, we have

$$\mathcal{Z}_3^*(h) = \mathrm{id}_{\mathcal{Z}_2(X)}.$$

If $h = \tau \otimes \mathrm{id}_X$, where τ has degree $p < 0$, then $\mathcal{Z}_3^*(h) = 0$.
(iv) Let $Y, X_1, \dots, X_n$ be in $\mathcal{A}_1(\mathcal{V})$, $Z_i \in \mathcal{Z}_1(X_i)$ for $i = 1, \dots, n$. Let

$$X = X_1 \times_S \dots \times_S X_n, \quad Z = Z_1 \times_{/S} \dots \times_{/S} Z_n.$$

If $f\colon \mathfrak{e}^{\otimes n} \otimes Y \to X \times Y$ is the morphism defined by the composition

$$\mathfrak{e}^{\otimes n} \otimes Y \xrightarrow{[Z_1]\otimes\dots\otimes[Z_n]\otimes\mathrm{id}_Y} X_1 \otimes \dots \otimes X_n \otimes Y \xrightarrow{\boxtimes_{X_1,\dots,X_n,Y}} X \times Y,$$

then, for $m \geq 0$, $\mathcal{Z}_3^*(\mathrm{id}_{\mathfrak{e}^{\otimes m}} \otimes f)\colon \mathcal{Z}_1(Y) \to \mathcal{Z}_1(X \times Y)$ is the map determined by the identity

$$\mathcal{Z}_3^*(\mathrm{id}_{\mathfrak{e}^{\otimes m} \otimes f})(W) = Z \times_{/S} W \text{ for all } W \in \mathcal{Z}_1(Y)$$

(by Appendix A, Remark 2.3.3, $Z \times_{/S} W$ is in $\mathcal{Z}_1(X \times_S Y)$). Moreover, the functor $\mathcal{Z}_3^*$ is uniquely determined by (i)-(iv).

PROOF. By (Part II, Chapter I, Proposition 2.5.2), the objects $\mathfrak{e}^{\otimes k} \otimes X$, $\mathfrak{e}^{\otimes k}$ and the morphisms of the form $h = \tau \otimes \mathrm{id}_X$ and $\mathrm{id}_{\mathfrak{e}^{\otimes m}} \otimes f$, together with the morphisms of $\mathcal{A}_2(\mathcal{V})$, generate $\mathcal{A}_3(\mathcal{V})^*$ as a graded additive category, whence the uniqueness of $\mathcal{Z}_3$.

For existence, we first note that, if Z_i is in $\mathcal{Z}_1(X_i)$ for $i = 1, 2$, then it follows immediately from the definitions that the cycle $Z_1 \times_{/S} Z_2$ is in $\mathcal{Z}_1(X_1 \times X_2)$. Thus, the expression for $\mathcal{Z}_3(\mathrm{id}_{\mathfrak{e}^{\otimes m} \otimes f})$ is well-defined.

It follows from (Part II, Chapter I, Proposition 2.5.2) that the formulas (i)-(iv) give, for each pair of objects X, Y of $\mathcal{A}_3(\mathcal{V})^*$, a well-defined homomorphism $\mathcal{Z}_3(X, Y)\colon \mathrm{Hom}_{\mathcal{A}_3(\mathcal{V})^*}(X, Y) \to \mathrm{Hom}_{\mathbf{GrAb}}(\mathcal{Z}_3(X), \mathcal{Z}_3(Y))$. The functoriality of the collection of maps $\mathcal{Z}_3(X, Y)$ is checked via the explicit form of the composition law in $\mathcal{A}_3(\mathcal{V})^*$, giving a graded additive functor $\mathcal{Z}_3^*\colon \mathcal{A}_3(\mathcal{V})^* \to \mathbf{GrAb}$. We then define $\mathcal{Z}_3$ as

$$\mathcal{Z}_3 = \mathcal{Z}_3^* \circ r_3,$$

completing the proof. □

3.2.3. DEFINITION. Recall (Definition 1.4.12) the graded tensor categories $\mathcal{A}_n^0(\mathcal{V})$ for $n = 4, 5$, and $n = \mathrm{mot}$, having the same objects as $\mathcal{A}_n(\mathcal{V})$. For $n = 4, 5$ and $n = \mathrm{mot}$, we let $\mathcal{A}_n^0(\mathcal{V})^*$ denote the full subcategory of $\mathcal{A}_n^0(\mathcal{V})$ generated by the objects of $\mathcal{A}_n(\mathcal{V})^*$.

3.2.4. The DG tensor functor (1.4.12.1) induces the DG tensor functor

$$H_n^*: \mathcal{A}_n(\mathcal{V})^* \to \mathcal{A}_n^0(\mathcal{V})^*.$$

The functors r_n, i_n of §3.1.5 induce functors

$$r_n^0: \mathcal{A}_n^0(\mathcal{V}) \to \mathcal{A}_n^0(\mathcal{V})^*,\ i_n^0: \mathcal{A}_n^0(\mathcal{V})^* \to \mathcal{A}_n^0(\mathcal{V}),$$

giving the commutative diagrams

$$\begin{array}{ccc} \mathcal{A}_n(\mathcal{V}) & \xrightarrow{H_n} & \mathcal{A}_n^0(\mathcal{V}) \\ {\scriptstyle r_n}\downarrow\uparrow{\scriptstyle i_n} & & {\scriptstyle i_n^0}\uparrow\downarrow{\scriptstyle r_n^0} \\ \mathcal{A}_n(\mathcal{V})^* & \xrightarrow[H_n^*]{} & \mathcal{A}_n^0(\mathcal{V})^* \end{array} \tag{3.2.4.1}$$

for $n = 4, 5, \mathrm{mot}$.

3.2.5. LEMMA. *There is an extension of $\mathcal{Z}_3: \mathcal{A}_3(\mathcal{V}) \to \mathbf{GrAb}$ to a functor of graded additive categories*

$$\mathcal{Z}_{\mathrm{mot}}: \mathcal{A}_{\mathrm{mot}}(\mathcal{V}) \to \mathbf{GrAb} \tag{3.2.5.1}$$

such that $\mathcal{Z}_{\mathrm{mot}}$ factors through $H^*_{\mathrm{mot}} \circ r_{\mathrm{mot}}$ as

$$\mathcal{A}_{\mathrm{mot}}(\mathcal{V}) \xrightarrow{H^*_{\mathrm{mot}} \circ r_{\mathrm{mot}}} \mathcal{A}^0_{\mathrm{mot}}(\mathcal{V})^* \xrightarrow{\mathcal{Z}^{0*}_{\mathrm{mot}}} \mathbf{GrAb}.$$

PROOF. One easily checks that the functor $\mathcal{Z}_3^*: \mathcal{A}_3(\mathcal{V})^* \to \mathbf{GrAb}$ (3.2.2.1) respects the relations of (1.4.7)(i)-(iv), giving the extension to $\mathcal{Z}_4^{0*}: \mathcal{A}_4^0(\mathcal{V})^* \to \mathbf{GrAb}$. Noting that $\mathcal{A}_4^0(\mathcal{V})^* = \mathcal{A}_5^0(\mathcal{V})^*$, we define $\mathcal{Z}_{\mathrm{mot}}$ as the restriction to $\mathcal{A}_{\mathrm{mot}}(\mathcal{V})$ of the composition $\mathcal{Z}_4^{0*} \circ H_5^* \circ r_5$. □

3.3. The motivic homotopy category

We now derive some basic properties of the triangulated tensor category $\mathbf{K}^b(\mathcal{V})$.

We let $\mathbf{C}^b_{\mathrm{mot}}(\mathcal{V})^*$ denote the full subcategory $\mathbf{C}^b(\mathcal{A}_{\mathrm{mot}}(\mathcal{V})^*)$ of $\mathbf{C}^b_{\mathrm{mot}}(\mathcal{V})$, Similarly, we let $\mathbf{K}^b_{\mathrm{mot}}(\mathcal{V})^*$ denote the full subcategory $\mathbf{K}^b(\mathcal{A}_{\mathrm{mot}}(\mathcal{V})^*)$ of $\mathbf{K}^b_{\mathrm{mot}}(\mathcal{V})$.

3.3.1. LEMMA. *(i) The functor of DG tensor categories $r_{\mathrm{mot}}: \mathcal{A}_{\mathrm{mot}}(\mathcal{V}) \to \mathcal{A}_{\mathrm{mot}}(\mathcal{V})^*$ and natural transformation $\boxtimes_{\mathrm{mot}}: \mathrm{id} \to i_{\mathrm{mot}} \circ r_{\mathrm{mot}}$ (cf. (3.1.5.1)) extend to the functor of DG tensor categories, and natural transformation of functors, compatible with the cone functors,*

$$\mathbf{C}^b(r_{\mathrm{mot}}): \mathbf{C}^b_{\mathrm{mot}}(\mathcal{V}) \to \mathbf{C}^b_{\mathrm{mot}}(\mathcal{V})^*,$$
$$\mathbf{C}^b(\boxtimes_{\mathrm{mot}}): \mathrm{id} \to \mathbf{C}^b(i_{\mathrm{mot}}) \circ \mathbf{C}^b(r_{\mathrm{mot}}).$$

These in turn extend to the functor of triangulated tensor categories (without unit), and natural transformation

$$\begin{aligned} &\mathbf{K}^b(r_{\mathrm{mot}}): \mathbf{K}^b_{\mathrm{mot}}(\mathcal{V}) \to \mathbf{K}^b_{\mathrm{mot}}(\mathcal{V})^*, \\ &\mathbf{K}^b(\boxtimes_{\mathrm{mot}}): \mathrm{id} \to \mathbf{K}^b(i_{\mathrm{mot}}) \circ \mathbf{K}^b(r_{\mathrm{mot}}). \end{aligned} \tag{3.3.1.1}$$

(ii) The functor (3.2.5.1) *extends to the functor of DG categories*

$$\mathcal{Z}_{\mathrm{mot}} := \mathbf{C}^b(\mathcal{Z}_{\mathrm{mot}}): \mathbf{C}^b_{\mathrm{mot}}(\mathcal{V}) \to \mathbf{C}^b(\mathbf{Ab}), \tag{3.3.1.2}$$

compatible with cones. This functor in turn extends to the functor of triangulated categories

$$\mathcal{Z}_{\mathrm{mot}} := \mathbf{K}^b(\mathcal{Z}_{\mathrm{mot}})\colon \mathbf{K}^b_{\mathrm{mot}}(\mathcal{V}) \to \mathbf{K}^b(\mathbf{Ab}). \tag{3.3.1.3}$$

PROOF. Apply the functors $\mathbf{C}^b(-)$ and $\mathbf{K}^b(-)$ to r_{mot}, $\boxtimes_{\mathrm{mot}}$ and $\mathcal{Z}_{\mathrm{mot}}$, and use the equivalences (Part II, Chapter II, §1.2.9)

$$\mathrm{Tot}\colon \mathbf{C}^b(\mathbf{C}^b(\mathbf{Ab})) \to \mathbf{C}^b(\mathbf{Ab}),$$
$$\mathrm{Tot}\colon \mathbf{K}^b(\mathbf{K}^b(\mathbf{Ab})) \to \mathbf{K}^b(\mathbf{Ab}).$$

□

3.3.2. Let (X, f) be in $\mathcal{L}(\mathcal{V})$, and let $\Gamma = \mathbb{Z}_X(q)_f$. For an integer $b \geq 0$, we let

$$i_{\Gamma,b}\colon \mathrm{Hom}_{\mathcal{A}_5(\mathcal{V})}(\mathfrak{e}^{\otimes a}, \Gamma)^* \to \mathrm{Hom}_{\mathcal{A}_5(\mathcal{V})}(\mathfrak{e}^{\otimes a+b}, \mathfrak{e}^{\otimes b} \otimes \Gamma)^* \tag{3.3.2.1}$$

be the map $i_{\Gamma,b}(f) = \mathrm{id}_{\mathfrak{e}^{\otimes b}} \otimes f$.

3.3.3. LEMMA. *The map* (3.3.2.1) *is a quasi-isomorphism.*

PROOF. Denote the complex $\mathrm{Hom}_{\mathcal{A}_5(\mathcal{V})}(\mathfrak{e}^{\otimes a+b}, \mathfrak{e}^{\otimes b} \otimes \Gamma)^*$ by $C^*_{a,b}$ and the complex $\mathrm{Hom}_{\mathcal{A}_5(\mathcal{V})}(\mathfrak{e}^{\otimes a}, \Gamma)^*$ by C^*_a. Let $\mathcal{H}$ denote the set of morphisms $h\colon \mathfrak{e}^{\otimes n} \to \mathbb{Z}_Y(n)_g$ adjoined to form the category $\mathcal{A}_5(\mathcal{V})$ from the category $\mathcal{A}_2(\mathcal{V})[\mathbb{E}]$ in Definition 1.4.6, Definition 1.4.8 and Definition 1.4.9. We may order the set $\mathcal{H}$ so that, if h and h' are in $\mathcal{H}$, and we adjoin h before adjoining h', then $h < h'$. We may then filter the two complexes via this ordering. Using a spectral sequence argument, it suffices to show that the map on the associated graded is a quasi-isomorphism. Let $\mathbf{gr}^h$ denote the term in the associated graded corresponding to the adjoined morphism h.

We refer to the description of the morphisms in C^*_a and $C^*_{a,b}$ given by (Part II, Chapter I, Proposition 2.5.2); each map in $C^*_{a,b}$ is a sum of compositions of the form

$$\begin{aligned}
&\mathfrak{e}^{\otimes a+b} \xrightarrow{\tau} \mathfrak{e}^{\otimes a+b} = \mathfrak{e}^{\otimes b} \otimes \mathfrak{e}^{\otimes a} = \mathfrak{e}^{\otimes b} \otimes \mathfrak{e}^{\otimes a_1} \otimes \ldots \otimes \mathfrak{e}^{\otimes a_1} \\
&\xrightarrow{\mathrm{id}_{\mathfrak{e}^{\otimes b}} \otimes h_1 \otimes \ldots \otimes h_s} \mathfrak{e}^{\otimes b} \otimes \Delta_1 \otimes \ldots \otimes \Delta_s \\
&\xrightarrow{\mathrm{id}_{\mathfrak{e}^{\otimes b}} \otimes \boxtimes_{\Delta_1, \ldots, \Delta_s}} \mathfrak{e}^{\otimes b} \otimes \Delta_1 \times \ldots \times \Delta_s \\
&\xrightarrow{\mathrm{id}_{\mathfrak{e}^{\otimes b}} \otimes p} \mathfrak{e}^{\otimes b} \otimes \Gamma.
\end{aligned} \tag{3.3.3.1}$$

Here $h_1 \leq \ldots \leq h_s$ is an increasing sequence of elements of $\mathcal{I}$, with $h_i\colon \mathfrak{e}^{\otimes a_i} \to \Delta_i$, τ is a morphism in $\mathbb{E}$, $\boxtimes$ is the external product, p is a morphism in $\mathcal{A}_1(\mathcal{V})$ and $a = a_1 + \ldots + a_s$. There is a similar description of the morphisms in C^*_a.

For an increasing sequence $h_* := h_1 \leq \ldots \leq h_s$, we let $S(h_*)$ denote the subgroup of the symmetric group S_s which preserves the order in the sequence h_*. We have the homomorphism $\rho_{h_*}\colon S(h_*) \to S_a$ gotten by letting a permutation in S_s act on $\{1, \ldots, a\}$ by permuting the blocks of size $a_1, \ldots, a_s$. We define a left $\mathbb{Z}[S(h_*)]$-module structure on $\mathrm{Hom}_{\mathbb{E}}(\mathfrak{e}^{\otimes a}, \mathfrak{e}^{\otimes a})$ by having $\sigma \in S(h_*)$ act by left composition with $\rho_{h_*}(\sigma)$. We give the group $\mathrm{Hom}_{\mathbb{E}}(\mathfrak{e}^{\otimes a+b}, \mathfrak{e}^{\otimes a+b})$ the left $\mathbb{Z}[S(h_*)]$-module structure defined by writing $\mathfrak{e}^{\otimes a+b} = \mathfrak{e}^{\otimes b} \otimes \mathfrak{e}^{\otimes a}$ and acting via $(\mathrm{id}_{\mathfrak{e}^{\otimes b}} \otimes \rho_{h_*}(\sigma))\circ$.

We let $\Delta(h_*)$ denote the object $\Delta_1 \times \ldots \times \Delta_s$ appearing in the composition (3.3.3.1). The group $\mathrm{Hom}_{\mathcal{A}_1(\mathcal{V})}(\Delta(h_*), \Gamma)$ is a right $\mathbb{Z}[S(h_*)]$-module, where σ acts by right composition with the symmetry isomorphism $\pm t^*_\sigma$, where the sign is given by the weighted sign representation determined by the degrees of the morphisms h_i.

The map (3.3.3.1) is in $F^{\le h}C^*_{a,b}$ if and only if $h_i \le h$ for each i. We may then filter $\mathbf{gr}^h C^*_{a,b}$ and $\mathbf{gr}^h C^*_b$ by the number of times h appears in the sequence $h_1 \le h_2 \le \ldots \le h_s$. Let $F^{\le m}\mathbf{gr}^h$ denote the subgroup for which h appears at most m times; since the differential of h is by construction in the category generated by the adjunction of the h' with $h' < h$, the Leibniz rule for differentiation implies that $F^{\le m}\mathbf{gr}^h C^*_{a,b}$ is a subcomplex of $\mathbf{gr}^h C^*_{a,b}$, and similarly for $F^{\le m}\mathbf{gr}^h C^*_b$. Again, we need only show that the map on the associated graded

$$\mathbf{gr}^m\mathbf{gr}^h i_{\Gamma,b}\colon \mathbf{gr}^m\mathbf{gr}^h C^*_b \to \mathbf{gr}^m\mathbf{gr}^h C^*_{a,b} \tag{3.3.3.2}$$

is a quasi-isomorphism.

Using the Leibniz rule again, we see that all the differentials in the complexes $\mathbf{gr}^m\mathbf{gr}^h C^*_b$ and $\mathbf{gr}^m\mathbf{gr}^h C^*_{a,b}$ are induced by the differentials in the category $\mathbb{E}$; using (Part II, Chapter I, Proposition 2.5.2), the complex $\mathbf{gr}^m\mathbf{gr}^h C^*_{a,b}$ is isomorphic to a direct sum of complexes of the following form

$$\mathbf{gr}^m\mathbf{gr}^h C^*_{a,b} \cong \oplus_{h_*}\mathrm{Hom}_{\mathcal{A}_1(\mathcal{V})}(\Delta(h_*),\Gamma)\otimes_{\mathbb{Z}[S(h_*)]}\mathrm{Hom}_{\mathbb{E}}(\mathfrak{e}^{\otimes a+b},\mathfrak{e}^{\otimes a+b})^*.$$

We have a similar description of $\mathbf{gr}^m\mathbf{gr}^h C^*_a$ as isomorphic to a direct sum of complexes of the form

$$\mathbf{gr}^m\mathbf{gr}^h C^*_{a,b} \cong \oplus_{h_*}\mathrm{Hom}_{\mathcal{A}_1(\mathcal{V})}(\Delta(h_*),\Gamma)\otimes_{\mathbb{Z}[S(h_*)]}\mathrm{Hom}_{\mathbb{E}}(\mathfrak{e}^{\otimes a},\mathfrak{e}^{\otimes a})^*,$$

where the two sums are over the same set of sequences h_*. The map (3.3.3.2) is the direct sum of the maps

$$\begin{aligned}\mathrm{id}\otimes i_{a,b}\colon &\mathrm{Hom}_{\mathcal{A}_1(\mathcal{V})}(\Delta(h_*),\Gamma)\otimes_{\mathbb{Z}[S(h_*)]}\mathrm{Hom}_{\mathbb{E}}(\mathfrak{e}^{\otimes a},\mathfrak{e}^{\otimes a})^*\\ &\to \mathrm{Hom}_{\mathcal{A}_1(\mathcal{V})}(\Delta(h_*),\Gamma)\otimes_{\mathbb{Z}[S(h_*)]}\mathrm{Hom}_{\mathbb{E}}(\mathfrak{e}^{\otimes a+b},\mathfrak{e}^{\otimes a+b})^*,\end{aligned} \tag{3.3.3.3}$$

where $i_{a,b}\colon \mathrm{Hom}_{\mathbb{E}}(\mathfrak{e}^{\otimes a},\mathfrak{e}^{\otimes a}) \to \mathrm{Hom}_{\mathbb{E}}(\mathfrak{e}^{\otimes a+b},\mathfrak{e}^{\otimes a+b})$ is the map gotten by writing $\mathfrak{e}^{\otimes a+b} = \mathfrak{e}^{\otimes b}\otimes\mathfrak{e}^{\otimes a}$ and defining $i_{a,b}(\tau) = \mathrm{id}_{\mathfrak{e}^{\otimes b}}\otimes f$.

Now let M be a right $\mathbb{Z}[S(h_*)]$-module, and let

$$i^M_{a,b}\colon M\otimes_{\mathbb{Z}[S(h_*)]}\mathrm{Hom}_{\mathbb{E}}(\mathfrak{e}^{\otimes a},\mathfrak{e}^{\otimes a})^* \to M\otimes_{\mathbb{Z}[S(h_*)]}\mathrm{Hom}_{\mathbb{E}}(\mathfrak{e}^{\otimes a+b},\mathfrak{e}^{\otimes a+b})^*$$

be the map $\mathrm{id}_M\otimes i_{a,b}$. The complex $\mathrm{Hom}_{\mathbb{E}}(\mathfrak{e}^{\otimes k},\mathfrak{e}^{\otimes k})^*$ is a free (left) $\mathbb{Z}[S_k]$-resolution of the trivial module $\mathbb{Z}$ (Part II, Chapter II, §3.1.12). Thus, if G is a sub-group of S_k, $\mathrm{Hom}_{\mathbb{E}}(\mathfrak{e}^{\otimes k},\mathfrak{e}^{\otimes k})^*$ is a free (left) $\mathbb{Z}[G]$-resolution of the trivial G-module $\mathbb{Z}$. From this we see that the map $i_{a,b}$ induces a map of $\mathbb{Z}[S(h_*)]$-free resolutions of the trivial $\mathbb{Z}[S(h_*)]$-module $\mathbb{Z}$, hence the map $i^M_{a,b}$ induces an isomorphism in cohomology. Taking $M = \mathrm{Hom}_{\mathcal{A}_1(\mathcal{V})}(\Delta(h_*),\Gamma)$ shows that (3.3.3.3) is a quasi-isomorphism, which proves the lemma. □

3.3.4. We have the cohomological functor on $\mathbf{K}^b(\mathbf{Ab})$

$$X \mapsto H^0(X) := \mathrm{Hom}_{\mathbf{K}^b(\mathbf{Ab})}(\mathbb{Z},X).$$

Let $B \ge 0$ be an integer, and let $\mathbf{K}^b_{\mathrm{mot}}(\mathcal{V})^*_B$ be the full triangulated subcategory of $\mathbf{K}^b_{\mathrm{mot}}(\mathcal{V})^*$ generated by the objects $\mathfrak{e}^{\otimes b}\otimes\mathbb{Z}_X(n)_f$, with (X,f) in $\mathcal{L}(\mathcal{V})$, n an integer, and b an integer with $0\le b\le B$. It is immediate that $\mathbf{K}^b_{\mathrm{mot}}(\mathcal{V})^*$ is the inductive limit

$$\mathbf{K}^b_{\mathrm{mot}}(\mathcal{V})^* = \varinjlim_B \mathbf{K}^b_{\mathrm{mot}}(\mathcal{V})^*_B.$$

For $\Gamma \in \mathbf{K}^b_{\mathrm{mot}}(\mathcal{V})^*$, we let B_Γ be defined by

$$B_\Gamma = \min\{B \mid \Gamma \text{ is in } \mathbf{K}^b_{\mathrm{mot}}(\mathcal{V})^*_B\}.$$

3.3.5. PROPOSITION. *Let $\Gamma = \mathbb{Z}_X(n)_f$, and let $a, b \geq 0$ be integers. Then*

(i) *If $n < 0$, or if $a < b$, then $\mathrm{Hom}_{\mathcal{A}_{\mathrm{mot}}(\mathcal{V})}(\mathfrak{e}^{\otimes a} \otimes 1, \mathfrak{e}^{\otimes b} \otimes \Gamma)^* = 0$. For all n, a, b, and all $q > 0$, we have $\mathrm{Hom}_{\mathcal{A}_{\mathrm{mot}}(\mathcal{V})}(\mathfrak{e}^{\otimes a} \otimes 1, \mathfrak{e}^{\otimes b} \otimes \Gamma)^q = 0$.*

(ii) *If $n \geq 0$, and $a > b$, then $\mathrm{Hom}_{\mathbf{K}^b_{\mathrm{mot}}(\mathcal{V})}(\mathfrak{e}^{\otimes a} \otimes 1, \mathfrak{e}^{\otimes b} \otimes \Gamma)^{2n+p} = 0$ for all $p \neq 0$.*

(iii) *Suppose that $a > b$. Then the map*

$$\mathrm{ev}_\Gamma \colon \mathrm{Hom}_{\mathbf{K}^b_{\mathrm{mot}}(\mathcal{V})}(\mathfrak{e}^{\otimes a} \otimes 1, \mathfrak{e}^{\otimes b} \otimes \Gamma)^{2n} \to \mathcal{Z}^n(X)_f$$

defined by $\mathrm{ev}_\Gamma(f) = \mathbf{K}^b(\mathcal{Z}_{\mathrm{mot}})(f)(1)$ (see (3.3.1.3)) is an isomorphism.

(iv) *Let Δ be an object of $\mathbf{K}^b_{\mathrm{mot}}(\mathcal{V})^*$. Then the map*

$$\mathcal{Z}_{\mathrm{mot}}(\mathfrak{e}^{\otimes a} \otimes 1, \Delta) \colon \mathrm{Hom}_{\mathbf{K}^b_{\mathrm{mot}}(\mathcal{V})}(\mathfrak{e}^{\otimes a} \otimes 1, \Delta) \to H^0(\mathcal{Z}_{\mathrm{mot}}(\Delta))$$

is an isomorphism for all $a > B_\Delta$.

PROOF. From the construction of the category $\mathcal{A}_5(\mathcal{V})$, together with the explicit description of the morphisms in $\mathcal{A}_5(\mathcal{V})$ given by (Part II, Chapter I, Proposition 2.5.2), the complex $\mathrm{Hom}_{\mathcal{A}_5(\mathcal{V})}(\mathfrak{e}^{\otimes a}, \mathfrak{e}^{\otimes b} \otimes \Gamma[2n])^*$ is zero in degrees $d > 0$ and if $n < 0$, this complex is the zero complex. Similarly, if $b > a$, then the complex is zero.

On the other hand, from Remark 1.4.11, the map

$$\mathrm{Hom}_{\mathcal{A}_5(\mathcal{V})}(\mathfrak{e}^{\otimes a}, \mathfrak{e}^{\otimes b} \otimes \Gamma)^* \to \mathrm{Hom}_{\mathcal{A}_5(\mathcal{V})}(\mathfrak{e}^{\otimes a} \otimes 1, \mathfrak{e}^{\otimes b} \otimes \Gamma)^*$$

which sends f to f^S is an isomorphism. This proves (i).

For (ii), we have the isomorphism

$$\mathrm{Hom}_{\mathbf{K}^b_{\mathrm{mot}}(\mathcal{V})}(\mathfrak{e}^{\otimes a} \otimes 1, \mathfrak{e}^{\otimes b} \otimes \Gamma)^q \cong H^q(\mathrm{Hom}_{\mathcal{A}_{\mathrm{mot}}(\mathcal{V})}(\mathfrak{e}^{\otimes a} \otimes 1, \mathfrak{e}^{\otimes b} \otimes \Gamma)^*),$$

and we have $\mathcal{Z}_{\mathrm{mot}}(f^S) = \mathcal{Z}_5(f)$.

Thus we need only show that the cohomology of the complex

$$\mathrm{Hom}_{\mathcal{A}_5(\mathcal{V})}(\mathfrak{e}^{\otimes a}, \mathfrak{e}^{\otimes b} \otimes \Gamma)^* \tag{3.3.5.1}$$

is zero in degrees $q \neq 2n$, and that $\mathcal{Z}_5$ induces an isomorphism

$$H^{2n}\mathcal{Z}_5 \colon H^{2n}(\mathrm{Hom}_{\mathcal{A}_5(\mathcal{V})}(\mathfrak{e}^{\otimes a}, \mathfrak{e}^{\otimes b} \otimes \Gamma)^*) \to \mathcal{Z}^n(X)_f.$$

If $a > b$, the cohomology of (3.3.5.1) is, by Lemma 3.3.3, the same as the cohomology in the complex $\mathrm{Hom}_{\mathcal{A}_5(\mathcal{V})}(\mathfrak{e}^{\otimes a-b}, \Gamma)^*$, i.e., we may assume that $b = 0$.

By the inductive construction of $\mathcal{A}_5(\mathcal{V})$, together with (Part II, Chapter I, Proposition 2.5.2), we have $H^q(\mathrm{Hom}_{\mathcal{A}_5(\mathcal{V})}(\mathfrak{e}^{\otimes a}, \Gamma)^*) = 0$ if $a > 0$ and $q \neq 2n$. This proves (ii). We now compute the cohomology H^{2n}.

By (Part II, Chapter I, Proposition 2.5.2), $\mathrm{Hom}_{\mathcal{A}_{\mathrm{mot}}(\mathcal{V})}(\mathfrak{e}^{\otimes a}, \Gamma)^{2n}$ is generated as an abelian group by maps of the form

$$f = p^* \circ \boxtimes \circ ([Z_1] \otimes \ldots \otimes [Z_a]) \circ \tau,$$

where $[Z_i] \colon \mathfrak{e} \to \mathbb{Z}_{Y_i}(e_i)_{g_i}$ are the maps of Definition 1.4.6 coming from elements $Z_i \in \mathcal{Z}^{e_i}(Y_i)_{g_i}$, $i = 1, \ldots, a$, $\boxtimes \colon \mathbb{Z}_{Y_1}(e_1)_{g_1} \otimes \ldots \otimes \mathbb{Z}_{Y_a}(e_a)_{g_a} \to \mathbb{Z}_Y(e)_g$ is the external product, with

$$Y = Y_1 \times_S \ldots \times_S Y_a; \quad g = g_1 \times \ldots \times g_a; \quad e = \Sigma_i e_i,$$

$\tau \colon \mathfrak{e}^{\otimes a} \to \mathfrak{e}^{\otimes a}$ is a symmetry isomorphism in the category $\mathbb{E}$, and $p \colon (X, f) \to (Y, g)$ is a map in $\mathcal{L}(\mathcal{V})$. By (Part II, Chapter II, Proposition 3.1.12), the map τ is homotopic in $\mathbb{E}$ to the identity, so we may assume that $\tau = \mathrm{id}$. Using the homotopies adjoined in Definition 1.4.8, there is a map h in $\mathcal{A}_5(\mathcal{V})$ with

$$dh = f - \boxtimes_{Y,1,\ldots,1} \circ ([W] \otimes [|S|] \otimes \ldots \otimes [|S|]),$$

where W is the cycle $p^*(Z_1 \times_{/S} \ldots \times_{/S} Z_a)$. We may therefore replace f with the map $F := \boxtimes_{Y,1,\ldots,1} \circ ([W] \otimes [|S|] \otimes \ldots \otimes [|S|])$.

As $W = \mathcal{Z}_5(F)(1)$, we find that the map $H^{2n}\mathcal{Z}_5$ is injective. The identity $W = \mathcal{Z}_5([W])(1)$ for $W \in \mathcal{Z}^n(X)_f$ also shows that $H^{2n}\mathcal{Z}_5$ is surjective, which completes the proof of (iii).

For (iv), suppose $a > B := B_\Delta$. The functor $\mathcal{Z}_{\mathrm{mot}}: \mathbf{K}^b_{\mathrm{mot}}(\mathcal{V}) \to \mathbf{K}^b(\mathbf{Ab})$ is exact, hence the functor $H^0 \circ \mathcal{Z}_{\mathrm{mot}}: \mathbf{K}^b_{\mathrm{mot}}(\mathcal{V}) \to \mathbf{K}^b(\mathbf{Ab})$ is a cohomological functor. By (ii), the cohomological functors $H^0 \circ \mathcal{Z}_{\mathrm{mot}}$ and $\mathrm{Hom}(\mathfrak{e}^{\otimes a} \otimes 1, -)$ agree on the objects $\mathfrak{e}^{\otimes b} \otimes \mathbb{Z}_X(n)[p]$ for all p as long as $a > b$; as the objects $\mathfrak{e}^{\otimes b} \otimes \mathbb{Z}_X(n)$ with $b \leq B$ generate $\mathbf{K}^b_{\mathrm{mot}}(\mathcal{V})^*_B$ as a triangulated category, we have

$$H^0 \circ \mathcal{Z}_{\mathrm{mot}} = \mathrm{Hom}(\mathfrak{e}^{\otimes a} \otimes 1, -)$$

on $\mathbf{K}^b_{\mathrm{mot}}(\mathcal{V})^*_B$. As Δ is in $\mathbf{K}^b_{\mathrm{mot}}(\mathcal{V})^*_B$, this proves (iv). □

3.3.6. Let

$$(3.3.6.1) \qquad \nu_{\Gamma,a}: \mathrm{Hom}_{\mathbf{K}^b_{\mathrm{mot}}(\mathcal{V})}(\mathfrak{e}^{\otimes a} \otimes 1, \Gamma) \to \mathrm{Hom}_{\mathbf{K}^b_{\mathrm{mot}}(\mathcal{V})}(\mathfrak{e}^{\otimes a+1} \otimes 1, \Gamma)$$

be the map sending $f: \mathfrak{e}^{\otimes a} \otimes 1 \to \Gamma$ to the composition

$$\mathfrak{e}^{\otimes a+1} \otimes 1 = \mathfrak{e}^{\otimes a} \otimes \mathfrak{e} \otimes 1 \xrightarrow{\mathrm{id} \otimes \nu_a} \mathfrak{e}^{\otimes a} \otimes 1. \xrightarrow{f} \Gamma.$$

3.3.7. LEMMA. *Let Δ be in $\mathbf{K}^b_{\mathrm{mot}}(\mathcal{V})^*$. Then*

$$\nu_{\Delta,a}: \mathrm{Hom}_{\mathbf{K}^b_{\mathrm{mot}}(\mathcal{V})}(\mathfrak{e}^{\otimes a} \otimes 1, \Delta) \to \mathrm{Hom}_{\mathbf{K}^b_{\mathrm{mot}}(\mathcal{V})}(\mathfrak{e}^{\otimes a+1} \otimes 1, \Delta)$$

is an isomorphism for all $a > B_\Delta$.

PROOF. This follows directly from Proposition 3.3.5(iv), and the fact that $\mathcal{Z}_{\mathrm{mot}}(\mathfrak{e}^{\otimes a} \otimes 1, \Delta)(f) = \mathcal{Z}_{\mathrm{mot}}(\mathfrak{e}^{\otimes a+1} \otimes 1, \Delta)(\nu_{\Delta,a} f)$. □

3.4. The triangulated motivic category

We now derive some information on the localization $\mathbf{D}^b_{\mathrm{mot}}(\mathcal{V})$ of $\mathbf{K}^b_{\mathrm{mot}}(\mathcal{V})$, and the full motivic category $\mathcal{DM}(\mathcal{V})$.

3.4.1. Form the triangulated tensor category without unit $\mathbf{D}^b_{\mathrm{mot}}(\mathcal{V})^*$ from the triangulated tensor category without unit $\mathbf{K}^b_{\mathrm{mot}}(\mathcal{V})^*$ by inverting the morphisms of Definition 2.1.4 (except for the Künneth isomorphism (c)).

The DG tensor functors and natural transformation (3.3.1.1) induce the functors

$$\mathbf{D}^b(i_{\mathrm{mot}}): \mathbf{D}^b_{\mathrm{mot}}(\mathcal{V})^* \to \mathbf{D}^b_{\mathrm{mot}}(\mathcal{V}),$$
$$\mathbf{D}^b(r_{\mathrm{mot}}): \mathbf{D}^b_{\mathrm{mot}}(\mathcal{V}) \to \mathbf{D}^b_{\mathrm{mot}}(\mathcal{V})^*,$$

and the natural transformation

$$\mathbf{D}^b(\boxtimes_{\mathrm{mot}}): \mathrm{id}_{\mathbf{D}^b_{\mathrm{mot}}(\mathcal{V})} \to \mathbf{D}^b(i_{\mathrm{mot}}) \circ \mathbf{D}^b(r_{\mathrm{mot}}).$$

We have $\mathbf{D}^b(r_{\mathrm{mot}}) \circ \mathbf{D}^b(i_{\mathrm{mot}}) = \mathrm{id}_{\mathbf{D}^b_{\mathrm{mot}}(\mathcal{V})^*}$.

3.4.2. THEOREM. *For each object X of $\mathbf{D}^b_{\mathrm{mot}}(\mathcal{V})$, the map $\mathbf{D}^b(\boxtimes_{\mathrm{mot}})(X): X \to \mathbf{D}^b(i_{\mathrm{mot}}) \circ \mathbf{D}^b(r_{\mathrm{mot}})(X)$ is an isomorphism, and the functors*

$$\begin{aligned} &\mathbf{D}^b(i_{\mathrm{mot}}): \mathbf{D}^b_{\mathrm{mot}}(\mathcal{V})^* \to \mathbf{D}^b_{\mathrm{mot}}(\mathcal{V}), \\ &\mathbf{D}^b(r_{\mathrm{mot}}): \mathbf{D}^b_{\mathrm{mot}}(\mathcal{V}) \to \mathbf{D}^b_{\mathrm{mot}}(\mathcal{V})^* \end{aligned} \tag{3.4.2.1}$$

are equivalences of triangulated tensor categories without unit.

PROOF. Suppose X is a tensor product of objects of $\mathcal{A}_1(\mathcal{V})$: $X = X_1 \otimes \ldots \otimes X_n$. Then $\mathbf{D}^b(\boxtimes_{\mathrm{mot}})(\mathfrak{e}^{\otimes k} \otimes X)$ is the map

$$\mathrm{id} \otimes \boxtimes_{X_1,\ldots X_n}: \mathfrak{e}^{\otimes k} \otimes X_1 \otimes \ldots \otimes X_n \to \mathfrak{e}^{\otimes k} \otimes X_1 \times \ldots \times X_n,$$

which is an isomorphism by the Künneth isomorphism (Definition 2.1.4(c)). As $\mathbf{D}^b_{\mathrm{mot}}(\mathcal{V})$ is generated as a triangulated category by objects of this form, this suffices to prove the theorem. □

3.4.3. COROLLARY. *Let R be a commutative ring, flat over $\mathbb{Z}$. The categories $\mathbf{D}^b_{\mathrm{mot}}(\mathcal{V})_R$ and $\mathcal{DM}(\mathcal{V})_R$ are triangulated R-tensor categories, with unit $1 = R_S(0)$.*

PROOF. By Lemma 3.1.6, the DG category $\mathcal{A}_{\mathrm{mot}}(\mathcal{V})^*$ has the structure of a DG tensor category with unit $1 = \mathbb{Z}_S(0)$. Applying the functor $\mathbf{K}^b(-)$, we see that $\mathbf{K}^b_{\mathrm{mot}}(\mathcal{V})^*$ has the structure of a triangulated tensor category with unit 1. This structure is preserved under localization (as a triangulated tensor category without unit), hence $\mathbf{D}^b_{\mathrm{mot}}(\mathcal{V})^*$ is a triangulated tensor category with unit 1. Applying the equivalence of triangulated tensor categories without unit $\mathbf{D}^b(r_{\mathrm{mot}}): \mathbf{D}^b_{\mathrm{mot}}(\mathcal{V}) \to \mathbf{D}^b_{\mathrm{mot}}(\mathcal{V})^*$ makes $\mathbf{D}^b_{\mathrm{mot}}(\mathcal{V})$ into a triangulated tensor category with unit 1. On easily checks that this structure is preserved by taking the pseudo-abelian hull, giving $\mathcal{DM}(\mathcal{V})$ the structure of a triangulated tensor category with unit 1. The proof for general R is the same. □

3.4.4. REMARK. For $\Gamma = \mathbb{Z}_X(n)_f$, the multiplication maps in $\mathcal{DM}(\mathcal{V})$

$$\mu^r_\Gamma: \Gamma \otimes 1 \to \Gamma; \quad \mu^l_\Gamma: 1 \otimes \Gamma \to \Gamma$$

are given by the external products: $\mu^r_\Gamma = \boxtimes_{\Gamma,1}$, $\mu^l_\Gamma = \boxtimes_{1,\Gamma}$. More generally, for each object Γ of $\mathbf{D}^b_{\mathrm{mot}}(\mathcal{V})$, we have the identity:

$$\mathbf{D}^b_{\mathrm{mot}}(r)(\Gamma \otimes 1) = \mathbf{D}^b_{\mathrm{mot}}(r)(1 \otimes \Gamma) = \mathbf{D}^b_{\mathrm{mot}}(r)(\Gamma);$$

the multiplication $\mu^l_\Gamma: 1 \otimes \Gamma \to \Gamma$ is given by the composition

$$\begin{aligned} 1 \otimes \Gamma &\xrightarrow{\mathbf{D}^b_{\mathrm{mot}}(\boxtimes_{\mathrm{mot}})(1 \otimes \Gamma)} \mathbf{D}^b_{\mathrm{mot}}(r_{\mathrm{mot}})(1 \otimes \Gamma) = \mathbf{D}^b_{\mathrm{mot}}(r_{\mathrm{mot}})(\Gamma) \\ &\xrightarrow{\mathbf{D}^b_{\mathrm{mot}}(\boxtimes_{\mathrm{mot}})(\Gamma)^{-1}} \Gamma. \end{aligned}$$

3.4.5. Form the triangulated category $\mathbf{D}^b_{\mathrm{mot}}(\mathcal{V})^*_{\mathrm{add}}$ as follows: Let $\mathcal{S}$ be the set of morphisms of the form $\mathrm{id}_{\mathfrak{e}^{\otimes a}} \otimes f$, where f is in the set of morphisms described in Definition 2.1.4(a), (b), (d), (e) and (f). Form the category $\mathbf{D}_{\mathrm{mot}}(\mathcal{V})^*_{\mathrm{add}}$ from the triangulated category $\mathbf{K}^b_{\mathrm{mot}}(\mathcal{V})^*$ by inverting the morphisms in $\mathcal{S}$ (as a triangulated category, *not* as a triangulated *tensor* category).

3.4.6. PROPOSITION. *The canonical exact functor $\mathbf{D}^b_{\mathrm{mot}}(\mathcal{V})^*_{\mathrm{add}} \to \mathbf{D}^b_{\mathrm{mot}}(\mathcal{V})^*$ is an equivalence of triangulated categories.*

PROOF. If (X, f) and (Y, g) are in $\mathcal{L}(\mathcal{V})$, and if $f' : Z \to X$ is a morphism in $\mathcal{V}$, then $\rho_{f,f'} \times \mathrm{id}_{\mathbb{Z}_Y(m)_g} : \mathbb{Z}_X(n)_f \times \mathbb{Z}_Y(n)_g \to \mathbb{Z}_X(n)_{f\cup f'} \times \mathbb{Z}_Y(n)_g$ is the morphism

$$\rho_{f\times g, f'\times g} : \mathbb{Z}_{X\times_S Y}(n+m)_{f\times g} \to \mathbb{Z}_{X\times_S Y}(n+m)_{f\cup f'\times g}.$$

Thus the set of morphisms of Definition 2.1.4(e) are closed under the operation $(-) \times \mathrm{id}_{\mathbb{Z}_Y(m)_g}$. Similarly, the set of morphisms of Definition 2.1.4(a), (b), (d) or (f) is closed under the operation $(-) \times \mathrm{id}_{\mathbb{Z}_Y(m)_g}$. Since the objects $\mathfrak{e}^{\otimes a} \otimes \mathbb{Z}_Y(m)_C$ generate $\mathbf{K}^b_{\mathrm{mot}}(\mathcal{V})^*$ as a triangulated category, the set of morphisms inverted in $\mathbf{K}^b_{\mathrm{mot}}(\mathcal{V})^*$ to form $\mathbf{D}^b_{\mathrm{mot}}(\mathcal{V})^*_{\mathrm{add}}$ is closed under the operation $(-) \times \mathrm{id}_Z$ for Z an arbitrary object in $\mathbf{K}^b_{\mathrm{mot}}(\mathcal{V})^*$. As $\times$ is the tensor operation on $\mathbf{K}^b_{\mathrm{mot}}(\mathcal{V})^*$, it follows that the canonical exact functor $\mathbf{D}^b_{\mathrm{mot}}(\mathcal{V})^*_{\mathrm{add}} \to \mathbf{D}^b_{\mathrm{mot}}(\mathcal{V})^*$ is an isomorphism. □

3.4.7. REMARKS. From Definition 1.4.12 and Definition 3.2.3, we have the graded tensor category $\mathcal{A}^0_{\mathrm{mot}}(\mathcal{V})$, the full subcategory $\mathcal{A}^0_{\mathrm{mot}}(\mathcal{V})^*$ of $\mathcal{A}^0_{\mathrm{mot}}(\mathcal{V})$, and the commutative diagram (3.2.4.1)

$$\begin{array}{ccc}
\mathcal{A}_{\mathrm{mot}}(\mathcal{V}) & \xrightarrow{H_{\mathrm{mot}}} & \mathcal{A}^0_{\mathrm{mot}}(\mathcal{V}) \\
{\scriptstyle r_{\mathrm{mot}}}\downarrow\uparrow{\scriptstyle i_{\mathrm{mot}}} & & {\scriptstyle i^0_{\mathrm{mot}}}\uparrow\downarrow{\scriptstyle r^0_{\mathrm{mot}}} \\
\mathcal{A}_{\mathrm{mot}}(\mathcal{V})^* & \xrightarrow[H^*_{\mathrm{mot}}]{} & \mathcal{A}^0_{\mathrm{mot}}(\mathcal{V})^*.
\end{array}$$

The natural transformation $\boxtimes_{\mathrm{mot}}$ induces the natural transformation

$$\boxtimes^0_{\mathrm{mot}} : \mathrm{id}_{\mathcal{A}^0_{\mathrm{mot}}(\mathcal{V})} \to i^0_{\mathrm{mot}} \circ r^0_{\mathrm{mot}};$$

the functor $\mathcal{Z}_{\mathrm{mot}} : \mathcal{A}_{\mathrm{mot}}(\mathcal{V}) \to \mathbf{GrAb}$ factors as

$$\mathcal{A}_{\mathrm{mot}}(\mathcal{V}) \xrightarrow{H^*_{\mathrm{mot}} \circ r_{\mathrm{mot}}} \mathcal{A}^0_{\mathrm{mot}}(\mathcal{V})^* \xrightarrow{\mathcal{Z}^{0*}_{\mathrm{mot}}} \mathbf{GrAb}.$$

Define the categories:

$$\mathbf{C}^{b0}_{\mathrm{mot}}(\mathcal{V}) := \mathbf{C}^b(\mathcal{A}^0_{\mathrm{mot}}(\mathcal{V})), \qquad \mathbf{C}^{b0}_{\mathrm{mot}}(\mathcal{V})^* := \mathbf{C}^b(\mathcal{A}^0_{\mathrm{mot}}(\mathcal{V})^*),$$
$$\mathbf{K}^{b0}_{\mathrm{mot}}(\mathcal{V}) := \mathbf{K}^b(\mathcal{A}^0_{\mathrm{mot}}(\mathcal{V})), \qquad \mathbf{K}^{b0}_{\mathrm{mot}}(\mathcal{V})^* := \mathbf{K}^b(\mathcal{A}^0_{\mathrm{mot}}(\mathcal{V})^*).$$

We let $\mathbf{D}^{b0}_{\mathrm{mot}}(\mathcal{V})$ be the triangulated tensor category gotten from $\mathbf{K}^{b0}_{\mathrm{mot}}(\mathcal{V})$ by inverting the maps of Definition 2.1.4, and define $\mathbf{D}^{b0}_{\mathrm{mot}}(\mathcal{V})^*$ similarly as a localization of $\mathbf{D}^{b0}_{\mathrm{mot}}(\mathcal{V})^*$. We have the triangulated category $\mathbf{D}^{b0}_{\mathrm{mot}}(\mathcal{V})^*_{\mathrm{add}}$ formed from $\mathbf{K}^{b0}_{\mathrm{mot}}(\mathcal{V})$ by inverting the maps of Definition 2.1.4 as triangulated category.

(i) The cycles functors

$$\mathbf{C}^b(\mathcal{Z}_{\mathrm{mot}}) : \mathbf{C}^b_{\mathrm{mot}}(\mathcal{V}) \to \mathbf{C}^b(\mathbf{Ab}),$$
$$\mathbf{K}^b(\mathcal{Z}_{\mathrm{mot}}) : \mathbf{K}^b_{\mathrm{mot}}(\mathcal{V}) \to \mathbf{K}^b(\mathbf{Ab})$$

factor as

$$\mathbf{C}^b_{\mathrm{mot}}(\mathcal{V}) \xrightarrow{\mathbf{C}^b(H^*_{\mathrm{mot}} \circ r_{\mathrm{mot}})} \mathbf{C}^{b0}_{\mathrm{mot}}(\mathcal{V})^* \xrightarrow{\mathbf{C}^b(\mathcal{Z}^{0*}_{\mathrm{mot}})} \mathbf{C}^b(\mathbf{Ab}),$$
$$\mathbf{K}^b_{\mathrm{mot}}(\mathcal{V}) \xrightarrow{\mathbf{K}^b(H^*_{\mathrm{mot}} \circ r_{\mathrm{mot}})} \mathbf{K}^{b0}_{\mathrm{mot}}(\mathcal{V})^* \xrightarrow{\mathbf{K}^b(\mathcal{Z}^{0*}_{\mathrm{mot}})} \mathbf{K}^b(\mathbf{Ab}).$$

(ii) Replace $?^b_{\mathrm{mot}}(\mathcal{V})$ with $?^{b0}_{\mathrm{mot}}(\mathcal{V})$, $?^b_{\mathrm{mot}}(\mathcal{V})^*$ with $?^{b0}_{\mathrm{mot}}(\mathcal{V})^*$, for $? = \mathbf{C}, \mathbf{K}$ and $\mathbf{D}$, and replace $\mathbf{D}^b_{\mathrm{mot}}(\mathcal{V})^*_{\mathrm{add}}$ with $\mathbf{D}^{b0}_{\mathrm{mot}}(\mathcal{V})^*_{\mathrm{add}}$. Then the analogs of all the results of this section remain valid, with similar proofs.

(iii) It follows from Proposition 3.3.5, together with the analog of Proposition 3.3.5 for the category $\mathbf{K}^{b0}_{\mathrm{mot}}(\mathcal{V})$, that the map

$$\mathbf{K}^b(H^*_{\mathrm{mot}})(\mathfrak{e}^{\otimes a}\otimes 1,\Delta):\mathrm{Hom}_{\mathbf{K}^b_{\mathrm{mot}}(\mathcal{V})}(\mathfrak{e}^{\otimes a}\otimes 1,\Delta)\to\mathrm{Hom}_{\mathbf{K}^{b0}_{\mathrm{mot}}(\mathcal{V})}(\mathfrak{e}^{\otimes a}\otimes 1,\Delta)$$

is an isomorphism for all Δ in $\mathbf{K}^b_{\mathrm{mot}}(\mathcal{V})^*_B$ (see §3.3.4), and all $a > B$.

(iv) We let $\mathcal{DM}^0(\mathcal{V})$ be the pseudo-abelian hull $\mathbf{D}^{b0}_{\mathrm{mot}}(\mathcal{V})_\#$ of $\mathbf{D}^{b0}_{\mathrm{mot}}(\mathcal{V})$. The functor $\mathbf{K}^b(H_{\mathrm{mot}})$ gives rise to the commutative diagram of exact tensor functors

$$\begin{array}{ccc}\mathbf{D}^b_{\mathrm{mot}}(\mathcal{V}) & \xrightarrow{\mathbf{D}^b(H_{\mathrm{mot}})} & \mathbf{D}^{b0}_{\mathrm{mot}}(\mathcal{V})\\ \downarrow & & \downarrow\\ \mathcal{DM}(\mathcal{V}) & \xrightarrow[\mathcal{DM}(H_{\mathrm{mot}})]{} & \mathcal{DM}^0(\mathcal{V}).\end{array}$$

We may view the categories $\mathcal{A}^0_{\mathrm{mot}}(\mathcal{V})$, $\mathbf{D}^{b0}_{\mathrm{mot}}(\mathcal{V})$, and $\mathcal{DM}^0(\mathcal{V})$ as the "naive" versions of $\mathcal{A}_{\mathrm{mot}}(\mathcal{V})$, $\mathbf{D}^b_{\mathrm{mot}}(\mathcal{V})$,and $\mathcal{DM}(\mathcal{V})$, as we have replaced the DG tensor structure in $\mathcal{A}_{\mathrm{mot}}(\mathcal{V})$ (which gives the structural identities only up to homotopy) with the graded tensor structure in $\mathcal{A}^0_{\mathrm{mot}}(\mathcal{V})$ (which gives the structural identities on the nose).

3.5. Cycles and cycle classes

In this section, we construct the cycle map and the cycle class map, and consider their basic properties.

3.5.1. *The cycle map and the cycle class map.* We have the cohomological functor

$$H^0:\mathbf{K}^b(\mathbf{Ab})\to\mathbf{Ab}$$
$$H^0(X):=\mathrm{Hom}_{\mathbf{K}^b(\mathbf{Ab})}(\mathbb{Z},X).$$

Let Γ be an object of $\mathbf{K}^b_{\mathrm{mot}}(\mathcal{V})^*_B$ (see §3.3.4). We have the functor (3.3.1.3), the corresponding object $\mathcal{Z}_{\mathrm{mot}}(\Gamma)$ of $\mathbf{K}^b(\mathbf{Ab})$, and the abelian group $H^0(\mathcal{Z}_{\mathrm{mot}}(\Gamma))$. By Proposition 3.3.5, we have the isomorphism

$$H^0\circ\mathcal{Z}_{\mathrm{mot}}(-):\mathrm{Hom}_{\mathbf{K}^b_{\mathrm{mot}}(\mathcal{V})}(\mathfrak{e}^{\otimes a}\otimes 1,\Gamma)\to H^0(\mathcal{Z}_{\mathrm{mot}}(\Gamma))\tag{3.5.1.1}$$

for all $a > B$. We define the map

$$\mathrm{cyc}_\Gamma:H^0(\mathcal{Z}_{\mathrm{mot}}(\Gamma))\to\varinjlim_a\mathrm{Hom}_{\mathbf{K}^b_{\mathrm{mot}}(\mathcal{V})}(\mathfrak{e}^{\otimes a}\otimes 1,\Gamma)\tag{3.5.1.2}$$

to be the inverse of the isomorphism (3.5.1.1); here the limit is with respect to the maps (3.3.6.1). For Γ in $\mathbf{K}^b_{\mathrm{mot}}(\mathcal{V})^*_B$, the limit is constant for $a > B$, by Lemma 3.3.7.

We have the unit isomorphism (2.2.4.1) $\nu_a:\mathfrak{e}^{\otimes a}\otimes 1\to 1$. Let Γ be an object of $\mathbf{K}^b_{\mathrm{mot}}(\mathcal{V})^*_B$, and let Z be an element of $H^0(\mathcal{Z}_{\mathrm{mot}}(\Gamma))$. We let $\mathrm{cl}_\Gamma(Z):1\to\Gamma$ be the morphism in $\mathbf{D}^b_{\mathrm{mot}}(\mathcal{V})$ defined by the composition

$$1\xrightarrow{\nu_a^{-1}}\mathfrak{e}^{\otimes a}\otimes 1\xrightarrow{\mathrm{cyc}_\Gamma(Z)}\Gamma$$

for any $a > B$. This is easily seen to be independent of the choice of a.

Let $H^0_{\mathrm{mot}}:\mathbf{K}^b_{\mathrm{mot}}(\mathcal{V})\to\mathbf{Ab}$ be the cohomological functor $\mathrm{Hom}_{\mathbf{D}^b_{\mathrm{mot}}(\mathcal{V})}(1,-)$. The assignment $Z\mapsto\mathrm{cl}_\Gamma(Z)$ defines the homomorphism

$$\mathrm{cl}_\Gamma:H^0(\mathcal{Z}_{\mathrm{mot}}(\Gamma))\to H^0_{\mathrm{mot}}(\Gamma).\tag{3.5.1.3}$$

3.5.2. *The cycle map and cycle class map for varieties.* Let X be in $\mathcal{V}$, (X, f) in $\mathcal{L}(\mathcal{V})$, and $Z \in \mathcal{Z}^d(X)_f$. The map (1.4.6.1) in $\mathcal{A}_3(\mathcal{V})$ determines the map in $\mathcal{A}_{\mathrm{mot}}(\mathcal{V})$, $[Z]^S : \mathfrak{e} \otimes 1 \to \mathbb{Z}_X(d)_f[2d]$ (see Remark 1.4.11). Since

$$\begin{aligned} \mathcal{Z}_{\mathrm{mot}}(\mathbb{Z}_X(d)_f[2d]) &= \mathcal{Z}^d(X)_f, \\ \mathcal{Z}_{\mathrm{mot}}(\mathfrak{e} \otimes 1) &= \mathbb{Z}, \\ \mathcal{Z}_{\mathrm{mot}}([Z]^S)(1) &= Z, \end{aligned} \tag{3.5.2.1}$$

it follows from Proposition 3.3.5 that sending Z to $[Z]^S$ gives the isomorphism

$$\mathcal{Z}^d(X)_f \xrightarrow{[-]^S} \mathrm{Hom}_{\mathbf{K}^b_{\mathrm{mot}}(\mathcal{V})}(\mathfrak{e} \otimes 1, \mathbb{Z}_X(d)_f[2d]).$$

Using the identities (3.5.2.1) we define

$$\mathrm{cyc}^d_{X,f} : \mathcal{Z}^d(X)_f \to \mathrm{Hom}_{\mathbf{K}^b_{\mathrm{mot}}(\mathcal{V})}(\mathfrak{e} \otimes 1, \mathbb{Z}_X(d)_f[2d]) \tag{3.5.2.2}$$

as the composition of $\mathrm{cyc}_{\mathbb{Z}_X(d)_f[2d]}$ with the canonical isomorphism

$$\mathcal{Z}^d(X)_f \cong H^0(\mathcal{Z}^d(X)_f) = H^0(\mathcal{Z}_{\mathrm{mot}}(\mathbb{Z}_X(d)_f[2d])). \tag{3.5.2.3}$$

It follows directly from the definitions that $\mathrm{cyc}^d_{X,f}(Z) = [Z]^S$.

Similarly, if $\hat{X}$ is a closed subset of X with complement $j : U \to X$, we have the subgroup $\mathcal{Z}^d_{\hat{X}}(X)_f$ of $\mathcal{Z}^d(X)_f$ defined by the exactness of

$$0 \to \mathcal{Z}^d_{\hat{X}}(X)_f \to \mathcal{Z}^d(X)_f \xrightarrow{j^*} \mathcal{Z}^d(U)_{j^*f}.$$

Since the map (3.5.2.2) is a functorial isomorphism, we have the canonically defined map

$$\mathrm{cyc}^d_{X,\hat{X},f} : \mathcal{Z}^d_{\hat{X}}(X)_f \to \mathrm{Hom}_{\mathbf{K}^b_{\mathrm{mot}}(\mathcal{V})}(\mathfrak{e} \otimes 1, \mathbb{Z}_{X,\hat{X}}(d)_f[2d]), \tag{3.5.2.4}$$

compatible with $\mathrm{cyc}^d_{X,f}$ via the canonical map

$$\mathrm{Hom}_{\mathbf{K}^b_{\mathrm{mot}}(\mathcal{V})}(\mathfrak{e} \otimes 1, \mathbb{Z}_{X,\hat{X}}(d)_f[2d]) \to \mathrm{Hom}_{\mathbf{K}^b_{\mathrm{mot}}(\mathcal{V})}(\mathfrak{e} \otimes 1, \mathbb{Z}_X(d)_f[2d]).$$

It follows similarly from the definitions and Proposition 3.3.5 that $\mathrm{cyc}^d_{X,\hat{X},f}(Z) = [Z]^S_{\hat{X}}$ in $\mathbf{K}^b_{\mathrm{mot}}(\mathcal{V})$, where $[Z]_{\hat{X}}$ is the map (2.1.3.3).

For $(X, f) \in \mathcal{L}(\mathcal{V})$, we have $H^{2d}(X, \mathbb{Z}(d)) = H^0_{\mathrm{mot}}(\mathbb{Z}_X(d)_f[2d])$ by definition. We define the homomorphism

$$\mathrm{cl}^d_{X,f} : \mathcal{Z}^d(X)_f \to H^{2d}(X, \mathbb{Z}(d)) \tag{3.5.2.5}$$

as the map $\mathrm{cl}_{\mathbb{Z}_X(d)_f[2d]}$, composed with the isomorphism (3.5.2.3). By the functoriality of the maps "change of f", the maps $\mathrm{cl}^d_{X,f}$ fit together to give a homomorphism

$$\mathrm{cl}^d_X : \mathcal{Z}^d(X/S) \to H^{2d}(X, \mathbb{Z}(d)),$$

which we call the *cycle class map*.

The cycle maps with support give similarly the map

$$\mathrm{cl}^d_{X,\hat{X},f} : \mathcal{Z}^d_{\hat{X}}(X)_f \to H^{2d}_{\hat{X}}(X, \mathbb{Z}(d)) \tag{3.5.2.6}$$

and the map

$$\mathrm{cl}^d_{X,\hat{X}} : \mathcal{Z}^d_{\hat{X}}(X/S) \to H^{2d}_{\hat{X}}(X, \mathbb{Z}(d)). \tag{3.5.2.7}$$

3.5.3. PROPOSITION. *Sending* $\Gamma \in \mathbf{K}^b_{\mathrm{mot}}(\mathcal{V})^*$ *to* cyc_Γ *defines an exact natural transformation of cohomological functors*

$$H^0(\mathcal{Z}_{\mathrm{mot}}(-)) \to \varinjlim_a \mathrm{Hom}_{\mathbf{K}^b_{\mathrm{mot}}(\mathcal{V})^*}(\mathfrak{e}^{\otimes a} \otimes 1, -)$$

from $\mathbf{K}^b_{\mathrm{mot}}(\mathcal{V})^*$ *to* $\mathbf{Ab}$. *Sending* $\Gamma \in \mathbf{K}^b_{\mathrm{mot}}(\mathcal{V})^*$ *to* cl_Γ *defines an exact natural transformation of cohomological functors* $H^0(\mathcal{Z}_{\mathrm{mot}}(-)) \to H^0_{\mathrm{mot}}(-)$ *from* $\mathbf{K}^b_{\mathrm{mot}}(\mathcal{V})^*$ *to* $\mathbf{Ab}$. *In particular, the following properties of the cycle class map hold:*

(i) *Let* $p: Y \to X$ *be a map in* $\mathcal{V}$, $\hat{X}$ *a closed subset of* X *and* $\hat{Y}$ *a closed subset of* Y *containing* $p^{-1}(\hat{X})$. *Then*

$$p^*(\mathrm{cl}^d_{X,\hat{X}}(Z)) = \mathrm{cl}^d_{Y,\hat{Y}}(p^*(Z)),$$

for Z *in* $\mathcal{Z}^d_{\hat{X}}(X)_{p\cup \mathrm{id}_X}$.

(ii) *Let* $i: X \to X \coprod Y$ *be the inclusion, with* X *and* Y *in* $\mathcal{V}$, *let* $\hat{X}$ *be a closed subset of* X *and* $\hat{Y}$ *a closed subset of* Y. *Then*

$$i_*(\mathrm{cl}^d_{X,\hat{X}}(Z)) = \mathrm{cl}^d_{X\sqcup Y,\hat{X}\sqcup\hat{Y}}(i_*(Z)),$$

for Z *in* $\mathcal{Z}^d_{\hat{X}}(X/S)$.

PROOF. This follows from the fact that $\mathcal{Z}_{\mathrm{mot}}(-)$ is an exact functor. □

3.5.4. LEMMA. *We have*

$$\mathrm{cl}^0_S(|S|) = \mathrm{id}_1.$$

PROOF. By definition (see Remark 1.4.11), the map $[|S|]^S: \mathfrak{e} \otimes 1 \to 1$ is the composition $\boxtimes_{1,1} \circ ([|S|] \otimes \mathrm{id}_1)$. This latter morphism is the isomorphism $\nu_1: \mathfrak{e} \otimes 1 \to 1$. As $\mathrm{cl}^0_S(|S|) = [|S|]^S \circ \nu_1^{-1}$ by definition, the lemma follows. □

Recall from §2.2.11 the definition of external products, and cup products, for motivic cohomology with support.

3.5.5. LEMMA. *Let* X *and* Y *be in* $\mathcal{V}$, *let* $\hat{X}$ *be a closed subset of* X *and* $\hat{Y}$ *a closed subset of* Y. *Take* A *in* $\mathcal{Z}^d_{\hat{X}}(X/S)$, *and* B *in* $\mathcal{Z}^e_{\hat{Y}}(Y/S)$. *Then the product cycle* $A \times_{/S} B$ *is in* $\mathcal{Z}^{e+d}_{\hat{X}\times_S\hat{Y}}(X \times_S Y/S)$, *and*

$$\mathrm{cl}^{d+e}_{X\times_S Y,\hat{X}\times_S\hat{Y}}(A \times_{/S} B) = \mathrm{cl}^d_{X,\hat{X}}(A) \cup^{\hat{X},\hat{Y}}_{X,Y} \mathrm{cl}^e_{Y,\hat{Y}}(B).$$

PROOF. It follows from (Appendix A, Remark 2.3.3(i)), that $A \times_{/S} B$ is in $\mathcal{Z}^{e+d}(X \times_S Y/S)$; clearly $A \times_{/S} B$ is supported in $\hat{X} \times \hat{Y}$. By Definition 1.4.8(ii), we have the identity in the homotopy category of $\mathcal{A}_5(\mathcal{V})$,

$$\boxtimes_{X\times_S Y,S} \circ ([A \times_{/S} B] \otimes [|S|]) = \boxtimes_{X,Y} \circ ([A] \otimes [B]),$$

as maps from $\mathfrak{e} \otimes \mathfrak{e}$ to $\mathbb{Z}_{X\times_S Y}(d+e)[2d+2e]$.

Using the notation of §2.2.11 we have the map

$$\theta^{\hat{X},\hat{Y}}_{X,Y}: \mathbb{Z}_{X\times_S Y,\hat{X}\times\hat{Y}}(d+e)_\Delta \to \mathbb{Z}_{X,\hat{X}}(d) \times \mathbb{Z}_{Y,\hat{Y}}(e).$$

By Proposition 3.3.5, the map

$$\mathrm{Hom}_{\mathbf{K}^b_{\mathrm{mot}}(\mathcal{V})}(\mathfrak{e} \otimes \mathfrak{e}, \mathbb{Z}_{X\times_S Y,\hat{X}\times\hat{Y}}(d+e)_\Delta[2d+2e]) \\ \to \mathrm{Hom}_{\mathbf{K}^b_{\mathrm{mot}}(\mathcal{V})}(\mathfrak{e} \otimes \mathfrak{e}, \mathbb{Z}_{X\times_S Y}(d+e)[2d+2e])$$

is injective, so we have the identity of maps in $\mathbf{K}^b_{\mathrm{mot}}(\mathcal{V})$,

$$\theta^{\hat{X},\hat{Y}}_{X,Y} \circ \boxtimes_{X\times_S Y,S} \circ ([A \times_{/S} B]_{\hat{X}\times_S\hat{Y}} \otimes [|S|]) = \boxtimes_{X,Y} \circ ([A]_{\hat{X}} \otimes [B]_{\hat{Y}}).$$

This in turn implies the identity of maps in $\mathcal{DM}$:

$$\mathrm{cl}^{d+e}_{X\times_S Y,\hat{X}\times_S\hat{Y}}(A \times_{/S} B) = \boxtimes^{\hat{X},\hat{Y}}_{X,Y} \circ ([A]_{\hat{X}} \otimes [B]_{\hat{Y}}) \circ \nu_2^{-1} : 1 \to \mathbb{Z}_{X\times_S Y}(d+e)[2d+2e].$$

From the definition of the tensor product in $\mathcal{DM}$, and the definition (2.2.11.3) of the product $\cup^{\hat{X},\hat{Y}}_{X,Y}$, we have the identity

$$\boxtimes^{\hat{X},\hat{Y}}_{X,Y} \circ ([A]_{\hat{X}} \otimes [B]_{\hat{Y}}) \circ \nu_2^{-1} = \mathrm{cl}^d_{\hat{X}}(A) \cup^{\hat{X},\hat{Y}}_{X,Y} \mathrm{cl}^e_{\hat{Y}}(B),$$

completing the proof. □

3.5.6. PROPOSITION. *Let X be in $\mathcal{V}$. Then*

(i) *$\oplus_{p,q} H^p(X, \mathbb{Z}(q))$, with product $\cup_X$, is an associative, bi-graded ring, graded-commutative with respect to p, with unit $1 \in H^0(X, \mathbb{Z}(0))$ given by the map $\mathrm{cl}^0_X(|X|) : 1 \to \mathbb{Z}_X(0)$.*

(ii) *Let $p: Y \to X$ be a map in $\mathcal{V}$. Then $p^* : \oplus_{p,q} H^p(X, \mathbb{Z}(q)) \to \oplus_{p,q} H^p(Y, \mathbb{Z}(q))$ is a ring homomorphism.*

PROOF. (i) Associativity and graded-commutativity of the product $\cup_X$ follow from the associativity and graded-commutativity of the tensor product in the tensor category $\mathcal{DM}$.

We now show that $\mathrm{cl}^0_X(|X|)$ acts as a unit. Let $p_X : X \to S$ be the structure morphism. We have the commutative diagram

$$\begin{array}{ccccc}
\mathbb{Z}_S(0) \otimes \mathbb{Z}_X(q)[p] & \xrightarrow{\boxtimes_{S,X}} & \mathbb{Z}_X(q)[p] & & \\
\downarrow{\scriptstyle p_X^* \otimes \mathrm{id}} & & \downarrow{\scriptstyle p_2^*} & & \\
\mathbb{Z}_X(0) \otimes \mathbb{Z}_X(q)[p] & \xrightarrow[\boxtimes_{X,X}]{} & \mathbb{Z}_{X\times_S X}(q)[p] & \xrightarrow[\Delta^*]{} & \mathbb{Z}_X(q)[p].
\end{array}$$

Let $f : 1 \to \mathbb{Z}_X(q)[p]$ be a map in $\mathcal{DM}$. By Corollary 3.4.3, $\mathcal{DM}$ is a tensor category with unit 1, hence we have the commutative diagram

$$\begin{array}{ccc}
1 \otimes 1 & \xrightarrow{\mu} & 1 \\
\downarrow{\scriptstyle \mathrm{id}\otimes f} & & \downarrow{\scriptstyle f} \\
1 \otimes \mathbb{Z}_X(q)[p] & \xrightarrow[\mu^l_X]{} & \mathbb{Z}_X(q)[p].
\end{array}$$

From the definition of the unit structure in $\mathcal{DM}$ (see Remark 3.4.4), we have

$$\mu^l_X = \boxtimes_{S,X} : \mathbb{Z}_S(0) \otimes \mathbb{Z}_X(q)[p] \to \mathbb{Z}_X(q)[p].$$

Thus, we may put the two commutative diagrams together, giving the identity

$$f = p_X^* \cup_X f. \tag{3.5.6.1}$$

By Proposition 3.5.3(i) and Lemma 3.5.4, we have

$$p_X^* = p_X^* \circ \mathrm{cl}^0_S(|S|) = \mathrm{cl}^0_X(|X|); \tag{3.5.6.2}$$

combining (3.5.6.1) and (3.5.6.2) shows that $\mathrm{cl}^0(|X|)$ is a unit.

The proof of (ii) is similar and is left to the reader. □

3.5.7. PROPOSITION. *Let X be in $\mathcal{V}$, let $\hat{X}_1$ and $\hat{X}_2$ be closed subsets of X, and let $A \in \mathcal{Z}^d_{\hat{X}_1}(X/S)$, $B \in \mathcal{Z}^e_{\hat{X}_2}(X/S)$ be cycles on X. Suppose that each component of the intersection $\operatorname{supp}(A) \cap \operatorname{supp}(B)$ intersects each fiber X_s of X over S in a subset of codimension at least $d+e$. Then the intersection product of A and B in X over S, $A \cdot_{X/S} B$, exists, $A \cdot_{X/S} B$ is in $\mathcal{Z}^{d+e}_{\hat{X}_1 \cap \hat{X}_2}(X/S)$, and*

$$\mathrm{cl}^{d+e}_{X,\hat{X}_1\cap\hat{X}_2}(A \cdot_{X/S} B) = \mathrm{cl}^d_{X,\hat{X}_1}(A) \cup_X^{\hat{X}_1,\hat{X}_2} \mathrm{cl}^e_{X,\hat{X}_2}(B).$$

PROOF. The intersection product of A and B in X over S is given by

$$A \cdot_{X/S} B = \Delta_X^*(A \times_{/S} B),$$

whenever $\Delta_X^*(A \times_{/S} B)$ is defined. By our assumptions on A, B and $\operatorname{supp}(A) \cap \operatorname{supp}(B)$, the cycle $A \times_{/S} B$ is in $\mathcal{Z}^{d+e}_{\hat{X}_1 \cap \hat{X}_2}(X \times_S X)_{\mathrm{id} \cup \Delta_X}$. By Lemma 1.2.2, the cycle $\Delta_X^*(A \times_{/S} B)$ is defined and is in $\mathcal{Z}^{d+e}(X/S)$. By Proposition 3.5.3, we have

$$\begin{aligned}\Delta_X^* \circ \mathrm{cl}^{d+e}_{X\times_S X,\hat{X}_1\times\hat{X}_2}(A \times_{/S} B) &= \mathrm{cl}^{d+e}_{X,\hat{X}_1\cap\hat{X}_2}(\Delta_X^*(A \times_{/S} B)) \\ &= \mathrm{cl}^{d+e}_{X,\hat{X}_1\cap\hat{X}_2}(A \cdot_{X/S} B);\end{aligned}$$

by Lemma 3.5.5, we have

$$\begin{aligned}\Delta_X^* \circ \mathrm{cl}^{d+e}_{X\times_S X,\hat{X}_1\times\hat{X}_2}(A \times_{/S} B) &= \Delta_X^*(\mathrm{cl}^d_{X,\hat{X}_1}(A) \cup_{X,X}^{\hat{X}_1,\hat{X}_2} \mathrm{cl}^e_{X,\hat{X}_2}(B)) \\ &= \mathrm{cl}^d_{X,\hat{X}_1}(A) \cup_X^{\hat{X}_1,\hat{X}_2} \mathrm{cl}^e_{X,\hat{X}_2}(B),\end{aligned}$$

completing the proof. □

CHAPTER II

Motivic Cohomology and Higher Chow Groups

In [**19**], Bloch defines his *higher Chow groups* as a candidate for a reasonable theory of motivic cohomology. In this chapter, we extend and modify Bloch's definition of higher Chow groups to give a theory of higher Chow groups for motives over a given base scheme S; in case $S = \operatorname{Spec} k$, for k a field, the higher Chow groups of the motive of a smooth k-variety agree with Bloch's original construction. We show that, if the motivic Chow groups satisfy certain natural conditions (see §3.2.1 and §3.3.1), then the motivic cohomology groups defined in Chapter I, §2.2.7 agree with the motivic Chow groups (Theorem 3.3.10). We are able to verify the axioms in case $S = \operatorname{Spec} k$, k a field, or if S is smooth and of dimension one over a field (Theorem 3.6.6), putting Bloch's higher Chow groups in a categorical framework.

The agreement of motivic cohomology with the motivic Chow groups gives an interpretation of motivic cohomology as Zariski hypercohomology, which enables us to prove some additional properties of motivic cohomology, such as a Gersten-type resolution, a local to global spectral sequence, and the like. These properties are treated in §3.4.

1. Hypercohomology in the motivic category

We begin by describing how to define Zariski hypercohomology for objects of $\mathbf{C}^b_{\mathrm{mot}}(\mathcal{V})$.

1.1. Čech resolutions for $\mathcal{A}_{\mathrm{mot}}$

1.1.1. Let (X, f) be in $\mathcal{L}(\mathcal{V})$, and let $\mathcal{U} := \{U_0, \ldots, U_m\}$ be a Zariski open cover of X. For an ordered index $I = (i_0 < \ldots < i_k)$, with $0 \le i_j \le m$, we let U_I denote the intersection $U_I = U_{i_0} \cap \ldots \cap U_{i_m}$. We have the augmented simplicial scheme $j_{\mathcal{U}}: \mathcal{U}_* \to X$ where $\mathcal{U}_*$ is the simplicial scheme with non-degenerate k-simplices $\mathcal{U}^{\mathrm{n.d.}}_* = \coprod_{I=(i_0<\ldots<i_k)} U_I$. Let $j_k: \mathcal{U}^{\mathrm{n.d.}}_k \to X$ be the union of the inclusions. This gives us the non-degenerate simplicial scheme $\mathcal{U}^{\mathrm{n.d.}}_*$, which is the empty scheme in degrees $n > m + 1$ (*cf.* Chapter I, §2.5.4).

We may then lift $\mathcal{U}^{\mathrm{n.d.}}_*$ to the non-degenerate simplicial object

$$(\mathcal{U}_*, j^* f)^{\mathrm{n.d.}}: \Delta^{\mathrm{op}}_{\mathrm{n.d.}} \to \mathcal{L}(\mathcal{V})$$

with $(j^* f)_k = j_k^* f$, as in Chapter I, *loc. cit.* We let $\mathbb{Z}_{\mathcal{U}^{\mathrm{n.d.}}_*}(q)_f$ be the corresponding object of $\mathbf{C}^b_{\mathrm{mot}}(\mathcal{V})$,

$$\mathbb{Z}_{\mathcal{U}^{\mathrm{n.d.}}_*}(q)^s_f := \mathbb{Z}_{\mathcal{U}^{\mathrm{n.d.}}_*([s])}(q)_{j^* f([s])},$$

with differential the usual alternating sum.

From the Mayer-Vietoris distinguished triangle (I.2.2.6.1), we see that the augmentation induces an isomorphism

$$j^*_{X,\mathcal{U}}: \mathfrak{e}^{\otimes a} \otimes \mathbb{Z}_X(q)_f[p] \to \mathfrak{e}^{\otimes a} \otimes \mathbb{Z}_{\mathcal{U}^{\text{n.d.}}_*}(q)_f[p] \tag{1.1.1.1}$$

in $\mathbf{D}^b_{\text{mot}}(\mathcal{V})$ for all $a \geq 0$.

We call the map (1.1.1.1) in $\mathbf{C}^b_{\text{mot}}(\mathcal{V})$ a *Čech resolution* of $\mathfrak{e}^{\otimes a} \otimes \mathbb{Z}_X(q)_f[p]$; we extend the notion of a Čech resolution to arbitrary objects of $\mathcal{A}_{\text{mot}}(\mathcal{V})^*$ by taking direct sums.

If we have a Čech resolution

$$j^*_{X,\mathcal{U}}: \mathfrak{e}^{\otimes a} \otimes \mathbb{Z}_X(q)_f[p] \to \mathfrak{e}^{\otimes a} \otimes \mathbb{Z}_{\mathcal{U}^{\text{n.d.}}_*}(q)_f[p]$$

coming from a cover $\mathcal{U}$ of X, each refinement $\rho: \mathcal{V} \to \mathcal{U}$ gives rise to a commutative diagram

$$\begin{array}{ccc} \mathfrak{e}^{\otimes a} \otimes \mathbb{Z}_X(q)_f[p] & \xrightarrow{j^*_{X,\mathcal{U}}} & \mathfrak{e}^{\otimes a} \otimes \mathbb{Z}_{\mathcal{U}^{\text{n.d.}}_*}(q)_f[p] \\ \| & & \downarrow{\scriptstyle \rho^*} \\ \mathfrak{e}^{\otimes a} \otimes \mathbb{Z}_X(q)_f[p] & \xrightarrow[j^*_{X,\mathcal{V}}]{} & \mathfrak{e}^{\otimes a} \otimes \mathbb{Z}_{\mathcal{V}^{\text{n.d.}}_*}(q)_f[p]. \end{array}$$

1.1.2. DEFINITION. Let $\Gamma = \oplus_{i=1}^m \mathfrak{e}^{\otimes a_i} \otimes \mathbb{Z}_{X_i}(q_i)[p_i]_{f_i}$ be an object of $\mathcal{A}_{\text{mot}}(\mathcal{V})^*$. A *Zariski open cover of* Γ consists of a finite Zariski open cover $\mathcal{U}_i = \{U_{0i}, \dots, U_{n_i i}\}$ of X_i for each $i = 1, \dots, m$. If $\mathcal{U}$ and $\mathcal{W}$ are Zariski open covers of Γ, a refinement of $\mathcal{U}$ by $\mathcal{W}$, $\rho: \mathcal{W} \to \mathcal{U}$, is a collection of refinements $\rho_i: \mathcal{W}_i \to \mathcal{U}_i$ for each $i = 1, \dots, m$.

If $\mathcal{U}$ is a Zariski open cover of $\Gamma \in \mathcal{A}_{\text{mot}}(\mathcal{V})^*$, the direct sum of the Čech resolutions for each component $\mathfrak{e}^{\otimes a_i} \otimes \mathbb{Z}_{X_i}(q_i)[p_i]_{f_i}$ gives a Čech resolution of Γ. We denote this map in $\mathbf{C}^b_{\text{mot}}(\mathcal{V})^*$ by $j^*_{\Gamma,\mathcal{U}}: \Gamma \to \Gamma_{\mathcal{U}}$.

1.2. A structural result

Before proceeding to extend the notions of §1.1 to $\mathbf{C}^b_{\text{mot}}(\mathcal{V})^*$, we need to examine the morphisms in the category $\mathcal{A}_{\text{mot}}(\mathcal{V})^*$ a bit more closely.

1.2.1. We recall that the category $\mathcal{A}_{\text{mot}}(\mathcal{V})$ is constructed from the tensor category $\mathcal{A}_2(\mathcal{V})$ by taking the coproduct with the DG tensor category $\mathbb{E}$, and then adjoining morphisms (see Chapter I, Definition 1.4.6, Definition 1.4.8 and Definition 1.4.9). By (Part II, Chapter II, Proposition 3.1.12(i) and (ii)) there is a graded symmetric semi-monoidal category $\mathcal{C}$, with objects generated by a single object $\mathfrak{e}$, such that, as a graded tensor category without unit, we have $\mathbb{E} = \mathcal{C}_{\mathbb{Z}}$, i.e., $\mathbb{E}$ is the graded additive category generated by $\mathcal{C}$ (with the relation $n(-f) = -nf$ for $n \in \mathbb{Z}$ and f a morphism in $\mathcal{C}$), and the tensor structure on $\mathbb{E}$ is induced by the symmetric monoidal structure of $\mathcal{C}$.

As $\mathcal{C}$ is a graded symmetric semi-monoidal category, there is the natural map $\{\pm 1\} \times S_n \to \text{Hom}_{\mathcal{C}}(\mathfrak{e}^{\otimes n}, \mathfrak{e}^{\otimes n})^0$ (see Part II, Chapter II, §3.1.5) which, when extended to $\mathbb{E}$, is the restriction of the canonical map $\mathbb{Z}[S_n] \to \text{Hom}_{\mathbb{E}}(\mathfrak{e}^{\otimes n}, \mathfrak{e}^{\otimes n})^0$ sending $\sigma \in S_n$ to the symmetry isomorphism τ_σ. In addition to the structure results described in the previous paragraph, a set of representatives in $\text{Hom}_{\mathcal{C}}(\mathfrak{e}^{\otimes n}, \mathfrak{e}^{\otimes n})^q \setminus \{*\}$ for the action of $\{\pm 1\} \times S_n$ forms a $\mathbb{Z}[S_n]$-basis of $\text{Hom}_{\mathbb{E}}(\mathfrak{e}^{\otimes n}, \mathfrak{e}^{\otimes n})^q$ (*cf.* Part II, Chapter II, Proposition 3.1.10(iii)).

1.2.2. Now suppose we have connected schemes X and Y in $\mathcal{V}$, and a morphism $q: \mathfrak{e}^{\otimes a} \otimes \mathbb{Z}_Y(b)_g \to \mathfrak{e}^{\otimes a'} \otimes \mathbb{Z}_X(b')$ in $\mathcal{A}_{\mathrm{mot}}(\mathcal{V})^*$. By (Chapter I, Lemma 3.1.3 and Part II, Chapter I, Proposition 2.5.2), we may write each such map q as a sum with $\mathbb{Z}$-coefficients of compositions of the form

$$\mathfrak{e}^{\otimes a} \otimes \mathbb{Z}_Y(b)_g \xrightarrow{\tau \otimes \mathrm{id}_Y} \mathfrak{e}^{\otimes a} \otimes \mathbb{Z}_Y(b)_g = \mathfrak{e}^{\otimes a'} \otimes \mathfrak{e}^{\otimes a-a'} \otimes \mathbb{Z}_Y(b)_g$$

$$\xrightarrow{\mathrm{id}_{\mathfrak{e}^{\otimes a'}} \otimes h_1 \otimes \ldots \otimes h_s \otimes \mathrm{id}}$$

$$\mathfrak{e}^{\otimes a'} \otimes \mathbb{Z}_{W_1}(b_1)_{g_1} \otimes \ldots \otimes \mathbb{Z}_{W_s}(b_s)_{g_s} \otimes \mathbb{Z}_Y(b)_g \tag{1.2.2.1}$$

$$\xrightarrow{\mathrm{id}_{\mathfrak{e}^{\otimes a'}} \otimes \boxtimes_{W_1, \ldots, W_s, Y}} \mathfrak{e}^{\otimes a'} \otimes \mathbb{Z}_{W_1 \times \ldots W_s \times Y}(\textstyle\sum_i b_i + b)_{g_1 \times \ldots g_s \times g}$$

$$\xrightarrow{\mathrm{id}_{\mathfrak{e}^{\otimes a'}} \otimes p^*} \mathfrak{e}^{\otimes a'} \otimes \mathbb{Z}_X(b')_f.$$

Here the h_i are maps $h_i: \mathfrak{e}^{\otimes e_i} \to \mathbb{Z}_{W_i}(b_i)_{g_i}$ adjoined in Chapter I, Definition 1.4.6, Definition 1.4.8 and Definition 1.4.9. We have

$$\sum_{i=1}^{s} e_i = a - a'; \qquad b_1 + \ldots + b_s + b = b',$$

and

$$p: (X, f) \to (W_1 \times_S \ldots \times_S W_s \times_S Y, g_1 \times \ldots g_s \times g)$$

is a map in $\mathcal{L}(\mathcal{V})$. The map $\tau: \mathfrak{e}^{\otimes a} \to \mathfrak{e}^{\otimes a}$ is a map in $\mathcal{C}$.

We may form a $\mathbb{Z}$-basis of $\mathrm{Hom}_{\mathcal{A}_{\mathrm{mot}}(\mathcal{V})^*}(\mathfrak{e}^{\otimes a} \otimes \mathbb{Z}_Y(b)_g, \mathfrak{e}^{\otimes a'} \otimes \mathbb{Z}_X(b'))$ consisting of compositions of the form (1.2.2.1) by ordering the set of adjoined maps h, taking $h_1 \leq \ldots \leq h_s$, and taking τ in a set of representative of $\mathrm{Hom}_{\mathcal{C}}(\mathfrak{e}^{\otimes n}, \mathfrak{e}^{\otimes n})^r \setminus \{*\}$ modulo the action of the $\{\pm 1\} \times S(h_*)$, where $S(h_*) \subset S_n$ is the group of order-preserving permutations of $\{h_1, \ldots, h_s\}$ (see Part II, *loc. cit.* for details).

For a triple (τ, h_*, p), with τ a morphism in $\mathcal{C}$, $h_* = (h_1 \leq \ldots \leq h_s)$, and p as in (1.2.2.1), we denote the morphism given by the composition (1.2.2.1) by $q(\tau, h_*, p)$. We let

$$\bar{q}(\tau, h_*, p): X \to Y \tag{1.2.2.2}$$

be the composition

$$X \xrightarrow{p} W_1 \times_S \ldots \times_S W_s \times_S Y \xrightarrow{p_Y} Y.$$

1.2.3. LEMMA. *(i) The map* (1.2.2.2) *depends only on* $q(\tau, h_*, p)$, *not on the choice of* τ, h_* *and* p.
(ii) When defined, the composition $q(\tau_2, h_{*2}, p_2) \circ q(\tau_1, h_{*1}, p_1)$ *is a map of the form* $q(\tau, h_*, p)$, *or is zero; if the composition is not zero, then*

$$\bar{q}(\tau_2, h_{*2}, p_2) \circ \bar{q}(\tau_1, h_{*1}, p_1) = \bar{q}(\tau, h_*, p).$$

PROOF. By (Part II, Chapter I, Proposition 2.5.2 and Chapter II, Proposition 3.1.10), the ambiguity in the choice of (τ, h_*, p) is given by

$$q(\pm\sigma \circ \tau, h_*, p) = \pm q(\tau, h_*, (t_\sigma \times \mathrm{id}_Y) \circ p),$$

where σ is in $S(h_*) \subset S_s$, and

$$t_\sigma : W_1 \times_S \ldots \times_S W_s \to W_{\sigma^{-1}(1)} \times_S \ldots \times_S W_{\sigma^{-1}(s)}$$

is the corresponding symmetry isomorphism. As $p_Y \circ (t_\sigma \times \mathrm{id}_Y) = p_Y$, (i) is proven.

For (ii), suppose the τ_i are maps $\tau_i : \mathfrak{e}^{\otimes a_i} \to \mathfrak{e}^{\otimes a_i}$. Using the notation of (1.2.2.1), in order that the composition in (ii) is defined, we have

$$a_1' = a_2,\ b_1' = b_2,\ X_1 = Y_2.$$

Let σ be the shuffle permutation which puts the sequence $h_{*2}h_{*1}$ in increasing order, and let h_* be the resulting increasing sequence. Let

$$W^i = W_1^i \times_S \ldots \times_S W_{s^i}^i; \qquad i = 1, 2,$$

and let W be the re-ordered version of $W^1 \times_S W^2$: $W = (W^1 \times_S W^2)^\sigma$. Let $p : X_2 \to W \times Y_1$ be the composition

$$\begin{aligned} X_2 &\xrightarrow{p_2} W^2 \times Y_2 = W^2 \times X_1 \\ &\xrightarrow{\mathrm{id}_{W^2} \times p_1} W^2 \times W^1 \times Y_1 \\ &\xrightarrow{t_\sigma^{-1} \times \mathrm{id}_Y} W \times Y_1. \end{aligned}$$

Then

$$q(\tau_2, h_{*2}, p_2) \circ q(\tau_1, h_{*1}, p_1) = \pm q(\sigma \circ (\tau_2 \otimes \mathrm{id}_{\mathfrak{e}^{\otimes a_1 - a_2}}) \circ \tau_1, h_*, p).$$

As $p_{Y_1} \circ p = (p_{Y_1} \circ p_1) \circ (p_{Y_2} \circ p_2)$, the proof of (ii) is complete. □

1.3. Čech resolutions for $\mathbf{C}^b_{\mathrm{mot}}(\mathcal{V})$

1.3.1. *Push-forward of open covers.* Let X and Y be in $\mathcal{V}$, and let $\mathcal{U} = \{U_0, \ldots, U_m\}$ be a Zariski open cover of Y. Suppose we have a map $q : \mathfrak{e}^{\otimes a} \otimes \mathbb{Z}_Y(b)_g \to \mathfrak{e}^{\otimes a'} \otimes \mathbb{Z}_X(b')$. We now define the Zariski open cover $q_*\mathcal{U}$ of X, and a commutative diagram in $\mathbf{C}^b_{\mathrm{mot}}(\mathcal{V})^*$:

(1.3.1.1)
$$\begin{CD} \mathfrak{e}^{\otimes a} \otimes \mathbb{Z}_Y(b)_g @>{j^*_{Y,\mathcal{U}}}>> \mathfrak{e}^{\otimes a} \otimes \mathbb{Z}_{\mathcal{U}_*^{\mathrm{n.d.}}}(b)_g \\ @V{q}VV @VV{q_{\mathcal{U}}}V \\ \mathfrak{e}^{\otimes a'} \otimes \mathbb{Z}_X(b')_f @>>{j^*_{X,q_*\mathcal{U}}}> \mathfrak{e}^{\otimes a'} \otimes \mathbb{Z}_{q_*\mathcal{U}_*^{\mathrm{n.d.}}}(b')_f. \end{CD}$$

Suppose at first that X and Y are connected, and that $q = q(\tau, h_*, p)$, giving the map $\bar{q} := \bar{q}(\tau, h_*, p) : X \to Y$ (1.2.2.2). For an open subset U of Y, we have the open subset $\bar{q}^{-1}(U)$ of X.

For each open subscheme $j_U : U \to Y$, let $k_V : V \to X$ denote the inclusion of $\bar{q}^{-1}(U)$ into X. If $j_{U,U'} : U' \to U$ is the inclusion of open subschemes of Y, we have the induced inclusion $k_{V,V'} : V' \to V$ where $V' = \bar{q}^{-1}(U')$. We have the map

(1.3.1.2)
$$q_U : \mathfrak{e}^{\otimes a} \otimes \mathbb{Z}_U(b)_{j_U^* g} \to \mathfrak{e}^{\otimes a'} \otimes \mathbb{Z}_V(b')_{k_V^* f}$$

defined as the composition

$$\begin{aligned}
\mathfrak{e}^{\otimes a} \otimes \mathbb{Z}_U(b)_{j_U^* g} &\xrightarrow{\tau \otimes \mathrm{id}_U} \mathfrak{e}^{\otimes a} \otimes \mathbb{Z}_U(b)_{j_U^* g} = \mathfrak{e}^{\otimes a'} \otimes \mathfrak{e}^{\otimes a-a'} \otimes \mathbb{Z}_U(b)_{j_U^* g} \\
&\xrightarrow{\mathrm{id}_{\mathfrak{e}^{\otimes a'}} \otimes h_1 \otimes \ldots \otimes h_s \otimes \mathrm{id}} \mathfrak{e}^{\otimes a'} \otimes \mathbb{Z}_{W_1}(b_1)_{g_1} \otimes \ldots \otimes \mathbb{Z}_{W_s}(b_s)_{g_s} \otimes \mathbb{Z}_U(b)_{j_U^* g} \\
&\xrightarrow{\mathrm{id}_{\mathfrak{e}^{\otimes a'}} \otimes \boxtimes_{W_1, \ldots, W_s, U}} \mathfrak{e}^{\otimes a'} \otimes \mathbb{Z}_{W_1 \times \ldots W_s \times U}(\textstyle\sum_i b_i + b)_{g_1 \times \ldots g_s \times j_U^* g} \\
&\xrightarrow{\mathrm{id}_{\mathfrak{e}^{\otimes a'}} \otimes p_{|V}^*} \mathfrak{e}^{\otimes a'} \otimes \mathbb{Z}_V(b')_{k_V^* f}.
\end{aligned}$$

The functoriality of the external products $\boxtimes_{**}$ implies the commutativity of

(1.3.1.3)
$$\begin{array}{ccc}
\mathfrak{e}^{\otimes a} \otimes \mathbb{Z}_U(b)_{j_U^* g} & \xrightarrow{q_U} & \mathfrak{e}^{\otimes a'} \otimes \mathbb{Z}_V(b')_{k_V^* f} \\
\downarrow{\scriptstyle j_{U,U'}^*} & & \downarrow{\scriptstyle k_{V,V'}^*} \\
\mathfrak{e}^{\otimes a} \otimes \mathbb{Z}_{U'}(b)_{j_{U'}^* g} & \xrightarrow[q_{U'}]{} & \mathfrak{e}^{\otimes a'} \otimes \mathbb{Z}_{V'}(b')_{k_{V'}^* f}
\end{array}$$

for $U' \subset U$, with $V = \bar{q}^{-1}(U)$, $V' = \bar{q}^{-1}(U')$.

If $\mathcal{U} = \{U_0, \ldots, U_m\}$ is an open cover of Y, we let $q_*\mathcal{U}$ be the open cover of X defined by

$$q_*\mathcal{U} = \{\bar{q}^{-1}(U_0), \ldots, \bar{q}^{-1}(U_m)\}.$$

Let $V_i = \bar{q}^{-1}(U_i)$.

Let $I = (i_0 < \ldots < i_s)$, and let $j_I : U_I \to Y$ and $k_I : V_I \to X$ be the inclusions; for $I \subset J$, we have the inclusions

$$j_{I \subset J} : U_J \to U_I; \qquad k_{I \subset J} : V_J \to V_I.$$

Using the commutativity of the diagram (1.3.1.3), the collection of maps q_{U_I} defines the map

(1.3.1.4)
$$q_{\mathcal{U}} : \mathfrak{e}^{\otimes a} \otimes \mathbb{Z}_{\mathcal{U}_*^{\text{n.d.}}}(b)_g \to \mathfrak{e}^{\otimes a'} \otimes \mathbb{Z}_{q_*\mathcal{U}_*^{\text{n.d.}}}(b')_f$$

giving the desired commutative diagram (1.3.1.1).

Suppose q is a sum of compositions (1.2.2.1)

$$q = \sum_{i=1}^{l} n_i q_i; \qquad n_i \in \mathbb{Z}, n_i \neq 0,$$

with the q_i basis elements as described in §1.2.2. Let $q_*\mathcal{U}$ be cover given by the open subsets

$$\bar{q}_1^{-1}(U_{i_1}) \cap \ldots \cap \bar{q}_l^{-1}(U_{i_l}); \quad 0 \leq i_j \leq m.$$

We then have the canonical refinement maps for each i $\rho_i : q_*\mathcal{U} \to q_{i*}\mathcal{U}$. Forming the maps $q_{i\mathcal{U}}$ (1.3.1.4) for each i, composing with the refinement map ρ_i^*, and summing, gives the desired map

$$q_{\mathcal{U}} : \mathfrak{e}^{\otimes a} \otimes \mathbb{Z}_{\mathcal{U}_*^{\text{n.d.}}}(b)_g \to \mathfrak{e}^{\otimes a'} \otimes \mathbb{Z}_{q_*\mathcal{U}_*^{\text{n.d.}}}(b')_f,$$

$$q_{\mathcal{U}} = \sum_{i=1}^{m} n_i(\rho_i^* \circ q_{i,\mathcal{U}}).$$

If q is the zero map, we define $q_*\mathcal{U}$ to be the trivial cover X, and $q_{\mathcal{U}}$ to be the zero map.

The formation of $q_{\mathcal{U}}$ and $q_*\mathcal{U}$ is compatible with refinement: each refinement $\rho: \mathcal{V} \to \mathcal{U}$ gives the refinement $q_*\rho: q_*\mathcal{V} \to q_*\mathcal{U}$ and we have the identity

$$(q_*\rho)^* \circ q_{\mathcal{U}} = q_{\rho^*\mathcal{U}} \circ \rho^*. \tag{1.3.1.5}$$

We might *not* have the identities

$$q'_* q_* \mathcal{U} = (q' \circ q)_* \mathcal{U}; \qquad (q' \circ q)_{\mathcal{U}} = q'_{q_*\mathcal{U}} \circ q_{\mathcal{U}}$$

due to possible cancellations in the expression for $q' \circ q$, however, it follows from Lemma 1.2.3 that there is a (non-canonical) refinement $\rho_{q',q}: q'_* q_* \mathcal{U} \to (q' \circ q)_* \mathcal{U}$, and for any such refinement, we have the relation

$$\rho^*_{q',q} \circ (q' \circ q)_{\mathcal{U}} = q'_{q_*\mathcal{U}} \circ q_{\mathcal{U}}. \tag{1.3.1.6}$$

For similar reasons, there is a (non-canonical) refinement $\rho_{dq,q}: q_*\mathcal{U} \to (dq)_*\mathcal{U}$ and for any such refinement we have the relation

$$d_1(q_{\mathcal{U}}) = \rho^*_{dq,q} \circ (dq)_{\mathcal{U}}, \tag{1.3.1.7}$$

where d_1 refers to the differential with respect to the category $\mathcal{A}_{\mathrm{mot}}(\mathcal{V})^*$, not the Čech differential.

We extend the definition of $q_{\mathcal{U}}$, $q_*\mathcal{U}$ and $\rho_{q',q}$ to arbitrary objects of $\mathcal{A}_{\mathrm{mot}}(\mathcal{V})^*$ by taking direct sums. The relations (1.3.1.5)-(1.3.1.7) continue to hold.

1.3.2. We recall from (Part II, Chapter II, §1.2.6), that for a DG category $\mathcal{A}$, we have the DG category Pre-Tr($\mathcal{A}$) with objects X being tuples of the form $(X_N, X_{N+1}, \ldots, X_M; q_{ij})$, where $N \le M$ are integers, the q_{ij} are morphisms $q_{ij}: X_j[-j] \to X_i[-i]$ in $\mathcal{A}$, and $\sum_k q_{ik} \circ q_{kj} = dq_{ij}$ for all i and j (including $i = j$). There is an operation of cone in Pre-Tr($\mathcal{A}$), and the category $\mathbf{C}^b(\mathcal{A})$ is the smallest full DG subcategory of Pre-Tr($\mathcal{A}$) containing $\mathcal{A}$ and closed under taking translations and cones. If $\mathcal{A}$ has trivial differential graded structure, $\mathbf{C}^b(\mathcal{A})$ is the usual DG category of bounded complexes in $\mathcal{A}$. The operation

$$(X_N, X_{N+1}, \ldots, X_M; q_{ij}) \mapsto \oplus_{i=N}^M X_i[-i]$$

defines the "forgetful functor" $FD: \mathbf{C}^b(\mathcal{A}) \to \mathcal{A}$; in case $\mathcal{A}$ has trivial differential structure, this is just the functor "forget the differential".

We recall from (Part II, Chapter II, §1.2.9 and Lemma 1.2.10), that taking the total complex defines the functor (see Part II, (II.1.2.9.1))

$$\mathrm{Tot}: \mathbf{C}^b(\mathbf{C}^b_{\mathrm{mot}}(\mathcal{V})^*) \to \mathbf{C}^b_{\mathrm{mot}}(\mathcal{V})^*.$$

1.3.3. DEFINITION. (i) Let Γ be an object of $\mathbf{C}^b_{\mathrm{mot}}(\mathcal{V})^*$. A *Zariski open cover of* Γ is a Zariski open cover of $FD(\Gamma)$. A refinement of a Zariski open cover of Γ is a refinement of the corresponding Zariski open cover of $FD(\Gamma)$.
(ii) Suppose we have $\Gamma = (\Gamma_N, \ldots, \Gamma_M; q_{ij})$ for objects Γ_i of $\mathcal{A}_{\mathrm{mot}}(\mathcal{V})^*$ and maps $q_{ij}: \Gamma_j[-j] \to \Gamma_i[-i]$ in $\mathcal{A}_{\mathrm{mot}}(\mathcal{V})^*$. A *Čech resolution of* Γ is a map $j: \Gamma \to \Gamma_{\mathcal{U}}$ in $\mathbf{C}^b(Z^0\mathbf{C}^b_{\mathrm{mot}}(\mathcal{A})^*)$ such that

(a) There are Zariski open covers $\mathcal{U}_i$ of Γ_i, and Čech resolutions $j_{\Gamma_i,\mathcal{U}_i}: \Gamma_i \to (\Gamma_i)_{\mathcal{U}_i}$ with associated open cover $\mathcal{U}_i$, $i = N, \ldots, M$.
(b) For each i and j with $q_{ij} \neq 0$, there is a refinement (on Γ_i) $\rho_{ij}: \mathcal{U}_i \to q_{ij*}\mathcal{U}_j$
(c) $\Gamma_{\mathcal{U}} = ((\Gamma_N)_{\mathcal{U}_N}, \ldots, (\Gamma_M)_{\mathcal{U}_M}; \tilde{q}_{ij})$, where $\tilde{q}_{ij}: (\Gamma_j)_{\mathcal{U}_j}[-j] \to (\Gamma_i)_{\mathcal{U}_i}[-i]$ is given by $\tilde{q}_{ij} = \rho^*_{ij} \circ (q_{ij})_{\mathcal{U}_j}$.

Letting $\mathcal{U}$ be the Zariski open cover of Γ determined by the $\mathcal{U}_i$, we say that the Čech resolution j has associated cover $\mathcal{U}$.
(iii) A map of Čech resolutions $\tilde{q}: (j: \Gamma \to \Gamma_{\mathcal{U}}) \to (j': \Gamma' \to \Gamma'_{\mathcal{U}'})$ over a map $q: \Gamma \to \Gamma'$ in $Z^0\mathbf{C}^b_{\mathrm{mot}}(\mathcal{V})^*$ is a map $\tilde{q}: \Gamma_{\mathcal{U}} \to \Gamma'_{\mathcal{U}'}$ in $\mathbf{C}^b(Z^0\mathbf{C}^b_{\mathrm{mot}}(\mathcal{V})^*)$ such that $\tilde{q} \circ j = j' \circ \tilde{q}$, and such that the map $FD(\tilde{q})$ is a map of the form $\rho^*_{\mathcal{U}', q_*\mathcal{U}} \circ q_{\mathcal{U}}$, where $\mathcal{U}$ and $\mathcal{U}'$ are the open covers of Γ and Γ' corresponding to j and j', for some choice of refinement mapping $\rho_{\mathcal{U}', q_*\mathcal{U}}: \mathcal{U}' \to q_*\mathcal{U}$.

1.3.4. REMARK. It follows directly from Definition 1.3.3 that, if $\tilde{q}: (j: \Gamma \to \Gamma_{\mathcal{U}}) \to (j': \Gamma' \to \Gamma'_{\mathcal{U}'})$ is a map of Čech resolutions over a map $q: \Gamma \to \Gamma'$ in $Z^0\mathbf{C}^b_{\mathrm{mot}}(\mathcal{V})^*$, then $(j[1], j'): \mathrm{cone}(q) \to \mathrm{cone}(\tilde{q})$ is a Čech resolution of $\mathrm{cone}(q)$, giving the commutative diagram

$$\begin{array}{ccccccc}
\Gamma & \xrightarrow{q} & \Gamma' & \longrightarrow & \mathrm{cone}(q) & \longrightarrow & \Gamma[1] \\
\downarrow{\scriptstyle j} & & \downarrow{\scriptstyle j'} & & \downarrow{\scriptstyle (j[1], j')} & & \downarrow{\scriptstyle j[1]} \\
\Gamma_{\mathcal{U}} & \xrightarrow[\tilde{q}]{} & \Gamma'_{\mathcal{U}'} & \longrightarrow & \mathrm{cone}(\tilde{q}) & \longrightarrow & \Gamma_{\mathcal{U}}[1]
\end{array}$$

with the columns standard cone sequences.

1.4. The category of hyper-resolutions

1.4.1. DEFINITION. (i) A sequence of maps

$$j_0: \Gamma \to \Gamma_{\mathcal{U}_1},\ j_1: \mathrm{Tot}(\Gamma_{\mathcal{U}_1}) \to \Gamma_{\mathcal{U}_2},\ \ldots,\ j_{m-1}: \mathrm{Tot}(\Gamma_{\mathcal{U}_{m-1}}) \to \Gamma_{\mathcal{U}_m},$$

in $\mathbf{C}^b(Z^0\mathbf{C}^b_{\mathrm{mot}}(\mathcal{A})^*)$ is called a *length m tower of Čech resolutions of* Γ if each map j_i is a Čech resolution. A map of length m towers over a map $q: \Gamma \to \Delta$ is a sequence of maps $\tilde{q}_i: \Gamma_{\mathcal{U}_i} \to \Delta_{\mathcal{W}_i}$ such that $\tilde{q}_1$ is a map of Čech resolutions over q, and $\tilde{q}_{i+1}$ is a map of Čech resolutions over $\mathrm{Tot}(\tilde{q}_i)$ for $1 \le i \le m-1$. We often write a tower of Čech resolutions as

$$\Gamma \xrightarrow{j_0} \Gamma_{\mathcal{U}_1} \to \ldots \to \Gamma_{\mathcal{U}_m}.$$

(ii) If we have a tower of Čech resolutions of Γ as in (i), we call the composition

$$j := j_{m-1} \circ \mathrm{Tot} j_{m-2} \circ \ldots \circ \mathrm{Tot} j_0 : \Gamma \to \Gamma_{\mathcal{U}_m}$$

a *hyper-resolution of* Γ.
(iii) Given two hyper-resolutions of Γ, $j: \Gamma \to \Gamma_{\mathcal{U}_m}$, $j': \Gamma' \to \Gamma'_{\mathcal{U}'_m}$, and a map $f: \Gamma \to \Gamma'$ in $Z^0\mathbf{C}^b_{\mathrm{mot}}(\mathcal{A})^*$, a map $\tilde{f}: \Gamma_{\mathcal{U}_m} \to \Gamma'_{\mathcal{U}'_m}$ in $\mathbf{C}^b(Z^0\mathbf{C}^b_{\mathrm{mot}}(\mathcal{A})^*)$ is a *map of hyper-resolutions over* f if there is an m and a map $(f_1, \ldots, f_m)$ over f of length m towers of Čech resolutions, such that $\tilde{f}$ is the map f_m. A map of hyper-resolutions of Γ is a map of hyper-resolutions over id_Γ.
(iv) We let $\mathbf{HR}$ be the sub-category of $\mathbf{C}^b(Z^0\mathbf{C}^b_{\mathrm{mot}}(\mathcal{A})^*)$ with objects the hyper-resolutions of objects of $\mathbf{C}^b_{\mathrm{mot}}(\mathcal{V})^*$, and maps the maps of hyper-resolutions; we let $\mathbf{HR}_\Gamma$ be the subcategory of $\mathbf{HR}$ with objects the hyper-resolutions of Γ and maps being maps over the identity.

1.4.2. LEMMA. *(i) Let Γ be in $\mathbf{C}^b_{\mathrm{mot}}(\mathcal{V})^*$, $\mathcal{U}$ a Zariski open cover of Γ. Then there is a refinement $\mathcal{W} \to \mathcal{U}$ of $\mathcal{U}$ and a Čech resolution $j: \Gamma \to \Gamma_{\mathcal{W}}$ of Γ with associated cover $\mathcal{W}$.*

(ii) Suppose we have a map $f:\Gamma \to \Gamma'$ in $Z^0\mathbf{C}^b_{\mathrm{mot}}(\mathcal{V})^*$, and Čech resolutions $j:\Gamma \to \Gamma_{\mathcal{W}}$, $j':\Gamma' \to \Gamma'_{\mathcal{W}'}$ with associated covers $\mathcal{W}$ and $\mathcal{W}'$. Then there is a refinement $\mathcal{U}'$ of $\mathcal{W}'$, a Čech resolution $j'':\Gamma' \to \Gamma'_{\mathcal{U}'}$ with associated cover $\mathcal{U}'$, a map of Čech resolutions $\tilde{f}:\Gamma_{\mathcal{W}} \to \Gamma_{\mathcal{U}'}$ over f, and a map of Čech resolutions over the identity $\tilde{\mathrm{id}}:\Gamma'_{\mathcal{W}'} \to \Gamma_{\mathcal{U}'}$.
(iii) Let Γ be in $\mathbf{C}^b_{\mathrm{mot}}(\mathcal{V})^*$, and let $j:\Gamma \to \Gamma_{\mathcal{U}}$ be a hyper-resolution of Γ. Then

$$\mathrm{Tot}(j):\Gamma \to \mathrm{Tot}(\Gamma_{\mathcal{U}})$$

is an isomorphism in $\mathbf{D}^b_{\mathrm{mot}}(\mathcal{V})$.

PROOF. Let $\Gamma = (\Gamma_N, \dots, \Gamma_M; q_{ij})$ be in $\mathbf{C}^b_{\mathrm{mot}}(\mathcal{V})^*$. As Γ is an iterated cone of objects of $\mathcal{A}_{\mathrm{mot}}(\mathcal{V})^*$, we may write each Γ_i as a direct sum

$$\Gamma_i = \oplus_k \Gamma_{ik}$$

such that the component $q_{ij}^{kk'}:\Gamma_{jk'}[-j] \to \Gamma_{ik}[-i]$ is zero if $k \le k'$. Let $\Gamma^k = \oplus_i \Gamma_{ik}[-i]$; we then have the collection of maps

$$q^{kk'}:\Gamma^{k'} \to \Gamma^k; \qquad N_1 \le k' < k \le M_1.$$

Given a Zariski open cover $\mathcal{U}$ of Γ, we have for each k a Zariski open cover $\mathcal{U}_k$ of Γ^k for all k. We may then inductively choose refinements $\mathcal{W}_k$ of $\mathcal{U}_k$ so that $\mathcal{W}_k$ is a refinement of $q_*^{kk'}\mathcal{W}_{k'}$ for each $k' < k$ via $\rho^{kk'}:\mathcal{W}_k \to q_*^{kk'}\mathcal{W}_{k'}$.

Let $\mathcal{W}_{ik}$ be the restriction of $\mathcal{W}_k$ to Γ_{ik}, let $\rho_{ij}^{kk'}$ be the restriction of $\rho^{kk'}$ to the refinement $\rho_{ij}^{kk'}:\mathcal{W}_{ik} \to q_{ij*}^{kk'}\mathcal{W}_{jk'}$, and let $\tilde{q}_{ij}^{kk'}:(\Gamma_{jk'}[-j])_{\mathcal{W}_{jk'}} \to (\Gamma_{ik}[-i])_{\mathcal{W}_{ik}}$ be the map

$$\tilde{q}_{ij}^{kk'} = (\rho_{ij}^{kk'})^* \circ (q_{ij}^{kk'})_{\mathcal{W}_{jk'}}.$$

The relations (1.3.1.5)-(1.3.1.7) imply that the tuple

$$(\dots, \oplus_k(\Gamma_{ik})_{\mathcal{W}_{ik}}, \dots; \oplus_{k,k'}\tilde{q}_{ij}^{kk'})$$

defines an object $\Gamma_{\mathcal{W}}$ of $\mathbf{C}^b(Z^0\mathbf{C}^b_{\mathrm{mot}}(\mathcal{V}))$, and that the collection of maps

$$j_{\mathcal{W}_{ik}}:\Gamma_{ik} \to (\Gamma_{ik}[-i])_{\mathcal{W}_{ik}}$$

define a Čech resolution $j_{\mathcal{W}}:\Gamma \to \Gamma_{\mathcal{W}}$. This proves (i).

If we already have a Čech resolution for the cover $\mathcal{W}$, $j_{\mathcal{U}}:\Gamma \to \Gamma_{\mathcal{W}}$ then the maps in $\Gamma_{\mathcal{W}}$ are completely determined by the maps in Γ and the choice of refinement mappings $\rho_{ij}^{\mathcal{W}}:\mathcal{W}_i \to q_{ij*}\mathcal{W}_j$.

Now, suppose we are given a map $f:\Gamma \to \Gamma'$, and Čech resolutions $j:\Gamma \to \Gamma_{\mathcal{W}}$, $j':\Gamma' \to \Gamma'_{\mathcal{W}'}$. Let Γ'' be the cone of the map f. We may order the summands Γ''_{ik} of Γ'' as above so that all the summands coming from Γ precede those coming from Γ'. Let $\mathcal{W}\coprod\mathcal{W}'$ be the Zariski open cover of Γ'' which gives $\mathcal{W}$ on $\Gamma[1]$ and $\mathcal{W}'$ on Γ'. We may then construct a refinement $\mathcal{W}''$ of $\mathcal{W}\coprod\mathcal{W}'$ as in the proof of (i), together with refinement maps $\rho_{ij}^{kk'}$ as above, which satisfies in addition

1. When restricted to $\Gamma[1]$, $(\mathcal{W}'', \rho_{**}^{**})$ is equal to the cover $\mathcal{W}$, together with the refinement mappings defining $\Gamma_{\mathcal{W}}[1]$.
2. When restricted to Γ', the refinement mapping $\mathcal{W}''_{|\Gamma'} \to \mathcal{W}'$ intertwines the refinement mappings in $\mathcal{W}''_{|\Gamma'}$ with those defining $\Gamma'_{\mathcal{W}'}$.

This gives us the Čech resolution $k:\Gamma'' \to \Gamma''_{\mathcal{W}''}$. The portion of $\Gamma''_{\mathcal{W}''}$ coming from the refinement of $\mathcal{W}'$ induced by $\mathcal{W}''$ then gives rise to the refinement of $\mathcal{W}'$, the Čech resolution $j'':\Gamma' \to \Gamma'_{\mathcal{U}'}$ of Γ', and the map of Čech resolutions $\Gamma'_{\mathcal{W}'} \to \Gamma'_{\mathcal{U}'}$

over the identity. The (degree 1) maps in $\Gamma_{\mathcal{W}''}$ which go from the terms in $\Gamma_{\mathcal{W}}[1]$ to the terms in $\Gamma'_{\mathcal{U}'}$ then define the (degree 0) map of Čech resolutions $\tilde{f}:\Gamma_{\mathcal{W}} \to \Gamma'_{\mathcal{U}'}$ over f, proving (ii).

For (iii), we have already noted in §1.1.1 that j is an isomorphism in $\mathbf{D}^b_{\mathrm{mot}}(\mathcal{V})$ in case Γ is an object of the form $\mathfrak{e}^{\otimes a} \otimes \mathbb{Z}_X(q)_f$. As such objects generate $\mathcal{A}_{\mathrm{mot}}(\mathcal{V})^*$ as an additive category, j is an isomorphism in $\mathbf{D}^b_{\mathrm{mot}}(\mathcal{V})$ for all Γ in $\mathcal{A}_{\mathrm{mot}}(\mathcal{V})^*$. In general, suppose Γ is a cone:

$$\Gamma_0 \xrightarrow{q} \Gamma_1 \to \Gamma = \mathrm{cone}(q) \to \Gamma_0[1].$$

By definition of a Čech resolution, we may fit the Čech resolution $j:\Gamma \to \Gamma_{\mathcal{U}}$ into a commutative diagram

$$\begin{array}{ccccccc}
\Gamma_{\mathcal{U}_0} & \xrightarrow{\tilde{q}} & \Gamma_{\mathcal{U}_1} & \longrightarrow & \Gamma_{\mathcal{U}} = \mathrm{cone}(\tilde{q}) & \longrightarrow & \Gamma_{\mathcal{U}_0}[1] \\
\uparrow{\scriptstyle j_0} & & \uparrow{\scriptstyle j_1} & & \uparrow{\scriptstyle j} & & \uparrow{\scriptstyle j_0[1]} \\
\Gamma_0 & \xrightarrow{q} & \Gamma_1 & \longrightarrow & \Gamma = \mathrm{cone}(q) & \longrightarrow & \Gamma_0[1]
\end{array}$$

with both rows cone sequences. As $\mathbf{C}^b_{\mathrm{mot}}(\mathcal{V})^*$ is generated by $\mathcal{A}_{\mathrm{mot}}(\mathcal{V})^*$ by taking cones, this proves (iii). □

1.4.3. PROPOSITION. *Let Γ be in $\mathbf{C}^b_{\mathrm{mot}}(\mathcal{V})^*$. Then the image of the category $\mathbf{HR}_\Gamma$ in the homotopy category $\mathbf{K}^b(Z^0\mathbf{C}^b_{\mathrm{mot}}(\mathcal{V}))$ is right-filtering.*

PROOF. As the identity map on Γ is a hyper-resolution of Γ, $\mathbf{HR}_\Gamma$ is non-empty. In addition, if we have a length m tower of Čech resolutions of Γ, we may extend the tower to length $m+1$ by adjoining an identity map at the end of the tower, so we may always consider two hyper-resolutions of Γ as coming from towers of Čech resolutions of Γ of the same length. Thus, it follows from Lemma 1.4.2(ii) that, given two hyper-resolutions of Γ, $j:\Gamma \to \tilde{\Gamma}$, $j':\Gamma \to \tilde{\Gamma}'_*$, there is a hyper-resolution $j'':\Gamma \to \tilde{\Gamma}''$ and maps of hyper-resolutions

$$\tilde{\Gamma} \to \tilde{\Gamma}''; \qquad \tilde{\Gamma}' \to \tilde{\Gamma}''.$$

Now suppose we have two maps of hyper-resolutions of Γ

$$\tilde{f}^1, \tilde{f}^2 : \tilde{\Gamma} \to \tilde{\Gamma}'.$$

As above, we may assume that the maps $\tilde{f}^1$, $\tilde{f}^2$ come from maps of length m towers of Čech resolutions

$$f^1_*, f^2_* : \Gamma_* \to \Gamma'_*.$$

From Definition 1.3.3 and Definition 1.4.1, the only choice that one has in forming a map of towers of Čech resolutions is the choice of the various refinement mappings, which must satisfy the compatibility requirement of Definition 1.3.3(iii).

Suppose we have a homotopy

$$dH_{m-1} = f^2_{m-1} - f^1_{m-1} : \Gamma_{m-1} \to \Gamma'_{m-1}$$

in $\mathbf{C}^b(Z^0\mathbf{C}^b_{\mathrm{mot}}(\mathcal{V}))$. Arguing as in the proof of Lemma 1.4.2(ii), we may extend $\mathrm{Tot}(H_{m-1})$ to a degree -1 map $H_m : \Gamma_m \to \Gamma'_m$. Replacing f^2_m with $dH_m + f^1_m$, we may assume that $f^2_{m-1} = f^1_{m-1}$. Applying Lemma 1.4.2(ii), we may assume that $f^2_{m-1} = f^1_{m-1} = \mathrm{id}$, reducing us to the case $m = 1$.

As is well known, if $\mathcal{U} = \{U_0, \dots, U_m\}$ and $\mathcal{V} = \{V_0, \dots, V_n\}$ are open covers of a topological space X and if we have two refinement mappings

$$\rho_1, \rho_2 : \mathcal{V} \to \mathcal{U},$$
$$V_i \subset U_{\rho_1(i)}; \quad V_i \subset U_{\rho_2(i)},$$

there is the homotopy $H = H(\rho_1, \rho_2)$ on the chain complexes (of nondegenerate simplices) associated to the maps of simplicial schemes

$$N(\rho_1), N(\rho_2) : N\mathcal{V} \to N\mathcal{U},$$
$$N(\rho_j) : V_{i_0} \cap \dots \cap V_{i_k} \to U_{\rho_j(i_0)} \cap \dots \cap U_{\rho_j(i_k)},$$

defined by sending $V_{i_0} \cap \dots \cap V_{i_k}$ to the sum

$$\sum_{j=0}^{k} (-1)^j [U_{\rho_1(i_0)} \cap \dots \cap U_{\rho_1(i_j)} \cap U_{\rho_2(i_j)} \cap \dots \cap U_{\rho_2(i_k)}].$$

Here $[U_{j_0} \cap \dots \cap U_{j_k}] = \operatorname{sgn}(j_0, \dots, j_k) \cdot (U_{j_0} \cap \dots \cap U_{j_k})$ and $\operatorname{sgn}(j_0, \dots, j_k)$ is the sign of the permutation which puts $j_0, \dots, j_k$ in increasing order if the j_i are distinct, and is zero if the j_i are not distinct.

The homotopy $H(\rho_1, \rho_2)$ is natural, in the following sense: Let $f : Y \to X$ be a continuous map of topological spaces, let $\rho_1^X, \rho_2^X : \mathcal{V}_X \to \mathcal{U}_X$ and $\rho_1^Y, \rho_2^Y : \mathcal{V}_Y \to \mathcal{U}_Y$ be refinements of open covers on X and Y, respectively, and let $f_{\mathcal{V}} : \mathcal{V}_Y \to f^{-1}(\mathcal{V}_X)$ and $f_{\mathcal{U}} : \mathcal{U}_Y \to f^{-1}(\mathcal{U}_X)$ be refinements of open covers on Y such that

$$f_{\mathcal{U}} \circ \rho_i^Y = f^{-1}(\rho_i^X) \circ f_{\mathcal{V}}.$$

Then $f_{\mathcal{V}}$ and $f_{\mathcal{U}}$ give rise to maps $N(f_{\mathcal{V}}) : N\mathcal{V}_Y \to N\mathcal{V}_X$ and $N(f_{\mathcal{U}}) : N\mathcal{U}_Y \to N\mathcal{U}_X$, and we have

$$H(\rho_1^X, \rho_2^X) \circ N(f_{\mathcal{V}}) = N(f_{\mathcal{U}}) \circ H(\rho_1^Y, \rho_2^Y)$$

Since the homotopy H has this naturality, we may apply H to the two compatible choices of refinement mappings determined by the maps $\tilde{f}^1$ and $\tilde{f}^2$, giving the desired homotopy between $\tilde{f}^1$ and $\tilde{f}^2$ in $\mathbf{C}^b(Z^0\mathbf{C}^b_{\mathrm{mot}}(\mathcal{V}))$. □

1.5. Hypercohomology

1.5.1. Let $h : \mathbf{C}^b_{\mathrm{mot}}(\mathcal{V})^* \to \mathbf{C}(\mathbf{Ab})$ be an DG functor, compatible with cones. We define the hypercohomology of Γ with respect to h as

$$\mathbb{H}^0_h(\Gamma) := \varinjlim_{\tilde{\mathcal{U}} \in \mathbf{HR}_\Gamma} H^0(h(\mathrm{Tot}(\Gamma_{\tilde{\mathcal{U}}}))). \tag{1.5.1.1}$$

By Proposition 1.4.3, the limit in (1.5.1.1) is equivalent to a filtered inductive limit.

The augmentation $\Gamma \to \Gamma_{\mathcal{U}}$ gives the natural map $H^0(h(\Gamma)) \to \mathbb{H}^0_h(\Gamma)$, hence sending Γ to $\mathbb{H}^0_h(\Gamma)$ defines a cohomological functor and exact natural transformation

$$\mathbb{H}^0_h : \mathbf{K}^b_{\mathrm{mot}}(\mathcal{V})^* \to \mathbf{Ab},$$
$$\mathbb{H}^0 : H^0 \circ h \to \mathbb{H}^0_h.$$

1.5.2. *Sheaves and hyper-resolutions.* We now relate the functor $\mathbb{H}_h^*$ to Zariski hypercohomology. For a scheme X, we have the full subcategory $\mathrm{Zar}(X)$ of $\mathbf{Sch}_X$ with objects the open subschemes of X.

Given (X, f) in $\mathcal{L}(\mathcal{V})$, an integer q, and a non-negative integer a, we may map $\mathrm{Zar}(X)^{\mathrm{op}}$ to $\mathcal{A}_{\mathrm{mot}}(\mathcal{V})$ by sending $(j_U: U \to X)$ to $\mathfrak{e}^{\otimes a} \otimes \mathbb{Z}_U(q)_{j_U^* f}$ (for $a = 0$, we send $(j_U: U \to X)$ to just $\mathbb{Z}_U(q)_{j_U^* f}$) and sending each inclusion $j_{U,V}: V \to U$ to $\mathrm{id} \otimes j_{U,V}^*: \mathbb{Z}_U(q)_{j_U^* f} \to \mathbb{Z}_V(q)_{j_V^* f}$.

We let $\mathcal{A}_{\mathrm{mot}}(\mathrm{Zar}(X, f))$ denote the additive subcategory of $\mathcal{A}_{\mathrm{mot}}(\mathcal{V})$ generated by image of $\{0, 1, \dots\} \times \mathrm{Zar}(X)^{\mathrm{op}} \times \mathbb{Z}$ under this functor. and $\mathbf{C}^b_{\mathrm{mot}}(\mathrm{Zar}(X, f))$ the category of bounded complexes over $\mathcal{A}_{\mathrm{mot}}(\mathrm{Zar}(X, f))$; $\mathbf{C}^b_{\mathrm{mot}}(\mathrm{Zar}(X, f))$ is naturally a DG subcategory of $\mathbf{C}^b_{\mathrm{mot}}(\mathcal{V})^*$.

Let $\mathrm{Zar}(X)_c$ denote the full subcategory of $\mathrm{Zar}(X)$ with objects the *connected* open subsets of X. It follows from (Chapter I, Lemma 3.1.2, Lemma 3.1.3 and Lemma 3.1.4) that $\mathcal{A}_{\mathrm{mot}}(\mathrm{Zar}(X, f))$ is isomorphic to the free additive category on $\{0, 1, \dots\} \times \mathrm{Zar}(X)_c^{\mathrm{op}} \times \mathbb{Z}$, with disjoint union of open subsets going over to the direct sum in $\mathcal{A}_{\mathrm{mot}}(\mathrm{Zar}(X, f))$. In particular, if P is a Zariski presheaf on X, with values in an additive category $\mathcal{A}$, and if P sends disjoint unions to direct sums, then P canonically defines an additive functor $P: \mathcal{A}_{\mathrm{mot}}(\mathrm{Zar}(X, f)) \to \mathcal{A}$, and thus gives the functor of DG categories $\mathbf{C}^b(P): \mathbf{C}^b_{\mathrm{mot}}(\mathrm{Zar}(X, f)) \to \mathbf{C}^b(\mathcal{A})$.

Note that, for each open subscheme U of X, the category $\mathbf{HR}_{\mathfrak{e}^{\otimes a} \otimes \mathbb{Z}_U(q)_f}$ of hyper-resolutions of $\mathfrak{e}^{\otimes a} \otimes \mathbb{Z}_U(q)_f$ (§1.1 and Definition 1.4.1) is a subcategory of $\mathbf{C}^b_{\mathrm{mot}}(\mathrm{Zar}(X, f))$.

1.5.3. LEMMA. *Let (X, f) be in $\mathcal{L}(\mathcal{V})$, and $a \geq 0$ an integer. Let S^* be a bounded above complex of presheaves on X which takes disjoint unions to direct sums, $\tilde{S}^*$ the sheafification of S^*, and let $j_U: U \to X$ be an open subscheme of X. Then there is a canonical functorial isomorphism*

$$\varinjlim_{\Gamma \in \mathbf{HR}_{\mathfrak{e}^{\otimes a} \otimes \mathbb{Z}_U(q)_f}} H^n(\mathbf{C}^b(S^*)(\Gamma)) \cong \mathbb{H}^n_{\mathrm{Zar}}(U, \tilde{S}^*)$$

(for $a = 0$, we take the limit over $\mathbf{HR}_{\mathbb{Z}_U(q)_f}$).

PROOF. We set $\Gamma_U = \mathfrak{e}^{\otimes a} \otimes \mathbb{Z}_U(q)_{j^* f}[p]$ (or $\mathbb{Z}_U(q)_{j^* f}[p]$ for $a = 0$). By the remarks in §1.5.2, we have the functor $\mathbf{C}^b(\mathcal{S}): \mathbf{C}^b_{\mathrm{mot}}(\mathrm{Zar}(X, f)) \to \mathbf{C}^b(\mathbf{Ab})$. In particular, for each open subscheme U of X, we have the functor $\mathbf{C}^b(\mathcal{S}): \mathbf{HR}_{\Gamma_U} \to \mathbf{C}^b(\mathbf{Ab})$, compatible with restriction maps $k^*: \mathbf{HR}_{\Gamma_U} \to \mathbf{HR}_{\Gamma_V}$ for inclusions $k: V \to U$ of open subschemes of X. We extend these constructions to complexes of presheaves by taking the associated total complex.

If $\mathcal{S}$ is a complex of presheaves, and $\Gamma_U \to \Gamma_{\mathcal{U}}$ is a Čech resolution, we have the natural isomorphism

$$\mathbf{C}^b(\mathcal{S})(\Gamma_{\mathcal{U}}) \to \mathcal{S}(\mathcal{U}), \tag{1.5.3.1}$$

where $\mathcal{S}(\mathcal{U})$ is the ordered non-degenerate Čech complex of $\mathcal{S}$ associated to the cover $\mathcal{U}$ of U. If $\mathcal{S}$ is a sheaf, we thus have have

$$H^0(\mathbf{C}^b(\mathcal{S})(\Gamma_{\mathcal{U}})) = \mathcal{S}(U), \tag{1.5.3.2}$$

and if $\mathcal{S}$ is an injective sheaf, we have

$$H^p(\mathbf{C}^b(\mathcal{S})(\Gamma_{\mathcal{U}})) = 0 \tag{1.5.3.3}$$

for $p > 0$.

Now let $\Gamma_U \to \tilde{\Gamma}$ be a hyper-resolution of Γ_U. By (1.5.3.1)-(1.5.3.3), induction and an elementary spectral sequence argument, we have

$$H^0(\mathbf{C}^b(\mathcal{S})(\tilde{\Gamma})) = \mathcal{S}(U) \tag{1.5.3.4}$$

if $\mathcal{S}$ is a sheaf, and if $\mathcal{S}$ is an injective sheaf, we have

$$H^p(\mathbf{C}^b(\mathcal{S})(\tilde{\Gamma})) = 0 \tag{1.5.3.5}$$

for $p > 0$.

Let $\mathcal{P}$ be a presheaf on X whose associated sheaf is zero, let $\tilde{\Gamma}$ be a hyper-resolution of Γ_X, and let α be a degree d element of $\mathcal{P}(\tilde{\Gamma})$. From Lemma 1.4.2(i), there is a map of hyper-resolutions of X $f:\tilde{\Gamma} \to \tilde{\Delta}$ over the identity such that $\mathbf{C}^b(\mathcal{P})(f)(\alpha) = 0$. From this and Proposition 1.4.3, it follows that setting

$$\mathbb{H}^*(\mathcal{S}) := \varinjlim_{\tilde{\Gamma}\in \mathbf{HR}_{\Gamma_X}} H^*(\mathbf{C}^b(\mathcal{S})(\tilde{\Gamma}))$$

defines a cohomological functor on the category of sheaves on X.

It follows then from (1.5.3.4) that there is a canonical map of cohomological functors (on the category of sheaves on X)

$$\mathbb{H}^*(-) \to H^*_{\mathrm{Zar}}(X, -); \tag{1.5.3.6}$$

it follows from (1.5.3.4) and (1.5.3.5) that the map (1.5.3.6) is an isomorphism. In addition, we have $\mathbb{H}^0(\mathcal{S}) \cong H^0_{\mathrm{Zar}}(X, \tilde{\mathcal{S}})$ for a presheaf $\mathcal{S}$ with associated sheaf $\tilde{\mathcal{S}}$; from this and the isomorphism (1.5.3.6) we have the canonical isomorphism $\mathbb{H}^*(\mathcal{S}) \to H^*_{\mathrm{Zar}}(X, \tilde{\mathcal{S}})$. □

1.5.4. Lemma. *(i) Let (X, f) be in $\mathcal{L}(\mathcal{V})$, and let $h: \mathbf{C}^b_{\mathrm{mot}}(\mathrm{Zar}(X, f)) \to \mathbf{C}(\mathbf{Ab})$ be a DG functor, compatible with cones. Let $\tilde{h}^X$ be the complex of Zariski sheaves on X associated to the presheaf h^X given by $h^X(j: U \to X) = h(\mathfrak{e}^{\otimes a} \otimes \mathbb{Z}_U(q)_{j^*f}[p])$ Let $\mathbb{H}^0_{\mathrm{Zar}}(X, \tilde{h}^X)$ denotes the Zariski hypercohomology. Then there is a natural isomorphism*

$$\mathbb{H}^0_h(\mathfrak{e}^{\otimes a} \otimes \mathbb{Z}_X(q)_f[p]) \cong \mathbb{H}^0_{\mathrm{Zar}}(X, \tilde{h}^X).$$

(ii) Let $h_n: \mathbf{C}^b_{\mathrm{mot}}(\mathcal{V})^* \to \mathbf{C}(\mathbf{Ab})$, $n = 0, 1, \ldots, \infty$ be DG functors, compatible with cones, together with a sequence of natural transformations

$$h_0 \xrightarrow{\pi_{10}} h_1 \xrightarrow{\pi_{21}} \ldots$$

and natural transformations $\pi_n: h_n \to h_\infty$, compatible with cones, such that

$$\pi_{n+1} \circ \pi_{n+1,n} = \pi_n.$$

Suppose that, for each pair of integers p and q, there is an integer $N_{p,q}$ such that $H^0(\pi_n): H^0(h_n(\mathbb{Z}_X(q)_f[m])) \to H^0(h_\infty(\mathbb{Z}_X(q)_f[m]))$ is an isomorphism for all $(X, f) \in \mathcal{L}(\mathcal{V})$, all $m \geq p$, and all $n \geq N_{p,q}$. Then, for each Γ in $\mathbf{C}^b_{\mathrm{mot}}(\mathcal{V})^*$, there is an integer N'_Γ such that the map $\mathbb{H}^0(\pi_N): \mathbb{H}^0_{h_N}(\Gamma) \to \mathbb{H}^0_{h_\infty}(\Gamma)$ is an isomorphism for all $N \geq N'_\Gamma$.

Proof. For (i), let $j: U \to X$ be a Zariski open subset of X. Suppose at first that h is a functor with values in $\mathbf{Ab}$. Then h^X is an abelian presheaf on X, and we have $\mathbf{C}^b(h^X)(\tilde{\Gamma}) = h(\tilde{\Gamma})$ for each hyper-resolution $\Gamma_U \to \tilde{\Gamma}$ of Γ_U.

This together with Lemma 1.5.3 proves (i) in case the functor h takes values in $\mathbf{Ab}$; the general case follows from this and a spectral sequence argument, noting that X has finite cohomological dimension by [**46**, Theorem 3.6.5].

We now prove (ii). The DG category $\mathbf{C}^b_{\mathrm{mot}}(\mathcal{V})^*$ is generated by $\mathcal{A}_{\mathrm{mot}}(\mathcal{V})^*$ by taking translations and cones. Since the functors $\mathbb{H}^0_{h_n}$ are cohomological functors, it suffices to prove the result for Γ a translate of an object in $\mathcal{A}_{\mathrm{mot}}(\mathcal{V})^*$. As $\mathcal{A}_{\mathrm{mot}}(\mathcal{V})^*$ is generated by the objects $\mathfrak{e}^{\otimes a} \otimes \mathbb{Z}_X(q)_f$, it suffices to proof the result for Γ of the form $\Gamma = \mathfrak{e}^{\otimes a} \otimes \mathbb{Z}_X(q)_f[p]$.

Let $\tilde{h}^X_n$ be the complex of Zariski sheaves on X associated to the presheaf

$$(j: U \to X) \mapsto h_n(\mathfrak{e}^{\otimes a} \otimes \mathbb{Z}_U(q)_{j^*f}[p]).$$

Suppose X has Krull dimension M. Then, by [**46**, *loc. cit.*], for all Zariski sheaves $\mathcal{F}$ on X, we have

$$H^n(X, \mathcal{F}) = 0; \qquad n > M.$$

By our assumption on the sequence of functors h_n, if $n \geq N_{p-M-1,q}$, the map of sheaves $\tilde{h}^X_n \to \tilde{h}^X_\infty$ induces an isomorphism on the cohomology sheaves

$$\mathcal{H}^m(\tilde{h}^X_n) \to \mathcal{H}^m(\tilde{h}^X_\infty)$$

for all $m \geq -M-1$. Applying this to the local to global spectral sequence for hypercohomology, we find that the natural map $\mathbb{H}^0_{\mathrm{Zar}}(X, \tilde{h}^X_n) \to \mathbb{H}^0_{\mathrm{Zar}}(X, \tilde{h}^X_\infty)$ is an isomorphism for $n \geq N_{p-M-1,q}$. This, together with (i), completes the proof. □

2. Higher Chow groups

We recall Bloch's construction of the higher Chow groups, and give an extension to motives over an arbitrary base. We also define the motivic cycle map from the motivic Chow group to motivic cohomology.

2.1. Bloch's higher Chow groups

We review the constructions of [**19**]. Fix a field k.

2.1.1. *The simplicial scheme* Δ^*_X. Let $\Delta^n_{\mathbb{Z}}$ be the affine space $\mathbb{A}^n_{\mathbb{Z}}$, given as the scheme

$$\Delta^n_{\mathbb{Z}} := \operatorname{Spec} \mathbb{Z}[t_0, \dots, t_n]/\sum_{i=0}^{n} t_i - 1.$$

$\Delta^n_{\mathbb{Z}}$ has the vertices v^n_i defined by $t_i = 1$, $t_j = 0$ for $j \neq i$. Each map $g: [n] \to [m]$ in Δ gives the map $g := \Delta^*_k(g): \Delta^n_{\mathbb{Z}} \to \Delta^m_{\mathbb{Z}}$, which sends v^n_i to $v^m_{g(i)}$, and is affine-linear. This defines the cosimplicial scheme $\Delta^*_{\mathbb{Z}}$. If X is a scheme, taking the product with X over $\mathbb{Z}$ defines the cosimplicial scheme Δ^*_X.

2.1.2. *Bloch's cycle complex.* A *face* of Δ^n_X is a subscheme of the form $\Delta^*_X(g)(\Delta^m_X)$ for some $g: [m] \to [n]$. Let $z^q(X, n)$ be the subgroup of $\mathcal{Z}^q(\Delta^n_X)$ generated by the codimension q subvarieties W of Δ^n_X such that, for each face F of Δ^n_X, each component of $W \cap F$ has codimension at least q on F. As each face F is a complete intersection in Δ^n_X, this implies that, for each $Z \in z^q(X, n)$, and each map $g: [m] \to [n]$ in Δ, the cycle pull-back $g^*(Z)$ is defined, and is in $z^q(X, *)$. This gives us the simplicial abelian group $n \mapsto z^q(X, n)$, and the associated (homological) complex of abelian groups $z^q(X, *)$, called *Bloch's cycle complex for* X.

2.1.3. *Functoriality.* Let $f: X \to Y$ be a morphism of k-varieties. If f is flat, then pull-back of cycles via the induced map $\Delta^n_X \to \Delta^n_Y$ gives rise to the map of complexes $f^*: z^q(Y,*) \to z^q(X,*)$.

If f is proper of relative dimension d, then push-forward of cycles gives the maps of complexes $f_*: z^{q+d}(X,*) \to z^q(Y,*)$.

2.1.4. DEFINITION. Let X be a reduced k-scheme, essentially of finite type over k. The *higher Chow groups of* X, $\mathrm{CH}^q(X,p)$, $p,q \geq 0$, are defined as

$$\mathrm{CH}^q(X,p) := H_p(z^q(X,*)).$$

2.1.5. Let $f: X \to Y$ be a map of varieties. The pull-back and push-forward operations of §2.1.3 give rise to the pull-back map $f^*: \mathrm{CH}^q(Y,p) \to \mathrm{CH}^q(X,p)$ if f is flat, and the push-forward map $f_*: \mathrm{CH}^{q+d}(X,p) \to \mathrm{CH}^q(Y,p)$ if f is a proper of relative dimension d. These are functorial, when the composition is defined.

2.1.6. *Properties of the higher Chow groups.* We give a list of the properties of $\mathrm{CH}^q(X,p)$; we take X to be quasi-projective over k.
(1) *Homotopy.* Let $p_X: X \times \mathbb{A}^1 \to X$ be the projection. Then

$$p_X^*: \mathrm{CH}^q(X,p) \to \mathrm{CH}^q(X \times \mathbb{A}^1, p)$$

is an isomorphism.
(2) *Localization and Mayer-Vietoris.* Let $i: Z \to X$ be a closed codimension d subscheme of a quasi-projective variety X, and $j: U \to X$ the complement. Then the sequence

$$z^{q-d}(Z,*) \xrightarrow{i_*} z^q(X,*) \xrightarrow{j^*} z^q(U,*)$$

defines a quasi-isomorphism $z^{q-d}(Z,*) \to \mathrm{cone}(j^*)[-1]$, giving rise to the long exact *localization sequence*

$$\ldots \to \mathrm{CH}^{q-d}(Z,p) \xrightarrow{i_*} \mathrm{CH}^q(X,p) \xrightarrow{j^*} \mathrm{CH}^q(U,p) \xrightarrow{\delta} \mathrm{CH}^{q-d}(Z,p-1) \to \ldots .$$

Similarly, if $X = U \cup V$, with $j_U: U \to X$ and $j_V: V \to X$ open subschemes, then the sequence

$$z^q(X,*) \xrightarrow{(j_U^*, j_V^*)} z^q(U,*) \oplus z^q(V,*) \xrightarrow{j^*_{U\cap V,U} - j^*_{U\cap V,V}} z^q(U \cap V,*)$$

defines a quasi-isomorphism $z^q(X,*) \to \mathrm{cone}(j^*_{U\cap V,U} - j^*_{U\cap V,V})[-1]$, giving rise to the long exact *Mayer-Vietoris sequence*

$$\ldots \to \mathrm{CH}^q(X,p) \xrightarrow{(j_U^*, j_V^*)} \mathrm{CH}^q(U,p) \oplus \mathrm{CH}^q(V,p) \xrightarrow{j^*_{U\cap V,U} - j^*_{U\cap V,V}}$$
$$\mathrm{CH}^q(U \cap V,p) \xrightarrow{\delta} \mathrm{CH}^q(X,p-1) \to \ldots .$$

(3) *Contravariant functoriality.* The functor $X \mapsto z^q(X,*)$ on the category of localizations of smooth quasi-projective k-varieties, with flat maps, extends to a functor

$$z^q(-,*): \mathbf{Sm}_k^{\mathrm{ess\ op}} \to \mathbf{D}^-(\mathbf{Ab}).$$

(4) *Products.* There are functorial maps of complexes in $\mathbf{D}^-(\mathbf{Ab})$

$$\boxtimes_{X,Y}: z^q(X,*) \otimes_{\mathbb{Z}} z^{q'}(Y,*) \to z^{q+q'}(X \times_k Y,*)$$

which are associative and (graded) commutative. Following $\boxtimes_{X,X}$ by the pull-back via the diagonal makes the bi-graded group $\oplus_{p,q}\mathrm{CH}^q(X,p)$ into a bi-graded ring, graded-commutative in p. If $f: X \to Y$ is a projective map, then

$$f_*(\alpha \cup_X f^*\beta) = f_*(\alpha) \cup_Y \beta$$

for $\alpha \in \mathrm{CH}^q(X,p)$ and $\beta \in \mathrm{CH}^{q'}(Y,p')$.

(5) *Comparison with K-theory.* Let X be a smooth quasi-projective variety. There are natural isomorphisms

$$\mathrm{CH}^q(X,p) \otimes \mathbb{Q} \cong K_p(X)^{(q)},$$

where $K_p(X)^{(q)}$ is the weight q subspace (for the Adams operations) in the rational higher algebraic K-theory of X, $K_p(X) \otimes \mathbb{Q}$.

2.1.7. REMARKS. (i) These properties were first listed in [**19**]; (1) was proved there and the construction of the products in (4) was given as well. There was an error in the proof of (2) in [**19**]; a correct proof of (2) was given in [**18**]. There was also an error in the proof of (3) in [**19**], which was fixed by Bloch [**16**]. We also give a proof of the contravariant functoriality in §3.5 of this chapter, for affine X; together with (2), this gives a proof of contravariant functoriality for arbitrary X. A proof of (5) (relying on (2) and (3)) was given in [**19**]; we have also given a proof of (5) in [**84**] which makes no use of (2) or (3). We give a proof of (5), following the argument of [**19**], in Chapter III, §3.6.
(ii) One consequence of (2) which we will use later is a comparison of $\mathrm{CH}^q(-,p)$ with Zariski hypercohomology. Let X be a quasi-projective k-variety. The functoriality of the complexes $z^q(-,*)$ allows us to sheafify these complexes on X, forming the complex of Zariski sheaves $\tilde{z}^q(*)_X$ associated to the complex of presheaves $U \mapsto z^q(U,*)$. An immediate consequence of (2) is that the natural map $\mathrm{CH}^q(X,p) \to \mathbb{H}^{-p}_{\mathrm{Zar}}(X, \tilde{z}^q(*)_X)$ is an isomorphism.

2.2. Suspension and the motivic cycle complex

We use the technique of "relative cycles" to give a first approximation to the Chow groups of motives. We fix a base scheme S.

We recall from Chapter I, §2.4, how to assign a motive to a "very smooth" cosimplicial scheme (see Chapter I, §2.4.1)

2.2.1. EXAMPLE. Recall the cosimplicial scheme $\Delta^* := \Delta^*_S$ of §2.1.1. One easily sees that Δ^* is a very smooth cosimplicial scheme in $\mathcal{V}$. We denote the maps (see Chapter I, §2.4.1) $f^n_{\Delta^*}: \Delta^{\leq n} \to \Delta^n$ by δ^n, giving the objects (see (I.2.4.1.1) and (I.2.4.2.2))

$$(\Delta^*, \delta^*): \Delta \to \mathcal{L}(\mathcal{V}),$$
$$\mathbb{Z}^{\leq N}_{\Delta^*}(0) \in \mathbf{C}^b_{\mathrm{mot}}(\mathcal{V})^*, \tag{2.2.1.1}$$

and the sequence of maps in $\mathbf{C}^b_{\mathrm{mot}}(\mathcal{V})^*$ (see (I.2.4.2.3))

$$\ldots \xrightarrow{\chi^{N,N-1}} \mathbb{Z}^{\leq N}_{\Delta^*}(0) \xrightarrow{\chi^{N+1,N}} \mathbb{Z}^{\leq N+1}_{\Delta^*}(0) \xrightarrow{\chi^{N+2,N+1}} \ldots. \tag{2.2.1.2}$$

We let

$$\mathbb{Z}_{\Delta^*}(0)_{\delta^*}: \Delta^{\mathrm{op}} \to \mathcal{A}_{\mathrm{mot}}(\mathcal{V})^*$$

denote the simplicial object $\mathbb{Z}(0)((\Delta^*, \delta^*))$, where $\mathbb{Z}(0): \mathcal{L}(\mathcal{V})^{\mathrm{op}} \to \mathcal{A}_{\mathrm{mot}}(\mathcal{V})^*$ is the functor $(X, f) \mapsto \mathbb{Z}_X(0)_f$.

2.2.2. DEFINITION. Let Γ be an object of $\mathbf{C}^b_{\mathrm{mot}}(\mathcal{V})^*$. We let $\Sigma^N\Gamma$ denote the object $\Gamma \times \mathbb{Z}^{\leq N}_{\Delta^*}(0)[-N]$, where $\times$ is the tensor operation in $\mathbf{C}^b_{\mathrm{mot}}(\mathcal{V})^*$. Sending Γ to $\Sigma^N\Gamma$ gives the cone-preserving functor

$$\Sigma^N : \mathbf{C}^b_{\mathrm{mot}}(\mathcal{V})^* \to \mathbf{C}^b_{\mathrm{mot}}(\mathcal{V})^*;$$

we have as well the extension of Σ^N to exact functors

$$\Sigma^N : \mathbf{K}^b_{\mathrm{mot}}(\mathcal{V})^* \to \mathbf{K}^b_{\mathrm{mot}}(\mathcal{V})^*,$$

and

$$\Sigma^N : \mathbf{D}^b_{\mathrm{mot}}(\mathcal{V})^* \to \mathbf{D}^b_{\mathrm{mot}}(\mathcal{V})^*.$$

2.2.3. Recall the cycles functor (I.3.3.1.2). For a simplicial object $[n] \mapsto \Gamma_n$ of $Z^0\mathbf{C}^b_{\mathrm{mot}}(\mathcal{V})^*$, let $\mathcal{Z}_{\mathrm{mot}}(\Gamma_*)$ be the object of $\mathbf{C}^-(\mathbf{Ab})$ defined as the total complex of the double complex

$$\ldots \to \mathcal{Z}_{\mathrm{mot}}(\Gamma_n) \xrightarrow{d^{-n}} \mathcal{Z}_{\mathrm{mot}}(\Gamma_{n-1}) \to \ldots \xrightarrow{d^{-1}} \mathcal{Z}_{\mathrm{mot}}(\Gamma_0),$$

where d^{-n} is the usual alternating sum

$$d^{-n} = \sum_{i=0}^{n} (-1)^i \mathcal{Z}_{\mathrm{mot}}(\Gamma(\delta_i^{n-1})),$$

and $\mathcal{Z}_{\mathrm{mot}}(\Gamma_p^q)$ is in total degree $q - p$.

2.2.4. DEFINITION. Let Γ be an object of $\mathbf{C}^b_{\mathrm{mot}}(\mathcal{V})^*$. Define the *motivic cycle complex* $\mathcal{Z}_{\mathrm{mot}}(\Gamma, *)$ by

$$\mathcal{Z}_{\mathrm{mot}}(\Gamma, *) := \mathcal{Z}_{\mathrm{mot}}(\Gamma \times \mathbb{Z}_{\Delta^*}(0)_{\delta^*})$$

(see Example 2.2.1).

2.2.5. PROPOSITION. *Suppose $S = \operatorname{Spec} k$, and X is a smooth quasi-projective variety over k. Then $\mathcal{Z}_{\mathrm{mot}}(\mathbb{Z}_X(q)[2q], *)$ is naturally isomorphic to Bloch's cycle complex $z^q(X, -*)$ (see §2.1.2).*

PROOF. We have the identity

$$\mathbb{Z}_X(q)_f[2q] \times \mathbb{Z}_{\Delta^p}(0)_{\delta^p} = \mathbb{Z}_{X\times_k\Delta^p}(q)_{\mathrm{id}_X\times\delta^p}[2q],$$

giving the identification of $\mathcal{Z}_{\mathrm{mot}}(\mathbb{Z}_X(q)[2q], -p)$ with the subgroup $\mathcal{Z}^q(X \times_k \Delta^p)_{\delta^p}$ of $\mathcal{Z}^q(X \times_k \Delta^p/k)$ generated by effective cycles W such that $(\mathrm{id}_X \times f)^*(W)$ is defined for all face maps $f: \Delta^m \to \Delta^p$. This is the same as the group $z^q(X, p)$ described in §2.1.2. With the shift $[2q]$, the graded group $\mathcal{Z}_{\mathrm{mot}}(\mathbb{Z}_X(q)[2q], -p)$ is concentrated in degree $-p$. The coboundary map

$$d^{-p}: \mathcal{Z}_{\mathrm{mot}}(\mathbb{Z}_X(q)[2q], -p) \to \mathcal{Z}_{\mathrm{mot}}(\mathbb{Z}_X(q)[2q], -p+1)$$

is given as the alternating sum of the restrictions to the codimension one faces of $X \times_k \Delta^p$, which is the same as the boundary map $d_p: z^q(X, p) \to z^q(X, p-1)$. $\square$

2.2.6. *Comparison maps.* The sequence of maps (2.2.1.2) gives the sequences of natural transformations

(2.2.6.1) $$\ldots \xrightarrow{\chi^{N,N-1}} \Sigma^N[N] \xrightarrow{\chi^{N+1,N}} \Sigma^{N+1}[N+1] \xrightarrow{\chi^{N+2,N+1}} \ldots .$$

We let

(2.2.6.2) $$i_N : \mathrm{id} \to \Sigma^N[N]$$

be the composition $\chi^{N,N-1} \circ \ldots \circ \chi^{1,0}$.

2.2.7. Sending Γ to $\mathcal{Z}_{\mathrm{mot}}(\Gamma, *)$ defines the DG functor

$$\mathcal{Z}_{\mathrm{mot}}(*): \mathbf{C}^b_{\mathrm{mot}}(\mathcal{V})^* \to \mathbf{C}^-(\mathbf{Ab}), \tag{2.2.7.1}$$

and extends to the exact functor $\mathcal{Z}_{\mathrm{mot}}(*): \mathbf{K}^b_{\mathrm{mot}}(\mathcal{V})^* \to \mathbf{K}^-(\mathbf{Ab})$. The natural maps (Part II, (III.1.1.4.2)) give the natural maps

$$\mathrm{id}_\Gamma \times \pi_N : \Gamma \times \mathbb{Z}^{\leq N}_{\Delta^*}(0) \to \Gamma \times \mathbb{Z}_N(j^*_N \mathbb{Z}_{\Delta^*}(0)_{\delta^*}),$$

which in turn give the natural transformation

$$\Pi_N : \mathcal{Z}_{\mathrm{mot}} \circ \Sigma^N(-)[N] \to \mathcal{Z}_{\mathrm{mot}}(-, *) \tag{2.2.7.2}$$

by applying the functor $\mathcal{Z}_{\mathrm{mot}}$ to the natural maps $\mathrm{id}_\Gamma \times \pi_N$, and composing with the natural inclusion

$$\mathcal{Z}_{\mathrm{mot}}(\Gamma \times \mathbb{Z}_N(j^*_N \mathbb{Z}_{\Delta^*}(0)_{\delta^*})) \subset \mathcal{Z}_{\mathrm{mot}}(\Gamma, *),$$

which is defined by identifying $\mathcal{Z}_{\mathrm{mot}}(\Gamma \times \mathbb{Z}_N(j^*_N \mathbb{Z}_{\Delta^*}(0)_{\delta^*}))$ with the total complex of the truncation

$$\mathcal{Z}_{\mathrm{mot}}(\Gamma \times \mathbb{Z}_{\Delta^N}(0)_{\delta^N}) \xrightarrow{d^{-N}} \dots \xrightarrow{d^{-1}} \mathcal{Z}_{\mathrm{mot}}(\Gamma \times \mathbb{Z}_{\Delta^0}(0)_{\delta^0})$$

of the double complex defining $\mathcal{Z}_{\mathrm{mot}}(\Gamma, *)$.

Applying $\mathcal{Z}_{\mathrm{mot}}$ to the sequence (2.2.6.1) gives us the sequence of natural transformations

$$\dots \xrightarrow{\mathcal{Z}_{\mathrm{mot}}(\chi^{N,N-1})} \Sigma^N \mathcal{Z}_{\mathrm{mot}}[N] \xrightarrow{\mathcal{Z}_{\mathrm{mot}}(\chi^{N+1,N})} \Sigma^{N+1} \mathcal{Z}_{\mathrm{mot}}[N+1] \xrightarrow{\mathcal{Z}_{\mathrm{mot}}(\chi^{N+2,N+1})} \dots . \tag{2.2.7.3}$$

The commutativity of the diagram (Part II, (III.1.1.4.3)) gives the relation

$$\Pi_{N+1} \circ \mathcal{Z}_{\mathrm{mot}}(\chi^{N+1,N}) = \Pi_N. \tag{2.2.7.4}$$

2.2.8. LEMMA. *(i) For each Γ in $\mathbf{C}^b_{\mathrm{mot}}(\mathcal{V})^*$, there is an integer N_Γ such that the maps*

$$H^0(\Pi_N(\Gamma)): H^0(\mathcal{Z}_{\mathrm{mot}}(\Sigma^N(\Gamma)[N])) \to H^0(\mathcal{Z}_{\mathrm{mot}}(\Gamma, *)),$$
$$H^0(\mathcal{Z}_{\mathrm{mot}}(\chi^{N+1,N})): H^0(\mathcal{Z}_{\mathrm{mot}}(\Sigma^N(\Gamma)[N])) \to H^0(\mathcal{Z}_{\mathrm{mot}}(\Sigma^{N+1}(\Gamma)[N+1]))$$

are isomorphisms for all $N \geq N_\Gamma$. In addition, if we take N_Γ minimal, we have $N_{\Gamma[-1]} = N_\Gamma + 1$.
(ii) For each pair of integers (p, q), there is an integer $N_{p,q}$ such that the maps

$$H^0(\Pi_N(\Gamma)): H^0(\mathcal{Z}_{\mathrm{mot}}(\Sigma^N(\Gamma)[N])) \to H^0(\mathcal{Z}_{\mathrm{mot}}(\Gamma, *)),$$
$$H^0(\mathcal{Z}_{\mathrm{mot}}(\chi^{N+1,N})): H^0(\mathcal{Z}_{\mathrm{mot}}(\Sigma^N(\Gamma)[N])) \to H^0(\mathcal{Z}_{\mathrm{mot}}(\Sigma^{N+1}(\Gamma)[N+1]))$$

are isomorphisms for all $N \geq N_{p,q}$, and for all Γ of the form $\Gamma = \mathbb{Z}_X(q)_f[m]$ with $m \geq p$.

PROOF. The assertion (i) for $\Gamma = \mathfrak{e}^{\otimes a} \otimes \mathbb{Z}_X(q)_f[p]$ follows from (Part II, Chapter III, Lemma 1.1.5), with $\mathcal{C} = \mathcal{A}_{\mathrm{mot}}(\mathcal{V})^*$, $F_*(\Gamma) = j^*_N \mathcal{Z}_{\mathrm{mot}}(\Gamma \times \mathbb{Z}_{\Delta^*}(0)_{\delta^*})$; we may take $N_\Gamma = \max(0, 2q - p + 1)$. Thus, taking $N_{p,q} = 2q - p + 1$ proves (ii).

As the extension of both functors from $\mathcal{A}_{\mathrm{mot}}(\mathcal{V})^*$ to $\mathbf{C}^b_{\mathrm{mot}}(\mathcal{V})^*$ preserves the operation of taking cones, and as $\mathbf{C}^b_{\mathrm{mot}}(\mathcal{V})^*$ is generated by translates of $\mathcal{A}_{\mathrm{mot}}(\mathcal{V})^*$ via the operation of taking cones, the assertion (i) is also true for arbitrary Γ in $\mathbf{C}^b_{\mathrm{mot}}(\mathcal{V})^*$. □

2.3. The naive Chow groups of a motive

2.3.1. DEFINITION. Let Γ be an object of $\mathbf{C}^b_{\mathrm{mot}}(\mathcal{V})^*$. The *naive higher Chow groups of* Γ, $\mathrm{CH}_{naif}(\Gamma, p)$, are defined by

$$\mathrm{CH}_{naif}(\Gamma, p) := H^{-p}(\mathcal{Z}_{\mathrm{mot}}(\Gamma, *)).$$

We often write $\mathrm{CH}_{naif}(\Gamma)$ for $\mathrm{CH}_{naif}(\Gamma, 0)$.

2.3.2. From Lemma 2.2.8, we have a natural isomorphism

$$\mathrm{CH}_{naif}(\Gamma, p) \cong H^{-p}(\mathcal{Z}_{\mathrm{mot}}(\Sigma^N(\Gamma)[N]))$$

for all $N \geq N_\Gamma + p$.

2.3.3. *Cohomological functors.* Sending Γ to $\mathrm{CH}_{naif}(\Gamma)$ defines a cohomological functor $\mathrm{CH}_{naif}(-): \mathbf{K}^b_{\mathrm{mot}}(\mathcal{V})^* \to \mathbf{Ab}$. The sequence of natural transformations (2.2.6.1) defines the cohomological functor (for each $a \geq 0$)

$$\varinjlim_N \mathrm{Hom}(\mathfrak{e}^{\otimes a} \otimes 1, \Sigma^N[N](-)): \mathbf{K}^b_{\mathrm{mot}}(\mathcal{V})^* \to \mathbf{Ab},$$

$$\Gamma \mapsto \varinjlim_N \mathrm{Hom}_{\mathbf{K}^b_{\mathrm{mot}}(\mathcal{V})}(\mathfrak{e}^{\otimes a} \otimes 1, \Sigma^N(\Gamma)[N]).$$

Applying the natural maps $\nu_{\Sigma^N(\Gamma)[N],a}$ (I.3.3.6.1) allows us to form the limit

$$\varinjlim_{N,a} \mathrm{Hom}_{\mathbf{K}^b_{\mathrm{mot}}(\mathcal{V})}(\mathfrak{e}^{\otimes a} \otimes 1, \Sigma^N(\Gamma)[N]).$$

2.3.4. PROPOSITION. *There is a natural exact isomorphism of cohomological functors from* $\mathbf{K}^b_{\mathrm{mot}}(\mathcal{V})^*$ *to* $\mathbf{Ab}$:

$$\Sigma^*[*]\mathrm{cyc}: \mathrm{CH}_{naif}(-) \to \varinjlim_{N,a} \mathrm{Hom}_{\mathbf{K}^b_{\mathrm{mot}}(\mathcal{V})}(\mathfrak{e}^{\otimes a} \otimes 1, \Sigma^N[N](-)).$$

The limit on the right is constant after a finite stage for each Γ *in* $\mathbf{K}^b_{\mathrm{mot}}(\mathcal{V})^*$.

PROOF. It follows from Chapter I, Proposition 3.3.5 that the functor $\mathcal{Z}_{\mathrm{mot}}$ gives an isomorphism

$$\mathrm{Hom}_{\mathbf{K}^b_{\mathrm{mot}}(\mathcal{V})}(\mathfrak{e}^{\otimes a} \otimes 1, \Sigma^N(\Gamma)[N]) \xrightarrow{\mathcal{Z}_{\mathrm{mot}}(\mathfrak{e}^{\otimes a} \otimes 1, \Sigma^N(\Gamma)[N])} H^0(\mathcal{Z}_{\mathrm{mot}}(\Sigma^N(\Gamma)[N]))$$

for all a sufficiently large. By Lemma 2.2.8 and Chapter I, Lemma 3.3.7, this, combined with the natural transformation of Lemma 2.2.8, gives the natural isomorphism

$$\varinjlim_{N,a} \mathrm{Hom}(\mathfrak{e}^{\otimes a} \otimes 1, \Sigma^N(\Gamma)[N]) \to \mathrm{CH}_{naif}(\Gamma);$$

we take $\Sigma^*[*]\mathrm{cyc}(\Gamma)$ to be the inverse of this isomorphism. □

2.3.5. LEMMA. *The sequence of natural transformations* (2.2.6.1) *composed with the functor* $\mathbf{K}^b_{\mathrm{mot}}(\mathcal{V})^* \to \mathbf{D}^b_{\mathrm{mot}}(\mathcal{V})^*$ *is a sequence of natural isomorphisms. In particular, for each* Γ *in* $\mathbf{K}^b_{\mathrm{mot}}(\mathcal{V})^*$, *the map*

$$i_N(\Gamma): \Gamma \to \Sigma^N\Gamma[N] \tag{2.3.5.1}$$

induced by the natural transformation (2.2.6.2) *is an isomorphism in* $\mathbf{D}^b_{\mathrm{mot}}(\mathcal{V})^*$.

PROOF. Let X be an object in an additive category $\mathcal{A}$. We have the constant functor $X^{\leq N}: \Delta^{\leq N \text{op}} \to \mathcal{A}$. We have the natural map (see Part II, (III.1.1.4.1))

$$\chi_X^{N+1,N}: \mathbb{Z}_N^{\oplus}(X^{\leq N}) \to \mathbb{Z}_{N+1}^{\oplus}(X^{\leq N+1});$$

let $B^{N+1}(X) := \text{cone}(\chi_X^{N+1,N})$ and set $B^0(X) = X$. In $\mathbf{K}^b(\mathcal{A})$, we have the natural isomorphism $B^{N+1}(X) \cong \text{cone}(\text{id}: B^N(X) \to B^N(X))[-1]$, showing that $B^N(X)$ is isomorphic to zero for all $N \geq 1$. This implies that the map $\chi_X^{N+1,N}$ is an isomorphism in $\mathbf{K}^b(\mathcal{A})$ for all $N \geq 0$.

The homotopy axiom (see Chapter I, Definition 2.1.4(a)), combined with the moving lemma axiom (Chapter I, Definition 2.1.4(e)), shows that the map

$$p_{\Delta^n}^*: \mathbb{Z}_S(0) \to \mathbb{Z}_{\Delta^n}(0)_{\delta^n}$$

is an isomorphism for each n in $\mathbf{D}^b_{\text{mot}}(\mathcal{V})^*$. Thus we have the isomorphism

$$p_{\Delta^*}^*: \mathbb{Z}_N^{\oplus}(\mathbb{Z}_S(0)^{\leq N}) \to \mathbb{Z}_{\Delta^*}^{\leq N}(0)$$

in $\mathbf{D}^b_{\text{mot}}(\mathcal{V})^*$. The remarks of the previous paragraph then show that the map $\chi^{N+1,N}: \mathbb{Z}_{\Delta^*}^{\leq N}(0) \to \mathbb{Z}_{\Delta^*}^{\leq N+1}(0)$ is an isomorphism in $\mathbf{D}^b_{\text{mot}}(\mathcal{V})^*$, as claimed. □

2.3.6. *The naive cycle class.* For Γ in $\mathbf{C}^b_{\text{mot}}(\mathcal{V})^*$, we have the map (2.3.5.1), which by Lemma 2.3.5 is an isomorphism in $\mathbf{D}^b_{\text{mot}}(\mathcal{V})$. We define the *naive cycle class map*

$$\text{cl}_{naif}(\Gamma): \text{CH}_{naif}(\Gamma) \to \text{Hom}_{\mathbf{D}^b_{\text{mot}}(\mathcal{V})}(1, \Gamma) \tag{2.3.6.1}$$

as the composition

$$\begin{aligned} \text{CH}_{naif}(\Gamma) &\xrightarrow{\Sigma^*[*]\text{cyc}(\Gamma)} \varinjlim_{N,a} \text{Hom}(\mathfrak{e}^{\otimes a} \otimes 1, \Sigma^N(\Gamma)[N]) \\ &= \text{Hom}_{\mathbf{K}^b_{\text{mot}}(\mathcal{V})}(\mathfrak{e}^{\otimes a} \otimes 1, \Sigma^N(\Gamma)[N]) \to \text{Hom}_{\mathbf{D}^b_{\text{mot}}(\mathcal{V})}(\mathfrak{e}^{\otimes a} \otimes 1, \Sigma^N(\Gamma)[N]) \\ &\xrightarrow{\nu_a^{-1} \circ (-) \circ i_N(\Gamma)^{-1}} \text{Hom}_{\mathbf{D}^b_{\text{mot}}(\mathcal{V})}(1, \Gamma), \end{aligned}$$

where N is any integer $\geq N_\Gamma$, a is sufficiently large (depending only on Γ) and ν_a is the isomorphism (I.2.2.4.1).

2.4. The naive higher Chow groups of a variety

2.4.1. It follows from Proposition 2.2.5 that there is a natural isomorphism

$$\text{CH}_{naif}(\mathbb{Z}_X(q)[2q], p) \cong \text{CH}^q(X, p) \tag{2.4.1.1}$$

for X a smooth quasi-projective k-variety, in case $S = \text{Spec}\, k$, k a field.

2.4.2. REMARKS. (i) Let X be in $\mathcal{V}$. The map $\delta^1: \Delta^{\leq 1} \to \Delta^1$ is the union $\text{id}_{\Delta^1} \cup i_1 \cup i_0$, where $i_0: S \to \Delta^1$, $i_1: S \to \Delta^1$ are the sections with value v_0 and v_1. We have the commutative diagram with exact columns

$$\begin{array}{ccc}
\mathcal{Z}_{\mathrm{mot}}(\mathbb{Z}_X(q)[2q] \times \mathbb{Z}_{\Delta^1}(0)_{\delta^1}) & = & \mathcal{Z}^q(X \times \Delta^1)_{\mathrm{id}_X \times \delta^1} \\
\downarrow{\scriptstyle d^{-1}} & & \downarrow{\scriptstyle i_1^* - i_0^*} \\
\mathcal{Z}_{\mathrm{mot}}(\mathbb{Z}_X(q)[2q]) & = & \mathcal{Z}^q(X/S) \\
\downarrow & & \\
\mathrm{CH}_{naif}(\mathbb{Z}_X(q)[2q]) & & \\
\downarrow & & \\
0. & &
\end{array} \tag{2.4.2.1}$$

If $S = \operatorname{Spec} k$, we have, via Proposition 2.2.5, the identification of the naive Chow group $\mathrm{CH}_{naif}(\mathbb{Z}_X(q)[2q])$ with the classical Chow group $\mathrm{CH}^q(X)$; the left-hand column in (2.4.2.1) is the standard sequence defining $\mathrm{CH}^q(X)$. We may use this as a definition for arbitrary base schemes: $\mathrm{CH}^q(X/S) := \mathrm{CH}_{naif}(\mathbb{Z}_X(q)[2q])$.
(ii) The cycle map cyc_Γ (I.3.5.1.2) and the cycle map $\Sigma^*[*]\mathrm{cyc}(\Gamma)$ of Proposition 2.3.4 are compatible in the following way: We have the commutative diagram

$$\begin{array}{ccc}
H^0(\mathcal{Z}_{\mathrm{mot}}(\Gamma)) & \xrightarrow{\mathrm{cyc}_\Gamma} & \mathrm{Hom}_{\mathbf{K}^b_{\mathrm{mot}}(\mathcal{V})}(\mathfrak{e}^{\otimes a} \otimes 1, \Gamma) \\
\downarrow{\scriptstyle \chi^{N,0}} & & \downarrow{\scriptstyle \chi^{N,0}} \\
H^0(\mathcal{Z}_{\mathrm{mot}}(\Sigma^N\Gamma[N])) & \xrightarrow[\mathrm{cyc}_{\Sigma^N\Gamma[N]}]{} & \mathrm{Hom}_{\mathbf{K}^b_{\mathrm{mot}}(\mathcal{V})}(\mathfrak{e}^{\otimes a} \otimes 1, \Sigma^N\Gamma[N]).
\end{array}$$

For $N \geq N_\Gamma$, we have the isomorphism $\Pi_N : H^0(\mathcal{Z}_{\mathrm{mot}}(\Sigma^N\Gamma[N])) \to \mathrm{CH}_{naif}(\Gamma)$ and the identity $\Sigma^*[*]\mathrm{cyc}(\Gamma) = \mathrm{cyc}_{\Sigma^N\Gamma[N]} \circ (\Pi_N)^{-1}$. For $\Gamma = \mathbb{Z}_X(q)[2q]$, this gives us the commutative diagram

$$\begin{array}{ccc}
\mathcal{Z}^q(X/S) & & \\
\| & & \\
\mathcal{Z}_{\mathrm{mot}}(\mathbb{Z}_X(q)[2q]) & \xrightarrow{\mathrm{cyc}^q_X} & \mathrm{Hom}_{\mathbf{K}^b_{\mathrm{mot}}(\mathcal{V})}(\mathfrak{e}^{\otimes a} \otimes 1, \mathbb{Z}_X(q)[2q]) \\
\downarrow & & \downarrow{\scriptstyle \chi^{1,0}} \\
\mathrm{CH}^q_{naif}(X/S) & \xrightarrow[\Sigma^*[*]\mathrm{cyc}(\mathbb{Z}_X(q)[2q])]{} & \mathrm{Hom}_{\mathbf{K}^b_{\mathrm{mot}}(\mathcal{V})}(\mathfrak{e}^{\otimes a} \otimes 1, \Sigma^1\mathbb{Z}_X(q)[2q+1]),
\end{array}$$

where the left-hand vertical arrows is the surjection of (i).
(iii) We define the *naive higher Chow groups of* X, for $X \in \mathcal{V}$, as

$$\mathrm{CH}^q_{naif}(X/S, p) = \mathrm{CH}_{naif}(\mathbb{Z}_X(q)[2q-p]).$$

By the isomorphism (2.4.1.1), this agrees with Bloch's higher Chow groups in case $S = \operatorname{Spec} k$.

2.4.3. *Products.* For the remainder of this subsection, we assume that $S = \operatorname{Spec} k$, k a field. Via the isomorphism (2.4.1.1), the naive cycle class map gives the map

$$\mathrm{cl}^{q,p}_{X,naif} : \mathrm{CH}^q(X, 2q-p) \to H^p(X, \mathbb{Z}(q)). \tag{2.4.3.1}$$

As both $\oplus_{p,q}\mathrm{CH}^q(X, 2q-p)$ and $\oplus_{p,q}H^p(X, \mathbb{Z}(q))$ are bi-graded rings, one may ask if the maps (2.4.3.1) give a ring homomorphism. We proceed to show that this is the case.

If S and T are partially ordered sets, we give $S \times T$ the product partial order: $(s,t) \leq ('s,t')$ if $s \leq s'$ and $t \leq t'$. Let $g = (g_1, g_2)\colon [k] \to [n] \times [m]$ be an order-preserving map. We let $\Delta(g)\colon \Delta^k \to \Delta^n \times \Delta^m$ be the affine-linear map with $\Delta(g)(v_i^k) = v^n_{g_1(i)} \times v^m_{g_2(i)}$. A *face* of $\Delta^n \times \Delta^m$ is a subscheme of the form $\Delta(g)(\Delta^k)$. Let F_g denote the face corresponding to g, and let

$$\delta^{n,m}\colon \coprod_{g\colon [k]\to[n]\times[m]} F_g \to \Delta^n \times \Delta^m$$

be the union of the inclusion maps, where g runs over injective, order-preserving maps.

We let $\mathbb{Z}^{\leq N\times M}_{\Delta^*\times\Delta^*}(0)^{-k}$ be the sum

$$\bigoplus_{\substack{f_1\colon [n]\to[N],\ f_2\colon [m]\to[M]\\ n+m=k}} \mathbb{Z}_{\Delta^n\times\Delta^m}(0)_{\delta^{n,m}},$$

where the sum is over injective maps f_1, f_2 in Δ. For $0 \leq i \leq n$, we map $\mathbb{Z}_{\Delta^n\times\Delta^m}(0)_{\delta^{n,m}}$ in the factor (f_1, f_2) to $\mathbb{Z}_{\Delta^{n-1}\times\Delta^m}(0)_{\delta^{n,m}}$ in the factor $(f_1\circ\delta_i^{n-1}, f_2)$ by the map $(\Delta(\delta_i^{n-1}) \times \mathrm{id})^*$. The sum of these maps gives the map

$$(\Delta(\delta_i^{n-1}) \times \mathrm{id})^*\colon \mathbb{Z}^{\leq N\times M}_{\Delta^*\times\Delta^*}(0)^{-k} \to \mathbb{Z}^{\leq N\times M}_{\Delta^*\times\Delta^*}(0)^{-k+1}.$$

For $0 \leq j \leq m$, we have the map

$$(\mathrm{id} \times \Delta(\delta_j^{m-1}))^*\colon \mathbb{Z}^{\leq N\times M}_{\Delta^*\times\Delta^*}(0)^{-k} \to \mathbb{Z}^{\leq N\times M}_{\Delta^*\times\Delta^*}(0)^{-k+1}$$

defined similarly. We let

$$d^{-k}_{N,M}\colon \mathbb{Z}^{\leq N\times M}_{\Delta^*\times\Delta^*}(0)^{-k} \to \mathbb{Z}^{\leq N\times M}_{\Delta^*\times\Delta^*}(0)^{-k+1}$$

be the map $\sum_{n+m=k}\sum_{i=0}^n(-1)^i(\Delta(\delta_i^{n-1})\times\mathrm{id})^* + (-1)^n\sum_{j=0}^m(-1)^j(\mathrm{id}\times\Delta(\delta_j^{m-1}))^*$, giving the complex

$$\ldots \to \mathbb{Z}^{\leq N\times M}_{\Delta^*\times\Delta^*}(0)^{-k} \xrightarrow{d^{-k}_{N,M}} \mathbb{Z}^{\leq N\times M}_{\Delta^*\times\Delta^*}(0)^{-k+1} \to \ldots,$$

which we denote by $\mathbb{Z}^{\leq N\times M}_{\Delta^*\times\Delta^*}(0)$.

The inclusion of $\mathbb{Z}_S(0)$ into $\mathbb{Z}^{\leq N\times M}_{\Delta^*\times\Delta^*}(0)^0$ as the summand corresponding to the vertex $v_0^N \times v_0^M$ defines the map

$$i_{N,M}\colon \mathbb{Z}_S \to \mathbb{Z}^{\leq N\times M}_{\Delta^*\times\Delta^*}(0).$$

The collection of identity maps on $\Delta^n \times \Delta^m$ defines the map in $\mathbf{C}^b_{\mathrm{mot}}(\mathbf{Sm}_k)$

$$\kappa_{N,M}\colon \mathbb{Z}^{\leq N\times M}_{\Delta^*\times\Delta^*}(0) \to \mathbb{Z}^{\leq N}_{\Delta^*}(0) \times \mathbb{Z}^{\leq M}_{\Delta^*}(0),$$

where $\times$ is the tensor operation in $\mathbf{C}^b_{\mathrm{mot}}(\mathbf{Sm}_k)^*$. By the moving lemma isomorphism (Chapter I, §2.2.2), $\kappa_{N,M}$ is an isomorphism in $\mathbf{D}^b_{\mathrm{mot}}(\mathbf{Sm}_k)$. In addition, we have

$$\kappa_{N,M} \circ i_{N,M} = i_N \times i_M.$$

2.4.4. *Triangulations.* We have the standard triangulation of $[N] \times [M]$, defined as the formal sum $\sum_g \operatorname{sgn}(g)g$, where the sum is over injective order-preserving maps $g: [N+M] \to [N] \times [M]$, and $\operatorname{sgn}(g)$ is defined as in (Part II, Chapter III, §3.4.5).

Let $f_1: [n] \to [N]$, $f_2: [m] \to [M]$, and $h: [n+m] \to [n] \times [m]$ be injective order-preserving maps. As each maximal totally ordered subset of $[N] \times [M]$ has $N+M+1$ elements, the composition $(f_1 \times f_2) \circ h$ can be extended to an injective order-preserving map $g: [N+M] \to [N] \times [M]$, i.e., there is an injective order-preserving map

$$\delta^h_{f_1,f_2}: [n+m] \to [N+M]$$

such that $g \circ \delta^h_{f_1,f_2} = (f_1 \times f_2) \circ h$. In fact, the map $\delta^h_{f_1,f_2}$ is independent of the choice of g, and is characterized by the identity

$$\delta^h_{f_1,f_2}(i+1) - \delta^h_{f_1,f_2}(i) = d((f_1 \times f_2) \circ h(i), (f_1 \times f_2) \circ h(i+1)).$$

Here $d(x,y)$ is the distance from x to y, for $x \leq y$ in the partially ordered set $[N] \times [M]$, i.e., the maximal r such that there is a string of strict inequalities

$$x = x_0 < x_1 < \ldots < x_r = y.$$

For each injective, order-preserving map $h: [n+m] \to [n] \times [m]$, we have the map $\Delta(h): \Delta^{n+m} \to \Delta^n \times \Delta^m$, giving the map

$$\Delta(h)^*: \mathbb{Z}_{\Delta^n \times \Delta^m}(0)_{\delta^{n,m}} \to \mathbb{Z}_{\Delta^{n+m}}(0)_{\delta^{n+m}}$$

in $\mathcal{A}_{\text{mot}}(\mathbf{Sm}_k)$. Define the map

$$\Delta(h)^{*,-k}: \mathbb{Z}^{\leq N \times M}_{\Delta^* \times \Delta^*}(0)^{-k} \to \mathbb{Z}^{\leq N+M}_{\Delta^*}(0)^{-k}$$

by sending the summand $\mathbb{Z}_{\Delta^n \times \Delta^m}(0)_{\delta^{n,m}}$ indexed by (f_1, f_2) to the summand $\mathbb{Z}_{\Delta^{n+m}}(0)_{\delta^{n+m}}$ indexed by $\delta^h_{f_1,f_2}$, via the map $\Delta(h)^*$. One checks that the maps $\sum_h \operatorname{sgn}(h)\Delta(h)^{*,-k}$, where the sum is over the injective, order-preserving maps $h: [n+m] \to [n] \times [m]$, $n+m=k$, define the map of complexes

$$T_{N,M}: \mathbb{Z}^{\leq N \times M}_{\Delta^* \times \Delta^*}(0) \to \mathbb{Z}^{\leq N+M}_{\Delta^*}(0).$$

By a direct computation, we have

$$T_{N,M} \circ i_{N,M} = i_{N+M}.$$

2.4.5. LEMMA. *The maps $T_{N,M}$ and $i_{N,M}$ are isomorphisms in $\mathbf{D}^b_{\text{mot}}(\mathbf{Sm}_k)$.*

PROOF. The maps i_M, i_N and i_{N+M} are isomorphisms in $\mathbf{D}^b_{\text{mot}}(\mathbf{Sm}_k)$ by Lemma 2.3.5 of Chapter II; we have already seen that $\kappa_{N,M}$ is an isomorphism in $\mathbf{D}^b_{\text{mot}}(\mathbf{Sm}_k)$. Since $\times = \boxtimes \circ \otimes$, it follows from the Künneth isomorphism (Chapter I, Definition 2.1.4(c)) and from (Chapter I, Theorem 3.4.2) that the map $i_N \times i_M$ is an isomorphism in $\mathbf{D}^b_{\text{mot}}(\mathbf{Sm}_k)$, hence $i_{N,M}$ is an isomorphism in $\mathbf{D}^b_{\text{mot}}(\mathbf{Sm}_k)$. Since $T_{N,M} \circ i_{N,M} = i_{N+M}$, $T_{N,M}$ is an isomorphism as well. □

2.4.6. PROPOSITION. *The map*

$$\oplus_{q,p}\mathrm{cl}^{q,p}_{X,naif}: \oplus_{q,p} \mathrm{CH}^q(X, 2q-p) \to \oplus_{q,p} H^p(X, \mathbb{Z}(q))$$

is a ring homomorphism.

PROOF. Since both products are gotten by taking external products and pulling back by the diagonal, it suffices to show that the maps $\mathrm{cl}^{q,p}_{naif}$ are compatible with the external products. Take $Z_1 \in Z_N(z^q(X,*))$, $Z_2 \in Z_M(z^{q'}(Y,*))$, giving the cycle $Z_1 \times Z_2$ on $X \times Y \times \Delta^N \times \Delta^M$. By [**19**, Theorem 5.1], changing Z_1 and Z_2 in $H_N(z^q(X,*))$ and $H_M(z^{q'}(Y,*))$, we may assume that $Z_1 \times Z_2$ intersects $X \times Y \times F$ properly, for all faces F of $\Delta^N \times \Delta^M$. Replacing $z^q(X,*)$ and $z^{q'}(Y,*)$ with the normalized subcomplexes, we may assume that $Z_1 \cdot (X \times F) = 0$ for each dimension $N-1$ face of Δ^N, and similarly for Z_2.

The appropriate cycle maps $\mathrm{cyc}^q(Z_1)$, $\mathrm{cyc}^{q'}(Z_2)$ thus define the cycle class maps in $\mathbf{D}^b_{\mathrm{mot}}(\mathbf{Sm}_k)$

$$\mathrm{cl}^q(Z_1): 1 \to \mathbb{Z}_X \times \mathbb{Z}^{\le N}_{\Delta^*}(q)[2q-N],\ \mathrm{cl}^{q'}(Z_2): 1 \to \mathbb{Z}_Y \times \mathbb{Z}^{\le M}_{\Delta^*}(q')[2q'-M].$$

The appropriate cycle map $\mathrm{cyc}^{q+q'}(Z_1 \times Z_2)$ similarly defines the cycle class map in $\mathbf{D}^b_{\mathrm{mot}}(\mathbf{Sm}_k)$

$$\mathrm{cl}^{q+q'}(Z_1 \times Z_2): 1 \to \mathbb{Z}_{X\times Y} \times \mathbb{Z}^{\le N\times M}_{\Delta^*\times\Delta^*}(q+q')[2(q+q')-(N+M)],$$

and we have

$$(\mathrm{id} \times \kappa_{N,M}) \circ \mathrm{cl}^{q+q'}(Z_1 \times Z_2) = \boxtimes \circ (\mathrm{cl}^q(Z_1) \otimes \mathrm{cl}^{q'}(Z_2)),$$

by (Chapter I, Lemma 3.5.5), after identifying $1 \otimes 1$ with 1 via $\mu: 1 \otimes 1 \to 1$.

Applying the map $\sum_g \mathrm{sgn}(g)[\mathrm{id}_{X\times Y} \times \Delta(g)]^*$ to the cycle $Z_1 \times Z_2$ gives the cycle

$$Z_1 \cup_{X,Y} Z_2 := \sum_g \mathrm{sgn}(g)[\mathrm{id}_{X\times Y} \times \Delta(g)]^*(Z_1 \times Z_2)$$

on $X \times Y \times \Delta^{N+M}$. By definition of the external product on the higher Chow groups given in [**19**, §5], the class of $Z_1 \cup_{X,Y} Z_2$ in

$$H_{N+M}(z^{q+q'}(X \times Y, *)) = CH^{q+q'}(X \times Y, N+M)$$

is the product of the classes defined by Z_1 and Z_2. On the other hand, letting $p = 2q - N$, $p' = 2q' - M$, we have (using Proposition 3.5.3 and Lemma 3.5.5 of Chapter I, and the definition of cl_{naif})

$$\begin{aligned}
\mathrm{cl}^{q,p}_{X,naif}(Z_1) \cup_{X,Y} \mathrm{cl}^{q',p'}_{Y,naif}(Z_2) &= (\mathrm{id} \times i_N)^{-1}(\mathrm{cl}^q(Z_1)) \cup_{X,Y} (\mathrm{id} \times i_M)^{-1}(\mathrm{cl}^{q'}(Z_2)) \\
&= (\mathrm{id} \times i_N \times i_M)^{-1}(\mathrm{cl}^q(Z_1) \times \mathrm{cl}^{q'}(Z_2)) \\
&= (\mathrm{id} \times i_{N,M})^{-1}(\mathrm{cl}^{q+q'}(Z_1 \times Z_2)) \\
&= (\mathrm{id} \times i_{N+M})^{-1}(T_{N,M} \circ \mathrm{cl}^{q+q'}(Z_1 \times Z_2)) \\
&= (\mathrm{id} \times i_{N+M})^{-1}(\mathrm{cl}^{q+q'}(Z_1 \cup_{X,Y} Z_2)) \\
&= \mathrm{cl}^{q+q',p+p'}_{X\times Y,naif}(Z_1 \cup_{X,Y} Z_2).
\end{aligned}$$

□

2.5. Motivic Chow groups

As described in Remark 2.1.7, for a base scheme S of the form $\mathrm{Spec}\, k$, the localization theorem for the higher Chow groups shows that the naive Chow groups $\mathrm{CH}^q_{naif}(X,p)$ may be also defined as the hypercohomology on X of the complex of sheaves associated to the presheaf $U \mapsto z^q(U,*)$. For a general base scheme S, the analogous statement is possibly not true for arbitrary X; we must therefore pass from the complex $\mathcal{Z}_{\rm mot}(\mathbb{Z}_X(q)_f[2q],*)$ to the associated complex of sheaves on X and take hypercohomology to get the proper definition of the higher Chow groups of X over S. In order to have a reasonable understanding of this operation and how it affects the maps in the category $\mathbf{D}^b_{\rm mot}(\mathcal{V})$, we use the notion of hypercohomology for the functor $\mathcal{Z}_{\rm mot}(-)$, developed in §1.5.

2.5.1. *Suspension.* We have the DG functor (2.2.7.1), compatible with cones,

$$\begin{aligned}\mathcal{Z}_{\rm mot}(*) \colon \mathbf{C}^b_{\rm mot}(\mathcal{V})^* &\to \mathbf{C}^-(\mathbf{Ab})\\ \Gamma &\mapsto \mathcal{Z}_{\rm mot}(\Gamma,*).\end{aligned}$$

For each N, we have the DG functor, compatible with cones

$$\Sigma^N\mathcal{Z}_{\rm mot}[N] \colon \mathbf{C}^b_{\rm mot}(\mathcal{V})^* \to \mathbf{C}^-(\mathbf{Ab}) \tag{2.5.1.1}$$

defined as the composition $\Sigma^N\mathcal{Z}_{\rm mot}[N] := \mathcal{Z}_{\rm mot}\circ\Sigma^N(-)[N]$ We have the natural transformation (2.2.7.2)

$$\Pi_N \colon \Sigma^N\mathcal{Z}_{\rm mot}[N] \to \mathcal{Z}_{\rm mot}(*), \tag{2.5.1.2}$$

inducing the natural transformation

$$\mathbb{H}^0(\Pi_N) \colon \mathbb{H}^0_{\Sigma^N\mathcal{Z}_{\rm mot}[N]} \to \mathbb{H}^0_{\mathcal{Z}_{\rm mot}(*)}. \tag{2.5.1.3}$$

2.5.2. Definition. Let Γ be in $\mathbf{C}^b_{\rm mot}(\mathcal{V})^*$. Define the higher Chow groups of Γ by

$$\mathcal{CH}(\Gamma,p) = \mathbb{H}^0_{\mathcal{Z}_{\rm mot}(*)}(\Gamma[-p]).$$

(*cf.* (1.5.1.1)). We write $\mathcal{CH}(\Gamma)$ for $\mathcal{CH}(\Gamma,0)$. The natural transformation

$$\mathbb{H}^0 \colon \mathcal{Z}_{\rm mot}(*) \to \mathbb{H}^0_{\mathcal{Z}_{\rm mot}(*)}$$

gives the natural map $\mathrm{CH}_{naif}(\Gamma,p) \to \mathcal{CH}(\Gamma,p)$.

2.5.3. Proposition. *Let Γ be in $\mathbf{C}^b_{\rm mot}(\mathcal{V})^*$. Then there is an integer N''_Γ such that, for all $N \geq N''_\Gamma$, the natural transformation* (2.5.1.3) *defines an isomorphism*

$$\mathbb{H}^0(\Pi_N)(\Gamma) \colon \mathbb{H}^0_{\Sigma^N\mathcal{Z}_{\rm mot}[N]}(\Gamma) \to \mathcal{CH}(\Gamma).$$

Proof. This follows from Lemma 1.5.4 and Lemma 2.2.8. □

2.5.4. *The motivic cycle class map.* We extend the naive cycle class map (2.3.6.1) to the cycle class map

$$\mathrm{cl}(\Gamma) \colon \mathcal{CH}(\Gamma) \to \mathrm{Hom}_{\mathbf{D}^b_{\rm mot}(\mathcal{V})}(1,\Gamma).$$

For this, let $j \colon \Gamma \to \Gamma_{\mathcal{U}}$ be a hyper-resolution of Γ. We have the naive cycle class map (2.3.6.1)

$$\mathrm{cl}_{naif}(\Gamma_{\mathcal{U}}) \colon \mathrm{CH}_{naif}(\mathrm{Tot}(\Gamma_{\mathcal{U}})) \to \mathrm{Hom}_{\mathbf{D}^b_{\rm mot}(\mathcal{V})}(1,\mathrm{Tot}(\Gamma_{\mathcal{U}})).$$

By Lemma 1.4.2(iii), $\mathrm{Tot}j$ is an isomorphism in $\mathbf{D}^b_{\rm mot}(\mathcal{V})$; composing $\mathrm{cl}_{naif}(\Gamma_{\mathcal{U}})$ with $(\mathrm{Tot}j)^{-1}$ gives the map

$$(\mathrm{Tot}j)^{-1}\circ\mathrm{cl}_{naif}(\mathrm{Tot}\Gamma_{\mathcal{U}}) \colon \mathrm{CH}_{naif}(\Gamma_{\mathcal{U}}) \to \mathrm{Hom}_{\mathbf{D}^b_{\rm mot}(\mathcal{V})}(1,\Gamma).$$

If we have another tower of Čech resolution of Γ, giving the hyper-resolution $j' \colon \Gamma \to \Gamma_{\mathcal{U}'}$ and a map $\eta \colon \Gamma_{\mathcal{U}} \to \Gamma_{\mathcal{U}'}$ over the identity, we have $j' = \eta\circ j$. As $\mathrm{cl}_{naif}(-)$ is natural, the maps $j^{-1}\circ\mathrm{cl}_{naif}(\Gamma_{\mathcal{U}})$ give a well-defined map on the limit

$$\mathrm{cl}(\Gamma) \colon \mathcal{CH}(\Gamma) \to \mathrm{Hom}_{\mathbf{D}^b_{\rm mot}(\mathcal{V})}(1,\Gamma). \tag{2.5.4.1}$$

By Lemma 1.4.2, we see that sending Γ to $\mathrm{cl}(\Gamma)$ defines a natural transformation of cohomological functors from $\mathbf{K}^b_{\mathrm{mot}}(\mathcal{V})$ to $\mathbf{Ab}$.

2.5.5. DEFINITION. Let (X, f) be in $\mathcal{L}(\mathcal{V})$. We let $\mathcal{Z}^q(X/S, *)_f$ denote the complex $\mathcal{Z}_{\mathrm{mot}}(\mathbb{Z}_X(q)_f[2q], *)$. We denote $\mathcal{CH}(\mathbb{Z}_X(q)_f[2q-p])$ by $\mathcal{CH}^q(X/S, p)_f$, and the map

$$\mathrm{cl}(\mathbb{Z}_X(q)[p]): \mathcal{CH}(\mathbb{Z}_X(q)[p])_f \to \mathrm{Hom}_{\mathbf{D}^b_{\mathrm{mot}}(\mathcal{V})}(1, \mathbb{Z}_X(q)_f[2q-p]) = H^p(X, \mathbb{Z}(q))$$

by

$$\mathrm{cl}^{q,p}_X: \mathcal{CH}^q(X/S, 2q-p)_f \to H^p(X, \mathbb{Z}(q)).$$

We write $\mathcal{CH}^q(X/S, p)$ for $\mathcal{CH}^q(X/S, p)_{\mathrm{id}_X}$.

3. The motivic cycle map

In this last section, we give criteria for the injectivity and surjectivity of the motivic cycle map, derive some consequences for motivic cohomology when these criteria are satisfied, and verify the criteria if the base scheme has dimension at most one over a field.

3.1. Sheafification

We relate the motivic Chow groups to Zariski hypercohomology.

3.1.1. Sending $(X, f) \in \mathcal{L}(\mathcal{V})$ to $\mathcal{Z}^q(X/S, *)_f$ defines the functor

$$\mathcal{Z}^q(-/S, *)_-: \mathcal{L}(\mathcal{V})^{\mathrm{op}} \to \mathbf{C}^-\mathbf{Ab};$$

in particular, sending an open subscheme $j: U \to X$ to $\mathcal{Z}^q(U/S, *)_{j^*f}$ defines a complex of presheaves on X. We let

$$\mathfrak{Z}^q_{X/S}(*)_f \tag{3.1.1.1}$$

denote the associated complex of Zariski sheaves.

3.1.2. Let $\mathrm{Sh}^{\mathbf{Ab}}(\mathrm{Zar}_S)$ be the category of Zariski sheaves of abelian groups on S-schemes: An object is a sheaf $\mathcal{F}$ on an S-scheme X, and a morphism $(X, \mathcal{F}) \to (Y, \mathcal{G})$ is a pair $(p, \tilde{p})$ consisting of a map $p: Y \to X$ and a map $\tilde{p}: \mathcal{F} \to p_*(\mathcal{G})$. Composition is given by $(q, \tilde{q}) \circ (p, \tilde{p}) = (q \circ p, \tilde{q} \circ q_*(\tilde{p}))$.

Sending (X, f) to $\mathfrak{Z}^q_{X/S}(*)_f$ then gives the functor

$$\mathfrak{Z}^q_{/S}(*): \mathcal{L}(\mathcal{V})^{\mathrm{op}} \to \mathbf{C}^-(\mathrm{Sh}^{\mathbf{Ab}}(\mathrm{Zar}_S)), \tag{3.1.2.1}$$

where we send a morphism $p: (X, f) \to (Y, g)$ to the pair (p, p^*), where p^* is the map $p^*: \mathfrak{Z}^q_{Y/S}(*)_g \to p_*(\mathfrak{Z}^q_{X/S}(*)_f)$. We write $\mathfrak{Z}^q_{X/S}(*)$ for $\mathfrak{Z}^q_{X/S}(*)_{\mathrm{id}_X}$.

3.1.3. PROPOSITION. *For each X in $\mathcal{V}$, there is a canonical identification*

$$\mathcal{CH}^q(X/S, p)_f \cong \mathbb{H}^{-p}_{\mathrm{Zar}}(X, \mathfrak{Z}^q_{X/S}(*)_f).$$

PROOF. This follows from Lemma 1.5.4(i). □

3.1.4. Sending (X, f) to $\mathcal{CH}^q(X/S, p)_f$ defines the functor

$$\mathcal{CH}^q(-, p)_f : \mathcal{L}(\mathcal{V})^{\rm op} \to \mathbf{Ab};$$

the cycle class maps $\mathrm{cl}_X^{q,p}$ define the natural transformation

$$\mathrm{cl}^{q,p} : \mathcal{CH}^q(-, 2q-p)_f \to H^p(-, \mathbb{Z}(q)). \tag{3.1.4.1}$$

Let $i: Z \to X$ be a closed embedding of smooth S-schemes in $\mathcal{V}$ of relative codimension d. If Y is in $\mathcal{V}$ and $W \in \mathcal{Z}^{q-d}(Z \times_S Y/S)$ is a cycle, we may consider W as a cycle on $X \times_S Y$; this defines the natural transformation

$$i_* : \mathcal{Z}^{q-d}(Z \times_S (-)/S) \to \mathcal{Z}^q(X \times_S (-)/S).$$

This extends in the obvious way to a natural map of complexes $i_* : \mathcal{Z}^{q-d}(Z/S, *) \to \mathcal{Z}^q(X/S, *)$, and to the natural map of complexes of sheaves on X,

$$i_* : i_* \mathfrak{Z}^{q-d}_{Z/S}(*) \to \mathfrak{Z}^q_{X/S}(*).$$

If $j^* : U \to X$ is the complement $X \setminus Z$, we have $j^* \circ i_* = 0$; giving the natural map of complexes of sheaves on X,

$$i_* : i_* \mathfrak{Z}^{q-d}_{Z/S}(*) \to \mathrm{cone}\big(j^* : \mathfrak{Z}^q_{X/S}(*) \to \mathfrak{Z}^q_{U/S}(*)\big)[-1]. \tag{3.1.4.2}$$

3.2. Surjectivity of the cycle map

We give a general criterion for the surjectivity of the cycle map. To simplify the notation, we take the coefficient ring to be $\mathbb{Z}$; making the obvious changes, the discussion goes through for a commutative ring, flat over $\mathbb{Z}$, as coefficient ring.

For a Zariski sheaf of abelian groups $\mathcal{F}$ on a scheme X, we have the classical *Godement resolution* [**49**] of $\mathcal{F}$, $\mathcal{F} \to \mathfrak{G}^*\mathcal{F}$, defined by letting $\mathfrak{G}^0\mathcal{F}$ be the sheaf on X with $\mathfrak{G}^0\mathcal{F}(U) := \prod_{x \in U} \mathcal{F}_x$, with inclusion $\mathcal{F} \to \mathfrak{G}^0\mathcal{F}$, and defining $\mathfrak{G}^n(\mathcal{F})$ inductively as

$$\mathfrak{G}^n(\mathcal{F}) := \mathfrak{G}^0(\mathfrak{G}^{n-1}(\mathcal{F})/\mathrm{Im}(\mathfrak{G}^{n-2}(\mathcal{F})),$$

with $\mathfrak{G}^{-1}(\mathcal{F}) = \mathcal{F}$.

For a complex of Zariski sheaves $\mathcal{F}$ of $\mathbb{Z}$-modules on a scheme X, we let $\mathfrak{G}\mathcal{F}$ denote the total complex of the Godement resolution, and $\mathfrak{R}_X\mathcal{F}$ the global sections $\Gamma(X, \mathfrak{G}\mathcal{F})$. If $\hat{X}$ is a closed subset of X, with complement $j: U \to X$, we let $\mathfrak{R}_X^{\hat{X}}\mathcal{F}$ denote the cone

$$\mathfrak{R}_X^{\hat{X}}\mathcal{F} := \mathrm{cone}\big(\mathfrak{R}j^* : \mathfrak{R}_X\mathcal{F} \to \mathfrak{R}_U(j^*\mathcal{F})\big)[-1].$$

The complex $\mathfrak{R}_X\mathcal{F}$ gives a functorial representative in $\mathbf{C}^+(\mathbf{Ab})$ for the object $R\Gamma(X, \mathcal{F})$ of $\mathbf{D}^+(\mathbf{Ab})$. Similarly, he complex $\mathfrak{R}_X^{\hat{X}}\mathcal{F}$ gives a functorial representative in $\mathbf{C}^+(\mathbf{Ab})$ for the object $R\Gamma^W(X, \mathcal{F})$ of $\mathbf{D}^+(\mathbf{Ab})$, where $\Gamma^W(X, -)$ is the functor "global sections with support in W".

3.2.1. *The surjectivity conditions.* Consider the following conditions:

(i) *Homotopy.* Let X be in $\mathcal{V}$. Then the map

$$\mathfrak{R}p^* : \mathfrak{R}_X \mathfrak{Z}^q_{X/S}(*) \to \mathfrak{R}_{X \times \mathbb{A}^1} \mathfrak{Z}^q_{X \times \mathbb{A}^1/S}(*)$$

induced by the projection $p: X \times \mathbb{A}^1 \to X$ is a quasi-isomorphism for all q.

(ii) *Moving lemma.* Let (X, f) be in $\mathcal{L}(\mathcal{V})$. Then the natural map $\mathfrak{Z}^q_{X/S}(*)_f \to \mathfrak{Z}^q_{X/S}(*)$ is an quasi-isomorphism for all q.

(iii) *Gysin isomorphism.* Let $i: Z \to X$ be a closed codimension d embedding in $\mathcal{V}$. Then the map $\mathfrak{R}i_*: \mathfrak{R}_Z \mathfrak{Z}^{q-d}_{Z/S}(*) \to \mathfrak{R}^Z_X \mathfrak{Z}^q_{X/S}(*)$ induced by the map (3.1.4.2) is a quasi-isomorphism for all q.

In this section we will show that, assuming these conditions, the cycle map (2.5.4.1) is surjective.

3.2.2. *The Chow realization.* We begin the construction of a functor from $\mathbf{D}^b_{\mathrm{mot}}(\mathcal{V})$ to $\mathbf{D}^-(\mathbf{Ab})$ which extends the assignment $X \mapsto \mathfrak{R}_X \mathfrak{Z}^q_{X/S}(*)$ for X in $\mathcal{V}$.

We denote the category $\mathbf{D}^b_{\mathrm{mot}}(\mathcal{V})$ by $\mathcal{D}$. We assume throughout this section that the conditions of §3.2.1 are satisfied.

We have the functor (2.2.7.1) and the functor (3.1.2.1).

Composing the functor $\mathfrak{Z}_{/S}(*)$ with $\mathfrak{R}(-)$ gives the functor

$$\mathfrak{R}\mathfrak{Z}^q_{/S}(*): \mathcal{L}(\mathcal{V})^{\mathrm{op}} \to \mathbf{C}^-(\mathbf{Ab}). \tag{3.2.2.1}$$

We now extend (3.2.2.1) to $\mathcal{A}_{\mathrm{mot}}(\mathcal{V})^*$.

For an object of $\mathcal{A}_{\mathrm{mot}}(\mathcal{V})^*$ of the form $\mathfrak{e}^{\otimes a} \otimes \mathbb{Z}_X(n)_f$, we define

$$\mathfrak{R}\mathfrak{Z}_{\mathrm{mot}}(\mathfrak{e}^{\otimes a} \otimes \mathbb{Z}_X(n)_f, *) := \mathfrak{R}\mathfrak{Z}^n_{X/S}(*)_f[-2n];$$

we extend the definition of $\mathfrak{R}\mathfrak{Z}_{\mathrm{mot}}(\Gamma, *)$ to arbitrary objects of $\mathcal{A}^b_{\mathrm{mot}}(\mathcal{V})^*$ by taking direct sums.

To define $\mathfrak{R}\mathfrak{Z}_{\mathrm{mot}}(q, *)$ for a morphism $q: \mathfrak{e}^{\otimes a'} \otimes \mathbb{Z}_Y(b')_g \to \mathfrak{e}^{\otimes a} \otimes \mathbb{Z}_X(b)_f$ we use the representation of q as a sum of compositions of the form (1.2.2.1). If q is one such composition, say $q = q(\tau, h_*, p)$, we have the associated map of S-schemes $\bar{q} := \bar{q}(\tau, h_*, p): X \to Y$ (1.2.2.2). For each open subscheme $j: U \to Y$, we have the inclusion $k: V \to X$, where $V = \bar{q}^{-1}(U)$, and the map (1.3.1.2)

$$q_U: \mathfrak{e}^{\otimes a'} \otimes \mathbb{Z}_U(b')_{j^*g} \to \mathfrak{e}^{\otimes a} \otimes \mathbb{Z}_V(b)_{k^*f}.$$

Let $\underline{\mathcal{Z}}_{\mathrm{mot}}(\mathfrak{e}^{\otimes a'} \otimes \mathbb{Z}_Y(b')_g, *)$ denote the complex of presheaves on Y defined by

$$\underline{\mathcal{Z}}_{\mathrm{mot}}(\mathfrak{e}^{\otimes a'} \otimes \mathbb{Z}_X(b')_g, *)(j: U \to Y) = \mathcal{Z}_{\mathrm{mot}}(\mathfrak{e}^{\otimes a'} \otimes \mathbb{Z}_U(b')_{j^*g}, *),$$

and define the complex of presheaves on X, $\underline{\mathcal{Z}}_{\mathrm{mot}}(\mathfrak{e}^{\otimes a} \otimes \mathbb{Z}_X(b)_f, *)$, similarly.

The commutativity of the diagram (1.3.1.3) implies that the maps

$$\mathcal{Z}_{\mathrm{mot}}(q_U, *): \mathcal{Z}_{\mathrm{mot}}(\mathfrak{e}^{\otimes a'} \otimes \mathbb{Z}_U(b')_{j^*g}, *) \to \mathcal{Z}_{\mathrm{mot}}(\mathfrak{e}^{\otimes a} \otimes \mathbb{Z}_V(b)_{k^*f}, *)$$

define a map of complexes of presheaves

$$\underline{\mathcal{Z}}_{\mathrm{mot}}(q, *): \underline{\mathcal{Z}}_{\mathrm{mot}}(\mathfrak{e}^{\otimes a'} \otimes \mathbb{Z}_X(b')_g, *) \to \underline{\mathcal{Z}}_{\mathrm{mot}}(\mathfrak{e}^{\otimes a} \otimes \mathbb{Z}_X(b)_f, *)$$

over the map $\bar{q}$.

Taking the map of associated sheaves, and noting that

$$\mathcal{Z}_{\mathrm{mot}}(\mathfrak{e}^{\otimes \alpha} \otimes \mathbb{Z}_W(q)_h, *) = \mathcal{Z}_{\mathrm{mot}}(\mathbb{Z}_W(q)_h, *) = \mathcal{Z}^q(W, *)_h[-2q]$$

(see Chapter I, Lemma 3.2.2(ii)), we have the map of sheaves over $\bar{q}$,

$$\mathfrak{Z}(q, *): \mathfrak{Z}^{b'}_{Y/S}(*)_g[-2b'] \to \mathfrak{Z}^b_{X/S}(*)_f[-2b].$$

We let

$$\mathfrak{R}\mathfrak{Z}_{\mathrm{mot}}(q, *): \mathfrak{R}\mathfrak{Z}_{\mathrm{mot}}(\mathfrak{e}^{\otimes a'} \otimes \mathbb{Z}_Y(b')_g, *) \to \mathfrak{R}\mathfrak{Z}_{\mathrm{mot}}(\mathfrak{e}^{\otimes a} \otimes \mathbb{Z}_X(b)_f, *) \tag{3.2.2.2}$$

be the map induced by $\mathfrak{Z}(q, *)$ on the global sections of the Godement resolution. We extend the definition of $\mathfrak{R}\mathfrak{Z}_{\mathrm{mot}}(q, *)$ to finite sums of compositions (1.2.2.1) by linearity, and extend to arbitrary maps between arbitrary objects of $\mathcal{A}_{\mathrm{mot}}(\mathcal{V})^*$ by taking direct sums.

The relations (1.3.1.5)-(1.3.1.7) imply that the maps (3.2.2.2) define a DG functor

$$\mathfrak{RZ}_{\mathrm{mot}}(*)\colon \mathcal{A}_{\mathrm{mot}}(\mathcal{V})^* \to \mathbf{C}^-(\mathbf{Ab}). \tag{3.2.2.3}$$

We may then extend (3.2.2.3) to the DG functor, compatible with cones, $\mathfrak{RZ}_{\mathrm{mot}}(*)\colon \mathbf{C}^b_{\mathrm{mot}}(\mathcal{V})^* \to \mathbf{C}^-(\mathbf{Ab})$, and the exact functor, $\mathfrak{RZ}_{\mathrm{mot}}(*)\colon \mathbf{K}^b_{\mathrm{mot}}(\mathcal{V})^* \to \mathbf{K}^-(\mathbf{Ab})$, by applying the functor $\mathrm{Tot}\circ\mathbf{C}^b$ (Part II, Chapter II, §1.2 and §1.2.9), and passing to the homotopy category.

We have the identity

$$\mathfrak{RZ}_{\mathrm{mot}}(\mathfrak{e}^{\otimes a}\otimes \mathbb{Z}_X(q)_f[2q], *) = \mathfrak{RZ}^q_{X/S}(*)_f, \tag{3.2.2.4}$$

and the canonical isomorphism (see Proposition 3.1.3)

$$H^0(\mathfrak{RZ}_{\mathrm{mot}}(\mathfrak{e}^{\otimes a}\otimes \mathbb{Z}_X(q)_f[2q-p], *)) \cong \mathcal{CH}^q(X,p).$$

As the functors

$$H^0(\mathfrak{RZ}_{\mathrm{mot}}(-,*))\colon \mathbf{K}^b_{\mathrm{mot}}(\mathcal{V})^* \to \mathbf{Ab},$$
$$\mathcal{CH}(-)\colon \mathbf{K}^b_{\mathrm{mot}}(\mathcal{V})^* \to \mathbf{Ab}$$

are cohomological functors, and the identity map on $\mathcal{Z}_{\mathrm{mot}}(*)$ gives the natural transformation $\mathcal{CH}(-)\to H^0(\mathfrak{RZ}_{\mathrm{mot}}(-,*))$, we have, for Γ in $\mathbf{C}^b_{\mathrm{mot}}(\mathcal{V})^*$, the canonical isomorphism

$$H^0(\mathfrak{RZ}_{\mathrm{mot}}(\Gamma,*)) \cong \mathcal{CH}(\Gamma). \tag{3.2.2.5}$$

3.2.3. PROPOSITION. *Under the conditions of §3.2.1, the functor*

$$\mathfrak{RZ}_{\mathrm{mot}}(*)\colon \mathbf{K}^b_{\mathrm{mot}}(\mathcal{V})^* \to \mathbf{K}^-(\mathbf{Ab})$$

extends to a functor of triangulated categories

$$\mathfrak{R}_{\mathcal{CH}}\colon \mathcal{D} \to \mathbf{D}^-(\mathbf{Ab}).$$

PROOF. The condition (i), together with the identity (3.2.2.4), implies that the morphisms of Chapter I, Definition 2.1.4(a) get sent to quasi-isomorphisms; similarly, the condition (ii) of §3.2.1 implies the functor $\mathfrak{RZ}_{\mathrm{mot}}(*)$ sends the morphisms of Definition 2.1.4(e) to quasi-isomorphisms.

Suppose we have a codimension d inclusion $i\colon Z\to P$, split by a smooth projection $p\colon P\to Z$. Then the map $i_*\colon \mathcal{Z}^{n-d}(Z,*)\to \mathcal{Z}^n(P,*)$ is the same as the composition $\cup[i(Z)]\circ p^*$. Applying the remarks of Chapter I, §2.2.5, we see that the condition (iii) of §3.2.1 implies that the morphisms of Chapter I, Definition 2.1.4(d) get sent to quasi-isomorphisms.

The excision property (Chapter I, Definition 2.1.4(b)) is a general property of the functor $\mathfrak{R}$. For each connected X, the complex $\mathcal{Z}_{\mathrm{mot}}(\mathbb{Z}_X(0),*)$ is the complex

$$\ldots\to\mathbb{Z}\to\mathbb{Z}\to\ldots\to\mathbb{Z}$$

with the maps alternatively the identity map and the zero map; thus the canonical map $\mathbb{Z}\to\mathcal{Z}_{\mathrm{mot}}(\mathbb{Z}_X(0),*)$ is a homotopy equivalence. From this, it is easy to verify that the morphism of Chapter I, Definition 2.1.4(f) is sent to a quasi-isomorphism.

By Chapter I, Proposition 3.4.6, this implies that we have the extension of $\mathfrak{RZ}_{\mathrm{mot}}(*)$ to the exact functor $\mathfrak{RZ}_{\mathrm{mot}}(*)\colon \mathbf{D}^b_{\mathrm{mot}}(\mathcal{V})^*\to\mathbf{D}^-(\mathbf{Ab})$. Composing with the retraction (Chapter I, Theorem 3.4.2) $\mathbf{D}^b(r_{\mathrm{mot}})\colon \mathbf{D}^b_{\mathrm{mot}}(\mathcal{V}) \to \mathbf{D}^b_{\mathrm{mot}}(\mathcal{V})^*$ gives the desired extension. □

3.2.4. We denote the category $\mathbf{D}^b_{\mathrm{mot}}(\mathcal{V})^*$ by $\mathcal{D}^*$, the category $\mathbf{K}^b_{\mathrm{mot}}(\mathcal{V})^*$ by $\mathcal{K}^*$ and the category $\mathbf{C}^b_{\mathrm{mot}}(\mathcal{V})^*$ by $\mathcal{C}^*$.

The identity $X = X \times \mathbb{Z}_S(0) = X \times \mathbb{Z}_{\Delta^0}(0)_{\delta^0}$ for $X \in \mathcal{C}^*$ gives the identity $\mathcal{Z}_{\mathrm{mot}}(-) = \mathcal{Z}_{\mathrm{mot}}(-,0)$ and thus gives us the natural transformation $\sigma_0\colon \mathcal{Z}_{\mathrm{mot}}(-) \to \mathcal{Z}_{\mathrm{mot}}(-,*)$. Following σ_0 with the natural transformation (presheaf to associated sheaf to Godement resolution to complex of global sections):

$$\mathcal{Z}_{\mathrm{mot}}(-,*) \to \mathfrak{RZ}_{\mathrm{mot}}(-,*) \tag{3.2.4.1}$$

gives the natural transformation

$$\mathfrak{RZ}\sigma_0\colon \mathcal{Z}_{\mathrm{mot}}(-) \to \mathfrak{RZ}_{\mathrm{mot}}(-,*). \tag{3.2.4.2}$$

We denote the category $\mathbf{K}^b_{\mathrm{mot}}(\mathcal{V})^*_B$ (see Chapter I, §3.3.4) by $\mathcal{K}^*_B$. For Δ in $\mathcal{K}^*$, we denote the map $\mathcal{Z}_{\mathrm{mot}}(\mathfrak{e}^{\otimes a} \otimes 1, \Delta)$ of Chapter I, Proposition 3.3.5(iv) by

$$\mathrm{ev}^{\Delta}_a\colon \mathrm{Hom}_{\mathcal{K}^*}(\mathfrak{e}^{\otimes a} \otimes 1, \Delta) \to H^0(\mathcal{Z}_{\mathrm{mot}}(\Delta)). \tag{3.2.4.3}$$

If Δ is in $\mathcal{K}^*_B$, then the map (3.2.4.3) is an isomorphism for all $a > B$, again by Chapter I, Proposition 3.3.5(iv).

3.2.5. LEMMA. *Let Γ and Ξ be objects of $\mathcal{K}^*_B$, $f\colon \Gamma \to \Xi$ a map in $\mathcal{K}^*$ which becomes an isomorphism in $\mathcal{D}^*$, and let $g\colon \mathfrak{e}^{\otimes a} \otimes 1 \to \Xi$ be a map in $\mathcal{K}^*$, with $a > B$. Then there are hyper-resolutions (see Definition 1.4.1)*

$$j_{\mathcal{U}}\colon \Gamma \to \Gamma_{\mathcal{U}}$$
$$j_{\mathcal{W}}\colon \Xi \to \Xi_{\mathcal{W}},$$

a map of hyper-resolutions over f, $\tilde{f}\colon \Gamma_{\mathcal{U}} \to \Xi_{\mathcal{W}}$, and an integer N such that, for each $n \geq N$, there is a map $h_n\colon \mathfrak{e}^{\otimes a} \otimes 1 \to \Sigma^n\Gamma_{\mathcal{U}}[n]$ in $\mathcal{K}^*$ satisfying

$$\Sigma^n(\mathrm{Tot}\tilde{f})[n] \circ h_n = i_n(\mathrm{Tot}\Xi_{\mathcal{W}}) \circ \mathrm{Tot}j_{\mathcal{W}} \circ g$$

in $\mathcal{K}^*$ (see Definition 2.2.2 and (2.2.6.2) for the notation).

PROOF. To simplify the notation, we omit the mention of the functor Tot. Let $\iota\colon \mathcal{K}^* \to \mathcal{D}^*$ and $\iota'\colon \mathbf{K}^-(\mathbf{Ab}) \to \mathbf{D}^-(\mathbf{Ab})$ be the natural maps.

Since the map f becomes an isomorphism in $\mathcal{D}^*$, the map in $\mathbf{D}^-(\mathbf{Ab})$,

$$\mathfrak{R}_{\mathcal{CH}}(f)\colon \mathfrak{R}_{\mathcal{CH}}(\Gamma) \to \mathfrak{R}_{\mathcal{CH}}(\Xi),$$

is an isomorphism. As $\mathfrak{R}_{\mathcal{CH}}(-) \circ \iota = \iota' \circ \mathfrak{RZ}_{\mathrm{mot}}(-,*)$, there is an element η of $H^0(\mathfrak{RZ}_{\mathrm{mot}}(\Gamma,*))$ such that

$$\mathfrak{RZ}_{\mathrm{mot}}(f,*)(\eta) = \mathfrak{RZ}\sigma_0(\mathrm{ev}^{\Xi}_a(g)) \tag{3.2.5.1}$$

in $H^0(\mathfrak{RZ}_{\mathrm{mot}}(\Xi,*))$ (see (3.2.4.2)).

As

$$H^0(\mathfrak{RZ}_{\mathrm{mot}}(\Gamma,*)) = \mathcal{CH}(\Gamma)$$
$$= \mathbb{H}^0_{\mathcal{Z}_{\mathrm{mot}}(*)}(\Gamma),$$

(see (1.5.1.1), Definition 2.5.2 and (3.2.2.5)) there is a hyper-resolution $j_{\mathcal{U}}\colon \Gamma \to \Gamma_{\mathcal{U}}$, and an element η' of $H^0(\mathcal{Z}_{\mathrm{mot}}(\Gamma_{\mathcal{U}},*))$ mapping to η under the natural map $H^0(\mathcal{Z}_{\mathrm{mot}}(\Gamma_{\mathcal{U}},*)) \to \mathbb{H}^0_{\mathcal{Z}_{\mathrm{mot}}(*)}(\Gamma)$.

By applying Lemma 1.4.2 repeatedly, we may assume that we have a hyper-resolution $j_{\mathcal{W}}:\Xi \to \Xi_{\mathcal{W}}$ of Ξ, and a map of hyper-resolutions over f, $\tilde{f}:\Gamma_{\mathcal{U}} \to \Xi_{\mathcal{W}}$. By (3.2.5.1), the difference $\mathcal{Z}_{\mathrm{mot}}(\tilde{f},*)(\eta') - \sigma_0(\mathrm{ev}_a^{\Xi_{\mathcal{W}}}(j_{\mathcal{W}} \circ g))$ goes to zero in $\mathbb{H}^0_{\mathcal{Z}_{\mathrm{mot}}(*)}(\Xi)$; using Lemma 1.4.2 again, we may assume we have the identity

$$\mathcal{Z}_{\mathrm{mot}}(\tilde{f},*)(\eta') = \sigma_0(\mathrm{ev}_a^{\Xi_{\mathcal{W}}}(j_{\mathcal{W}} \circ g)) \tag{3.2.5.2}$$

in $H^0(\mathcal{Z}_{\mathrm{mot}}(\Xi_{\mathcal{W}},*))$.

By Lemma 2.2.8, there is an integer N such that the natural maps (2.2.7.2)

$$\Pi_n(\Gamma_{\mathcal{U}}):\mathcal{Z}_{\mathrm{mot}}(\Sigma^n\Gamma_{\mathcal{U}}[n]) \to \mathcal{Z}_{\mathrm{mot}}(\Gamma_{\mathcal{U}},*)$$
$$\Pi_n(\Xi_{\mathcal{W}}):\mathcal{Z}_{\mathrm{mot}}(\Sigma^n\Xi_{\mathcal{W}}[n]) \to \mathcal{Z}_{\mathrm{mot}}(\Xi_{\mathcal{W}},*)$$

induce an isomorphism on H^0 for all $n \geq N$. Take $\eta_n \in H^0(\mathcal{Z}_{\mathrm{mot}}(\Sigma^n\Gamma_{\mathcal{U}}[n]))$ with $\Pi_n(\eta_n) = \eta'$ in $H^0(\mathcal{Z}_{\mathrm{mot}}(\Gamma_{\mathcal{U}},*))$. The relation (3.2.5.2) then gives the identity

$$\mathcal{Z}_{\mathrm{mot}}(\Sigma^n(\tilde{f})[n])(\eta_n) = \mathrm{ev}_a^{\Sigma^n\Xi_{\mathcal{W}}[n]}(i_n(\Xi_{\mathcal{W}}) \circ j_{\mathcal{W}} \circ g) \tag{3.2.5.3}$$

in $H^0(\mathcal{Z}_{\mathrm{mot}}(\Sigma^n\Xi_{\mathcal{W}}[n]))$.

By Chapter I, Proposition 3.3.5(iv), there is a unique map $h_n:\mathfrak{e}^{\otimes a} \otimes 1 \to \Sigma^n\Gamma_{\mathcal{U}}[n]$ in $\mathcal{K}^*$ such that

$$\mathrm{ev}_a^{\Sigma^n\Gamma_{\mathcal{U}}[n]}(h_n) = \eta_n$$

in $H^0(\mathcal{Z}_{\mathrm{mot}}(\Sigma^n\Gamma_{\mathcal{U}}[n]))$. The identity (3.2.5.3) implies the identity

$$\begin{aligned}\mathrm{ev}_a^{\Sigma^n\Xi_{\mathcal{W}}[n]}(\Sigma^n(\tilde{f})[n](h_n)) &= \mathcal{Z}_{\mathrm{mot}}(\Sigma^n(\tilde{f})[n])(\eta_n)\\ &= \mathrm{ev}_a^{\Sigma^n\Xi_{\mathcal{W}}[n]}(i_n(\Xi_{\mathcal{W}}) \circ j_{\mathcal{W}} \circ g)\end{aligned}$$

in $H^0(\mathcal{Z}_{\mathrm{mot}}(\Sigma^n\Xi_{\mathcal{U}}[n]))$; applying Chapter I, Proposition 3.3.5 again, we have the identity of maps in $\mathcal{K}^*$,

$$\Sigma^n(\tilde{f})[n](h_n) = i_n(\Xi_{\mathcal{W}}) \circ j_{\mathcal{W}} \circ g,$$

completing the proof. □

3.2.6. We have the equivalence of triangulated categories (I.3.4.2.1)

$$\mathbf{D}^b_{\mathrm{mot}}(r):\mathbf{D}^b_{\mathrm{mot}}(\mathcal{V}) \to \mathbf{D}^b_{\mathrm{mot}}(\mathcal{V})^*.$$

For $\Gamma \in \mathbf{D}^b_{\mathrm{mot}}(\mathcal{V})$, we define $\mathcal{CH}(\Gamma) := \mathcal{CH}(\mathbf{D}^b_{\mathrm{mot}}(r)(\Gamma))$, and define the map

$$\mathrm{cl}(\Gamma):\mathcal{CH}(\Gamma) \to \mathrm{Hom}_{\mathbf{D}^b_{\mathrm{mot}}(\mathcal{V})}(1,\Gamma) \tag{3.2.6.1}$$

as the composition (see (2.5.4.1))

$$\mathcal{CH}(\Gamma) = \mathcal{CH}(\Gamma^*) \xrightarrow{\mathrm{cl}(\Gamma^*)} \mathrm{Hom}_{\mathbf{D}^b_{\mathrm{mot}}(\mathcal{V})^*}(1,\Gamma^*) \xrightarrow{\mathbf{D}^b_{\mathrm{mot}}(r)(1,\Gamma)^{-1}} \mathrm{Hom}_{\mathbf{D}^b_{\mathrm{mot}}(\mathcal{V})}(1,\Gamma),$$

where $\Gamma^* = \mathbf{D}^b_{\mathrm{mot}}(r)(\Gamma)$.

3.2.7. THEOREM. *Suppose the conditions of §3.2.1 hold, and let Γ be in $\mathbf{C}^b_{\mathrm{mot}}(\mathcal{V})$. Then the map (3.2.6.1)*

$$\mathrm{cl}(\Gamma):\mathcal{CH}(\Gamma) \to \mathrm{Hom}_{\mathbf{D}^b_{\mathrm{mot}}(\mathcal{V})}(1,\Gamma)$$

is surjective. In particular, the map (3.1.4.1)

$$\mathrm{cl}_X^{q,p}:\mathcal{CH}^q(X,p) \to H^{2q-p}(X,\mathbb{Z}(q))$$

is surjective for all X in $\mathcal{V}$.

PROOF. To simplify the notation, we omit the mention of the functor Tot.

Using the equivalence $\mathbf{D}^b_{\mathrm{mot}}(r)$ (Chapter I, Theorem 3.4.2), we may replace $\mathbf{D}^b_{\mathrm{mot}}(\mathcal{V})$ with $\mathcal{D}^*$, and assume that Γ is in $\mathcal{D}^*$.

We have the isomorphism (I.2.2.4.1) in $\mathcal{D}^*$, $\nu_a\colon \mathfrak{e}^{\otimes a}\otimes 1 \to 1$. Since $\mathcal{D}^*$ is a localization of $\mathcal{K}^*$, each map $\phi\colon 1\to\Gamma$ in $\mathcal{D}^*$ may be factored as a composition

$$1 \xrightarrow{(\nu_a)^{-1}} \mathfrak{e}^{\otimes a}\otimes 1 \xrightarrow{g} \Xi \xrightarrow{f^{-1}} \Gamma,$$

where $g\colon \mathfrak{e}^{\otimes a}\otimes 1\to\Xi$ and $f\colon\Gamma\to\Xi$ are maps in $\mathcal{K}^*$, and f is invertible in $\mathcal{D}^*$ (see Part II, Chapter II, §2.3.3). Since the diagram

$$\begin{array}{ccc}
\mathfrak{e}^{\otimes a}\otimes\mathfrak{e}^{\otimes b} & = & \mathfrak{e}^{\otimes a+b} \\
\downarrow{\scriptstyle \mathrm{id}_{\mathfrak{e}^{\otimes a}}\otimes\nu_b} & & \downarrow{\scriptstyle \nu_{a+b}} \\
\mathfrak{e}^{\otimes a}\otimes 1 & \xrightarrow[\nu_a]{} & 1
\end{array}$$

commutes in $\mathcal{K}^*$, we may assume that Γ and Ξ are in $\mathcal{K}^*_B$ (see §3.2.4) with $a > B$.

Applying Lemma 3.2.5, there are hyper-resolutions

$$j_{\mathcal{U}}\colon\Gamma\to\Gamma_{\mathcal{U}},$$
$$j_{\mathcal{W}}\colon\Xi\to\Xi_{\mathcal{W}},$$

an integer n, and maps

$$h_n\colon \mathfrak{e}^{\otimes a}\otimes 1\to\Sigma^n\Gamma_{\mathcal{U}}[n],$$
$$\tilde f\colon\Gamma_{\mathcal{W}}\to\Xi_{\mathcal{U}}$$

in $\mathcal{K}^*$ such that

(3.2.7.1) $$\Sigma^n(\tilde f)[n]\circ h_n = i_n(\Xi_{\mathcal{W}})\circ j_{\mathcal{W}}\circ g.$$

In addition, the diagram

(3.2.7.2)
$$\begin{array}{ccc}
\Gamma & \xrightarrow{f} & \Xi \\
\downarrow{\scriptstyle j_{\mathcal{U}}} & & \downarrow{\scriptstyle j_{\mathcal{W}}} \\
\Gamma_{\mathcal{U}} & \xrightarrow{\tilde f} & \Xi_{\mathcal{W}} \\
\downarrow{\scriptstyle i_n(\Gamma_{\mathcal{U}})} & & \downarrow{\scriptstyle i_n(\Xi_{\mathcal{W}})} \\
\Sigma^n\Gamma_{\mathcal{U}}[n] & \xrightarrow{\Sigma^n(\tilde f)[n]} & \Sigma^n\Xi_{\mathcal{W}}[n]
\end{array}$$

commutes in $\mathcal{K}^*$. Since $i_n(\Gamma_{\mathcal{U}})$, $j_{\mathcal{U}}$, $i_n(\Xi_{\mathcal{W}})$ and $j_{\mathcal{W}}$ are isomorphisms in $\mathcal{D}^*$ (Lemma 2.3.5 and Lemma 1.4.2(iii)), the relation (3.2.7.1) and the commutativity of (3.2.7.2) gives us the identity

(3.2.7.3)
$$\begin{aligned}
f^{-1}\circ g\circ(\nu_a)^{-1} &= f^{-1}\circ(j_{\mathcal{W}})^{-1}\circ(i_n(\Xi_{\mathcal{W}}))^{-1}\circ\Sigma^n(\tilde f)[n]\circ h_n\circ(\nu_a)^{-1} \\
&= (j_{\mathcal{U}})^{-1}\circ(i_n(\Gamma_{\mathcal{U}}))^{-1}\circ h_n\circ(\nu_a)^{-1}.
\end{aligned}$$

Let $\tilde\eta$ be the image of $\mathrm{ev}_a^{\Sigma^n\Gamma_{\mathcal{U}}[n]}(h_n)$ (see (3.2.4.3)) in $H^0(\mathcal{Z}_{\mathrm{mot}}(\Gamma_{\mathcal{U}},*))$, under the map (2.2.7.2),

$$\Pi_n(\Sigma^n\Gamma_{\mathcal{U}}[n])\colon\mathcal{Z}_{\mathrm{mot}}(\Sigma^n\Gamma_{\mathcal{U}}[n])\to\mathcal{Z}_{\mathrm{mot}}(\Gamma_{\mathcal{U}},*).$$

By Definition 2.5.2, we have

$$\mathcal{CH}(\Gamma) = \mathbb{H}^0_{\mathcal{Z}_{\mathrm{mot}}(*)}(\Gamma) = \varinjlim_{\Gamma_{\mathcal{U}} \in \mathbf{HR}_\Gamma} \mathcal{Z}_{\mathrm{mot}}(\Gamma_{\mathcal{U}}, *),$$

hence $\tilde{\eta}$ has a well-defined image $\eta \in \mathcal{CH}(\Gamma)$. By definition of the map (2.5.4.1), we have

$$\mathrm{cl}(\Gamma)(\eta) = (j_{\mathcal{U}})^{-1} \circ (i_n(\Gamma_{\mathcal{U}}))^{-1} \circ h_n \circ (\nu_a)^{-1};$$

as this is the map ϕ by (3.2.7.3), surjectivity is proved. □

3.3. Injectivity of the cycle map

We give a criterion for the injectivity of the cycle map.

3.3.1. *Cohomology vanishing.* In order to prove injectivity, we need, in addition to the conditions of §3.2.1, the following hypothesis: Let X be in $\mathcal{V}$, let $p: X \times \mathbb{A}^n \to X$ be the projection, and let $(\mathbb{A}^n, g)$ and (X, f) be liftings of $\mathbb{A}^n$ and X to objects of $\mathcal{L}(\mathbf{Sm}_{\mathrm{Spec}\,\mathbb{Z}})$ and $\mathcal{L}(\mathcal{V})$, respectively. Then the map

$$p^*: \mathfrak{Z}^q_{X/S}(*)_f \to p_*(\mathfrak{Z}^q_{X \times \mathbb{A}^n/S}(*)_{f \times g})$$

is a quasi-isomorphism of complexes of sheaves on X.

3.3.2. *A double cycle complex.* Recall from Example 2.2.1 the cosimplicial scheme $\Delta^*: \Delta \to \mathcal{V}$, the cosimplicial object of $\mathcal{L}(\mathcal{V})$, $(\Delta^*, \delta^*): \Delta \to \mathcal{L}(\mathcal{V})$, and the associated simplicial object $\mathbb{Z}_{\Delta^*}(0)_{\delta^*}$ (Definition 2.2.2(iii)) of $\mathcal{A}_{\mathrm{mot}}(\mathcal{V})^*$. For $\Gamma \in \mathcal{A}_{\mathrm{mot}}(\mathcal{V})^*$, the complex $\mathcal{Z}_{\mathrm{mot}}(\Gamma, *)$ is the complex associated to the simplicial object $\mathcal{Z}_{\mathrm{mot}}(\Gamma \times \mathbb{Z}_{\Delta^*}(0)_{\delta^*}): \Delta^{\mathrm{op}} \to \mathbf{C}^b(\mathbf{Ab})$ (see Definition 2.2.4). We now form the bi-simplicial object

$$\mathcal{Z}_{\mathrm{mot}}(\Gamma \times \mathbb{Z}_{\Delta^*}(0)_{\delta^*} \times \mathbb{Z}_{\Delta^*}(0)_{\delta^*}): \Delta^{\mathrm{op}} \times \Delta^{\mathrm{op}} \to \mathbf{C}^b(\mathbf{Ab}).$$

and let $\mathcal{Z}_{\mathrm{mot}}(\Gamma, *, *)$ be the associated double complex. Since $(\mathbb{Z}_{\Delta^*}(0)_{\delta^*})_0 = \mathbb{Z}_S(0)$, and $\mathbb{Z}_S(0)$ is the unit for the tensor operation $\times$, the sub-complexes $\mathcal{Z}_{\mathrm{mot}}(\Gamma, *, 0)$ and $\mathcal{Z}_{\mathrm{mot}}(\Gamma, 0, *)$ are canonically isomorphic to the complex $\mathcal{Z}_{\mathrm{mot}}(\Gamma, *)$. This defines the two inclusions

$$i_1, i_2: \mathcal{Z}_{\mathrm{mot}}(\Gamma, *) \to \mathrm{Tot}(\mathcal{Z}_{\mathrm{mot}}(\Gamma, *, *)). \tag{3.3.2.1}$$

We let $\mathcal{Z}^q(X, *)_f$ denote the complex $\mathcal{Z}_{\mathrm{mot}}(\mathbb{Z}_X(q)_f[2q], *)$, and we define the double complex $\mathcal{Z}^q(X, *, *)_f$ by

$$\mathcal{Z}^q(X, *, *)_f = \mathcal{Z}_{\mathrm{mot}}(\mathbb{Z}_X(q)[2q], *, *). \tag{3.3.2.2}$$

The inclusions (3.3.2.1) give the natural maps

$$i_1, i_2: \mathcal{Z}^q(X, *)_f \to \mathrm{Tot}(\mathcal{Z}^q(X, *, *)_f). \tag{3.3.2.3}$$

3.3.3. We may sheafify the construction of §3.3.2 over X. Let $\mathfrak{Z}^q_{X/S}(*, *)_f$ be the double complex of sheaves on X associated to the presheaf $(j: U \to X) \mapsto \mathcal{Z}^q(U, *, *)_{j^* f}$; the maps (3.3.2.3) define the maps

$$i_1, i_2: \mathfrak{Z}^q_{X/S}(*)_f \to \mathrm{Tot}(\mathfrak{Z}^q_{X/S}(*, *)_f). \tag{3.3.3.1}$$

3.3.4. LEMMA. *Assuming the condition of §3.3.1, the maps (3.3.3.1) are quasi-isomorphisms.*

PROOF. We consider one of the two convergent spectral sequences associated to the double complex of sheaves $\mathfrak{Z}^q_{X/S}(*,*)_f$. The E_1-terms are given by

$$E_1^{a,b} = \mathcal{H}^a(\mathfrak{Z}^q_{X/S}(*,b)_f),$$

where $\mathfrak{Z}^q_{X/S}(*,-b)_f$ is the complex of sheaves on X associated to the presheaf $U \mapsto \mathcal{Z}^q(X \times \Delta^b, *)_{f\times\delta^b}$, and $\mathcal{H}^a$ is the sheaf of cohomology groups on X.

Let $p^n: X \times_S \Delta^n \to X$ be the projection. We have the identity

$$\mathfrak{Z}^q_{X/S}(*,-b)_f = p^b_*(\mathfrak{Z}^q_{X\times_S\Delta^b/S}(*)_{f\times\delta^b}).$$

By our assumption of §3.3.1, the map $p^{b*}: \mathfrak{Z}^q_{X/S}(*)_f \to \mathfrak{Z}^q_{X/S}(*,-b)_f$ is a quasi-isomorphism. This implies that the complex of E_1-terms is

$$\ldots \to \mathcal{H}^a(\mathfrak{Z}^q_{X/S}(*)_f) \to \mathcal{H}^a(\mathfrak{Z}^q_{X/S}(*)_f) \to \ldots \to \mathcal{H}^a(\mathfrak{Z}^q_{X/S}(*)_f),$$

where the maps alternate between the zero map and the identity map, with the last map being the zero map. Thus, the spectral sequence degenerates at E_2, and the inclusion $i_1: \mathfrak{Z}^q_{X/S}(*)_f = \mathfrak{Z}^q_{X/S}(*,0)_f \to \mathrm{Tot}(\mathfrak{Z}^q_{X/S}(*,*)_f)$ is a quasi-isomorphism. The other inclusion i_2 is handled by using the other spectral sequence. □

3.3.5. We now return to the cosimplicial object $(X,f)\times(\Delta^*,\delta^*): \Delta \to \mathcal{L}(\mathcal{V})$. We may view the double complex $\mathcal{Z}^q(X,*,*)_f$ as the double complex associated to the simplicial object

$$\begin{aligned} \mathcal{Z}^q(X\times_S\Delta^*,*)_{f\times\delta^*}&: \Delta^{\mathrm{op}} \to \mathbf{C}^-\mathbf{Ab} \\ n &\mapsto \mathcal{Z}^q_{\mathrm{mot}}(X\times_S\Delta^n,*)_{f\times\delta^n}. \end{aligned} \tag{3.3.5.1}$$

We may apply the natural transformation (3.2.4.1), $\iota_{Y,g}: \mathcal{Z}^q(Y,*)_g \to \mathfrak{R}\mathfrak{Z}^q_{Y/S}(*)_g$, to (3.3.5.1), giving the simplicial object

$$\begin{aligned} \mathfrak{R}\mathfrak{Z}^q_{X\times_S\Delta^*/S}(*)_{f\times\delta^*}&: \Delta^{\mathrm{op}} \to \mathbf{C}^-\mathbf{Ab} \\ n &\mapsto \mathfrak{R}\mathfrak{Z}^q_{X\times_S\Delta^n/S}(*)_{f\times\delta^n}, \end{aligned}$$

and the natural map of simplicial objects

$$\iota_{X\times_S\Delta^*,\delta^*}: \mathcal{Z}^q(X\times_S\Delta^*/S,*)_{f\times\delta^*} \to \mathfrak{R}\mathfrak{Z}^q_{X\times_S\Delta^*/S}(*)_{f\times\delta^*}.$$

We let

$$\iota_X(*)(*)_f: \mathcal{Z}^q(X,*,*)_f \to \mathfrak{R}\mathfrak{Z}^q_{X/S}(*)(*)_f \tag{3.3.5.2}$$

denote the induced map on the associated double complexes; here the indices in the double complex $\mathfrak{R}\mathfrak{Z}^q_{X/S}(*)(*)_f$ are arranged so that

$$\mathfrak{R}\mathfrak{Z}^q_{X/S}(m)(-n)_f = \mathfrak{R}\mathfrak{Z}^q_{X\times_S\Delta^n/S}(m)_{f\times\delta^n}.$$

One easily sees that the map (3.3.5.2) factors canonically through the natural map $\mathcal{Z}^q(X,*,*)_f \to \mathfrak{R}(\mathfrak{Z}^q_{X/S}(*,*)_f)$, giving the map

$$\mathfrak{R}\mathfrak{Z}\iota_X: \mathfrak{R}(\mathfrak{Z}^q_{X/S}(*,*)_f) \to \mathfrak{R}\mathfrak{Z}^q_{X/S}(*)(*)_f. \tag{3.3.5.3}$$

3.3.6. LEMMA. *Assume the conditions of §3.2.1 and §3.3.1 hold. Then the map (3.3.5.3) induces a quasi-isomorphism on the associated total complexes.*

PROOF. By §3.3.1 and §3.2.1(i) and (ii), the map

$$p^{n*}:\mathfrak{R}\mathfrak{Z}^q_{X/S}(*)_f \to \mathfrak{R}\mathfrak{Z}^q_{X\times_S\Delta^n/S}(*)_{f\times\delta^n}$$

is a quasi-isomorphism for each n. The same spectral sequence argument as in the proof of Lemma 3.3.4 then shows that the inclusion

$$\iota_1:\mathfrak{R}\mathfrak{Z}^q_{X/S}(*)_f = \mathfrak{R}\mathfrak{Z}^q_{X/S}(*)(0)_f \to \mathrm{Tot}(\mathfrak{R}\mathfrak{Z}^q_{X/S}(*)(*)_f)$$

is a quasi-isomorphism. By Lemma 3.3.4, the inclusion

$$\mathfrak{R}(i_1):\mathfrak{R}\mathfrak{Z}^q_{X/S}(*)_f = \mathfrak{R}(\mathfrak{Z}^q_{X/S}(*,0)_f) \to \mathrm{Tot}(\mathfrak{R}(\mathfrak{Z}^q_{X/S}(*,*)_f))$$

is a quasi-isomorphism. As $(\mathrm{Tot}\mathfrak{R}\mathfrak{Z}\iota_X)\circ\mathfrak{R}(i_1) = \iota_1$, the lemma is proved. □

3.3.7. Let Γ be in $\mathbf{C}^b_{\mathrm{mot}}(\mathcal{V})^*$. We may form the functor $\mathcal{Z}^\Gamma_{\mathrm{mot}}:\mathbf{C}^b_{\mathrm{mot}}(\mathcal{V})^* \to \mathbf{C}^b(\mathbf{Ab})$ defined by

$$\mathcal{Z}^\Gamma_{\mathrm{mot}}(-) = \mathcal{Z}_{\mathrm{mot}}((-)\times\Gamma).$$

For $(X,f)\in\mathcal{L}(\mathcal{V})$, we may form the presheaf on X,

$$(j:U\to X)\mapsto \mathcal{Z}^\Gamma_{\mathrm{mot}}(\mathbb{Z}_U(q)_{j^*f}[2q]);$$

we let $\mathfrak{Z}^{q,\Gamma}_{X/S,f}$ denote the associated sheaf. The natural transformation (3.2.4.2) defines the natural map

$$\phi_\Gamma:\mathfrak{Z}^{q,\Gamma}_{X/S,f} \to \mathfrak{R}\mathfrak{Z}_{\mathrm{mot}}(\Gamma\times\mathbb{Z}_X(q)_f[2q],*). \tag{3.3.7.1}$$

3.3.8. We have the object $\mathbb{Z}^{\le N}_{\Delta^*}(0)$ of $\mathbf{C}^b_{\mathrm{mot}}(\mathcal{V})^*$ (see (2.2.1.1)), the functor

$$\Sigma^N(-)[N] := (-)\times\mathbb{Z}^{\le N}_{\Delta^*}(0):\mathbf{C}^b_{\mathrm{mot}}(\mathcal{V})^* \to \mathbf{C}^b_{\mathrm{mot}}(\mathcal{V})^*$$

(Definition 2.2.2), and the functor (2.5.1.1)

$$\Sigma^N\mathcal{Z}_{\mathrm{mot}}[N] = \mathcal{Z}_{\mathrm{mot}}\circ\Sigma^N(-)[N]:\mathbf{C}^b_{\mathrm{mot}}(\mathcal{V})^* \to \mathbf{C}^b(\mathbf{Ab}).$$

Using the notation of §3.3.7, we may write $\Sigma^N\mathcal{Z}_{\mathrm{mot}}[N]$ as $\mathcal{Z}^{\mathbb{Z}^{\le N}_{\Delta^*}(0)}_{\mathrm{mot}}(-)$.

Denote $\Sigma^N\mathcal{Z}_{\mathrm{mot}}[N](\mathbb{Z}_X(q)_f[2q])$ by $\Sigma^N\mathcal{Z}^q_{\mathrm{mot}}(X)_f[N]$. As in §3.3.7, we may sheafify the Zariski presheaf $(j:U\to X)\mapsto\Sigma^N\mathcal{Z}^q_{\mathrm{mot}}(U)_{j^*f}[N]$ over X, giving the complex of sheaves $\Sigma^N\mathfrak{Z}^q_{X/S,f}[N]$ on X, and the functor

$$\Sigma^N\mathfrak{Z}^q_{-/S,-}[N]:\mathcal{L}(\mathcal{V})^{\mathrm{op}} \to \mathbf{C}^-(\mathrm{Sh}^{\mathbf{Ab}}(\mathrm{Zar}_S)).$$

We have the identity

$$\Sigma^N\mathfrak{Z}^q_{X/S,f}[N] = \mathfrak{Z}^{q,\mathbb{Z}^{\le N}_{\Delta^*}(0)}_{X/S,f};$$

applying (3.3.7.1) gives the natural map

$$\phi_{X,N}:\mathfrak{R}_X\Sigma^N\mathfrak{Z}^q_{X/S,f}[N] \to \mathfrak{R}\mathfrak{Z}_{\mathrm{mot}}(\mathbb{Z}_X(q)_f\times\mathbb{Z}^{\le N}_{\Delta^*}(0)[2q],*). \tag{3.3.8.1}$$

3.3.9. LEMMA. *Let p be an integer. For fixed X, f and q, there is an integer N_p such that the map (3.3.8.1) induces an isomorphism in cohomology $H^m(-)$ for all $m\ge -p$ if $N\ge N_p$.*

PROOF. The natural transformation (2.5.1.2) gives the natural transformation (on the category of open subschemes $j:U\to X$ of X)

$$\mathfrak{Z}\Pi_N(j:U\to X):\Sigma^N\mathfrak{Z}^q_{U/S,j^*f}[N] \to \mathfrak{Z}^q_{U/S}(*)_{j^*f},$$

which in turn gives the natural map

$$\mathfrak{R}\mathfrak{Z}\Pi_N : \mathfrak{R}_X\Sigma^N\mathfrak{Z}^q_{X/S,f}[N] \to \mathfrak{R}\mathfrak{Z}^q_{X/S}(*)_f. \tag{3.3.9.1}$$

We recall from Example 2.2.1 and Chapter I, §2.4.2, that $\mathbb{Z}^{\leq N}_{\Delta^*}(0)$ is the complex

$$\mathbb{Z}^{\leq N}_{\Delta^*}(0)^{-N} \to \ldots \to \mathbb{Z}^{\leq N}_{\Delta^*}(0)^0,$$

where $\mathbb{Z}^{\leq N}_{\Delta^*}(0)^{-p}$ is the direct sum $\mathbb{Z}^{\leq N}_{\Delta^*}(0)^{-p} = \oplus_{g\colon[p]\to[N]}\mathbb{Z}_{\Delta^p}(0)_{\delta^p}$, with the sum being over injective ordered maps g. The map Π_N in each degree p is the map induced on $\mathcal{Z}_{\mathrm{mot}}(-)$ by the sum map $\Sigma_{N,p}\colon \oplus_{g\colon[p]\to[N]}\mathbb{Z}_{\Delta^p}(0)_{\delta^p} \to \mathbb{Z}_{\Delta^p}(0)_{\delta^p}$. Applying the functor $\mathfrak{R}\mathfrak{Z}_{/S}(-,*)$ to $\mathrm{id}\times\Sigma_{N,p}$ gives the natural map

$$\mathfrak{R}\mathfrak{Z}\Pi_N(*) : \mathfrak{R}\mathfrak{Z}_{\mathrm{mot}}(\mathbb{Z}_X(q)_f \times \mathbb{Z}^{\leq N}_{\Delta^*}(0)[2q], *) \to \mathrm{Tot}\mathfrak{R}\mathfrak{Z}^q_{X/S}(*)(*)_f, \tag{3.3.9.2}$$

where $\mathfrak{R}\mathfrak{Z}\Pi_N(m)$ maps $\mathfrak{R}\mathfrak{Z}_{\mathrm{mot}}(\mathbb{Z}_X(q)_f \times \mathbb{Z}^{\leq N}_{\Delta^*}(0)[2q], *)$ to $\mathfrak{R}\mathfrak{Z}^q_{X/S}(m)(*)_f$.

By Lemma 2.2.8, we may apply Lemma 1.5.4, with $h_N = \mathcal{Z}_{\mathrm{mot}}(\Sigma^N[N])$, $h_\infty = \mathcal{Z}_{\mathrm{mot}}$, $\pi_{n+1,n}$ the natural transformation $\mathcal{Z}_{\mathrm{mot}}(\chi_{n+1,n})$ (2.2.7.3) and π_n the natural transformation (2.2.7.2). Thus, there is an N_p such that, for all $N \geq N_p$, the map (3.3.9.1) gives an isomorphism on H^m for all $m \geq -p$.

We have the natural transformation (2.2.6.2), inducing the natural map

$$\mathfrak{R}\mathfrak{Z}_{\mathrm{mot}}(\mathbb{Z}_X(q)_f[2q], *) \xrightarrow{\mathfrak{R}\mathfrak{Z}_{\mathrm{mot}}(i_N(\mathbb{Z}_X(q)_f[2q]))} \mathfrak{R}\mathfrak{Z}_{\mathrm{mot}}(\mathbb{Z}_X(q)_f \times \mathbb{Z}^{\leq N}_{\Delta^*}(0)[2q], *). \tag{3.3.9.3}$$

By Lemma 2.3.5, the map

$$i_N(\mathbb{Z}_X(q)_f[2q]) : \mathbb{Z}_X(q)_f[2q] \to \mathbb{Z}_X(q)_f \times \mathbb{Z}^{\leq N}_{\Delta^*}(0)[2q]$$

is an isomorphism in $\mathbf{D}^b_{\mathrm{mot}}(\mathcal{V})$. By Proposition 3.2.3, the map (3.3.9.3) is thus a quasi-isomorphism.

We have the identity $\mathfrak{R}\mathfrak{Z}_{\mathrm{mot}}(\mathbb{Z}_X(q)_f[2q], *) = \mathfrak{R}\mathfrak{Z}^q_{X/S}(*)_f$. Let

$$\iota_{X,j} : \mathfrak{R}\mathfrak{Z}_{\mathrm{mot}}(\mathbb{Z}_X(q)_f[2q], *) \to \mathrm{Tot}\mathfrak{R}\mathfrak{Z}^q_{X/S}(*)(*)_f; \qquad j = 1, 2,$$

be the composition

$$\mathfrak{R}\mathfrak{Z}_{\mathrm{mot}}(\mathbb{Z}_X(q)_f[2q], *) \xrightarrow{\mathfrak{R}(i_j)} \mathrm{Tot}(\mathfrak{R}(\mathfrak{Z}^q_{X/S}(*, *)_f)) \xrightarrow{\mathfrak{R}\mathfrak{Z}\iota_X} \mathrm{Tot}\mathfrak{R}\mathfrak{Z}^q_{X/S}(*)(*)_f$$

(*cf.* (3.3.3.1) and (3.3.5.3)).

We have the commutative diagram

$$\begin{array}{ccc}
\mathfrak{R}\mathfrak{Z}_{\mathrm{mot}}(\mathbb{Z}_X(q)_f[2q], *) & = & \mathfrak{R}\mathfrak{Z}_{\mathrm{mot}}(\mathbb{Z}_X(q)_f[2q], *) \\
{\scriptstyle \mathfrak{R}\mathfrak{Z}_{\mathrm{mot}}(i_N(\mathbb{Z}_X(q)_f[2q]))}\downarrow & & \downarrow{\scriptstyle \iota_{X,1}} \\
\mathrm{Tot}\mathfrak{R}\mathfrak{Z}_{\mathrm{mot}}(\mathbb{Z}_X(q)_f \times \mathbb{Z}^{\leq N}_{\Delta^*}(0)[2q], *) & \xrightarrow[\mathfrak{R}\mathfrak{Z}\Pi_N(*)]{} & \mathrm{Tot}\mathfrak{R}\mathfrak{Z}^q_{X/S}(*)(*)_f.
\end{array}$$

By Lemma 3.3.4 and Lemma 3.3.6, the map

$$\iota_{X,1} : \mathfrak{R}\mathfrak{Z}_{\mathrm{mot}}(\mathbb{Z}_X(q)_f[2q], *) \to \mathrm{Tot}\mathfrak{R}\mathfrak{Z}^q_{X/S}(*)(*)_f$$

is a quasi-isomorphism. As the map (3.3.9.3) is a quasi-isomorphism, the map (3.3.9.2) is a quasi-isomorphism.

By Lemma 3.3.4 and Lemma 3.3.6, the map

$$\iota_{X,2} : \mathfrak{R}\mathfrak{Z}_{\mathrm{mot}}(\mathbb{Z}_X(q)_f[2q], *) \to \mathrm{Tot}\mathfrak{R}\mathfrak{Z}^q_{X/S}(*)(*)_f$$

is also a quasi-isomorphism. One easily sees that the diagram

$$\begin{array}{ccc} \mathfrak{R}_X\Sigma^N\mathfrak{Z}^q_{X/S,f}[N] & \xrightarrow{\phi_{X,N}} & \mathfrak{R}\mathfrak{Z}_{\mathrm{mot}}(\mathbb{Z}_X(q)_f\times\mathbb{Z}^{\le N}_{\Delta^*}(0)[2q],*) \\ {\scriptstyle \mathfrak{R}\Pi_N}\downarrow & & \downarrow{\scriptstyle \mathfrak{R}\mathfrak{Z}\Pi_N(*)} \\ \mathfrak{R}\mathfrak{Z}^q_{X/S}(*)_f & \xrightarrow[\iota_{X,2}]{} & \mathrm{Tot}\mathfrak{R}\mathfrak{Z}^q_{X/S}(*)(*)_f \end{array}$$

commutes.

Take $N\ge N_p$. As the map $\mathfrak{R}\Pi_N$ gives an isomorphism on H^m for all $m\ge -p$, and the maps $\iota_{X,2}$, $\mathfrak{R}\mathfrak{Z}\Pi_N(*)$, and $\mathfrak{R}\Pi_N$ are quasi-isomorphisms, the map $\phi_{X,N}$ gives an isomorphism on H^m for all $m\ge -p$. □

3.3.10. THEOREM. *Suppose the conditions of §3.2.1 and §3.3.1 are satisfied. Then the map*

$$\mathrm{cl}(\Gamma)\colon\mathcal{CH}(\Gamma)\to\mathrm{Hom}_{\mathcal{DM}(\mathcal{V})}(1,\Gamma)$$

is an isomorphism for all Γ in $\mathcal{DM}(\mathcal{V})$.

PROOF. It suffices to prove the result for Γ in $\mathbf{D}^b_{\mathrm{mot}}(\mathcal{V})$; using the equivalence (I.3.4.2.1) $\mathbf{D}^b_{\mathrm{mot}}(r)\colon\mathbf{D}^b_{\mathrm{mot}}(\mathcal{V})\to\mathbf{D}^b_{\mathrm{mot}}(\mathcal{V})^*$, we may assume Γ is in $\mathbf{D}^b_{\mathrm{mot}}(\mathcal{V})^*$. As $\mathbf{D}^b_{\mathrm{mot}}(\mathcal{V})^*$ is generated as a triangulated category by the objects $\mathfrak{e}^{\otimes a}\otimes\mathbb{Z}_X(q)_f$, and since $\mathrm{cl}(-)$ is an exact natural transformation of cohomological functors we may take Γ to be a translate of $\mathfrak{e}^{\otimes a}\otimes\mathbb{Z}_X(q)_f$; as $\mathfrak{e}^{\otimes a}\otimes\mathbb{Z}_X(q)_f$ is isomorphic to $\mathbb{Z}_X(q)_f$ in $\mathbf{D}^b_{\mathrm{mot}}(\mathcal{V})^*$, we may take Γ to be a translate of $\mathbb{Z}_X(q)_f$. By Theorem 3.2.7, we need only prove injectivity.

By Proposition 2.5.3, there is an N_1 such that the map $\mathbb{H}^0(\Pi_N)$ gives an isomorphism

$$\mathcal{CH}(\mathbb{Z}_X(q)_f[2q-p])\cong\mathbb{H}^0_{\Sigma^N\mathcal{Z}_{\mathrm{mot}}[N]}(\mathbb{Z}_X(q)_f[2q-p]) \tag{3.3.10.1}$$

for all $N\ge N_1$. By Lemma 1.5.4(i), we may identify the hypercohomology with respect to the functor $\Sigma^N\mathcal{Z}_{\mathrm{mot}}[N]$ as Zariski hypercohomology:

$$\mathbb{H}^0_{\Sigma^N\mathcal{Z}_{\mathrm{mot}}[N]}(\mathbb{Z}_X(q)_f[2q-p]) = H^{-p}(\mathfrak{R}\Sigma^N\mathfrak{Z}^q_{X/S,f}[N]). \tag{3.3.10.2}$$

It follows directly from the construction of $\mathfrak{R}_{\mathcal{CH}}$ in Proposition 3.2.3 that the composition

(3.3.10.3)

$$\mathfrak{R}_{\mathcal{CH}}\circ\mathrm{cl}(\mathbb{Z}_X(q)_f[2q-p])\colon\mathcal{CH}(\mathbb{Z}_X(q)_f[2q-p])\to H^0(\mathfrak{R}_{\mathcal{CH}}(\mathbb{Z}_X(q)_f[2q-p]))$$

is the map induced on H^{-p} by the map (3.3.8.1)

$$\phi_{X,N}\colon\mathfrak{R}\Sigma^N\mathfrak{Z}^q_{X/S,f}[N]\to\mathfrak{R}\mathfrak{Z}_{\mathrm{mot}}(\mathbb{Z}_X(q)_f\times\mathbb{Z}^{\le N}_{\Delta^*}(0)[2q],*)$$

once we identify $\mathcal{CH}(\mathbb{Z}_X(q)_f[2q-p])$ with $H^{-p}(\mathfrak{R}\Sigma^N\mathfrak{Z}^q_{X/S,f}[N])$ via (3.3.10.1) and (3.3.10.2). By Lemma 3.3.9, the map (3.3.10.3) is an isomorphism, once we take N large enough. Thus $\mathrm{cl}(\mathbb{Z}_X(q)_f[2q-p])$ is injective, completing the proof. □

We recall the triangulated tensor category $\mathcal{DM}^0(\mathcal{V})$, and the exact tensor functor $\mathcal{DM}(H_{\mathrm{mot}})\colon\mathcal{DM}(\mathcal{V})\to\mathcal{DM}^0(\mathcal{V})$ (see Chapter I, Remark 3.4.7).

3.3.11. THEOREM. *Suppose the conditions of §3.2.1 and §3.3.1 are satisfied. Then the functor $\mathcal{DM}(H_{\mathrm{mot}})$ induces an isomorphism*

$$\mathcal{DM}(H_{\mathrm{mot}})\colon\mathrm{Hom}_{\mathcal{DM}(\mathcal{V})}(1,\Gamma)\to\mathrm{Hom}_{\mathcal{DM}^0(\mathcal{V})}(1,\mathcal{DM}(H_{\mathrm{mot}})(\Gamma))$$

for all Γ *in* $\mathcal{DM}(\mathcal{V})$.

PROOF. We use throughout the notation of Chapter I, Remark 3.4.7. The arguments of §2.2-§3.3 can be applied, replacing the categories $\mathcal{A}_{\rm mot}(\mathcal{V})^*$, $\mathbf{C}^b_{\rm mot}(\mathcal{V})^*$, $\mathbf{K}^b_{\rm mot}(\mathcal{V})^*$, and $\mathbf{D}^b_{\rm mot}(\mathcal{V})^*$ with $\mathcal{A}^0_{\rm mot}(\mathcal{V})^*$, $\mathbf{C}^{b0}_{\rm mot}(\mathcal{V})^*$, $\mathbf{K}^{b0}_{\rm mot}(\mathcal{V})^*$, and $\mathbf{D}^{b0}_{\rm mot}(\mathcal{V})^*$, respectively, to prove the analog of Theorem 3.3.10 for the category $\mathcal{DM}^0(\mathcal{V})$, i.e., that there is a natural cycle class map $\mathrm{cl}(\Gamma): \mathcal{CH}(\Gamma) \to \mathrm{Hom}_{\mathcal{DM}^0(\mathcal{V})}(1, \Gamma)$, which is an isomorphism for all Γ in $\mathcal{DM}(\mathcal{V})$.

Noting that the construction of $\mathrm{cl}(-)$ is compatible with the functor $H_{\rm mot}$ proves the results. □

3.4. Some consequences

In the previous section, we have given criteria (§3.2.1 and §3.3.1) for the cycle class map $\mathrm{cl}(\Gamma): \mathcal{CH}(\Gamma) \to \mathrm{Hom}_{\mathcal{DM}(\mathcal{V})}(1, \Gamma)$ to be an isomorphism for all Γ in $\mathcal{DM}(\mathcal{V})$. In this section, we suppose these criteria to be satisfied for $\mathbf{Sm}_S^{\rm ess}$, and deduce some consequences.

The first is the interpretation of motivic cohomology in terms of Zariski hypercohomology.

For X in $\mathcal{V}$, we have the complex of Zariski sheaves (3.1.1.1), $\mathfrak{Z}^q_{X/S}(*)$.

3.4.1. THEOREM. *Suppose the conditions of §3.2.1 and §3.3.1 are satisfied for* $\mathcal{V} = \mathbf{Sm}_S^{\rm ess}$. *Let* $\mathcal{V}$ *be a full subcategory of* $\mathbf{Sm}_S^{\rm ess}$ *such that the conditions of Chapter I, Definition 2.1.4 are satisfied. Then for* X *in* $\mathcal{V}$, *with closed subset* $\hat{X}$, *we have the natural isomorphism*

$$\mathbb{H}^{p-2q}_{\hat{X}}(X, \mathfrak{Z}^q_{X/S}(*)) \cong \mathrm{Hom}_{\mathcal{DM}(\mathcal{V})}(1, \mathbb{Z}_{X,\hat{X}}(q)[p]),$$

where $\mathbb{H}^*_{\hat{X}}$ *is the Zariski hypercohomology with support.*

PROOF. Let Γ be in $\mathbf{C}^b_{\rm mot}(\mathcal{V})$. From (3.2.2.5), we have the natural isomorphism $H^0(\mathfrak{R}\mathfrak{Z}_{\rm mot}(\Gamma, *)) \cong \mathcal{CH}(\Gamma)$. Taking $\Gamma = \mathbb{Z}_{X,\hat{X}}(q)[p]$ and noting that the Godement resolution $\mathfrak{R}\mathfrak{Z}_{\rm mot}(\mathbb{Z}_{X,\hat{X}}(q)[p], *)$ represents the object $R\Gamma_{\hat{X}}(X, \mathfrak{Z}^q_{X/S}(*)[p-2q])$ in $\mathbf{D}^-(\mathbf{Ab})$ gives the isomorphism

$$\mathcal{CH}(\mathbb{Z}_{X,\hat{X}}(q)[p]) \cong \mathbb{H}^{p-2q}_{\hat{X}}(X, \mathfrak{Z}^q_{X/S}(*)).$$

By Theorem 3.3.10, we have the natural isomorphism

$$\mathrm{cl}(\mathbb{Z}_{X,\hat{X}}(q)[p]): \mathcal{CH}(\mathbb{Z}_{X,\hat{X}}(q)[p]) \to \mathrm{Hom}_{\mathcal{DM}(\mathcal{V})}(1, \mathbb{Z}_{X,\hat{X}}(q)[p]).$$

□

The next result is the independence of motivic cohomology on the choice of category $\mathcal{V}$ in $\mathbf{Sm}_S^{\rm ess}$. If we have a full subcategory $\mathcal{V}$ of $\mathbf{Sm}_S^{\rm ess}$ for which the conditions of Chapter I, Definition 2.1.4 are satisfied, then the inclusion $i: \mathcal{V} \to \mathbf{Sm}_S^{\rm ess}$ induces the exact tensor functor $i_*: \mathcal{DM}(\mathcal{V}) \to \mathcal{DM}(\mathbf{Sm}_S^{\rm ess})$.

3.4.2. COROLLARY. *Suppose the conditions of §3.2.1 and §3.3.1 are satisfied for* $\mathcal{V} = \mathbf{Sm}_S^{\rm ess}$. *Let* $i: \mathcal{V} \to \mathbf{Sm}_S^{\rm ess}$ *be a full subcategory of* $\mathbf{Sm}_S^{\rm ess}$ *such that the conditions of Chapter I, Definition 2.1.4 are satisfied. Then the functor* $i_*: \mathcal{DM}(\mathcal{V}) \to \mathcal{DM}(\mathbf{Sm}_S^{\rm ess})$ *induces an isomorphism*

$$\mathrm{Hom}_{\mathcal{DM}(\mathcal{V})}(1, \Gamma) \to \mathrm{Hom}_{\mathcal{DM}(\mathbf{Sm}_S^{\rm ess})}(1, i_*(\Gamma))$$

for all Γ *in* $\mathcal{DM}(\mathcal{V})$.

PROOF. It follows easily from the definition of the functor $\mathcal{Z}_{\mathrm{mot}}$ in Chapter I, §3.2 that $\mathcal{Z}_{\mathrm{mot}}(\mathbb{Z}_X(q)_f[p]) = \mathcal{Z}_{\mathrm{mot}}(i_*(\mathbb{Z}_X(q)_f[p]))$ for all (X,f) in $\mathcal{L}(\mathcal{V})$. From this, it follows that we have the identity $\mathcal{Z}_{\mathrm{mot}}(\mathbb{Z}_X(q)_f[p],*) = \mathcal{Z}_{\mathrm{mot}}(i_*(\mathbb{Z}_X(q)_f[p]),*)$ (see Definition 2.2.4), from which it follows that the complex $\mathfrak{Z}^q_{X/S}(*)$ is independent of the choice of the category $\mathcal{V}$ containing X.

Applying Theorem 3.4.1, it follows that i_* induces the isomorphism

$$\mathrm{Hom}_{\mathcal{DM}(\mathcal{V})}(1,\mathbb{Z}_X(q)[p]) \cong \mathrm{Hom}_{\mathcal{DM}(\mathbf{Sm}_S^{\mathrm{ess}})}(1, i_*(\mathbb{Z}_X(q)[p]))$$

for all X in $\mathcal{V}$. As $\mathcal{DM}(\mathcal{V})$ is generated as a triangulated category by the objects $\mathbb{Z}_X(q)[p]$, and taking direct summands, the map

$$\mathrm{Hom}_{\mathcal{DM}(\mathcal{V})}(1,\Gamma) \to \mathrm{Hom}_{\mathcal{DM}(\mathbf{Sm}_S^{\mathrm{ess}})}(1, i_*(\Gamma))$$

i an isomorphism for all Γ in $\mathcal{DM}(\mathcal{V})$. □

We have as well a compatibility of motivic cohomology with filtered projective limits.

3.4.3. COROLLARY. *Let $\{S_\alpha \mid \alpha \in A\}$ be a filtered inverse of reduced schemes, with projective limit S. Suppose the conditions of §3.2.1 and §3.3.1 are satisfied for $\mathbf{Sm}_{S_\alpha}^{\mathrm{ess}}$, for each α. Let X in $\mathbf{Sm}_S^{\mathrm{ess}}$ be a filtered projective limit in* **Sch**: *$X = \lim_{\leftarrow} X_\alpha$, with X_α in $\mathbf{Sm}_{S_\alpha}^{\mathrm{ess}}$ for each $\alpha \in A$, such that the canonical maps $\pi_\alpha : X \to S \times_{S_\alpha} X_\alpha$ are flat. Let $\hat{X}$ be a closed subset of X, and suppose we have closed subsets $\hat{X}_\alpha$ of X_α, compatible with the transition maps in the inverse system, and with $\hat{X} = \lim_{\leftarrow} X_\alpha$. Then the natural map*

$$\lim_{\to} H^p_{\hat{X}_\alpha}(X_\alpha, \mathbb{Z}(q)) \to H^p_{\hat{X}}(X, \mathbb{Z}(q))$$

is an isomorphism. In addition, the conditions of §3.2.1 and §3.3.1 are satisfied for $\mathbf{Sm}_S^{\mathrm{ess}}$

PROOF. Let Y be in $\mathbf{Sm}_S^{\mathrm{ess}}$, then for large enough α, Y is a localization of a scheme of the form $S \times_{S_\alpha} Y_\alpha$ for Y_α in $\mathbf{Sm}_{S_\alpha}^{\mathrm{ess}}$. In particular, Y is a projective limit of an inverse system $\alpha \mapsto Y_\alpha \in \mathbf{Sm}_{S_\alpha}^{\mathrm{ess}}$, with the canonical map $\pi_\alpha : Y \to S \times_{S_\alpha} Y_\alpha$ being flat. It follows from the definition of the complexes $\mathfrak{Z}^q_{Y/S}(*)$ (which are functorial in Y for flat maps, and functorial in S for arbitrary maps) that the natural map

$$\mathfrak{Z}^q_{Y/S}(*) \to \lim_{\to}(p_1 \circ \pi_\alpha)^* \mathfrak{Z}^q_{Y_\alpha/S_\alpha}(*)$$

is an isomorphism. This shows that the conditions of §3.2.1 and §3.3.1 are satisfied for $\mathbf{Sm}_S^{\mathrm{ess}}$.

In addition, taking $Y = X$, we have the isomorphism

$$\mathbb{H}^{p-2q}_{\hat{X}}(X, \mathfrak{Z}^q_{X/S}(*)) \cong \lim_{\to} \mathbb{H}^{p-2q}_{\hat{X}_\alpha}(X_\alpha, \mathfrak{Z}^q_{X_\alpha/S}(*));$$

applying Theorem 3.4.1 completes the proof. □

3.4.4. *Local to global spectral sequence.* For X in $\mathbf{Sm}_S^{\mathrm{ess}}$, we have the presheaf of motivic cohomology groups on X_{Zar}, $\mathcal{H}^p(\mathbb{Z}(q))$, gotten by sheafifying the presheaf $U \mapsto H^p(U, \mathbb{Z}(q))$. From Corollary 3.4.3, we have the natural isomorphism

$$\mathcal{H}^p(\mathbb{Z}(n))_x \cong H^p(\mathrm{Spec}\, \mathcal{O}_{X,x}, \mathbb{Z}(n)). \tag{3.4.4.1}$$

Combining (3.4.4.1) with Theorem 3.4.1 and the local to global hypercohomology spectral sequence

$$E_2^{p,q} := H^q_{\mathrm{Zar}}(X, \mathcal{H}^p(\mathfrak{Z}^n_{X/S}(*))) \Longrightarrow \mathbb{H}^{p+q}_{\mathrm{Zar}}(X, \mathfrak{Z}^n_{X/S}(*))$$

gives the local to global spectral sequence for motivic cohomology

$$E_2^{p,q} : H^q_{\mathrm{Zar}}(X, \mathcal{H}^p(\mathbb{Z}(n)) \Longrightarrow H^{p+q}(X, \mathbb{Z}(n)), \tag{3.4.4.2}$$

assuming that the conditions of §3.2.1 and §3.3.1 are satisfied for $\mathcal{V} = \mathbf{Sm}_S^{\mathrm{ess}}$.

3.4.5. *Quillen spectral sequence.* The results of this paragraph rely in part on some material in Chapter III and Chapter IV; we will not be using any of the results proved here in Chapter III or in Chapter IV.

We now suppose that the base scheme S is of the form $S = \operatorname{Spec} k$, where k is a perfect field. We suppose in addition that the conditions of §3.2.1 and §3.3.1 are satisfied for $\mathcal{V} = \mathbf{Sm}_k^{\mathrm{ess}}$.

Let X be in $\mathbf{Sm}_k^{\mathrm{ess}}$, and suppose we have a filtration of X by closed subsets

$$X = X^0 \supset X^1 \supset \ldots \supset X^n \supset X^{n+1} = \emptyset, \tag{3.4.5.1}$$

such that

1. For $j = 0, \ldots, n$, X^j has pure codimension j on X
2. For $j = 0, \ldots, n$, $X^j \setminus X^{j+1}$ is smooth over k.

Taking the motive of X with support in X^j, we have the distinguished triangles (I.2.2.10.1)

$$\mathbb{Z}_{X,X^{j+1}} \to \mathbb{Z}_{X,X^j} \to \mathbb{Z}_{X\setminus X^{j+1},X^j\setminus X^{j+1}} \to \mathbb{Z}_{X,X^{j+1}}[1].$$

We have as well the *Gysin isomorphism* (III.2.1.2.2)

$$i_{j*} : \mathbb{Z}_{X^j\setminus X^{j+1}} \to \mathbb{Z}_{X\setminus X^{j+1},X^j\setminus X^{j+1}}(j)[2j],$$

giving the linked distinguished triangles

$$\mathbb{Z}_{X,X^{j+1}} \to \mathbb{Z}_{X,X^j} \to \mathbb{Z}_{X^j\setminus X^{j+1}}(-j)[-2j] \to \mathbb{Z}_{X,X^{j+1}}[1].$$

This gives the strongly convergent spectral sequence

$$E_1^{p,q}(X^*) := H^{q-p}(X^p \setminus X^{p+1}, \mathbb{Z}(n-p)) \Longrightarrow H^{p+q}(X, \mathbb{Z}(n))$$

If we then pass to the limit over filtrations (3.4.5.1), and use Corollary 3.4.3, we have the strongly convergent spectral sequence

$$E_1^{p,q}(X^*) := \oplus_{x\in X^{(p)}} H^{q-p}(\operatorname{Spec} k(x), \mathbb{Z}(n-p)) \Longrightarrow H^{p+q}(X, \mathbb{Z}(n)) \tag{3.4.5.2}$$

where $X^{(p)}$ is the set of codimension p points of X.

3.4.6. *Gersten complex.* We let

$$\begin{aligned} H^q(X, \mathbb{Z}(n)) &\to \oplus_{x\in X^{(1)}} H^{q-1}(k(x), \mathbb{Z}(n-1)) \to \ldots \\ &\to \oplus_{x\in X^{(\dim X)}} H^{q-\dim_k X}(k(x), \mathbb{Z}(n-\dim X)) \end{aligned} \tag{3.4.6.1}$$

be the complex of E_1-terms in the spectral sequence (3.4.5.2), where $\dim X$ is the Krull dimension of X. As the spectral sequence is natural in X (for open immersions), we may sheafify over X, giving the *Gersten complex*

$$\begin{aligned} \mathcal{H}^q(\mathbb{Z}(n)) &\to \coprod_{x\in X^{(1)}} i_{x*} H^{q-1}(k(x), \mathbb{Z}(n-1)) \to \ldots \\ &\to \coprod_{x\in X^{(\dim X)}} i_{x*} H^{q-\dim_k X}(k(x), \mathbb{Z}(n-\dim X)). \end{aligned} \tag{3.4.6.2}$$

Quillen's proof of Gersten's conjecture gives the analogous result for motivic cohomology.

3.4.7. LEMMA [Gersten's conjecture for motivic cohomology]. *Let k be a perfect field, and suppose that the conditions of §3.2.1 and §3.3.1 are satisfied for $\mathcal{V} = \mathbf{Sm}_k^{\mathrm{ess}}$. Take X in $\mathbf{Sm}_k^{\mathrm{ess}}$, and let x be a finite set of points of X. Let $X_x := \operatorname{Spec} \mathcal{O}_{X,x}$, let X_x^j be a closed, codimension $j > 0$ subset of X_x, and take $\eta \in H^q_{X_x^j}(X_x, \mathbb{Z}(n))$. Then there is a codimension $j-1$ closed subset X_x^{j-1} of X_x with $X_x^j \subset X_x^{j-1}$ and with η going to zero in $H^q_{X_x^{j-1}}(X_x, \mathbb{Z}(n))$.*

PROOF. By Corollary 3.4.3, we may assume that η is the restriction to X_x of an element $\tilde{\eta} \in H^q_{X^j}(X, \mathbb{Z}(n))$ for some codimension j closed subset X^j of X (shrinking X if necessary). We may similarly assume that X is affine, and of finite type over k; let $d = \dim_k X$. Take a codimension one closed subset D of X containing X^j. As in [**102**, §7, Lemma 5.12], there is a morphism $\pi: X \to \mathbb{A}_k^{d-1}$, with $\pi(x) = 0$, such that

1. the restriction of π to D is finite
2. π is smooth with fiber dimension one over an open neighborhood U of $D \cap \pi^{-1}(0)$.

Shrinking U, we may assume that the image $\pi(U \cap D)$ is an open neighborhood V of 0, and $U \cap D$ is finite over V. Let $\{x_1, \ldots, x_r\} = \pi^{-1}(0) \cap D$, and let $X_U^j = X^j \cap U$.

Form the pull-back diagram

$$\begin{array}{ccc} U \times_V U & \xrightarrow{p_2} & U \\ {\scriptstyle p_1}\downarrow & & \downarrow{\scriptstyle \pi} \\ U & \xrightarrow[\pi]{} & V \end{array}$$

and let $s: U \to U \times_V U$ be the diagonal section. Since $\pi: U \to V$, is smooth, the diagonal $s(U)$ in $U \times_V U$ is a Cartier divisor [**5**, II 4.15], hence $s(U)$ is defined by a single equation $t = 0$ in a neighborhood of $\{\ldots, x_i \times x_j, \ldots\} \subset U \times_V U$. Shrinking U and V, we may assume that $s(U)$ is a principle divisor in $U \times_V U$. Thus

$$\mathrm{cl}^1_{U \times_V U}(|s(U)|) = 0$$

in $H^2(U \times_V U, \mathbb{Z}(1))$.

Let $\tilde{X}^{j-1} := p_1^{-1}(X_U^j)$. Since $\pi: D \cap U \to V$ is finite, $\tilde{X}^{j-1}$ is finite over U, hence X^{j-1} is closed in U and $\tilde{X}^{j-1}$ is finite over X^{j-1}.

We have the Gysin map $s_*: \mathbb{Z}_{U, X_U^j} \to \mathbb{Z}_{U \times_V U, \tilde{X}^{j-1})}(1)[2]$ (III.2.1.2.3).

Let $\tilde{i}: s(X_U^j) \to \tilde{X}^{j-1}$ and $i: X_U^j \to X^{j-1}$ be the inclusions. Restricting p_2 gives the maps $p_2^j: s(X_U^j) \to X^j$ and $p_2^{j-1}: \tilde{X}^{j-1} \to X^{j-1}$. By the functoriality of the Borel-Moore motive (Chapter IV, §2.4.6), we have the commutative diagram of pushforward morphisms

$$\begin{array}{ccc} \mathbb{Z}_{U \times_V U, s(X_U^j)}(1)[2] & \xrightarrow{p_{2*}^j} & \mathbb{Z}_{U, X_U^j} \\ {\scriptstyle \tilde{i}_*}\downarrow & & \downarrow{\scriptstyle i_*} \\ \mathbb{Z}_{U \times_V U, \tilde{X}^{j-1}}(1)[2] & \xrightarrow[p_{2*}^{j-1}]{} & \mathbb{Z}_{U, X^{j-1}}, \end{array}$$

and in addition $p_{2*}^{j-1} \circ s_* = i_*$.

On the other hand, by (Chapter III, Lemma 2.2.7, with the closed subset F taken to be $U \times_V U$), we have

$$s_* = \cup \mathrm{cl}^1_{U\times_V U}(|s(U)|) \circ p_1^* = 0,$$

hence $i_*: \mathbb{Z}_{U,X^j_U} \to \mathbb{Z}_{U,X^{j-1}}$ is the zero map, as desired. □

Gersten's conjecture yields the following result:

3.4.8. THEOREM [Gersten resolution]. *Let k be a perfect field, and suppose that the conditions of §3.2.1 and §3.3.1 are satisfied for $\mathcal{V} = \mathbf{Sm}_k^{\mathrm{ess}}$.*
(i) Let y be a finite set of points on a scheme Y in $\mathbf{Sm}_k^{\mathrm{ess}}$. Then the complex (3.4.6.1) for $X = \operatorname{Spec} \mathcal{O}_{Y,y}$ is exact.
(ii) Let X be in $\mathbf{Sm}_k^{\mathrm{ess}}$. Then the Gersten complex (3.4.6.2) forms an acyclic resolution of the sheaf $\mathcal{H}^q(\mathbb{Z}(n))$ on X.

PROOF. The argument is the same as in [**102**]. Let (Y, y) be as in (i). The vanishing proved in §3.4.7 implies that the spectral sequence (3.4.5.2) for $X = \operatorname{Spec} \mathcal{O}_{Y,y}$ has

$$E_2^{p,q} = \begin{cases} H^q(\operatorname{Spec} \mathcal{O}_{Y,y}, \mathbb{Z}(n)); & \text{for } p = 0 \\ 0; & \text{otherwise,} \end{cases}$$

which proves (i). The assertion (ii) follows from (i). □

For arbitrary X in $\mathbf{Sm}_k^{\mathrm{ess}}$, Theorem 3.4.8 identifies the E_2-term of the spectral sequence (3.4.5.2) as the cohomology

$$E_2^{p,q} = H^p_{\mathrm{Zar}}(X, \mathcal{H}^q(\mathbb{Z}(n))).$$

From this, it follows by a standard argument that the Quillen spectral sequence agrees with the local to global spectral sequence (3.4.4.2) from E_2 on.

3.4.9. *Bloch's formula.* If we suppose that the cycle class map $\mathrm{cl}_X^{q,p}: \mathrm{CH}^q(X, p) \to H^{2q-p}(X, \mathbb{Z}(q))$ is an isomorphism for all X in $\mathbf{Sm}_k^{\mathrm{ess}}$ (e.g, if the conditions of §3.2.1 and §3.3.1 are satisfied for $\mathcal{V} = \mathbf{Sm}_k^{\mathrm{ess}}$), then in particular, we have $H^p(X, \mathbb{Z}(q)) = 0$ for $q < 0$, $H^0(F, \mathbb{Z}(0)) = \mathbb{Z}$, and $H^1(F, \mathbb{Z}(1)) = F^\times$ for all fields F of finite type over k. Thus, the Gersten resolution (3.4.6.2) for $\mathcal{H}^q(\mathbb{Z}(q))$ ends with

$$\coprod_{x\in X^{(q-1)}} i_{x*}k(x)^\times \to \coprod_{x\in X^{(q)}} i_{x*}\mathbb{Z}.$$

As in [**102**], this gives the isomorphism

$$H^q_{\mathrm{Zar}}(X, \mathcal{H}^q(\mathbb{Z}(q))) \cong \mathrm{CH}^q(X). \tag{3.4.9.1}$$

Indeed, it suffices to show that the connecting homomorphism in the Gersten resolution is given by the divisor map. We will show this in Chapter VI, Proposition 1.1.11, when we discuss Milnor K-theory and motivic cohomology. We will also show in Chapter VI, Theorem 1.1.16 that there is a natural isomorphism of sheaves $\mathcal{H}^q(\mathbb{Z}(q)) \cong \mathcal{K}_q^M$, where $\mathcal{K}_q^M$ is the qth *Milnor K-sheaf*, defined as the kernel of the tame symbol map for Milnor K-theory

$$\coprod_{x\in X^{(0)}} i_{x*}K_q^M(k(x)) \to \coprod_{x\in X^{(1)}} i_{x*}K_{q-1}^M(k(x))$$

(see [**7**], [**108**]). Bloch's formula (3.4.9.1) thus gives us the isomorphism

$$H^q_{\mathrm{Zar}}(X, \mathcal{K}_q^M) \cong \mathrm{CH}^q(X)$$

(this result is not new, see [**108**] and [**76**]).

3.5. Some moving lemmas

In this section, we verify a version of the classical moving lemma for the complexes $\mathcal{Z}^q(X,*)_f$ in case the base S is $\operatorname{Spec} k$ for a field k, and X is affine. This helps in the next section, where we verify the criteria of §3.2.1 and §3.3.1 in case S is smooth, essentially of finite type, and of dimension ≤ 1 over a field k.

The main results of this section have been also proved by Bloch [**16**] by essentially the same method; as this work has not appeared in published form, we give the details here.

3.5.1. As in §2.1.2, we call a subvariety of Δ^p of the form $\Delta^*(h)(\Delta^m)$ for some $h\colon [m] \to [p]$ in Δ a *face* of Δ^p; all faces F of Δ^p are given by equations of the form $t_{i_1} = \ldots = t_{i_s} = 0$, where

$$\Delta^p = \operatorname{Spec} k[t_0, \ldots, t_p]/\sum_{i=0}^{p} t_i - 1.$$

Let X be a smooth k-variety, and $\mathcal{C} = \{C_1, \ldots, C_s\}$ a finite collection of irreducible locally closed subsets of X; let $i_j\colon C_j \to X$ be the inclusion. Let $m = (m_1, \ldots, m_s)$ be a sequence of integers such that $m_j \leq q$, $j = 1, \ldots, s$, and let $\mathcal{Z}^q_{\mathcal{C},m}(X,p)$ be the subgroup of $\mathcal{Z}^q(X,p)$ generated by the codimension q subvarieties W of $X \times \Delta^p$ such that

1. W is in $\mathcal{Z}^q(X,p)$
2. for each face F of Δ^p and each i, we have

$$\operatorname{codim}_{C_i \times F}(W \cap (C_i \times F)) \geq m_i$$

   or the intersection is empty.

One easily sees that $\mathcal{Z}^q_{\mathcal{C},m}(X,*)$ forms a subcomplex of $\mathcal{Z}^q(X,*)$.

3.5.2. LEMMA. *Let (X,f) be in $\mathcal{L}(\mathbf{Sm}_k)$. Then the complex $\mathcal{Z}^q(X,*)_f$ is equal to $\mathcal{Z}^q_{\mathcal{C},m}(X,*)$ for some finite set of locally closed irreducible subsets $\mathcal{C}$, and some sequence m.*

PROOF. Write f as $f\colon X' \to X$. Write X' as a union of connected components $X' = \coprod_{i=1}^{s} X'_i$, and let $f_i\colon X'_i \to X$ be the restriction of f to X'_i. As X' is smooth over k, each X'_i is irreducible; let $n_i = \dim_k(X'_i)$.

Let $C_{i,j}$ be the subset of X defined as the set of points x such that each irreducible component of $f_i^{-1}(x)$ of maximal dimension has dimension j (over $k(x)$). The sets $C_{i,j}$ are constructible subsets of X, and form a filtration of the constructible subset $f_i(X'_i)$ of X. Write each $C_{i,j}$ as a finite union of irreducible subsets $C^l_{i,j}$, with each $C^l_{i,j}$ locally closed in X, and let $d^l_{i,j} = \dim_k(C^l_{i,j})$. Clearly, we have

$$d^l_{i,j} + j \leq n_i. \tag{3.5.2.1}$$

Now let W be a reduced irreducible codimension q closed subset of $X \times \Delta^p$, and let $F \cong \Delta^m$ be a face of Δ^p with inclusion $g\colon F \to \Delta^p$. Let W' be an irreducible component of $(f_i \times g)^{-1}(W)$; then there is a j and an irreducible component $C^l_{i,j}$

of $C_{i,j}$ such that

$$(f_i \times g)(W') \subset \bar{C}^l_{i,j} \times F$$
$$(f_i \times g)(W') \not\subset C_{i,j+1} \times F.$$

From this it follows that

$$\text{(3.5.2.2)} \qquad \dim_k(W') \le j + \dim_k((f_i \times g)(W')) \le j + \dim_k(W \cap (C^l_{i,j} \times F)).$$

Now suppose that $\operatorname{codim}_{X'_i \times F}(W') < q$. Then (3.5.2.2) implies

$$\begin{aligned} n_i + m - q &< \dim_k(W') \\ &\le j + \dim_k(W \cap (C^l_{i,j} \times F)) \\ &= j + d^l_{i,j} + m - \operatorname{codim}_{C^l_{i,j} \times F}(W \cap (C^l_{i,j} \times F)), \end{aligned}$$

or

$$\text{(3.5.2.3)} \qquad \operatorname{codim}_{C^l_{i,j} \times F}(W \cap (C^l_{i,j} \times F)) < j + d^l_{i,j} - n_i + q.$$

Conversely, suppose that (3.5.2.3) holds for some i, j, l. Take an irreducible component Z of the intersection $W \cap (C^l_{i,j} \times F)$ of maximal dimension; then

$$\begin{aligned} \dim_k((f_i \times g)^{-1}(W)) &\ge \dim_k((f_i \times g)^{-1}(Z)) \\ &\ge j + \dim_k(Z) \\ &> j + d^l_{i,j} + m - (j + d^l_{i,j} - n_i + q) \\ &= n_i + m - q. \end{aligned}$$

Thus, if we let $m^l_{i,j}$ be defined by $m^l_{i,j} = j + d^l_{i,j} - n_i + q$, then

$$\begin{gathered} \operatorname{codim}_{X' \times F}((f \times g)^{-1}(W)) \ge q \quad \text{for all faces } g\colon F \to \Delta^p \\ \Updownarrow \\ \operatorname{codim}_{C^l_{i,j} \times F}(W \cap (C^l_{i,j} \times F)) \ge m^l_{i,j} \quad \text{for all } i, j, l \text{ and all faces } F. \end{gathered}$$

In addition, by (3.5.2.1), we have $m^l_{i,j} \le q$.

This gives the equality $\mathcal{Z}^q(X, *)_f = \mathcal{Z}^q_{\mathcal{C},m}(X, *)$ for

$$\mathcal{C} = \{\dots, C^l_{i,j}, \dots\}; \qquad m = (\dots, m^l_{i,j}, \dots).$$

□

3.5.3. *Generic projections.* We take k to be an infinite field. Let X be a smooth affine k-variety of dimension n, embedded as a closed subset of $\mathbb{A}^N$, with $N > n$. We let $\bar{X}$ be the closure of X in $\mathbb{P}^N \supset \mathbb{A}^N$. Let $\mathbb{P}^{N-1}_\infty$ denote the complement $\mathbb{P}^N - \mathbb{A}^N$, and $\bar{X}_\infty$ the intersection $\bar{X} \cap \mathbb{P}^{N-1}_\infty$.

For a linear subvariety $L \subset \mathbb{P}^N$ of dimension $N-n-1$, we let $\pi_L\colon \mathbb{P}^N - L \to \mathbb{P}^n$ denote the projection with center L; the projection with center $L \subset \mathbb{P}^{N-1}_\infty$ gives the affine-linear map $\pi^0_L\colon \mathbb{A}^N \to \mathbb{A}^n$. The restriction of π^0_L to X: $\pi_{L,X}\colon X \to \mathbb{A}^n$ is finite if and only if $L \cap \bar{X} = \emptyset$. We let $\mathcal{U}_X$ denote the subset of the Grassmannian $\mathbf{Gr}_{\mathbb{P}^{N-1}_\infty}(N-n-1)$ consisting of those L with $L \cap \bar{X} = \emptyset$.

If we have constructible subsets A and C of X, we let $e(A, C)$ denote the maximum among the irreducible components C_i of C and irreducible components Z of $A \cap C_i$ of the expression

$$\max(\operatorname{codim}_X(A) - \operatorname{codim}_{C_i}(Z), 0).$$

For an irreducible locally closed subset A of X, and an $L \in \mathcal{U}_X$, let $L^+(A)$ be the closure in $\pi_{L,X}^{-1}(\pi_{L,X}(A))$ of $\pi_{L,X}^{-1}(\pi_{L,X}(A)) \setminus A$; for general A, we define $L^+(A)$ to be the union of the $L^+(A_i)$, over the irreducible components A_i of A. We let $R_L \subset X$ denote the ramification locus of the map $\pi_{L,X}$.

The following result is a version of the classical moving lemma for algebraic cycles.

3.5.4. LEMMA [see [**106**], [**29**]]. *Let $X \subset \mathbb{A}_k^N$ be a smooth k-variety of dimension n, embedded as a closed subset of $\mathbb{A}_k^N$. Let A be an irreducible, locally closed subset of X, and C a locally closed subset of X. Then there is a non-empty open subset U of $\mathcal{U}_X$ such that R_L contains no irreducible component of A, $A \cap C$, or C, and*

$$e(L^+(A), C) \leq \max(e(A, C) - 1, 0)$$

for all $L \in U$.

PROOF. We may assume that C is irreducible. Let $\ell(A, C)$ be the set of lines l in $\mathbb{P}_k^N$ such that there are points $p \in A$, $q \in C$ with $p \neq q$ and with $p, q \in l$. Let $S(A, C)$ be the *secant space* of A and C, i.e., the subset of $\mathbb{P}_k^N$

$$S(A, C) := \bigcup_{l \in \ell(A,C)} l.$$

For a locally closed subset Y of X, let $T(X; Y)$ be the set of lines l in $\mathbb{P}_k^N$ which are tangent to X at some point $p \in Y$, and let

$$R(X; Y) := \bigcup_{l \in T(X;Y)} l.$$

For a constructible subset Y of $\mathbb{P}_k^N$, let $\dim_k(Y)$ stands for the maximum of the dimension of the irreducible components of Y. By Chevalley's theorem, $S(A, C)$ and $R(X; Y)$ are constructible subsets of $\mathbb{P}_k^N$, with

$$\dim_k(S(A, C)) \leq \dim_k(A) + \dim_k(C) + 1,$$
$$\dim_k(R(X; Y)) \leq \dim_k(Y) + n.$$

Since both $S(A, C)$ and $R(X; Y)$ have no irreducible component contained in $\mathbb{P}_\infty^{N-1}$, we have

$$\begin{aligned} \dim_k(S(A, C) \cap \mathbb{P}_\infty^{N-1}) &\leq \dim_k(A) + \dim_k(C) \\ \dim_k(R(X; Y) \cap \mathbb{P}_\infty^{N-1}) &\leq \dim_k(Y) + n - 1. \end{aligned} \tag{3.5.4.1}$$

If a point x of X is in $R_L \cap Y$, then there is an $l \in T(X; Y)$ with $x \in l$ and $l \cap L \neq \emptyset$. For $L \in \mathcal{U}_X$, this implies that l is not contained in X, hence

$$\dim_k(R_L \cap Y) \leq \dim_k(L \cap R(X; Y)). \tag{3.5.4.2}$$

Similarly, a point x is in $L^+(A) \cap C \setminus R_L \cap A \cap C$ if and only if there is an $l \in \ell(A, C)$ with $x \in l$ and $l \cap L \neq \emptyset$. For $L \in \mathcal{U}_X$, this gives

$$\dim_k(L^+(A) \cap C \setminus R_L \cap A \cap C) \leq \dim_k(L \cap S(A, C)). \tag{3.5.4.3}$$

Let U be the subset of $\mathcal{U}_X$ consisting of those L which intersect $S(A, C)$ properly, and $R(X; Y)$ properly for all irreducible components Y of A, $A \cap C$ and C, and take L in U. By (3.5.4.2), R_L contains no component of A, C or $A \cap C$. Combining (3.5.4.1), (3.5.4.2) and (3.5.4.3), we have

$$\begin{aligned}\dim_k(L^+(A) \cap C) \leq \max(&\dim_k(A \cap C) + n - 1 + (N - n - 1) - (N - 1),\\ &\dim_k(A) + \dim_k(C) + (N - n - 1) - (N - 1))\\ = \max(&\dim_k(A \cap C) - 1, \dim_k(A) + \dim_k(C) - n),\end{aligned}$$

which is the desired result. □

3.5.5. *Finite pull-back.* Let $f: X \to Y$ be a finite surjective morphism of smooth k-varieties of finite type over k. Then the maps $(f \times \mathrm{id}_{\Delta^p})_*$, $(f \times \mathrm{id}_{\Delta^p})^*$ give maps of complexes

$$f_*: \mathcal{Z}^q(X, *) \to \mathcal{Z}^q(Y, *),$$
$$f^*: \mathcal{Z}^q(Y, *) \to \mathcal{Z}^q(X, *).$$

Suppose we have a finite surjective map $f: X \to Y$, a collection of locally closed subsets $\mathcal{C} = \{C_1, \dots, C_s\}$ of X, and a sequence of integers $m_1, \dots, m_s$ with $0 \leq m_j \leq q$. We let $\mathcal{Z}^q_{\mathcal{C},m,f}(Y, p)$ be the subgroup of $\mathcal{Z}^q(Y, p)$ generated by the irreducible codimension q subvarieties W of $Y \times \Delta^p$ such that

$$\mathrm{codim}_{C_j \times F}\big((C_j \times F) \cap (f \times \mathrm{id})^{-1}(W)\big) \geq m_j$$

for all faces F of Δ^p. The $\mathcal{Z}^q_{\mathcal{C},m,f}(Y, p)$ form a subcomplex of $\mathcal{Z}^q(Y, *)$, and the map $f^*: \mathcal{Z}^q(Y, *) \to \mathcal{Z}^q(X, *)$ restricts to the map $f^*: \mathcal{Z}^q_{\mathcal{C},m,f}(Y, *) \to \mathcal{Z}^q_{\mathcal{C},m}(X, *)$.

For a sequence of integers $m = (m_1, \dots, m_s)$ with $m_i \leq q$, we let $m - 1$ be the sequence $(m_1 - 1, \dots, m_s - 1)$ and $m + 1$ the sequence $(m'_1, \dots, m'_s)$, where

$$m'_j = \begin{cases} m_j + 1 & \text{if } m_j < q, \\ q & \text{if } m_j = q. \end{cases}$$

We let $m_{\max}$ denote the constant sequence $m_{\max} = (q, \dots, q)$.

3.5.6. LEMMA. *Let W be an irreducible closed subvariety of $X \times \Delta^p$ such that W is in $\mathcal{Z}^q_{\mathcal{C},m-1}(X, p)$. Then there is an open subset $U_{W,\mathcal{C},m}$ of $\mathcal{U}_X$ such that, for each $L \in U_{W,\mathcal{C},m}$, we have*

(a) $(\pi_{L,X} \times \mathrm{id}_{\Delta^p})_*(|W|)$ *is in* $\mathcal{Z}^q_{\mathcal{C},m-1,\pi_{L,X}}(\mathbb{A}^n, p)$.
(b) $(\pi_{L,X} \times \mathrm{id}_{\Delta^p})^*((\pi_{L,X} \times \mathrm{id}_{\Delta^p})_*(|W|)) = |W| + W'$, *with W' effective.*
(c) W' *is in* $\mathcal{Z}^q_{\mathcal{C},m}(X, p)$.

PROOF. Let $f: X \to Y$ be a finite surjective morphism of smooth k-varieties, and let Z be in $\mathcal{Z}^*(X, *)$. Then $f^*(f_*(Z))$ is equal to $Z + Z'$, with Z' effective if Z is effective. This proves (b), and shows that (c) implies (a).

To prove (a), let F be a face of Δ^p. Write $W \cap (X \times F)$ as a union of irreducible components $W \cap (X \times F) = W_F^1 \cup \dots W_F^t$ and let $W_F^{i,j}$ be the locally closed subset of X defined by

$$x \in W_F^{i,j} \iff \dim_k((x \times F) \cap W_F^i) = j$$
$$j = 0, 1, \dots, \dim_k(F); \qquad i = 1, \dots, t.$$

Let $C_0 = X$, $m_0 = q$. We note that $|W|$ is in $\mathcal{Z}^q_{\mathcal{C},m-1}(X, p)$ if and only if the inequalities

$$\mathrm{codim}_{C_l}(W_F^{i,j} \cap C_l) \geq m'_l + j - \dim_k(F) \tag{3.5.6.1}$$

hold for all i, j and l, where

$$m'_l = \begin{cases} m_l - 1 & l = 1, \dots, s, \\ q & l = 0. \end{cases}$$

Define $L^+(|W|)$ to be the support of the cycle

$$W' := (\pi_{L,X} \times \mathrm{id}_{\Delta^p})^*((\pi_{L,X} \times \mathrm{id}_{\Delta^p})_*(|W|)) - |W|.$$

Write $L^+(|W|) \cap (X \times F)$ as a union of irreducible components, $L^+(|W|) \cap (X \times F) = \cup_{i'} L^+(|W|)^{i'}_F$, and define $L^+(|W^i_F|)$ similarly to $L^+(|W|)$. Then each $L^+(|W|)^{i'}_F$ is an irreducible component of $L^+(|W^i_F|)$ for some i; we write this as $i = \nu(i')$.

By Lemma 3.5.4, there is an open subset U_0 of $\mathcal{U}_X$ such that, for $L \in U_0$, the ramification locus R_L contains no component of any $W^{i,j}_F$ or $W^{i,j}_F \cap C_l$, and that $R_L \times \Delta^p$ contains no component of W or $W \cap C_l \times \Delta^p$.

Take L in U_0. If we define the locally closed subsets $L^+(|W|)^{i',j}_F$ for $L^+(|W|)$ in a similar fashion to the definition of $W^{i,j}_F$, then $L^+(|W|)^{i',j}_F$ is a union of irreducible components of $L^+(W^{\nu(i'),j}_F)$.

We now apply Lemma 3.5.4 with $A = W^{i,j}_F$; we find that, for each F, i and j, there is a non-empty open subset $U^{i,j}_F$ of U_0 such that, for $L \in U^{i,j}_F$, we have

(3.5.6.2)
$$\mathrm{codim}_{C_l}(L^+(|W|)^{i',j}_F \cap C_l) \geq \min(m'_l + j - \dim_k(F) + 1, \mathrm{codim}_X(W^{i,j}_F))$$

for all i' with $\nu(i') = i$, and for all l. On the other hand, by (3.5.6.1) for $l = 0$, we have

$$\mathrm{codim}_X(W^{i,j}_F) \geq q + j - \dim_k(F),$$

so (3.5.6.2) is equivalent to

(3.5.6.3)
$$\mathrm{codim}_{C_l}(L^+(|W|)^{i',j}_F \cap C_l) \geq \min(m'_l + j - \dim_k(F) + 1, q + j - \dim_k(F)).$$

Noting that $m_l = \min(m'_l + 1, q)$, we see that (3.5.6.3) is equivalent to

(3.5.6.4) $$\mathrm{codim}_{C_l}(L^+(|W|)^{i',j}_F \cap C_l) \geq m_l + j - \dim_k(F); \quad l = 0, \dots, s.$$

Now take L in the intersection of all the $U^{i,j}_F$. As (3.5.6.4) implies

$$\mathrm{codim}_{C_l}(L^+(|W|)^{i',j}_F \cap C_l) \geq m_l + j - \dim_k(F),$$

for all i', j, l and F. and as $L^+(|W|)$ is the support of W', we see that W' is in $\mathcal{Z}^q_{\mathcal{C},m}(X, p)$, as desired. □

3.5.7. *A triangulation.* We have the vertices $v^p_0, \dots, v^p_p$ of Δ^p, where the vertex v^p_j is given by $t_j = 1$, $t_i = 0$, $i \neq j$. For $i = 0, 1$, $j = 0, \dots, p$, we let $v^p_{i,j}$ be the point of $\Delta^1 \times \Delta^p$

$$v_{i,j} = v^1_i \times v^p_j; \qquad i = 0, 1;\ j = 0, \dots, p.$$

We let $[n]$ denote the set $\{0, \dots, n\}$. For each $i = 0, \dots, p$, we let $f^p_i \colon [p+1] \to [1] \times [p]$ be given by

$$f^p_i(j) = \begin{cases} (0, j) & \text{if } 0 \leq j \leq i, \\ (1, j-1) & \text{if } i + 1 \leq j \leq p + 1. \end{cases}$$

If we let $h^p = \sum_{i=0}^p (-1)^i f_i^p$, the h^p form a triangulation of $\Delta^1 \times \Delta^p$:

$$h^p \circ (\sum_{i=0}^{p+1} (-1)^i \delta_i^p) + (\sum_{i=0}^{p} (-1)^i (\mathrm{id} \times \delta_i^{p-1})) \circ h_{p-1} = i_1^p - i_0^p,$$

where the i_j^p are the maps

$$i_j^p : [p] \to [1] \times [p]$$
$$i_j^p(k) = (j, k).$$

Write $f_i^p = (f_{i,1}^p, f_{i,2}^p)$. We let

$$F_i^p : \Delta^{p+1} = \mathbb{A}^{p+1} \to \Delta^1 \times \Delta^p = \mathbb{A}^{p+1}; \qquad i = 0, \ldots, p,$$

be the affine-linear map with $F_i^p(v_j^{p+1}) = v^1_{f_{i,1}^p(j)} \times v^p_{f_{i,2}^p}$. We call a linear subset F of $\Delta^1 \times \Delta^p$ a *face* if $F = F_i^p(F')$ for some i and some face F' of Δ^{p+1}.

3.5.8. DEFINITION. Let $f: X \to Y$ be a finite surjective morphism, $\mathcal{C}$ a finite set of locally closed subsets of X. Let $\mathcal{Z}^q_{\mathcal{C},m,f,h}(Y \times \Delta^1, p)$ be the subgroup of $\mathcal{Z}^q(Y \times \Delta^1, p)$ generated by the codimension q subvarieties W of $Y \times \Delta^1 \times \Delta^p$ such that $(\mathrm{id}_Y \times F_i^p)^*(W)$ is in $\mathcal{Z}^q_{\mathcal{C},m,f}(Y, p+1)$ for all $i = 0, \ldots, p$.

3.5.9. We now take $Y = \mathbb{A}^n$. Let F be a face of $\Delta^1 \times \Delta^p$, let $\mathcal{C} := \{C_1, \ldots, C_s\}$ be a set of locally closed subsets of X, and let C be the disjoint union $C = \coprod_{i=1}^s C_i$. Let W be an irreducible codimension q subvariety of $Y \times \Delta^p$ such that W is in $\mathcal{Z}^q(Y, p)$, and let W_F be the intersection

$$W_F := p_{13}^{-1}(W) \cap (Y \times F) \subset Y \times \Delta^1 \times \Delta^p.$$

Let $(G, 1)$ be the pointed affine space $(\mathbb{A}^n_k, 0)$, considered as an algebraic group under addition, and acting on $Y = \mathbb{A}^n$ via translation. We use coordinates $x_1, \ldots,$ x_n for G and $y_1, \ldots, y_n$ for Y. Let $\pi: G \setminus \{1\} \to \mathbb{P}^{n-1}$ be the canonical map

$$\pi(x_1, \ldots, x_n) = (x_1 : \ldots : x_n).$$

For $x = (x_1, \ldots, x_n) \in G \setminus \{1\}$, the closure in G of fiber $\pi^{-1}(x)$ is canonically isomorphic to Δ^1 via the unique linear map which sends 0 to v_0^1, and sends x to v_1^1. We write this isomorphism as $\phi_x : \Delta^1 \to G$.

Let $i_F: W_F \to Y \times F$ be the inclusion, let $f_C: C \to Y$ be the composition of f with the natural map $C \to X$, and let $T: G \times C \to Y$ be the map

$$T(g, c) = g + f_C(c).$$

Let $p_{F,1}: F \to \Delta^1$ and $p_{F,2}: F \to \Delta^p$ be the projections.

Consider the diagram

$$\begin{array}{ccc} & & p_{13}^{-1}(W) \\ & & \downarrow \\ G \times C \times \Delta^1 \times \Delta^p & \xrightarrow[q]{} & Y \times \Delta^1 \times \Delta^p, \end{array}$$

where q is the map

$$q(g, c, t, \lambda) = (g + f_C(c), t, \lambda).$$

Let $\phi:\Delta^1 \to G$ be an affine-linear map sending v_0^1 to 0, and let Φ be the map

$$\Phi: C \times \Delta^1 \times \Delta^p \to G \times C \times \Delta^1 \times \Delta^p$$
$$\Phi(c,t,\lambda) = (\phi(t), c, t, \lambda).$$

This gives the diagram

$$\begin{array}{ccc} \phi_C^* W & \longrightarrow & p_{13}^{-1}(W) \\ \downarrow & & \downarrow \\ C \times \Delta^1 \times \Delta^p & \xrightarrow[q\circ\Phi]{} & Y \times \Delta^1 \times \Delta^p, \end{array}$$

where $\phi_C^* W$ and the maps $\phi_C^* W \to p_{13}^{-1}(W)$ and $\phi_C^* W \to C \times \Delta^1 \times \Delta^p$ are defined to make the diagram cartesian. For a face F, we let $\phi_C^* W_F$ be the intersection $\phi_C^* W \cap (C \times F)$. We let F^0 be the open subset $F \setminus v_0^1 \times \Delta^p$ of F, and $\phi_C^* W_F^0$ the open subset $\phi_C^* W_F \cap C \times F^0$ of $\phi_C^* W_F$.

3.5.10. LEMMA. *There is a non-empty Zariski open subset $U_{W,C}$ of $G \setminus \{1\}$ such that, for each $x \in U_{W,C}$, for each face F of $\Delta^1 \times \Delta^p$, and for $\phi = \phi_x$, $\phi_C^* W_F^0$ has codimension q on $C \times F^0$, or is empty.*

PROOF. It suffices to show the existence, for each face F, of a non-empty open subset $U_{W,C,F}$ of U such that, for each $x \in U_{W,C,F}$, if we take $\phi = \phi_x$, then $\phi_C^* W_F^0$ has codimension q on $C \times F^0$. We may assume that F is not contained in $v_0^1 \times \Delta^p$. Let $F' = p_{F,2}(F)$. We consider three cases:

(a) $p_{F,2}: F \to F'$ is an isomorphism, and $p_{F,1}: F \to \Delta^1$ is surjective.
(b) $F = v_1^1 \times F'$.
(c) $F = \Delta^1 \times F'$.

It suffices to handle the case of a single locally closed subset C of X; we consider the case (a) first.

In case (a), we may identify F with the transpose of the graph of a surjective affine linear map $L: F' \to \mathbb{A}^1$, where we identify $\mathbb{A}^1$ with Δ^1 via the affine-linear map sending 0 to v_0^1 and 1 to v_1^1. Let $F'^0 = L^{-1}(\mathbb{A}^1 \setminus \{0\})$, and let $\Psi: G \times C \times F'^0 \to Y \times F^0$ be the map

$$\Psi(x,c,\lambda) = (L(\lambda)\cdot x + f(c), L(\lambda), \lambda).$$

We claim that Ψ is surjective, with fibers of dimension $\dim_k(C)$. Indeed, for $(y, L(\lambda), \lambda) \in Y \times F^0$, we have $L(\lambda) \neq 0$. Thus the translates of $f(C)$ by elements of the form $L(\lambda)\cdot x$ cover all of Y, and the projection $p_2: \Psi^{-1}\big((y, L(\lambda), \lambda)\big) \to C$ is a bijection, proving the claim.

Since W is in $\mathcal{Z}^q(Y,p)$, $W \cap Y \times F'$ has codimension q on $Y \times F'$. Thus $W_F := p_{13}^{-1}(W) \cap Y \times F$ has codimension q on $Y \times F$, and hence $\Psi^{-1}(W_F)$ has codimension q on $G \times C \times F'^0$.

Let $\Pi: G \setminus \{1\} \times C \times F'^0 \to \mathbb{P}^{n-1}$ be the map induced by the canonical projection $\pi: G \setminus \{1\} \to \mathbb{P}^{n-1}$. Since $\Psi^{-1}(W_F)$ has codimension q on $\mathbb{A}^n \times C \times F'^0$, it follows that $\Psi^{-1}(W_F) \cap \Pi^{-1}(z)$ has codimension q on $\pi^{-1}(z) \times C \times F'^0$ for all z in an open subset V of $\mathbb{P}^{n-1}$. For $x \in G - \{0\}$, the map $\phi := \phi_x$ gives an isomorphism of $\Delta^1 \setminus \{v_0^1\}$ with $\pi^{-1}(\pi(x))$, and identifies $\Psi^{-1}(W_F) \cap \Pi^{-1}(z)$ with $\phi_C^* W_F^0$, completing the proof in case (a).

The case (b) is similar; we write $F = v_1^1 \times F$, which identifies F with the transpose of the graph of the constant map $L\colon F' \to \mathbb{A}^1$ with value 1. The same proof as in (a) gives the desired conclusion.

For (c), we have $F^0 = (\Delta^1 \setminus \{v_0^1\}) \times F'$; let Ψ be the map

$$\Psi\colon G \times C \times F^0 \to Y \times F^0$$
$$\Psi(x, c, t, \lambda) = (t \cdot x + f(c), t, \lambda).$$

The argument of (a) shows that Ψ is surjective with fibers of dimension $\dim_k(C)$; continuing the argument by considering the projection $\Pi\colon G \setminus \{1\} \times C \times F^0 \to \mathbb{P}^{n-1}$ leads to the desired conclusion. $\square$

For a pointed map $\phi\colon (\mathbb{A}^1, 0) \to (G, 1)$, we have the automorphism

$$T^\phi\colon Y \times \Delta^1 \to Y \times \Delta^1$$
$$T^\phi(x, t) = (\phi(t) + x, t),$$

where we identify (Δ^1, v_0^1, v_1^1) with $(\mathbb{A}^1, 0, 1)$ as above. For a cycle W on $Y \times \Delta^p$, we let $T_\phi^*(W)$ be the cycle $(T_\phi \times \mathrm{id})^*(p_{13}^*(W))$ on $Y \times \Delta^1 \times \Delta^p$.

Let i_1 be the inclusion

$$i_1\colon Y \to Y \times \Delta^1$$
$$i_1(y) = (y, v_1^1).$$

3.5.11. LEMMA. *Let W be an subvariety of $Y \times \Delta^p$ which is in $\mathcal{Z}^q_{\mathcal{C},m,f}(Y, p)$. Then*
(i) For each $x \in U_{W,\mathcal{C}}$, $T_{\phi_x}^*(W)$ is in $\mathcal{Z}^q_{\mathcal{C},m,f,h}(Y \times \Delta^1, p)$.
(ii) The cycle $i_1^*(W)$ is in $\mathcal{Z}^q_{\mathcal{C},m_{\max},f}(Y, p)$.

PROOF. Let F be a face of $\Delta^1 \times \Delta^p$. If F is contained in $v_0^1 \times \Delta^p$, then, for each $C \in \mathcal{C}$, or for $C = X$, we have

$$(f \times \mathrm{id})^*(T_{\phi_x}^*(W)) \cap (C \times F) = f^*(W) \cap C \times (p_2(F)).$$

As W is in $\mathcal{Z}^q_{\mathcal{C},m,f}(Y, p)$, this intersection has the required codimension on $C \times F$. For all other faces F, it follows from the definition of $U_{W,\mathcal{C}}$ in the statement of Lemma 3.5.10 that $(f \times \mathrm{id})^*(T_{\phi_x}^*(W)) \cap (C \times F^0)$ has codimension q on $C \times F^0$. Taking F to be of the form $v_1^1 \times F'$ proves (ii). For the other F, $C \times F \setminus C \times F^0 = C \times F'$ for some face F' contained in $v_0^1 \times \Delta^p$. As we have already shown that the intersection with $C \times F'$ has the required codimension on $C \times F'$, we have an even better bound for the codimension of $(f \times \mathrm{id})^*(T_{\phi_x}^*(W)) \cap (C \times F)$. This completes the proof. $\square$

3.5.12. *The homotopy.* We now take $k \subset K \subset F$ to be transcendental extensions $K = k(t_{11}, \ldots, t_{Nn})$, $F = K(s_1, \ldots, s_n)$. Let $\pi_t\colon \mathbb{A}^N_K \to \mathbb{A}^n_K$ the linear map with matrix

$$\begin{pmatrix} t_{11}, \ldots, t_{1.n} \\ \vdots, \ldots, \vdots \\ t_{N1}, \ldots, t_{Nn} \end{pmatrix},$$

and let $\pi_{t,X}\colon X_K \to \mathbb{A}^n_K = Y_K$ be the restriction to X_K. Let $\phi_s\colon (\mathbb{A}^1_F, 0) \to (G_F, 1)$ be the map

$$\phi_s(z) = z \cdot s.$$

Let $H_p\colon \mathcal{Z}^q_{\mathcal{C},m-1,f,h}(Y\times\Delta^1,p)\to\mathcal{Z}^q_{\mathcal{C},m-1,f}(Y,p+1)$ be the map

$$H_p=\sum_{i=0}^{p}(-1)^iF_i^{p*}.$$

Let $H_p^{X,\mathcal{C},m}$ denote the composition

$$\mathcal{Z}^q_{\mathcal{C},m-1}(X,p)\xrightarrow{\pi_{t,X*}\circ p_F^*}\mathcal{Z}^q_{\mathcal{C},m-1,\pi_{t,X}}(Y_F,p)\xrightarrow{T^*_{\phi_s}}\mathcal{Z}^q_{\mathcal{C},m-1,\pi_{t,X},h}(Y_F\times\Delta^1,p)$$
$$\xrightarrow{H_p}\mathcal{Z}^q_{\mathcal{C},m-1,f}(Y_F,p+1)\xrightarrow{\pi^*_{t,X}}\mathcal{Z}^q_{\mathcal{C},m-1}(X_F,p+1),$$

where p_F^* is induced by the projection $p_F\colon X_F\to X$. It follows from Lemma 3.5.6 and Lemma 3.5.11 that $H_p^{X,\mathcal{C},m}$ is well-defined, and that the maps $H_p^{X,\mathcal{C},m}$ give a homotopy between the maps

$$\pi^*_{t,X}\circ\pi_{t,X*}\circ p_F^*\colon\mathcal{Z}^q_{\mathcal{C},m-1}(X,*)\to\mathcal{Z}^q_{\mathcal{C},m-1}(X_F,*),$$
$$\pi^*_{t,X}\circ i_1^*\circ T^*_{\phi_s}\circ\pi_{t,X*}\circ p_F^*\colon\mathcal{Z}^q_{\mathcal{C},m-1}(X,*)\to\mathcal{Z}^q_{\mathcal{C},m-1}(X_F,*).$$

In addition, by Lemma 3.5.11, the map $\pi^*_{t,X}\circ i_1^*\circ T^*_{\phi_s}\circ\pi_{t,X*}\circ p_F^*$ factors through the inclusion $\mathcal{Z}^q_{\mathcal{C},m_{\max}}(X_F,*)\to\mathcal{Z}^q_{\mathcal{C},m-1}(X_F,*)$. Finally, it follows from Lemma 3.5.6 that the map $p_F^*\circ\pi^*_{t,X}\circ\pi_{t,X*}\circ p_F^*-p_F^*$ factors through the inclusion $\mathcal{Z}^q_{\mathcal{C},m}(X_F,*)\to\mathcal{Z}^q_{\mathcal{C},m-1}(X_F,*)$. Thus, we have shown

3.5.13. LEMMA. *The base extension from k to F induces a homotopically trivial map*

$$\frac{\mathcal{Z}^q_{\mathcal{C},m-1}(X,*)}{\mathcal{Z}^q_{\mathcal{C},m}(X,*)}\to\frac{\mathcal{Z}^q_{\mathcal{C},m-1}(X_F,*)}{\mathcal{Z}^q_{\mathcal{C},m}(X_F,*)}.$$

3.5.14. THEOREM. *Let k be a field, not necessarily infinite. Let (X,f) be in $\mathbf{Sm}_k^{\mathrm{ess}}$, with X affine. Then the inclusion*

$$\mathcal{Z}^q(X,*)_f\to\mathcal{Z}^q(X,*)$$

is a quasi-isomorphism.

PROOF. As the complexes $\mathcal{Z}^q(X,*)_f$ and $\mathcal{Z}^q(X,*)$ transform filtered projective limits in (X,f) to filtered inductive limits, we may assume that X is in $\mathbf{Sm}_k$, i.e, that X is of finite type over k.

Let K be a finite extension of k; we then have the base-extension and norm maps

$$p_K^*\colon\mathcal{Z}^q(X,*)_f\to\mathcal{Z}^q(X_K,*)_{f_K},$$
$$p_{K*}\colon\mathcal{Z}^q(X_K,*)_{f_K}\to\mathcal{Z}^q(X_K,*)_f,$$

with

(3.5.14.1) $$p_{K*}\circ p_K^*=[K:k]\cdot\mathrm{id}.$$

If k is a finite field, there exist infinite pro-l extensions of k for each prime l different from $\mathrm{char}(k)$; using (3.5.14.1), we may assume that k is infinite.

From Lemma 3.5.2 and an elementary induction, it suffices to show that

(3.5.14.2) $$\frac{\mathcal{Z}^q_{\mathcal{C},m-1}(X,*)}{\mathcal{Z}^q_{\mathcal{C},m}(X,*)}$$

is acyclic for all choices of $\mathcal{C}$ and m. By Lemma 3.5.13, the map

$$\frac{\mathcal{Z}^q_{\mathcal{C},m-1}(X,*)}{\mathcal{Z}^q_{\mathcal{C},m}(X,*)} \to \frac{\mathcal{Z}^q_{\mathcal{C},m-1}(X_F,*)}{\mathcal{Z}^q_{\mathcal{C},m}(X_F,*)}$$

is zero on homology. On the other hand, since F is a pure transcendental extension of the infinite field k, an elementary specialization argument shows that the above map is injective on homology. Indeed, if W is an element of $\mathcal{Z}^q_{\mathcal{C},m-1}(X,p)$, and if we have elements B_F of $\mathcal{Z}^q_{\mathcal{C},m-1}(X_F,p+1)$ and Z_F of $\mathcal{Z}^q_{\mathcal{C},m}(X_F,p)$ with $W\times_k F = Z_F + dB_F$, then there is an open subset U of an affine space over k, and elements

$$B_U \in \mathcal{Z}^q_{\mathcal{C},m-1}(X\times_k U,p+1),$$
$$Z_U \in \mathcal{Z}^q_{\mathcal{C},m}(X\times_k U,p)$$

such that B_F and Z_F are the restrictions of B_U and Z_U to the generic point of U, and such that $W\times_k U = Z_U + dB_U$. We may then find a k-point s of U such that restrictions of B_U and Z_U to $X\times s$ are all defined, giving the relation $W = i_s^*(W\times_k U) = i_s^*(Z_U) + d(i_s^*(B_U))$. Thus, the complex (3.5.14.2) is acyclic, as desired. □

3.5.15. COROLLARY. *Let (X,f) and (Y,g) be in $\mathcal{L}(\mathbf{Sm}_k^{\mathrm{ess}})$, with Y affine. Let $p\colon X\times_k Y\to X$ be the projection. Then the natural map*

$$\mathrm{id}^*\colon p_*(\mathfrak{Z}^q_{X\times_k Y/k}(*)_{f\times g}) \to p_*(\mathfrak{Z}^q_{X\times_k Y/k}(*))$$

is a quasi-isomorphism of complexes of sheaves on X.

PROOF. As in the proof of Theorem 3.5.14, we may assume that (X,f) and (Y,g) are in $\mathcal{L}(\mathbf{Sm}_k)$. Let x be a point of X. The stalk $[p_*\mathfrak{Z}^q_{X\times_k Y/k}(*)_{f\times g})]_x$ is the inductive limit of the complexes $\mathcal{Z}^q(U\times_k Y,*)_{j^*f\times g}$ over affine open neighborhoods $j\colon U\to X$; we have the similar description of the stalk $[p_*(\mathfrak{Z}^q_{X\times_k Y/k}(*))]_x$. By Theorem 3.5.14, the map $\mathcal{Z}^q(U\times_k Y,*)_{j^*f\times g}\to \mathcal{Z}^q(U\times_k Y,*)$ is a quasi-isomorphisms for all affine U, whence the result. □

3.6. Motivic cohomology and the higher Chow groups

We now verify the criteria of §3.2.1 and §3.3.1 in case S is smooth and of dimension ≤ 1 over a field k.

3.6.1. We first consider the case $S=\operatorname{Spec} k$. Let X in $\mathbf{Sm}_k^{\mathrm{ess}}$. From the Mayer-Vietoris property of the higher Chow groups (§2.1.6(2), see also Remark 2.1.7), the natural map

$$\mathcal{Z}^q(X,*) \to \mathfrak{R}\mathfrak{Z}^q_{X/k}(*) \tag{3.6.1.1}$$

is a quasi-isomorphism. From the homotopy property (§2.1.6(1)), the map

$$p_1^*\colon \mathcal{Z}^q(X,*) \to \mathcal{Z}^q(X\times_k \mathbb{A}^1,*) \tag{3.6.1.2}$$

is a quasi-isomorphism.

For $\hat{X}$ a closed subset of X, with complement $j\colon U\to X$, we let $\mathcal{Z}^q_{\hat{X}}(X,*)$ denote the cone

$$\mathcal{Z}^q_{\hat{X}}(X,*) := \mathrm{cone}\big(j^*\colon \mathcal{Z}^q(X,*)\to \mathcal{Z}^q(X-Z,*)\big)[-1].$$

If Z is a smooth subvariety of X, of codimension d, and $\hat{Z}$ is a closed subset of Z, the localization property (§2.1.6(2)) implies that the inclusion $i_Z: Z \to X$ induces a quasi-isomorphism

$$i_{Z*}: \mathcal{Z}^{q-d}_{\hat{Z}}(Z,*) \to \mathcal{Z}^q_{\hat{Z}}(X,*). \tag{3.6.1.3}$$

3.6.2. PROPOSITION. *Let k be a field. Then the conditions of §3.2.1 and §3.3.1 are satisfied for $S = \operatorname{Spec} k$, $\mathcal{V} = \mathbf{Sm}_k^{\mathrm{ess}}$.*

PROOF. The Gysin morphism condition §3.2.1(iii) follows from (3.6.1.1) and (3.6.1.3). We reduce the homotopy condition §3.2.1(i) to the usual homotopy property (3.6.1.2), using (3.6.1.1) and (3.6.1.3). The conditions of §3.2.1(ii) and §3.3.1 follow in a similar fashion from (3.6.1.1)-(3.6.1.3), together with Corollary 3.5.15. □

3.6.3. *The case of curves.* Let k be a field, let $p_S: S \to \operatorname{Spec} k$ be in $\mathbf{Sm}_k^{\mathrm{ess}}$ and of dimension one over k. The functor "compose with p_S" gives the functor $p_{S*}: \mathbf{Sm}_S^{\mathrm{ess}} \to \mathbf{Sm}_k^{\mathrm{ess}}$ inducing the functor $p_{S*}: \mathcal{L}(\mathbf{Sm}_S^{\mathrm{ess}}) \to \mathcal{L}(\mathbf{Sm}_k^{\mathrm{ess}})$. We usually ignore the p_{S*} in the notion, and simply consider an object (X,f) of $\mathcal{L}(\mathbf{Sm}_S^{\mathrm{ess}})$ as an object of $\mathcal{L}(\mathbf{Sm}_k^{\mathrm{ess}})$. In particular, we have the natural inclusion of complexes of sheaves on X

$$\iota_{S/k}: \mathfrak{Z}^q_{X/S}(*)_f \to \mathfrak{Z}^q_{X/k}(*)_f.$$

More generally, for each (Y,g) in $\mathcal{L}(\mathbf{Sm}_k^{\mathrm{ess}})$, we have the inclusion of complexes of sheaves on X:

$$\iota_{S/k}: p_*(\mathfrak{Z}^q_{X\times_k Y/S}(*)_{f\times_k g}) \to p_*(\mathfrak{Z}^q_{X\times_k Y/k}(*)_{f\times_k g}), \tag{3.6.3.1}$$

where $p: X\times_k Y \to X$ is the projection.

3.6.4. LEMMA. *The map (3.6.3.1) is a quasi-isomorphism.*

PROOF. By taking limits, we may replace $\mathbf{Sm}_S^{\mathrm{ess}}$ and $\mathbf{Sm}_k^{\mathrm{ess}}$ with $\mathbf{Sm}_S$ and $\mathbf{Sm}_k$.

Let x be a point of X, and take $W \in p_*(\mathfrak{Z}^q_{X\times_k Y/k}(*)_{f\times_k g})(U)$ for some affine open neighborhood $j: U \to X$ of x. Let $p_X: X \to S$ be the structure map, and let $s = p_X(x)$. We have the auxiliary maps $f: X' \to X$ and $g: Y' \to Y$; let $j^*f_s: U'_s \to U$ be the restriction of f to the fiber of $X' \times_X U$ over s. If W is in the subcomplex $p_*(\mathfrak{Z}^q_{X\times_k Y/S}(*)_{f\times_k g})(U)$, then $(j^*f_s \times g)^*(W)$ is defined and is in $\mathcal{Z}^q(U'_s \times_k Y', *)$, since $(j^*f_s \times g)^*(W)$ is the fiber of $(j^*f\times g)^*(W)$ over s. Conversely, suppose that the cycle $(j^*f_s \times g)^*(W)$ is defined and is in $\mathcal{Z}^q(U'_s \times_k Y', *)$. Since S has dimension one, this implies that $(j^*f \times g)^*W$ is equi-dimensional over an open neighborhood of s in S; similarly, all the necessary intersections of $(j^*f \times g)^*W$ with faces are equi-dimensional over a neighborhood of s in S. Thus W is in $p_*(\mathfrak{Z}^q_{X\times_k Y/S}(*)_{f\times_k g})(V)$ for some neighborhood $V \subset U$ of x. We have just shown the identity on the stalks

$$[p_*(\mathfrak{Z}^q_{X\times_k Y/k}(*)_{(f\cup f_s)\times_k g})]_x = [p_*(\mathfrak{Z}^q_{X\times_k Y/S}(*)_{f\times_k g})]_x.$$

As the inclusion $[p_*(\mathfrak{Z}^q_{X\times_k Y/k}(*)_{(f\cup f_s)\times_k g})]_x \to [p_*(\mathfrak{Z}^q_{X\times_k Y/k}(*)_{f\times_k g})]_x$ is a quasi-isomorphism by Corollary 3.5.15, the proof is complete. □

3.6.5. PROPOSITION. *Let k be a field, and let S be in $\mathbf{Sm}_k^{\mathrm{ess}}$ and of dimension one over k. Then the conditions of §3.2.1 and §3.3.1 are satisfied for $\mathcal{V} = \mathbf{Sm}_S^{\mathrm{ess}}$.*

PROOF. A smooth, dimension one k-scheme $S \to \operatorname{Spec} k$, is a projective limit of smooth dimension one k schemes $Y_\alpha \to \operatorname{Spec} k$, of finite type over k. Using Corollary 3.4.3, we reduce to the case of S of finite type over k. The result then follows from Lemma 3.6.4 and Proposition 3.6.2. □

3.6.6. THEOREM. *Let S be scheme which is filtered projective limit of schemes S_α, such that each S_α is a smooth k_α-scheme of finite type for some field k_α, with S_α of dimension at most one over k_α. Let $\mathcal{V}$ be a full subcategory of $\mathbf{Sm}_S^{\mathrm{ess}}$ such that the conditions of Chapter I, Definition 2.1.4 are satisfied. Then*
(i) The cycle class map

$$\mathrm{cl}(\Gamma) : \mathcal{CH}(\Gamma) \to \mathrm{Hom}_{\mathcal{DM}(\mathcal{V})}(1, \Gamma)$$

is an isomorphism for all Γ in $\mathcal{DM}(\mathcal{V})$.
(ii) The cycle class map

$$\mathrm{cl}_X^{q,p} : \mathcal{CH}^q(X, 2q-p) \to H^p(X, \mathbb{Z}(q))$$

is an isomorphism for all X in $\mathcal{V}$.
(iii) Suppose S is smooth of dimenison at most one over a field k. Then the natural map $\mathcal{Z}^q(X/k,*) \to \mathfrak{R}\mathfrak{Z}^q(X/S,*)$ induces an isomorphism

$$i_X^{q,p} : \mathrm{CH}^q_{naif}(X/k, p) \to \mathcal{CH}^q(X/S, p)$$

for all X in $\mathcal{V}$. The group $\mathrm{CH}^q_{naif}(X/k,p)$ is Bloch's higher Chow group $\mathrm{CH}^q(X,p)$.
(iv) For X in $\mathcal{V}$ there is a natural isomorphism

$$K_{2q-p}(X)^{(q)} \to H^p(X, \mathbb{Q}(q)),$$

where $K_n(X)^{(q)}$ is the weight q Adams eigenspace of $K_n(X) \otimes \mathbb{Q}$.
(v) Suppose $S = \operatorname{Spec} k$, k a field. Then the map

$$\oplus_{p,q}\mathrm{cl}_X^{q,p} : \oplus_{p,q} \mathcal{CH}^q(X/S, 2q-p) \to \oplus_{p,q} H^p(X, \mathbb{Z}(q))$$

is an isomorphism of rings, where we make $\oplus_{p,q}\mathcal{CH}^q(X/S, 2q-p)$ a ring using the products on $\mathrm{CH}^q(X,p)$ defined in [**19**, §5] *and the isomorphism $\mathcal{CH}^q(X/S,p) \cong \mathrm{CH}^q(X,p)$ from (iii).*

PROOF. By Corollary 3.4.2 and Corollary 3.4.3, it suffices to prove (i) and (ii) in case of S is a smooth finite type k-scheme of dimension at most one over k. The statement (i) then follows from Theorem 3.3.10, Proposition 3.6.2 and Proposition 3.6.5. The assertion (ii) follows from (i) and the definition of the map $\mathrm{cl}_X^{q,p}$, the higher Chow group $\mathcal{CH}^q(X,p)$, and the motivic cohomology group $H^p(X, \mathbb{Z}(q))$ as

$$\begin{aligned} H^p(X, \mathbb{Z}(q)) &= \mathrm{Hom}_{\mathcal{DM}(\mathcal{V})}(1, \mathbb{Z}_X(q)[p]), \\ \mathcal{CH}^q(X,p) &= \mathcal{CH}(\mathbb{Z}_X(q)[2q-p]), \\ \mathrm{cl}_X^{q,p} &= \mathrm{cl}(\mathbb{Z}_X(q)[2q-p]). \end{aligned}$$

The first part of (iii) follows from the Mayer-Vietoris property of the higher Chow groups (see Remark 2.1.7(ii)), together with Proposition 3.1.3. The second part follows from Proposition 2.2.5.

The statement (iv) for S smooth of dimension at most one over a field follows from (iii) and §2.1.6(5) (see also Theorem 3.6.12 of Chapter III). The general case follows from this, Corollary 3.4.2, Corollary 3.4.3, and compatibility of K-theory with filtered inductive limits of exact categories [**102**, §2].

Proposition 2.4.6, (ii), (iii), and the fact that $\mathrm{cl}^{q,p}_{X,naif} = \mathrm{cl}^{q,p}_X \circ i^{q,2q-p}_X$ prove (v). □

3.6.7. REMARK. In order to extend a version of Theorem 3.6.6 to more general base schemes, it may be necessary to modify the definition of $\mathcal{CH}(\Gamma)$. One such modification would be to form the inductive limit over the maps

$$\mathcal{CH}(\Gamma) \xrightarrow{\mathrm{id}\otimes p^*} \mathcal{CH}(\Gamma \otimes \mathbb{Z}_{\mathbb{A}^1}) \xrightarrow{\mathrm{id}\otimes p^*} \dots \xrightarrow{\mathrm{id}\otimes p^*} \mathcal{CH}(\Gamma \otimes \mathbb{Z}_{\mathbb{A}^1}^{\otimes n}) \xrightarrow{\mathrm{id}\otimes p^*} \dots .$$

Clearly the cycle map to $\mathrm{Hom}_{\mathcal{DM}}(1, \Gamma)$ factors through this inductive limit. It is not difficult to show that the evident revision of the conditions §3.2.1 and §3.3.1 imply that the modified cycle map is an isomorphism. The advantage would be that one could then increase the *dimension* of cycles of fixed codimension at will.

CHAPTER III

K-Theory and Motives

In this chapter, we describe the fundamental constructions relating K-theory and motivic cohomology: Chern classes and the Riemann-Roch theorem. In order to state Riemann-Roch, one needs the operation of push-forward, so we construct this as well. Most of our arguments are adaptations of standard constructions for a Bloch-Ogus cohomology theory, or for the Chow ring, but, as we are working in the motivic category, we occasionally need to modify an argument to rely entirely on either formal properties of triangulated categories, or on purely geometric considerations. Our heavy debt to Grothendieck *et al.* [**57**] and [**2**], Gillet [**46**], Baum-Fulton-MacPherson [**8**] and Fulton [**44**] will be readily apparent.

We will only prove the Riemann-Roch theorem for K-theory; the G-theory version (Riemann-Roch for singular varieties) will have to wait until we have defined motivic Borel-Moore homology in Chapter IV.

1. Chern classes

The first Chern class of a line bundle is given by the cycle class of the zero section. We prove the projective bundle formula, and then use the classic method of Grothendieck [**57**] to define Chern classes of vector bundles on simplicial schemes by means of the splitting principle. Grothendieck's geometric proof of the Whitney product formula translates directly into the motivic setting. A modification of Gillet's construction gives us Chern classes for higher K-theory. We treat the case of diagrams of schemes as well.

1.1. Cycles on simplicial schemes

We extend the definition of cycles to simplicial schemes; we will use the constructions and notation of Chapter I, §2.5 for the motives associated to simplicial schemes.

We have the inclusion $j_n : \Delta^{\leq n} \to \Delta$, and for $n' < n$ the inclusion $j_{n',n} : \Delta^{n'} \to \Delta^n$. If $X : \Delta^{\mathrm{op}} \to \mathcal{C}$ is a simplicial object in a category $\mathcal{C}$, we write $X^{\leq n}$ for $j_n^* X$; if $X : \Delta^{n\mathrm{op}} \to \mathcal{C}$ is an n-truncated simplicial object, and $n' < n$, we write $X^{\leq n'}$ for $j_{n',n}^* X$. We use a similar notation for morphisms. This notation conflicts with the notation of Chapter I, §2.4.1 for cosimplicial objects, but the context will make clear which notation is being used.

1.1.1. Let $(X, f) : \Delta^{\leq n\mathrm{op}} \to \mathcal{L}(\mathcal{V})$ be a truncated simplicial object of $\mathcal{L}(\mathcal{V})$. We have the sequence of maps in $\mathbf{C}^b_{\mathrm{mot}}(\mathcal{V})$

$$\mathbb{Z}_X(q)_f^{m\leq * \leq n}[m] \xrightarrow{\pi_m} \mathbb{Z}_{X_m}(q)_{f_m} \xrightarrow{d^m} \mathbb{Z}_{X_{m+1}}(q)_{f_{m+1}}, \tag{1.1.1.1}$$

where π_m is the canonical map of complexes (I.2.5.3.4). From Chapter I, Proposition 3.3.5, the sequence (1.1.1.1) gives the exact sequence

$$(1.1.1.2)\quad 0 \to H^0(\mathcal{Z}_{\mathrm{mot}}(\mathbb{Z}_X(q)^{m\le*\le n}[2q+m])) \to \mathcal{Z}^q(X_m)_{f_m} \xrightarrow{\mathcal{Z}^q(d^m)} \mathcal{Z}^q(X_{m+1})_{f_{m+1}}.$$

We set

$$\mathcal{Z}^q(X)_f^{m\le*\le n} := H^0(\mathcal{Z}_{\mathrm{mot}}(\mathbb{Z}_X(q)_f^{m\le*\le n}[2q+m]);$$

the exact sequence (1.1.1.2) thus gives the exact sequence

$$(1.1.1.3)\qquad 0 \to \mathcal{Z}^q(X)_f^{m\le*\le n} \to \mathcal{Z}^q(X_m)_{f_m} \xrightarrow{\mathcal{Z}^q(d^m)} \mathcal{Z}^q(X_{m+1})_{f_{m+1}}.$$

By Chapter I, Proposition 3.3.5 again, we have the canonical isomorphism

$$(1.1.1.4)\quad \mathrm{cyc}^q_{(X,f)^{m\le*\le n}}: \mathcal{Z}^q(X)_f^{m\le*\le n} \to \mathrm{Hom}_{\mathbf{K}^b_{\mathrm{mot}}}(\mathfrak{e}\otimes 1, \mathbb{Z}_X(q)_f^{m\le*\le n}[2q+m]);$$

one easily checks that, for $m=n$, this agrees with the cycle class map (I.3.5.2.2) (after a shift).

For $W \in \mathcal{Z}^q(X)_f^{m\le*\le n}$, we let

$$\mathrm{cl}^q_{(X,f)^{m\le*\le n}}(W) \in \mathrm{Hom}_{\mathbf{D}^b_{\mathrm{mot}}(\mathcal{V})}(1, \mathbb{Z}_X(q)^{m\le*\le n}[2q+m]) = H^{2q+m}_{\mathrm{mot}}(\mathbb{Z}_X(q)^{m\le*\le n})$$

be the composition in $\mathbf{D}^b_{\mathrm{mot}}(\mathcal{V})$

$$1 \xrightarrow{\nu_1^{-1}} \mathfrak{e}\otimes 1 \xrightarrow{\mathrm{cyc}^q_{(X,f)^{m\le*\le n}}(W)} \mathbb{Z}_X(q)_f^{m\le*\le n}[2q+m],$$

where ν_1 is the map (I.2.2.4.1). This defines the homomorphism

$$(1.1.1.5)\qquad \mathrm{cl}^q_{(X,f)^{m\le*\le n}}: \mathcal{Z}^q(X)_f^{m\le*\le n} \to H^{2q+m}_{\mathrm{mot}}(\mathbb{Z}_X(q)_f^{m\le*\le n});$$

as above, this agrees with the shifted cycle class map (I.3.5.2.5) for $m=n$.

For $m \le n' < n$, we have the canonical map of complexes

$$(1.1.1.6)\qquad \rho_{m;n',n}: \mathbb{Z}_X(q)_f^{m\le*\le n} \to \mathbb{Z}_X(q)_f^{m\le*\le n'};$$

in particular, we have the map (I.2.5.3.3)

$$(1.1.1.7)\qquad \rho_{n',n} := \rho_{0;n',n}: \mathbb{Z}_X(q)_f \to \mathbb{Z}_{X^{\le n'}}(q)_{f^{\le n'}}.$$

We may take $m=0$ in (1.1.1.1)-(1.1.1.6); we write

$$\mathcal{Z}^q(X)_f := \mathcal{Z}^q(X)_f^{0\le*\le n},$$
$$\mathrm{cyc}^q_{(X,f)} := \mathrm{cyc}^q_{(X,f)^{0\le*\le n}},$$

etc.

For a truncated simplicial object $X: \Delta^{\le n\mathrm{op}} \to \mathcal{V}$ we have the lifting (I.2.5.2.1) to the truncated simplicial object (X, f_X) of $\mathcal{L}(\mathcal{V})$; we write

$$\mathcal{Z}^q(X)^{m\le*\le n} := \mathcal{Z}^q(X)_{f_X}^{m\le*\le n},$$
$$\mathcal{Z}^q(X/S) := \mathcal{Z}^q(X)_{f_X},$$
$$\mathrm{cyc}^q_X := \mathrm{cyc}^q_{(X,f_X)},$$

etc.

For $0 \leq n' < n$, the map (1.1.1.7) induces the injective map

$$\mathcal{Z}^q(\rho_{n',n}) : \mathcal{Z}^q(X)_f \to \mathcal{Z}^q(X^{\leq n'})_{f^{\leq n'}}.$$

1.1.2. *Products.* We recall the construction of cup products from Chapter I, §2.5.6. We have the map $f_0^{m,0}: [0] \to [m]$, $f_0^{m,0}(0) = m$ (see Part II, (III.1.2.1.1)).

1.1.3. PROPOSITION. *(i) The maps (1.1.1.4) and (1.1.1.5) define natural transformations of functors from* $\mathrm{s.}^{\leq n}\mathcal{L}$ *to* $\mathbf{Ab}$.
(ii) Let

$$(X, f) : \Delta^{\leq n\mathrm{op}} \to \mathcal{L}(\mathcal{V}),$$
$$(Y, g) : \Delta^{\leq n\mathrm{op}} \to \mathcal{L}(\mathcal{V})$$

be truncated simplicial objects of $\mathcal{L}(\mathcal{V})$*, and take*

$$W_X \in \mathcal{Z}^q(X)_f \subset \mathcal{Z}^q(X_0)_{f_0},$$
$$W_Y \in \mathcal{Z}^{q'}(Y)_g^{m\leq * \leq n} \subset \mathcal{Z}^{q'}(Y_m)_{g_m}.$$

Then the cycle $W_Y \times_{/S} X(f_0^{m,0})^*(W_X)$ *is in* $\mathcal{Z}^{q+q'}(Y \times_S X)_{g\times f}^{m\leq * \leq n}$ *and we have*

$$\begin{aligned}\mathrm{cl}^q_{(Y\times_S X, g\times f)^{m\leq *\leq n}}(W_Y \times_{/S} X(f_0^{m,0})^*(W_X)) \\ = \mathrm{cl}^q_{(Y,g)^{m\leq *\leq n}}(W_Y) \cup_{Y,X} \mathrm{cl}^q_{(X,f)}(W_X).\end{aligned}$$

(iii) Let $X : \Delta^{\leq n\mathrm{op}} \to \mathcal{V}$ *be a truncated simplicial object of* $\mathcal{V}$*, and take*

$$W \in \mathcal{Z}^q(X) \subset \mathcal{Z}^q(X_0)_{f_{X_0}},$$
$$W' \in \mathcal{Z}^{q'}(X)^{m\leq * \leq n} \subset \mathcal{Z}^q(X_m)_{f_{X_m}}.$$

Suppose that $X(f_0^{m,0})^*(W)$ *and* W' *intersect properly on* X_m*. Then the intersection product* $W' \cdot_{X_m} X(f_0^{m,0})^*(W)$ *is defined, is in* $\mathcal{Z}^{q+q'}(X)^{m\leq * \leq n}$*, and we have*

$$\mathrm{cl}^q_{X^{m\leq *\leq n}}(W' \cdot_{X_m} X(f_0^{m,0})^*(W)) = \mathrm{cl}^{q'}_{X^{m\leq *\leq n}}(W') \cup_X \mathrm{cl}^q_X(W).$$

PROOF. The assertion (i) follows from Chapter I, Proposition 3.5.3 and the definitions, and (iii) follows from (ii), (i) and the definition (see Chapter I, §2.5.6 and Part II, (III.1.2.3.6)) of $\cup_X$. The assertion (ii) follows from (Chapter I, Lemma 1.2.2 and Proposition 3.5.3), and the definition of $\cup_{X,Y}$ (see Chapter I, §2.5.6 and Part II, (III.1.2.1.4)). □

1.1.4. *Simplicial closed subsets.* Let $X : \Delta^{\leq n\mathrm{op}} \to \mathcal{V}$ be a truncated simplicial object of $\mathcal{V}$. Suppose we have, for $0 \leq k \leq n$, a closed subset $\hat{X}_k$ of $X([k])$. We say that the collection $\{\hat{X}_k\}$ defines a *simplicial closed subset* $\hat{X}$ *of* X if $X(g)^{-1}(\hat{X}_k) = \hat{X}_m$ for each $g : [k] \to [m]$ in $\Delta^{\leq n}$.

Let $X : \Delta^{\leq n\mathrm{op}} \to \mathcal{V}$ be a truncated simplicial scheme, and $\hat{X}$ a simplicial closed subset of X. Let U_k be the complement, $U_k := X([k]) \setminus \hat{X}_k$. Then, for each $g : [k] \to [m]$ in $\Delta^{\leq n}$, we have $X(g)(U_m) \subset U_k$, so the open subschemes U_k define an open simplicial subscheme $j : U \to X$ of X. We write this as $U := X \setminus \hat{X}$, and call U the complement of $\hat{X}$. In particular, we may define the *twisted motive of* X*, with support in* $\hat{X}$, $\mathbb{Z}_{X,\hat{X}}(q)$, by

$$\mathbb{Z}_{X,\hat{X}}(q) := \mathrm{cone}\big(j^* : \mathbb{Z}_X(q) \to \mathbb{Z}_U(q)\big)[-1].$$

1.1.5. *Cycles associated to simplicial subschemes.* Suppose we have a truncated simplicial object $(X, f): \Delta^{\leq n\mathrm{op}} \to \mathcal{L}(\mathcal{V})$ of $\mathcal{L}(\mathcal{V})$, together with a closed simplicial subscheme $Z \subset X$ such that

(a) For each $m \leq n$, Z_m is a pure codimension q subscheme of X_m.

(b) For each $m \leq n$, the codimension q cycle $|Z_m|$ determined by Z_m is in $\mathcal{Z}^q(X_m)_{f_m}$.

(c) For each map $g: [m] \to [k]$ in $\Delta^{\leq n}$, we have $X(g)^*(|Z_m|) = |Z_k|$ (note that, by (b) and Chapter I, Lemma 1.2.2, $X(g)^*(|Z_m|)$ is defined).

It follows directly from (a)-(c) that the cycle $|Z_0|$ is in $\mathcal{Z}^q(X)_f$. We write

$$|Z| \in \mathcal{Z}^q(X)_f \tag{1.1.5.1}$$

for the cycle $|Z_0|$ considered as an element of $\mathcal{Z}^q(X)_f$. We call a subscheme Z of X satisfying (a)-(c) a *codimension q closed subscheme of* (X, f), and the cycle (1.1.5.1) the *codimension q cycle on (X, f) determined by Z.*

If Z is a codimension q closed subscheme of (X, f), then the collection of closed subsets $\{\mathrm{supp}(Z([k]))\}$ forms a simplicial closed subset of X.

1.1.6. *Simplicial vector bundles.* Let $X: \Delta^{\leq N\mathrm{op}} \to \mathbf{Sch}$ be an N-truncated simplicial scheme. A *vector bundle of rank r on X* is a map of truncated simplicial schemes $p: E \to X$ together with the structure of a vector bundle of rank r on the n-simplices $p_n: E_n \to X_n$ for each $n \leq N$, such that, for each $g: [n] \to [m]$ in $\Delta^{\leq N}$, the map $E(g): E_m \to E_n$ is a map of vector bundles over $X(g)$, and in addition, the map $E_m \to X(g)^*(E_n)$ induced by $E(g)$ is an isomorphism. A line bundle on X is as usual a vector bundle of rank 1.

A map $f: E \to E'$ of vector bundles on X is a map over X of truncated simplicial schemes such that the map of n-simplices $f_n: E_n \to E'_n$. is a map of vector bundles on X_n (i.e., fiber-wise linear). A sequence of maps of vector bundles on X

$$E' \to E \to E''$$

is *exact* if the sequence of n-simplices

$$E'_n \to E_n \to E''_n$$

is exact for each n.

This defines the category $\mathcal{P}_X$ of vector bundles on X; we have the *Grothendieck group* $K_0(X) := K_0(\mathcal{P}_X)$ defined as usual as the free abelian group on the isomorphism classes of objects in $\mathcal{P}_X$, modulo relations

$$[E] = [E'] + [E'']$$

for each exact sequence

$$0 \to E' \to E \to E'' \to 0$$

in $\mathcal{P}_X$.

Ignoring the truncation at N, these notions are defined for a simplicial scheme as well.

There is a more sophisticated notion of $K_0(X)$ for X an N-truncated simplicial scheme, involving a homotopy limit over $\Delta^{\leq N}$, as in the definition in Appendix B of the K-theory of a functor X. We will ignore the question of whether the two definitions of $K_0(X)$ agree; we will always use the definition of K_0 given in this section when X is a (truncated) simplicial scheme.

1.1.7. LEMMA. *Let $X:\Delta^{\leq n\mathrm{op}} \to \mathcal{V}$ be a truncated simplicial object of $\mathcal{V}$, and let $(X, f_X):\Delta^{\leq n\mathrm{op}} \to \mathcal{L}(\mathcal{V})$ be the lifting (I.2.5.2.1). Suppose we have a rank N vector bundle $\pi: E \to X$ on the truncated simplicial scheme X, together with a section $s: X \to E$. Let Z_m be the subscheme of X_m determined by $s_m = 0$. Suppose that*

(a) *The subscheme Z_0 of X_0 defined by $s_0 = 0$ has pure codimension N on X_0.*
(b) *The cycle $|Z_0|$ on X_0 determined by Z_0 is in $\mathcal{Z}^q(X_0)_{f_{X_0}}$.*

Then the collection of subschemes $Z_m \subset X_m$ determines a codimension N closed subscheme Z of (X, f_X). In particular, the cycle $|Z_0|$ determines the element $|Z|$ of $\mathcal{Z}^q(X)$.

PROOF. Let $g: [m] \to [k]$ be a map in $\Delta^{\leq n}$. We have the commutative diagram

$$\begin{array}{ccc} E_k & \xrightarrow{E(g)} & E_m \\ s_k \uparrow\downarrow \pi_k & & \pi_m \downarrow\uparrow s_m \\ X_k & \xrightarrow[X(g)]{} & X_m. \end{array}$$

As the map $E(g)$ induces an isomorphism $\tilde{E}(g): E_k \to X(g)^*(E_m)$, we have the identity of subschemes of X_k:

$$Z_k = X(g)^{-1}(Z_m). \tag{1.1.7.1}$$

By our hypotheses (a) and (b), and the definition of f_X, it follows from (1.1.7.1), with $m = 0$, that each Z_k has pure codimension N on X_k. Since each Z_l is a local complete intersection in X_l, and X_l is smooth over S, it follows from (1.1.7.1) that

$$\mathrm{Tor}_p^{\mathcal{O}_{X_m}}(\mathcal{O}_{Z_m}, \mathcal{O}_{X_k}) = 0 \tag{1.1.7.2}$$

for all $p > 0$, where $\mathcal{O}_{X_k}$ is an $\mathcal{O}_{X_m}$-module via the morphism $X(g)$. Thus, from (1.1.7.1) and (1.1.7.2), we have the identity of cycles $|Z_k| = X(g)^*(|Z_m|)$. Taking $m = 0$, the assumptions (a) and (b) together with Lemma 1.2.2 of Chapter I imply that the cycle $|Z_k|$ is in $\mathcal{Z}^N(X_k)_{f_{X_k}}$ for each k. From the definitions in §1.1.5, this completes the proof. $\square$

We conclude this section with an elementary but useful extension of the homotopy property.

1.1.8. LEMMA. *Let X be an N-truncated simplicial object of $\mathcal{V}$, and let $p: E \to X$ be a vector bundle on X. Then $p^*: \mathbb{Z}_X \to \mathbb{Z}_E$ is an isomorphism in $\mathbf{D}^b_{\mathrm{mot}}(\mathcal{V})$.*

PROOF. The map p^* defines the map of distinguished triangles in $\mathbf{D}^b_{\mathrm{mot}}(\mathcal{V})$

$$(\mathbb{Z}_{X_N}[N] \to \mathbb{Z}_X \to \mathbb{Z}_{X^{\leq N-1}}) \xrightarrow{(p^*[N], p^*, p^*_{\leq N-1})} (\mathbb{Z}_{E_N}[N] \to \mathbb{Z}_E \to \mathbb{Z}_{E^{\leq N-1}}).$$

Induction on N reduces us to the case $N = 0$, i.e., X in $\mathcal{V}$. The Mayer-Vietoris property (Chapter I, §2.2.6) reduces to the case of a trivial bundle; the result then follows from the homotopy property (Chapter I, §2.2.1). $\square$

1.2. Chern classes of line bundles

We define the motivic first Chern class of a line bundle on a truncated simplicial scheme in $\mathcal{V}$.

1.2.1. *Line bundles.* Let $X:\Delta^{\leq n\mathrm{op}} \to \mathcal{V}$ be a truncated simplicial object in $\mathcal{V}$, and $p:L \to X$ a line bundle on X. By Lemma 1.1.8 the map $p^*:\mathbb{Z}_X(q) \to \mathbb{Z}_L(q)$ is an isomorphism in $\mathbf{D}^b_{\mathrm{mot}}(\mathcal{V})$. Applying Lemma 1.1.7 to the tautological section of p^*L over L, the zero subscheme of L determines the element $0_L \in \mathcal{Z}^1(L/S)$. We may then take the cycle class map (1.1.1.5) in $\mathbf{D}^b_{\mathrm{mot}}(\mathcal{V})$, $\mathrm{cl}^1_L(0_L):1 \to \mathbb{Z}_L(1)[2]$.

1.2.2. DEFINITION. Let $X:\Delta^{\leq n\mathrm{op}} \to \mathcal{V}$ be a truncated simplicial object in $\mathcal{V}$, and $p:L \to X$ a line bundle on X. The *first Chern class of* L,

$$c_1(L) \in H^2(X, \mathbb{Z}(1)),$$

is the element corresponding to the morphism $(p^*)^{-1} \circ \mathrm{cl}^1_L(0_L):1 \to \mathbb{Z}_X(1)[2]$ in $\mathbf{D}^b_{\mathrm{mot}}(\mathcal{V})$.

1.2.3. PROPOSITION. *The first Chern class satisfies*
(i) Functoriality: For $f:Y \to X$ *a morphism in* $\mathrm{s}.^{\leq n}\mathcal{V}$, *and* L *a line bundle on* X, *we have*

$$c_1(f^*(L)) = f^*(c_1(L)).$$

In addition, the simplicial first Chern class is stable in n, *i.e., for* $n' \leq n$, *we have*

$$\rho_{n',n}(c_1(L)) = c_1(L^{\leq n'}).$$

(ii) Additivity: For L_1 *and* L_2 *line bundles on* $X \in \mathrm{s}.^{\leq n}\mathcal{V}$, *we have*

$$c_1(L_1 \otimes L_2) = c_1(L_1) + c_1(L_2).$$

(iii) Compatibility with divisors: Let L *be a line bundle on* $X \in \mathrm{s}.^{\leq n}\mathcal{V}$, *and let* $s:X \to L$ *be a section such that the divisor* D_0 *of* $s_0:X_0 \to L_0$ *is in* $\mathcal{Z}^1(X_0)_{f_{X_0}}$. *Let* D *be the divisor on* (X, f_X) *determined by the codimension one subscheme* $s = 0$ *of* (X, f_X) *(see §1.1.5 and Lemma 1.1.7). Then*

$$c_1(L) = \mathrm{cl}^1_X(D).$$

PROOF. For (i), let $f_L:f^*(L) \to L$ be the canonical map of line bundles over the map f, giving the commutative diagram

$$\begin{array}{ccc} f^*(L) & \xrightarrow{f_L} & L \\ {\scriptstyle p_Y}\downarrow & & \downarrow{\scriptstyle p_X} \\ Y & \xrightarrow[f]{} & X. \end{array} \tag{1.2.3.1}$$

We have the identity of cycles $f_L^*(0_L) = 0_{f^*(L)}$, which, from Proposition 1.1.3(i), gives the identity $f_L^*(\mathrm{cl}^1_L(0_L)) = \mathrm{cl}^1_{f^*(L)}(0_{f^*(L)})$. This, together with the commutativity of (1.2.3.1) and the definition of c_1, proves the first part of (i). The second part follows by a similar argument.

For (ii), we have the map over X, $\pi:L_1 \times_X L_2 \to L_1 \otimes L_2$, defined on a fiber over a point x of X^m by sending (s,t) to the product st. We also have the projections $p_1:L_1 \times_X L_2 \to L_1$, $p_2:L_1 \times_X L_2 \to L_2$, and the maps $p:L_1 \times_X L_2 \to X$, $q:L_1 \otimes L_2 \to X$.

Let 0_1, 0_2 and 0_{12} denote the zero sections on L_1, L_2 and $L_1 \otimes L_2$, respectively. One easily checks that

$$\begin{aligned} \pi^*(0_{12}) &= 0_1 \times_X L_2 + L_1 \times_X 0_2, \\ p_1^*(0_2) &= L_1 \times_X 0_2, \\ p_2^*(0_1) &= 0_1 \times_X L_2. \end{aligned} \tag{1.2.3.2}$$

It follows immediately from the definition of c_1 and Proposition 1.1.3(i) that

$$c_1(L_1) = (p^*)^{-1}(p_2^*(\mathrm{cl}^1_{L_1}(0_1))); \ c_1(L_2) = (p^*)^{-1}(p_1^*(\mathrm{cl}^1_{L_2}(0_1))).$$

Since $c_1(L_1 \otimes L_2) = (q^*)^{-1}(\mathrm{cl}^1_{L_1\otimes L_2}(0_{12}))$, and the cycle class map cl^1 is additive and functorial, the relations (1.2.3.2) prove (ii).

Finally, for (iii), it suffices to prove that $\mathrm{cl}^1_L(p^*(D)) = \mathrm{cl}^1_L(0_L)$ in $H^2(L, \mathbb{Z}(1))$, where $p: L \to X$ is the structure map for the line bundle L.

We may form the sheaf $\mathcal{O}_X(D)$ on X with $[\mathcal{O}_X(D)]_m = \mathcal{O}_{X_m}(D_m)$, where D_m is the divisor of the section $s_m: X_m \to L_m$. We have the canonical map of sheaves on X

$$i_D: \mathcal{O}_X \to \mathcal{O}_X(D);$$

the resulting section s_D of L defines by Lemma 1.1.7 and the hypothesis of (iii) a codimension one subscheme $s_D = 0$ of (X, f_X) with divisor on (X, f_X) equal to D. Pulling back by p, we have the section s_1 of $p^*(L)$ over L with divisor $p^*(D)$ on (L, f_L). On the other hand the identity map on L determines the tautological section s_2 of $p^*(L)$ over L with divisor 0_L on (L, f_L).

Let $q: L \times_S \mathbb{A}^1_S \to L$ be the projection, and form the section $s_3 := tq^*(s_1) + (1-t)q^*(s_2)$ of $q^*(p^*(L))$ over $L \times_S \mathbb{A}^1_S$, where t is the coordinate on $\mathbb{A}^1_S$.

Let E be the divisor of the section s_3, and take a geometric point a of S. For a geometric point $b \neq 0$ of $\mathbb{A}^1$ and for $m \leq n$, the restriction of E_a to $L^m_a \times b$ is locally isomorphic to the graph of a function on X^m_a; in particular E^m_a is reduced, locally irreducible and pure codimension one on $(L^m \times_S \mathbb{A}^1_S)_a$. Let $i_0: L \to L \times_S \mathbb{A}^1_S$ and $i_1: L \to L \times_S \mathbb{A}^1_S$ be the 0 and 1 sections. From (Appendix A, Remark 2.3.4), and the identities

$$i_0^*(E_m) = 0_{L_m}, \qquad i_1^*(E_m) = p^*(D_m) \tag{1.2.3.3}$$

of divisors on L_m, it follows that E_m is in $\mathcal{Z}^1(L_m \times_S \mathbb{A}^1_S)_{\mathrm{id}\cup i_0 \cup i_1}$ for each m. From this it follows that the subscheme of $L \times_S \mathbb{A}^1_S$ defined by the section s_3 is a codimension one subscheme of $(L \times_S \mathbb{A}^1_S, f_{L\times_S \mathbb{A}^1_S} \cup i_0 \cup i_1)$, with corresponding cycle the divisor E.

By (1.2.3.3) and Proposition 1.1.3(i), we have the identity of divisors on (L, f_L)

$$i_0^*(E) - i_1^*(E) = 0_L - p^*(D). \tag{1.2.3.4}$$

By the homotopy axiom (Chapter I, Definition 2.1.4(a)), (1.2.3.4) implies that

$$\mathrm{cl}^1_L(p^*(D)) = \mathrm{cl}^1_L(0_L).$$

This gives the desired identity. □

1.3. Projective bundle formula and Chern classes

We use the splitting principle to define the motivic Chern classes of vector bundles, following the classic method of Grothendieck [**57**].

1.3.1. Let $X:\Delta^{\leq n\text{op}} \to \mathcal{V}$ be a truncated simplicial object of $\mathcal{V}$, let $p: E \to X$ be a rank $N+1$ vector bundle on X, and $q: \mathbb{P}(E) \to X$ the associated $\mathbb{P}^N$-bundle. We have the tautological surjection on $\mathbb{P}(E)$, $q^*(E) \to L_E$, where L_E is the line bundle associated to the invertible sheaf $\mathcal{O}(1)$ on $\mathbb{P}(E)$.

Let $\hat{X}$ be a simplicial closed subset of X, $\hat{P}$ the inverse image $q^{-1}(\hat{X})$. For each integer $i \geq 0$, we have the map

$$\alpha_i^E: \mathbb{Z}_{X,\hat{X}}(q-i)[-2i] \to \mathbb{Z}_{\mathbb{P}(E),\hat{P}}(q) \tag{1.3.1.1}$$

defined as the composition

$$\begin{aligned} \mathbb{Z}_{X,\hat{X}}(q-i)[-2i] &\cong \mathbb{Z}_{X,\hat{X}}(q-i)[-2i] \otimes 1 \\ &\xrightarrow{\mathrm{id}\otimes c_1(L_E)^i} \mathbb{Z}_{X,\hat{X}}(q-i)[-2i] \otimes \mathbb{Z}_{\mathbb{P}(E)}(i)[2i] \\ &\xrightarrow{\cup_{\mathbb{P}(E),X}} \mathbb{Z}_{X\times_S\mathbb{P}(E),\hat{X}\times_S\mathbb{P}(E)}(q), \\ &\xrightarrow{\Delta_E^*} \mathbb{Z}_{\mathbb{P}(E),\hat{P}}(q), \end{aligned}$$

where $\Delta_E: \mathbb{P}(E) \to \mathbb{P}(E)\times_S X$ is the map (id, q), and $\cup_{\mathbb{P}(E),X}$ is the map (I.2.5.6.4).

1.3.2. THEOREM [projective bundle formula]. *The map*

$$\sum_{i=0}^N \alpha_i^E : \oplus_{i=0}^N \mathbb{Z}_{X,\hat{X}}(q-i)[-2i] \to \mathbb{Z}_{\mathbb{P}(E),\hat{P}}(q)$$

is an isomorphism in $\mathbf{D}^b_{\mathrm{mot}}(\mathcal{V})$, natural in $(X,\hat{X},E)$.

PROOF. By the naturality of c_1 (Proposition 1.2.3(i)), the maps α_i^E are natural in the triple $(X,\hat{X},E)$; using the definition of $\mathbb{Z}_{X,\hat{X}}$ and $\mathbb{Z}_{\mathbb{P}(E),\hat{P}}$ as shifted cones (I.2.1.3.1), we reduce to the case $\hat{X} = \emptyset$.

We now reduce to the case of an object of $\mathcal{V}$ rather than a simplicial object, i.e., to the case $n = 0$. Suppose $n > 0$. We have the distinguished triangles in $\mathbf{D}^b_{\mathrm{mot}}(\mathcal{V})$

$$\mathbb{Z}_{X_n}(q)[n] \to \mathbb{Z}_X(q) \to \mathbb{Z}_{X^{\leq n-1}}(q) \to \mathbb{Z}_{X_n}(q)[n+1],$$
$$\mathbb{Z}_{\mathbb{P}(E)_n}(q)[n] \to \mathbb{Z}_{\mathbb{P}(E)}(q) \to \mathbb{Z}_{\mathbb{P}(E)^{\leq n-1}}(q) \to \mathbb{Z}_{\mathbb{P}(E)_n}(q)[n+1].$$

From the definition of the product maps (I.2.5.6.3), we see that the map α_i^E induces the map $\alpha_{i,n}^E: \mathbb{Z}_{X_n}(q-i)[n-2i] \to \mathbb{Z}_{\mathbb{P}(E)_n}(q)[n]$. By Proposition 1.1.3(ii), the definition Definition 1.2.2 of the first Chern class, and the naturality of c_1 (Proposition 1.2.3(i)), we have

$$\alpha_{i,n}^E = \alpha_i^{E_n}[n]. \tag{1.3.2.1}$$

Similarly, α_i^E induces the map $\alpha_i^{E,\leq n-1}: \mathbb{Z}_{X^{\leq n-1}}(q-i)[-2i] \to \mathbb{Z}_{\mathbb{P}(E)^{\leq n-1}}(q)$; the naturality of c_1 (Proposition 1.2.3(i)) implies

$$\alpha_i^{E,\leq n-1} = \alpha_i^{E^{\leq n-1}}. \tag{1.3.2.2}$$

By (1.3.2.1) and (1.3.2.2), we have the map of distinguished triangles

$$\begin{array}{ccccc} \mathbb{Z}_{X_n}(q)[n] & \longrightarrow & \mathbb{Z}_X(q) & \longrightarrow & \mathbb{Z}_{X^{\leq n-1}}(q) \\ \downarrow{\scriptstyle \sum_{i=0}^N \alpha_i^{E_n}[n]} & & \downarrow{\scriptstyle \sum_{i=0}^N \alpha_i^E} & & \downarrow{\scriptstyle \sum_{i=0}^N \alpha_i^{E^{\leq n-1}}} \\ \mathbb{Z}_{\mathbb{P}(E)_n}(q)[n] & \longrightarrow & \mathbb{Z}_{\mathbb{P}(E)}(q) & \longrightarrow & \mathbb{Z}_{\mathbb{P}(E)^{\leq n-1}}(q). \end{array}$$

By induction, this reduces us to the case $n = 0$, $X \in \mathcal{V}$.

Using the Mayer-Vietoris distinguished triangle (I.2.2.6.1), and the naturality of c_1, we reduce to the case of trivial E:

$$E \cong \operatorname{Spec}_{\mathcal{O}_X}(\mathcal{O}_X[X_0, \dots, X_n]),$$
$$\mathbb{P}(E) \cong \operatorname{Proj}_{\mathcal{O}_X}(\mathcal{O}_X[X_0, \dots, X_n]).$$

We let $i: 0_E \to \mathbb{P}(E)$ denote the subscheme of $\mathbb{P}(E)$ defined by $X_1 = \dots = X_N = 0$, and let $j: U \to \mathbb{P}(E)$ be the complement of 0_E.

We have the projection $\pi: U \to \mathbb{P}_X^{N-1}$ defined by

$$\pi(x_0 : \dots : x_N) = (x_1 : \dots : x_N);$$

this gives U the structure of a line bundle over $\mathbb{P}_X^{N-1}$. By Lemma 1.1.8, the map $\pi^*: \mathbb{Z}_{\mathbb{P}_X^{N-1}}(q) \to \mathbb{Z}_U(q)$ is an isomorphism; by induction, we have the isomorphism

$$\sum_{i=0}^{N-1} \alpha_i^{N-1}: \oplus_{i=0}^{N-1} \mathbb{Z}_X(q-i)[-2i] \to \mathbb{Z}_{\mathbb{P}_X^{N-1}}(q).$$

The naturality of c_1 implies the identity $j^* \circ \alpha_i = \pi^* \circ \alpha_i^{N-1}$, giving us the isomorphism

$$j^* \circ \sum_{i=0}^{N-1} \alpha_i: \oplus_{i=0}^{N-1} \mathbb{Z}_X(q-i)[-2i] \to \mathbb{Z}_U(q).$$

The homogeneous functions X_i define sections of L_E which are smooth over S; in fact, each subscheme of $\mathbb{P}(E)$ defined by an equation of the form $X_{i_1} = \dots = X_{i_s} = 0$ for $i_1 < \dots < i_s$ is smooth over S. Thus, by (Appendix A, Remark 2.3.4), and Proposition 1.2.3(iii), we have the identity

(1.3.2.3) $$c_1(L_E)^N = \mathrm{cl}_{\mathbb{P}(E)}^N(0_E).$$

We have the object $\mathbb{Z}_{\mathbb{P}(E),0_E}(q)$ (I.2.1.3.1) of $\mathbf{C}^b_{\mathrm{mot}}(\mathcal{V})$, defined as the shifted cone of the morphism $j^*: \mathbb{Z}_{\mathbb{P}(E)}(q) \to \mathbb{Z}_U(q)$; the cone sequence thus gives the distinguished triangle in $\mathbf{D}^b_{\mathrm{mot}}(\mathcal{V})$

$$\mathbb{Z}_{\mathbb{P}(E),0_E}(q) \xrightarrow{i_{\mathbb{P}(E),0_E}} \mathbb{Z}_{\mathbb{P}(E)}(q) \xrightarrow{j^*} \mathbb{Z}_U(q).$$

We have the Gysin isomorphism (I.2.2.5.1)

$$\cup[0_E] \circ q^*: \mathbb{Z}_X(q-N)[-2N] \to \mathbb{Z}_{\mathbb{P}(E),0_E}(q);$$

the identity $i_{\mathbb{P}(E),0_E} \circ (\cup[0_E] \circ q^*) = \alpha_N$ follows from (1.3.2.3). This gives us the map of distinguished triangles:

$$\begin{array}{ccccc} \mathbb{Z}_{\mathbb{P}(E),0_E}(q) & \xrightarrow{i_{\mathbb{P}(E),0_E}} & \mathbb{Z}_{\mathbb{P}(E)}(q) & \xrightarrow{j^*} & \mathbb{Z}_U(q) \\ \uparrow{\scriptstyle \cup[0_E]\circ q^*} & & \uparrow{\scriptstyle \sum_{i=0}^N \alpha_i} & & \uparrow{\scriptstyle j^*\circ\sum_{i=0}^{N-1} \alpha_i} \\ \mathbb{Z}_X(q-N)[-2N] & \longrightarrow & \oplus_{i=0}^N \mathbb{Z}_X(q-i)[-2i] & \longrightarrow & \oplus_{i=0}^{N-1} \mathbb{Z}_X(q-i)[-2i]. \end{array}$$

As the two maps on the ends are isomorphisms, the map in the middle is an isomorphism as well, completing the proof. □

1.3.3. *Splitting principle.* Let X be an n-truncated simplicial object of $\mathcal{V}$, and $p: E \to X$ be a vector bundle on X. We have the *flag variety* $q: \mathcal{F}l(E) \to X$, gotten by forming the projective bundle $q_1: \mathbb{P}(E) \to X$, taking the kernel E_1 of the canonical surjection $q_1^*: E \to \mathcal{O}(1)$, forming $\mathbb{P}(E_1)$, and so on, until the resulting kernel has rank 1. The pull-back q^*E then has the canonical filtration

$$E = E^0 \supset E^1 \supset \ldots \supset E^N \supset E^{N+1} = 0 \tag{1.3.3.1}$$

with E^i/E^{i+1} a line bundle on $\mathcal{F}(E)$ for each i. We may then pull back further, to the bundle of splittings of (1.3.3.1), $\tilde{q}: \mathcal{S}p(E) \to X$, giving the isomorphism $\tilde{q}^*(E) \cong \oplus_{i=1}^N L_i$, with the L_i line bundles on $\mathcal{S}p(E)$.

The bundle $r: \mathcal{S}p(E) \to \mathcal{F}l(E)$ is a sequence of Zariski torsors for the vector bundle $\mathcal{H}om(E^{i-1}/E^i, E^i)$; using Mayer-Vietoris and the homotopy property, one proves that the map $r^*: \mathbb{Z}_{\mathcal{F}l(E)} \to \mathbb{Z}_{\mathcal{S}p(E)}$ is an isomorphism. From the projective bundle formula, the map $q^*: \mathbb{Z}_X \to \mathbb{Z}_{\mathcal{F}l(E)}$ is *injective*, hence, so is the map $\tilde{q}^*: \mathbb{Z}_X \to \mathbb{Z}_{\mathcal{S}p(E)}$. This enables us to reduce proofs of identities among characteristic classes of vector bundles to the case of sums of line bundles. We may use a similar construction to replace any finite collection of exact sequences with split exact sequences among direct sums of line bundles.

1.3.4. DEFINITION. Let $X: \Delta^{\leq n\mathrm{op}} \to \mathcal{V}$ be a truncated simplicial object of $\mathcal{V}$, let $\hat{X}$ be a closed simplicial subscheme of X, and let $E \to X$ be a vector bundle of rank N on X. Let $q: \mathbb{P}(E) \to X$ be the associated projective bundle with tautological quotient line bundle L_E, and let $\zeta = c_1(L_E)$. The *Chern classes of E* are the elements $c_i(E) \in H^{2i}(X, \mathbb{Z}(i))$ satisfying

$$\sum_{i=0}^{N} (-1)^i q^*(c_i(E))\zeta^{N-i} = 0, \ c_0(E) = 1. \tag{1.3.4.1}$$

By Theorem 1.3.2, the $c_i(E)$ exist and are uniquely determined by the identity (1.3.4.1). We define the total Chern class $c(E)$ to be the sum

$$c(E) = \sum_{i=0}^{N} c_i(E).$$

1.3.5. THEOREM. *The Chern classes satisfy*
(i) Naturality: Let $f: Y \to X$ be a morphism in $\mathrm{s}.^{\leq n}\mathcal{V}$, E a vector bundle on X. Then

$$f^*(c(E)) = c(f^*(E)).$$

Similarly, if we have X in $\mathrm{s}.^{\leq n}\mathcal{V}$, E a vector bundle on X, and $0 \leq n' < n$, then

$$\rho_{n',n}(c(E)) = c(E^{\leq n'}),$$

where $\rho_{n',n}: \mathbb{Z}_X(q) \to \mathbb{Z}_{X^{\leq n'}}(q)$ is the map (1.1.1.7).
(ii) Normalization: The two definitions (Definition 1.2.2 and Definition 1.3.4) of the first Chern class of a line bundle agree.

PROOF. The first part of (i) follows from the naturality of the first Chern class Proposition 1.2.3(i), and the naturality of the projective bundle isomorphism of Theorem 1.3.2; the second part follows by using Proposition 1.2.3(i), and noting that the projective bundle isomorphism of Theorem 1.3.2 is compatible with truncation. The statement (ii) follows from the defining relation $c_1(L) - \zeta = 0$ for c_1 of a line bundle $L \to X$ in Definition 1.3.4, and the identification of the tautological line bundle L_L on $\mathbb{P}(L) = X$ with L. $\square$

1.3.6. REMARK. Suppose we have a morphism of base schemes $p: T \to S$ as in (Chapter I, §2.3), an object X of $\mathrm{s.}^{\leq n}\mathcal{V}$ and a vector bundle $E \to X$. Suppose that $\mathcal{W}$ is a subcategory of $\mathbf{Sm}_T$ for which $\mathcal{DM}(\mathcal{W})$ is defined and with $\mathcal{W} \supset p^*\mathcal{V}$. Essentially the same proof as for Theorem 1.3.5(i), using the properties of pull-back $p^*: \mathcal{DM}(\mathcal{V}) \to \mathcal{DM}(\mathcal{W})$ given in (Chapter I, §2.3), shows that the Chern classes are functorial in this setting: $p^*(c_q(E)) = c_q(p^*(E))$, where p^*E is the pull-back bundle $E \times_S T \to X \times_S T$.

1.3.7. THEOREM [Whitney product formula]. *Let X be an n-truncated simplicial object in $\mathcal{V}$, and*

$$0 \to E_1 \to E \to E_2 \to 0$$

an exact sequence of vector bundles on X. Then

$$c(E) = c(E_1)c(E_2).$$

PROOF. Using the splitting principle of §1.3.3, we may assume that $E = E_1 \oplus E_2$ and that E_1 and E_2 are direct sums of line bundles. This reduces us to showing, for line bundles $L_1, \ldots, L_N$ on X, that

$$c(\oplus_{k=1}^N L_k) = \prod_{k=1}^N (1 + c_1(L_k)).$$

Let $E = \oplus_{k=1}^N L_k$, and let $q: \mathbb{P} \to X$ be the projective bundle $\mathbb{P}(E)$. We have the canonical surjection $\pi: q^*E \to \mathcal{O}(1)$, giving the maps $p_k: q^*L_k \to \mathcal{O}(1)$, $i = 1, \ldots, N$, defined as the composition

$$q^*L_k \hookrightarrow q^*E \xrightarrow{\pi} \mathcal{O}(1).$$

Twisting by $q^*L_k^{-1}$ gives the sections $s_k: \mathcal{O}_{\mathbb{P}} \to \mathcal{O}(1) \otimes q^*L_k^{-1}$, $i = 1, \ldots, N$. Let D_k be the subscheme of $\mathbb{P}$ defined by the vanishing of s_k.

Locally on X, the divisors $D_1, \ldots D_N$ are independent hyperplanes in $\mathbb{P}$ (which is a Zariski locally trivial $\mathbb{P}^{N-1}$-bundle); in particular, the D_k are smooth S-schemes, hence in $\mathcal{V}$. Thus, the cycles D_k are in $\mathcal{Z}^1(\mathbb{P}/S)$; it follows from Proposition 1.2.3(iii) that $\mathrm{cl}^1_{\mathbb{P}}(D_k) = \zeta - q^*(c_1(L_k))$, where $\zeta = c_1(\mathcal{O}(1))$. Since the intersection $D_1 \cap \ldots \cap D_N$ is empty on $\mathbb{P}$, we have by Proposition 1.1.3(iii)

$$\begin{aligned} 0 &= \mathrm{cl}^N(D_1 \cap \ldots \cap D_N) \\ &= \mathrm{cl}^1(D_1) \cup \ldots \cup \mathrm{cl}^1(D_N) \\ &= \prod_{k=1}^N (\zeta - q^*(c_1(L_k))) \\ &= \zeta^N + \sum_{k=1}^N (-1)^k \zeta^{N-k} q^*(\sigma_k), \end{aligned}$$

where σ_k is the kth symmetric function in the Chern classes $c_1(L_1), \ldots, c_1(L_N)$. By the defining relation Definition 1.3.4 for the Chern classes of E, this shows

$c_k(E) = \sigma_k$, i.e.

$$c(\oplus_{k=1}^N L_k) = \prod_{k=1}^N (1 + c_1(L_k)),$$

as desired. □

We give a few immediate consequences of the product formula.

Recall from §1.1.6 the definition of $K_0(X)$ for X an n-truncated simplicial scheme.

1.3.8. COROLLARY. *Let X be an n-truncated simplicial object of $\mathcal{V}$. Then sending a vector bundle E on X to $c_q(E) \in H^{2q}(X, \mathbb{Z}(q))$ descends to a map (of sets)*

$$c_q : K_0(X) \to H^{2q}(X, \mathbb{Z}(q)).$$

PROOF. Form the group

$$1 + \widehat{H^{2*}(X, \mathbb{Z}(*))}^+ := 1 \times \prod_{q \geq 1} H^{2q}(X, \mathbb{Z}(q))$$

with group law

$$(1 + \sum_q x_q) + (1 + \sum_q y_q) = (1 + \sum_q x_q)(1 + \sum_q y_q),$$

where the multiplication is as formal series. For a vector bundle E, let

$$\hat{c}(E) := 1 + \sum_q c_q(E) \in 1 + \widehat{H^{2*}(X, \mathbb{Z}(*))}^+.$$

The Whitney product formula implies that $\hat{c}(E) = \hat{c}(E') + \hat{c}(E'')$ if there is an exact sequence

$$0 \to E' \to E \to E'' \to 0,$$

hence $\hat{c}$ descends to a group homomorphism

$$\hat{c} : K_0(X) \to 1 + \widehat{H^{2*}(X, \mathbb{Z}(*))}^+.$$

□

We have an extension of Proposition 1.2.3(iii) to vector bundles of arbitrary rank.

1.3.9. COROLLARY. *Let $p : E \to X$ be a vector bundle of rank r on a truncated simplicial object X in $\mathcal{V}$, and let $0_E \subset E$ denote the 0-section. Let $s : X \to E$ be a section satisfying the conditions (a) and (b) of Lemma 1.1.7. Then*

$$\mathrm{cl}^r_X(s^*(|0_E|)) = c_r(E)$$

in $H^{2r}(X, \mathbb{Z}(r))$.

PROOF. Since $s^*(|0_E|) = |s^{-1}(0_E)|$, it follows from Lemma 1.1.7 that $s^*(|0_E|)$ is in $\mathcal{Z}^r(X/S)$. By the splitting principle, we may pull back to the flag bundle $\mathcal{F}(E)$ over X, and, by homotopy, we may pull back further to the affine bundle of splittings of the canonical flag in E over $\mathcal{F}(E)$, so we may assume that E is a direct sum of line bundles, $E \cong \oplus_{i=1}^r L_i$, $p_i : L_i \to X$. Let $q_i : E \to L_i$ be the projection.

Let $\tilde{s}$ be the tautological section of p^*E over E. As in the proof of Proposition 1.2.3(iii), we have

$$p^*(\mathrm{cl}^r_X(s^*(|0_E|))) = \mathrm{cl}^r_E(\tilde{s}^*(|0_{p^*E}|))).$$

Letting s_i be the tautological section of $p_i^*L_i$ over L_i, we have

$$\tilde{s}^*(|0_{p^*E}|) = \cap_{i=1}^r q_i^*(s_i^*(|0_{p_i^*L_i}|)).$$

By Proposition 1.2.3, Theorem 1.3.5, and Theorem 1.3.7, together with Proposition 3.5.7 of Chapter I, we thus have

$$\begin{aligned} p^*(\mathrm{cl}^r_X(s^*(|0_E|))) &= \mathrm{cl}^r_E(\tilde{s}^*(|0_{p^*E}|)) \\ &= \mathrm{cl}^1_E(q_1^*s_1^*(|0_{p_1^*L_1}|) \cup \ldots \cup \mathrm{cl}^1_E(q_r^*s_r^*(|0_{p_r^*L_r}|) \\ &= c_1(q_1^*p_1^*L_1) \cup \ldots \cup c_1(q_r^*p_r^*L_1) \\ &= c_r(p^*E) \\ &= p^*(c_r(E)). \end{aligned}$$

Since $p^*: \mathbb{Z}_X \to \mathbb{Z}_E$ is an isomorphism in $\mathcal{DM}(\mathcal{V})$ by homotopy and Mayer-Vietoris, we thus have

$$\mathrm{cl}^r_X(s^*(|0_E|)) = c_r(E).$$

□

1.4. Chern classes for higher K-theory

We use the method of Gillet [**46**] to define motivic Chern classes for higher K-theory.

1.4.1. *Representable sheaves.* We refer to the constructions, notations, and results of Chapter II, §1.5.2 and Lemma 1.5.3; in particular, for X in $\mathcal{V}$, we have the category of Zariski open subsets of X, $\mathrm{Zar}(X)$, and the subcategory $\mathbf{C}^b_{\mathrm{mot}}(\mathrm{Zar}(X)) := \mathbf{C}^b_{\mathrm{mot}}(\mathrm{Zar}(X, \mathrm{id}_X))$ of $\mathbf{C}^b_{\mathrm{mot}}(\mathcal{V})$, which contains the category of hyper-resolutions $\mathbf{HR}_{\mathbb{Z}_U(q)}$ for all open subschemes U of X. For an abelian presheaf S on X which takes disjoint unions to direct sums, we have the functor

$$\mathbf{C}^b(S): \mathbf{C}^b_{\mathrm{mot}}(\mathrm{Zar}(X)) \to \mathbf{C}^b(\mathbf{Ab}).$$

As a special case of this construction, we may take the presheaf S to be the restriction to $\mathrm{Zar}(X)$ of the free abelian group on a representable functor, $S(U) := \mathbb{Z}[\mathrm{Hom}_{\mathcal{L}(\mathcal{V})}((U, \mathrm{id}_U), (Z, g))]$, which we denote by $H^X_{(Z,g)}$. Sending a morphism of $h: (U, \mathrm{id}_U) \to (Z, g)$ to the map $h^*: \mathbb{Z}_Z(q)_g \to \mathbb{Z}_U(q)$ defines the map

$$(1.4.1.1) \qquad \xi(Z, g)(U): H^X_{(Z,g)}(U) \to \mathrm{Hom}_{\mathcal{A}_{\mathrm{mot}}}(\mathbb{Z}_Z(q)_g, \mathbb{Z}_U(q))$$

which is natural in both (Z, g) and in $U \in \mathrm{Zar}(X)$.

We have as well the representable functor $H^X_Z(U) := \mathbb{Z}[\mathrm{Hom}_{\mathcal{V}}(U, Z)]$. The subcategory of $\mathcal{L}(\mathcal{V})$ of maps $(Z, g) \to (Z, g')$ over the identity on Z is filtering; indeed $(Z, g \cup g')$ dominates (Z, g) and (Z, g'). We have as well the identity

$$(1.4.1.2) \qquad H^X_Z = \varinjlim_g H^X_{(Z,g)}.$$

Suppose we have non-degenerate simplicial object $(Z, g): \Delta^{\mathrm{op}}_{\mathrm{n.d.}} \to \mathcal{L}(\mathcal{V})$ of $\mathcal{L}(\mathcal{V})$. We may then form the complex of presheaves on X, $C^*_X((Z, g); \mathbb{Z})$, by setting

$C^k_X((Z,g);\mathbb{Z})(U) := H^X_{(Z([-k]),g([-k]))}(U)$, with differential the alternating sum induced by the maps in Z. We may make a similar construction for Z an n-truncated non-degenerate simplicial object of $\mathcal{L}(\mathcal{V})$. If Z is a non-degenerate simplicial object, or an n-truncated non-degenerate simplicial object of $\mathcal{V}$, we have the complex of presheaves on X, $C^*_X(Z;\mathbb{Z})$, with $C^k_X(Z;\mathbb{Z})(U) := H^X_{Z([-k])}(U)$. The identity (1.4.1.2) gives

$$C^*_X(Z;\mathbb{Z}) = \varinjlim_g C^*_X((Z,g);\mathbb{Z}). \tag{1.4.1.3}$$

Let Z be a non-degenerate simplicial object of $\mathcal{V}$, and let $(Z,g)\colon \Delta^{\mathrm{op}}_{\mathrm{n.d.}} \to \mathcal{L}(\mathcal{V})$ be a lifting to a non-degenerate simplicial object of $\mathcal{L}(\mathcal{V})$, giving the associated motive $\mathbb{Z}_Z(q)^{*\leq n}_g$ for each $n \geq 0$, as in (Chapter I, §2.5.4).

If Γ is in $\mathbf{C}^b_{\mathrm{mot}}(\mathrm{Zar}(X),q)$, the natural transformation (1.4.1.1) gives the natural map of complexes

$$\xi_n(Z,g)(\Gamma)\colon \mathbf{C}^b(C^{*\geq -n}_X((Z,g);\mathbb{Z}))(\Gamma) \to \mathrm{Hom}_{\mathbf{C}^b_{\mathrm{mot}}(\mathcal{V})}(\mathbb{Z}_Z(q)^{*\leq n}_g, \Gamma).$$

Now suppose that Γ is in the subcategory $\mathbf{HR}_{\mathbb{Z}_X(q)}$ of $\mathbf{C}^b_{\mathrm{mot}}(\mathrm{Zar}(X),q)$. Since the augmentation $\epsilon\colon \mathbb{Z}_X(q) \to \Gamma$ is an isomorphism in $\mathbf{D}^b_{\mathrm{mot}}(\mathcal{V})$ (see Chapter II, Lemma 1.4.2(iii)), we have the natural map

(1.4.1.4)

$$H^n(\mathbf{C}^b(C^{*\geq -n}_X((Z,g);\mathbb{Z}))(\Gamma)) \xrightarrow{\epsilon^{-1}\circ H^n(\xi(Z,g)(\Gamma))} \mathrm{Hom}_{\mathbf{D}^b_{\mathrm{mot}}(\mathcal{V})}(\mathbb{Z}_Z(q)^{*\leq n}_g, \mathbb{Z}_X(q)[n]).$$

As the map $\mathrm{id}^*_Z : \mathbb{Z}_Z(q)^{*\leq n}_g \to \mathbb{Z}_Z(q)^{*\leq n}_{g'}$ induced by a map $\mathrm{id}_Z\colon (Z,g') \to (Z,g)$ is an isomorphism in $\mathbf{D}^b_{\mathrm{mot}}(\mathcal{V})$, the map (1.4.1.4) defines via (1.4.1.3) the natural map

$$H^n(\mathbf{C}^b(C^{*\geq -n}_X(Z;\mathbb{Z}))(\Gamma)) \to \mathrm{Hom}_{\mathbf{D}^b_{\mathrm{mot}}(\mathcal{V})}(\mathbb{Z}_Z(q)^{*\leq n}_g, \mathbb{Z}_X(q)[n]).$$

Taking the limit over $\mathbf{HR}_{\mathbb{Z}_X(q)}$ and applying (Chapter II, Lemma 1.5.3) gives the natural map

$$\Xi_n(Z)\colon \mathbb{H}^m_{\mathrm{Zar}}(X, \tilde{C}^{*\geq -n}_X(Z;\mathbb{Z})) \to \mathrm{Hom}_{\mathbf{D}^b_{\mathrm{mot}}(\mathcal{V})}(\mathbb{Z}_Z(q)^{*\leq n}, \mathbb{Z}_X(q)[m]). \tag{1.4.1.5}$$

Let $c\colon 1 \to \mathbb{Z}_Z(q)^{*\leq n}[a]$ be a morphism in $\mathbf{D}^b_{\mathrm{mot}}(\mathcal{V})$. Composing with the map (1.4.1.5) (suitably shifted) gives the natural map

$$\Xi_n(Z)\circ c\colon \mathbb{H}^m_{\mathrm{Zar}}(X, \tilde{C}^{*\geq -n}_X(Z;\mathbb{Z})) \to \mathrm{Hom}_{\mathbf{D}^b_{\mathrm{mot}}(\mathcal{V})}(1, \mathbb{Z}_X(q)[a+m]).$$

This gives us the natural map

$$\mathbb{H}^m_{\mathrm{Zar}}(X, \tilde{C}^{*\geq -n}_X(Z;\mathbb{Z})) \otimes \mathrm{Hom}_{\mathbf{D}^b_{\mathrm{mot}}(\mathcal{V})}(1, \mathbb{Z}_Z(q)^{*\leq n}[a]) \xrightarrow{\Psi_n(Z)} H^{a+m}(X, \mathbb{Z}(q)). \tag{1.4.1.6}$$

Since X has finite Zariski cohomological dimension, we have the identity

$$\mathbb{H}^m_{\mathrm{Zar}}(X, \tilde{C}^{*\geq -n}_X(Z;\mathbb{Z})) = \mathbb{H}^m_{\mathrm{Zar}}(X, \tilde{C}^*_X(Z;\mathbb{Z}))$$

for all n sufficiently large (depending on m). As

$$H^a(Z,\mathbb{Z}(q)) = \varprojlim_{(\mathbb{N},\leq)^{\mathrm{op}}} [n \mapsto \mathrm{Hom}_{\mathbf{D}^b_{\mathrm{mot}}(\mathcal{V})}(1, \mathbb{Z}_Z(q)^{*\leq n}[a])]$$

by definition (see Chapter I, §2.5.5), the map (1.4.1.6) gives us the natural map

$$\Psi(Z)\colon \mathbb{H}^m_{\mathrm{Zar}}(X, \tilde{C}^*_X(Z;\mathbb{Z})) \otimes H^a(Z,\mathbb{Z}(q)) \to H^{a+m}(X,\mathbb{Z}(q)). \tag{1.4.1.7}$$

1.4.2. *Homology and motivic cohomology.* We refer the reader to §1.1.1-§1.1.3 of Appendix B for notions related to classifying schemes and the general linear group.

For a scheme X, we have the sheaf of groups $\mathcal{GL}_N := \mathcal{GL}_N/X$ defined as the sheafification of the presheaf $U \mapsto \mathrm{GL}_N(\Gamma(U, \mathcal{O}_X))$, and the sheaf of simplicial sets $\mathcal{BGL}_N/X$ defined similarly. For a ring A, we have the simplicial abelian group $\mathbb{Z}\mathrm{BGL}_N(A)$ with k-simplices being the free abelian on $\mathrm{BGL}_N(A)([k])$; applying this construction to the sheaf $\mathcal{BGL}_N/X$ gives the presheaf of simplicial abelian groups $\mathbb{Z}\mathcal{BGL}_N/X$, and the associated complex of presheaves $C^*(\mathcal{BGL}_N/X; \mathbb{Z})$, $C^k(\mathcal{BGL}_N/X; \mathbb{Z})(U) = \mathbb{Z}[\mathcal{BGL}_N/X([-k])(U)]$.

The stalk $C^*(\mathcal{BGL}_N/X; \mathbb{Z})_x$ is the complex computing the homology of the discrete group $\mathrm{GL}_N(\mathcal{O}_{X,x})$:

$$H^{-p}(C^*(\mathcal{BGL}_N/X; \mathbb{Z})_x) = H_p(\mathrm{GL}_N(\mathcal{O}_{X,x}); \mathbb{Z}).$$

We define $H_p(X, \mathcal{GL}_N; \mathbb{Z})$ by

$$H_p(X, \mathcal{GL}_N; \mathbb{Z}) := \mathbb{H}^{-p}_{\mathrm{Zar}}(X, \tilde{C}^*(\mathcal{BGL}_N/X; \mathbb{Z})).$$

Suppose that X is an S-scheme. We have the simplicial S-scheme BGL_N/S, which satisfies the flatness conditions of §1.4.1. We have in addition the identity of complexes of presheaves on X, $C^*(\mathcal{BGL}_N/X; \mathbb{Z}) = C^*_X(\mathrm{BGL}_N/S; \mathbb{Z})$, so the map (1.4.1.7) gives us the natural map

$$\Psi_N : H_p(X, \mathcal{GL}_N; \mathbb{Z}) \otimes H^a(\mathrm{BGL}_N/S, \mathbb{Z}(q)) \to H^{a-p}(X, \mathbb{Z}(q)). \tag{1.4.2.1}$$

1.4.3. *Stabilization.* We have the stabilization map $i_N : \mathrm{GL}_N/S \to \mathrm{GL}_{N+1}/S$ defined by

$$i_N(g) = \begin{pmatrix} g & 0 \\ 0 & 1 \end{pmatrix}.$$

This induces stabilization maps $\mathrm{B}i_N : \mathrm{BGL}_N/S \to \mathrm{BGL}_{N+1}/S$. For X in $\mathcal{V}$, this gives stabilization maps $C^*(\mathcal{BGL}_N/X; \mathbb{Z}) \to C^*(\mathcal{BGL}_{N+1}/X; \mathbb{Z})$, and stabilization maps on hypercohomology, $H_*(X, \mathcal{GL}_N; \mathbb{Z}) \to H_*(X, \mathcal{GL}_{N+1}; \mathbb{Z})$.

We set

$$H^a(\mathrm{BGL}/S, \mathbb{Z}(q)) := \varprojlim H^a(\mathrm{BGL}_N/S, \mathbb{Z}(q)),$$
$$H_p(X, \mathcal{GL}; \mathbb{Z}) := \varinjlim H_p(X, \mathcal{GL}_N; \mathbb{Z}).$$

The maps (1.4.2.1) for varying N thus give the map

$$\Psi : H_p(X, \mathcal{GL}; \mathbb{Z}) \otimes H^a(\mathrm{BGL}/S, \mathbb{Z}(q)) \to H^{a-p}(X, \mathbb{Z}(q)). \tag{1.4.3.1}$$

1.4.4. *Universal Chern classes.* From Appendix B, §1.1.3, we have the universal rank N vector bundle $p_n : E_N \to \mathrm{BGL}_N/S$.

We have $\mathrm{B}i_N^*(E_{N+1}) \cong E_N \oplus 1$, where 1 denotes the trivial line bundle. Thus, by the Whitney product formula (Theorem 1.3.7) and the stability of Chern classes (Theorem 1.3.5(i)) we have

$$\mathrm{B}i_N^*(c(E_{N+1}^{\leq n})) = c(E_N^{\leq n}),$$
$$\rho_{n',n}(c(E_N^{\leq n})) = c(E_N^{\leq n'}),$$

for all $n \geq n' \geq 0$. Thus, the qth Chern class $c_q(E_N^{\leq n})$ for $n = 1, 2, \ldots$ determines the element $c_q(E_N) \in H^{2q}(\mathrm{BGL}_N/S, \mathbb{Z}(q))$, and the classes $c_q(E_N)$ for $N = 1, 2 \ldots$ determines the element $c_q(E) \in H^{2q}(\mathrm{BGL}/S, \mathbb{Z}(q))$.

1.4.5. We apply the map (1.4.3.1) to the universal Chern class $c_q(E)$; we denote the map $\Psi(-\otimes c_q(E))$ by

(1.4.5.1) $$Hc^{q,2q-p}: H_p(X, \mathcal{GL}; \mathbb{Z}) \to H^{2q-p}(X, \mathbb{Z}(q)).$$

From Appendix B, §2.2.2, we have the Hurewicz map

(1.4.5.2) $$h_p^X : K_p(X) \to H_p(X, \mathcal{GL}; \mathbb{Z}).$$

Composing (1.4.5.1) with (1.4.5.2) gives the Chern class map

$$c^{q,2q-p}: K_p(X) \to H^{2q-p}(X, \mathbb{Z}(q)).$$

1.4.6. REMARK. Let $E \to X$ be a rank r-vector bundle on a scheme $X \in \mathcal{V}$. We may take a trivializing open cover $\mathcal{U} = \{U_0, \dots, U_N\}$ for E; a choice of trivializing isomorphisms $\psi_i : E_{|U_i} \to U_i \times \mathbb{A}^r$ gives the transition maps $g_{ij} := \psi_i \circ \psi_j^{-1} : U_i \cap U_j \to \mathrm{GL}_r/S$ which extend to give the map of simplicial schemes over S:

$$g: N\mathcal{U} \to \mathrm{BGL}_r/S$$
$$g_{i_0,\dots,i_n} = (g_{i_0,i_1}, g_{i_1,i_2}, \dots, g_{i_{n-1},i_n})_{|U_{i_0}\cap\dots\cap U_{i_n}}.$$

The isomorphisms ψ_i then give the isomorphism $p_{\mathcal{U}}^* E \cong g^* E_r$, where $p_{\mathcal{U}} : N\mathcal{U} \to X$ is the augmentation.

We have the truncated Chern classes $c_i(E_r^{\leq n}) \in H^{2i}(\mathbb{Z}_{\mathrm{BGL}_r/S}(i)^{*\leq n})$. The map g defines the map $g^* : \mathbb{Z}^{*\leq n}_{\mathrm{BGL}_r/S} \to \mathbb{Z}^{*\leq n}_{\mathcal{U}}$. We may then pull back the c_i via g^* to give classes $g^*(c_i(E_r^{\leq n})) \in H^{2i}(\mathbb{Z}_{\mathcal{U}}(i)^{*\leq n})$. On the other hand, the map $p_{\mathcal{U}}$ induces the isomorphism in $\mathcal{DM}(\mathcal{V})$, $p_{\mathcal{U}}^* : \mathbb{Z}_X(i) \to \mathbb{Z}_{\mathcal{U}}(i)^{*\leq n} = \mathbb{Z}_{\mathcal{U}}(i)$, for all $n \geq N+1$. Thus, we get the elements

$$(p_{\mathcal{U}}^*)^{-1} \circ g^*(c_i(E_r^{\leq n})) \in H^{2i}(\mathbb{Z}_X(i)) = H^{2i}(X, \mathbb{Z}(i)).$$

It follows from the naturality of the Chern classes that

$$(p_{\mathcal{U}}^*)^{-1} \circ g^*(c_i(E_r^{\leq n})) = c_i(E); \qquad i = 0, 1, \dots$$

for all $n \geq N+1$.

From this it follows that $c^{q,2q}$ agrees with the Chern class c_q.

1.4.7. *Chern classes for diagrams.* We proceed to extend the construction of Chern classes given in §1.4.5 to diagrams in $\mathcal{V}$. We use the notions and notations of (Appendix B, §2.1.3 and Remark 2.2.3) and Chapter I, §2.7.

Let I be the category associated to a finite partially ordered set and let $X : I \to \mathcal{V}^+$ be a functor. We have the lifting $(X, f_X) : I \to \mathcal{L}(\mathcal{V}^+)$, and the motive of X, $\mathbb{Z}_X$, defined as in (Chapter I, §2.7.2) as the non-degenerate homotopy limit

$$\mathbb{Z}_X = \operatorname*{holim}_{I,\mathrm{n.d.}}(i \mapsto \mathbb{Z}_{X(i)}(0)_{f_X(i)}).$$

Define $H_p(X, \mathcal{GL}_N; \mathbb{Z})$ as the hypercohomology

$$H_p(X, \mathcal{GL}_N; \mathbb{Z}) := \mathbb{H}^{-p}_{\mathrm{Zar}}(X, \tilde{C}^*(\mathcal{BGL}_N/X; \mathbb{Z})),$$

and set

$$H_p(X, \mathcal{GL}; \mathbb{Z}) := \varinjlim_N H_p(X, \mathcal{GL}_N; \mathbb{Z}).$$

We have the Hurewicz map (from Appendix B, Remark 2.2.3)

(1.4.7.1) $$h_{X,N} : K_p(X) \to H_p(X, \mathcal{GL}_N; \mathbb{Z})$$

for all N sufficiently large (stable in N).

We note that a bounded above complex of presheaves S^* on X (which takes disjoint unions to direct sums) defines the functor $\mathbf{C}^b(S^*)\colon \mathbf{HR}_{\mathbb{Z}_X(q)} \to \mathbf{C}^-(\mathbf{Ab})$, just as in (Chapter II, §1.5.2). Using the distinguished triangles of (Chapter I, §2.7.3) and (Part II, (III.3.3.1.3)), we extend Lemma 1.5.3 of Chapter II to give a canonical isomorphism

$$\varinjlim_{\Gamma\in\mathbf{HR}_{\mathbb{Z}_X(q)}} H^n(\mathbf{C}^b(S^*)(\Gamma)) \cong \mathbb{H}^n_{\mathrm{Zar}}(X, \tilde{S}^*).$$

For a simplicial scheme Z satisfying the flatness conditions of §1.4.1, we may then use the construction of that section to give a natural map

$$\Psi(Z)\colon \mathbb{H}^m_{\mathrm{Zar}}(X, \tilde{C}^*_X(Z;\mathbb{Z})) \otimes H^a(Z, \mathbb{Z}(q)) \to H^{a+m}(X, \mathbb{Z}(q)) \tag{1.4.7.2}$$

extending the natural map (1.4.1.7).

Taking $Z = \mathrm{BGL}_N/S$, stabilizing and evaluating at the universal Chern classes $c_q(E)$ gives us the map (as in (1.4.5.1))

$$Hc^{q,2q-p}\colon H_p(X, \mathcal{GL};\mathbb{Z}) \to H^{2q-p}(X, \mathbb{Z}(q)). \tag{1.4.7.3}$$

Composing with the Hurewicz map (1.4.7.1) gives the Chern class

$$c^{q,2q-p}\colon K_p(X) \to H^{2q-p}(X, \mathbb{Z}(q)). \tag{1.4.7.4}$$

1.4.8. EXAMPLES. (i) Take I to be the category $* > 0 < 1$, and U be an open subscheme of X, with complement $\hat{X}$. We have the functor $(X, \hat{X})\colon I \to \mathcal{V}^+$ with

$$(X,\hat{X})(0) = U,\ (X,\hat{X})(1) = X,\ (X,\hat{X})(*) = *, (X,\hat{X})(0<1) = j_U\colon U \to X.$$

Then $\mathbb{Z}_{(X,\hat{X})}(q)$ is the motive with support $\mathbb{Z}_{X,\hat{X}}(q)$ and the K-group $K_n\big((X,\hat{X})\big)$ is the *K-group with support* $K_n^{\hat{X}}(X)$, defined as the homotopy group π_{n+1} of the homotopy fiber of the map $j_U^*\colon \mathrm{BQ}\mathcal{P}_X \to \mathrm{BQ}\mathcal{P}_U$

The Chern classes (1.4.7.4) give the Chern classes with support:

$$c_{\hat{X}}^{q,2q-p}\colon K_p^{\hat{X}}(X) \to H_{\hat{X}}^{2q-p}(X, \mathbb{Z}(q)),$$

compatible with the Chern classes without support via the "forget the support" maps $K_p^{\hat{X}}(X) \to K_p(X)$, $H_{\hat{X}}^{2q-p}(X,\mathbb{Z}(q)) \to H^{2q-p}(X,\mathbb{Z}(q))$.

(ii) We have the *n-cube*, $<n>$ (see Chapter I, §2.6.1), the opposite of the category of subsets of $\{1,\ldots,n\}$; take I to be the category $<n>* := <n> \cup *$, (the pointed n-cube) with $* > J$ for each non-empty $J \subset \{1,\ldots,n\}$.

Given X in $\mathcal{V}$ and a collection of closed subschemes $D_1,\ldots,D_n$ of X, such that each intersection $D_J := \cap_{j\in J\subset\{1,\ldots,n\}} D_j$ is in $\mathcal{V}$, we then have the functor

$$\begin{aligned}(X; D_1,\ldots,D_n)\colon &<n>* \to \mathcal{V}^+\\ &J \mapsto D_J;\quad * \mapsto *.\end{aligned}$$

The resulting object $\mathbb{Z}_{(X;D_1,\ldots,D_n)}(q)$ of $\mathcal{DM}(\mathcal{V})$ is isomorphic to the *motive of X relative to* $D_1,\ldots,D_n$ (see Chapter I, §2.6.6 and §1.5.1 below). The *motivic cohomology of X relative to* $D_1,\ldots,D_n$ is defined as

$$H^{2q-p}(X; D_1,\ldots,D_n, \mathbb{Z}(q)) := \mathrm{Hom}_{\mathcal{DM}(\mathcal{V})}(1, \mathbb{Z}_{(X;D_1,\ldots,D_n)}(q)[2q-p]).$$

We have the *K-groups of X relative to* $D_1,\ldots,D_n$, $K_n(X; D_1,\ldots,D_n)$, defined as

$$K_n(X; D_1,\ldots,D_n) := \pi_{n+1}(\operatorname*{holim}_{<n>*} J \mapsto \mathrm{BQ}P_{(X;D_1,\ldots,D_n)(J)}).$$

The Chern classes (1.4.7.4) give the Chern classes for relative K-groups:

$$c^{q,2q-p}_{(X;D_1,\dots,D_n)}\colon K_p(X;D_1,\dots,D_n)\to H^{2q-p}(X;D_1,\dots,D_n,\mathbb{Z}(q)).$$

(iii) We may combine (i) and (ii): suppose we have a closed subset $\hat{X}$ of X with complement $j:U\to X$, and closed subschemes $D_1,\dots,D_n$ such that each intersection D_I is in $\mathcal{V}$. This gives the pointed $n+1$-cube $(X;D_1,\dots,D_n)^{\hat{X}}$ defined via the map of pointed n-cubes $j\colon(U;D_1\cap U,\dots,D_n\cap U)\to(X;D_1,\dots,D_n)$. The homotopy limit of $\mathrm{B}QP_{(X;D_1,\dots,D_n)^{\hat{X}}}$ over $<n+1>*$ then has homotopy groups

$$K_n^{\hat{X}}(X;D_1,\dots,D_n):=\pi_{n+1}(\operatorname*{holim}_{<n+1>*} J\mapsto \mathrm{B}QP_{(X;D_1,\dots,D_n)^{\hat{X}}(J)}),$$

the *K-groups of X relative to $D_1,\dots,D_n$, with support in $\hat{X}$*. The *motive of X relative to $D_1,\dots,D_n$, with support in $\hat{X}$* is the object $\mathbb{Z}_{(X;D_1,\dots,D_n)^{\hat{X}}}(q)$ of $\mathcal{DM}(\mathcal{V})$. The Chern classes (1.4.7.4) then give the Chern classes for relative K-groups with support:

$$c^{q,2q-p}_{(X;D_1,\dots,D_n),\hat{X}}\colon K_p^{\hat{X}}(X;D_1,\dots,D_n)\to H^{2q-p}_{\hat{X}}(X;D_1,\dots,D_n,\mathbb{Z}(q)).$$

(iv) Mod-n Chern classes (see [**26**]). Let S^m/n be the *mod n Moore space*: S^2/n is the CW-complex gotten by attaching the boundary of a 2-disk to S^1 by the n to 1 cover $S^1\to S^1$, and S^m/n is the $m-2$-fold suspension of S^2/n. For a pointed space $(X,*)$ one defines the *mod-n homotopy groups of X* by

$$\pi_p(X;\mathbb{Z}/n):=[(S^p/n,*),(X,*)];\quad n\geq 3,$$

where $[-,-]$ means pointed homotopy classes of pointed maps. The addition is given by a co-H-space structure on S^p/n, similar to that of S^p. If X is an H-space, the addition in H gives the same group structure, so one may extend the definition to $\pi_2(X;\mathbb{Z}/n)$. One has the fundamental short exact sequence

$$0\to\pi_p(X)/n\to\pi_p(X;\mathbb{Z}/n)\to{}_n\pi_{p-1}(X)\to 0;\quad p\geq 2,\tag{1.4.8.1}$$

where ${}_n\pi_{p-1}(X)$ is the n-torsion subgroup of $\pi_{p-1}(X)$.

One defines the mod-n K-groups of a scheme X by

$$K_p(X;\mathbb{Z}/n):=\pi_{p+1}(\mathrm{B}QP_X;\mathbb{Z}/n)$$

for $p\geq 1$. For $p=0$, one deloops $\mathrm{B}QP_X$ and takes $\pi_2(-,\mathbb{Z}/n)$.

For a simplicial abelian group S, the Dold-Kan isomorphism (see e.g. [**95**, Chapter V]) gives a natural isomorphism $\pi_p(S;\mathbb{Z}/n)\cong H^{-p}(C^*(S)\otimes^L\mathbb{Z}/n)$, where $C^*(S)$ is the chain complex associated to S, and $C^*(S)\otimes^L\mathbb{Z}/n$ is the cone

$$C^*(S)\otimes^L\mathbb{Z}/n:=\operatorname{cone}(C^*(S)\xrightarrow{\times n}C^*(S)).$$

The sequence (1.4.8.1) is then just the sequence one gets from breaking up the cohomology sequence for the distinguished triangle

$$C^*(S)\otimes^L\mathbb{Z}/n[-1]\to C^*(S)\xrightarrow{\times n}C^*(S)\to C^*(S)\otimes^L\mathbb{Z}/n$$

into short exact sequences. Thus, the construction of the Hurewicz map (1.4.5.2) gives the mod-n Hurewicz map

$$K_p(X;\mathbb{Z}/n)\to H_p(X,\mathcal{GL}_N;\mathbb{Z}/n)\tag{1.4.8.2}$$

(at least for $p\geq 2$); the extension to diagrams of schemes (1.4.7.1) gives rise to the mod-n Hurewicz map for X a functor as in §1.4.7.

For Γ in $\mathbf{C}^b_{\mathrm{mot}}(\mathcal{V})$, we have the object $\Gamma \otimes^L \mathbb{Z}/n$ defined in Chapter I, §2.2.8 as

$$\Gamma \otimes^L \mathbb{Z}/n := \mathrm{cone}(\Gamma \xrightarrow{\times n} \Gamma),$$

and the mod-n motivic cohomology

$$H^p(\Gamma, \mathbb{Z}/n(q)) := H^p(\Gamma \otimes^L \mathbb{Z}/n, \mathbb{Z}(q)).$$

Taking $\Gamma = \mathbb{Z}_X(0)$ defines the mod-n motivic cohomology of X

$$H^p(X, \mathbb{Z}/n(q)) := H^p(\mathbb{Z}_X(0) \otimes^L \mathbb{Z}/n, \mathbb{Z}(q)).$$

The map (1.4.1.7) thus gives the mod-n version

$$\Psi(Z) \otimes^L \mathbb{Z}/n \colon \mathbb{H}^m_{\mathrm{Zar}}(X, \tilde{C}^*_X(Z;\mathbb{Z}) \otimes^L \mathbb{Z}/n) \otimes H^a(Z, \mathbb{Z}(q)) \to H^{a+m}(X, \mathbb{Z}/n(q)).$$

Thus, the construction of the map (1.4.5.1) (or (1.4.7.3) for X a diagram of schemes) gives the map $Hc^{q,2q-p} \colon H_p(X, \mathcal{GL}; \mathbb{Z}/n) \to H^{2q-p}(X, \mathbb{Z}/n(q))$. Composing with the Hurewicz map gives the mod-n Chern class

$$c^{q,2q-p} \colon K_p(X; \mathbb{Z}/n) \to H^{2q-p}(X, \mathbb{Z}/n(q)); \quad p \geq 2,$$

for X in $\mathcal{V}$, as well as for $X \colon I \to \mathcal{V}^+$ a functor as in §1.4.7. The constructions of (i), (ii) and (iii) thus also have their mod-n versions.

1.4.9. Proposition. *Let I be the category associated to a finite partially ordered set and let $X \colon I \to \mathcal{V}^+$ be a functor.*
(i) For $p \geq 1$, the Chern class maps (1.4.7.4) are additive.
(ii) The Chern class maps $c_q := c^{q,2q} \colon K_0(X) \to H^{2q}(X, \mathbb{Z}(q))$ satisfy the Whitney product formula

$$c(x+y) = c(x) \cup c(y).$$

(iii) Let J be the category associated to a finite partially ordered set, let $\iota \colon J \to I$ be a functor, and let $Y \colon J \to \mathcal{V}^+$ be a functor. Let $f \colon Y \to X \circ \iota$ be a map of functors, inducing the pull-back maps $f^* \colon K_p(X) \to K_p(Y)$ and $f^* \colon H^{2q-p}(X, \mathbb{Z}(q)) \to H^{2q-p}(Y, \mathbb{Z}(q))$. Then the diagram

$$\begin{array}{ccc} K_p(X) & \xrightarrow{c^{q,2q-p}} & H^{2q-p}(X, \mathbb{Z}(q)) \\ {\scriptstyle f^*}\downarrow & & \downarrow{\scriptstyle f^*} \\ K_p(Y) & \xrightarrow[c^{q,2q-p}]{} & H^{2q-p}(Y, \mathbb{Z}(q)) \end{array}$$

commutes.
(iv) Let $g \colon T \to S$ be a map of reduced schemes, and let $\mathcal{W}$ be a full subcategory of $\mathbf{Sm}^{\mathrm{ess}}_T$ containing $g^*\mathcal{V}$, such that $\mathcal{DM}(\mathcal{W})$ is defined, giving the pull-back functor $g^* \colon \mathcal{DM}(\mathcal{V}) \to \mathcal{DM}(\mathcal{W})$ (I.2.3.1.1). Let $g^*X \colon I \to \mathcal{W}$ be the functor $g^*X(i) := X \times_S T$, giving the map of functors from I to $\mathbf{Sch}^+$, $p_1 \colon g^*X \to X$. Then the diagram

$$\begin{array}{ccc} K_p(X) & \xrightarrow{c^{q,2q-p}_X} & H^{2q-p}(X, \mathbb{Z}(q)) \\ {\scriptstyle p_1^*}\downarrow & & \downarrow{\scriptstyle g^*} \\ K_p(g^*X) & \xrightarrow[c^{q,2q-p}_{g^*X}]{} & H^{2q-p}(g^*X, \mathbb{Z}(q)) \end{array}$$

commutes.

PROOF. The addition in $K_p(X)$ is induced by the direct sum operation in $\mathcal{P}_X$, which gives rise to the H-space structure in ΩBQP_X. Thus, the addition in $K_p(X)$ agrees with the group law as a homotopy group $K_p(X) = \pi_p(\Omega BQP_X)$ for $p > 0$. The Hurewicz map (1.4.7.1) is a group homomorphism for $p > 0$, and the map (1.4.7.3) is a group homomorphism for all p, which proves (i).

Formula (ii) follows from the Whitney product formula for the Chern classes of the universal direct sum bundle $p_1^*E_N \oplus p_2^*E_M$ on $\mathrm{BGL}_N/S \times_S \mathrm{BGL}_M/S$ (see Theorem 1.3.7).

The functoriality (iii) follows directly from the definitions and the functoriality of the Hurewicz map. The functoriality (iv) is proved similarly, using the functoriality of the universal Chern class with respect to base-change described in Remark 1.3.6. □

1.4.10. REMARK. The mod-n Chern classes

$$c^{q,2q-p}: K_p(X;\mathbb{Z}/n) \to H^{2q-p}(X, \mathbb{Z}/n(q))$$

satisfy the functorialities of Proposition 1.4.9; they are also additive for $p \geq 3$ by the same reasoning as in Proposition 1.4.9. For $p = 2$, $c^{q,2q-p}$ is additive if n is odd. If n is even, then $c^{q,2q-2}$ is not in general additive; this is due to the fact that the mod n Hurewicz map is *not* in general a group homomorphism for even n! This phenomenon and its consequences is discussed in [**127**], where the consequences for the étale Chern classes are given in detail. Exactly the same consequences hold for the motivic Chern classes. For instance, the motivic mod 2 Chern classes $c^{q,2q-2}: K_2(X;\mathbb{Z}/2) \to H^{2q-2}(X, \mathbb{Z}/2(q))$ satisfy

$$c^{q,2q-2}(a+b) = c^{q,2q-2}(a) + c^{q,2q-2}(b) + c^{q,2q-2}(\partial a \cup \partial b),$$

where $\partial: K_2(X;\mathbb{Z}/2) \to K_1(X)$ is the map in the universal coefficient sequence

$$0 \to K_2(X)/2 \to K_2(X;\mathbb{Z}/2) \xrightarrow{\partial} {}_2K_1(X) \to 0$$

[**127**, Proposition 2.4].

1.5. Localization and relativization

The relative K-theory with support, and the relative motivic cohomology with support give rise to the fundamental *relativization sequences* and *localization sequences*; we now show that they are compatible via the Chern classes described in Example 1.4.8. To describe these sequences, we first require a few generalities on iterated homotopy fibers and iterated cones.

1.5.1. *Homotopy fiber sequences.* We refer the reader to (Part II, Chapter III, Section 3) for the notions in this paragraph related to homotopy limits, and to [**115**] and [**95**] for the basic notions of algebraic topology and simplicial sets.

Recall the category $I := * > 0 < 1$ of Example 1.4.8(i). Let $f: (X, *) \to (Y, *)$ be a map of pointed simplicial sets. We may then form the functor $\tilde{f}: I^{\mathrm{op}} \to \mathbf{Top}^*$ by

$$\tilde{f}(0) = Y, \ \tilde{f}(1) = X, \ \tilde{f}(*) = *;$$

$\tilde{f}(0 < 1)$ is the map f and $\tilde{f}(0 < *)$ is the inclusion of $*$ as the base-point of Y.

Let $[0,1]$ denote the simplicial set $\mathrm{Hom}(-, [1]): \Delta^{\mathrm{op}} \to \mathbf{Sets}$, with inclusions

$$0: * \to [0,1]; \quad 1: * \to [0,1]$$

given by the inclusions $0 \mapsto 0$ and $0 \mapsto 1$ of $[0]$ in $[1]$. The *simplicial path space* $P(f)$ of the map f is defined as the fiber product

$$P(f) := X \times_Y \mathcal{H}om([0,1], Y),$$

over the diagram

$$\begin{array}{ccc} & & \mathcal{H}om([0,1],Y) \\ & & \downarrow{\scriptstyle 1^*} \\ X & \xrightarrow[f]{} & Y, \end{array}$$

and the *simplicial homotopy fiber* of f, defined as the fiber product

$$\mathrm{Fib}(f) := P(f) \times_Y \mathcal{H}om\big(([0,1],0),(Y,*)\big),$$

over the diagram

$$\begin{array}{ccc} & & \mathcal{H}om\big(([0,1],0),(Y,*)\big) \\ & & \downarrow{\scriptstyle 1^*} \\ P(f) & \xrightarrow[0^* \circ p_2]{} & Y. \end{array}$$

One constructs directly from the definition of the homotopy limit a natural isomorphism of $\mathrm{holim}_{I^{\mathrm{op}}} \tilde{f}$ with $\mathrm{Fib}(f)$.

Suppose X and Y are *fibrant* (see e.g. [**25**, V,§3]). As the functor holim transforms pointwise weak equivalences of fibrant simplicial sets to weak equivalences [**25**, XI, 5.6], one can replace X and Y with the singular complex of their geometric realizations, without changing the weak equivalence class of $\mathrm{holim}_{I^{\mathrm{op}}} \tilde{f}$. Thus, it follows that the geometric realization of $\mathrm{holim}_{I^{\mathrm{op}}} \tilde{f}$ is weakly equivalent to the usual homotopy fiber of the map induced by f on the geometric realizations of X and Y (see [**93**, Chapter III]).

Similarly, suppose we have an n-cube of (fibrant) pointed simplicial sets

$$X\colon <n> \to \mathbf{s.Sets}^*.$$

We may form the *iterated homotopy fiber of* X, $\mathrm{Fib}^n(X)$, inductively, by writing X as a map of $n-1$-cubes $f\colon X^+ \to X^-$ as in Chapter I, §2.6.4, taking the induced map on the iterated homotopy fibers

$$\mathrm{Fib}^{n-1} f\colon \mathrm{Fib}^{n-1} X^+ \to \mathrm{Fib}^{n-1} X^-,$$

and then taking the homotopy fiber. Sending $*$ to the one-point space $*$ extends the functor X to the functor

$$X* : <n>* \to \mathbf{s.Sets}^*,$$

and one has the natural isomorphism of $\mathrm{holim}_{<n>*} X*$ with $\mathrm{Fib}^n(X)$.

If we let

$$\Omega X^- \colon <n> \to \mathbf{s.Sets}^*$$

be the functor which is the one-point simplicial set $*$ on all $I \subset \{1, \ldots, n\}$ with $n \in I$, and $X^-(I)$ on all $I \subset \{1, \ldots, n-1\}$, then

$$(\Omega X^-)^+ \colon <n-1> \to \mathbf{s.Sets}^*$$

is the constant functor with value $*$, and $(\Omega X^-)^- = X^-$. Thus, we have the isomorphism

$$\operatorname*{holim}_{<n>*} \Omega X^- \cong \Omega \operatorname*{holim}_{<n-1>*} X^-,$$

where the *loop space* ΩY of a pointed simplicial set Y is the homotopy fiber of the inclusion of the base-point $* \to Y$.

The maps $* \to X^+$ and $(\Omega X^-)^- = X^- \xrightarrow{\text{id}} X^-$ give the map $\iota_X : (\Omega X^-)^- \to X$, yielding the sequence of functors

$$(\Omega X^-)^- * \circ i_n \to X * \circ i_n \to X^+ * \to X^- *, \tag{1.5.1.1}$$

where $i_n : <n-1>* \to <n>*$ is the inclusion functor $i_n(I) := I \cup \{n\}$.

Applying holim to (1.5.1.1), we have the sequence of simplicial sets

$$\operatorname*{holim}_{<n>*} \Omega X^- * \to \operatorname*{holim}_{<n>*} X* \to \operatorname*{holim}_{<n-1>*} X^+ * \to \operatorname*{holim}_{<n-1>*} X^- *, \tag{1.5.1.2}$$

which is isomorphic to the (weak) fibration sequence

$$\Omega \operatorname*{holim}_{<n-1>*} X^- * \to \operatorname*{holim}_{<n>*} X* \to \operatorname*{holim}_{<n-1>*} X^+ * \to \operatorname*{holim}_{<n-1>*} X^- * .$$

Now let $\mathcal{A}$ be an additive category, and $X : <n> \to \mathbf{C}^b(\mathcal{A})$ a functor. Extend X to $X* : <n>* \to \mathbf{C}^b(\mathcal{A})$ by sending $*$ to 0. We may inductively form the iterated shifted cone of X, $\operatorname{cone}^n(X)[-n]$, by viewing X as a map $f : X^+ \to X^-$, and taking the shifted cone

$$\operatorname{cone}^n(X)[-n] :=$$
$$\operatorname{cone}\big(\operatorname{cone}^{n-1}(X^+)[-(n-1)] \xrightarrow{\operatorname{cone}^{n-1}(f)[-(n-1)]} \operatorname{cone}^{n-1}(X^-)[-(n-1)]\big)[-1].$$

If we take the non-degenerate homotopy limit $\operatorname{holim}_{<n>* \text{ n.d.}} X*$, we construct as above a natural isomorphism in $\mathbf{K}^b(\mathcal{A})$

$$\operatorname*{holim}_{<n>* \text{ n.d.}} X* \cong \operatorname{cone}^n(X)[-n].$$

Similarly, the sequence

(1.5.1.3)

$$\operatorname*{holim}_{<n>* \text{ n.d.}} \Omega X^- * \to \operatorname*{holim}_{<n>* \text{ n.d.}} X* \to \operatorname*{holim}_{<n-1>* \text{ n.d.}} X^+ * \to \operatorname*{holim}_{<n-1>* \text{ n.d.}} X^- *$$

is isomorphic in $\mathbf{K}^b(\mathcal{A})$ to the shifted cone sequence

$$\operatorname{cone}^{n-1}(X^-)[-(n-1)][-1] \to \operatorname{cone}^n(X)[-n]$$
$$\to \operatorname{cone}^{n-1}(X^+)[-(n-1)] \to \operatorname{cone}^{n-1}(X^-)[-(n-1)].$$

1.5.2. *Localization and relativization sequences.* We apply the sequences (1.5.1.2) and (1.5.1.3) of §1.5.1 to relative K-theory and relative motivic cohomology.

Let X be in $\mathcal{V}$, let $Y_1, \dots, Y_n$ be subschemes of X with $Y_I := \cap_{i \in I} Y_i$ in $\mathcal{V}$ for each subset I of $\{1, \dots, n\}$. Let W be a closed subset of X, giving the relative K-theory with support $K_p^W(X; Y_1, \dots, Y_n)$ defined in Example 1.4.8 as the homotopy groups of a homotopy limit over the pointed $n+1$-cube $<n+1>*$ of a certain functor to simplicial sets. Similarly, we have the relative motivic cohomology with support $H^*_W(X; Y_1, \dots, Y_n, \mathbb{Z}(q))$ defined either via an iterated shifted cone (Chapter I, §2.6.2-§2.6.6), or via a homotopy limit over $<n+1>*$ using the isomorphism mentioned in §1.5.1.

The fibration sequence (1.5.1.2) gives the long exact *relativization sequence*

$$(1.5.2.1)\quad \to K_{p+1}^{W\cap Y_n}(Y_n;Y_1\cap Y_n,\dots,Y_{n-1}\cap Y_n)$$
$$\to K_p^W(X;Y_1,\dots,Y_n)\to K_p^W(X;Y_1,\dots,Y_{n-1})$$
$$\xrightarrow{i_n^*} K_p^{W\cap Y_n}(Y_n;Y_1\cap Y_n,\dots,Y_{n-1}\cap Y_n)\to,$$

where the map i_n^* is induced by pull-back with respect to the inclusion $i_n\colon Y_n\to X$.

Choosing a different face of $n+1$-cube for the "last variable" gives the long exact *localization sequence*

$$(1.5.2.2)\quad \to K_{p+1}(X\setminus W;Y_1\setminus W,\dots,Y_n\setminus W)\to K_p^W(X;Y_1,\dots,Y_n)$$
$$\to K_p(X;Y_1,\dots,Y_n)\xrightarrow{j^*} K_p(X\setminus W;Y_1\setminus W,\dots,Y_n\setminus W)\to,$$

where j^* is induced by pull-back with respect to the open immersion $j\colon X\setminus W\to X$.

More generally, if W and F are closed subsets of X, the same construction gives the localization sequence with support

$$(1.5.2.3)\quad \to K_{p+1}^{F\setminus W}(X\setminus W;Y_1\setminus W,\dots,Y_n\setminus W)\to K_p^{W\cup F}(X;Y_1,\dots,Y_n)$$
$$\to K_p^F(X;Y_1,\dots,Y_n)\xrightarrow{j^*} K_p^{F\setminus W}(X\setminus W;Y_1\setminus W,\dots,Y_n\setminus W)\to .$$

We have the analogous sequences for motivic cohomology: The relativization sequence

$$(1.5.2.4)\quad \to H^{p-1}_{W\cap Y_n}(Y_n;Y_1\cap Y_n,\dots,Y_{n-1}\cap Y_n,\mathbb{Z}(q))$$
$$\to H^p_W(X;Y_1,\dots,Y_n,\mathbb{Z}(q))\to H^p_W(X;Y_1,\dots,Y_{n-1},\mathbb{Z}(q))$$
$$\xrightarrow{i_n^*} H^p_{W\cap Y_n}(Y_n;Y_1\cap Y_n,\dots,Y_{n-1}\cap Y_n,\mathbb{Z}(q))\to,$$

and the localization sequence

$$(1.5.2.5)$$
$$\to H^{p-1}_{F\setminus W}(X\setminus W;Y_1\setminus W,\dots,Y_n\setminus W,\mathbb{Z}(q))\to H^p_{W\cup F}(X;Y_1,\dots,Y_n,\mathbb{Z}(q))$$
$$\to H^p_F(X;Y_1,\dots,Y_n,\mathbb{Z}(q))\xrightarrow{j^*} H^p_{F\setminus W}(X\setminus W;Y_1\setminus W,\dots,Y_n\setminus W,\mathbb{Z}(q))\to$$

(the localization sequence without support in F is obtained by taking $F=X$).

As the sequences (1.5.2.1)-(1.5.2.5) are constructed by taking the long exact homotopy, resp. motivic cohomology, associated to the fibration sequence (1.5.1.2), resp. cone sequence (1.5.1.3), the functoriality of the Chern classes described in Proposition 1.4.9(iii) and (iv) imply that the K-theory sequences are compatible with the corresponding motivic cohomology sequences via the appropriate Chern classes defined in Example 1.4.8. For example, we have the commutative ladder

$$\begin{array}{ccc}
\downarrow & & \downarrow \\
K_{p+1}^{W\cap Y_n}(Y_n; Y_{1\le *<n}\cap Y_n) & \xrightarrow{c^{q,2q-p-1}} & H^{2q-p-1}_{W\cap Y_n}(Y_n; Y_{1\le *<n}\cap Y_n, \mathbb{Z}(q)) \\
\downarrow & & \downarrow \\
K_p^W(X; Y_{1\le *\le n}) & \xrightarrow{c^{q,2q-p}} & H^{2q-p}_W(X; Y_{1\le *\le n}, \mathbb{Z}(q)) \\
\downarrow & & \downarrow \\
K_p^W(X; Y_{1\le *<n}) & \xrightarrow{c^{q,2q-p}} & H^{2q-p}_W(X; Y_{1\le *<n}) \\
{\scriptstyle i_n^*}\downarrow & & \downarrow{\scriptstyle i_n^*} \\
K_p^{W\cap Y_n}(Y_n; Y_{1\le *<n}\cap Y_n) & \xrightarrow{c^{q,2q-p}} & H^{2q-p}_{W\cap Y_n}(Y_n; Y_{1\le *<n}\cap Y_n, \mathbb{Z}(q)) \\
\downarrow & & \downarrow
\end{array}$$

Here $Y_{1\le *\le n}$ denotes the collection $Y_1, \ldots, Y_n$, and $Y_{1\le *<n}$ stands for the collection $Y_1, \ldots, Y_{n-1}$.

2. Push-forward

In this section, we define the push-forward morphism in $\mathcal{DM}(\mathcal{V})$ associated to a projective morphism in $\mathcal{V}$, and verify the properties normally satisfied by projective push-forward in a reasonable cohomology theory: functoriality, compatibility with pull-back in cartesian squares, and the projection formula. We verify the compatibility with cycle classes for the case of a closed embedding; the compatibility of projective push-forward with cycle classes for an arbitrary projective morphism is also valid for $S = \operatorname{Spec} k$, where k is a field. We conclude the section with an extension of push-forward to diagrams of schemes in $\mathcal{V}$.

2.1. The Gysin morphism

We use the method of "deformation to the normal bundle" from [**8**] to define the Gysin morphism associated to a closed embedding.

2.1.1. *The split case.* Let $f: P \to Z$ be a smooth map in $\mathcal{V}$, with section $s: Z \to P$, let $\hat{Z}$ be a closed subset of Z, and let $\hat{P} = s(\hat{Z})$.

We have the cycle class with support $\mathrm{cl}^d_{P,s(Z)}(|s(Z)|) \in H^{2d}_{s(Z)}(P, \mathbb{Z}(d))$, *cf.* (I.3.5.2.7). Using the cup products with support (Chapter I, §2.2.11) we have the map

$$(2.1.1.1) \qquad \cup[s(Z)]_{s(Z)} \circ f^*: \mathbb{Z}_{Z,\hat{Z}}(-d)[-2d] \to \mathbb{Z}_{P,\hat{P}}$$

defined as the composition

$$\mathbb{Z}_{Z,\hat{Z}}(-d)[-2d] \xrightarrow{f^*} \mathbb{Z}_{P,f^{-1}(\hat{Z})}(-d)[-2d] \xrightarrow{(-)\cup_P^{s(Z),f^{-1}(\hat{Z})}\mathrm{cl}^d_{P,s(Z)}(|s(Z)|)} \mathbb{Z}_{P,\hat{P}}.$$

Let $j: Z \setminus \hat{Z} \to Z$ and $j_P: P \setminus f^{-1}(\hat{Z}) \to P$ be the inclusions. The maps (I.2.2.5.1)

$$\cup[s(Z)] \circ f^*: \mathbb{Z}_Z(-d)[-2d] \to \mathbb{Z}_{P,s(Z)}$$
$$\cup[s(Z \setminus \hat{Z})] \circ f^*: \mathbb{Z}_{Z\setminus\hat{Z}}(-d)[-2d] \to \mathbb{Z}_{P\setminus f^{-1}(\hat{Z}),s(Z\setminus\hat{Z})}$$

are isomorphisms by the remarks of Chapter I, §2.2.5. The triple

$$\alpha := (\cup[s(Z)]_{s(Z)} \circ f^*, \cup[s(Z)] \circ f^*, \cup[s(Z \setminus \hat{Z})] \circ f^*)$$

gives the map of the distinguished localization triangles in $\mathcal{DM}(\mathcal{V})$ (see Chapter I, §2.2.10)

$$(\mathbb{Z}_{Z,\hat{Z}}(-d)[-2d] \to \mathbb{Z}_Z(-d)[-2d] \xrightarrow{j^*} \mathbb{Z}_{Z\setminus\hat{Z}}(-d)[-2d] \to \mathbb{Z}_{Z,\hat{Z}}(-d)[1-2d])$$
$$\xrightarrow{\alpha} (\mathbb{Z}_{P,\hat{P}} \to \mathbb{Z}_{P,s(Z)} \xrightarrow{j_P^*} \mathbb{Z}_{P\setminus f^{-1}(\hat{Z}),s(Z\setminus\hat{Z})} \to \mathbb{Z}_{P,\hat{P}}[1]),$$

hence the map (2.1.1.1) is an isomorphism.

We often omit the support ${}_{s(\hat{Z})}$ from the notation; the meaning will be clear from the context.

2.1.2. *The deformation diagram.* Let $i: Z \to X$ be a closed codimension d embedding in $\mathcal{V}$, and let $\hat{Z}$ a be a closed subset of Z. Let $q: Y \to X \times_S \mathbb{A}^1_S$ be the blow-up of $X \times_S \mathbb{A}^1_S$ along $Z \times 1$, $\hat{Y}$ the proper transform of $\hat{Z} \times_S \mathbb{A}^1_S$ to Y, P the full inverse image of $Z \times 1$ in Y, $\hat{P}$ the closed subset $P \cap \hat{Y}$ of P. Let $i_0: X \to Y$ be the composition of the inclusion

$$\mathrm{id}_X \times j_0: X \to X \times_S \mathbb{A}^1_S$$
$$x \mapsto (x, 0),$$

with the inverse of the blow-up $q: Y \to X \times_S \mathbb{A}^1_S$, $i_1: P \to Y$ the inclusion. P is isomorphic to the projectivization of the normal bundle of $Z \times 1$ in $X \times_S \mathbb{A}^1_S$; let $f: P \to Z$ be the resulting projection. Let $[Z \times \mathbb{A}^1]$ denote the proper transform of $Z \times \mathbb{A}^1$ to Y. We note that the restriction of f to $[Z \times \mathbb{A}^1]$ gives an isomorphism $[Z \times \mathbb{A}^1] \to Z \times \mathbb{A}^1$, determining sections s' to q over $Z \times \mathbb{A}^1$, and $s : Z \to P$ to f over Z, with $s(\hat{Z}) = \hat{P}$. We encapsulate the above discussion in the following diagram:

(2.1.2.1)

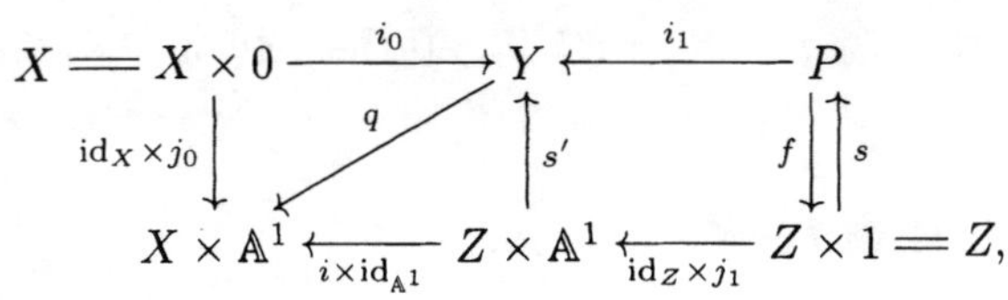

where $j_0: 0 \to \mathbb{A}^1$, $j_1: 1 \to \mathbb{A}^1$ are the inclusions.

By the results of §2.1.1, the map (2.1.1.1)

$$\cup[s(Z)]_{s(Z)} \circ f^*: \mathbb{Z}_{Z,\hat{Z}}(-d)[-2d] \to \mathbb{Z}_{P,\hat{P}}$$

is an isomorphism; the maps $i_1^*: \mathbb{Z}_{Y,\hat{Y}} \to \mathbb{Z}_{P,\hat{P}}$ and $i_0^*: \mathbb{Z}_{Y,\hat{Y}} \to \mathbb{Z}_{X,\hat{Z}}$ are isomorphisms by the homotopy axiom (Chapter I, Definition 2.1.4(a)). This gives the sequence of isomorphisms in $\mathcal{DM}(\mathcal{V})$:

$$\mathbb{Z}_{Z,\hat{Z}}(-d)[-2d] \xrightarrow{\cup[s(Z)]_{s(Z)} \circ f^*} \mathbb{Z}_{P,\hat{P}} \xrightarrow{(i_1^*)^{-1}} \mathbb{Z}_{Y,\hat{Y}} \xrightarrow{i_0^*} \mathbb{Z}_{X,\hat{Z}};$$

we denote the composition by

$$i_*: \mathbb{Z}_{Z,\hat{Z}}(-d)[-2d] \to \mathbb{Z}_{X,\hat{Z}}. \tag{2.1.2.2}$$

For $\hat{Z} = Z$, this gives the isomorphism $i_*: \mathbb{Z}_Z(-d)[-2d] \to \mathbb{Z}_{X,Z}$.

More generally, if $\hat{X}$ is a closed subset of X containing $\hat{Z}$, we denote the composition

$$\mathbb{Z}_{Z,\hat{Z}}(-d)[-2d] \xrightarrow{\cup[s(Z)]_{s(Z)}\circ f^*} \mathbb{Z}_{P,\hat{P}} \xrightarrow{(i_1^*)^{-1}} \mathbb{Z}_{Y,\hat{Y}} \xrightarrow{i_0^*} \mathbb{Z}_{X,\hat{X}}$$

by

$$i_*: \mathbb{Z}_{Z,\hat{Z}}(-d)[-2d] \to \mathbb{Z}_{X,\hat{X}}. \tag{2.1.2.3}$$

2.1.3. *The Gysin distinguished triangle.* Combining the localization distinguished triangle (I.2.2.10.2)

$$\mathbb{Z}_{X,\hat{Z}} \to \mathbb{Z}_X \to \mathbb{Z}_{X-\hat{Z}} \to \mathbb{Z}_{X,\hat{Z}}[1]$$

with the isomorphism (2.1.2.2) gives the Gysin distinguished triangle

$$\mathbb{Z}_{Z,\hat{Z}}(-d)[-2d] \xrightarrow{i_*} \mathbb{Z}_X \xrightarrow{j^*} \mathbb{Z}_{X-\hat{Z}} \xrightarrow{\partial} \mathbb{Z}_{Z,\hat{Z}}(-d)[1-2d]; \tag{2.1.3.1}$$

the long exact cohomology sequence associated to (2.1.3.1) (after twisting by $\mathbb{Z}(q)$) gives the exact Gysin sequence

$$\begin{aligned} \ldots \to H^{p-2d}_{\hat{Z}}(Z,\mathbb{Z}(q-d)) \to H^p(X,\mathbb{Z}(q)) \to H^p(X-\hat{Z},\mathbb{Z}(q)) \\ \to H^{p-2d-1}_{\hat{Z}}(Z,\mathbb{Z}(q-d)) \to \ldots. \end{aligned}$$

In particular, for $\hat{Z} = Z$, we have the Gysin distinguished triangle

$$\mathbb{Z}_Z(-d)[-2d] \xrightarrow{i_*} \mathbb{Z}_X \xrightarrow{j^*} \mathbb{Z}_{X-Z} \xrightarrow{\partial} \mathbb{Z}_Z(-d)[1-2d]$$

and the exact Gysin sequence

$$\begin{aligned} \ldots \to H^{p-2d}(Z,\mathbb{Z}(q-d)) \to H^p(X,\mathbb{Z}(q)) \to H^p(X-\hat{Z},\mathbb{Z}(q)) \\ \to H^{p-2d-1}(Z,\mathbb{Z}(q-d)) \to \ldots. \end{aligned}$$

2.2. Properties of the Gysin morphism

2.2.1. PROPOSITION. *Suppose we have subschemes $W \xrightarrow{i} Z \xrightarrow{j} X$ of a scheme X, with X, Z and W in $\mathcal{V}$, and with $\operatorname{codim}_X(Z) = d$ and $\operatorname{codim}_Z(W) = e$. Let $\hat{W}$ be a closed subset of W, $\hat{Z}$ a closed subset of Z, with $\hat{Z} \subset \hat{W}$. Then the diagrams*

(1)
$$\begin{array}{ccc} \mathbb{Z}_{W,\hat{W}}(-d-e)[-2d-2e] & \xrightarrow{j_*} & \mathbb{Z}_{Z,\hat{W}}(-d)[-2d] \\ & \searrow^{(i\circ j)_*} & \downarrow{i_*} \\ & & \mathbb{Z}_{X,\hat{W}} \end{array}$$

and

(2)
$$\begin{array}{ccc} \mathbb{Z}_{Z,\hat{W}}(-d)[-2d] & \xrightarrow{i_*} & \mathbb{Z}_{X,\hat{W}} \\ \downarrow{\scriptstyle \mathrm{id}_Z^*} & & \downarrow{\scriptstyle \mathrm{id}_X^*} \\ \mathbb{Z}_{Z,\hat{Z}}(-d)[-2d] & \xrightarrow[i_*]{} & \mathbb{Z}_{X,\hat{Z}} \end{array}$$

commute in $\mathcal{DM}(\mathcal{V})$. *In addition, if* $i = \mathrm{id}_X$, *then* $i_* = \mathrm{id}$.

PROOF. For the commutativity of (1), let

$$Y_W,\ \hat{Y}_W,\ P_W,\ \hat{P}_W,\ i_{W0}\colon X \to Y_W,\ i_{W1}\colon P_W \to Y_W,$$
$$f_W\colon P_W \to W,\ [W \times \mathbb{A}^1]_W \text{ and } s_W\colon W \to P_W$$

be as in §2.1.2 with $\hat{W}$ replacing $\hat{Z}$, and W replacing Z. Similarly, let

$$Y_W^Z,\ \hat{Y}_W^Z,\ P_W^Z,\ \hat{P}_W^Z,\ i_{W0}^Z\colon Z \to Y_W^Z,\ i_{W1}^Z\colon P_W^Z \to Y_W^Z,$$
$$f_W^Z\colon P_W^Z \to W,\ [W \times \mathbb{A}^1]_W^Z, \text{ and } s_W^Z\colon W \to P_W^Z$$

be as in §2.1.2, after replacing $\hat{Z}$ with $\hat{W}$, replacing X with Z and replacing Z with W. Finally, let

$$Y,\ \hat{Y}_Z,\ P,\ \hat{P}_Z,\ i_0\colon X \to Y,\ i_1\colon P \to Y,$$
$$f\colon P \to Z,\ [Z \times \mathbb{A}^1], \text{ and } s\colon Z \to P$$

be as in §2.1.2, with $\hat{W}$ replacing $\hat{Z}$ (and leaving X and Z the same). We let $[W \times \mathbb{A}^1]$ denote the proper transform of $W \times \mathbb{A}^1$ to Y.

We have the subscheme $T := s(W)$ of P, and closed subset $\hat{T} := \hat{P}_Z = s(\hat{W})$ of T. Let $i_T\colon T \to P$ be the inclusion. Let

$$Y_T, \hat{Y}_T, P_T, \hat{P}_T, i_{T0}\colon P \to Y_T, i_{T1}\colon P_T \to Y_T,$$
$$f_T\colon P_T \to T, [T \times \mathbb{A}^1]_T, \text{ and } s_T\colon T \to P_T$$

be as in §2.1.2, after replacing X with P, replacing Z with T and replacing $\hat{Z}$ with $\hat{T}$.

Let $[f^{-1}(W)]$ be the proper transform of $f^{-1}(W)$ to Y_T, let $Y_T^0 := Y_T \setminus (P_T \cap [f^{-1}(W)])$, and let $P_T^0 = P_T \setminus [f^{-1}(W)]$. Then the rational map $Y_T \to Y_W^Z$ induced by $f \times \mathrm{id}_{\mathbb{A}^1}\colon P \times \mathbb{A}^1 \to Z \times \mathbb{A}^1$ restricts to a morphism

$$f_{Z/W}\colon Y_T^0 \to Y_W^Z.$$

The section $s\colon Z \to P$ to f gives the section $s \times \mathrm{id}\colon Z \times \mathbb{A}^1 \to P \times \mathbb{A}^1$ to $f \times \mathrm{id}_{\mathbb{A}^1_S}$; blowing up $W \times 1$ and $s(W) \times 1$ gives the section $s_{Z/W}\colon Y_W^Z \to Y_T^0$ to $f_{Z/W}$. In particular, Y_T^0 contains $\hat{Y}_T$; by the excision isomorphism the inclusion $Y_T^0 \to Y_T$ induces the isomorphism $\mathbb{Z}_{Y_T,\hat{Y}_T} \cong \mathbb{Z}_{Y_T^0,\hat{Y}_T}$. Similarly, we have the isomorphism $\mathbb{Z}_{P_T,\hat{P}_T} \cong \mathbb{Z}_{P_T^0,\hat{P}_T}$.

Restricting $s_{Z/W}$ to P_W^Z gives the commutative diagram

$$\begin{array}{ccc} P_W^Z & \xrightarrow{s_{Z/W}} & P_T \\ {\scriptstyle f_W^Z}\downarrow\uparrow{\scriptstyle s_W^Z} & & {\scriptstyle s_T}\uparrow\downarrow{\scriptstyle f_T} \\ W & \xrightarrow[s_{|W}]{} & T. \end{array}$$

In addition, we have the identity of cycles $f^*_{Z/W}(|s^Z_W(W)|)\cdot|s_{Z/W}(P^Z_W)| = |s_T(T)|$, the intersection product taking place on P^0_T. This gives us the commutative diagram of isomorphisms

$$
\begin{array}{ccccc}
 & & \mathbb{Z}_{P,\hat{T}} & \xleftarrow{\ \cup[s(Z)]\circ f^*\ } & \mathbb{Z}_{Z,\hat{W}}(-d)[-2d] \\
 & \swarrow{\scriptstyle (i^*_{T0})^{-1}} & \downarrow{\scriptstyle (i^*_{T0})^{-1}} & & \downarrow{\scriptstyle (i^*_{W0})^{-1}} \\
\mathbb{Z}_{Y_T,\hat{Y}_T} & \xrightarrow{\ \sim\ } & \mathbb{Z}_{Y^0_T,\hat{Y}_T} & \xleftarrow{\ \cup[s_{Z/W}(Y^Z_W)]\circ f^*_{Z/W}\ } & \mathbb{Z}_{Y^Z_W,\hat{Y}^Z_W}(-d)[-2d] \\
{\scriptstyle i^*_{T1}}\downarrow & & {\scriptstyle i^*_{T1}}\downarrow & & \downarrow{\scriptstyle i^{Z*}_{W1}} \\
\mathbb{Z}_{P_T,\hat{P}_T} & \xrightarrow{\ \sim\ } & \mathbb{Z}_{P^0_T,\hat{P}_T} & \xleftarrow{\ \cup[s_{Z/W}(P^Z_W)]\circ f^*_{Z/W}\ } & \mathbb{Z}_{P^Z_W,\hat{P}^Z_W}(-d)[-2d] \\
 & \nwarrow{\scriptstyle \cup[s_T(T)]\circ f^*_T} & \uparrow{\scriptstyle \cup[s_T(T)]\circ f^*_T} & & \uparrow{\scriptstyle \cup[s^Z_W(W)]\circ f^{Z*}_W} \\
 & & \mathbb{Z}_{T,\hat{T}}(-d-e)[-2d-2e] & \xrightarrow[\ s^*_{|W}\ ]{} & \mathbb{Z}_{W,\hat{W}}(-d-e)[-2d-2e].
\end{array}
\tag{2.2.1.1}
$$

Let $h_1\colon Q_1 \to X\times\mathbb{A}^1\times\mathbb{A}^1$ be the blow-up of $X\times\mathbb{A}^1\times\mathbb{A}^1$ along the subscheme $Z\times\mathbb{A}^1\times 1$, let $[W\times 1\times\mathbb{A}^1]_1 \subset Q_1$ be the proper transform of $W\times 1\times\mathbb{A}^1$, and let $h_2\colon Q\to Q_1$ be the blow-up along $[W\times 1\times\mathbb{A}^1]_1$. Let $h\colon Q\to X\times\mathbb{A}^1\times\mathbb{A}^1$ the composition $h_1\circ h_2$. Then we have isomorphisms (as $X\times\mathbb{A}^1$-schemes)

$$
h^{-1}(X\times\mathbb{A}^1\times 0)\cong Y_W;\quad h^{-1}(X\times 0\times\mathbb{A}^1)\cong Y. \tag{2.2.1.2}
$$

We identify $h^{-1}(X\times\mathbb{A}^1\times 0)$ with Y_W, and $h^{-1}(X\times 0\times\mathbb{A}^1)$ with Y via these isomorphisms.

Let E_1 be the exceptional divisor of h_1, E the exceptional divisor of h_2, and $[E_1]$ the proper transform of E_1 to Q. SInce E is isomorphic to $P\times\mathbb{A}^1$, and $E\cap[W\times 1\times\mathbb{A}^1]$ goes to $T\times 1$ under this isomorphism, we have an isomorphism

$$
[E_1]\cong Y_T \tag{2.2.1.3}
$$

as a $Z\times\mathbb{A}^1$-scheme; we identify $[E_1]$ with Y_T via this isomorphism.

Let $[W\times\mathbb{A}^1\times\mathbb{A}^1]$ and $[\hat{W}\times\mathbb{A}^1\times\mathbb{A}^1]$ be the proper transforms of $W\times\mathbb{A}^1\times\mathbb{A}^1$ and $\hat{W}\times\mathbb{A}^1\times\mathbb{A}^1$ to Q. Then the map $h_{|[W\times\mathbb{A}^1\times\mathbb{A}^1]}\colon[W\times\mathbb{A}^1\times\mathbb{A}^1]\to W\times\mathbb{A}^1\times\mathbb{A}^1$ is an isomorphism, and we have

$$
\begin{aligned}
h^{-1}(X\times\mathbb{A}^1\times 0)\cap[W\times\mathbb{A}^1\times\mathbb{A}^1] &= [W\times\mathbb{A}^1]_W,\\
h^{-1}(X\times 0\times\mathbb{A}^1)\cap[W\times\mathbb{A}^1\times\mathbb{A}^1] &= [W\times\mathbb{A}^1],\\
[E_1]\cap[W\times\mathbb{A}^1\times\mathbb{A}^1] &= [T\times\mathbb{A}^1]_T.
\end{aligned}
\tag{2.2.1.4}
$$

We have similar equalities with $\hat{W}$ replacing W, and with $\hat{T}$ replacing T.

Let $\hat{Q} = [\hat{W}\times\mathbb{A}^1\times\mathbb{A}^1]$ and $\hat{E} = E\cap[\hat{W}\times\mathbb{A}^1\times\mathbb{A}^1]$. As $h_1\colon[W\times\mathbb{A}^1\times 1]_1\to W\times\mathbb{A}^1\times 1$ is an isomorphism, we have the projection $h_E\colon E\to W\times\mathbb{A}^1$. The restriction of h_E to $E\cap[W\times\mathbb{A}^1\times\mathbb{A}^1]$ gives isomorphisms $E\cap[W\times\mathbb{A}^1\times\mathbb{A}^1]\to W\times\mathbb{A}^1$ and $\hat{E}\to\hat{W}\times\mathbb{A}^1$, which thus defines the section $s_E\colon W\times\mathbb{A}^1\to E$ to h_E.

We have $Y_W \cap E = P_W$, $Y_W \cap \hat{E} = \hat{P}_W$, $Y_T \cap E = P_T$, and $Y_T \cap \hat{E} = \hat{P}_T$. The isomorphisms (2.2.1.2), (2.2.1.3) and (2.2.1.4) give us inclusions

$$\begin{gathered} i_{\#0}\colon Y_W \to Q;\ i_{0\#}\colon Y \to Q;\ i_{\#1}\colon Y_T \to Q;\ i_{1\#}\colon E \to Q; \\ i_{00}\colon X \to Q;\ i_{10}\colon P_W \to Q;\ i_{01}\colon P \to Q;\ i_{11}\colon P_T \to Q; \\ i_{E0}\colon P_W \to E;\ i_{E1}\colon P_T \to E. \end{gathered} \tag{2.2.1.5}$$

We use the convention that the image of i_{ab} lies in the fiber of Q over $X \times (a, b)$, that of $i_{\#b}$ lies in the fiber over $X \times \mathbb{A}^1 \times b$, etc.

Putting (2.2.1.2)-(2.2.1.5) together and using the homotopy axiom gives the commutative diagram of isomorphisms in $\mathcal{DM}(\mathcal{V})$,

$$\begin{array}{ccccc} \mathbb{Z}_{X,\hat{W}} & \xleftarrow{i_0^*} & \mathbb{Z}_{Y,\hat{Y}_Z} & \xrightarrow{i_1^*} & \mathbb{Z}_{P,\hat{P}_Z} \\ \uparrow{\scriptstyle i_{W0}^*} & \nwarrow^{i_{00}^*} & \uparrow{\scriptstyle i_{0\#}^*} & \nearrow^{i_{01}^*} & \uparrow{\scriptstyle i_{T0}^*} \\ \mathbb{Z}_{Y_W,\hat{Y}_W} & \xleftarrow{i_{\#0}^*} & \mathbb{Z}_{Q,\hat{Q}} & \xrightarrow{i_{\#1}^*} & \mathbb{Z}_{Y_T,\hat{Y}_T} \\ \downarrow{\scriptstyle i_{W1}^*} & \swarrow^{i_{10}^*} & \downarrow{\scriptstyle i_{1\#}^*} & \searrow^{i_{11}^*} & \downarrow{\scriptstyle i_{T1}^*} \\ \mathbb{Z}_{P_W,\hat{P}_W} & \xleftarrow[i_{E0}^*]{} & \mathbb{Z}_{E,\hat{E}} & \xrightarrow[i_{E1}^*]{} & \mathbb{Z}_{P_T,\hat{P}_T}. \end{array} \tag{2.2.1.6}$$

Let $i_0\colon W \to W \times \mathbb{A}^1$, $i_1\colon W \to W \times \mathbb{A}^1$ be the inclusions $i_0(w) = (w, 0)$, $i_1(w) = (w, 1)$. We have the commutative diagram of isomorphisms in $\mathcal{DM}(\mathcal{V})$,

(2.2.1.7)

$$\begin{array}{ccc} \mathbb{Z}_{W,\hat{W}}(-d-e)[-2d-2e] \xleftarrow{s_{|W}^*} \mathbb{Z}_{T,\hat{T}}(-d-e)[-2d-2e] & \xrightarrow{\cup[s_T(W)]\circ f_T^*} & \mathbb{Z}_{P_T,\hat{P}_T} \\ \uparrow{\scriptstyle i_1^*} & & \uparrow{\scriptstyle i_{E1}^*} \\ \mathbb{Z}_{W\times\mathbb{A}^1,\hat{W}\times\mathbb{A}^1}(-d-e)[-2d-2e] & \xrightarrow{\cup[s_E(W\times\mathbb{A}^1)]\circ h_E^*} & \mathbb{Z}_{E,\hat{E}} \\ \downarrow{\scriptstyle i_0^*} & & \downarrow{\scriptstyle i_{E0}^*} \\ \mathbb{Z}_{W,\hat{W}}(-d-e)[-2d-2e]. & \xrightarrow[\cup[s_W(W)]\circ f_W^*]{} & \mathbb{Z}_{P_W,\hat{P}_W}. \end{array}$$

Since the composition $(i_0^*)^{-1} \circ i_1^*$ is the identity, the composition

$$\begin{aligned} \mathbb{Z}_{W,\hat{W}}(-d-e)[-2d-2e] &\xrightarrow{\cup[s_W(W)]\circ f_W^*} \mathbb{Z}_{P_W,\hat{P}_W} \xrightarrow{(i_{E0}^*)^{-1}} \mathbb{Z}_{E,\hat{E}} \\ &\xrightarrow{i_{E1}^*} \mathbb{Z}_{P_T,\hat{P}_T} \xrightarrow{(\cup[s_T(W)]\circ f_T^*)^{-1}} \mathbb{Z}_{T,\hat{T}}(-d-e)[-2d-2e] \\ &\xrightarrow{s_{|W}^*} \mathbb{Z}_{W,\hat{W}}(-d-e)[-2d-2e] \end{aligned}$$

is the identity.

We thus have

$$\begin{aligned}(i\circ j)_* &= i^*_{W0}\circ(i^*_{W1})^{-1}\circ\cup[s_W(W)]\circ f^*_W && \text{(by definition)}\\ &= i^*_{W0}\circ(i^*_{W1})^{-1}\circ i^*_{E0}\circ(i^*_{E1})^{-1}\circ\\ &\qquad\cup[s_T(T)]\circ f^*_T\circ(s^*_W)^{-1} && (2.2.1.7)\\ &= i^*_0\circ(i^*_1)^{-1}\circ i^*_{T0}\circ(i^*_{T1})^{-1}\circ\\ &\qquad\cup[s_T(T)]\circ f^*_T\circ s^*_W && (2.2.1.6)\\ &= (i^*_0\circ(i^*_1)^{-1}\circ\cup[s(Z)]\circ f^*)\circ\\ &\qquad(i^{Z*}_{W0}\circ(i^{Z*}_{W1})^{-1}\circ\cup[s^Z_W(W)]\circ f^{Z*}_W) && (2.2.1.1)\\ &= i_*\circ j_*. && \text{(by definition)}\end{aligned}$$

The commutativity of (2) follows directly from the functoriality of the cycle map, Chapter I, Proposition 3.5.3(i).

To prove the assertion $\mathrm{id}_{X*} = \mathrm{id}$, we note that the blow-up of $X\times\mathbb{A}^1$ along $X\times 1$ is isomorphic to $X\times\mathbb{A}^1$, hence id_{X*} is the composition

$$\mathbb{Z}_{X,\hat{X}} \xrightarrow{(i_1^*)^{-1}} \mathbb{Z}_{X\times\mathbb{A}^1,\hat{X}\times\mathbb{A}^1} \xrightarrow{i_0^*} \mathbb{Z}_{X,\hat{X}},$$

where $i_0\colon X\to X\times\mathbb{A}^1$ and $i_1\colon X\to X\times\mathbb{A}^1$ are the 0 and 1 sections. As $p_1\circ i_0 = p_1\circ i_1 = \mathrm{id}_X$, the above composition is the identity. □

2.2.2. PROPOSITION [projection formula]. *Let $i\colon Z\to X$ be a closed embedding in $\mathcal{V}$ of codimension d and let $\hat{X}_i$ be closed subsets of X, $i=1,2$. Let $\hat{Z}_i = Z\cap\hat{X}_i$, $i=1,2$. Then the diagram*

$$\begin{array}{ccc}\mathbb{Z}_{Z,\hat{Z}_1}(-d)[-2d]\otimes\mathbb{Z}_{X,\hat{X}_2} & \xrightarrow{\mathrm{id}^*_Z\otimes i^*} & \mathbb{Z}_{Z,\hat{Z}_1}(-d)[-2d]\otimes\mathbb{Z}_{Z,\hat{Z}_2}\\ & & \big\downarrow{\scriptstyle\cup_Z}\\ \big\downarrow{\scriptstyle i_*\otimes\mathrm{id}} & & \mathbb{Z}_{Z,\hat{Z}_1\cap\hat{Z}_2}(-d)[-2d]\\ & & \big\downarrow{\scriptstyle i_*}\\ \mathbb{Z}_{X,\hat{X}_1}\otimes\mathbb{Z}_{X,\hat{X}_2} & \xrightarrow[\cup_X]{} & \mathbb{Z}_{X,\hat{X}_1\cap\hat{X}_2}\end{array}$$

commutes in $\mathcal{DM}(\mathcal{V})$.

PROOF. To simplify the notation, we give the proof in the case $Z=\hat{Z}_1=\hat{Z}_2$, $X=\hat{X}_1=\hat{X}_2$.

Via the diagram (2.1.2.1), we have the definition of the map $i_*\colon\mathbb{Z}_Z(-d)[-2d]\to\mathbb{Z}_X$ as the composition

(2.2.2.1) $$\mathbb{Z}_Z(-d)[-2d]\xrightarrow{f^*}\mathbb{Z}_P(-d)[-2d]\xrightarrow{\cup[s(Z)]}\mathbb{Z}_{P,s(Z)}$$
$$\xrightarrow{(i_1^*)^{-1}}\mathbb{Z}_{Y,Z\times\mathbb{A}^1}\xrightarrow{i_0^*}\mathbb{Z}_X.$$

Taking the product of (2.1.2.1) with X yields the diagram ($X_X := X \times_S X$, $P_X := P \times_S X$, etc.)

(2.2.2.2)
$$
\begin{array}{ccccc}
X_X = X_X \times 0 & \xrightarrow{i_{X0}} & Y_X & \xleftarrow{i_{X1}} & P_X \\
\downarrow{\scriptstyle \mathrm{id}_{X_X}\times j_0} & \swarrow{\scriptstyle q_X} & \uparrow{\scriptstyle s'_X} & & {\scriptstyle f_X}\downarrow\uparrow{\scriptstyle s_X} \\
X_X \times \mathbb{A}^1 & \xleftarrow[i_X\times \mathrm{id}_{\mathbb{A}^1}]{} & Z_X \times \mathbb{A}^1 & \xleftarrow[\mathrm{id}_{Z_X}\times j_1]{} & Z_X \times 1 = Z_X,
\end{array}
$$

which gives the definition of the map $(i \times \mathrm{id}_X)_*\colon \mathbb{Z}_{Z\times X}(-d)[-2d] \to \mathbb{Z}_{X\times X, Z\times X}$ as the composition

(2.2.2.3)
$$
\begin{aligned}
\mathbb{Z}_{Z\times X}(-d)[-2d] \xrightarrow{f_X^*} \mathbb{Z}_{P_X}(-d)[-2d] &\xrightarrow{\cup[s_X(Z_X)]} \mathbb{Z}_{P_X, s_X(Z_X)} \\
&\xrightarrow{(i_{X1}^*)^{-1}} \mathbb{Z}_{Y_X, Z_X\times\mathbb{A}^1} \xrightarrow{i_{X0}^*} \mathbb{Z}_{X\times X}.
\end{aligned}
$$

We have the commutative diagram

$$
\begin{array}{ccc}
\mathbb{Z}_Z(-d)[-2d] \otimes \mathbb{Z}_X & \xrightarrow{\boxtimes_{Z,X}} & \mathbb{Z}_{Z\times X}(-d)[-2d] \\
\downarrow{\scriptstyle f^*\otimes \mathrm{id}} & & \downarrow{\scriptstyle f_X^*} \\
\mathbb{Z}_P(-d)[-2d] \otimes \mathbb{Z}_X & \xrightarrow{\boxtimes_{P,X}} & \mathbb{Z}_{P_X}(-d)[-2d] \\
\downarrow{\scriptstyle \cup[s(Z)]\otimes \mathrm{id}} & & \downarrow{\scriptstyle \cup[s_X(Z_X)]} \\
\mathbb{Z}_{P,s(Z)} \otimes \mathbb{Z}_X & \xrightarrow{\boxtimes_{P,X}} & \mathbb{Z}_{P_X, s_X(Z_X)} \\
\uparrow{\scriptstyle i_0^*\otimes \mathrm{id}} & & \uparrow{\scriptstyle i_{X0}^*} \\
\mathbb{Z}_X \otimes \mathbb{Z}_X & \xrightarrow{\boxtimes_{X,X}} & \mathbb{Z}_{X\times X},
\end{array}
$$

which, together with (2.2.2.1) and (2.2.2.3), yields the commutativity of the diagram

$$
\begin{array}{ccc}
\mathbb{Z}_Z(-d)[-2d] \otimes \mathbb{Z}_X & \xrightarrow{\boxtimes_{Z,X}} & \mathbb{Z}_{Z\times X}(-d)[-2d] \\
\downarrow{\scriptstyle i_*\otimes \mathrm{id}} & & \downarrow{\scriptstyle i_{X*}} \\
\mathbb{Z}_X \otimes \mathbb{Z}_X & \xrightarrow[\boxtimes_{X,X}]{} & \mathbb{Z}_{X\times X}.
\end{array}
$$

The naturality of the external products $\boxtimes_{*,*}$ implies that the diagram

$$
\begin{array}{ccc}
\mathbb{Z}_Z(-d)[-2d] \otimes \mathbb{Z}_X & \xrightarrow{\mathrm{id}_Z^*\otimes i^*} & \mathbb{Z}_Z(-d)[-2d] \otimes \mathbb{Z}_Z \\
\downarrow{\scriptstyle \boxtimes_{Z,X}} & & \downarrow{\scriptstyle \boxtimes_{Z,Z}} \\
\mathbb{Z}_{Z\times X}(-d)[-2d] & \xrightarrow[(\mathrm{id}_Z\times i)^*]{} & \mathbb{Z}_{Z\times Z}(-d)[-2d]
\end{array}
$$

commutes. Thus, we need only check the commutativity of the diagram

(2.2.2.4)
$$
\begin{array}{ccc}
\mathbb{Z}_{Z\times X}(-d)[-2d] & \xrightarrow{(\mathrm{id}_Z, i)^*} & \mathbb{Z}_Z(-d)[-2d] \\
\downarrow{\scriptstyle (i\times \mathrm{id}_X)_*} & & \downarrow{\scriptstyle i_*} \\
\mathbb{Z}_{X\times X} & \xrightarrow[\Delta_X^*]{} & \mathbb{Z}_X.
\end{array}
$$

We have the commutative diagram

$$\begin{array}{ccc} Z & \xrightarrow{(i,\mathrm{id}_Z)} & Z \times X \\ {\scriptstyle i}\downarrow & & \downarrow{\scriptstyle i\times \mathrm{id}_X} \\ X & \xrightarrow[\Delta_X]{} & X \times X. \end{array}$$

We take the product with $\mathbb{A}^1$, and blow up along $Z \times 1$ and $Z \times X \times 1$, which, together with the diagram (2.1.2.1) and (2.2.2.2), gives us the commutative diagram

$$\text{(2.2.2.5)} \qquad \begin{array}{ccccccc} Z & \underset{f}{\overset{s}{\rightleftarrows}} & P & \xrightarrow{i_1} & Y & \xleftarrow{i_0} & X \\ {\scriptstyle (i,\mathrm{id}_Z)}\downarrow & & {\scriptstyle \delta_P}\downarrow & & {\scriptstyle \delta_Y}\downarrow & & \downarrow{\scriptstyle \Delta_X} \\ Z_X & \underset{f_X}{\overset{s_X}{\rightleftarrows}} & P_X & \xrightarrow[i_{X1}]{} & Y_X & \xleftarrow{i_{X0}} & X_X. \end{array}$$

In addition, we have

$$\text{(2.2.2.6)} \qquad \delta_Y^*(|[Z_X \times \mathbb{A}^1]|) = |[Z \times \mathbb{A}^1]|; \quad \delta_P^*(|s_X([Z_X])|) = |s([Z])|.$$

Putting (2.2.2.5) and (2.2.2.6) together gives the commutative diagram

$$\begin{array}{ccc} \mathbb{Z}_{X_X} & \xrightarrow{\Delta_X^*} & \mathbb{Z}_X \\ {\scriptstyle i_{X0}^*}\uparrow & & \uparrow{\scriptstyle i_0^*} \\ \mathbb{Z}_{Y_X, Z_X\times\mathbb{A}^1} & \xrightarrow{\delta_Y^*} & \mathbb{Z}_{Y, Z\times\mathbb{A}^1} \\ {\scriptstyle i_{X1}^*}\downarrow & & \downarrow{\scriptstyle i_1^*} \\ \mathbb{Z}_{P_X, s_X(Z_X)} & \xrightarrow{\delta_P} & \mathbb{Z}_{P, s(Z)} \\ {\scriptstyle \cup[s_X(Z_X)]}\uparrow & & \uparrow{\scriptstyle \cup[s(Z)]} \\ \mathbb{Z}_{P_X}(-d)[-2d] & \xrightarrow{\delta_P} & \mathbb{Z}_P(-d)[-2d] \\ {\scriptstyle f_X^*}\uparrow & & \uparrow{\scriptstyle f^*} \\ \mathbb{Z}_{Z\times X}(-d)[-2d] & \xrightarrow{(i,\mathrm{id}_Z)^*} & \mathbb{Z}_Z(-d)[-2d]. \end{array}$$

As this implies the commutativity of (2.2.2.4), the proof is complete. □

2.2.3. THEOREM. *Let $i\colon Z \to X$ be a closed embedding in $\mathcal{V}$, of codimension d, and let W be in $\mathcal{Z}^p(Z/S)$, supported on a closed subset $\hat{Z}$ of Z. Then*

$$i_*(\mathrm{cl}^p_{Z,\hat{Z}}(W)) = \mathrm{cl}^{p+q}_{X,\hat{Z}}(i_*(W)),$$

where $\mathrm{cl}^p_{Z,\hat{Z}}$, $\mathrm{cl}^{p+q}_{X,\hat{Z}}$ are the cycle classes with support (I.3.5.2.6).

PROOF. We use the notation from §2.1.2. From Chapter I, Proposition 3.5.3, we have

$$\begin{aligned}(\cup[s(Z)]\circ f^*)(\mathrm{cl}^p_{Z,\hat Z}(W)) &= \mathrm{cl}^{p+q}_{P,s(\hat Z)}(s(Z)\cup f^*(W))\\ &= \mathrm{cl}^{p+q}_{P,s(\hat Z)}(s_*(W)).\end{aligned}$$

$$i_1^*(\mathrm{cl}^{p+q}_{Y,\hat Z\times\mathbb{A}^1}(i_*(W)\times\mathbb{A}^1)) = \mathrm{cl}^{p+q}_{P,s(\hat Z)}(s_*(W)).$$

$$i_0^*(\mathrm{cl}^{p+q}_{Y,\hat Z\times\mathbb{A}^1}(i_*(W)\times\mathbb{A}^1)) = \mathrm{cl}^{p+q}_{X,\hat Z}(i_*(W)).$$

These identities, together with the definition of i_*, prove the theorem. □

2.2.4. LEMMA. *Let $i\colon Z\to X$ be a closed embedding in $\mathcal{V}$, $p\colon W\to X$ a morphism in $\mathcal{V}$, giving the cartesian diagram*

$$\begin{array}{ccc} W\times_X Z & \xrightarrow{p_1} & W\\ {\scriptstyle p_2}\downarrow & & \downarrow{\scriptstyle p}\\ Z & \xrightarrow[i]{} & X. \end{array}$$

Suppose that $\mathrm{Tor}_p^{\mathcal{O}_X}(\mathcal{O}_Z,\mathcal{O}_W)=0$ *for all* $p>0$. *Then*

$$p^*\circ i_* = p_{1*}\circ p_2^*.$$

PROOF. We use the notation from §2.1.2, and give the proof without closed support to simplify the notation.

Since i is a closed embedding in $\mathcal{V}$, Z is a local complete intersection in X. The vanishing of the Tors implies that the closed embedding $p_1\colon W\times_X Z\to W$ identifies $W\times_X Z$ with a local complete intersection of codimension equal to the codimension d of Z in X.

Applying the product $W\times_X(-)$ to the diagram (2.1.2.1) gives the diagram

$$\begin{array}{ccccc} W = W\times 0 & \xrightarrow{i_{W0}} & W\times_X Y & \xleftarrow{i_{W1}} & W\times_X P\\ {\scriptstyle \mathrm{id}_W\times j_0}\downarrow & \swarrow{\scriptstyle q_W} & \uparrow{\scriptstyle s'_W} & & {\scriptstyle f_W}\downarrow\uparrow{\scriptstyle s_W}\\ W\times\mathbb{A}^1 & \xleftarrow[p_W\times\mathrm{id}_{\mathbb{A}^1}]{} & W\times_X Z\times\mathbb{A}^1 & \xleftarrow[\mathrm{id}_{W\times_X Z}\times j_1]{} & W\times_X Z\times 1 = W\times_X Z. \end{array}$$

Since $W\times_X Z$ is a codimension d local complete intersection, this diagram is the same as the deformation diagram for the closed embedding p_1.

We have the commutative diagram

$$\begin{array}{ccccccc} W\times_X Z & \underset{f_W}{\overset{s_W}{\rightleftarrows}} & W\times_X P & \xrightarrow{i_{W1}} & W\times_X Y & \xleftarrow{i_{W0}} & W\\ {\scriptstyle p_2}\downarrow & & {\scriptstyle p_2}\downarrow & & {\scriptstyle p_2}\downarrow & & \downarrow{\scriptstyle p_2}\\ Z & \underset{f}{\overset{s}{\rightleftarrows}} & P & \xrightarrow[i_1]{} & Y & \xleftarrow[i_0]{} & X. \end{array}$$

In addition, we have

$$p_2^*(|[Z\times\mathbb{A}^1]|) = |[W\times_X Z\times\mathbb{A}^1]|;\quad p_2^*(|s(Z)|) = |s_W(W\times_X Z)|.$$

This gives us the commutative diagram

$$\text{(2.2.4.1)}\qquad \begin{array}{ccc}
\mathbb{Z}_{W,W\times_X Z} & \xleftarrow{\;p^*\;} & \mathbb{Z}_{X,Z} \\
{\scriptstyle i^*_{W0}}\uparrow & & \uparrow{\scriptstyle i^*_0} \\
\mathbb{Z}_{W\times_X Y, W\times_X Z\times\mathbb{A}^1} & \xleftarrow{\;p_2^*\;} & \mathbb{Z}_{Y,Z\times\mathbb{A}^1} \\
{\scriptstyle i^*_{W1}}\downarrow & & \downarrow{\scriptstyle i^*_1} \\
\mathbb{Z}_{W\times_X P, s(W\times_X Z)} & \xleftarrow{\;p_2^*\;} & \mathbb{Z}_{P,s(Z)} \\
{\scriptstyle \cup[s_W(W\times_X Z)]}\uparrow & & \uparrow{\scriptstyle \cup[s(Z)]} \\
\mathbb{Z}_{W\times_X P, s(W\times_X Z)}(-d)[-2d] & \xleftarrow{\;p_2^*\;} & \mathbb{Z}_{P,s(Z)}(-d)[-2d] \\
{\scriptstyle f^*_W}\uparrow & & \uparrow{\scriptstyle f^*} \\
\mathbb{Z}_{W\times_X Z}(-d)[-2d] & \xleftarrow{\;p_2^*\;} & \mathbb{Z}_Z(-d)[-2d].
\end{array}$$

By definition, p_{1*} is the composition

$$\mathbb{Z}_{W\times_X Z}(-d)[-2d] \xrightarrow{\cup[s_W(W\times_X Z)]\circ f^*_W} \mathbb{Z}_{W\times_X P, s(W\times_X Z)}$$
$$\xrightarrow{(i^*_{W1})^{-1}} \mathbb{Z}_{W\times_X Y, W\times_X Z\times\mathbb{A}^1} \xrightarrow{i^*_{W0}} \mathbb{Z}_{W,W\times_X Z}.$$

This, together with the definition of i_* and the diagram (2.2.4.1), completes the proof. □

2.2.5. THEOREM [semi-purity]. *Suppose that the base scheme S is* $\operatorname{Spec} k$ *for a field k. Let X be in $\mathcal{V}$, and $\hat{X}$ a closed subset of codimension $\geq d$. Then*

$$H^{2q-p}_{\hat{X}}(X, \mathbb{Z}(q)) = 0, \textit{ if } q = d \textit{ and } p \neq 0, \textit{ or if } q < d.$$

PROOF. Each X in $\mathcal{V}$ is a projective limit of schemes X_α in $\mathbf{Sm}_{k_\alpha}$, with k_α finitely generated over the prime field k_0. By (Chapter II, Corollary 3.4.3 and Theorem 3.6.6), we may assume that k is finitely generated over k_0, and that X is in $\mathbf{Sm}_k$.

Let f_*X denote X, considered as an object of $\mathbf{Sm}^{\mathrm{ess}}_{k_0}$. Since we have the identity of complexes $\mathcal{Z}_{\mathrm{mot}}(X/k,*) = \mathcal{Z}_{\mathrm{mot}}(f_*X/k_0,*)$, Theorem 3.6.6 of Chapter II gives us the isomorphism of motivic cohomology, $H^{2q-p}_{f_*\hat{X}}(f_*X, \mathbb{Z}(q)) \cong H^{2q-p}_{\hat{X}}(X, \mathbb{Z}(q))$. Thus, we may assume that k is perfect.

We proceed by downward induction on d, the case $d = \dim_k X+1$ being trivially true. Let $\hat{Y} \subset \hat{X}$ be the singular locus of $\hat{X}$, together with all components of $\hat{X}$ which have codimension $> d$, so $\operatorname{codim}_X(\hat{Y}) \geq d+1$. We have the distinguished triangle (I.2.2.10.1)

$$\mathbb{Z}_{X,\hat{Y}}(q) \to \mathbb{Z}_{X,\hat{X}}(q) \to \mathbb{Z}_{X\setminus\hat{Y},\hat{X}\setminus\hat{Y}} \to \mathbb{Z}_{X,\hat{Y}}(q)[1].$$

Applying the induction hypothesis to the long exact cohomology sequence associated to this triangle, we reduce to the case of a smooth $\hat{X}$, of pure codimension d.

In this case, we have the isomorphism $i_*: \mathbb{Z}_{\hat{X}}(q-d) \to \mathbb{Z}_{X,\hat{X}}(q)[2d]$, giving the isomorphism $i_*: H^{-p}(\hat{X}, \mathbb{Z}(q-d)) \to H^{2d-p}_{\hat{X}}(X, \mathbb{Z}(q))$. By Theorem 3.6.6 of

Chapter II, we have $H^{-p}(\hat{X}, \mathbb{Z}(q-d)) = \mathrm{CH}^{q-d}(\hat{X}, 2(q-d)+p)$, which is zero if $q < d$, or if $q = d$ and $p \neq 0$. Thus, the induction goes through, completing the proof. □

2.2.6. We now check that, in case of a section $i: Z \to X$ to a smooth projection $q: X \to Z$, with $s(Z)$ codimension d on X, the definition (2.1.2.3) of i_* agrees with the map (2.1.1.1)

$$\cup[i(Z)]_{i(Z)} \circ q^* : \mathbb{Z}_{Z,\hat{Z}}(-d)[-2d] \to \mathbb{Z}_{X,i(\hat{Z})}.$$

We consider a somewhat more general situation.

Let $\hat{Z}$ be a closed subset of Z, and let $\hat{X}$ be a closed subset of X containing $i(\hat{Z})$. Let F be a closed subset of X with $i(Z) \subset F$ and with $F \cap q^{-1}(\hat{Z}) \subset \hat{X}$. We have the cycle class with support $\mathrm{cl}^d_{X,F}(|i(Z)|) \in H^{2d}_F(X, \mathbb{Z}(d))$; let $\cup[i(Z)]_F \circ q^* : \mathbb{Z}_{Z,\hat{Z}}(-d)[-2d] \to \mathbb{Z}_{X,\hat{X}}$ be the composition

$$\mathbb{Z}_{Z,\hat{Z}}(-d)[-2d] \xrightarrow{q^*} \mathbb{Z}_{X,q^{-1}(\hat{Z})} \xrightarrow{(-)\cup_X^{F,q^{-1}(\hat{Z})} \mathrm{cl}^d_{X,F}(|i(Z)|)} \mathbb{Z}_{X,q^{-1}(\hat{Z})\cap F} \to \mathbb{Z}_{X,\hat{X}}.$$

2.2.7. LEMMA. *The two maps*

$$\cup[i(Z)]_F \circ q^* : \mathbb{Z}_{Z,\hat{Z}}(-d)[-2d] \to \mathbb{Z}_{X,\hat{X}},$$
$$i_* : \mathbb{Z}_{Z,\hat{Z}}(-d)[-2d] \to \mathbb{Z}_{X,\hat{X}}$$

agree in $\mathcal{DM}(\mathcal{V})$.

PROOF. By the functoriality of the cycle class with respect to change of support, we have $\cup[i(Z)]_F \circ q^* = \cup[i(Z)]_{i(Z)} \circ q^*$, so we may assume that $F = i(Z)$.

We use the notation of §2.1.2. The map q induces maps $q_Y: Y \to Z \times \mathbb{A}^1$ and $q_P: P \to Z$, with s' a section to q_Y, and $q_P = f$. Letting $i_{Z0}: Z \to Z \times \mathbb{A}^1$ and $i_{Z1}: Z \to Z \times \mathbb{A}^1$ be the 0 and 1 sections, respectively, we have the commutative diagram

$$\begin{array}{ccccc} X & \xrightarrow{i_0} & Y & \xleftarrow{i_1} & P \\ {\scriptstyle q}\downarrow\uparrow{\scriptstyle i} & & {\scriptstyle q_Y}\downarrow\uparrow{\scriptstyle s'} & & {\scriptstyle f}\downarrow\uparrow{\scriptstyle s} \\ Z & \xrightarrow[i_{Z0}]{} & Z \times \mathbb{A}^1 & \xleftarrow[i_{Z1}]{} & Z. \end{array}$$

In addition, we have the identity of cycles $|i(Z)| = i_0^*(|s'(Z \times \mathbb{A}^1)|)$. This gives the commutative diagram

$$(2.2.7.1)\qquad \begin{array}{ccc} \mathbb{Z}_{Z,\hat{Z}} & \xrightarrow{\cup[i(Z)]_{i(Z)} \circ q^*} & \mathbb{Z}_{X,\hat{X}} \\ {\scriptstyle i_{Z0}^*}\uparrow & & \uparrow{\scriptstyle i_0^*} \\ \mathbb{Z}_{Z\times\mathbb{A}^1, \hat{Z}\times\mathbb{A}^1} & \xrightarrow{\cup[s'(Z\times\mathbb{A}^1)]_{s'(Z\times\mathbb{A}^1)} \circ q_Y^*} & \mathbb{Z}_{Y, s'(\hat{Z}\times\mathbb{A}^1)} \\ {\scriptstyle i_{Z1}^*}\downarrow & & \downarrow{\scriptstyle i_1^*} \\ \mathbb{Z}_{Z,\hat{Z}} & \xrightarrow{\cup[s(Z)]_{s(Z)} \circ f^*} & \mathbb{Z}_{P,s(\hat{Z})}. \end{array}$$

As $i_{Z0}^* = (p_1^*)^{-1} = i_{Z1}^*$, (2.2.7.1), together with the definition of i_*, completes the proof. □

Finally, we check the compatibility of the Gysin morphism with the external products.

2.2.8. PROPOSITION. *Let $i\colon Z \to X$ be a closed embedding in $\mathcal{V}$ of codimension d, $\hat{Z}$ a closed subset of Z and $\hat{X}$ a closed subset of X with $\hat{Z} \subset \hat{X}$. Let W be in $\mathcal{V}$, $\hat{W}$ a closed subset of W. Then the diagram*

$$\begin{array}{ccc}
\mathbb{Z}_{Z,\hat{Z}}(-d)[-2d] \otimes \mathbb{Z}_{W,\hat{W}} & \xrightarrow{i_*\otimes \mathrm{id}} & \mathbb{Z}_{X,\hat{X}} \otimes \mathbb{Z}_{W,\hat{W}} \\
\downarrow{\scriptstyle \boxtimes_{Z,W}} & & \downarrow{\scriptstyle \boxtimes_{X,W}} \\
\mathbb{Z}_{Z\times_S W,\hat{Z}\times_S \hat{W}}(-d)[-2d] & \xrightarrow[(i\times \mathrm{id}_W)_*]{} & \mathbb{Z}_{X\times_S W,\hat{X}\times_S\hat{W}}
\end{array}$$

commutes in $\mathcal{DM}(\mathcal{V})$.

PROOF. Taking the diagram (2.1.2.1) and forming the product $(-)\times_S W$ gives the diagram defining the map $(i\times \mathrm{id}_W)_*$. We have the commutative diagram

$$\begin{array}{ccc}
\mathbb{Z}_{X,\hat{X}} \otimes \mathbb{Z}_{W,\hat{W}} & \xrightarrow{\boxtimes_{X,W}} & \mathbb{Z}_{X\times_S W,\hat{X}\times_S \hat{W}} \\
\uparrow{\scriptstyle i_0^*\otimes\mathrm{id}} & & \uparrow{\scriptstyle (i_0\times\mathrm{id}_W)^*} \\
\mathbb{Z}_{Y,\hat{Z}\times\mathbb{A}^1} \otimes \mathbb{Z}_{W,\hat{W}} & \xrightarrow{\boxtimes_{Y,W}} & \mathbb{Z}_{Y\times_S W,\hat{Z}\times_S\hat{W}\times\mathbb{A}^1} \\
\downarrow{\scriptstyle i_1^*\otimes\mathrm{id}} & & \downarrow{\scriptstyle (i_1\times\mathrm{id}_W)^*} \\
\mathbb{Z}_{P,s(\hat{Z})} \otimes \mathbb{Z}_{W,\hat{W}} & \xrightarrow{\boxtimes_{P,W}} & \mathbb{Z}_{P\times_S W,(s\times\mathrm{id}_W)(\hat{Z}\times_S\hat{W})} \\
\uparrow{\scriptstyle \cup[s(Z)]\otimes\mathrm{id}} & & \uparrow{\scriptstyle \cup[(s\times\mathrm{id}_W)(Z\times_S W)]} \\
\mathbb{Z}_{P,s(\hat{Z})}(-d)[-2d] \otimes \mathbb{Z}_{W,\hat{W}} & \xrightarrow{\boxtimes_{P,W}} & \mathbb{Z}_{P\times_S W,(s\times\mathrm{id}_W)(\hat{Z}\times_S\hat{W})}(-d)[-2d] \\
\uparrow{\scriptstyle f^*\otimes\mathrm{id}} & & \uparrow{\scriptstyle (f\times\mathrm{id}_W)^*} \\
\mathbb{Z}_{Z,\hat{Z}}(-d)[-2d] \otimes \mathbb{Z}_{W,\hat{W}} & \xrightarrow{\boxtimes_{Z,W}} & \mathbb{Z}_{Z\times_S W,\hat{Z}\times_S\hat{W}}(-d)[-2d].
\end{array}$$

This, together with the definition of i_* and $(i\times \mathrm{id}_W)_*$, completes the proof. □

2.3. Push-forward for a projection

We use the projective bundle formula to define the push-forward q_* for $q\colon \mathbb{P}(E) \to X$ the projective space bundle associated to a vector bundle $E \to X$.

2.3.1. *The definition of push-forward for a projection.* Let X be in $\mathcal{V}$, let $p\colon E \to X$ be a vector bundle of rank $N+1$, and let $q\colon \mathbb{P}(E) \to X$ the associated $\mathbb{P}^N$-bundle with tautological bundle $\mathcal{O}_E(1)$. Let $\hat{X}$ be a closed subset of X, and $\hat{P}_E$ the inverse image $q^{-1}(\hat{X})$. We let $\zeta = \mathrm{cl}^1(\mathcal{O}(1))$. By Theorem 1.3.2, we have the isomorphism

$$\alpha^E_{X,\hat{X}} := \sum_{i=0}^N \alpha_i^E\colon \oplus_{i=0}^N \mathbb{Z}_{X,\hat{X}}(N-i)[2N-2i] \to \mathbb{Z}_{\mathbb{P}(E),\hat{P}_E}(N)[2N].$$

We define the map $q_*\colon \mathbb{Z}_{\mathbb{P}(E),\hat{P}_E}(N)[2N] \to \mathbb{Z}_{X,\hat{X}}$ to be the composition $\pi_N \circ (\alpha^E_{X,\hat{X}})^{-1}$, where $\pi_N\colon \oplus_{i=0}^N \mathbb{Z}_{X,\hat{X}}(N-i)[2N-2i] \to \mathbb{Z}_{X,\hat{X}}$ is the projection on the summand $i = N$.

2.3.2. LEMMA. *Let X be in $\mathcal{V}$, and let $j: E \to F$ be a surjection of vector bundles on X, giving the closed embedding over X, $\bar{j}: \mathbb{P}(F) \to \mathbb{P}(E)$. Let $q_E: \mathbb{P}(E) \to X$, $q_F: \mathbb{P}(F) \to X$ be the structure morphisms. Then*

$$q_{F*} = q_{E*} \circ \bar{j}_*.$$

PROOF. Let $\zeta_E = c_1(\mathcal{O}_E(1))$ and $\zeta_F = c_1(\mathcal{O}_F(1))$. By the naturality of the first Chern class (Proposition 1.2.3(i)), and of the tautological bundle $\mathcal{O}(1)$, we have

$$\bar{j}^*(\zeta_E) = \zeta_F. \tag{2.3.2.1}$$

Let

$$\alpha_i^E : \mathbb{Z}_{X,\hat{X}}(-i)[-2i] \to \mathbb{Z}_{\mathbb{P}(E),\hat{P}_E},$$
$$\alpha_i^F : \mathbb{Z}_{X,\hat{X}}(-i)[-2i] \to \mathbb{Z}_{\mathbb{P}(F),\hat{P}_F}$$

be the maps (1.3.1.1), i.e., the respective compositions

$$\mathbb{Z}_{X,\hat{X}}(-i)[-2i] \xrightarrow{q_E^*} \mathbb{Z}_{\mathbb{P}(E),\hat{P}_E}(-i)[-2i] \xrightarrow{\cup\zeta_E^i} \mathbb{Z}_{\mathbb{P}(E),\hat{P}_E},$$
$$\mathbb{Z}_{X,\hat{X}}(-i)[-2i] \xrightarrow{q_F^*} \mathbb{Z}_{\mathbb{P}(F),\hat{P}_F}(-i)[-2i] \xrightarrow{\cup\zeta_F^i} \mathbb{Z}_{\mathbb{P}(F),\hat{P}_F}.$$

Let $N+1 = \operatorname{rnk} E$, $M+1 = \operatorname{rnk} F$, and $d = N - M$. Let $\bar{j}_*|\mathbb{P}(F)|$ denote the cycle on $\mathbb{P}(E)$ determined by the subscheme $\mathbb{P}(F)$; by (Appendix A, Remark 2.3.4), $\bar{j}_*|\mathbb{P}(F)|$ is an element of $\mathcal{Z}^d(\mathbb{P}(E)/S)$. Since $q_F = q_E \circ \bar{j}$, we have

$$\bar{j}_*(q_F^*(-) \cup \zeta_F^i) = \bar{j}_*(\mathrm{cl}^0_{\mathbb{P}(F)}(|\mathbb{P}(F)|) \cup \bar{j}^*(\zeta_E^i \cup q_E^*(-))) \qquad \text{by Proposition 3.5.6 and Proposition 3.5.3, Chapter I}$$

$$= \mathrm{cl}^d_{\mathbb{P}(E)}(\bar{j}_*|\mathbb{P}(F)|) \cup (\zeta_E^i \cup q_E^*(-)) \qquad \text{by Proposition 2.2.2 and Theorem 2.2.3.}$$

Thus

$$\bar{j}_* \circ \alpha_i^F(-) = \mathrm{cl}^d_{\mathbb{P}(E)}(\bar{j}_*|\mathbb{P}(F)|) \cup \alpha_i^E(-). \tag{2.3.2.2}$$

We claim that there are elements $a_i \in H^{2i}(X, \mathbb{Z}(i))$ such that

$$\mathrm{cl}^d_{\mathbb{P}(E)}(\bar{j}_*|\mathbb{P}(F)|) = \zeta_E^d + \sum_{i=1}^{d} q_E^*(a_i)\zeta_E^{d-i}. \tag{2.3.2.3}$$

Indeed, by the projective bundle formula (Theorem 1.3.2) we have

$$\mathrm{cl}^d_{\mathbb{P}(E)}(\bar{j}_*|\mathbb{P}(F)|) = \sum_{i=0}^{N} q_E^*(b_i)\zeta_E^i$$

for unique elements $b_i \in H^{2d-2i}(X, \mathbb{Z}(d-i))$. By the splitting principle (§1.3.3), we may assume that the kernel of the surjection $j: E \to F$ is a direct sum of line bundles: $\ker j = \oplus_{i=1}^d L_i$.

Let $t_i: q_E^* L_i \to \mathcal{O}_{\mathbb{P}(E)}(1)$ be the composition

$$q_E^* L_i \hookrightarrow q_E^* E \xrightarrow{\pi} \mathcal{O}_{\mathbb{P}(E)}(1),$$

where π is the canonical surjection. Twisting by $q_E^* L_i^{-1}$, t_i determines the section $s_i: \mathcal{O}_{\mathbb{P}(E)} \to \mathcal{O}_{\mathbb{P}(E)}(1) \otimes q_E^* L_i^{-1}$; let H_i be the zero subscheme of s_i. One checks by a local computation that H_i is smooth over X, hence the cycle $|H_i|$ is in $\mathcal{Z}^1(\mathbb{P}(E)/S)$. By Proposition 1.2.3(iii), we have $\mathrm{cl}^1(H_i) = \zeta_E - q_E^*(c_1(L_i))$.

One also checks by a local computation that $\bar{j}(\mathbb{P}(F))$ is the scheme-theoretic complete intersection $\bar{j}(\mathbb{P}(F)) = \cap_{i=1}^d H_i$. Thus

$$\begin{aligned}\mathrm{cl}^d_{\mathbb{P}(E)}(\bar{j}_*|\mathbb{P}(F)|) &= \mathrm{cl}^d_{\mathbb{P}(E)}(|H_1 \cap \ldots \cap H_d|)\\ &= \mathrm{cl}^1_{\mathbb{P}(E)}(|H_1|) \cup \ldots \cup \mathrm{cl}^1_{\mathbb{P}(E)}(|H_d|)\\ &= \prod_{i=1}^d (\zeta_E - q_E^*(c_1(L_i)))\\ &= \zeta_E^d + \sum_{i=1}^d q_E^*(a_i)\zeta_E^{d-i},\end{aligned}$$

as claimed.

Combining (2.3.2.3) with the identity (2.3.2.2), we have the identity

$$\bar{j}_* \circ \alpha_j^F = \alpha_{d+j}^E + \sum_{i=1}^d \alpha_{d+j-i}^E \circ (a_i \cup_X (-)), \text{ for } 0 \le j \le M.$$

Thus there is an $N+1 \times M+1$ matrix

$$P := (p_{ij})\,; \quad p_{ij} \in H^*(X, \mathbb{Z}(*)),$$

with

$$\begin{aligned} p_{d+j,j} &= 1; \quad j = 0, \ldots, M\\ p_{ij} &= 0; \quad \text{for } i > d+j,\end{aligned}$$

such that the diagram

$$\begin{array}{ccc} \mathbb{Z}_{\mathbb{P}(F),\hat{P}_F}(-d)[-2d] & \xrightarrow{\bar{j}_*} & \mathbb{Z}_{\mathbb{P}(E),\hat{P}_E}\\ \uparrow{\scriptstyle \alpha^F_{X,\hat{X}}} & & \uparrow{\scriptstyle \alpha^E_{X,\hat{X}}}\\ \oplus_{i=0}^M \mathbb{Z}_{X,\hat{X}}(-i-d)[-2i-2d] & \xrightarrow[P\cup(-)]{} & \oplus_{i=0}^N \mathbb{Z}_{X,\hat{X}}(-i)[-2i] \end{array}$$

commutes. This implies the desired identity $q_{F*} = q_{E*} \circ \bar{j}_*$. □

2.3.3. PROPOSITION [projection formula]. *Let X be in $\mathcal{V}$. Let $p\colon E \to X$ be a vector bundle of rank $N+1$, $q\colon P = \mathbb{P}(E) \to X$ the associated $\mathbb{P}^N$-bundle, $\hat{X}_i$, $i = 1, 2$ closed subsets of X, and $\hat{P}_i = q^{-1}(\hat{X}_i)$. Then the diagram*

$$\begin{array}{ccc} \mathbb{Z}_{P,\hat{P}_1}(N)[2N] \otimes \mathbb{Z}_{X,\hat{X}_2} & \xrightarrow{\mathrm{id}_P^* \otimes q^*} & \mathbb{Z}_{P,\hat{P}_1}(N)[2N] \otimes \mathbb{Z}_{P,\hat{P}_2}\\ & & \downarrow{\scriptstyle \cup_P}\\ \downarrow{\scriptstyle q_*\otimes \mathrm{id}} & & \mathbb{Z}_{P,\hat{P}_1\cap\hat{P}_2}(N)[2N]\\ & & \downarrow{\scriptstyle q_*}\\ \mathbb{Z}_{X,\hat{X}_1} \otimes \mathbb{Z}_{X,\hat{X}_2} & \xrightarrow[\cup_X]{} & \mathbb{Z}_{X,\hat{X}_1\cap\hat{X}_2} \end{array}$$

commutes.

PROOF. We give the proof in case $\hat{X}_1 = \hat{X}_2 = X$ to simplify the notation. The associativity and commutativity of products implies the commutativity of the diagram

$$\begin{array}{ccc} \mathbb{Z}_X \otimes \mathbb{Z}_X & \xrightarrow{\alpha_i^E \otimes q^*} & \mathbb{Z}_{P,\hat{P}_1}(i)[2i] \otimes \mathbb{Z}_P \\ \downarrow{\scriptstyle \cup_X} & & \downarrow{\scriptstyle \cup_P} \\ \mathbb{Z}_X & \xrightarrow[\alpha_i^E]{} & \mathbb{Z}_P(i)[2i]. \end{array}$$

This, together with the definition of q_*, proves the proposition. □

2.3.4. LEMMA. *Let $q: \mathbb{P}(E) \to X$ be the projective bundle associated to a rank $N+1$ vector bundle $E \to X$ on X, $p: W \to X$ a morphism in $\mathcal{V}$, giving the cartesian diagram*

$$\begin{array}{ccccc} \mathbb{P}(p^*E) & = & W \times_X \mathbb{P}(E) & \xrightarrow{p_2} & \mathbb{P}(E) \\ & & \downarrow{\scriptstyle p_1} & & \downarrow{\scriptstyle q} \\ & & W & \xrightarrow[p]{} & X. \end{array}$$

Then

$$p^* \circ q_* = p_{1*} \circ p_2^*.$$

PROOF. Let $\zeta = c_1\mathcal{O}_E(1)$ and $\zeta_W = c_1\mathcal{O}_{p^*E}(1)$. Then $p_2^*(\zeta) = \zeta_W$. This implies the relation $p_2^* \circ \alpha_i^E = \alpha_i^{p^*E} \circ p^*$, which in turn implies the desired result. □

2.3.5. REMARK. In the case of the dimension 0 projective bundle, $q: \mathbb{P}_X^0 = X \to X$, the projective bundle isomorphism is the identity map, hence $q_* = \mathrm{id}$.

2.3.6. PROPOSITION. *Let $p: E \to X$ be a vector bundle of rank $N+1$, $q: P := \mathbb{P}(E) \to X$ the associated $\mathbb{P}^N$-bundle, $\hat{X}$ a closed subset of X, $\hat{P} = q^{-1}(\hat{X})$. Let W be in $\mathcal{V}$, $\hat{W}$ a closed subset of W giving us the projective bundle $q \times \mathrm{id}_W : P \times_S W \to X \times_S W$ associated to the vector bundle $p_1^*E \to X \times_S W$. Then the diagram*

$$\begin{array}{ccc} \mathbb{Z}_{P,\hat{P}}(N)[2N] \otimes \mathbb{Z}_{W,\hat{W}} & \xrightarrow{q_* \otimes \mathrm{id}} & \mathbb{Z}_{X,\hat{X}} \otimes \mathbb{Z}_{W,\hat{W}} \\ \downarrow{\scriptstyle \boxtimes_{P,W}} & & \downarrow{\scriptstyle \boxtimes_{X,W}} \\ \mathbb{Z}_{P \times_S W, \hat{P} \times_S \hat{W}}(N)[2N] & \xrightarrow[(q \times \mathrm{id}_W)_*]{} & \mathbb{Z}_{X \times_S W, \hat{X} \times_S \hat{W}} \end{array}$$

commutes.

PROOF. Since $p_1^*(c_1(\mathcal{O}_{\mathbb{P}(E)}(1))) = c_1(\mathcal{O}_{\mathbb{P}(p_1^*E)}(1))$, we have the commutative diagram

$$\begin{array}{ccc} \mathbb{Z}_{X,\hat{X}}(N-i)[2N-2i] \otimes \mathbb{Z}_{W,\hat{W}} & \xrightarrow{\alpha_i^E \otimes \mathrm{id}} & \mathbb{Z}_{P,\hat{P}}(N)[2N] \otimes \mathbb{Z}_{W,\hat{W}} \\ \downarrow{\scriptstyle \boxtimes_{X,W}} & & \downarrow{\scriptstyle \boxtimes_{P,W}} \\ \mathbb{Z}_{X \times_S W, \hat{X} \times_S \hat{W}}(N-i)[2N-2i] & \xrightarrow[\alpha_i^{p_1^*E}]{} & \mathbb{Z}_{P \times_S W, \hat{P} \times_S \hat{W}}(N)[2N]. \end{array}$$

This, together with the definition of q_* and $(q \times \mathrm{id}_W)_*$, proves the result. □

2.4. Push-forward for a projective morphism

In this section, we assume that the base-scheme S admits an ample family of line bundles. Since each scheme X in $\mathcal{V} \subset \mathbf{Sm}_S$ is a quasi-projective scheme over S, this implies that X admits an ample family of line bundles as well, i.e., for each vector bundle $E \to X$ there is a line bundle L on X such that $E \otimes L$ is generated by global sections.

2.4.1. Let $p: Y \to X$ be a projective morphism in $\mathcal{V}$. By definition, there is a vector bundle $E \to X$, with associated projective bundle $q: \mathbb{P}(E) \to X$, and a closed embedding $i: Y \to \mathbb{P}(E)$ such that $p = q \circ i$.

2.4.2. LEMMA. *Suppose X and Y are of pure dimension d and e over S, respectively. Let $\hat{X}$ be a closed subset of X, $\hat{Y}$ a closed subset of Y such that $p(\hat{Y}) \subset \hat{X}$. Let $p: Y \to X$ be a projective morphism, and let $Y \xrightarrow{i} \mathbb{P}(E) \xrightarrow{q} X$ be a factorization of p as a closed embedding followed by a projection. Then the composition*

$$\mathbb{Z}_{Y,\hat{Y}}(e)[2e] \xrightarrow{i_*} \mathbb{Z}_{\mathbb{P}(E),q^{-1}(\hat{X})}(a)[2a] \xrightarrow{q_*} \mathbb{Z}_{X,\hat{X}}(d)[2d]$$

($a = \dim_S(\mathbb{P}(E))$) depends only on the morphism p.

PROOF. Suppose we have another factorization of p as $q' \circ i'$, with $i': Y \to \mathbb{P}(E')$ a closed embedding, and $q': \mathbb{P}(E') \to X$ the projection. The projections $E \oplus E' \to E$, $E \oplus E' \to E'$ induce the closed embeddings $j: \mathbb{P}(E) \to \mathbb{P}(E \oplus E')$ and $j': \mathbb{P}(E') \to \mathbb{P}(E \oplus E')$. Letting $r: \mathbb{P}(E \oplus E') \to X$ be the structure morphism, we have the commutative diagrams

$$\begin{array}{ccc} Y & \xrightarrow{i} & \mathbb{P}(E) \\ {\scriptstyle p}\downarrow & \swarrow{\scriptstyle q} & \downarrow{\scriptstyle j} \\ X & \xleftarrow[r]{} & \mathbb{P}(E \oplus E') \end{array} \qquad \begin{array}{ccc} Y & \xrightarrow{i'} & \mathbb{P}(E') \\ {\scriptstyle p}\downarrow & \swarrow{\scriptstyle q'} & \downarrow{\scriptstyle j'} \\ X & \xleftarrow[r]{} & \mathbb{P}(E \oplus E'). \end{array}$$

This reduces us to considering the case of a projection $j: E = F \oplus F' \to F$, giving the induced closed embedding $\bar{j}$:

$$\begin{array}{ccc} \mathbb{P}(F) & \xrightarrow{\bar{j}} & \mathbb{P}(E) \\ & {\scriptstyle q'}\searrow \quad \swarrow{\scriptstyle q} & \\ & X & \end{array}$$

and closed embedding $i': Y \to \mathbb{P}(F)$ with $i = \bar{j} \circ i'$.

By Proposition 2.2.1 and Lemma 2.3.2, we have

$$\begin{aligned} q_* \circ i_* &= q_* \circ \bar{j}_* \circ i'_* \\ &= q'_* \circ i'_*, \end{aligned}$$

completing the proof. □

2.4.3. DEFINITION. Let $p: Y \to X$ be a projective morphism in $\mathcal{V}$. Suppose X and Y are of pure dimension d and e over S, respectively. Let $\hat{X}$ be a closed subset of X, $\hat{Y}$ a closed subset of Y such that $p(\hat{Y}) \subset \hat{X}$. Choose a vector bundle $E \to X$, with associated projective bundle $q: \mathbb{P}(E) \to X$, and a closed embedding $i: Y \to \mathbb{P}(E)$

such that $p = q \circ i$. Define

$$p_*: \mathbb{Z}_{Y,\hat{Y}}(e)[2e] \to \mathbb{Z}_{X,\hat{X}}(d)[2d]$$

to be the composition $q_* \circ i_*$. By Lemma 2.4.2, p_* is well-defined.

2.4.4. Let X be in $\mathcal{V}$. For integers $N, M \geq 0$, we let $i_{N,M}: \mathbb{P}^N_X \times_X \mathbb{P}^M_X \to \mathbb{P}^{NM+N+M}_X$ be the Segre embedding. We let $q^n: \mathbb{P}^n_X \to X$ denote the structure morphism, and

$$p_1: \mathbb{P}^N_X \times_X \mathbb{P}^M_X \to \mathbb{P}^N_X,$$
$$p_2: \mathbb{P}^N_X \times_X \mathbb{P}^M_X \to \mathbb{P}^M_X$$

the projections.

2.4.5. LEMMA. *We have*

$$q^N_* \circ p_{1*} = q^M_* \circ p_{2*} = q^{NM+N+M}_* \circ i_{N,M*}.$$

PROOF. We give the proof without closed support to simplify the notation.

Let ζ_1 be the first Chern class of the tautological line bundle on $\mathbb{P}^N_X$, ζ_2 the first Chern class of the tautological line bundle on $\mathbb{P}^M_X$, and ζ the first Chern class of the tautological line bundle on $\mathbb{P}^{NM+N+M}_X$.

Two applications of Theorem 1.3.2 give the isomorphism

$$\alpha^{N,M} := \sum_{i=0}^{N}\sum_{j=0}^{M} \alpha_{i,j}: \oplus_{i=0}^N \oplus_{j=0}^M \mathbb{Z}_X(N+M-i-j)[2N+2M-2i-2j] \\ \to \mathbb{Z}_{\mathbb{P}^N_X \times_X \mathbb{P}^M_X}(N+M)[2N+2M],$$

where $\alpha_{i,j}$ is the composition

$$\mathbb{Z}_X(N+M-i-j)[2N+2M-2i-2j] \\ \xrightarrow{(q^N\circ p_1)^*} \mathbb{Z}_{\mathbb{P}^N_X \times_X \mathbb{P}^M_X}(N+M-i-j)[2N+2M-2i-2j] \\ \xrightarrow{\cup(p_1^*(\zeta_1^i)\cup p_2^*(\zeta_2^j))} \mathbb{Z}_{\mathbb{P}^N_X \times_X \mathbb{P}^M_X}(N+M)[2N+2M].$$

It is easy to see that the composition

$$q^N_* \circ p_{1*} \circ \alpha^{N,M}: \oplus_{i=0}^N \oplus_{j=0}^M \mathbb{Z}_X(N+M-i-j)[2N+2M-2i-2j] \to \mathbb{Z}_X$$

is projection on the factor $\mathbb{Z}_X$ $(i = N,\ j = M)$.

Let $K = NM + M + N$. We denote the standard homogeneous coordinates on $\mathbb{P}^N_X$, $\mathbb{P}^M_X$ and $\mathbb{P}^K_X$ by $Y^1_0, \ldots, Y^1_N$, $Y^2_0, \ldots, Y^2_M$, and $X_0, \ldots, X_K$, respectively.

By Proposition 1.2.3(iii), ζ_1 and ζ_2 are the classes of respective hyperplanes $Y^1_i = 0$, $Y^2_j = 0$ in $\mathbb{P}^N_X$ and $\mathbb{P}^M_X$ (for any choice of i or j). Thus $p_1^*(\zeta_1^i) \cup p_2^*(\zeta_2^j) = \mathrm{cl}^{i+j}(L^i_1 \times L^j_2)$, where L^i_1 is any codimension i linear subspace of $\mathbb{P}^N_X$ defined by equations of the form $Y^1_{n_1} = \ldots = Y^1_{n_i} = 0$, $0 \leq n_1 < \ldots < n_i \leq N$, and L^j_2 is any codimension j linear subspace of $\mathbb{P}^M_X$ defined by equations of the form $Y^2_{m_1} = \ldots = Y^2_{m_j} = 0$, $0 \leq m_1 < \ldots < m_j \leq M$. By Theorem 2.2.3, we have

$$i_{M,N*}(p_1^*(\zeta_1^i) \cup p_2^*(\zeta_2^j)) = \mathrm{cl}^{NM+i+j}(i_{M,N*}(L^i_1 \times_X L^j_2)).$$

Let $H \subset \mathbb{P}^{NM+N+M}_X$ be the hyperplane $X_0 = 0$. Then $\zeta = \mathrm{cl}^1(H)$ and the cycle $H \cdot (\mathbb{P}^N_X \times_X \mathbb{P}^M_X)$ is rationally equivalent to $L^1_1 \times \mathbb{P}^M_X + \mathbb{P}^N_X \times L^1_2$. Counting intersection multiplicities, and using the classical projection formula for algebraic cycles, we have the rational equivalence

$$i_{M,N*}(|L_1^i \times L_2^j|) \sim_r \binom{N+M-i-j}{N-i} H^{(NM+i+j)}$$

for $i \leq N$ and $j \leq M$. Thus, applying Chapter I, Proposition 3.5.3, we have

$$\mathrm{cl}^{NM+i+j}(i_{M,N*}(|L_1^i \times L_2^j|)) = \binom{N+M-i-j}{N-i} \zeta^{NM+i+j}$$

if $i \leq N$ and $j \leq M$. Therefore

$$i_{M,N*}(p_1^*(\zeta_1^i) \cup p_2^*(\zeta_2^j)) = \binom{N+M-i-j}{N-i} \zeta^{NM+i+j} \tag{2.4.5.1}$$

if $i \leq N$ and $j \leq M$.

We have the isomorphism of Theorem 1.3.2

$$\alpha^K = \sum_{k=0}^{K} \alpha_k : \oplus_{k=0}^K \mathbb{Z}_X(K-k)[2K-2k] \to \mathbb{Z}_{\mathbb{P}_X^K}(K)[2K];$$

from the projection formula Proposition 2.2.2, and (2.4.5.1), we have the identity

$$i_{M,N*} \circ \alpha_{i,j} = \binom{N+M-i-j}{N-i} \alpha_{NM+i+j}$$

if $i \leq N$ and $j \leq M$. In addition $q_*^K \circ \alpha^K$ is the projection on the factor $\mathbb{Z}_X$ (i.e., the summand $k = K$).Thus $q_*^K \circ i_{M,N*} \circ \alpha^{N,M}$ is the projection on the factor $\mathbb{Z}_X$ (i.e., the summand $i = N$, $j = M$), hence $q_*^K \circ i_{M,N*} = q_*^N \circ p_{1*}$. The identity $q_*^K \circ i_{M,N*} = q_*^M \circ p_{2*}$ is proved similarly. □

2.4.6. LEMMA. *Let $E \to X$ be a vector bundle, $i: Z \to X$ a closed embedding in $\mathcal{V}$, $i^*E \to Z$ the restriction of E to Z, giving the cartesian diagram*

$$\begin{array}{ccc} \mathbb{P}(i^*E) & \xrightarrow{j} & \mathbb{P}(E) \\ \downarrow{\scriptstyle q'} & & \downarrow{\scriptstyle q} \\ Z & \xrightarrow[i]{} & X. \end{array}$$

Then

$$q_* \circ j_* = i_* \circ q'_*.$$

PROOF. We give the proof without closed support. Suppose E has rank $N+1$. We have the isomorphisms of Theorem 1.3.2

$$\alpha^E = \sum_{k=0}^{N} \alpha_k^E : \oplus_{k=0}^N \mathbb{Z}_X(-k)[-2k] \to \mathbb{Z}_{\mathbb{P}(E)}$$

$$\alpha^{i^*E} = \sum_{k=0}^{N} \alpha_k^{i^*E} : \oplus_{k=0}^N \mathbb{Z}_Z(-k)[-2k] \to \mathbb{Z}_{\mathbb{P}(i^*E)},$$

with

$$\alpha_k^E = \cup\zeta_E^k \circ q^*,$$
$$\alpha_k^{i^*E} = \cup\zeta_{i^*E}^k \circ q'^*.$$

By the functoriality of c_1 (Proposition 1.2.3(i)), we have $i^*(\zeta_E) = \zeta_{i^*E}$; thus, by Lemma 2.2.4 and the projection formula (Proposition 2.2.2), we have $\alpha_k^E \circ i_* = j_* \circ \alpha_k^{i^*E}$. This, together with the definition of q_* and q'_*, proves the lemma. □

2.4.7. THEOREM. *Suppose we have a sequence of projective morphisms in $\mathcal{V}$,*

$$Z \xrightarrow{p'} Y \xrightarrow{p} X,$$

with X, Y and Z of pure dimension d, e and f over S, respectively, together with closed subsets $\hat{X}$ of X, $\hat{Y}$ of Y and $\hat{Z}$ of Z, such that $p'(\hat{Z}) \subset \hat{Y}$ and $p(\hat{Y}) \subset \hat{X}$. Then the diagram

$$\begin{array}{ccc} \mathbb{Z}_{Z,\hat{Z}}(f)[2f] & \xrightarrow{p'_*} & \mathbb{Z}_{Y,\hat{Y}}(e)[2e] \\ & {\scriptstyle (p\circ p')_*} \searrow & \downarrow {\scriptstyle p_*} \\ & & \mathbb{Z}_{X,\hat{X}}(d)[2d] \end{array}$$

commutes. In addition $\mathrm{id}_{X*} = \mathrm{id}$.

PROOF. As we may factor id_X as a composition of the identity closed embedding into $X = \mathbb{P}^0_X$, followed by the identity projection $q\colon \mathbb{P}^0_X \to X$, the assertion $\mathrm{id}_{X*} = \mathrm{id}$ follows from Proposition 2.2.1 and Remark 2.3.5.

For the first assertion, let E be a vector bundle on X. Since, by assumption, X has an ample family of line bundles, there is a line bundle L on X and a surjection $1^N \to E \otimes L$ for $N >> 0$, where 1^N is the trivial rank N vector bundle on X. This gives the closed embedding $\mathbb{P}(E) \cong \mathbb{P}(E \otimes L) \to \mathbb{P}^N_X$. Thus, we may factor the projective morphism $p\colon Y \to X$ as $Y \xrightarrow{i} \mathbb{P}^N_X \xrightarrow{q^N} X$ with i a closed embedding and q the projection. Similarly, we may factor p' as $Z \xrightarrow{i'} \mathbb{P}^M_Y \xrightarrow{q^M} Y$.

We therefore have the closed embeddings

$$j\colon Z \to \mathbb{P}^N_X \times_X \mathbb{P}^M_X$$
$$j'\colon \mathbb{P}^M_Y \to \mathbb{P}^N_X \times_X \mathbb{P}^M_X$$

with

$$p \circ p' = q^N \circ p_1 \circ j$$
$$j = j' \circ i',$$
$$p_1 \circ j' = i \circ q^M.$$

This gives the factorization of $p \circ p'$ as

$$p \circ p' = q^{NM+N+M} \circ i_{N,M} \circ j' \circ i'.$$

Thus, we have

$$\begin{aligned} (p \circ p')_* &= q_*^{NM+N+M} \circ (i_{N,M} \circ j' \circ i')_*, && \text{by Definition 2.4.3} \\ &= q_*^{NM+N+M} \circ i_{N,M*} \circ j'_* \circ i'_*, && \text{by Proposition 2.2.1} \\ &= q_*^N \circ p_{1*} \circ j'_* \circ i'_*, && \text{by Lemma 2.4.5} \\ &= q_*^N \circ i_* \circ q_*^M \circ i'_*, && \text{by Lemma 2.4.6} \\ &= p_* \circ p'_* && \text{by Definition 2.4.3.} \end{aligned}$$

□

The projection formula for a closed embedding and for a projection give the general version.

2.4.8. THEOREM [projection formula]. *Let $p: Y \to X$ be a projective morphism in $\mathcal{V}$, with X and Y pure dimension d and e, respectively. Let $\hat{X}_i$, $i = 1, 2$, be closed subsets of X, and let $\hat{Y}_i = p^{-1}(\hat{X}_i)$, $i = 1, 2$. Then the diagram*

$$
\begin{array}{ccc}
\mathbb{Z}_{Y,\hat{Y}_1}(e)[2e] \otimes \mathbb{Z}_{X,\hat{X}_2} & \xrightarrow{\mathrm{id}\otimes p^*} & \mathbb{Z}_{Y,\hat{Y}_1}(e)[2e] \otimes \mathbb{Z}_{Y,\hat{Y}_2} \\
 & & \big\downarrow{\scriptstyle \cup_Y} \\
{\scriptstyle p_*\otimes \mathrm{id}}\big\downarrow & & \mathbb{Z}_{Y,\hat{Y}_1\cap\hat{Y}_2}(e)[2e] \\
 & & \big\downarrow{\scriptstyle p_*} \\
\mathbb{Z}_{X,\hat{X}_1}(d)[2d] \otimes \mathbb{Z}_{X,\hat{X}_2} & \xrightarrow[\cup_X]{} & \mathbb{Z}_{X,\hat{X}_1\cap\hat{X}_2}(d)[2d]
\end{array}
$$

commutes.

PROOF. This follows from the definition of p_*, together with Proposition 2.2.2 and Proposition 2.3.3. □

The usual compatibility of push-forward with pull-back in cartesian square is valid as well. We call a cartesian square

$$
\begin{array}{ccc}
Y \times_X Z & \xrightarrow{p_2} & Z \\
{\scriptstyle p_1}\big\downarrow & & \big\downarrow{\scriptstyle f} \\
Y & \xrightarrow[p]{} & X
\end{array}
$$

in $\mathcal{V}$ *transverse* if $\mathrm{Tor}_p^{\mathcal{O}_X}(\mathcal{O}_Z, \mathcal{O}_Y) = 0$ for all $p > 0$.

2.4.9. THEOREM. *Let*

$$
\begin{array}{ccc}
Y \times_X Z & \xrightarrow{p_2} & Z \\
{\scriptstyle p_1}\big\downarrow & & \big\downarrow{\scriptstyle f} \\
Y & \xrightarrow[p]{} & X
\end{array}
$$

be a transverse cartesian square in $\mathcal{V}$, with p a projective morphism. Then

$$f^* \circ p_* = p_{2*} \circ p_1^*.$$

PROOF. Let

$$Y \xrightarrow{i} X \times_S \mathbb{P}_S^N \xrightarrow{q} X$$

be a factorization of p, with i a closed embedding and q the projection. We have the isomorphism

$$Y \times_X Z \cong Y \times_{X \times_S \mathbb{P}_S^N} Z \times_S \mathbb{P}_S^N.$$

Let $j: Y \times_X Z \to Z \times_S \mathbb{P}_S^N$ be the map induced by the projection $Y \times_{X\times_S\mathbb{P}_S^N} Z \times_S \mathbb{P}_S^N \to Z \times_S \mathbb{P}_S^N$, and let $r: Z \times_S \mathbb{P}_S^N \to Z$ be the projection. We have the

commutative diagram

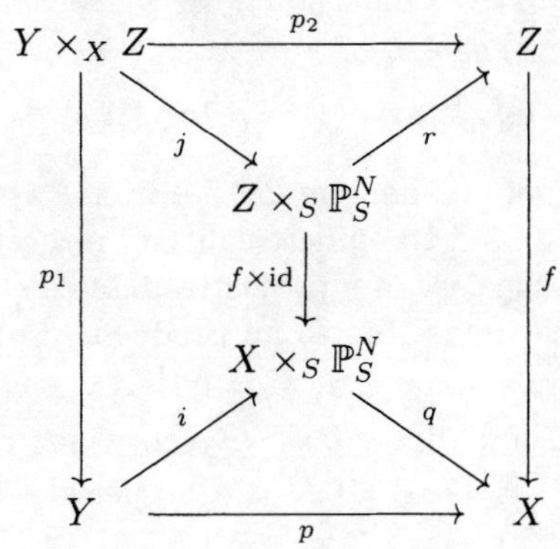

with the two trapezoids cartesian; in particular, j is a closed embedding. By definition, we have $p_{2*} = r_* \circ i_*$ and $p_* = q_* \circ i_*$. Since q is faithfully flat, the transversality hypothesis implies that $\mathrm{Tor}_p^{\mathcal{O}_{X\times_S\mathbb{P}^N}}(\mathcal{O}_Y, \mathcal{O}_{Z\times_S\mathbb{P}^N}) = 0$ for all $p > 0$. The result then follows from Lemma 2.2.4 and Lemma 2.3.4. □

2.4.10. THEOREM. *Let $p: X \to Y$ be a projective morphism in $\mathcal{V}$ of relative dimension d, $\hat{X}$ a closed subset of X and $\hat{Y}$ a closed subset of Y with $f(\hat{X}) \subset \hat{Y}$. Let W be in $\mathcal{V}$, with closed subset $\hat{W}$. Then the diagram*

$$\begin{array}{ccc}
\mathbb{Z}_{X,\hat{X}}(d)[2d] \otimes \mathbb{Z}_{W,\hat{W}} & \xrightarrow{p_*\otimes\mathrm{id}} & \mathbb{Z}_{Y,\hat{Y}} \otimes \mathbb{Z}_{W,\hat{W}} \\
\downarrow{\scriptstyle \boxtimes_{X,W}} & & \downarrow{\scriptstyle \boxtimes_{Y,W}} \\
\mathbb{Z}_{X\times_S W, \hat{X}\times_S\hat{W}}(d)[2d] & \xrightarrow[(p\times\mathrm{id}_W)_*]{} & \mathbb{Z}_{Y\times_S W, \hat{Y}\times_S\hat{W}}
\end{array}$$

commutes.

PROOF. This follows from Proposition 2.3.6, Proposition 2.2.8, and the definition (Definition 2.4.3) of p_* and $(p \times \mathrm{id}_W)_*$. □

2.5. Some useful results

We collect some miscellaneous results on projective push-forward.

2.5.1. *Naturality.* Let $i: Z \to X$ be a closed codimension d embedding in $\mathbf{Sm}_S^{\mathrm{ess}}$, and let $f: T \to S$ be a map of reduced schemes. Then, applying f^* to the deformation diagram (2.1.2.1) for i gives us the deformation diagram for the closed embedding $f^*(i)$. This, together with the definition of the pull-back functor (Chapter I, §2.3) $f^*: \mathcal{DM}(\mathbf{Sm}_S^{\mathrm{ess}}) \to \mathcal{DM}(\mathbf{Sm}_T^{\mathrm{ess}})$, implies that $f^*(i_*) = f^*(i)_*$. Similarly, we have $f^*(q_*) = f^*(q)_*$ for $q: \mathbb{P}(E) \to X$ a projective bundle, hence the push-forward for a projective morphism $p: X \to Y$, is natural with respect to f^*:

$$f^*(p_*) = f^*(p)_*.$$

2.5.2. *Push-forward of cycles.* We have shown in Theorem 2.2.3 that cycle classes are compatible with the Gysin morphism. The situation for a general projective morphism is not as clear; however, in case the base scheme is $\mathrm{Spec}\, k$, k a field, the semi-purity of motivic cohomology enables us to prove compatibility in general.

2.5.3. THEOREM. *Suppose $S = \operatorname{Spec} k$, where k is a field. Let $p: Y \to X$ be a projective morphism in $\mathcal{V}$ of relative dimension d, W an element of $\mathcal{Z}^q(Y/S)$, supported on a closed subset $\hat{Y}$ of Y. Then*

$$\mathrm{cl}^{q-d}_{p(\hat{Y})}(p_*(W)) = p_*(\mathrm{cl}^q_{\hat{Y}}(W)).$$

PROOF. As in the proof of Theorem 2.2.5, we may assume that k is a perfect field. Using Theorem 2.2.3, and the functoriality of projective push-forward (Theorem 2.4.7), we reduce to the case of a projective bundle $q: \mathbb{P}^N_X \to X$. We may also assume that W is the cycle associated to an irreducible subscheme A of $\mathbb{P}^N_X$.

Let $B = q(A)$; we may assume that $\hat{Y} = q^{-1}(A)$. By Theorem 2.2.5, we have $H^{2(q-d)}_B(X, \mathbb{Z}(q-d)) = 0$ if $\operatorname{codim}_X(B) > q - d$, which proves the result in this case. Suppose $\operatorname{codim}_X(B) = q - d$. If $\hat{B}$ is a proper closed subset of B, we have the exact sequence

$$H^{2(q-d)}_{\hat{B}}(X, \mathbb{Z}(q-d)) \to H^{2(q-d)}_B(X, \mathbb{Z}(q-d)) \to H^{2(q-d)}_{B\setminus\hat{B}}(X\setminus\hat{B}, \mathbb{Z}(q-d))$$
$$\to H^{2(q-d)+1}_{\hat{B}}(X, \mathbb{Z}(q-d)).$$

Applying Theorem 2.2.5, we arrive at the isomorphism

$$H^{2(q-d)}_B(X, \mathbb{Z}(q-d)) \to H^{2(q-d)}_{B\setminus\hat{B}}(X\setminus\hat{B}, \mathbb{Z}(q+d));$$

we may therefore remove from X any proper closed subset of B. In particular, we may assume that B is smooth over k, and that A is finite over B, hence A has codimension N in $\mathbb{P}^N_B$.

We have the cartesian square

(2.5.3.1)
$$\begin{array}{ccc} \mathbb{P}^N_B & \xrightarrow{j} & \mathbb{P}^N_X \\ {\scriptstyle q'}\downarrow & & \downarrow{\scriptstyle q} \\ B & \xrightarrow[i]{} & X. \end{array}$$

By Theorem 2.4.7 and Theorem 2.2.3, we need only show that $\mathrm{cl}^0_B(q'_*(|A|)) = q'_*(\mathrm{cl}^N_{\mathbb{P}^N_B}(|A|))$.

Let $s: B \to \mathbb{P}^N_B$ be a constant section, and let $\zeta = c_1\mathcal{O}_{\mathbb{P}^N_B}(1)$. Since A is finite over B, we have the rational equivalence $|A| \sim_r a \cdot s(B) + W'$, where W' is a cycle supported over a proper closed subset $\hat{B}$ of B, and a is the degree of A over B. Removing $\hat{B}$, we may assume that $W' = 0$. Since the cycle class respects rational equivalence, and since $\mathrm{cl}^N(s(|B|)) = \zeta^N$, we have

$$\mathrm{cl}^N_{\mathbb{P}^N_B}(|A|) = a \cdot \zeta^N$$
$$q'_*(|A|) = a \cdot |B|.$$

Since the composite

$$\mathbb{Z}_B \xrightarrow{\cup\zeta^N \circ q^*} \mathbb{Z}_{\mathbb{P}^N_B} \xrightarrow{q'_*} \mathbb{Z}_B$$

is the identity, we have

$$\begin{aligned} \mathrm{cl}^0(|B|) &= q'_*(\zeta^N \cup q'^*(\mathrm{cl}^0(|B|))) \\ &= q'_*(\zeta^N). \end{aligned}$$

Thus $q'_*(\mathrm{cl}^N_{\mathbb{P}^N_B}(|A|)) = a \cdot \mathrm{cl}^0_B(|B|)$, verifying (2.5.3.1) and completing the proof. □

2.5.4. COROLLARY. *Suppose $S = \operatorname{Spec} k$, with k a field. Let $f: Y \to X$ be a finite morphism in $\mathcal{V}$ of degree d. Then*

$$f_* \circ f^* = d \cdot \mathrm{id}_{\mathbb{Z}_X}$$

in $\mathbf{D}^b_{\mathrm{mot}}(\mathcal{V})$.

PROOF. Let $p_Y: Y \to S$, $p_X: X \to S$ be the structure morphisms. Since $p_Y = p_X \circ f$, and $\mathrm{cl}^0_Y(|Y|) = p_Y^*(\mathrm{cl}^0_S(|S|)) = p_Y^*$, (Chapter I, Proposition 3.5.3 and Lemma 3.5.4), the projection formula (Theorem 2.4.8) implies that $f_* \circ f^*$ is equal to the composition

$$\mathbb{Z}_X \cong 1 \otimes \mathbb{Z}_X \xrightarrow{f_*(\mathrm{cl}^0_Y(|Y|)) \otimes \mathrm{id}} \mathbb{Z}_X \otimes \mathbb{Z}_X \xrightarrow{\cup_X} \mathbb{Z}_X.$$

By Theorem 2.5.3, we have

$$f_*(\mathrm{cl}^0_Y(|Y|)) = d \cdot \mathrm{cl}^0_X(|X|) = d \cdot p_X^*,$$

whence the result. □

2.5.5. *Borel-Moore motives.* We will examine the Borel-Moore motive and extend its definition to certain singular varieties in Chapter IV, §2.2, §2.3 and §2.4; we content ourselves here with a brief introduction.

2.5.6. DEFINITION. (i) Let $\mathcal{V}_{\mathrm{proj}}$ be the subcategory of $\mathcal{V}$ with the same objects, and with morphisms being the projective morphisms. Let $\mathbf{P}\mathcal{V}_{\mathrm{proj}}$ be the category of pairs in $\mathcal{V}_{\mathrm{proj}}$, i.e., objects are pairs (X, W), with X in $\mathcal{V}$ and W a closed subset of X, and a morphism $f:(X,W) \to (Y,T)$ is a projective map $f: X \to Y$ with $f(W) \subset T$.
(ii) Let X be in $\mathcal{V}$. If X is connected, then X is equi-dimensional of dimension d_X over some connected component of S; we define the *Borel-Moore motive of* X, $\mathbb{Z}_X^{\mathrm{B.M.}}$, by

$$\mathbb{Z}_X^{\mathrm{B.M.}} = \mathbb{Z}_X(d_X)[2d_X].$$

We extend the definition of $\mathbb{Z}_X^{\mathrm{B.M.}}$ to general $X \in \mathcal{V}$ by taking direct sums over the connected components of X.
(iii) Let X be in $\mathcal{V}$, W a closed subset of X, and $j: U \to X$ the complement of W. Define the *Borel-Moore motive of X with support in W*, $\mathbb{Z}_{X,W}^{\mathrm{B.M.}}$, by

$$\mathbb{Z}_{X,W}^{\mathrm{B.M.}} := \mathrm{cone}(j^*: \mathbb{Z}_X^{\mathrm{B.M.}} \to \mathbb{Z}_U^{\mathrm{B.M.}})[-1].$$

2.5.7. THEOREM. *Sending $(X, W) \in \mathbf{P}\mathcal{V}$ to $\mathbb{Z}_{X,W}^{\mathrm{B.M.}}$, and sending a projective morphism $f:(X,W) \to (Y,T)$ to $f_*: \mathbb{Z}_{X,W}^{\mathrm{B.M.}} \to \mathbb{Z}_{Y,T}^{\mathrm{B.M.}}$ defines a functor*

$$\mathbb{Z}^{\mathrm{B.M.}}: \mathbf{P}\mathcal{V}_{\mathrm{proj}} \to \mathcal{DM}(\mathcal{V}).$$

PROOF. The map f_* is defined in Definition 2.4.3; the functoriality follows from Theorem 2.4.7. □

2.6. Push-forward for diagrams

We will now explain how to extend the construction of the push-forward morphism to diagrams of schemes.

2.6.1. *Projective morphisms of diagrams.* Let I be a small category, and let $Z: I \to \mathbf{Sm}_S^{\mathrm{ess}}$ and $X: I \to \mathbf{Sm}_S^{\mathrm{ess}}$ be functors. We call a morphism of functors $f: Z \to X$ a *codimension d closed embedding* if

1. For each $i \in I$, the morphism $f(i): Z(i) \to X(i)$ is a codimension d closed embedding.
2. For each morphism $s: i \to j$ in I, the canonical map $f/s: Z(i) \to Z(j) \times_{X(j)} X(i)$ induced by the diagram

$$\begin{array}{ccc} Z(i) & \xrightarrow{f(i)} & X(i) \\ {\scriptstyle Z(s)}\downarrow & & \downarrow{\scriptstyle X(s)} \\ Z(j) & \xrightarrow[f(j)]{} & X(j) \end{array}$$

   is an isomorphism.

One easily verifies that the composition of a codimension d closed embedding with a codimension e closed embedding is a codimension $d + e$ closed embedding.

A *rank N vector bundle* $p: E \to X$ is a morphism of functors $E: I \to \mathbf{Sm}_S^{\mathrm{ess}}$, $X: I \to \mathbf{Sm}_S^{\mathrm{ess}}$, such that

1. For each $i \in I$, $p(i): E(i) \to X(i)$ is a rank N vector bundle on $X(i)$.
2. For each morphism $s: i \to j$ in I, the natural map $E(i) \to X(s)^*(E(j))$ is an isomorphism of vector bundles on $X(i)$.

A map of vector bundles $f: E \to F$ over X is a morphism of functors such that $f(i): E(i) \to F(i)$ is a map of vector bundles over $X(i)$ for each $i \in I$. This makes the category of vector bundles over X into an exact category, where a sequence is exact if it is exact on each $X(i)$. The usual operations of vector bundles, e.g., direct sum, tensor product, pull-back, extend in the obvious way to this setting.

We say a vector bundle $E \to X$ is *generated by global sections* if there is a trivial vector bundle 1_X^M and a surjection $1_X^M \to E$.

Given a rank N vector bundle $p: E \to X$, we may form the associated projective bundle $q: \mathbb{P}(E) \to X$ by taking the functor $i \mapsto [q(i): \mathbb{P}(E(i)) \to X(i)]$. We have the canonical surjection $q^*E \to \mathcal{O}_E(1)$ induced by the canonical surjections $q(i)^*(E(i)) \to \mathcal{O}_{E(i)}(1)$.

A morphism of functors $f: Y \to X$ is call a *projective morphism* if f admits a factorization

$$Y \xrightarrow{i} \mathbb{P}(E) \xrightarrow{q} X$$

with i a closed, codimension d embedding for some d, and $q: \mathbb{P}(E) \to X$ the projective bundle associated to a vector bundle E on X which is generated by global sections.

From this latter requirement, one sees that each projective morphism $f: Y \to X$ may be factored as $Y \xrightarrow{i} \mathbb{P}_X^N \xrightarrow{q} X$; from this one easily verifies that the composition of two projective morphisms is a projective morphism.

2.6.1.1. REMARK. Suppose that S is quasi-projective over a noetherian ring. Let $X: I \to \mathbf{Sm}_S^{\mathrm{ess}}$ be a functor. If I is the category associated to a finite partially ordered set, if all the morphisms $X(i) \to X(j)$ are *affine*, and if each $X(i)$ is quasi-projective over S, then, for each vector bundle on X, there is a line bundle L on X with $E \otimes L$ generated by global sections. In particular, we may dispense with the requirement that the vector bundle E be generated by global sections in the definition of a projective morphism.

2.6.2. *Adjoining a disjoint base-point.* We refer the reader to §2.7.1 of Chapter I for the construction of the category $\mathbf{Sm}_S^{\mathrm{ess}+}$.

Suppose we have functors $Y, X: I \to \mathbf{Sm}_S^{\mathrm{ess}+}$ and a map of functors $f: Y \to X$. We call f a codimension d closed embedding (resp. a projective morphism) if

1. $Y(i) = *$ if and only if $X(i) = *$.
2. Let I_0 be the full subcategory of I with objects those $i \in I$ such that $Y(i) \neq *$, giving us the map of functors $f_0: Y_0 \to X_0$, where $Y_0, X_0: I_0 \to \mathbf{Sm}_S^{\mathrm{ess}}$ are the restrictions of Y and X to I_0. Then f_0 is a codimension d closed embedding (resp. a projective morphism).

We extend the other notions described above (vector bundles, projective bundles, etc.) to functors from I to $\mathbf{Sm}_S^{\mathrm{ess}+}$ similarly by requiring that a vector bundle or projective bundle over $*$ be just $*$.

2.6.3. *Blow-ups.* Let $f: Z \to X$ be a closed, codimension $d > 1$ embedding of functors $Z, X: I \to \mathbf{Sm}_S^{\mathrm{ess}+}$. For each i with $Z(i)$ and $X(i)$ different from $*$, we may form the blow-up $Y(i)$ of $X(i)$ along $Z(i)$. If $Z(i) = X(i) = *$, define $Y(i) = *$ as well; this gives us the functor $Y: I \to \mathbf{Sm}_S^{\mathrm{ess}+}$ and the diagram of functors

$$\begin{array}{ccc} & & Y \\ & & \downarrow{\scriptstyle \mu} \\ Z & \xrightarrow[f]{} & X. \end{array}$$

Similarly, letting $E(i)$ be the exceptional divisor of $\mu(i): Y(i) \to X(i)$ (or $*$ if $X(i) = Z(i) = Y(i) = *$), we have the diagram

$$\begin{array}{ccc} E & \xrightarrow{g} & Y \\ {\scriptstyle q}\downarrow & & \downarrow{\scriptstyle \mu} \\ Z & \xrightarrow[f]{} & X \end{array}$$

with g a codimension one closed embedding, and q the projective bundle associated to the normal bundle of Z in X.

2.6.4. *Closed subdiagrams.* Let $X: I \to \mathbf{Sm}_S^{\mathrm{ess}+}$ be a functor. We let I_0 be the full subcategory of I with objects i such that $X(i) \neq *$.

A collection of closed subsets $W(i) \subset X(i)$ for $i \in I_0$ forms *a closed subdiagram of* X if, for each $s: i \to j$ in I_0 we have $X(s)^{-1}(W(j)) = W(i)$. For example, if $f: Z \to X$ is a codimension d closed embedding, then the support $\mathrm{supp}(Z(i))$ form a closed subdiagram of X.

If $W \subset X$ is a closed subdiagram of X, let $j(i): U(i) \to X(i)$ be the complement of $W(i)$ (we set $U(i) = *$ if $X(i) = *$). Then the $U(i)$ form a functor $U: I \to X$, giving us morphism of functors $j: U \to X$. We call U the *complement* of W in X, and call $j: U \to X$ an open immersion.

2.6.5. *Motives.* Suppose that I is a finite category, and $X: I \to \mathcal{V}^+$ a functor. We then have the motive of X, $\mathbb{Z}_X$, in $\mathbf{C}^b_{\mathrm{mot}}(\mathcal{V})$, defined as in (Chapter I, §2.7), by taking the canonical lifting of X, $(X, f_X): I \to \mathcal{L}(\mathcal{V})^+$ (Chapter I, §2.7.2), and forming the non-degenerate homotopy limit $\operatorname{holim}_{I^{\mathrm{op}},\mathrm{n.d.}} \mathbb{Z}_{(X,f_X)}$ (Part II, Chapter III, §3.2.7).

If $W \subset X$ is a closed subdiagram of X, with complement $U: I \to \mathcal{V}^+$, we define the *motive of X with support in W* by

$$\mathbb{Z}_{X,W} := \operatorname{cone}(j^*: \mathbb{Z}_X \to \mathbb{Z}_U)[-1].$$

If $i \in I$ is a maximal element, then i is a minimal element of I^{op}, hence we have the distinguished triangle in $\mathbf{D}^b_{\mathrm{mot}}(\mathcal{V})$

$$\mathbb{Z}_{X,W} \to \mathbb{Z}_{X(i),W(i)} \oplus \mathbb{Z}_{X_{|I\setminus\{i\}},W_{|I\setminus\{i\}}} \to \mathbb{Z}_{X^{i/},W^{i/}} \to \mathbb{Z}_{X,W}[1] \tag{2.6.5.1}$$

(see (I.2.7.3.1) and Part II, (III.3.2.9.1)).

2.6.6. REMARK. Via the distinguished triangle (2.6.5.1), the properties of motives with support described in (Chapter I, §2.2) extend to the setting of diagrams by using induction on $\dim I$ and $|\mathcal{N}(I)_{\mathrm{n.d.}}([\dim I])|$; for example, we have the homotopy property, moving lemma, localization, and Mayer-Vietoris for the motives of diagrams with support in a closed subdiagram.

2.6.7. *Cup products and cycle classes.* Let $X: I \to \mathcal{V}^+$ be a functor, with I a finite category, let W and W' be closed subdiagrams of X. We have the canonical lifting $(X, f_X): I \to \mathcal{L}(\mathcal{V})^+$.

Suppose we have for each $i \in I$ a cycle $Z_i \in \mathcal{Z}^q_{W(i)}(X(i))_{f_X(i)}$ (we set $\mathcal{Z}^q(*) := \{0\}$) such that, for each $s: i \to j$ in I, we have $X(s)^*(Z_j) = Z_i$. We call the collection of cycles $Z(i)$ a codimension q cycle on X, supported in W, and we denote this group of cycles by $\mathcal{Z}^q_W(X)$.

If Z is in $\mathcal{Z}^q_W(X)$, we have the morphism

$$\cup[Z]: \mathbb{Z}_{X,W'} \to \mathbb{Z}_{X,W\cap W'}(q)[2q];$$

to define this, we first note the following construction:

Let I be a finite category, $\mathcal{A}$ a DG category with translation structure, and $X, Y: I \to \mathcal{A}$ two functors. Suppose we have maps $f(i): X(i) \to Y(i)$ in $Z^0\mathcal{A}$, for each $i \in I$, and in addition, for each non-degenerate t-simplex

$$i_0 \xrightarrow{s_1} i_1 \cdots i_{t-1} \xrightarrow{s_t} i_t$$

in I, a map $h_{s_t,\ldots,s_1}: X(i_0) \to Y(i_t)[-t]$ of degree 0, with

$$dh_{s_t,\ldots,s_1} + h_{s_t,\ldots,s_2} \circ X(s_1) + \sum_{i=1}^{t-1} (-1)^i h_{s_t,\ldots,s_{i+1}s_i,\ldots,s_1} + (-1)^t Y(s_t)[-t+1] \circ h_{s_{t-1},\ldots,s_1} = 0,$$

where we set $h_\emptyset$ to be f (evaluated at the appropriate element of I), and we identify degree zero maps $A \to B[-t+1]$ with degree one maps $A \to B[-t]$.

Let

$$(f,h): \operatorname*{holim}_{I,\mathrm{n.d.}} X \to \operatorname*{holim}_{I,\mathrm{n.d.}} Y$$

be the map gotten by sending $X(j_r)[-r]$ in the component of $\operatorname{holim}^{\mathcal{A}}_{I,\mathrm{n.d.}} X$ indexed by the r-simplex

$$j_0 \xrightarrow{s_1} j_1 \cdots j_{r-1} \xrightarrow{s_r} j_r$$

to $\operatorname{holim}_{I,\text{n.d.}} Y$ by taking the sum of all the maps $h_{s'_t,\dots,s'_{r+1}}[-r]: X(i'_r)[-r] \to Y(i_t)[-t]$, where $Y(i_t)$ is in the component of $\operatorname{holim}_{I,\text{n.d.}} Y$ indexed by the simplex

$$i'_0 \xrightarrow{s'_1} i'_1 \cdots i'_{t-1} \xrightarrow{s'_t} i'_t,$$

with $s_a = s'_a$ for $a = 1, \dots, r < t$. For $r = t$, we assume $s_a = s'_a$ for $a = 1, \dots, t$, and map $X(i_t)[-t]$ to $Y(i_t)[-t]$ via $f(i_t)[-t]$.

By a direct computation, we have

2.6.7.1. LEMMA. *The map*

$$(f, h)\colon \operatorname*{holim}_{I,\text{n.d.}} X \to \operatorname*{holim}_{I,\text{n.d.}} Y$$

is a map in $Z^0\mathbf{C}^b(\mathcal{A})$.

We now define the map $\cup[Z]$. For each $i \in I$, we have the map (Chapter I, Definition 1.4.6) $[Z_i]\colon \mathfrak{e} \to \mathbb{Z}_{X(i),W(i)}(d)_{fX(i)}[2d]$ in $\mathcal{A}_5$. For each map $s\colon i \to j$ in I^{op}, the morphisms adjoined in (Chapter I, Definition 1.4.8) give us the map $h_s\colon \mathfrak{e} \to \mathbb{Z}_{X(j),W(j)}(d)_{fX(j)}[2d-1]$ of degree 0, with $dh_s = X(s)^* \circ [Z_i] - [Z_j]$.

Let

$$i_0 \xrightarrow{s_1} i_1 \cdots i_{t-1} \xrightarrow{s_t} i_t$$

be a non-degenerate t-simplex in I. Using the morphisms adjoined in (Chapter I, Definition 1.4.9), it follows by an elementary induction that we have morphisms of degree 0, $h_{s_t,\dots,s_1}\colon \mathfrak{e} \to \mathbb{Z}_{X(i_t),W(i_t)}(d)_{fX(i_t)}[2d-t]$, with

$$dh_{s_t,\dots,s_1} = h_{s_t,\dots,s_2} + \sum_{i=1}^{t-1}(-1)^i h_{s_t,\dots,s_{i+1}s_i,\dots,s_1} + (-1)^t X^*(s_t) \circ h_{s_{t-1},\dots,s_1}.$$

By the above lemma, the maps

$$\mathfrak{e} \otimes \mathbb{Z}_{X(i_t),W'(i_t)}(0)_{fX(i_t)} \xrightarrow{h_{s_t,\dots,s_1}\otimes \mathrm{id}} \mathbb{Z}_{X(i_t),W(i_t)}(d)_{fX(i_t)}[2d-t] \otimes \mathbb{Z}_{X(i_t),W'(i_t)}(0)_{fX(i_t)}$$

give a well-defined map

$$[Z] \otimes \mathrm{id}\colon \operatorname*{holim}_{I^{\text{op}},\text{n.d.}} \mathfrak{e} \otimes \mathbb{Z}_{X,W'}(0)_{fX} \to \operatorname*{holim}_{I^{\text{op}},\text{n.d.}} \mathbb{Z}_{X,W}(d)_{fX}[2d] \otimes \mathbb{Z}_{X,W'}(0)_{fX}.$$

One similarly shows that $[Z] \otimes \mathrm{id}$ is independent (modulo homotopy) of the choice of the maps $h_{s_t,\dots,s_1}$.

We compose $[Z] \otimes \mathrm{id}$ with the map

$$\operatorname*{holim}_{I^{\text{op}},\text{n.d.}} \mathbb{Z}_{X,W}(d)_{fX}[2d] \otimes \mathbb{Z}_{X,W'}(0)_{fX} \xrightarrow{\operatorname{holim}\boxtimes} \operatorname*{holim}_{I^{\text{op}},\text{n.d.}} \mathbb{Z}_{X\times_S X, W\times W'}(d)_{fX\times fX}[2d]$$

induced by the external products. We have the isomorphisms in $\mathbf{D}^b_{\text{mot}}(\mathcal{V})$

$$\operatorname*{holim}_{I^{\text{op}},\text{n.d.}} \mathbb{Z}_{X\times_S X, W\times W'}(d)_{fX\times fX}[2d] \cong \mathbb{Z}_{X\times_S X, W\times W'}(d)[2d]$$

$$\operatorname*{holim}_{I^{\text{op}},\text{n.d.}} \mathfrak{e} \otimes \mathbb{Z}_{X,W'}(0)_{fX} \cong \mathbb{Z}_{X,W'};$$

pulling back by the diagonal (using the moving lemma for diagrams; see Remark 2.6.6) gives us the desired map in $\mathbf{D}^b_{\mathrm{mot}}(\mathcal{V})$

$$\cup[Z]:\mathbb{Z}_{X,W'} \to \mathbb{Z}_{X,W\cap W'}(d)[2d].$$

It is easy to check that the map $\cup[Z]$ is natural in the functor X; in particular, if $i \in I^{\mathrm{op}}$ is a minimal element, the triple $(\cup[Z], \cup[Z_i] \oplus \cup[Z_{I\setminus\{i\}}], \cup[Z^{i/}])$ gives a map of distinguished triangles

$$(2.6.7.2)\quad (\mathbb{Z}_{X,W} \to \mathbb{Z}_{X_{|I\setminus\{i\}},W_{|I\setminus\{i\}}} \oplus \mathbb{Z}_{X(i),W(i)} \to \mathbb{Z}_{X^{i/},W^{i/}}) \xrightarrow{(\cup[Z],\cup[Z_i]\oplus\cup[Z_{I\setminus\{i\}}],\cup[Z^{i/}])}$$

$$(\mathbb{Z}_{X,W}(d)[2d] \to \mathbb{Z}_{X_{|I\setminus\{i\}},W_{|I\setminus\{i\}}}(d)[2d] \oplus \mathbb{Z}_{X(i),W(i)}(d)[2d] \to \mathbb{Z}_{X^{i/},W^{i/}}(d)[2d]).$$

2.6.8. *Gysin morphism.* Let I be a finite category, $Z, X: I \to \mathcal{V}^+$ functors, and $i: Z \to X$ a codimension d closed embedding. Let W be a closed subdiagram of Z. We have the cycle $i_*(|Z|)$ in $\mathcal{Z}^d(X)$.

2.6.8.1. LEMMA. *Suppose the map i is split by a smooth map $p: X \to Z$, giving the map $p^*: \mathbb{Z}_{Z,W} \to \mathbb{Z}_{X,p^{-1}(W)}$. Then the map $\cup[i_*(|Z|)] \circ p^*: \mathbb{Z}_{Z,W}(-d)[2d] \to \mathbb{Z}_{X,W}$ is an isomorphism.*

PROOF. The result follows from (Chapter III, §2.1.1) in case I is the one-point category; in general, we use induction on the number of elements of I, the distinguished triangle (2.6.5.1), and the naturality (2.6.7.2) of the map $\cup[i_*(|Z|)]$. □

We now define the Gysin isomorphism

$$(2.6.8.2)\qquad i_*: \mathbb{Z}_{Z,W}(-d)[2d] \to \mathbb{Z}_{X,W}$$

using the deformation diagram

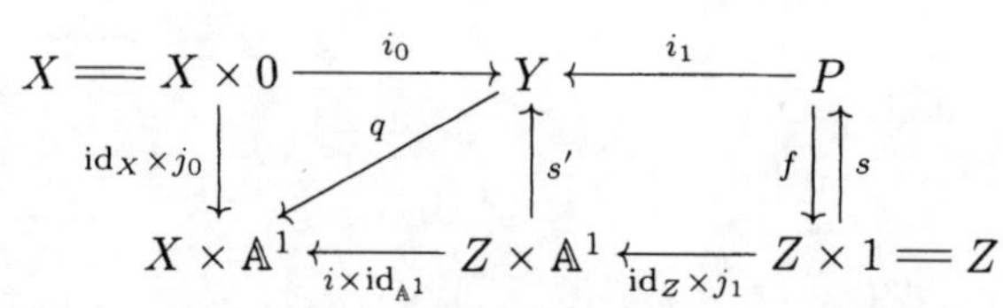

defined just as the diagram (2.1.2.1). Indeed, the maps $i_0^*: \mathbb{Z}_{Y,s'(W\times\mathbb{A}^1)} \to \mathbb{Z}_{X,W}$ and $i_1^*: \mathbb{Z}_{Y,s'(W\times\mathbb{A}^1)} \to \mathbb{Z}_{P,s(W)}$ are isomorphisms by the homotopy property for diagrams (see Remark 2.6.6), and $\cup[s_*(|Z|)] \circ f^*: \mathbb{Z}_{Z,W}(-d)[-2d] \to \mathbb{Z}_{P,s(W)}$ is an isomorphism by Lemma 2.6.8.1. Thus, we may define i_* by

$$i_* := i_0^* \circ (i_1^*)^{-1} \circ (\cup[s_*(|Z|)] \circ f^*).$$

If W_X be a closed subdiagram of X containing $i(W)$, we define the Gysin morphism $i_*: \mathbb{Z}_{Z,W}(-d)[-2d] \to \mathbb{Z}_{X,W_X}$ as the composition of the Gysin isomorphism (2.6.8.2) with the map $\mathbb{Z}_{X,W} \to \mathbb{Z}_{X,W_X}$ induced by the identity on X.

2.6.9. *Pushforward for a projection.* Let I be a finite category, $X: I \to \mathcal{V}^+$ a functor, and $p: E \to X$ a rank $N+1$ vector bundle on X. This gives us the projective bundle $q: \mathbb{P}(E) \to X$ and the tautological line bundle $r: L \to \mathbb{P}(E)$.

Let W be a closed subdiagram of X. The zero section of L gives the cycle $[0_L]$ in $\mathcal{Z}^1(\mathcal{L})$; for each i, define the map

$$\alpha_i^E: \mathbb{Z}_{X,W}(-i)[-2i] \to \mathbb{Z}_{\mathbb{P}(E),q^{-1}(W)}$$

as the composition of the map

$$(\cup[0_L])^i \circ (q\circ r)^* : \mathbb{Z}_{X,W}(-i)[-2i] \to \mathbb{Z}_{L,(q\circ r)^{-1}(W)}$$

with the inverse of the isomorphism $r^* : \mathbb{Z}_{\mathbb{P}(E),q^{-1}(W)} \to \mathbb{Z}_{L,(q\circ r)^{-1}(W)}$ given by the homotopy property for diagrams (see Remark 2.6.6).

2.6.9.1. LEMMA [projective bundle formula]. *The map*

$$\alpha := \sum_{i=0}^{N} \alpha_i : \oplus_{i=0}^{N} \mathbb{Z}_{X,W}(-i)[-2i] \to \mathbb{Z}_{\mathbb{P}(E),q^{-1}(W)}$$

is an isomorphism.

PROOF. The proof follows from the case of the one-point category, the naturality of the maps $\cup[0_L]$ (2.6.7.2), and the distinguished triangle (2.6.5.1). □

We may then define the pushforward map

$$q_* : \mathbb{Z}_{\mathbb{P}(E),q^{-1}(W)}(N)[2N] \to \mathbb{Z}_{X,W}$$

as the inverse of the map $\alpha^E(N)[2N]$, followed by the projection on the factor $\mathbb{Z}_{X,W}$.

2.6.10. *Pushforward for a projective morphism.* Let $f : Y \to X$ be a projective morphism diagrams of relative dimension d, let W_X be a closed subdiagram of X and let W_Y be a closed subdiagram of Y with $W_Y \subset f^{-1}(W_X)$. Factor f as

$$Y \xrightarrow{i} \mathbb{P}(E) \xrightarrow{q} X$$

where i is a codimension e closed embedding, for some e, and $\mathbb{P}(E)$ is the projective bundle associated to a vector bundle E which is generated by global sections. We then define $f_* : \mathbb{Z}_{Y,W_Y}(d)[2d] \to \mathbb{Z}_{X,W_X}$ by

$$f_* = q_* \circ i_*.$$

The proofs of most of the main results of (Chapter III, Section 2) go through without change, using the machinery we have developed in this section. We list these results in an omnibus theorem for future reference.

2.6.11. THEOREM. *Let I be a finite category, $X, Y : I \to \mathcal{V}^+$ functors, W_X a closed subdiagram of X, W_Y a closed subdiagram of Y, and $f : Y \to X$ a projective morphism of relative dimension d with $W_Y \subset f^{-1}(W_X)$.*
(i) The map

$$f_* : \mathbb{Z}_{Y,W_Y}(d)[2d] \to \mathbb{Z}_{X,W_X}$$

is well-defined, independent of the choice of factorization of f (see Lemma 2.4.2).
(ii) We have the functoriality

$$(f\circ g)_* = f_* \circ g_*$$

for composable projective morphisms of diagrams (see Theorem 2.4.7).

(iii) The projection formula (see Theorem 2.4.8) holds: Let $\hat{X}_i$, $i = 1, 2$, be closed subdiagrams of X, and let $\hat{Y}_i = f^{-1}(\hat{X}_i)$, $i = 1, 2$. Then the diagram

$$\begin{array}{ccc}
\mathbb{Z}_{Y,\hat{Y}_1}(d)[2d] \otimes \mathbb{Z}_{X,\hat{X}_2} & \xrightarrow{\mathrm{id}\otimes f^*} & \mathbb{Z}_{Y,\hat{Y}_1}(d)[2d] \otimes \mathbb{Z}_{Y,\hat{Y}_2} \\
 & & \downarrow{\scriptstyle \cup_Y} \\
\downarrow{\scriptstyle f_*\otimes\mathrm{id}} & & \mathbb{Z}_{Y,\hat{Y}_1\cap\hat{Y}_2}(d)[2d] \\
 & & \downarrow{\scriptstyle f_*} \\
\mathbb{Z}_{X,\hat{X}_1} \otimes \mathbb{Z}_{X,\hat{X}_2} & \xrightarrow[\cup_X]{} & \mathbb{Z}_{X,\hat{X}_1\cap\hat{X}_2}
\end{array}$$

commutes.
(iv) Let

$$\begin{array}{ccc}
Y' & \xrightarrow{g'} & Y \\
\downarrow{\scriptstyle f'} & & \downarrow{\scriptstyle f} \\
X' & \xrightarrow[g]{} & X
\end{array}$$

be a transverse cartesian diagram of functors $I \to \mathcal{V}^+$, i.e., for each i with $Y'(i)$, $Y(i)$, $X'(i)$ and $X(i)$ in $\mathcal{V}$, the diagram

$$\begin{array}{ccc}
Y'(i) & \xrightarrow{g'(i)} & Y(i) \\
\downarrow{\scriptstyle f'(i)} & & \downarrow{\scriptstyle f(i)} \\
X'(i) & \xrightarrow[g(i)]{} & X(i)
\end{array}$$

is a transverse cartesian diagram in $\mathcal{V}$, and if one of $Y'(i)$, $X'(i)$, $Y(i)$ or $X(i)$ is $*$, then they are all $*$.

Then f' is a projective morphism, and

$$g^* \circ f_* = f'_* \circ g'^*$$

(see Theorem 2.4.9).
(v) Let $T: I \to \mathcal{V}^+$ be a diagram, with closed subdiagram W_T. Then the diagram

$$\begin{array}{ccc}
\mathbb{Z}_{Y,W_Y}(d)[2d] \otimes \mathbb{Z}_{T,W_T} & \xrightarrow{f_*\otimes\mathrm{id}} & \mathbb{Z}_{X,W_X} \otimes \mathbb{Z}_{T,W_T} \\
\downarrow{\scriptstyle \boxtimes_{Y,T}} & & \downarrow{\scriptstyle \boxtimes_{X,T}} \\
\mathbb{Z}_{Y\times_S T, W_Y\times_S W_T}(d)[2d] & \xrightarrow[(f\times\mathrm{id}_T)_*]{} & \mathbb{Z}_{X\times_S T, W_X\times_S W_T}
\end{array}$$

commutes (see Theorem 2.4.10).

□

We remind the reader that the product of pointed diagrams $X, Y: I \to \mathcal{V}^+$ is the *pointwise smash product*, i.e.,

$$(X \times_S Y)(i) = \begin{cases} X(i) \times_S Y(i); & \text{if } X(i) \neq * \text{ and } Y(i) \neq * \\ *; & \text{otherwise.} \end{cases}$$

2.6.12. REMARK. The projective bundle formula for diagrams (Lemma 2.6.9.1) implies that we have the *splitting principle* available for Chern classes of vector bundles on a functor $X: I \to \mathcal{V}^+$.

3. Riemann-Roch

The machinery is now in place for a proof of the motivic Riemann-Roch theorem; the argument proceeds along the lines laid out in [**8**] and [**45**].

As an application, we show in §3.6 that the Chern character gives an isomorphism of weight-graded K-theory with rational motivic cohomology, when the base scheme is a field or a smooth curve over a field.

3.1. Lambda rings

3.1.1. Recall from [**2**, exposé 0, Appendix] that a *lambda ring* is a commutative ring R, together with operations

$$\lambda^k: R \to R; \quad k = 0, 1, \ldots$$

such that

1. $\lambda^0(x) = 1$, $\lambda^1(x) = x$ for all $x \in R$.
2. For x and y in R, we have

$$\lambda^n(x+y) = \sum_{i+j=n} \lambda^i(x)\lambda^j(y).$$

3.1.2. *Universal polynomials.* For symbols $x_1, x_2, \ldots$, we have the *symmetric functions* $\sigma_k(x_1, x_2, \ldots)$ defined by the formal identity

$$\prod_{i=1}^{\infty}(1 + x_i t) = 1 + \sum_{i=1}^{\infty} \sigma_k(x_1, x_2, \ldots)t^k.$$

Given elements $a_1, \ldots, a_n$ in a commutative ring R, we define $\sigma_k(a_1, a_2, \ldots, a_n) \in R$ by setting

$$x_i = \begin{cases} a_i & i = 1, \ldots, n, \\ 0 & i > n \end{cases}$$

in $\sigma_k(x_1, x_2, \ldots)$. The following result is well-known (see for example [**82**, V, §9]):

3.1.2.1. THEOREM. *Let R be a commutative ring, and let the symmetric group S_n act on the polynomial ring $R[X_1, \ldots, X_n]$ by*

$$\sigma(f)(X_1, \ldots, X_n) = f(X_{\sigma(1)}, \ldots, X_{\sigma(n)}).$$

Then the subring of invariants $R[X_1, \ldots, X_n]^{S_n}$ is equal to the polynomial ring over R in $\sigma_1(X_1, \ldots, X_n), \ldots, \sigma_n(X_1, \ldots, X_n)$.

Define polynomials

$$P_k(X_1, \ldots, X_k; Y_1, \ldots, Y_k) \in \mathbb{Z}[X_1, \ldots, X_k; Y_1, \ldots, Y_k]$$
$$P_{k,j}(X_1, \ldots, X_{kj}) \in \mathbb{Z}[X_1, \ldots, X_{kj}]$$

by setting $X_i = \sigma_i(U_1, U_2, \dots)$, $Y_j = \sigma_j(V_1, V_2, \dots)$, and

$$\sum_{k\geq 0} P_k(X_1, \dots, X_k; Y_1, \dots, Y_k)t^k = \prod_{i,j}(1 + U_iV_jt)$$
$$\sum_{k\geq 0} P_{k,j}(X_1, \dots, X_{kj})t^k = \prod_{i_1<\dots<i_j} (1 + U_{i_1}\cdots U_{i_j}t).$$

3.1.3. *Special lambda rings.* A lambda ring (R, λ^*) is called a *special* lambda ring if

$$\lambda^n(xy) = P_n(\lambda^1(x), \dots, \lambda^n(x); \lambda^1(y), \dots, \lambda^n(y)),$$
$$\lambda^k(\lambda^j(x)) = P_{k,j}(\lambda^1(x), \dots, \lambda^{kj}(x)),$$

for all $x, y \in R$.

3.1.4. EXAMPLES [see [**2**], exposé 0]. (i) Let $A = \oplus_{q=1}^{\infty} A^q$ be a graded ring (without identity). Set

$$\tilde{A} := \mathbb{Z} \times [1 \times \prod_{q=1}^{\infty} A^q]$$

and make $\tilde{A}$ a group by

$$(n, 1 + \sum_{i=1}^{\infty} x_i) + (m, 1 + \sum_{i=1}^{\infty} y_i) = (n + m, (1 + \sum_{i=1}^{\infty} x_i)(1 + \sum_{i=1}^{\infty} y_i)).$$

Define the product

$$(n, 1 + \sum_{i=1}^{\infty} x_i)\bigstar(m, 1 + \sum_{i=1}^{\infty} y_i)$$

by

$$(n, 1 + \sum_{i=1}^{\infty} x_i)\bigstar(m, 1 + \sum_{i=1}^{\infty} y_i)$$
$$= (nm, [(1 + \sum_{i=1}^{\infty} x_i) * (1 + \sum_{i=1}^{\infty} y_i)](1 + \sum_{i=1}^{\infty} x_i)^m (1 + \sum_{i=1}^{\infty} y_i)^n),$$

where

$$(1 + \sum_{i=1}^{\infty} x_i) * (1 + \sum_{i=1}^{\infty} y_i) := 1 + \sum_{i=1}^{\infty} P_i(x_1, \dots, x_i; y_1, \dots, y_k).$$

This makes $\tilde{A}$ into a commutative ring.

The ring $\mathbb{Z}$ is a lambda ring with

$$\lambda^k(n) = \begin{cases} \begin{pmatrix} n \\ k \end{pmatrix}; & \text{for } n \geq 0, \\ (-1)^k \begin{pmatrix} k - n - 1 \\ k \end{pmatrix}; & \text{for } n < 0. \end{cases}$$

Define the operations λ^k on $\tilde{A}$ by

$$\lambda^k(0, 1 + \sum_{i=1}^{\infty} x_i) := (0, 1 + \sum_{i=1}^{\infty} P_{k,i}(x_1, \dots, x_{ki})),$$

and

$$\lambda^k(n, 1 + X) := \sum_{i=0}^{k} (\lambda^i(n), 1) \bigstar \lambda^{k-i}(0, 1 + X).$$

This gives $\tilde{A}$ the structure of a special lambda ring.
(ii) Let (R, λ^*) be a commutative ring with operations λ^k, $k = 0, 1, \dots$ such that $\lambda^0(x) = 1$. We have the graded polynomial ring $R[t]$; let $\lambda_t : R \to \widetilde{R[t]}$ be the map

$$\lambda_t(r) = (0, 1 + \sum_{i=1}^{\infty} \lambda^i(r) t^i).$$

Then λ_t is additive if and only if (R, λ^*) is a lambda ring, and λ_t is a lambda ring homomorphism if and only if (R, λ^*) is a special lambda ring.
(iii) Suppose we have a graded ring A as in (i), and a graded A-module M, $M := \oplus_{q \geq 1} M^q$. Let $\tilde{M} := \prod_{q \geq 1} M^q$. Form the ring $A \oplus M$ with

$$(a, m)(a', m') := (aa', am' + a'm).$$

Then we have the inclusions

$$\tilde{A} \hookrightarrow \widetilde{A \oplus M} \hookleftarrow \tilde{M}$$

identifying $\tilde{A}$ with a subring of $\widetilde{A \oplus M}$ and $\tilde{M}$ with an ideal in $\widetilde{A \oplus M}$. This allows us to define the product $\bigstar : \tilde{A} \otimes \tilde{M} \to \tilde{M}$, making $\tilde{M}$ into a $\tilde{A}$-module.

Similarly, if M and M' and M'' are (positively) graded abelian groups with a graded product $M \otimes M' \to M''$, we have the (non-unital) ring $M \oplus M' \oplus M''$ with

$$(m, m', m'')(n, n', n'') = (0, 0, mn')$$

As above, this gives the product $\bigstar : \tilde{M} \otimes \tilde{M}' \to \tilde{M}''$.

3.2. K_0 and Chern classes

3.2.1. *Lambda ring structure on* K_0. Let $X : \Delta^{\leq N\mathrm{op}} \to \mathbf{Sch}$ be a truncated simplicial scheme. Tensor product of vector bundles makes $K_0(X)$ a commutative ring. If

$$0 \to E' \to E \to E'' \to 0$$

is a short exact sequences of vector bundles on a scheme X, the images of the maps $\Lambda^i E' \otimes \Lambda^{k-i} E \to \Lambda^k E$ give an increasing filtration on $\Lambda^k E$ with ith graded quotient canonically isomorphic to $\Lambda^i E' \otimes \Lambda^{k-i} E''$. This gives the identity in $K_0(X)$

$$[\Lambda^k E] = \sum_{i=0}^{k} [\Lambda^i E'][\Lambda^{k-i} E''].$$

This identity implies that $[\Lambda^k E]$ depends only on the K_0-class $[E]$, giving natural operations

$$\lambda^k : K_0(X) \to K_0(X),$$

which make $K_0(X)$ into a lambda ring.

3.2.2. *Splitting principle for* K_0. Let X be a scheme and $q: \mathcal{F}l(E) \to X$ the flag bundle over X associated to a vector bundle $E \to X$. By [**2**, exposé VI, Théorème 1.1] the map $q^*: K_0(X) \to K_0(\mathcal{F}(E))$ is injective, so identities in $K_0(X)$ may be checked in $K_0(\mathcal{F}(E))$, or even in K_0 of a product of flag bundles. As an application, we have

3.2.3. THEOREM. *Let X be a scheme. Then $(K_0(X), \lambda^*)$ is a special lambda ring.*

PROOF. Using the splitting principle, we need only consider elements x, y in $K_0(X)$ with

$$x = \sum_{i=1}^{n} [L_i]; \quad y = \sum_{j=1}^{m} [M_j]$$

and the L_i and M_j line bundles on X. We then have

$$\lambda^k(x) = \sum_{1 \le i_1 < \ldots < i_k \le n} [L_{i_1}] \cdots [L_{i_k}] = \sigma_k([L_1], \ldots, [L_n]),$$

and similarly $\lambda^j(y) = \sigma_j([M_1], \ldots, [M_m])$. The special lambda ring identities then follow from the definition of the polynomials P_k and $P_{k,j}$ in terms of symmetric functions. □

We will ignore the question of whether $K_0(X)$ is a special lambda ring for X a truncated simplicial scheme.

3.3. Chern classes and Chern character for higher K-theory

3.3.1. *Chern classes.* Let X be a connected truncated simplicial object of $\mathcal{V}$. We have the (non-unital) ring $H^{2*}(X, \mathbb{Z}(*)) := \oplus_{q \ge 1} H^{2q}(X, \mathbb{Z}(q))$. Define the *augmented total Chern class*

$$\tilde{c}_{X,0}: K_0(X) \to H^{2*}\widetilde{(X, \mathbb{Z}(*))}$$

by

$$\tilde{c}_{X,0} := (\mathrm{rnk}, 1 + \sum_q c_q),$$

where rnk is the rank function.

For non-connected X with connected components $X_1, \ldots, X_s$, define $\tilde{c}_{X,0}$ to be the product

$$\prod_i \tilde{c}_{X_i,0}: K_0(X) = \prod_i K_0(X_i) \to \prod_i H^{2*}\widetilde{(X_i, \mathbb{Z}(*))}.$$

Fix an integer $p > 0$, and let $H^{2*-p}(X, \mathbb{Z}(*)) := \oplus_{q=1}^{\infty} H^{2q-p}(X, \mathbb{Z}(q))$. The cup product in motivic cohomology makes $H^{2*-p}(X, \mathbb{Z}(*))$ a graded module over the (non-unital) graded ring $H^{2*}(X, \mathbb{Z}(*))$, and gives the graded products

$$H^{2*-p}(X, \mathbb{Z}(*)) \otimes H^{2*-p'}(X, \mathbb{Z}(*)) \to H^{2*-p-p'}(X, \mathbb{Z}(*)).$$

Following Example 3.1.4, we have the $H^{2*}\widetilde{(X, \mathbb{Z}(*))}$-module $H^{2*-p}\widetilde{(X, \mathbb{Z}(*))}$, and the products

$$\bigstar: H^{2*-p}\widetilde{(X, \mathbb{Z}(*))} \otimes H^{2*-p'}\widetilde{(X, \mathbb{Z}(*))} \to H^{2*-p-p'}\widetilde{(X, \mathbb{Z}(*))}.$$

Let

$$\tilde{c}_{X,p}: K_p(X) \to \widetilde{H^{2*-p}(X, \mathbb{Z}(*))}$$

be the total Chern class , $\tilde{c}_{X,p} = \sum_{q \geq 1} c^{q,2q-p}$.

Similarly, if I is the category associated to a finite partially ordered set, and if $X: I \to \mathcal{V}^+$ is a functor, we have the Chern classes

$$c^{q,2q-p}: K_p(X) \to H^{2q-p}(X, \mathbb{Z}(q)),$$

and the total Chern class (augmented for $p = 0$)

$$\tilde{c}_{X,p}: K_p(X) \to \widetilde{H^{2*-p}(X, \mathbb{Z}(*))}$$

(see §1.4.7).

3.3.2. LEMMA. *Let X be a truncated simplicial object of $\mathcal{V}$. Then $\tilde{c}_{X,0}$ is a homomorphism of lambda rings.*

PROOF. By the projective bundle formula for motivic cohomology of X (Theorem 1.3.2), we may check the various identities on a product of flag bundles over X; this reduces us to the case of elements which are sums of line bundles. In this case, the necessary identities all follows from the Whitney product formula (Theorem 1.3.7), and the definition of the polynomials P_k and $P_{k,i}$ in terms of symmetric functions (§3.1.2). □

3.3.3. Suppose we have an element $C_q \in H^{2q}(\mathrm{BGL}, \mathbb{Z}(q))$. If one applies the construction of the Chern classes for a functor $X: I \to \mathcal{V}^+$ in §1.4.7, replacing the universal Chern class $c_q(E)$ with the given class C_q, we get the natural map

$$C^{q,2q-p}: K_p(X) \to H^{2q-p}(X, \mathbb{Z}(q)).$$

As the product of Chern classes $c_q(E_N^{\leq n}) \cup c_{q'}(E_N^{\leq n})$ for varying N and n gives the "product" $c_q(E) \cup c_{q'}(E) \in H^{2(q+q')}(\mathrm{BGL}, \mathbb{Z}(q+q'))$, we thus have the corresponding map

$$(c_q \cup c_{q'})^{q+q',2q+2q'-p}: K_p(X) \to H^{2q+2q'-p}(X, \mathbb{Z}(q+q')). \tag{3.3.3.1}$$

Suppose we have functors $X, Y: I \to \mathcal{V}^+$. We then have the pointed product functor $X \times_S Y: I \to \mathcal{V}^+$ with

$$X \times_S Y(i) := \begin{cases} X(i) \times_S Y(i); & \text{if } X(i) \neq * \text{ and } Y(i) \neq * \\ *; & \text{otherwise.} \end{cases}$$

The natural products

$$\mathrm{B}Q\mathcal{P}_{X(i)} \wedge \mathrm{B}Q\mathcal{P}_{Y(i)} \to \mathrm{B}Q\mathcal{P}_{X(i) \times_S Y(i)}$$

give us the product

$$\boxtimes_{X,Y}: \operatorname*{holim}_I \mathrm{B}Q\mathcal{P}_X \wedge \operatorname*{holim}_I \mathrm{B}Q\mathcal{P}_Y \to \operatorname*{holim}_I \mathrm{B}Q\mathcal{P}_{X \times_S Y},$$

which in turn induces the external product

$$\boxtimes_{X,Y}: K_p(X) \otimes K_{p'}(Y) \to K_{p+p'}(X \times_S Y)$$

(see Appendix B, Remark 2.2.6). Taking the motives of X and Y, and applying the non-degenerate homotopy limit, we have the external product (see §2.7.4 and Part II, Chapter III, Section 3)

$$\boxtimes_{X,Y}: H^p(X; \mathbb{Z}(q)) \otimes H^{p'}(Y; \mathbb{Z}(q')) \to H^{p+p'}(X \times_S Y; \mathbb{Z}(q+q')).$$

This induces the external product

$$\bigstar_{X,Y}: H^{2*-p}\widetilde{(X;\mathbb{Z}(*))} \otimes H^{2*-p'}\widetilde{(Y;\mathbb{Z}(*))} \to H^{2*-p-p'}\widetilde{(X \times_S Y;\mathbb{Z}(*))}.$$

3.3.4. LEMMA. *Let I be the category associated to a finite partially ordered set, and let $X, Y: I \to \mathcal{V}^+$ be functors.*
(i) For $x \in K_p(X)$, $y \in K_{p'}(Y)$, we have

$$\tilde{c}_{X\times_S Y,p+p'}(x \boxtimes_{X,Y} y) = \tilde{c}_{X,p}(x) \bigstar_{X,Y} \tilde{c}_{Y,p'}(y),$$

and for $x \in K_p(X)$, $y \in K_{p'}(X)$, we have

$$\tilde{c}_{X,p+p'}(xy) = \tilde{c}_{X,p}(x) \bigstar \tilde{c}_{X,p'}(y).$$

(ii) The map (3.3.3.1) is zero for all $p > 0$.
(iii) For $x, y \in K_p(X)$, we have

$$c^{q,2q-p}(x+y) = c^{q,2q-p}(x) + c^{q,2q-p}(y)$$

for all $p \geq 1$.

PROOF. The second assertion in (i) follows from the first by taking $X = Y$ and pulling back by the diagonal. For the first assertion in (i), recall from (Appendix B, §2.2.4 and Remark 2.2.6) the construction of products

$$\mathbb{H}^{-p}(X, \tilde{\mathbb{Z}} \oplus \tilde{C}^*(\mathcal{BGL};\mathbb{Z})) \otimes \mathbb{H}^{-q}(Y, \tilde{\mathbb{Z}} \oplus \tilde{C}^*(\mathcal{BGL};\mathbb{Z}))$$
$$\xrightarrow{\boxtimes^H_{X,Y}} \mathbb{H}^{-p-q}(X \times_S Y, \tilde{\mathbb{Z}} \oplus \tilde{C}^*(\mathcal{BGL};\mathbb{Z})),$$

compatible with the external products $\boxtimes_{X,Y}: K_*(X) \otimes K_*(Y) \to K_*(X \times_S Y)$ via the Hurewicz map. Letting

$$H\tilde{c}_{X,p}: \mathbb{H}^{-p}(X, \tilde{\mathbb{Z}} \oplus \tilde{C}^*(\mathcal{BGL};\mathbb{Z})) \to H^{2*-p}\widetilde{(X, \mathbb{Z}(*))}$$

be the total homology Chern class (augmented for $p = 0$), constructed as $\tilde{c}_{p,X}$ by using the map (1.4.7.3), we reduce to showing

$$H\tilde{c}_{X\times_S Y,p+p'}(\boxtimes^H_{X,Y}(x \otimes y)) = H\tilde{c}_{X,p}(x) \bigstar_{X,Y} H\tilde{c}_{Y,p'}(y)$$

for $x \in H_p(X, \mathcal{GL};\mathbb{Z})$, $y \in H_{p'}(Y, \mathcal{GL};\mathbb{Z})$ (or $x \in H^0(X, \tilde{\mathbb{Z}}) \times H_0(X, \mathcal{GL};\mathbb{Z})$ for $p = 0$).

Let $\mathrm{BGL}_{N,M}/S$ be the product simplicial scheme $\mathrm{BGL}_N/S \times_S \mathrm{BGL}_M/S$. By Lemma 3.3.2, we have

$$\tilde{c}_{\mathrm{BGL}^{\leq n}_{N,M}/S,0}(p_1^* E_N \otimes p_2^* E_M) = p_1^* \tilde{c}_{\mathrm{BGL}^{\leq n}_N/S,0}(E_N) \bigstar p_2^* \tilde{c}_{\mathrm{BGL}^{\leq n}_M/S,0}(E_M)$$

for all n. Thus, letting 1^k stand for the trivial rank k vector bundle, we have

$$c_{\mathrm{BGL}^{\leq n}_{N,M}/S,0}(p_1^* E_N \otimes p_2^* E_M - p_1^* E_N \otimes p_2^* 1^M - p_1^* 1^N \otimes p_2^* E_M)$$
$$= p_1^*(c_{\mathrm{BGL}^{\leq n}_N/S}(E_N)) * p_2^*(c_{\mathrm{BGL}^{\leq n}_M/S,0}(E_M))$$

where $*$ is the operation

$$(1 + \sum_q x_q) * (1 + \sum_q y_q) = 1 + \sum_q P_q(x_1, \ldots, x_q; y_1, \ldots, y_q)$$

(see §3.1.2 and Example 3.1.4(i)). The result then follows from the definition of the external product $\boxtimes^H_{X,Y}$, and of the Chern classes $c^{q,p}$.

We now prove (ii); we refer the reader to Appendix B for the notation. The elements x and y are represented by maps $x: S^p \to K(X)$ and $y: S^p \to K(X)$, where $K(X)$ is the space $\operatorname{holim}_I \Omega \mathrm{B}P_X$. Let $h^X: K_p(X) \to \mathbb{H}^{-p}(X, \mathcal{GL}; \mathbb{Z})$ denote the Hurewicz map.

An element x of $K_p(X)$ is "trivialized" on an open cover $\mathcal{U} := \{U_0, \ldots, U_n\}$ of X (see Appendix B, §2.1.3 and (B.2.1.1.2)), and hence comes from an element

$$\tilde{x} \in \pi_p(\operatorname*{holim}_{(I\times\underline{[n]})^{\mathrm{op}}} \mathbb{Z} \times \mathrm{BGL}_N(\Gamma(\underline{\mathcal{U}}, \mathcal{O}_X))^+).$$

The Hurewicz map h_X for x factors through the map on π_p induced by the map

$$\tilde{h}_X\colon \operatorname*{holim}_{(I\times\underline{[n]})^{\mathrm{op}}} \mathbb{Z} \times \mathrm{BGL}_N(\Gamma(\underline{\mathcal{U}}, \mathcal{O}_X))^+ \to \operatorname*{holim}_{(I\times\underline{[n]})^{\mathrm{op}}} \mathbb{Z} \times \mathbb{Z}\mathrm{BGL}_N(\Gamma(\underline{\mathcal{U}}, \mathcal{O}_X))^+,$$

(for N sufficiently large) which in turn is induced by the point-wise Hurewicz maps $\mathbb{Z} \times \mathrm{BGL}_N(\Gamma(\underline{\mathcal{U}}, \mathcal{O}_X))^+(i) \to \mathbb{Z} \times \mathbb{Z}\mathrm{BGL}_N(\Gamma(\underline{\mathcal{U}}, \mathcal{O}_X))^+(i)$.

Take $p > 0$ and let $\iota_p\colon S^p \to S^p \wedge S^p$ be the diagonal embedding. If

$$\gamma\colon S^p \to \operatorname*{holim}_{(I\times\underline{[n]})^{\mathrm{op}}} \mathbb{Z}\mathrm{BGL}_N(\Gamma(\underline{\mathcal{U}}, \mathcal{O}_X))^+$$

is a map, we have the map

$$S^p \xrightarrow{(\gamma\wedge\gamma)\circ\iota_p} \operatorname*{holim}_{(I\times\underline{[n]})^{\mathrm{op}}} \mathbb{Z}\mathrm{BGL}_N(\Gamma(\underline{\mathcal{U}}, \mathcal{O}_X))^+ \wedge \operatorname*{holim}_{(I\times\underline{[n]})^{\mathrm{op}}} \mathbb{Z}\mathrm{BGL}_N(\Gamma(\underline{\mathcal{U}}, \mathcal{O}_X))^+.$$

By the Dold-Kan equivalence of the homotopy category of simplicial abelian groups with the homotopy category of complexes of abelian groups (see e.g. [**95**, Chapter V]), we may consider $(\gamma \wedge \gamma) \circ \iota_p$ as an element of

$$H^{-p}(C^{*\geq -n}(\mathcal{U}, \mathrm{BGL}_N; \mathbb{Z}) \otimes C^{*\geq -n}(\mathcal{U}, \mathrm{BGL}_N; \mathbb{Z}))$$

for all n sufficiently large.

Let $\epsilon\colon \mathbb{Z}_X \to \Gamma_{\mathcal{U}}$ be the Čech resolution of $\mathbb{Z}_X$ coming from the open cover $\mathcal{U}$ (see Chapter II, §1.3). As in §1.4.1, we have the natural map of complexes

$$\xi\colon C^{*\geq -n}(\mathcal{U}, \mathrm{BGL}_N; \mathbb{Z}) \to \varinjlim_{g} \mathrm{Hom}_{\mathbf{C}^b_{\mathrm{mot}}(\mathcal{V})}(\mathbb{Z}_{\mathrm{BGL}_N^{\leq n}}(0)_g, \Gamma_{\mathcal{U}}).$$

Using the tensor structure in $\mathbf{C}^b_{\mathrm{mot}}(\mathcal{V})$, this gives us the map of complexes

$$\begin{aligned} &C^{*\geq -n}(\mathcal{U}, \mathrm{BGL}_N; \mathbb{Z}) \otimes C^{*\geq -n}(\mathcal{U}, \mathrm{BGL}_N; \mathbb{Z}) \\ &\qquad \xrightarrow{\xi\otimes\xi} \varinjlim_{g} \mathrm{Hom}_{\mathbf{C}^b_{\mathrm{mot}}(\mathcal{V})}(\mathbb{Z}_{\mathrm{BGL}_N^{\leq n}}(0)_g \otimes \mathbb{Z}_{\mathrm{BGL}_N^{\leq n}}(0)_g, \Gamma_{\mathcal{U}} \otimes \Gamma_{\mathcal{U}}). \end{aligned}$$

Taking cohomology, composing with $\epsilon(q)^{-1} \otimes \epsilon(q')^{-1}$, applying the moving lemma isomorphism, and using the product structure on $\mathbb{Z}_X$ (see Chapter I, §2.7.4) gives the map

$$\begin{aligned} &H^{-p}(C^{*\geq -n}(\mathcal{U}, \mathrm{BGL}_N; \mathbb{Z}) \otimes C^{*\geq -n}(\mathcal{U}, \mathrm{BGL}_N; \mathbb{Z})) \\ &\qquad \xrightarrow{\Xi\otimes\Xi} \mathrm{Hom}_{\mathbf{D}^b_{\mathrm{mot}}(\mathcal{V})}(\mathbb{Z}_{\mathrm{BGL}_N^{\leq n}}(q) \otimes \mathbb{Z}_{\mathrm{BGL}_N^{\leq n}}(q'), \mathbb{Z}_X(q+q')[-p]). \end{aligned}$$

We may then compose $\Xi \otimes \Xi[2q + 2q']$ with the map

$$1 \cong 1 \otimes 1 \xrightarrow{c_q(E_N^{\leq n}) \otimes c_{q'}(E_N^{\leq n})} \mathbb{Z}_{\mathrm{BGL}_N^{\leq n}}(q) \otimes \mathbb{Z}_{\mathrm{BGL}_N^{\leq n}}(q')[2q + 2q'],$$

giving the map

$$H^{-p}(C^{*\geq -n}(\mathcal{U},\mathrm{BGL}_N;\mathbb{Z})\otimes C^{*\geq -n}(\mathcal{U},\mathrm{BGL}_N;\mathbb{Z})) \xrightarrow{(\Xi\otimes\Xi)(c_q\otimes c_{q'})} H^{2q+2q'-p}(X,\mathbb{Z}(q+q')).$$

It follows from the definition of (3.3.3.1) that

$$(c_q\cup c_{q'})^{q+q',2q+2q'-p}(x)=(\Xi\otimes\Xi)(c_q\otimes c_{q'})\big((\gamma\wedge\gamma)\circ\iota_p\big),$$

with $\gamma=\tilde{h}_X(x)$. Since the map ι_p is homotopically trivial for $p>0$, we have $(c_q\cup c_{q'})^{q+q',2q+2q'-p}(x)=0$.

The assertion (iii) follows from the Whitney product formula (Theorem 1.3.7) for the universal total Chern classes $c(E_N)^{\leq n}$, the construction of the Chern classes $c^{q,2q-p}$, and (ii). □

3.3.5. Example. As an example, we consider the case of K-theory with support. Let X be in $\mathcal{V}$, with closed subset W, and open complement $j\colon U\to X$, $U:=X\setminus W$. As explained in Example 1.4.8, the K-theory with support $K_*^W(X)$ is gotten by taking the holim of $\mathrm{B}Q\mathcal{P}_{(-)}$ over the diagram

$$(X,W):=\begin{array}{ccc} X & \xleftarrow{\;j\;} & U\\ & & \downarrow\\ & & * \end{array}$$

and the motive with support $\mathbb{Z}_{X,W}$ is given similarly. We may also consider the "constant" diagram

$$(X):=\begin{array}{ccc} X & = & X\\ & & \|\\ & & X. \end{array}$$

There is the canonical map $\mathrm{B}Q\mathcal{P}_X\to\mathrm{holim}_{(X)}\mathrm{B}Q\mathcal{P}_{(-)}$, defined by taking the constant maps $\mathcal{N}(I/-)\to\mathrm{B}Q\mathcal{P}_{(-)}$. Similarly, we have the canonical map $\mathbb{Z}_X\to\mathbb{Z}_{(X)}$. The appropriate diagonal maps give the morphism of diagrams

$$\delta\colon(X,U)\to(X)\times_S(X,U).$$

If we then take the external product, followed by the pull-back by δ, we have the action of $K(X)$ on $K^W(X)$, and the action of $\mathbb{Z}_X$ on $\mathbb{Z}_{X,W}$:

$$K_p(X)\otimes K_{p'}^W(X)\to K_{p+p'}^W(X)$$
$$H^p(X,\mathbb{Z}(q))\otimes H_W^{p'}(X,\mathbb{Z}(q'))\to H_W^{p+p'}(X,\mathbb{Z}(q+q')).$$

Thus, Lemma 3.3.4(i) gives the identity

$$\tilde{c}^W_{X,p+p'}(xy)=\tilde{c}_{X,p}(x)\bigstar\tilde{c}^W_{X,p'}(y)$$

for $x\in K_p(X)$, $y\in K_{p'}^W(X)$, where $\tilde{c}^W_{X,*}$ is the total Chern class with support.

We have a similar description of the action of $K_*(X)$ on the relative K-theory with support, $K_*^W(X;Y_1,\dots,Y_n)$, the action of $\mathbb{Z}_X$ on the relative motive with support $\mathbb{Z}_{(X;Y_1,\dots,Y_n),W}$, and a similar identity for the total Chern classes.

3.3.6. *Chern character.* For each $k = 1, 2, \ldots$, let $S_k(t_1, \ldots, t_k) \in \mathbb{Z}[t_1, \ldots, t_k]$ be the polynomial such that

$$S_k(\sigma_1(x_1, x_2, \ldots), \ldots, \sigma_k(x_1, x_2, \ldots)) = \sum_i x_i^k.$$

We have the universal Chern character

$$\mathrm{ch}(E_N)^{\leq n} \in \oplus_{q\geq 0} H^{2q}(\mathrm{BGL}_N/S^{\leq n}, \mathbb{Q}(q)),$$

given as

$$\mathrm{ch}(E_N)^{\leq n} := N \cdot \mathrm{cl}^0(|\mathrm{BGL}_N^{\leq n}|) + \sum_{q\geq 1} \frac{1}{q!} S_q(c_1, \ldots, c_q)(E_N^{\leq n}).$$

Write $\mathrm{ch}(E_N)^{\leq n} = \sum_{q\geq 0} \mathrm{ch}(E_N)_q^{\leq n}$.

The classes $\mathrm{ch}(E_N)_q^{\leq n}$ are stable in n for all q, and are stable in N for $q > 0$, giving the element

$$\mathrm{ch}(E)^+ := \sum_{q\geq 1} \mathrm{ch}(E)_q \in \prod_{q\geq 1} H^{2q}(\mathrm{BGL}/S, \mathbb{Z}(q)).$$

Let I be the category associated to a finite partially ordered set, and let $X: I \to \mathcal{V}^+$ be a functor, as in §1.4.7. We may then pair the classes $\mathrm{ch}(E)_q$ with $H_p(X, \mathcal{GL}; \mathbb{Z})$ via the map (1.4.7.2), and compose with the Hurewicz map, as in the construction of the Chern classes for $K_*(X)$ in §1.4.7, giving the maps

$$\mathrm{ch}_{X,p,q}: K_p(X) \to H^{2q-p}(X, \mathbb{Q}(q)) \tag{3.3.6.1}$$

for $q > 0$. For $q = p = 0$, we have the rank function $\mathrm{rnk}: K_0(X) \to H^0_{\mathrm{Zar}}(X, \mathbb{Z})$; sending $i \in I$ and $n \in \mathbb{Z}$ to $n \cdot \mathrm{cl}^0_{X(i)}(|X_i|) \in H^0(X(i), \mathbb{Z}(0))$ extends to the map $\mathrm{cl}^0: H^0_{\mathrm{Zar}}(X, \mathbb{Z}) \to H^0(X, \mathbb{Z}(0))$. We then define $\mathrm{ch}_{X,0,0}$ to be the composition

$$K_0(X) \xrightarrow{\mathrm{rnk}} H^0_{\mathrm{Zar}}(X, \mathbb{Z}) \xrightarrow{\mathrm{cl}^0} H^0(X, \mathbb{Z}(0)) \to H^0(X, \mathbb{Q}(0)).$$

By the construction of the Chern classes $c^{q,2q-p}$, it follows that, for each element $x \in K_p(X)$, we have $c^{q,2q-p}(x) = 0$ for all q sufficiently large (the Chern class $Hc^{q,2q-p}$ comes from a collection of maps into BGL_N for some N, and $c_q(E_N) = 0$ for $q > N$, see Appendix B, §2.2.2). By Lemma 3.3.4(ii), we have $\mathrm{ch}_{X,p,q}(x) = \frac{(-1)^{q-1}}{(q-1)!} c^{q,2q-p}(x)$, so we have the map

$$\mathrm{ch}_{X,p} := \sum_{q\geq 1} \mathrm{ch}_{X,p,q}: K_p(X) \to \oplus_q H^{2q-p}(X, \mathbb{Q}(q)).$$

For $q = 0$, we have the map

$$\mathrm{ch}_{X,0} := \sum_{q\geq 0} \mathrm{ch}_{X,0,q}: K_0(X) \to \prod_{q\geq 0} H^{2q}(X, \mathbb{Q}(q)).$$

3.3.7. REMARKS. (i) The functorialities for the Chern classes described in Proposition 1.4.9(iii) and (iv) give similar functorialities for the Chern character. Similarly, the compatibility of the localization/relativization sequences via Chern classes, described in §1.5.2, gives a similar compatibility of the localization/relativization sequences via the Chern character.

(ii) Let X be in $\mathcal{V}$ and let E be a vector bundle on X. Suppose that the base scheme S is a k-scheme of essentially finite type for k an infinite field, and has an ample family of line bundles. Let $d_X < \infty$ be the dimension of X over k. One can show that each polynomial of weighted homogeneous degree N in the Chern classes $c_q(E)$ (where we give $c_q(E)$ degree q) vanishes in $H^{2N}(X, \mathbb{Z}(N))$ for $N > d_X$.

In fact, since X has an ample family of line bundles, there is a Grassmann variety $\mathbf{Gr}_k(r, N)$, and k-morphism $f: X \to H$, with $H = \mathbb{P}^N_k \times_k \mathbf{Gr}_k(r, N)$, such that

$$f^*(p_1^*\mathcal{O}(-1) \otimes p_2^*U(r, N)) = E,$$

where $U(r, N)$ is the universal bundle on $\mathbf{Gr}_S(r, N)$. Write U for the bundle $p_1^*\mathcal{O}(-1) \otimes p_2^*U(r, N)$.

Let $f_S: X \to H \times_k S$ be the S-morphism induced by f. and let $p: S \to \operatorname{Spec} k$ be the structure morphism. We have the bundle $p^*U \to H \times_k S$.

From Remark 1.3.6, the Chern classes are functorial with respect to the motivic pull-back of (Chapter I, §2.3); we thus have $c_q(p^*U) = p^*(c_q(U))$. By functoriality of the Chern classes, we have as well

$$c_q(E) = f_S^*(c_q(p^*U)).$$

Since k is a field, we know by (Chapter II, Theorem 3.6.6) that $H^{2q}(H, \mathbb{Z}(q)) = \mathrm{CH}^q(H)$, so there are codimension q cycles σ_q on H such that $c_q(U) = \mathrm{cl}^q(\sigma_q)$. Thus

$$c_q(p^*U) = p^*\mathrm{cl}^q(\sigma_q) = \mathrm{cl}^q(p^*\sigma_q).$$

The automorphism group of H acts on the cycle σ_q, and leaves the cycle class invariant. By Kleiman's transversality result [**78**], one may translate any finite number of the cycles σ_q so that they intersect properly on H, and so that the cycle pull-back by f is defined. In particular, if we take a finite collection of cycles σ^i such that the sum of their codimensions is more that d_X, then there are translates $g_i \cdot \sigma^i$ such that the cycle intersection $\cap_i g_i \cdot \sigma^i$ is well defined, and $f(X) \cap \cap_i \mathrm{supp}(g_i \cdot \sigma^i) = \emptyset$. Thus $f_S(X) \cap \cap_i \mathrm{supp}(p^*(g_i \cdot \sigma^i)) = \emptyset$, and hence

$$P(\ldots, c_q(E), \ldots) = f_S^*(P(\ldots, \mathrm{cl}^q(p^*\sigma_q), \ldots) = 0$$

if P has weighted degree $N > d_X$.

This implies that the Chern character $\mathrm{ch}_{X,0}: K_0(X) \to \prod_q H^{2q}(X, \mathbb{Q}(q))$ has image in the direct sum $\oplus_q H^{2q}(X, \mathbb{Q}(q))$.

3.3.8. *Properties of the Chern character.* We recall from (Appendix B, §2.2.4 and Remark 2.2.6) and (Part II, Chapter III, §3.4.4) the construction of products and external products for $K_*(X)$, and the construction of products and external products for $H^*(X, \mathbb{Z}(*))$ in (Chapter I, §2.7.4).

3.3.9. PROPOSITION. *Let I be the category associated to a finite partially ordered set, and let $X, Y: I \to \mathcal{V}^+$ be functors. Then*
(i) The Chern character

$$\oplus_p \mathrm{ch}_{X,p}: \oplus_p K_p(X) \to \oplus_p \prod_q H^{2q-p}(X, \mathbb{Q}(q))$$

is a ring homomorphism.
(ii) For $x \in K_p(X)$, $y \in K_{p'}(Y)$, we have

$$\mathrm{ch}_{X \times_S Y, p+p'}(x \boxtimes_{X,Y} y) = \mathrm{ch}_{X,p}(x) \boxtimes_{X,Y} \mathrm{ch}_{Y,p'}(y).$$

PROOF. This follows from Lemma 3.3.4, and the fact (see [**3**, exposé V, §6]), for a positively graded ring A, the map $\mathrm{ch}_A : \tilde{A} \to \mathbb{Q} \oplus \prod_{q\geq 1} A^q \otimes \mathbb{Q}$ given by

$$\mathrm{ch}_A(n, 1 + \sum_q x_q) := n + \sum_q \frac{1}{q!} S_q(x_1, \dots, x_q)$$

is a ring homomorphism. □

3.3.10. REMARKS. (i) Via the discussion in Example 1.4.8, the Chern character for diagrams constructed in §3.3.6 give the Chern character for relative K-theory with support,

$$\mathrm{ch}^W_{X;Y_1,\dots,Y_n,p} : K^W_p(X; Y_1, \dots, Y_n) \to \prod_{q\geq 0} H^{2q-p}_W(X; Y_1, \dots, Y_n, \mathbb{Z}(q)).$$

(ii) As in Example 3.3.5, if W is a closed subset of X, with X in $\mathbf{Sm}_S$, we have

$$\mathrm{ch}^W_{X,p+q}(a \cup^W_X b) = \mathrm{ch}_{X,p}(a) \cup^W_X \mathrm{ch}^W_{X,q}(b),$$

where a is in $K_p(X)$, b is in $K^W_q(X)$, and the $\cup^W_X$ are the products

$$K_p(X) \otimes K^W_q(X) \to K^W_{p+q}(X)$$
$$H^p(X, \mathbb{Q}(q)) \otimes H^{p'}_W(X, \mathbb{Q}(q')) \to H^{p+p'}_W(X, \mathbb{Q}(q + q')).$$

3.4. Riemann-Roch without denominators

We now proceed to give the proof of the Riemann-Roch formula without denominators; we use the arguments from [**2**], [**8**], [**45**], and [**46**].

3.4.1. For a scheme X, let $\mathcal{H}_X$ denote the category of coherent sheaves of finite homological dimension on X. If X has finite Krull dimension and an ample family of line bundles, the natural map $K_p(X) \to K_p(\mathcal{H}_X)$ is an isomorphism (this follows from Quillen's resolution theorem [**102**], §4, Corollary 1); in particular, each coherent sheaf $\mathcal{F}$ on X of finite homological dimension has a class $[\mathcal{F}]$ in $K_0(X)$, and a class in $K^W_0(X)$ if $\mathcal{F}$ is supported on a closed subset W of X. If X is in $\mathcal{V}$, we may then take the augmented total Chern class with support $\tilde{c}^W_X([\mathcal{F}]) \in H^{2*}_W(X, \mathbb{Z}(*))$.

3.4.2. *Push-forward in K-theory.* For a projective morphism $f : X \to Y$ in $\mathcal{V}$, we may factor f as a composition $X \xrightarrow{i} \mathbb{P}^N_Y \xrightarrow{q} Y$, where i is a closed embedding, and q is the projection. Since X and Y are smooth over S, i is a regular embedding [**55**, 16.9.2] hence f is a morphism of finite Tor-dimension. If we let $\mathcal{P}_f$ be the full subcategory of $\mathcal{P}_X$ of locally free coherent sheaves P on X such that $R^q f_* P = 0$ for $q > 0$, then, by the resolution theorem of Quillen [**102**, §4, Corollary 1], the natural map $K_p(\mathcal{P}_f) \to K_p(\mathcal{P}_X)$ is an isomorphism. As the functor $f_* : \mathcal{P}_f \to \mathcal{H}_Y$ is exact, we have the push-forward map

$$K_p(X) = K_p(\mathcal{P}_f) \xrightarrow{f_*} K_p(\mathcal{H}_Y) = K_p(Y).$$

This extends directly to push-forward for K-theory with support

$$f_* : K^W_p(X) \to K^{W'}_p(Y),$$

where W is a closed subset of X, W' is a closed subset of Y, and $f(W) \subset W'$. Indeed, let $j_U : U \to X$ and $j_V : V \to Y$ be the complements of W and W', and let

$j^*f\colon U \to V$ be the restriction of f. We then have the commutative diagram of exact functors of exact categories

$$\begin{array}{ccc} \mathcal{P}_f & \xrightarrow{j_U^*} & \mathcal{P}_{j^*f} \\ {\scriptstyle f_*}\downarrow & & \downarrow{\scriptstyle (j^*f)_*} \\ \mathcal{P}_Y & \xrightarrow[j_V^*]{} & \mathcal{P}_V \end{array}$$

giving the map on the homotopy fibers

$$\mathrm{Fib}(\mathrm{B}Q\mathcal{P}_f \xrightarrow{j_U^*} \mathrm{B}Q\mathcal{P}_{j^*f}) \to \mathrm{Fib}(\mathrm{B}Q\mathcal{P}_Y \xrightarrow{j_V^*} \mathrm{B}Q\mathcal{P}_V).$$

One can easily check (using results of [**102**]) that the projective push-forward is well-defined, functorial, satisfies the projection formula, and commutes with pull-back in cartesian squares.

3.4.3. *More universal polynomials.* Let $x := x_1, x_2, \ldots, y := y_1, \ldots, y_r$ be variables, and form the formal series (with integral coefficients)

$$G_r(x;y) := \prod_{i=1}^{\infty}\prod_{j=1}^{r}\prod_{k_1<\ldots<k_j}(1 + x_i - (y_{k_1} + \ldots + y_{k_j}))^{(-1)^j}.$$

Clearly, $G_r(x;y)$ is symmetric in the variables x and in the variables y. In addition,

$$G_r(x,y) - 1 \text{ is divisible by } y_1 \cdots y_r. \tag{3.4.3.1}$$

Indeed, by symmetry, we need only check that y_1 divides $G_r(x,y) - 1$. If we set $y_1 = 0$, then we may pair each term

$$(1 + x_i - (y_{k_1} + \ldots + y_{k_j}))^{(-1)^j}$$

with $k_1 > 1$ with the term

$$(1 + x_i - (y_1 + y_{k_1} + \ldots + y_{k_j}))^{(-1)^{j+1}},$$

so that the resulting product is 1, verifying (3.4.3.1).

Thus there is a unique power series $Q_r(s_1, s_2, \ldots; t_1, \ldots, t_r)$, with integral coefficients, such that

$$G_r(x;y) = 1 + \sigma_r(y)Q_r(\sigma_1(x), \sigma_2(x), \ldots; \sigma_1(y), \ldots, \sigma_r(y)). \tag{3.4.3.2}$$

Giving the x_i and y_j degree 1, and giving s_i and t_i degree i, we may decompose (3.4.3.2) into homogeneous terms, giving the identities

$$G_r(x;y)^{(d)} = \sigma_r(y)Q_r(\sigma_1(x), \ldots, \sigma_{d-r}(x); \sigma_1(y), \ldots, \sigma_r(y))^{(d-r)}; \quad d = 1, 2, \ldots$$

with $Q_r(s_1, \ldots, s_{d-r}; t_1, \ldots, t_r)^{(d-r)}$ a polynomial with integral coefficients, of weighted degree $d - r$.

3.4.4. For x in a lambda ring R, we have the formal sum

$$\lambda_t(x) := \sum_i \lambda^i(x)t^i,$$

which we may evaluate at $t = r \in R$ if $\lambda^k(x)$ is non-zero for only finitely many k.

Now let $A = \oplus_{i\geq 1}$ be a graded (non-unital) ring, and let

$$s := (n, 1 + \sum_i s_i); \quad t := (r, 1 - t_1 + \ldots + (-1)^r t_r)$$

be elements of $\tilde{A}$, where s is arbitrary and t has the given special form. We thus have

$$\begin{aligned}&s\bigstar\lambda_{-1}t\\&=(nr,(1+t_r\sum_d Q_r(s_1,\dots,s_{d-r};t_1,\dots,t_r)^{(d-r)})(1+\sum_i s_i)^r(1+\sum_j(-1)^jt_j)^n).\end{aligned}$$

Thus, there are universal polynomials $Q_{r,j-r}(V,X_1,\dots,X_{j-r};Y_1,\dots,Y_r)$ with integral coefficients such that

$$\begin{aligned}(3.4.4.1)\quad &[(n,1+\sum_i s_i)\bigstar(r,1-t+\ldots+(-1)^rt_r)]_j\\&\qquad\qquad=t_rQ_{r,j-r}(n,s_1,\dots,s_{j-r};t_1,\dots,t_r)\end{aligned}$$

for $j\geq 1$. Here $[(a,1+\sum_i y_i)]_j:=y_j$. Note that

$$Q_{r,j-r}(V,X_1,\dots,X_{j-r};Y_1,\dots,Y_r)\equiv 0$$

if $1\leq j<r$.

Also, for a graded A-module M, we may evaluate $Q_{r,j-r}$ at elements $n\in\mathbb{Z}$, $s_i\in M^i$ and $t_j\in A^j$, using the product $\bigstar\colon\tilde{M}\otimes\tilde{A}\to\tilde{M}$. This gives the element $Q_{r,j-r}(n,s_1,\dots,s_{j-r};t_1,\dots,t_r))\in M^{j-r}$.

3.4.5. LEMMA. *Let X be in $\mathcal{V}$, q a vector bundle of rank r on X, W a closed subset of X, and let e be in $K_p^W(X)$. Then*

$$\begin{aligned}&c_W^{j,2j-p}(e\lambda_{-1}q^\vee)\\&\qquad=c_r(q)Q_{r,j-r}(\mathrm{rnk}(e),c_W^{1,2-p}(e),\dots,c_W^{j-r,2j-2r-p}(e);c_1(q),\dots,c_r(q))\end{aligned}$$

in $H_W^{2j-p}(X,\mathbb{Z}(j))$.

PROOF. It follows from Lemma 3.3.4 that

$$\tilde{c}^W_{X,p}(e\lambda_{-1}q^\vee)=\tilde{c}^W_{X,p}(e)\bigstar\lambda_{-1}\tilde{c}_{X,0}(q^\vee).$$

It follows easily from the splitting principle that $\tilde{c}(q^\vee)=(r,1-c_1(q)+\ldots+(-1)^rc_r(q))$. The identity then follows from (3.4.4.1). □

3.4.6. LEMMA. *Let $p\colon E\to Y$ be a rank r vector bundle over Y in $\mathcal{V}$, and let $i\colon Y\to E$ be the zero section, and let W be a closed subset of Y. Let $i_*\colon K_p^W(Y)\to K_p^{i(W)}(E)$ and $i_*\colon H_W^{*-2r}(Y,\mathbb{Z}(*-r))\to H^*_{i(W)}(E,\mathbb{Z}(*))$ be the push-forward maps. Then for $x\in K_p^W(Y)$, we have*

$$\begin{aligned}&c_{i(W)}^{j,2j-p}(i_*x)\\&\qquad=i_*(Q_{r,j-r}(\mathrm{rnk}(x),c_W^{1,2-p}(x),\dots,c_W^{j-r,2j-2r-p}(x);c_1(E),\dots,c_r(E))).\end{aligned}$$

PROOF. Let $q\colon\bar{E}\to Y$ be the projective bundle $\mathbb{P}(E^\vee\oplus 1_Y)$, where $^\vee$ denotes dual, and 1_Y is the trivial line bundle on Y. We have the canonical open immersion $j\colon E\to\bar{E}$. By excision, the restriction map $j^*\colon\mathbb{Z}_{\bar{E},Y}\to\mathbb{Z}_{E,Y}$ is an isomorphism in $\mathcal{DM}(\mathcal{V})$, hence we may replace E with $\bar{E}$. As $q\circ i=\mathrm{id}_Y$ the composition

$$\mathbb{Z}_{Y,W}(-r)[-2r]\xrightarrow{i_*}\mathbb{Z}_{\bar{E},W}\to\mathbb{Z}_{\bar{E},q^{-1}(W)}$$

is split by $q_*\colon\mathbb{Z}_{\bar{E},q^{-1}(W)}\to\mathbb{Z}_{Y,q^{-1}(W)}(-r)[-2r]$ hence we may work with support in $q^{-1}(W)$.

Let Q be the kernel of the canonical surjection $q^*(E^\vee \oplus 1_Y) \to \mathcal{O}_{\bar{E}}(1)$. The projection of $q^*(E^\vee \oplus 1_Y)$ onto $q^*(1_Y) = \mathcal{O}_{\bar{E}}$ gives the map $\pi: Q \to O_{\bar{E}}$ and thus the section $s: \mathcal{O}_{\bar{E}} \to Q^\vee$. The projection of $q^*(E^\vee \oplus 1_Y)$ onto $q^*E^\vee$ gives the map $\pi': Q \to q^*E^\vee$; restricting to $E \subset \bar{E}$ gives the isomorphism $\pi'': j^*Q \to p^*E^\vee$. In particular, we have

$$i^*Q \cong E^\vee.$$

An elementary computation shows that the $i(Y)$ is the zero-subscheme of s, hence, by Corollary 1.3.9, we have

$$c_r(Q^\vee) = \mathrm{cl}^r_{\bar{E}}(i_*(|Y|)). \tag{3.4.6.1}$$

Additionally, the cokernel of π is isomorphic to $\mathcal{O}_{i(Y)}$. Thus the augmented Koszul complex

$$0 \to \Lambda^r Q \to \ldots \to Q \xrightarrow{\pi} \mathcal{O}_{\bar{E}} \to \mathcal{O}_{i(Y)} \to 0$$

is a locally free resolution of $\mathcal{O}_{i(Y)}$. This gives the identity in $K_0(\bar{E})$

$$[\mathcal{O}_{i(Y)}] = \lambda_{-1}([Q]).$$

As the map $i_*: K^W_*(Y) \to K^{q^{-1}(W)}_*(\bar{E})$ is equal to the composition

$$K^W_*(Y) \xrightarrow{q^*} K^{q^{-1}(W)}_*(\bar{E}) \xrightarrow{\otimes[\mathcal{O}_{i(Y)}]} K^{q^{-1}(W)}_*(\bar{E}),$$

we have the identity

$$i_*(x) = q^*(x)\lambda_{-1}([Q]) \tag{3.4.6.2}$$

for all $x \in K^W_*(Y)$.

Applying Lemma 3.4.5, (3.4.6.1), (3.4.6.2), and the projection formula, we have

$$\begin{aligned}
c^{j,2j-p}(i_*x) &= c_r(Q^\vee)Q_{r,j-r}(\mathrm{rnk}(x), q^*c^{1,2-p}(x), \ldots; c_1(Q^\vee), \ldots, c_r(Q^\vee)) \\
&= i_*(\mathrm{cl}^0_Y(|Y|))Q_{r,j-r}(\mathrm{rnk}(x), q^*c^{1,2-p}(x), \ldots; c_1(Q^\vee), \ldots, c_r(Q^\vee)) \\
&= i_*Q_{r,j-r}(\mathrm{rnk}(x), c^{1,2-p}(x), \ldots; i^*c_1(Q^\vee), \ldots, i^*c_r(Q^\vee)) \\
&= i_*Q_{r,j-r}(\mathrm{rnk}(x), c^{1,2-p}(x), \ldots; c_1(E), \ldots, c_r(E)).
\end{aligned}$$

In the last two lines we use the naturality of the Chern classes (Theorem 1.3.5), and the fact that $\mathrm{cl}^0_Y(|Y|)$ is the identity in $H^*(Y, \mathbb{Z}(*)$ (Chapter I, Proposition 3.5.6). □

3.4.7. THEOREM. *Let $i: Z \to X$ be a closed codimension r embedding of schemes in $\mathcal{V}$, with normal bundle N, and let W be a closed subset of Z. Then*

$$\begin{aligned}
&c^{q,2q-p}_{i(W)}(i_*x) \\
&\quad = i_*(Q_{r,q-r}(\mathrm{rnk}(x), c^{1,2-p}_W(x), \ldots, c^{q-r,2q-2r-p}_W(x); c_1(N), \ldots, c_r(N))),
\end{aligned}$$

for $x \in K^W_p(Z)$, $q \geq 1$.

PROOF. We have the deformation diagram

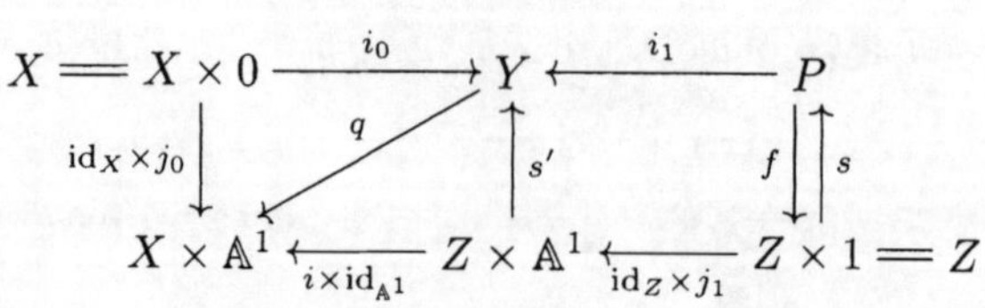

as in §2.1.2. By the Thomason-Trobaugh theorem [**121**], the maps

$$\begin{aligned} s'_*: K_p^{W\times\mathbb{A}^1}(Z\times\mathbb{A}^1) &\to K_p^{[W\times\mathbb{A}^1]}(Y)\\ i_*: K_p^W(Z) &\to K_p^W(X)\\ s_*: K_p^W(Z) &\to K_p^{s(W)}(P)\end{aligned}$$

are isomorphisms.

Furthermore, the restriction maps $j_1^*, j_0^*: K_p^{W\times\mathbb{A}^1}(Z\times\mathbb{A}^1)\to K_p^W(Z)$ are surjections, split by the pull-back map $p^*: K_p^W(Z)\to K_p^{W\times\mathbb{A}^1}(Z\times\mathbb{A}^1)$. We have as well the maps $i_1^*: K_p^{[W\times\mathbb{A}^1]}(Y)\to K_p^{s(W)}(P)$ and $i_0^*: K_p^{[W\times\mathbb{A}^1]}(Y)\to K_p^{i(W)}(X)$, satisfying

$$i_1^*\circ s'_* = s_*\circ j_1^*;\quad i_0^*\circ s'_* = i_*\circ j_0^*. \tag{3.4.7.1}$$

The above maps all have their counterparts for motivic cohomology, which, by Theorem 2.4.9 satisfy the relation (3.4.7.1). In addition, the maps

$$\begin{aligned} i_1^*: H^{2q-p}_{[W\times\mathbb{A}^1]}(Y,\mathbb{Z}(q)) &\to H^{2q-p}_{s(W)}(P,\mathbb{Z}(q)),\\ i_0^*: H^{2q-p}_{[W\times\mathbb{A}^1]}(Y,\mathbb{Z}(q)) &\to H^{2q-p}_{W}(X,\mathbb{Z}(q)),\end{aligned}$$

are isomorphisms by the homotopy property.

We have the diagrams

$$\begin{array}{ccccc} K_p^Z(X) & \xleftarrow{\ i_0^*\ } & K_p^{[W\times\mathbb{A}^1]}(Y) & \xrightarrow{\ i_1^*\ } & K_p^{s(W)}(P)\\ \downarrow{\scriptstyle c^{q,2q-p}} & & \downarrow{\scriptstyle c^{q,2q-p}} & & \downarrow{\scriptstyle c^{q,2q-p}}\\ H_Z^{2q-p}(X,\mathbb{Z}(q)) & \xleftarrow{\ i_0^*\ } & H^{2q-p}_{[W\times\mathbb{A}^1]}(Y,\mathbb{Z}(q)) & \xrightarrow{\ i_1^*\ } & H^{2q-p}_{s(W)}(P,\mathbb{Z}(q))\end{array}$$

and

$$\begin{array}{ccccc} K_p^W(Z) & \xleftarrow{\ j_0^*\ } & K_p^{W\times\mathbb{A}^1}(Z\times\mathbb{A}^1) & \xrightarrow{\ j_1^*\ } & K_p^W(Z)\\ \downarrow{\scriptstyle c^{q',2q'-p}} & & \downarrow{\scriptstyle c^{q',2q'-p}} & & \downarrow{\scriptstyle c^{q',2q-p}}\\ H_W^{2q'-p}(Z,\mathbb{Z}(q')) & \xleftarrow{\ i_0^*\ } & H^{2q'-p}_{W\times\mathbb{A}^1}(Z\times\mathbb{A}^1,\mathbb{Z}(q')) & \xrightarrow{\ i_1^*\ } & H_W^{2q'-p}(Z,\mathbb{Z}(q')),\end{array}$$

which commute by the naturality of the Chern classes. This reduces us to the case of the inclusion $s: Z\to P$.

Since an open neighborhood of $s(Z)$ in P is isomorphic to the normal bundle N, $j: N\to P$, with s going over to the zero section, and since

$$\begin{aligned} j^*: H^*_{s(W)}(P,\mathbb{Z}(*)) &\to H^*_{0_W}(N,\mathbb{Z}(*))\\ j^*: K_*^{s(W)}(P) &\to K_*^{0_W}(N)\end{aligned}$$

are isomorphisms (the isomorphism for K-theory uses results of [**121**] in the nonregular case), we may finish the proof by applying Lemma 3.4.6. □

3.5. Riemann-Roch

3.5.1. *The Todd class.* For a graded ring A, let

$$\hat{A} := \prod_{q \geq 0} A^q,$$

and write

$$1 + \hat{A}^+ \subset \hat{A}^\times$$

for the group of power series $1 + \sum_{q \geq 1} x_q$ with $x_q \in A^q$.

If A is a graded $\mathbb{Q}$-algebra, we have the *Todd character*

$$\mathrm{Todd}\colon 1 + \hat{A}^+ \to 1 + \hat{A}^+$$

defined by

$$\mathrm{Todd}(1 + X) := \frac{X}{1 - e^{-X}}.$$

Note that Todd is *multiplicative*:

$$\mathrm{Todd}((1+X)(1+Y)) = \mathrm{Todd}(1+X)\mathrm{Todd}(1+Y).$$

If X is in $\mathcal{V}$, and x is an element of $K_0(X)$, we have the (reduced) total Chern class $c(x) \in 1 + \widehat{H^{2*}(X, \mathbb{Z}(*))}^+$; we write $\mathrm{Todd}(x)$ for $\mathrm{Todd}(c(x)) \in \widehat{H^{2*}(X, \mathbb{Z}(*))}$. The multiplicativity of Todd and the Whitney product formula gives the relation

$$\mathrm{Todd}(x + y) = \mathrm{Todd}(x)\mathrm{Todd}(y).$$

Since X is smooth over S, we have the *relative tangent bundle* $T_{X/S} \to S$ which has the $K_0(X)$-class $[T_{X/S}]$. Define

$$\mathrm{Todd}(X/S) := \mathrm{Todd}([T_{X/S}]).$$

The naturality of the total Chern class implies that $\mathrm{Todd}(X/S)$ is functorial in X.

Before stating the Riemann-Roch theorem, we prove two preliminary results.

3.5.2. LEMMA. *Let X be in $\mathcal{V}$, and let E be a vector bundle of rank n on X. Then*

$$\mathrm{ch}(\lambda_{-1}E) = c_n(E^\vee)\mathrm{Todd}(E^\vee)^{-1} \text{ in } \widehat{H^{2*}(X, \mathbb{Z}(*))}.$$

PROOF. Since $\lambda_t(x + y) = \lambda_t(x)\lambda_t(y)$, as formal power series, the function $\lambda_{-1}(x)$ is multiplicative, and thus so is $\mathrm{ch}(\lambda_{-1}x)$. By the Whitney product formula, the map $F \mapsto c_{\mathrm{rnk} F}(F^\vee)$ is multiplicative, hence so is $F \mapsto c_{\mathrm{rnk} F}(F^\vee)\mathrm{Todd}(F^\vee)^{-1}$. By the splitting principle, we may assume that E has rank 1.

In this case, we have

$$\lambda_{-1}(E) = 1 - E,$$
$$c_n(E^\vee) = c_1(E^\vee) = -c_1(E),$$
$$c_q(E) = 0 \text{ for } q > 1.$$

The result then follows from the power series identity

$$1 - e^X = -X\left(\frac{-X}{1 - e^X}\right)^{-1}.$$

□

3.5.3. LEMMA. *Let $\phi_{N,k}$ be the power series*

$$\phi_{N,k}(x) := e^{kx}(\frac{x}{1-e^{-x}})^{N+1}.$$

The coefficient of x^N in $\phi_{N,k}$ is zero for $-N \leq k < 0$ and is 1 for $k = 0$.

PROOF. The coefficient of X^N in $\phi_{N,k}$ is the residue (at 0) of the differential form $\frac{e^{kx}dx}{(1-e^{-x})^{N+1}}$. Setting $y = 1 - e^{-x}$, this is the same as the residue of $\frac{(1-y)^{-k-1}dy}{y^{N+1}}$. The coefficient of y^N in $(1-y)^{-k-1}$ is zero for $-N \leq k < 0$ and 1 for $k = 0$, whence the result. □

3.5.4. THEOREM [Riemann-Roch]. *Let $f: X \to Y$ be a projective morphism in $\mathcal{V}$, W a closed subset of X and T a closed subset of Y with $f(W) \subset T$. Then for $x \in K_p^W(X)$, we have*

$$f_*(\mathrm{ch}^W_{X,p}(x) \cup \mathrm{Todd}(X/S)) = \mathrm{ch}^T_{Y,p}(f_*(x)) \cup \mathrm{Todd}(Y/S)$$

in $\prod_{q \geq 0} H^{2q-p}_{W'}(Y, \mathbb{Z}(q))$.

PROOF. Factor f as $X \xrightarrow{i} \mathbb{P}^N_Y \xrightarrow{q} Y$ where i is a closed embedding, and q is the projection. Since projective pushforward is functorial for both motivic cohomology (Theorem 2.4.7) and for K-theory [**102**], we reduce to proving the theorem for i and for q.

RR for a closed embedding: We use the notation of the proof of Lemma 3.4.6. As in the proof of Theorem 3.4.7, we may assume that the embedding is the zero-section into the projective closure of a rank r vector bundle $E \to X$: $i: X \to \bar{E} := \mathbb{P}(E \oplus 1_Y)$, which is split by the projection $q: \bar{E} \to X$. From the proof of Lemma 3.4.6, we have

$$\tilde{c}^{q^{-1}(W)}_{\bar{E},p}(i_*(x)) = \tilde{c}^{q^{-1}(W)}_{\bar{E},p}(q^*(x)) \bigstar \tilde{c}_{\bar{E},0}(\lambda_{-1}Q).$$

As taking the Chern character transforms the multiplication $\bigstar$ to the cup product in the ring $\oplus_p \prod_q H^{2q-p}(-, \mathbb{Z}(q))$, (*cf.* the proof of Proposition 3.3.9) we thus have

$$\mathrm{ch}^{q^{-1}(W)}_{\bar{E},p}(i_*(x)) = \mathrm{ch}^{q^{-1}(W)}_{\bar{E},p}(q^*(x)) \cup \mathrm{ch}(\lambda_{-1}Q). \tag{3.5.4.1}$$

From the proof of Lemma 3.4.6, we have $c_r(Q^\vee) = i_*(\mathrm{cl}^0_X(|X|))$, and $i^*Q = E^\vee$. Applying Lemma 3.5.2, (3.5.4.1) becomes

$$\begin{aligned}\mathrm{ch}^{q^{-1}(W)}_{\bar{E},p}(i_*(x)) &= q^*\mathrm{ch}^W_{X,p}(x) \cup i_*(|X|) \cup \mathrm{Todd}(Q^\vee)^{-1} \\ &= i_*(\mathrm{ch}_{X,p}(x) \cup \mathrm{Todd}(E)^{-1})\end{aligned}$$

in $H^*_{q^{-1}(W)}(\bar{E}, \mathbb{Z}(*))$. Since q_* splits i_*, we thus have the identity

$$\mathrm{ch}^W_{\bar{E},p}(i_*(x)) = i_*(\mathrm{ch}_{X,p}(x) \cup \mathrm{Todd}(E)^{-1})$$

in $H^*_W(\bar{E}, \mathbb{Z}(*))$.

We have the exact sequence of vector bundles on Y,

$$0 \to T_{X/S} \to i^*T_{\bar{E}/S} \to E = N_{X:\bar{E}} \to 0.$$

Since the Todd character is multiplicative in exact sequences, we have

$$\mathrm{Todd}(E)^{-1} = i^*\mathrm{Todd}(\bar{E}/S)^{-1}\mathrm{Todd}(X/S),$$

giving us

$$\begin{aligned}\mathrm{ch}^W_{\bar{E},p}(i_*(x)) &= i_*(\mathrm{ch}^W_{X,p}(x) \cup i^*\mathrm{Todd}(\bar{E}/S)^{-1}\mathrm{Todd}(X/S)) \\ &= i_*(\mathrm{ch}^W_{X,p}(x) \cup \mathrm{Todd}(X/S)) \cup \mathrm{Todd}(\bar{E}/S)^{-1},\end{aligned}$$

or

$$i_*(\mathrm{ch}^W_{X,p}(x) \cup \mathrm{Todd}(Y/S)) = \mathrm{ch}^W_{\bar{E},p}(i_*(x)) \cup \mathrm{Todd}(\bar{E}/S).$$

RR for a projection: Let $q\colon \mathbb{P}^N_Y \to Y$ be the projection. By the projective bundle formula for K-theory [**102**, §8, Theorem 2.1], $K_*^{q^{-1}(T)}(\mathbb{P}^N_Y)$ is a free $K_*^T(Y)$-module with basis $\{[\mathcal{O}_{\mathbb{P}^N_Y}(-i)] \mid i = 0, \dots, N\}$. If we express an element x of $K_p^{q^{-1}(T)}(\mathbb{P}^N_Y)$ in terms of this basis,

$$x = \sum_{i=0}^N q^*(y_i)[\mathcal{O}_{\mathbb{P}^N_Y}(-i)]; \quad y_i \in K_p^T(Y),$$

we have, by the projection formula for K-theory,

$$\begin{aligned}q_*(\mathrm{ch}^{q^{-1}(T)}(x)\mathrm{Todd}(\mathbb{P}^N_Y)) &= q_*(\sum_{i=0}^N q^*\mathrm{ch}^T(y_i)\mathrm{ch}[\mathcal{O}_{\mathbb{P}^N_Y}(-i)]\mathrm{Todd}(\mathbb{P}^N_Y)) \\ &= \sum_{i=0}^N \mathrm{ch}^T(y_i)q_*(\mathrm{ch}[\mathcal{O}_{\mathbb{P}^N_Y}(-i)]\mathrm{Todd}(\mathbb{P}^N_Y)).\end{aligned}$$

In addition, since $Rq^p_*(\mathcal{O}_{\mathbb{P}^N_Y}(-i)) = 0$ for all p if $0 < i \leq N$, and for all $p > 0$ if $i = 0$, and since $q_*\mathcal{O}_{\mathbb{P}^N_Y} = \mathcal{O}_Y$, we have

$$q_*([\mathcal{O}_{\mathbb{P}^N_Y}(-i)]) = \begin{cases} 0 & \text{for } 0 < i \leq N, \\ [\mathcal{O}_Y] = 1 & \text{for } i = 0. \end{cases}$$

Thus

$$\mathrm{ch}^T(q_*(x))\mathrm{Todd}(Y) = \mathrm{ch}^T(y_0)\mathrm{Todd}(Y).$$

This reduces us to showing that

$$q_*(\mathrm{ch}[\mathcal{O}_{\mathbb{P}^N_Y}(-i)]\mathrm{Todd}(\mathbb{P}^N_Y)q^*\mathrm{Todd}(Y)^{-1}) = \begin{cases} 0 & \text{for } 0 < i \leq N, \\ |Y| = 1 & \text{for } i = 0. \end{cases}$$

We have the *relative tangent bundle* $T_{\mathbb{P}^N_Y/Y}$, defined by the exact sequence

$$0 \to T_{\mathbb{P}^N_Y/Y} \to T_{\mathbb{P}^N_Y/S} \xrightarrow{dq} q^*T_{Y/S} \to 0,$$

giving $\mathrm{Todd}(\mathbb{P}^N_Y)q^*\mathrm{Todd}(Y)^{-1} = \mathrm{Todd}(T_{\mathbb{P}^N_Y/Y})$. In addition, we have the exact sequence

$$0 \to \mathcal{O}_{\mathbb{P}^N_Y} \to \mathcal{O}_{\mathbb{P}^N_Y}(1)^{N+1} \to T_{\mathbb{P}^N_Y/Y} \to 0,$$

hence $\mathrm{Todd}(T_{\mathbb{P}^N_Y/Y}) = \mathrm{Todd}(\mathcal{O}_{\mathbb{P}^N_Y}(1))^{N+1}$.

Write $\zeta = c_1(\mathcal{O}_{\mathbb{P}^N_Y}(1))$. Then, by the projective bundle formula (Theorem 1.3.2), $H^*(\mathbb{P}^N_Y, \mathbb{Z}(*))$ is a free $H^*(Y, \mathbb{Z}(*))$-module, with basis $1, \zeta, \dots \zeta^N$; the pushforward on motivic cohomology is defined by

$$q_*(\sum_{i=0}^N q^*(t_i)\zeta^i) := t_N$$

(see §2.3.1). In addition, $\mathbb{P}^N_Y$ is $\mathbb{P}(\mathcal{O}_Y^{N+1})$; the defining relation for the Chern classes of the trivial bundle $\mathcal{O}_Y^{N+1}$ gives

$$\zeta^{N+1} = \zeta^{N+1} + \sum_{i=1}^{N} c_i(\mathcal{O}_Y^{N+1})\zeta^{N+1-i} = 0.$$

Since $c_1(\mathcal{O}(-i)) = -i\zeta$, we have

$$\mathrm{ch}(\mathcal{O}(-i)) = e^{-i\zeta}.$$

Similarly, we have (formally)

$$\mathrm{Todd}(T_{\mathbb{P}^N_Y/Y}) = (\frac{\zeta}{1-e^{-\zeta}})^{N+1}.$$

Thus $q_*(\mathrm{ch}(\mathcal{O}(-i))\mathrm{Todd}(T_{\mathbb{P}^N_Y/Y}))$ is given by the coefficient of ζ^N in the formal expression $e^{-i\zeta}(\frac{\zeta}{1-e^{-\zeta}})^{N+1}$. By Lemma 3.5.3, this coefficient is zero for $0 < i \leq N$ and 1 for $i = 0$, completing the proof. □

3.5.5. REMARK. One could also deduce Riemann-Roch for a closed embedding directly from the Riemann-Roch theorem without denominators (Theorem 3.4.7) by a formal power series identity.

3.6. The Chern character isomorphism

We conclude the chapter with a sketch of Bloch's argument in [**19**], which shows that the Chern character

$$\mathrm{ch}_{X,p}\colon K_p(X)_\mathbb{Q} \to \oplus_q H^{2q-p}(X, \mathbb{Q}(q))$$

is an isomorphism in case the base scheme S is $\mathrm{Spec}\, k$, for k a field, or smooth and of dimension at most one over a field.

We refer to §1.5.2 for the fundamental localization and relativization sequences we will need. In §3.6.1-§3.6.3, we list some well-known facts about the gamma filtration, Adams operations, and the weight space decomposition for lambda rings; for details and proofs, we direct the reader to [**6**] and [**64**]. We give the main argument in §3.6.4-§3.6.10, where we assume that $S = \mathrm{Spec}\, k$, with k a perfect field.

3.6.1. *Lambda operations on K-theory.* Functorial lambda operations

$$\lambda^k, \quad k = 1, 2, \dots,$$

on the higher K-groups of a commutative ring were first constructed by Quillen (the construction was described in an article of Hiller's [**64**]) and Kratzer [**80**]. These operations satisfy the special lambda ring identities in the following sense: Make $K_0(A) \oplus K_p(A)$ into a ring with

$$(e, x)(e', x') = (ee', ex' + e'x)$$

where the product in the second factor uses the graded product structure on $K_*(A)$. The lambda operations on $K_0(A)$ and $K_p(A)$ give lambda operations on $K_0(A) \oplus K_p(A)$ by

$$\lambda^k(e, x) := (\lambda^k(e), \sum_{i=1}^{k} \lambda^{k-i}(e)\lambda^i(x)).$$

Then $K_0(A) \oplus K_p(A)$ is a special lambda ring. Kratzer refers to this structure on $K_p(X)$ as a *special* $K_0(A)$ *lambda algebra.*

The lambda operations for $K_p(A)$ were extended to give lambda operations for $K_p(X)$ for a scheme X by Soulé [**114**]; these make $K_p(X)$ into a special $K_0(X)$ lambda algebra. This was further extended by Gillet-Soulé in [**48**] to the case of an object in a Grothendieck topos, satisfying a certain cohomological finiteness condition. Another treatment of lambda operations, this time for the K-theory of a functor $X: I \to \mathbf{Sch}^+$, where I is a finite category, appears in [**83**]. In any case, the relative K-theory with support described in Example 1.4.8 has functorial lambda operations, compatible with the boundary maps in the localization and relativization sequences (see §1.5.2), which make $K_p^W(X; Y_1, \ldots, Y_n)$ into a special $K_0(X)$ lambda algebra.

3.6.2. *The gamma filtration.* Let (R, λ^*) be a special lambda ring with augmentation $\epsilon: R \to \mathbb{Z}$ of lambda rings (where $\mathbb{Z}$ has its uniquely defined lambda ring structure), and set

$$F_\gamma^1 R := \ker \epsilon.$$

We call (R, ϵ) an *augmented* lambda ring.

Define the operations

$$\gamma^k : R \to R; \quad k = 1, 2, \ldots$$

by

$$\gamma^k(x) := \lambda^k(x + k - 1).$$

Define $F_\gamma^k R$, $k = 2, 3, \ldots$, as the subgroup of R generated by elements of the form $\gamma^{i_1}(x_1) \cdot \ldots \cdot \gamma^{i_s}(x_s)$, with $x_j \in F_\gamma^1 R$ and $\sum_{j=1}^s i_j \geq k$. This gives the *gamma filtration*

$$R = F_\gamma^0 R \supset F_\gamma^1 R \supset F_\gamma^2 R \supset \ldots .$$

3.6.3. *Adams operations.* Let (R, λ^*) be a special lambda ring. The *Adams operations* $\psi^k : R \to R$ are defined as the polynomial in the lambda operations

$$\psi^k = S_k(\lambda_1, \ldots, \lambda_k),$$

where S_k is the polynomial of weighted degree k in $X_1, \ldots, X_k$ (with X_i having degree i) such that

$$S_k(\sigma_1(x_1, \ldots), \ldots, \sigma_k(x_1, \ldots)) = \sum_i x_i^k.$$

The ψ^k are ring homomorphisms, and satisfy

$$\psi^k \circ \psi^l = \psi^{kl}. \tag{3.6.3.1}$$

The main result on the Adams operations is

THEOREM. *Let R be an augmented lambda ring. Then each ψ^k preserves the gamma filtration, and ψ^k acts on $\mathrm{gr}_\gamma^q R$ as multiplication by k^q.*

In particular, let $R^{(q)}$ be the weight q eigenspace of ψ^k $(k \geq 2)$, acting on $R_{\mathbb{Q}}$:

$$R^{(q)} := \{x \in R_{\mathbb{Q}} \mid \psi^k(x) = k^q \cdot x\}.$$

If R is generated by elements x such that $\lambda^M(x) = 0$ for some M (depending on x), then the relation (3.6.3.1) implies that $R^{(q)}$ is independent of the choice of

$k \geq 2$. If we make the stronger assumption that $F_\gamma^{N+1}R = 0$ for some N, then $R_{\mathbb{Q}} = \oplus_{q=0}^{N} R^{(q)}$.

Let W be a closed subset of a smooth quasi-projective k-scheme X, and let $D_1, \ldots, D_n$ be closed subschemes of X forming a normal crossing divisor. In [**83**] we have shown that $K_p^W(X; D_1, \ldots, D_n)_{\mathbb{Q}}$ has the finite, functorial direct sum decomposition

$$K_p^W(X; D_1, \ldots, D_n)_{\mathbb{Q}} = \oplus_{q=\alpha}^{\dim_k X+p} K_p^W(X; D_1, \ldots, D_n)^{(q)}, \tag{3.6.3.2}$$

where

$$\alpha = \begin{cases} 0; & \text{for } p = 0 \\ 1; & \text{for } p = 1 \\ 2; & \text{for } p \geq 2, \end{cases}$$

relying on the analogous result of Soulé [**114**], for $n = 0$.

As the lambda ring structure on $K_*^W(X; D_1, \ldots, D_n)$ is functorial and compatible with the localization and relativization sequences, the same holds for the weight decomposition (3.6.3.2).

3.6.4. *Relative cycles.* We now require that k be perfect. Let X be a smooth quasi-projective k-scheme, $D_1, \ldots, D_n$ subschemes forming a normal crossing divisor in X, and W a closed subset of X. For a subset I of $\{1, \ldots, n\}$, let D_I be the intersection $\cap_{i \in I} D_i$, and let $f : \coprod_{I \subset \{1, \ldots, n\}} D_I \to X$ be the union of the inclusions. For each i, intersection with D_i defines the map

$$\mathcal{Z}^q(X)_f \xrightarrow{\cdot D_i} \mathcal{Z}^q(D_i);$$

let $\mathcal{Z}^q(X; D_1, \ldots, D_n)$ be the subgroup of $\mathcal{Z}^q(X)_f$ defined by the exactness of

$$0 \to \mathcal{Z}^q(X; D_1, \ldots, D_n) \to \mathcal{Z}^q(X)_f \xrightarrow{\oplus_i \cdot D_i} \oplus_{i=1}^n \mathcal{Z}^q(D_i).$$

We let $\mathcal{Z}_W^q(X; D_1, \ldots, D_n)$ be the subgroup of $\mathcal{Z}^q(X; D_1, \ldots, D_n)$ consisting of those cycles with support in W.

3.6.5. LEMMA. *Let X be a smooth quasi-projective k-scheme, $D_1, \ldots, D_n$ subschemes forming a normal crossing divisor in X, and W is a closed subset of X of codimension $\geq q$. Suppose that $W \cap D_I$ has codimension $\geq q$ on D_I for each $I \subset \{1, \ldots, n\}$. Then there are natural isomorphisms*

$$H\mathrm{cyc}_W^q : \mathcal{Z}_W^q(X; D_1, \ldots, D_n)_{\mathbb{Q}} \to H_W^{2q}(X; D_1, \ldots, D_n, \mathbb{Q}(q)),$$
$$K\mathrm{cyc}_W^q : \mathcal{Z}_W^q(X; D_1, \ldots, D_n)_{\mathbb{Q}} \to K_0^W(X; D_1, \ldots, D_n)^{(q)};$$

the isomorphism $H\mathrm{cyc}_W^q$ being induced by the cycle class map $\mathrm{cl}_{X,W}^q : \mathcal{Z}_W^q(X)_f \to H_W^{2q}(X, \mathbb{Z}(q))$ (Chapter I, §3.5.2).

PROOF. We first prove the assertion for K-theory. As a preliminary result, we claim that

$$K_p^W(X; D_1, \ldots, D_n)^{(q)} = 0$$

for all $p > 0$. We prove this by induction on n.

For $n = 0$, let F be a closed subset of X contained in W. We have the exact localization sequence

$$\to K_p^F(X)^{(q)} \to K_p^W(X)^{(q)} \to K_p^{W \setminus F}(X \setminus F)^{(q)} \to;$$

if $U \subset X$ is an open subscheme containing W, we have the excision isomorphism

$$K_p^W(X)^{(q)} \cong K_p^W(U)^{(q)}.$$

Thus, by noetherian induction, we may assume that W is smooth over k, that X is affine, and that W is defined (as a reduced subscheme) by q equations. In this case, Soulé [**114**] has shown that the push-forward isomorphism $i_{W*}: K_p(W) \to K_p^W(X)$ gives isomorphisms $i_{W*}^{(r)}: K_p(W)^{(r)} \to K_p^W(X)^{(r+q)}$. Since $p > 0$, we have $K_p(W)^{(0)} = 0$ by (3.6.3.2), so

$$K_p^W(X)^{(q)} = 0.$$

For general n, we have the relativization sequence

$$\to K_{p+1}^{W\cap D_n}(D_n; D_1 \cap D_n, \dots, D_{n-1} \cap D_n) \to K_p^W(X; D_1, \dots, D_n) \\ \to K_p^W(X; D_1, \dots, D_{n-1}) \to$$

which gives the exact sequence on the weight q subspaces

$$\to K_{p+1}^{W\cap D_n}(D_n; D_1 \cap D_n, \dots, D_{n-1} \cap D_n)^{(q)} \to K_p^W(X; D_1, \dots, D_n)^{(q)} \to \\ K_p^W(X; D_1, \dots, D_{n-1})^{(q)} \to .$$

Then our induction hypothesis implies

$$K_p^W(X; D_1, \dots, D_n)^{(q)} = 0.$$

Taking $p = 0$ in the relativization sequence and applying our preliminary result thus gives the exact sequence

$$0 \to K_0^W(X; D_1, \dots, D_n)^{(q)} \to K_0^W(X; D_1, \dots, D_{n-1})^{(q)} \\ \to K_0^{W\cap D_n}(D_n; D_1 \cap D_n, \dots, D_{n-1} \cap D_n)^{(q)};$$

by an elementary induction, we thus have the exact sequence

$$(3.6.5.1) \qquad 0 \to K_0^W(X; D_1, \dots, D_n)^{(q)} \to K_0^W(X)^{(q)} \to \oplus_{i=1}^n K_0^{W\cap D_i}(D_i)^{(q)}.$$

Arguing by localization as above, we have the natural isomorphisms

$$K_0^W(X)^{(q)} \cong K_0(W')^{(0)}$$
$$K_0^{W\cap D_i}(D_i)^{(q)} \cong K_0(W' \cap D_i)^{(0)},$$

where W' is any open subset of W which is smooth over k, contains all generic points of W and $W \cap D_i$, and such that $W' \cap D_i$ (with reduced scheme structure) is smooth over k for each i. Localizing further, we reduce the computation of $K_0(W')^{(0)}$ to the case of Spec of a product of fields; as it easily seen that $K_0(F) = \mathbb{Z}$ by rank (which is the augmentation for K_*), we thus have the natural isomorphism

$$K_0^W(X)^{(q)} \cong \oplus_{w\in W^{(0)}} \mathbb{Q},$$

where $W^{(0)}$ is the set of generic points for W. We have a similar computation for $K_0^{W\cap D_i}(D_i)^{(q)}$. We may therefore identify $K_0^W(X; D_1, \dots, D_n)^{(q)}$ via the exact sequence (3.6.5.1) with the kernel of

$$\mathcal{Z}_W^q(X)_\mathbb{Q} \xrightarrow{\oplus_i \cdot D_i} \oplus_{i=1}^n \mathcal{Z}^q(D_i)_\mathbb{Q}.$$

This gives the desired isomorphism

$$K_0^W(X; D_1, \dots, D_n)^{(q)} \cong \mathcal{Z}_W^q(X; D_1, \dots, D_n)_\mathbb{Q}.$$

The proof for $H^{2q}_W(X; D_1, \dots, D_n, \mathbb{Q}(q))$ is essentially the same, using the correspondence

$$K_p^{?_2}(?_1)^{(q)} \longleftrightarrow H^{2q-p}_{?_2}(?_1, \mathbb{Q}(q)).$$

We use as well the Gysin isomorphism $i_*: H^a(W, \mathbb{Z}(b)) \to H^{a+2q}_W(X, \mathbb{Z}(q+b))$ (2.1.2.2) for $i: W \to X$ a closed embedding in $\mathbf{Sm}_k$ (this is where we need k to be perfect), the isomorphism $H^{2q-p}(Y, \mathbb{Z}(q)) \cong \mathrm{CH}^q(Y, p)$ for Y in $\mathbf{Sm}_k$ (Chapter II, Theorem 3.6.6), and the identity

$$\mathrm{CH}^0(Y, p) = \begin{cases} 0; & \text{for } p > 0, \\ H^0_{\mathrm{Zar}}(Y, \mathbb{Z}); & \text{for } p = 0. \end{cases}$$

We leave the details to the reader. □

3.6.6. *The cycle class map to K-theory.* For a smooth k-scheme X, with subschemes $D_1, \dots, D_n$ forming a normal crossing divisor, define

$$\mathcal{Z}^q_{(q)}(X; D_1, \dots, D_n) := \varinjlim_W \mathcal{Z}^q_W(X; D_1, \dots, D_n),$$

where W ranges over the closed codimension q subsets of X such that $W \cap D_I$ has codimension q on D_I (or is empty) for each $I \subset \{1, \dots, \}$. Define

$$K_0^{(q)}(X; D_1, \dots, D_n)^{(q)},\ H^{2q}_{(q)}(X; D_1, \dots, D_n, \mathbb{Z}(q))$$

similarly. The isomorphisms of Lemma 3.6.5 give the isomorphisms

$$H\mathrm{cyc}^q_{(q)}: \mathcal{Z}^q_{(q)}(X; D_1, \dots, D_n)_{\mathbb{Q}} \to H^{2q}_{(q)}(X; D_1, \dots, D_n, \mathbb{Q}(q)),$$
$$K\mathrm{cyc}^q_{(q)}: \mathcal{Z}^q_{(q)}(X; D_1, \dots, D_n)_{\mathbb{Q}} \to K_0^{(q)}(X; D_1, \dots, D_n)^{(q)}.$$

Let

$$H\iota: H^{2q}_{(q)}(X; D_1, \dots, D_n, \mathbb{Q}(q)) \to H^{2q}(X; D_1, \dots, D_n, \mathbb{Q}(q)),$$
$$K\iota: K_0^{(q)}(X; D_1, \dots, D_n)^{(q)} \to K_0(X; D_1, \dots, D_n)^{(q)},$$

be the "forget the support" maps.

Define the map

$$\gamma^q_{X; D_1, \dots, D_n}: \mathcal{Z}^q_{(q)}(X; D_1, \dots, D_n)_{\mathbb{Q}} \to K_0(X; D_1, \dots, D_n)^{(q)}$$

as the composition

$$\mathcal{Z}^q_{(q)}(X; D_1, \dots, D_n)_{\mathbb{Q}} \xrightarrow{K\mathrm{cyc}^q_{(q)}} K_0^{(q)}(X; D_1, \dots, D_n)^{(q)} \xrightarrow{K\iota} K_0(X; D_1, \dots, D_n)^{(q)}.$$

We have the cycle class map

$$\mathrm{cl}^q_{X; D_1, \dots, D_n}: \mathcal{Z}^q_{(q)}(X; D_1, \dots, D_n)_{\mathbb{Q}} \to H^{2q}(X; D_1, \dots, D_n, \mathbb{Q}(q))$$

defined similarly.

3.6.7. LEMMA. *We have*

$$\mathrm{ch}_{X; D_1, \dots, D_n} \circ \gamma^q_{X; D_1, \dots, D_n} = \mathrm{cl}^q_{X; D_1, \dots, D_n} \mod \prod_{r > q} H^{2r}(X; D_1, \dots, D_n, \mathbb{Q}(r)),$$

where

$$\mathrm{ch}_{X;D_1,\dots,D_n}: K_0(X;D_1,\dots,D_n) \to \prod_{r\geq 0} H^{2r}(X;D_1,\dots,D_n,\mathbb{Q}(r))$$

is the Chern character for relative K-theory (Remark 3.3.10).

PROOF. Let W be a closed codimension q subset of X which intersects each D_I in codimension q. We write D_* for $D_1,\dots,D_n$.

We consider the Chern character with support

$$\mathrm{ch}^W: K_0^W(X;D_*)_{\mathbb{Q}} \to \prod_{q\geq 0} H^{2q}_W(X;D_*,\mathbb{Q}(q))$$

(see Remark 3.3.10). From the commutative diagram (§1.5.2)

$$\begin{array}{ccc}
K_0^W(X;D_*)^{(q)} & \xrightarrow{\mathrm{ch}_q^W} & H^{2q}_W(X;D_*,\mathbb{Q}(q)) \\
\downarrow{\scriptstyle K\iota} & & \downarrow{\scriptstyle H\iota} \\
K_0(X;D_*)^{(q)} & \xrightarrow[\mathrm{ch}_q]{} & H^{2q}(X;D_*,\mathbb{Q}(q))
\end{array}$$

it suffices to prove

$$\mathrm{ch}^W \circ K\mathrm{cyc}^q_W = H\mathrm{cyc}^q_W \mod \prod_{r>q} H^{2r}_W(X;D_1,\dots,D_n,\mathbb{Q}(r)). \tag{3.6.7.1}$$

We note that $H^p_W(X;D_*,\mathbb{Z}(r)) = 0$ for $r < q$; indeed, for $n=0$, this follows from the semi-purity theorem (Theorem 2.2.5), and in general by induction, together with the long exact relativization sequence

$$\to H^{p-1}_W(D_n;D_{*<n}\cap D,\mathbb{Z}(r)) \to H^p_W(X;D_*,\mathbb{Z}(r)) \to H^p_W(X;D_{*<n},\mathbb{Z}(r)) \to .$$

Thus $\mathrm{ch}_r^W = 0$ for $r = 0,\dots,q-1$, so we need only compute ch_q^W.

Let $\bar{W}$ be a closed subset of W. We write W^0 for $W\setminus\bar{W}$, X^0 for $X\setminus\bar{W}$, etc.

We have the commutative diagrams

$$\begin{array}{ccccc}
K_0^W(X;D_*)^{(q)} & \xleftarrow{K\mathrm{cyc}^q_W} & \mathcal{Z}^q_W(X;D_*)_{\mathbb{Q}} & \xrightarrow{H\mathrm{cyc}^q_W} & H^{2q}_W(X;D_*,\mathbb{Q}(q)) \\
\downarrow{\scriptstyle j^*} & & \downarrow{\scriptstyle j^*} & & \downarrow{\scriptstyle j^*} \\
K_0^{W^0}(X^0;D^0_*)^{(q)} & \xleftarrow[K\mathrm{cyc}^q_{W^0}]{} & \mathcal{Z}^q_{W^0}(X^0;D^0_*)_{\mathbb{Q}} & \xrightarrow[H\mathrm{cyc}^q_{W^0}]{} & H^{2q}_{W^0}(X^0;D^0_*,\mathbb{Q}(q))
\end{array}$$

and

$$\begin{array}{ccc}
K_0^W(X;D_*)^{(q)} & \xrightarrow{\mathrm{ch}_q^W} & H^{2q}_W(X;D_*,\mathbb{Q}(q)) \\
\downarrow{\scriptstyle j^*} & & \downarrow{\scriptstyle j^*} \\
K_0^{W^0}(X^0;D^0_*)^{(q)} & \xrightarrow[\mathrm{ch}_q^{W^0}]{} & H^{2q}_{W^0}(X^0;D^0_*,\mathbb{Q}(q))
\end{array}$$

(see §1.5.2) with j^* being the appropriate restriction map.

As the map $j^*: \mathcal{Z}^q_W(X;D_*) \to \mathcal{Z}^q_{W^0}(X^0;D^0_*)$ is injective if $\bar{W}$ contains no generic point of W, so are the two other maps j^*. Thus, we may pass to a neighborhood of the generic points of W, so we may assume that $n=0$ and the W is a disjoint union of smooth k varieties of codimension q in X with trivial normal bundle in X. We may also assume W irreducible.

By the Riemann-Roch theorem (Theorem 3.5.4), we have

(3.6.7.2)
$$\mathrm{ch}^W(i_*x) = i_*(\mathrm{ch}(x)) \equiv i_*(\mathrm{rnk}(x)\cdot|W|) \mod \prod_{r>q} H^{2r}_W(X; D_1, \ldots, D_n, \mathbb{Q}(r))$$

for $x \in K_0(W)$, where the i_* are the push-forward isomorphisms

$$i_*: H^0(W, \mathbb{Q}(0)) \to H^{2q}_W(X, \mathbb{Q}(q))$$
$$i_*: K_0(W)_{\mathbb{Q}} \to K_0^W(X).$$

We know that the i_* in K-theory sends $K_0(W)^{(0)}_{\mathbb{Q}}$ isomorphically onto $K_0^W(X)^{(q)}$. In addition, the rank function maps $K_0(W)^{(0)}_{\mathbb{Q}}$ isomorphically onto $\mathbb{Q}$; by the definition of the map $K\mathrm{cyc}^q_W$, we have

$$K\mathrm{cyc}^q_W(1\cdot|W|) = i_*(\mathrm{rnk}^{-1}(1) \in K_0(W)^{(0)}_{\mathbb{Q}}).$$

This, together with (3.6.7.2), proves (3.6.7.1).

□

In fact, the higher degree terms in the Chern character on $K_0(-)^{(q)}$ vanish as well; as we won't need this result, we omit the proof.

3.6.8. *Cycles and higher K-theory.* Take Y in $\mathbf{Sm}_k^{\mathrm{ess}}$. We have the cosimplicial scheme (Chapter II, §2.1.1) $\Delta^*_Y := \Delta^* \times Y$. We apply the constructions of the previous section to

$$X := \Delta^N_Y; D_* = \partial\Delta^N_{Y*} := \{\Delta^N_{Y0}, \ldots, \Delta^N_{YN}\},$$

where Δ^N_{Yi} is the face $t_i = 0$. We consider as well $(\Delta^{N+1}_Y; \partial_0\Delta^{N+1}_{Y*})$, where

$$\partial_0\Delta^{N+1}_{Y*} = \{\Delta^{N+1}_{Y0}, \ldots, \Delta^{N+1}_{YN}\}.$$

Let

$$f\colon \coprod_{I\subset\{0,\ldots,N+1\}} \partial\Delta^{N+1}_{YI} \to \Delta^{N+1}_Y$$

be the union of the inclusion maps. We let $\mathcal{Z}^q(\Delta^{N+1}_Y; \partial_0\Delta^{N+1}_Y)'$ be the subgroup of $\mathcal{Z}^q(\Delta^{N+1}_Y; \partial_0\Delta^{N+1}_Y)$ defined by the exactness of

$$0 \to \mathcal{Z}^q(\Delta^{N+1}_Y; \partial_0\Delta^{N+1}_Y)' \to \mathcal{Z}^q(\Delta^{N+1}_Y)_f \xrightarrow{\oplus_{i=0}^N\cdot|\Delta^{N+1}_{Yi}|} \oplus_{i=0}^N \mathcal{Z}^q(\Delta^{N+1}_{Yi});$$

this is just the subgroup of $\mathcal{Z}^q(\Delta^{N+1}_Y; \partial_0\Delta^{N+1}_Y)$ of cycles z which have proper intersection with each face of $\Delta^N_Y = \Delta^{N+1}_{YN+1}$. The intersection with $|\Delta^{N+1}_{YN+1}|$ thus gives the map

$$i^*_{N+1}\colon \mathcal{Z}^q(\Delta^{N+1}_Y; \partial_0\Delta^{N+1}_Y)' \to \mathcal{Z}^q(\Delta^N_Y; \partial\Delta^N_Y).$$

We have as well the inclusions

$$\mathcal{Z}^q(\Delta^{N+1}_Y; \partial_0\Delta^{N+1}_Y)' \hookrightarrow z^q(Y, N+1),$$
$$\mathcal{Z}^q(\Delta^N_Y; \partial\Delta^N_Y) \hookrightarrow z^q(Y, N),$$

where $z^q(Y,*)$ is Bloch's cycle complex (Chapter II, §2.1.2); by the Dold-Kan equivalence, these inclusions induce an isomorphism

$$\text{(3.6.8.1)} \qquad \operatorname{coker} i^*_{N+1} \cong H_N(z^q(Y,*)) =: \mathrm{CH}^q(Y,N).$$

We write γ^q_N for

$$\gamma^q_{\Delta^N_Y;\partial\Delta^N_Y} : \mathcal{Z}^q(\Delta^N_Y;\partial\Delta^N_Y)_{\mathbb{Q}} \to K_0(\Delta^N_Y;\partial\Delta^N_Y)^{(q)},$$

and cl^q_N for

$$\mathrm{cl}^q_{\Delta^N_Y;\partial\Delta^N_Y} : \mathcal{Z}^q_{(q)}(\Delta^N_Y;\partial\Delta^N_Y)_{\mathbb{Q}} \to H^{2q}(\Delta^N_Y;\partial\Delta^N_Y,\mathbb{Q}(q))$$

3.6.9. LEMMA. *We have $\gamma^q_N \circ d^{N+1} = 0$ and $\mathrm{cl}^q_N \circ d^{N+1} = 0$.*

PROOF. Denote the restriction of $\gamma^q_{\Delta^{N+1}_Y;\partial_0\Delta^{N+1}_Y}$ to $\mathcal{Z}^q(\Delta^{N+1}_Y;\partial_0\Delta^{N+1}_Y)'_{\mathbb{Q}}$ by $\gamma^{q\prime}_{N+1}$. We have the commutative diagram

$$\text{(3.6.9.1)} \qquad \begin{array}{ccc} \mathcal{Z}^q(\Delta^{N+1}_Y;\partial_0\Delta^{N+1}_Y)'_{\mathbb{Q}} & \xrightarrow{\gamma^{q\prime}_{N+1}} & K_0(\Delta^{N+1}_Y;\partial_0\Delta^{N+1}_Y)^{(q)} \\ {\scriptstyle i^*_{N+1}}\downarrow & & \downarrow{\scriptstyle i^*_{N+1}} \\ \mathcal{Z}^q(\Delta^N_Y;\partial\Delta^N_Y)_{\mathbb{Q}} & \xrightarrow[\gamma^q_N]{} & K_0(\Delta^N_Y;\partial\Delta^N_Y)^{(q)}, \end{array}$$

where the map i^*_{N+1} on K_0 is given by the restriction to the face $\partial\Delta^{N+1}_{YN+1}$.

Since Y is regular, the pull-back map $p^* : K_*(Y) \to K_*(\Delta^m_Y)$ is an isomorphism for all $m \ge 0$. This implies $K_p(\Delta^m_Y, \partial\Delta^m_{Yi}) = 0$ for each i. We have the relativization sequence, gotten by identifying Δ^{m-1}_Y with the face Δ^m_{Yn}, $n < m$,

$$\text{(3.6.9.2)} \qquad K_{p+1}(\Delta^{m-1}_Y, \partial\Delta^{m-1}_{Y*<n}) \to K_p(\Delta^m_Y, \partial\Delta^m_{Y*\le n}) \to K_p(\Delta^m_Y, \partial\Delta^m_{Y*<n}).$$

The inclusion $\Delta^m_{Yn} \to \Delta^m_Y$ is split by projection

$$\pi : \Delta^m_Y \to \Delta^m_{Yn}$$
$$(t_0, \dots, t_m) \mapsto (t_0, \dots, t_n + t_{n+1}, \dots, t_m)$$

for $0 \le n < m$. As π sends the face Δ^m_{Yj} to Δ^{m-1}_{Yj} for $0 \le j < n$, π induces a splitting in the sequence (3.6.9.2). Thus, by induction, we have $K_*(\Delta^m_Y, \partial\Delta^m_{Y*\le n}) = 0$ for all $n < m$. In particular, we have $K_0(\Delta^{N+1}_Y;\partial_0\Delta^{N+1}_Y) = 0$. The result for γ^q_N then follows from the commutativity of (3.6.9.1).

A similar argument proves the result for cl^q_N. □

Via the isomorphism (3.6.8.1), Lemma 3.6.9 shows that the map γ^q_N descends to $\gamma^q_N : \mathrm{CH}^q(Y,N) \to K_0(\Delta^N_Y;\partial\Delta^N_{Y*})^{(q)}$, and cl^q_N descends to $\mathrm{cl}^q_N : \mathrm{CH}^q(Y,N)_{\mathbb{Q}} \to H^{2q}(\Delta^N_Y;\partial\Delta^N_{Y*},\mathbb{Q}(q))$.

By Lemma 3.6.7, we have

$$\text{(3.6.9.3)} \qquad \mathrm{ch}_{\Delta^N_Y;\partial\Delta^N_{Y*},0,q} \circ \gamma^q_N = \mathrm{cl}^q_{N\mathbb{Q}}.$$

3.6.10. *Suspension isomorphisms.* We have a natural isomorphism

$$K\Sigma^N : K_p(\Delta_Y^N; \partial\Delta_Y^N) \to K_{N+p}(Y).$$

Indeed, from the proof of Lemma 3.6.9, we have $K_p(\Delta_Y^N; \partial_0\Delta_{Y*}^N) = 0$ for all p. The relativization sequence

$$K_{p+1}(\Delta_Y^{N-1}; \partial\Delta_{Y*}^{N-1}) \to K_p(\Delta_Y^N; \partial\Delta_{Y*}^N) \to K_p(\Delta_Y^N; \partial_0\Delta_{Y*}^N) \to$$

thus gives the isomorphisms

$$K_p(\Delta_Y^N; \partial\Delta_{Y*}^N) \cong K_{p+1}(\Delta_Y^{N-1}; \partial\Delta_{Y*}^{N-1}) \cong \ldots \cong K_{N+p}(Y).$$

The relativization sequence being compatible with the weight decomposition, we have the isomorphisms

$$K\Sigma^N : K_p(\Delta_Y^N; \partial\Delta_Y^N)^{(q)} \to K_{N+p}(Y)^{(q)}$$

as well.

Similarly, we have the natural isomorphism

$$H\Sigma^N : H^p(\Delta_Y^N; \partial\Delta_Y^N, \mathbb{Q}(q)) \to H^{p-N}(Y, \mathbb{Q}(q)).$$

As the relativization sequences for $H^*(-, \mathbb{Z}(*)$ and for $K_*(-)$ are compatible via the appropriate Chern classes (see §1.5.2 and Remark 3.3.10), we have the commutative diagram

$$\begin{array}{ccc} K_p(\Delta_Y^N; \partial\Delta_Y^N) & \xrightarrow{K\Sigma^N} & K_{N+p}(Y) \\ \downarrow{\scriptstyle \mathrm{ch}_{p,q}} & & \downarrow{\scriptstyle \mathrm{ch}_{N+p,q}} \\ H^{2q-p}(\Delta_Y^N; \partial\Delta_Y^N, \mathbb{Q}(q)) & \xrightarrow[H\Sigma^N]{} & H^{2q-p-N}(Y, \mathbb{Q}(q)). \end{array} \tag{3.6.10.1}$$

The $\mathbb{Q}$-cycle class map, which gives the isomorphism $\mathrm{cl}_Y^{q,2q-N} : \mathrm{CH}^q(Y, N)_\mathbb{Q} \to H^{2q-N}(Y, \mathbb{Q}(q))$ of (Chapter III, Theorem 3.6.6) is none other than the composition

$$H\Sigma^N \circ \mathrm{cl}_N^q : \mathrm{CH}^q(Y, N)_\mathbb{Q} \to H^{2q-N}(Y, \mathbb{Q}(q))$$

(at least up to sign). This follows from an elementary comparison of the isomorphism (II.2.2.6.2) used to define cl_{naif} (see (II.2.3.6.1)) with the map $H\Sigma^N$ defined above via the linked relativization sequences. Thus, combining the identity (3.6.9.3) with the commutativity of (3.6.10.1), we have shown

3.6.11. LEMMA. *The Chern character*

$$\mathrm{ch}_{Y,N} : K_N(Y)_\mathbb{Q} \to \oplus_{q\geq 0} H^{2q-N}(Y, \mathbb{Q}(q))$$

is a split surjection.

To show the injectivity of $\mathrm{ch}_{Y,N,q}$, it suffices to show that the map

$$\gamma_N^q : \mathrm{CH}^q(Y, N)_\mathbb{Q} \to K_0(\Delta_Y^N; \partial\Delta_Y^N)^{(q)}$$

is surjective. As the map with support

$$\gamma_N^q : \mathcal{Z}^q(\Delta_Y^N; \partial\Delta_Y^N)_\mathbb{Q} \to K_0^{(q)}(\Delta_Y^N; \partial\Delta_Y^N)^{(q)}$$

is an isomorphism by Lemma 3.6.5, it suffices to show that the map

$$K_0^{(q)}(\Delta_Y^N; \partial\Delta_Y^N)^{(q)} \to K_0(\Delta_Y^N; \partial\Delta_Y^N)^{(q)}$$

is surjective.

The proof of this latter fact would take us rather far from the main thread of this text; we therefore refer the reader to [**85**, Theorem 2.3], where the following result is proven

LEMMA. *Let k be an infinite field, let X be in $\mathbf{Sm}_k$, and let $Y_1, \ldots, Y_n$ be subschemes which form a normal crossing divisor. Then the map*

$$K_0^{(q)}(X;Y_*)^{(q)} \to K_0(X;Y_*)^{(q)}$$

is surjective.

Actually, in the result referred to in [**85**], the superscript $^{(q)}$ means the k^q-characteristic $\mathbb{Q}$-subspace with respect to ψ^k; the fact that ψ^k acts by $k^q \times \mathrm{id}$ on gr^q_γ implies that the k^q-characteristic subspace is the same as the k^q-eigenspace.

As both $K_0^{(q)}(X;Y_*)^{(q)}$ and $K_0(X;Y_*)^{(q)}$ send filtered projective limits to filtered inductive limits, the above result extends directly to X in $\mathbf{Sm}_k^{\mathrm{ess}}$.

We collect our results in

3.6.12. THEOREM. *Let k be a field (not necessarily perfect). Then the Chern character*

$$\mathrm{ch}_{Y,N}: K_N(Y)_\mathbb{Q} \to \oplus_{q\geq 0} H^{2q-N}(Y, \mathbb{Q}(q))$$

is an isomorphism for all $Y \in \mathbf{Sm}_k^{\mathrm{ess}}$. If S smooth and of dimension at most one over k, the same holds (using the motivic cohomology and Chern character for $\mathcal{DM}(\mathbf{Sm}_S^{\mathrm{ess}})$) for all Y in $\mathbf{Sm}_S^{\mathrm{ess}}$.

PROOF. If k is an infinite perfect field, the result for $\mathbf{Sm}_k^{\mathrm{ess}}$ follows from the above discussion. If k is not infinite or not perfect, we may pass to the algebraic closure $\bar{k}$; the pull-back maps on K-theory and on the higher Chow groups,

$$K_N(Y)_\mathbb{Q} \to K_N(Y_{\bar{k}})_\mathbb{Q},\ \mathrm{CH}^q(Y,N)_\mathbb{Q} \to \mathrm{CH}^q(Y_{\bar{k}},N)_\mathbb{Q},$$

are injective. As $\mathrm{CH}^q(-,N)_\mathbb{Q}$ and $H^{2q-N}(-,\mathbb{Q}(q))$ are naturally isomorphic, the pull-back map on motivic cohomology $H^{2q-N}(Y,\mathbb{Q}(q)) \to H^{2q-N}(Y_{\bar{k}},\mathbb{Q}(q))$ is injective as well. The Chern character is compatible with change of base scheme (Remark 3.3.7), hence the result for k follows from that for $\bar{k}$.

Now take $f: S \to \mathrm{Spec}\, k$ smooth and of dimension one over k, and take Y in $\mathbf{Sm}_S^{\mathrm{ess}}$. Then S is an inductive limit of smooth dimension one k-schemes $S_\alpha \to \mathrm{Spec}\, k$ in $\mathbf{Sm}_k$, and Y is an inductive limit of schemes Y_α in $\mathbf{Sm}_{S_\alpha}$, with the canonical maps $\pi_\alpha: Y \to S \times_{S_\alpha} Y_\alpha$ being flat. By (Chapter II, Corollary 3.4.3 and Theorem 3.6.6), we have

$$H^{2q-N}(Y,\mathbb{Q}(q)) = \varinjlim H^{2q-N}(Y_\alpha,\mathbb{Q}(q)).$$

By [**102**, §2] we have

$$K_N(Y) = \varinjlim K_N(Y_\alpha);$$

by the functoriality of the Chern character (Remark 3.3.7), we may assume that S is of finite type over k.

Take $p_Y: Y \to S$ in $\mathbf{Sm}_S^{\mathrm{ess}}$. Let f_*Y denote Y considered as an object of $\mathbf{Sm}_k^{\mathrm{ess}}$, let $p_1: f^*f_*Y = f_*Y \times_k S \to f_*Y$ be the projection, and let $\delta: Y \to f_*Y \times_k S$ be the map (id, p_Y). Then the compositions

$$K_p(Y) = K_p(f_*Y) \xrightarrow{p_1^*} K_p(f^*f_*Y) \xrightarrow{\delta^*} K_p(Y)$$

$$\mathrm{CH}^q(Y, 2q-p) = \mathrm{CH}^q(f_*Y, 2q-p) \xrightarrow{p_1^*} \mathrm{CH}^q(f^*f_*Y, 2q-p) \xrightarrow{\delta^*} \mathrm{CH}^q(Y, 2q-p)$$

are the identity maps, hence (by Theorem 3.6.6 of Chapter II) the map

$$H^p(f_*Y, \mathbb{Z}(q)) \xrightarrow{p_1^*} H^p(f^*f_*Y, \mathbb{Z}(q)) \xrightarrow{\delta^*} H^p(Y, \mathbb{Z}(q))$$

is an isomorphism. Thus, the result for S follows from that for k, together with the naturality of the Chern character with respect to the pull-back f^*. □

CHAPTER IV

Homology, Cohomology, and Duality

The first section of this chapter begins with some general results on duality in a tensor category, followed by some results on duality in certain triangulated tensor categories. In our main result of the section, Theorem 1.4.2, we show that the full sub-category $\mathcal{DM}(\mathcal{V})^{\mathrm{pr}}$ of $\mathcal{DM}(\mathcal{V})$, gotten by taking the pseudo-abelian hull of the triangulated sub-category generated by the motives of smooth projective S-schemes in $\mathcal{V}$, admits a duality involution. In particular, if $S = \operatorname{Spec} k$, and if one has resolution of singularities for k-varieties, then the category $\mathcal{DM}(k)$ has a duality involution, making $\mathcal{DM}(k)$ a rigid triangulated tensor category.

We begin the second section by embedding the category of Chow motives over a field k into the triangulated motivic category $\mathcal{DM}(\operatorname{Spec} k)$. We then examine motivic versions of the classical theories of homology, Borel-Moore homology, and compactly supported cohomology.

We give an extension of these theories to certain non-smooth S-schemes (for all quasi-projective k-schemes in case $S = \operatorname{Spec} k$, k a perfect field), and we prove a Riemann-Roch theorem for the K-theory of coherent sheaves on these S-schemes (Riemann-Roch for singular varieties). We conclude with a brief discussion of the Tate motivic category.

In the third and final section, we restrict our attention to a base scheme of the form $\operatorname{Spec} k$, with k a perfect field for which resolution of singularities holds for k-schemes of finite type. We apply the methods of [**60**] and [**61**] to extend the construction of the motive of a smooth quasi-projective k-scheme to arbitrary reduced finite type k-schemes. We extend the homological, Borel-Moore, and compactly supported motive as well.

We assume in this chapter that the base scheme S admits an ample family of line bundles.

1. Duality

1.1. Duality in tensor categories

We recall some basic facts about duality in tensor categories. This material is taken from [**109**], [**37**], and [**31**]; we give the treatment here mainly to fix notation and to keep our presentation self-contained.

We give an applications of the duality involution in §1.5.2 and Theorem 1.5.5, showing that, in case the base scheme $S = \operatorname{Spec} k$ where k is a field which admits resolution of singularities, the motivic category $\mathcal{DM}(k)$ can be constructed from the "naive" version $\mathcal{A}^0_{\mathrm{mot}}(k)$, i.e., we may replace all the homotopy identities in the construction of the motivic DG tensor category $\mathcal{A}_{\mathrm{mot}}(\mathbf{Sm}_S)$ with strict identities. Combining this with Chapter I, Theorem 3.4.2, we arrive at a construction of $\mathbf{D}^b_{\mathrm{mot}}(k)$ as a localization of the homotopy category of the usual category of complexes in the additive category $\mathcal{A}^0_{\mathrm{mot}}(\mathbf{Sm}_S)^*$, with the tensor structure induced by the product in the category $\mathcal{L}(\mathbf{Sm}_S)$, similar to the classical Grothendieck construction.

1.1.1. Let $\mathcal{A}$ be an tensor category, X and X' objects of $\mathcal{A}$, $\iota: 1 \to X \otimes X'$ a morphism. For objects A and B of $\mathcal{A}$, we have the homomorphisms

$$\begin{aligned} &\iota'(A,B): \mathrm{Hom}_{\mathcal{A}}(X' \otimes A, B) \to \mathrm{Hom}_{\mathcal{A}}(A, X \otimes B), \\ &\iota''(A,B): \mathrm{Hom}_{\mathcal{A}}(A \otimes X, B) \to \mathrm{Hom}_{\mathcal{A}}(A, B \otimes X'), \end{aligned} \tag{1.1.1.1}$$

where $\iota'(A,B)(f)$ is the composition

$$A \cong 1 \otimes A \xrightarrow{\iota \otimes \mathrm{id}_f} X \otimes X' \otimes A \xrightarrow{\mathrm{id}_X \otimes f} X \otimes B,$$

and $\iota''(A,B)(g)$ is the composition

$$A \cong A \otimes 1 \xrightarrow{\mathrm{id}_A \otimes \iota} A \otimes X \otimes X' \xrightarrow{g \otimes \mathrm{id}_X} B \otimes X'.$$

In case $A = 1$, or $B = 1$, we will often make the identifications

$$X \otimes 1 \cong 1 \otimes X \cong X, \quad X' \otimes 1 \cong 1 \otimes X' \cong X',$$

giving the maps

$$\begin{aligned} &\iota'(1,B): \mathrm{Hom}_{\mathcal{A}}(X', B) \to \mathrm{Hom}_{\mathcal{A}}(1, X \otimes B) \\ &\iota'(A,1): \mathrm{Hom}_{\mathcal{A}}(X' \otimes A, 1) \to \mathrm{Hom}_{\mathcal{A}}(A, X), \\ &\iota'(1,1): \mathrm{Hom}_{\mathcal{A}}(X', 1) \to \mathrm{Hom}_{\mathcal{A}}(1, X), \end{aligned}$$

and similarly for ι''.

Clearly, the maps (1.1.1.1) define natural transformations

$$\begin{aligned} &\iota': \mathrm{Hom}_{\mathcal{A}}(X' \otimes ?_1, ?_2) \to \mathrm{Hom}_{\mathcal{A}}(?_1, X \otimes ?_2) \\ &\iota'': \mathrm{Hom}_{\mathcal{A}}(?_1 \otimes X, ?_2) \to \mathrm{Hom}_{\mathcal{A}}(?_1, ?_2 \otimes X'), \end{aligned} \tag{1.1.1.2}$$

of the functors

$$\begin{aligned} &\mathrm{Hom}_{\mathcal{A}}(X' \otimes ?_1, ?_2), \mathrm{Hom}_{\mathcal{A}}(?_1, X \otimes ?_2): \mathcal{A}^{\mathrm{op}} \otimes \mathcal{A} \to \mathcal{A}, \\ &\mathrm{Hom}_{\mathcal{A}}(?_1 \otimes X, ?_2), \mathrm{Hom}_{\mathcal{A}}(?_1, ?_2 \otimes X'): \mathcal{A}^{\mathrm{op}} \otimes \mathcal{A} \to \mathcal{A}. \end{aligned}$$

1.1.2. DEFINITION. Let X be an object of $\mathcal{A}$. A *dual* to X is a pair (X^D, ι_X), with X^D an object of $\mathcal{A}$, and $\iota_X: 1 \to X \otimes X^D$ a morphism, such that the natural transformations (1.1.1.2) are isomorphisms.

Clearly, the relation of duality is symmetric: If (X^D, ι_X) is a dual to X, then (X, ι_{X^D}) is a dual to X^D, where $\iota_{X^D} = \tau_{X,X^D} \circ \iota_X$.

1.1.3. LEMMA. *Let X be an object of $\mathcal{A}$, (X^D, ι_X) and (X^{*D}, ι^*_X) two duals to X. Then there is a unique morphism $f: X^{*D} \to X^D$ such that $(\mathrm{id}_X \otimes f)(\iota^*_X) = \iota_X$. In addition, f is an isomorphism.*

PROOF. We have the isomorphism

$$\iota^{*\prime}_X := \iota^{*\prime}_X(1, X^D): \mathrm{Hom}_{\mathcal{A}}(X^{*D}, X^D) \to \mathrm{Hom}_{\mathcal{A}}(1, X \otimes X^D).$$

Letting $f = (\iota^{*\prime}_X)^{-1}(\iota_X)$ gives the desired morphism $f: X^{*D} \to X^D$.

If $g: X^{*D} \to X^D$ satisfies $(\mathrm{id}_X \otimes g)(\iota^*_X) = \iota_X$, then

$$\iota^{*\prime}_X(g) = \iota_X = \iota^{*\prime}_X(f);$$

since $\iota^{*\prime}_X$ is an isomorphism, we have $g = f$, hence f is unique.

By symmetry, there is an $h: X^D \to X^{*D}$ such that $(\mathrm{id}_X \otimes h)(\iota_X) = \iota_X^*$, hence

$$(\mathrm{id}_X \otimes h \circ f)(\iota_X^*) = \iota_X^*, \quad (\mathrm{id}_X \otimes f \circ h)(\iota_X) = \iota_X.$$

By the uniqueness just proven, we have $h \circ f = \mathrm{id}_{X^{*D}}$ and $f \circ h = \mathrm{id}_{X^D}$, hence f is an isomorphism. □

By Lemma 1.1.3, we may speak of *the* dual (X^D, ι_X) to X.

1.1.4. *The dual of a morphism.* If X and Y are objects of $\mathcal{A}$, with duals (X^D, ι_X) and (Y^D, ι_Y), we have the isomorphism

$$(-)^D: \mathrm{Hom}_{\mathcal{A}}(X, Y) \to \mathrm{Hom}_{\mathcal{A}}(Y^D, X^D) \tag{1.1.4.1}$$

given as the composition

$$\mathrm{Hom}_{\mathcal{A}}(X, Y) \xrightarrow{\iota''_X} \mathrm{Hom}_{\mathcal{A}}(1, Y \otimes X^D) \xrightarrow{(\iota'_Y)^{-1}} \mathrm{Hom}_{\mathcal{A}}(Y^D, X^D).$$

1.1.5. LEMMA. *(i) If X, Y and Z are objects of $\mathcal{A}$, with duals (X^D, ι_X),(Y^D, ι_Y) and (Z^D, ι_Z), and if $f: Y \to Z$ and $g: X \to Y$ are morphisms, then*

$$(f \circ g)^D = g^D \circ f^D.$$

The dual of the identity map id_X is id_{X^D}.
(ii) Let (X^{*D}, ι_X^*) and (Y^{*D}, ι_Y^*) be another choice for the duals of X and Y, $F: X^{*D} \to X^D$ and $G: Y^{*D} \to Y^D$ the canonical isomorphisms. Let

$$(-)^{*D}: \mathrm{Hom}_{\mathcal{A}}(X, Y) \to \mathrm{Hom}_{\mathcal{A}}(Y^{*D}, X^{*D})$$

be the isomorphism (1.1.4.1), formed using the duals (X^{*D}, ι_X^*) and (Y^{*D}, ι_Y^*). Then, for $f: X \to Y$, we have

$$F \circ f^{*D} = f^D \circ G.$$

(iii) Let $f: X \to Y$ be a map in $\mathcal{A}$. Take $(X, \tau_{X,X^D} \circ \iota_X)$, $(Y, \tau_{Y,Y^D} \circ \iota_Y)$ for duals to X^D and Y^D. Then

$$(f^D)^D = f.$$

(iv) Let $f: X \to Y$ be a morphism in $\mathcal{A}$, and take duals as in (iii). Let A and B be in $\mathcal{A}$. Then the diagrams

$$\begin{array}{ccc} \mathrm{Hom}_{\mathcal{A}}(X^D \otimes A, B) & \xrightarrow{\iota'_X(A,B)} & \mathrm{Hom}_{\mathcal{A}}(A, X \otimes B) \\ \downarrow{\scriptstyle (f^D \otimes \mathrm{id}_A)^*} & & \downarrow{\scriptstyle (f \otimes \mathrm{id}_B)_*} \\ \mathrm{Hom}_{\mathcal{A}}(Y^D \otimes A, B) & \xrightarrow[\iota'_Y(A,B)]{} & \mathrm{Hom}_{\mathcal{A}}(A, Y \otimes B) \end{array}$$

and

$$\begin{array}{ccc} \mathrm{Hom}_{\mathcal{A}}(A \otimes Y, B) & \xrightarrow{\iota''_Y(A,B)} & \mathrm{Hom}_{\mathcal{A}}(A, B \otimes Y^D) \\ \downarrow{\scriptstyle (\mathrm{id}_A \otimes f)^*} & & \downarrow{\scriptstyle (\mathrm{id}_B \otimes f^D)^*} \\ \mathrm{Hom}_{\mathcal{A}}(A \otimes X, B) & \xrightarrow[\iota''_X(A,B)]{} & \mathrm{Hom}_{\mathcal{A}}(A, B \otimes X^D) \end{array}$$

commute.

PROOF. All four assertions follow easily from the definitions; we leave the details to the reader. □

1.1.6. THEOREM. *Let $\mathcal{A}$ be a tensor category. Suppose each object X of $\mathcal{A}$ has a dual (X^D, ι_X).*
(i) Sending X to X^D, $f: X \to Y$ to $f^D: Y^D \to X^D$ defines a functor (of additive categories)

$$(-)^D: \mathcal{A}^{\mathrm{op}} \to \mathcal{A}.$$

(ii) The functor $(-)^D$ is independent, up to canonical isomorphism, of the choice of duals.
(iii) Suppose we have $1^{\otimes nD} = 1^{\otimes n}$, with $\iota_{1^{\otimes n}}: 1^{\otimes n} \to 1^{\otimes n} \otimes 1^{\otimes n}$ the inverse to the multiplication $\mu: 1^{\otimes n} \otimes 1^{\otimes n} \to 1^{\otimes n}$. Then the functor $(-)^D$ is a pseudo-tensor functor, i.e.:

(a) *There is a natural isomorphism $\rho: (- \otimes -)^D \to (-)^D \otimes (-)^D$ of functors $(- \otimes -)^D \to (-)^D \otimes (-)^D: \mathcal{A}^{\mathrm{op}} \otimes \mathcal{A}^{\mathrm{op}} \to \mathcal{A}$, such that*

$$(\rho_{X,Y} \otimes \mathrm{id}_Z) \circ \rho_{X \otimes Y, Z} = (\mathrm{id}_X \otimes \rho_{Y,Z}) \circ \rho_{X, Y \otimes Z}$$

for all X, Y and Z in $\mathcal{A}$.

(b) $\rho(1^{\otimes a}, 1^{\otimes b}) = \mathrm{id}$.

(c) *The maps $\rho_{X,Y}$ intertwine the symmetry isomorphisms $\tau^D_{X,Y}$ and τ_{X^D,Y^D}, and the maps $\rho_{1,X}$ (resp. $\rho_{X,1}$) intertwine the multiplication isomorphisms $\mu^D_{X,l}$ (resp. $\mu^D_{X,r}$) and $\mu_{X^D,l}$ (resp. $\mu_{X^D,r}$) (cf. Part II, Chapter I, §1.3.7).*

(iv) There is a canonical natural isomorphism of pseudo-tensor functors

$$\mathrm{id}_{\mathcal{A}} \to ((-)^D)^D.$$

PROOF. The assertion (i) follows directly from Lemma 1.1.5(i), and (ii) follows from Lemma 1.1.3 and Lemma 1.1.5(ii). Lemma 1.1.3 and Lemma 1.1.5(iii) imply (iv). For (iii), let (X^D, ι_X) be a dual to X, (Y^D, ι_Y) a dual to Y, and let $\iota^*_{X \otimes Y} = \tau \circ (\iota_X \otimes \iota_Y)$, where

$$\tau: X \otimes X^D \otimes Y \otimes Y^D \to X \otimes Y \otimes X^D \otimes Y^D$$

is the symmetry isomorphism. We note that $(X^D \otimes Y^D, \iota^*_{X \otimes Y})$ is a dual to $X \otimes Y$. Indeed, for objects A and B of $\mathcal{A}$, the maps

$$\iota'_Y(A \otimes X^D, B): \mathrm{Hom}_{\mathcal{A}}(A \otimes X^D \otimes Y^D, B) \to \mathrm{Hom}_{\mathcal{A}}(A \otimes X^D, B \otimes Y),$$
$$\iota'_X(A, B \otimes Y): \mathrm{Hom}_{\mathcal{A}}(A \otimes X^D, B \otimes Y) \to \mathrm{Hom}_{\mathcal{A}}(A, B \otimes Y \otimes X)$$

are isomorphisms. This implies that $\iota^{*\prime}_{X \otimes Y}(A, B)$ is an isomorphism. Similarly, $\iota^{*\prime\prime}_{X \otimes Y}(A, B)$ is an isomorphism. Via Lemma 1.1.3, we have the canonical isomorphism $\rho_{X,Y}: (X \otimes Y)^D \to X^D \otimes Y^D$. The relation of (iii) follows from the uniqueness portion of Lemma 1.1.3. □

1.1.7. REMARKS. (i) A tensor category such that each object has a dual is called a *rigid* tensor category.
(ii) If $\mathcal{A}$ is a graded tensor category with translation structure (see Part II, Chapter II, Definition 1.1.4), and if $X \in \mathcal{A}$ has a dual (X^D, ι), then $((X^D)^{[-1]}, \iota_1)$ is a dual to $X[1]$, where $\iota_1: 1 \to X[1] \otimes (X^D)^{[-1]}$ is the image of ι under composition with the canonical isomorphism $X \otimes X^D \to X[1] \otimes (X^D)^{[-1]}$. Similarly, $((X^D)[-1], \iota_2)$ is a dual to $X^{[1]}$, via the isomorphism $X \otimes X^D \to X^{[1]} \otimes (X^D)[-1]$, where ι_2 is induced from ι via this isomorphism. Since, in a graded tensor category with translation structure, we usually identify $X[1]$ and $X^{[1]}$ via the symmetry isomorphism $\tau_{T,X}: X[1] \to X^{[1]}$ (see Part II, Chapter II, §1.1.3), we have the dual $(X^D[1], \iota')$ to

X, with ι' induced from ι_1 via the isomorphism τ_{T,X^D}. We call this choice of dual on $X[1]$ the *canonical* dual.

If we have a morphism $f: X \to Y$, where X and Y have duals (X^D, ι_X), (Y^D, ι_Y), and if we make the canonical choice of dual for $X[1]$ and $Y[1]$, we have the identity $f[1]^D = f^D[-1]$. Thus, if $\mathcal{A}$ is a graded tensor category such that each object has a translation which has a dual, then each object of $\mathcal{A}$ has a dual, and, we may assume that the dual of $X[1]$ is the canonical dual $(X^D[-1], \iota')$ for each X in $\mathcal{A}$. In this case, Theorem 1.1.6 extends to a version for a graded tensor category with translation structure, in which the duality functor of (i) is a graded functor.

1.1.8. *A duality criterion.* We now give a criterion for a given morphism $\iota: 1 \to X \otimes X^D$ to give a dual (X^D, ι) to X. We prove a somewhat more general statement, for later use. We refer the reader to (Part II, Chapter I, §1.3.7) for the notion of a pseudo-tensor functor.

Let $\iota: 1 \to X \otimes X^D$ be a morphism in $\mathcal{A}$, $(F, \theta): \mathcal{A} \to \mathcal{B}$ a pseudo-tensor functor, and let A and B be objects of $\mathcal{B}$. Define the map

$$\iota'_F(A, B): \mathrm{Hom}_{\mathcal{B}}(F(X^D) \otimes A, B) \to \mathrm{Hom}_{\mathcal{B}}(A, F(X) \otimes B) \tag{1.1.8.1}$$

by setting $\iota'_F(A, B)(f)$ equal to the composition

$$A \cong 1 \otimes A \xrightarrow{F(\iota)\otimes \mathrm{id}_A} F(X \otimes X^D) \otimes A$$
$$\xrightarrow{\theta(X, X^D)^{-1} \otimes \mathrm{id}_A} F(X) \otimes F(X^D) \otimes A \xrightarrow{\mathrm{id}_{F(X)} \otimes f} F(X) \otimes B.$$

Define the map

$$\iota''_F(A, B): \mathrm{Hom}_{\mathcal{B}}(A \otimes F(X), B) \to \mathrm{Hom}_{\mathcal{B}}(A, B \otimes F(X^D)) \tag{1.1.8.2}$$

similarly by setting $\iota''_F(A, B)(f)$ equal to the composition

$$A \cong A \otimes 1 \xrightarrow{\mathrm{id}_A \otimes F(\iota)} A \otimes F(X \otimes X^D)$$
$$\xrightarrow{\mathrm{id}_A \otimes \theta(X, X^D)^{-1}} A \otimes F(X) \otimes F(X^D) \xrightarrow{f \otimes \mathrm{id}_{F(X^D)}} B \otimes F(X^D).$$

If (F, θ) is the identity pseudo-tensor functor on $\mathcal{A}$, these maps are just the maps ι' and ι'' defined in (1.1.1.1).

1.1.9. PROPOSITION. *Suppose there is a map* $\epsilon: X^D \otimes X \to 1$ *in* $\mathcal{A}$ *such that the compositions*

$$X \cong 1 \otimes X \xrightarrow{\iota \otimes \mathrm{id}_X} X \otimes X^D \otimes X \xrightarrow{\mathrm{id}_X \otimes \epsilon} X \otimes 1 \cong X \tag{1.1.9.1}$$
$$X^D \cong X^D \otimes 1 \xrightarrow{\mathrm{id}_{X^D} \otimes \iota} X^D \otimes X \otimes X^D \xrightarrow{\epsilon \otimes \mathrm{id}_{X^D}} 1 \otimes X^D \cong X^D,$$

are the respective identity maps. Then the maps (1.1.8.1) *and* (1.1.8.2) *are isomorphisms for all* A *and* B *in* $\mathcal{B}$. *In particular,* (X^D, ι) *is a dual to* X.

PROOF. Fix A and B in $\mathcal{B}$, and write ι'_F for $\iota'_F(A, B)$. Define the map

$$\sigma'_F: \mathrm{Hom}_{\mathcal{A}}(A, F(X) \otimes B) \to \mathrm{Hom}_{\mathcal{A}}(F(X^D) \otimes A, B)$$

by sending $g\colon A \to F(X)\otimes B$ to the composition

$$F(X^D)\otimes A \xrightarrow{\mathrm{id}_{F(X^D)}\otimes g} F(X^D)\otimes F(X)\otimes B \xrightarrow{\theta(X,X^D)\otimes \mathrm{id}_B} F(X^D\otimes X)\otimes B \xrightarrow{F(\epsilon_X)\otimes \mathrm{id}_B} 1\otimes B\cong B.$$

Let $f\colon F(X^D)\otimes A\to B$ be a morphism in $\mathcal{B}$. Then $\sigma'_F(\iota'_F(f))$ is given by the composition

$$F(X^D)\otimes A\cong F(X^D)\otimes 1\otimes A \xrightarrow{\mathrm{id}_{F(X^D)}\otimes F(\iota)\otimes \mathrm{id}_A} F(X^D)\otimes F(X\otimes X^D)\otimes A \xrightarrow{[\theta(X^D\otimes X,X^D)^{-1}\circ\theta(X^D,X\otimes X^D)]\otimes\mathrm{id}} F(X^D\otimes X)\otimes F(X^D)\otimes A \xrightarrow{\mathrm{id}\otimes f} F(X^D\otimes X)\otimes B \xrightarrow{F(\epsilon)\otimes\mathrm{id}_B} 1\otimes B\cong B.$$

We may commute $\mathrm{id}\otimes f$ and $F(\epsilon)\otimes\mathrm{id}$, hence $\sigma'_F(\iota'_F(f))$ is equal to the composition

$$F(X^D)\otimes A\cong F(X^D)\otimes 1\otimes A \xrightarrow{\theta(X^D,1)\otimes\mathrm{id}_A} F(X^D\otimes 1)\otimes A \xrightarrow{F(\mathrm{id}_{X^D}\otimes\iota)\otimes\mathrm{id}_A} F(X^D\otimes X\otimes X^D)\otimes A \xrightarrow{F(\epsilon\otimes\mathrm{id}_{X^D})\otimes\mathrm{id}_A} F(1\otimes X^D)\otimes A \xrightarrow{F(\mu)\otimes\mathrm{id}} F(X^D)\otimes A \xrightarrow{f} B.$$

By assumption, the composition $F(\epsilon\otimes\mathrm{id}_{X^D})\circ F(\mathrm{id}_{X^D}\otimes\iota)$ is the canonical isomorphism $F(X^D\otimes 1)\cong F(X^D)\cong F(1\otimes X^D)$. Since 1 is the unit in $\mathcal{A}$, we therefore have $\sigma'_F(\iota'_F(f))=f$.

Now let $g\colon A\to F(X)\otimes B$ be a map in $\mathcal{B}$. Then $\iota'_F(\sigma'_F(g))$ is given by the composition

$$A\cong 1\otimes A\xrightarrow{F(\iota)\otimes\mathrm{id}_A} F(X\otimes X^D)\otimes A\xrightarrow{\mathrm{id}\otimes g} F(X\otimes X^D)\otimes F(X)\otimes B \xrightarrow{[\theta(X,X^D\otimes X)^{-1}\circ\theta(X\otimes X^D,X)]\otimes\mathrm{id}} F(X)\otimes F(X^D\otimes X)\otimes B \xrightarrow{\mathrm{id}\otimes F(\epsilon)\otimes\mathrm{id}} F(X)\otimes 1\otimes B\cong F(X)\otimes B.$$

We may commute $\mathrm{id}\otimes g$ and $F(\iota)\otimes\mathrm{id}_A$, and rewrite this as the composition

$$A\cong 1\otimes A\xrightarrow{\mathrm{id}\otimes g} 1\otimes F(X)\otimes B\xrightarrow{\theta(1,X)\otimes\mathrm{id}} F(1\otimes X)\otimes B \xrightarrow{F(\iota\otimes\mathrm{id}_X)\otimes\mathrm{id}_B} F(X\otimes X^D\otimes X)\otimes B\xrightarrow{F(\mathrm{id}_X\otimes\epsilon)\otimes\mathrm{id}_B} F(X\otimes 1)\otimes B \xrightarrow{F(\mu)\otimes\mathrm{id}} F(X)\otimes B.$$

By assumption, the composition $F(\mathrm{id}_X\otimes\epsilon)\circ F(\iota\otimes\mathrm{id}_X)$ is the canonical isomorphism $F(1\otimes X)\cong F(X)\cong F(X\otimes 1)$. Since 1 is the unit in $\mathcal{A}$, we have $\iota'_F(\sigma'_F(g))=g$. Thus σ'_F is the inverse to ι'_F, hence $\iota'_F=\iota'_F(A,B)$ is an isomorphism. The proof that $\iota''_F(A,B)$ is an isomorphism is essentially the same. $\square$

There is a converse to Proposition 1.1.9, namely,

1.1.10. PROPOSITION. *Suppose (X^D,ι) is a dual to X. Then there is a unique map $\epsilon\colon X^D\otimes X\to 1$ such that the compositions* (1.1.9.1) *are the respective identity maps.*

PROOF. If such an ϵ exists, then the first composition in (1.1.9.1) is equal to $\iota'(X,X)(\epsilon)$, after making the canonical identification of $X \otimes 1$ with X. Similarly, the second composition in (1.1.9.1) is $\iota''(X^D, X^D)(\epsilon)$, after making a similar identification. Since $\iota'(X,X)$ is an isomorphism, ϵ is unique.

To show existence, we use the canonical structure for the dual of $X \otimes X^D$, i.e., $(X \otimes X^D)^D = X^D \otimes X$, with map $\iota_{X\otimes X^D}: 1 \to X \otimes X^D \otimes X^D \otimes X$ being the composition

$$1 \cong 1 \otimes 1 \xrightarrow{\iota\otimes\iota} X \otimes X^D \otimes X \otimes X^D \xrightarrow{\tau_{3,4}} X \otimes X^D \otimes X^D \otimes X \xrightarrow{\tau_{2,3}} X \otimes X^D \otimes X^D \otimes X. \tag{1.1.10.1}$$

Taking the dual of ι gives the map $\iota^D: X^D \otimes X \to 1$; we claim that $\epsilon = \iota^D$ is the desired map. It suffices to show that $\iota'(X,X)(\iota^D)$ is the identity on X and that $\iota''(X^D, X^D)(\iota^D)$ is the identity on X^D, after making the identifications as above.

We use the canonical dual for 1: $1^D = 1$ with $\iota_1: 1 \to 1 \otimes 1$ the inverse of the multiplication. By the definition of duality, ι^D is characterized by the fact that the composition

$$1 \xrightarrow{\iota_{X\otimes X^D}} X \otimes X^D \otimes X^D \otimes X \xrightarrow{\mathrm{id}\otimes\iota^D} X \otimes X^D \otimes 1 \cong X \otimes X^D$$

is the map ι. By definition, the map $\iota'(X,X)(\iota^D): X \to X$ is the composition

$$X \cong 1 \otimes X \xrightarrow{\iota\otimes\mathrm{id}_X} X \otimes X^D \otimes X \xrightarrow{\mathrm{id}_X\otimes\iota^D} X \otimes 1 \cong X.$$

From this, it follows that the map $\iota''(1,X)(\iota'(X,X)(\iota^D))$ is the composition

$$1 \xrightarrow{\iota} X \otimes X^D \cong 1 \otimes X \otimes X^D \xrightarrow{\iota\otimes\mathrm{id}_X\otimes\mathrm{id}_{X^D}} X \otimes X^D \otimes X \otimes X^D \xrightarrow{\mathrm{id}_X\otimes\iota^D\otimes\mathrm{id}_{X^D}} X \otimes 1 \otimes X^D \cong X \otimes X^D. \tag{1.1.10.2}$$

Using the above characterization of ι^D, together with the definition (1.1.10.1) of $\iota_{X\otimes X^D}$, it follows easily from (1.1.10.2) that $\iota''(1,X)(\iota'(X,X)(\iota^D)) = \iota$. Since $\iota''(1,X)(\mathrm{id}_X) = \iota$ as well, and since $\iota''(1,X)$ is an isomorphism, we have

$$\iota'(X,X)(\iota^D) = \mathrm{id}_X.$$

The identity $\iota''(X^D, X^D)(\iota^D) = \mathrm{id}_{X^D}$ is verified similarly. □

The dual maps $\epsilon = \iota^D: X^D \otimes X \to 1$ give a description of the composition law in $\mathcal{A}$, as follows:

1.1.11. PROPOSITION. *Let X and Y be in $\mathcal{A}$, with respective duals (X^D, ι_X), (Y^D, ι_Y). Let $\epsilon_Y: Y^D \otimes Y \to 1$ be the map given by Proposition 1.1.10. Suppose we have maps $f: X \to Y$ and $g: Y \to Z$ in $\mathcal{A}$. Then $\iota''_X(1,Z)(g \circ f): 1 \to Z \otimes X^D$ is the composition*

$$1 \cong 1 \otimes 1 \xrightarrow{\iota''_Y(1,Z)(g)\otimes\iota''_X(1,Y)(f)} Z \otimes Y^D \otimes Y \otimes X^D \xrightarrow{\mathrm{id}_Z\otimes\epsilon_Y\otimes\mathrm{id}_{X^D}} Z \otimes 1 \otimes X^D \cong Z \otimes X^D. \tag{1.1.11.1}$$

PROOF. We may expand the composition (1.1.11.1) as

$$1 \cong 1 \otimes 1 \xrightarrow{\mathrm{id}_1 \otimes \iota_X} 1 \otimes X \otimes X^D \xrightarrow{\iota_Y \otimes \mathrm{id}_{X \otimes X^D}} Y \otimes Y^D \otimes X \otimes X^D$$
$$\xrightarrow{\mathrm{id}_{Y \otimes Y^D} \otimes f \otimes \mathrm{id}_{X^D}} Y \otimes Y^D \otimes Y \otimes X^D$$
$$\xrightarrow{g \otimes \mathrm{id}_{Y^D} \otimes \mathrm{id}_Y \otimes \mathrm{id}_{X^D}} Z \otimes Y^D \otimes Y \otimes X^D$$
$$\xrightarrow{\mathrm{id}_Z \otimes \epsilon_Y \otimes \mathrm{id}_{X^D}} Z \otimes 1 \otimes X^D \cong Z \otimes X^D.$$

We may then commute $g \otimes \mathrm{id}_{Y^D} \otimes \mathrm{id}_Y \otimes \mathrm{id}_{X^D}$ with $\mathrm{id}_Z \otimes \epsilon_Y \otimes \mathrm{id}_{X^D}$, and $\iota_Y \otimes \mathrm{id}_{X \otimes X^D}$ with $\mathrm{id}_{Y \otimes Y^D} \otimes f \otimes \mathrm{id}_{X^D}$ to give the composition

$$1 \cong 1 \otimes 1 \xrightarrow{\mathrm{id}_1 \otimes \iota_X} 1 \otimes X \otimes X^D \xrightarrow{\mathrm{id}_1 \otimes f \otimes \mathrm{id}_{X^D}} 1 \otimes Y \otimes X^D$$
$$\xrightarrow{\iota_Y \otimes \mathrm{id}_Y \otimes \mathrm{id}_{X^D}} Y \otimes Y^D \otimes Y \otimes X^D \xrightarrow{\mathrm{id}_Y \otimes \epsilon_Y \otimes \mathrm{id}_{X^D}} Y \otimes 1 \otimes X^D$$
$$\cong Y \otimes X^D \xrightarrow{g \otimes \mathrm{id}_{X^D}} Z \otimes X^D.$$

Since $(\mathrm{id}_Y \otimes \epsilon_Y) \circ (\iota_Y \otimes \mathrm{id}_Y)$ is the canonical identification $1 \otimes Y \cong Y \otimes 1$, we may rewrite this composition as

$$1 \cong 1 \otimes 1 \xrightarrow{\mathrm{id}_1 \otimes \iota_X} 1 \otimes X \otimes X^D \xrightarrow{\mathrm{id}_1 \otimes f \otimes \mathrm{id}_{X^D}} 1 \otimes Y \otimes X^D$$
$$\cong Y \otimes X^D \xrightarrow{g \otimes \mathrm{id}_{X^D}} Z \otimes X^D.$$

Eliminating the superfluous 1's gives the composition

$$1 \xrightarrow{\iota_X} X \otimes X^D \xrightarrow{f \otimes \mathrm{id}_{X^D}} Y \otimes X^D \xrightarrow{g \otimes \mathrm{id}_{X^D}} Z \otimes X^D; \tag{1.1.11.2}$$

since

$$(g \otimes \mathrm{id}_{X^D}) \circ (f \otimes \mathrm{id}_{X^D}) = (g \circ f) \otimes \mathrm{id}_{X^D},$$

the composition (1.1.11.2) is $\iota''_X(1, Z)(g \circ f)$. □

1.1.12. REMARK. One can phrase duality in a tensor category $\mathcal{A}$ as in [**109**] in terms of *internal Hom objects*, where $\mathcal{H}om(X, Y)$ is a representing object for the functor $\mathrm{Hom}_{\mathcal{A}}(-\otimes X, Y): \mathcal{A}^{\mathrm{op}} \to \mathbf{Ab}$. In our setting, we may define $\mathcal{H}om(X, Y) := Y \otimes X^D$, with the necessary isomorphism $\mathrm{Hom}_{\mathcal{A}}(- \otimes X, Y) \to \mathrm{Hom}_{\mathcal{A}}(-, \mathcal{H}om(X, Y))$ being given by the isomorphism $\iota''(-, Y)$.

1.2. Duality in triangulated tensor categories

We show how the existence of duals for generating objects in certain triangulated tensor categories gives rise to an exact duality on the entire category.

1.2.1. Let $\mathcal{A}$ be a DG tensor category without unit. We may then form the category of complexes, $\mathbf{C}^b(\mathcal{A})$ and the homotopy category $\mathbf{K}^b(\mathcal{A})$ (see Part II, Chapter II, §1.2). The tensor product on $\mathcal{A}$ induces the structure of a DG tensor category without unit on $\mathbf{C}^b(\mathcal{A})$, and the structure of a triangulated tensor category without unit on $\mathbf{K}^b(\mathcal{A})$. We may form a localization $\mathcal{D}$ of $\mathbf{K}^b(\mathcal{A})$ with respect to a thick tensor subcategory; $\mathcal{D}$ is then a triangulated tensor category without unit.

If we have two distinguished triangles in $\mathcal{D}$:

$$X_1 \to Y_1 \to Z_1 \to X_1[1],$$
$$Z_2 \to Y_2 \to X_2 \to Z_2[1],$$

we may then form the commutative square

$$
\begin{array}{ccccccc}
X_1 \otimes Z_2 & \longrightarrow & Y_1 \otimes Z_2 & \longrightarrow & Z_1 \otimes Z_2 & \xrightarrow{a'} & X_1[1] \otimes Z_2 \\
\downarrow & & \downarrow & & \downarrow{\scriptstyle c'} & & \downarrow \\
X_1 \otimes Y_2 & \longrightarrow & Y_1 \otimes Y_2 & \xrightarrow{b'} & Z_1 \otimes Y_2 & \longrightarrow & X_1[1] \otimes Y_2 \\
\downarrow & & \downarrow{\scriptstyle b} & & \downarrow & & \downarrow \\
X_1 \otimes X_2 & \xrightarrow{c} & Y_1 \otimes X_2 & \longrightarrow & Z_1 \otimes X_2 & \longrightarrow & X_1[1] \otimes X_2 \\
{\scriptstyle a}\downarrow & & \downarrow & & \downarrow & & \downarrow \\
X_1 \otimes Z_2[1] & \longrightarrow & Y_1 \otimes Z_2[1] & \longrightarrow & Z_1 \otimes Z_2[1] & \longrightarrow & X_1[1] \otimes Z_2[1].
\end{array}
\tag{1.2.1.1}
$$

Identifying $X_1[1] \otimes Z_2$ with $X_1 \otimes Z_2[1]$ by the canonical isomorphism, we have the map

$$X_1 \otimes X_2 \oplus Z_1 \otimes Z_2 \xrightarrow{a+a'} X_1 \otimes Z_2[1],$$

and we may form the distinguished triangle

$$K \xrightarrow{q} X_1 \otimes X_2 \oplus Z_1 \otimes Z_2 \xrightarrow{a+a'} X_1 \otimes Z_2[1] \to K[1] \tag{1.2.1.2}$$

in $\mathcal{D}$. In addition, we have the maps

$$Y_1 \otimes Y_2 \xrightarrow{(b,b')} Y_1 \otimes X_2 \oplus Z_1 \otimes Y_2, \tag{1.2.1.3}$$

and

$$X_1 \otimes X_2 \oplus Z_1 \otimes Z_2 \xrightarrow{c \oplus c'} Z_1 \otimes Y_2 \oplus Y_1 \otimes X_2. \tag{1.2.1.4}$$

Putting the maps (1.2.1.3) and (1.2.1.4) together gives the diagram

$$
\begin{array}{ccc}
 & & X_1 \otimes X_2 \oplus Z_1 \otimes Z_2 \\
 & & \downarrow{\scriptstyle c\oplus c'} \\
Y_1 \otimes Y_2 & \xrightarrow[(b,b')]{} & Y_1 \otimes X_2 \oplus Z_1 \otimes Y_2.
\end{array}
\tag{1.2.1.5}
$$

1.2.2. LEMMA. *There is a morphism $\beta \colon K \to Y_1 \otimes Y_2$ so that the diagram* (1.2.1.5) *fills in to a commutative diagram*

$$
\begin{array}{ccccccc}
K & \xrightarrow{q} & X_1 \otimes X_2 \oplus Z_1 \otimes Z_2 & \xrightarrow{a+a'} & X_1 \otimes Z_2[1] & \longrightarrow & K[1] \\
{\scriptstyle \beta}\downarrow & & \downarrow{\scriptstyle c\oplus c'} & & & & \\
Y_1 \otimes Y_2 & \xrightarrow[(b,b')]{} & Y_1 \otimes X_2 \oplus Z_1 \otimes Y_2, & & & &
\end{array}
\tag{1.2.2.1}
$$

with the top row the distinguished triangle (1.2.1.2).

PROOF. Let X and Y be objects of $\mathcal{D}$, and $f \colon X \to Y$ a morphism in $\mathcal{D}$. As $\mathcal{D}$ is a localization of $\mathbf{K}^b(\mathcal{A})$, the morphism f can be factored as a composition $X \xrightarrow{i} Y' \xrightarrow{j^{-1}} Y$ with i and j morphisms in $\mathbf{K}^b(\mathcal{A})$, and j invertible in $\mathcal{D}$. Let

$$X \xrightarrow{f} Y \to Z \to X[1]$$

be a completion of f to a distinguished triangle in $\mathcal{D}$. As the diagram

$$\begin{array}{ccc} X & \xrightarrow{f} & Y \\ & \searrow^{i} & \downarrow j \\ & & Y' \end{array}$$

commutes, there is a map of triangles

$$\begin{array}{ccccccc} X & \xrightarrow{f} & Y & \longrightarrow & Z & \longrightarrow & X[1] \\ \| & & \downarrow j & & \downarrow k & & \| \\ X & \xrightarrow[i]{} & Y' & \longrightarrow & \mathrm{cone}(i) & \longrightarrow & X[1] \end{array}$$

in $\mathcal{D}$, where the bottom row is the image of the cone sequence. Since j is an isomorphism in $\mathcal{D}$, so is k (the "five lemma" for triangulated categories), hence each distinguished triangle in $\mathcal{D}$ is isomorphic to the image of a cone sequence from $\mathbf{C}^b(\mathcal{A})$.

Thus, it suffices to prove the lemma in the case of shifted standard distinguished triangles from $\mathbf{C}^b(\mathcal{A})$:

$$X_1 \xrightarrow{g_1} Y_1 = \mathrm{cone}(f_1) \xrightarrow{h_1} Z_1 \xrightarrow{-f_1[1]} X_1[1]$$

$$Z_2 \xrightarrow{g_2} Y_2 = \mathrm{cone}(f_2) \xrightarrow{h_2} X_2 \xrightarrow{-f_2[1]} Z_2[1]$$

We may then take K to be given by

$$K := \mathrm{cone}\big(-\mathrm{id}_{X_1} \otimes f_2[1] - f_1[1] \otimes \mathrm{id}_{Z_2} : X_1 \otimes X_2 \oplus Z_1 \otimes Z_2 \to X_1 \otimes Z_2[1]\big)[-1],$$

and the sequence (1.2.1.2) to be the shifted standard cone sequence, with $q: K \to X_1 \otimes X_2 \oplus Z_1 \otimes Z_2$ the canonical projection (the map $X_1 \otimes Z_2[1] \to K[1]$ is then *minus* the canonical inclusion, so that we have a distinguished triangle). Explicitly, K is the total complex of the double complex

$$X_1 \otimes X_2[-1] \oplus Z_1[-1] \otimes Z_2 \xrightarrow{\mathrm{id} \otimes f_2 + f_1 \otimes \mathrm{id}} X_1 \otimes Z_2.$$

The definition of Y_1 and Y_2 as cones give the description of $Y_1 \otimes Y_2$ as the total complex of the double complex

$$Z_1[-1] \otimes X_2[-1] \xrightarrow{(f_1 \otimes \mathrm{id}, -\mathrm{id} \otimes f_2)} X_1 \otimes X_2[-1] \oplus Z_1[-1] \otimes Z_2 \xrightarrow{\mathrm{id} \otimes f_2 + f_1 \otimes \mathrm{id}} X_1 \otimes Z_2.$$

The total complex of the subcomplex gotten by omitting the term $Z_1[-1] \otimes X_2[-1]$ is just K, so the inclusion of K as this subcomplex of $Y_1 \otimes Y_2$ gives us the map $\beta: K \to Y_1 \otimes Y_2$. One then verifies the commutativity of (1.2.2.1) by inspection. □

We suppose for the remainder of this section that the category $\mathcal{D}$ is a triangulated tensor category with unit 1.

1.2.3. LEMMA. *Let $X \xrightarrow{f} Y \xrightarrow{g} Z \xrightarrow{h} X[1]$ be a distinguished triangle in $\mathcal{D}$. Suppose that X has a dual (X^D, ι_X), and Z has a dual (Z^D, ι_Z). Then Y has a dual (Y^D, ι_Y). In addition, the sequence*

$$X^D[-1] \xrightarrow{h^D} Z^D \xrightarrow{g^D} Y^D \xrightarrow{f^D} X^D \tag{1.2.3.1}$$

is a distinguished triangle in $\mathcal{D}$.

PROOF. We use the canonical dual $X[1]^D = X^D[-1]$, with $\iota_{X[1]}$ the image of ι_X under the canonical isomorphism $X \otimes X^D \to X[1] \otimes X^D[-1]$. The map $h: Z \to X[1]$ gives rise to the dual map $h^D: X^D[-1] \to Z^D$; we define Y^D and the maps $i: Z^D \to Y^D$ and $j: Y^D \to X^D$ by requiring that the sequence

$$X^D[-1] \xrightarrow{h^D} Z^D \xrightarrow{i} Y^D \xrightarrow{j} X^D \tag{1.2.3.2}$$

be a distinguished triangle in $\mathcal{D}$. By axiom (TR2) (Part II, Chapter II, §2.1.1), the triangle

$$Z^D \xrightarrow{i} Y^D \xrightarrow{j} X^D \xrightarrow{-h^D[1]} X^D$$

is distinguished. Let $p: X \otimes X^D \oplus Z \otimes Z^D \to X \otimes Z^D[1]$ be the map $h \otimes \mathrm{id} - \mathrm{id} \otimes h^D[1]$, and let

$$K \xrightarrow{q} X \otimes X^D \oplus Z \otimes Z^D \xrightarrow{p} X \otimes Z^D[1] \to K[1]$$

be the extension of p to a distinguished triangle in $\mathcal{D}$.

We identify $Z \otimes Z^D$ with $Z[-1] \otimes Z^D[1] = Z[-1] \otimes Z[-1]^D$ by the canonical isomorphism and let $\iota_{Z[-1]}: 1 \to Z[-1] \otimes Z[-1]^D$ be the map corresponding to ι_Z. Then, by definition, we have

$$\begin{aligned}
&(h \otimes 1) \circ \iota_Z = (h[-1] \otimes 1) \circ \iota_{Z[-1]} = \iota''_{Z[-1]}(h[-1]),\\
&(1 \otimes h^D[1]) \circ \iota_X = (1 \otimes h[-1]^D) \circ \iota_X = \iota'_X(h[-1]^D),\\
&h[-1]^D = (\iota'_X)^{-1}(\iota''_{Z[-1]}(h[-1])).
\end{aligned}$$

Thus, we have $p \circ (\iota_X, \iota_Z) = 0$, hence there is a map

$$\iota_K: 1 \to K \tag{1.2.3.3}$$

with

$$q \circ \iota_K = (\iota_X, \iota_Z). \tag{1.2.3.4}$$

From Lemma 1.2.2 we have the commutative diagram (1.2.2.1):

$$\begin{array}{ccc}
K & \xrightarrow{\beta} & Y \otimes Y^D \\
{\scriptstyle q}\downarrow & & \downarrow{\scriptstyle (\mathrm{id}\otimes j,\, g\otimes \mathrm{id})} \\
X \otimes X^D \oplus Z \otimes Z^D & \xrightarrow[f\otimes\mathrm{id}\oplus\mathrm{id}\otimes i]{} & Y \otimes X^D \oplus Z \otimes Y^D;
\end{array} \tag{1.2.3.5}$$

let $\iota_Y = \beta \circ \iota_K$. Then we have

$$\begin{aligned}\iota'_Y(j) &= (\mathrm{id} \otimes j) \circ \iota_Y \\ &= (f \otimes \mathrm{id}) \circ \iota_X \\ &= \iota''_X(f),\end{aligned}$$

(1.2.3.6)

$$\begin{aligned}\iota''_Y(g) &= (g \otimes \mathrm{id}) \circ \iota_Y \\ &= (\mathrm{id} \otimes i) \circ \iota_Z \\ &= \iota'_Z(i).\end{aligned}$$

Let A and B be objects of $\mathcal{D}$, and consider the diagram

(1.2.3.7)
$$\begin{array}{ccc}\mathrm{Hom}_{\mathcal{D}}(X^D \otimes A, B) & \xrightarrow{\iota''_X} & \mathrm{Hom}_{\mathcal{D}}(A, X \otimes B) \\ \downarrow{\scriptstyle (j\otimes\mathrm{id})^*} & & \downarrow{\scriptstyle (f\otimes\mathrm{id})_*} \\ \mathrm{Hom}_{\mathcal{D}}(Y^D \otimes A, B) & \xrightarrow[\iota''_Y]{} & \mathrm{Hom}_{\mathcal{D}}(A, Y \otimes B).\end{array}$$

For a map $\alpha: X^D \otimes A \to B$, the map $(f \otimes \mathrm{id})_*(\iota''_X(\alpha))$ is the composition

$$A \cong 1 \otimes A \xrightarrow{\iota_X \otimes \mathrm{id}_A} X \otimes X^D \otimes A \xrightarrow{\mathrm{id}_X \otimes \alpha} X \otimes B \xrightarrow{f \otimes \mathrm{id}_B} Y \otimes B.$$

We may commute the last two maps in this composition, giving the identity

$$(f \otimes \mathrm{id})_*(\iota''_X(\alpha)) = (\mathrm{id}_Y \otimes \alpha) \circ ([(f \otimes \mathrm{id}_{X^D}) \circ \iota_X] \otimes \mathrm{id}_A).$$

(we ignore the identification of A and $1 \otimes A$). By (1.2.3.6), this gives the identity

$$(f \otimes \mathrm{id})_*(\iota''_X(\alpha)) = [\mathrm{id}_Y \otimes (\alpha \circ (j \otimes \mathrm{id}_A))] \circ (\iota_Y \otimes \mathrm{id}_A),$$

which shows that the diagram (1.2.3.7) commutes.

One shows that the diagram

(1.2.3.8)
$$\begin{array}{ccc}\mathrm{Hom}_{\mathcal{D}}(A \otimes Y, B) & \xrightarrow{\iota'_Y} & \mathrm{Hom}_{\mathcal{D}}(A, B \otimes Y^D) \\ \downarrow{\scriptstyle (\mathrm{id}\otimes f)^*} & & \downarrow{\scriptstyle (\mathrm{id}\otimes j)_*} \\ \mathrm{Hom}_{\mathcal{D}}(A \otimes X, B) & \xrightarrow[\iota'_X]{} & \mathrm{Hom}_{\mathcal{D}}(A, B \otimes X^D)\end{array}$$

commutes, using a similar argument.

Using the second identity of (1.2.3.6), the same argument gives the commutativity of the diagrams

(1.2.3.9)
$$\begin{array}{ccc}\mathrm{Hom}_{\mathcal{D}}(Y^D \otimes A, B) & \xrightarrow{\iota''_Y} & \mathrm{Hom}_{\mathcal{D}}(A, Y \otimes B) \\ \downarrow{\scriptstyle (i\otimes\mathrm{id})^*} & & \downarrow{\scriptstyle (g\otimes\mathrm{id})_*} \\ \mathrm{Hom}_{\mathcal{D}}(Z^D \otimes A, B) & \xrightarrow[\iota''_Z]{} & \mathrm{Hom}_{\mathcal{D}}(A, Z \otimes B)\end{array}$$

and

$$(1.2.3.10)\qquad \begin{CD} \mathrm{Hom}_{\mathcal{D}}(A\otimes Z,B) @>{\iota'_Z}>> \mathrm{Hom}_{\mathcal{D}}(A,B\otimes Z^D)\\ @V{(\mathrm{id}\otimes g)^*}VV @VV{(\mathrm{id}\otimes i)_*}V \\ \mathrm{Hom}_{\mathcal{D}}(A\otimes Y,B) @>>{\iota'_Y}> \mathrm{Hom}_{\mathcal{D}}(A,B\otimes Y^D). \end{CD}$$

Since we are using the canonical duals of Remark 1.1.7, the diagrams (1.2.3.7)-(1.2.3.10) remain commutative after applying a shift. Thus, the commutativity of diagrams (1.2.3.7)-(1.2.3.10), together with Lemma 1.1.5(iv), gives the commutativity of the diagrams

$$(1.2.3.11)\qquad \begin{CD} \mathrm{Hom}_{\mathcal{D}}(A\otimes X[1],B) @>{\iota'_{X[1]}}>> \mathrm{Hom}_{\mathcal{D}}(A,B\otimes X^D[-1])\\ @V{(\mathrm{id}\otimes h)^*}VV @VV{(\mathrm{id}\otimes h^D)_*}V \\ \mathrm{Hom}_{\mathcal{D}}(A\otimes Z,B) @>{\iota'_Z}>> \mathrm{Hom}_{\mathcal{D}}(A,B\otimes Z^D)\\ @V{(\mathrm{id}\otimes g)^*}VV @VV{(\mathrm{id}\otimes i)_*}V \\ \mathrm{Hom}_{\mathcal{D}}(A\otimes Y,B) @>{\iota'_Y}>> \mathrm{Hom}_{\mathcal{D}}(A,B\otimes Y^D)\\ @V{(\mathrm{id}\otimes f)^*}VV @VV{(\mathrm{id}\otimes j)_*}V \\ \mathrm{Hom}_{\mathcal{D}}(A\otimes X,B) @>{\iota'_X}>> \mathrm{Hom}_{\mathcal{D}}(A,B\otimes X^D)\\ @V{(\mathrm{id}\otimes h[-1])^*}VV @VV{(\mathrm{id}\otimes h^D[1])_*}V \\ \mathrm{Hom}_{\mathcal{D}}(A\otimes Z[-1],B) @>{\iota'_{Z[-1]}}>> \mathrm{Hom}_{\mathcal{D}}(A,B\otimes Z^D[1]) \end{CD}$$

and

$$(1.2.3.12)\qquad \begin{CD} \mathrm{Hom}_{\mathcal{D}}(Z^D[1]\otimes A,B) @>{\iota''_{Z[-1]}}>> \mathrm{Hom}_{\mathcal{D}}(A,Z[-1]\otimes B)\\ @V{(h^D[1]\otimes\mathrm{id})^*}VV @VV{(h[-1]\otimes\mathrm{id})_*}V \\ \mathrm{Hom}_{\mathcal{D}}(X^D\otimes A,B) @>{\iota''_X}>> \mathrm{Hom}_{\mathcal{D}}(A,X\otimes B)\\ @V{(j\otimes\mathrm{id})^*}VV @VV{(f\otimes\mathrm{id})_*}V \\ \mathrm{Hom}_{\mathcal{D}}(Y^D\otimes A,B) @>{\iota''_Y}>> \mathrm{Hom}_{\mathcal{D}}(A,Y\otimes B)\\ @V{(i\otimes\mathrm{id})^*}VV @VV{(g\otimes\mathrm{id})_*}V \\ \mathrm{Hom}_{\mathcal{D}}(Z^D\otimes A,B) @>{\iota''_Z}>> \mathrm{Hom}_{\mathcal{D}}(A,Z\otimes B)\\ @V{(h^D\otimes\mathrm{id})^*}VV @VV{(h\otimes\mathrm{id})_*}V \\ \mathrm{Hom}_{\mathcal{D}}(X^D[-1]\otimes A,B) @>{\iota''_{X[1]}}>> \mathrm{Hom}_{\mathcal{D}}(A,X[1]\otimes B). \end{CD}$$

As the columns of (1.2.3.11) and (1.2.3.12) are Hom sequences arising from distinguished triangles, they are exact; the five lemma then implies that the maps ι'_Y and ι''_Y are isomorphisms. Thus, (Y^D, ι_Y) is a dual to Y.

Finally, the identities (1.2.3.6) show that $g^D = i$ and $f^D = j$; the distinguished triangle (1.2.3.2) is thus the desired triangle (1.2.3.1). □

1.2.4. REMARK. Consider the diagram (1.2.3.5). Suppose we have chosen duals (X^D, ι_X) to X and (Z^D, ι_Z) to Z; this gives the object Y^D by the sequence (1.2.3.2). From Lemma 1.2.2 we have the diagram (1.2.2.1)

$$\begin{array}{ccc} K & \xrightarrow{\quad\beta\quad} & Y \otimes Y^D \\ {\scriptstyle q}\downarrow & & \downarrow \\ X \otimes X^D \oplus Z \otimes Z^D & \longrightarrow & Y \otimes Y^D \oplus Z \otimes Y^D. \end{array}$$

It follows from the proof of Lemma 1.2.3 that, if we have a map as in (1.2.3.3) $\iota: 1 \to K$ in $\mathcal{D}$, satisfying the identity $q \circ \iota = (\iota_X, \iota_Z)$, then $(Y^D, \beta \circ \iota)$ is dual to Y.

1.2.5. THEOREM. *Suppose $\mathcal{D}$ is generated (as a triangulated category) by a set of objects $\mathcal{S}$ such that each $X \in \mathcal{S}$ has a dual (X^D, ι_X). Then*
(i) Every object Y of $\mathcal{D}$ has a dual (Y^D, ι_Y).
(ii) If we assume that the choice of dual for $X[1]$ is the canonical one (see Remark 1.1.7) for each X in $\mathcal{D}$, then, sending X to its dual X^D and a morphism $f: X \to Y$ to the dual morphism $f^D: Y^D \to X^D$ defines an exact functor

$$(-)^D : \mathcal{D}^{\mathrm{op}} \to \mathcal{D}.$$

The functor $(-)^D$ is a pseudo-tensor functor (see Theorem 1.1.6(iii)).

PROOF. Part (i) follows directly from Lemma 1.2.3. From (i), Theorem 1.1.6 and Remark 1.1.7, sending X to its dual X^D and f to its dual f^D defines a graded functor $(-)^D: \mathcal{D}^{\mathrm{op}} \to \mathcal{D}$ which is a pseudo-tensor functor. Thus, we need only show that $(-)^D$ is exact.

Let $X \xrightarrow{f} Y \xrightarrow{g} Z \xrightarrow{h} X[1]$ be a distinguished triangle in $\mathcal{D}$. By Lemma 1.2.3, there is a choice of dual (Y^{*D}, ι_{*Y}) for Y such that the sequence

$$X^D[-1] \xrightarrow{h^D} Z^D \xrightarrow{g^{*D}} Y^{*D} \xrightarrow{f^{*D}} X^D$$

is a distinguished triangle, where g^{*D} and f^{*D} are the maps defined with respect to the choice (Y^{*D}, ι_{*Y}) of dual for Y. Let $F: Y^{*D} \to Y^D$ be the canonical isomorphism given by Lemma 1.1.3. By Lemma 1.1.5, we have $F \circ g^{*D} = g^D$, $f^{*D} = f^D \circ F$. Thus, we have a commutative diagram

$$\begin{array}{ccccccc} X^D[-1] & \xrightarrow{h^D} & Z^D & \xrightarrow{g^{*D}} & Y^{*D} & \xrightarrow{f^{*D}} & X^D \\ \| & & \| & & \downarrow{\scriptstyle F} & & \| \\ X^D[-1] & \xrightarrow[h^D]{} & Z^D & \xrightarrow[g^D]{} & Y^D & \xrightarrow[f^D]{} & X^D. \end{array}$$

As F is an isomorphism, the sequence

$$X^D[-1] \xrightarrow{h^D} Z^D \xrightarrow{g^D} Y^D \xrightarrow{f^D} X^D$$

is a distinguished triangle, completing the proof. □

1.3. The diagonal and co-diagonal

We examine the diagonal morphism for a smooth projective S-scheme.

1.3.1. Let $p_X : X \to S$ be a smooth projective S-scheme in $\mathcal{V}$ of dimension d over S. We have the diagonal $\delta_X : X \to X \times_S X$, giving the maps $\delta_X^* : \mathbb{Z}_{X\times_S X}(d)[2d] \to \mathbb{Z}_X(d)[2d]$ and $\delta_*^X : \mathbb{Z}_X \to \mathbb{Z}_{X\times_S X}(d)[2d]$. We also have the external products

$$\begin{aligned}\boxtimes_{X,X} &: \mathbb{Z}_X(a) \otimes \mathbb{Z}_X(d-a)[2d] \to \mathbb{Z}_{X\times_S X}(d)[2d],\\ \boxtimes_{X,X} &: \mathbb{Z}_X(d-a)[2d] \otimes \mathbb{Z}_X(a) \to \mathbb{Z}_{X\times_S X}(d)[2d].\end{aligned}$$

We define the maps in $\mathcal{DM}(\mathcal{V})$

$$\begin{aligned}\iota_X &: \mathbb{Z}_S = 1 \to \mathbb{Z}_X(a) \otimes \mathbb{Z}_X(d-a)[2d],\\ \epsilon_X &: \mathbb{Z}_X(d-a)[2d] \otimes \mathbb{Z}_X(a) \to 1,\end{aligned} \tag{1.3.1.1}$$

by

$$\begin{aligned}\iota_X &= \boxtimes_{X,X}^{-1} \circ \delta_*^X \circ p_X^*,\\ \epsilon_X &= p_{X*} \circ \delta_X^* \circ \boxtimes_{X,X}.\end{aligned}$$

We form the composition

$$\begin{aligned}\mathbb{Z}_X(a) \cong 1 \otimes \mathbb{Z}_X(a) &\xrightarrow{\iota_X\otimes\mathrm{id}} \mathbb{Z}_X(a) \otimes \mathbb{Z}_X(d-a)[2d] \otimes \mathbb{Z}_X(a)\\ &\xrightarrow{\mathrm{id}\otimes\epsilon_X} \mathbb{Z}_X(a) \otimes 1 \cong \mathbb{Z}_X(a)\end{aligned} \tag{1.3.1.2}$$

and the composition

$$\begin{aligned}\mathbb{Z}_X(d-a)[2d] &\cong \mathbb{Z}_X(d-a)[2d] \otimes 1\\ &\xrightarrow{\mathrm{id}\otimes\iota_X} \mathbb{Z}_X(d-a)[2d] \otimes \mathbb{Z}_X(a) \otimes \mathbb{Z}_X(d-a)[2d]\\ &\xrightarrow{\epsilon_X\otimes\mathrm{id}} 1 \otimes \mathbb{Z}_X(d-a)[2d] \to \mathbb{Z}_X(d-a)[2d].\end{aligned} \tag{1.3.1.3}$$

1.3.2. LEMMA. *The compositions* (1.3.1.2) *and* (1.3.1.3) *are identity maps.*

PROOF. Let $\delta_X^{12} : X \times X \to X \times X \times X$ and $\delta_X^{23} : X \times X \to X \times X \times X$ be the maps $\delta_X^{12} = \delta_X \times \mathrm{id}_X$ and $\delta_X^{23} = \mathrm{id}_X \times \delta_X$, respectively.

We recall from Chapter I, Remark 3.4.4 that the multiplication isomorphism $\mu_l : 1 \otimes \mathbb{Z}_X(a) \to \mathbb{Z}_X(a)$ is the external product

$$\boxtimes_{S,X} : \mathbb{Z}_S \otimes \mathbb{Z}_X(a) \to \mathbb{Z}_{S\times_S X}(a) \xrightarrow{p_2^*} \mathbb{Z}_X(a).$$

We have the isomorphism

$$\boxtimes_{X,X,X} : \mathbb{Z}_X(a) \otimes \mathbb{Z}_X(d-a)[2d] \otimes \mathbb{Z}_X(a) \to \mathbb{Z}_{X\times X\times X}(d+a)[2d];$$

applying Theorem 2.4.10 of Chapter III, we find that (1.3.1.2) is equal to the composition

$$\begin{aligned}\mathbb{Z}_X(a) \xrightarrow{p_2^*} \mathbb{Z}_{X\times X}(a) &\xrightarrow{\delta_*^{12}} \mathbb{Z}_{X\times X\times X}(d+a)[2d]\\ &\xrightarrow{\delta_X^{23*}} \mathbb{Z}_{X\times X}(d+a)[2d] \xrightarrow{p_{1*}} \mathbb{Z}_X(a).\end{aligned}$$

We have the transverse cartesian diagram

$$\begin{array}{ccc} X & \xrightarrow{\delta_X} & X \times_S X \\ {\scriptstyle \delta_X}\downarrow & & \downarrow{\scriptstyle \delta_X^{23}} \\ X \times_S X & \xrightarrow[\delta_X^{12}]{} & X \times_S X \times_S X. \end{array}$$

By Chapter III, Theorem 2.4.9, we have $\delta_X^{23*} \circ \delta_*^{12} = \delta_{X*} \circ \delta_X^*$. Thus (1.3.1.2) is equal to the composition

(1.3.2.1)
$$\begin{aligned} &\mathbb{Z}_X(a) \xrightarrow{p_2^*} \mathbb{Z}_{X\times X}(a) \xrightarrow{\delta_X^*} \mathbb{Z}_X(a) \\ &\xrightarrow{\delta_{X*}} \mathbb{Z}_{X\times X}(d+a)[2d] \xrightarrow{p_{1*}} \mathbb{Z}_X(a). \end{aligned}$$

Since $p_2 \circ \delta_X = \mathrm{id}_X$, we have $\delta_X^* \circ p_2^* = \mathrm{id}$. Since $p_1 \circ \delta_X = \mathrm{id}_X$, it follows from Chapter III, Theorem 2.4.7 that $p_{1*} \circ \delta_{X*} = \mathrm{id}$. Thus, the composition (1.3.2.1) is the identity, completing the proof that (1.3.1.2) is the identity. The proof that (1.3.1.3) is the identity is similar, and is left to the reader. □

1.4. The duality involution

We describe the duality structure on the motivic category.

1.4.1. *The dual for projective X.* We let $\mathcal{DM}(\mathcal{V})^{\mathrm{pr}}$ denote the smallest strictly full triangulated subcategory of $\mathcal{DM}(\mathcal{V})$ containing the objects $\mathbb{Z}_X(a)$, with X in $\mathcal{V}$ smooth and projective over S, and closed under taking summands. Since $\mathbb{Z}_X(a) \otimes \mathbb{Z}_Y(b)$ is isomorphic to $\mathbb{Z}_{X\times_S Y}(a+b)$, $\mathcal{DM}(\mathcal{V})^{\mathrm{pr}}$ is a triangulated tensor subcategory of $\mathcal{DM}(\mathcal{V})$.

For X in $\mathcal{V}$, smooth and projective over S, we set

(1.4.1.1)
$$\mathbb{Z}_X(a)[b]^D := \mathbb{Z}_X(d-a)[2d-b].$$

We have the morphism (1.3.1.1) $\iota_X : 1 \to \mathbb{Z}_X(a)[b] \otimes \mathbb{Z}_X(a)[b]^D$; by Lemma 1.3.2 and Proposition 1.1.9, $(\mathbb{Z}_X(a)^D, \iota_X)$ is a dual to $\mathbb{Z}_X(a)$. Thus, for X and Y in $\mathcal{V}$, smooth and projective over S, we have the isomorphism (1.1.4.1)

(1.4.1.2)
$$(-)^D : \mathrm{Hom}_{\mathcal{DM}(\mathcal{V})}(\mathbb{Z}_X(a)[b], \mathbb{Z}_Y(a')[b']) \to \mathrm{Hom}_{\mathcal{DM}(\mathcal{V})}(\mathbb{Z}_Y(a')[b']^D, \mathbb{Z}_X(a)[b]^D).$$

1.4.2. THEOREM. *The operation $(-)^D$ defined for projective X by (1.4.1.1) and (1.4.1.2):*

$$\begin{gathered} \mathbb{Z}_X(a)[b] \mapsto \mathbb{Z}_X(a)[b]^D, \\ (f : \mathbb{Z}_X(a)[b] \to \mathbb{Z}_Y(a')[b']) \mapsto (f^D : \mathbb{Z}_Y(a')[b']^D \to \mathbb{Z}_X(a)[b]^D), \end{gathered}$$

extends to an exact pseudo-tensor functor (see Theorem 1.1.6(iii))

$$(-)^D : (\mathcal{DM}(\mathcal{V})^{\mathrm{pr}})^{\mathrm{op}} \to \mathcal{DM}(\mathcal{V})^{\mathrm{pr}}$$

defining an exact duality on $\mathcal{DM}(\mathcal{V})^{\mathrm{pr}}$, i.e., for A, B and C in $\mathcal{DM}(\mathcal{V})^{\mathrm{pr}}$, there are natural isomorphisms

$$\begin{aligned} \mathrm{Hom}_{\mathcal{DM}(\mathcal{V})}(A \otimes B^D, C) &\to \mathrm{Hom}_{\mathcal{DM}(\mathcal{V})}(A, C \otimes B), \\ \mathrm{Hom}_{\mathcal{DM}(\mathcal{V})}(A \otimes B, C) &\to \mathrm{Hom}_{\mathcal{DM}(\mathcal{V})}(A, C \otimes B^D), \end{aligned}$$

which are exact in the variables A, B *and* C. *In addition, there is a natural isomorphism* $\mathrm{id} \to ((-)^D)^D$.

PROOF. Define the subcategory $\mathbf{D}^b_{\mathrm{mot}}(\mathcal{V})^{\mathrm{pr}}$ of $\mathbf{D}^b_{\mathrm{mot}}(\mathcal{V})$ to be the full triangulated subcategory generated by the objects $\mathbb{Z}_X(a)$ for X in $\mathcal{V}$, smooth and projective over S. Then the extension of the operation $(-)^D$ to an exact, pseudo-tensor functor $(-)^D: (\mathbf{D}^b_{\mathrm{mot}}(\mathcal{V})^{\mathrm{pr}})^{\mathrm{op}} \to \mathbf{D}^b_{\mathrm{mot}}(\mathcal{V})^{\mathrm{pr}}$, with a natural isomorphism $\mathrm{id} \to ((-)^D)^D$, follows directly from Theorem 1.2.5.

We have the functor $_\#$ (see Part II, Chapter II, §2.4) on the category of tensor categories, where $\mathcal{A}_\#$ is the pseudo-abelian hull of a category $\mathcal{A}$. Applying Theorem 1.2.5 again, it suffices to show that, if $\mathcal{A}$ is a tensor category having a duality involution $(-)^D$, if A is an object of $\mathcal{A}$, and if B is the summand of A in $\mathcal{A}_\#$ corresponding to an idempotent endomorphism $p: A \to A$, then B has a dual (B^D, ι_B) in $\mathcal{A}_\#$.

To see this, the idempotent endomorphism $p: A \to A$ gives rise to the endomorphism $p^D: A^D \to A^D$. Since $(-)^D$ is a functor on $\mathcal{A}$, p^D is an idempotent endomorphism of A^D. Let B^D be the summand of A^D in $\mathcal{A}_\#$ corresponding to p^D. The idempotent endomorphism $p \otimes p^D$ then defines the summand $B \otimes B^D$ of $A \otimes A^D$. We let $\iota_B: 1 \to B \otimes B^D$ be the map gotten by projecting $\iota_A: 1 \to A \otimes A^D$ onto the summand $B \otimes B^D$:

$$\iota_B = (p \otimes p^D) \circ \iota_A \in (p \otimes p^D) \circ \mathrm{Hom}_{\mathcal{A}}(1, A \otimes A^D) := \mathrm{Hom}_{\mathcal{A}_\#}(1, B \otimes B^D).$$

It is then an elementary exercise to show that (B^D, ι_B) is a dual to B, which completes the proof of the theorem. □

1.5. An application

The duality involution on $\mathcal{DM}(\mathcal{V})^{\mathrm{pr}}$ implies that the morphisms in $\mathcal{DM}(\mathcal{V})^{\mathrm{pr}}$ are determined by the motivic cohomology, i.e., the functor $\mathrm{Hom}_{\mathcal{DM}(\mathcal{V})}(1, -)$. We formalize this principle, and give an application.

1.5.1. THEOREM. *Let* $\mathcal{A}$ *be a triangulated* R*-tensor category, and let*

$$F: \mathcal{DM}(\mathcal{V})_R \to \mathcal{A}$$

be an exact R*-pseudo-tensor functor (cf. Part II, Chapter I, §1.3.7). Suppose the map* $F(1, \Gamma): \mathrm{Hom}_{\mathcal{DM}(\mathcal{V})}(1, \Gamma) \to \mathrm{Hom}_{\mathcal{A}}(1, F(\Gamma))$ *is an isomorphism for each* Γ *in* $\mathcal{DM}(\mathcal{V})_R$. *Then, for each* Δ *in* $\mathcal{DM}(\mathcal{V})_R^{\mathrm{pr}}$, *and each* Γ *in* $\mathcal{DM}(\mathcal{V})_R$, *the map* $F(\Delta, \Gamma): \mathrm{Hom}_{\mathcal{DM}(\mathcal{V})_R}(\Delta, \Gamma) \to \mathrm{Hom}_{\mathcal{A}}(F(\Delta), F(\Gamma))$ *is an isomorphism, hence the restriction of* F *to* $\mathcal{DM}(\mathcal{V})_R^{\mathrm{pr}}$ *fully faithful. In particular, if* $\mathcal{DM}(\mathcal{V})_R^{\mathrm{pr}} = \mathcal{DM}(\mathcal{V})_R$, *then* F *is fully faithful.*

PROOF. We give the proof for $R = \mathbb{Z}$. Each object of $\mathcal{DM}(\mathcal{V})^{\mathrm{pr}}$ is a summand of an iterated cone of objects of the form $\mathbb{Z}_X(a)[b]$, with X smooth and projective over S. Since F is exact, it suffices to show that $F(\Delta, \Gamma)$ is an isomorphism for $\Delta = \mathbb{Z}_X(a)[b]$.

This follows from the hypothesis on F, and Theorem 1.4.2. □

1.5.2. We recall the graded tensor category $\mathcal{A}^0_{\mathrm{mot}}(\mathbf{Sm}_S)$ (see Chapter I, Definition 1.4.12), and the DG tensor functor (I.1.4.12.1)

$$H_{\mathrm{mot}}: \mathcal{A}_{\mathrm{mot}}(\mathbf{Sm}_S) \to \mathcal{A}^0_{\mathrm{mot}}(\mathbf{Sm}_S).$$

The category $\mathcal{A}^0_{\rm mot}(\mathbf{Sm}_S)$ and functor $H_{\rm mot}$ are characterized by the identity

$$\begin{aligned}&\mathrm{Hom}_{\mathcal{A}^0_{\rm mot}(\mathbf{Sm}_S)}(\mathfrak{e}^{\otimes a}\otimes\mathbb{Z}_X,\mathfrak{e}^{\otimes b}\otimes\mathbb{Z}_Y(n))\\ &\qquad\qquad = H^{2n}(\mathrm{Hom}_{\mathcal{A}_{\rm mot}(\mathbf{Sm}_S)}(\mathfrak{e}^{\otimes a}\otimes\mathbb{Z}_X,\mathfrak{e}^{\otimes b}\otimes\mathbb{Z}_Y(n))).\end{aligned}$$

We have the triangulated tensor category $\mathbf{K}^{b0}_{\rm mot}(\mathbf{Sm}_S) := \mathbf{K}^b(\mathcal{A}^0_{\rm mot}(\mathbf{Sm}_S))$ (Chapter I, Remark 3.4.7), and the exact tensor functor

$$\mathbf{K}^b(H_{\rm mot})\colon \mathbf{K}^b_{\rm mot}(\mathbf{Sm}_S)\to\mathbf{K}^{b0}_{\rm mot}(\mathbf{Sm}_S).$$

The category $\mathbf{D}^{b0}_{\rm mot}(\mathbf{Sm}_S)$ is gotten from $\mathbf{K}^{b0}_{\rm mot}(\mathbf{Sm}_S)$ by inverting the morphisms of Chapter I, Definition 2.1.4, and the category $\mathcal{DM}^0(S)$ is gotten from $\mathbf{D}^{b0}_{\rm mot}(\mathbf{Sm}_S)$ by forming the pseudo-abelian hull. This gives the exact tensor functors

$$\mathbf{D}^b(H_{\rm mot})\colon \mathbf{D}^b_{\rm mot}(\mathbf{Sm}_S)\to\mathbf{D}^{b0}_{\rm mot}(\mathbf{Sm}_S)$$
$$\mathcal{DM}(H_{\rm mot})\colon\mathcal{DM}(S)\to\mathcal{DM}^0(S)$$

1.5.3. DEFINITION. Let Y be in $\mathbf{Sm}_S$, and $Z_1,\ldots,Z_N$ closed subschemes. We say that $Z_1,\ldots,Z_N$ form a *normal crossing subscheme of* Y for each collection of indices $1\le i_1<\ldots<i_s\le N$,

$$\mathrm{codim}_Y(Z_{i_1}\cap\ldots\cap Z_{i_s})=\sum_{j=1}^{s}\mathrm{codim}_Y(Z_{i_j}),$$

and the closed subscheme $Z_{i_1}\cap\ldots\cap Z_{i_s}$ is smooth over S (or is empty). We call the union $\cup_{i=1}^N Z_i$ a *normal crossing subscheme* of Y.

1.5.4. LEMMA. *Let X be in $\mathbf{Sm}_S$. Suppose there is an open immersion $j\colon X\to\bar{X}$ with $\bar{X}$ smooth and projective over S, such that*

(i) *The complement $Z:=\bar{X}\setminus X$ is a union of smooth projective S-schemes, $Z=\cup_{i=1}^N Z_i$, with each Z_i a union of irreducible components of Z.*
(ii) *For each collection of indices $i_1,\ldots,i_s$, the closed subset $Z_{i_1}\cap\ldots\cap Z_{i_s}$ of $\bar{X}$ is smooth over S*

(This is the case if, for instance Z is a normal crossing subscheme of $\bar{X}$). Then $\mathbb{Z}_X$ is in $\mathcal{DM}(\mathbf{Sm}_S)^{\rm pr}$.

PROOF. Let $U=\bar{X}\setminus Z_1$, $Z_U=Z\cap U$. We have the distinguished triangles (I.2.2.10.1) in $\mathcal{DM}(S)$:

$$\begin{aligned}&\mathbb{Z}_{\bar{X},Z}(a)\to\mathbb{Z}_{\bar{X}}(a)\to\mathbb{Z}_X(a)\to\mathbb{Z}_{\bar{X},Z}(a)[1],\\ &\mathbb{Z}_{\bar{X},Z_1}(a)\to\mathbb{Z}_{\bar{X},Z}(a)\to\mathbb{Z}_{U,Z_U}(a)\to\mathbb{Z}_{\bar{X},Z_1}(a)[1],\\ &\mathbb{Z}_{\bar{X},Z_1}(a)\to\mathbb{Z}_{\bar{X}}(a)\to\mathbb{Z}_U(a)\to\mathbb{Z}_{\bar{X},Z_1}(a)[1].\end{aligned}$$

Since Z_1 is smooth, say of codimension d, we have the isomorphism (III.2.1.2.2) $\mathbb{Z}_{\bar{X},Z_1}(a)\cong\mathbb{Z}_{Z_1}(a-d)[-2d]$, hence $\mathbb{Z}_U(a)$ is in $\mathcal{DM}(S)^{\rm pr}$. Similarly, $\mathbb{Z}_{Z_i\cap U}(b)$ is in $\mathcal{DM}(S)^{\rm pr}$ for each $i=2,\ldots,N$; by induction, this implies $\mathbb{Z}_{U,Z_U}(a)$ is in $\mathcal{DM}(S)^{\rm pr}$. Thus $\mathbb{Z}_{\bar{X},Z}(a)$ is in $\mathcal{DM}(S)^{\rm pr}$, hence $\mathbb{Z}_X(a)$ is in $\mathcal{DM}(S)^{\rm pr}$. □

1.5.5. THEOREM. *Suppose $S=\mathrm{Spec}\,k$ for a field k, and that, for each X in $\mathbf{Sm}_k$, there is an open immersion $j\colon X\to\bar{X}$ with $\bar{X}$ smooth and projective over k, such that*

(i) *The complement $Z:=\bar{X}\setminus X$ is a union of smooth projective irreducible k-schemes: $Z=\cup_{i=1}^N Z_i$.*

(ii) *For each collection of indices* $i_1, \ldots, i_s$, *the closed subset* $Z_{i_1} \cap \ldots \cap Z_{i_s}$ *of* $\bar{X}$ *is smooth over* k.

Then the functors $\mathbf{D}^b(H_{\text{mot}})$ *and* $\mathcal{DM}(H_{\text{mot}})$ *are equivalences.*

PROOF. It suffices to show that $\mathbf{D}^b(H_{\text{mot}})$ is an equivalence of categories. By Lemma 1.5.4, $\mathbf{D}^b_{\text{mot}}(\mathbf{Sm}_k)^{\text{pr}} = \mathbf{D}^b_{\text{mot}}(\mathbf{Sm}_k)$. Since $\mathbf{D}^{b0}_{\text{mot}}(\mathbf{Sm}_k)$ is generated by the objects in the image of $\mathbf{D}^b_{\text{mot}}(\mathbf{Sm}_k)$, it suffices to show that $\mathbf{D}^b(H_{\text{mot}})$ is fully faithful.

By (Chapter II, Theorem 3.3.11 and Theorem 3.6.6), the functor $\mathbf{D}^b(H_{\text{mot}})$ gives an isomorphism

$$\text{Hom}_{\mathbf{D}^b_{\text{mot}}(\mathbf{Sm}_k)}(1, \mathbb{Z}_X(a)[b]) \to \text{Hom}_{\mathbf{D}^{b0}_{\text{mot}}(\mathbf{Sm}_k)}(1, \mathbb{Z}_X(a)[b])$$

for each X in $\mathbf{Sm}_k$. Since $\mathbf{D}^b_{\text{mot}}(\mathbf{Sm}_k)$ is generated as a triangulated category by the objects $\mathbb{Z}_X(a)$, it follows that $\mathbf{D}^b(H_{\text{mot}})$ gives an isomorphism

$$\text{Hom}_{\mathbf{D}^b_{\text{mot}}(\mathbf{Sm}_k)}(1, \Gamma) \to \text{Hom}_{\mathbf{D}^{b0}_{\text{mot}}(\mathbf{Sm}_k)}(1, \Gamma)$$

for each Γ in $\mathbf{D}^b_{\text{mot}}(\mathbf{Sm}_k)$. Since $\mathbf{D}^b_{\text{mot}}(\mathbf{Sm}_k)^{\text{pr}} = \mathbf{D}^b_{\text{mot}}(\mathbf{Sm}_k)$, it follows from Theorem 1.5.1 that $\mathbf{D}^b(H_{\text{mot}})$ is fully faithful, completing the proof. □

If, for example, the field k has characteristic zero, then the hypotheses of Theorem 1.5.5 are satisfied, by Hironaka's resolution of singularities [**66**].

2. Classical constructions

We begin this section by showing in §2.1 how the morphisms in $\mathcal{DM}(\mathcal{V})^{\text{pr}}$ can be realized as correspondences; for $S = \text{Spec}\, k$, this allows us to embed the category of R-Chow motives into $\mathcal{DM}(k)_R$.

We then proceed to define motivic versions of homology, Borel-Moore homology, and compactly supported cohomology. In §2.2, we define the motive with compact support, $\mathbb{Z}_X^{c/S}$, as well as motivic homology, motivic compactly supported cohomology and motivic Borel-Moore homology, and verify the standard properties of these theories. In §2.3, we give a detailed discussion of duality for "open relative motives", reminiscent of the Hodge-theory yoga of forms with log poles at infinity. As a special case, for a smooth S-scheme X which admits a compactification $\bar{X}$ which is smooth over S and has a normal crossing complement $\bar{X} \setminus X = D_1 \cup \ldots \cup D_n$, we identify the motive of X with compact support as the motive of $\bar{X}$ relative to $D_1, \ldots, D_n$.

In §2.4, we identify the Borel-Moore motive as the motive with support in a closed subset, and use this to extend the Borel-Moore motive to "smoothly decomposable" S-schemes (see Definition 2.4.1(i)). Similarly, we extend the definition of the motive with compact support to S-schemes which admit a "compactifiable closed embedding" into a smooth quasi-projective S scheme (Definition 2.4.1(ii)). We show that the resulting motivic cohomology with compact support/motivic Borel-Moore homology have the usual properties of classical cohomology with compact support/Borel-Moore homology.

For $S = \text{Spec}\, k$, where k is a perfect field, every quasi-projective k-scheme is smoothly decomposable; if resolution of singularities holds for quasi-projective k-schemes, then every quasi-projective k-scheme admits a compactifiable embedding, so our construction gives a Borel-Moore motive, and a motive with compact support for arbitrary quasi-projective k-schemes, as well as the resulting motivic Borel-Moore homology and motivic cohomology with compact support.

In §2.6, we define the Tate motivic category, and discuss some of its basic properties.

2.1. Correspondences

Via the duality isomorphism of §1.4, we may interpret maps in $\mathcal{DM}(S)$ between motives of projective varieties X and Y as classes in the motivic cohomology of the product. In this section, we show how the one generalizes the construction of the category of Chow motives over a field k to arbitrary base-schemes, and how the generalized category of Chow motives embeds into $\mathcal{DM}$.

2.1.1. LEMMA. *Let X be a smooth projective S-scheme, of dimension d over S, let Y be in $\mathbf{Sm}_S$, and let $f:\mathbb{Z}_X \to \mathbb{Z}_Y(a)[b]$ be a morphism in $\mathcal{DM}(S)$.*
(i) Let $\zeta: 1 \to \mathbb{Z}_{Y\times_S X}(a+d)[b+2d]$ be the map $\boxtimes_{Y,X}\circ \iota''_X(f)$. Then f is equal to the composition

$$\mathbb{Z}_X \xrightarrow{p_2^*} \mathbb{Z}_{Y\times_S X} \xrightarrow{\cup_{Y\times_S X}\zeta} \mathbb{Z}_{Y\times_S X}(a+d)[b+2d] \xrightarrow{p_{1*}} \mathbb{Z}_Y(a)[b].$$

(ii) Suppose that Y is projective over S of dimension e, and write the map $\boxtimes_{X,Y}\circ \iota''_Y(f^D)$ as $\zeta^D: 1 \to \mathbb{Z}_{X\times_S Y}(a+d)[b+2d]$. Then

$$\zeta^D = t^*_{X,Y}(\zeta),$$

where $t_{X,Y}: X\times_S Y \to Y\times_S X$ is the exchange of factors. Additionally, f^D is equal to the composition

$$\mathbb{Z}_Y(e-a)[2e-b] \xrightarrow{p_1^*} \mathbb{Z}_{Y\times_S X}(e-a)[2e-b]$$
$$\xrightarrow{\cup_{Y\times_S X}\zeta} \mathbb{Z}_{Y\times_S X}(d+e)[2d+2e] \xrightarrow{p_{2*}} \mathbb{Z}_X(d)[2d].$$

PROOF. We first prove (i). If we denote the composition in (i) by g, it suffices to show that $\iota''_X(f) = \iota''_X(g)$. The map $\iota''_X(g)$ is the composition

$$1 \xrightarrow{\iota_X} \mathbb{Z}_X \otimes \mathbb{Z}_X(d)[2d] \xrightarrow{p_2^*\otimes \mathrm{id}} \mathbb{Z}_{Y\times_S X}\otimes \mathbb{Z}_X(d)[2d]$$
$$\xrightarrow{\cup_{Y\times_S X}\zeta\otimes\mathrm{id}} \mathbb{Z}_{Y\times_S X}(a+d)[b+2d]\otimes \mathbb{Z}_X(d)[2d]$$
$$\xrightarrow{p_{1*}\otimes\mathrm{id}} \mathbb{Z}_Y(a)[b]\otimes \mathbb{Z}_X(d)[2d].$$

We may rewrite this as

$$\boxtimes_{Y,X}\circ \iota''_X(g) = p_{13*}\circ \cup_{Y\times_S X\times_S X}(p_{12}^*\zeta)\circ p_{23}^*\circ \delta_{X*}\circ p_X^*.$$

We have the transverse cartesian diagram

$$\begin{array}{ccc} Y\times_S X & \xrightarrow{\mathrm{id}_Y\times\delta_X} & Y\times_S X\times_S X \\ \Big\downarrow{\scriptstyle p_2} & & \Big\downarrow{\scriptstyle p_{23}} \\ X & \xrightarrow[\delta_X]{} & X\times_S X; \end{array}$$

applying Theorem 2.4.9 of Chapter III, we have the identity $p_{23}^*\circ\delta_{X*} = (\mathrm{id}_Y\times \delta_X)_*\circ p_2^*$. Using this and the projection formula (Chapter III, Theorem 2.4.8), we may rewrite $\boxtimes_{Y,X} \circ \iota''_X(g)$ as

$$\begin{aligned}\boxtimes_{Y,X} \circ \iota''_X(g) &= p_{13*} \circ (\mathrm{id}_Y \times \delta_X)_* \circ \cup_{Y\times_S X}(\zeta) \circ p^*_{Y\times_S X} \\ &= \mathrm{id}_{Y\times_S X*} \circ \zeta \\ &= \zeta.\end{aligned}$$

Thus $\iota''_X(g) = \boxtimes^{-1}_{Y,X} \circ \zeta = \iota''_X(f)$, completing the proof of (i).

For (ii), we note the $\iota'_Y(f^D) = \iota''_X(f)$, by definition of f^D. On the other hand, it follows from the symmetry of the diagonal map that the maps

$$\begin{aligned}\iota_{\mathbb{Z}_Y(a)[b]} &: 1 \to \mathbb{Z}_Y(a)[b] \otimes \mathbb{Z}_Y(e-a)[2e-b], \\ \iota_{\mathbb{Z}_Y(e-a)[2e-b]} &: 1 \to \mathbb{Z}_Y(e-a)[2e-b] \otimes \mathbb{Z}_Y(a)[b]\end{aligned}$$

are related by $t^*_{Y,Y} \circ \boxtimes_{Y,Y} \circ \iota_{\mathbb{Z}_Y(a)[b]} = \boxtimes_{Y,Y} \circ \iota_{\mathbb{Z}_Y(e-a)[2e-b]}$. From this, it follows that $\boxtimes_{X,Y} \circ \iota''_Y(f^D) = t^*_{X,Y} \circ \boxtimes_{Y,X} \circ \iota'_Y(f^D)$, i.e., $\zeta^D = t^*_{X,Y} \circ \zeta$, proving the first part of (ii); the second part follows from the first and (i). □

2.1.2. LEMMA. *Suppose X, Y and Z are in $\mathbf{Sm}_S$, with X and Y projective over S, of relative dimensions d and e, respectively. Let $f: \mathbb{Z}_X \to \mathbb{Z}_Y(a)[b]$, $g: \mathbb{Z}_Y(a)[b] \to \mathbb{Z}_Z(a'+a)[b'+b]$ be morphisms in $\mathcal{DM}(S)$, and let*

$$\begin{aligned}\zeta_f &: 1 \to \mathbb{Z}_{Y\times_S X}(a+d)[b+2d], \\ \zeta_g &: 1 \to \mathbb{Z}_{Z\times_S Y}(a'+e)[b'+2e], \\ \zeta_{g\circ f} &: 1 \to \mathbb{Z}_{Z\times_S X}(a+a'+d)[b+b'+2d]\end{aligned}$$

be the respective morphisms

$$\boxtimes_{Y,X} \circ \iota''_X(f), \quad \boxtimes_{Z,Y} \circ \iota''_Y(g), \quad \boxtimes_{Z,X} \circ \iota''_X(g\circ f).$$

Let

$$\begin{aligned}p_{12} &: Z\times_S Y \times_S X \to Z \times_S Y, \\ p_{13} &: Z\times_S Y \times_S X \to Z \times_S X, \\ p_{23} &: Z\times_S Y \times_S X \to Y \times_S X\end{aligned}$$

be the projections. Then

$$\zeta_{g\circ f} = p_{13*}(p^*_{12}(\zeta_g) \cup p^*_{23}(\zeta_f)).$$

PROOF. By Proposition 1.1.11 and the uniqueness (Proposition 1.1.10) of the co-diagonal $\epsilon: \mathbb{Z}_Y(e-a)[2e-b] \otimes \mathbb{Z}_Y(a)[b] \to 1$, $\zeta_{g\circ f}$ is the composition

$$\begin{aligned}&1 \cong 1 \otimes 1 \\ &\xrightarrow{\iota''_Y(g)\otimes\iota''_X(f)} \mathbb{Z}_Z(a+a')[b+b'] \otimes \mathbb{Z}_Y(e-a)[2e-b] \otimes \mathbb{Z}_Y(a)[b] \otimes \mathbb{Z}_X(d)[2d] \\ &\xrightarrow{\mathrm{id}\otimes\epsilon_Y\otimes\mathrm{id}} \mathbb{Z}_Z(a+a')[b+b'] \otimes 1 \otimes \mathbb{Z}_X(d)[2d] \\ &\cong \mathbb{Z}_Z(a+a')[b+b'] \otimes \mathbb{Z}_X(d)[2d] \\ &\xrightarrow{\boxtimes_{Z,X}} \mathbb{Z}_{Z\times_S X}(a+a'+d)[b+b'+2d].\end{aligned}$$

Since $\epsilon_Y = p_{Y*} \circ \delta_Y^* \circ \boxtimes_{Y,Y}$ we may rewrite this as the composition

$$\begin{aligned}
1 &\cong 1 \otimes 1 \\
&\xrightarrow{\iota_Y \otimes \iota_X} \mathbb{Z}_Y(a)[b] \otimes \mathbb{Z}_Y(e-a)[2e-b] \otimes \mathbb{Z}_X \otimes \mathbb{Z}_X(d)[2d] \\
&\xrightarrow{g \otimes \mathrm{id} \otimes f \otimes \mathrm{id}} \mathbb{Z}_Z(a+a')[b+b'] \otimes \mathbb{Z}_Y(e-a)[2e-b] \otimes \mathbb{Z}_Y(a)[b] \otimes \mathbb{Z}_X(d)[2d] \\
&\xrightarrow{\mathrm{id} \otimes (\delta_Y^* \circ \boxtimes_{Y,Y}) \otimes \mathrm{id}} \mathbb{Z}_Z(a+a')[b+b'] \otimes \mathbb{Z}_Y(e)[2e] \otimes \mathbb{Z}_X(d)[2d] \\
&\xrightarrow{\mathrm{id} \otimes p_{Y*} \otimes \mathrm{id}} \mathbb{Z}_Z(a+a')[b+b'] \otimes 1 \otimes \mathbb{Z}_X(d)[2d] \\
&\cong \mathbb{Z}_Z(a+a')[b+b'] \otimes \mathbb{Z}_X(d)[2d] \\
&\xrightarrow{\boxtimes_{Z,X}} \mathbb{Z}_{Z \times_S X}(a+a'+d)[b+b'+2d].
\end{aligned}$$

Using the definition of ζ_f and ζ_g, and Theorem 2.4.10 of Chapter III, this in turn may be rewritten as the composition

(2.1.2.1)
$$\begin{aligned}
1 &\cong 1 \otimes 1 \\
&\xrightarrow{\boxtimes_{Z \times_S Y, Y \times_S X} \circ (\zeta_g \otimes \zeta_f)} \mathbb{Z}_{Z \times_S Y \times_S Y \times_S X}(a+a'+e+d)[b+b'+2e+2d] \\
&\xrightarrow{(\mathrm{id}_Z \times \delta_Y \times \mathrm{id}_X)^*} \mathbb{Z}_{Z \times_S Y \times_S X}(a+a'+e+d)[b+b'+2e+2d] \\
&\xrightarrow{p_{13*}} \mathbb{Z}_{Z \times_S X}(a+a'+d)[b+b'+2d].
\end{aligned}$$

Since the composition

$$\begin{aligned}
1 &\cong 1 \otimes 1 \\
&\xrightarrow{\boxtimes_{Z \times_S Y, Y \times_S X} \circ (\zeta_g \otimes \zeta_f)} \mathbb{Z}_{Z \times_S Y \times_S Y \times_S X}(a+a'+e+d)[b+b'+2e+2d]
\end{aligned}$$

is the same as the cup product $p_{12}^*(\zeta_g) \cup p_{34}^*(\zeta_f)$, the composition (2.1.2.1) is the same as the composition

$$\begin{aligned}
1 &\xrightarrow{p_{12}^*(\zeta_g) \cup p_{23}^*(\zeta_f)} \mathbb{Z}_{Z \times_S Y \times_S X}(a+a'+e+d)[b+b'+2e+2d] \\
&\xrightarrow{p_{13*}} \mathbb{Z}_{Z \times_S X}(a+a'+d)[b+b'+2d],
\end{aligned}$$

completing the proof. □

2.1.3. LEMMA. *Let X be a smooth projective S-scheme, of dimension d over S.*

(i) *Let Y be in $\mathbf{Sm}_S$, and let $f: Y \to X$ be a morphism in $\mathbf{Sm}_S$, giving the morphism $f^*: \mathbb{Z}_X \to \mathbb{Z}_Y$. Let $\zeta_{f^*} = \boxtimes_{Y,X} \circ \iota_X''(f^*)$, and let $\Gamma_f \subset Y \times_S X$ be the graph of f. Then*
$$\zeta_{f^*} = \mathrm{cl}^d_{Y \times_S X}(|\Gamma_f|).$$

(ii) *Let Y be in $\mathbf{Sm}_S$, and let $f: X \to Y$ be a morphism in $\mathbf{Sm}_S$ of relative codimension a, giving the morphism $f_*: \mathbb{Z}_X \to \mathbb{Z}_Y(a)[2a]$. Let $\zeta_{f_*} = \boxtimes_{Y,X} \circ \iota_X''(f_*)$, and let $\Gamma_f^t \subset Y \times_S X$ be the transpose $t_{X,Y}(|\Gamma_f|)$ of the graph of f. Then*
$$\zeta_{f_*} = \mathrm{cl}^{a+d}_{Y \times_S X}(\Gamma_f^t).$$

(iii) *Let* $Z \in \mathcal{Z}^q(Y \times_S X/S)$ *be a codimension* q *cycle on* $Y \times_S X$*. Let* $\gamma_Z : \mathbb{Z}_X \to \mathbb{Z}_Y(q-d)[2q-2d]$ *be the composition*

$$\mathbb{Z}_X \xrightarrow{p_2^*} \mathbb{Z}_{X \times_S Y} \xrightarrow{\cup \mathrm{cl}^q(Z)} \mathbb{Z}_{X \times_S Y}(q)[2q] \xrightarrow{p_{1*}} \mathbb{Z}_Y(q-d)[2q-2d],$$

and let $\zeta_{\gamma_Z} = \boxtimes_{Y,X} \circ \iota''_X(\gamma_Z)$*. Then*

$$\zeta_{\gamma_Z} = \mathrm{cl}^q_{Y \times_S X}(Z).$$

(iv) *Let* Y *be in* $\mathbf{Sm}_S$*, and let* $Z \in \mathcal{Z}^q(Y/S)$ *be a codimension* q *cycle on* Y*, giving the map* $\mathrm{cl}^q_Y(Z): 1 \to \mathbb{Z}_Y(q)[2q]$*. Let* $\zeta_Z = \boxtimes_{Y,S} \circ \iota''_S(\mathrm{cl}^q_Y(Z))$*. Then*

$$\zeta_Z = \mathrm{cl}^q_Y(Z).$$

(v) *Let* $Z \in \mathcal{Z}^q(X/S)$ *be a codimension* q *cycle on* X*; cup product with* $\mathrm{cl}^q_X(Z)$ *gives the map* $\cup \mathrm{cl}^q_X(Z): \mathbb{Z}_X \to \mathbb{Z}_X(q)[2q]$*. Let* $\zeta_{\cup Z} = \boxtimes_{X,X} \circ \iota''_X(\cup \mathrm{cl}^q(Z))$*. Then*

$$\zeta_{\cup Z} = \mathrm{cl}^{d+q}_{X \times_S X}(\delta_{X*}(Z)).$$

PROOF. For (i), we use the relation

$$\begin{aligned}
\boxtimes_{Y,X} \circ \iota_X &= \delta_{X*} \circ p_X^* \\
&= \delta_{X*} \circ p_X^* \circ \mathrm{cl}^0_S(|S|) && \text{by Chapter I, Lemma 3.5.4} \\
&= \delta_{X*} \circ \mathrm{cl}^0_X(|X|) && \text{by Chapter I, Proposition 3.5.3} \\
&= \mathrm{cl}^d_{X \times_S X}(\delta_{X*}(|X|)) && \text{by Chapter III, Theorem 2.2.3} \\
&= \mathrm{cl}^d_{X \times_S X}(|\Delta_X|).
\end{aligned}$$

From this it follows that

$$\begin{aligned}
\boxtimes_{Y,X} \circ \iota''_X(f^*) &= (f \times \mathrm{id}_X)^*(\mathrm{cl}^d_{X \times_S X}(|\Delta_X|)) \\
&= \mathrm{cl}^d_{Y \times_S X}(|\Gamma_f|) && \text{by Proposition 3.5.3, Chapter I,}
\end{aligned}$$

proving (i).

For (ii), we have

$$\begin{aligned}
\boxtimes_{Y,X} \circ \iota''_X(f_*) &= (f \times \mathrm{id}_X)_* \circ \delta_{X*} \circ p_X^* \\
&= (f \times \mathrm{id}_X)_* \circ \delta_{X*} \circ p_X^* \circ \mathrm{cl}^0_S(|S|) && \text{by Lemma 3.5.4, Chapter I} \\
&= (f, \mathrm{id}_X)_* \circ \mathrm{cl}^0_X(|X|) && \text{by Theorem 2.4.7, Chapter III} \\
&= \mathrm{cl}^d_{Y \times_S X}(|\Gamma^t_f|) && \text{by Theorem 2.2.3, Chapter III.}
\end{aligned}$$

The assertion (iii) follows directly from Lemma 2.1.1. As the cycle class map $\mathrm{cl}^q_Y(Z)$ may be factored as the composition

$$1 = \mathbb{Z}_S \xrightarrow{p_2^*} \mathbb{Z}_{Y \times_S S} \xrightarrow{\cup \mathrm{cl}^q(Z)} \mathbb{Z}_{Y \times_S S}(q)[2q] \xrightarrow{p_{1*}} \mathbb{Z}_Y(q)[2q],$$

(iv) follows from (iii). For (v), we have

$$\begin{aligned} &p_{1*}\circ(p_2^*(-)\cup[\delta_{X*}\circ \mathrm{cl}^q_X(Z)]) \\ &\qquad = p_{1*}\circ\delta_{X*}(\delta_X^*\circ p_2^*(-)\cup \mathrm{cl}^q_X(Z)) && \text{by Theorem 2.4.8, Chapter III} \\ &\qquad = \cup\mathrm{cl}^q_X(Z) && \text{by Theorem 2.4.7, Chapter III.}\end{aligned}$$

By Theorem 2.2.3 of Chapter III, $\delta_{X*}\circ\mathrm{cl}^q_Y(Z)=\mathrm{cl}^{d+q}_{X\times_S X}(\delta_{X*}(Z))$; this and (iii) proves (v). □

2.1.4. REMARK. Let X and Y be smooth projective S-schemes.
(i)Let $f{:}\,X\to Y$ a morphism. Suppose X has dimension d over S and Y has dimension e over S. Then the morphism $f^*:\mathbb{Z}_Y(a)\to\mathbb{Z}_X(a)$ has dual the morphism $f_*{:}\,\mathbb{Z}_X(d-a)[2d]\to\mathbb{Z}_Y(e-a)[2e]$. The morphism $f_*{:}\,\mathbb{Z}_X(a)\to\mathbb{Z}_Y(e-d+a)[2e-2d]$ has dual $f^*{:}\,\mathbb{Z}_Y(d-a)[2d]\to\mathbb{Z}_X(d-a)[2d]$.
(ii) Let $Z\in\mathcal{Z}^q(X/S)$ be a cycle. Then the morphism $\cup_X\mathrm{cl}^q_X(Z){:}\,\mathbb{Z}_X(a)\to\mathbb{Z}_X(a+q)[2q]$ has dual the morphism $\cup_X\mathrm{cl}^q_X(Z){:}\,\mathbb{Z}_X(d-a-q)[2d-2q]\to\mathbb{Z}_X(d-a)[2d]$. The morphism $\mathrm{cl}^q_X(Z){:}\,1\to\mathbb{Z}_X(q)[2q]$ has dual the composition

$$\mathbb{Z}_X(d-q)[2d-2q]\xrightarrow{\cup_X\mathrm{cl}^q_X(Z)}\mathbb{Z}_X(d)[2d]\xrightarrow{p_{X*}}1.$$

Indeed, the computations of the dual of f^*, f_* and $\cup_X\mathrm{cl}^q_X(Z)$ follow easily from Lemma 2.1.1 and Lemma 2.1.3. For $\mathrm{cl}^q_X(Z)$, we have $\mathrm{cl}^q_X(Z)=(\cup\mathrm{cl}^q_X(Z))\circ p_X^*$, hence

$$(\mathrm{cl}^q_X(Z))^D=(p_X^*)^D\circ(\cup\mathrm{cl}^q_X(Z))^D=p_{X*}\circ(\cup\mathrm{cl}^q_X(Z)).$$

2.1.5. *Chow motives.* We recall the construction of the category of *graded Chow motives over* k with R-coefficients, $\mathrm{Mot}_R(k)$ (see [**79**] or [**94**] for a slightly different, but equivalent, definition).

Let k be a field. The category Pre-$\mathrm{Mot}_R(k)$ has objects $\mathcal{M}(X)(a)$, where X is a smooth projective k-scheme, and a an integer. The morphisms are given by

$$\mathrm{Hom}_{\text{Pre-Mot}_R(k)}(\mathcal{M}(X)(a),\mathcal{M}(Y)(b))=\mathrm{CH}^{d_X+b-a}(Y\times X)\otimes_{\mathbb{Z}}R$$

if X has dimension d_X over k. The composition law is given by

$$W'\circ W=p_{ZX*}(p_{ZY}^*(W')\cup p_{YX}^*(W)),$$

for $W\in\mathrm{CH}^{d_X+n}(Y\times X)$ and $W'\in\mathrm{CH}^{d_Y+m}(Z\times Y)$. Pre-$\mathrm{Mot}_R(k)$ is an R-tensor category, with direct sum being disjoint union, and tensor product induced by the product over k. The duality involution on Pre-$\mathrm{Mot}_R(k)$ is given by the interchange of factors in $Y\times X$. The category $\mathrm{Mot}_R(k)$ is the R-tensor category Pre-$\mathrm{Mot}_{R\#}$ gotten from Pre-$\mathrm{Mot}_R(k)$ by taking the pseudo-abelian hull.

Let R be a commutative ring, flat over $\mathbb{Z}$, and let $S=\mathrm{Spec}\,k$. It follows immediately from Lemma 2.1.1, Lemma 2.1.2, and Lemma 2.1.3 that sending $\mathcal{M}(X)(a)$ to $R_X(a)[2a]$ and

$$Z\in\mathrm{Hom}_{\text{Pre-Mot}_R(k)}(\mathcal{M}(X)(a),\mathcal{M}(Y)(b))=\mathrm{CH}^{d_X+b-a}(Y\times X)\otimes_{\mathbb{Z}}R$$

to

$$(\boxtimes_{Y,X}\circ\iota''_X)^{-1}(\mathrm{cl}^{d_X+b-a}_{Y\times_S X}(Z)){:}\,R_X(a)[2a]\to R_Y(b)[2b]$$

extends to an R-tensor functor $\phi\colon \mathrm{Mot}_R(k) \to \mathcal{DM}(S)_R$ compatible with the respective duality involutions. From Chapter II, Theorem 3.6.6, ϕ induces an isomorphism $\mathrm{Hom}_{\mathrm{Pre\text{-}Mot}_R(k)}(1, \mathcal{M}(X)(a)) \to \mathrm{Hom}_{\mathcal{DM}_R}(1, R_X(a)[2a])$ for all smooth projective k-schemes X. Using the duality isomorphisms

$$\begin{aligned}&\mathrm{Hom}_{\mathrm{Pre\text{-}Mot}_R(k)}(\mathcal{M}(X)(a), \mathcal{M}(Y)(b))\\ &\qquad\cong \mathrm{Hom}_{\mathrm{Pre\text{-}Mot}_R(k)}(1, \mathcal{M}(X \times Y)(b - a + d_X))\end{aligned}$$

$$\begin{aligned}&\mathrm{Hom}_{\mathcal{DM}_R}(R_X(a)[2a], R_Y(b)[2b])\\ &\qquad\cong \mathrm{Hom}_{\mathcal{DM}_R}(1, R_{X\times Y}(b - a + d_X)[2(b - a + d_X)])\end{aligned}$$

we see that ϕ is a fully faithful embedding.

2.2. Homology and compactly supported cohomology

We use the results of §1.4 and §2.1 to define and relate the homological motive, the Borel-Moore motive, and the motive with compact support. The resulting cohomology theories yield motivic homology, motivic Borel-Moore homology, and motivc cohomology with compact support.

2.2.1. We have the full subcategory $\mathbf{Sm}_S^{\mathrm{pr}}$ of $\mathbf{Sm}_S$ with objects being those X in $\mathbf{Sm}_S$ for which $\mathbb{Z}_X$ is in $\mathcal{DM}(S)^{\mathrm{pr}}$; let $\mathbf{Sm}_{S,\mathrm{proj}}^{\mathrm{pr}}$ be the subcategory of $\mathbf{Sm}_S^{\mathrm{pr}}$ with the same objects, but with only projective morphisms. Similarly, let $\mathbf{Sm}_{S,\mathrm{proj}}$ be the subcategory of $\mathbf{Sm}_S$ with the same objects, but with only the projective morphisms.

If $X \in \mathbf{Sm}_S$ is projective over S, then X is in $\mathbf{Sm}_S^{\mathrm{pr}}$; more generally, if U is an open subset of a smooth projective S-scheme $X \in \mathbf{Sm}_S$, and if we can write the complement of U as $X - U = \cup_{i=1}^n D_i$ such that the closed subsets $D_{i_1} \cap \ldots \cap D_{i_s}$ are smooth over S for each collection of indices $i_1, \ldots, i_s$, then, by Lemma 1.5.4, U is in $\mathbf{Sm}_S^{\mathrm{pr}}$.

2.2.2. DEFINITION. (i) Let X be in $\mathbf{Sm}_S$. The *homological motive* of X, $\mathbb{Z}_X^h$, is the dual $\mathbb{Z}_X^D$ of $\mathbb{Z}_X$. The *motivic homology* $H_p(X, \mathbb{Z}(q))$ of X is defined by

$$H_p(X, \mathbb{Z}(q)) := \mathrm{Hom}_{\mathcal{DM}(S)}(1, \mathbb{Z}_X^h(-q)[-p]).$$

(ii) Let X be in $\mathbf{Sm}_S$. We have the Borel-Moore motive $\mathbb{Z}_X^{\mathrm{B.M.}}$ defined in Chapter III, §2.5.5. The *Borel-Moore homology* of X, $H_p^{\mathrm{B.M.}}(X, \mathbb{Z}(q))$, is defined by

$$H_p^{\mathrm{B.M.}}(X, \mathbb{Z}(q)) = \mathrm{Hom}_{\mathcal{DM}(S)}(1, \mathbb{Z}_X^{\mathrm{B.M.}}(-q)[-p]).$$

(iii) Let X be in $\mathbf{Sm}_S^{\mathrm{pr}}$. We define the *motive of X with compact support over S*, $\mathbb{Z}_X^{c/S}$, by

$$\mathbb{Z}_X^{c/S} := (\mathbb{Z}_X^{\mathrm{B.M.}})^D.$$

The *compactly supported motivic cohomology* of X, $H^p_{c/S}(X, \mathbb{Z}(q))$, is defined by

$$H^p_{c/S}(X, \mathbb{Z}(q)) := \mathrm{Hom}_{\mathcal{DM}(S)}(1, \mathbb{Z}_X^{c/S}(q)[p]).$$

2.2.3. *Functorialities and dualities.* Dualizing the restriction to $\mathbf{Sm}_S^{\mathrm{pr}}$ of the basic motivic functor (I.2.2.9.1), $\mathbb{Z}_{(-)}\colon \mathbf{Sm}_S^{\mathrm{op}} \to \mathcal{DM}(S)$, gives the *homological motivic functor*

$$\mathbb{Z}_{(-)}^h\colon \mathbf{Sm}_S^{\mathrm{pr}} \to \mathcal{DM}(S). \tag{2.2.3.1}$$

Similarly, dualizing the Borel-Moore functor $\mathbb{Z}^{\mathrm{B.M.}}_{(-)}:\mathbf{Sm}_{S,\mathrm{proj}}\to\mathcal{DM}(S)$ (Chapter III, §2.5.5) gives the compactly supported motivic functor

$$\mathbb{Z}^{c/S}_{(-)}:(\mathbf{Sm}^{\mathrm{pr}}_{S,\mathrm{proj}})^{\mathrm{op}}\to\mathcal{DM}(S). \tag{2.2.3.2}$$

For a map $f:X\to Y$ in $\mathbf{Sm}^{\mathrm{pr}}_S$, we write $\mathbb{Z}^h(f)$ as $f_*:\mathbb{Z}^h_X\to\mathbb{Z}^h_Y$ and for f projective, write $\mathbb{Z}^{c/S}(f)$ as $f^!:\mathbb{Z}^{c/S}_Y\to\mathbb{Z}^{c/S}_X$.

Composing the above functors with the cohomological functor $\mathrm{Hom}_{\mathcal{DM}(S)}(1,-)$ gives the appropriate functoriality for the (co)homology theories described in Definition 2.2.2. Applying the duality involution gives the natural isomorphisms (for X in $\mathbf{Sm}^{\mathrm{pr}}_S$)

$$\begin{aligned}H^{\mathrm{B.M.}}_p(X,\mathbb{Z}(q))&\cong\mathrm{Hom}_{\mathcal{DM}(S)}(\mathbb{Z}^{c/S}_X(q)[p],1),\\ H_p(X,\mathbb{Z}(q))&\cong\mathrm{Hom}_{\mathcal{DM}(S)}(\mathbb{Z}_X(q)[p],1).\end{aligned}\tag{2.2.3.3}$$

2.2.4. *The homological motive.* The homotopy and Künneth isomorphisms, and the Mayer-Vietoris and Gysin distinguished triangles (Chapter I, §2.2) dualize to yield the corresponding properties of the homological motive.

1. *Homotopy.* The map $p_*:\mathbb{Z}^h_{X\times\mathbb{A}^1}\to\mathbb{Z}^h_X$ is an isomorphism
2. *Products.* Taking the dual of the inverse of the external products $\boxtimes_{X,Y}:\mathbb{Z}_X\otimes\mathbb{Z}_Y\to\mathbb{Z}_{X\times_SY}$ gives natural associative and commutative external products $\boxtimes^h_{X,Y}:\mathbb{Z}^h_X\otimes\mathbb{Z}^h_Y\to\mathbb{Z}^h_{X\times_SY}$, which are isomorphisms.
3. *Mayer-Vietoris.* Write X as a union of open subschemes $X=U\cup V$, with X, U and V in $\mathbf{Sm}^{\mathrm{pr}}_S$. Then $U\cap V$ is in $\mathbf{Sm}^{\mathrm{pr}}_S$ and we have the Mayer-Vietoris distinguished triangle
$$\mathbb{Z}^h_{U\cap V}\xrightarrow{(j_{U\cap V,U*},-j_{U\cap V,V*})}\mathbb{Z}^h_U\oplus\mathbb{Z}^h_V\xrightarrow{j^*_U+j^*_V}\mathbb{Z}^h_X\to\mathbb{Z}^h_{U\cap V}[1].$$
4. *Localization.* Let $j:U\to X$ be an open immersion, with X and U in $\mathbf{Sm}^{\mathrm{pr}}_S$, and let W be the complement of U in X. Define the *homological motive of X relative to U*, $\mathbb{Z}^h_{X/U}$, as the dual of the motive with support $\mathbb{Z}_{X,W}$. Dualizing the localization distinguished triangle of (Chapter I, §2.2.10) gives the distinguished triangle
$$\mathbb{Z}^h_U\xrightarrow{j_{U*}}\mathbb{Z}^h_X\xrightarrow{p_*}\mathbb{Z}^h_{X/U}\to\mathbb{Z}^h_U[1].$$

The Gysin isomorphism and projective push-forward morphisms of Chapter III, Section 2 dualize to give functorial pull-back for projective morphisms $f:Y\to X$ of relative dimension d: $f^*:\mathbb{Z}^h_X\to\mathbb{Z}^h_Y(-d)[-2d]$ and the Thom isomorphism $i^*:\mathbb{Z}^h_{X/U}\to\mathbb{Z}^h_W(d)[2d]$ for $i:W\to X$ a closed embedding in $\mathbf{Sm}^{\mathrm{pr}}_S$ of codimension d, with complement U. The above properties have the obvious translations to properties of the motivic homology groups: homotopy property, external products, long exact Mayer-Vietoris and localization sequences, projective pull-back and Thom isomorphism.

The composition operation, combined with the Tate twist isomorphism,

$$\begin{aligned}&\mathrm{Hom}_{\mathcal{DM}(S)}(1,\mathbb{Z}_X(q)[p])\otimes\mathrm{Hom}_{\mathcal{DM}(S)}(\mathbb{Z}_X(q')[p'],1)\\ &\qquad\to\mathrm{Hom}_{\mathcal{DM}(S)}(\mathbb{Z}_X(q')[p'],\mathbb{Z}_X(q)[p])\\ &\qquad\qquad\cong\mathrm{Hom}_{\mathcal{DM}(S)}(\mathbb{Z}_X(q'-q)[p'-p],1)\end{aligned}$$

gives, via the isomorphism (2.2.3.3), the cap product pairing

$$\cap^X : H^p(X, \mathbb{Z}(q)) \otimes H_{p'}(X, \mathbb{Z}(q')) \to H_{p'-p}(X, \mathbb{Z}(q'-q)).$$

Let $f: Y \to X$ be a morphism in $\mathbf{Sm}_S^{\mathrm{pr}}$; one easily verifies the identity

$$f^* \circ (\mathrm{id}_X \cup \alpha) = (\mathrm{id}_Y \cup f^*(\alpha)) \circ f^*$$

for $\alpha: 1 \to \mathbb{Z}_X(p')[q']$. This, together with the associativity of composition, immediately implies the projection formula

$$f_*(f^*(\alpha) \cap^Y \beta) = \alpha \cap^X f_*(\beta)$$

for elements $\alpha \in H^p(X, \mathbb{Z}(q))$, $\beta \in H_{p'}(Y, \mathbb{Z}(q'))$.

2.2.5. *The motive with compact support and the Borel-Moore motive.* For a morphism $f: X \to Y$ in $\mathbf{Sm}_S^{\mathrm{pr}}$ of relative dimension d_f, we let

$$f_!: \mathbb{Z}_X^{c/S} \to \mathbb{Z}_Y^{c/S}(-d_f)[-2d_f] \tag{2.2.5.1}$$

denote the (shifted and twisted) dual of the pull-back map $f^*: \mathbb{Z}_Y \to \mathbb{Z}_X$.

The properties of the motive $\mathbb{Z}_X$ (homotopy, external products and Künneth isomorphism, Mayer-Vietoris, localization, Gysin isomorphism) translate directly to the analogous properties of the Borel-Moore motive, after making the appropriate twist and shift. Dualizing gives the same properties for the motive with compact support.

For example, taking the inverse of the dual of the Künneth isomorphism gives the Künneth isomorphism $\boxtimes_{X,Y}^{c/S}: \mathbb{Z}_X^{c/S} \otimes \mathbb{Z}_Y^{c/S} \to \mathbb{Z}_{X\times Y}^{c/S}$, giving external products in compactly supported cohomology

$$\cup_{X,Y}^{c/S} : H^p_{c/S}(X, \mathbb{Z}(q)) \otimes H^{p'}_{c/S}(Y, \mathbb{Z}(q')) \to H^{p+p'}_{c/S}(X \times_S Y, \mathbb{Z}(q+q')).$$

If $X = Y$, we may then pull-back by $\Delta_X^!$, giving cup product in compactly supported cohomology:

$$\cup_X^{c/S} : H^p_{c/S}(X, \mathbb{Z}(q)) \otimes H^{p'}_{c/S}(X, \mathbb{Z}(q')) \to H^{p+p'}_{c/S}(X, \mathbb{Z}(q+q')),$$
$$\cup_X^{c/S} = \Delta_X^! \circ \cup_{X,X}^{c/S},$$

which makes $H^*_{c/S}(X, \mathbb{Z}(*)) := \oplus_{p,q} H^p_{c/S}(X, \mathbb{Z}(q))$ into a bi-graded ring (in general, without unit). We have external products in Borel-Moore homology, defined similarly.

Let $H_*^{\mathrm{B.M.}}(X, \mathbb{Z}(*)) := \oplus_{p,q} H_p^{\mathrm{B.M.}}(X, \mathbb{Z}(q))$. Composition defines as above the cap product

$$\cap_X^c : H^p_{c/S}(X, \mathbb{Z}(q)) \otimes H_{p'}^{\mathrm{B.M.}}(X, \mathbb{Z}(q')) \to H_{p'-p}^{\mathrm{B.M.}}(X, \mathbb{Z}(q'-q))$$

satisfying $f_{!*}(f_!(\alpha) \cap_Y^c \beta) = \alpha \cap_X^c f_!^*(\beta)$ for $\alpha \in H^{p'}_{c/S}(X, \mathbb{Z}(q'))$, $\beta \in H_p^{\mathrm{B.M.}}(Y, \mathbb{Z}(q))$, and $f: X \to Y$ a morphism in $\mathbf{Sm}_S$. This gives $H_*^{\mathrm{B.M.}}(X, \mathbb{Z}(*))$ the structure of a bi-graded module over $H^*_{c/S}(X, \mathbb{Z}(*))$.

2.2.6. *Poincaré duality.* For X smooth and projective of dimension d over S, the identity (1.4.1.1)

$$\mathbb{Z}_X^D = \mathbb{Z}_X(d)[2d],$$

the identifications (Remark 2.1.4) $f^{*D} = f_*$, $f_*^D = f^*$ for a morphism $f: X \to Y$ of smooth projective S-schemes, and the fact that duality is an exact involution, gives the functorial isomorphisms

$$\begin{gathered} H_p(X, \mathbb{Z}(q)) \cong H^{2d-p}(X, \mathbb{Z}(2d-q)) \\ H^p_{c/S}(X, \mathbb{Z}(q)) \cong H^p(X, \mathbb{Z}(q)) \\ H_p^{\mathrm{B.M.}}(X, \mathbb{Z}(q)) \cong H_p(X, \mathbb{Z}(q)) \cong H^{2d-p}(X, \mathbb{Z}(2d-q)). \end{gathered}$$

Via these isomorphisms, the cap products defined above are identified with the cup product in motivic cohomology.

2.3. Relative cohomology and homology

In this section, we discuss duality for the complement of a normal crossing subscheme in a smooth projective S-scheme, relative to another normal crossing subscheme. As a special case, we identify the compactly supported cohomology with the cohomology of a projective compactification over S, relative to a "normal crossing complement at infinity" in case such exists; we identify the Borel-Moore homology with a similarly defined relative homology group. In particular, if $S = \operatorname{Spec} k$ for a perfect field k, and if one has resolution of singularities for quasi-projective k schemes, then one has this interpretation of compactly supported motivic cohomology, and Borel-Moore homology, for all quasi-projective k-schemes.

2.3.1. *Relative cycle classes.* Let X be in $\mathbf{Sm}_S$, $F_1, \ldots, F_k$ closed subschemes of X forming a normal crossing subscheme (Definition 1.5.3), and let W be a closed subset of X, disjoint from all the F_i. Then, as $F_i \cap (X \setminus W) = F_i$, it follows from Chapter I, Lemma 2.6.5, that the canonical map $\mathbb{Z}_{(X;F_1,\ldots,F_k),W} \to \mathbb{Z}_{X,W}$ is an isomorphism. In particular, the cycle class map with support (I.3.5.2.7) $\mathrm{cl}^q_{X,W}: \mathcal{Z}^q_W(X/S) \to \mathbb{Z}_{X,W}(q)[2q]$ defines the *relative cycle class map*

$$\mathrm{cl}^q_{(X;F_1,\ldots,F_k),W}: \mathcal{Z}^q_W(X/S) \to \mathbb{Z}_{(X;F_1,\ldots,F_k),W}.$$

2.3.2. *Push-forward for relative cycles.* Let X be in $\mathbf{Sm}_S$, $F_1, \ldots, F_k, Y$ closed subschemes of X forming a normal crossing scheme, and let W be a closed subset of Y, disjoint from all the F_i. Let $i: Y \to X$ denote the inclusion, and suppose that Y has codimension e on X; write F_i^Y for $F^i \cap Y$.

We have the push-forward

$$i_*: \mathbb{Z}_{(Y;F_1^Y,\ldots,F_k^Y)} \to \mathbb{Z}_{(X;F_1,\ldots,F_k)}(e)[2e]$$

(see Chapter III, §2.6).

2.3.3. LEMMA. *Let W be a closed subset of Y, disjoint from $F_1, \ldots, F_k$, and let z be in $\mathcal{Z}^q_W(Y/S)$. Then*

$$\mathrm{cl}_{(X;F_1,\ldots,F_k)}(i_*(z)) = i_* \circ \mathrm{cl}_{(Y;F_1^Y,\ldots,F_k^Y)}(z).$$

PROOF. We have the push-forward for the relative motives with support in W (Chapter III, *loc. cit.*)

$$i_*: \mathbb{Z}_{(Y;F_1^Y,\ldots,F_k^Y),W} \to \mathbb{Z}_{(X;F_1,\ldots,F_k),W}(e)[2e],$$

the push-forward for the motive with support in W (III.2.1.2.3)

$$i_*:\mathbb{Z}_{Y,W}\to\mathbb{Z}_{X,W}(e)[2e],$$

and the commutative diagram

$$\begin{array}{ccccc}\mathbb{Z}_{Y,W} & \longleftarrow & \mathbb{Z}_{(Y;F_1^Y,\ldots,F_k^Y),W} & \longrightarrow & \mathbb{Z}_{(Y;F_1^Y,\ldots,F_k^Y)}\\ \downarrow{\scriptstyle i_*} & & \downarrow{\scriptstyle i_*} & & \downarrow{\scriptstyle i_*}\\ \mathbb{Z}_{X,W}(e)[2e] & \longleftarrow & \mathbb{Z}_{(X;F_1,\ldots,F_k),W}(e)[2e] & \longrightarrow & \mathbb{Z}_{(X;F_1,\ldots,F_k)}(e)[2e].\end{array}$$

As the left-hand horizontal arrows are isomorphisms, we reduce to the case $k=0$, which is just Theorem 2.2.3 of Chapter II. □

2.3.4. *Relative motives and duality.* Let X be in $\mathbf{Sm}_S$ of dimension d over S, $D_1,\ldots,D_n$ closed subschemes of X which form a normal crossing subscheme (Definition 1.5.3) of X. Let $U:=X\setminus(D_1\cup\ldots\cup D_i)$, $V:=X\setminus(D_{i+1}\cup\ldots\cup D_n)$, and let $\delta_{U\cap V}:U\cap V\to V\times_S U$ be the diagonal inclusion. We suppose that X is equi-dimensional over S; let d be the dimension of X over S.

Let $D_j^V:=V\cap D_j$ and $D_j^U:=D_j\cap U$. We have the relative motives

$$\mathbb{Z}_{(V;D^V_{*\le i})}:=\mathbb{Z}_{(V;D_1^V,\ldots,D_i^V)},\quad \mathbb{Z}_{(U;D^U_{i<*})}:=\mathbb{Z}_{(U;D^U_{i+1},\ldots,D^U_n)},$$

and

$$\mathbb{Z}_{(V\times_SU;D^V_{*\le i}\times_SU,V\times_SD^U_{i<*})}:=\mathbb{Z}_{(V\times_SU;D_1^V\times_SU,\ldots,D_i^V\times_SU,V\times_SD^U_{i+1},\ldots,V\times_SD^U_n)}$$

in $\mathbf{D}^b_{\mathrm{mot}}(\mathbf{Sm}_S)$ (see Chapter I, §2.6.6).

Let $\Delta_{U\cap V}$ be the image of $\delta_{U\cap V}$, and $|\Delta_{U\cap V}|$ the corresponding cycle.

By the method of §2.3.1, we have the cycle class map

$$\mathrm{cl}(|\Delta_{U\cap V}|):1\to\mathbb{Z}_{(V\times_SU;D^V_{*\le i}\times_SU,V\times_SD^U_{i<*})}(d)[2d].$$

We let

$$\delta_{U,V}:1\to\mathbb{Z}_{(V;D^V_{*\le i})}\otimes\mathbb{Z}_{(U;D^U_{i<*})}(d)[2d] \tag{2.3.4.1}$$

be the map in $\mathbf{D}^b_{\mathrm{mot}}(\mathbf{Sm}_S)$ defined by composing $\mathrm{cl}(|\Delta_{U\cap V}|)$ with the inverse of the Künneth isomorphism

$$\mathbb{Z}_{(V;D^V_{*\le i})}\otimes\mathbb{Z}_{(U;D^U_{i<*})}(d)[2d]\xrightarrow{\boxtimes_{*,*}}\mathbb{Z}_{(V\times_SU;D^V_{*\le i}\times_SU,V\times_SD^U_{i<*})}(d)[2d].$$

Suppose we have closed subschemes $E,D_1,\ldots,D_n$ of X, with transverse intersection; we suppose that E has codimension $d_{E:X}$ in X. Let $E_V:=E\cap V$, $D^V_{Ei}:=D_i\cap E_V$, and let $\mathbb{Z}_{(E_V;D^V_{E,*\le i})}$ denote the relative motive $\mathbb{Z}_{(E_V;D^V_{E1},\ldots,D^V_{Ei})}$. The collection of inclusions $D^V_{Ei}\hookrightarrow D^V_i$ defines the morphism

$$i^*_{E_V}:\mathbb{Z}_{(V;D^V_{*\le i})}\to\mathbb{Z}_{(E_V;D^V_{E,*\le i})}. \tag{2.3.4.2}$$

Similarly, using the Gysin morphism for diagrams (see Chapter III, §2.6), we have the push-forward

$$i_{E_U*}:\mathbb{Z}_{(E_U;D^U_{E,i<*})}(-d_E)[-2d_{E:X}]\to\mathbb{Z}_{(U;D^U_{i<*})}. \tag{2.3.4.3}$$

2.3.5. LEMMA. *(i) The pair* $(\mathbb{Z}_{(V;D^V_{*\le i})},\delta_{U,V})$ *is the dual of* $\mathbb{Z}_{(U;D^U_{i<*})}(d)[2d]$.
(ii) The map (2.3.4.3) *is the dual of the map* (2.3.4.2).

PROOF. To simplify the notation, we suppress the auxiliary maps $_f$ in expressions of the form $\mathbb{Z}_Y(n)_f$. We will use the notation for relative motives and relative motives with support employed in the previous few paragraphs.

We prove (i) and (ii) together by induction on n, the case $n = 0$ for (i) being the definition (1.4.1.1) of the dual of $\mathbb{Z}_X(d)[2d]$ for X smooth and projective over S, and for (ii) being Remark 2.1.4. We may suppose that each D_i has pure codimension d_i on X.

Let $V' := X \setminus (D_{i+1} \cup \ldots \cup D_{n-1})$. Let d' be the dimension of $D_n^{V'}$ over S, and let $j: V \to V'$ and $i: D_n^{V'} \to V'$ be the inclusions. We use the notation $D_{n,j}^{V'} := D_n^{V'} \cap D_j$, $D_{D_{n,*\le i}}^{V'} := \{D_{n,1}^{V'}, \ldots, D_{n,i}^{V'}\}$.

The Gysin isomorphism

$$i_*: \mathbb{Z}_{(D_n^{V'}; D_{D_{n,*\le i}}^{V'})}(d')[2d'] \to \mathbb{Z}_{(V', D_{*\le i}^{V'}), D_n^{V'}}(d)[2d]$$

(III.2.6.8.2), together with the localization sequence for the relative motive with support (I.2.6.6.2), gives us the Gysin distinguished triangle

$$\text{(2.3.5.1)}\quad \mathbb{Z}_{(V', D_{*\le i}^{V'})}(d)[2d] \xrightarrow{j^*} \mathbb{Z}_{(V, D_{*\le i}^{V})}(d)[2d]$$
$$\to \mathbb{Z}_{(D_n^{V'}; D_{D_{n,*\le i}}^{V'})}(d')[2d'+1] \xrightarrow{i_*} \mathbb{Z}_{(V', D_{*\le i}^{V'})}(d)[2d+1].$$

Applying our induction hypothesis, the dual of the map i_* is the map

$$i^*: \mathbb{Z}_{(U; D_{i<*\le n-1}^U)}[-1] \to \mathbb{Z}_{(D_n^U; D_{D_{n,i<*\le n-1}}^U)}[-1];$$

this latter map fits into the relativization sequence (I.2.6.6.1)

$$\text{(2.3.5.2)}\quad \mathbb{Z}_{(U; D_{i<*\le n-1}^U)}[-1] \xrightarrow{i^*} \mathbb{Z}_{(D_n^U; D_{D_{n,i<*\le n-1}}^U)}[-1]$$
$$\xrightarrow{i_n} \mathbb{Z}_{(U; D_{i<*}^U)} \xrightarrow{j_n} \mathbb{Z}_{(U; D_{i<*\le n-1}^U)},$$

with $\mathbb{Z}_{(U; D_{i<*}^U)}$ the cone of i^*, i_n the canonical inclusion, and j_n the canonical projection.

We have the relative motive with support on $D_n^{V'}$:

$$\mathbb{Z}_{(V; D_{*\le i}^{V'}), D_n^{V'}} := \mathrm{cone}\big(\mathbb{Z}_{(V', D_{*\le i}^{V'})} \xrightarrow{j^*} \mathbb{Z}_{(V, D_{*\le i}^{V})}\big)[-1];$$

the distinguished triangle (2.3.5.1) is by definition isomorphic to the cone sequence

$$\text{(2.3.5.3)}\quad \mathbb{Z}_{(V', D_{*\le i}^{V'})}(d)[2d] \xrightarrow{j^*} \mathbb{Z}_{(V, D_{*\le i}^{V})}(d)[2d]$$
$$\xrightarrow{i_0} \mathbb{Z}_{(V; D_{*\le i}^{V'}), D_n^{V'}}(d)[2d+1] \xrightarrow{j_0} \mathbb{Z}_{(V', D_{*\le i}^{V'})}(d)[2d+1],$$

with i_0 the canonical inclusion, and j_0 the canonical projection.

Let

$$X_1 = \mathbb{Z}_{(D_n^U; D_{D_{n,i<*\le n-1}}^U)}[-1],\ Y_1 = \mathbb{Z}_{(U; D_{i<*}^U)},\ Z_1 = \mathbb{Z}_{(U; D_{i<*\le n-1}^U)};$$
$$X_2 = \mathbb{Z}_{(V'; D_{*\le i}^{V'}), D_n^{V'}}(d)[2d+1],\ Y_2 = \mathbb{Z}_{(V, D_{*\le i}^{V})}(d)[2d],\ Z_2 = \mathbb{Z}_{(V', D_{*\le i}^{V'})}(d)[2d].$$

We may rewrite the cone sequences (2.3.5.2) and (2.3.5.3) as

$$\begin{gathered} X_1 \xrightarrow{i_n} Y_1 \xrightarrow{j_n} Z_1 \xrightarrow{-i^*[1]} X_1[1], \\ Z_2 \xrightarrow{j^*} Y_2 \xrightarrow{i_0} X_2 \xrightarrow{j_0} Z_2[1]. \end{gathered} \tag{2.3.5.4}$$

We form the 4×4 diagram (1.2.1.1) by tensoring the two sequences of (2.3.5.4) together, using the tensor product $\times$ in the category $\mathbf{C}^b(\mathbf{Sm}_S)^*$ (see Chapter I, §3.3), and let

$$K = \operatorname{cone}\bigl(\mathrm{id} \times j_0 - i^*[1] \times \mathrm{id} \colon X_1 \times X_2 \oplus Z_1 \times Z_2 \to X_1 \times Z_2[1]\bigr)[-1].$$

From §1.2, (see Lemma 1.2.2 and (1.2.1.5)) we have the commutative diagram in $\mathbf{C}^b_{\mathrm{mot}}(\mathbf{Sm}_S)$:

$$\begin{array}{ccc} K & \xrightarrow{\quad\beta\quad} & Y_1 \times Y_2 \\ {\scriptstyle q}\downarrow & & \downarrow{\scriptstyle (\mathrm{id}\times i_1, j_n \times \mathrm{id})} \\ X_1 \times X_2 \oplus Z_1 \times Z_2 & \xrightarrow[i_n \times \mathrm{id} \oplus \mathrm{id} \times j^*]{} & Y_1 \times X_2 \oplus Z_1 \times Y_2 \\ {\scriptstyle \mathrm{id}\times j_0 - i^* \times \mathrm{id}}\downarrow & & \\ X_1 \times Z_2[1] & & \end{array} \tag{2.3.5.5}$$

where the left-hand column is the defining cone sequence.

For a smooth S-scheme X with smooth closed subschemes $W, F_1, \ldots, F_k$, let $\mathcal{Z}^d_W(X; F_1, \ldots, F_k)$ be the subgroup of $\mathcal{Z}^d(X/S)$ consisting of those cycles z which are supported in W, intersect each F_i properly, and have zero intersection with each F_i.

By Proposition 3.3.5 of Chapter I, the cohomological functors $H^0(\mathcal{Z}_{\mathrm{mot}}(-))$ and $\mathrm{Hom}_{\mathbf{K}^b_{\mathrm{mot}}(\mathcal{V})}(\mathfrak{e} \otimes 1, -)$ are isomorphic on the full triangulated subcategory of $\mathbf{K}^b_{\mathrm{mot}}(\mathcal{V})$ generated by the objects $\mathbb{Z}_X(a)_f[b]$, with (X, f) in $\mathcal{L}(\mathcal{V})$.

In addition, we have

$$\begin{gathered} H^0(\mathcal{Z}_{\mathrm{mot}}(X_1 \times X_2)) = \mathcal{Z}^d_{D^n_U \times D^{V'}_n}(D^U_n \times_S V'; D^U_{D_n, i<*\le n-1} \times V', D^U_n \times D^{V'}_{*\le i}), \\ H^0(\mathcal{Z}_{\mathrm{mot}}(X_1 \times Z_2[1])) = \mathcal{Z}^d(D^U_n \times_S V'; D^U_{D_n, i<*\le n-1} \times V', D^U_n \times D^{V'}_{*\le i}), \\ H^0(\mathcal{Z}_{\mathrm{mot}}(Y_1 \times Y_2)) = \mathcal{Z}^d(U \times_S V; D^U_{i<*} \times V, U \times D^V_{*\le i}), \\ H^0(\mathcal{Z}_{\mathrm{mot}}(Z_1 \times Z_2)) \subset \mathcal{Z}^d(U \times_S V'; D^U_{i<*\le n-1} \times V', U \times D^{V'}_{*\le i}), \\ H^0(\mathcal{Z}_{\mathrm{mot}}(Z_1 \times Y_2)) \subset \mathcal{Z}^d(U \times_S V; D^U_{i<*\le n-1} \times V, U \times D^V_{*\le i}). \end{gathered}$$

In the last two lines, the H^0 is the subgroup given by those cycles which have proper intersection with *all* the subschemes $(D^U_{i_1} \cap \ldots \cap D^U_{i_p}) \times V'$, $i \le i_j \le n$, not just those with $1 \le i_j \le n-1$. It also follows from (Chapter I, *loc. cit.*) that $H^{-1}(\mathcal{Z}_{\mathrm{mot}}(X_1 \times Z_2[1])) = 0$.

Thus, applying the functor $H^0(\mathcal{Z}_{\mathrm{mot}}(-))$ to the diagram (2.3.5.5) gives the commutative diagram with exact columns

$$\begin{array}{ccc} 0 & & \\ \downarrow & & \\ H^0(\mathcal{Z}_{\mathrm{mot}}(K)) & \xrightarrow{\beta} & H^0(\mathcal{Z}_{\mathrm{mot}}(Y_1 \times Y_2)) \\ \downarrow{\scriptstyle q} & & \downarrow{\scriptstyle (\mathrm{id}\times i_1, j_n\times\mathrm{id})} \\ H^0(\mathcal{Z}_{\mathrm{mot}}(X_1 \times X_2)) & & H^0(\mathcal{Z}_{\mathrm{mot}}(Y_1 \times X_2)) \\ \oplus & \xrightarrow{i_n\times\mathrm{id}\oplus\mathrm{id}\times j^*} & \oplus \\ H^0(\mathcal{Z}_{\mathrm{mot}}(Z_1 \times Z_2)) & & H^0(\mathcal{Z}_{\mathrm{mot}}(Z_1 \times Y_2)) \\ \downarrow{\scriptstyle \mathrm{id}\times j_0 - i^*\times\mathrm{id}} & & \\ H^0(\mathcal{Z}_{\mathrm{mot}}(X_1 \times Z_2[1])). & & \end{array} \tag{2.3.5.6}$$

Now let $\Delta_{U,V'}$ be the diagonal in $U \times V'$, and Δ_D the diagonal in $D_n^U \times D_n^{V'}$. Since $\Delta_{U,V'}$ avoids the subschemes $U \times D_j^{V'}$, $1 \le j \le i$, and the subschemes $D_j^U \times V'$, $j = i+1, \dots, n-1$, the cycle $|\Delta_{U,V'}|$ is an element of $H^0(\mathcal{Z}_{\mathrm{mot}}(Z_1 \times Z_2))$. Similarly, the cycle $|\Delta_D|$ is an element of $H^0(\mathcal{Z}_{\mathrm{mot}}(X_1 \times X_2))$. Since j_0 is the map "forget the support", we have $(\mathrm{id} \times j_0)(|\Delta_D|) = (i^* \times \mathrm{id})(|\Delta_{U,V'}|)$, hence there is a unique element η of $H^0(\mathcal{Z}_{\mathrm{mot}}(K))$ lifting the pair $(|\Delta_D|, |\Delta_{U,V'}|)$. By the commutativity of the diagram (2.3.5.6), we have

$$(j_n \times \mathrm{id})(\beta(\eta)) = (\mathrm{id} \times j^*)(|\Delta_{U,V'}|) = |\Delta_{U,V}|.$$

Since $H^0(\mathcal{Z}_{\mathrm{mot}}(Y_1 \times Y_2))$ and $H^0(\mathcal{Z}_{\mathrm{mot}}(Z_1 \times Y_2))$ are both subgroups of $\mathcal{Z}^d(U \times_V)$, the map

$$j_n \times \mathrm{id} \colon H^0(\mathcal{Z}_{\mathrm{mot}}(Y_1 \times Y_2)) \to H^0(\mathcal{Z}_{\mathrm{mot}}(Z_1 \times Y_2))$$

is injective, which gives us the identity

$$\beta(\eta) = |\Delta_{U,V}|. \tag{2.3.5.7}$$

From the definition of the cycle class map (Chapter I, §3.5.1) the elements

$$\eta \in H^0(\mathcal{Z}_{\mathrm{mot}}(K)),\ |\Delta_D| \in H^0(\mathcal{Z}_{\mathrm{mot}}(X_1 \times X_2)),$$
$$|\Delta_{U,V'}| \in H^0(\mathcal{Z}_{\mathrm{mot}}(Z_1 \times Z_2)),\ |\Delta_{U,V}| \in H^0(\mathcal{Z}_{\mathrm{mot}}(Y_1 \times Y_2))$$

determine the maps in $\mathbf{D}^b_{\mathrm{mot}}(\mathcal{V})$

$$\mathrm{cl}_K(\eta) \colon 1 \to K,\ \mathrm{cl}_{X_1 \times X_2}(|\Delta_D|) \colon 1 \to X_1 \times X_2$$
$$\mathrm{cl}_{Z_1 \times Z_2}(|\Delta_{U,V'}|) \colon 1 \to Z_1 \times Z_2,\ \mathrm{cl}_{Y_1 \times Y_2}(|\Delta_{U,V}|) \colon 1 \to Y_1 \times Y_2$$

with

$$q \circ \mathrm{cl}(\eta) = (\mathrm{cl}_{X_1 \times X_2}(|\Delta_D|), \mathrm{cl}_{Z_1 \times Z_2}(|\Delta_{U,V'}|)), \tag{2.3.5.8}$$

and, from (2.3.5.7),

$$\beta \circ \mathrm{cl}(\eta) = \mathrm{cl}_{Y_1 \times Y_2}(|\Delta_{U,V}|). \tag{2.3.5.9}$$

Additionally, these cycle class maps agree with the maps constructed by the method of §2.3.1.

Write $X_2' := \mathbb{Z}_{(D_n^{V'};D_{D_{n,*\le i}}^{V'})}(d')[2d']$. Then, as above, the diagonal Δ_D gives an element $|\Delta_D'| \in H^0(\mathcal{Z}_{\mathrm{mot}}(X_1 \times X_2'))$, hence we have the cycle class

$$\mathrm{cl}_{X_1\times X_2'}(|\Delta_D'|): 1 \to X_1 \times X_2'.$$

By Lemma 2.3.3, we have

$$(\mathrm{id} \times i_*) \circ \mathrm{cl}_{X_1\times X_2'}(|\Delta_D'|) = \mathrm{cl}_{X_1\times X_2}(|\Delta_D|). \tag{2.3.5.10}$$

Write $\boxtimes$ for the Künneth isomorphism for $X_1 \otimes X_2$, $Z_1 \otimes Z_2$, etc. By our induction hypothesis, the cycle class maps in $\mathbf{D}^b_{\mathrm{mot}}(\mathbf{Sm}_S)$

$$\mathrm{cl}_{Z_1\times Z_2}(|\Delta_{U,V'}|): 1 \to Z_1 \times Z_2,$$
$$\mathrm{cl}_{X_1\times X_2'}(|\Delta_D'|): 1 \to X_1 \times X_2',$$

gives the duals

$$(Z_1, \boxtimes^{-1} \circ \mathrm{cl}_{Z_1\times Z_2}(|\Delta_{U,V'}|)), \quad (X_1, \boxtimes^{-1} \circ \mathrm{cl}_{X_1\times X_2'}(|\Delta_D'|))$$

to Z_2 and X_2', respectively.

By (2.3.5.10), and the fact that the Gysin map $i_{D_n^{V'}*}: X_2' \to X_2$ is an isomorphism, it follows that $(X_1, \boxtimes^{-1} \circ \mathrm{cl}_{X_1\times X_2}(|\Delta_D|))$ is the dual to X_2. By (2.3.5.8), (2.3.5.9), and Remark 1.2.4 this implies that $(Y_1, \boxtimes^{-1}\mathrm{cl}_{Y_1\times Y_2}(|\Delta_{U,V}|))$ is the dual of Y_2, which proves part (i).

For part (ii), we need only show that

$$(\mathrm{id} \times i_{E_U*}) \circ \mathrm{cl}(|\Delta_{E_U\cap E_V}|) = (i^*_{E_V} \times \mathrm{id}) \circ \mathrm{cl}(|\Delta_{U\cap V}|) \tag{2.3.5.11}$$

as maps $1 \to \mathbb{Z}_{(E_V\times_S U; D^V_{E_V,*\le i}\times_S U, E_V\times_S D^U_{i<*})}$. Indeed, we may apply (i) to the relative motives

$$\mathbb{Z}_{(E_V;D^V_{E_V,*\le i})}, \quad \mathbb{Z}_{(E_U;D^U_{E_U,i<*})}, \tag{2.3.5.12}$$

showing that the map

$$\mathrm{cl}(|\Delta_{E_U\cap E_V}|): 1 \to \mathbb{Z}_{(E_V\times_S E_U; D^V_{E_V,*\le i}\times_S E_U, E_V\times_S D^U_{E_U,i<*})}$$

(followed by the inverse of the Künneth isomorphism) gives the duality between the two relative motives (2.3.5.12). We then apply (i) as it stands in the statement of the lemma, from which the identity (2.3.5.11) is equivalent to $\iota''(i^*_{E_V}) = \iota'(i_{E_U*})$. This in turn implies (ii) by the definition of the duality involution.

The identity (2.3.5.11) follows from the identity of cycles

$$(\mathrm{id}_{E_U} \times i_{E_U})_*(|\Delta_{E_U\cap E_V}|) = (i_{E_V} \times \mathrm{id}_V)^*(|\Delta_{U\cap V}|),$$

using the trick of lifting to the relative motives with support, as in Lemma 2.3.3. □

The description of the morphism i_{E_U*} given above can be extended as follows.

2.3.6. LEMMA. *Let X and Y be smooth, projective and equi-dimensional over S, let $D_1, \ldots, D_n$ form a normal crossing subscheme of X and $E_1, \ldots, E_n$ a normal crossing subscheme of Y. Let*

$$U_X := X \setminus (D_1 \cup \ldots \cup D_i), \quad V_X := X \setminus (D_{i+1} \cup \ldots \cup D_n),$$
$$U_Y := Y \setminus (E_1 \cup \ldots \cup E_j), \quad V_Y := Y \setminus (E_{j+1} \cup \ldots \cup E_m).$$

Suppose we have a morphism $f: X \to Y$ such that f restricts to a proper map $f_U: U_X \to U_Y$, and f restricts to a map $f_V: V_X \to V_Y$. Suppose in addition that

$\operatorname{codim}_X(D_j) = \operatorname{codim}_Y(E_j)$ *and* $f(D_j \cap U_X) \subset E_j \cap U_Y$, *for* $j = i+1, \dots, n$. *Then the dual of*

$$f_{U*}: \mathbb{Z}_{(U_X; D^U_{i+1}, \dots, D^U_n)}(d_X)[2d_X] \to \mathbb{Z}_{(U_Y; E^U_{j+1}, \dots, E^U_n)}(d_Y)[2d_Y]$$

is

$$f^*_V: \mathbb{Z}_{(V_Y; E^V_1, \dots, E^V_j)} \to \mathbb{Z}_{(V_X; D^V_1, \dots, D^V_i)}.$$

PROOF. Note that the first two hypotheses on f imply that f maps each D_k to some E_l, with $1 \le l \le j$ for $1 \le k \le i$, hence the map f^*_V is defined (see Chapter I, §2.6.7); the remaining hypotheses on f imply that f defines a projective morphism of $n-i$-cubes $f_U: (U_X; D^U_{i+1}, \dots, D^U_n)_* \to (U_Y; E^U_{j+1}, \dots, E^U_n)_*$, hence the push-forward f_{U*} is defined following (Chapter III, §2.6).

The proof is then the same as the proof of Lemma 2.3.5(ii). □

2.3.7. *Good compactifications.* Let X be smooth and projective over S, $D_1, \dots, D_n$ closed subschemes of X which form a normal crossing subscheme, and let $U = X \setminus \cup_{i=1}^n D_i$. We call the collection $(X; D_1, \dots, D_n)$ a *good compactification of* U *over* S. If U admits a good compactification, then, by Lemma 1.5.4, U is in $\mathbf{Sm}^{\mathrm{pr}}_S$.

2.3.8. PROPOSITION. *(i) Let* $(X; D_1, \dots, D_n)$ *be a good compactification of* U *over* S. *Then there is a canonical isomorphism*

$$\mathbb{Z}_{(X; D_1, \dots, D_n)} \to \mathbb{Z}^{c/S}_U.$$

In particular, if $(X; D_1, \dots, D_n)$ *and* $(X'; D'_1, \dots, D'_m)$ *are two good compactifications of* U *over* S, *then there is a canonical isomorphism*

$$\mathbb{Z}_{(X; D_1, \dots, D_n)} \to \mathbb{Z}_{(X'; D'_1, \dots, D'_m)}$$

(ii) Let $(X; D_1, \dots, D_n)$ *and* $(Y; E_1, \dots, E_n)$ *be good compactifications of* U *and* V *over* S, *with* $D_1, \dots, D_n$ *and* $E_1, \dots, E_n$ *normal crossing divisors. Let* $f: X \to Y$ *be a morphism such that* $f(D_i) \subset E_i$, $i = 1, \dots, n$, *inducing the push-forward map*

$$f_*: \mathbb{Z}_{(X; D_1, \dots, D_n)}(d_X)[2d_X] \to \mathbb{Z}_{(Y; E_1, \dots, E_n)}(d_Y)[2d_Y]$$

(see Chapter III, §2.6). Then the isomorphism of (i) identifies f_* *with* $f_!: \mathbb{Z}^{c/S}_U \to \mathbb{Z}^{c/S}_V$, *and the dual of* f_* *with* $f^*: \mathbb{Z}_V \to \mathbb{Z}_U$.
(iii) Let $(X; D_1, \dots, D_n)$ *be a good compactification of* U, *and* $(Y; E_1, \dots, E_m)$ *a good compactification of* V, *and let* $\bar{g}: X \to Y$ *be a morphism such that* $\bar{g}$ *restricts to a proper morphism* $g: U \to V$. *Then the isomorphism of (i) identifies* $f^!: \mathbb{Z}^{c/S}_V \to \mathbb{Z}^{c/S}_U$ *with* $g^*: \mathbb{Z}_{(Y; E_1, \dots, E_n)} \to \mathbb{Z}_{(X; D_1, \dots, D_n)}$.

PROOF. The first assertion follows from Lemma 1.1.3, Lemma 2.3.5, and the definition of $\mathbb{Z}^{c/S}_U$ (Definition 2.2.2).

For (ii), the assertion that $f_* = f_!$ follows from the assertion that $f^D_* = f^*$, and the definition of $f_!$ (2.2.5.1). The identity $f^D_* = f^*$ is a special case of Lemma 2.3.6.

For (iii), the condition that g is proper implies that $\bar{g}^{-1}(\cup_{i=1}^m E_i) \subset \cup_{j=1}^n D_j$, so the map of relative motives $g^*: \mathbb{Z}_{(Y; E_1, \dots, E_n)} \to \mathbb{Z}_{(X; D_1, \dots, D_n)}$ is defined (see Chapter I, §2.6.7). As $g^!$ is the dual of $g_*: \mathbb{Z}_U(d_U)[2d_U] \to \mathbb{Z}_V(d_V)[2d_V]$ by definition, (iii) follows from Lemma 2.3.6.

□

2.3.9. *Relative homology and cohomology.* We define the relative motivic homology $H_p((X; D_1, \dots, D_n), \mathbb{Z}(q))$ by

$$H_p(X; D_1, \dots, D_n, \mathbb{Z}(q)) := \mathrm{Hom}_{\mathcal{DM}(S)}(\mathbb{Z}_{(X;D_1,\dots,D_n)}(q)[p], \mathbb{Z}_S);$$

this is compatible with our earlier definition in Chapter I, §2.6.6 of relative motivic cohomology as $\mathrm{Hom}_{\mathcal{DM}(S)}(\mathbb{Z}_S, \mathbb{Z}_{(X;D_1,\dots,D_n)}(q)[p])$. Via Lemma 2.3.5 and Proposition 2.3.8, we may identify the Borel-Moore homology, respectively the compactly supported cohomology, of an S-scheme U which admits a good compactification $(X; D_1, \dots, D_n)$ with relative motivic (co)homology:

$$\begin{aligned} H_p^{\mathrm{B.M.}}(U, \mathbb{Z}(q)) &\cong H_p(X; D_1, \dots, D_n, \mathbb{Z}(q)) \\ H^p_{c/S}(U, \mathbb{Z}(q)) &\cong H^p(X; D_1, \dots, D_n, \mathbb{Z}(q)). \end{aligned}$$

2.4. The Borel-Moore motive for singular schemes

We show how to interpret the Borel-Moore homology as homology with support in a "smoothly decomposable" closed subscheme (Definition 2.4.1(i)). This enables us to extend the definition of the Borel-Moore motive and Borel-Moore homology to such S-schemes. We also consider the extension of the motive with compact support to certain S-schemes which are not smooth over S: those which are smoothly decomposable and admit a "compactifiable" closed embedding into a smooth S-scheme (see Definition 2.4.1(ii)). For such S-schemes, we define the motive with compact support and the resulting motivic cohomology with compact support.

2.4.1. DEFINITION. (i) Let W be a reduced quasi-projective S-scheme. A sequence of closed subsets of W:

$$\emptyset = W_0 \subset W_1 \subset \dots \subset W_{n-1} \subset W_n = W$$

is an *S-smooth stratification of W* if $W_{i+1} \setminus W_i$ is smooth and equi-dimensional over S for each $i = 0, \dots, n-1$. We call W *smoothly decomposable over S* if W has an S-smooth stratification.
(ii) Let W be a smoothly decomposable S-scheme, $i: W \to X$ a closed embedding of W into a smooth S-scheme X, with complement U. We call the embedding i *compactifiable* if X and U are in $\mathbf{Sm}_S^{\mathrm{pr}}$.

2.4.2. *Push-forward for Borel-Moore motives with support.* Let $i: W \to X$ be a closed embedding, with X in $\mathbf{Sm}_S$. Recall from Chapter III, Definition 2.5.6, the Borel-Moore motive with support in $i(W)$, $\mathbb{Z}^{\mathrm{B.M.}}_{X,i(W)}$, which we abbreviate by $\mathbb{Z}_i^{\mathrm{B.M.}}$.

Let $g: W \to W'$ be a projective S-morphism of smoothly decomposable S-schemes, and take closed embeddings $i: W \to X$ and $i': W' \to X'$. Suppose we have a morphism $f: X \to X'$ which makes the diagram

$$\begin{array}{ccc} W & \xrightarrow{i} & X \\ {\scriptstyle g}\downarrow & & \downarrow{\scriptstyle f} \\ W' & \xrightarrow[i']{} & X' \end{array}$$

commute. Factor f as a composition

$$X \xrightarrow{i_X} U_X \xrightarrow{j_X} \mathbb{P}^N_{X'} \xrightarrow{q_X} X',$$

where i_X is a closed embedding, j_X is an open immersion, and q_X is the projection; since g is proper, the composition $j_X \circ i_X \circ i : W \to \mathbb{P}^N_{X'}$ is a closed embedding. We then define $(g,f)_*: \mathbb{Z}^{\mathrm{B.M.}}_i \to \mathbb{Z}^{\mathrm{B.M.}}_{i'}$ as the composition

$$\mathbb{Z}^{\mathrm{B.M.}}_i \xrightarrow{i_{X*}} \mathbb{Z}^{\mathrm{B.M.}}_{i_X \circ i} \xrightarrow{(j_X^*)^{-1}} \mathbb{Z}^{\mathrm{B.M.}}_{j_X \circ i_X \circ i} \xrightarrow{q_{X*}} \mathbb{Z}^{\mathrm{B.M.}}_{i'}.$$

2.4.3. LEMMA. *(i) Suppose that $f: W \to W'$ is an isomorphism of S-schemes. Then*

$$(g,f)_*: \mathbb{Z}^{\mathrm{B.M.}}_i \to \mathbb{Z}^{\mathrm{B.M.}}_{i'}$$

is an isomorphism in $\mathcal{DM}(S)$.
(ii) For all (g,f), the map

$$(g,f)_*: \mathbb{Z}^{\mathrm{B.M.}}_i \to \mathbb{Z}^{\mathrm{B.M.}}_{i'}$$

is independent of the choice of factorization.

PROOF. For (i), let $W = W_X = W_Y$. The reader will easily verify that the maps i_{X*}, j_X^* and q_{X*} are all compatible with localization on W, i.e., suppose $\bar{W}$ is a closed subset of W, with complement W_0. Let X_0 and Y_0 be the complement of $\bar{W}$ in X and Y, and let U_0 be the complement of $q_X^{-1}(W')$ in U. Let

$$\bar{i}: \bar{W} \to X, \quad \bar{i}': \bar{W} \to Y,$$
$$i_0: W_0 \to X_0, \quad i'_0: W_0 \to Y_0$$

be the inclusions; the maps i_X, j_X and q_X thus induce maps

$$\overline{(f,g)}_*: \mathbb{Z}^{\mathrm{B.M.}}_{\bar{i}} \to \mathbb{Z}^{\mathrm{B.M.}}_{\bar{i}'},$$
$$(f_0,g_0)_*: \mathbb{Z}^{\mathrm{B.M.}}_{i_0} \to \mathbb{Z}^{\mathrm{B.M.}}_{i'_0},$$

and give the map of localization distinguished triangles

$$(\mathbb{Z}^{\mathrm{B.M.}}_{\bar{i}} \xrightarrow{i_{\bar{W}*}} \mathbb{Z}^{\mathrm{B.M.}}_{i} \xrightarrow{j^*_{W_0}} \mathbb{Z}^{\mathrm{B.M.}}_{i_0} \to)$$
$$\xrightarrow{(\overline{(f,g)}_*, (f,g)_*, (f_0,g_0)_*)} (\mathbb{Z}^{\mathrm{B.M.}}_{\bar{i}'} \xrightarrow{i_{\bar{W}*}} \mathbb{Z}^{\mathrm{B.M.}}_{i'} \xrightarrow{j^*_{W_0}} \mathbb{Z}^{\mathrm{B.M.}}_{i'_0} \to).$$

Taking the filtration of W into locally closed subsets, smooth over S and using noetherian induction, we reduce to the case in which W is smooth over S.

In this case, the maps $i_*: \mathbb{Z}_W \to \mathbb{Z}^{\mathrm{B.M.}}_i$ and $i'_*: \mathbb{Z}_W \to \mathbb{Z}^{\mathrm{B.M.}}_{i'}$ are isomorphisms. From (Chapter III, Theorem 2.4.7 and Theorem 2.4.9), it follows that

$$i'_* = (g,f)_* \circ i_*,$$

whence (i).

For (ii), it follows easily from the excision isomorphism (Chapter I, Definition 2.1.4(b)) that the map $(g,f)_*$ is independent of the choice of open subscheme U_X.

Suppose we have another factorization of f as

$$X \xrightarrow{i'_X} U'_X \xrightarrow{j'_X} \mathbb{P}^M_{X'} \xrightarrow{q'_X} X'.$$

Form the diagonal embedding $(i_X, j'_X \circ i'_X): X \to U_X \times_{X'} \mathbb{P}^M_{X'}$. As $i_X(X)$ is closed in U_X, and $U_X \times_{X'} \mathbb{P}^M_{X'}$ is proper over U_X, $(i_X, i'_X)(X)$ is closed in $U_X \times_{X'} \mathbb{P}^M_{X'}$. By the functoriality of proper push-forward (Chapter III, Theorem 2.4.7), we have

$$p_{1*} \circ (i_X, j'_X \circ i'_X)_* = i_{X*}: \mathbb{Z}^{\mathrm{B.M.}}_i \to \mathbb{Z}^{\mathrm{B.M.}}_{i_X \circ i}.$$

By the compatibility of push-forward and pull-back in transverse cartesian squares (Chapter III, Theorem 2.4.9), we have

$$p_{1*} \circ [(j_X \times j'_X)^*]^{-1} \circ (i_X, i'_X)_* = (j_X^*)^{-1} \circ i_{X*} : \mathbb{Z}_i^{\mathrm{B.M.}} \to \mathbb{Z}_{j_X \circ i_X \circ i}^{\mathrm{B.M.}}.$$

Similarly, we have

$$p_{2*} \circ [(j_X \times j'_X)^*]^{-1} \circ (i_X, i'_X)_* = (j_X'^*)^{-1} \circ i'_{X*} : \mathbb{Z}_i^{\mathrm{B.M.}} \to \mathbb{Z}_{j'_X \circ i'_X \circ i}^{\mathrm{B.M.}}.$$

Since $q_{X*} \circ p_{1*} = q'_{X*} \circ p_{2*}$, by the functoriality of push-forward (*loc. cit.*), we have the desired result. □

2.4.4. *Borel-Moore motives and motives with compact support.* Form the category $\mathbf{E}$ with objects the closed embeddings $i: W \to X$, $X \in \mathbf{Sm}_S$, where a morphism $i \to i'$ is a commutative diagram

$$\begin{array}{ccc} W & \xrightarrow{i} & X \\ {\scriptstyle g}\downarrow & & \downarrow{\scriptstyle f} \\ W' & \xrightarrow[i']{} & X' \end{array}$$

with g proper.

Let $\mathbf{SDS}_S$ be the full subcategory of $\mathbf{Sch}_S$ with objects the smoothly decomposable S-schemes, and let $\mathbf{SDS}_{S\mathrm{proj}}$ the sub-category of $\mathbf{SDS}_S$ with the same objects as $\mathbf{SDS}_S$, and with morphisms being the projective morphisms. We have the functor $s: \mathbf{E} \to \mathbf{SDS}_{S\mathrm{proj}}$ sending $i: W \to X$ to W.

For a proper map $g: W \to W'$ in $\mathbf{SDS}_S$, let $\mathbf{E}(g)$ be the fiber of s over g, i.e., the category of maps $(g, f): i \to i'$, where a morphism $(g, f_1) \to (g, f_2)$, with $(g, f_j): i_j \to i'_j$, $j = 1, 2$, is a commutative diagram

$$\begin{array}{ccc} i_1 & \xrightarrow{(\mathrm{id}_W, s_1)} & i_2 \\ {\scriptstyle (g,f_1)}\downarrow & & \downarrow{\scriptstyle (g,f_2)} \\ i'_1 & \xrightarrow[(\mathrm{id}_{W'}, s_2)]{} & i'_2. \end{array}$$

Let $\mathrm{Ho}\mathbf{E}$ be the category gotten from $\mathbf{E}$ by inverting all the morphisms in $\mathbf{E}(\mathrm{id}_W)$, for all W in $\mathbf{SDS}_S$.

By Lemma 2.4.3, sending $i: W \to X$ to $\mathbb{Z}_i^{\mathrm{B.M.}}$, and a map of maps $(g, f): i \to i'$ to $(g, f)_*: \mathbb{Z}_i^{\mathrm{B.M.}} \to \mathbb{Z}_{i'}^{\mathrm{B.M.}}$ defines the functor

$$\mathbb{Z}^{\mathrm{B.M.}}: \mathrm{Ho}\mathbf{E} \to \mathbf{D}^b_{\mathrm{mot}}(\mathbf{Sm}_S). \tag{2.4.4.1}$$

We note the following result:

2.4.5. LEMMA. *The functor $s: \mathbf{E} \to \mathbf{SDS}_{S\mathrm{proj}}$ induces an equivalence of categories $\mathrm{Ho}(s): \mathrm{Ho}\mathbf{E} \to \mathbf{SDS}_{S\mathrm{proj}}$.*

PROOF. Let $g: W \to W'$ be a proper morphism in $\mathbf{SDS}_S$, and let $\mathbf{E}(g)^*$ be the image of $\mathbf{E}(g)$ in $\mathrm{Ho}\mathbf{E}$, i.e., we identify two morphism in $\mathbf{E}(g)$ if they have the same image in $\mathrm{Ho}\mathbf{E}$. It suffices to show that $\mathbf{E}(g)^*$ is left filtering.

In fact, suppose this is the case. It then follows easily that each morphism $\eta: i \to i'$, in $\mathrm{Ho}\mathbf{E}$ ($i: W \to X$, $i': W' \to X'$) can be factored as $\eta = s \circ t^{-1}$, where $t: i'' \to i$ and $s: i'' \to i'$ are maps in $\mathbf{E}$ and t is a map over id_W. From this, and the fact that $\mathbf{E}(\mathrm{id}_W)^*$ is left filtering, we see that the fiber of $\mathrm{Ho}(s)$ over id_W is a connected and simply connected groupoid. Since the fiber of $\mathrm{Ho}(s)$ over a given map g in $\mathbf{SDS}_{S\mathrm{proj}}$ is non-empty, it is then an easy exercise to show that $\mathrm{Ho}(s)$ is an equivalence.

Take a closed embedding $i'\colon W' \to X'$, and a projective morphism $g\colon W \to W'$. We will show there is a closed embedding $i\colon W \to X$, and a map $(g, f)\colon i \to i'$. Indeed, since g is projective, we may factor g as

$$W \xrightarrow{i_g} \mathbb{P}^N_{W'} \xrightarrow{q_g} W'$$

with i_g a closed embedding, and q_g the projection. The embedding $W' \to X'$ induces the embedding $\mathbb{P}^N_{W'} \to \mathbb{P}^N_{X'}$, giving the commutative diagram

$$\begin{array}{ccc} W & \xrightarrow{i} & \mathbb{P}^N_{X'} \\ {\scriptstyle g}\downarrow & & \downarrow{\scriptstyle q} \\ W' & \xrightarrow[i']{} & X'. \end{array}$$

If we have embeddings

$$\begin{aligned} i'_1\colon W' \to X'_1, \quad & i'_2\colon W' \to X'_2, \\ i_1\colon W \to X_1, \quad & i_2\colon W \to X_2, \end{aligned}$$

and maps $(g, f_j)\colon i_j \to i'_j$, $j = 1, 2$, then we have the product embeddings

$$(i_1, i_2)\colon W \to X_1 \times_S X_2, \quad (i'_1, i'_2)\colon W' \to X'_1 \times_S X'_2,$$

and the map $(g, f_1 \times f_2)\colon (i_1, i_2) \to (i'_1, i'_2)$, which dominates both (g, f_1) and (g, f_2) by taking the first and second projections.

Finally, suppose we have embeddings $i\colon W \to X$, $i'\colon W' \to X'$, and two maps $(g, f_1), (g, f_2)\colon i \to i'$. We have the two maps of X to $X \times_S X'$,

$$g_i := (\mathrm{id}_X, f_i)\colon X \to X \times_S X',$$

with $g_1 \circ i = g_2 \circ i = (i, i' \circ g)$, giving the commutative diagram

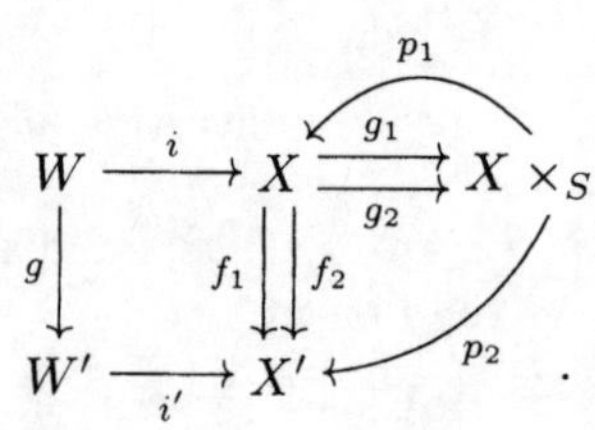

This shows that, in $\mathrm{Ho}\mathbf{E}$, the maps $g_1, g_2\colon (i\colon W \to X) \to ((i, i' \circ g)\colon W \to X \times_S X')$ are equal. Thus, in $\mathbf{E}(g)^*$,

$$(g, f_1 \circ p_1) = (g, f_2 \circ p_1)\colon (i, i' \circ g) \to i',$$

so $p_1\colon (i, i' \circ g) \to i$ equalizes the maps f_1 and f_2. □

2.4.6. By Lemma 2.4.5, the functor (2.4.4.1) descends to the functor

$$\mathbb{Z}^{\mathrm{B.M.}}_{(-)}\colon \mathbf{SDS}_{S\mathrm{proj}} \to \mathbf{D}^b_{\mathrm{mot}}(\mathbf{Sm}_S). \tag{2.4.6.1}$$

For W in $\mathbf{SDS}_S$, we call $\mathbb{Z}^{\mathrm{B.M.}}_W$ the *Borel-Moore motive* of W; for a proper map $g\colon W \to W'$, we write the map $\mathbb{Z}^{\mathrm{B.M.}}(g)$ as $g_*\colon \mathbb{Z}^{\mathrm{B.M.}}_W \to \mathbb{Z}^{\mathrm{B.M.}}_{W'}$.

Let $\mathbf{SDS}^{\mathrm{pr}}_{S\mathrm{proj}}$ be the full subcategory of $\mathbf{SDS}_{S\mathrm{proj}}$ with objects those W which admit a compactifiable closed embedding. Restricting $\mathbb{Z}^{\mathrm{B.M.}}_{(-)}$ to $\mathbf{SDS}^{\mathrm{pr}}_{S\mathrm{proj}}$ and dualizing gives the functor

$$\mathbb{Z}^{c/S}_{(-)}: (\mathbf{SDS}^{\mathrm{pr}}_{S\mathrm{proj}})^{\mathrm{op}} \to \mathbf{D}^b_{\mathrm{mot}}(\mathbf{Sm}_S). \tag{2.4.6.2}$$

We call $\mathbb{Z}^{c/S}_W$ the *motive of W with compact support over S*; for a proper map $g: W \to W'$, we write $\mathbb{Z}^{c/S}(g)$ as $g^!: \mathbb{Z}^{c/S}_{W'} \to \mathbb{Z}^{c/S}_W$.

2.4.7. *Open immersions.* Let $g: W \to W'$ be an open immersion with complement F. Take a closed embedding $i': W' \to X'$ with X' of dimension N over S, and let $i: W \to X := X' \setminus i'(F)$ be the restriction of i'. Let $j: X \to X'$ be the inclusion. Then j gives the map $j^*: \mathbb{Z}_{i'} \to \mathbb{Z}_i$. It follows from the compatibility of push-forward with transverse pull-back (Chapter III, Theorem 2.4.9), together with an argument similar to that of Lemma 2.4.5, that the map j^* induces a well-defined map $g^*: \mathbb{Z}^{\mathrm{B.M.}}_{W'} \to \mathbb{Z}^{\mathrm{B.M.}}_W$, independent of the choice of closed embeddings, and choice of extension of g to the open immersion j. In addition, we have

$$(g \circ g')^* = g'^* \circ g^*; \tag{2.4.7.1}$$

the same compatibility of push-forward with transverse pull-back gives the identity

$$f_* \circ g^* = g'^* \circ f'_* \tag{2.4.7.2}$$

for a cartesian diagram

$$\begin{array}{ccc} T' & \xrightarrow{g'} & T \\ {\scriptstyle f'}\downarrow & & \downarrow{\scriptstyle f} \\ W' & \xrightarrow[g]{} & W \end{array}$$

with g an open immersion, and f proper.

We let

$$g_!: \mathbb{Z}^{c/S}_W \to \mathbb{Z}^{c/S}_{W'}$$

be the dual of g^* (when defined). Dualizing the relations (2.4.7.1) and (2.4.7.2) gives functoriality

$$(g \circ g')_! = g_! \circ g'_!;$$

and compatibility in cartesian squares

$$g^! \circ f_! = f'^! \circ g'_!.$$

2.4.8. Remarks. (i) If W is already smooth and equi-dimensional over S, we may take the embedding $i: W \to X$ to be the identity map. From this, one sees that $\mathbb{Z}^{\mathrm{B.M.}}_W$ agrees with the definition of $\mathbb{Z}^{\mathrm{B.M.}}_W$ given in Chapter III, Definition 2.5.6, and the functor (2.4.6.1) is an extension of the functor $\mathbb{Z}^{\mathrm{B.M.}}$ of Theorem 2.5.7. Similarly, for W in $\mathbf{Sm}^{\mathrm{pr}}_S$, the definition (2.4.6.2) of $\mathbb{Z}^{c/S}_W$ agrees with that of Definition 2.2.2(ii).

(ii) Let $i: W \to X$ be a compactifiable embedding of a smoothly decomposable S-scheme with complement $j: U \to X$. Then $\mathbb{Z}^{\mathrm{B.M.}}_W$ is canonically isomorphic to the Borel-Moore motive with support, $\mathbb{Z}^{\mathrm{B.M.}}_{X,W}$, giving the distinguished triangle

$$\mathbb{Z}^{\mathrm{B.M.}}_W \xrightarrow{i_*} \mathbb{Z}^{\mathrm{B.M.}}_X \xrightarrow{j^*} \mathbb{Z}^{\mathrm{B.M.}}_U \to \mathbb{Z}^{\mathrm{B.M.}}_W[1].$$

Now $(j^*)^D = j_!$ (2.2.5.1), $i^! = i_*^D$ by definition, and duality is an exact involution (Theorem 1.2.5), so the motive $\mathbb{Z}_W^{c/S}$ fits into a distinguished triangle

$$\mathbb{Z}_U^{c/S} \xrightarrow{j_!} \mathbb{Z}_X^{c/S} \xrightarrow{i^!} \mathbb{Z}_W^{c/S} \to \mathbb{Z}_U^{c/S}[1].$$

We collect the results of the preceding paragraphs in the following omnibus theorem for future reference.

2.4.9. THEOREM. *(i) We have functors*

$$\mathbb{Z}^{\text{B.M.}}: \mathbf{SDS}_{S\text{proj}} \to \mathcal{DM}(S),$$
$$\mathbb{Z}^{c/S}: \mathbf{SDS}^{\text{prop}}_{S\text{proj}} \to \mathcal{DM}(S).$$

If W is in **SDS**, *and $i: W \to X$ is a closed embedding with X smooth of dimension d over S, there is a canonical isomorphism*

$$\mathbb{Z}_W^{\text{B.M.}} \cong \mathbb{Z}_{X,i(W)}(d)[2d].$$

If i is a compactifiable closed embedding, there is a canonical isomorphism

$$\mathbb{Z}_W^{c/S} \cong \mathbb{Z}^D_{X,i(W)}(-d)[-2d].$$

(ii) The functor $\mathbb{Z}^{\text{B.M.}}$ of (i) is an extension of the functor $\mathbb{Z}^{\text{B.M.}}: \mathbf{Sm}_{S\text{proj}} \to \mathcal{DM}(S)$ defined in Chapter III, Theorem 2.5.7, and the functor $\mathbb{Z}^{c/S}$ of (i) is an extension of the functor (2.2.3.2) $\mathbb{Z}^{c/S}: \mathbf{Sm}^{\text{op}}_{S,\text{proj}} \to \mathcal{DM}(S)$.
(iii) Let $j: W \to W'$ be an open immersion in $\mathbf{SDS}_S$. We have functorial pull-back maps

$$j^*: \mathbb{Z}_{W'}^{\text{B.M.}} \to \mathbb{Z}_W^{\text{B.M.}},$$

compatible with the proper push-forward maps in cartesian squares. If j is a map in $\mathbf{SDS}_S^{\text{pr}}$, we have functorial push-forward maps

$$j_!: \mathbb{Z}_W^{c/S} \to \mathbb{Z}_{W'}^{c/S},$$

compatible with the proper pullback maps in cartesian squares.

2.4.10. DEFINITION. (i) Let W be a smoothly decomposable S-scheme. The *motivic Borel-Moore homology of W* is defined by

$$H_p^{\text{B.M.}}(W, \mathbb{Z}(q)) = \text{Hom}_{\mathcal{DM}(S)}(1, \mathbb{Z}_W^{\text{B.M.}}(-q)[-p]).$$

(ii) Let W be a smoothly decomposable S-scheme which has a compactifiable closed embedding $i: W \to X$ into a smooth quasi-projective S-scheme X. The *motivic cohomology of W with compact support* is defined by

$$H^p_{c/S}(W, \mathbb{Z}(q)) = \text{Hom}_{\mathcal{DM}(S)}(1, \mathbb{Z}_W^{c/S}(q)[p]).$$

Since we have the duality isomorphism

$$\text{Hom}_{\mathcal{DM}(S)}(1, \mathbb{Z}_X^{\text{B.M.}}(-q)[-p]) \cong \text{Hom}_{\mathcal{DM}(S)}(\mathbb{Z}_X^{c/S}(q)[p], 1)$$

for X in $\mathbf{Sm}^{\text{pr}}$, the definition Definition 2.4.10 of Borel-Moore homology and compactly supported cohomology extends that given in Definition 2.2.2. It follows from Theorem 2.4.9 that the Borel-Moore homology is covariantly functorial for projective maps, and contravariantly functorial for open immersions; in addition, the pull-back and push-forward are compatible in cartesian squares. The dual statements follows for the compactly supported cohomology via the duality involution, using Remark 2.4.8(ii).

The cup and cap products for Borel-Moore homology and compactly supported cohomology defined in §2.2.5 extend in the obvious way to the singular case whenever all the groups are defined; one applies Mayer-Vietoris and the Künneth isomorphism to give canonical isomorphisms $\mathbb{Z}_W^{\mathrm{B.M.}} \otimes \mathbb{Z}_{W'}^{\mathrm{B.M.}} \to \mathbb{Z}_{W\times_S W'}^{\mathrm{B.M.}}$, and then takes the inverse of the dual to give canonical isomorphisms $\mathbb{Z}_W^{c/S} \otimes \mathbb{Z}_{W'}^{c/S} \to \mathbb{Z}_{W\times_S W'}^{c/S}$; the remainder of the construction of cup and cap product then proceeds formally the same way as the smooth case, as does the external products. The various properties: functoriality, projection formula, etc. described in §2.2 also extend without trouble. In particular, there is a functorial bi-graded ring structure on the compactly supported cohomology, and Borel-Moore homology is a bi-graded module for the compactly supported cohomology ring.

Additionally, suppose we have an S-morphism $f: W \to Y$, with Y smooth over S, and W smoothly decomposable over S. If we embed W as a closed subscheme of some X in $\mathbf{Sm}_S$, $i: W \hookrightarrow X$, then we have the embedding $(i, f): W \hookrightarrow X \times_S Y$. Via Theorem 2.4.9, we have the canonical identification of $\mathbb{Z}_W^{\mathrm{B.M.}}$ with the motive with support, $\mathbb{Z}_W^{\mathrm{B.M.}} \cong \mathbb{Z}_{W, X\times_S Y}(N)[2N]$, where N is the dimension of $X \times_S Y$ over S. Combining the cup product

$$\mathbb{Z}_{X\times_S Y}(q) \otimes \mathbb{Z}_{W, X\times_S Y}(q' + N)[2N] \to \mathbb{Z}_{W, X\times_S Y}(q + q'N)[2N]$$

with the pull-back $f^*: \mathbb{Z}_Y \to \mathbb{Z}_{X\times_S Y}$ gives the cap product

$$\cap_f : H^p(Y, \mathbb{Z}(q)) \otimes H_{p'}^{\mathrm{B.M.}}(W, \mathbb{Z}(q')) \to H_{p'-p}^{\mathrm{B.M.}}(W, \mathbb{Z}(q' - q)). \tag{2.4.10.1}$$

One shows, as in the proof of Lemma 2.4.5, that $\cap_f$ is independent of the choice of embedding i.

2.5. Riemann-Roch for singular schemes

Using the constructions of §2.4, we give a version of the Riemann-Roch theorem for the K-theory of coherent sheaves, á la [**8**]. In this section, we assume that the base scheme S is a *regular* scheme.

2.5.1. *K-theory of coherent sheaves.* Let X be a scheme. We have the exact category $\mathcal{M}_X$ of coherent sheaves on X, giving the Quillen K-theory space $\mathrm{B}Q\mathcal{M}_X$ (see Appendix B, §1.2). We write $G_p(X)$ for the homotopy groups $G_p(X) := \pi_p(\mathrm{B}Q\mathcal{M}_X)$.

We will require the following properties of $G_*(X)$, for proofs, see [**102**]:
(i) *Functoriality:* Let $f: X \to Y$ be a map of schemes. If f is flat, the pull-back functor $f^*: \mathcal{M}_Y \to \mathcal{M}_X$ is exact, hence induces the map $f^*: G_p(Y) \to G_p(X)$. This makes $G_p(-)$ a contravariant functor from the category of schemes, with morphisms being flat maps, to abelian groups. Similarly, if f is finite, the push-forward $f_*: \mathcal{M}_X \to \mathcal{M}_Y$ is exact, giving the map $f_*: G_p(X) \to G_p(Y)$, making $G_p(-)$ a functor from schemes with finite morphisms, to abelian groups. One extends this covariant functoriality from finite morphisms to projective morphisms by the methods of [**102**].
(ii) *Products:* Tensor product $\otimes: \mathcal{P}_X \otimes \mathcal{M}_X \to \mathcal{M}_X$ is an exact pairing of exact categories, hence Waldhausen's products (see [**126**, p. 235], [**128**, §3], as well as Appendix B, §2.2.4) give the graded group $G_*(X)$ the structure of a graded $K_*(X)$-module; we write the resulting product as $\cap_X$. One has the *projection formula:*

$$f_*(f^*(\alpha) \cap_X \beta) = \alpha \cap_Y f_*(\beta)$$

for $f: X \to Y$ a projective morphisms, $\alpha \in K_p(Y)$ and $\beta \in G_q(X)$.

Similarly, for flat S-schemes X and Y, we have the exact pairing of exact categories $\boxtimes_{X,Y}: \mathcal{M}_X \otimes \mathcal{M}_Y \to \mathcal{M}_{X\times_S Y}$ defined by sending (M, N) to $p_1^*M \otimes_{\mathcal{O}_{X\times_S Y}} p_2^*N$. This induces as above the *external products*

$$\boxtimes_{X,Y}: G_p(X) \otimes G_q(Y) \to G_{p+q}(X \times_S Y).$$

(iii) *Localization:* Let $i: Z \to X$ be the inclusion of a closed subscheme, with complement $j: U \to X$. Let $BQ\mathcal{M}_{X,U}$ denote the homotopy fiber of the map $j^*: BQ\mathcal{M}_X \to BQ\mathcal{M}_U$. The composition $j^* \circ i_*: BQ\mathcal{M}_Z \to BQ\mathcal{M}_U$ is canonically contractible, hence we have the lifting of $i_*: BQ\mathcal{M}_Z \to BQ\mathcal{M}_X$ to the map $i_*: BQ\mathcal{M}_Z \to BQ\mathcal{M}_{X,U}$. This latter map is a weak homotopy equivalence, giving the isomorphism

$$i_*: G_p(Z) \to K_p^Z(X), \tag{2.5.1.1}$$

and the long exact *localization* sequence

$$\ldots \to G_p(Z) \xrightarrow{i_*} \mathbb{G}_p(X) \xrightarrow{j^*} G_p(U) \xrightarrow{\delta} G_{p-1}(Z) \to \ldots .$$

(iv) *Poincaré duality:* If X is a regular scheme, then the inclusion $\mathcal{P}_X \to \mathcal{M}_X$ induces an isomorphism $K_p(X) \cong G_p(X)$.

(v) *Projective bundle formula:* Let $E \to X$ be a rank N vector bundle, with associated projective bundle $q: \mathbb{P}(E) \to X$. Let $\mathcal{O}(1)$ be the tautological quotient bundle, and $\mathcal{O}(-1)$ its dual. Let $\alpha_i: G_p(X) \to G_p(\mathbb{P}(E))$ be the map $\alpha_i(x) := [\mathcal{O}(-i)] \cap_{\mathbb{P}(E)} q^*(x)$, where $[\mathcal{O}(-i)]$ is the K_0-class. Then

$$\sum_{i=0}^{N} \alpha_i : \oplus_{i=0}^N G_p(X) \to G_p(\mathbb{P}(E))$$

is an isomorphism.

2.5.2. *The map* τ. We recall from §2.4.4 the category $\mathbf{SDS}_S$ of smoothly decomposable S-schemes, and the subcategory $\mathbf{SDS}_{S\text{proj}}$ with the same objects, and with morphisms being the projective morphisms over S. For X in $\mathbf{SDS}_S$, we let $\prod_q' H_{p+2q}^{\text{B.M.}}(-, \mathbb{Q}(q))$ denote the subgroup of the full product consisting of sequences

$$(\ldots, \alpha_q \in H_{p+2q}^{\text{B.M.}}(-, \mathbb{Q}(q)), \ldots)$$

such that there is a q_0 with $\alpha_q = 0$ for all $q \geq q_0$.

Take W in $\mathbf{SDS}_S$, and choose an embedding $i: W \to X$ with X in $\mathbf{Sm}_S$, of dimension d over S. We then have the isomorphism (2.5.1.1) $i_*: G_p(W) \to K_p^W(X)$, and the isomorphism

$$\mathbb{Z}_W^{\text{B.M.}} \cong \mathbb{Z}_{X,W}(d)[2d]. \tag{2.5.2.1}$$

We have as well the Chern character for K-theory with support

$$\text{ch}_{p,X}^W: K_p^W(X) \to \prod_{q\geq 0} H_W^{2q-p}(X, \mathbb{Q}(q)),$$

and the Todd class $\text{Todd}(X/S) \in \prod_{q\geq 0} H^{2q}(X, \mathbb{Q}(q))$ (see Chapter III, §3.3.6 and §3.5.1).

We define

$$\tau_p^X: G_p(W) \to \prod_q' H_{p+2q}(W, \mathbb{Q}(q)) \tag{2.5.2.2}$$

to be the cap product (2.4.10.1)

$$\mathrm{Todd}(X/S) \cap_{i:W\to X} \mathrm{ch}^W_{p,X} : G_p(W) \to \prod_{q\le d} H_{p+2q}(W, \mathbb{Q}(q)),$$

where we identify $\prod_{q\ge 0} H_W^{2q-p}(X, \mathbb{Q}(q))$ with $\prod_{q\le d} H_{p+2q}(W, \mathbb{Q}(q))$ via the canonical isomorphism (2.5.2.1), and identify $K_p^W(X)$ with $G_p(W)$ via the isomorphism (2.5.1.1).

2.5.3. THEOREM [Riemann-Roch for singular varieties]. *Suppose that S is a regular scheme. Then the maps (2.5.2.2) for W in* $\mathbf{SDS}_S$ *are independent of the choice of embedding, and define a natural transformation*

$$\tau_p : G_p(-) \to {\prod_q}' H^{\mathrm{B.M.}}_{p+2q}(-, \mathbb{Q}(q))$$

of functors from $\mathbf{SDS}_{S\mathrm{proj}}$ *to* $\mathbf{Ab}$*, satisfying the following conditions:*

1. *Let $f: X \to Y$ be a morphism in* $\mathbf{Sch}_S$*, with X smoothly decomposable over S, and with Y in* $\mathbf{Sm}_S$*. Then for $\alpha \in K_p(Y)$ and $\beta \in G_q(X)$, we have*
$$\tau_{p+q}(f^*(\alpha) \cap_X \beta) = \mathrm{ch}(\alpha) \cap_f \tau_q(\beta).$$
2. *For X and Y in* $\mathbf{SDS}_S$*, $\alpha \in G_p(X)$ and $\beta \in G_q(Y)$, we have*
$$\tau_{p+q}(\alpha \boxtimes_{X,Y} \beta) = \tau_p(\alpha) \boxtimes_{X,Y} \tau_q(\beta).$$
3. *If $j: U \to X$ is an open immersion of smoothly decomposable S-schemes, then, for $\alpha \in G_p(X)$, we have*
$$j^*\tau_p(\alpha) = \tau_p(j^*\alpha).$$
4. *If X is in* $\mathbf{Sm}_S$ *of dimension d over S, then we have the class $[\mathcal{O}_X]$ in $G_0(X)$ and the fundamental cycle class $\mathrm{cl}^0_X(|X|)$ in $H^{\mathrm{B.M.}}_{2d}(X, \mathbb{Z}(d)) = H^0(X, \mathbb{Z}(0))$. Then*
$$\tau_0([\mathcal{O}_X]) = \mathrm{Todd}(X/S) \cap_X \mathrm{cl}^0_X(|X|)$$
(see Chapter III, §3.5.1).

PROOF. Once we have shown that the maps τ_p^X are independent of the choice of X, the naturality of τ_p follows directly from the Riemann-Roch theorem. Indeed, given a projective morphism in $\mathbf{SDS}_S$ $f: W \to T$, we may assume that W is a closed subscheme of $\mathbb{P}^N_T$, and that f is the restriction of the projection $q: \mathbb{P}^N_T \to T$. If we embed T as a closed subscheme of some X in $\mathbf{Sm}_S$, then we have the embedding of W as a closed subscheme of $\mathbb{P}^N_X$, as well as the commutative diagram

$$\begin{array}{ccc} W & \subset & \mathbb{P}^N_X \\ {\scriptstyle f}\downarrow & & \downarrow{\scriptstyle q} \\ T & \subset & X. \end{array}$$

From Theorem 3.5.4, we have

$$q_* \circ (\mathrm{Todd}(\mathbb{P}^N_X/S) \cup \mathrm{ch}^W_{\mathbb{P}^N_X}(-)) = \mathrm{Todd}(X/S) \cup \mathrm{ch}^T(q_*(-))$$

as maps from $K^W_*(\mathbb{P}^N_X))$ to $H^*_T(X, \mathbb{Z}(*))$, which implies the desired naturality.

On the other hand, the argument of Lemma 2.4.3, together with the above naturality statement, implies the independence on the choice of embedding.

Indeed, suppose we have another closed embedding $i': W \to X'$. If i' factors through a closed embedding $s: X \to X'$, then the Riemann-Roch theorem for K-theory with support (Chapter III, Theorem 3.5.4) gives the identity

$$s_*(\mathrm{Todd}(X/S) \cup \mathrm{ch}^W_{p,X}(-)) = \mathrm{Todd}(X'/S) \cup \mathrm{ch}^W_{p,X'}(s_*(-))$$

in $\prod_{q \geq 0} H^{2q-p}_W(X', \mathbb{Z}(q))$; from Lemma 2.4.3, this implies that $\tau^X_p = \tau^{X'}_p$. Similarly, if i' factors through an open immersion $j: X \to X'$, then excision, the naturality of Todd and ch, and Lemma 2.4.3 imply $\tau^X_p = \tau^{X'}_p$.

We may therefore assume that X and X' are open subschemes of projective spaces over S, say X' is an open subscheme of $\mathbb{P}^n_S$. We then have the diagonal embedding

$$W \to X \times_S X' \hookrightarrow X \times_S \mathbb{P}^n_S = \mathbb{P}^n_X,$$

and W is still closed in $\mathbb{P}^n_X$. Riemann-Roch for the projection $q: \mathbb{P}^n_X \to X$ gives the commutativity of the diagram

$$\begin{array}{ccc} K^W_p(\mathbb{P}^n_X) & \xrightarrow{q_*} & K^W_*(X) \\ \downarrow{\scriptstyle \tau^{\mathbb{P}^n_X}_p} & & \downarrow{\scriptstyle \tau^X_p} \\ \prod_{q \geq 0} H^{2q-p}_W(\mathbb{P}^n_X, \mathbb{Q}(q)) & \xrightarrow[q_*]{} & \prod_{q \geq 0} H^{2q-p}_W(\mathbb{P}^n_X, \mathbb{Q}(q)). \end{array}$$

Combining this with the case of the open immersion $X \times_S X' \to \mathbb{P}^N_X$, Lemma 2.4.3, and symmetry, gives the desired independence of the choice of embedding.

For (2), we may assume, as in the definition of $\cap_f$, that f is a closed embedding. Then, by the definition of τ_*, we have

$$\begin{aligned} \tau_{p+q}(f^*(\alpha) \cap_X \beta) &= \mathrm{Todd}(Y/S) \cup \mathrm{ch}^X_{Y,p+q}(\alpha \cup^X_Y \beta) \\ &= \mathrm{Todd}(Y/S) \cup [\mathrm{ch}_{Y,p}(\alpha) \cap_X \mathrm{ch}^X_{Y,q}(\beta)] \\ &= \mathrm{ch}_{Y,p}(\alpha) \cup^X_Y [\mathrm{Todd}(Y/S) \cup^X_Y \mathrm{ch}^X_{Y,q}(\beta)] \\ &= \mathrm{ch}_{Y,p}(\alpha) \cap_X \tau_q(\beta). \end{aligned}$$

Here $\cup^X_Y$ is the cup product $H^p(Y, \mathbb{Q}(q)) \otimes H^{p'}_X(Y, \mathbb{Q}(q')) \to H^{p+p'}_X(Y, \mathbb{Q}(q+q'))$; we identify Borel-Moore homology with cohomology with support in X (as modules over $H^*(Y, \mathbb{Q}(*))$) via the isomorphism (2.5.2.1), and we use the multiplicativity of the Chern character (see Chapter III, Remark 3.3.10(ii)). This proves (2). The proof of (3) is similar, where we reduce to the case of Riemann-Roch for K-theory with support by first embedding X and Y in smooth S-schemes.

The assertion (4) follows from the definition of τ_*, using the naturality of ch and Todd, and (5) follows from the fact that $\mathrm{ch}_{X,0}([\mathcal{O}_X]) = 1$, and that, as $\mathrm{cl}^0_X(|X|)$ is the unit in $H^*(X, \mathbb{Z}(*))$ (Chapter I, Proposition 3.5.6), the cap product with $\mathrm{cl}^0_X(|X|)$ is just the identification of motivic cohomology of X with Borel-Moore homology, via (2.5.2.1). □

2.6. The triangulated Tate motivic category

We give the definition of the triangulated Tate motivic category $\mathcal{DTM}(S)_R$ as a subcategory of $\mathcal{DM}(S)_R$. If the base scheme S is $\mathrm{Spec}\, k$ for k a field, we show that $\mathcal{DTM}(S)_R$ is equivalent to a subcategory of both $\mathbf{D}^b_{\mathrm{mot}}(\mathbf{Sm}_k)_R$ and $\mathbf{D}^b_{\mathrm{mot}}(\mathbf{Sm}_k)^0_R$.

In addition, the duality involution on $\mathcal{DM}(\mathbf{Sm}_S)_R$ restricts to a duality involution on $\mathcal{DTM}(S)_R$.

2.6.1. DEFINITION. Let S be a reduced scheme and R a commutative ring, flat over $\mathbb{Z}$. The *triangulated Tate motivic category* $\mathcal{DTM}(S)_R$ is the strictly full triangulated tensor subcategory of $\mathcal{DM}(S)_R$ generated by the objects $R_S(q)$, $q = 0, \pm 1$.

2.6.2. LEMMA. *The category $\mathcal{DTM}(S)_R$ is equal to the strictly full triangulated subcategory of $\mathcal{DM}(S)_R$ generated by the objects $R_S(q)$, $q \in \mathbb{Z}$.*

PROOF. This follows immediately from the exactness of the tensor product operation in $\mathcal{DM}(S)$, and the Künneth isomorphism $R_S(a) \otimes R_S(b) \cong R_S(a+b)$ (see Chapter I, Definition 2.1.4(c)). □

2.6.3. PROPOSITION. *The duality involution $(-)^D : \mathcal{DM}(S)_R^{\mathrm{op}} \to \mathcal{DM}(S)_R$ (Theorem 1.4.2) restricts to an involution*

$$(-)^D : \mathcal{DTM}(S)_R^{\mathrm{op}} \to \mathcal{DTM}(S)_R.$$

PROOF. We have $\mathbb{Z}_S^D = 1^D = 1 = \mathbb{Z}_S$, hence $\mathbb{Z}_S(q)^D = \mathbb{Z}_S(-q)$ for each integer q. Since the involution $(-)^D$ is exact (Theorem 1.4.2), this implies that the strictly full triangulated subcategory of $\mathcal{DM}(S)^{\mathrm{op}}$ is mapped into its opposite by $(-)^D$. Applying Lemma 2.6.2 completes the proof. □

2.6.4. We recall the graded tensor category $\mathcal{A}^0_{\mathrm{mot}}(\mathcal{V})$ (Chapter I, Definition 1.4.12), and the DG tensor functor $H_{\mathrm{mot}} : \mathcal{A}_{\mathrm{mot}}(\mathcal{V}) \to \mathcal{A}^0_{\mathrm{mot}}(\mathcal{V})(I.1.4.12.1)$. We have as well the triangulated tensor categories $\mathbf{D}^{b0}_{\mathrm{mot}}(\mathcal{V})$ and $\mathcal{DM}^0(\mathcal{V})_R$ formed from $\mathcal{A}^0_{\mathrm{mot}}(\mathcal{V})$ in a way paralleling the construction of $\mathbf{D}^b_{\mathrm{mot}}(\mathcal{V})$ and $\mathcal{DM}(\mathcal{V})$ from $\mathcal{A}_{\mathrm{mot}}(\mathcal{V})$ (see §1.5.2 and Chapter I, Remark 3.4.7). In particular, we have the commutative diagram

$$\begin{array}{ccccc}
Z^0\mathcal{A}_{\mathrm{mot}}(\mathcal{V}) & \longrightarrow & \mathbf{D}^b_{\mathrm{mot}}(\mathcal{V}) & \longrightarrow & \mathcal{DM}(\mathcal{V}) \\
\Big\downarrow {\scriptstyle Z^0 H_{\mathrm{mot}}} & & \Big\downarrow {\scriptstyle \mathbf{D}^b_{\mathrm{mot}}(H_{\mathrm{mot}})} & & \Big\downarrow {\scriptstyle \mathcal{DM}(H_{\mathrm{mot}})} \\
Z^0\mathcal{A}^0_{\mathrm{mot}}(\mathcal{V}) & \longrightarrow & \mathbf{D}^{b0}_{\mathrm{mot}}(\mathcal{V}) & \longrightarrow & \mathcal{DM}^0(\mathcal{V})_R.
\end{array}$$

If we take $\mathcal{V} = \mathbf{Sm}_k$ for k a field, then, if, e.g., we have resolution of singularities for k-varieties, it follows from Theorem 1.5.5 that the functors $\mathbf{D}^b_{\mathrm{mot}}(H_{\mathrm{mot}})$ and $\mathcal{DM}(H_{\mathrm{mot}})$ are equivalences.

Let $\mathcal{DT}(S)_R$ be the full triangulated tensor subcategory of $\mathbf{D}^b_{\mathrm{mot}}(\mathbf{Sm}_S)_R$ generated by the objects $R_S(q)$, $q = 0, \pm 1$, and let $\mathcal{DT}(S)^0_R$ be the full triangulated tensor subcategory of $\mathbf{D}^b_{\mathrm{mot}}(\mathbf{Sm}_S)^0_R$ generated by the objects $R_S(q)$, $q = 0, \pm 1$. Similarly, let $\mathcal{DTM}(S)^0_R$ be the full triangulated tensor subcategory of $\mathcal{DM}(S)^0_R$ generated by the objects $R_S(q)$, $q = 0, \pm 1$.

2.6.5. THEOREM. *The functors*

$$\mathbf{D}^b_{\mathrm{mot}}(\mathbf{Sm}_S)_R \to \mathcal{DM}(S)_R,$$
$$\mathbf{D}^{b0}_{\mathrm{mot}}(\mathbf{Sm}_S)_R \to \mathcal{DM}(S)^0_R$$

induce equivalences

$$\mathcal{DT}(S)_R \to \mathcal{DTM}(S)_R,$$
$$\mathcal{DT}(S)^0_R \to \mathcal{DTM}(S)^0_R.$$

Under the hypothesis of Theorem 1.5.5, the functors $\mathbf{D}^b_{\mathrm{mot}}(H_{\mathrm{mot}})$ *and* $\mathcal{DM}(H_{\mathrm{mot}})$ *induce equivalences*

$$\mathcal{DT}(S)_R \to \mathcal{DT}(S)^0_R,$$
$$\mathcal{DTM}(S)_R \to \mathcal{DTM}(S)^0_R.$$

PROOF. As the objects $R_S(q)$ generating $\mathcal{DTM}(S)_R$ are in $\mathcal{DT}(S)_R$, and as the functor $\mathbf{D}^b_{\mathrm{mot}}(\mathbf{Sm}_S)_R \to \mathcal{DM}(S)_R$ is a fully faithful embedding, the categories $\mathcal{DT}(S)_R$ and $\mathcal{DTM}(S)_R$ are equivalent. The second pair of equivalences follows from Theorem 1.5.5. □

2.6.6. *Functoriality.* We recall from Chapter I, §2.3 that the formation of the category $\mathcal{DM}(S)$ is functorial in S. If $p: T \to S$ is a map of reduced schemes, the functor $\mathcal{DM}(p^*): \mathcal{DM}(S)_R \to \mathcal{DM}(T)_R$ induces the functor $\mathcal{DTM}(p^*): \mathcal{DTM}(S) \to \mathcal{DTM}(T)$. This determines the functor

$$\mathcal{DTM}(-)_R: \mathbf{Sch} \to \mathbf{TT}_R$$
$$S \mapsto \mathcal{DTM}(S_{\mathrm{red}})_R$$

from the category of schemes to the category of triangulated rigid R-tensor categories.

2.6.7. REMARKS. (i) As the Tate motives $\mathbb{Z}_S(q)$ are in $\mathcal{DTM}(S)$, one can recover the motivic cohomology of S entirely from within $\mathcal{DTM}(S)$:

$$H^p(S, \mathbb{Z}(q)) = \mathrm{Hom}_{\mathcal{DTM}(S)}(1, \mathbb{Z}_S(q)[p]).$$

For $S = \mathrm{Spec}\, F$, F a field, we thus have the natural isomorphism

$$\mathrm{Hom}_{\mathcal{DTM}(S)_{\mathbb{Q}}}(1, \mathbb{Q}_S(q)[2q-p]) \cong K_p(F)^{(q)}. \tag{2.6.7.1}$$

This follows from Theorem 3.6.6 and Theorem 3.6.12 of Chapter II.
(ii) Beilinson and Soulé (see [**114**]) have conjectured that, for a field F, one has the following vanishing of the weight-graded pieces of the K-theory of F:

$$K_p(F)^{(q)} = 0; \quad \text{for } 2q \le p,\ p > 0. \tag{2.6.7.2}$$

It follows from (i) and [**85**] that, assuming the vanishing (2.6.7.2), the triangulated category $\mathcal{DTM}(\mathrm{Spec}\, F)_{\mathbb{Q}}$ has a canonical t-structure, with heart $\mathcal{MTM}(F)_{\mathbb{Q}}$ containing the Tate motives $\mathbb{Q}(q)$; in fact, an abelian subcategory of $\mathcal{MTM}(F)_{\mathbb{Q}}$ containing the $\mathbb{Q}(q)$ and closed under extensions is all of $\mathcal{MTM}(F)_{\mathbb{Q}}$. The duality involution makes $\mathcal{MTM}(F)_{\mathbb{Q}}$ into a rigid tensor category, and the natural map $\mathrm{Ext}^1_{\mathcal{MTM}(F)_{\mathbb{Q}}}(1, \mathbb{Q}(q)) \to \mathrm{Hom}_{\mathcal{DTM}(F)_{\mathbb{Q}}}(1, \mathbb{Q}(q)[1])$ is an isomorphism. There is a natural weight filtration on $\mathcal{MTM}(F)_{\mathbb{Q}}$, and a fiber functor to graded $\mathbb{Q}$-vector spaces, making $\mathcal{MTM}(F)_{\mathbb{Q}}$ a Tannakian category (again, assuming the Beilinson-Soulé vanishing conjectures).

It is not known if the Beilinson-Soulé vanishing conjectures suffice to imply that the natural map

$$\mathrm{Ext}^p_{\mathcal{MTM}(F)_{\mathbb{Q}}}(1, \mathbb{Q}(q)) \to \mathrm{Hom}_{\mathcal{DTM}(F)_{\mathbb{Q}}}(1, \mathbb{Q}(q)[p]) \tag{2.6.7.3}$$

is an isomorphism for all p, however, surjectivity in (2.6.7.3) for all p implies injectivity. If F is a subfield of $\bar{\mathbb{Q}}$, then the vanishing of $K_{2q-p}(F)^{(q)}$ for for $q > 0$ and $p \ne 1$ (see [**22**] and [**23**]) and (i) imply that the Beilinson-Soulé vanishing conjectures are valid for F, and that both sides of (2.6.7.3) are zero for $q > 0$ and $p > 1$. This gives a construction of the category of mixed Tate motives over F, for F a subfield of $\mathbb{Q}$. It is not known, even in this case, if the inclusion of $\mathcal{MTM}(F)_{\mathbb{Q}}$ into $\mathcal{DTM}(F)_{\mathbb{Q}}$ induces an equivalence of the bounded derived category of $\mathcal{MTM}(F)_{\mathbb{Q}}$ with $\mathcal{DTM}(F)_{\mathbb{Q}}$.

3. Motives over a perfect field

In this section, we take the base scheme S to be of the form $S = \operatorname{Spec} k$, with k a perfect field admitting resolution of singularities for reduced k-schemes of finite type. We take this to mean:

1. Let X be a reduced k-scheme of finite type. Then there is a sequence of blow-ups with smooth center $Y \to X$ such that Y is smooth.
2. Let D be a closed codimension one subset of a smooth finite type k-scheme X. Then there is a sequence of blow-ups with smooth center $\mu: Y \to X$ such that $\mu^{-1}(D)_{\mathrm{red}}$ is a divisor with normal crossing.
3. Let $f: X \to Z$ a rational map of reduced finite type k-schemes, i.e., there is an open subscheme U of X, containing each generic point of X, and a morphism $f_U: U \to Z$. Then there is a sequence of blow-ups of X with smooth centers lying over $X \setminus U$, $\mu: Y \to X$, such that the rational map $f \circ \mu: Y \to Z$ is a morphism.

A strong version of Chow's lemma is an immediate consequence of resolution of singularities, in other words, if X is a reduced finite type k-scheme, there is a sequence of blow-ups with smooth centers $Y \to X$ such that Y is smooth and quasi-projective over k. By [**66**], a field of characteristic zero admits resolution of singularities.

Using the method of hyperresolutions constructed in [**60**], Hanamura [**61**] has shown how to extend the definition of the motive (in the sense of [**61**]) of smooth quasi-projective varieties to finite type k-schemes, assuming one has resolution of singularities for finite type k-schemes. In this section, we show how to apply his methods to give the motive of a finite type k-scheme as an object of $\mathbf{D}^b_{\mathrm{mot}}(\mathbf{Sm}_k)$. We then give a description of the fundamental properties of (cohomological) motives of singular schemes, a construction of the dual homological motive, and a comparison with the motive with compact support and the Borel-Moore motive.

3.1. Hyperresolutions

We first prove a result about the motive of a blow-up along a smooth center. We let $\mathbf{Sch}_k^{\mathrm{fin}}$ denote the category of reduced finite type k-schemes.

3.1.1. *Blow-up distinguished triangle.* Let X be in $\mathbf{Sm}_S$, and C a smooth closed subscheme of X. Let $\mu: Y \to X$ be the blow-up of X along C, with exceptional divisor E. The map μ induces the commutative diagram of pull-back maps

$$\begin{array}{ccc} \mathbb{Z}_X & \xrightarrow{i_C^*} & \mathbb{Z}_C \\ {\scriptstyle \mu^*}\downarrow & & \downarrow{\scriptstyle \mu_{|E}^*} \\ \mathbb{Z}_Y & \xrightarrow[i_E^*]{} & \mathbb{Z}_E, \end{array}$$

which induces the map

$$\mathbb{Z}_X \xrightarrow{(\mu^*, i_C^*)} \operatorname{cone}\left(\mathbb{Z}_Y \oplus \mathbb{Z}_C \xrightarrow{i_E^* - \mu_{|E}^*} \mathbb{Z}_E\right)[-1]. \tag{3.1.1.1}$$

3.1.2. LEMMA. *The map (3.1.1.1) is an isomorphism in* $\mathbf{D}^b(\mathbf{Sm}_S)$.

PROOF. Using the Mayer-Vietoris property on X, we may assume that C has trivial normal bundle in X.

Let $U := X \setminus C = Y \setminus E$, and let d be the codimension of C in X. We have the identification of $\mu_{|E}: E \to C$ with the projectivized normal bundle $q: \mathbb{P}(N_{C/X}) \to C$, i.e., E is a projective space $\mathbb{P}^{d-1}_C$.

Let $\zeta \in H^2(E, \mathbb{Z}(1))$ be the first Chern class of the tautological bundle $\mathcal{O}(1)$ on E. We have the isomorphism $\mathcal{O}(1) \cong \mathcal{O}_E(-E)$.

We claim that the diagram

$$\begin{array}{ccc} \mathbb{Z}_C(-d)[-2d] & \xrightarrow{i_{C*}} & \mathbb{Z}_X \\ {\scriptstyle \cup\zeta^{d-1}\circ\mu_{|E}^*}\downarrow & & \downarrow{\scriptstyle \mu^*} \\ \mathbb{Z}_E(-1)[-2] & \xrightarrow{i_{E*}} & \mathbb{Z}_Y \end{array}$$

commutes.

Indeed, by taking the deformation to the normal bundle, we may assume that

$$X = \bar{N}_{C/X} := \mathbb{P}(N_{C/X} \oplus 1_X) \cong \mathbb{P}^d_C,$$

with i_C the section $i_0: C \to \mathbb{P}^d_C$ with constant value $(1 : 0 : \ldots : 0)$.

The projection $p: \mathbb{P}^d_C \to C$ induces the splitting $p_Y : Y \to E$ to the inclusion of E in Y. Thus, by (Chapter III, Lemma 2.2.7), we have

$$\begin{aligned} i_{C*} &= \cup \mathrm{cl}^d_{\mathbb{P}^d_C}(|i_0(C)|) \circ p^*, \\ i_{E*} = \cup \mathrm{cl}^1(|E|) \circ p_Y^* &= \cup(\mathrm{cl}^1_Y(c_1(\mathcal{O}(E)))), \end{aligned}$$

hence, by the functoriality of cycle classes,

$$\begin{aligned} \mu^* \circ i_{C*} &= \mu^* \circ (\cup \mathrm{cl}^d_{\mathbb{P}^d_C}(|i_0(C)|) \circ p^*) \\ &= \cup(\mu^*(\mathrm{cl}^d_{\mathbb{P}^d_C}(|i_0(C)|))) \circ p_Y^* \circ \mu_{|E}^*. \end{aligned}$$

On the other hand, $i_0(C)$ is the zero subscheme of a section of the vector bundle $\mathcal{O}_{\mathbb{P}^d_C}(1)^d$, hence by Corollary 1.3.9 of Chapter III, we have $\mathrm{cl}^d_{\mathbb{P}^d_C}(|i_0(C)|) = c_d(\mathcal{O}_{\mathbb{P}^d_C}(1)^d)$. Letting $i_1: C \to \mathbb{P}^d_C$ be the section with constant value $(0 : 1 : \ldots : 0)$, we have $\mathrm{cl}^d_{\mathbb{P}^d_S}(|i_1(C)|) = c_d(\mathcal{O}_{\mathbb{P}^d_C}(1)^d)$ for the same reason, hence $\mathrm{cl}^d_{\mathbb{P}^d_S}(|i_1(C)|) = \mathrm{cl}^d_{\mathbb{P}^d_S}(|i_0(C)|)$. As μ is an isomorphism over a neighborhood of $i_1(C)$, we have a lifting of i_1 to a section $\tilde{i}_1: C \to Y$, and thus have

$$\mu^* \circ i_{C*} = \cup \mathrm{cl}^d_Y(|\tilde{i}_1(C)|) \circ p_Y^* \circ \mu_{|E}^*.$$

Let l be the $\mathbb{P}^1_C$ in $\mathbb{P}^d_C$ through the points $(0 : 1 : \ldots : 0)$ and $(1 : 0 : \ldots : 0)$, and let l' be the proper transform of l to Y. The intersection $l' \cap E$ determines a section $p: C \to l'$. We may view $\tilde{i}_1$ as a section $\tilde{i}_1: C \to l'$. Since $p(C)$ and $\tilde{i}_1(C)$ are both zero subschemes (on l') of the tautological line bundle $\mathcal{O}_{l'}(1)$, we have $\mathrm{cl}^1_{l'}(|p(C)|) = \mathrm{cl}^1_{l'}(|\tilde{i}_1(C)|)$ by (*loc. cit.*). From Theorem 2.2.3 of Chapter III, it follows that $\mathrm{cl}^d_Y(|\tilde{i}_1(C)|) = \mathrm{cl}^d_Y(|p(C)|)$, so

$$\mu^* \circ i_{C*} = \cup \mathrm{cl}^d_Y(|p(C)|) \circ p_Y^* \circ \mu_{|E}^*.$$

Since $\zeta = c_1(\mathcal{O}_E(1))$, we have $\mathrm{cl}_E^{d-1}(|p(C)|) = \zeta^{d-1}$. Thus, using Theorem 2.2.3 of Chapter III and the projection formula, we have

$$\begin{aligned}\mu^* \circ i_{C*} &= \cup[i_{E*}(\zeta^{d-1})] \circ p_Y^* \circ \mu^*_{|E} \\ &= i_{E*} \circ (\cup(\zeta^{d-1}) \circ \mu^*_{|E}).\end{aligned}$$

This verifies the claim.

Thus we have the distinguished triangle

$$\mathbb{Z}_C(-d)[-2d] \xrightarrow{(i_{C*},(-1)^{d-1}\zeta^{d-1}\cup\circ\mu^*_{|E})} \mathbb{Z}_X \oplus \mathbb{Z}_E \to \mathbb{Z}_Y \to$$

Via the projective bundle formula, we have $\mathbb{Z}_E \cong \oplus_{i=0}^{d-1}\mathbb{Z}_C(-i)[-2i]$, which gives us the isomorphism

$$\mathbb{Z}_Y \cong \mathbb{Z}_X \bigoplus \oplus_{i=1}^{d-1}\mathbb{Z}_C(-i)[-2i].$$

Via this isomorphism, the map $i_E^*: \mathbb{Z}_Y \to \mathbb{Z}_E$ becomes

$$i_C^* \oplus \mathrm{id}: \mathbb{Z}_X \bigoplus \oplus_{i=1}^{d-1}\mathbb{Z}_C(-i)[-2i] \to \mathbb{Z}_C \bigoplus \oplus_{i=1}^{d-1}\mathbb{Z}_C(-i)[-2i],$$

which proves the result. □

3.1.3. We recall some notions from [**60**]. Let I be the category associated to a finite partially ordered set. As in Chapter I, §2.7, an I-diagram of k-schemes is just a functor $X: I \to \mathbf{Sch}_k$. We call a map $f: X \to Y$ of I-diagrams a closed embedding, an open immersion, proper, etc., if $f(i): X(i) \to Y(i)$ is a closed embedding, etc., for each $i \in I$. Call an I-diagram X *smooth* if each $X(i)$ is smooth over k.

We work throughout with *reduced* k-schemes; for instance, we call a diagram cartesian if it is the diagram of reduced schemes associated to a (usual) cartesian diagram. As above, this generalizes to the notion of a cartesian diagram of I-diagrams.

The *discriminant* of a map $f: X \to Y$ of I-diagrams is the I-diagram

$$i \mapsto \mathrm{disc} f(i),$$

where $\mathrm{disc} f(i)$ is the complement of the largest open subset $U(i)$ of $Y(i)$ over which $f(i)$ is an isomorphism.

Following the notation of [**60**], we denote the opposite of the category associated to the partially ordered set $\{0 < 1\}^{n+1}$ by $\square_n^+$, and the full subcategory gotten by deleting the object $(0, \dots, 0)$ by $\square_n$. $\square_n^+$ is isomorphic to the $n+1$-cube $<n+1>$.

3.1.4. DEFINITION. Let X be an I-diagram of (reduced) finite type k-schemes. A *2-resolution* of X is a cartesian $\square_1^+ \times I$-diagram of the form

$$\begin{array}{ccc} Z_{11} & \longrightarrow & Z_{01} \\ \downarrow & & \downarrow f \\ Z_{10} & \longrightarrow & Z_{00} \end{array}$$

where
(3.1.4.1)

1. $Z_{00} = X$,
2. Z_{01} is in $\mathbf{Sm}_k$,
3. the horizontal arrows are closed embeddings,
4. f is proper,
5. Z_{10} contains the discriminant locus of f.

We call the 2-resolution *strict* if, for each $i \in I$, $\dim_k Z_{01}(i) = \dim_k Z_{00}(i)$, and the restriction of f to the components of $Z_{01}(i)$ and $Z_{00}(i)$ of maximal dimension is birational.

3.1.5. We now recall the *reduction* operation. Suppose we have $\square_n^+ \times I$-diagrams X_*^n for $1 \le n \le r$ such that the $\square_{n-1}^+ \times I$-diagrams X_{00*}^{n+1} and X_{1*}^n are the same for all n, $1 \le n < r$. Define the $\square_r^+ \times I$-diagram

$$Z_* := rd(X_*^1, \dots, X_*^r)$$

inductively as follows: For $r = 1$, set $Z_* := X_*^1$. For $r = 2$, define

$$Z_{ab} := \begin{cases} X_{0b}^1; & \text{if } a = (0,0), \\ X_{ab}^2; & \text{for } a \in \square_1, \end{cases}$$

for all $b \in \square_0^+$, with the evident morphisms. For $r > 2$, define $rd(X_*^1, \dots, X_*^r) := rd(rd(X_*^1, \dots, X_*^{r-1}), X_r^*)$, where we identify $\square_r^+ \times I$ with $\square_2^+ \times (\square_{r-2}^+ \times I)$ and $\square_{r-1}^+ \times I$ with $\square_1^+ \times (\square_{r-2}^+ \times I)$ by the evident isomorphism.

We now present the main definition of this section:

3.1.6. DEFINITION. Let X be an I-diagram of reduced finite type k-schemes. A *cubical hyperresolution* (or hyperresolution, for short) of X is a $\square_r \times I$-diagram of reduced finite type k-schemes Z_*, such that Z_* is the restriction to $\square_r \times I$ of $rd(X_*^1, \dots, X_*^r)$, where

1. X_*^1 is a 2-resolution of X
2. For $1 \le n < r$, X_*^{n+1} is a 2-resolution of X_*^n
3. Z_a is in $\mathbf{Sm}_k$ for all $a \in \square_r \times I$.

We call the hyperresolution *strict* if the 2-resolutions in (1) and (2) are strict.

If Z_* is a hyperresolution of X, we have the $\square_r^+ \times I$ diagram Z_*^+ gotten by "remembering" the component $X = rd(X_*^1, \dots, X_*^r)_{(0,\dots,0)}$. We call Z_*^+ an *augmented* hyperresolution, and write Z_*^+ as $Z_* \to X$.

3.1.7. A map of hyperresolutions $Z_* \to Z'_*$ over a map $f\colon X \to X'$ is given by taking an inclusion functor $i_{r,r'}\colon \square_r^+ \to \square_{r'}^+$, $r \le r'$, by filling in a given r-tuple with 0's in fixed spots $i_1, \dots, i_{r'-r}$, and taking a map of $\square_r^+ \times I$-diagrams $f_*\colon Z_*^+ \to Z_*\prime + \circ\, i_{r,r'}$ such that the map $f_{(0,\dots,0)}$ is f. This gives us the functor w from the category of hyperresolutions of I-diagrams of reduced finite type k-schemes to $\mathbf{Sch}_k^{\mathrm{fin}}$ by taking the $(0,\dots,0)$ component of the associated augmented hyperresolution.

Let $\mathbf{Hr}$ denote the category of hyperresolutions, and $\mathbf{Hrs}$ the category of strict hyperresolutions. We let Ho$\mathbf{Hr}$ and Ho$\mathbf{Hrs}$ denote the localizations of $\mathbf{Hr}$ and $\mathbf{Hrs}$ with respect to the maps over some identity map in $\mathbf{Sch}_k^{\mathrm{fin}}$.

The following facts are shown in [**60**] (here I is as above the category associated to a finite partially ordered set).

(3.1.7.1)

1. Each I-diagram of k-schemes admits a strict hyperresolution.
2. The functors

$$\mathrm{Ho}(w)\colon \mathrm{Ho}\mathbf{Hr} \to \mathbf{Sch}_k^{\mathrm{fin}}$$

and

$$\mathrm{Ho}(w)\colon \mathrm{Ho}\mathbf{Hrs} \to \mathbf{Sch}_k^{\mathrm{fin}}$$

are equivalences of categories.

3.1.8. REMARK. In fact, the notion of a *strict* hyperresolution is not explicitly formulated in [**60**], but it is implicit throughout the discussion there, and the arguments apply to strict hyperresolutions without modification. Also, in [**60**], a 2-resolution is somewhat less restrictive, as one does not require that Z_{01} be in $\mathbf{Sm}_k$ (smooth and quasi-projective over k) but only smooth over k. Using the remark on resolution of singularities at the beginning of this section, the arguments of [**60**] give the results stated above.

3.1.9. *The motive of a hyperresolution.* Let Z_* be a $\square_r$-diagram in $\mathbf{Sm}_k$. We form the *motive* of Z_*, $\mathbb{Z}_{Z_*}$, exactly as we formed the motive of an n-cube (see Chapter I, §2.6), except that we use 0 for the missing spot $(0,\dots,0)$. In particular, if X is a reduced finite type k-scheme, and X_* is a strict hyperresolution of X, we have the motive $\mathbb{Z}_{X_*}$.

We may (and will) adjust the auxiliary maps in the definition of $\mathbb{Z}_{Z_*}$ without explicitly indicating this in the notation, so that each morphism $f\colon Z_* \to W_*$ of hyperresolutions gives a resulting morphism of motives $f^*\colon \mathbb{Z}_{W_*} \to \mathbb{Z}_{Z_*}$ in $\mathbf{C}^b_{\mathrm{mot}}(\mathbf{Sm}_k)$, and similarly for all finite diagrams of morphisms. The different objects of $\mathbf{C}^b_{\mathrm{mot}}(\mathbf{Sm}_k)$ are all canonically isomorphic in $\mathbf{D}^b_{\mathrm{mot}}(\mathbf{Sm}_k)$, so the change of representing object in $\mathbf{C}^b_{\mathrm{mot}}(\mathbf{Sm}_k)$ has no effect in $\mathbf{D}^b_{\mathrm{mot}}$. This does allow us, however, to construct the cone of the morphism $f^*\colon \mathbb{Z}_{W_*} \to \mathbb{Z}_{Z_*}$, which is thus uniquely defined up to canonical isomorphism in $\mathbf{D}^b_{\mathrm{mot}}(\mathbf{Sm}_k)$.

3.2. The motive of a k-scheme

The main result concerning the motive $\mathbb{Z}_{Z_*}$ is

3.2.1. THEOREM. *Let U be a reduced finite type k-scheme, and Z_*, Z'_* two strict hyperresolutions of U. Then the motives $\mathbb{Z}_{Z_*}$ and $\mathbb{Z}_{Z'_*}$ are canonically isomorphic.*

In order to prove Theorem 3.2.1, we need an auxiliary statement, useful in its own right.

3.2.2. THEOREM. *Let*

$$\begin{array}{ccc} Z' & \xrightarrow{\alpha} & U' \\ {\scriptstyle g}\downarrow & & \downarrow{\scriptstyle f} \\ Z & \xrightarrow[\beta]{} & U \end{array}$$

be a cartesian diagram of reduced finite type k-schemes such that α *and* β *are closed embeddings,* f *is proper, and* f *induces an isomorphism* $U' \setminus Z' \to U \setminus Z$. *Take strict hyperresolutions of* U, U', Z *and* Z' *which fit into a commutative square*

$$\begin{array}{ccc} Z'_* & \xrightarrow{\alpha} & U'_* \\ {\scriptstyle g}\downarrow & & \downarrow{\scriptstyle f} \\ Z_* & \xrightarrow[\beta]{} & U_*. \end{array}$$

Then the square

$$\begin{array}{ccc} \mathbb{Z}_{Z'_*} & \xleftarrow{\alpha^*} & \mathbb{Z}_{U'_*} \\ {\scriptstyle g^*}\uparrow & & \uparrow{\scriptstyle f^*} \\ \mathbb{Z}_{Z_*} & \xleftarrow[\beta^*]{} & \mathbb{Z}_{U_*} \end{array}$$

is distinguished, i.e., the induced map $(\alpha^*, \beta^*)\colon \mathrm{cone}(f^*) \to \mathrm{cone}(g^*)$ *is an isomorphism in* $\mathbf{D}^b_{\mathrm{mot}}(\mathbf{Sm}_k)$, *or, equivalently, the induced map* $(f^*, g^*)\colon \mathrm{cone}(\beta^*) \to \mathrm{cone}(\alpha^*)$ *is an isomorphism in* $\mathbf{D}^b_{\mathrm{mot}}(\mathbf{Sm}_k)$.

3.2.3. To prove Theorem 3.2.1 and Theorem 3.2.2, we use the argument of [**61**], proceeding by noetherian induction. We let Theorem $3.2.1_n$ be the statement of Theorem 3.2.1 for U of dimension at most n, and Theorem $3.2.2_n$ be the statement of Theorem 3.2.2 for U and U' of dimension at most n. Let Theorem $3.2.1_{n,l}$ be the statement of Theorem 3.2.1 for U of dimension at most n and having at most l irreducible components of dimension n, and Theorem $3.2.2_{n,l}$ be the statement of Theorem 3.2.2 for U and U' of dimension at most n, and having at most l irreducible components of dimension n.

We first show

Induction Step 1: *Theorem* $3.2.1_{n,l}$ *implies Theorem* $3.2.2_{n,l}$

PROOF. Using Theorem $3.2.1_{n,l}$, we are free to choose our strict hyperresolutions.

Suppose first of all that we know Theorem $3.2.2_{n,l}$ when the map f is birational. In particular, if V is a k-scheme of dimension at most n, with irreducible components $V_1, \ldots, V_r$, having at most l components of dimension n, we may apply Theorem $3.2.2_{n,l}$ to the cartesian diagram

$$\begin{array}{ccc} Z' & \xrightarrow{\alpha} & V' \\ {\scriptstyle g}\downarrow & & \downarrow{\scriptstyle f} \\ Z & \xrightarrow[\beta]{} & V, \end{array}$$

where $V' := \coprod_{i=1}^r V_i$, $Z = \cup_{1\le i<j\le r} V_i \cap V_j$. This gives the *Mayer-Vietoris* property for unions of closed subschemes; using this Mayer-Vietoris property, we reduce to the case of irreducible U and U'. Thus, it suffices to prove Theorem $3.2.2_{n,l}$ in case Z contains no generic point of U, and Z' contains no generic point of U'.

Assuming this, let $\tilde{U} \to U'$ be a resolution of singularities of U' with $\tilde{U}$ in $\mathbf{Sm}_k$, let $\tilde{Z}$ be the inverse image of Z'. We may take strict hyperresolutions of Z, Z' and

$\tilde{Z}$ so that there is a commutative diagram

$$\begin{array}{ccccc} \tilde{Z}_* & \longrightarrow & \tilde{Z} & \longrightarrow & \tilde{U} \\ \downarrow & & \downarrow & & \downarrow \\ Z'_* & \longrightarrow & Z' & \longrightarrow & U' \\ \downarrow & & \downarrow & & \downarrow \\ Z_* & \longrightarrow & Z & \longrightarrow & U. \end{array}$$

Then

$$\begin{array}{ccc} \tilde{Z}_* & \longrightarrow & \tilde{U} \\ \downarrow & & \\ Z_* & & \end{array}$$

is a strict hyperresolution of U' and

$$\begin{array}{ccc} \tilde{Z}_* & \longrightarrow & \tilde{U} \\ \downarrow & & \\ Z'_* & & \end{array}$$

is a strict hyperresolution of U, from which Theorem 3.2.2 for $f: U' \to U$ follows directly. □

Of course, it follows directly from the above that Theorem $3.2.1_n$ implies Theorem $3.2.2_n$.

3.2.3.1. LEMMA. *Let*

$$\begin{array}{ccc} Z' & \xrightarrow{\alpha} & U' \\ {\scriptstyle g}\downarrow & & \downarrow{\scriptstyle f} \\ Z & \xrightarrow[\beta]{} & U \end{array}$$

be a cartesian diagram of quasi-projective k-schemes such that α and β are closed embeddings, f is projective, and f induces an isomorphism $U' \backslash Z' \to U \backslash Z$. Suppose that U and U' are smooth, that the restriction of f to the maximal dimension components of U and U' is birational, and that Z does not contain all the maximal dimensional components of U. Take strict hyperresolutions Z_* of Z and Z'_* of Z' with a map $g: Z'_* \to Z_*$ over $g: Z \to Z'$. Assume that $\dim_k U \leq n+1$, that either U is irreducible and Theorem $3.2.1_n$ holds, or U has at most $l+1$ components of dimension $n+1$, and Theorem $3.2.1_{n+1,l}$ holds. Then the square

$$\begin{array}{ccc} \mathbb{Z}_{Z'_*} & \xleftarrow{\alpha^*} & \mathbb{Z}_{U'} \\ {\scriptstyle g^*}\uparrow & & \uparrow{\scriptstyle f^*} \\ \mathbb{Z}_{Z_*} & \xleftarrow[\beta^*]{} & \mathbb{Z}_U \end{array}$$

is distinguished.

PROOF. By our hypotheses, the motives $\mathbb{Z}_{Z_*}$ and $\mathbb{Z}_{Z'_*}$ are independent of the choice of strict hyperresolution. Suppose first that $U' \to U$ is the inclusion of a union of components of U, necessarily containing all the components of maximal dimension. Then Z must contain all the components of U not occurring in U'. As the components of U are all smooth and disjoint, the fact that the $\mathbb{Z}_{Z_*}$ and $\mathbb{Z}_{Z'_*}$ are independent of the choice of strict hyperresolution, and that forming the motive takes disjoint union to direct sum, we reduce to the case in which all the components of U and U' have dimension $n+1$. The map f is thus birational.

Suppose first that f is the blow-up of U along a smooth center $C \subset Z$. Let E be the exceptional divisor, giving us the commutative diagram

$$\begin{array}{ccccc} E & \longrightarrow & Z' & \xrightarrow{\alpha} & U' \\ \downarrow & & \downarrow{\scriptstyle g} & & \downarrow{\scriptstyle f} \\ C & \longrightarrow & Z & \xrightarrow[\beta]{} & U \end{array}$$

with all squares cartesian. By our hypotheses, the result is independent of the choice of strict hyperresolutions, and we may apply Theorem 3.2.2 to the left square. Applying Lemma 3.1.2 to the outside square proves the result in this case.

Thus, the result is true for a map f which is a sequence of blow-ups with smooth centers lying over Z.

In general, we may dominate f with a map $f': U'' \to U$ which is a sequence of blow-ups with smooth centers lying over Z. Thus, the map $(g^*, f^*): \mathrm{cone}(\beta^*) \to \mathrm{cone}(\alpha^*)$ admits a left splitting, say h. We may then dominate the map $U'' \to U'$ by a sequence of blow-ups with smooth centers, which shows that h admits a left splitting as well, hence (g^*, f^*) is an isomorphism. □

The following result completes the inductive argument and the proof of Theorem 3.2.1.

Theorem 3.2.2$_n$ implies Theorem 3.2.1$_{n+1,1}$ and Theorem 3.2.2$_{n+1,l}$ implies Theorem 3.2.1$_{n+1,l+1}$

PROOF. By (3.1.7.1), it suffices to show that a map of strict hyperresolutions of U induces an isomorphism of the motives.

A strict hyperresolution of U is gotten by first forming a strict 2-resolution

$$\begin{array}{ccc} Z' & \longrightarrow & U' \\ \downarrow & & \downarrow \\ Z & \longrightarrow & U, \end{array}$$

and then taking a strict hyperresolution

$$\begin{array}{ccc} Z'_* & \longrightarrow & Z' \\ \downarrow & & \downarrow \\ Z_* & \longrightarrow & Z \end{array}$$

of the diagram $Z' \to Z$, giving the strict hyperresolution

$$\begin{array}{ccc} Z'_* & \longrightarrow & U' \\ \downarrow & & \\ Z_* & & \end{array}$$

of U. We first compare two strict hyperresolutions of U with the same U'. We may assume that the second strict hyperresolution is of the form

$$\begin{array}{ccccc} W'_* & \longrightarrow & W' & \longrightarrow & U' \\ \downarrow & & \downarrow & & \downarrow \\ W_* & \longrightarrow & W & \longrightarrow & U \end{array}$$

with $Z \subset W$, that $Z'_* \to Z_*$ maps to $W'_* \to W_*$, and that W is a proper closed subscheme of U.

Assume at first that U is irreducible and has dimension $n+1$. Since the hyperresolution is strict, U' is also irreducible and of dimension $n+1$. We may then apply Theorem $3.2.2_n$ to the square

$$\begin{array}{ccc} Z' & \longrightarrow & W' \\ \downarrow & & \downarrow \\ Z & \longrightarrow & W, \end{array}$$

giving the distinguished square

$$\begin{array}{ccc} \mathbb{Z}_{Z'} & \longleftarrow & \mathbb{Z}_{W'} \\ \uparrow & & \uparrow \\ \mathbb{Z}_Z & \longleftarrow & \mathbb{Z}_W. \end{array}$$

This proves the result in this case. In general, suppose U has $l+1$ components of dimension $n+1$. Then W has at most l components of dimension $n+1$, and since $U' \to U$ is birational on the components of maximal dimension, W' has at most l components of dimension $n+1$ as well. Arguing as above, Theorem $3.2.2_{n+1,l}$ implies Theorem $3.2.1_{n+1,l+1}$ for hyperresolutions of this type.

Now suppose we have two strict hyperresolutions arising from the diagrams

$$\begin{array}{ccccc} Z''_* & \longrightarrow & Z'' & \longrightarrow & U'' \\ \downarrow & & \downarrow & & \downarrow \\ W_* & \longrightarrow & W & \longrightarrow & U \end{array} \qquad \begin{array}{ccccc} Z'_* & \longrightarrow & Z' & \longrightarrow & U' \\ \downarrow & & \downarrow & & \downarrow \\ Z_* & \longrightarrow & Z & \longrightarrow & U. \end{array}$$

By the result we have already proven, we may assume that $Z = W$; we may also assume that U'' maps to U'. This gives us the cartesian square

$$\begin{array}{ccc} Z'' & \longrightarrow & U'' \\ \downarrow & & \downarrow{\scriptstyle g} \\ Z' & \longrightarrow & U'. \end{array}$$

Since the hyperresolution is strict, the map g is birational on the components of maximal dimension. We then apply Lemma 3.2.3.1 to this square, which proves Theorem 3.2.1 for U. □

3.2.4. *Cohomological motives.* Once we have the independence on the choice of hyperresolution, the rest is easy. For a reduced finite type k-scheme X, we define the motive $\mathbb{Z}_X$ to be $\mathbb{Z}_{X_*}$, where X_* is a strict hyperresolution of X.

3.2.5. THEOREM. *Let k be a perfect field admitting resolution of singularities. The functor $\mathbb{Z}_{(-)}: \mathbf{Sm}_k \to \mathbf{D}^b_{\mathrm{mot}}(\mathbf{Sm}_k)$ extends to the functor $\mathbb{Z}_{(-)}: \mathbf{Sch}_k^{\mathrm{fin}} \to \mathbf{D}^b_{\mathrm{mot}}(\mathbf{Sm}_k)$.*

PROOF. By Theorem 3.2.1, the motive $\mathbb{Z}_X$ of X in $\mathbf{Sch}_k^{\mathrm{fin}}$ is well-defined, independent of the choice of hyperresolution. In particular, if X is in $\mathbf{Sm}_k$, we may use the identity hyperresolution

$$\begin{array}{ccc} \emptyset & \longrightarrow & X \\ \downarrow & & \| \\ \emptyset & \longrightarrow & X, \end{array}$$

so $\mathbb{Z}_X$ agrees with the old definition.

Let $f: X \to Y$ be a morphism of k-schemes. By the equivalence of categories (3.1.7.1), we may find a morphism of hyperresolutions $f: X_* \to Y_*$ over f, and the resulting map $f^*: \mathbb{Z}_{Y_*} \to \mathbb{Z}_{X_*}$ canonically induces the map $f^*: \mathbb{Z}_Y \to \mathbb{Z}_X$. The functoriality $(f \circ g)^* = g^* \circ f^*$, follows similarly, as does the fact that f^* is the old f^* if X is in $\mathbf{Sm}_k$. □

The restriction to strict hyperresolutions is somewhat awkward, and in fact unnecessary.

3.2.6. THEOREM. *Let X be a reduced finite type k-scheme. Then all motives $\mathbb{Z}_{X_*}$ of hyperresolutions X_* of X are canonically isomorphic in $\mathbf{D}^b_{\mathrm{mot}}(\mathbf{Sm}_k)$.*

PROOF. Let

$$\begin{array}{ccc} Z_{11} & \longrightarrow & Z_{01} \\ \downarrow & & \downarrow \\ Z_{10} & \longrightarrow & X \end{array}$$

be a 2-resolution of X. By Theorem 3.2.2, the induced map $\mathbb{Z}_X \to \mathbb{Z}_{Z_*}$ is an isomorphism. By using the linked distinguished triangles of an n-cube (see Chapter I, §2.6.4), the same holds true if X is a $\square_n^+$-diagram of reduced finite type k-schemes. Thus, if $X_* \to X$ is a hyperresolution, the induced map $\mathbb{Z}_X \to \mathbb{Z}_{X_*}$ is an isomorphism. □

Let I be a full subcategory of $\square_n^+$, such that, if i is in I, and $j \to i$ is a map in $\square_n^+$, then j is in I; we call such a subcategory *left closed.* If $X: I \to \mathbf{Sch}_k$ is an I-diagram, then, using the equivalence of categories (3.1.7.1) and Theorem 3.2.6, we have the I^{op}-diagram of motives (uniquely defined as a pro-object, with transition maps isomorphisms in $\mathbf{D}^b_{\mathrm{mot}}$) $\mathbb{Z}_{X/I}: I^{\mathrm{op}} \to \mathbf{C}^b_{\mathrm{mot}}(\mathbf{Sm}_k)$. We may then extend $\mathbb{Z}_{X/I}$ to all of $(\square_n^+)^{\mathrm{op}}$, taking the value 0 at those a not in I, and then forming the resulting motive $\mathbb{Z}_X$ as the associated total complex (or iterated shifted cone) as in (Chapter I, §2.6, §2.6.2-§2.6.4).

3.2.7. COROLLARY. *Let I be a left closed subcategory of $\square_n^+$, let $X: I \to \mathbf{Sch}_k^{\text{fin}}$ be an I-diagram of reduced finite type k-schemes, and let $X_* \to X$ be a hyperresolution. Then the induced map $\mathbb{Z}_X \to \mathbb{Z}_{X_*}$ is an isomorphism in $\mathbf{D}^b_{\text{mot}}(\mathbf{Sm}_k)$.*

PROOF. This follows from Theorem 3.2.6 and the linked distinguished triangles of an n-cube (*loc. cit.*). □

We call a commutative diagram in $\mathbf{Sch}_k^{\text{fin}}$

$$\begin{array}{ccc} Z_{11} & \longrightarrow & Z_{01} \\ \downarrow & & \downarrow \\ Z_{10} & \longrightarrow & X \end{array}$$

a *weak 2-resolution* of X if all the properties of a 2-resolution (3.1.4.1) are satisfied except (2), i.e., we do not require that Z_{01} be in $\mathbf{Sm}_k$. A weak hyperresolution $X_* \to X$ of X is then defined as a hyperresolution (§3.1), with weak 2-resolutions replacing 2-resolutions, and omitting the condition (3) that all the terms X_a, $a \neq (0, \dots, 0)$, be in $\mathbf{Sm}_k$. The proof of Theorem 3.2.6 and Corollary 3.2.7 gives the generalization

3.2.8. THEOREM. *Let I be a left closed subcategory of $\square_n^+$, $X: I \to \mathbf{Sch}_k^{\text{fin}}$ an I-diagram of reduced finite type k-schemes, and let $X_* \to X$ be a weak hyperresolution. Then the induced map $\mathbb{Z}_X \to \mathbb{Z}_{X_*}$ is an isomorphism in $\mathbf{D}^b_{\text{mot}}(\mathbf{Sm}_k)$.*

3.2.9. REMARK. One can abstract the argument above. Let $\mathcal{A}$ be a DG category, and $F: \mathbf{Sm}_k^{\text{op}} \to \mathbf{C}^b(\mathcal{A})$ a functor (in our case, we take $\mathcal{A}$ to be the category of pro-objects in $\mathcal{A}_{\text{mot}}(\mathbf{Sm}_k)$ such that the transition maps are isomorphisms in $\mathbf{D}^b_{\text{mot}}(\mathbf{Sm}_k)$, and the functor F to be the map sending X to the system of motives $\mathbb{Z}_X(0)_f$). Let $\mathbf{K}^b(\mathcal{A})$ be the homotopy category of $\mathbf{C}^b(\mathcal{A})$, and let $G: \mathbf{K}^b(\mathcal{A}) \to \mathcal{D}$ be an exact functor of triangulated categories. Since F is a functor to complexes, we may extend $G \circ F$ to hyperresolutions Z_* by taking the total complex of the associated multi-dimensional complex $F(Z_*)$, and then applying G.

Suppose $G \circ F$ satisfies "descent for blow-ups with smooth center": Given X in $\mathbf{Sm}_k$, and $C \subset X$ a smooth closed subscheme, form the blow-up $\mu: Y \to X$ of X along X, with exceptional divisor E. Then the square

$$\begin{array}{ccc} (G \circ F)(E) & \xleftarrow{(G\circ F)(i_E)} & (G \circ F)(Y) \\ {\scriptstyle (G\circ F)(\mu_{|E})}\uparrow & & \uparrow{\scriptstyle (G\circ F)(\mu)} \\ (G \circ F)(C) & \xleftarrow[(G\circ F)(i_C)]{} & (G \circ F)(X) \end{array}$$

is distinguished.

We then have descent for hyperresolutions: The functor $G \circ F: \mathbf{Hrs} \to \mathcal{D}$ extends to the homotopy category $\mathbf{HoHrs}$, and therefore descends to the functor $G \circ F: \mathbf{Sch}_k^{\text{fin}} \to \mathcal{D}$. Furthermore, this functor is an extension of the functor $G \circ F: \mathbf{Sm}_k \to \mathcal{D}$, and satisfies the property given by Theorem 3.2.2, namely, $G \circ F$ transforms cartesian diagrams (with the horizontal maps closed embeddings, and the vertical maps proper) to distinguished squares. The various extensions described above hold in the abstract situation as well.

One may vary the above data, replacing $\mathcal{A}$ with an additive category and using various other categories of complexes, $\mathbf{C}^*(\mathcal{A})$ $(* = +, -, \emptyset)$, or one can replace $\mathbf{C}^b(\mathcal{A})$ with a closed simplicial model category $\mathcal{C}$, replace the total complex with an iterated homotopy fiber, and replace $\mathcal{D}$ with the homotopy category.

Similar results are discussed in the paper [**59**].

3.2.10. *Products.* Suppose we have reduced finite type k-schemes X and Y, and a hyperresolution $Y_* \to Y$. Then $X \times_k Y_* \to X \times_k Y$ is a weak hyperresolution of $X \times_k Y$, where $X \times_k Y_*$ is the diagram

$$i \mapsto X \times_k Y_i.$$

Similarly, if $X_* \to X$ is hyperresolution of X, then the augmented $\square_n^{\mathrm{op}} \times \square_m^{\mathrm{op}}$ diagram (for appropriate n and m) $X_* \times_k Y_* \to X \times_k Y_*$ is a hyperresolution of $X \times_k Y_*$. By Theorem 3.2.8, the map $\mathbb{Z}_{X \times Y} \to \mathbb{Z}_{X_* \times_k Y_*}$ induced by the two augmentations is an isomorphism in $\mathbf{D}^b_{\mathrm{mot}}(k)$.

The external products $\boxtimes_{X_i, Y_j} : \mathbb{Z}_{X_i} \otimes \mathbb{Z}_{Y_j} \to \mathbb{Z}_{X_i \times_k Y_j}$ give the external product

$$\boxtimes_{X_*, Y_*} : \mathbb{Z}_{X_*} \otimes \mathbb{Z}_{Y_*} \to \mathbb{Z}_{X_* \times_k Y_*},$$

which is an isomorphism in $\mathbf{D}^b_{\mathrm{mot}}(\mathbf{Sm}_k)$. Composing this external product with the isomorphisms $\mathbb{Z}_X \otimes \mathbb{Z}_Y \to \mathbb{Z}_{X_*} \otimes \mathbb{Z}_{Y_*}$ and $\mathbb{Z}_{X_* \times_k Y_*} \to \mathbb{Z}_{X \times_k Y}$ gives the external product

$$\boxtimes_{X,Y} : \mathbb{Z}_X \otimes \mathbb{Z}_Y \to \mathbb{Z}_{X \times_k Y}. \tag{3.2.10.1}$$

It follows from the equivalence of categories (3.1.7.1) that the map $\boxtimes_{X,Y}$ is independent of the choices made, and that we get the same map if we first resolve X and then resolve Y. This implies that the external products are commutative and associative; the linked distinguished triangles of an n-cube imply that the product $\boxtimes_{X,Y}$ is an isomorphism.

3.2.11. *Properties of the cohomological motive.* The fundamental properties of motives described in Chapter I extend to the singular case as well, using the properties of smooth motives, and the distinguished triangles associated to the n-cube defining the motive of a hyperresolution. In particular, we have

1. *Homotopy.* For each reduced finite type k-scheme X, the pull-back $p^* : \mathbb{Z}_X \to \mathbb{Z}_{\mathbb{A}^1 \times X}$ is an isomorphism in $\mathbf{D}^b_{\mathrm{mot}}(\mathbf{Sm}_k)$.
2. *Künneth isomorphism.* For X and Y reduced finite type k-schemes, we have the natural commutative external product $\boxtimes_{X,Y} : \mathbb{Z}_X \otimes \mathbb{Z}_Y \to \mathbb{Z}_{X \times_k Y}$, which is an isomorphism in $\mathbf{D}^b_{\mathrm{mot}}(\mathbf{Sm}_k)$.
3. *Mayer-Vietoris.* Let X be a reduced finite type k-scheme, $j_U : U \to X$ and $j_V : V \to X$ open subschemes with $X = U \cup V$. Then the sequence

$$\mathbb{Z}_X \xrightarrow{(j_U^*, j_V^*)} \mathbb{Z}_U \oplus \mathbb{Z}_V \xrightarrow{j_{U\cap V,U}^* - j_{U\cap V,V}^*} \mathbb{Z}_{U\cap V}$$

induces an isomorphism in $\mathbf{D}^b_{\mathrm{mot}}(\mathbf{Sm}_k)$

$$\mathbb{Z}_X \to \mathrm{cone}\big(\mathbb{Z}_U \oplus \mathbb{Z}_V \xrightarrow{j_{U\cap V,U}^* - j_{U\cap V,V}^*} \mathbb{Z}_{U\cap V}\big)[-1].$$

4. *Blow-up distinguished triangle.* Let

$$\begin{array}{ccc} Z' & \xrightarrow{i'} & X' \\ {\scriptstyle g}\downarrow & & \downarrow{\scriptstyle f} \\ Z & \xrightarrow[i]{} & X \end{array}$$

be a cartesian diagram of reduced finite type k-schemes, with i and i' closed embeddings, f proper, and $f: X' \setminus Z' \to X \setminus Z$ an isomorphism. Then the map

$$\mathbb{Z}_X \xrightarrow{(i^*,f^*)} \mathrm{cone}(\mathbb{Z}_Z \oplus \mathbb{Z}_{X'} \xrightarrow{g^*-i'^*} \mathbb{Z}_{Z'})[-1]$$

is an isomorphism in $\mathbf{D}^b_{\mathrm{mot}}(\mathbf{Sm}_k)$.

Indeed, the homotopy property follows from the homotopy property for smooth k-schemes, together with the fact that $\mathbb{A}^1 \times X_*$ is a hyperresolution of $\mathbb{A}^1 \times X$ if X_* is a hyperresolution of X. The Künneth isomorphism was discussed in §3.2.10.

Define the motivic cohomology by

$$H^p(X, \mathbb{Z}(q)) := \mathrm{Hom}_{\mathcal{DM}(k)}(1, \mathbb{Z}_X(q)[p]).$$

As in the case of smooth k-schemes, the Künneth isomorphism, followed by pull-back by the diagonal, gives the bi-graded motivic cohomology $H^*(X, \mathbb{Z}(*)) := \oplus_{p,q} H^p(X, \mathbb{Z}(q))$ the functorial structure of a bi-graded ring (graded-commutative in p), with identity given by $p_X^*: \mathbb{Z}_{\mathrm{Spec}\, k} \to \mathbb{Z}_X$.

For the Mayer-Vietoris property, take a hyperresolution $X_* \to X$ of X, with X_* a $\square_r$-diagram. Then the pull-backs $U_* := X_* \times_X U$, $V_* := X_* \times_X V$, and $(U \cap V)_* := X_* \times_X (U \cap V)$ form hyperresolutions of U, V and $U \cap V$, respectively, with $U_\alpha \cup V_\alpha = X_\alpha$ and $U_\alpha \cap V_\alpha = (U \cap V)_\alpha$ for each $\alpha \in \square_r$. Thus, the map

$$Z_{X_\alpha} \to \mathrm{cone}(\mathbb{Z}_{U_\alpha} \oplus \mathbb{Z}_{V_\alpha} \to \mathbb{Z}_{(U \cap V)_\alpha})[-1]$$

are isomorphisms for each α, which gives the result for $\mathbb{Z}_X$.

The blow-up triangle is a consequence of Theorem 3.2.2; note as in the proof of Induction Step 1, §3.1.9, that as a special case, we have the Mayer-Vietoris distinguished triangle for the union of closed subschemes.

3.2.12. *Homological motives of singular schemes.* Dualizing the motive $\mathbb{Z}_X$ gives the *homological motive*

$$\mathbb{Z}^h_X := \mathbb{Z}^D_X.$$

By duality, Theorem 3.2.5 gives

3.2.13. THEOREM. *Let k be a perfect field, admitting resolution of singularities. Then sending X in $\mathbf{Sch}^{\mathrm{fin}}_k$ to $\mathbb{Z}^h_X$ extends to a functor*

$$\mathbb{Z}^h_{(-)}: \mathbf{Sch}^{\mathrm{fin}}_k \to \mathbf{D}^b_{\mathrm{mot}}(k)$$

whose restriction to $\mathbf{Sm}_k$ is the functor $\mathbb{Z}^D_{(-)}$.

In particular, the restriction of $\mathbb{Z}^h$ to $\mathbf{Sm}_k$ agrees with the homological motive functor (2.2.3.1).

For a morphism $f: X \to Y$, we denote the morphism $\mathbb{Z}^h(f): \mathbb{Z}^h_X \to \mathbb{Z}^h_Y$ by f_*.

The properties of the cohomological motive listed in §3.2.11 dualize to give

1. *Homotopy.* The map $p_*: \mathbb{Z}^h_{\mathbb{A}^1 \times_k X} \to \mathbb{Z}^h_X$ is an isomorphism for all X in $\mathbf{Sch}_k^{\mathrm{fin}}$.
2. *Künneth isomorphism.* There are natural commutative external products
$$\boxtimes^h_{X,Y}: \mathbb{Z}^h_X \otimes \mathbb{Z}^h_Y \to \mathbb{Z}^h_{X \times_k Y}$$
which are isomorphisms.
3. Define the *motivic homology* of X by
$$H_p(X, \mathbb{Z}(q)) := \mathrm{Hom}_{\mathcal{DM}(k)}(1, \mathbb{Z}^h_X(-q)[-p]).$$
The external products $\boxtimes^h_{X,Y}$ induce commutative and associative external products
$$\boxtimes_{X,Y}: H_p(X, \mathbb{Z}(q)) \otimes H_{p'}(Y, \mathbb{Z}(q')) \to H_{p+p'}(X \times_k Y, \mathbb{Z}(q+q')).$$
4. *Mayer-Vietoris* Write X as a union of open subschemes, $X = U \cup V$. Then the map
$$\mathrm{cone}\big(\mathbb{Z}^h_{U\cap V} \xrightarrow{(j_{U\cap V,U*}, -j_{U\cap V,V*})} \mathbb{Z}^h_U \oplus \mathbb{Z}^h_V\big) \xrightarrow{j_{U*}+j_{V*}} \mathbb{Z}^h_X$$
is an isomorphism in $\mathbf{D}^b_{\mathrm{mot}}(k)$.
5. *Blow-up distinguished triangle* Let
$$\begin{array}{ccc} Z' & \xrightarrow{i'} & X' \\ {\scriptstyle g}\downarrow & & \downarrow{\scriptstyle f} \\ Z & \xrightarrow[i]{} & X \end{array}$$
be a cartesian diagram of reduced finite type k-schemes, with i and i' closed embeddings, f proper, and $f: X' \setminus Z' \to X \setminus Z$ an isomorphism. Then the map
$$\mathrm{cone}\big(\mathbb{Z}_{Z'} \xrightarrow{(g_*, -i'_*)} \mathbb{Z}_Z \oplus \mathbb{Z}_{X'}\big) \xrightarrow{(i_*+f_*)} \mathbb{Z}_X$$
is an isomorphism in $\mathbf{D}^b(k)$.

In addition, the cap products for motivic cohomology and homology for smooth varieties extend to the cap product
$$\cap_X: H^p(X, \mathbb{Z}(q)) \otimes H_{p'}(X, \mathbb{Z}(q')) \to H_{p'-p}(X, \mathbb{Z}(q'-q)),$$
using the same construction as in the smooth case: Identify $H_{p'}(X, \mathbb{Z}(q')$ with $\mathrm{Hom}_{\mathcal{DM}(k)}(\mathbb{Z}_X(p')[q'], 1)$ by duality, compose with an element of $H^p(X, \mathbb{Z}(q)) := \mathrm{Hom}_{\mathcal{DM}(k)}(1, \mathbb{Z}_X(p)[q])$ to get to $\mathrm{Hom}_{\mathcal{DM}(k)}(\mathbb{Z}_X(p')[q'], \mathbb{Z}_X(p)[q])$, and then twist and shift to get to $\mathrm{Hom}_{\mathcal{DM}(k)}(\mathbb{Z}_X(p'-p)[q'-q], 1) \cong H_{p'-p}(X, \mathbb{Z}(q'-q))$.

3.3. Comparison of motives

For $S = \mathrm{Spec}\, k$, with k a perfect field, all reduced quasi-projective k-schemes are smoothly decomposable, hence the Borel-Moore motive, and Borel-Moore homology are defined for all reduced quasi-projective k-schemes. If, in addition, resolution of singularities holds for reduced quasi-projective k-schemes, then, by Lemma 1.5.4, all reduced quasi-projective k-schemes admit a compactifiable closed embedding into a smooth quasi-projective k-scheme. Thus the compactly supported cohomology is defined for all reduced quasi-projective k-schemes. We conclude the discussion of the motives of k-schemes by comparing the cohomological motive with the cohomological motive with compact support, and the homological motive with the Borel-Moore motive.

3.3.1. *Cohomology of normal crossing schemes.* Suppose $D_0, \dots, D_n$ define a normal crossing scheme in some $X \in \mathbf{Sm}_k$, and let $D := \cup_{i=1}^n D_i$. We have the $\square_n$-diagram $D_*: \square_n \to \mathbf{Sm}_k$ defined by

$$D_{\alpha_0,\dots,\alpha_n} = \cap_{i,\ \alpha_i=1} D_i,$$

with maps being the inclusions. We may extend D_* to the $\square_n^+$-diagram $(X, D)_*$ with value X at $(0, \dots, 0)$. By the evident isomorphism of $\square_n^+$ with the $n+1$-cube $<n+1>$, the diagram $(X, D)_*$ agrees with the $n+1$-cube $(X; D_0, \dots, D_n)_*$ defined in Chapter I, §2.6.6.

One sees by an elementary induction that D_* is a hyperresolution of D, hence we have the canonical isomorphism in $\mathbf{D}^b_{\mathrm{mot}}(\mathbf{Sm}_k)$ $\mathbb{Z}_D \cong \mathbb{Z}_{D_*}$. Similarly, we may view the $\square_n^+$-diagram $(X, D)_*$ as a hyperresolution of the diagram $D \to X$, giving us the identification of the relative motive $\mathbb{Z}_{(X; D_0, \dots, D_n)}$ with the shifted cone

$$\mathbb{Z}_{(X; D_0, \dots, D_n)} \cong \mathrm{cone}\big(i_D^* : \mathbb{Z}_X \to \mathbb{Z}_D\big)[-1].$$

We denote the cone of i_D^* by $\mathbb{Z}_{(X;D)}$.

3.3.2. *Hyperresolutions of compactifications.* Form the category $\mathbf{Sch}_k^{\mathrm{cpt}}$ as the category of open immersions $j: X \to \bar{X}$ with X and $\bar{X}$ in $\mathbf{Sch}_k^{\mathrm{fin}}$ and $\bar{X}$ projective over k; a map $(j_1: X_1 \to \bar{X}_1) \to (j_2: X_2 \to \bar{X}_2)$ is a pair of proper maps

$$f: X_1 \to X_2; \quad \bar{f}: \bar{X}_1 \to \bar{X}_2$$

making the evident diagram commute. Letting $\mathbf{Sch}_{k,\mathrm{pr}}^{\mathrm{fin}}$ be the category with the same objects as $\mathbf{Sch}_k^{\mathrm{fin}}$, but with proper maps, we have the functor $F: \mathbf{Sch}_k^{\mathrm{cpt}} \to \mathbf{Sch}_k^{\mathrm{fin}}$ gotten by ignoring the compactification $\bar{X}$.

Form the category $\mathrm{Ho}\mathbf{Sch}_k^{\mathrm{cpt}}$ by inverting all maps of the form $(\mathrm{id}, \bar{f})$.

3.3.3. LEMMA. *The functor F induces an equivalence of categories*

$$\mathrm{Ho}F: \mathrm{Ho}\mathbf{Sch}_k^{\mathrm{cpt}} \to \mathbf{Sch}_{k,\mathrm{pr}}^{\mathrm{fin}}.$$

PROOF. The proof is essentially the same as the proof of Lemma 2.4.5, but easier, and is left to the reader. □

3.3.4. *Compactly supported motives.* For $j: X \to \bar{X}$ in $\mathbf{Sch}_k^{\mathrm{cpt}}$, we write $D(j) := \bar{X} \setminus X$, and let $i_{D(j)}: D(j) \to \bar{X}$ be the inclusion. Define the *motive with compact support* $\mathbb{Z}_j^c$ by

$$\mathbb{Z}_j^c := \mathrm{cone}\big(i_{D(j)}^* : \mathbb{Z}_X \to \mathbb{Z}_{D(j)}\big)[-1].$$

To be precise, we take a hyperresolution Z_* of the diagram $D(j) \xrightarrow{i_{D(j)}} \bar{X}$ and set $\mathbb{Z}_j^c := \mathbb{Z}_{Z_*}$. From Theorem 3.2.6 and (3.1.7.1), $\mathbb{Z}_j^c$ is well-defined, up to canonical isomorphism, and each map $(f, g): j \to j'$ in $\mathbf{Sch}_k^{\mathrm{cpt}}$ induces a map $(f, g)^*: \mathbb{Z}_{j'}^c \to \mathbb{Z}_j^c$, giving the functor

$$\mathbb{Z}^c: (\mathbf{Sch}_k^{\mathrm{cpt}})^{\mathrm{op}} \to \mathbf{D}^b_{\mathrm{mot}}(\mathbf{Sm}_k).$$

From Theorem 3.2.2, each map in $\mathbf{Sch}_k^{\mathrm{cpt}}$ of the form $(\mathrm{id}, \bar{f})$ induces an isomorphism $(\mathrm{id}, \bar{f})^*: \mathbb{Z}_{j'}^c \to \mathbb{Z}_j^c$, hence, by Lemma 3.3.3, we may descend $\mathbb{Z}^c$ to the functor

$$\mathbb{Z}^c\colon (\mathbf{Sch}^{\mathrm{fin}}_{k,\mathrm{pr}})^{\mathrm{op}} \to \mathbf{D}^b_{\mathrm{mot}}(\mathbf{Sm}_k). \tag{3.3.4.1}$$

Recall the category $\mathbf{Sm}^{\mathrm{pr}}_{k,\mathrm{proj}}$ (§2.2.1); by the comments at the beginning of this section, we have $\mathbf{Sm}^{\mathrm{pr}}_{k,\mathrm{proj}} = \mathbf{Sm}_{k,\mathrm{proj}}$, the category with the same objects as $\mathbf{Sm}_k$, but with only projective morphisms.

3.3.5. LEMMA. *The functor* (3.3.4.1) *defines an extension of the functor* (2.2.3.2)

$$\mathbb{Z}^{c/k}\colon \mathbf{Sm}^{\mathrm{op}}_{k,\mathrm{proj}} \to \mathbf{D}^b_{\mathrm{mot}}(\mathbf{Sm}_k).$$

PROOF. Let X be in $\mathbf{Sm}_k$, and take a compactification $j\colon X \to \bar{X}$ with complement $D(j)$ a normal crossing divisor. Write $D(j)$ as the union of its irreducible components $D(j) = \cup_{i=1}^n D_i$. By the identification in §3.3.1 of the relative motive $\mathbb{Z}_{\bar{X},D(j)}$ with the relative motive $\mathbb{Z}_{(\bar{X};D_1,\dots,D_n)}$, together with the definition of $\mathbb{Z}^c_X$, we have the canonical and functorial identification $\mathbb{Z}^c_X \cong \mathbb{Z}_{(\bar{X};D_1,\dots,D_n)}$. The result then follows from Proposition 2.3.8. □

By duality, the functor $\mathbb{Z}^{\mathrm{B.M.}} := (\mathbb{Z}^c)^D\colon \mathbf{Sch}^{\mathrm{fin}}_{k,\mathrm{pr}} \to \mathbf{D}^b_{\mathrm{mot}}(\mathbf{Sm}_k)$ is an extension of the functor (see Chapter III, Theorem 2.5.7) $\mathbb{Z}^{\mathrm{B.M.}}\colon \mathbf{Sm}_{k,\mathrm{proj}} \to \mathbf{D}^b_{\mathrm{mot}}(\mathbf{Sm}_k)$.

We let $H^p_c(X,\mathbb{Z}(q))$ denote the *cohomology with compact support*

$$H^p_c(X,\mathbb{Z}(q)) := \mathrm{Hom}_{\mathcal{DM}(k)}(1, \mathbb{Z}^c_X(q)[p]),$$

and $H^{\mathrm{B.M.}}_p(X,\mathbb{Z}(q))$ the *Borel-Moore homology*

$$H^{\mathrm{B.M.}}_p(X,\mathbb{Z}(q)) := \mathrm{Hom}_{\mathcal{DM}(k)}(1, \mathbb{Z}^{\mathrm{B.M.}}_X(-q)[-p]).$$

3.3.6. *Products.* Let $j_1\colon X_1 \to \bar{X}_1$ and $j_2\colon X_2 \to \bar{X}_2$ be compactifications, and write D_i for $D(j_i)$. We have the Mayer-Vietoris distinguished triangle (see §3.2.11)

$$\mathbb{Z}_{D_1\times_k\bar{X}_2\cup\bar{X}_1\times_k D_2} \to \mathbb{Z}_{D_1\times_k\bar{X}_2} \oplus \mathbb{Z}_{\bar{X}_1\times_k D_2} \to \mathbb{Z}_{D_1\times_k D_2} \to .$$

Using this, we see that the external products for the cohomological motive (3.2.10.1) define the external product $\boxtimes\colon \mathbb{Z}_{(\bar{X}_1;D_1)} \otimes \mathbb{Z}_{(\bar{X}_2;D_2)} \to \mathbb{Z}_{(\bar{X}_1\times_k\bar{X}_2;D_1\times_k X_2\cup X_1\times_k D_2)}$, which is an isomorphism. This defines the external product

$$\boxtimes^c_{X_1,X_2}\colon \mathbb{Z}^c_{X_1} \otimes \mathbb{Z}^c_{X_2} \to \mathbb{Z}^c_{X_1\times_k X_2},$$

which is an isomorphism. Taking the pull-back by the diagonal defines the cup product $\cup_X\colon \mathbb{Z}^c_X \otimes \mathbb{Z}^c_X \to \mathbb{Z}^c_X$, giving the compactly supported cohomology $H^*_c(X,\mathbb{Z}(*))$ a natural ring structure (without unit, in general).

Dualizing the external products gives external products for Borel-Moore motive and for Borel-Moore homology. Cap products are defined as in §2.2.5 by identifying $H^{\mathrm{B.M.}}_{p'}(X,\mathbb{Z}(q'))$ with $\mathrm{Hom}_{\mathcal{DM}(k)}(\mathbb{Z}^c_X(q')[p'],1)$ and using the composition

$$\begin{aligned}
&\mathrm{Hom}_{\mathcal{DM}(k)}(1,\mathbb{Z}^c_X(q)[p]) \otimes \mathrm{Hom}_{\mathcal{DM}(k)}(\mathbb{Z}^c_X(q')[p'],1)\\
&\qquad\to \mathrm{Hom}_{\mathcal{DM}(k)}(\mathbb{Z}^c_X(q')[p'],\mathbb{Z}^c_X(q)[p])\\
&\qquad\qquad\cong \mathrm{Hom}_{\mathcal{DM}(k)}(\mathbb{Z}^c_X(q'-q)[p'-p],1).
\end{aligned}$$

3.3.7. *Open push-forward.* Let $j_U : U \to X$ be an open subscheme, and let $j : X \to \bar{X}$ be a compactification, giving us the compactification $j \circ j_U : U \to \bar{X}$. The natural map $\mathbb{Z}_{\bar{X}, D(j \circ j_U)} \to \mathbb{Z}_{\bar{X}, D(j)}$ induced by the commutative diagram

$$\begin{array}{ccc} D(j) & \longrightarrow & \bar{X} \\ \downarrow & & \| \\ D(j \circ j_U) & \longrightarrow & \bar{X} \end{array}$$

defines the map $j_{U!} : \mathbb{Z}^c_U \to \mathbb{Z}^c_X$. These push-forward maps are evidently functorial and extend the push-forward maps defined in §2.2.5 (*cf.* the proof of Lemma 3.3.5).

Dualizing gives the functorial pull-back maps $j_U^* : \mathbb{Z}_X^{\mathrm{B.M.}} \to \mathbb{Z}_U^{\mathrm{B.M.}}$.

3.3.8. Let $j : X \to \bar{X}$ be a compactification. The canonical map $j^* : \mathbb{Z}_{\bar{X}, D(j)} \to \mathbb{Z}_X$ defines the natural map $\iota_X : \mathbb{Z}^c_X \to \mathbb{Z}_X$. The maps ι_X and their duals define the natural transformations $\iota : \mathbb{Z}^c \to \mathbb{Z}$, $\iota^D : \mathbb{Z}^h \to \mathbb{Z}^{\mathrm{B.M.}}$ of the functors

$$\mathbb{Z}^c, \mathbb{Z} : (\mathbf{Sch}_k^{\mathrm{fin}})^{\mathrm{op}} \to \mathbf{D}^b_{\mathrm{mot}}(\mathbf{Sm}_k),$$
$$\mathbb{Z}^h, \mathbb{Z}^{\mathrm{B.M.}} : (\mathbf{Sch}_k^{\mathrm{fin}})^{\mathrm{op}} \to \mathbf{D}^b_{\mathrm{mot}}(\mathbf{Sm}_k).$$

It is easy to see that ι and ι^D are compatible with the respective products.

CHAPTER V

Realization of the Motivic Category

In this chapter, we describe a mapping property satisfied by the category $\mathcal{DM}(\mathcal{V})$. The main theorem of this chapter, Theorem 1.3.1, gives a criterion for a cohomology theory defined by a complex of sheaves $\mathcal{F}$ on a Grothendieck site to define the "$\mathcal{F}$-realization" of $\mathcal{DM}(\mathcal{V})$. One should view this more as a prototype than a final result; many interesting cohomology theories have been defined in a somewhat more general setting than the one covered by our result, but it seems difficult to give an all-encompassing result covering all the known cases. We will consider various important examples of cohomology theories in the Section 2, where we give the realizations corresponding to singular cohomology, étale cohomology, Hodge (Deligne) cohomology, and Jannsen's motivic cohomology of mixed absolute Hodge complexes. Some of these theories do not quite satisfy the criterion we give in Theorem 1.3.1, but minor modifications allow the construction to go through.

1. Realization for geometric cohomology

1.1. Geometric cohomology theories

We give axioms for a cohomology theory which suffice to give a realization functor from $\mathcal{DM}$.

1.1.1. Let $\mathcal{C}$ be a full subcategory of $\mathbf{Sch}_S$, closed under finite fiber products, and taking open subsets. Following Bloch-Ogus [**20**] and Gillet [**46**], a graded cohomology theory $\Gamma(*)$ on $\mathcal{C}$ is a graded complex of sheaves of R-modules $\Gamma^*(*)$ on the big Zariski site $\mathcal{C}_{\mathrm{Zar}}$ of $\mathcal{C}$, together with a pairing in the derived category of graded complexes of sheaves of R-modules on $\mathcal{C}_{\mathrm{Zar}}$, $\Gamma^*(*)\otimes^L\Gamma^*(*)\to\Gamma^*(*)$, which is associative with unit and graded-commutative, and satisfies certain additional axioms. We give here a slightly different version of this notion.

1.1.2. *Basic assumptions.* We begin with a Grothendieck site $(\mathfrak{S},\mathfrak{T})$ (see Part II, Chapter IV, §1.1) which has a final object $*$ and initial object $\emptyset$, and admits finite products over $*$ and finite coproducts under $\emptyset$.
(1) We assume we have a functor $\alpha\colon\mathcal{V}\to\mathfrak{S}$ which is *cocontinuous*, i.e., if $U\to X$ is the inclusion of a Zariski open subset of X in $\mathcal{V}$, then $\alpha(U)\to\alpha(X)$ is an open in $(\mathfrak{S},\mathfrak{T})$. We suppose that $\alpha(\mathrm{Spec}\,S)=*$, $\alpha(\emptyset)=\emptyset$, and that α sends finite products over $\mathrm{Spec}\,S$ to finite products over $*$, and similarly for coproducts under $\emptyset$. In particular, the operation of product over S (resp. product over $*$), together with the canonical isomorphisms $X\times_S Y\to Y\times_S X$ and $A\times_* B\to B\times_* A$, defines the structure of a symmetric monoidal category on $\mathcal{V}$ (resp. $\mathfrak{S}$), and makes α into a symmetric monoidal functor.

We will usually omit mention of the functor α when the distinction between $\mathcal{V}$ and $\mathfrak{S}$ is clear from the context.

(2) We assume that the topos of sheaves of sets on $\mathfrak{S}$, $\tilde{\mathfrak{S}}$, has a conservative family of points (see Part II, Chapter IV, §1.3 and Definition 1.3.5).
(3) We fix a commutative ring R, and let $\mathcal{A}$ denote the abelian tensor category $\mathbf{Mod}_R$. We denote the category of sheaves on $\mathfrak{S}$ with values in $\mathcal{A}$ by $\mathrm{Sh}^{\mathcal{A}}_{\mathfrak{S},\mathfrak{T}}$; for an object X of $\mathfrak{S}$, we let $\mathrm{Sh}^{\mathcal{A}}_{\mathfrak{S},\mathfrak{T}}(X)$ denote the category of $\mathcal{A}$-valued sheaves on X, for the induced topology.
(4) We say a sheaf $\mathcal{F} \in \mathrm{Sh}^{\mathcal{A}}_{\mathfrak{S},\mathfrak{T}}$ is *flat* if the functor $-\otimes\mathcal{F}\colon \mathrm{Sh}^{\mathcal{A}}_{\mathfrak{S},\mathfrak{T}} \to \mathrm{Sh}^{\mathcal{A}}_{\mathfrak{S},\mathfrak{T}}$ is exact. A presheaf $\mathcal{F}$ on $\mathfrak{S}$ is flat if for each U in $\mathfrak{S}$, the functor $-\otimes\mathcal{F}(U)\colon \mathcal{A} \to \mathcal{A}$ is exact. We assume that the functor $i_*i^*\colon \mathrm{Sh}^{\mathcal{A}}_{\mathfrak{S},\mathfrak{T}} \to \mathrm{Sh}^{\mathcal{A}}_{\mathfrak{S},\mathfrak{T}}$ sends flat sheaves to flat presheaves. By (Part II, Chapter IV, Proposition 2.4.3 and Remark 2.4.4) this condition is satisfied if R is noetherian.

1.1.3. From [**4**, II 6.9], $\mathrm{Sh}^{\mathcal{A}}_{\mathfrak{S},\mathfrak{T}}$ and $\mathrm{Sh}^{\mathcal{A}}_{\mathfrak{S},\mathfrak{T}}(X)$ are abelian categories with enough injectives.

We have the category of bounded below complexes of sheaves $\mathbf{C}^+(\mathrm{Sh}^{\mathcal{A}}_{\mathfrak{S},\mathfrak{T}})$, the homotopy category $\mathbf{K}^+(\mathrm{Sh}^{\mathcal{A}}_{\mathfrak{S},\mathfrak{T}})$, and the derived category $\mathbf{D}^+(\mathrm{Sh}^{\mathcal{A}}_{\mathfrak{S},\mathfrak{T}})$.

For an object X of $\mathfrak{S}$, the categories

$$\mathbf{C}^+(\mathrm{Sh}^{\mathcal{A}}_{\mathfrak{S},\mathfrak{T}}(X)),\ \mathbf{K}^+(\mathrm{Sh}^{\mathcal{A}}_{\mathfrak{S},\mathfrak{T}}(X)),\ \mathbf{D}^+(\mathrm{Sh}^{\mathcal{A}}_{\mathfrak{S},\mathfrak{T}}(X))$$

are similarly defined.

We let $p_X\colon X \to S$ denote the structure morphism.

1.1.4. For X in $\mathcal{V}$ and W a closed subset of X, we have the functors

$$p_{X*}\colon \mathrm{Sh}^{\mathcal{A}}_{\mathfrak{S},\mathfrak{T}}(X) \to \mathrm{Sh}^{\mathcal{A}}_{\mathfrak{S},\mathfrak{T}}(*)$$
$$p^W_{X*}\colon \mathrm{Sh}^{\mathcal{A}}_{\mathfrak{S},\mathfrak{T}}(X) \to \mathrm{Sh}^{\mathcal{A}}_{\mathfrak{S},\mathfrak{T}}(*)$$

where p^W_{X*} is the functor "sections with support in W". This gives the derived functors

$$Rp_{X*}\colon \mathbf{D}^+(\mathrm{Sh}^{\mathcal{A}}_{\mathfrak{S},\mathfrak{T}}(X)) \to \mathbf{D}^+(\mathrm{Sh}^{\mathcal{A}}_{\mathfrak{S},\mathfrak{T}}(*))$$
$$Rp^W_{X*}\colon \mathbf{D}^+(\mathrm{Sh}^{\mathcal{A}}_{\mathfrak{S},\mathfrak{T}}(X)) \to \mathbf{D}^+(\mathrm{Sh}^{\mathcal{A}}_{\mathfrak{S},\mathfrak{T}}(*))$$

and the natural transformation

$$Ri^W_* : Rp^W_{X*} \to Rp_{X*}. \tag{1.1.4.1}$$

We have the subgroup $\mathcal{Z}^q_W(X/S)$ of $\mathcal{Z}^q(X/S)$ consisting of cycles with support in W.

1.1.5. If $\mathcal{B}$ is a tensor category, and $\mathcal{F}$ is in $\mathrm{Sh}^{\mathcal{B}}_{\mathfrak{S},\mathfrak{T}}$, we recall from (Part II, Chapter IV, §2.3.3) the notion of a (associative, commutative) multiplication

$$\mu\colon p_1^*\mathcal{F} \otimes p_2^*\mathcal{F} \to \mathcal{F} \circ \times,$$

i.e., a collection of natural maps of sheaves for X and Y in $\mathfrak{S}$,

$$\mu_{X,Y}\colon p_1^*(\mathcal{F}_{|X}) \otimes p_2^*(\mathcal{F}_{|Y}) \to \mathcal{F}_{|X\times Y}.$$

If $\mathcal{F} = \oplus_q \mathcal{F}(q)$ is a graded object in $\mathrm{Sh}^{\mathcal{B}}_{\mathfrak{S},\mathfrak{T}}$, a multiplication μ is said to be *graded* if μ restricts to natural transformations $\mu_{p,q}\colon p_1^*\mathcal{F}(p) \otimes p_2^*\mathcal{F}(q) \to \mathcal{F}(p+q) \circ \times$.

1.1.6. DEFINITION. Let $\mathcal{F} = \oplus_{q=0}^{\infty}\mathcal{F}(q) \in \mathbf{C}^+(\mathrm{Sh}^{\mathcal{A}}_{\mathfrak{S},\mathfrak{T}})$ be a graded complex of flat sheaves, with an associative, commutative, graded multiplication

$$\mu\colon p_1^*\mathcal{F} \otimes p_2^*\mathcal{F} \to \mathcal{F} \circ \times \tag{1.1.6.1}$$

We say that $\mathcal{F}$ defines a *geometric cohomology theory on* $\mathcal{V}$ if the $\mathcal{F}$ has the following properties:

(i) *Homotopy.* Let $p\colon X \to Y$ be the inclusion of a closed codimension one subscheme. Let $T \subset Y$ be a reduced closed subscheme, and let $W = p^{-1}(T)$. Suppose that the inclusion $p\colon X \to Y$ is a map in $\mathcal{V}$. Suppose further that $T \cong \mathbb{A}^1_W$, and that, via this isomorphism, $p\colon W \to T$ is the inclusion of $W \times 0$ into $\mathbb{A}^1_W$. Then the map $p^*\colon Rp^T_{Y*}\mathcal{F}_Y \to Rp^W_{X*}\mathcal{F}_X$ is an isomorphism in $\mathbf{D}^+(\mathrm{Sh}^{\mathcal{A}}_{\mathfrak{S},\mathfrak{T}}(*))$.

(ii) *Cycle classes.* Let X be in $\mathcal{V}$, and $W \subset X$ a closed subset such that W is the support of an effective cycle in $\mathcal{Z}^q(X/S)$. Then there is a homomorphism

$$\mathrm{cl}^q_{X,W}\colon \mathcal{Z}^q_W(X/S) \to \mathrm{Hom}_{\mathbf{D}^+(\mathrm{Sh}^{\mathcal{A}}_{\mathfrak{S},\mathfrak{T}}(*))}(\tilde{1}, Rp^W_{X*}\mathcal{F}_X(q)[2q]).$$

Here $\tilde{1}$ is the constant sheaf on $* = \alpha(S)$ with value the unit $1 \in \mathcal{A}$. The maps $\mathrm{cl}^q_{X,W}$ are functorial in the following sense:

(a) If $f\colon Y \to X$ is a map in $\mathcal{V}$, and if $f^{-1}(W)$ is contained in the support W' of some effective cycle in $\mathcal{Z}^q(Y/S)$, then the diagram

$$\begin{array}{ccc} \mathcal{Z}^q_W(X/S) & \xrightarrow{\mathrm{cl}^q_{X,W}} & \mathrm{Hom}_{\mathbf{D}^+(\mathrm{Sh}^{\mathcal{A}}_{\mathfrak{S},\mathfrak{T}}(*))}(\tilde{1}, Rp^W_{X*}\mathcal{F}_X(q)[2q]) \\ \downarrow{\scriptstyle f^*} & & \downarrow{\scriptstyle f^*} \\ \mathcal{Z}^q_{W'}(Y/S) & \xrightarrow[\mathrm{cl}^q_{Y,W'}]{} & \mathrm{Hom}_{\mathbf{D}^+(\mathrm{Sh}^{\mathcal{A}}_{\mathfrak{S},\mathfrak{T}}(*))}(\tilde{1}, Rp^{W'}_{Y*}\mathcal{F}_Y(q)[2q]) \end{array}$$

commutes (by our assumption on W and f, the cycle $f^*(Z)$ is defined for all $Z \in \mathcal{Z}^q_W(X/S)$).

(b) If $T \subset Y$ is the support of an effective cycle in $\mathcal{Z}^{q'}(Y/S)$, then

$$\mathrm{cl}^q_{X,W}(Z) \boxtimes \mathrm{cl}^{q'}_{Y,T}(Z') = \mathrm{cl}^{q+q'}_{X\times_S Y, W\times_S T}(Z \times_{/S} Z')$$

for all $Z \in \mathcal{Z}^q_W(X)$, $Z' \in \mathcal{Z}^{q'}_T(Y)$. Here $\boxtimes$ is the external product induced by μ.

(iii) *Semi-purity.* Let X be in $\mathcal{V}$, and $W \subset X$ a closed subset which is the support of an effective cycle in $\mathcal{Z}^q(X/S)$. Then

$$\mathrm{Hom}_{\mathbf{D}^+(\mathrm{Sh}^{\mathcal{A}}_{\mathfrak{S},\mathfrak{T}}(*))}(\tilde{1}, Rp^W_{X*}\mathcal{F}_X(q)[2q-p]) = 0$$

for $p > 0$.

(iv) *Künneth isomorphism.* For all X, Y in $\mathcal{V}$, the external products

$$\boxtimes^{q_1,q_2}_{\mathcal{F}}(X,Y)\colon Rp_{X*}\mathcal{F}(q_1) \otimes^L Rp_{Y*}\mathcal{F}_Y(q_2) \to Rp_{X\times_S Y*}\mathcal{F}_{X\times_S Y}(q_1+q_2)$$

induced by the product (1.1.6.1) are isomorphisms in $\mathbf{D}^+(\mathrm{Sh}^{\mathcal{A}}_{\mathfrak{S},\mathfrak{T}}(*))$.

(v) *Gysin isomorphism.* Let $p\colon P \to X$ be a smooth morphism in $\mathcal{V}$ of relative dimension d, with section $s\colon X \to P$, giving the map $\mathrm{cl}^d_{P,s(X)}(|s(X)|)\colon \tilde{1} \to Rp^{s(X)}_{P*}\mathcal{F}_P(d)[2d]$. Then the composition

$$Rp_{X*}\mathcal{F}_X(q) \xrightarrow{p^*} Rp_{P*}\mathcal{F}_P(q) \xrightarrow{(-)\cup\mathrm{cl}^d_{s(X),P}(|s(X)|)} Rp^{s(X)}_{P*}\mathcal{F}_P(q+d)[2d]$$

is an isomorphism in $\mathbf{D}^+(\mathrm{Sh}^{\mathcal{A}}_{\mathfrak{S},\mathfrak{T}}(*))$.

(vi) *Unit.* The cycle class map $\mathrm{cl}^0(|S|)\colon \tilde{1} \to \mathcal{F}_S(0)$ associated to the fundamental class on S is an isomorphism in $\mathbf{D}^+(\mathrm{Sh}^{\mathcal{A}}_{\mathfrak{S},\mathfrak{T}}(*))$.

1.1.7. REMARK. Suppose we have a *twisted duality theory* $\Gamma(*)$ on $\mathcal{V}$, in the sense of [**20**] or [**46**]. Then, for $p: X \to Y$ the inclusion of a closed codimension d subscheme, with X and Y smooth over S, we have the Poincaré duality isomorphism, $H^p(X, \Gamma(q)) \to H_X^{p+2d}(Y, \Gamma(q+d))$. This implies part (iv) above as a special case, and reduces part (i) to the usual form of the homotopy axiom:
(i)′ For X in $\mathcal{V}$, the map $p^*: Rp_{X*}\mathcal{F}_X \to Rp_{\mathbb{A}^1_X*}\mathcal{F}_{\mathbb{A}^1_X}$ is an isomorphism, where $p: \mathbb{A}^1_X \to X$ is the projection.

If the base is a perfect field k, the semi-purity condition (iii) reduces to the condition $H^p(X, \Gamma(0)) = 0$ for $p < 0$, and (ii) is implied by requiring $H^0(X, \Gamma(0))$ to be the free $H^0(S, \Gamma(0))$-module on the fundamental classes of the connected components of X, together with the projection formula. In particular, for $S = \operatorname{Spec} k$, k a perfect field, a twisted duality theory $\Gamma(*)$ gives rise to a geometric cohomology theory if $\Gamma(*)$ (with its product) is given as $\Gamma(*) = R\beta_*(\mathcal{F}(*))$, where $\beta: \mathcal{V}_{\mathfrak{T}} \to \mathcal{V}_{\mathrm{Zar}}$ is a map of a Grothendieck site $(\mathcal{V}, \mathfrak{T})$ with enough points to the Zariski site on $\mathcal{V}$, and $\mathcal{F}(*)$ is a graded complex of flat sheaves on $\mathcal{V}$ for the topology $\mathfrak{T}$ with an associative and commutative product.

1.2. Cohomology and cohomology with support

We show how to define canonical cochain complexes for cohomology and for "cohomology with support in codimension q".

1.2.1. Let (X, f, q) be in $\mathcal{L}(\mathcal{V}) \times \mathbb{Z}$, and let $(X, f)^{(q)}$ denote the set of closed subsets $W \subset X$ such that W is the support of an effective cycle in $\mathcal{Z}^q(X)_f$.

Recall the symmetric monoidal category $\mathcal{L}(\mathcal{V})$ (Chapter I, Definition 1.1.2). We have the faithful functor

$$\begin{aligned} i&: \mathcal{L}(\mathcal{V}) \times \mathbb{Z} \to \mathcal{V} \\ i&(X, f, q) = X. \end{aligned} \tag{1.2.1.1}$$

Via i we identify $\operatorname{Hom}_{\mathcal{L}(\mathcal{V})\times\mathbb{Z}}((X, f, q), (Y, f', q))$ with a subset of $\operatorname{Hom}_{\mathcal{V}}(Y, X)$; there are no morphisms from (X, f, q) to (Y, f', q') if $q \neq q'$.

1.2.2. LEMMA. *(i) Let $g: (Y, f') \to (X, f)$ be a map in $\mathcal{L}(\mathcal{V})$. Then for each W in $(X, f)^{(q)}$, $g^{-1}(W)$ is in $(Y, f')^{(q)}$.*
(ii) For $W \in (X, f)^{(q)}$, $W' \in (X', f')^{(q')}$, the subset $W \times_S W'$ of $X \times_S X'$ is in $(X \times_S X', f \times f')^{(q+q')}$.

PROOF. (i) Suppose W is the support of an effective cycle $Z \in \mathcal{Z}^q(X)_f$. By Chapter I, Lemma 1.2.2, $g^*(Z)$ is defined and is in $\mathcal{Z}^q(Y)_{f'}$. Since Z is effective, and the map g is a map of smooth S-schemes, it follows that $g^*(Z)$ is effective, and $g^{-1}(W)$ is the support of $g^*(Z)$, hence $g^{-1}(W)$ is in $(Y, f')^{(q)}$.

For (ii), write f and f' as $f: Y \to X$ and $f': Y' \to X'$. Then by assumption, $\operatorname{codim}_Y(f^{-1}(W)) \geq q$ and $\operatorname{codim}_{Y'}(f'^{-1}(W')) \geq q'$, hence

$$\operatorname{codim}_{Y\times_S Y'}(f \times f')^{-1}(W \times_S W')) \geq q + q',$$

i.e., $W \times_S W'$ is in $(X \times_S X', f \times f')^{(q+q')}$. □

1.2.3. Let $\mathcal{B}$ be the category $\mathbf{C}^+(\mathcal{A} \times \mathbb{Z})$, i.e., the category of bounded below, graded complexes in $\mathcal{A}$; we will refer to the grading coming from the factor $\mathbb{Z}$ as the *Adams degree.* The tensor structure on $\mathcal{A}$ induces the structure of a tensor category on $\mathcal{B}$. Let $\operatorname{Sh}^{\mathcal{B}}_{\mathfrak{S},\mathfrak{T}}(*)_{\mathrm{fl}}$ denote the full subcategory of $\operatorname{Sh}^{\mathcal{B}}_{\mathfrak{S},\mathfrak{T}}(*)$ with objects the flat sheaves. It follows from §1.1.2(2) and (Part II, Chapter IV, Lemma 2.4.2) that $\mathcal{F}$ is flat if and only if $\mathcal{F}_p$ is a flat R-module for all points p of $\tilde{\mathfrak{S}}$; since $(\mathcal{F}\otimes\mathcal{G})_p \cong \mathcal{F}_p\otimes\mathcal{G}_p$ (Part II, Chapter IV, Lemma 2.4.1), $\mathrm{Sh}^{\mathcal{B}}_{\mathfrak{S},\mathfrak{T}}(*)_{\mathrm{fl}}$ is a tensor subcategory of $\mathrm{Sh}^{\mathcal{B}}_{\mathfrak{S},\mathfrak{T}}(*)$.

We recall from (Part II, Chapter IV, §2.2 and (IV.2.2.1.1)) the construction of the cosimplicial Godement resolution of $\mathcal{F}$ as the augmented cosimplicial object $\mathcal{F} \xrightarrow{\iota} G_{\mathcal{B}}\mathcal{F}$, and the associated augmented cochain complex (*loc. cit.* (IV.2.2.1.2)) $\mathcal{F} \xrightarrow{\iota} \mathrm{cc}G_{\mathcal{B}}\mathcal{F}$.

We now define the functors

$$p_*G\mathcal{F}\colon \mathcal{L}(\mathcal{V})^{\mathrm{op}} \times \mathbb{Z} \to \mathrm{c.s.Sh}^{\mathcal{B}}_{\mathfrak{S},\mathfrak{T}}(*)_{\mathrm{fl}}, \tag{1.2.3.1}$$

$$p_*^{(*)}G\mathcal{F}\colon \mathcal{L}(\mathcal{V})^{\mathrm{op}} \times \mathbb{Z} \to \mathrm{c.s.Sh}^{\mathcal{B}}_{\mathfrak{S},\mathfrak{T}}(*)_{\mathrm{fl}}, \tag{1.2.3.2}$$

and natural transformation

$$i_*^{(*)}G\colon p_*^{(*)}G\mathcal{F} \to p_*G\mathcal{F}. \tag{1.2.3.3}$$

Here "c.s." is the category of cosimplicial objects.

To define the functor (1.2.3.1), we start with the functor

$$p_*G\mathcal{F}_{\mathcal{V}}\colon \mathcal{V}^{\mathrm{op}} \times \mathbb{Z} \to \mathrm{c.s.Sh}^{\mathcal{B}}_{\mathfrak{S},\mathfrak{T}}(*)_{\mathrm{fl}}$$

gotten by sending X in $\mathcal{V}$ to $p_{X*}G_{\mathcal{B}}\mathcal{F}(q)(X)$, concentrated in Adams degree q, and similarly for morphisms; it follows from §1.1.2(2) and (4) that $p_{X*}G_{\mathcal{B}}\mathcal{F}(q)(X)$ is in fact a flat sheaf on S in each degree.

The projection on the first and last factors defines the functor $p_{13}\colon \mathcal{L}(\mathcal{V})^{\mathrm{op}}\times\mathbb{Z} \to \mathcal{V}^{\mathrm{op}} \times \mathbb{Z}$; we then define $p_*G\mathcal{F} := p_*G\mathcal{F}_{\mathcal{V}} \circ p_{13}$.

For the functor (1.2.3.2), let (X, f, q) be in $\mathcal{L}(\mathcal{V})^{\mathrm{op}}\times\mathbb{Z}$, let W be in $(X, f, q)^{(q)}$, and define

$$p_*^W G\mathcal{F} := \ker[p_*G\mathcal{F}(j^*)\colon p_*G\mathcal{F}(X, f, q) \to p_*G\mathcal{F}(X \setminus W, j^*f, q)],$$

where $j\colon X\setminus W \to X$ is the inclusion; by (Part II, Chapter IV, Lemma 2.2.3) $p_*^W G\mathcal{F}$ is also flat. We then set

$$p_*^{(*)}G\mathcal{F}(X, f, q) := \varinjlim_{W\in(X,f,q)^{(q)}} p_*^W G\mathcal{F},$$

which is flat by §1.1.2(2), since the functor i^* preserves inductive limits, and an inductive limit of flat R-modules is flat.

By Lemma 1.2.2, the subcosimplicial object $p_*^{(*)}G\mathcal{F}(X, f, q)$ of $p_*G\mathcal{F}(X, f, q)$ is functorial with respect to the morphisms in $\mathcal{L}(\mathcal{V})^{\mathrm{op}}\times\mathbb{Z}$, giving the functor (1.2.3.2). The inclusions $p_*^{(*)}G\mathcal{F}(X, f, q) \subset p_*G\mathcal{F}(X, f, q)$ define the natural transformation (1.2.3.3).

Taking the total complex of the cochain complex associated to the cosimplicial object defines the functors

$$\begin{aligned} &\check{\mathcal{F}}\colon \mathcal{L}(\mathcal{V})^{\mathrm{op}} \times \mathbb{Z} \to \mathbf{C}^+(\mathrm{Sh}^{\mathcal{B}}_{\mathfrak{S},\mathfrak{T}}(*)_{\mathrm{fl}}) \\ &\check{\mathcal{F}} := \mathrm{Tot}(\mathrm{cc}(p_*G^*\mathcal{F})), \\ &\check{\mathcal{F}}^{(*)}\colon \mathcal{L}(\mathcal{V})^{\mathrm{op}} \times \mathbb{Z} \to \mathbf{C}^+(\mathrm{Sh}^{\mathcal{B}}_{\mathfrak{S},\mathfrak{T}}(*)_{\mathrm{fl}}) \\ &\check{\mathcal{F}}^{(*)} := \mathrm{Tot}(\mathrm{cc}(p_*^{(*)}G^*\mathcal{F})), \end{aligned} \tag{1.2.3.4}$$

and natural transformation

$$(1.2.3.5)\qquad \begin{aligned} &\check{\imath}_*:\check{\mathcal{F}}^{(*)}\to\check{\mathcal{F}}\\ &\check{\imath}_* := \mathrm{Tot}(\mathrm{cc}(i_*^{(*)}G)). \end{aligned}$$

Since $\check{\mathcal{F}}(X,f,q)$ only depends on (X,q), we often write $\check{\mathcal{F}}(X,q)$ for $\check{\mathcal{F}}(X,f,q)$.

We define $Rp_{X*}^{(q)}\mathcal{F}(X)(q)_f$ as the inductive limit:

$$Rp_{X*}^{(q)}\mathcal{F}(X)(q)_f := \varinjlim_{W\in(X,f)^{(q)}} Rp_{X*}^W\mathcal{F}(X)(q)_f.$$

Sending (X,f,q) to $Rp_{X*}^{(q)}\mathcal{F}_X(q)_f$ or $Rp_{X*}\mathcal{F}_X(q)$ (concentrated in Adams degree q) defines functors

$$\begin{aligned} Rp_*^{(-)}\mathcal{F}&:\mathcal{L}(\mathcal{V})^*\to\mathbf{D}^+(\mathrm{Sh}^{\mathcal{B}}_{\mathfrak{S},\mathfrak{T}}(*)),\\ Rp_*\mathcal{F}&:\mathcal{L}(\mathcal{V})^*\to\mathbf{D}^+(\mathrm{Sh}^{\mathcal{B}}_{\mathfrak{S},\mathfrak{T}}(*)). \end{aligned}$$

The natural transformation (1.1.4.1) defines the natural transformation

$$Ri_*\mathcal{F}: Rp_*^{(-)}\mathcal{F}\to Rp_*\mathcal{F}.$$

1.2.4. LEMMA. *(i) For $(X,f,q)\in\mathcal{L}(\mathcal{V})^*$, the complexes $\check{\mathcal{F}}^{(*)}(X,f,q)$ and $\check{\mathcal{F}}(X,f,q)$ are complexes of acyclic sheaves on S.*
(ii) For $\mathcal{G}\in\mathbf{C}^+(\mathrm{Sh}^{\mathcal{B}}_{\mathfrak{S},\mathfrak{T}}(*))$, we let $R\mathcal{G}$ denote the image of $\mathcal{G}$ in $\mathbf{D}^+(\mathrm{Sh}^{\mathcal{B}}_{\mathfrak{S},\mathfrak{T}}(*))$. There are canonical isomorphisms of functors

$$\begin{aligned} R\check{\mathcal{F}}^{(-)}&\to Rp_*^{(-)}\mathcal{F},\\ R\check{\mathcal{F}}&\to Rp_*\mathcal{F}. \end{aligned}$$

In addition, the diagram

$$\begin{array}{ccc} R\check{\mathcal{F}}^{(-)} & \longrightarrow & Rp_*^{(-)}\mathcal{F}\\ {\scriptstyle R\check{\imath}_*}\downarrow & & \downarrow{\scriptstyle Ri_*}\\ R\check{\mathcal{F}} & \longrightarrow & Rp_*\mathcal{F} \end{array}$$

commutes.

PROOF. This follows from Part II, Chapter IV, Lemma 2.2.2, Lemma 2.2.3 and Remark 2.2.4. □

1.3. The construction of the realization functor

We now give the construction of the realization functor

$$\Re_{\mathcal{F}}:\mathcal{DM}(\mathcal{V})\to\mathbf{D}^+(\mathrm{Sh}^{\mathcal{B}}_{\mathfrak{S},\mathfrak{T}}(*))$$

associated to a geometric cohomology theory $\mathcal{F}$ on $\mathcal{V}$. Except for one point, the construction would be an essentially straightforward step-by-step extension of the functor

$$\begin{aligned} \check{\mathcal{F}}&:\mathcal{V}^{\mathrm{op}}\times\mathbb{Z}\to\mathbf{C}^+(\mathrm{Sh}^{\mathcal{B}}_{\mathfrak{S},\mathfrak{T}}(*)_{\mathfrak{A}})\\ (X,q)&\mapsto\check{\mathcal{F}}(X,f,q), \end{aligned}$$

to the DG tensor category $\mathcal{A}_{\mathrm{mot}}(\mathcal{V})$, and from there, a direct extension to the category of complexes $\mathbf{C}^b_{\mathrm{mot}}(\mathcal{V})$, the homotopy category $\mathbf{K}^b_{\mathrm{mot}}(\mathcal{V})$, and the localization $\mathbf{D}^b_{\mathrm{mot}}(\mathcal{V})$. One then applies (Part II, Chapter II, Corollary 2.4.10), to give the extension to $\mathcal{DM}(\mathcal{V})$. The problem is that in general one cannot define the external product on the complexes $\check{\mathcal{F}}(X, f, q)$ to be associative and graded-commutative (on the level of complexes), but only graded-commutative up to homotopy (although still associative). We must then replace the DG category $\mathcal{A}_{\mathrm{mot}}(\mathcal{V})$ with an up-to-homotopy commutative model $\mathcal{A}^{\mathfrak{sh}}_{\mathrm{mot}}(\mathcal{V})$, and use (Part II, Chapter II, Theorem 2.2.2), to get back to the homotopy category $\mathbf{K}^b_{\mathrm{mot}}(\mathcal{V})$. The extension to $\mathcal{DM}(\mathcal{V})$ then proceeds as outlined above. We now give the details of this construction.

1.3.1. THEOREM. *Let $\mathcal{F} = \oplus_{q=0}^{\infty}\mathcal{F}(q) \in \mathbf{C}^+(\mathrm{Sh}^{\mathcal{B}}_{\mathfrak{S},\mathfrak{T}})$ be a graded complex of flat sheaves on the site $(\mathfrak{S}, \mathfrak{T})$, with values in $\mathcal{A} := \mathbf{Mod}_R$, having an associative, graded-commutative product* (1.1.6.1). *Suppose $\mathcal{F}$ defines a geometric cohomology theory on $\mathcal{V}$ (see Definition 1.1.6). Let A be a commutative ring, flat over $\mathbb{Z}$, and suppose R is an A-algebra. Then the functor*

$$\check{\mathcal{F}}\colon \mathcal{V}^{\mathrm{op}} \times \mathbb{Z} \to \mathbf{C}^+(\mathrm{Sh}^{\mathcal{B}}_{\mathfrak{S},\mathfrak{T}}(*))$$
$$(X, q) \mapsto \check{\mathcal{F}}(X, q)$$

has a canonical extension to an exact functor

$$\Re_{\mathcal{F}}\colon \mathcal{DM}(\mathcal{V})_A \to \mathbf{D}^+(\mathrm{Sh}^{\mathcal{B}}_{\mathfrak{S},\mathfrak{T}}(*)).$$

The functor $\Re_{\mathcal{F}}$ is natural in the geometric cohomology theory $\mathcal{F}$, and is natural, up to canonical isomorphism, in the category $\mathcal{V}$. In addition, there is a full tensor triangulated subcategory $\mathcal{DM}_{\mathfrak{sh}}(\mathcal{V})_A$ of $\mathcal{DM}(\mathcal{V})_A$, containing the motives $A_X(n)$ for all X in $\mathcal{V}$, and with essential image all of $\mathcal{DM}(\mathcal{V})_A$, such that the restriction of $\Re_{\mathcal{F}}$ to $\mathcal{DM}_{\mathfrak{sh}}(\mathcal{V})_A$ is an exact, pseudo-tensor functor (Part II, Chapter I, §1.3.7). The subcategory $\mathcal{DM}_{\mathfrak{sh}}(\mathcal{V})_A$ is independent of $\mathcal{F}$, and is natural in $\mathcal{V}$.

The proof proceeds in a series of steps:
Step 1. The extension to $\mathcal{A}_1(\mathcal{V})$:

1.3.2. Using the additive structure of $\mathbf{C}^+(\mathrm{Sh}^{\mathcal{B}}_{\mathfrak{S},\mathfrak{T}}(*)_{\mathrm{fl}})$, and the fact that sheaves transform disjoint unions to direct sums, the functors (1.2.3.4) and natural transformation (1.2.3.5) canonically extend to functors

$$\begin{aligned}\check{\mathcal{F}}_1^*&\colon \mathcal{A}_1(\mathcal{V}) \to \mathbf{C}^+(\mathrm{Sh}^{\mathcal{B}}_{\mathfrak{S},\mathfrak{T}}(*)_{\mathrm{fl}}),\\ \check{\mathcal{F}}_1&\colon \mathcal{A}_1(\mathcal{V}) \to \mathbf{C}^+(\mathrm{Sh}^{\mathcal{B}}_{\mathfrak{S},\mathfrak{T}}(*)_{\mathrm{fl}}),\end{aligned} \tag{1.3.2.1}$$

with $\check{\mathcal{F}}_1^*(\mathbb{Z}_X(a)_f) = \check{\mathcal{F}}^{(*)}(X, a, f)$ and $\check{\mathcal{F}}_1(\mathbb{Z}_X(a)_f) = \check{\mathcal{F}}(X, a, f)$, and natural transformation

$$\check{i}_*\mathcal{F}_1\colon \check{\mathcal{F}}_1^* \to \check{\mathcal{F}}_1 \tag{1.3.2.2}$$

(see Chapter I, Definition 1.4.1).
Step 2. The category $\mathcal{A}_2^{\mathfrak{sh}}(\mathcal{V})$, and the extension to $\mathcal{A}_2^{\mathfrak{sh}}(\mathcal{V})$:

1.3.3. $\mathcal{A}_2^{\mathfrak{sh}}(\mathcal{V})$. We now apply the constructions of (Part II, Chapter III, §2.1). Using the notation of (Part II, Chapter I, §2.4), and referring to Chapter I, Definition 1.4.4, the category $\mathcal{A}_2(\mathcal{V})$ is the universal commutative external product, $\mathcal{A}_1(\mathcal{V})^{\otimes,c}$, on $\mathcal{A}_1(\mathcal{V})$. Using the construction of (Part II, Chapter III, §2.1.5), and applying (Part II, Chapter III, Theorem 2.1.7), we have the DG tensor category $\mathcal{A}_1(\mathcal{V})^{\otimes,\mathfrak{sh}}$, and the DG tensor functor $\mathfrak{c}\colon \mathcal{A}_1(\mathcal{V})^{\otimes,\mathfrak{sh}} \to \mathcal{A}_1(\mathcal{V})^{\otimes,c}$, which is the identity on objects, surjective on morphisms and a homotopy equivalence. We have as well the additive functor $i^{\mathfrak{sh}}: \mathcal{A}_1(\mathcal{V}) \to \mathcal{A}_1(\mathcal{V})^{\otimes,\mathfrak{sh}}$ with $\mathfrak{c} \circ i^{\mathfrak{sh}}$ the canonical functor (Part II, (I.2.4.3.1)) $i^c: \mathcal{A}_1(\mathcal{V}) \to \mathcal{A}_1(\mathcal{V})^{\otimes,c}$. We set $\mathcal{A}_2^{\mathfrak{sh}}(\mathcal{V}) := \mathcal{A}_1(\mathcal{V})^{\otimes,\mathfrak{sh}}$.

1.3.4. *Multiplicative structure.* We extend the functors (1.2.3.1) and (1.2.3.2) and natural transformation (1.2.3.3) to functors

$$p_*G\mathcal{F}: \mathcal{L}(\mathcal{V})^* \to \text{c.s.Sh}^{\mathcal{B}}_{\mathfrak{S},\mathfrak{T}}(*)_{\mathfrak{A}},$$
$$p_*^{(*)}G\mathcal{F}: \mathcal{L}(\mathcal{V})^* \to \text{c.s.Sh}^{\mathcal{B}}_{\mathfrak{S},\mathfrak{T}}(*)_{\mathfrak{A}},$$

and natural transformation

$$i_*^{(*)}G: p_*^{(*)}G\mathcal{F} \to p_*G\mathcal{F},$$

using the additivity of $\mathcal{F}$ with respect to disjoint union.

By (Part II, Chapter IV, Proposition 2.3.7), the graded product $\mu: p_1^*\mathcal{F} \otimes p_2^*\mathcal{F} \to \mathcal{F} \circ \times$ induces the natural associative, commutative product $G\mu: p_1^*G\mathcal{F} \otimes p_2^*G\mathcal{F} \to G\mathcal{F} \circ \times$. This product induces the natural transformation

$$p_*G\mu: p_*G\mathcal{F} \otimes p_*G\mathcal{F} \to p_*G\mathcal{F} \circ \times, \tag{1.3.4.1}$$

where $\times$ is the symmetric monoidal product on $\mathcal{L}(\mathcal{V})^*$ (see Part II, Chapter III, §2.2.2). It follows from Lemma 1.2.2(ii) that the natural transformation (1.3.4.1) maps $p_*^{(*)}G\mathcal{F} \otimes p_*^{(*)}G\mathcal{F}$ into $p_*^{(*)}G\mathcal{F} \circ \times$, giving the natural transformation

$$p_*^{(*)}G\mu: p_*^{(*)}G\mathcal{F} \otimes p_*^{(*)}G\mathcal{F} \to p_*^{(*)}G\mathcal{F} \circ \times,$$

compatible with (1.3.4.1) via the natural transformation (1.2.3.3). As these products are associative and commutative, they define compatible multiplications on the functors $p_*G\mathcal{F}$ and $p_*^{(*)}G\mathcal{F}$, in the sense of (Part II, Chapter III, §2.2.2).

We have the additive category $\mathbb{Z}\mathcal{L}(\mathcal{V})^*$ freely generated by $\mathcal{L}(\mathcal{V})^*$. Using the results of (Part II, Chapter III, Theorem 2.2.4), and applying the total complex functor Tot, we have the functors of DG tensor categories without unit

$$\begin{aligned} \check{\mathcal{F}}^*: (\mathbb{Z}\mathcal{L}(\mathcal{V})^*)^{\otimes,\mathfrak{sh}} &\to \mathbf{C}^+(\text{Sh}^{\mathcal{B}}_{\mathfrak{S},\mathfrak{T}}(*)_{\mathfrak{A}}), \\ \check{\mathcal{F}}: (\mathbb{Z}\mathcal{L}(\mathcal{V})^*)^{\otimes,\mathfrak{sh}} &\to \mathbf{C}^+(\text{Sh}^{\mathcal{B}}_{\mathfrak{S},\mathfrak{T}}(*)_{\mathfrak{A}}) \end{aligned} \tag{1.3.4.2}$$

extending the functors (1.3.2.1). The natural transformation (1.3.2.2) similarly extends to the natural transformation

$$\check{\mathcal{F}}^* \to \check{\mathcal{F}}. \tag{1.3.4.3}$$

It follows directly from the definition (Part II, Chapter III, §2.1.5) of the functor $(-)^{\otimes,\mathfrak{sh}}$ that $\mathcal{A}_2^{\mathfrak{sh}}(\mathcal{V}) := (\mathcal{A}_1(\mathcal{V}))^{\otimes,\mathfrak{sh}}$ is isomorphic to the DG tensor category gotten by imposing the relations of Chapter I, Definition 1.4.1 on $(\mathbb{Z}\mathcal{L}(\mathcal{V})^*)^{\otimes,\mathfrak{sh}}$. Thus, the functors (1.3.4.2) and the natural transformation (1.3.4.3) extend to the functors of DG tensor categories without unit

$$\begin{aligned} \check{\mathcal{F}}_2^*: \mathcal{A}_2^{\mathfrak{sh}}(\mathcal{V}) &\to \mathbf{C}^+(\text{Sh}^{\mathcal{B}}_{\mathfrak{S},\mathfrak{T}}(*)_{\mathfrak{A}}), \\ \check{\mathcal{F}}_2: \mathcal{A}_2^{\mathfrak{sh}}(\mathcal{V}) &\to \mathbf{C}^+(\text{Sh}^{\mathcal{B}}_{\mathfrak{S},\mathfrak{T}}(*)_{\mathfrak{A}}), \end{aligned} \tag{1.3.4.4}$$

and natural transformation

$$\check{\mathcal{F}}_2^* \to \check{\mathcal{F}}_2. \tag{1.3.4.5}$$

Before proceeding further with our construction, we note the following result; the proof is elementary and is left to the reader:

1.3.5. LEMMA. *Let $F: \mathcal{A} \to \mathcal{B}$ be a DG tensor functor of DG tensor categories without unit. Suppose that*

(i) *F is an isomorphism on objects,*
(ii) *F is surjective on morphisms,*
(iii) *F is a homotopy equivalence.*

Let $f: F(X) \to F(Y)$ be a map of degree a in $\mathcal{B}$ such that $df = 0$. Then there is a map $s: X \to Y$ of degree a in $\mathcal{A}$ such that $F(s) = f$ and $ds = 0$. In addition, let $\mathcal{A}[h_s]$ be the DG tensor category without unit gotten by adjoining a morphism $h_s: X \to Y$ of degree $a-1$ with $dh_s = s$, let $\mathcal{B}[h_f]$ the DG tensor category without unit defined by adjoining a morphism $h_f: F(X) \to F(Y)$ of degree $a-1$ with $dh_f = f$, and let $F': \mathcal{A}[h_s] \to \mathcal{B}[h_f]$ be the extension of F with $F'(h_s) = h_f$. Then F' satisfies (i) and (ii).

Step 3. The category $\mathcal{A}^{\mathfrak{sh}}_{\mathrm{mot}}(\mathcal{V})$ and the extension to $\mathcal{A}_{\mathrm{mot}}(\mathcal{V})$:

1.3.6. *The category $\mathcal{A}^{\mathfrak{sh}}_{\mathrm{mot}}(\mathcal{V})$.* We now form a sequence of DG tensor categories without unit

$$\begin{array}{c} \mathcal{A}_2^{\mathfrak{sh}}(\mathcal{V}) \to \mathcal{A}_3^{\mathfrak{sh}}(\mathcal{V}) \to \mathcal{A}_4^{\mathfrak{sh}}(\mathcal{V}) \to \mathcal{A}_5^{\mathfrak{sh}}(\mathcal{V}) \\ \qquad\qquad\qquad\qquad\qquad\qquad \cup \\ \qquad\qquad\qquad\qquad\qquad\qquad \mathcal{A}^{\mathfrak{sh}}_{\mathrm{mot}}(\mathcal{V}) \end{array}$$

analogous to the sequence of DG tensor categories formed in Chapter I, §1.4.

We recall the homotopy one-point category $\mathbb{E}$ constructed in (Part II, Chapter II, §3.1). $\mathbb{E}$ is a DG tensor category without unit, with the generating object $\mathfrak{e}$. $\mathbb{E}$ has no morphisms of positive degree, no morphisms from $\mathfrak{e}^{\otimes m}$ to $\mathfrak{e}^{\otimes n}$ if $n \neq m$, and

$$H^q(\mathrm{Hom}(\mathfrak{e}^{\otimes n}, \mathfrak{e}^{\otimes n})^*) = \begin{cases} \mathbb{Z} \cdot \mathrm{id} & \text{for q=0,} \\ 0 & \text{otherwise.} \end{cases} \tag{1.3.6.1}$$

We have the coproduct of DG tensor categories without unit $\mathcal{A}_2(\mathcal{V})[\mathbb{E}]$; the DG tensor category $\mathcal{A}_3(\mathcal{V})$ (Chapter I, Definition 1.4.6) is formed from the DG tensor category $\mathcal{A}_2(\mathcal{V})[\mathbb{E}]$ by adjoining morphisms $[Z]: \mathfrak{e} \to \mathbb{Z}_X(n)_f$ of degree $2n$ for each non-zero $Z \in \mathcal{Z}^n(X)_f$. We form the DG tensor category $\mathcal{A}_3^{\mathfrak{sh}}(\mathcal{V})$ from the coproduct $\mathcal{A}_2^{\mathfrak{sh}}(\mathcal{V})[\mathbb{E}]$ by adjoining morphisms $[Z]^{\mathfrak{sh}}: \mathfrak{e} \to \mathbb{Z}_X(n)_f$ of degree $2n$, with $d[Z]^{\mathfrak{sh}} = 0$ for each non-zero $Z \in \mathcal{Z}^n(X)_f$. We extend the functor $\mathfrak{c}[\mathrm{id}_{\mathbb{E}}]: \mathcal{A}_2^{\mathfrak{sh}}(\mathcal{V})[\mathbb{E}] \to \mathcal{A}_2(\mathcal{V})[\mathbb{E}]$ to

$$\mathfrak{c}_3: \mathcal{A}_3^{\mathfrak{sh}}(\mathcal{V}) \to \mathcal{A}_3(\mathcal{V}) \tag{1.3.6.2}$$

by setting $\mathfrak{c}_3([Z]^{\mathfrak{sh}}) = [Z]$. By (Part II, Chapter II, Proposition 2.2.4), the functor (1.3.6.2) is a homotopy equivalence.

For each pair of objects Γ, Δ of $\mathcal{A}_1(\mathcal{V})$, we have the external product $\boxtimes_{\Gamma,\Delta}: \Gamma \otimes \Delta \to \Gamma \times \Delta$ in $\mathcal{A}_2(\mathcal{V})$, and the lifting of $\boxtimes_{\Gamma,\Delta}$ to the external product $\boxtimes^{\mathfrak{sh}}_{\Gamma,\Delta}: \Gamma \otimes \Delta \to \Gamma \times \Delta$ in $\mathcal{A}_2^{\mathfrak{sh}}(\mathcal{V})$ (see (Part II, (III.2.1.6.1) and Chapter III, Theorem 2.1.7). We note that $d\boxtimes^{\mathfrak{sh}}_{\Gamma,\Delta} = 0$.

The DG tensor category $\mathcal{A}_4(\mathcal{V})$ is formed from graded tensor category $\mathcal{A}_3(\mathcal{V})$ by selecting certain morphisms f in $\mathcal{A}_3(\mathcal{V})$, and adjoining morphisms h_f with $dh_f = f$ (see Chapter I, Definition 1.4.8). The morphisms f are all constructed from the morphisms $[Z]$, $\boxtimes_{**}$ and $\otimes$, together with morphisms of the category $\mathcal{A}_1(\mathcal{V})$. Given such an expression for a morphism f, we let $f^{\mathfrak{sh}}$ be the morphism in $\mathcal{A}_3^{\mathfrak{sh}}(\mathcal{V})$ gotten by replacing each occurrence of the morphism $[Z]$ with the morphism $[Z]^{\mathfrak{sh}}$, and replacing $\boxtimes_{**}$ with $\boxtimes_{**}^{\mathfrak{sh}}$. Since $d[Z]^{\mathfrak{sh}} = 0$ and $d\boxtimes_{**}^{\mathfrak{sh}} = 0$, we have $df^{\mathfrak{sh}} = 0$ as well. We then adjoin, for each such f, a morphism $h_f^{\mathfrak{sh}}$ to $\mathcal{A}_3^{\mathfrak{sh}}(\mathcal{V})$ with $dh_f^{\mathfrak{sh}} = f^{\mathfrak{sh}}$, forming the DG tensor category without unit $\mathcal{A}_4^{\mathfrak{sh}}(\mathcal{V})$.

We extend (1.3.6.2) to $\mathfrak{c}_4\colon \mathcal{A}_4^{\mathfrak{sh}}(\mathcal{V}) \to \mathcal{A}_4(\mathcal{V})$ by setting $\mathfrak{c}_4(h_f^{\mathfrak{sh}}) = h_f$. By (Part II, Chapter II, Proposition 2.2.4), $\mathfrak{c}_4$ is a homotopy equivalence; by Lemma 1.3.5, $\mathfrak{c}_4$ is the identity on objects and surjective on morphisms.

The category $\mathcal{A}_5(\mathcal{V})$ is formed from $\mathcal{A}_4(\mathcal{V})$ by forming a succession of categories

$$\mathcal{A}_5(\mathcal{V})^{(0)} = \mathcal{A}_4(\mathcal{V}) \subset \ldots \subset \mathcal{A}_5(\mathcal{V})^{(r,k-1)} \subset \mathcal{A}_5(\mathcal{V})^{(r,k)} \subset \ldots$$

and letting $\mathcal{A}_5(\mathcal{V})$ be the inductive limit. The category $\mathcal{A}_5(\mathcal{V})^{(r,k)}$ is formed from $\mathcal{A}_5(\mathcal{V})^{(r,k-1)}$ by adjoining morphisms $h_f\colon \mathfrak{e}^{\otimes k} \to \mathbb{Z}_X(n)_f$ of degree $2n-r-1$, with $dh_f = f$ for each non-zero morphism

$$f\colon \mathfrak{e}^{\otimes k} \to \mathbb{Z}_X(n)_f \tag{1.3.6.3}$$

of degree r in $\mathcal{A}_5(\mathcal{V})^{(r-1)}$ with $df = 0$. This is done successively for $k = 1, 2, \ldots$, which gives us the category $\mathcal{A}_5(\mathcal{V})^{(r+1,0)}$; we then take the inductive limit of the categories $\mathcal{A}_5(\mathcal{V})^{(r,0)}$ to form $\mathcal{A}_5(\mathcal{V})$.

Using Lemma 1.3.5 and (Part II, Chapter II, Proposition 2.2.4), we may inductively construct the sequence of DG tensor categories

$$\mathcal{A}_5^{\mathfrak{sh}}(\mathcal{V})^{(0)} = \mathcal{A}_4^{\mathfrak{sh}}(\mathcal{V}) \subset \ldots \subset \mathcal{A}_5^{\mathfrak{sh}}(\mathcal{V})^{(r,k-1)} \subset \mathcal{A}_5^{\mathfrak{sh}}(\mathcal{V})^{(r,k)} \subset \ldots$$

and DG tensor functors

$$\mathfrak{c}_5^{(r)}\colon \mathcal{A}_5^{\mathfrak{sh}}(\mathcal{V})^{(r,k)} \to \mathcal{A}_5(\mathcal{V})^{(r,k)} \tag{1.3.6.4}$$

which are homotopy equivalences, the identity on objects and surjective on morphisms as follows: Assuming we have constructed the sequence up to $(r, k-1)$, we may lift each morphism (1.3.6.3) to a morphism $f^{\mathfrak{sh}}\colon \mathfrak{e}^{\otimes k} \to \mathbb{Z}_X(n)_f$ in $\mathcal{A}_5^{\mathfrak{sh}}(\mathcal{V})^{(r,k-1)}$ with $df^{\mathfrak{sh}} = 0$. We may then adjoin morphisms $h_f^{\mathfrak{sh}}\colon \mathfrak{e}^{\otimes k} \to \mathbb{Z}_X(n)_f$ with $dh_f^{\mathfrak{sh}} = f^{\mathfrak{sh}}$, forming the DG tensor category $\mathcal{A}_5^{\mathfrak{sh}}(\mathcal{V})^{(r,k)}$. The extension of $\mathfrak{c}_5^{(r,k-1)}$ to $\mathfrak{c}_5^{(r,k)}$ is defined by $\mathfrak{c}_5^{(r,k)}(h_f^{\mathfrak{sh}}) = h_f$.

Taking the inductive limit over (r, k) of (1.3.6.4) gives the DG tensor functor

$$\mathfrak{c}_5\colon \mathcal{A}_5^{\mathfrak{sh}}(\mathcal{V}) \to \mathcal{A}_5(\mathcal{V}), \tag{1.3.6.5}$$

which is a homotopy equivalence.

Finally, the category $\mathcal{A}_{\mathrm{mot}}(\mathcal{V})$ is defined as the full DG tensor subcategory of $\mathcal{A}_5(\mathcal{V})$ generated by objects of the form $\mathbb{Z}_X(n)_f$ or $\mathfrak{e}^{\otimes a} \otimes \mathbb{Z}_X(n)_f$, $a \geq 1$ (see Chapter I, Definition 1.4.10). We let $\mathcal{A}_{\mathrm{mot}}^{\mathfrak{sh}}(\mathcal{V})$ be the full DG tensor subcategory of $\mathcal{A}_5^{\mathfrak{sh}}(\mathcal{V})$ generated by objects of the form $\mathbb{Z}_X(n)_f$ or $\mathfrak{e}^{\otimes a} \otimes \mathbb{Z}_X(n)_f$, $a \geq 1$. Since (1.3.6.5) is a homotopy equivalence, the identity on objects and surjective on morphisms, the same is true for the restriction $\mathfrak{c}_{\mathrm{mot}}\colon \mathcal{A}_{\mathrm{mot}}^{\mathfrak{sh}}(\mathcal{V}) \to \mathcal{A}_{\mathrm{mot}}(\mathcal{V})$.

1.3.7. *The extension to $\mathcal{A}_{\mathrm{mot}}^{\mathfrak{sh}}(\mathcal{V})$.* It is now a straightforward matter to extend the functors (1.3.4.4) and the natural transformation (1.3.4.5) to $\mathcal{A}_{\mathrm{mot}}^{\mathfrak{sh}}(\mathcal{V})$.

We consider $\mathcal{A}$ as a DG tensor category with all morphisms in degree zero, and all differentials zero. By (1.3.6.1) and the universal mapping property of the category $\mathbb{E}$ (see Part II, Chapter II, Proposition 3.1.13), there is a unique functor of DG tensor categories $I:\mathbb{E}\to\mathcal{A}$ with $I(\mathfrak{e})=R$.

The sheaf $\tilde{1}^{\otimes n}$ is the constant sheaf on $*$ corresponding to $R^{\otimes n}\cong R$. Let

$$(1.3.7.1)\qquad \begin{aligned}\tilde{I}:\mathbb{E}&\to \mathrm{Sh}^{\mathcal{B}}_{\mathfrak{S},\mathfrak{T}}(*)_{\mathfrak{A}}\\ \tilde{I}(\mathfrak{e}^{\otimes n})&=\tilde{1}^{\otimes n}\end{aligned}$$

be the sheafification of the functor I (where the objects $\tilde{1}^{\otimes n}$ are in Adams degree 0).

Taking the coproduct of $\tilde{I}$ with the functors (1.3.4.4) gives the DG tensor functors

$$(1.3.7.2)\qquad \begin{aligned}\check{\mathcal{F}}_2^*[\tilde{I}]:\mathcal{A}_2^{\mathfrak{sh}}(\mathcal{V})[\mathbb{E}]&\to \mathbf{C}^+(\mathrm{Sh}^{\mathcal{B}}_{\mathfrak{S},\mathfrak{T}}(*)_{\mathfrak{A}}),\\ \check{\mathcal{F}}_2[\tilde{I}]:\mathcal{A}_2^{\mathfrak{sh}}(\mathcal{V})[\mathbb{E}]&\to \mathbf{C}^+(\mathrm{Sh}^{\mathcal{B}}_{\mathfrak{S},\mathfrak{T}}(*)_{\mathfrak{A}}).\end{aligned}$$

The natural transformation (1.3.4.5) extends similarly to the natural transformation

$$(1.3.7.3)\qquad \check{\mathcal{F}}_2^*[\tilde{I}]\to\check{\mathcal{F}}_2[\tilde{I}].$$

It follows from Lemma 1.2.4 that we have the isomorphism

$$(1.3.7.4)\qquad \begin{aligned}&\mathrm{Hom}_{K^+(\mathrm{Sh}^{\mathcal{B}}_{\mathfrak{S},\mathfrak{T}}(*)_{\mathfrak{A}})}(\tilde{1},\check{\mathcal{F}}_X^n(n)_f)\\ &\quad\cong \varinjlim_{W\in(X,f)^{(q)}} \mathrm{Hom}_{D^+(\mathrm{Sh}^{\mathcal{B}}_{\mathfrak{S},\mathfrak{T}}(*)_{\mathfrak{A}})}(\tilde{1},Rp^W_{X*}\mathcal{F}_X(n)).\end{aligned}$$

For each non-zero Z in $\mathcal{Z}^n(X)_f$, define $\check{\mathcal{F}}_3^*([Z]^{\mathfrak{sh}}):\tilde{1}\to\check{\mathcal{F}}_X^n(n)_f[2n]$ to be a choice of a map in $\mathbf{C}^+(\mathrm{Sh}^{\mathcal{B}}_{\mathfrak{S},\mathfrak{T}}(*)_{\mathfrak{A}})$ representing the map

$$\mathrm{cl}^q_{X,W}(Z):\tilde{1}\to Rp^W_{X*}\mathcal{F}_X(n)[2n]$$

in $\mathbf{D}^+(\mathrm{Sh}^{\mathcal{B}}_{\mathfrak{S},\mathfrak{T}}(*)_{\mathfrak{A}})$ given by Definition 1.1.6(ii). We define $\check{\mathcal{F}}_3([Z]^{\mathfrak{sh}}):\tilde{1}\to\check{\mathcal{F}}_X(n)_f$ to be the composition of $\check{\mathcal{F}}_3^*([Z]^{\mathfrak{sh}})$ with the natural map $\check{\mathcal{F}}_X^*(n)_f\to\check{\mathcal{F}}_X(n)_f$. This gives the extension of the functors (1.3.7.2) to functors

$$(1.3.7.5)\qquad \begin{aligned}\check{\mathcal{F}}_3^*:\mathcal{A}_3^{\mathfrak{sh}}(\mathcal{V})&\to\mathbf{C}^+(\mathrm{Sh}^{\mathcal{B}}_{\mathfrak{S},\mathfrak{T}}(*)_{\mathfrak{A}}),\\ \check{\mathcal{F}}_3:\mathcal{A}_3^{\mathfrak{sh}}(\mathcal{V})&\to\mathbf{C}^+(\mathrm{Sh}^{\mathcal{B}}_{\mathfrak{S},\mathfrak{T}}(*)_{\mathfrak{A}}),\end{aligned}$$

and the natural transformation (1.3.7.3) extends to the natural transformation

$$(1.3.7.6)\qquad \check{\mathcal{F}}_3^*\to\check{\mathcal{F}}_3.$$

The extension of (1.3.7.5) to the category $\mathcal{A}_4^{\mathfrak{sh}}(\mathcal{V})$ is accomplished using the functoriality of the cycle classes in Definition 1.1.6(ii). For example, let $f:(Y,g)\to(X,f)$ be a morphism in $\mathcal{L}(\mathcal{V})$, giving the map $f^*:\mathbb{Z}_X(n)_f\to\mathbb{Z}_Y(n)_g$ in $\mathcal{A}_1(\mathcal{V})$. Take $Z\in\mathcal{Z}^n(X)_f$. Then we have the map $h^{\mathfrak{sh}}_{X,Y,[Z],f^*}:\mathfrak{e}\to\mathbb{Z}_Y(n)_g$ in $\mathcal{A}_4^{\mathfrak{sh}}(\mathcal{V})$ with $dh^{\mathfrak{sh}}_{X,Y,[Z],f^*}=f^*\circ[Z]^{\mathfrak{sh}}-[f^*(Z)]^{\mathfrak{sh}}$. The functoriality of the cycle classes gives the relation $\check{\mathcal{F}}_3^*(f^*\circ[Z]^{\mathfrak{sh}})-\check{\mathcal{F}}_3^*([f^*(Z)])=d\beta$ for some map $\beta:\tilde{1}\to\check{\mathcal{F}}_3^*(\mathbb{Z}_Y(n)_g)$ of degree $2n-1$. We then define $\check{\mathcal{F}}_4^*(h^{\mathfrak{sh}}_{X,Y,[Z],f^*})=\beta$. The definition of $\check{\mathcal{F}}_4^*$ for the other types of maps adjoined to form $\mathcal{A}_4^{\mathfrak{sh}}(\mathcal{V})$ is similar; we let $\check{\mathcal{F}}_4$ be the composition of $\check{\mathcal{F}}_4^*$ with the natural transformation (1.3.7.6).

The extension to $\mathcal{A}_5^{\mathfrak{sh}}(\mathcal{V})$ is accomplished in a similar manner, relying on the semi-purity hypothesis Definition 1.1.6(iii) for the cohomology theory $\mathcal{F}$. Restricting to the subcategory $\mathcal{A}^{\mathfrak{sh}}_{\text{mot}}(\mathcal{V})$ and extending scalars to A gives the DG tensor functor

$$\check{\mathcal{F}}^{\mathfrak{sh}}_{\text{mot}}: \mathcal{A}^{\mathfrak{sh}}_{\text{mot}}(\mathcal{V}) \otimes A \to \mathbf{C}^+(\mathrm{Sh}^{\mathcal{B}}_{\mathfrak{S},\mathfrak{T}}(*)_{\mathfrak{A}}). \tag{1.3.7.7}$$

Step 4. The extension to $\mathcal{DM}(\mathcal{V})$:

1.3.8. Applying the functor $\mathbf{C}^b$ (Part II, Chapter II, Definition 1.2.7) to the DG tensor functor (1.3.7.7) gives the functor

$$\mathbf{C}^b(\check{\mathcal{F}}^{\mathfrak{sh}}_{\text{mot}}): \mathbf{C}^b(\mathcal{A}^{\mathfrak{sh}}_{\text{mot}}(\mathcal{V}) \otimes A) \to \mathbf{C}^b(\mathbf{C}^+(\mathrm{Sh}^{\mathcal{B}}_{\mathfrak{S},\mathfrak{T}}(*)_{\mathfrak{A}}));$$

composing with the equivalence (see Part II, Chapter II, §1.2.9)

$$\mathrm{Tot}: \mathbf{C}^b(\mathbf{C}^+(\mathrm{Sh}^{\mathcal{B}}_{\mathfrak{S},\mathfrak{T}}(*)_{\mathfrak{A}})) \to \mathbf{C}^+(\mathrm{Sh}^{\mathcal{B}}_{\mathfrak{S},\mathfrak{T}}(*)_{\mathfrak{A}})$$

gives the DG tensor functor

$$\tilde{\mathbf{C}}^b(\check{\mathcal{F}}^{\mathfrak{sh}}_{\text{mot}}): \mathbf{C}^b(\mathcal{A}^{\mathfrak{sh}}_{\text{mot}}(\mathcal{V}) \otimes A) \to \mathbf{C}^+(\mathrm{Sh}^{\mathcal{B}}_{\mathfrak{S},\mathfrak{T}}(*)_{\mathfrak{A}}).$$

Passing to the homotopy category gives the exact tensor functor

$$\tilde{\mathbf{K}}^b(\check{\mathcal{F}}^{\mathfrak{sh}}_{\text{mot}}): \mathbf{K}^b(\mathcal{A}^{\mathfrak{sh}}_{\text{mot}}(\mathcal{V}) \otimes A) \to \mathbf{K}^+(\mathrm{Sh}^{\mathcal{B}}_{\mathfrak{S},\mathfrak{T}}(*)_{\mathfrak{A}}). \tag{1.3.8.1}$$

By (Part II, Chapter II, Theorem 2.2.2), the functor

$$\mathbf{K}^b(\mathfrak{c}_{\text{mot}}): \mathbf{K}^b(\mathcal{A}^{\mathfrak{sh}}_{\text{mot}}(\mathcal{V}) \otimes A) \to \mathbf{K}^b_{\text{mot}}(\mathcal{V})_A := \mathbf{K}^b(\mathcal{A}_{\text{mot}}(\mathcal{V}) \otimes A) \tag{1.3.8.2}$$

is an equivalence of triangulated categories; the functor (1.3.8.1) thus gives the exact functor

$$\mathbf{K}^b(\check{\mathcal{F}}_{\text{mot}}): \mathbf{K}^b_{\text{mot}}(\mathcal{V})_A \to \mathbf{K}^+(\mathrm{Sh}^{\mathcal{B}}_{\mathfrak{S},\mathfrak{T}}(*)_{\mathfrak{A}}). \tag{1.3.8.3}$$

In addition, if we let $\mathbf{K}^b_{\text{mot}\mathfrak{sh}}(\mathcal{V})_A$ be the full image of $\mathbf{K}^b(\mathfrak{c}_{\text{mot}})$, then $\mathbf{K}^b(\mathfrak{c}_{\text{mot}})$ gives a pseudo-tensor equivalence

$$\mathbf{K}^b(\mathfrak{c}_{\text{mot}}): \mathbf{K}^b(\mathcal{A}^{\mathfrak{sh}}_{\text{mot}}(\mathcal{V}) \otimes A) \to \mathbf{K}^b_{\text{mot}\mathfrak{sh}}(\mathcal{V})_A$$

of A-triangulated tensor categories, and the essential image of $\mathbf{K}^b_{\text{mot}\mathfrak{sh}}(\mathcal{V})_A$ in $\mathbf{K}^b_{\text{mot}}(\mathcal{V})_A$ is all of $\mathbf{K}^b_{\text{mot}}(\mathcal{V})_A$.

Let X be in $\mathcal{V}$, and suppose X is a union of two open subschemes in $\mathcal{V}$: $X = U \cup V$. Let $j: \check{\mathcal{F}}_U(n) \oplus \check{\mathcal{F}}_V(n) \to \check{\mathcal{F}}_{U\cap V}(n)$ be the difference of the two restriction maps. Since the image of $\check{\mathcal{F}}_X(n)$ in the derived category is isomorphic to $Rp_{X*}\mathcal{F}_X(n)$, the natural map $\check{\mathcal{F}}_X(n) \to \mathrm{cone}(j)[-1]$ is a quasi-isomorphism. This, together with Definition 1.1.6(i), (iii), (vi), implies that the composition of (1.3.8.3) with the canonical map $\mathbf{K}^+(\mathrm{Sh}^{\mathcal{B}}_{\mathfrak{S},\mathfrak{T}}(*)_{\mathfrak{A}}) \to \mathbf{D}^+(\mathrm{Sh}^{\mathcal{B}}_{\mathfrak{S},\mathfrak{T}}(*))$. factors through the localization $\mathbf{D}^b_{\text{mot}}(\mathcal{V})_A$ of $\mathbf{K}^b_{\text{mot}}(\mathcal{V})_A$ (Chapter I, §2.1.5), giving the exact functor

$$\mathbf{D}^b(\check{\mathcal{F}}_{\text{mot}}): \mathbf{D}^b_{\text{mot}}(\mathcal{V})_A \to \mathbf{D}^+(\mathrm{Sh}^{\mathcal{B}}_{\mathfrak{S},\mathfrak{T}}(*)), \tag{1.3.8.4}$$

and the restriction to the full image $\mathbf{D}^b_{\text{mot}\mathfrak{sh}}(\mathcal{V})_A$ of $\mathbf{K}^b_{\text{mot}\mathfrak{sh}}(\mathcal{V})_A$ is an exact pseudo-tensor functor.

Finally, by (Part II, Chapter II, Theorem 2.4.8.2), the functor (1.3.8.4) extends canonically to $\mathcal{DM}(\mathcal{V})_A$, giving the desired realization functor

$$\mathfrak{R}_{\mathcal{F}}: \mathcal{DM}(\mathcal{V})_A \to \mathbf{D}^+(\mathrm{Sh}^{\mathcal{B}}_{\mathfrak{S},\mathfrak{T}}(*)). \tag{1.3.8.5}$$

Let $\mathcal{DM}_{\mathfrak{sh}}(\mathcal{V})_A$ be the full triangulated tensor subcategory of $\mathcal{DM}(\mathcal{V})_A$ gotten by forming the pseudo-abelian hull of $\mathbf{D}^b_{\mathrm{mot}\mathfrak{sh}}(\mathcal{V})_A$; it follows from the theorem just mentioned that the restriction of $\Re_{\mathcal{F}}$ to $\mathcal{DM}_{\mathfrak{sh}}(\mathcal{V})_A$ is an exact pseudo-tensor functor.

This completes the construction of the realization functor. We now show that the functor (1.3.8.5) is canonical. For this, we first note that the extension from $\mathcal{A}_2^{\mathfrak{sh}}$ to $\mathcal{A}_{\mathrm{mot}}^{\mathfrak{sh}}$ is independent of the various choices of the maps in Step 3, up to a homotopy of the resulting functors; this follows directly from the semi-purity hypothesis Definition 1.1.6(iii), and the fact that the category $\mathcal{A}_{\mathrm{mot}}^{\mathfrak{sh}}$ is freely generated from $\mathcal{A}_2^{\mathfrak{sh}}$ by the adjoined maps. The only other choice is the choice of the functor giving the inverse to (1.3.8.2). We make one choice of such an inverse, independent of $\mathcal{F}$, which shows that $\Re_{\mathcal{F}}$ is canonical. The choice of the inverse is unique up to unique isomorphism, which verifies the remaining assertions of Theorem 1.3.1.

1.3.9. REMARK. Applying $\Re_{\mathcal{F}}$ to $A_X(q)[p]$ for X in $\mathcal{V}$ gives the realization map on cohomology

$$\Re_{\mathcal{F}}\colon H^p(X, A(q)) \to H^p_{\mathfrak{T}}(\alpha(X), \mathcal{F}(q))$$

and similarly for cohomology with support. The realization functor $\Re_{\mathcal{F}}$ sends the Chern classes defined in Chapter III to the Chern classes defined in [**46**] for the twisted duality theory defined by $\mathcal{F}$.

2. Concrete realizations

We show how the theory of Section 1, with some modifications, gives the realizations corresponding to singular cohomology, étale cohomology, and Hodge cohomology. We also give some extensions of these constructions for more general base schemes, as well as the "motivic" realizations to the category of compatible realizations.

2.1. The Betti realizations

2.1.1. *Points for the classical site.* For a topological space T with point $p \in T$, we have the fiber functor $\phi_p\colon \mathrm{Sh}_T \to \mathbf{Sets}$ defined by

$$\phi_p(\mathcal{F}) = \varinjlim_{p\in U} F(U),$$

where the limit is over open neighborhoods U of p in T. This gives us the point $i_p\colon \mathbf{Sets} \to \mathrm{Sh}_T$ of the topos Sh_T. The inclusion the category of open subsets of T into $\mathbf{Top}$ defines the morphism of topoi $i_T\colon \mathrm{Sh}_T \to \mathrm{Sh}_{\mathbf{Top}}$. Composing these two gives the point $i_{p,T}\colon \mathbf{Sets} \to \mathrm{Sh}_{\mathbf{Top}}$ of $\mathrm{Sh}_{\mathbf{Top}}$. As the set of points $\{i_p \mid p \in T\}$ forms a conservative family of points for the topos Sh_T, it follows from [**3**, Chapter IV, Proposition 6.5(b)] that a set of points of the form $i_{p,T}$ forms a conservative family of points for $\mathrm{Sh}_{\mathbf{Top}}$.

2.1.2. *The $\mathbb{C}$-Betti realizations.* Take $S = \operatorname{Spec}\mathbb{C}$ and $\mathcal{V} = \mathbf{Sm}_{\mathbb{C}}$. We take the cohomology theory defined on the full subcategory $\mathbf{CW}$ of $\mathbf{Top}$, with objects the topological spaces having the homotopy type of a CW complex, by the graded sheaf $\mathbb{Z}(*) = \oplus_{q=0}^{\infty}(2\pi i)^q\mathbb{Z}$. We let $\mathrm{an}\colon \mathcal{V} \to \mathbf{CW}$ be the functor sending a $\mathbb{C}$-scheme X to the topological space X_{an} formed by the $\mathbb{C}$-points $X(\mathbb{C})$ with the classical topology. The well known properties of singular cohomology show that $(\mathbb{Z}(*), \alpha)$ defines a geometric cohomology theory on $\mathbf{Sm}_{\mathbb{C}}$; we then apply Theorem 1.3.1 to give the $\mathbb{C}$-Betti realization functor

$$\mathfrak{R}_{\mathfrak{B}_{\mathbb{C}}}: \mathcal{DM}(\mathbf{Sm}_{\mathbb{C}}) \to \mathbf{D}^{+}(\mathbf{Ab}).$$

More generally, for an arbitrary base scheme S, each $\mathbb{C}$-valued point of S, $\sigma: \operatorname{Spec}\mathbb{C} \to S$, defines the realization functor

$$\mathfrak{R}_{\mathfrak{B}_{\mathbb{C},\sigma}}: \mathcal{DM}(\mathbf{Sm}_{S}) \to \mathbf{D}^{+}(\mathbf{Ab})$$

by composing $\mathfrak{R}_{\mathfrak{B}_{\mathbb{C}}}$ with $\sigma^*: \mathcal{DM}(\mathbf{Sm}_S) \to \mathcal{DM}(\mathbf{Sm}_{\mathbb{C}})$.

Using the same method, we may form the Betti realization over a smooth base scheme. Let B be a smooth $\mathbb{C}$-scheme of finite type, with associated topological space B_{an}. Let $\mathbf{CW}/B_{\mathrm{an}}$ be the category of maps $E \to B_{\mathrm{an}}$ in $\mathbf{CW}$, with the topology induced from $\mathbf{CW}$. The conservative family of points for $\mathrm{Sh}_{\mathbf{CW}}$ described in §2.1.1 induces a conservative family of points for $\mathrm{Sh}_{\mathbf{CW}/B_{\mathrm{an}}}$. Using the cohomology theory on $\mathbf{Sm}_B$ induced by the graded sheaf $\mathbb{Z}(*) = \oplus_{q=0}^{\infty}(2\pi i)^q\mathbb{Z}$ and the functor $\mathrm{an}: \mathbf{Sm}_B \to \mathbf{CW}_{B_{\mathrm{an}}}$ gives via Theorem 1.3.1 the B-Betti realization functor

$$\mathfrak{R}_{\mathfrak{B}_{B,\mathbb{C}}}: \mathcal{DM}(\mathbf{Sm}_B) \to \mathbf{D}^{+}(\mathrm{Sh}^{\mathbf{Ab}}_{B_{\mathrm{an}}}).$$

2.1.3. *The $\mathbb{R}$-Betti realizations.* The $\mathbb{R}$-Betti realization is defined by adding the data of a "real Frobenius" to the $\mathbb{C}$-Betti realization.

Take $S = \operatorname{Spec}\mathbb{R}$ and $\mathcal{V} = \mathbf{Sm}_{\mathbb{R}}$, the category of smooth, quasi-projective $\mathbb{R}$-schemes. We have the Grothendieck site $\mathbb{Z}/2 - \mathbf{CW}$, where an object is a CW-complex X together with a continuous involution $F_\infty: X \to X$.

We replace the functor $\mathrm{an}: \mathcal{V} \to \mathbf{CW}$ with the functor $\mathrm{an}_{\mathbb{R}}: \mathcal{V} \to \mathbb{Z}/2 - \mathbf{CW}$ by $X_{\mathrm{an}_{\mathbb{R}}} := (X_{\mathrm{an}}, F_\infty)$, where F_∞ is the continuous involution induced by complex conjugation $^{-} \otimes \mathrm{id}_X: \mathbb{C} \otimes_{\mathbb{R}} X \to \mathbb{C} \otimes_{\mathbb{R}} X$.

As the forgetful functor $\mathbb{Z}/2 - \mathbf{CW} \to \widetilde{\mathbf{CW}}$ is faithful, the conservative family of points for $\mathbf{CW}$ gives a conservative family of points for $\mathbb{Z}/2 - \mathbf{CW}$.

Using the same cohomology theory as for the $\mathbb{C}$-Betti realization, we then apply Theorem 1.3.1 to give the $\mathbb{R}$-Betti realization functor

$$\mathfrak{R}_{\mathfrak{B}_{\mathbb{R}}}: \mathcal{DM}(\mathbf{Sm}_{\mathbb{R}}) \to \mathbf{D}^{+}(\mathbf{Mod}_{\mathbb{Z}[\mathbb{Z}/2]}).$$

If B is a smooth $\mathbb{R}$-scheme, we have the category $\mathbb{Z}/2 - \mathbf{CW}/B_{\mathrm{an}_{\mathbb{R}}}$ of maps $(E, \sigma) \to B_{\mathrm{an}_{\mathbb{R}}}$ in $\mathbb{Z}/2 - \mathbf{CW}/B_{\mathrm{an}_{\mathbb{R}}}$. This gives us the Grothendieck site of $\mathbb{Z}/2$-CW complexes over $B_{\mathrm{an}_{\mathbb{R}}}$, and the realization functor

$$\mathfrak{R}_{\mathfrak{B}_{B,\mathbb{R}}}: \mathcal{DM}(\mathbf{Sm}_{\mathbb{R}}) \to \mathbf{D}^{+}(\mathrm{Sh}^{\mathbf{Ab}}_{B_{\mathrm{an}_{\mathbb{R}}}}).$$

As above, the restriction of these functors to the full triangulated subcategory $\mathcal{DM}_{\mathfrak{gh}}(\mathbf{Sm}_B)$ are exact pseudo-tensor functors.

2.2. The étale realization

For simplicity, we assume that the base scheme S is smooth and essentially of finite type over a ring R, where R is either an algebraically closed field, a global field, a local field, a finite field, or a ring of integers in a global field or a local field.

2.2.1. *Points for the étale site.* For each point s of S, fix an algebraic closure $\overline{k(s)}$ of the residue field $k(s)$ of s. If $X \to S$ is a quasi-projective S-scheme, a *geometric point* of X is an equivalence class of maps over S, $x\colon \mathrm{Spec}\, \overline{k(s)} \to X$, where x is equivalent to x' if x and x' differ by an automorphism of $\mathrm{Spec}\, \overline{k(s)}$ over X. We let X_{geom} denote the set of geometric points of X.

A *pointed étale map* $f\colon (U, u) \to (X, x)$ is an étale map $f\colon U \to X$, with geometric points u of U, x of X such that $f(u) = x$. We form the fiber functor $\phi_x\colon \mathrm{Sh}_{(X,\text{ét})} \to \mathbf{Sets}$ by

$$\phi_x(F) = \varinjlim_{(U,u)\to(X,x)} F(U),$$

where the limit is over pointed étale maps $(U, u) \to (X, x)$. As in §2.1.1, this gives the point $i_x\colon \mathbf{Sets} \to \mathrm{Sh}_{\text{ét}}(X)$ the point $i_{x,X}\colon \mathbf{Sets} \to \mathrm{Sh}_{S,\text{ét}}$ of the topos of sheaves on the big étale site over S, and a conservative family of points of $\mathrm{Sh}_{S,\text{ét}}$.

2.2.2. *The category of inverse systems.* We sketch Ekedahl's construction [**41**] of the category of sheaves of modules over the inverse system

$$\mathbb{Z}/l^* := (\mathbb{Z}/l \leftarrow \mathbb{Z}/l^2 \leftarrow \ldots \leftarrow \mathbb{Z}/l^n \leftarrow \ldots).$$

Let X be in $\mathbf{Sm}_S^{\mathrm{ess}}$. Let $\mathrm{Sh}_{\text{ét}}^{\mathbb{Z}/l^*}(X)$ be the category of inverse systems of sheaves on X,

$$\mathcal{F} := (\mathcal{F}_0 \leftarrow \mathcal{F}_1 \leftarrow \ldots)$$

which form a module over the inverse system of rings $\mathbb{Z}/l^*$; maps being commutative diagrams

$$\begin{array}{ccccc}
\mathcal{F}_0 & \longleftarrow & \mathcal{F}_1 & \longleftarrow & \cdots \\
\downarrow & & \downarrow & & \\
\mathcal{G}_0 & \longleftarrow & \mathcal{G}_1 & \longleftarrow & \cdots
\end{array}$$

of maps of sheaves of $\mathbb{Z}/l^*$-modules.

Forming the sheaf projective limit of the projective system defines the functor $\pi_*\colon \mathrm{Sh}_{\text{ét}}^{\mathbb{Z}/l^*}(X) \to \mathrm{Sh}_{\text{ét}}^{\mathbb{Z}_l}(X)$, where $\mathrm{Sh}_{\text{ét}}^{\mathbb{Z}_l}(X)$ is the category of sheaves of continuous $\mathbb{Z}_l$-modules. We have the functor $\pi^*\colon \mathrm{Sh}_{\text{ét}}^{\mathbb{Z}_l}(X) \to \mathrm{Sh}_{\text{ét}}^{\mathbb{Z}/l^*}(X)$ defined by

$$\pi^*(\mathcal{G}) := \mathcal{G} \otimes \mathbb{Z}/l^* := (\mathcal{G} \otimes \mathbb{Z}/l \leftarrow \mathcal{G} \otimes \mathbb{Z}/l^2 \leftarrow \ldots).$$

The categories $\mathrm{Sh}_{\text{ét}}^{\mathbb{Z}/l^*}(X)$ and $\mathrm{Sh}_{\text{ét}}^{\mathbb{Z}_l}(X)$ are abelian categories with enough injectives; the functors π_* and π^* extend to exact functors

$$R\pi_*\colon \mathbf{D}(\mathrm{Sh}_{\text{ét}}^{\mathbb{Z}/l^*}(X)) \to \mathbf{D}(\mathrm{Sh}_{\text{ét}}^{\mathbb{Z}_l}(X)),$$
$$L\pi^*\colon \mathbf{D}(\mathrm{Sh}_{\text{ét}}^{\mathbb{Z}_l}(X)) \to \mathbf{D}(\mathrm{Sh}_{\text{ét}}^{\mathbb{Z}/l^*}(X)).$$

For $\mathcal{F}$ in $\mathbf{D}(\mathrm{Sh}_{\text{ét}}^{\mathbb{Z}/l^*}(X))$, define $\hat{\mathcal{F}} := L\pi^*(R\pi_*(\mathcal{F}))$; we have the canonical map $\hat{\mathcal{F}} \to \mathcal{F}$.

2.2.3. DEFINITION [[**41**], Definition 2.1]. Let $\mathcal{F}$ be in $\mathbf{D}(\mathrm{Sh}_{\text{ét}}^{\mathbb{Z}/l^*}(X))$. Call $\mathcal{F}$ *a normalized $\mathbb{Z}/l^*$-complex* if the map $\hat{\mathcal{F}} \to \mathcal{F}$ is a quasi-isomorphism.

2.2.4. By [**41**, Proposition 2.2], $\mathcal{F}$ is normalized if and only if

$$\mathbb{Z}/l^n \otimes^L_{\mathbb{Z}/l^{n+1}} \mathcal{F}_{n+1} \to \mathcal{F}_n$$

is a quasi-isomorphism for all n. In particular, if the stalks of $\mathcal{F}_n$ are flat $\mathbb{Z}/l^n$-modules, and the maps $\mathcal{F}_{n+1} \to \mathcal{F}_n$ are all surjective, then $\mathcal{F}$ is normalized.

2.2.5. EXAMPLE. The system of Tate sheaves

$$\mathbb{Z}_{l,X}(q) := (\mathbb{Z}/l(q) \leftarrow \mathbb{Z}/l^2(q) \leftarrow \dots)$$

is a normalized $\mathbb{Z}/l^*$-complex.

2.2.6. The full subcategory $\mathbf{K}^* \lim \mathrm{Sh}^{\mathbb{Z}_l}_{\text{ét}}(X)$ of $\mathbf{K}^*(\mathrm{Sh}^{\mathbb{Z}/l^*}_{\text{ét}}(X))$ ($* = +, -, b, \emptyset$) with objects the normalized $\mathbb{Z}/l^*$-complexes is closed under cones. Let $\mathbf{D}^* \lim \mathrm{Sh}^{\mathbb{Z}_l}_{\text{ét}}(X)$ be the localization of $\mathbf{K}^*(\mathrm{Sh}^{\mathbb{Z}/l^*}_{\text{ét}}(X))$ with respect to quasi-isomorphisms.

In [**41**], a more general notion, that of a $\mathbb{Z}/l^*$-complex, is defined, as well as a thick subcategory containing the acyclic complexes, the *negligible* complexes. By [**41**, Proposition 2.7] the inclusion functor from the homotopy category of normalized $\mathbb{Z}/l^*$-complexes to $\mathbb{Z}/l^*$-complexes induces an equivalence from $\mathbf{D}^* \lim \mathrm{Sh}^{\mathbb{Z}_l}_{\text{ét}}(X)$ to the localization of the homotopy category of $\mathbb{Z}/l^*$-complexes with respect to the negligible complexes.

It is shown in [**41**] that the tensor product and internal Hom of complexes define a tensor structure and internal Hom for $\mathbf{D}^* \lim \mathrm{Sh}^{\mathbb{Z}_l}_{\text{ét}}(X)$. This makes $\mathbf{D}^* \lim \mathrm{Sh}^{\mathbb{Z}_l}_{\text{ét}}(X)$ a triangulated tensor category, with internal Hom's, functorially in X.

A map $f\colon X \to Y$ in $\mathbf{Sm}^{\text{ess}}_S$ induces the functor $f_*\colon \mathrm{Sh}^{\mathbb{Z}/l^*}_{\text{ét}}(X) \to \mathrm{Sh}^{\mathbb{Z}/l^*}_{\text{ét}}(Y)$, which extends to the exact functor $Rf_*\colon \mathbf{D}^+ \lim \mathrm{Sh}^{\mathbb{Z}_l}_{\text{ét}}(X) \to \mathbf{D}^+ \lim \mathrm{Sh}^{\mathbb{Z}_l}_{\text{ét}}(Y)$.

We have the isomorphism $\Gamma(X, \pi_*(-)) \cong \mathrm{Hom}_{\mathrm{Sh}^{\mathbb{Z}/l^*}_{\text{ét}}(X)}(\mathbb{Z}/l^*, -)$; define the *continuous hypercohomology* of an object $\mathcal{F}$ of $\mathbf{D}^+ \lim \mathrm{Sh}^{\mathbb{Z}_l}_{\text{ét}}(X)$ by

$$\mathbb{H}^p_{\text{cont}}(X, \mathcal{F}) := \mathrm{Hom}_{\mathbf{D}^+ \lim \mathrm{Sh}^{\mathbb{Z}_l}_{\text{ét}}(X)}(\mathbb{Z}_{l,X}, \mathcal{F}[p]).$$

The continuous hypercohomology with support in a closed subset $i_W\colon W \to X$ is defined similarly as

$$\mathbb{H}^p_{\text{cont},W}(X, \mathcal{F}) := \mathrm{Hom}_{\mathbf{D}^+ \lim \mathrm{Sh}^{\mathbb{Z}_l}_{\text{ét}}(X)}(i_{W*}\mathbb{Z}_{l,W}, \mathcal{F}[p]).$$

The continuous cohomology of an $\mathcal{F}$ in $\mathrm{Sh}^{\mathbb{Z}_l}_{\text{ét}}(X)$ which satisfies the conditions in §2.2.4 is thus defined, in particular, we have $H^p_{\text{cont}}(X, \mathbb{Z}_{l,X}(q))$. It is easy to see that the definition given above agrees with the continuous cohomology of Jannsen [**72**].

The triangulated category $\mathbf{D}^+ \lim \mathrm{Sh}^{\mathbb{Z}_l}_{S,\text{ét}}$, constructed in a similar manner by localizing the homotopy category of normalized $\mathbb{Z}/l^*$-complexes on the big étale site over S with respect to quasi-isomorphisms, has the analogous formal properties.

2.2.7. REMARK. The construction given in [**41**] is more general than what we have presented here; in particular, one may replace the étale topology with another Grothendieck topology, and the same construction goes through without change, as long as one assumes that X has finite cohomological dimension in the new topology.

2.2.8. *The construction of the étale realization.* Take a prime l which is invertible on S, and let X be an S-scheme.

We have the objects $\mathbb{Z}_{l,X}(q)$ defined in Example 2.2.5, sending X to $\mathbb{Z}_{l,X}(q)$ gives us the objects $\mathbb{Z}_l(q)$ of $\mathbf{D}^+\mathrm{limSh}^{\mathbb{Z}_l}_{S,\text{ét}}$ with value $\mathbb{Z}_{l,X}(q)$ on X. The continuous l-adic cohomology is then defined as the cohomology theory associated to the graded object $\mathbb{Z}_{l,\text{ét}}(*) := \oplus_{q=0}^{\infty}\mathbb{Z}_l(q)$ of $\mathbf{D}^+\mathrm{limSh}^{\mathbb{Z}_l}_{S,\text{ét}}$.

Although Theorem 1.3.1 does not directly apply to this situation, the same proof, replacing the homotopy category of sheaves over S with the homotopy category $\mathbf{K}^+\lim\mathrm{Sh}^{\mathbb{Z}_l}_{\text{ét}}(S)$, yields the analogous result. We give a sketch of the necessary changes.

The work of Jannsen [**72**] (see also [**98**]) verifies that $\mathbb{Z}_{l,\text{ét}}(*)$ satisfies the axioms of a geometric cohomology theory, suitably interpreted, in case the base scheme S is smooth over a field. If S is smooth over a number ring, or a ring of integers in a global field, we need to use Thomason's purity result [**120**] for $\mathbb{Q}_l$ étale cohomology, so we will only get a $\mathbb{Q}_l$-realization (see §2.2.11 below).

One calls an inverse system $\mathcal{F}$ *flat* if $\mathcal{F}_n$ has stalks which are flat $\mathbb{Z}/l^n$-modules, for all n; call $\mathcal{F}$ *strongly normalized* if $\mathcal{F}$ is flat, and in addition, the maps $\mathbb{Z}/l^n\otimes\mathcal{F}_{n+1}\to\mathcal{F}_n$ are isomorphisms (i.e., $\mathcal{F}_{n+1}\to\mathcal{F}_n$ is surjective for all n). The sheaves $\mathbb{Z}_l(q)$ are strongly normalized.

Call a strongly normalized sheaf $\mathcal{F}$ *strongly acyclic* if the sheaves $\mathcal{F}_n$ and $\ker(\mathcal{F}_{n+1}\to\mathcal{F}_n)$ are acyclic for all n.

Using (Part II, Chapter IV, Lemma 2.1.5 and Lemma 2.2.2), we see that the Godement resolution takes strongly normalized complexes to complexes of strongly acyclic sheaves. In addition, if $f\colon X\to Y$ is a map in $\mathbf{Sm}_S^{\text{ess}}$, then $f_*G\mathcal{F}$ is a complex of strongly acyclic sheaves if $\mathcal{F}$ is strongly normalized. Similarly, the tensor product of strongly normalized complexes is again strongly normalized.

The construction of the realization functor then goes through without change until the middle of Step 3, where we have the isomorphism (1.3.7.4). We alter the construction at this point as follows: Let $\mathcal{F}$ be a strongly acyclic sheaf on S. Then

$$R^q\pi_*\mathcal{F}=\begin{cases}0 & \text{for } q>0,\\ \varprojlim_n\mathcal{F}_n & \text{for } q=0.\end{cases}$$

Since $\mathcal{F}$ is strongly acyclic, and the maps $\mathcal{F}_{n+1}\to\mathcal{F}_n$, are surjective, we have the inverse system

$$H^0(S,\mathcal{F}_0)\leftarrow H^0(S,\mathcal{F}_1)\leftarrow\dots$$

with all maps begin surjective. Thus, $\varprojlim^1 H^0(S,\mathcal{F}_*)=0$, hence

$$H^p(S,\pi_*(\mathcal{F}))=\begin{cases}0 & \text{for } p>0,\\ \varprojlim_n H^0(S,\mathcal{F}_n) & \text{for } p=0.\end{cases}$$

This gives us

$$\mathbb{H}^p_{\text{cont}}(S,\mathcal{F})=\begin{cases}0 & \text{for } p\neq 0,\\ \varprojlim_n H^0(S,\pi_*\mathcal{F}) & \text{for } p=0;\end{cases}$$

as

$$\mathrm{Hom}_{\mathbf{K}^+\mathrm{limSh}^{\mathbb{Z}_l}_{\text{ét}}(S)}(\mathbb{Z}_{l,S}, \mathcal{F}[p]) = \begin{cases} 0 & \text{for } p \neq 0, \\ \varprojlim_n H^0(S, \pi_* \mathcal{F}) & \text{for } p = 0 \end{cases}$$

as well, we have

$$\mathrm{Hom}_{\mathbf{K}^+\mathrm{limSh}^{\mathbb{Z}_l}_{\text{ét}}(S)}(\mathbb{Z}_{l,S}, \mathcal{F}[p]) = \mathrm{Hom}_{\mathbf{D}^+\mathrm{limSh}^{\mathbb{Z}_l}_{\text{ét}}(S)}(\mathbb{Z}_{l,S}, \mathcal{F}[p])$$

for all p. By devissage, we have

$$\mathrm{Hom}_{\mathbf{K}^+\mathrm{limSh}^{\mathbb{Z}_l}_{\text{ét}}(S)}(\mathbb{Z}_{l,S}, \mathcal{F}) = \mathrm{Hom}_{\mathbf{D}^+\mathrm{limSh}^{\mathbb{Z}_l}_{\text{ét}}(S)}(\mathbb{Z}_{l,S}, \mathcal{F}) \tag{2.2.8.1}$$

for a complex of strongly acyclic sheaves $\mathcal{F}$ on S.

We then use the isomorphism (2.2.8.1) instead of (1.3.7.4), and the construction of the realization functor goes through without further change.

This gives us

2.2.9. THEOREM. *Let S be the localization of a smooth scheme over a finite, local, global, or algebraically closed field, with l invertible on S. Then sending (X, q) to $Rp_{X*}\mathbb{Z}_{\text{ét},X,l}(q)$ extends canonically to the exact l-adic realization functor*

$$\mathfrak{R}_{\text{ét},l,S}: \mathcal{DM}(\mathbf{Sm}_S) \to \mathbf{D}^+\mathrm{limSh}^{\mathbb{Z}_l}_{\text{ét}}(S).$$

The restriction of $\mathfrak{R}_{\text{ét},l,S}$ to $\mathcal{DM}_{\mathfrak{sh}}(\mathbf{Sm}_S)$ is an exact pseudo-tensor functor.

2.2.10. *The mod-n realization.* Suppose n is invertible on S. To form the mod-n realization, first take the product of the l-adic realization functors for all l dividing n. If Γ is an object of $\mathbf{C}^b_{\text{mot}}(\mathcal{V})$, define $\Gamma \otimes \mathbb{Z}/n$ to be the object $\mathrm{cone}(\Gamma \xrightarrow{\times n} \Gamma)$. Let $\mathcal{DM}(\mathcal{V}; \mathbb{Z}/n)$ be the full subcategory of $\mathcal{DM}(\mathbf{Sm}_S)$ generated by the objects $\Gamma \otimes \mathbb{Z}/n$. Restricting the product of the l-adic realizations to $\mathcal{DM}(\mathbf{Sm}_S; \mathbb{Z}/n)$ defines the mod n realization

$$\mathfrak{R}_{\text{ét},\mathbb{Z}/n}: \mathcal{DM}(\mathbf{Sm}_S; \mathbb{Z}/n) \to \mathbf{D}^+\mathrm{limSh}^{\mathbb{Z}_l}_{\text{ét}}(S).$$

Using the quasi-isomorphism

$$\mu_n^{\otimes q} \to \mathrm{cone}\Big(\prod_{l|n} \mathbb{Z}_l(q) \xrightarrow{\times n} \prod_{l|n} \mathbb{Z}_l(q)\Big),$$

we may form an equivalent realization in the usual category of étale sheaves of $\mathbb{Z}/n$-modules

$$\mathfrak{R}_{\text{ét},\mathbb{Z}/n}: \mathcal{DM}(\mathbf{Sm}_S; \mathbb{Z}/n) \to \mathbf{D}^+(\mathrm{Sh}^{\mathbb{Z}/n}_{\text{ét}}(S)).$$

2.2.11. *The $\mathbb{Q}_l$ realization.* We tensor the l-adic realization with $\mathbb{Q}$, and use the argument of (Part II, Chapter II, Theorem 2.4.8.2) to give the $\mathbb{Q}_l$ realization

$$\mathfrak{R}_{\text{ét},l}: \mathcal{DM}(\mathbf{Sm}_S)_{\mathbb{Q}} \to \mathbf{D}^+\mathrm{limSh}^{\mathbb{Z}_l}_{\text{ét}}(S)_{\mathbb{Q}}.$$

The restriction of the various étale realization functors to the full triangulated subcategory $\mathcal{DM}_{\mathfrak{sh}}(\mathbf{Sm}_S)$ are exact pseudo-tensor functors. Using Thomason's purity theorem [**120**], we have the $\mathbb{Q}_l$ realization for S essentially smooth over a number ring, or the ring of integers in a number field, as well as for S essentially smooth over an algebraically closed field, a number field, or a local field.

2.3. The Hodge realizations

We first construct the Hodge realization over a point via a modification of Beilinson's category of Hodge complexes.

2.3.1. *Absolute Hodge complexes.* Let $\mathcal{H}_R$ denote the category of polarizable R-mixed Hodge structures (R a noetherian subring of $\mathbb{R}$). We begin by recalling Beilinson's construction of the category of absolute Hodge complexes. This consists of a subcategory of the category of diagrams

$$\mathcal{F} = \begin{array}{ccccccc} & & \mathcal{F}'_{\mathbb{Q}} & & (\mathcal{F}'_{\mathbb{C}}, W'_{\mathbb{C}}) & & \\ & \nearrow & & \nwarrow \quad \nearrow & & \nwarrow & \\ \mathcal{F}_R & & & (\mathcal{F}_{\mathbb{Q}}, W_{\mathbb{Q}}) & & & (\mathcal{F}_{\mathbb{C}}, W_{\mathbb{C}}, F) \end{array} \tag{2.3.1.1}$$

Here, $\mathcal{F}_R$, $\mathcal{F}'_{\mathbb{Q}}$, $\mathcal{F}_{\mathbb{Q}}$, $\mathcal{F}_{\mathbb{C}}$ and $\mathcal{F}'_{\mathbb{C}}$ are complexes of R-modules, $R\otimes\mathbb{Q}$-modules, $R\otimes\mathbb{Q}$-modules, $\mathbb{C}$-vector spaces and $\mathbb{C}$-vector spaces, resp., W, W' denotes an increasing filtration, and F denotes a decreasing filtration. The arrows in the diagram denote the following maps:

- The arrow $\mathcal{F}_R \to \mathcal{F}'_{\mathbb{Q}}$ is a quasi-isomorphism $\mathcal{F}_R \otimes \mathbb{Q} \to \mathcal{F}'_{\mathbb{Q}}$.
- The arrow $(\mathcal{F}_{\mathbb{Q}}, W_{\mathbb{Q}}) \to \mathcal{F}'_{\mathbb{Q}}$ is a quasi-isomorphism $\mathcal{F}_{\mathbb{Q}} \to \mathcal{F}'_{\mathbb{Q}}$.
- The arrow $(\mathcal{F}_{\mathbb{Q}}, W_{\mathbb{Q}}) \to (\mathcal{F}'_{\mathbb{C}}, W'_{\mathbb{C}})$ is a filtered quasi-isomorphism
$$(\mathcal{F}_{\mathbb{Q}} \otimes \mathbb{C}, W_{\mathbb{Q}} \otimes \mathbb{C}) \to (\mathcal{F}'_{\mathbb{C}}, W'_{\mathbb{C}}).$$
- The arrow $(\mathcal{F}_{\mathbb{C}}, W_{\mathbb{C}}, F) \to (\mathcal{F}'_{\mathbb{C}}, W'_{\mathbb{C}})$ is a filtered quasi-isomorphism
$$(\mathcal{F}_{\mathbb{C}}, W_{\mathbb{C}}) \to (\mathcal{F}'_{\mathbb{C}}, W'_{\mathbb{C}}).$$

The category $C^*_{\mathcal{H}_R}$ ($*$ = a boundedness condition) of mixed Hodge complexes are those diagrams as above for which the following conditions are satisfied (see [**11**, Definition 3.2]):

(i) The $H^p(\mathcal{F}_R)$ are finitely generated R-modules.
(ii) For a in $\mathbb{Z}$, consider the filtered complex $(\mathrm{gr}_a^{W_{\mathbb{C}}}\mathcal{F}_{\mathbb{C}}, \mathrm{gr}_a^{W_{\mathbb{C}}} F)$. The differential of this complex is strictly compatible with the filtration.
(iii) The filtration on $H^*(\mathrm{gr}_a^{W_{\mathbb{C}}}\mathcal{F}_{\mathbb{C}})$ induced by $\mathrm{gr}_a^{W_a} F$, together with the isomorphism $H^*(\mathrm{gr}_a^{W_{\mathbb{Q}}}\mathcal{F}_{\mathbb{Q}}) \otimes \mathbb{C} \to H^*(\mathrm{gr}_a^{W_{\mathbb{C}}}\mathcal{F}_{\mathbb{C}})$ that comes from the diagram, defines on $H^*(\mathrm{gr}_a^{W_{\mathbb{Q}}}\mathcal{F}_{\mathbb{Q}})$ a pure, polarizable $R\otimes\mathbb{Q}$-Hodge structure of weight a.

In particular, taking the cohomology of the complexes in (2.3.1.1) defines a mixed Hodge structure on $H^*(\mathcal{F}_R)$; let $\underline{H}^*(\mathcal{F})$ denote the resulting mixed Hodge structure.

The category $C^*_{\mathcal{H}_R}$ is closed under taking cones, and thus the homotopy category $K^*_{\mathcal{H}_R}$ is a triangulated category; the functor $\underline{H}^*(-)$ defines a cohomological functor from $K^*_{\mathcal{H}_R}$ to $\mathcal{H}_R$. Localizing $K^*_{\mathcal{H}_R}$ with respect to $\underline{H}^*(-)$ gives the category $D^*_{\mathcal{H}_R}$; Beilinson shows [**11**, Theorem 3.4] that the resulting functor

$$D^b_{\mathcal{H}_R} \to D^b(\mathcal{H}_R) \tag{2.3.1.2}$$

is an equivalence of triangulated categories.

2.3.2. *Tensor structure.* Let $\mathcal{D}$ be a diagram of the form

$$\begin{array}{ccccccccc} & & B_1 & & & & B_2 & & \\ & {}^{f_1}\nearrow & & \nwarrow^{g_2} & & {}^{f_2}\nearrow & & \nwarrow^{g_2} & & {}^{f_3}\nearrow \\ A_1 & & & & A_2 & & & & A_3 \end{array}$$

and let $\mathcal{D}'$ be similarly defined, replacing A_i with A'_i, etc. The tensor product $\mathcal{D}\otimes\mathcal{D}'$ is the diagram

$$\mathcal{D}\otimes\mathcal{D}' := \begin{array}{ccccccc} & & B_1\otimes B'_1 & & & B_2\otimes B'_2 & & \\ & \overset{f_1\otimes f'_1}{\nearrow} & & \overset{g_1\otimes g'_1}{\nwarrow}\ \overset{f_2\otimes f'_2}{\nearrow} & & \overset{g_2\otimes g'_2}{\nwarrow}\ \overset{f_3\otimes f'_3}{\nearrow} & \\ A_1\otimes A'_1 & & & A_2\otimes A'_2 & & & A_3\otimes A'_3 \end{array}.$$

This makes $C^*_{\mathcal{H}_R}$ a DG tensor category, and the equivalence (2.3.1.2) a tensor equivalence [**11**, Theorem 3.4].

2.3.3. *Enlarged diagrams.* We consider a slight modification of the above constructions. Form the category $C^*_{\mathcal{H}'_R}$ as the category of diagrams

$$\text{(2.3.3.1)}\qquad \mathcal{F} = \begin{array}{ccccccc} & \mathcal{F}'_{\mathbb{Q}} & & (\mathcal{F}'_{\mathbb{C}}, W'_{\mathbb{C}}) & & (\mathcal{F}''_{\mathbb{C}}, W''_{\mathbb{C}}) & \\ \nearrow & & \nwarrow\ \nearrow & & \nwarrow\ \nearrow & & \nwarrow \\ \mathcal{F}_R & & (\mathcal{F}_{\mathbb{Q}}, W_{\mathbb{Q}}) & & (\mathcal{F}_{\mathbb{C}}, W_{\mathbb{C}}) & & (\mathcal{F}_{\mathbb{C}}, W_{\mathbb{C}}, F) \end{array}$$

where the arrows are as in §2.3.1 quasi-isomorphisms of the appropriate objects in the appropriate category, and the conditions of §2.3.1 are satisfied, where we use the diagram of maps

$$\begin{array}{ccccc} & (\mathcal{F}'_{\mathbb{C}}, W'_{\mathbb{C}}) & & (\mathcal{F}''_{\mathbb{C}}, W''_{\mathbb{C}}) & \\ \nearrow & & \nwarrow\ \nearrow & & \nwarrow \\ (\mathcal{F}_{\mathbb{Q}}, W_{\mathbb{Q}}) & & (\mathcal{F}_{\mathbb{C}}, W_{\mathbb{C}}) & & (\mathcal{F}_{\mathbb{C}}, W_{\mathbb{C}}, F) \end{array}$$

to give the $\mathbb{Q}$-weight filtration on $H^*(\mathcal{F}_{\mathbb{C}})$. Taking the resulting mixed Hodge structure gives the functor $\underline{H}^*(\mathcal{F})'$.

We map $C^*_{\mathcal{H}'_R}$ to $C^*_{\mathcal{H}_R}$ by replacing the portion

$$\begin{array}{ccc} (\mathcal{F}'_{\mathbb{C}}, W'_{\mathbb{C}}) & & (\mathcal{F}''_{\mathbb{C}}, W''_{\mathbb{C}}) \\ & {}^{f}\nwarrow\ \nearrow^{g} & \\ & (\mathcal{F}_{\mathbb{C}}, W_{\mathbb{C}}) & \end{array}$$

of the diagram (2.3.3.1) with

$$\overline{(\mathcal{F}'_{\mathbb{C}}, W'_{\mathbb{C}})} := \operatorname{cone}\big((f,g)\colon \mathcal{F}_{\mathbb{C}} \to \mathcal{F}'_{\mathbb{C}} \oplus \mathcal{F}''_{\mathbb{C}}\big),$$

forming the diagram

$$\overline{\mathcal{F}} := \begin{array}{ccccc} & \mathcal{F}'_{\mathbb{Q}} & & \overline{(\mathcal{F}'_{\mathbb{C}}, W'_{\mathbb{C}})} & \\ \nearrow & & \nwarrow\ \nearrow & & \nwarrow \\ \mathcal{F}_R & & (\mathcal{F}_{\mathbb{Q}}, W_{\mathbb{Q}}) & & (\mathcal{F}_{\mathbb{C}}, W_{\mathbb{C}}, F). \end{array}$$

We map $C^*_{\mathcal{H}_R}$ to $C^*_{\mathcal{H}'_R}$ by adding two identity maps, forming the diagram

$$\begin{array}{ccccccc} & \mathcal{F}'_{\mathbb{Q}} & & (\mathcal{F}'_{\mathbb{C}}, W'_{\mathbb{C}}) & & (\mathcal{F}'_{\mathbb{C}}, W'_{\mathbb{C}}) & \\ \nearrow & & \nwarrow\ \nearrow & & {}_{\mathrm{id}}\nwarrow\ \nearrow_{\mathrm{id}} & & \nwarrow \\ \mathcal{F}_R & & (\mathcal{F}_{\mathbb{Q}}, W_{\mathbb{Q}}) & & (\mathcal{F}'_{\mathbb{C}}, W'_{\mathbb{C}}) & & (\mathcal{F}_{\mathbb{C}}, W_{\mathbb{C}}, F). \end{array}$$

These maps give a tensor equivalence of the homotopy categories

$$\text{(2.3.3.2)}\qquad K^*_{\mathcal{H}_R} \to K^*_{\mathcal{H}'_R}.$$

The functor $\underline{H}^*(\mathcal{F})'$ is compatible with the functor $\underline{H}^*(\mathcal{F})$ of §2.3.1 via the equivalence (2.3.3.2). Defining $D^*_{\mathcal{H}'_R}$ to be the localization of $K^*_{\mathcal{H}'_R}$ with respect to $\underline{H}^*(\mathcal{F})'$, we thus have the equivalence of categories $D^*_{\mathcal{H}'_R} \to D^*_{\mathcal{H}_R} \to D^*(\mathcal{H}_R)$.

2.3.4. *Godement resolutions.* The set of points of the topological space X_{an} allows us to define the cosimplicial Godement resolution (see Part II, Chapter IV, §2.1) of a sheaf of abelian groups S on X_{an}: $S \to G_X S$. By (Part II, Chapter IV, Lemma 2.2.2), the induced augmented cochain complex $\epsilon_{X,S}: S \to \mathrm{cc}G_X S$ is a quasi-isomorphism of S with the acyclic complex $\mathrm{cc}G_X S$. The formation of $G_X S$ is natural in S and in X; as G_X is an exact functor, $\mathrm{cc}G_X$ thus preserves filtered quasi-isomorphisms. We often write G^*_X for $\mathrm{cc}G_X$.

Each product $\mu: p_1^* S \otimes p_2^* S' \to S''$ on $X \times_{\mathbb{C}} Y$ induces a product of augmented cosimplicial sheaves $G_X\mu: p_1^* G_X S \otimes p_2^* G_Y S' \to G_{X\times_{\mathbb{C}} Y} S''$ (see Part II, Chapter IV, Proposition 2.3.7). If μ comes from a multiplication of sheaves on the big analytic site, and is associative and commutative, then so is $G_{(-)}\mu$. The analogous results hold for complexes of sheaves of abelian groups.

2.3.5. *Thom-Sullivan cochains.* We have taken this material from [**65**].

Let $|\Delta_n|$ be the real n-simplex

$$|\Delta_n| := \{(t_0, \dots, t_n) \in \mathbb{R}^{n+1} \mid \sum_{i=0}^{n} t_i = 1, 0 \le t_i\},$$

and let Δ_n be the simplicial set $\mathrm{Hom}_\Delta(-, [n]): \Delta^{\mathrm{op}} \to \mathbf{Sets}$. For $f: [m] \to [n]$ in Δ, let $|f|: |\Delta_m| \to |\Delta_n|$ denote the corresponding affine-linear map.

For a cosimplicial abelian group G, we have the associated complex $\mathrm{cc}G$, and the *normalized* subcomplex, $\mathrm{Norm}(G)$, with $\mathrm{Norm}(G)^p$ the subgroup of those $g \in G([p])$ such $G(f)(g) = 0$ for all $f: [p] \to [q]$ in Δ which are not injective. The inclusion of $\mathrm{Norm}(G) \hookrightarrow \mathrm{cc}G$ is a homotopy equivalence. In particular, we have the complex of (simplicial) cochains of Δ_n, and the subcomplex of normalized cochains $Z^*(\Delta_n)$.

Let $\Omega^*(|\Delta_n|)$ denote the complex of $\mathbb{Q}$-polynomial differential forms on $|\Delta_n|$

$$\Omega^*(|\Delta_n|) := \Omega^*_{\mathbb{Q}[t_0,\dots,t_n]/\sum_{i=0}^n t_i - 1}.$$

Sending $[n]$ to $Z^*(\Delta_n)$, $\Omega^*(|\Delta_n|)$ determines functors

$$Z^*: \Delta^{\mathrm{op}} \to \mathbf{C}^{\ge 0}(\mathbf{Ab}), \quad \Omega^*: \Delta^{\mathrm{op}} \to \mathbf{C}^{\ge 0}(\mathbf{Mod}_{\mathbb{Q}}),$$

where $\mathbf{C}^{\ge 0}(\mathbf{Mod}_{\mathbb{Q}})$ is the category of complexes of $\mathbb{Q}$-vector spaces concentrated in degrees ≥ 0, and $\mathbf{C}^{\ge 0}(\mathbf{Ab})$ is the integral version. There is a natural homotopy equivalence

$$\int : \Omega^* \to Z^* \otimes \mathbb{Q}$$

defined by

$$\int(\omega)(\sigma) = \int_{|\sigma|} \omega$$

for $\omega \in \Omega^m(|\Delta_n|)$ and σ an m-simplex of Δ_n.

We have the category $\mathrm{Mor}(\Delta)$, with objects the morphisms in Δ, where a morphism $f \to g$ is a commutative diagram

$$\begin{array}{ccc} \bullet & \xrightarrow{f} & \bullet \\ \uparrow & & \downarrow \\ \bullet & \xrightarrow[g]{} & \bullet \end{array}.$$

For functors $F\colon \Delta^{\mathrm{op}} \to \mathbf{C}^{\geq 0}(\mathbf{Ab})$ and $G\colon \Delta \to \mathbf{Ab}$, we have the functor

$$\begin{aligned} F(\text{domain}) \otimes G(\text{range})\colon \mathrm{Mor}(\Delta) &\to \mathbf{C}^{\geq 0} \\ f &\mapsto F(\text{domain}(f)) \otimes G(\text{range}(f)); \end{aligned}$$

let $F \underset{\leftarrow}{\otimes} G$ be the projective limit

$$F \underset{\leftarrow}{\otimes} G := \underset{\substack{\longleftarrow \\ \mathrm{Mor}(\Delta)}}{\lim} F(\text{domain}) \otimes G(\text{range}).$$

Explicitly, an element ϵ of $(F \underset{\leftarrow}{\otimes} G)^m$ is given by a collection $p \mapsto \epsilon_p \in F^m([p]) \otimes G([p])$ such that

$$F(f) \otimes G(\mathrm{id})(\epsilon_p) = F(\mathrm{id}) \otimes G(f)(\epsilon_q)$$

for each $f\colon [q] \to [p]$ in Δ. The operation $\underset{\leftarrow}{\otimes}$ is functorial and respects homotopy equivalence.

For a simplicial abelian group G, we have the well-defined map $e\colon Z^* \underset{\leftarrow}{\otimes} G \to \mathrm{Norm}(G)$ defined by sending $\epsilon := (\ldots \epsilon_p \ldots) \in (Z^* \underset{\leftarrow}{\otimes} G)^q$ to $\epsilon_q(\mathrm{id}_{[q]}) \in G([q])$. In [**65**, Lemma 3.1] it is shown that this map is well-defined, lands in $\mathrm{Norm}(G)$, and gives an isomorphism of complexes. Thus, we have the natural homotopy equivalences

$$\int \underset{\leftarrow}{\otimes} \mathrm{id}\colon \Omega^* \underset{\leftarrow}{\otimes} G \to \mathrm{Norm}(G) \otimes \mathbb{Q} \to G \otimes \mathbb{Q}.$$

The operation of wedge product makes Ω^* into a simplicial differential graded algebra. If G is a cosimplicial commutative ring without unit, i.e., if we have a product $\mu_G\colon G \otimes G \to G$ which is commutative and associative, then the cochain complex $\Omega^* \underset{\leftarrow}{\otimes} G$ has the commutative and associative product

$$\Omega^* \mu_G\colon (\Omega^* \underset{\leftarrow}{\otimes} G) \otimes (\Omega^* \underset{\leftarrow}{\otimes} G) \to \Omega^* \underset{\leftarrow}{\otimes} G$$

induced by the map $(\omega \otimes g) \otimes (\omega' \otimes g') \mapsto \omega \wedge \omega' \otimes \mu(g \otimes g')$, for $\omega \otimes g \in \Omega^q(|\Delta_p|) \otimes G([p])$, $\omega' \otimes g' \in \Omega^{q'}(|\Delta_p|) \otimes G([p])$. It is easy to check that this gives a well-defined functorial product of cochain complexes.

In addition, suppose we have a commutative ring without unit A, which we may consider as a constant cosimplicial ring without unit, and an augmentation $\iota\colon A \to G$ of cosimplicial rings without unit. Then $\Omega^* \underset{\leftarrow}{\otimes} A = A$, and the product $\Omega^* \mu_A$ agrees with μ_A. Thus the products μ_A and $\Omega^* \mu_G$ are compatible via the augmentation

$$A = \Omega^* \underset{\leftarrow}{\otimes} A \xrightarrow{\Omega^* \underset{\leftarrow}{\otimes} \iota} \Omega^* \underset{\leftarrow}{\otimes} G.$$

We extend the functor $\Omega^* \underset{\leftarrow}{\otimes}(-)$ to the functor $G^*: \Delta^{\mathrm{op}} \to \mathbf{C}^+(\mathbf{Ab})$ by taking the total complex of the double complex

$$\ldots \to \Omega^* \underset{\leftarrow}{\otimes} G^0 \to \Omega^* \underset{\leftarrow}{\otimes} G^1 \to \ldots .$$

All the properties of $\Omega^* \underset{\leftarrow}{\otimes}(-)$ described above extend to the case of complexes.

We may replace the category **Ab** with any abelian tensor category for which projective limits and filtered inductive limits are representable; in particular, we the above construction is valid for the category of sheaves of abelian groups on a Grothendieck site.

2.3.6. *Compactifications.* Let X be a smooth quasi-projective $\mathbb{C}$-scheme. A *compactification* of X is a birational inclusion $j: X \to \overline{X}$ of X as an open subscheme of a smooth projective $\mathbb{C}$-scheme $\overline{X}$, such that the complement $D := \overline{X}\setminus X$ is a normal crossing divisor.

For as complex of sheaves S on X, we have the *canonical filtration* $\tau_X^{\leq}$ of S

$$(\tau_X^{\leq p} S)^q = \begin{cases} S^q; & \text{if } q < p \\ \ker(d^q: S^q \to S^{q+1}); & \text{if } q = p \\ 0; & \text{if } q > p. \end{cases}$$

We have the similar notion for a complex of sheaves on $\overline{X}$.

Form the category $\mathcal{C}^*(X, \overline{X})$ ($* = b,+,-$ or $\emptyset$ is a boundedness condition) of diagrams

(2.3.6.1)
$$\begin{array}{ccccccccc} & \mathcal{F}'_{\mathbb{Q}} & & (\mathcal{F}_{\mathbb{C}}, W_{\mathbb{C}}) & & (\mathcal{F}''_{\mathbb{C}}, W''_{\mathbb{C}}) & & \\ \nearrow & & \nwarrow \quad \nearrow & & \nwarrow \quad \nearrow & & \nwarrow & \\ \mathcal{F}_R & & (\mathcal{F}_{\mathbb{Q}}, W_{\mathbb{Q}}) & & (\mathcal{F}'_{\mathbb{C}}, W'_{\mathbb{C}}) & & (\mathcal{G}_{\mathbb{C}}, W_{\mathbb{C}}, F). \end{array}$$

Here $\mathcal{F}_R$ is a complex of sheaves of R-modules on X, $\mathcal{F}'_{\mathbb{Q}}$ is a complex of sheaves of $R\otimes\mathbb{Q}$-vector spaces on X, $(\mathcal{F}_{\mathbb{Q}}, W_{\mathbb{Q}})$ is a filtered complex of sheaves of $R\otimes\mathbb{Q}$-vector spaces on $\overline{X}$, $(\mathcal{F}_{\mathbb{C}}, W_{\mathbb{C}})$, $(\mathcal{F}'_{\mathbb{C}}, W'_{\mathbb{C}})$ and $(\mathcal{F}''_{\mathbb{C}}, W''_{\mathbb{C}})$ are filtered complexes of sheaves of $\mathbb{C}$-vector spaces on $\overline{X}$, and $(\mathcal{G}_{\mathbb{C}}, W_{\mathbb{C}}, F)$ is a bi-filtered complex of sheaves of $\mathbb{C}$-vector spaces on $\overline{X}$. The W filtrations are all increasing and the F filtration is decreasing.

The arrows in the diagram are as follows:

- The arrow $\mathcal{F}_R \to \mathcal{F}'_{\mathbb{Q}}$ is a quasi-isomorphism $\mathcal{F}_R \otimes \mathbb{Q} \to \mathcal{F}'_{\mathbb{Q}}$.
- The arrow $(\mathcal{F}_{\mathbb{Q}}, W_{\mathbb{Q}}) \to \mathcal{F}'_{\mathbb{Q}}$ is a map $j^*\mathcal{F}_{\mathbb{Q}} \to \mathcal{F}'_{\mathbb{Q}}$ which by adjunction induces an isomorphism $\mathcal{F}_{\mathbb{Q}} \to Rj_*\mathcal{F}'_{\mathbb{Q}}$ in $\mathbf{D}^+(\mathrm{Sh}^{\mathbb{Q}}_{\overline{X}_{\mathrm{an}}})$.
- The arrow $\mathcal{F}_{\mathbb{Q}} \to (\mathcal{F}_{\mathbb{C}}, W_{\mathbb{C}})$ is a filtered quasi-isomorphism $(\mathcal{F}_{\mathbb{Q}}, W_{\mathbb{Q}}) \otimes \mathbb{C} \to (\mathcal{F}_{\mathbb{C}}, W_{\mathbb{C}})$.
- The arrows $(\mathcal{F}'_{\mathbb{C}}, W'_{\mathbb{C}}) \to (\mathcal{F}_{\mathbb{C}}, W_{\mathbb{C}})$ and $(\mathcal{F}'_{\mathbb{C}}, W'_{\mathbb{C}}) \to (\mathcal{F}''_{\mathbb{C}}, W''_{\mathbb{C}})$ are filtered quasi-isomorphisms.
- The arrow $(\mathcal{G}_{\mathbb{C}}, W_{\mathbb{C}}, F) \to (\mathcal{F}''_{\mathbb{C}}, W''_{\mathbb{C}})$ is a filtered quasi-isomorphism

$$(\mathcal{G}_{\mathbb{C}}, W_{\mathbb{C}}) \to (\mathcal{F}''_{\mathbb{C}}, W''_{\mathbb{C}}).$$

Maps in $\mathcal{C}^*(X, \overline{X})$ are component-wise; $\mathcal{C}^*(X, \overline{X})$ forms a DG tensor category, with the tensor product given component-wise, and the cone functors on each component defines the cone functor for $\mathcal{C}^*(X, \overline{X})$.

If D is a diagram (2.3.6.1), we have the diagram

(2.3.6.2)

$$p_{(X,\overline{X})*}G^*D :=$$

$$\begin{array}{ccccccc}
& \Gamma(X, G_X^*\mathcal{F}'_{\mathbb{Q}}) & & \Gamma(\overline{X}, G_{\overline{X}}^*(\mathcal{F}_{\mathbb{C}}, W_{\mathbb{C}})) & & \Gamma(\overline{X}, G_{\overline{X}}^*(\mathcal{F}''_{\mathbb{C}}, W''_{\mathbb{C}})) & \\
\nearrow & & \nwarrow \quad \nearrow & & \nwarrow \quad \nearrow & & \nwarrow \\
\Gamma(X, G_X^*\mathcal{F}_R) & & \Gamma(\overline{X}, G_{\overline{X}}^*(\mathcal{F}_{\mathbb{Q}}, W_{\mathbb{Q}})) \quad \Gamma(\overline{X}, G_{\overline{X}}^*(\mathcal{F}'_{\mathbb{C}}, W'_{\mathbb{C}})) & & & & \Gamma(\overline{X}, G_{\overline{X}}^*(\mathcal{G}_{\mathbb{C}}, W_{\mathbb{C}}, F)).
\end{array}$$

We have the full DG tensor subcategory $\mathcal{C}^*_{\mathcal{H}}(X,\overline{X})$ of $\mathcal{C}^*(X,\overline{X})$ consisting of diagrams D as in (2.3.6.1) for which the diagram $p_{(X,\overline{X})*}G^*D$ is in $C^*_{\mathcal{H}'_R}$.

2.3.7. *Décalage.* The most obvious weight filtration to put on the analytic De Rham complex of a smooth projective variety X is to put everything in weight zero. This would not agree with the natural weights on the cohomology $H^*(X)$, as $H^n(X)$ has a Hodge structure of weight n. The operation of décalage gives the necessary shift.

Let C be a complex with an increasing filtration W. Define the filtration $\mathrm{Dec}(W)$ by

$$\mathrm{Dec}(W)_p(C^n) := \ker\left(W_{p-n}(C^n) \xrightarrow{d} C^{n+1}/W_{p-n-1}(C^{n+1})\right).$$

The effect of replacing W with $\mathrm{Dec}(W)$ is

$$\mathrm{gr}_p^{\mathrm{Dec}(W)}H^n(C) = \mathrm{gr}^W_{p-n}H^n(C) \tag{2.3.7.1}$$

(see [**38**, 1.3.4]).

2.3.8. *The representing diagram.* Let $j: X \to \overline{X}$ be a compactification. Let Ω_X^* denote the analytic de Rham complex, and let $\Omega^*_{\overline{X}}(\log D)$ be the complex of forms with log poles. Let $(\Omega^*_{\overline{X}}(\log D), W)$ denote the filtration by order of pole, and $(\Omega^*_{\overline{X}}(\log D), F)$ the stupid filtration.

We have the following natural maps

$$\mathbb{Z}_X \otimes \mathbb{Q} \to \Omega^* \underset{\leftarrow}{\otimes} G_X\mathbb{Q}_X$$

$$j^*(\Omega^* \underset{\leftarrow}{\otimes} j_*G_X\mathbb{Q}_X) \to \Omega^* \underset{\leftarrow}{\otimes} G_X\mathbb{Q}_X$$

$$(\Omega^* \underset{\leftarrow}{\otimes} j_*G_X\mathbb{Q}_X, \tau^{\leq}_{\overline{X}}) \otimes \mathbb{C} \to (\Omega^* \underset{\leftarrow}{\otimes} j_*G_X\Omega^*_X, \tau^{\leq}_{\overline{X}})$$

$$(\Omega^*_{\overline{X}}(\mathrm{log}D), \tau^{\leq}_{\overline{X}}) \to (\Omega^* \underset{\leftarrow}{\otimes} j_*G_X\Omega^*_X, \tau^{\leq}_{\overline{X}})$$

$$(\Omega^*_{\overline{X}}(\log D), \tau^{\leq}_{\overline{X}}) \to (\Omega^*_{\overline{X}}(\log D), W).$$

The first is the composition

$$\mathbb{Z}_X \otimes \mathbb{Q} \cong \mathbb{Q}_X = \Omega^* \underset{\leftarrow}{\otimes} \mathbb{Q}_X \xrightarrow{\Omega^* \underset{\leftarrow}{\otimes} \iota_{\mathbb{Q}_X}} \Omega^* \underset{\leftarrow}{\otimes} G_X\mathbb{Q}_X$$

with $\iota_{\mathbb{Q}_X}$ the canonical augmentation. The second is the isomorphism

$$j^*(\Omega^* \underset{\leftarrow}{\otimes} j_*(G_X\mathbb{Q}_X)) \cong \Omega^* \underset{\leftarrow}{\otimes} G_X\mathbb{Q}_X.$$

The third is the map induced by applying $\Omega^* \underset{\leftarrow}{\otimes} j_*G_X$ to the augmentation $\mathbb{Q}_X \otimes \mathbb{C} \cong \mathbb{C}_X \hookrightarrow \Omega^*_X$ and taking the canonical filtration on $\overline{X}$. The fourth is given by the composition

$$\Omega^*_{\overline{X}}(\log D) \to j_*\Omega^*_X = \Omega^* \underset{\leftarrow}{\otimes} j_*\Omega^*_X \to \Omega^* \underset{\leftarrow}{\otimes} j_*G_X\Omega^*_X$$

and taking the canonical filtration, and the last line induced by the identity map on $\Omega^*_{\overline{X}}(\log D)$ (the canonical filtration is finer than the weight filtration).

It follows from the discussion in §2.3.4 and §2.3.5 that the first, third and fourth maps are quasi-isomorphisms and that the second map induces the sequence of isomorphisms

$$\Omega^* \underset{\leftarrow}{\otimes} j_*G_X\mathbb{Q}_X \cong j_*G^*_X\mathbb{Q}_X \cong Rj_*G^*_X\mathbb{Q}_X \cong Rj_*\Omega^* \underset{\leftarrow}{\otimes} G_X\mathbb{Q}_X$$

in $\mathbf{D}^+(\mathrm{Sh}^{\mathbb{Q}}_{\mathrm{an}}(X))$. Since we are taking the canonical filtration, the third and fourth maps are filtered quasi-isomorphisms. The fifth map is a filtered quasi-isomorphism by [**38**, Prop. 3.1.8].

This gives us the diagram

(2.3.8.1)

$$D[X, \overline{X}] :=$$

$$\begin{array}{ccccccccc} & & \Omega^* \underset{\leftarrow}{\otimes} G_X\mathbb{Q}_X & & (\Omega^* \underset{\leftarrow}{\otimes} j_*G_X\Omega^*_X, \mathrm{Dec}\tau^{\leq}_{\overline{X}}) & & (\Omega^*_{\overline{X}}(\log D), \mathrm{Dec}W) & & \\ & \nearrow & & \nwarrow \quad \nearrow & & \nwarrow \quad \nearrow & & \nwarrow & \\ \mathbb{Z}_X & & & (\Omega^* \underset{\leftarrow}{\otimes} j_*G_X\mathbb{Q}_X, \mathrm{Dec}\tau^{\leq}_{\overline{X}}) & & (\Omega^*_{\overline{X}}(\log D), \mathrm{Dec}\tau^{\leq}_{\overline{X}}) & & & (\Omega^*_{\overline{X}}(\log D), \mathrm{Dec}W, F) \end{array}$$

in $\mathcal{C}^+(X, \overline{X})$. It follows from the results of [**11**, §4] that this diagram is in fact in $\mathcal{C}^+_{\mathcal{H}'}(X, \overline{X})$.

2.3.9. *The Tate object.* For an abelian group A, let $W(q)_A$ be the filtration on A given by

$$W(q)_{A,p} = \begin{cases} A; & \text{if } p \geq -2q \\ 0; & \text{if } p < -2q. \end{cases}$$

We let $F(q)$ be the filtration on $\mathbb{C}$ given by

$$F(q)^p = \begin{cases} \mathbb{C}; & \text{if } p \leq -q \\ 0; & \text{if } p > -q. \end{cases}$$

We have the *R-Tate mixed Hodge structure*

$$R(q) :=$$

$$\begin{array}{ccccccc} & (2\pi i)^q R_{\mathbb{Q}} & & (\mathbb{C}, W(q)_{\mathbb{C}}) & & (\mathbb{C}, W(q)_{\mathbb{C}}) & \\ \nearrow & & \nwarrow \quad \nearrow & & \nwarrow \quad \nearrow & & \nwarrow \\ (2\pi i)^q R & & ((2\pi i)^q R_{\mathbb{Q}}, W(q)_{(2\pi i)^q R_{\mathbb{Q}}}) \quad (\mathbb{C}, W(q)_{\mathbb{C}}) & & & & (\mathbb{C}, W(q)_{\mathbb{C}}, F(q)) \end{array}$$

(where $R_{\mathbb{Q}} = R \otimes \mathbb{Q}$). The object $R(0)$ is the unit for the tensor structure on $C^+_{\mathcal{H}'_R}$.

Define $R^{\mathrm{Hdg}}_{(X,\bar{X})}(q)$ to be the object $R(q) \otimes p_{X*}G^*D[X, \bar{X}]$ of $C^+_{\mathcal{H}'_R}$, and let $R^{\mathrm{Hdg}}_X(q)$ be the image in $D^+_{\mathcal{H}'}$ of the inductive limit of the objects $R^{\mathrm{Hdg}}_{(X,\bar{X})}(q)$ over compactifications $\bar{X}$ of X.

2.3.10. THEOREM. *Sending (X, q) to $R_X^{\mathrm{Hdg}}(q)$ extends canonically to an exact functor*

$$\mathfrak{R}_{\mathrm{Hdg}}: \mathcal{DM}(\mathcal{V})_R \to D^+_{\mathcal{H}'_R}.$$

The restriction of $\mathfrak{R}_{\mathrm{Hdg}}$ to $\mathcal{DM}_{\mathfrak{sh}}(\mathcal{V})_R$ is an exact pseudo-tensor functor.

PROOF. The proof is essentially the same as the proof of Theorem 1.3.1 with $\mathcal{H}$ playing the role of the abelian tensor category $\mathcal{A}$; there are four main differences: (i) We replace the categories $\mathbf{C}^+(\mathcal{H}_R)$, $\mathbf{K}^+(\mathcal{H}_R)$ and $\mathbf{D}^+(\mathcal{H}_R)$ with the more convenient categories $C^+_{\mathcal{H}'_R}$, $K^+_{\mathcal{H}'_R}$ and $D^+_{\mathcal{H}'_R}$.

Flatness is not a problem, as all the complexes involved are either complexes of torsion-free abelian groups, or complexes of vector spaces over a field.
(ii) After applying $p_{(X,\overline{X})*}G^*$ to the diagram $D[X, \overline{X}]$, we then form the inductive limit (in $C^+_{\mathcal{H}'_R}$) with respect to compactifications $\overline{X}$ of X; as the category of compactifications is filtering, and the resulting Hodge structure is independent of the compactification, the inductive limit

$$p_{X*}G^*D[X, -] := \varinjlim_{X \hookrightarrow \overline{X}} p_{X*}G^*D[X, \overline{X}]$$

is still in $C^+_{\mathcal{H}'}$.

For a diagram D in $\mathcal{C}^*(X, \overline{X})$, let $p_{(X,\overline{X})*}GD$ be defined as in (2.3.6.2), only using the cosimplicial Godement resolutions G_X, $G_{\overline{X}}$ instead of the associated complex G^*_X, $G^*_{\overline{X}}$.

Let $p_{X*}GD[X, -]$ be the inductive limit

$$p_{X*}GD[X, -] := \varinjlim_{X \hookrightarrow \overline{X}} p_{X*}GD[X, \overline{X}].$$

$p_{X*}G^*D[X, -]$ is thus the diagram of complexes associated to the cosimplicial diagram $p_{X*}GD[X, -]$.

The discussion of §2.3.4 and §2.3.5 shows that the compatible associative and commutative multiplications on the functorial complexes of sheaves

$$(X, \overline{X}) \mapsto \mathbb{Z}_X,\ \mathbb{Q}_X,\ \Omega^*_X,\ \Omega_{\overline{X}}(\log D)$$

give functorial associative, commutative multiplications

$$\mu_{X,Y}: p_{X*}GD[X, -] \otimes p_{Y*}GD[Y, -] \to p_{X \times_{\mathbb{C}} Y*}GD[X \times_{\mathbb{C}} Y, -]$$

in the category of cosimplicial diagrams of complexes.
(iii) We need to give a somewhat different construction for the functor (1.3.7.1), and the isomorphism (1.3.7.4). We give the construction for $R = \mathbb{Z}$ for simplicity.

Consider the enlarged mixed Hodge complex P_* which in degree 0 is

$$\begin{array}{ccccccccccc} & & \mathbb{Q} \oplus \mathbb{Q} & & (\mathbb{C} \oplus \mathbb{C}, W(0)_{\mathbb{C}} \oplus W(0)_{\mathbb{C}}) & & & & (\mathbb{C} \oplus \mathbb{C}, W(0)_{\mathbb{C}} \oplus W(0)_{\mathbb{C}}) & & \\ & \nearrow & & \nwarrow \quad \nearrow & & \nwarrow & & \nearrow & & \nwarrow & \\ \mathbb{Z} & & & (\mathbb{Q}, W(0)_{\mathbb{Q}}) & & & (\mathbb{C}, W(0)_{\mathbb{C}}) & & & & (\mathbb{C}, W(0)_{\mathbb{C}}, F(0)) \end{array}$$

and in degree -1 is

$$\begin{array}{ccccccccccc} & & \mathbb{Q} & & & (\mathbb{C}, W(0)_{\mathbb{C}}) & & & (\mathbb{C}, W(0)_{\mathbb{C}}) & & \\ & \nearrow & & \nwarrow & \nearrow & & \nwarrow & \nearrow & & \nwarrow & \\ 0 & & & & 0 & & & 0 & & & 0. \end{array}$$

The maps in the degree 0 portion are the obvious ones, with the left-hand side from the lower row going into the left-hand summand, and the right-hand side from the lower row going into the right-hand summand. The differential $P_{-1} \to P_0$ is the diagonal map. We map P_0 to $\mathbb{Z}(0)$ by the identity on the lower row and the difference map on the upper row. This determines the map of enlarged mixed Hodge complexes

$$P_* \to \mathbb{Z}(0) \tag{2.3.10.1}$$

which is an isomorphism in $K^+_{\mathcal{H}'_R}$. It is easy to see that the map (2.3.10.1) gives an isomorphism of functors

$$\mathrm{Hom}_{K^+_{\mathcal{H}'_R}}(P_*^{\otimes n}, -) \to \mathrm{Hom}_{D^+_{\mathcal{H}'_R}}(\mathbb{Z}(0)^{\otimes n}, -) \cong \mathrm{Hom}_{D^+_{\mathcal{H}'_R}}(\mathbb{Z}(0), -) \tag{2.3.10.2}$$

for all $n \geq 0$.

In addition, the complex of morphisms of complexes $\mathrm{Hom}_{C^+_{\mathcal{H}'_R}}(P_*^{\otimes n}, P_*^{\otimes m})$ is free of two-torsion for all $n, m \geq 1$; by (2.3.10.2), we have

$$H^p(\mathrm{Hom}_{C^+_{\mathcal{H}'_R}}(P_*^{\otimes n}, P_*^{\otimes m})) = \begin{cases} 0; & \text{for } p \neq 0, \\ \mathbb{Z} \cdot \mathrm{id}; & \text{for } p = 0. \end{cases}$$

By (1.3.6.1) and the universal mapping property of the category $\mathbb{E}$ (see Part II, Chapter II, Proposition 3.1.13), there is a functor of DG tensor categories

$$I_{\mathcal{H}} : \mathbb{E} \to C^+_{\mathcal{H}'_R} \tag{2.3.10.3}$$

with $I_{\mathcal{H}}(\mathfrak{e}^{\otimes n}) = P_*^{\otimes n}$. $I_{\mathcal{H}}$ is unique up to homotopy equivalence.

Using the functor (2.3.10.3) instead of the functor (1.3.7.1), the isomorphism (2.3.10.2) instead of the isomorphism (1.3.7.4), and using the categories $C^+_{\mathcal{H}'_R}$, $K^+_{\mathcal{H}'_R}$ and $D^+_{\mathcal{H}'_R}$ instead of $\mathbf{C}^+(\mathcal{H}_R)$, $\mathbf{K}^+(\mathcal{H}_R)$ and $\mathbf{D}^+(\mathcal{H}_R)$, the proof of Theorem 1.3.1 goes through without essential change. □

We may compose the functor $\Re_{\mathrm{Hdg}}$ with the derived functor $R\mathrm{Hdg} : D^+_{\mathcal{H}'_R} \to \mathbf{D}^+(\mathbf{Mod}_R)$ of the absolute Hodge cohomology functor $H^0_{\mathrm{Hdg}}(-) := F^0(-) \cap W_0(-)$ (more precisely, $H^0_{\mathrm{Hdg}}(D)$ is the R-submodule of D_R of elements which land in $W^0(D_{\mathbb{Q}})$ and in $F^0(D_{\mathbb{C}})$ under the comparison maps in the diagram D) to give the absolute Hodge realization

$$\Re_{AHdg} : \mathcal{DM}(\mathbf{Sm}_{\mathbb{C}})_R \to \mathbf{D}^+(\mathbf{Mod}_R).$$

For $R = \mathbb{Z}$, the resulting cohomology groups $H^p(\Re_{AHdg}(\mathbb{Z}_X(q)))$ are the absolute Hodge cohomology groups $H^p_{\mathrm{AH}}(X, \mathbb{Z}(q))$; when X is projective, these agree with the Deligne cohomology groups $H^p_{\mathcal{D}}(X, \mathbb{Z}(q))$. The restriction of these realization functors to $\mathcal{DM}_{\mathfrak{sh}}(\mathbf{Sm}_{\mathbb{C}})_R$ gives, as above, exact, pseudo-tensor functors.

2.3.11. *The* $\mathbb{R}$*-Hodge realization.* One can extend the $\mathbb{R}$-Betti realization to the $\mathbb{R}$-*Hodge realization* by adding the data of a real Frobenius (i.e., an involution) in each component of a diagram in $C^+_{\mathcal{H}'_R}$, forming the category $C^{+\infty}_{\mathcal{H}'_R}$. The comparison maps between the components $\mathcal{F}_R$, $\mathcal{F}'_{\mathbb{Q}}$ and $(\mathcal{F}_{\mathbb{Q}}, W)$, as well as the comparison maps between the components $(\mathcal{F}_{\mathbb{C}}, W)$, $(\mathcal{F}'_{\mathbb{C}}, W)$, $(\mathcal{F}''_{\mathbb{C}}, W)$ and $(\mathcal{G}_{\mathbb{C}}, W, F)$, are required to be F_∞ equivariant. The comparison map $f : (\mathcal{F}_{\mathbb{Q}}, W) \otimes \mathbb{C} \to (\mathcal{F}_{\mathbb{C}}, W)$ satisfies $F_\infty \circ f = f \circ (F_\infty \otimes \sigma)$ where $\sigma : \mathbb{C} \to \mathbb{C}$ is complex conjugation.

For an $\mathbb{R}$-scheme X with compactification $\overline{X}$, the diagram $D[X, \overline{X}]$ carries the required involution, where the F_∞ on the $\mathbb{Z}$ and $\mathbb{Q}$ portions is induced by complex conjugation on the space X_{an} as in §2.1.3, and the F_∞ on the $\mathbb{C}$ portion is induced by $\eta \mapsto \overline{F_\infty^* \eta}$. We let $R_X^{\mathrm{Hdg}\infty}(q)$ be the object of $D^{+\infty}_{\mathcal{H}'_R}$ determined by the object $R_X^{\mathrm{Hdg}}(q)$ together with the real Frobenius defined above.

The construction of the Hodge realization in Theorem 2.3.10 gives the $\mathbb{R}$-Hodge version:

2.3.12. THEOREM. *Sending (X, q) to $R_X^{\mathrm{Hdg}\infty}(q)$ extends canonically to an exact functor*

$$\mathfrak{R}_{\mathbb{R}\mathrm{Hdg}} : \mathcal{DM}(\mathbf{Sm}_{\mathbb{R}})_R \to D^{+\infty}_{\mathcal{H}'_R}.$$

The restriction of $\mathfrak{R}_{\mathbb{R}\mathrm{Hdg}}$ to $\mathcal{DM}_{\mathfrak{sh}}(\mathbf{Sm}_{\mathbb{R}})_R$ is an exact pseudo-tensor functor.

2.3.13. *Hodge realization over a smooth base scheme.* Let S be a smooth quasi-projective $\mathbb{C}$-scheme. The construction of the Hodge realization over $\mathbb{C}$ extends to give a realization to M. Saito's category $\mathrm{MHM}(S)$ of *algebraic mixed Hodge modules over S*. To describe this, we first briefly recall some facts about $\mathrm{MHM}(S)$; for further details, we refer the reader to [**111**], especially §2 and §5, and [**110**].

Let X be a complex manifold. We have the sheaf of linear holomorphic differential operators $\mathcal{D}_X$, with (decreasing) filtration F by order. A (right) filtered $\mathcal{D}_X$-module (M, F) is *holonomic* if $\mathrm{gr}^F_m M$ is a coherent $\mathcal{O}_X$-module; in this case, F is a *good filtration*. This gives us the category $\mathrm{MF}_h(\mathcal{D}_X)$ of holonomic $\mathcal{D}_X$-modules with good filtration.

For a filtered holonomic $\mathcal{D}_X$-module (M, F), we have the object in the derived category of constructible $\mathbb{C}$-sheaves $\mathrm{DR}(M) := M \otimes^L_{\mathcal{D}_X} \mathcal{O}_X$. In fact, $\mathrm{DR}(M)$ is in the abelian subcategory $\mathrm{Perv}_{\mathbb{C}}(X)$ of perverse $\mathbb{C}$-sheaves. There is an explicit complex representing $\mathrm{DR}(M)$, which we now describe. Suppose X has dimension d. We have the Koszul complex (Θ_X is the holomorphic tangent bundle)

$$\mathcal{D}_X \otimes_{\mathcal{O}_X} \Lambda^{-*}\Theta_X := (\ldots \to \mathcal{D}_X \otimes_{\mathcal{O}_X} \Lambda^{-p}\Theta_X \to \ldots \to \mathcal{D}_X),$$

which gives a free $\mathcal{D}_X$ resolution of the left $\mathcal{D}_X$-module $\mathcal{O}_X$. We give $\mathcal{D}_X \otimes_{\mathcal{O}_X} \Lambda^{-*}\Theta_X$ the filtration

$$F_p(\mathcal{D}_X \otimes_{\mathcal{O}_X} \Lambda^{-i}\Theta_X) := F_{p+i}\mathcal{D}_X \otimes \Lambda^{-i}\Theta.$$

Then

$$\widetilde{\mathrm{DR}}(M, F) := (M, F) \otimes_{\mathcal{D}} (\mathcal{D}_X \otimes_{\mathcal{O}_X} \Lambda^{-*}\Theta_X, F)$$

is a filtered complex, which represents $\mathrm{DR}(M, F)$.

2.3.13.1. EXAMPLE. Let ω_X be the filtered holonomic $\mathcal{D}_X$-module (Ω^d_X, F), $d = \dim_{\mathbb{C}} X$, with $F_p = 0$ for $p \neq -d$ ($\mathcal{D}_X$ acts through the quotient $\mathcal{O}_X$). Then $\widetilde{\mathrm{DR}}(\omega_X, F)$ is isomorphic to the shifted de Rham complex $\Omega^*_X[d]$, with filtration the (shifted) Hodge filtration. The canonical map $\mathbb{C}_X[d] \to \Omega^*_X[d]$ gives a quasi-isomorphism $\mathbb{C}_X[d] \to \widetilde{\mathrm{DR}}(\omega_X)$.

For a filtered $\mathcal{O}_X$-module (L, F), set $\mathrm{DR}^{-1}(L, F) := (L, F) \otimes_{\mathcal{O}} (\mathcal{D}_X, F)$.

Let X be a smooth quasi-projective $\mathbb{C}$-scheme. We have the category $\mathrm{Perv}_{\mathbb{Q}}(X)$ of perverse $\mathbb{Q}$-sheaves on X_{an} which have algebraic stratifications such that the restrictions of their cohomology sheaves are local systems (and similarly the category $\mathrm{Perv}_{\mathbb{C}}(X)$ of perverse $\mathbb{C}$ sheaves with the same condition).

The category $\mathrm{MF}_h\mathrm{W}(\mathcal{D}_X,\mathbb{Q})$ has objects $((M,F,W),(K,W),\alpha)$, where (M,F) is in $\mathrm{MF}_h(\mathcal{D}_{X_{\mathrm{an}}})$, K is in $\mathrm{Perv}_{\mathbb{Q}}(X)$, W is a locally finite increasing filtration, and $\alpha\colon \widetilde{\mathrm{DR}}(M,F)\to K\otimes_{\mathbb{Q}}\mathbb{C}$ is an isomorphism respecting W (i.e. a filtered quasi-isomorphism). $\mathrm{MHM}(X)$ is a certain abelian subcategory of $\mathrm{MF}_h\mathrm{W}(\mathcal{D}_X,\mathbb{Q})$ (see [**111**, Proposition 5.1.14] and [**110**, §4]). If X has dimension d, there are objects $\mathbb{Q}_X(q)[d]$, $q=0,\pm1,\pm2,\dots$ in $\mathrm{MHM}(X)$.

There are external products $\boxtimes_{X,Y}\colon \mathrm{MHM}(X)\otimes\mathrm{MHM}(Y)\to\mathrm{MHM}(X\times Y)$, defined by

$$\begin{aligned}((M_X,F_X,W_X),(K_X,W_X),\alpha_X)&\boxtimes((M_Y,F_Y,W_Y),(K_Y,W_Y),\alpha_Y)\\ =\big((p_X^{-1}M_X\otimes_{\mathbb{C}}p_Y^{-1}M_Y)&\otimes_{p_X^{-1}\mathcal{O}_X\otimes p_Y^{-1}\mathcal{O}_Y}\mathcal{O}_{X\times Y},W_X\otimes W_Y,F_X\otimes F_Y),\\ &(p_X^{-1}K_X\otimes_{\mathbb{C}}p_Y^{-1}K_Y,W_X\otimes W_Y),\alpha_X\boxtimes\alpha_Y\big),\end{aligned}$$

where the tensor product filtration $W_X\otimes W_Y$ is, in degree n, the submodule generated by $p_X^{-1}W_X^a\otimes p_Y^{-1}W_Y^b$ with $a+b=n$, and similarly for the F-filtration. $\alpha_X\boxtimes\alpha_Y$ is the isomorphism induced by α_X and α_Y. We note that the functor $\widetilde{\mathrm{DR}}$ admits a commutative, associative multiplication with respect to the external products: the natural map

$$\widetilde{\mathrm{DR}}(M_X,F_X)\boxtimes_{X,Y}\widetilde{\mathrm{DR}}(M_Y,F_Y)\to\widetilde{\mathrm{DR}}\big((M_X,F_X)\boxtimes_{X,Y}(M_Y,F_Y)\big)$$

induced by the usual product on the Koszul complex.

Each morphism $p\colon X\to Y$ induces the push-forward

$$p_*\colon \mathbf{D}^b(\mathrm{MHM}(X))\to\mathbf{D}^b(\mathrm{MHM}(Y)) \tag{2.3.13.2}$$

[**110**, §2.c, §4]. We now describe a representing complex for p_*, assuming p is proper.

Let $((M,F,W),(K,W),\alpha)$ be in $\mathrm{MHM}(X)$; we suppose that α is given by a filtered isomorphism of filtered complexes $\alpha'\colon\widetilde{\mathrm{DR}}(M,F,W)\to(K',W')$, with (K',W') isomorphic to $(K,W)\otimes\mathbb{C}$ in the filtered derived category of $\mathbb{C}$-sheaves on X. Take the complexes associated to the cosimplicial Godement resolutions, $G_X^*\widetilde{\mathrm{DR}}(M,F,W)$, $G_X^*(K,W)$, and $G_X^*(K',W')$, take push-forward, and take DR^{-1} for the first term:

$$(\mathrm{DR}^{-1}p_*G_X^*\widetilde{\mathrm{DR}}(M,F,W),p_*G_X^*(K,W)).$$

Since $\mathcal{D}_X\otimes_{\mathcal{O}_X}\Lambda^{-*}\Theta_X$ is a resolution of $\mathcal{O}_X$, we have the natural map

$$\widetilde{\mathrm{DR}}\mathrm{DR}^{-1}p_*G_X^*\widetilde{\mathrm{DR}}(M,F,W)\xrightarrow{\theta}p_*G_X^*\widetilde{\mathrm{DR}}(M,F,W),$$

which is a filtered quasi-isomorphism. Let $p_*\tau\colon p_*G_X^*(K',W')\to p_*G_X^*(K,W)\otimes\mathbb{C}$ be the filtered quasi-isomorphism induced by the given filtered quasi-isomorphism on X. Then

$$(\mathrm{DR}^{-1}p_*G_X^*\widetilde{\mathrm{DR}}(M,F,W),p_*G_X^*(K,W),\tau\circ p_*G_X^*\alpha'\circ\theta)$$

represents $p_*((M,F,W),(K,W),\alpha)$.

We now describe a representative for $j_*\mathbb{Q}_X(0)$, where $j\colon X\to\bar{X}$ is a compactification.

2.3.13.3. Example. (see [**110**], proof of Theorem 3.27) Let $\bar{X}$ be a compactification of X with normal crossing divisor D at infinity, let $d=\dim_{\mathbb{C}}(X)$. We have the filtered $\mathcal{D}_{\bar{X}}$-module $(\Omega^d_{\bar{X}}(\log D),F)$, which we denote by $\omega_{\bar{X}}(*D)$. The usual weight filtration gives the filtration W on $\omega_{\bar{X}}(*D)[-d]$. Then there is a canonical filtered isomorphism β of $\mathrm{DR}(\omega_{\bar{X}}(*D)[-d], W)$ with $(\Omega^*_{\bar{X}}(\log D), W)$, and the filtration F on $\omega_{\bar{X}}(*D)[-d]$ induces the Hodge filtration F on $\Omega^*_{\bar{X}}(\log D)$. The diagram

$$\mathrm{MW}[X, \bar{X}] :=$$

(2.3.13.4)
$$\begin{array}{ccccccc} & (\Omega^* \underset{\leftarrow}{\otimes} j_* G_X \Omega^*_X, \tau^{\leq}_{\bar{X}}) & & & (\Omega^*_{\bar{X}}(\log D), W) & & \\ \nearrow & & \nwarrow & \nearrow & & \nwarrow & \\ (\Omega^* \underset{\leftarrow}{\otimes} j_* G_X \mathbb{Q}_X, \tau^{\leq}_{\bar{X}}) & & (\Omega^*_{\bar{X}}(\log D), \tau^{\leq}_{\bar{X}}) & & & & (\Omega^*_{\bar{X}}(\log D), W, F) \end{array}$$

together with the isomorphism β defines a quasi-isomorphism α of $(\omega_{\bar{X}}(*D)[-d], W)$ with $(Rj_*\mathbb{Q}_X, \tau^{\leq}_{\bar{X}})$; this defines an object of $\mathbf{D}^b(\mathrm{MHM}(\bar{X}))$ canonically isomorphic to $j_*\mathbb{Q}_X$.

2.3.14. We write $\mathrm{MH}[X, \bar{X}]$ for the diagram $((\omega_{\bar{X}}(*D)[-d], W), \mathrm{MW}[X, \bar{X}], \alpha)$ representing $j_*\mathbb{Q}_X$. Suppose that X is an S-scheme, with S a smooth quasi-projective $\mathbb{C}$-scheme, $p_X : X \to S$. Extend p_X to $\bar{p}_X : \bar{X} \to \bar{S}$, where $i : S \to \bar{S}$ is a smooth compactification of S with normal crossing divisor at infinity.

We let $\bar{p}_{X*}G^*X\mathrm{MH}[X, \bar{X}]$ be the diagram of complexes

$$(\mathrm{DR}^{-1} p_* G^*_{\bar{X}} \widetilde{\mathrm{DR}}(\omega_{\bar{X}}(*D), W), p_{X*}G^*\mathrm{MW}[X, \bar{X}], \alpha_S),$$

which by the discussion above represents $\bar{p}_{X*}j_*\mathbb{Q}_X$ in $\mathbf{D}^b(\mathrm{MHM}(\bar{S}))$. We then restrict to S, giving $\Re_{\mathrm{MHM}}(X) \in \mathbf{D}^b(\mathrm{MHM}(S))$. Let $\Re_{\mathrm{MHM}}(X, q) := \Re_{\mathrm{MHM}}(X) \otimes \mathbb{Q}_S(q)$.

We can now apply the argument for Theorem 2.3.10; in fact, since we are working with $\mathbb{Q}$-vector spaces, we may systematically use the Thom-Sullivan cochains in place of the categorical cohomology operations used in the proof of Theorem 1.3.1, so that our representing complexes admit strictly associative and commutative external products. The various properties required of the cohomology theory

$$H^p_{\mathrm{MHM}}(X, \mathbb{Q}(q)) := \mathrm{Ext}^p_{\mathrm{MHM}(X)}(\mathbb{Q}_X(0), \mathbb{Q}_X(q))$$

follow from the fact (see e.g. [**68**, Appendix A, Corollary A.1.10]) that the groups agree with Beilinson's absolute Hodge cohomology. Thus, we have

2.3.15. THEOREM. *Let S be a smooth quasi-projective $\mathbb{C}$-scheme. Sending (X, q) to $\Re_{\mathrm{MHM}}(X, q)$ extends canonically to the exact tensor functor*

$$\Re_{\mathrm{MHM}} : \mathcal{DM}(S)_{\mathbb{Q}} \to \mathbf{D}^b(\mathrm{MHM}(S)),$$

natural in S.

2.3.16. REMARKS. (i) If S is a smooth $\mathbb{R}$-scheme, we have a version with the additional data of a real Frobenius, as in Theorem 2.3.12.
(ii) We may replace $\mathbb{Q}$ with a subfield R of $\mathbb{R}$, giving the exact realization functor

$$\Re_{\mathrm{MHM},R} : \mathcal{DM}(S)_R \to \mathbf{D}^b(\mathrm{MHM}_R(S)).$$

2.4. The motivic realization

Let k be a field of characteristic zero, finitely generated over $\mathbb{Q}$. We conclude this section with an extension of the construction of the Hodge realization to give a realization of $\mathcal{DM}(\mathbf{Sm}_k)$ into a version of Jannsen's category of mixed absolute Hodge complexes. This is constructed similarly to Beilinson's category $D^+_{\mathcal{H}}$; the essential difference is the addition of l-adic data to the Betti-Hodge data encoded in $D^+_{\mathcal{H}}$.

We give a brief review of this construction, somewhat modified as in the previous section. For details see [**71**, §6], especially pages 97-104; for a more detailed version, see [**67**]. The version we will give includes integral data, as in [**71**], but takes some features from [**67**] as well.

2.4.1. DEFINITION. Let G_k denote the Galois group of $\bar{k}$ over k. A *polarizable mixed absolute Hodge complex* (MAH-complex) over k is a commutative diagram $\mathcal{D}$ of the following form:

$\mathcal{D} :=$

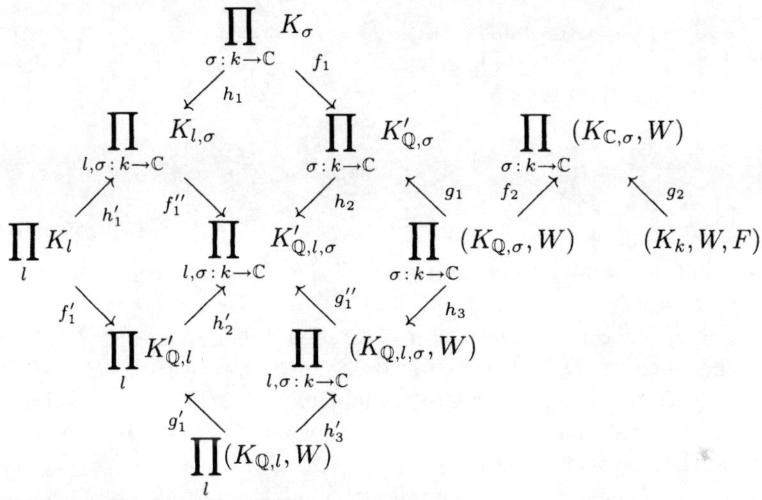

where $\sigma : k \to \mathbb{C}$ denotes an embedding, l is a prime number, and

(i) For each l, K_l, $K_{\mathbb{Q},l}$ and $K'_{\mathbb{Q},l}$ are bounded below complexes of continuous G_k-modules. K_l is a complex of $\mathbb{Z}_l$-modules such that the homology groups are finitely generated $\mathbb{Z}_l$-modules, $K_{\mathbb{Q},l}$ and $K'_{\mathbb{Q},l}$ are complexes of $\mathbb{Q}_l$-vector spaces such that the cohomology groups are finite dimensional $\mathbb{Q}_l$-vector spaces; W is a finite increasing filtration on $K_{\mathbb{Q},l}$.

(ii) For each l and each $\sigma : k \to \mathbb{C}$, $K'_{\mathbb{Q},l,\sigma}$ and $K_{\mathbb{Q},l,\sigma}$ are bounded below complexes of $\mathbb{Q}_l$-vector spaces with finite dimensional cohomology, and $K'_{l,\sigma}$ is a bounded below complex of $\mathbb{Z}_l$-modules with cohomology finitely generated over $\mathbb{Z}_l$. W is a finite increasing filtration.

(iii) For each $\sigma : k \to \mathbb{C}$, $K_{\mathbb{Q},\sigma}$ and $K'_{\mathbb{Q},\sigma}$ are bounded below complexes of $\mathbb{Q}$-vector spaces, and $K_{\mathbb{C},\sigma}$ are bounded below complexes of $\mathbb{C}$-vector spaces, all with finite dimensional cohomology. For each $\sigma : k \to \mathbb{C}$, K_σ is a bounded below complex of abelian groups, with finitely generated cohomology. W denotes a finite increasing filtration on the various complexes.

(iv) K_k is a complex of k-vector spaces, bounded below, with finite dimensional cohomology, an increasing filtration W, and a decreasing filtration F, both finite.

(v) $f_1' = \prod_l f_{1,l}'$, $g_1' = \prod_l g_{1,l}'$, where $f_{1,l}' : K_l \otimes_{\mathbb{Z}_l} \mathbb{Q}_l \to K_{\mathbb{Q},l}'$ and $g_{1,l}' : K_{\mathbb{Q},l} \to K_{\mathbb{Q},l}'$ are quasi-isomorphisms.

(vi) $h_1' = \prod_{l,\sigma} h_{1,l,\sigma}'$, where each $h_{1,l,\sigma}'$ is a family of quasi-isomorphisms

$$h_{1,l,\overline{\sigma}} : K_l \to K_{l,\sigma}$$

indexed by the set of extensions $\overline{\sigma} : \overline{k} \to \mathbb{C}$ of σ, with $h_{1,l,\overline{\sigma}\rho}'$ homotopic to $h_{1,l,\overline{\sigma}}'$ for each $\rho \in G_k$. h_2 and h_3 are similarly defined, with $h_{2,l,\overline{\sigma}} : K_{\mathbb{Q},\sigma}' \otimes \mathbb{Z}_l \to K_{\mathbb{Q},l,\sigma}'$ a quasi-isomorphism, and $h_{3,l,\overline{\sigma}} : (K_{\mathbb{Q},\sigma}, W) \otimes \mathbb{Z}_l \to (K_{\mathbb{Q},l,\sigma}, W)$ a filtered quasi-isomorphism; in addition, for each l, $\overline{\sigma}$ and ρ, the homotopies for the $h_{i,l,\overline{\sigma}}$, $i = 1, 2, 3$, are compatible via the maps f_1', f_1'', g_1' and g_1''.

(vii) The maps h_1, h_2 and h_3 are families of maps indexed by l and σ, with $h_{1,l,\sigma} : K \to K_{l,\sigma}$ and $h_{2,l,\sigma}' : K_{\mathbb{Q},\sigma}' \to K_{\mathbb{Q},l,\sigma}'$ quasi-isomorphisms, and

$$h_{3,l,\sigma} : (K_{\mathbb{Q},\sigma}, W) \to (K_{\mathbb{Q},l,\sigma}, W)$$

a filtered quasi-isomorphism.

(viii) $f_1 = \prod_\sigma f_{1,\sigma}$, $g_1 = \prod_\sigma g_{1,\sigma}$, $f_2 = \prod_\sigma f_{2,\sigma}$, $g_2 = \prod_\sigma g_{2,\sigma}$, where, for each $\sigma : k \to \mathbb{C}$, the diagram

$$\begin{array}{ccccc} & K_{\mathbb{Q},\sigma}' & & (K_{\mathbb{C},\sigma}, W) & \\ {}^{f_{1,\sigma}}\nearrow & & \nwarrow^{g_{1,\sigma}} \quad {}^{f_{2,\sigma}}\nearrow & & \nwarrow^{g_{2,\sigma}} \\ K_\sigma & & (K_{\mathbb{Q},\sigma}, W) & & (K_k, W, F) \otimes_{k,\sigma} \mathbb{C} \end{array}$$

is in $C_{\mathcal{H},\mathbb{Q}}^b$, and defines on $H^i(\mathrm{gr}_m^W K_\sigma)$ a pure, polarizable $\mathbb{Z}$-Hodge structure of weight m.

(ix) Let $\underline{H}$ denote the collection of graded cohomologies $\mathrm{gr}_m^W H^i$ arising from the diagram $\mathcal{D}$. Then there are bilinear forms $\underline{H}_? \otimes \underline{H}_? \to \mathbb{Q}_?(-m)$ for each component, which are compatible under the various comparison isomorphisms, and which give a polarization of the real Hodge structure given by the diagram in (viii).

(x) The cohomology H^* of complexes K_l, $K_{\mathbb{Q},l}'$, and $(K_{\mathbb{Q},l}, W)$ defines (via the comparison maps in the diagram) a constructible filtered continuous $\mathbb{Z}_l[G_k]$-module with $\mathrm{gr}_m^W H^*$ having weight m (see [**67**, Part I, Chapter 9]).

(xi) For each real embedding $\sigma : k \to \mathbb{R} \subset \mathbb{C}$, the factors involving σ all have the additional data of a real Frobenius, i.e., the structure of a complex of $G_\mathbb{R} = \mathrm{Gal}(\mathbb{C}/\mathbb{R})$-modules. For such σ, all the comparison maps, except for $f_{2\sigma}$, are $G_\mathbb{R}$ equivariant; the map $f_{2\sigma}$ is anti-equivariant: $F_\infty \circ f_{2\sigma} = f_{2\sigma} \circ (F_\infty \otimes \overline{(-)})$.

2.4.2. Let $C_{\mathcal{M}AH,k}^b$ denote the category of polarizable mixed absolute Hodge complexes over k, where a morphism $\mathcal{D}_1 \to \mathcal{D}_2$ consists of a collection of maps on each component, in the appropriate category of complexes, such that each of the resulting squares commutes. We have the Tate object $\mathbb{Z}_{\mathcal{M}AH,k}(q)$ in $C_{\mathcal{M}AH,k}^+$, defined by putting the Tate object from the appropriate category in the appropriate spot, i.e., use the Galois module $\mathbb{Z}_l(q)$ for K_l, $\mathbb{Q}_l(q)$ for $K_{\mathbb{Q},l}'$, and $(\mathbb{Q}_l(q), W(q))$ for $(K_{\mathbb{Q},l}, W)$. We use $\mathbb{Z}(q) = (2\pi i)^q \mathbb{Z}$ for K, $\mathbb{Q}(q) = (2\pi i)^q \mathbb{Q}$ for $K_\mathbb{Q}'$ and the remaining data is given as in §2.3.9.

2.4.3. $C_{\mathcal{M}AH,k}^b$ is a DG tensor category ($\otimes$ given component-wise) with a cone functor; the unit is the Tate object $\mathbb{Z}_{\mathcal{M}AH,k}(0)$. This gives the homotopy category $K_{\mathcal{M}AH,k}^b$ the structure of a triangulated tensor category. Taking the 0th cohomology of each component of a diagram $\mathcal{D}$ in $C_{\mathcal{M}AH,k}^b$ gives an object $\underline{H}^0$ in the category of diagrams equivalent to a version $\underline{MR}_k^p$ of Jannsen's category [**71**] of polarizable mixed realizations (the requirement Definition 2.4.1(x) means we land in the refined version of Jannsen's category given in [**67**, Part II, Chapter 11]); one can give a quick definition of this category as the category of diagrams given by Definition 2.4.1, where we require that all the complexes are concentrated in degree zero. The category $\underline{MR}_k^p$ is similar to Deligne's category of systems of realizations described in [**32**, §1]; the main difference being that Deligne's category uses only $\mathbb{Q}$-data, but adds a "crystalline realization" to the Betti, étale, and de Rham components of $\underline{MR}_k^p$.

The functor $\underline{H}^0$ extends to a cohomological functor from $K^b_{\mathcal{M}AH,k}$ to $\underline{MR}_k^p$.

A map f in $K^b_{\mathcal{M}AH,k}$ is called a quasi-isomorphism if f induces a quasi-isomorphism (not necessarily filtered) in each component. Localizing $K^b_{\mathcal{M}AH,k}$ with respect to quasi-isomorphisms defines the triangulated tensor category $D^b_{\mathcal{M}AH,k}$; the cohomological functor $\underline{H}^0$ extends to $D^b_{\mathcal{M}AH,k}$.

2.4.4. *Enlarged diagrams.* As for the Hodge realization, it will be useful to enlarge the basic diagram of Definition 2.4.1 by adding zigzag diagrams of (filtered) quasi-isomorphisms between (filtered) complexes in the same category. We will make various enlargements without further comment; as in §2.3, these enlargements lead to equivalent homotopy categories, and equivalent "derived" categories, by adding in zigzag diagrams of identity maps to enlarge a small diagram or by taking the appropriate cone to shrink an enlarged diagram.

2.4.5. *Tempered Thom-Sullivan cochains.* We will want to apply the construction of §2.3.5 to continuous $\mathbb{Q}_l$-modules, and to $\mathbb{Z}/l^*$-modules; for this we need to consider *tempered* differential forms.

For a variable x, we let $\gamma^a(x)$ denote $\frac{x^a}{a!}$. We have the subgroup $\Omega^{p,q}(|\Delta_n|)$ of $\Omega^p(|\Delta_n|)$ generated by the differential p-forms $\gamma^{a_0}(t_0)\cdot\ldots\cdot\gamma^{a_n}(t_n)dt_{i_1}\ldots dt_{i_p}$ with $\sum_{i=0}^n a_i + p = q$. One checks that the graded group $\Omega^{*,q}(|\Delta_n|)$ of $\Omega^*(|\Delta_n|)$ forms a subcomplex; sending n to $\Omega^{*,q}(|\Delta_n|)$ thus defines a simplicial subcomplex $\Omega^{*,q}$ of the simplicial complex Ω^*. Sending $n \in \mathbb{Z}$ to $n\gamma^q(t_0) = \frac{n}{q!}$ in $\Omega^{0,q}$ define the augmentation $\frac{1}{q!}\mathbb{Z} \xrightarrow{\epsilon_q} \Omega^{*,q}$. Clearly we have $\Omega^{p,q} = 0$ for $p > q$.

The following facts about $\Omega^{*,q}$ are proved in [**97**] and [**27**]:

(2.4.5.1)

1. The $\Omega^{*,q}$ form an increasing filtration of Ω^*, and $\Omega^* = \cup_{q\geq 0}\Omega^{*,q}$.
2. The integration map $\int : \Omega^* \to Z^* \otimes \mathbb{Q}$ restricts to the integration map $\int_q : \Omega^{*,q} \to \frac{1}{q!}Z^*$, i.e., $q!\int$ maps $\Omega^{*,q}$ into Z^*.
3. $H^p(\Omega^{*,q}(|\Delta_n|)) = 0$ for $p > 0$, and the augmentation ϵ_q induces an isomorphism $\frac{1}{q!}\mathbb{Z} \cong H^0(\Omega^{*,q}(|\Delta_n|))$.

We now consider the simplicial abelian group $\Omega^{p,q}$. For a simplicial set X, let $\Omega^{p,q}(X)$ be the group of maps of simplicial sets $X \to \Omega^{p,q}$; if $|X|$ is the geometric realization of X, we sometimes write $\Omega^{p,q}(|X|)$ for $\Omega^{p,q}(X)$. Note that restricting to the non-degenerate simplex of Δ_n defines an isomorphism of $\Omega^{p,q}(\Delta_n)$ with $\Omega^{p,q}(|\Delta_n|)$, so the notation is unambiguous.

2.4.5.2. LEMMA. *The simplicial abelian group $\Omega^{p,q}$ is acyclic for $0 \le p < q$:*

$$\pi_m(\Omega^{p,q}) = 0; \quad \textit{for } 0 \le p < q.$$

PROOF. It suffices to prove the following *extension property*: Let $m \geq 0$ be an integer, and let $i: S^m \to |\Delta_m|$ be the union of the faces $t_i = 0$, $i = 0, \dots, m$. Suppose we have an element $\tau \in \Omega^{p,q}(S^m)$. Then there is an $\omega \in \Omega^{p,q}(|\Delta_m|)$ with $i^*\omega = \tau$.

Let v_i be the vertex $t_i = 1$ of $|\Delta_m|$. If σ is the simplex of $|\Delta_m|$ spanned by vertices $v_{i_0}, \dots, v_{i_s}$, with $i_0 < \dots < i_s$, then $\Omega^{p,q}(\sigma)$ is the free $\mathbb{Z}$-module on generators

$$g = \gamma^{a_{i_0}}(t_{i_0}) \cdot \ldots \cdot \gamma^{a_{i_s}}(t_{i_s}) dt_{j_1} \dots dt_{j_p} \tag{2.4.5.3}$$

with $\sum_j a_{i_j} = q - p$, and with $\{j_1 < \dots < j_p\}$ a subset of $\{i_1 < \dots < i_s\}$ (see [**97**, §2]). We define the operators

$$h_{r,a}: \Omega^{p,q}(\sigma) \to \Omega^{p,q}(\sigma), \quad r = 0, \dots, m, \ a = 1, \dots, q-p,$$

on basis elements g as in (2.4.5.3) by setting $h_{r,a}(g) = g$ if $a_r = a$ and $a_{i_j} = 0$ for $i_j < r$; we define $h_{r,a}(g) = 0$ otherwise. We then extend by linearity. Since $p < q$, we have

$$\sum_{r,a} h_{r,a} = \mathrm{id}. \tag{2.4.5.4}$$

One checks that the maps $h_{r,a}$ are compatible with the restriction maps to subfaces, hence they extend to gives well-defined operators $h_{r,a}: \Omega^{p,q}(L) \to \Omega^{p,q}(L)$ for each subcomplex L of $|\Delta_m|$; the $h_{r,a}$ are natural in L, and satisfy the relation (2.4.5.4).

Let Λ_r be the closed star neighborhood of the vertex v_r in S_m, i.e., Λ_r is the closure of the complement in S_m of the face $t_r = 0$.

Now take $\tau \in \Omega^{p,q}(S^m)$. Using (2.4.5.4), it suffices to extend an element of the form $\tau := \gamma^a(t_r)\tau_r$, with $a > 0$ and $\tau_r \in \Omega^{p,q-a}(\Lambda_r)$, where we extend τ to S_m by zero on the face $t_r = 0$. Since $\Omega^{p,q-a}$ is a simplicial abelian group, $\Omega^{p,q}$ is a Kan complex, hence the element τ_r of $\Omega^{p,q-a}(\Lambda_r)$ extends to an element ω_r of $\Omega^{p,q-a}(|\Delta_m|)$. Then $\omega := \gamma^a(t_r)\omega_r$ is the desired extension of τ. □

For a complex C^*, we let $\sigma^{\le N}C^*$ denote the "stupid" truncation, i.e., $\sigma^{\le N}C^p = C^p$ for $p \le N$, and $\sigma^{\le N}C^p = 0$ for $p > N$.

2.4.5.5. LEMMA. *There is a map of complexes (over the identity on $\frac{1}{q!}\mathbb{Z}$)*

$$T_q: \sigma^{\le q-1}\frac{1}{q!}Z^* \to \sigma^{\le q-1}\Omega^{*,q}$$

and maps

$$h_p: \frac{1}{q!}Z^p \to \frac{1}{q!}Z^{p-1}; \quad g_n: \Omega^{p,q} \to \Omega^{p-1,q}$$

for $p = 0, \dots, q-1$ which define homotopies $\mathrm{id}_{Z^*} \sim \int_q \circ T_q$, $\mathrm{id}_{\Omega^{*,q}} \sim T_q \circ \int_q$ in degrees $\le q-2$.

PROOF. This follows by applying the arguments of [**24**, §2]; we give a brief description of how these methods apply, using the notations from *loc. cit.* It follows from (2.4.5.1)(3) that $\Omega^{*,q}$ is *acyclic on models*. From Lemma 2.4.5.2, the simplicial abelian groups $\Omega^{p,q}$ are *corepresentable* for $p < q$. In *loc. cit.*, it is shown that Z^* is acyclic on models, and Z^p is corepresentable for all p. The result then follows from the method of acyclic models, as described in *loc. cit.* □

Now let G be a cosimplicial abelian group. From Lemma 2.4.5.5, the map

$$\int_q \underset{\leftarrow}{\otimes} \mathrm{id}\colon \Omega^{*,q} \underset{\leftarrow}{\otimes} G \to \frac{1}{q!} Z^* \underset{\leftarrow}{\otimes} G$$

is a cohomology isomorphism in degrees $\leq q-2$. Setting

$$\Omega^{**} \underset{\leftarrow}{\otimes} G := \varinjlim_q \Omega^{*,q} \underset{\leftarrow}{\otimes} G,$$

we thus have the quasi-isomorphism

$$\varinjlim_q \int_q \underset{\leftarrow}{\otimes} \mathrm{id}\colon \Omega^{**} \underset{\leftarrow}{\otimes} G \to \varinjlim_q \frac{1}{q!} Z^* \underset{\leftarrow}{\otimes} G \cong \mathbb{Q} \otimes Z^* \underset{\leftarrow}{\otimes} G \cong \mathbb{Q} \otimes \mathrm{Norm}(G).$$

In particular, the inclusion

$$\Omega^{**} \underset{\leftarrow}{\otimes} G \to \Omega^* \underset{\leftarrow}{\otimes} G$$

is a quasi-isomorphism. In addition, the wedge product of forms gives the product $\Omega^{p,q} \otimes \Omega^{p',q'} \to \Omega^{p+p',q+q'}$. Thus, using the construction of §2.3.5, a commutative multiplication on G gives $\Omega^{**} \underset{\leftarrow}{\otimes} G$ the structure of a DG algebra over $\mathbb{Q}$, and makes the above inclusion a quasi-isomorphism of DG algebras.

2.4.6. We now proceed to define the motivic realization

$$\mathfrak{R}_{\mathrm{mot}}\colon \mathcal{DM}_{\mathrm{mot}}(\mathbf{Sm}_k) \to D^b_{\mathcal{M}AH,k}.$$

2.4.7. *Sites.* Let ét$_k$ denote the big étale site over k; We use a slightly different site for the classical topology than the usual one. For an embedding $\sigma\colon k \to \mathbb{C}$ and for $X \in \mathbf{Sm}_k$, let X^σ denote the complex manifold associated to the $\mathbb{C}$-scheme $X \times_{k,\sigma} \mathbb{C}$. Let an$_k$ denote the site where an open cover of $X \in \mathbf{Sm}_k$ is a collection of maps $f^\sigma\colon U^\sigma \to X^\sigma$, where f^σ is surjective, and f^σ is locally (on U^σ) a homeomorphism.

We have the map of sites $\alpha\colon \text{ét}_k \to \text{an}_k$ given by sending $U \to X$ to

$$\coprod_\sigma U^\sigma \to X^\sigma.$$

Similarly, letting Zar$_k$ denote the big Zariski site over k, we have the map of sites $\beta\colon \mathrm{Zar}_k \to \mathrm{an}_k$. Each of these site has a conservative family of points gotten by taking the points of the topological space X (for the Zariski or classical topology), the geometric points of X (for the étale topology), or the points of the topological space X^σ (for the classical topology). These collections of points are compatible with respect to the change of topology maps.

2.4.8. Let X be in $\mathbf{Sm}_k$. There is a compactification $X \to \overline{X}$ defined over k, and the category of compactifications of X over k forms a directed filtering category. For a compactification $X \to \overline{X}$, with normal crossing divisor D at infinity, we form the diagram of complexes of sheaves $\mathcal{D}[X, \overline{X}]$ as in Definition 2.4.1 by

1. K_σ is $\mathbb{Z}^{\mathrm{an}}_{X^\sigma}$. K_l is the inverse system $\mathbb{Z}^{\text{ét}}_{l,X}$ in $\mathrm{Sh}^{\mathbb{Z}/l^*}_{\text{ét}}(X)$ described in Example 2.2.5. Via Remark 2.2.7, we have the category $\mathrm{Sh}^{\mathbb{Z}/l^*}_{\mathrm{an}}(X^\sigma)$ for each embedding $\sigma\colon k \to \mathbb{C}$. We take $K_{l,\sigma}$ to be the inverse system $\mathbb{Z}^{\mathrm{an}}_{l,X^\sigma}$ in $\mathrm{Sh}^{\mathbb{Z}/l^*}_{\mathrm{an}}(X^\sigma)$ gotten from $\mathbb{Z}^{\text{ét}}_{l,X}$ by change of topology with respect to σ.

2. We set

$$K'_{\mathbb{Q},\sigma} := \Omega^{**} \underset{\leftarrow}{\otimes} j_* G_{X^\sigma} \mathbb{Z}^{\mathrm{an}}_{X^\sigma}$$

where $\Omega^{**} \underset{\leftarrow}{\otimes}$ is the tempered Thom-Sullivan cochain construction described in §2.4.5, and G_{X^σ} is the cosimplicial Godement resolution for the analytic topology.

We define $K'_{\mathbb{Q},l}$ and $K'_{\mathbb{Q},l,\sigma}$ similarly by

$$K'_{\mathbb{Q},l} := \Omega^{**} \underset{\leftarrow}{\otimes} j_* G_X \mathbb{Z}^{\text{ét}}_{l,X},$$
$$K'_{\mathbb{Q},l,\sigma} := \Omega^{**} \underset{\leftarrow}{\otimes} j_* G_{X^\sigma} \mathbb{Z}^{\mathrm{an}}_{l,X^\sigma},$$

which are objects in the $\mathbb{Q}$-localizations $\mathrm{Sh}^{\mathbb{Z}/l^*}_{\text{ét}}(X) \otimes \mathbb{Q}$ and $\mathrm{Sh}^{\mathbb{Z}/l^*}_{\mathrm{an}}(X^\sigma) \otimes \mathbb{Q}$ of $\mathrm{Sh}^{\mathbb{Z}/l^*}_{\text{ét}}(X)$ and $\mathrm{Sh}^{\mathbb{Z}/l^*}_{\mathrm{an}}(X^\sigma)$, respectively.

3. We take $K_{\mathbb{Q},\sigma} = K'_{\mathbb{Q},\sigma}$, $K_{\mathbb{Q},l} = K'_{\mathbb{Q},l}$, and $K_{\mathbb{Q},l,\sigma} = K'_{\mathbb{Q},l,\sigma}$. The weight filtrations are given by taking $\mathrm{Dec}\tau^{\leq}_{\bar{X}}$ (or $\mathrm{Dec}\tau^{\leq}_{\bar{X}^\sigma}$), where $\tau^{\leq}_{\bar{X}}$ and $\tau^{\leq}_{\bar{X}^\sigma}$ are the canonical filtrations, in the appropriate topology.

4. $(K_k, W, F) := (\Omega^{\mathrm{Zar}}_{\bar{X}/k}(\log(D)), \mathrm{Dec}W^{\mathrm{Zar}}, F)$, and $(K_{\mathbb{C},\sigma}, W)$ is the zigzag diagram

$$\begin{array}{ccc} (\Omega^* \underset{\leftarrow}{\otimes} j_* G_{X^\sigma} \Omega^{\mathrm{an}}_{X^\sigma}, \mathrm{Dec}\tau^{\leq}_{\bar{X}^\sigma}) & & (\Omega^{\mathrm{an}}_{\bar{X}^\sigma}(\log(D)), \mathrm{Dec}W^{\mathrm{an}}) \\ & \nwarrow \quad \nearrow & \\ & (j_* \Omega^{\mathrm{an}}_{X^\sigma}, \mathrm{Dec}\tau^{\leq}_{\bar{X}^\sigma}) & \end{array}$$

defined as in §2.3.8: "Zar" and "an" refer to the Zariski and analytic topologies, $\Omega^{\mathrm{an}}_{X^\sigma}$, (resp. $\Omega^{\mathrm{an}}_{\bar{X}^\sigma}(\log(D))$) is the complex of sheaves of holomorphic forms (resp. holomorphic forms with log poles), $\Omega^{\mathrm{Zar}}_{\bar{X}/k}(\log(D))$ is the complex of sheaves of algebraic forms with log poles. W^{an} and W^{Zar} are the filtrations by order of pole, and F is the stupid filtration.

5. The map f_1 is induced by the natural inclusion $\iota_G : G \to \Omega^{**} \underset{\leftarrow}{\otimes} G$, which gives the natural map of sheaves

$$\mathbb{Z}^{\mathrm{an}}_{X^\sigma} \to \Omega^{**} \underset{\leftarrow}{\otimes} G_{X^\sigma} \mathbb{Z}^{\mathrm{an}}_{X^\sigma},$$

and similarly for f'_1 and f''_1.

6. The maps g_1, g'_1 and g''_1 are "forget the filtration".
7. The maps h_i are gotten by passing from the analytic sheaf $\mathbb{Z}_{X^\sigma}$ to the analytic sheaf of $\mathbb{Z}/l^*$-modules $\mathbb{Z}_{X^\sigma} \otimes \mathbb{Z}/l^*$.
8. For each $\sigma : k \to \mathbb{C}$, we have the map of sites $\alpha_\sigma : X^\sigma_{\mathrm{an}} \to X_{\text{ét}}$. The maps h'_i are induced by the product over l and σ of the change of topology morphism $\alpha^*_\sigma \mathbb{Z}^{\text{ét}}_{l,X} \to \mathbb{Z}^{\mathrm{an}}_{l,X^\sigma}$.
9. For each $\sigma : k \to \mathbb{C}$, we have the map of sites $\beta_\sigma : X^\sigma_{\mathrm{an}} \to X_{\mathrm{Zar}}$. The arrow g_2 is the product over σ of the change of topology maps

$$\beta^*(\Omega^{\mathrm{Zar}}_{\bar{X}/k}(\log(D)), W^{\mathrm{Zar}}) \to (\Omega^{\mathrm{an}}_{\bar{X}^\sigma}(\log(D)), W^{\mathrm{an}}),$$

and the remainder of the diagram is the product over σ of the portion of the Hodge realization diagram (2.3.8.1).

We form the push-forward diagram $p_{(X,\bar{X})*}G\mathcal{D}[X,\bar{X}]$ as a cosimplicial object by first taking the cosimplicial Godement resolution of each term (either on $X_{\text{ét}}$, X^σ_{an}, $\overline{X}^\sigma_{\text{an}}$ or $\overline{X}_{\text{Zar}}$, filtered or bifiltered, as necessary) and taking global sections; we then take the projective limit in the l-adic components.

We identify a sheaf on $k_{\text{ét}}$ with a G_k-module. As in the construction of the l-adic realization, the inverse systems involved are all normalized sheaves on the étale site over k, so taking the projective limit preserves quasi-isomorphisms and the resulting projective limit is a continuous G_k-module [**72**, Theorem 2.2 and Theorem 3.2]).

Let $\mathbb{Z}_{\mathcal{M}AH,k,(X,\bar{X})}(q)_\Delta$ be the cosimplicial diagram

$$\mathbb{Z}_{\mathcal{M}AH,k,(X,\bar{X})}(q)_\Delta := \mathbb{Z}_{\mathcal{M}AH,k}(q) \otimes p_{(X,\bar{X})*}G\mathcal{D}[X,\bar{X}],$$

and let $\mathbb{Z}_{\mathcal{M}AH,k,(X,\bar{X})}(q)$ be the diagram of associated total complexes.

We let $\mathbb{Z}_{\mathcal{M}AH,k,X}(q)_\Delta$ be the inductive limit of the diagrams $\mathbb{Z}_{\mathcal{M}AH,k,(X,\bar{X})}(q)_\Delta$, over compactifications $\bar{X}$ of X, and let $\mathbb{Z}_{\mathcal{M}AH,k,X}(q)$ be the diagram of associated total complexes; equivalently, $\mathbb{Z}_{\mathcal{M}AH,k,X}(q)$ is the inductive limit of the diagrams $\mathbb{Z}_{\mathcal{M}AH,k,(X,\bar{X})}(q)$.

It follows from [**72**, Theorem 2.2 and Theorem 3.2] that $p_{(X,\bar{X})*}G\mathcal{D}[X,\bar{X}]$ and $\mathbb{Z}_{\mathcal{M}AH,k,(X,\bar{X})}(q)$ are in fact objects of $C^+_{\mathcal{M}AH,k}$, and that the maps in the inductive limit are quasi-isomorphisms in each component, hence $\mathbb{Z}_{\mathcal{M}AH,k,X}(q)$ is an object of $C^+_{\mathcal{M}AH,k}$, canonically isomorphic to each $\mathbb{Z}_{\mathcal{M}AH,k,(X,\bar{X})}(q)$ in $D^+_{\mathcal{M}AH,k}$.

The results of [**72**, Part I] and [**67**, §15] imply that the conditions of Theorem 1.3.1 are satisfied for the objects $\mathbb{Z}_{\mathcal{M}AH,k,(X,\bar{X})}(q)$ of $D^+_{\mathcal{M}AH,k}$ (as objects of the triangulated tensor category $D^+_{\mathcal{M}AH,k}$, rather than the derived category of sheaves). In fact, each natural map which gives a quasi-isomorphism in each of the individual cohomology theories gives an isomorphism in $D^+_{\mathcal{M}AH,k}$, which gives all the properties except for cycle classes and semi-purity. As the cohomology group $\mathrm{Hom}_{D^+_{\mathcal{M}AH,k}}(1, \mathbb{Z}_{\mathcal{M}AH,k,(X,\bar{X})}(q)[p])$ can be computed from the cohomologies in the individual theories by a cone construction, the compatible cycle classes in each theory, together with semi-purity in each theory, gives cycle classes and semi-purity for $\mathcal{M}AH$.

Additionally, the compatible external products in the various sheaves

$$\mathbb{Z}^{\text{ét}}_{l,X},\ \mathbb{Z}^{\text{an}}_{X^\sigma},\ \mathbb{Z}^{\text{an}}_{l,X^\sigma},\ \mathbb{Q}^{\text{an}}_{X^\sigma},$$
$$j_*\Omega^{\text{an}}_{X^\sigma},\ \Omega^{\text{an}}_{\overline{X}^\sigma}(\log(D)),\ \Omega^{\text{Zar}}_{\overline{X}/k}(\log(D)),$$

together with their various filtrations, give, via §2.3.5 and §2.3.4, a natural associative and commutative external product for the diagrams $\mathbb{Z}_{\mathcal{M}AH,k,X}(q)_\Delta$, as cosimplicial objects.

If we now repeat the construction of the realization functor in the proof of Theorem 1.3.1, making the modifications we used to construct the étale realization and the Hodge realization, we have the following result:

2.4.9. THEOREM. *Let R be a noetherian subring of $\mathbb{R}$. Sending (X,q) in $\mathbf{Sm}_k \times \mathbb{Z}$ to $R_{\mathcal{M}AH,k,X}(q)$ extends canonically to an exact functor*

$$\mathfrak{R}_{\mathcal{M}AH,k} : \mathcal{DM}(\mathbf{Sm}_k)_R \to D^+_{\mathcal{M}AH,k,R}.$$

The restriction of $\mathfrak{R}_{\mathcal{M}AH,k}$ to $\mathcal{DM}_{\mathfrak{sh}}(\mathbf{Sm}_k)_R$ is an exact, pseudo-tensor functor.

Here $D^+_{\mathcal{M}AH,k,R}$ is constructed as $D^+_{\mathcal{M}AH,k}$, replacing $\mathbb{Z}$ with R, $\mathbb{Q}$ with $R\otimes\mathbb{Q}$, etc., and $R_{\mathcal{M}AH,k,X}(q)$ is the image of $\mathbb{Z}_{\mathcal{M}AH,k,X}(q)$ in $D^+_{\mathcal{M}AH,k,R}$.

2.5. Questions, projects, and open problems

2.5.1. *Singular base-schemes.* One should have Betti, étale, and Hodge realizations over a general base scheme. Do these various cohomology theories satisfy the properties required by Theorem 1.3.1, for objects smooth over a singular base?

2.5.2. *L-functions.* Suppose that the base field k is a number field, and fix an integer $m \geq 0$. The motivic realization gives us, for each motive X over k, a packet of (virtual) l-adic G_k-representations of weight m, $\mathrm{gr}^W_m \mathfrak{R}_{\mathcal{M}AH,k}(X)_{\text{ét},l}$, as well as a pure Hodge structure of weight m, $\mathrm{gr}^W_m \mathfrak{R}_{\mathcal{M}AH,k}(X)_{\mathrm{Hdg},\sigma}$, for each embedding σ of k, with real Frobenius for each real embedding. Suppose that the characteristic function of Frobenius F_p, for each prime p of $\mathcal{O}_k$ for which $\mathrm{gr}^W_m \mathfrak{R}_{\mathcal{M}AH,k}(X)_{\text{ét},l}$ is unramified, is independent of $l \neq p$. Suppose that one has a good notion of inertia invariants, and that the resulting characteristic function of F_p on $\mathrm{gr}^W_m \mathfrak{R}_{\mathcal{M}AH,k}(X)^{I_p}_{\text{ét},l}$ is independent of the choice of $l \neq p$. Then one can define the L-function $L^{(m)}(X,s)$ as the Artin L-function of this packet of virtual G_k-representations. The Hodge structures $\mathrm{gr}^W_m \mathfrak{R}_{\mathcal{M}AH,k}(X)_{\mathrm{Hdg},\sigma}$ define the factor at infinity $L^{(m)}_\infty(X,s)$, as in [**36**] and [**113**]; see also [**34**, §5.3]. This gives the completed L-function $\Lambda(X,s) := L^{(m)}_\infty(X,s)L^{(m)}(X,s)$.

If E is a number field, and X is in $\mathcal{DM}(X)_E$, one should be able to define an $E\otimes\mathbb{C}$-valued series $\Lambda(X,s)$ as in [**34**, §2.2].

Suppose now that X is in $\mathbf{D}^b_{\mathrm{mot}}(k)_E$. Then the characteristic function of Frobenius F_p on $\mathrm{gr}^W_m \mathfrak{R}_{\mathcal{M}AH,k}(X)_{\text{ét},l}$ is independent of l for all l outside a finite set of primes. Indeed, by Lemma 1.5.4 of Chapter IV, X is expressible in terms of finitely many motives of the form $\mathbb{Z}_X(q)$, with X smooth and projective over k. As the G_k representations $H^{m+2q}(X,\mathbb{Q}_l(q))$ have the desired independence property, the same holds for $\mathrm{gr}^W_m \mathfrak{R}_{\mathcal{M}AH,k}(X)_{\text{ét},l}$. Thus, the L-function $L^{(m)}(X,s)$ is defined, except for finitely many factors. Is the above independence on l valid for all X in $\mathcal{DM}(X)_E$, at least for all but finitely many p?

2.5.3. *Mixed motives.* It would be nice to add the crystalline data, in some form, to the realization $\mathfrak{R}_{\mathcal{M}AH,k}$. Let us assume we have done this, and are working in $D^b_{\mathcal{M}AH,k}\otimes\mathbb{Q}$. In [**32**, Définition 1.11] the category of mixed motives is "defined" as the subcategory of the category of systems of realizations generated (by $\oplus$, $\otimes$, dual, and subquotient) by systems of geometric origin, where this latter term is left undefined. A reasonable definition would be the objects $\underline{H}^0(\mathfrak{R}_{\mathcal{M}AH,k}(X))$ for X in $\mathcal{DM}(k)_\mathbb{Q}$. Let $\mathcal{MM}_{gm}(k)$ denote the full subcategory generated by the objects of geometric origin. Then $\mathcal{MM}_{gm}(k)$ is closed under $\oplus$, $\otimes$ and dual. Is $\mathcal{MM}_{gm}(k)$ already abelian?

In any case, let $\mathcal{MM}(k)$ be the closure of $\mathcal{MM}_{gm}(k)$ under $\oplus$, $\otimes$, dual, and subquotient, and let $\mathcal{DMM}(k)$ denote the full subcategory of $D^b_{\mathcal{M}AH,k}$ of objects with $\underline{H}^p$ in $\mathcal{MM}(k)$ for all p. Is $\mathcal{DMM}(k)$ equivalent to the bounded derived category $\mathbf{D}^b(\mathcal{MM}(k))$? Is the restricted realization functor

$$\mathfrak{R}_{\mathcal{M}AH,k}\colon \mathcal{DM}(k)_\mathbb{Q} \to \mathcal{DMM}(k)$$

an equivalence of categories? What if k is a number field?

CHAPTER VI

Motivic Constructions and Comparisons

We begin this chapter by interpreting Milnor K-theory in terms of the category $\mathcal{DM}$, including a motivic proof of the Steinberg relation (Proposition 1.1.7). We show in §1.2 how to construct Beilinson's polylogarithm as a motive; although we are not able to give an integral version, we do get an explicit bound on the denominators involved. We conclude by comparing our motivic category $\mathcal{DM}(\operatorname{Spec} k)$ with Voevodsky's category $DM_{gm}(k)$ [**124**]; we show that the two categories are equivalent if k is a perfect field admitting resolution of singularities for finite type k-varieties.

1. Motivic constructions

1.1. The group of units and Milnor K-theory

1.1.1. Let $i_0 : S \to \mathbb{A}^1_S$ and $i_1 : S \to \mathbb{A}^1_S$ be the 0-section and the 1-section, let $T := \mathbb{A}^1_S \setminus \{i_0(S)\}$, and let $\mathbb{Z}_{\mathbb{G}_m}$ be the image in $\mathcal{DM}$ of the object

$$\operatorname{cone}\big(i_1^* : \mathbb{Z}_T(0)_{(i_1 \cup \mathrm{id}_T)} \to \mathbb{Z}_S\big)[-1]$$

of $\mathbf{C}^b_{\mathrm{mot}}(\mathbf{Sm}_S)$. Let $\square^1$ be the open subscheme $\mathbb{P}^1_S \setminus \{i_1(S)\}$ of $\mathbb{P}^1_S$, let $i_0 : S \to \square^1$ and $i_\infty : S \to \square^1$ be the 0-section and the ∞-section, and let $\mathbb{Z}_{(\square^1;\partial\square^1)}$ be the image in $\mathcal{DM}$ of the object

$$\operatorname{cone}\big((i_0^*, i_\infty^*) : \mathbb{Z}_{\square^1}(0)_{(i_0 \cup i_\infty \cup \mathrm{id}_{\square^1})} \to \mathbb{Z}_S \oplus \mathbb{Z}_S\big)[-1]$$

of $\mathbf{C}^b_{\mathrm{mot}}(\mathbf{Sm}_S)$. Letting $\partial\square^1$ be the set of divisors $\{i_0(S), i_\infty(S)\}$, the notation $\mathbb{Z}_{(\square^1;\partial\square^1)}$ agrees with the notation for relative motives given in Chapter I, §2.6.6.

Let $\Delta_{\mathbb{P}^1_S}$ be the diagonal in $\mathbb{P}^1_S \times_S \mathbb{P}^1_S$, and let D be the intersection of $\Delta_{\mathbb{P}^1_S}$ with the open subscheme $T \times_S \square^1$. As in (Chapter IV, §2.3.4 and (IV.2.3.4.1)), the cycle class of D, $\mathrm{cl}(|D|) : 1 \to \mathbb{Z}_{(\square^1 \times T; \partial\square^1 \times T, \square^1 \times 1)}(1)[2]$, gives us the map

$$\delta : 1 \to \mathbb{Z}_{(\square^1;\partial\square^1)}(1)[2] \otimes \mathbb{Z}_{\mathbb{G}_m}.$$

Let $t := X_1/X_0$ be the canonical rational function on $\mathbb{P}^1_S$. Let $\square^n$ be the n-fold product of $\square^1$ over S, with the rational coordinate functions $t_1, \ldots, t_n$, and let $\partial\square^n$ be the collection of divisors $\partial\square^n := \{t_1 = 0, t_1 = \infty, \ldots, t_n = 0, t_n = \infty\}$. This gives us the relative motive $\mathbb{Z}_{(\square^n;\partial\square^n)}$. The Künneth isomorphism gives the canonical isomorphism

$$\mathbb{Z}_{(\square^n;\partial\square^n)} \cong \mathbb{Z}^{\otimes n}_{(\square^1;\partial\square^1)}. \tag{1.1.1.1}$$

Taking the nth tensor power of δ, and using the identification (1.1.1.1) gives us the map

$$\delta^{\otimes n} : 1 \to \mathbb{Z}_{(\square^n;\partial\square^n)}(n)[2n] \otimes \mathbb{Z}^{\otimes n}_{\mathbb{G}_m}.$$

1.1.2. LEMMA. *For each $n = 1, 2, \dots,$ the pair $(\delta^{\otimes n}, \mathbb{Z}_{(\square^n;\partial\square^n)}(n)[2n])$ is the dual of $\mathbb{Z}_{\mathbb{G}_m}^{\otimes n}$.*

PROOF. We reduce immediately to the case $n = 1$. The result is then a special case of Lemma 2.3.5 of Chapter IV. $\square$

1.1.3. Now let $u \in \Gamma(X, \mathcal{O}_X^*)$ be a unit on a smooth S-scheme X. We view u as a map $u: X \to \mathbb{A}^1_S \setminus \{0\}$; sending u to the composition $\mathbb{Z}_{\mathbb{G}_m} \xrightarrow{\pi_0} \mathbb{Z}_{\mathbb{A}^1_S\setminus\{0\}} \xrightarrow{u^*} \mathbb{Z}_X$ gives the map of sets

$$\mathrm{cl}_{\mathbb{G}_m,X} : \Gamma(X, \mathcal{O}_X^*) \to \mathrm{Hom}_{\mathcal{DM}}(\mathbb{Z}_{\mathbb{G}_m}, \mathbb{Z}_X).$$

We compose with the duality isomorphism

$$\mathrm{Hom}_{\mathcal{DM}}(\mathbb{Z}_{\mathbb{G}_m}, \mathbb{Z}_X) \cong \mathrm{Hom}_{\mathcal{DM}}(1, \mathbb{Z}_X \otimes \mathbb{Z}^D_{\mathbb{G}_m}),$$

the isomorphism of Lemma 1.1.2,

$$\mathrm{Hom}_{\mathcal{DM}}(1, \mathbb{Z}_X \otimes \mathbb{Z}^D_{\mathbb{G}_m}) \cong \mathrm{Hom}_{\mathcal{DM}}(1, \mathbb{Z}_X \otimes \mathbb{Z}_{(\square^1;\partial\square^1)}(1)[2]),$$

and the Künneth isomorphism

$$\mathrm{Hom}_{\mathcal{DM}}(1, \mathbb{Z}_X \otimes \mathbb{Z}_{(\square^1;\partial\square^1)}(1)[2]) \cong \mathrm{Hom}_{\mathcal{DM}}(1, \mathbb{Z}_{(X\times\square^1;X\times\partial\square^1)}(1)[2]),$$

to give the map

$$\mathrm{cl}_{\mathbb{G}^D_m,X} : \Gamma(X, \mathcal{O}_X^*) \to \mathrm{Hom}_{\mathcal{DM}}(1, \mathbb{Z}_{(X\times\square^1;X\times\partial\square^1)}(1)[2]).$$

Similarly, if $u: X \to \mathbb{A}^1_S \setminus \{0\}$ is a unit, the cycle class of the graph Γ_u in $X \times (\mathbb{A}^1_S \setminus \{0\})$ gives a well-defined cycle class map

$$\mathrm{cl}_{[\Gamma_u]} : 1 \to \mathbb{Z}_{(X\times\square^1;X\times\partial\square^1)}(1)[2]$$

as in Chapter IV, §2.3.1.

1.1.4. LEMMA. *(i) For u in $\Gamma(X, \mathcal{O}_X^*)$, we have*

$$\mathrm{cl}_{[\Gamma_u]} = \mathrm{cl}_{\mathbb{G}^D_m,X}(u).$$

(ii) The map $\mathrm{cl}_{\mathbb{G}_m,X} : \Gamma(X, \mathcal{O}_X^*) \to \mathrm{Hom}_{\mathcal{DM}}(\mathbb{Z}_{\mathbb{G}_m}, \mathbb{Z}_X)$ is a group homomorphism, and is natural with respect to morphisms $f: Y \to X$ in $\mathbf{Sm}_S$.

PROOF. The map $\mathrm{cl}_{\mathbb{G}^D_m,X}(u)$ is given by the composition

$$1 \xrightarrow{\delta} \mathbb{Z}_{\mathbb{G}_m} \otimes \mathbb{Z}_{(\square^1;\partial\square^1)}(1)[2] \xrightarrow{u^*\otimes\mathrm{id}} \mathbb{Z}_X \otimes \mathbb{Z}_{(\square^1;\partial\square^1)}(1)[2] \cong \mathbb{Z}_{(X\times\square^1;X\times\partial\square^1)}(1)[2].$$

This is the same as the composition

$$1 \xrightarrow{\mathrm{cl}_{\partial,D}(1\cdot D)} \mathbb{Z}_{(\mathbb{G}_m\times\square^1;\mathbb{G}_m\times\partial\square^1),D}(1)[2] \xrightarrow{(u\times\mathrm{id})^*} \mathbb{Z}_{(X\times\square^1;X\times\partial\square^1),\Gamma_u}(1)[2] \to \mathbb{Z}_{(X\times\square^1;X\times\partial\square^1)}(1)[2].$$

Since $(u \times \mathrm{id})^*(D) = \Gamma_u$, this in turn is the same as the composition

$$1 \xrightarrow{\mathrm{cl}_{\partial,\Gamma_u}(1\cdot\Gamma_u)} \mathbb{Z}_{(X\times\square^1;X\times\partial\square^1),\Gamma_u}(1)[2] \to \mathbb{Z}_{(X\times\square^1;X\times\partial\square^1)}(1)[2],$$

which is exactly $\mathrm{cl}_{[\Gamma_u]}$.

For (ii), let Δ_2 be the open subscheme of $\mathbb{P}^2_S$ (with projective coordinates X_0, X_1, X_2) defined by $X_0 - X_1 - X_2 \neq 0$, and let $\partial\Delta^2$ be the set of closed subschemes $\{D_0, D_1, D_2\}$, with D_i defined by $X_i = 0$. We identify each D_i with $\square^1$ by the respective coordinate functions X_1/X_0 on D_2, X_2/X_0 on D_1, and $-X_2/X_1$ on D_0, giving the three inclusions $i_j : \square^1 \to \Delta_2$, $j = 0, 1, 2$.

Use the standard coordinate functions t_1, t_2 on T^2, and let W be the subscheme of $T^2 \times \Delta_2$ defined by $t_2X_1 + t_1X_2 = t_1t_2X_0$. Let $p_j : T^2 \to T$, $j = 1, 2$, be the projections, and let $q : T^2 \to T$ be the map $q(t_1, t_2) = t_2/t_1$. Then

$$(1.1.4.1)\quad \begin{aligned}&(\mathrm{id} \times i_2)^*(W) - (\mathrm{id} \times i_1)^*(W) + (\mathrm{id} \times i_0)^*(W)\\ &\quad = (p_1 \times \mathrm{id})^*(D) - (p_2 \times \mathrm{id})^*(D) + (q \times \mathrm{id})^*(D)\\ &\quad = \Gamma_{t_1} - \Gamma_{t_2} + \Gamma_{t_2/t_1}.\end{aligned}$$

We have the standard affine n-simplex

$$\Delta^n := \mathrm{Spec}\, \mathcal{O}_S[t_0, \dots, t_n]/\sum_{i=0}^n t_i - 1.$$

We identify $\square^1$ with Δ^1 by sending 0 to $(0,1)$ and ∞ to $(1,0)$, and identify Δ_2 with Δ^2 by sending D_i to the face $t_i = 0$ of Δ^2, $i = 0, 1, 2$, both via affine linear maps.

Under these identifications, if Y is a smooth S-scheme, then a divisor on $Y \times \square^1$ which is disjoint from $Y \times \{0, \infty\}$ defines an element of $\mathrm{CH}_{naif}(\mathbb{Z}_Y(1)[2], 1)$ (see Chapter II, Definition 2.3.1). The relation (1.1.4.1) shows that the cycle $\Gamma_{t_1} - \Gamma_{t_2} + \Gamma_{t_2/t_1}$ is zero in $\mathrm{CH}_{naif}(\mathbb{Z}_{(\mathbb{A}^1\setminus\{0\})^2}(1)[2], 1)$. As the cycle class map factors through the class in CH_{naif} via the naive cycle class map (see (II.2.3.6.1)), it follows that

$$\mathrm{cl}_{[\Gamma_{t_2/t_1}]} = \mathrm{cl}_{[\Gamma_{t_2}]} - \mathrm{cl}_{[\Gamma_{t_1}]}.$$

By (i), and the duality isomorphism, this proves that $\mathrm{cl}_{\mathbb{G}_m, X}$ is a group homomorphism.

The naturality of $\mathrm{cl}_{\mathbb{G}_m, X}$ follows easily from the definition. □

1.1.5. *The Steinberg relation.* We give a proof of the Steinberg relation, based solely on formal properties in $\mathcal{DM}$.

We first note that the *Leibniz rule* for relative motivic cohomology holds with respect to localization sequences: Let X be in $\mathbf{Sm}_S$, and $j : U \to X$ an open subscheme with reduced closed complement $i : W \to X$. Suppose that W is smooth over S. The Gysin isomorphism $\mathbb{Z}_W(-1)[-2] \cong \mathbb{Z}_{X,W}$ gives us the distinguished triangle in $\mathcal{DM}$,

$$(1.1.5.1)\qquad \mathbb{Z}_W(-1)[-2] \xrightarrow{i_*} \mathbb{Z}_X \xrightarrow{j^*} \mathbb{Z}_U \xrightarrow{\partial} \mathbb{Z}_W(-1)[-1].$$

The appropriate diagonal maps composed with the external products give rise to the cup products

$$\begin{aligned}\cup_X &: \mathbb{Z}_X \otimes \mathbb{Z}_X \to \mathbb{Z}_X,\\ \cup_{U,X} &: \mathbb{Z}_U \otimes \mathbb{Z}_X \to \mathbb{Z}_U,\\ \cup_{W,X} &: \mathbb{Z}_W \otimes \mathbb{Z}_X \to \mathbb{Z}_W,\\ \cup_U &: \mathbb{Z}_U \otimes \mathbb{Z}_U \to \mathbb{Z}_U,\\ \cup_W &: \mathbb{Z}_W \otimes \mathbb{Z}_W \to \mathbb{Z}_W,\end{aligned}$$

with

$$\cup_{W,X} = \cup_W \circ (\mathrm{id} \otimes i^*), \tag{1.1.5.2}$$
$$\cup_{U,X} = \cup_U \circ (\mathrm{id} \otimes j^*).$$

1.1.6. LEMMA. *Let Γ be an object of $\mathcal{DM}$, $f:\Gamma \to \mathbb{Z}_U$ and $g:\Gamma \to \mathbb{Z}_X$ morphisms in $\mathcal{DM}$. Then*

$$\partial(f \cup_U j^* g) = \partial(f) \cup_W i^*(g).$$

PROOF. Tensoring the distinguished triangle (1.1.5.1) with $\mathbb{Z}_{(X;D)}$ gives the distinguished triangle

(1.1.6.1)

$$\mathbb{Z}_W \otimes \mathbb{Z}_X(-1)[-2] \xrightarrow{i_*\otimes \mathrm{id}} \mathbb{Z}_X \otimes \mathbb{Z}_X \xrightarrow{j^*\otimes \mathrm{id}} \mathbb{Z}_U \otimes \mathbb{Z}_X \xrightarrow{\partial\otimes \mathrm{id}} \mathbb{Z}_W \otimes \mathbb{Z}_X(-1)[-1].$$

The maps $\cup_{W,X}$, $\cup_X$ and $\cup_{U,X}$ define the map of the distinguished triangle (1.1.5.1) to the distinguished triangle (1.1.6.1). This, together with the identity (1.1.5.2), completes the proof. □

We let t_1, t_2 be the standard coordinate functions on T^2, giving elements $x_i := \mathrm{cl}_{\mathbb{G}_m,T^2}(t_i)$, $i = 1, 2$, of $\mathrm{Hom}_{\mathcal{DM}}(\mathbb{Z}_{\mathbb{G}_m}, \mathbb{Z}_{T^2})$, and the cup product $x_1 \cup_{T^2} x_2 : \mathbb{Z}_{\mathbb{G}_m}^{\otimes 2} \to \mathbb{Z}_{T^2}$. Let $L_0 = \mathbb{A}^1 \setminus \{0, 1\}$, and let $\iota: L_0 \to T^2$ be the map

$$\iota(t) = (t, 1 - t).$$

We also have the map

$$\rho: T \to T^2$$
$$\rho(t) = (t, -t).$$

1.1.7. PROPOSITION [The Steinberg relation]. *(i) The map $\iota^* \circ (x_1 \cup_X x_2): \mathbb{Z}_{\mathbb{G}_m}^{\otimes 2} \to \mathbb{Z}_{L_0}$ is zero.*
(ii) The map $\rho^* \circ (x_1 \cup_X x_2): \mathbb{Z}_{\mathbb{G}_m}^{\otimes 2} \to \mathbb{Z}_T$ is zero.

PROOF. (i) Let $Y = \mathbb{A}^2 \setminus \mathbb{A}^1 \times 0 = \mathbb{A}^1 \times T$, let $\mu: \tilde{Y} \to Y$ be the blow-up of Y at the point $(0, 1)$, and let E be the exceptional curve $E := \mu^{-1}((0, 1))$. The inclusion of $\mathbb{A}^1 \times 1$ in Y lifts to the inclusion $i_1: \mathbb{A}^1 \to \tilde{Y}$ and the inclusion of $0 \times T$ into Y lifts uniquely to the inclusion $i_0: T \to \tilde{Y}$.

Let F_1 be the image of i_1 and F_0 the image of i_0 and let $X = \tilde{Y} \setminus F_0$. Let $W = X \cap E$, and let $j: U \to X$ be the inclusion of $X \setminus W$. Then μ gives an isomorphism $\mu_0: U \to Y \setminus 0 \times T = T^2$.

Let z be the rational function $(1 - t_2)/t_1$ on X; then z is a regular function and (z, t_1) are global coordinates on X. In addition, F_1 is contained in X, and F_1 and W are defined respectively by the equations $z = 0$, $t_1 = 0$ The function $t_2 = 1 - t_1 z$ is a unit on X, and t_1 is a unit on U. This gives us the identity $\mu_0^*(x_1 \cup x_2) = \mu_0^*(x_1) \cup j^* \mu^*(x_2)$.

The exact triangle

$$\mathbb{Z}_W(-1)[-2] \to \mathbb{Z}_X \xrightarrow{j^*} \mathbb{Z}_U \xrightarrow{\partial} \mathbb{Z}_W(-1)[-1]$$

gives the exact Hom-sequences

$$\mathrm{Hom}_{\mathcal{DM}}(\mathbb{Z}_{\mathbb{G}_m}^{\otimes i}, \mathbb{Z}_X) \xrightarrow{j^*} \mathrm{Hom}_{\mathcal{DM}}(\mathbb{Z}_{\mathbb{G}_m}^{\otimes i}, \mathbb{Z}_U) \xrightarrow{\partial_i} \mathrm{Hom}_{\mathcal{DM}}(\mathbb{Z}_{\mathbb{G}_m}^{\otimes i}, \mathbb{Z}_W(-1)[-1])$$

for $i = 1, 2, \ldots$. By the Leibniz rule (Lemma 1.1.6), we have

$$\begin{aligned}\partial_2(\mu_0^*(x_1 \cup x_2)) &= \partial_2(\mu_0^*(x_1) \cup j^*\mu^*(x_2)) \\ &= \partial_1(\mu_0^*(x_1)) \cup i_W^*(\mu^*(x_2)) \\ &= \partial_1(\mu_0^*(x_1)) \cup \mathrm{cl}_{\mathbb{G}_m,W}(i_W^*(\mu^*(t_2))) \\ &= \partial_1(\mu_0^*(x_1)) \cup \mathrm{cl}_{\mathbb{G}_m,W}(1) \\ &= 0.\end{aligned}$$

Thus $\mu_0^*(x_1 \cup x_2)$ lifts to a map $\gamma: \mathbb{Z}_{\mathbb{G}_m}^{\otimes 2} \to \mathbb{Z}_X$.

Both F_1 and W are isomorphic to $\mathbb{A}^1_S$, hence by the homotopy axiom there are maps $\gamma_{F_1}: \mathbb{Z}_{\mathbb{G}_m}^{\otimes 2} \to \mathbb{Z}_S$, $\gamma_W: \mathbb{Z}_{\mathbb{G}_m}^{\otimes 2} \to \mathbb{Z}_S$, such that

$$i_{F_1}^* \circ \gamma = p_{F_1}^* \circ \gamma_{F_1}; \quad i_W^* \circ \gamma = p_W^* \circ \gamma_W,$$

where $p_W: W \to S$, $p_{F_1}: F_1 \to S$ are the structure maps. Since F_1 and W intersect (at the point $(z, t_1) = (0, 0)$), we have $\gamma_{F_1} = \gamma_W$. Additionally, $F_1 \cap U$ maps to $\mathbb{A}^1 \times 1$ via μ_0; let $\bar{\mu}_0: F_1 \cap U \to \mathbb{A}^1 \times 1$ be the induced map. Then

$$i_{F_1 \cap U}^* \circ \gamma = \bar{\mu}_0^* \circ i_{A^1 \times 1}^* \circ (x_1 \cup x_2) = 0,$$

since $t_2 = 1$ on $\mathbb{A}^1 \times 1$. Thus $\gamma_{F_1} = 0$, hence

$$i_W^* \circ \gamma = p_W^* \circ \gamma_W = p_W^* \circ \gamma_{F_1} = 0.$$

Now blow-up $\mathbb{A}^2$ at $(1, 0)$ and $(0, 1)$, $\tau: Y_1 \to \mathbb{A}^2$, with exceptional curve $E_1 \coprod E_2$, $E_1 := \tau^{-1}((0, 1))$, $E_2 := \tau^{-1}((1, 0))$. The inclusion of $\mathbb{A}^1 \times 0 \cup 0 \times \mathbb{A}^1$ in $\mathbb{A}^2$ lifts to the inclusion $i: \mathbb{A}^1 \times 0 \cup 0 \times \mathbb{A}^1 \to Y_1$. Let F be the image of i, and let $W_i = E_i \setminus F$. Let $V = \tau^{-1}(T^2)$, $X_1 = Y_1 \setminus F$, and let $\tau_0: V \to T^2$ be the restriction of τ. A neighborhood of W_1 in Y_1 is isomorphic to X; repeating the above argument for the corresponding neighborhood of W_2 in Y_1, we find that $\tau_0^*(x_1 \cup x_2)$ extends to a map $\gamma_1: \mathbb{Z}_{\mathbb{G}_m}^{\otimes 2} \to \mathbb{Z}_{X_1}$ with

$$i_{W_1}^* \circ \gamma_1 = 0 = i_{W_2}^* \circ \gamma_1.$$

The inclusion $\iota: L_0 \to T^2$ extends uniquely to the inclusion $\iota_1: \mathbb{A}^1_S \to X_1$ with $\iota_1(0) \in W_1$ and $\iota_1(1) \in W_2$. Letting $j: L_0 \to \mathbb{A}^1_S$ be the inclusion, we have

$$j^* \circ \iota_1^* \circ \gamma_1 = \iota^* \circ (x_1 \cup c_2).$$

On the other hand, by the homotopy property there is a map $\gamma_0: \mathbb{Z}_{\mathbb{G}_m}^{\otimes 2} \to \mathbb{Z}_S$ with $\iota_1^* \circ \gamma_1 = p^* \circ \gamma_0$, where $p: \mathbb{A}^1_S \to S$ is the structure map. Since $i_{W_1}^* \circ \gamma_1 = 0$, it follows that $\gamma_0 = 0$.

For (ii), let $\rho_0: L_0 \to T^2$ be the restriction of ρ to L_0, and let $\sigma: L_0 \to L_0$ be the map $\sigma(t) = t^{-1}$. Then

(1.1.7.1)

$$\begin{aligned}0 = (\iota \circ \sigma)^* \circ (x_1 \cup x_2) &= \mathrm{cl}_{\mathbb{G}_m}(1/t) \cup \mathrm{cl}_{\mathbb{G}_m}((t-1)/t) \\ &= \mathrm{cl}_{\mathbb{G}_m}(t) \cup \mathrm{cl}_{\mathbb{G}_m}(-t) - \mathrm{cl}_{\mathbb{G}_m}(t) \cup \mathrm{cl}_{\mathbb{G}_m}(1-t) \\ &= \mathrm{cl}_{\mathbb{G}_m}(t) \cup \mathrm{cl}_{\mathbb{G}_m}(-t) \\ &= \rho_0^*(x_1 \cup x_2).\end{aligned}$$

Now let $x: S \to \mathbb{A}^1_S$ be a section, and $p: \mathbb{A}^1_S \to S$ the structure morphism; by the homotopy property, the map $x^*: \mathbb{Z}_{\mathbb{A}^1_S} \to \mathbb{Z}_S$ is inverse to $p^*: \mathbb{Z}_S \to \mathbb{A}^1_S$. As the sections $x+1$ and x are disjoint, it follows from Theorem 2.4.9 of Chapter III that $(x+1)^* \circ x_* = 0$, hence $x_*: \mathbb{Z}_S(-1)[-2] \to \mathbb{A}^1_S$ is the zero map. Let $j: T \to \mathbb{A}^1_S$ be the inclusion and let $y: S \to T$ be a section. Then $y_* = j^* \circ (j \circ y)_* = 0$. In particular, for the section $i_1: S \to T$ with value one, the Gysin map $i_{1*}: \mathbb{Z}_S(-1)[-2] \to \mathbb{Z}_T$ is the zero map.

Letting $k: L_0 \to T$ be the inclusion, we have seen that $k^* \circ (\rho^* \circ (x_1 \cup x_2)) = 0$. From the Gysin distinguished triangle

$$\mathbb{Z}_S(-1)[-2] \xrightarrow{i_{1*}} \mathbb{Z}_T \xrightarrow{k^*} \mathbb{Z}_{L_0} \to \mathbb{Z}_S(-1)[-1],$$

there is a map $\alpha: \mathbb{Z}_{T^2} \to \mathbb{Z}_S(-1)[-2]$ with $\rho^* \circ (x_1 \cup x_2) = i_{1*} \circ \alpha = 0$. □

1.1.8. *Milnor K-groups.* For a field F, the graded ring $K_*^M(F)$ is defined to be the tensor algebra (over $\mathbb{Z}$) of the multiplicative group $F^\times$ of F, modulo the two-sided ideal generated by elements of the form $a \otimes (1-a)$, with $a, 1-a \in F^\times$. From the identity used in (1.1.7.1), this ideal also contains the elements $a \otimes -a$, $a \in F^\times$, hence $K_*^M(F)$ is graded commutative.

For an arbitrary commutative ring, one can mimic this definition defining the graded ring $K_*^M(R)$ by

$$K_*^M(R) := \oplus_{n=0}^\infty (R^\times)^{\otimes_{\mathbb{Z}} n}/\mathcal{I},$$

with $\mathcal{I}$ the two-sided ideal generated by elements $a \otimes (1-a)$, with $a, 1-a \in R^\times$, and $a \otimes -a$, with $a \in R^\times$. It follows from Lemma 1.1.4 and Proposition 1.1.7 that the map $\mathrm{cl}_{\mathbb{G}_m,X}$ descends uniquely to a graded ring homomorphism

$$\mathrm{cl}^M_{\mathbb{G}_m,X}: K_*^M(\Gamma(X, \mathcal{O}_X)) \to \oplus_{n=0}^\infty \mathrm{Hom}_{\mathcal{DM}}(\mathbb{Z}_{\mathbb{G}_m}^{\otimes n}, \mathbb{Z}_X).$$

Via the duality isomorphism from Lemma 1.1.2, $\mathrm{cl}^M_{\mathbb{G}_m,X}$ induces the graded ring homomorphism

$$\mathrm{cl}^M_{\mathbb{G}_m^D,X}: K_*^M(\Gamma(X, \mathcal{O}_X)) \to \oplus_{n=0}^\infty \mathrm{Hom}_{\mathcal{DM}}(1, \mathbb{Z}_{(X\times\square^n, X\times\partial\square^n)}(n)[2n]).$$

As in (Chapter II, Lemma 2.3.5), the identification of $\mathbb{Z}_X(n)$ with the summand of the complex $\mathbb{Z}_{(X\times\square^n, X\times\partial\square^n)}(n)$ corresponding to the vertex $(0, \dots, 0)$ of $\square^n$ determines an isomorphism in $\mathcal{DM}$:

$$i_{X,n}: \mathbb{Z}_X(n)[-n] \to \mathbb{Z}_{(X\times\square^n, X\times\partial\square^n)}(n).$$

Composing $\mathrm{cl}^M_{\mathbb{G}_m^D,X}$ with $i_{X,n}[n]^{-1}$ gives the natural graded ring homomorphism

$$\mathrm{cl}^M_X: K_*^M(\Gamma(X, \mathcal{O}_X)) \to \oplus_{n=0}^\infty H^n(X, \mathbb{Z}(n)). \tag{1.1.8.1}$$

1.1.9. THEOREM. *Suppose that $S = \operatorname{Spec} k$, where k is a field, and let F be a finitely generated field over k. Then the map (1.1.8.1) (for $X = \operatorname{Spec} F$) is an isomorphism.*

PROOF. Let $u_1, \dots, u_n$ be units in a commutative ring R and let $X = \operatorname{Spec} R$. It follows from Lemma 1.1.4(i) that $\mathrm{cl}^M_{\mathbb{G}_m^D,X}(\{u_1, \dots, u_n\})$ is represented by the cycle class map associated to the graph of the morphism $(u_1, \dots, u_n): X \to \square^n$.

Let $\partial_0\square^n$ be the subset of $\partial\square^n$ gotten by deleting the divisor $t_n = 0$, and write $\mathcal{Z}^q(X, n)^c$ for the cycle group $\mathcal{Z}^q(\mathbb{Z}_{(X\times\square^n; X\times\partial_0\square^n)}(q)[2q])$. Explicitly $\mathcal{Z}^q(X, n)^c$ can be defined as follows: Let $\mathcal{Z}^q(X, n)'$ be the subgroup of $\mathcal{Z}^q(X \times \square^n)$ generated by irreducible codimension q subschemes W of $X \times \square^n$ such that W intersects each "face" $X \times F$ in codimension q, where F is a subvariety of $\square^n$ defined by an equation of the form $t_{i_1} = \epsilon_1, \dots, t_{i_s} = \epsilon_s$, $\epsilon_i \in \{0, 1\}$. For each i, we have the restriction to the face $X \times \partial_i^\epsilon \square^n$, defined by $t_i = \epsilon$, $\epsilon \in \{0,1\}$, $\delta_{i,\epsilon}^*: \mathcal{Z}^q(X,n)' \to \mathcal{Z}^q(X,n-1)'$, and

$$\mathcal{Z}^q(X,n)^c = \cap_{i=0}^n \ker(\delta_{i,1}^*) \cap \cap_{i=0}^{n-1} \ker(\delta_{i,0}^*).$$

The map $\delta_{n,0}^*$ gives us the map $d_n: \mathcal{Z}^q(X,n)^c \to \mathcal{Z}^q(X,n-1)^c$, forming the complex $\mathcal{Z}^q(X,*)^c$.

In [**85**], we have constructed a natural quasi-isomorphism

$$\phi_X: \mathcal{Z}^q(X,*)^c \to \mathcal{Z}^q(X,*);$$

additionally, one can repeat the arguments of Chapter II, replacing $\mathcal{Z}^q(X,*)^c$ with $\mathcal{Z}^q(X,*)$ throughout, and replacing the truncated simplicial object $\mathbb{Z}_{X\times\Delta^*}^{\leq n}$ with the cubical object $\mathbb{Z}_{(X\times\square^n; X\times\partial_0\square^n)}$, to show that sending a cycle $Z \in Z_p(\mathcal{Z}^q(X,*)^c)$ to the corresponding morphism in $\mathcal{DM}$, $\mathrm{cl}_{[Z]}: 1 \to \mathbb{Z}_X(q)[2q-p]$, gives an isomorphism

$$H_p(\mathcal{Z}^q(X,*)^c) \cong H^{2q-p}(X,\mathbb{Z}(q)). \tag{1.1.9.1}$$

In [**122**] it is shown that the map sending $(u_1,\dots,u_n) \in F^{\times n}$ to the point $(u_1,\dots,u_n)$ of $\operatorname{Spec} F \times \square^n$ gives an isomorphism $K_n^M(F) \to H_n(\mathcal{Z}^q(X,*)^c)$; putting this together with the explicit description of the map $\mathrm{cl}^M_{\mathbb{G}_m^D,X}$, and the isomorphism (1.1.9.1) proves the theorem. □

1.1.10. *The localization connecting homomorphism.* For the discussion of the connecting homomorphism, we assume the base scheme S is normal. Let U be a dense open subscheme of some $X \in \mathbf{Sm}_S^{\mathrm{ess}}$, and let f be a unit on U. We extend f to a rational function on X, which we may view as a morphism $f: X^0 \to \mathbb{P}^1$, where X^0 is the complement of a codimension two closed subset of X. Let $W := X^0 \setminus U$.

We have the element $\mathrm{cl}_U^M(f) \in H^1(U,\mathbb{Z}(1))$; we want to compute the image of $\mathrm{cl}_U^M(f)$ in $H^2(X^0,\mathbb{Z}(1))$ under the connecting homomorphism in the localization sequence (Chapter I, §2.2.10)

$$\dots \to H^1(X^0,\mathbb{Z}(1)) \to H^1(U,\mathbb{Z}(1)) \xrightarrow{\partial} H^2_W(X^0,\mathbb{Z}(1)) \to \dots$$

1.1.11. PROPOSITION. *Suppose that* $\mathrm{Div}(f)$ *is in* $\mathcal{Z}^1(X^0/S)$. *Then*

$$\partial(\mathrm{cl}_U^M(f)) = \mathrm{cl}^1_{X^0,W}(\mathrm{Div}(f)).$$

PROOF. Since f is a morphism on X^0, the zero locus and infinity locus of f are disjoint; it follows from excision that

$$\mathbb{Z}_{X^0, f^{-1}(0)\amalg f^{-1}(\infty)} \cong \mathbb{Z}_{X^0\setminus f^{-1}(\infty), f^{-1}(0)} \oplus \mathbb{Z}_{X^0\setminus f^{-1}(0), f^{-1}(\infty)}.$$

Thus, we may compute the boundary terms due to the zero's and poles of f separately, hence we may assume that f is a regular function on X^0. By functoriality, we reduce to the case $U = T := \mathbb{A}^1 \setminus \{0\}$, $X^0 = \mathbb{A}^1$, $f = t$, with t the standard coordinate on $\mathbb{A}^1$.

We now construct various relative motives and relative motives with support; to simplify the notation, we omit the auxiliary h from the notation $\mathbb{Z}_X(q)_h$.

We have the relative motive

$$\mathbb{Z}_{(T\times\square^1; T\times\{0,\infty\})} := \mathrm{cone}\big(\mathbb{Z}_{T\times\square^1} \xrightarrow{(i_0^*, i_\infty^*)} \mathbb{Z}_{T\times 0} \oplus \mathbb{Z}_{T\times\infty}\big)[-1].$$

Let Γ be the intersection of the diagonal $\Delta_{\mathbb{P}^1} \subset \mathbb{P}^1 \times \mathbb{P}^1$ with the open subscheme $T \times \square^1$; Γ is then Γ_t in the notation of §1.1.3.

Let $U := T \times \square^1 \setminus \Gamma$. We have the relative motive $\mathbb{Z}_{(U;T\times\{0,\infty\})}$ and the relative motive with support,

$$\mathbb{Z}_{(T\times\square^1;T\times\{0,\infty\}),\Gamma} := \mathrm{cone}\big(\mathbb{Z}_{(T\times\square^1;T\times\{0,\infty\})} \xrightarrow{(j_U^*,\mathrm{id})} \mathbb{Z}_{(U;T\times\{0,\infty\})}\big)[-1].$$

Let $\bar{\Gamma}$ be the closure of Γ in $\mathbb{A}^1 \times \square^1$, and let $V := \mathbb{A}^1 \times \square^1 \setminus \bar{\Gamma}$. We have the relative motive

$$\mathbb{Z}_{(\mathbb{A}^1\times\square^1;T\times 0,\mathbb{A}^1\times\infty)} := \mathrm{cone}\big(\mathbb{Z}_{\mathbb{A}^1\times\square^1} \xrightarrow{(i_0^*,i_\infty^*)} \mathbb{Z}_{T\times 0} \oplus \mathbb{Z}_{\mathbb{A}^1\times\infty}\big)[-1],$$

the relative motive with support

$$\mathbb{Z}_{(\mathbb{A}^1\times\square^1;T\times 0,\mathbb{A}^1\times\infty),\bar{\Gamma}} := \mathrm{cone}\big(\mathbb{Z}_{(\mathbb{A}^1\times\square^1;T\times 0,\mathbb{A}^1\times\infty)} \xrightarrow{(j_V^*,\mathrm{id})} \mathbb{Z}_{(V;T\times 0,\mathbb{A}^1\times\infty)}\big)[-1],$$

and the relative motive

$$\mathbb{Z}_{(\mathbb{A}^1\times\square^1;\mathbb{A}^1\times\{0,\infty\})} := \mathrm{cone}\big(\mathbb{Z}_{\mathbb{A}^1\times\square^1} \xrightarrow{(i_0^*,i_\infty^*)} \mathbb{Z}_{T\times 0} \oplus \mathbb{Z}_{\mathbb{A}^1\times\infty}\big)[-1].$$

We have the commutative diagram

$$\begin{array}{ccc}
\mathbb{Z}_{(\mathbb{A}^1\times\square^1;T\times 0,\mathbb{A}^1\times\infty),\bar{\Gamma}} & \xrightarrow{j^*} & \mathbb{Z}_{(T\times\square^1;T\times\{0,\infty\}),\Gamma} \\
\Big\downarrow{\scriptstyle\pi} & & \Big\downarrow{\scriptstyle\pi} \\
\mathbb{Z}_{\mathbb{A}^1\times\square^1,\bar{\Gamma}} & \xrightarrow{j^*} & \mathbb{Z}_{T\times\square^1,\Gamma}
\end{array}$$

where the maps j^* are given by the appropriate collection of restriction maps (to open subschemes), and the maps π are the projections onto the quotient complexes

$$\mathbb{Z}_{\mathbb{A}^1\times\square^1,\bar{\Gamma}} := \mathrm{cone}\big(\mathbb{Z}_{\mathbb{A}^1\times\square^1} \xrightarrow{j_V^*} \mathbb{Z}_V\big)[-1]$$
$$\mathbb{Z}_{T\times\square^1,\Gamma} := \mathrm{cone}\big(\mathbb{Z}_{T\times\square^1} \xrightarrow{j_U^*} \mathbb{Z}_U\big)[-1].$$

The maps π are isomorphisms in the homotopy category; from this, we see that the class of the cycle $|\Gamma|$, $\mathrm{cl}_{\partial,\Gamma}(|\Gamma|)\colon 1 \to \mathbb{Z}_{(T\times\square^1;T\times\{0,\infty\}),\Gamma}(1)[2]$ (see §1.1.3), lifts canonically to the class of the cycle $|\bar{\Gamma}|$, $\mathrm{cl}_{\partial,\bar{\Gamma}}(|\bar{\Gamma}|)\colon 1 \to \mathbb{Z}_{(\mathbb{A}^1\times\square^1;T\times 0,\mathbb{A}^1\times\infty),\bar{\Gamma}}(1)[2]$.

Let

$$\mathrm{cl}_{[\Gamma]}\colon 1 \to \mathbb{Z}_{(T\times\square^1;T\times\{0,\infty\})}(1)[2],$$
$$\mathrm{cl}_{[\bar{\Gamma}]}\colon 1 \to \mathbb{Z}_{(\mathbb{A}^1\times\square^1;T\times 0,\mathbb{A}^1\times\infty)}(1)[2]$$

be the maps induced by $\mathrm{cl}_{\partial,\Gamma}(|\Gamma|)$ and $\mathrm{cl}_{\partial,\bar{\Gamma}}(|\bar{\Gamma}|)$ by forgetting the support.

We have the commutative diagram, with the columns being the localization distinguished triangles, and horizontal maps the isomorphisms in $\mathcal{DM}$ induced by the inclusion of $\mathbb{Z}_?$ as the summand $\mathbb{Z}_{?\times 0}$:

$$\begin{array}{ccc}
\mathbb{Z}_{\mathbb{A}^1}(1)[1] & \xrightarrow{i} & \mathbb{Z}_{(\mathbb{A}^1\times\square^1;\mathbb{A}^1\times\{0,\infty\})}(1)[2] \\
\Big\downarrow & & \Big\downarrow{\scriptstyle k^*(1)[2]} \\
\mathbb{Z}_T(1)[1] & \xrightarrow{i} & \mathbb{Z}_{(T\times\square^1;T\times\{0,\infty\})}(1)[2] \\
\Big\downarrow{\scriptstyle\partial} & & \Big\downarrow{\scriptstyle\partial} \\
\mathbb{Z}_{\mathbb{A}^1,\{0\}}(1)[2] & \xrightarrow{i} & \mathrm{cone}(k^*)(1)[2].
\end{array}$$

We have as well the commutative diagram

$$\begin{CD}
\mathbb{Z}_{(\mathbb{A}^1\times\square^1;\mathbb{A}^1\times\{0,\infty\})} @= \mathbb{Z}_{(\mathbb{A}^1\times\square^1;\mathbb{A}^1\times\{0,\infty\})}\\
@V{\tilde{k}^*}VV @VV{k^*}V\\
\mathbb{Z}_{(\mathbb{A}^1\times\square^1;T\times\{0\},\mathbb{A}^1\times\{\infty\})} @>{\tilde{j}}>> \mathbb{Z}_{(T\times\square^1;T\times\{0,\infty\})}\\
@V{\rho}VV @VV{\partial}V\\
\mathrm{cone}(\tilde{k}^*) @>{j}>> \mathrm{cone}(k^*)
\end{CD}$$

with ρ the canonical map; it follows directly from the homotopy property that $\tilde{j}$ is an isomorphism in $\mathcal{DM}$, hence the map j is an isomorphism as well. The map i lifts in the evident manner to the map $\tilde{i}\colon \mathbb{Z}_{\mathbb{A}^1,\{0\}}(1)[1] \to \mathrm{cone}(\tilde{k}^*)(1)[2]$ with $j(1)[2]\circ\tilde{i} = i$.

Let $i_0\colon \mathbb{A}^1 \to \mathbb{A}^1\times\square^1$ be the section with value 0. We have the map

$$\Pi\colon \mathrm{cone}(\tilde{k}^*) \to \mathbb{Z}_{\mathbb{A}^1,\{0\}},$$

defined by taking the sum of the maps

$$\begin{aligned}
i_0^*\colon \mathbb{Z}_{\mathbb{A}^1\times\square^1} &\to \mathbb{Z}_{\mathbb{A}^1},\\
-i_0^*\colon \mathbb{Z}_{\mathbb{A}^1\times 0} &\to \mathbb{Z}_{A^1},\\
i_0^*\colon \mathbb{Z}_{T\times 0} &\to \mathbb{Z}_T,
\end{aligned}$$

and mapping all the remaining terms in $\mathrm{cone}(\tilde{k}^*)$ to zero. Let $\tau\colon \mathbb{Z}_{\mathbb{A}^1\times\square^1,\bar{\Gamma}} \to \mathbb{Z}_{(\mathbb{A}^1\times\square^1;T\times\{0\},\mathbb{A}^1\times\{\infty\})}$ be the standard homotopy inverse to π, followed by the map "forget the support". Let $\tilde{i}_0^*\colon \mathbb{Z}_{(\mathbb{A}^1\times\square^1;T\times\{0\},\mathbb{A}^1\times\{\infty\})} \to \mathbb{Z}_{\mathbb{A}^1,\{0\}}$ be the map on the cones given by the maps

$$\begin{aligned}
\mathbb{Z}_{\mathbb{A}^1\times\square^1} &\xrightarrow{i_0^*} \mathbb{Z}_{\mathbb{A}^1},\\
\mathbb{Z}_{T\times 0}\oplus\mathbb{Z}_{\mathbb{A}^1\times\infty} &\xrightarrow{(i_0^*+0)} \mathbb{Z}_T.
\end{aligned}$$

This gives the commutative diagram

$$\begin{CD}
\mathbb{Z}_{(\mathbb{A}^1\times\square^1;\mathbb{A}^1\times\{0,\infty\})} @= \mathbb{Z}_{(\mathbb{A}^1\times\square^1;\mathbb{A}^1\times\{0,\infty\})} @.\\
@V{\tilde{k}^*}VV @VV{\tilde{k}^*}V @.\\
\mathbb{Z}_{(\mathbb{A}^1\times\square^1;T\times\{0\},\mathbb{A}^1\times\{\infty\})} @= \mathbb{Z}_{(\mathbb{A}^1\times\square^1;T\times\{0\},\mathbb{A}^1\times\{\infty\})} @<{\tau}<< \mathbb{Z}_{\mathbb{A}^1\times\square^1,\bar{\Gamma}}\\
@V{\rho}VV @V{\tilde{i}_0^*}VV @VV{i_0^*}V\\
\mathrm{cone}(\tilde{k}^*) @>>{\Pi}> \mathbb{Z}_{\mathbb{A}^1,\{0\}} @= \mathbb{Z}_{\mathbb{A}^1,\{0\}}.
\end{CD}$$

We have in addition $\Pi(1)[2]\circ\tilde{i} = \mathrm{id}_{\mathbb{Z}_{\mathbb{A}^1,\{0\}}(1)[2]}$.

Thus, we have

$$\partial(i^{-1}(\mathrm{cl}_{[\Gamma]})) = i_0^*(\mathrm{cl}_{\partial,\bar{\Gamma}}(|\bar{\Gamma}|)) = \mathrm{cl}_{\mathbb{A}^1,\{0\}}(|0|).$$

Since $\mathrm{cl}^M(t) = i^{-1}(\mathrm{cl}_{[\Gamma]})$ (by Lemma 1.1.4), and since $|0| = \mathrm{Div}(t)$, the proof is complete. $\square$

1.1.12. REMARK. Suppose that X is in $\mathbf{Sm}_k$, for k a field. By semi-purity (Chapter III, Theorem 2.2.5), and the fact that the complement of X^0 in X has codimension two, it follows that the restriction map $H^2_{\bar{W}}(X, \mathbb{Z}(1)) \to H^2_W(X^0, \mathbb{Z}(1))$ is an isomorphism, for all closed subsets W of X^0. Thus, Proposition 1.1.11 implies the identity $\partial(\mathrm{cl}^M(f)) = \mathrm{cl}^1_{X,\bar{W}}(\mathrm{Div}(f))$, where ∂ is the connecting homomorphism in the localization sequence

$$\to H^1(X, \mathbb{Z}(1)) \to H^1(U, \mathbb{Z}(1)) \xrightarrow{\partial} H^2_{\bar{W}}(X, \mathbb{Z}(1)) \to .$$

1.1.13. *The tame symbol.* Let $\mathcal{O}$ be a DVR with residue field k and quotient field K. We have the *tame symbol homomorphism* $T_{\mathcal{O}}: K^M_n(K) \to K^M_{n-1}(k)$, which is characterized by the following properties:

1. $T_{\mathcal{O}}$ vanishes on the image of $K^M_n(\mathcal{O})$.
2. Let π be a generator of the maximal ideal of $\mathcal{O}$, and let $u_2, \ldots, u_n$ be units in $\mathcal{O}$. Then $T_{\mathcal{O}}(\{\pi, u_2, \ldots, u_n\}) = \{\bar{u}_2, \ldots, \bar{u}_n\}$, with $\bar{u}_i$ the image of u_i in k.

For details on Milnor K-theory, and the tame symbol map, we refer the reader to [**7**] and [**99**].

Now take $\mathcal{O} = \mathcal{O}_{X,D}$, where X is in $\mathbf{Sm}_S$ and D is a reduced irreducible codimension one closed subscheme of X, with generic point a regular point of X. Let $Y := \mathrm{Spec}\,\mathcal{O}$, $W := \mathrm{Spec}\,k$, and $U := Y \setminus W$. We have the connecting homomorphism ∂ in the Gysin sequence

$$H^n(Y, \mathbb{Z}(n)) \to H^n(U, \mathbb{Z}(n)) \xrightarrow{\partial} H^{n-1}(W, \mathbb{Z}(n-1)),$$

and the diagram

$$\begin{array}{ccc} H^n(U, \mathbb{Z}(n)) & \xrightarrow{\partial} & H^{n-1}(W, \mathbb{Z}(n-1)) \\ \uparrow{\scriptstyle \mathrm{cl}^M_U} & & \uparrow{\scriptstyle \mathrm{cl}^M_W} \\ K^M_n(K) & \xrightarrow[T_{\mathcal{O}}]{} & K^M_{n-1}(k). \end{array} \tag{1.1.13.1}$$

1.1.14. LEMMA. *The diagram* (1.1.13.1) *is commutative.*

PROOF. From the exactness of the Gysin sequence

$$H^n(Y, \mathbb{Z}(n)) \to H^n(U, \mathbb{Z}(n)) \xrightarrow{\partial} H^{n-1}(W, \mathbb{Z}(n-1)),$$

and the functoriality of $\mathrm{cl}^M_?$, we see that we see that ∂ vanishes on cl^M_U of the image of $K^M_n(\mathcal{O})$. The Leibniz rule (Lemma 1.1.6) implies that

$$\begin{aligned} \partial(\mathrm{cl}^M_U(\{\pi, u_2, \ldots, u_n\})) &= \partial(\mathrm{cl}^M_U(\pi) \cup_U j^*_U \mathrm{cl}^M_Y(\{u_2, \ldots, u_n\})) \\ &= \partial(\mathrm{cl}^M_U(\pi)) \cup_W i^*_W \mathrm{cl}^M_Y(\{u_2, \ldots, u_n\}). \end{aligned}$$

By Proposition 1.1.11, and the functoriality of $\mathrm{cl}^M_?$, we have

$$\partial(\mathrm{cl}^M_U(\pi)) \cup_W i^*_W \mathrm{cl}^M_Y(\{u_2, \ldots, u_n\}) = \mathrm{cl}^0(|W|) \cup_W \mathrm{cl}^M_W(\{\bar{u}_2, \ldots, u_n\});$$

since $\mathrm{cl}^0(|W|)$ is the unit in the motivic cohomology ring of W (Chapter I, Proposition 3.5.6), the proof is complete. □

1.1.15. *The Milnor K-sheaf.* Suppose now that the base scheme S is Spec of a perfect field, and let X be in $\mathbf{Sm}_S^{\mathrm{ess}}$. One may define the Zariski sheaf $\mathcal{K}_n^M$ on X as the kernel of the map

$$\coprod_{x\in X^{(0)}} i_{x*}K_n^M(k(x)) \xrightarrow{\amalg_{x\in X^{(1)}} T_x} \coprod_{x\in X^{(1)}} i_{x*}K_{n-1}^M(k(x)),$$

where T_x stands for the tame symbol map $T_{\mathcal{O}_{X,x}}$. We have the motivic cohomology sheaf $\mathcal{H}^n(\mathbb{Z}(n))$ on X defined as the Zariski sheaf associated to the presheaf $U \mapsto H^n(U, \mathbb{Z}(n))$; it follows from (Chapter II, Theorem 3.4.8 and Proposition 3.6.2) that the sequence

$$0 \to \mathcal{H}^n(\mathbb{Z}(n)) \xrightarrow{\amalg_{x\in X^{(0)}} i_x^*} \coprod_{x\in X^{(0)}} i_{x*}H^n(k(x), \mathbb{Z}(n))$$
$$\xrightarrow{\amalg_{x\in X^{(1)}} \partial_x} \coprod_{x\in X^{(1)}} i_{x*}H^{n-1}(k(x), \mathbb{Z}(n-1))$$

is exact, where ∂_x is the connecting homomorphism in the localization sequence

$$\to H^n(\mathrm{Spec}\, \mathcal{O}_{X,x}, \mathbb{Z}(n)) \to H^n(\mathrm{Spec}\, k(X)_x, \mathbb{Z}(n)) \xrightarrow{\partial} H^{n-1}(k(x), \mathbb{Z}(n-1)) \to,$$

with $k(X)_x$ the function field of the irreducible component of X containing x.

It follows directly from Lemma 1.1.14 that the maps

$$\mathrm{cl}_{k(x)}^M : K_n^M(k(x)) \to H^n(k(x), \mathbb{Z}(n))$$

for $x \in X^{(0)}$ induce the map of Zariski sheaves

$$\mathrm{cl}_X^M : \mathcal{K}_n^M \to \mathcal{H}^n(\mathbb{Z}(n)). \tag{1.1.15.1}$$

The following is thus a direct consequence of Theorem 1.1.9:

1.1.16. THEOREM. *Let k be a perfect field, X in $\mathbf{Sm}_k^{\mathrm{ess}}$. Then the map (1.1.15.1) is an isomorphism of Zariski sheaves on X.*

1.2. Motivic polylogarithm

We give a version of Beilinson's construction of the rational motivic polylogarithm [**9**], keeping track of the denominators. For a detailed description of Beilinson's construction, and Zagier's conjecture on polylogarithms and values of L-functions, see [**13**], [**68**], and [**129**].

1.2.1. *The cube of schemes $Y^{(n)}$.* As in §1, we have the S-scheme $T := \mathbb{A}_S^1 \setminus \{i_0(S)\}$. This gives us the S-scheme T^{n+1}, with coordinate functions $x_0, \dots, x_n$. We have as well the coordinates $y_0, y_1, \dots, y_n$ with

$$y_i = \begin{cases} x_i/x_{i+1} & 0 \le i < n, \\ x_n & i = n. \end{cases}$$

We view T^{n+1} as a T-scheme via the coordinate function $x_0 : T^{n+1} \to T$.

For $i = 0, \dots, n$ let $Y_i^{(n)}$ be the subscheme $y_i = 1$ of T^{n+1}. More generally, for a subset I of $\{0, \dots, n\}$, let $Y_I^{(n)}$ be the subscheme of T^{n+1} defined by

$$Y_I^{(n)} = \begin{cases} \cap_{i\in I} Y_i^{(n)} & \text{for } I \neq \{0, \dots, n\}, \\ \emptyset & \text{for } I = \{0, \dots, n\}. \end{cases}$$

This defines the $n+1$-cube $Y^{(n)}$, $I \mapsto Y_I^{(n)}$, where $Y_{J\supset I}^{(n)}: Y_J^{(n)} \to Y_I^{(n)}$ is the inclusion.

Form the object $\mathbb{Z}_{Y^{(n)}}$ in $\mathbf{C}^b(\mathbf{Sm}_S)$ as the complex

$$
(1.2.1.1)\quad \mathbb{Z}_{T^{n+1}}(0)_{f^\emptyset} \xrightarrow{d^0} \oplus_{i=0}^n \mathbb{Z}_{Y_i^{(n)}}(0)_{f^{\{i\}}} \xrightarrow{d^1} \dots
$$

$$
\dots \xrightarrow{d^{k-1}} \oplus_{|I|=k} \mathbb{Z}_{Y_I^{(n)}}(0)_{f^I} \xrightarrow{d^k} \dots \xrightarrow{d^{n-1}} \oplus_{|I|=n} \mathbb{Z}_{Y_I^{(n)}}(0)_{f^I},
$$

where d^k is the sum of the maps d_I^k, $I = \{i_0 < \dots < i_k\}$,

$$
d_I^k := \sum_{j=0}^{k} (-1)^j Y_{I\setminus\{i_j\}\subset I}^{(n)*} : \oplus_{j=0}^k \mathbb{Z}_{Y_{I\setminus\{i_j\}}^{(n)}} \to \mathbb{Z}_{Y_I^{(n)}},
$$

and the maps f^I are given by

$$
f^I = \bigcup_{\substack{J\supset I\\ J\neq\{0,\dots,n\}}} Y_{J\supset I}^{(n)} : \coprod_{\substack{J\supset I\\ J\neq\{0,\dots,n\}}} Y_J^{(n)} \to Y_I^{(n)}.
$$

Here the term $\mathbb{Z}_{T^{n+1}}$ is in degree 0. Note that we may construct $\mathbb{Z}_{Y^{(n)}}$ from the relative motive $\mathbb{Z}_{(T^{n+1}; Y_0^{(n)}, \dots, Y_n^{(n)})}$ (see Chapter I, §2.6.6) by deleting the degree $n+1$ term $\mathbb{Z}_{Y^{(n)}(0)_{\{0,\dots,n\}}}$, and deleting the component

$$
Y_{\{0,\dots,n\}\supset I}^{(n)} : Y_{\{0,\dots,n\}}^{(n)} \to Y_I^{(n)}
$$

from the maps giving the lifting of $Y^{(n)}$ to an $n+1$-cube in $\mathcal{L}(\mathbf{Sm}_S)$, as in Chapter I, §2.6.2.

If $t: U \to T$ is a T-scheme, smooth over S, we have the $n+1$-cube of U-schemes $I \mapsto Y_{tU,I}^{(n)} := U \times_T Y_I^{(n)}$, giving us the complex $\mathbb{Z}_{Y_{tU}^{(n)}}$ constructed as in (1.2.1.1).

We denote by $tU(-n)$ the n-cube $I \mapsto U \times_T Y_I^{(n)}$, $I \subset \{1, \dots, n\}$, giving the relative motive $\mathbb{Z}_{tU(-n)}$ in $\mathbf{C}^b(\mathbf{Sm}_S)$ (Chapter I, §2.6.6).

The T-isomorphism

$$
\begin{aligned}
j_n : T^{n+1} &\to T^{n+1} \\
(x_0, \dots, x_n) &\mapsto (x_0, y_1, \dots, y_n)
\end{aligned}
$$

sends $Y_i^{(n)}$ to the subscheme $y_i = 1$ for $i = 1, \dots, n$. The inclusions

$$
\begin{aligned}
i_n : T^n &\to T^{n+1} \\
(x_0, \dots, x_{n-1}) &\mapsto (x_0, x_0, x_1, \dots, x_{n-1})
\end{aligned}
$$

give the isomorphism $i_n : Y_i^{(n-1)} \to Y_{0,i+1}^{(n)}$. Pulling back by $t: U \to T$ gives the map $i_{tU,n}^* : \mathbb{Z}_{tU(-n)} \to \mathbb{Z}_{Y_{tU}^{(n-1)}}$ and the isomorphism

$$
(1.2.1.2)\qquad j_{tU,n}^* : \mathbb{Z}_{U\times_S \mathbb{G}_m^n} \to \mathbb{Z}_{tU(-n)}.
$$

From the construction of $\mathbb{Z}_{Y_{tU}^{(n)}}$, we have the identity in $\mathbf{C}^b(\mathbf{Sm}_S)$:

$$
(1.2.1.3)\qquad \mathbb{Z}_{Y_{tU}^{(n)}} = \operatorname{cone}(i_{tU,n}^*)[-1].
$$

1.2.2. *The spectral sequence.* The isomorphism (1.2.1.2), together with the Künneth isomorphism in $\mathcal{DM}(S)$, $\mathbb{Z}_{U\times\mathbb{G}_m^n} \cong \mathbb{Z}_U \otimes \mathbb{Z}_{\mathbb{G}_m}^{\otimes n}$, and the isomorphism $\mathbb{Z}_{\mathbb{G}_m} \cong \mathbb{Z}_S(-1)[-1]$, gives the natural isomorphism

$$H^p(\mathbb{Z}_{tU(-n)}, \mathbb{Z}(q)) \cong H^{p-n}(U, \mathbb{Z}(q-n)). \tag{1.2.2.1}$$

From (1.2.1.3), we have the linked distinguished triangles

$$\begin{gathered}\mathbb{Z}_{Y_{tU}^{(n-1)}}[-1] \to \mathbb{Z}_{Y_{tU}^{(n)}} \to \mathbb{Z}_{tU(-n)} \to \mathbb{Z}_{Y_{tU}^{(n-1)}},\\ \mathbb{Z}_{Y_{tU}^{(n-2)}}[-1] \to \mathbb{Z}_{Y_{tU}^{(n-1)}} \to \mathbb{Z}_{tU(-n+1)} \to \mathbb{Z}_{Y_{tU}^{(n-2)}},\\ \vdots\\ \mathbb{Z}_{Y_{tU}^{(0)}}[-1] \to \mathbb{Z}_{Y_{tU}^{(1)(-1)^{n-1}}} \to \mathbb{Z}_{tU(-1)} \to \mathbb{Z}_{Y_{tU}^{(0)}}.\end{gathered} \tag{1.2.2.2}$$

As $Y_{tU}^{(0)} = U$, (1.2.2.1) and (1.2.2.2) give us the convergent spectral sequence
(1.2.2.3)

$$E_1^{p,q}(tU,n) = H^q(U(p), \mathbb{Z}(N)) \cong H^{p+q}(U, \mathbb{Z}(N+p)) \Longrightarrow H^{p+q+n}(\mathbb{Z}_{Y_{tU}^{(n)}}, \mathbb{Z}(N)).$$

with $E_1^{p,q} = 0$ for $p > 0$, $p < -n$. We now examine this spectral sequence.

Let $a \geq 1$ be an integer. and let $t_a : T^{p+1} \to T^{p+1}$ be the morphism

$$(x_0, \dots, x_p) \mapsto (x_0^a, \dots, x_p^a).$$

The morphism t_a maps $Y_I^{(p)}$ to $Y_I^{(p)}$ for each subset I of $\{0, \dots, n\}$, giving the morphisms $\mathbb{Z}_{Y_{tU}^{(p)}} \xrightarrow{t_a^*} \mathbb{Z}_{Y_{t\circ t_aU}^{(p)}}$ and $\mathbb{Z}_{U(-p)} \xrightarrow{t_a^*} \mathbb{Z}_{U(-p)}$. As these preserve the linked distinguished triangles (1.2.2.2), the morphism t_a thus induces a map of the spectral sequences (1.2.2.3) $t_a^* : E(tU,n) \to E(t\circ t_aU, n)$.

1.2.3. LEMMA. *(i) The differential*

$$d_1^{p,q}(tU,n) : H^{p+q}(U, \mathbb{Z}(N+p)) \to H^{p+q+1}(U, \mathbb{Z}(N+p+1))$$

is $\cup\mathrm{cl}_{\mathbb{G}_m}((-1)^{p-1}t)$.

(ii) Let $i_W : W \to U$ *be the inclusion of a closed codimension one subscheme* W *in* $\mathbf{Sm}_S$*, giving the open complement* $j_V : V \to U$*. Then we have a canonical distinguished triangle in* $\mathcal{DM}(S)$*:*

$$\mathbb{Z}_{Y_{t\circ i_W W}^{(n)}}(-1)[-2] \xrightarrow{i_{W*}} \mathbb{Z}_{Y_{tU}^{(n)}} \xrightarrow{j_V^*} \mathbb{Z}_{Y_{t\circ j_V V}^{(n)}} \xrightarrow{\partial_W} \mathbb{Z}_{Y_{t\circ i_W W}^{(n)}}(-1)[-1].$$

This gives the localization sequence

$$\begin{aligned}\dots \to H^{p-2}(\mathbb{Z}_{Y_{t\circ i_W W}^{(n)}}, \mathbb{Z}(N-1)) &\xrightarrow{i_{W*}} H^p(\mathbb{Z}_{Y_{tU}^{(n)}}, \mathbb{Z}(N))\\ &\xrightarrow{j_V^*} H^p(\mathbb{Z}_{Y_{t\circ j_V V}^{(n)}}, \mathbb{Z}(N)) \xrightarrow{\partial_W} H^{p-1}(\mathbb{Z}_{Y_{t\circ i_W W}^{(n)}}, \mathbb{Z}(N-1)) \to \dots,\end{aligned}$$

and induces a long exact sequence of E_r*-terms for the spectral sequence* (1.2.2.3)*:*

$$\to E_r^{p-1,q-1}(W) \xrightarrow{i_{W*}} E_r^{p,q}(U) \xrightarrow{j_V^*} E_r^{p,q}(V) \xrightarrow{\partial_W} E_r^{p-1,q}(W) \to . \tag{1.2.3.1}$$

For $r = 1$*, the sequence* (1.2.3.1) *agrees with the usual localization sequence*

$$\begin{aligned}\dots \to H^{p+q-2}(W, \mathbb{Z}(p+N-1)) &\xrightarrow{i_{W*}} H^{p+q}(U, \mathbb{Z}(p+N))\\ &\xrightarrow{j_V^*} H^{p+q}(V, \mathbb{Z}(p+N)) \xrightarrow{\partial_W} H^{p+q-1}(W, \mathbb{Z}(p+N-1)) \to \dots .\end{aligned}$$

(iii) Let $i_\zeta : S \to T$ be the section with value ζ, with $\zeta^N = 1$, and let $a \geq 1$ be an integer with $a \equiv 1 \mod N$. Then $i_\zeta \circ t_a = i_\zeta$, and the induced map $t_a^* : E_r^{p,q}(i_\zeta, n) \to E_r^{p,q}(i_\zeta, n)$ is multiplication by a^{-p}.

PROOF. The composition $j_p \circ i_p \circ j_{p-1}^{-1}$ is the map

$$(x_0, y_1, \ldots, y_{p-1}) \mapsto (x_0, x_0 / \prod_{i=1}^{p-1} y_i, y_1, \ldots, y_{p-1}).$$

The isomorphism $H^{p+q}(U, \mathbb{Z}(N+p)) \to H^q(U(-p), \mathbb{Z}(N))$ (1.2.2.1) sends a class α to $\alpha \cup \mathrm{cl}_{\mathbb{G}_m}(y_1) \cup \ldots \cup \mathrm{cl}_{\mathbb{G}_m}(y_p)$. The relation $\mathrm{cl}_{\mathbb{G}_m}(y) \cup \mathrm{cl}_{\mathbb{G}_m}(-y) = 0$ of Proposition 1.1.7(ii) gives

$$\begin{aligned} (1.2.3.2)\quad & (j_p \circ i_p \circ j_{p-1}^{-1})^*(\mathrm{cl}_{\mathbb{G}_m}(y_1) \cup \ldots \cup \mathrm{cl}_{\mathbb{G}_m}(y_p)) \\ &= \mathrm{cl}_{\mathbb{G}_m}((-1)^{p-1} x_0 / \prod_{i=1}^{p-1} -y_i) \cup \mathrm{cl}_{\mathbb{G}_m}(y_1) \cup \ldots \cup \mathrm{cl}_{\mathbb{G}_m}(y_{p-1}) \\ &= \mathrm{cl}_{\mathbb{G}_m}((-1)^{p-1} x_0) \cup \mathrm{cl}_{\mathbb{G}_m}(y_1) \cup \ldots \cup \mathrm{cl}_{\mathbb{G}_m}(y_{p-1}). \end{aligned}$$

As the differential $d_1^{-p,q}(tU, n)$ is given by the composition

$$\mathbb{Z}_U(-p)[-p] \cong \mathbb{Z}_{U(-p)} \xrightarrow{(j_p \circ i_p \circ j_{p-1}^{-1})^*} \mathbb{Z}_{U(-p+1)} \cong \mathbb{Z}_U(-p+1)[-p+1],$$

(i) follows from (1.2.3.2) and the naturality of the spectral sequence.

The assertion (ii) follows from the localization sequence for motives with support, together with the Gysin isomorphism for diagrams (III.2.6.8.2).

For (iii), we need only prove the case $r = 1$. The result then follows directly from the identity

$$\begin{aligned} (1.2.3.3)\quad t_a^*(\mathrm{cl}_{\mathbb{G}_m}(y_1) \cup \ldots \cup \mathrm{cl}_{\mathbb{G}_m}(y_p)) &= \mathrm{cl}_{\mathbb{G}_m}(y_1^a) \cup \ldots \cup \mathrm{cl}_{\mathbb{G}_m}(y_p^a) \\ &= a^p \mathrm{cl}_{\mathbb{G}_m}(y_1) \cup \ldots \cup \mathrm{cl}_{\mathbb{G}_m}(y_p). \end{aligned}$$

□

1.2.4. *Some numerology.* For $p = 1, 2, 3, \ldots$, $r = 1, \ldots, p$, let $N_{p,r}$ be the integer defined as

$$N_{p,r} = \gcd_{a=2,3,\ldots} (a^p - a^{p-r}).$$

We note that $N_{p,p} = 1$ and $N_{p,1} = 2$ for $p > 1$. If $r < p$, if q is a odd prime, and $k \geq 1$, then q^k exactly divides $N_{p,r}$ if and only if k is the maximal number with $1 \leq k \leq p - r$ such that $q^{k-1}(q-1)$ divides r; if $q - 1$ does not divide r, then q does not divide $N_{p,r}$. Similarly, suppose that $r < p$ and that $r > 1$ and $p - r > 1$. Then 2^k exactly divides $N_{p,r}$ if and only if k is the maximal number with $2 \leq k \leq p - r$ such that 2^{k-2} divides r If $p - r = 1$, then 2 exactly divides $N_{p,r}$. In particular, only primes $q \leq r + 1$ divide $N_{p,r}$; more precisely: $N_{p,r} \mid 4(r+1)!$. On the other hand, if r is prime, then $N_{p,r} = 2$ for all $p > r$. Also, $N_{p+N,r}$ is eventually constant in N. We set $N_{p,r} = 1$ for $r > p$ or for $p \leq 0$.

1.2.5. LEMMA. *Let $i_1 : S \to T$ be the section with value 1. Then the spectral sequence $E(i_1 S, n)$ satisfies $N_{-p,r} d_r^{p,q} = 0$ for all p and q, and all $r \geq 1$. In addition, $d_1^{p,q} = 0$ for p odd.*

PROOF. From Lemma 1.2.3(i), the differential $d_1^{p,q}$ is given by

$$d_1^{p,q}(x) = \mathrm{cl}_{\mathbb{G}_m}((-1)^{p-1}) \cup x.$$

From Lemma 1.2.3(iii), we have $a^{-p}d_r^{p,q}(x) = d_r^{p,q}(a^{-p}x) = a^{-p-r}d_r^{p,q}(x)$ for all $x \in E_r^{p,q}(i_1, n)$, all q and all $p = 0, -1, \dots, -n$. □

We now take $t: U \to T$ to be the inclusion of $T^0 := T \setminus \{i_1(S)\}$, and write $Y_t^{(n)}$ for $Y_{tT^0}^{(n)}$. We write the motivic cohomology by $H^p(Y_t^{(n)}, \mathbb{Z}(q)) := H^p(\mathbb{Z}_{Y_t^{(n)}}, \mathbb{Z}(q))$.

The localization sequence on T^0 gives the canonical isomorphism

$$\mathbb{Z}_S \oplus \mathbb{Z}_S(-1)[-1] \oplus \mathbb{Z}_S(-1)[-1] \xrightarrow{p^*+p^*\cup\mathrm{cl}_{\mathbb{G}_m}(t)+p^*\cup\mathrm{cl}_{\mathbb{G}_m}(1-t)} \mathbb{Z}_{T^0},$$

where $p: T^0 \to S$ is the structure morphism. We denote the summand coming from the image of $p^* \cup \mathrm{cl}_{\mathbb{G}_m}(t)$ as $\mathbb{Z}_S(-1)[-1] \cup t$, and similarly for $\mathbb{Z}_S(-1)[-1] \cup (1-t)$. The edge homomorphisms

$$E_1^{0,1}(t, n+1) \to H^{n+1}(Y_t^{(n)}, \mathbb{Z}(n+1)) \to E_1^{-n,n+1}(t, n+1)$$

give the complex

$$(1.2.5.1) \qquad 0 \to H^1(S, \mathbb{Z}(n+1)) \xrightarrow{\alpha_n} H^{n+1}(Y_t^{(n)}, \mathbb{Z}(n+1)) \xrightarrow{\beta_n} H^1(T^0, \mathbb{Z}(1)).$$

For $n \geq 1$, let N_n be the integer

$$N_n := \begin{cases} 1 & \text{for } n = 1 \\ 2^n \prod_{r=2,\dots,n-1} N_{n,r} & \text{for } n \geq 2 \text{ odd}, \\ 2^{n+1} \prod_{r=2,\dots,n-1} N_{n,r} & \text{for } n \geq 2 \text{ even}. \end{cases}$$

1.2.6. PROPOSITION. *Let $N > 0$ be an integer.*
(i) If $N \cdot H^0(S, \mathbb{Z}(r)) = 0$ and $N \cdot H^{-1}(S, \mathbb{Z}(0)) = 0$ for all $r = 1, \dots, n$, then homology of the complex (1.2.5.1) is killed by N.
(ii) If $H^0(S, \mathbb{Z}(r))[1/N] = 0$ and $H^{-1}(S, \mathbb{Z}(0))[1/N] = 0$ for all $r = 1, \dots, n$, then homology of the complex (1.2.5.1) is killed by inverting N.
(iii) If $n \geq 2$, we suppose that 2 is invertible on S. For all $n \geq 1$, the image of β_n contains the subgroup $N_n \cdot H^0(S, \mathbb{Z}(0)) \cup (1-t)$.

PROOF. For all $-n \leq p \leq 0$, we have the complex of E_1-terms

$$\to E_1^{p-1,q} \xrightarrow{\cup\mathrm{cl}_{\mathbb{G}_m}((-1)^p t)} E_1^{p,q} \xrightarrow{\cup\mathrm{cl}_{\mathbb{G}_m}((-1)^{p-1} t)} E_1^{p+1,q} \to;$$

this gives the E_2-terms as follows: For $-n > p > 0$, let

$$M^{\mathrm{odd}}(-p, q) := [H^{q-p-1}(S, \mathbb{Z}(n-p-1))/H^{q-p-2}(S, \mathbb{Z}(n-p-2)) \cup \mathrm{cl}_{\mathbb{G}_m}(-1)],$$

$$M^{\mathrm{ev}}(-p, q) := \ker[H^{q-p-1}(S, \mathbb{Z}(n-p-1)) \xrightarrow{\cup\mathrm{cl}_{\mathbb{G}_m}(-1)} H^{q-p}(S, \mathbb{Z}(n-p))].$$

Then

$$E_2^{-p,q} = \begin{cases} M^{\mathrm{odd}}(-p, q) \cup (1-t) & \text{for } p \text{ odd}, \\ M^{\mathrm{ev}}(-p, q) \cup (1-t) & \text{for } p \text{ even}. \end{cases}$$

Let M^{odd} be the subgroup of $H^1(S, \mathbb{Z}(1)) \oplus H^0(S, \mathbb{Z}(0)) \cup t$ consisting of elements of the form $(x \cup \mathrm{cl}_{\mathbb{G}_m}(-1), x)$, and let M^{ev} be the kernel of $\cup\mathrm{cl}_{\mathbb{G}_m}(-1): H^0(S, \mathbb{Z}(0)) \to H^1(S, \mathbb{Z}(1))$. Then

$$E_2^{-n,n+1} = \begin{cases} M^{\mathrm{odd}} \oplus H^0(S, \mathbb{Z}(0)) \cup (1-t) & \text{for } n \text{ odd}, \\ H^0(S, \mathbb{Z}(0)) \cup t \oplus M^{\mathrm{ev}} \cup (1-t) & \text{for } n \text{ even}. \end{cases}$$

We have a similar description of $E_2^{-n,n}$, shifting the cohomology degree down by one. Finally,

$$E_2^{0,q} = H^q(S,\mathbb{Z}(n+1)) \oplus [H^{q-1}(S,\mathbb{Z}(n))/H^{q-2}(S,\mathbb{Z}(n-1)) \cup \mathrm{cl}_{\mathbb{G}_m}(-1)] \cup (1-t).$$

All these computations are easy consequences of the relations of Proposition 1.1.7, and the additivity of $\mathrm{cl}_{\mathbb{G}_m}$.

Suppose as in (i) that $N \cdot H^0(S,\mathbb{Z}(r)) = 0$ and $N \cdot H^{-1}(S,\mathbb{Z}(0)) = 0$ for all $r = 1,\dots,n$. Then $E_2^{-p,1+p}$ is N-torsion for $0 < p \le n$, and the summand $H^1(S,\mathbb{Z}(n+1))$ of $E_2^{0,1}$ is isomorphic to $E_2^{0,1}$, modulo to N-torsion. Thus $E_\infty^{-p,1+p}$ is N-torsion for $0 < p \le n$, and the summand $H^1(S,\mathbb{Z}(n+1))$ of $E_2^{0,1}$ is isomorphic to $E_\infty^{0,1}$, modulo N-torsion. This proves (i). The proof of (ii) is similar.

For (iii), first take $n = 1$. One can check directly that $1\cdot(1-t)$ is in the image of β_1. Indeed, this is just saying that $x_1 \cup (1-t)$ lifts to an element of $H^2(Y_t^{(1)},\mathbb{Z}(2))$. From the distinguished triangle relating $Y_t^{(1)}$, $Y_t^{(0)} = T^0$ and $T^0(-1)$, this comes down to the Steinberg relation $t \cup (1-t) = 0$ of Proposition 1.1.7(i).

Now take $n \ge 2$, and let $\partial_1 : H^m(T^0,\mathbb{Z}(q)) \to H^{m-1}(S,\mathbb{Z}(q-1))$ be the boundary in the localization sequence coming from the section i_1 of T. By Lemma 1.2.3(ii), we have $d_r(\partial_1(x)) = \partial_1(d_r(x))$; as ∂_1 gives an isomorphism

$$\partial_1 : E_2^{-p,2+p}(tT^0, n+1) \to E_2^{-p,2+p}(i_1S, n)$$

for $-n < p < 0$, it follows from Lemma 1.2.5 that

$$N_{n,r} d_r^{-n,n+1} = 0 \tag{1.2.6.1}$$

for $2 \le r < n$.

Similarly, this argument shows that $N_{r,r} d_n^{-r,r+1} = d_r^{-r,r+1}$ maps $E_r^{-r,r+1}$ into the kernel of ∂_1 on $E_r^{0,2}$; denote this kernel by $E_r^{0,2+}$. We now consider the section $i_{-1}: S \to T^0$ with constant value -1, and the resulting map of spectral sequences $i_{-1}^*: E(tT^0, n+1) \to E(i_{-1}S, n+1)$. For $a \ge 2$ an even integer, the maps t_a^* act on $E(i_{-1}S, n+1)$; arguing as in Lemma 1.2.5, the maps

$$d_r^{-r,r+1} : E_r^{-r,r+1}(i_{-1}, n+1) \to E_r^{0,2}(i_{-1}, n+1)$$

are killed by the g.c.d. of the numbers $a^r - 1$, as a runs over positive odd integers. Taking $a = q$, for q an odd prime, we see that this g.c.d. is 2, hence the surjection $E_1^{0,2}(i_{-1}) \to E_n^{0,2}(i_{-1})$ has kernel killed by 2^{n-1}. But as $E_1^{0,2}(i_{-1}, n+1) = H^2(S,\mathbb{Z}(n+1))$, and $E_2^{0,2}(tT^0, n+1) = H^2(S,\mathbb{Z}(n+1)) \oplus H^2(S,\mathbb{Z}(n)) \cup (1-t)$, it follows that $E_2^{0,2+}$ is equal to the summand $H^2(S,\mathbb{Z}(n+1))$, and the map $i_{-1}^* : E_n^{0,2+} \to E_n^{0,2}(i_{-1}, n+1)$ has kernel which is 2^{n-1}-torsion. As $2d_n^{-n,n+1}(i_{-1}) = 0$, it follows that

$$2^n d_n^{-n,n+1} = 0. \tag{1.2.6.2}$$

Since $d_1^{-n,n+1} = 0$ for n odd and $2d_1^{-n,n+1} = 0$ for n even, (iii) follows from (1.2.6.1) and (1.2.6.2). □

1.2.7. REMARKS. (i) Suppose that $S = \operatorname{Spec} F$, where F is a field. Then by Theorem 3.6.6 of Chapter II, we have the isomorphism $H^p(S,\mathbb{Z}(q)) \cong \mathrm{CH}^q(S, 2q-p)$.

In particular, we have

$$H^p(S,\mathbb{Z}(0)) = \begin{cases} 0 & \text{for } p \neq 0, \\ \mathbb{Z} & \text{for } p = 0, \end{cases}$$

$$H^p(S,\mathbb{Z}(1)) = \begin{cases} 0 & \text{for } p \neq 1, \\ F^\times & \text{for } p = 1. \end{cases}$$

In addition, by [**83**, Theorem 8.1], we have the isomorphisms

$$H^p(S,\mathbb{Z}[\frac{1}{(2q-p-1)!}](q)) \cong \mathrm{gr}^q_\gamma K_{2q-p}(S)[\frac{1}{(2q-p-1)!}].$$

Suppose now that F is a number field. Then (see [**22**] and [**23**]) $\mathrm{gr}^q_\gamma K_{2q-p}(F)$ is a finite group for $p \neq 1$. Thus, there is an integer N satisfying the condition of Proposition 1.2.6(ii).

(ii) We may consider $\mathbb{Z}_{Y_t^{(n)}}$ as an object of $\mathcal{DM}(T^0)$; the linked distinguished triangles (1.2.2.2) and the isomorphism (1.2.1.2) are then valid in $\mathcal{DM}(T^0)$.

1.2.8. Now choose an element $\pi \in H^{n+1}(Y_t^{(n)},\mathbb{Z}(n+1))$ with $\beta_n(\pi) = N_n\cdot 1\cup(1-t)$, where 1 is the image in $H^0(S,\mathbb{Z}(0))$ of the identity map on S (for $n \geq 2$, we suppose that S is a scheme over $\mathbb{Z}[\frac{1}{2}]$). We may view π as a map

$$\mathbb{Z}_S(-(n+1)) \xrightarrow{\pi} \mathbb{Z}_{Y_t^{(n)}}[n][1]$$

in $\mathcal{DM}(S)$. Let $p: T^0 \to S$ be the structure map, let $\delta: Y_t^{(n)} \to Y_t^{(n)} \times_S T^0$ be the canonical map induced by the structure map $Y_t^{(n)} \to T^0$, and let

$$\Pi: \mathbb{Z}_{T^0}(-(n+1)) \to \mathbb{Z}_{Y_t^{(n)}}[n][1]$$

be the map in $\mathcal{DM}(T^0)$ defined as the composition

$$\mathbb{Z}_{T^0}(-(n+1)) \xrightarrow{p^*(\pi)} \mathbb{Z}_{Y_t^{(n)}\times_S}[n][1] \xrightarrow{\delta^*} \mathbb{Z}_{Y_t^{(n)}}[n][1].$$

Complete the map Π to a distinguished triangle

$$\mathbb{Z}_{T^0}(-(n+1)) \xrightarrow{\Pi} \mathbb{Z}_{Y_t^{(n)}}[n][1] \to \Phi \to \mathbb{Z}(-(n+1))[1]$$

and let

$$N_n\mathrm{pol}^S_{n+1} := \Phi[-1].$$

1.2.9. PROPOSITION. *$N_n\mathrm{pol}^S_{n+1}$ is in the triangulated Tate subcategory $\mathcal{DTM}(T^0)$ of $\mathcal{DM}(T^0)$. More precisely, we have the distinguished triangles in $\mathcal{DM}(T^0)$:*

$$\begin{gathered}\mathbb{Z}_{Y_t^{(n)}}[n] \to N_n\mathrm{pol}^S_{n+1} \to \mathbb{Z}_{T^0}(-(n+1)) \to \mathbb{Z}_{Y_t^{(n)}}[n+1] \\ \mathbb{Z}_{Y_t^{(n-1)}}[n-1] \to \mathbb{Z}_{Y_t^{(n)}}[n] \to \mathbb{Z}_{T^0}(-n) \to \mathbb{Z}_{Y_t^{(n-1)}}[n] \\ \vdots \\ \mathbb{Z}_{Y_t^{(0)}} = \mathbb{Z}_{T^0} \to \mathbb{Z}_{Y_t^{(1)}}[1] \to \mathbb{Z}_{T^0}(-1) \to \mathbb{Z}_{Y_t^{(0)}}[1].\end{gathered}$$

PROOF. This follows from the definition of $N_n\mathrm{pol}^S_{n+1}$, the linked distinguished triangles (1.2.2.2), after shifting, and the isomorphism (1.2.1.2), together with Remark 1.2.7(ii). □

1.2.10. REMARKS. (i) If one passes to the *rational* motivic category, one can define $\mathrm{pol}^S_{n+1\mathbb{Q}}$ as the object corresponding to an element $\pi_\mathbb{Q} \in H^{n+1}(Y_t^{(n)}, \mathbb{Q}(n+1))$ with $\beta_n(\pi_\mathbb{Q}) = 1 \cup (1-t)$. For $S = \mathrm{Spec}\, F$, with F a number field, this agrees with Beilinson's construction of the motivic polylogarithm in [**9**]. We thus have

$$N_n \mathrm{pol}^S_{n+1} = N_n \cdot \mathrm{pol}^S_{n+1\mathbb{Q}}$$

in $\mathcal{DM}(S)_\mathbb{Q}$ (up to the image of α_n, see (ii)).
(ii) Taking $S = \mathbb{Q}$, the rational version $\mathrm{pol}^S_{n+1\mathbb{Q}}$ is *unique* for n odd. Indeed, the lack of uniqueness comes from $H^1(\mathrm{Spec}\,\mathbb{Q}, \mathbb{Q}(n+1)) \cong K_{2n+1}(\mathbb{Q})^{(n+1)}$, which, by Borel [**22**], is zero for n odd. For n even, we have $H^1(\mathrm{Spec}\,\mathbb{Q}, \mathbb{Q}(n+1)) \cong K_{2n+1}(\mathbb{Q})^{(n+1)} \cong \mathbb{Q}$, so there is a real lack of uniqueness.

Following [**68**], we may normalize our choice of π (for arbitrary S). For each subset I of $\{0, \dots, n\}$ of size n, we have the identification of T^0 with the term $Y_I^{(n)}$ in the $n+1$-cube $Y^{(n)}$, giving the composition

$$\mathbb{Z}_S \xrightarrow{p^*} \mathbb{Z}_{T^0} \to \mathbb{Z}_{Y_t^{(n)}}[n],$$

which we call $a_I : \mathbb{Z}_S \to \mathbb{Z}_{Y_t^{(n)}}[n]$. The map

$$\alpha_n : H^1(S, \mathbb{Z}(n+1)) \to H^{n+1}(Y_t^{(n)}, \mathbb{Z}(n+1))$$

is just the map on motivic cohomology induced by a_I with $I = \{0, 1, \dots, n-1\}$; on the other hand, it is easy to see that all the maps a_I induce the same map on motivic cohomology up to sign for n odd, and the *same* map for n even.

The symmetric group S_n acts on $T^0 \times T^n$, by permuting the factors in T^n; this extends to an action of S_n on $Y_t^{(n)}$. As $Y^{(n)}_{\{1,\dots,n\}}$ is defined by the equations $x_1 = x_2 = \dots = x_n = 1$, which is fixed by S_n, α_n maps into the S_n-invariants of $H^{n+1}(Y_t^{(n)}, \mathbb{Z}(n+1))$ for n even. For n odd, S_n acts by even permutations on the set of indices I as above, so α_n maps into the S_n-invariants of $H^{n+1}(Y_t^{(n)}, \mathbb{Z}(n+1))$ for n odd as well.

On the other hand, S_n acts by the sign representation on $H^p(T^0(-n), \mathbb{Z}(q))$, so we have a splitting in the exact sequence of Proposition 1.2.6 after inverting $n!$ and N. This gives a unique choice of $Nn!\pi$ by taking the alternating projection of π.
(iii) We have the étale realization of $N_n \mathrm{pol}^S_{n+1}$ in $\mathbf{D}^* \lim \mathrm{Sh}^{\mathbb{Z}_l}_{\text{ét}}(X)$ (Chapter V, §2.2) and the Hodge realization in $\mathbf{D}^b(\mathrm{MHM}(T^0))$ (Chapter V, §2.3.13). This recovers Beilinson's étale and Hodge realization of the motivic polylogarithm.

2. Comparison with the category $DM_{gm}(k)$

Voevodsky has defined a triangulated tensor category of motives over a perfect field k, $DM_{gm}(k)$; in case the field k admits resolution of singularities (e.g., in characteristic zero), the resulting motivic cohomology groups agree with Bloch's higher Chow groups. We show here (Theorem 2.5.5) that our category $\mathcal{DM}(\mathrm{Spec}\, k)$ is equivalent to $DM_{gm}(k)$. We begin by giving a brief description of the construction of the category $DM_{gm}(k)$; for details, we refer the reader to [**124**]. We write $\mathcal{DM}(k)$ for $\mathcal{DM}(\mathrm{Spec}\, k)$.

2.1. The category $DM_{gm}(k)$

2.1.1. *SmCor*(k). Let k be a field, and let Sm/k denote the category of smooth k-schemes of finite type. For X and Y in Sm/k, let $c(X,Y)$ be the free abelian group on irreducible subschemes W of $X \times_k Y$ which are *finite and dominant* over an irreducible component of X. Let $\circ\colon c(Y,Z)\times c(X,Y) \to c(X,Z)$ be the operation defined by

$$Z \circ Z' := p_{XZ*}(p_{YZ}^*(Z) \cdot p_{XY}(Z')),$$

where $\cdot$ is the intersection product; the projection p_{XZ*} is well-defined since each component of $p_{YZ}^*(Z)\cdot p_{XY}^*(Z')$ is finite over $X\times_k Z$. One checks that the operations $\circ$ are associative, giving the pre-additive category $SmCor(k)$, with Hom-groups $\mathrm{Hom}_{SmCor(k)}(X,Y) := c(X,Y)$, and composition $\circ$. In fact, the operation of disjoint union is a direct sum, making $SmCor(k)$ into an additive category, and the operation of product over k defines the structure of a tensor category on $SmCor(k)$.

Sending a morphism $f\colon X \to Y$ in Sm/k to the graph $\Gamma_f \subset X \times_k Y$ defines a faithful functor $Sm/k \to SmCor(k)$; we consider Sm/k as a subcategory of $SmCor(k)$ via this functor.

2.1.2. *Nisnevic sheaves with transfer.* Recall that the *Nisnevic topology* on Sm/k is the Grothendieck topology on Sm/k for which a cover $U \to X$ is a cover in the étale topology such that, for each field K containing k, the map on K-valued points, $U(K) \to X(K)$, is surjective. A *Nisnevic sheaf with transfers on* $SmCor(k)$ is a functor $S\colon SmCor(k)^{\mathrm{op}} \to \mathbf{Ab}$, such that the restriction of S to Sm/k is a Nisnevic sheaf. For example, for X in Sm/k, the "representing" presheaf $L(X)$, $L(X)(Y) = c(Y,X)$, is a Nisnevic sheaf with transfers. Denote the category of Nisnevic sheaf with transfers on $SmCor(k)$ by $Sh_{Nis}(SmCor(k))$.

Note that sending X to $L(X)$, and $f\colon X \to X'$ to the map of functors

$$f_*\colon c(-,X) \to c(-,X')$$

defines the functor

$$L\colon Sm/k \to Sh_{Nis}(SmCor(k)).$$

Voevodsky [**124**, Theorem 3.1.4 and Lemma 3.1.6] shows that the category $Sh_{Nis}(SmCor(k))$ is an abelian category with enough injectives. Let

$$D^-(Sh_{Nis}(SmCor(k)))$$

be the derived category of bounded above complexes.

2.1.3. *Homotopy invariant sheaves.* Let F be in $Sh_{Nis}(SmCor(k))$. Call F *homotopy invariant* if the projection $X \times \mathbb{A}^1 \to X$ induces an isomorphism $F(X) \to F(\mathbb{A}^1 \times X)$ for all X in Sm/k.

Now assume that k is a perfect field. Let $HI(k)$ be the full subcategory of $Sh_{Nis}(SmCor(k))$ consisting of the homotopy invariant sheaves. Then $HI(k)$ is abelian, and the inclusion $HI(k) \to Sh_{Nis}(SmCor(k))$ is exact [**124**, Proposition 3.1.12]. Let $DM_-^{eff}(k)$ be the full subcategory of $D^-(Sh_{Nis}(SmCor(k)))$ consisting of complexes whose cohomology sheaves are in $HI(k)$.

2.1.4. *The homotopy localization.* Let Δ^* be the cosimplicial scheme of affine spaces

$$\Delta^n := \mathrm{Spec}\, k[t_0,\dots,t_n]/\sum_i t_i - 1.$$

For F in $Sh_{Nis}(SmCor(k))$, let $C^*(F)$ be the complex of sheaves associated to the simplicial sheaf $F(\Delta^*)$, $C^{-n}(F) = F(\Delta^n)$. Then, for *all* F, the cohomology sheaves of $C^*(F)$ are homotopy invariant [**124**, Lemma 3.2.1], giving the element $C^*(F)$ of $DM_-^{eff}(k)$; sending F to $C^*(F)$ extends to an exact functor

$$\mathbf{R}C^* : D^-(Sh_{Nis}(SmCor(k))) \to DM_-^{eff}(k),$$

which is left adjoint to the embedding $DM_-^{eff}(k) \to D^-(Sh_{Nis}(SmCor(k)))$ [**124**, Proposition 3.2.3].

One can also consider the localization $D^-(Sh_{Nis}(SmCor(k)))/Htp$ of the triangulated category $D^-(Sh_{Nis}(SmCor(k)))$ with respect to the thick subcategory Htp generated by the objects $\mathrm{cone}(L(p_X) : L(X \times \mathbb{A}^1) \to L(X))$, where $p_X : X \times \mathbb{A}^1 \to X$ is the projection. Let

$$\alpha : D^-(Sh_{Nis}(SmCor(k))) \to D^-(Sh_{Nis}(SmCor(k)))/Htp$$

be the canonical map. By [**124**, Proposition 3.2.3], the restriction of α to $DM_-^{eff}(k)$ is an equivalence of triangulated categories, and the functor $\mathbf{R}C^*$ descends to give the inverse equivalence.

2.1.5. *The tensor structure.* Voevodsky defines a tensor structure on the triangulated category $DM_-^{eff}(k)$ [**124**, §3.2] by first defining a tensor structure on the sheaf category $Sh_{Nis}(SmCor(k))$. For this, let F be in $Sh_{Nis}(SmCor(k))$. One has the canonical surjection

$$\oplus_{\phi, X} L(X) \to F, \tag{2.1.5.1}$$

where (ϕ, X) runs over elements ϕ of $F(X)$, with X in Sm/k. Applying this operation to the kernel of (2.1.5.1) and repeating gives the canonical resolution $\mathcal{L}(F) \to F$, with each term in $\mathcal{L}(F)$ a direct sum of the representing sheaves $L(X)$. Define $L(X) \otimes L(Y) := L(X \times_k Y)$, so that the tensor product $\mathcal{L}(F) \otimes \mathcal{L}(F')$ is thus defined. This gives the definition of $F \otimes F'$ as

$$F \otimes F' := \mathcal{H}_0(\mathcal{L}(F) \otimes \mathcal{L}(F')),$$

where $\mathcal{H}_0$ is the sheaf homology. This extends to a well-defined operation on $D^-(Sh_{Nis}(SmCor(k)))$, making $D^-(Sh_{Nis}(SmCor(k)))$ into a triangulated tensor category.

The thick subcategory Htp turns out to be a thick *tensor* subcategory [**124**, Lemma 3.2.4], so the triangulated category $D^-(Sh_{Nis}(SmCor(k)))/Htp$ inherits a the structure of a triangulated tensor category from $D^-(Sh_{Nis}(SmCor(k)))$. The equivalence of $D^-(Sh_{Nis}(SmCor(k)))/Htp$ and $DM_-^{eff}(k)$ via the functor $\mathbf{R}C^*$ makes $DM_-^{eff}(k)$ into a triangulated tensor category.

2.1.6. REMARK. One can find representatives in $C^-(Sh_{Nis}(SmCor(k)))$ for the tensor product of objects of $DM_-^{eff}(k)$ by using the functor C^*. If $F_i = C^*(Z_i)$ for objects Z_i of $Sh_{Nis}(SmCor(k))$, $i = 1, \ldots, n$, then $F_1 \otimes \ldots \otimes F_n$ is represented by $C^*(Z_1 \otimes \ldots \otimes Z_n)$. Since, for F in $DM_-^{eff}(k)$, the canonical map $F \to C^*(F)$ in isomorphism, we can also apply C^* as many times as we like, and still get a representative of $F_1 \otimes \ldots \otimes F_n$ in $C^-(Sh_{Nis}(SmCor(k)))$.

2.1.7. *Motives and Hom-objects.* We write $M(X)$, the "motive" of X, for the object $\mathbf{R}C^*(L(X))$ of $DM_-^{eff}(k)$. For X in Sm/k, let $\tilde{L}(X)$ be the kernel of the map $L(p)\colon L(X) \to L(\operatorname{Spec} k)$, where p is the structure morphism, and let $\tilde{M}(X)$ be the object $\mathbf{R}C^*(\tilde{L}(X))$ of $DM_-^{eff}(k)$: the *reduced* motive of X. For $n \geq 1$ we denote $(\tilde{M}(\mathbb{P}^1)[-2])^{\otimes n}$ by $\mathbb{Z}(n)$, and let $\mathbb{Z} = \mathbb{Z}(0) = M(\operatorname{Spec} k)$; for an object F of $DM_-^{eff}(k)$, write $F(n)$ for $F \otimes \mathbb{Z}(n)$.

We write $DM_{gm}^{eff}(k)$ for the strictly full triangulated subcategory of $DM_-^{eff}(k)$ generated by first taking the triangulated subcategory generated by the objects $M(X)$ for X in Sm/k, and then taking the pseudo-abelian hull. This definition differs from Voevodsky's (see [**124**, §2]), but it is an immediate consequence of [**124**, Theorem 3.2.6] that the two definitions agree up to equivalence. In addition, $DM_{gm}^{eff}(k)$ is closed under the tensor operation, so it is a triangulated tensor subcategory of $DM_-^{eff}(k)$.

For objects A and B of $D(Sh_{Nis}(SmCor(k)))$ (the unbounded derived category), we have the internal Hom object $\underline{RHom}(A, B)$. By [**124**, Proposition 3.2.8] $\underline{RHom}(A, B)$ is in $DM_-^{eff}(k)$ if A is in $DM_{gm}^{eff}(k)$ and B is in $DM_-^{eff}(k)$; in this case, denote this object of $DM_-^{eff}(k)$ by $\underline{Hom}_{DM^{eff}}(A, B)$. If both A and B are in $DM_{gm}^{eff}(k)$, then $\underline{Hom}_{DM^{eff}}(A, B \otimes \mathbb{Z}(n))$ is in $DM_{gm}^{eff}(k)$ for all sufficiently large n.

2.1.8. *The category of geometric motives.* We suppose for the remainder of the discussion that k admits resolution of singularities. Let $DM_{gm}(k)$ be the category gotten from $DM_{gm}^{eff}(k)$ by inverting the Tate object $\mathbb{Z}(1)$, i.e., objects are pairs (A, a), with $a \in \mathbb{Z}$, and

$$\operatorname{Hom}_{DM_{gm}(k)}((A, a), (B, b)) = \varinjlim_n \operatorname{Hom}_{DM_{gm}^{eff}(k)}(A \otimes \mathbb{Z}(n + a), B \otimes \mathbb{Z}(n + b)).$$

Denoting (A, a) by $A(a)$, the category $DM_{gm}(k)$ inherits from $DM_{gm}^{eff}(k)$ the structure of a triangulated tensor category, and the natural functor $DM_{gm}^{eff}(k) \to DM_{gm}(k)$ is a full embedding (see [**124**, §2.1, and Theorem 4.3.1]). In addition, for objects A and B of $DM_{gm}(k)$, the internal Hom-object $\underline{Hom}_{DM_{gm}}(A, B)$ exists in $DM_{gm}(k)$ [**124**, Corollary 4.3.5], in the sense that there are natural isomorphisms

$$\operatorname{Hom}_{DM_{gm}}(A, \underline{Hom}_{DM_{gm}}(B, C)) \cong \operatorname{Hom}_{DM_{gm}}(A \otimes B, C)$$

for objects A, B and C of DM_{gm}. In fact, one may take $\underline{Hom}_{DM_{gm}}(A, B)$ to be the object $\underline{Hom}_{DM^{eff}}(A, B \otimes \mathbb{Z}(n)) \otimes \mathbb{Z}(-n)$ for n sufficiently large.

We form the triangulated tensor category $DM_-(k)$ from $DM_-^{eff}(k)$ by inverting $\mathbb{Z}(1)$.

2.1.9. *Motivic cohomology.* For X in Sm/k, define the cohomology $H_V^p(X, \mathbb{Z}(q))$ by

$$H_V^p(X, \mathbb{Z}(q)) := \operatorname{Hom}_{DM_{gm}}(M(X), \mathbb{Z}(q)[p]).$$

Then [**124**, Corollary 4.2.7, Proposition 4.2.9 and Theorem 4.3.7] there are natural isomorphisms

$$H_V^p(X, \mathbb{Z}(q)) \cong \mathrm{CH}^q(X, 2q - p)$$

for X in Sm/k. On the other hand, we have (Chapter II, Theorem 3.6.6) the natural isomorphism

$$H^p(X, \mathbb{Z}(q)) := \operatorname{Hom}_{\mathcal{DM}_k}(\mathbb{Z}, \mathbb{Z}_X(q)[p]) \cong \mathrm{CH}^q(X, 2q - p)$$

for X smooth and quasi-projective over k. Thus, one should expect that sending $\mathbb{Z}_X(q)$ to the internal Hom object

(2.1.9.1) $$\underline{Hom}_{DM_{gm}}(M(X), \mathbb{Z}(q))$$

should extend to an equivalence of $\mathcal{DM}_k$ with $DM_{gm}(k)$. We now proceed to realize this construction.

2.2. A complex of correspondences

We will be interested in understanding the morphisms among the various internal Hom-objects (2.1.9.1). As it is difficult to get a handle on all morphisms, we construct a complex of correspondences which give enough morphisms for our purposes, and which is concrete enough to compute with. In this section, we study this complex of correspondences.

2.2.1. *Quasi-finite cycles.* For F in $Sh_{Nis}(SmCor(k))$, and Y in Sm/k, let F^Y be the presheaf on $SmCor(k)$ with $F^Y(U) := F(Y \times_k U)$. It is immediate that F^Y is a Nisnevic sheaf on Sm/k, hence an object of the category $Sh_{Nis}(SmCor(k))$.

For X in Sm/k, let $L^c(X)$ be the presheaf on Sm/k with $L^c(X)(Y)$ the free abelian group on the set of reduced irreducible subschemes W of $Y \times X$ which are *quasi-finite* over Y and dominate a component of Y; this is in fact an object of $Sh_{Nis}(SmCor(k))$. If X is proper over k, then $L^c(X) = L(X)$.

2.2.2. *Higher correspondences.* Let X_i and Y_i be in Sm/k for $i = 1, 2$. We will describe an explicit subgroup of the group of morphisms $L^c(X_1)^{Y_1} \to L^c(X_2)^{Y_2}$.

Let $c(X_1^{Y_1}, X_2^{Y_2})$ be the free abelian group generated by subvarieties W of $X_1 \times Y_1 \times X_2 \times Y_2$ which satisfy

(2.2.2.1)

1. W is proper over $X_2 \times Y_2$.
2. W is quasi-finite and dominant over a component of $X_1 \times Y_2$.
3. The closure of the projection of W to $Y_1 \times Y_2$ is finite over Y_2.

2.2.3. LEMMA. *(i) Let Z be in $c(X_1^{Y_1}, X_2^{Y_2})$. Then, for U in Sm/k, the correspondence $Z \times \Delta_U$ from $X_1 \times Y_1 \times U$ to $X_2 \times Y_2 \times U$ defines a map*

$$(Z \times \Delta_U)_*: L^c(X_1)^{Y_1}(U) \to L^c(X_2)^{Y_2}(U),$$

giving the map of sheaves with transfer $Z_*: L^c(X_1)^{Y_1} \to L^c(X_2)^{Y_2}$.
(ii) Take Z in $c(X_1^{Y_1}, X_2^{Y_2})$ and W in $c(X_2^{Y_2}, X_3^{Y_3})$. Then the cycle-theoretic intersection product

$$(Z \times X_3 \times Y_3) \cdot (X_1 \times Y_1 \times W)$$

is defined, has support which proper over $X_1 \times Y_1 \times X_3 \times Y_3$, and the projection

$$W \circ Z := p_{X_1 \times Y_1 \times X_3 \times Y_3}(Z \times (X_3 \times Y_3) \cdot (X_1 \times Y_1) \times W)$$

is in $c(X_1^{Y_1}, X_3^{Y_3})$
(iii) With Z and W as in (ii), we have

$$(W \circ Z)_* = W_* \circ Z_*: L^c(X_1)^{Y_1} \to L^c(X_3)^{Y_3}.$$

(iv) If W is in $c(X_1^{Y_1}, X_2^{Y_2})$, and X_1 is irreducible, then each component of W is equi-dimensional over Y_2.

PROOF. Indeed,

$$Z \in c(X_1^{Y_1}, X_2^{Y_2}) \Longrightarrow Z \times \Delta_U \in c(X_1^{Y_1 \times U}, X_2^{Y_2 \times U}),$$

hence we may assume $U = \operatorname{Spec} k$; we may also assume $Z = 1 \cdot W$ for W irreducible. We may assume the X_i and Y_i are irreducible. If $C \subset X_1 \times Y_1$ is irreducible and in $L^c(X_1)^{Y_1}$, then $\dim(C) = \dim(Y_1)$, hence

$$\operatorname{codim}(C \times X_2 \times Y_2) + \operatorname{codim}(W) = \dim X_1 + \dim X_2 + \dim Y_1,$$

using (2.2.2.1)(2). Thus

$$\dim(C \times X_2 \times Y_2 \cap W) \geq \dim(Y_2).$$

On the other hand, take a point y in Y_2. Then, by (2.2.2.1)(3), there is a finite subset S of Y_1 such that

$$C \times X_2 \times y \cap W \subset (X_1 \times S \cap C) \times X_2 \times y.$$

Since C is in $L^c(X_1)(Y_1)$, there is a finite subset S' of X_1 with $X_1 \times S \cap C \subset S' \times S$, hence

$$C \times X_2 \times y \cap W \subset S' \times Y_1 \times X_2 \times y \cap W.$$

Since W is quasi-finite over $X_1 \times Y_2$, $C \times X_2 \times y \cap W$ is finite. Thus

$$\dim(C \times X_2 \times Y_2 \cap W) = \dim(p_{Y_2}(C \times X_2 \times Y_2 \cap W)) \leq \dim(Y_2),$$

or $C \times X_2 \times Y_2 \cap W = \emptyset$, and thus the intersection product $C \times X_2 \times Y_2 \cdot W$ is defined. Since W is proper over $X_2 \times Y_2$, the cycle

$$W_*(C) := p_{X_2 \times Y_2 *}(C \times X_2 \times Y_2 \cap W)$$

is defined.

In addition, $p_{X_2 \times Y_2}(C \times X_2 \times Y_2 \cap W)$ is quasi-finite over Y_2. Since the dimension of each component of $W_*(C)$ and Y_2 are the same, each component of $W_*(C)$ must dominate a component of Y_2, hence $W_*(C)$ is in $L^c(X_2)^{Y_2}$.

For (ii), one verifies (using e.g. the valuative criterion for properness) that for $X \subset A \times B$, $Y \subset B \times C$ with X proper over B and Y proper over C, that $X \times C \cap A \times Y$ is proper over C. Thus $|Z| \times (X_3 \times Y_3) \cap (X_1 \times Y_1) \times |W|$ is proper over $X_3 \times Y_3$, hence also proper over $X_1 \times Y_1 \times X_3 \times Y_3$.

Now fix $y_3 \in Y_3$ and $x_1 \in X_1$. By (2.2.2.1)(3), there is a finite set $S_2 \subset Y_2$ such that

$$|W| \cap X_2 \times Y_2 \times X_3 \times y_3 \subset X_2 \times S_2 \times X_3 \times y_3,$$

and there is a finite set $S_1 \subset Y_1$ such that

$$|Z| \cap (X_1 \times Y_1 \times X_2 \times S_2) \subset X_1 \times S_1 \times X_2 \times S_2.$$

By (2.2.2.1)(2), there is a finite set $T_2 \subset X_2$ such that

$$|Z| \cap (x_1 \times Y_1 \times X_2 \times S_2) \subset x_1 \times S_1 \times T_2 \times S_2,$$

and a finite subset $T_3 \subset X_3$ such that

$$\begin{aligned}[][|Z| \times (X_3 \times Y_3) \cap (X_1 \times Y_1) \times |W|] \cap x_1 \times Y_1 \times X_2 \times Y_2 \times X_3 \times y_3 \\ \subset x_1 \times S_1 \times T_2 \times S_2 \times T_3 \times y_3,\end{aligned}$$

i.e., $|Z| \times (X_3 \times Y_3) \cap (X_1 \times Y_1) \times |W|$ is quasi-finite over $X_1 \times Y_3$. Counting dimensions as in (i), this implies that the intersection product in (ii) is defined and that $W \circ Z$ has support which is quasi-finite over $X_1 \times Y_3$.

Let $\bar{Z}$ be the closure of the projection of $|Z|$ to $Y_1 \times Y_2$, and define $\bar{W} \subset Y_2 \times Y_3$ similarly. Since $|Z|$ and $|W|$ are finite over Y_2 and Y_3, the intersection $Y_1 \times |W| \cap |Z| \times Y_3$ is defined and is finite over Y_3. Letting $\overline{W \circ Z}$ be the closure of the projection of $|W \circ Z|$ to $Y_1 \times Y_3$, we clearly have

$$\overline{W \circ Z} \subset p_{13}(Y_1 \times |W| \cap |Z| \times Y_3),$$

hence $\overline{W \circ Z}$ is finite over Y_3, completing the proof of (ii).

The assertion (iii) follows from the fact that, for the action of correspondences on cycles, correspondence product is compatible with composition .

For (iv), take $y \in Y_2$, let W be a subvariety of $X_1 \times Y_1 \times X_2 \times Y_2$ which is quasi-finite and dominant over a component of $X_1 \times Y_2$, and assume X_1 is irreducible. Then $W \cap (X_1 \times X_2 \times Y_1 \times y)$ is empty or is quasi-finite and dominant over X_1. Thus W is equi-dimensional of dimension $\dim X_1$ over Y_2. □

2.2.4. Sending W to W_* gives us the injective map

$$Tr: c(X_1^{Y_1}, X_2^{Y_2}) \to \mathrm{Hom}_{Sh_{Nis}(SmCor(k))}(L^c(X_1)^{Y_1}, L^c(X_2)^{Y_2}).$$

By Lemma 2.2.3 we may define an additive subcategory $Sh^{Tr}_{Nis}(SmCor(k))$ of $Sh_{Nis}(SmCor(k))$ with objects the direct sums of sheaves $L^c(X)^Y$, and morphisms given as the image of Tr.

For example, the map $L^c(f)^g: L^c(X)^Y \to L^c(X')^{Y'}$ associated to a proper map $f: X \to X'$ and a $g \in c(Y', Y)$ is in the subcategory $Sh^{Tr}_{Nis}(SmCor(k))$.

2.2.5. *Operations.* The groups $c(X_1^{Y_1}, X_2^{Y_2})$ admit the following operations:
(1) **Push-forward in $\mathbf{Y_1}$.** Let $f: Y_1 \to Y_1'$ be a morphism in Sm/k, and take W in $X_1 \times Y_1 \times X_2 \times Y_2$ satisfying (2.2.2.1)(1)-(3). As W is proper over $X_2 \times Y_2$, W is also proper over $X_1 \times Y_1' \times X_2 \times Y_2$, hence the cycle $f_*(W)$ is defined. It is easy to check that this gives

$$f_*: c(X_1^{Y_1}, X_2^{Y_2}) \to c(X_1^{Y_1'}, X_2^{Y_2}),$$

with $(fg)_* = f_* g_*$.
(2) **Pull-back in $\mathbf{Y_2}$.** Let g be in $c(Y_2', Y_2)$, giving the correspondence

$$G := \Delta_{X_1} \times \Delta_{X_2} \times \Delta_{Y_1} \times g$$

from $X_1 \times X_2 \times Y_1 \times Y_2$ to $X_1 \times X_2 \times Y_1 \times Y_2'$. It follows from Lemma 2.2.3(iv) that $G_*(W)$ is defined for all W in $c(X_1^{Y_1}, X_2^{Y_2})$; it is easy to check that $G_*(W)$ is in fact in $c(X_1^{Y_1}, X_2^{Y_2'})$, giving the homomorphism

$$g^*: c(X_1^{Y_1}, X_2^{Y_2}) \to c(X_1^{Y_1}, X_2^{Y_2'}),$$

with $(g \circ f)^* = f^* \circ g^*$.
(3) **Products.** Sending W to $W \times \Delta_Y$ (and making the appropriate exchange of factors) defines

$$^{\times Y}: c(X_1^{Y_1}, X_2^{Y_2}) \to c(X_1^{Y_1 \times Y}, X_2^{Y_2 \times Y}).$$

This operation is compatible, via Tr, with the map

$$\mathrm{Hom}_{Sh_{Nis}(SmCor(k))}(L^c(X_1)^{Y_1}, L^c(X_2)^{Y_2}) \to \mathrm{Hom}_{Sh_{Nis}(SmCor(k))}(L^c(X_1)^{Y_1 \times Y}, L^c(X_2)^{Y_2 \times Y})$$

gotten by restricting the presheaves $L^c(X_1)^{Y_1}$ and $L^c(X_2)^{Y_2}$ to the k-schemes of the form $Y \times U$, U in Sm/k.

Similarly, sending W to $W \times \Delta_X$ (and making the appropriate exchange of factors) defines

$$\times X : c(X_1^{Y_1}, X_2^{Y_2}) \to c((X_1 \times X)^{Y_1 \times Y}, (X_2 \times X)^{Y_2 \times Y}).$$

We have

$$f_* \circ g^* = g^* \circ f_*.$$

In addition, the maps f_* and g^* are compatible, via Tr, with the maps

$$\mathrm{Hom}_{Sh_{Nis}(SmCor(k))}(L^c(X_1)^{Y_1}, L^c(X_2)^{Y_2}) \xrightarrow{f_*} \mathrm{Hom}_{Sh_{Nis}(SmCor(k))}(L^c(X_1)^{Y_1}, L^c(X_2)^{Y_2'}),$$

$$\mathrm{Hom}_{Sh_{Nis}(SmCor(k))}(L^c(X_1)^{Y_1}, L^c(X_2)^{Y_2}) \xrightarrow{g^*} \mathrm{Hom}_{Sh_{Nis}(SmCor(k))}(L^c(X_1)^{Y_1}, L^c(X_2)^{Y_2'})$$

defined using the functoriality in $L^c(-)^-$.

2.2.6. REMARK. Let X_i, Y_i, Z_i, W_i, $i = 1, 2$ be in Sm/k. Taking the "product" of cycles defines the map

$$\otimes : c(X_1^{Y_1}, X_2^{Y_2}) \otimes c(Z_1^{W_1}, Z_2^{W_2}) \to c([X_1 \times Z_1]^{Y_1 \times W_1}, [X_2 \times Z_2]^{Y_2 \times W_2}).$$

Similarly, the exchange of factors gives the isomorphism

$$\tau : c([X_1 \times Z_1]^{Y_1 \times W_1}, [X_2 \times Z_2]^{Y_2 \times W_2}) \to c([Z_1 \times X_1]^{W_1 \times Y_1}, [Z_2 \times X_2]^{W_2 \times Z_2}).$$

One checks that, via Tr, this gives $Sh^{Tr}_{Nis}(SmCor(k))$ the structure of a tensor category.

2.2.7. *The Hom-complex.* Let X_i, Y_i, $i = 1, 2$ be in Sm/k. Applying the constructions of §2.2.1 to $(X_1, Y_1 \times \Delta^*, X_2, Y_2 \times \Delta^*)$, and using the functoriality described in §2.2.5 gives the complex $c(X_1^{Y_1 \times \Delta^*}, X_2^{Y_2 \times \Delta^*})$, defined as the extended total complex associated to the functor from $\Delta \times \Delta^{\mathrm{op}}$ to $\mathbf{Ab}$:

$$([n], [m]) \mapsto c(X_1^{Y_1 \times \Delta^n}, X_2^{Y_2 \times \Delta^m})$$

(see Part II, Chapter III, §2.1.1). We also have the map of complexes

$$Tr : c(X_1^{Y_1 \times \Delta^*}, X_2^{Y_2 \times \Delta^*}) \to \mathrm{Hom}_{Sh_{Nis}(SmCor(k))}(L^c(X_1)^{Y_1 \times \Delta^*}, L^c(X_2)^{Y_2 \times \Delta^*}).$$

2.2.8. LEMMA. *Let X_i, Y_i be in Sm/k. Then the map*

$$\times \mathbb{A}^1 : c(X_1^{Y_1 \times \Delta^*}, X_2^{Y_2 \times \Delta^*}) \to c((X_1 \times \mathbb{A}^1)^{Y_1 \times \Delta^*}, (X_2 \times \mathbb{A}^1)^{Y_2 \times \Delta^*})$$

is a homotopy equivalence.

PROOF. Let

$$[(X_1 \times \mathbb{A}^1) \times Z_1 \times (X_2 \times \mathbb{A}^1) \times Z_2]_\Delta \subset (X_1 \times \mathbb{A}^1) \times Z_1 \times (X_2 \times \mathbb{A}^1) \times Z_2$$

be the closed subscheme with points of the form $(x_1, t, z_1, x_2, t, z_2)$, and let

$$c((X_1 \times \mathbb{A}^1)^{Z_1}, (X_2 \times \mathbb{A}^1)^{Z_2})_\Delta$$

be the subgroup of $c((X_1 \times \mathbb{A}^1)^{Z_1}, (X_2 \times \mathbb{A}^1)^{Z_2})$ consisting of those cycles supported on $[(X_1 \times \mathbb{A}^1) \times Z_1 \times (X_2 \times \mathbb{A}^1) \times Z_2]_\Delta$. The map

$$\times \mathbb{A}^1 : c(X_1^{Z_1}, X_2^{Z_2}) \to c((X_1 \times \mathbb{A}^1)^{Z_1}, (X_2 \times \mathbb{A}^1)^{Z_2}) \tag{2.2.8.1}$$

of §2.2.5 sends $c(X_1^{Z_1}, X_2^{Z_2})$ into $c((X_1 \times \mathbb{A}^1)^{Z_1}, (X_2 \times \mathbb{A}^1)^{Z_2})_\Delta$.

Taking $Z_i = Y_i \times \Delta^*$, we first show that the inclusion of complexes

$$c((X_1 \times \mathbb{A}^1)^{Y_1 \times \Delta^*}, (X_2 \times \mathbb{A}^1)^{Y_2 \times \Delta^*})_\Delta \xrightarrow{i} c((X_1 \times \mathbb{A}^1)^{Y_1 \times \Delta^*}, (X_2 \times \mathbb{A}^1)^{Y_2 \times \Delta^*})$$

is a homotopy equivalence.

Let

$$(X_1 \times \mathbb{A}^1) \times Z_1 \times (X_2 \times \mathbb{A}^1) \times (Z_2 \times \mathbb{A}^1) \xrightarrow{\phi} (X_1 \times \mathbb{A}^1) \times Z_1 \times (X_2 \times \mathbb{A}^1) \times (Z_2 \times \mathbb{A}^1)$$

be the map

$$(x_1, s_1, z_1, x_2, s_2, z_2, t) \mapsto (x_1, t(s_1 - s_2) + s_2, z_1, x_2, s_2, z_2, t).$$

For each cycle W on $(X_1 \times \mathbb{A}^1) \times Z_1 \times (X_2 \times \mathbb{A}^1) \times (Z_2 \times \mathbb{A}^1)$ which is finite over $X_1 \times Z_1 \times (X_2 \times \mathbb{A}^1) \times (Z_2 \times \mathbb{A}^1)$, the pushforward $\phi_*(W)$ is defined. One easily checks that this defines

$$\phi_*: c((X_1 \times \mathbb{A}^1)^{Z_1}, (X_2 \times \mathbb{A}^1)^{Z_2 \times \mathbb{A}^1}) \to c((X_1 \times \mathbb{A}^1)^{Z_1}, (X_2 \times \mathbb{A}^1)^{Z_2 \times \mathbb{A}^1}).$$

Letting $p: Z_2 \times \mathbb{A}^1 \to Z_2$ be the projection, we define the map

$$\psi: c((X_1 \times \mathbb{A}^1)^{Z_1}, (X_2 \times \mathbb{A}^1)^{Z_2}) \to c((X_1 \times \mathbb{A}^1)^{Z_1}, (X_2 \times \mathbb{A}^1)^{Z_2 \times \mathbb{A}^1})$$

by $\psi = \phi_* \circ p^*$. Letting $i_0: Z_2 \to Z_2 \times \mathbb{A}^1$ and $i_1: Z_2 \to Z_2 \times \mathbb{A}^1$ be the 0-section and 1-section, we then have

$$i_1^* \circ \psi = \mathrm{id}.$$

In addition,

$$i_0^* \circ \psi(c((X_1 \times \mathbb{A}^1)^{Z_1}, (X_2 \times \mathbb{A}^1)^{Z_2})) \subset c((X_1 \times \mathbb{A}^1)^{Z_1}, (X_2 \times \mathbb{A}^1)^{Z_2})_\Delta,$$

and the restriction of $i_0^* \circ \psi$ to $c((X_1 \times \mathbb{A}^1)^{Z_1}, (X_2 \times \mathbb{A}^1)^{Z_2})_\Delta$ is the inclusion i.

Taking Z_i to be the cosimplicial scheme $Y_i \times \Delta^*$ gives the map

$$c((X_1 \times \mathbb{A}^1)^{Y_1 \times \Delta^*}, (X_2 \times \mathbb{A}^1)^{Y_2 \times \Delta^*}) \xrightarrow{\psi_*} c((X_1 \times \mathbb{A}^1)^{Y_1 \times \Delta^*}, (X_2 \times \mathbb{A}^1)^{Y_2 \times \Delta^* \times \mathbb{A}^1}).$$

Now, if $F: Sm/k^{\mathrm{op}} \to \mathbf{C}^+(\mathbf{Ab})$ is a functor, let $C^*(F): Sm/k^{\mathrm{op}} \to \mathbf{C}(\mathbf{Ab})$ be the functor given by defining $C^*(F)(Z)$ to be the extended total complex associated to the simplicial object $F(Z \times \Delta^*)$. Then the map $p^*: C^*(F) \to C^*(F \times \mathbb{A}^1)$ is a homotopy equivalence. Indeed, one applies F to the standard triangulation of $\Delta^n \times \mathbb{A}^1$, $n = 0, 1, \ldots$ (see Chapter II, §3.5.7), to show that $i_0^*: C^*(F \times \mathbb{A}^1) \to C^*(F)$ is a homotopy inverse to p^*, where i_0 is the zero section.

In particular, the maps

$$c((X_1 \times \mathbb{A}^1)^{Y_1 \times \Delta^*}, (X_2 \times \mathbb{A}^1)^{Y_2 \times \Delta^*}) \xrightarrow{i_0^* \circ \psi_*, i_1^* \circ \psi_*} c((X_1 \times \mathbb{A}^1)^{Y_1 \times \Delta^*}, (X_2 \times \mathbb{A}^1)^{Y_2 \times \Delta^*})$$

are homotopic; the properties of $i_0^* \circ \psi_*$, $i_1^* \circ \psi_*$ described above thus show that the inclusion i is a homotopy equivalence.

We now show that the map (2.2.8.1) is a homotopy equivalence. Let

$$[(X_1 \times \mathbb{A}^1) \times Z_1 \times (X_2 \times \mathbb{A}^1) \times (Z_2 \times \mathbb{A}^1)]_\Delta \xrightarrow{\pi} [(X_1 \times \mathbb{A}^1) \times Z_1 \times (X_2 \times \mathbb{A}^1) \times (Z_2 \times \mathbb{A}^1)]_\Delta$$

be the map

$$(x_1, s, z_1, x_2, s, z_2, t) \mapsto (x_1, ts, z_1, x_2, ts, z_2, t).$$

The pullback $\pi^*(W)$ is defined for each cycle W on $[(X_1 \times \mathbb{A}^1) \times Z_1 \times (X_2 \times \mathbb{A}^1) \times (Z_2 \times \mathbb{A}^1)]_\Delta$ which intersects $(X_1 \times 0) \times Z_1 \times (X_2 \times 0) \times (Z_2 \times \mathbb{A}^1)$ properly.

Let W be in $c((X_1 \times \mathbb{A}^1)^{Z_1}, (X_2 \times \mathbb{A}^1)^{Z_2 \times \mathbb{A}^1})$. By (2.2.2.1)(2), W intersects $(X_1 \times 0) \times Z_1 \times (X_2 \times \mathbb{A}^1) \times (Z_2 \times \mathbb{A}^1)$ properly on $(X_1 \times \mathbb{A}^1) \times Z_1 \times (X_2 \times \mathbb{A}^1) \times (Z_2 \times \mathbb{A}^1)$. Thus, if W is in $c((X_1 \times \mathbb{A}^1)^{Z_1}, (X_2 \times \mathbb{A}^1)^{Z_2 \times \mathbb{A}^1})_\Delta$, W intersects

$$(X_1 \times 0) \times Z_1 \times (X_2 \times 0) \times (Z_2 \times \mathbb{A}^1)$$

properly on

$$[(X_1 \times \mathbb{A}^1) \times Z_1 \times (X_2 \times \mathbb{A}^1) \times (Z_2 \times \mathbb{A}^1)]_\Delta,$$

hence the cycle $\pi^*(W)$ is defined for all $W \in c((X_1 \times \mathbb{A}^1)^{Z_1}, (X_2 \times \mathbb{A}^1)^{Z_2 \times \mathbb{A}^1})_\Delta$. The properties of (2.2.2.1) are preserved by taking π^*, hence we have the pull-back homomorphism

$$\pi^*: c((X_1 \times \mathbb{A}^1)^{Z_1}, (X_2 \times \mathbb{A}^1)^{Z_2 \times \mathbb{A}^1})_\Delta \to c((X_1 \times \mathbb{A}^1)^{Z_1}, (X_2 \times \mathbb{A}^1)^{Z_2 \times \mathbb{A}^1})_\Delta.$$

Define the map

$$\xi: c((X_1 \times \mathbb{A}^1)^{Z_1}, (X_2 \times \mathbb{A}^1)^{Z_2}) \to c((X_1 \times \mathbb{A}^1)^{Z_1}, (X_2 \times \mathbb{A}^1)^{Z_2 \times \mathbb{A}^1})_\Delta$$

by $\xi = \pi^* \circ p^*$.

We have $i_1^* \circ \xi = \mathrm{id}$. In addition,

$$i_0^* \circ \pi^* \big(c((X_1 \times \mathbb{A}^1)^{Z_1}, (X_2 \times \mathbb{A}^1)^{Z_2})_\Delta\big) \subset \times\mathbb{A}^1(c(X_1^{Z_1}, X_2^{Z_2})),$$

and the restriction of $i_0^* \circ \pi^*$ to $\times\mathbb{A}^1(c(X_1^{Z_1}, X_2^{Z_2}))$ is the inclusion

$$i' : \times\mathbb{A}^1(c(X_1^{Z_1}, X_2^{Z_2})) \to c((X_1 \times \mathbb{A}^1)^{Z_1}, (X_2 \times \mathbb{A}^1)^{Z_2})_\Delta.$$

As above, if we take $Z_i = Y_i \times \Delta^*$, $i = 1, 2$, this shows that i' is a homotopy equivalence. Since the map

$$\times\mathbb{A}^1: c(X_1^{Y_1 \times \Delta^*}, X_2^{Y_2 \times \Delta^*}) \to c((X_1 \times \mathbb{A}^1)^{Y_1 \times \Delta^*}, (X_2 \times \mathbb{A}^1)^{Y_2 \times \Delta^*})_\Delta$$

is injective, the lemma is proved. □

2.2.9. *Multiplicative structure.* For $f \in L(V')(V)$ and $s \in L^c(X)(Y \times V')$, we have the map

$$\phi_{f,s}: L(V) \to \oplus_{U \in Sm/k, s \in L^c(X)(Y \times U)} L(U),$$

defined by taking the map $L(f): L(V) \to L(V')$ and including $L(V')$ as the summand corresponding to $(V', s \in L^c(X)(Y \times U))$.

For X, Y, V in Sm/k, let

$$\mathcal{R}(X,Y,V) := \bigcup_{n=1}^{\infty}\{(f_1,\dots,f_n;s_1,\dots,s_n) \mid f_i \in L(U_i)(V), s_i \in L^c(X)^Y(U_i), \\ U_i \in Sm/k,\ i=1,\dots,n, \\ \text{and } \sum_i f_i^*(s_i)=0\}.$$

For $(f_1,\dots,f_n;s_1,\dots,s_n) \in \mathcal{R}(X,Y,V)$, we have the map

$$\phi_{f_*,s_*}: L(V) \to \oplus_{U,s\in L^c(X)(Y\times U)} L(U)$$

by taking the sum of the maps ϕ_{f_i,s_i}.

The resolution $\mathcal{L}(L^c(X)^Y)$ of §2.1.5 starts out as

$$\bigoplus_{\substack{V\in Sm/k\\(f_*;s_*)\in\mathcal{R}(X,Y,V)}} L(V) \xrightarrow{\oplus\phi_{f_*,s_*}} \bigoplus_{U,s\in L^c(X)(Y\times U)} L(U) \to L^c(X)^Y.$$

This gives the presentation of $L^c(X)^Y \otimes L^c(X')^{Y'}$ as

$$\bigoplus_{\substack{V,U'\in Sm/k\\(f_*;s_*)\in\mathcal{R}(X,Y,V)}} L(V\times U') \oplus \bigoplus_{\substack{V',U\in Sm/k\\(f'_*;s'_*)\in\mathcal{R}(X',Y',V')}} L(V'\times U)$$

$$\longrightarrow \bigoplus_{\substack{U,U'\in Sm/k\\ s\in L^c(X)(Y\times U), s'\in L^c(X')(Y'\times U')}} L(U\times U')$$

$$\longrightarrow L^c(X)^Y \otimes L^c(X')^{Y'}.$$

Now take V, U, U' in Sm/k, $s \in L^c(X)(Y\times U)$, $s' \in L^c(X')(Y'\times U')$, and $t \in L(U\times U')(V)$. We consider s as a cycle on $X\times Y\times U$, s' as a cycle on $X'\times Y'\times U'$, giving the "product" cycle $s\times' s' \in L^c(X\times X')^{Y\times Y'}(U\times U')$. This give the cycle $L(t)(s\times' s') \in L^c(X\times X')^{Y\times Y'}(V)$. One checks that this operation respects the relations defining $L^c(X)^Y \otimes L^c(X')^{Y'}$, and gives the map of sheaves with transfer

$$\boxtimes^{Y,Y'}_{X,X'}: L^c(X)^Y \otimes L^c(X')^{Y'} \to L^c(X\times X')^{Y\times Y'}.$$

The maps $\boxtimes^{Y,Y'}_{X,X'}$ are natural in X, X', Y, Y', and satisfy the obvious associativity and commutativity conditions.

In addition, the maps $\boxtimes^{Y,Y'}_{X,X'}$ define a commutative external product (see Part II, Chapter I, §2.4) for the inclusion functor $Sh^{Tr}_{Nis}(SmCor(k)) \to Sh_{Nis}(SmCor(k))$.

2.3. Representing complexes

We now describe some good representatives in $C^-(Sh_{Nis}(SmCor(k)))$ for the internal Hom-object (2.1.9.1).

2.3.1. For X, Y in Sm/k, we write $\mathfrak{z}(Y,X)$ for the complex of sheaves with transfer $C^*(L^c(X)^Y)$. Explicitly, $\mathfrak{z}(Y,X)^{-n}$ is the sheaf defined by

$$\mathfrak{z}(Y,X)^{-n}(U) = L^c(X)(U\times Y\times\Delta^n);$$

the cosimplicial structure on Δ^* giving the differential for the complex $\mathfrak{z}(Y,X)$. More generally, let $\mathfrak{z}^m(Y,X)$ denote the total complex associated to the m-simplicial object

$$(n_1,\dots,n_m)\mapsto L^c(X)^{Y\times\Delta^{n_1}\times\ldots\times\Delta^{n_m}}.$$

The external products $\boxtimes^{*,*}_{X,X'}$ then give the map of complexes

$$\boxtimes^{Y,Y';n,n'}_{X,X'}:\mathfrak{z}^n(Y,X)\otimes\mathfrak{z}^{n'}(Y',X')\to\mathfrak{z}^{n+n'}(Y\times Y',X\times X'). \tag{2.3.1.1}$$

For Y in Sm/k we write

$$C^*(Y):=C^*(L(Y)),$$
$$C^*_c(Y):=C^*(L^c(Y)).$$

The complex $\mathfrak{z}(Y,X)$ is the complex $\mathfrak{z}(Y,X,0)$ of [**124**] (with cohomological notation rather than the homological notation of [**124**]).

We record the following properties of $\mathfrak{z}^n(Y,X)$:

(2.3.1.2)

(1) For an integer $n\geq 0$, let δ_n be the diagonal in $\Delta^n\times\Delta^n$. For $y\in L(Y)(\Delta^m)$, the cycle $y\times\delta_n$ defines an element $y(n)$ of $L(Y\times\Delta^n)(\Delta^m\times\Delta^n)$. For $x\in L^c(X)(Y\times\Delta^n)$, we have the element $x\circ y(n)$ of $L^c(X)(\Delta^m\times\Delta^n)$. Sending y to $x\circ y(n)$ gives the map $\phi_{n,m}(x): L(Y)(\Delta^m)\to L^c(X)(\Delta^m\times\Delta^n)$. As $\mathfrak{z}(Y,X)^{-n}(\mathrm{Spec}\,k)=L^c(X)(Y\times\Delta^n)$, sending x to $\phi_{n,m}(x)$ gives the map of simplicial abelian groups

$$\phi_{Y,X}(k): L^c(X)(Y\times\Delta^*)\to\mathrm{Hom}_{\mathbf{C}^-(Sh_{Nis}(SmCor(k)))}(C^*(Y),C^*_c(X)(\Delta^*)),$$

and thus the map

$$H^i(\phi_{Y,X}(k)): H^i(\mathfrak{z}(Y,X)(\mathrm{Spec}\,k))\to\mathrm{Hom}_{DM^{eff}}(C^*(Y),C^*_c(X)[i]).$$

By [**43**, Theorem 8.1] (for $r=0$) and [**124**, Theorem 4.1.10], $H^i(\phi_{Y,X}(k))$ is an isomorphism.

(2) Replacing Δ^m with $U\times\Delta^m$ gives the map

$$L^c(X)(Y\times U\times\Delta^*)\xrightarrow{\phi_{Y,X}(U)}\mathrm{Hom}_{\mathbf{C}^-(Sh_{Nis}(SmCor(k)))}(C^*(Y)(U),C^*_c(X)(U\times\Delta^*))$$

defined as in (1), which induces the map in $DM^{eff}_-(k)$

$$\phi_{Y,X}:\mathfrak{z}(Y,X)\to\underline{Hom}_{DM^{eff}}(C^*(Y),C^*_c(X)).$$

By [**124**, Corollary 4.2.7], $\phi_{Y,X}$ is an isomorphism. Similarly, we have the natural isomorphism

$$\phi_{Y,X,n}:\mathfrak{z}^n(Y,X)\to\underline{Hom}_{DM^{eff}}((C^*)^n(Y),C^*_c(X)),$$

where $(C^*)^n(Y)$ is gotten by applying C^* n times to $L(Y)$. As restriction map

$$\underline{Hom}_{DM^{eff}}((C^*)^n(Y),C^*_c(X))\to\underline{Hom}_{DM^{eff}}(C^*(Y),C^*_c(X))$$

is an isomorphism, the restriction to Δ^* in the first factor gives the isomorphism $\mathfrak{z}^n(Y,X)\to\mathfrak{z}(Y,X)$ in DM^{eff}_-.

Taking $X=\mathbb{A}^m$ and applying [**124**, Corollary 4.1.8] gives the isomorphisms $\mathfrak{z}^n(Y,\mathbb{A}^m)\to\underline{Hom}_{DM^{eff}}(C^*(Y),\mathbb{Z}(m)[2m])$. By [**124**, Corollary 4.3.6], the internal

Hom $\underline{Hom}_{DM_{gm}}(M(Y),\mathbb{Z}(m))$ is in DM_{-}^{eff} for $m \geq \dim_k Y$, and is isomorphic to $\underline{Hom}_{DM^{eff}}(C^*(Y),\mathbb{Z}(m))$. Thus, we get the natural isomorphism

$$\phi_{Y,m,n}\colon \mathfrak{z}^n(Y,\mathbb{A}^m) \to \underline{Hom}_{DM_{gm}}(M(Y),\mathbb{Z}(m)[2m])$$

for all $m \geq \dim_k Y$.

(3) The maps $\phi_{Y,m,n}$ are compatible with products, i.e., the diagram

$$\begin{array}{ccc}
\mathfrak{z}^n(Y,\mathbb{A}^m)\otimes\mathfrak{z}^{n'}(Y',\mathbb{A}^{m'}) & \xrightarrow{\phi_{Y,m,n}\otimes\phi_{Y',m',n'}} & \underline{Hom}_{DM_{gm}}(M(Y),\mathbb{Z}(m)[2m])\otimes\underline{Hom}_{DM_{gm}}(M(Y'),\mathbb{Z}(m')[2m']) \\
\downarrow \boxtimes & & \downarrow \boxtimes \\
\mathfrak{z}^{n+n'}(Y\times Y',\mathbb{A}^{m+m'}) & \xrightarrow{\phi_{Y\times Y',m+m',n+n'}} & \underline{Hom}_{DM_{gm}}(M(Y\times Y'),\mathbb{Z}(m+m')[2(m+m')])
\end{array}$$

commutes; here the tensor product is the tensor product in DM_{-}^{eff} induced from that of $D^-(Sh_{Nis}(SmCor(k)))$ via the localizing functor $\mathbf{R}C^*$. Since the product on the right-hand side is an isomorphism, so is the product

$$\boxtimes\colon \mathfrak{z}^n(Y,\mathbb{A}^m)\otimes\mathfrak{z}^{n'}(Y',\mathbb{A}^{m'}) \to \mathfrak{z}^{n+n'}(Y\times Y',\mathbb{A}^{m+m'})$$

(in DM_{-}^{eff}) for $m \geq \dim_k Y$ and $m' \geq \dim_k Y'$.

2.4. Homotopy commutativity

The technically most difficult part of defining our equivalence comes from the lack of commutativity in the product for our representing complexes $\mathfrak{z}(X,\mathbb{A}^m)$. We now have the tools to deal with this problem.

2.4.1. For an additive category $\mathcal{A}$, let $\mathrm{Gr}_+C^-\mathcal{A}$ denote the category of non-negatively graded, uniformly bounded below complexes. We use the constructions and notations of Part II, Chapter I, §2.3.

We proceed to define a functor

$$\Phi_0\colon \Omega_0 \to \mathrm{Gr}_+C^-(Sh^{Tr}_{Nis}(SmCor(k))). \tag{2.4.1.1}$$

On objects, Φ_0 is given by

$$\Phi_0(n) = \bigoplus_{\substack{(m_1,\dots,m_n)\\ m_i\geq 0}} \mathfrak{z}^n(\mathrm{Spec}\,k,\mathbb{A}^{\Sigma_i m_i})$$

with $\sum_i m_i$ giving the grading.

To define Φ_0 on morphisms, we first recall the standard triangulation of $\Delta^{n_1}\times\Delta^{n_2}$. We give $[n_1]\times[n_2]$ the product partial order

$$(a,b)\leq(a',b') \Leftrightarrow a\leq a' \text{ and } b\leq b'.$$

For an injective, order-preserving map $g := (g_1,g_2)\colon [n_1+n_2]\to[n_1]\times[n_2]$, define $\mathrm{sgn}(g)\in\{\pm1\}$ as in the sign of the shuffle permutation determined by g, as in

(Part II, Chapter III, §3.4.5). Then $\sum_{g:[n_1+n_2]\to[n_1]\times[n_2]}\operatorname{sgn}(g)\cdot g$ is the standard triangulation of $\Delta^{n_1}\times\Delta^{n_2}$.

Each order-preserving map $g=(g_1,g_2)\colon[n_1+n_2]\to[n_1]\times[n_2]$ defines the affine-linear map of schemes

$$\Delta(g)\colon\Delta^{n_1+n_2}=\mathbb{A}^{n_1+n_2}\to\Delta^{n_1}\times\Delta^{n_2}=\mathbb{A}^{n_1+n_2}$$

by sending the vertex v_i of $\Delta^{n_1+n_2}$ to the point $(v_{g_1(i)},v_{g_2(i)})$ of $\Delta^{n_1}\times\Delta^{n_2}$.

Let

$$\cup^{n_1,n_2}_{m_1,m_2}\colon L^c(\mathbb{A}^{m_1+m_2})(\Delta^{n_1}\times\Delta^{n_2})\to L^c(\mathbb{A}^{m_1+m_2})(\Delta^{n_1+n_2})$$

be the map $\sum_g\operatorname{sgn}(g)L^c(\mathbb{A}^{m_1+m_2})(\Delta(g))$, where the sum is over the injective order-preserving maps $g\colon[n_1+n_2]\to[n_1]\times[n_2]$. We then let

$$\cup_{m_1,m_2}\colon\mathfrak{z}^2(\operatorname{Spec}k,\mathbb{A}^{m_1+m_2})\to\mathfrak{z}^1(\operatorname{Spec}k,\mathbb{A}^{m_1+m_2})$$

be the cup product, i.e., the product of maps $\cup^{n_1,n_2}_{m_1,m_2}$. Let $\cup\colon\Phi_0(2)\to\Phi_0(1)$ be the product of the maps $\cup_{m_1,m_2}$.

The map $\cup$ are easily seen to be associative, hence, for each ordered surjective map $f\colon\underline{n}\to\underline{m}$, there is a unique morphism $\cup_f\colon\Phi_0(n)\to\Phi_0(m)$ such that

1. $\cup_f\circ\cup_g=\cup_{fg}$,
2. $\cup_{f_{21}}=\cup$,

where $f_{21}\colon\underline{2}\to\underline{1}$ is the unique surjection.

For $\sigma\in S_n$, we have symmetry isomorphism

$$L^c(\mathbb{A}^{m_1}\times\ldots\times\mathbb{A}^{m_n})(\Delta^{a_1}\times\ldots\times\Delta^{a_n})\xrightarrow{\tau^{m_1,\ldots,m_n}_{\sigma;a_1,\ldots,a_n}}L^c(\mathbb{A}^{m'_1}\times\ldots\times\mathbb{A}^{m'_n})(\Delta^{a'_1}\times\ldots\times\Delta^{a'_n})$$

gotten by permuting the factors $\mathbb{A}^{m_i}$ via σ and the factors Δ^{a_i} via σ^{-1}. Twisting $\tau^{m_1,\ldots,m_n}_{\sigma;a_1,\ldots,a_n}$ by the weighted sign map (with weights $(a_1,\ldots,a_n)$) gives the symmetry isomorphism $\tau_\sigma\colon\Phi_0(n)\to\Phi_0(n)$. One easily checks that $d\tau_\sigma=0$.

If we now send a morphism $(f,\sigma)\colon\underline{n}\to\underline{m}$ in Ω_0 to the composition $\cup_f\circ\tau_\sigma$, we have a well-defined functor Φ_0.

2.4.2. LEMMA. *(i) Let σ be in S_n. Then the map $\tau_\sigma\colon\Phi_0(n)\to\Phi_0(n)$ is homotopic to the identity.*
(ii) Each map in $\mathrm{Gr}_+C^-(Sh^{Tr}_{Nis}(SmCor(k)))$ of non-zero degree, $\psi\colon\Phi_0(n)\to\Phi_0(n)$, with $d\psi=0$, is homotopic to zero.

PROOF. It follows directly from the definition of the groups $c(X^Y,Z^W)$ (see (2.2.2.1)) that

$$\begin{aligned}c(\operatorname{Spec}k^{\Delta^*},\operatorname{Spec}k^{\Delta^*})&=L(\Delta^*)(\Delta^*)\\&=C^*(\Delta^*).\end{aligned}$$

Since $p_{2*}\colon C^*(\mathbb{A}^1\times X)\to C^*(X)$ is a homotopy equivalence for each X, it follows that the projection $C^*(\Delta^*)\to C^*(\operatorname{Spec}k)$ is a homotopy equivalence. As $L(\operatorname{Spec}k)$ is the constant sheaf $\mathbb{Z}$, we have the homotopy equivalence $c(\operatorname{Spec}k^{\Delta^*},\operatorname{Spec}k^{\Delta^*})\cong\mathbb{Z}$. As the map

$$\times\mathbb{A}^m\colon c(\operatorname{Spec}k^{\Delta^*},\operatorname{Spec}k^{\Delta^*})\to c(\mathbb{A}^{m\Delta^*},\mathbb{A}^{m\Delta^*})$$

is a homotopy equivalence (Lemma 2.2.8), we see that the complex $c(\mathbb{A}^{m\Delta^*}, \mathbb{A}^{m\Delta^*})$ is acyclic in non-zero degrees. An elementary spectral sequence argument then implies that the extended total complex $c(m)_*$ associated to the functor

$$c(m)\colon \Delta^n \times \Delta^{n\mathrm{op}} \to \mathbf{Ab}$$

$$c(m)(a_1, \dots, a_n; b_1, \dots, b_n) = c(\mathbb{A}^{m\Delta^{a_1}\times\Delta^{a_n}}, \mathbb{A}^{m\Delta^{b_1}\times\Delta^{b_n}})$$

is acyclic in negative degrees. As the Hom-complex

$$\mathrm{Hom}_{\mathrm{Gr}_+C^-(Sh^{Tr}_{Nis}(SmCor(k)))}(\Phi_0(n), \Phi_0(n))$$

is a product over $m = 0, 1, 2, \dots$ of direct sums of $c(m)_*$, the assertion (ii) is proved.

For (i), it suffices to prove the case $n = 2$, with $\sigma \in S_2$ the non-trivial permutation. Let $\tau'_\sigma\colon \Phi_0(2) \to \Phi_0(2)$ be the map defined similarly to τ_σ, except that we only permute the factors in $\Delta^{a_1} \times \Delta^{a_2}$. Let

$$W_{a_1,a_2} \subset (\mathbb{A}^1 \times \mathbb{A}^1) \times \Delta^{a_1} \times \Delta^{a_2} \times (\mathbb{A}^1 \times \mathbb{A}^1) \times \Delta^{a_2} \times \Delta^{a_1} \times \mathbb{A}^1$$

be the subvariety with points $((x_1, x_2), (s_1, s_2), t(x_2, x_1) + (1-t)(x_1, x_2), (s_2, s_1), t)$. Let

$$(\mathbb{A}^1 \times \mathbb{A}^1) \times \Delta^{a_1} \times \Delta^{a_2} \times (\mathbb{A}^1 \times \mathbb{A}^1) \times \Delta^{a_2} \times \Delta^{a_1}$$
$$\xrightarrow{i_0, i_1} (\mathbb{A}^1 \times \mathbb{A}^1) \times \Delta^{a_1} \times \Delta^{a_2} \times (\mathbb{A}^1 \times \mathbb{A}^1) \times \Delta^{a_2} \times \Delta^{a_1} \times \mathbb{A}^1$$

be the 0-section and 1-section. One sees immediately that

$$W_{a_1,a_2} \in c(\mathbb{A}^{2\Delta^{a_1}\times\Delta^{a_2}}, \mathbb{A}^{2\Delta^{a_2}\times\Delta^{a_1}\times\mathbb{A}^1}),$$

and that the collection of cycles $i_0^*[(-1)^{a_1a_2}W_{a_1,a_2}]$, $i_1^*[(-1)^{a_1a_2}W_{a_1,a_2}]$, define τ'_σ and τ_σ, respectively. It follows as in the proof of Lemma 2.2.8 that the collection of cycles $(-1)^{a_1a_2}W_{a_1,a_2}$ define a homotopy between τ'_σ and τ_σ.

Set

$$\mathrm{Hom}([p]\coprod[q], [r]\coprod[s]) :=$$
$$\mathrm{Hom}_\Delta([p], [r]) \times \mathrm{Hom}_\Delta([q], [s]) \coprod \mathrm{Hom}_\Delta([p], [s]) \times \mathrm{Hom}_\Delta([q], [r]),$$

giving the functor

$$\mathrm{Hom}(-\coprod-, -\coprod-)\colon \Delta^{\mathrm{op}2} \times \Delta^2 \to \mathbf{Sets}.$$

Let $\mathbb{Z}\mathrm{Hom}(\Delta^* \times \Delta^*, \Delta^* \times \Delta^*)$ denote the extended total complex of the free abelian group on $\mathrm{Hom}(-\coprod-, -\coprod-)$. Sending an element $g \in \mathrm{Hom}([p]\coprod[q], [r]\coprod[s])$ to the graph of the affine linear map

$$\mathbb{A}^2 \times \Delta^p \times \mathbb{A}^2 \times \Delta^q \to \mathbb{A}^2 \times \Delta^r \times \mathbb{A}^2 \times \Delta^s$$

or

$$\mathbb{A}^2 \times \Delta^p \times \mathbb{A}^2 \times \Delta^q \to \mathbb{A}^2 \times \Delta^s \times \mathbb{A}^2 \times \Delta^r$$

given on the vertices by g (and the identity on the factors $\mathbb{A}^2$), gives a map of complexes

$$\mathbb{Z}\mathrm{Hom}(\Delta^* \times \Delta^*, \Delta^* \times \Delta^*) \xrightarrow{\rho} \mathrm{Hom}_{\mathrm{Gr}_+C^-(Sh^{Tr}_{Nis}(SmCor(k)))}(\Phi_0(2), \Phi_0(2)).$$

On the other hand, the collection of symmetry isomorphisms $\sigma_{a,b}\colon [a]\coprod[b] \to [b]\coprod[a]$ gives the element

$$\sigma_{*,*} := [(a, b) \mapsto (-1)^{ab}\sigma_{a,b}] \in Z^0(\mathbb{Z}\mathrm{Hom}(\Delta^* \times \Delta^*, \Delta^* \times \Delta^*)),$$

with $\rho(\sigma_{*,*}) = \tau'_\sigma$. By (Part II, Chapter III, Lemma 2.1.3.1), $\sigma_{*,*}$ is determined up to homotopy by the term $\sigma_{0,0}: [0] \times [0] \to [0] \times [0]$. As $\sigma_{0,0} = \mathrm{id}$, this implies that τ'_σ is homotopic to the identity, proving (i). □

We refer the reader to (Part II, Chapter III, §3.2.1) for the construction of the DG tensor category $\mathbb{Z}\Omega^{\mathfrak{h}}$. Using Lemma 2.4.2, we extend the functor (2.4.1.1) to a DG tensor functor

$$\Phi^{\mathfrak{h}}: \mathbb{Z}\Omega^{\mathfrak{h}} \to \mathrm{Gr}_+ C^-(Sh^{Tr}_{Nis}(SmCor(k)))$$

(recall that $Sh^{Tr}_{Nis}(SmCor(k))$ is a tensor category, via Remark 2.2.6). For this, suppose we have defined $\Phi^{\mathfrak{h}}$ on morphisms $\mathrm{Hom}_{\mathbb{Z}\Omega^{\mathfrak{h}}}(a,b)$ for $1 \le b \le a < n$, so that $\Phi^{\mathfrak{h}}$ satisfies the axioms for a DG tensor functor, when the appropriate operations are defined.

First suppose that $n = 2$. By Lemma 2.4.2(i), we may extend Φ_0 to the complex $\mathrm{Hom}_{\mathbb{Z}\Omega^{\mathfrak{h}}}(2,1)^{*\ge -1}$, and then use Lemma 2.4.2(ii) to extend to all of the complex $\mathrm{Hom}_{\Omega^{\mathfrak{h}}}(2,1)$. Now suppose $n > 2$. The tensor structure on the category of complexes $\mathrm{Gr}_+ C^-(Sh^{Tr}_{Nis}(SmCor(k)))$ gives the extension of $\Phi^{\mathfrak{h}}$ to $\mathrm{Hom}_{\mathbb{Z}\Omega^{\mathfrak{h}}}(n,1)_0$, and we then extend to $\mathrm{Hom}_{\mathbb{Z}\Omega^{\mathfrak{h}}}(n,1)$ by using Lemma 2.4.2. By the construction of $\mathbb{Z}\Omega^h$, this process does indeed define a DG tensor functor.

2.5. The equivalence

2.5.1. *The category $\mathcal{DM}^{\mathfrak{h}}$*. We proceed to define a category $\mathcal{DM}^{\mathfrak{h}}$ equivalent to $\mathcal{DM}(\mathrm{Spec}\, k)$, and an exact tensor functor $\Psi: \mathcal{DM}^{\mathfrak{h}} \to DM_{gm}(k)$.

Let $\mathcal{A}_1^{eff}$ be the full subcategory of $\mathcal{A}_1(\mathbf{Sm}_k)$, with objects direct sums of $\mathbb{Z}_X(m)_f$ with $m \ge 0$. We let $\mathcal{A}_2^{eff}$ be the full tensor subcategory of $\mathcal{A}_2(\mathbf{Sm}_k)$ generated by $\mathcal{A}_1^{eff}$; for $n = 3,4,5$, let $\mathcal{A}_n^{eff}$ be the full DG tensor subcategory of $\mathcal{A}_n(\mathbf{Sm}_k)$ generated by $\mathfrak{e}$ and $\mathcal{A}_2^{eff}$. Let $\mathcal{A}_{\mathrm{mot}}^{eff}$ be the full DG tensor subcategory of $\mathcal{A}_{\mathrm{mot}}(\mathbf{Sm}_k)$ generated by the objects $\mathfrak{e}^{\otimes k} \otimes \mathbb{Z}_X(m)_f$, with $k \ge 0$ and with $\mathbb{Z}_X(m)_f$ in $\mathcal{A}_1^{eff}$.

We begin with the additive functor

$$\phi_1: \mathcal{A}_1^{eff} \to C^-(Sh^{Tr}_{Nis}(SmCor(k))),$$

which send $\mathbb{Z}_X(m)_f$ to $\mathfrak{z}(X, \mathbb{A}^m)[-2m]$. For a map $g: Y \to X$, ϕ_1 sends the morphism $g^*: \mathbb{Z}_X(m)_f \to \mathbb{Z}_Y(m)_{f'}$ to $g^*: \mathfrak{z}(X,\mathbb{A}^m)[-2m] \to \mathfrak{z}(Y,\mathbb{A}^m)[-2m]$.

We now proceed as in Chapter V, §1.3, to define a sequence of DG tensor categories $\mathcal{A}_n^{eff\mathfrak{h}}$, which are homotopy equivalent to the categories $\mathcal{A}_n^{eff}$, and DG tensor functors

$$\phi_n: \mathcal{A}_n^{eff\mathfrak{h}} \to C^-(Sh_{Nis}(SmCor(k))),$$

$n = 2,3,4,5,\mathrm{mot}$.

The category $\mathcal{A}_2^{eff\mathfrak{h}}$ is $(\mathcal{A}_1^{eff})^{\otimes,\mathfrak{h}}$, defined in (Part II, Chapter III, §3.2.4). For an integer $n \ge 1$, and for objects $\mathbb{Z}_{X_i}(m_i)_{f_i}$ of $\mathcal{A}_1^{eff}$, define

$$\phi_2(\mathbb{Z}_{X_1}(m_1)_{f_1} \otimes \ldots \otimes \mathbb{Z}_{X_n}(m_n)_{f_n}) = \phi_1(\mathbb{Z}_{X_1}(m_1)_{f_1}) \otimes \ldots \otimes \phi_1(\mathbb{Z}_{X_n}(m_n)_{f_n}).$$

Write

$$X := X_1 \times \ldots \times X_n;\ m := \sum_i m_i;\ f := f_1 \times \ldots \times f_n,$$
$$\mathbb{Z}_{X_1}(m_1)_{f_1} \otimes \ldots \otimes \mathbb{Z}_{X_n}(m_n)_{f_n} = \Xi.$$

Each pair (g,h), with $g \in \mathrm{Hom}_{\mathbb{Z}\Omega^{\mathfrak{h}}}(n,1)$ and $h: X' \to X$, gives the element $g \otimes h^*$ of $\mathrm{Hom}_{\mathcal{A}_2^{eff\mathfrak{h}}}(\Xi, \mathbb{Z}_{X'}(m)_{f'})$ (see Part II, (III.3.2.4.1)). Define

$$\phi_2(g \otimes h^*): \phi_2(\Xi) \to \phi_2(\mathbb{Z}_{X'}(m')_{f'}) = \mathfrak{z}(X', \mathbb{A}^m)[-2m]$$

as the composition

$$\phi_2(\Xi) \xrightarrow{\boxtimes_\Xi} \mathfrak{z}^n(X, \mathbb{A}^m)[-2m]$$
$$\xrightarrow{(\Phi^{\mathfrak{h}}(g)_{m_1,\dots,m_n} \times X)[-2m]} \mathfrak{z}(X, \mathbb{A}^m)[-2m]$$
$$\xrightarrow{h^*} \mathfrak{z}(X', \mathbb{A}^m)[-2m],$$

where $\boxtimes$ is the external product (2.3.1.1), and $\times X$ is the operation described in §2.2.5(3).

Since the maps $\boxtimes$ form a commutative exterior product, this definition of $\phi_2(g \otimes h^*)$ gives a well-defined map of complexes

$$\phi_2: \mathrm{Hom}_{\mathcal{A}_2^{eff\mathfrak{h}}}(\Xi, \mathbb{Z}_X(m)_f) \to \mathrm{Hom}_{C^-(Sh_{Nis}(SmCor(k)))}(\phi_2(\Xi), \phi_2(\mathbb{Z}_X(m)_f)).$$

Using the tensor structure of $C^-(Sh_{Nis}(SmCor(k)))$, and the explicit description of the maps in $(\mathcal{A}_1^{eff})^{\otimes,\mathfrak{h}}$ from (Part II, Chapter III, §3.2.4), it follows that the formula for ϕ_2 described above extends uniquely to give the functor

$$\phi_2: \mathcal{A}_2^{eff\mathfrak{h}} \to C^-(Sh_{Nis}(SmCor(k))).$$

2.5.2. *Cycle classes and semi-purity.* For a smooth equi-dimensional k-scheme X, let $z^r_{\mathrm{eq}}(X)$ be the sheaf with $z^r_{\mathrm{eq}}(X)(Y)$ the codimension r cycles on $X \times Y$ which are equi-dimensional over Y (see Appendix A). By [**43**, Theorem 8.1], the evident inclusion

$$\mathfrak{z}(X, \mathbb{A}^q)(\mathrm{Spec}\, k) \hookrightarrow C_*(z^q_{\mathrm{eq}}(X \times \mathbb{A}^q))(\mathrm{Spec}\, k)$$

is a quasi-isomorphism.

For a closed subset $W \subset X$, let $\mathfrak{z}_W(X, \mathbb{A}^q)$ be defined as

$$\mathfrak{z}_W(X, \mathbb{A}^q) := \mathrm{cone}\big(\mathfrak{z}(X, \mathbb{A}^q) \xrightarrow{j^*} \mathfrak{z}(X \setminus W, \mathbb{A}^q)\big)[-1],$$

and let $C_*(z^q_{\mathrm{eq}}(X \times \mathbb{A}^q))_W$ be similarly defined as

$$C_*(z^q_{\mathrm{eq}}(X \times \mathbb{A}^q))_W := \mathrm{cone}\big(C_*(z^q_{\mathrm{eq}}(X \times \mathbb{A}^q)) \xrightarrow{j^*} C_*(z^q_{\mathrm{eq}}((X \setminus W \times \mathbb{A}^q))\big)[-1].$$

We then have the quasi-isomorphism

$$\mathfrak{z}_W(X, \mathbb{A}^q)(\mathrm{Spec}\, k) \to C_*(z^q_{\mathrm{eq}}(X \times \mathbb{A}^q))_W(\mathrm{Spec}\, k). \tag{2.5.2.1}$$

We have Bloch's cycle complex $z^q(Y,*)$ (see Chapter II, §2.1.2), for a closed subset F of Y, define $z^q_F(Y,*)$ to be the shifted cone, as above. We have the evident inclusion $C_*(z^q_{\mathrm{eq}}(X \times \mathbb{A}^q))(\mathrm{Spec}\, k) \to z^q(X \times \mathbb{A}^q, *)$; by [**124**, Proposition 4.2.9], this inclusion is a quasi-isomorphism, giving the quasi-isomorphism

$$C_*(z^q_{\mathrm{eq}}(X \times \mathbb{A}^q))_W(\mathrm{Spec}\, k) \to z^q_{W \times \mathbb{A}^q}(X \times \mathbb{A}^q, *). \tag{2.5.2.2}$$

We have the group of codimension q cycles on X with support in W, $\mathcal{Z}^q_W(X)$. Via (2.5.2.1) and (2.5.2.2), we have the canonical map

$$\mathrm{cl}^q_{X,W}(k): \mathcal{Z}^q_W(X) \to \mathrm{Hom}_{\mathbf{K}^-(\mathbf{Ab})}(\mathbb{Z}, \mathfrak{z}_W(X, \mathbb{A}^q)(\mathrm{Spec}\, k)) \tag{2.5.2.3}$$

satisfying the functorialities of Chapter V, Definition 1.1.6(ii). Also, if W has pure codimension $\geq q$ on X, we similarly have

$$H^p(\mathfrak{z}_W(X, \mathbb{A}^q)(\operatorname{Spec} k)) = 0; \quad \text{for } p < 0. \tag{2.5.2.4}$$

2.5.3. *The extension.* The categories $\mathcal{A}_n^{eff\mathfrak{h}}$, $n = 3, 4, 5$ and $n = \text{mot}$ are now constructed exactly as the categories $\mathcal{A}_n^{\mathfrak{s}\mathfrak{h}}(\mathcal{V})$ in Chapter V, §1.3.6; note that all the maps adjoined to $\mathcal{A}_2(\mathcal{V})$ to form $\mathcal{A}_n(\mathcal{V})$ in Chapter I, §1.4 are of the form $h: \mathfrak{e}^{\otimes k} \to \mathbb{Z}_X(n)_f[m]$ with $n \geq 0$.

We have the canonical isomorphism (for $\mathcal{F}$ in $C^-(Sh_{Nis}(SmCor(k)))$)

$$\operatorname{Hom}_{\mathbf{K}^-(\mathbf{Ab})}(\mathbb{Z}, \mathcal{F}(\operatorname{Spec} k)) \to \operatorname{Hom}_{K^-(Sh_{Nis}(SmCor(k)))}(\mathbb{Z}, \mathcal{F}). \tag{2.5.3.1}$$

The cycle class map (2.5.2.3) thus gives the map

$$\mathrm{cl}^q_{X,W}: \mathcal{Z}^q_W(X) \to \operatorname{Hom}_{K^-(Sh_{Nis}(SmCor(k)))}(\mathbb{Z}, \mathfrak{z}_W(X, \mathbb{A}^q)),$$

satisfying the functorialities of Chapter V, Definition 1.1.6(ii).

The extension of ϕ_2 to

$$\phi_n: \mathcal{A}_n^{eff\mathfrak{h}} \to C^-(Sh_{Nis}(SmCor(k))); \quad n = 3, 4, 5, \text{mot},$$

is then constructed as the extension to $\mathcal{A}_n^{\mathfrak{s}\mathfrak{h}}(\mathcal{V})$ described in Chapter V, §1.3.6, using the cohomology vanishing (2.5.2.4) and the isomorphism (2.5.3.1). One then takes complexes, and passes to the homotopy category, giving the exact tensor functor

$$\mathbf{K}^b(\phi_{\text{mot}}): \mathbf{K}^b(\mathcal{A}_{\text{mot}}^{eff\mathfrak{h}}) \to K^-(Sh_{Nis}(SmCor(k))),$$

as in Chapter V, §1.3.8.

Using (Part II, Chapter II, Theorem 2.2.2), passing to the derived category, and localizing with respect to the thick subcategory Htp, we have the exact functor

$$\mathbf{K}^b(\phi_{\text{mot}})': \mathbf{K}^b(\mathcal{A}_{\text{mot}}^{eff}) \to DM_-^{eff}(k),$$

and the full triangulated tensor subcategory $\mathbf{K}^b_{\mathfrak{h}}(\mathcal{A}_{\text{mot}}^{eff})$ of $\mathbf{K}^b(\mathcal{A}_{\text{mot}}^{eff})$, with essential image all of $\mathbf{K}^b(\mathcal{A}_{\text{mot}}^{eff})$; the restriction of $\mathbf{K}^b(\phi_{\text{mot}})'$ to $\mathbf{K}^b_{\mathfrak{s}\mathfrak{h}}(\mathcal{A}_{\text{mot}}^{eff})$ is an exact pseudo-tensor functor.

We now let $\mathcal{A}_{\text{mot}}^{eff+}$ be the full DG tensor subcategory of $\mathcal{A}_{\text{mot}}^{eff}$ generated by objects of the form $\mathfrak{e}^{\otimes k} \otimes \mathbb{Z}_X(m)_f$ with $m \geq \dim_k X$. Restricting $\mathbf{K}^b(\phi_{\text{mot}})'$ gives the exact functor

$$\mathbf{K}^b(\phi_{\text{mot}})^+: \mathbf{K}^b(\mathcal{A}_{\text{mot}}^{eff+}) \to DM_-^{eff}(k)$$

and the exact pseudo-tensor functor

$$\mathbf{K}^b(\phi_{\text{mot}})^+: \mathbf{K}^b_{\mathfrak{h}}(\mathcal{A}_{\text{mot}}^{eff+}) \to DM_-^{eff}(k).$$

Since

$$\mathbf{K}^b(\phi_{\text{mot}})^+(\mathfrak{e}^{\otimes k} \otimes \mathbb{Z}_X(m)_f) = \mathfrak{z}(X, \mathbb{A}^m)[-2m],$$

and $m \geq \dim_k X$, it follows from (2.3.1.2) that we have the isomorphism

$$\mathbf{K}^b(\phi_{\text{mot}})^+(\mathfrak{e}^{\otimes k} \otimes \mathbb{Z}_X(m)_f) \cong \underline{Hom}_{DM^{eff}}(M(X), \mathbb{Z}(m)) \cong M(X)^D \otimes \mathbb{Z}(m).$$

Since $m \geq \dim_k X$, it follows from [**124**, Corollary 4.3.4] that $M(X)^D \otimes \mathbb{Z}(m)$ is in the strictly full subcategory $DM_{gm}^{eff}(k)$ of $DM_-^{eff}(k)$. Thus we have exact functor

$$\mathbf{K}^b(\phi_{\text{mot}})^+: \mathbf{K}^b(\mathcal{A}_{\text{mot}}^{eff+}) \to DM_{gm}^{eff}(k),$$

and restriction to the exact pseudo-tensor functor

$$\mathbf{K}^b(\phi_{\mathrm{mot}})^+ : \mathbf{K}^b_{\mathfrak{h}}(\mathcal{A}^{eff+}_{\mathrm{mot}}) \to DM^{eff}_{gm}(k).$$

We may localize $\mathbf{K}^b(\mathcal{A}^{eff+}_{\mathrm{mot}})$ with respect to the morphisms of Chapter I, Definition 2.1.4, with the restriction that the morphisms inverted are in $\mathbf{C}^b(\mathcal{A}^{eff+}_{\mathrm{mot}})$, forming the category $\mathbf{D}^{b\ eff+}_{\mathrm{mot}}$; the basic properties of motivic cohomology listed in [**124**, §2], together with the properties of (2.3.1.2), imply that $\mathbf{K}^b(\phi_{\mathrm{mot}})^+$ extends to the exact functor

$$\Psi^{eff+} : \mathbf{D}^{b\ eff+}_{\mathrm{mot}} \to DM^{eff}_{gm}(k).$$

Letting $\mathbf{D}^{b\ eff+}_{\mathrm{mot}\mathfrak{h}}$ be the full image of $\mathbf{K}^b_{\mathfrak{h}}(\mathcal{A}^{eff+}_{\mathrm{mot}})$, we have the full triangulated tensor subcategory $\mathbf{D}^{b\ eff+}_{\mathrm{mot}\mathfrak{h}}$ of $\mathbf{D}^{b\ eff+}_{\mathrm{mot}}$, with essential image all of $\mathbf{D}^{b\ eff+}_{\mathrm{mot}}$, and the exact pseudo-tensor functor

$$\Psi^{eff+} : \mathbf{D}^{b\ eff+}_{\mathrm{mot}\mathfrak{h}} \to DM^{eff}_{gm}(k).$$

2.5.4. LEMMA. *(i) The symmetry isomorphism* $\tau : \mathbb{Z}_S(1) \otimes \mathbb{Z}_S(1) \to \mathbb{Z}_S(1) \otimes \mathbb{Z}_S(1)$ *in* $\mathbf{D}^{b\ eff+}_{\mathrm{mot}}$ *is the identity, so one may invert* $\mathbb{Z}_S(1)$ *to form the triangulated tensor category* $\mathbf{D}^{b\ \pm}_{\mathrm{mot}}$
(ii) The natural functor $\mathbf{D}^{b\ \pm}_{\mathrm{mot}\mathfrak{h}} \to \mathbf{D}^b_{\mathrm{mot}}$ *is an equivalence of triangulated tensor categories.*

PROOF. (i) follows from the fact that the commutative external product

$$\mathbb{Z}_S(1) \otimes \mathbb{Z}_S(1) \to \mathbb{Z}_S(2)$$

is an isomorphism in $\mathbf{D}^{b\ eff+}_{\mathrm{mot}}$ (see Chapter I, Definition 2.1.4). For (ii), we note that the equivalence of triangulated categories of Chapter I, Theorem 3.4.2, works as well for $\mathbf{D}^{b\ eff+}_{\mathrm{mot}}$, giving the subcategory $\mathbf{C}^{b\ eff+}_{\mathrm{mot}}(\mathbf{Sm}_k)^*$ of $\mathbf{C}^b_{\mathrm{mot}}(\mathbf{Sm}_k)^*$, the functors

$$\mathbf{C}^{b\ eff+}(r_{\mathrm{mot}}) : \mathbf{C}^{b\ eff+}_{\mathrm{mot}}(\mathbf{Sm}_k) \to \mathbf{C}^{b\ eff+}_{\mathrm{mot}}(\mathbf{Sm}_k)^*,$$
$$\mathbf{K}^{b\ eff}(r_{\mathrm{mot}}) : \mathbf{K}^{b\ eff+}_{\mathrm{mot}}(\mathbf{Sm}_k) \to \mathbf{K}^{b\ eff+}_{\mathrm{mot}}(\mathcal{V})^*,$$

and the equivalences of triangulated tensor categories

$$\mathbf{D}^b(r_{\mathrm{mot}}) : \mathbf{D}^b_{\mathrm{mot}}(\mathcal{V}) \to \mathbf{D}^b_{\mathrm{mot}}(\mathcal{V})^*, \tag{2.5.4.1}$$
$$\mathbf{D}^{b\ eff+}(r_{\mathrm{mot}}) : \mathbf{D}^{b\ eff+}_{\mathrm{mot}}(\mathcal{V}) \to \mathbf{D}^{b\ eff+}_{\mathrm{mot}}(\mathcal{V})^*.$$

We have the cone-preserving functor $s(n) : \mathbf{C}^b_{\mathrm{mot}}(\mathcal{V})^* \to \mathbf{C}^b_{\mathrm{mot}}(\mathcal{V})^*$ defined by

$$s(n)(\Gamma) = \mathbf{C}^b(r_{\mathrm{mot}})(\Gamma \otimes \mathbb{Z}_S(n)).$$

As $r_{\mathrm{mot}}(\mathbb{Z}_X(m)_f \otimes \mathbb{Z}_S(n)) = \mathbb{Z}_X(m+n)_f$, it follows that, for each Γ in $\mathbf{C}^b_{\mathrm{mot}}(\mathcal{V})^*$, $s(n)(\Gamma)$ is in $\mathbf{C}^{b\ eff+}_{\mathrm{mot}}(\mathcal{V})^*$ for all n sufficiently large. As $s(n)$ has the inverse $s(-n)$, we have the isomorphism

$$\mathrm{Hom}_{\mathbf{K}^b_{\mathrm{mot}}(\mathcal{V})^*}(\Gamma, \Gamma') \cong \mathrm{Hom}_{\mathbf{K}^b_{\mathrm{mot}}(\mathcal{V})^*}(s(n)(\Gamma), s(n)(\Gamma')) \tag{2.5.4.2}$$

for all Γ, Γ' in $\mathbf{C}^b_{\mathrm{mot}}(\mathcal{V})^*$.

We have the natural isomorphism

$$\mathbf{D}^b(\boxtimes_{\mathrm{mot}}) : \Gamma \otimes \mathbb{Z}_S(n) \to s(n)(\Gamma),$$

which gives us the identity

$$\text{(2.5.4.3)} \qquad \mathrm{Hom}_{\mathbf{D}^b_{\mathrm{mot}}(\mathcal{V})^*}(\Gamma,\Gamma') \cong \mathrm{Hom}_{\mathbf{D}^b_{\mathrm{mot}}(\mathcal{V})^*}(s(n)(\Gamma), s(n)(\Gamma'))$$

for all Γ, Γ' in $\mathbf{C}^b_{\mathrm{mot}}(\mathcal{V})^*$. As each morphism and each identity between morphisms in a localization of $\mathbf{K}^b_{\mathrm{mot}}(\mathcal{V})^*$ can be described by a finite diagram of morphisms in $\mathbf{K}^b_{\mathrm{mot}}(\mathcal{V})^*$, it follows from (2.5.4.2) and (2.5.4.3) that have the identity

$$\begin{aligned}\mathrm{Hom}_{\mathbf{D}^b_{\mathrm{mot}}(\mathcal{V})^*}(\Gamma,\Gamma') &\cong \varinjlim_n \mathrm{Hom}_{\mathbf{D}^{b\ eff+}_{\mathrm{mot}}(\mathcal{V})^*}(s(n)(\Gamma), s(n)(\Gamma')) \\ &\cong \varinjlim_n \mathrm{Hom}_{\mathbf{D}^{b\ eff+}_{\mathrm{mot}}(\mathcal{V})}(s(n_0)(\Gamma)\otimes \mathbb{Z}_S(n), s(n_0)(\Gamma')\otimes \mathbb{Z}_S(n)),\end{aligned}$$

where n_0 is chosen so that $s(n_0)(\Gamma)$ and $s(n_0)(\Gamma')$ are in $\mathbf{D}^{b\ eff+}_{\mathrm{mot}}$. Together with the equivalences (2.5.4.1), this proves (ii). □

Via Lemma 2.5.4, we have the exact functor

$$\Psi^b : \mathbf{D}^b_{\mathrm{mot}}(\mathcal{V}) \to DM_{gm}(k).$$

As projectors have a kernel and cokernel in $DM_{gm}(k)$ [**124**, Lemma 3.1.13], we may extend Ψ^b to the exact functor

$$\Psi : \mathcal{DM}(k) \to DM_{gm}(k).$$

Let $\mathbf{D}^b_{\mathrm{mot}\mathfrak{h}}(\mathcal{V})$ be the image of $\mathbf{D}^{b\ eff+}_{\mathrm{mot}\mathfrak{h}}$ in $\mathbf{D}^b_{\mathrm{mot}}(\mathcal{V})$, and let $\mathcal{DM}_{\mathfrak{h}}(k)$ be the pseudo-abelian hull of $\mathbf{D}^b_{\mathrm{mot}\mathfrak{h}}(\mathcal{V})$. Then $\mathcal{DM}_{\mathfrak{h}}(k)$ is naturally a full triangulated tensor subcategory of $\mathcal{DM}(k)$, with essential image all of $\mathcal{DM}(k)$, and Ψ restricts to the exact pseudo-tensor functor

$$\Psi_{\mathfrak{h}} : \mathcal{DM}_{\mathfrak{h}}(k) \to DM_{gm}(k).$$

2.5.5. THEOREM. *(i) The functor Ψ is an equivalence of triangulated categories. (ii) The functor $\Psi_{\mathfrak{h}}$ is a pseudo-tensor equivalence of the triangulated tensor category $\mathcal{DM}_{\mathfrak{h}}(k)$ with its full image in $DM_{gm}(k)$; the essential image of $\Psi_{\mathfrak{h}}$ is all of $DM_{gm}(k)$.*

PROOF. Let Γ be an object of $\mathbf{C}^b(\mathcal{A}_1(\mathbf{Sm}_k))$, i.e., a bounded complex with terms direct sums of motives $\mathbb{Z}_X(a)_f[b]$. We have the cycle class map (I.3.5.1.3)

$$\mathrm{cl}_\Gamma : H^0(\mathcal{Z}_{\mathrm{mot}}(\Gamma)) \to \mathrm{Hom}_{\mathcal{DM}(k)}(1,\Gamma).$$

Composing cl_Γ with the map induced by the functor Ψ gives us the map

$$\Psi\circ\mathrm{cl}_\Gamma : H^0(\mathcal{Z}_{\mathrm{mot}}(\Gamma)) \to \mathrm{Hom}_{DM_{gm}(k)}(1,\Psi(\Gamma)).$$

The natural quasi-isomorphisms (2.5.2.1) and (2.5.2.2), and the natural isomorphism

$$H^i(\phi_{Y,X}(k)) : H^i(\mathfrak{z}(Y,X)(\mathrm{Spec}\,k)) \to \mathrm{Hom}_{DM^{eff}}(C^*(Y), C^*_c(X)[i])$$

of (2.3.1.2)(1) induces the isomorphism

$$\xi_\Gamma : \mathrm{Hom}_{DM_{gm}(k)}(1,\Psi(\Gamma)) \to H^0(\mathcal{Z}_{\mathrm{mot}}(\Gamma,*))$$

(see Chapter II, Definition 2.2.4). It follows from the construction of the cycle class map (2.5.2.3) that the composition

$$\xi_\Gamma\circ\Psi\circ\mathrm{cl}_\Gamma : H^0(\mathcal{Z}_{\mathrm{mot}}(\Gamma)) \to H^0(\mathcal{Z}_{\mathrm{mot}}(\Gamma,*))$$

is just the map induced by the canonical inclusion

$$\mathcal{Z}_{\mathrm{mot}}(\Gamma) = \mathcal{Z}_{\mathrm{mot}}(\Gamma, 0) \hookrightarrow \mathcal{Z}_{\mathrm{mot}}(\Gamma, *). \tag{2.5.5.1}$$

We now take $\Gamma = \Sigma^N(\mathbb{Z}_X(q)[p])[N]$ (Chapter II, Definition 2.2.2). For N sufficiently large we have $H^0(\mathcal{Z}_{\mathrm{mot}}(\Sigma^N(\mathbb{Z}_X(q)[p])[N]) \cong \mathrm{CH}^q(X, 2q-p)$ (Chapter II, Proposition 2.2.5 and Lemma 2.2.8). Via this isomorphism, the map

$$\mathrm{cl}_\Gamma : \mathrm{CH}^q(X, 2q-p) \to H^p(X, \mathbb{Z}(q))$$

is the isomorphism of (Chapter II, Theorem 3.6.6), and the composition (see Chapter II, Proposition 3.2.3)

$$\mathrm{CH}^q(X, 2q-p) \xrightarrow{\mathrm{cl}_\Gamma} H^p(X, \mathbb{Z}(q)) \xrightarrow{\mathfrak{R}_{\mathcal{CH}}} \mathcal{CH}^q(X, 2q-p) \cong \mathrm{CH}^q(X, 2q-p)$$

is the map on cohomology induced by the inclusion (2.5.5.1). This latter composition is shown to be an isomorphism in the proof of Theorem 3.6.6 of Chapter II.

Putting this all together, we have shown that the map

$$\Psi : \mathrm{Hom}_{\mathcal{DM}(k)}(1, \Sigma^N(\mathbb{Z}_X(q)[p])[N]) \to \mathrm{Hom}_{DM_{gm}(k)}(1, \Psi(\Sigma^N(\mathbb{Z}_X(q)[p])[N]))$$

is an isomorphism. As the map $i_N : \mathbb{Z}_X(q)[p] \to \Sigma^N(\mathbb{Z}_X(q)[p])[N]$ (II.2.2.6.2) is an isomorphism in $\mathcal{DM}(k)$ (Chapter II, Lemma 2.3.5), we conclude that

$$\Psi : \mathrm{Hom}_{\mathcal{DM}(k)}(1, \mathbb{Z}_X(q)[p]) \to \mathrm{Hom}_{DM_{gm}(k)}(1, \Psi(\mathbb{Z}_X(q)[p]))$$

is an isomorphism.

Since $\mathcal{DM}_{\mathfrak{h}}(k)$ is generated as a triangulated category by the objects $\mathbb{Z}_X(n)$, and then taking the pseudo-abelian hull, it follows from Theorem 1.5.1 of Chapter IV that $\Psi_{\mathfrak{h}}$ is fully faithful; as each object of $\mathcal{DM}(k)$ is isomorphic to an object of $\mathcal{DM}_{\mathfrak{h}}(k)$, it follows that Ψ is fully faithful.

To see that the essential image of $\Psi_{\mathfrak{h}}$ (or equivalently, of Ψ) is all of $DM_{gm}(k)$, the category $DM_{gm}(k)$ has the duality involution $\Gamma \mapsto \Gamma^D := \underline{Hom}_{DM_{gm}}(\Gamma, \mathbb{Z})$. By (Chapter IV, Lemma 1.1.3, Proposition 1.1.9 and Proposition 1.1.10), we have $\Psi(\Delta^D) \cong \Psi(\Delta)^D$ for each Δ in $\mathcal{DM}(k)$. For each X in $\mathbf{Sm}_k$, it follows from (2.3.1.2) that

$$\Psi(\mathbb{Z}_X(m)_f) \cong \underline{Hom}_{DM_{gm}}(M(X), \mathbb{Z}) \otimes \mathbb{Z}(m) = M(X)^D \otimes \mathbb{Z}(m),$$

hence

$$\Psi(\mathbb{Z}_X(m)_f^D) \cong \Psi(\mathbb{Z}_X(m)_f)^D \cong M(X) \otimes \mathbb{Z}(-m).$$

As $DM_{gm}(k)$ is generated as a triangulated category by the objects $M(X) \otimes \mathbb{Z}(-m)$, as X runs over $\mathbf{Sm}_k$, and $m \in \mathbb{Z}$, and then taking the pseudo-abelian hull, the essential image of Ψ is all of $DM_{gm}(k)$.

Since $\Psi_{\mathfrak{h}}$ is fully faithful, we may define the inverse equivalence on the image of $\Psi_{\mathfrak{h}}$ by choosing a lifting on objects; one easily verifies that this canonically extends to a pseudo-tensor inverse

$$\Phi_{\mathfrak{h}} : \mathrm{Im}(\Psi_{\mathfrak{h}}) \to \mathbf{D}^b_{\mathrm{mot}\mathfrak{h}}(\mathcal{V}),$$

giving the desired pseudo-tensor equivalence. □

APPENDIX A

Equi-dimensional Cycles

In this appendix, we give a review of a part of the theory of equi-dimensional cycles as developed by Suslin-Voevodsky [**117**]. All the main ideas are from [**117**]; all errors are of course the responsibility of this author.

1. Cycles over a normal scheme

1.1. Relative cycles

1.1.1. *Dimension.* Let S be a scheme. Let k be an algebraically closed field, W a reduced, irreducible k-scheme, essentially of finite type over k. We define $\dim_k(W)$ to be the transcendence dimension over k of the function field $k(W)$; since W is essentially of finite type over k, $\dim_k(W)$ is finite. If s is a point of a scheme S, we say that S has dimension r at s if the Krull dimension of the local ring $\mathcal{O}_{S,s}$ is r; we say that a local scheme (S, s) has dimension r if S has dimension r at s. A *geometric point* s of a scheme S is an equivalence class of maps $s\colon \operatorname{Spec} k \to S$, where k is an algebraically closed field, and where s and s' are equivalent if there is a commutative triangle

$$\begin{array}{ccc} \operatorname{Spec} k & \xrightarrow{p} & \operatorname{Spec} k' \\ & {\scriptstyle s}\searrow \quad \swarrow{\scriptstyle s'} & \\ & S & \end{array}$$

with p an isomorphism. We let $k(s)$ denote a representative modulo isomorphism of the collection of the algebraically closed fields k associated to a geometric point s. We let S_{geom} denote the set of geometric points of a scheme S.

For a map of schemes $W \to S$, and a geometric point s of S, we denote $W \times_S \operatorname{Spec} k(s)$ by W_s; we also use the notation $W_s := W \times_S s$ when s is a point of S. If W is an irreducible S-scheme, essentially of finite type over S, s a geometric point of S, then $\dim_{k(s)}(Z)$ is independent of the choice of irreducible component Z of $(W_s)_{\text{red}}$; we denote the common value of such dimensions by $\dim_{k(s)}(W_s)$.

Let Z be a closed subset of a scheme W. We let $\mathcal{O}_{W,Z}$ denote the semi-local ring of Z in W; for a $\mathcal{O}_{W,Z}$-module M, we let $\operatorname{lng}_{\mathcal{O}_{W,Z}}(M)$ denote the length of M.

1.1.2. Definition/Lemma. (i) Let S be a reduced scheme, and let $p\colon W \to S$ be a reduced irreducible S-scheme, essentially of finite type over S. We say that W *has dimension d over S* if W dominates an irreducible component S' of S, and, for each $s \in S'_{\text{geom}}$, either W_s is empty, or $\dim_{k(s)}(W_s) = d$. We say a reduced S-scheme $Z \to S$, essentially of finite type over S, has dimension d over S if each irreducible component of Z has dimension d over S. We say that Z is *equi-dimensional over S* if Z has dimension d over S for some d.

(ii) Let S be a reduced scheme, $X \to S$ an S-scheme, essentially of finite type over S. A *relative dimension d cycle on X* is a finite sum $\sum_{i=1}^n n_i Z_i$ where the n_i are integers, and each Z_i is a closed, reduced irreducible subscheme of X, which has dimension d over S. We define a relative dimension d $\mathbb{Q}$-cycle on X similarly, allowing $\mathbb{Q}$-coefficients. We let $\mathcal{C}_d(X/S)_\mathbb{Z}$, (resp. $\mathcal{C}_d(X/S)_\mathbb{Q}$) denote the group of relative dimension d cycles on X (resp. the $\mathbb{Q}$-vector space of relative dimension d $\mathbb{Q}$-cycles on X).

Let $Z = \sum_{i=1}^N n_i Z_i$ be in $\mathcal{C}_d(X/S)_\mathbb{Z}$ or $\mathcal{C}_d(X/S)_\mathbb{Q}$, with the Z_i distinct and irreducible, and all $n_i \neq 0$. We call Z an *effective cycle* if $n_i > 0$ for $i = 1, \ldots, N$. The collection of effective relative dimension d cycles on X (together with 0) forms a sub-monoid $\mathcal{C}_d(X/S)_\mathbb{Z}^{\geq 0}$ of $\mathcal{C}_d(X/S)_\mathbb{Z}$; we have as well the $\mathbb{Q}$-sub-cone $\mathcal{C}_d(X/S)_\mathbb{Q}^{\geq 0}$ of $\mathcal{C}_d(X/S)_\mathbb{Q}$ consisting of effective relative dimension d $\mathbb{Q}$-cycles on X. The *support* of Z is the union $\cup_{i=1}^n Z_i$.

(iii) Let S be a reduced scheme and $X \to S$ be an S-scheme with connected components $X_1, \ldots, X_m$. Suppose each X_i has dimension n_i over S. We have the group $\mathcal{C}^d(X/S)_\mathbb{Z}$ of relative codimension d cycles on X, defined as

$$\mathcal{C}^d(X/S)_\mathbb{Z} = \oplus_{i=1}^m \mathcal{C}_{n_i-d}(X_i/S)_\mathbb{Z},$$

and the $\mathbb{Q}$-vector space $\mathcal{C}^d(X/S)_\mathbb{Q}$ of relative codimension d $\mathbb{Q}$-cycles on X:

$$\mathcal{C}^d(X/S)_\mathbb{Q} = \oplus_{i=1}^m \mathcal{C}_{n_i-d}(X_i/S)_\mathbb{Q}.$$

We have as well the monoid of effective cycles $\mathcal{C}^d(X/S)_\mathbb{Z}^{\geq 0}$, which generates the positive $\mathbb{Q}$-cone $\mathcal{C}^d(X/S)_\mathbb{Q}^{\geq 0}$.

(iv) If S is a normal scheme, $p: T \to S$ a map of schemes, and if $Z \to S$ has dimension d over S, then $(Z \times_S T)_{\text{red}} \to T$ has dimension d over T, or is empty.

PROOF. (of (iv)). We may assume that T, Z and S are irreducible and $Z \times_S T$ is not empty; as the geometric fibers of $(Z \times_S T)_{\text{red}} \to T$ form a subset of the geometric fibers of $f: Z \to S$, it suffices to show that each irreducible component of $(Z \times_S T)_{\text{red}}$ dominates T.

By Gruson-Raynaud [**105**], there is a blow-up $\mu: S' \to S$ of S such that, if $U \subset S$ is a non-empty open subset of S such that μ is an isomorphism over U, then the closure $Z_{S'}$ of $Z \times_S \mu^{-1}(U)$ in $Z \times_S S'$ is flat over S'. It suffices to show that

$$Z_{S'} = (Z \times_S S')_{\text{red}}. \tag{1.1.2.1}$$

Indeed, if this is the case, form the cartesian diagram

$$\begin{array}{ccc} T \times_S S' & \xrightarrow{q} & S' \\ \downarrow{\scriptstyle \mu_T} & & \downarrow{\scriptstyle \mu} \\ T & \xrightarrow[p]{} & S. \end{array}$$

As $S' \to S$ is proper and surjective, so is $T \times_S S' \to T$; thus there is an irreducible component T' of $T \times_S S'$ which maps surjectively to T. If $Z \times_S T$ has an irreducible component which does not dominate T, then $Z \times_S T'$ has an irreducible component which does not dominate T'. As we have $Z \times_S T' \cong (Z \times_S S') \times_{S'} T'$, as T'-schemes, (1.1.2.1) would imply that $(Z \times_S T')_{\text{red}} \cong (Z_{S'} \times_{S'} T')_{\text{red}}$ as T'-schemes. Since $Z_{S'}$ is flat over S', $Z_{S'} \times_{S'} T'$ is flat over T', hence each irreducible component of $Z_{S'} \times_{S'} T'$ dominates T'.

We now prove (1.1.2.1). Pick a point s of S; to prove (1.1.2.1), it suffices to show that, for each $s' \in S'$ with $\mu(s') = s$ we have

$$(1.1.2.2) \qquad (Z_{S'} \times_{S'} s')_{\mathrm{red}} = (Z \times_S s')_{\mathrm{red}}.$$

Let $Y = \mu^{-1}(s)_{\mathrm{red}}$, and let $\mu_Y: Y \to s$ be the restriction of μ. The map $q: Z_{S'} \to Z$ induced by the projection $p_1: Z \times_S S' \to Z$ is proper and dominant, hence the restriction of q to $Z_{S'} \times_{S'} Y$, $q_Y: Z_{S'} \times_{S'} Y \to Z \times_S s$, is also proper and dominant; in particular, q_Y is surjective.

Let $\overline{k(s)}$ be the algebraic closure of $k(s)$, $\bar{s} = \operatorname{Spec} \overline{k(s)}$, and W be an irreducible component of $Z \times_S \bar{s}$. Let $\bar{Y} = Y \times_S \bar{s}$. Since $q: Z_{S'} \to Z$ is proper and surjective, there is an irreducible component Z' of $Z_{S'} \times_{S'} \bar{Y}$ with $q_Y(Z') = W$.

Since $Z_{S'}$ is flat over S', and of dimension d over $\mu^{-1}(U)$, each fiber of $Z_{S'}$ over S' has dimension d, or is empty. If Z' is contained in $Z_{S'} \times_{S'} \bar{Y}_0$ for some irreducible component $\bar{Y}_0$ of $\bar{Y}$, the fact the generic fiber of Z' over $\bar{Y}_0$ has dimension d implies that $Z' = W \times_{\bar{s}} \bar{Y}_0$.

Since $Z_{S'} \times_{S'} \bar{Y}_i$ is flat over $\bar{Y}_i$ for each irreducible component $\bar{Y}_i$ of $\bar{Y}$, each irreducible component of $Z_{S'} \times_{S'} \bar{Y}_i$ dominates $\bar{Y}_i$. Suppose y is a point of $\bar{Y}_i \cap \bar{Y}_0$. Then $(Z_{S'} \times_{S'} \bar{Y}_i) \times_{\bar{Y}_i} y = (Z_{S'} \times_{S'} \bar{Y}_0) \times_{\bar{Y}_0} y$, hence, if Z'_y is an irreducible component of $Z' \times_{\bar{Y}_0} y$, there is an irreducible component Z'_i of $Z_{S'} \times_{S'} \bar{Y}_i$ containing Z'_y. As $Z' \times_{\bar{Y}_0} y = W \times_{\bar{s}} y$, we have $Z'_i = W \times_{\bar{s}} \bar{Y}_i$.

As S is normal, $\bar{Y}$ is connected by Zariski's connectedness theorem [**130**], [**56**, III, Théorème 4.3.1]. By the two preceding paragraphs, $Z_{S'} \times_{S'} \bar{Y}$ contains $(Z \times_S \bar{s})_{\mathrm{red}} \times_{\bar{s}} \bar{Y}$, which implies (1.1.2.2). This completes the proof. □

1.1.3. Remark. Given an S-scheme $f: X \to S$, the groups $\mathcal{C}^d(U/V)_{\mathbb{Z}}$ form a contravariant functor on the category of pairs (U, V), with U open in X, V open in S, and $f(U) \subset V$. This makes $U \mapsto \mathcal{C}^d(U/S)_{\mathbb{Z}}$ into a sheaf on X, and $V \mapsto \mathcal{C}^d(f^{-1}(V)/V)_{\mathbb{Z}}$ into a sheaf on S. In particular a cycle W on X is in $\mathcal{C}^d(X/S)_{\mathbb{Z}}$ if and only if the stalk of W at p is in $\mathcal{C}^d(\operatorname{Spec}(\mathcal{O}_{X,p})/\operatorname{Spec}(\mathcal{O}_{S,f(p)}))_{\mathbb{Z}}$ for all $p \in X$. Similar remarks hold for effective cycles, and for the $\mathbb{Q}$-versions.

1.1.4. *Intersection multiplicities.* As a general reference for this section, see [**112**].

Suppose S is a regular scheme, and W is an irreducible S-scheme of dimension d over S. Since S is regular, the $\mathcal{O}_S$-module $\mathcal{O}_W$ has finite Tor-dimension over $\mathcal{O}_S$, i.e., there is an N such that the coherent sheaf of $\mathcal{O}_{W_s}$-modules $\operatorname{Tor}_p^{\mathcal{O}_S}(\mathcal{O}_W, k(s))$ is zero for all $p > N$, and all points s of S. Let $C \to S$ be a reduced, irreducible S-scheme with generic point c, and let Z be an irreducible component of $(W \times_S C)_{\mathrm{red}}$ with generic point z. Let $m(Z; W \cdot_S C)$ be the integer

$$m(Z; W \cdot_S C) = \sum_{p=0}^{N} (-1)^p \operatorname{lng}_{\mathcal{O}_{W \times_S C, z}} (\operatorname{Tor}_p^{\mathcal{O}_S}(\mathcal{O}_W, k(c))_z).$$

Then $m(Z; W \cdot_S C)$ is well-defined; we let $W \cdot_S C$ be the d-dimensional cycle over C,

$$W \cdot_S C = \sum_Z m(Z; W \cdot_S C) \cdot Z,$$

where the sum runs over all irreducible components Z of $(W \times_S C)_{\mathrm{red}}$. We extend the definition of the operation $(-) \cdot_S C$ to all d-dimensional cycles over S, or all

d-dimensional $\mathbb{Q}$-cycles over S by linearity. If C is also regular, and $A \to C$ is a reduced irreducible C-scheme, we have the associativity relation

$$(W \cdot_S C) \cdot_C A = W \cdot_S A \tag{1.1.4.1}$$

for all d-dimensional cycles over S.

One may define $W \cdot_S C$ for an arbitrary base scheme S, using the same formula, if one assumes that each irreducible component of W has finite Tor dimension over S.

1.1.5. *Dominant pull-back.* Let $p: T \to S$ be a *dominant* morphism of irreducible normal schemes, let $X \to S$ be an S-scheme, essentially of finite type, and let Z be an irreducible closed subscheme of X, of dimension d over S. By Definition/Lemma 1.1.2, each irreducible component W of $Z \times_S T$ has dimension d over T. Let t be the generic point of T, s the generic point of S, and let

$$m(W, p^*(Z)) := m(W_t, Z_s \cdot_s t).$$

We set

$$p^*(Z) := \sum_W m(W, p^*(Z))W,$$

where the sum is over the irreducible components W of $Z \times_S T$. It follows from the associativity (1.1.4.1) that this makes the assignment $T \mapsto \mathcal{C}_d(X \times_S T/T)_\mathbb{Z}$ into a functor on the category of normal S-schemes, with maps being dominant morphisms. The same holds for the effective cycles, and cycles with rational coefficients. This pull-back is compatible with the restriction maps for open immersions described in Remark 1.1.3.

1.1.6. REMARK. Suppose that X and S are affine. We have the identity

$$\mathcal{C}_d(X/S)_\mathbb{Z} = \varinjlim \mathcal{C}_d(X_\alpha/S_\alpha)_\mathbb{Z},$$

where $X_\alpha \to S_\alpha$ is a morphism, essentially of finite type, S_α is a normal scheme, essentially of finite type over $\mathbb{Z}$, and the limit is over the category of commutative diagrams

$$\begin{array}{ccc} X & \longrightarrow & X_\alpha \\ \downarrow & & \downarrow \\ S & \xrightarrow[f]{} & S_\alpha \end{array}$$

which identify X with a localization of $X_\alpha \times_{S_\alpha} S$, and with f dominant; maps are $(X_\beta \to S_\beta) \to (X_\alpha \to S_\alpha)$ with $S_\beta \to S_\alpha$ dominant, and X_β a localization of $X_\alpha \times_{S_\alpha} S_\beta$.

Indeed, if $S = \operatorname{Spec} A$, then A is the inductive limit over finitely generated subrings; by [**100**, pg. 93, Theorem 3], the normalization B^N of a ring B finitely generated over $\mathbb{Z}$ is finite over B, so A is the inductive limit of normal finitely generated subrings. Since X is essentially of finite type over S, this shows that X is the projective limit of schemes X_α of the type considered above; similarly, each closed subscheme Z of X is the projective limit of closed subschemes Z_α of X_α.

If now Z is an irreducible closed subscheme of X of dimension d over S, and if Z is the restriction of $Z_\alpha \times_{S_\alpha} S$ to X for some irreducible subscheme Z_α of some $X_\alpha \to S_\alpha$, with morphism $f: (X \to S) \to (X_\alpha \to S_\alpha)$ as above, then the set

C of points X_α at which X_α fails to have dimension d over S_α is a constructible subset of X_α. Since X is the projective limit of the X_α, we may factor f as $(X \to S) \xrightarrow{f_1} (X_\beta \to S_\beta) \xrightarrow{f_2} (X_\alpha \to S_\alpha)$, with morphisms as above, and with C disjoint from the image of f_2. Taking Z_β to be the unique irreducible component of $Z_\alpha \times_{X_\alpha} X_\beta$ which dominates S_β, Z_β has dimension d over S_β, and $f_1^* Z_\beta$ is Z.

This allows us to replace an arbitrary normal base with one essentially of finite type over $\mathbb{Z}$ when making computations which are local on X.

1.2. Connectivity

The fundamental result on connectivity is Zariski's theorem on the connectedness of the fibers of a proper birational morphism to a normal scheme (see [**130**] and [**56**, III, Théorème 4.3.1]). In this section, we give a series of elaborations on this result.

1.2.1. Let E be a pure codimension p closed subset of a scheme Y. We say that E is *connected in codimension one* if there is a pure codimension one subset F of E such that the intersection of E with the semi-local scheme $\operatorname{Spec} \mathcal{O}_{Y,F}$ is connected. If this is the case, we say that E *is connected by* F.

We recall from [**100**, pg. 93, Theorem 3] that, for a scheme Y, essentially of finite type over $\mathbb{Z}$, the normalization $Y^N \to Y$ is *finite* over Y.

1.2.2. LEMMA. *Let $S = \operatorname{Spec} \mathcal{O}$ be a semi-local normal scheme, s a closed point of S. Suppose S is essentially of finite type over $\mathbb{Z}$. Then there is a projective morphism $f: Y \to S$ such that*

(i) *Y is normal and f is birational.*
(ii) *$f^{-1}(s)$ is the support of a Cartier divisor on Y.*
(iii) *$f^{-1}(s)$ is connected in codimension one.*

In addition, if $s_1, \dots, s_p$ are points of S different from s, we may assume that f is an isomorphism over each s_j.

PROOF. If S has dimension one at s, we may take f to be the identity; suppose then that S has dimension $r > 1$ at s. Let $\mathfrak{m}$ be the maximal ideal of $\mathcal{O}_{S,s}$. Since S has dimension r, we may find elements $x_1, \dots, x_r$ of $\mathfrak{m}$ such that the subscheme defined by any k of the x_i's has codimension k in S. We may also assume that all the x_i's are units at each of the points s_j, $j = 1, \dots, p$, and at each of the other closed points of S. In particular, the ideal $(x_1, \dots, x_r)$ contains $\mathfrak{m}^n$ for some n.

Form the sequence of blow-ups and normalizations:

$$S = Y_0 = Y_0^N \leftarrow Y_1 \leftarrow Y_1^N \leftarrow \dots \leftarrow Y_{r-1} \leftarrow Y_{r-1}^N,$$

where Y_j is the blow-up of Y_{j-1}^N along the subscheme defined by the pull-back of the ideal $(x_1, \dots, x_{j+1})$, and Y_j^N is the normalization of Y_j.

Let $\mu_j : Y_j^N \to S$ be the composite of the above maps; we take $Y = Y_{r-1}^N$, and $f: Y \to S$ to be μ_{r-1}. This verifies (i).

If $\mathcal{I}$ is an ideal sheaf on S, we write $\mu_j^* \mathcal{I}$ for the ideal sheaf generated by the pullback of $\mathcal{I}$ to Y_j^N.

Since Y_{r-1} is the blow-up of Y_{r-2}^N along the ideal sheaf $\mu_{r-2}^*(x_1, \dots, x_r)$, the pull-back of the ideal $(x_1, \dots, x_r)$ to Y_{r-1} is locally principal, hence the ideal sheaf $\mu_{r-1}^*(x_1, \dots, x_r)$ is locally principal. As $(x_1, \dots, x_r)$ contains $\mathfrak{m}^n$, this proves (ii)

Let $\mathcal{I}_j$ denote the sheaf of ideals $\mu_j^*(x_1, \ldots, x_{j+2})$, let Z_j be the subscheme of Y_j^N defined by $\mathcal{I}_j$, and let Z_j^k be the subscheme of Y_j^N defined by the ideal sheaf $\mu_j^*(x_1, \ldots, x_{k+2})$ for $k \geq j$. Let us assume by induction that Z_j^k has pure codimension $k-j$ on Y_j^N. As $\mu_j^*(x_1, \ldots, x_{j+1})$ is locally principal, it follows that $\mathcal{I}_j$ is locally generated by two elements. Let $\pi_{j+1,j}: Y_{j+1}^N \to Y_j^N$ be the normalization of the blow-up of Y_j^N along Z_j. As $\pi_{j+1,j}^{-1}(Z_j)$ is a Cartier divisor, and Z_j has pure codimension two, the generic fibers of $\pi_{j+1,j}$ have a component of positive dimension; by connectedness and upper semi-continuity of fiber dimension, each irreducible component of the fiber of $\pi_{j+1,j}$ over each point of Z_j has positive dimension. Since $\mathcal{I}_j$ is locally generated by two elements, the fibers of $\pi_{j+1,j}$ over Z_j are all pure dimension one. This shows that Z_{j+1}^k has pure codimension $k-j-1$ on Y_{j+1}^N, and the induction goes through. In particular, $E_j := \mu_j^{-1}(s)$ has pure codimension $r-j$ on Y_j^N.

We have seen that all fibers of $\pi_{j+1,j}$ are connected and of dimension one. We may suppose by induction that E_j is connected in codimension one; let F_j be a codimension one subset of E_j such that E_j is connected by F_j. For each point p of E_j, the fiber $\pi_{j+1,j}^{-1}(p)$ is connected by a finite set of points $F_{j+1}^1(p)$; let F_{j+1}^1 be the union of the closures of the points $F_{j+1}^1(p)$, as p runs over the generic points of E_j; clearly F_{j+1}^1 is a pure codimension one closed subset of E_{j+1}. Let

$$F_{j+1}^2 = \pi_{j+1,j}^{-1}(F_j),$$
$$F_{j+1} = F_{j+1}^1 \cup F_{j+1}^2.$$

If E^1 is an irreducible component of E_j, then clearly $\pi_{j+1,j}^{-1}(E^1)$ is connected by the components of F_{j+1}^1 lying over E^1. Similarly, if E^1 and E^2 are irreducible components of E_j, with $E^1 \cap E^2$ containing a component F^{12} of F_j, then there is an irreducible component E_{j+1}^1 of $\pi_{j+1,j}^{-1}(E^1)$ and an irreducible component E_{j+1}^2 of $\pi_{j+1,j}^{-1}(E^2)$ such that $E_{j+1}^2 \cap E_{j+1}^1$ contains an irreducible component of F_{j+1}^2. Thus E_{j+1} is connected by F_{j+1}, and (iii) follows by induction. □

1.2.3. Lemma. *Let S be a normal semi-local scheme with a closed point s. We suppose S is essentially of finite type over $\mathbb{Z}$. Let $f: Y \to S$ be a projective birational map, with Y normal. Let E_1 and E_2 be pure codimension one irreducible subsets of Y, with $f(E_i) = s$. Then there is a normal projective S-scheme $g: Z \to S$, and a pure codimension one closed subset E of Z such that*

(i) *g is a birational map and $g(E) = s$.*
(ii) *The rational map $h := f^{-1} \circ g: Z \to Y$ is a morphism, and $h(E)$ contains E_1 and E_2.*
(iii) *E is connected in codimension one.*

Proof. We may suppose S has dimension $r > 1$ at s. We may replace Y with the normalization of the blow-up of Y along $f^{-1}(s)$, hence we may assume that $f^{-1}(s)$ has pure codimension one on Y, and is the support of a Cartier divisor on Y. Since S is normal, $f^{-1}(s)$ is connected.

Write $f^{-1}(s)$ as a union of its irreducible components, $f^{-1}(s) = E_1 \cup \ldots \cup E_p$, and let F be the union of the pair-wise intersections, $F = \cup_{i \neq j}(E_i \cap E_j)$. Write F

as a union of irreducible components, $F = F_1 \cup \ldots \cup F_a \cup F_{a+1} \cup \ldots \cup F_b$, where

$$\operatorname{codim}_Y(F_i) > 2 \text{ for } i = 1, \ldots, a,$$
$$\operatorname{codim}_Y(F_i) = 2 \text{ for } i = a+1, \ldots, b.$$

Let Y^F be the semi-local scheme $\operatorname{Spec} \mathcal{O}_{Y,F}$; Y has the closed points $s_1, \ldots, s_b$, where $s_i = F_i \cap Y^F$ for $i = 1, \ldots, b$. The intersection E^F of $f^{-1}(s)$ with Y^F is still connected. We may apply Lemma 1.2.2 repeatedly, forming the blow-up $\pi_F : Z^F \to Y^F$ such that $\pi^{-1}(s_i)$ is the support of a Cartier divisor, and is connected in codimension one for each $i = 1, \ldots, a$. We may also assume that π_F is an isomorphism over each s_i, $i = a+1, \ldots, b$.

Let $E^F_{Z^F}$ be the proper transform of E^F to Z^F, and let E' be an irreducible component of $E^F_{Z^F}$. Since $\pi^{-1}(s_i)$ is the support of a Cartier divisor, the intersection $\pi^{-1}(s_i) \cap E'$ is pure codimension one on E'. This implies that $\pi_F^{-1}(E^F)$ is connected in codimension one.

The map $\pi_F : Z^F \to Y^F$ is the blow-up of a sheaf of ideals $\mathcal{I}^F$; let $\mathcal{I}$ be the maximal extension of $\mathcal{I}^F$ to a sheaf of ideals on Y, and let $h : Z \to Y$ be the normalization of the blow-up of Y along $\mathcal{I}$. This gives the commutative diagram

$$\begin{array}{ccc} Z^F & \longrightarrow & Z \\ {\scriptstyle \pi^F}\downarrow & & \downarrow{\scriptstyle h} \\ Y^F & \longrightarrow & Y, \end{array}$$

where the top horizontal arrow identifies Z^F with a localization of Z. Letting E be the closure of E^F in Z gives the desired result. □

1.2.4. LEMMA. *Let $S = \operatorname{Spec} \mathcal{O}$ be a local normal domain, essentially of finite type over $\mathbb{Z}$, with closed point s. Suppose we have discrete valuation rings $\mathcal{O}_1$, $\mathcal{O}_2$, and birational local inclusions $q_1 : \mathcal{O} \to \mathcal{O}_1$; $q_2 : \mathcal{O} \to \mathcal{O}_2$. Suppose that S has dimension $r > 1$, and that $\mathcal{O}_1$ and $\mathcal{O}_2$ are localizations of $\mathcal{O}$-algebras of finite type over $\mathcal{O}$. Then there is a regular, two-dimensional irreducible scheme X, a birational morphism $p : X \to S$, essentially of finite type over S, reduced irreducible codimension one closed subschemes C_1, C_2 of X, and a codimension one closed subset C of X such that*

(i) *the maps $p^* : \mathcal{O} \to \mathcal{O}_{X,C_1}$ and $p^* : \mathcal{O} \to \mathcal{O}_{X,C_2}$ are isomorphic to the inclusions q_1, q_2.*

(ii) *C is connected, each irreducible component of C is regular, $p(C) = s$ and C contains C_1 and C_2.*

PROOF. Let $T_i = \operatorname{Spec} \mathcal{O}_i$, with closed point t_i, for $i = 1, 2$. By assumption, $T_i = \operatorname{Spec} \mathcal{O}_{Y_i, D_i}$, for some finite type normal affine S-scheme $Y_i \to S$, with closed, codimension one subset $D_i \subset Y_i$, $i = 1, 2$. We may assume that each Y_i is reduced and irreducible; as the inclusions q_i are birational, the maps $Y_i \to S$ are also birational. We may embed Y_i as a closed subset of an affine space $\mathbb{A}_S^{N_i}$ over S; replacing Y_i with the closure of Y_i in $\mathbb{P}_S^{N_i}$ and changing notation, we may assume that Y_i is projective over S.

Let $f : Y \to S$ be the normalization of the S-scheme $(Y_1 \times_S Y_2)_{\text{red}}$. The projections $p_i : (Y_1 \times_S Y_2)_{\text{red}} \to Y_i$ induce the proper maps $\pi_i : Y \to Y_i$. Thus there are irreducible codimension one closed subsets E_i of Y such that $\pi_i(E_i) = D_i$; this gives the birational local maps $\pi_i^* : \mathcal{O}_i = \mathcal{O}_{Y_i, D_i} \to \mathcal{O}_{Y, E_i}$, which are necessarily isomorphisms as both rings are discrete valuation rings. Since $\pi_i(E_i) = D_i$, we have $f(E_i) = s$. Replacing Y_1 and Y_2 with Y, we may assume from the start that q_i is the map $f^*: \mathcal{O} \to \mathcal{O}_{Y,E_i}$.

By Lemma 1.2.3, we may assume that Y contains a pure codimension one closed subset E such that

(a) $E \supset E_i$; $i = 1, 2$,
(b) $f(E) = s$,
(c) E is connected in codimension one.

Suppose E is connected by a pure codimension one subset F. Let S' be the scheme $\operatorname{Spec} \mathcal{O}_{Y,F}$, and let E^F be the restriction of E to S'. Let E_i^F be the restriction of E_i to S'. Then S' is a normal two dimensional excellent domain, hence, by [**91**], there is a projective birational map $\pi: X \to S'$ such that X is regular. In addition, E^F is connected; as S' is normal, this implies that $C := \pi^{-1}(E^F)$ is connected. Since π is an isomorphism over each codimension one point of S', there are irreducible components C_1, C_2 of C with $\pi(C_i) = E_i^F$. Blowing up X further, we may assume that each component of C is regular. Taking $p: X \to S$ to be the composition $X \xrightarrow{\pi} S' \subset Y \xrightarrow{f} S$ completes the proof. □

1.3. Index of inseparability

1.3.1. Let k be a field, W a reduced irreducible k-scheme, $K \supset k$ an extension field. Let W' be an irreducible component of $(W \times_k K)_{\text{red}}$. We define the integer $m_{K:k}(W'/W)$ by

$$m_{K:k}(W'/W) := \operatorname{lng}_{\mathcal{O}_{W\times_k K,W'}}(\mathcal{O}_{W\times_k K,W'}) = m(W', W \cdot_{\operatorname{Spec} k} \operatorname{Spec} K).$$

The function $m_{-:-}(-/-)$is multiplicative in towers:

$$m_{L:k}(W''/W) = m_{L:K}(W''/W') \cdot m_{K:k}(W'/W), \tag{1.3.1.1}$$

for $L \supset K \supset k$, W' an irreducible component of $(W \times_k K)_{\text{red}}$ and W'' an irreducible component of $(W' \times_K L)_{\text{red}}$. In addition, if $\operatorname{char}(k) = 0$, then $m_{K:k}(W'/W) = 1$ for all K, W' and W; if $\operatorname{char}(k) = p > 0$, then $m_{K:k}(W'/W) = p^\alpha$ for some integer $\alpha \geq 0$.

1.3.2. LEMMA. *Let k be a field, and let W be an irreducible reduced k-scheme, of dimension d over k. Let $\mathcal{O}$ be a local k-algebra, essentially of finite type over k, and suppose that $\mathcal{O}$ is a discrete valuation ring. Let $C = \operatorname{Spec} \mathcal{O}$, with closed point a and generic point c. Let $W_C = W \times_k C$, let Z_a be an irreducible component of $(W_C \times_C a)_{\text{red}}$, let Z_C be an irreducible component of $(W_C)_{\text{red}}$ containing Z_a, and let $Z_c = Z_C \times_C c$. Then the multiplicity $m(Z_a; Z_C \cdot_C a)$ of Z_a in $Z_C \cdot_C a$ is given by*

$$m(Z_a; Z_C \cdot_C a) := \frac{m_{k(a):k}(Z_a/W)}{m_{k(c):k}(Z_c/W)}.$$

PROOF. We may assume that $W = \operatorname{Spec} K$ for some field extension K of k. We have $(Z_C \times_C a)_{\text{red}} = Z_a \coprod Z'_a$; let

$$W_C^0 := W \times_k C \setminus Z'_a,\ W_a^0 := W \times_k a \setminus Z'_a,\ Z_C^0 := Z_C \setminus Z'_a.$$

Let $\mathcal{M}^{Z_C^0}$ be the category of coherent sheaves on W_C^0, with support on Z_C^0, $\mathcal{M}^{Z_c}$ the category of coherent sheaves on $W \times_k c$ with support on Z_c, and $\mathcal{M}^{Z_a}$

the category of coherent sheaves on W^0_a with support on Z_a. The restriction and inclusion functors

$$j^*: \mathcal{M}^{Z^0_C} \to \mathcal{M}^{Z_c}$$
$$i_*: \mathcal{M}^{Z_a} \to \mathcal{M}^{Z^0_C}$$

give the exact localization sequence

$$K_0(\mathcal{M}^{Z_a}) \xrightarrow{i_*} K_0(\mathcal{M}^{Z^0_C}) \xrightarrow{j^*} K_0(\mathcal{M}^{Z_c}) \to 0. \tag{1.3.2.1}$$

The maps

$$\mathrm{lng}_{\mathcal{O}_{W\times_k c, Z_c}}: K_0(\mathcal{M}^{Z_c}) \to \mathbb{Z},\ \mathrm{lng}_{\mathcal{O}_{W^0_a, Z_a}}: K_0(\mathcal{M}^{Z_a}) \to \mathbb{Z},$$

are isomorphisms.

Let $(-)\otimes^L k(a)$ be the map from $\mathcal{M}^{Z^0_C}$ to $K_0(\mathcal{M}^{Z_a})$ given by

$$M \otimes^L k(a) = [M \otimes_{\mathcal{O}_{W^0_C, Z^0_C}} \mathcal{O}_{W^0_a, Z_a}] - [\mathrm{Tor}_1^{\mathcal{O}_{W^0_C, Z^0_C}}(M, \mathcal{O}_{W^0_a, Z_a})],$$

where $[-]$ denotes class in $K_0(\mathcal{M}^{Z_a})$. Then $(-)\otimes^L k(a)$ defines a map

$$(-)\otimes^L k(a): K_0(\mathcal{M}^{Z^0_C}) \to K_0(\mathcal{M}^{Z_c})$$

with $i_*(x)\otimes^L k(a) = 0$ for all $x \in K_0(\mathcal{M}^{Z_a})$. Also, we have the expression for the intersection multiplicity $m(Z_a; Z_C \cdot_C a)$:

$$m(Z_a; Z_C \cdot_C a) = \mathrm{lng}_{\mathcal{O}_{W^0_a, Z_a}}([\mathcal{O}_{Z^0_C}] \otimes^L k(a)). \tag{1.3.2.2}$$

Since

$$\mathrm{lng}_{\mathcal{O}_{W\times_k c, Z_c}}(\mathcal{O}_{W\times_k c, Z_c}) = m_{k(c):k}(Z_c/W),\ \mathrm{lng}_{\mathcal{O}_{W\times_k c, Z_c}}(\mathcal{O}_{Z_c}) = 1,$$
$$\mathrm{lng}_{\mathcal{O}_{W^0_a, Z_a}}(\mathcal{O}_{W^0_a, Z_a}) = m_{k(a):k}(Z_a/W),\ \mathrm{lng}_{\mathcal{O}_{W^0_a, Z_a}}(\mathcal{O}_{Z_a}) = 1,$$

we have

$$[\mathcal{O}_{W\times_k c, Z_c}] = m_{k(c):k}(Z_c/W)\cdot[\mathcal{O}_{Z_c}] \text{ in } K_0(\mathcal{M}^{Z_c}),$$
$$[\mathcal{O}_{W^0_a, Z_a}] = m_{k(a):k}(Z_a/W)\cdot[\mathcal{O}_{Z_a}] \text{ in } K_0(\mathcal{M}^{Z_a}),$$

where $[-]$ denotes the class in the appropriate K_0. Applying the localization sequence (1.3.2.1), and noting that $j^*(\mathcal{O}_{Z^0_C}) = \mathcal{O}_{Z_c}$, we have

$$[\mathcal{O}_{W^0_C, Z^0_C}] = m_{k(c):k}(Z_c/W)\cdot[\mathcal{O}_{Z^0_C}] + i_*(x) \text{ in } K_0(\mathcal{M}^{Z^0_C})$$

for some $x \in K_0(\mathcal{M}^{Z_a})$. This gives

$$\begin{aligned}[\mathcal{O}_{W^0_C, Z^0_C}]\otimes^L k(a) &= m_{k(c):k}(Z_c/W)\cdot[\mathcal{O}_{Z^0_C}]\otimes^L k(a) + i_*(x)\otimes^L k(a)\\ &= m_{k(c):k}(Z_c/W)\cdot[\mathcal{O}_{Z^0_C}]\otimes^L k(a).\end{aligned}$$

Applying the length homomorphism and using (1.3.2.2), we have

$$m_{k(c):k}(Z_c/W)\cdot m(Z_a; Z_C \cdot_C a) = \mathrm{lng}_{\mathcal{O}_{W^0_a, Z_a}}([\mathcal{O}_{W^0_C, Z_C}]\otimes^L k(a)). \tag{1.3.2.3}$$

Since k is a field, W is flat over k, hence W^0_C is flat over C. Thus

$$\begin{aligned}\mathrm{lng}_{\mathcal{O}_{W^0_a, Z_a}}([\mathcal{O}_{W^0_C, Z_C}]\otimes^L k(a)) &= \mathrm{lng}_{\mathcal{O}_{W^0_a, Z_a}}(\mathcal{O}_{W^0_C, Z_C}\otimes k(a))\\ &= \mathrm{lng}_{\mathcal{O}_{W^0_a, Z_a}}(\mathcal{O}_{W^0_a, Z_a})\\ &= m_{k(a):k}(Z_a/W).\end{aligned}$$

This, together with (1.3.2.3) gives $m(Z_a; Z_C \cdot_C a) = \dfrac{m_{k(a):k}(Z_a/W)}{m_{k(c):k}(Z_c/W)}$, as desired. □

1.4. Multiplicities over a normal base

1.4.1. PROPOSITION. *(i) Let S be a normal scheme, $p: W \to S$ an equi-dimensional, reduced irreducible S-scheme, s a point of S, and $\bar{W}$ an irreducible component of the reduced fiber $(W_s)_{\mathrm{red}}$. Let $\mathcal{O}$ be a discrete valuation ring, $T = \operatorname{Spec}\mathcal{O}$, and t the closed point of T. Suppose we have a birational morphism, essentially of finite type, $f: T \to S$, with $f(t) = s$. Let $W_T = (W \times_S T)_{\mathrm{red}}$, with induced maps $p_T: W_T \to T$ and $\tilde{f}: W_T \to W$. Let $\bar{W}_T$ be an irreducible component of the reduced fiber $p_T^{-1}(t)_{\mathrm{red}}$ with $\tilde{f}(\bar{W}_T) = \bar{W}$, and let m_T be the intersection multiplicity $m_T = m(\bar{W}_T, W_T \cdot_T t) = \mathrm{lng}_{\mathcal{O}_{W_T,\bar{W}_T}}(\mathcal{O}_{W_T,\bar{W}_T} \otimes k(t)) > 0$. Then the positive rational number*

$$m(\bar{W}; W, s) := \frac{m_T}{m_{k(t):k(s)}(\bar{W}_T/\bar{W})} \tag{1.4.1.1}$$

depends only on s, W and the choice of component $\bar{W}$.
(ii) If W has finite Tor-dimension over S, then

$$m(\bar{W}; W, s) = m(\bar{W}; W \cdot_S s).$$

PROOF. We first prove (i). We may assume that S is local, and W is affine over S. Since W and T are S-schemes essentially of finite type, we may assume that S is essentially of finite type over $\mathbb{Z}$.

Suppose we have two such maps, $f_i: T_i \to S$, $i = 1, 2$. By Lemma 1.2.4, we may find a regular two-dimensional scheme X, with a birational map $p: X \to S$, a connected curve C on X, contained in $p^{-1}(s)$, and irreducible components C_1, C_2 of C such that the maps f_i are the maps $p_i: \operatorname{Spec}\mathcal{O}_{X,C_i} \to S$ induced by p. In addition, all the components of C are regular. By induction, we may assume that C_1 and C_2 intersect at a point a of X. Let c_i be the generic point of C_i.

Let W_X be the reduced pull-back $(W \times_S X)_{\mathrm{red}}$. Since X, C_1 and C_2 are regular, the intersection cycles

$$W_1 := W_X \cdot_X C_1,\ W_2 := W_X \cdot_X C_2,\ W_X \cdot_X a,$$
$$W_1 \cdot_{C_1} a, \text{ and } W_2 \cdot_{C_2} a$$

are all defined. From the associativity relation (1.1.4.1), we have

$$W_X \cdot_X a = W_1 \cdot_{C_1} a = W_2 \cdot_{C_2} a. \tag{1.4.1.2}$$

Let Z_a be an irreducible component of $(W_X \times_X a)_{\mathrm{red}}$ lying over $\bar{W}$, let Z_1 be the reduced irreducible component of the cycle W_1 containing Z_a, and let Z_2 be the reduced irreducible component of the cycle W_2 containing Z_a. We have the intersection multiplicities

$$m_1 = m(Z_1; W_X \cdot_X C_1),\ m_2 = m(Z_2; W_X \cdot_X C_2),\ m_a = m(Z_a; W_X \cdot_X a).$$

By Lemma 1.3.2, Z_a appears with multiplicity

$$m(Z_a; Z_i \cdot_{C_i} a) = \frac{m_{k(a):k(s)}(Z_a/\bar{W})}{m_{k(c_i):k(s)}(Z_{c_i}/\bar{W})}$$

in $Z_i \cdot_{C_i} a$, hence, by (1.4.1.2), we have

$$m_1 \cdot \frac{m_{k(a):k(s)}(Z_a/\bar{W})}{m_{k(c_1):k(s)}(Z_{c_1}/\bar{W})} = m_a = m_2 \cdot \frac{m_{k(a):k(s)}(Z_a\bar{W})}{m_{k(c_2):k(s)}(Z_{c_2}/\bar{W})},$$

which gives the desired identity

$$\frac{m_1}{m_{k(c_1):k(s)}} = \frac{m_2}{m_{k(c_2):k(s)}}.$$

To prove (ii), we may assume that S is local, with closed point s, and that W is a closed subscheme of $Y := \mathbb{A}^N_S$ for some N. Since Y is smooth over S, the assumption that W has finite Tor dimension over S implies there is a finite resolution of $\mathcal{O}_W$ by locally free coherent $\mathcal{O}_Y$-modules, $\mathcal{P}_* \to \mathcal{O}_W$. We may pull this complex back to $Y_T := Y \times_S T$, forming the complex $p_1^*\mathcal{P}_* \to \mathcal{O}_{W\times_S T}$; composing with the surjection $\mathcal{O}_{W\times_S T} \to \mathcal{O}_{W_T}$ gives us the complex

$$p_1^*\mathcal{P}_* \to \mathcal{O}_{W_T} \tag{1.4.1.3}$$

on Y_T. Let Y_t and W_t be the fiber of Y and W_T over the closed point t of T and let Y_s be the fiber of Y over s. Let $\mathcal{M}_{Y_T}^{W_T}$ be the abelian category of coherent sheaves on Y_T with support in W_T, and let $\mathcal{M}_{Y_t}^{W_t}$ be the abelian category of coherent sheaves on Y_t with support in W_t. The inclusion $i: Y_t \to Y_T$ induces the functor $i_*: \mathcal{M}_{Y_t}^{W_t} \to \mathcal{M}_{Y_T}^{W_T}$. As the map $f: T \to S$ is birational, the complex (1.4.1.3) is exact away from Y_t. From this, we arrive at the identity in $K_0(\mathcal{M}_{Y_T}^{W_T})$:

$$[\mathcal{O}_{W_T}] = [H_0(p_1^*\mathcal{P}_*)] + i_*(x), \tag{1.4.1.4}$$

where $[\mathcal{F}]$ denotes the class in $K_0(\mathcal{M}_{Y_T}^{W_T})$ of a coherent sheaf $\mathcal{F}$, and x is some element of $K_0(\mathcal{M}_{Y_t}^{W_t})$. In addition, $H_p(p_1^*\mathcal{P}_*)$ is in $\mathcal{M}_{Y_t}^{W_t}$ for all $p > 0$. By Serre's intersection vanishing formula [**112**] (for the case of a DVR), we have

$$\sum_{q=0}^{\infty}(-1)^q \mathrm{lng}_{\mathcal{O}_{Y_t,\bar{W}_T}}(\mathrm{Tor}_q^{\mathcal{O}_T}(\mathcal{F}, k(t)) \otimes \mathcal{O}_{Y_t,\bar{W}_T}) = 0$$

for all $\mathcal{F}$ in $\mathcal{M}_{Y_t}^{W_t}$ and all components $\bar{W}_T$ of W_t; from this and (1.4.1.4) it follows that

$$\sum_{p=0}^{\infty}(-1)^p \mathrm{lng}_{\mathcal{O}_{Y_t,\bar{W}_T}}(H_p(p_1^*\mathcal{P}_* \otimes_{\mathcal{O}_T} k(t)) \otimes \mathcal{O}_{Y_t,\bar{W}_T}) = \mathrm{lng}_{\mathcal{O}_{Y_t,\bar{W}_T}}(\mathcal{O}_{W_T,\bar{W}_T} \otimes k(t)). \tag{1.4.1.5}$$

On the other hand, the extension $k(t) \supset k(s)$ is flat, hence

$$H_p(p_1^*\mathcal{P}_* \otimes_{\mathcal{O}_T} k(t)) = H_p(\mathcal{P}_* \otimes_{\mathcal{O}_S} k(s)) \otimes_{k(s)} k(t). \tag{1.4.1.6}$$

Also, if $\mathcal{F}$ is a coherent sheaf on Y_s, supported on $\bar{W}$, we have

$$\mathrm{lng}_{\mathcal{O}_{Y_t,\bar{W}_T}}(\mathcal{F} \otimes_{k(s)} k(t)) = m_{k(t):k(s)}(\bar{W}_T/\bar{W}) \cdot \mathrm{lng}_{\mathcal{O}_{Y_s,\bar{W}}}(\mathcal{F}). \tag{1.4.1.7}$$

From (1.4.1.5)-(1.4.1.7), we have

$$\begin{aligned}\mathrm{lng}_{\mathcal{O}_{Y_t,\bar{W}_T}}(\mathcal{O}_{W_T,\bar{W}_T}\otimes k(t))\\ &= \sum_{p=0}^{\infty} m_{k(t):k(s)}(\bar{W}_T/\bar{W})\cdot(-1)^p\mathrm{lng}_{\mathcal{O}_{Y_s,\bar{W}_T}}(H_p(\mathcal{P}_*\otimes_{\mathcal{O}_S}k(s))\otimes\mathcal{O}_{Y_t,\bar{W}_T})\\ &= m_{k(t):k(s)}(\bar{W}_T/\bar{W})\cdot m(\bar{W},W\cdot_S s).\end{aligned}$$

This proves (ii). □

1.4.2. DEFINITION. (i) Let S be a normal scheme, $p\colon W\to S$ an irreducible reduced S-scheme of dimension d over S, s a point of S. Let $W(s)$ be the dimension d cycle over $k(s)$ with positive rational coefficients,

$$W(s) := \sum_{\bar{W}} m(\bar{W};W,s)\cdot\bar{W},$$

where the sum is over all irreducible components $\bar{W}$ of $W\times_S s$, and the rational number $m(\bar{W};W,s)>0$ is given by (1.4.1.1). We extend the definition to $\mathbb{Q}$-linear sums of such W by linearity.
(ii) Let $f\colon T\to S$ be a map of reduced normal schemes, and $W\to S$ be a relative dimension d reduced irreducible scheme over S. Let t be the generic point of T, and let $s=f(t)$. For an irreducible component W^i_T of $(W\times_S T)_{\mathrm{red}}$, let W^i_s be the irreducible component $f(W^i_T)\cap W_s$ of the fiber W_s. Define the multiplicity $m(W^i_T;W,f)>0$ by

$$m(W^i_T;W,f) = m(W^i_s;W,s)\cdot m_{k(t):k(s)}(W^i_T/W^i_s).$$

We set $f^*(W)$ to be the effective dimension d $\mathbb{Q}$-cycle over T:

$$f^*(W) := \sum_i m(W^i_T;W,f)\cdot W^i_T,$$

where the sum is over all irreducible components W^i_T of $(W\times_S T)_{\mathrm{red}}$.
(iii) Let $p\colon T\to S$ be a map of normal schemes, $X\to S$ an S-scheme, essentially of finite type over S. Sending W to $p^*(W)$ defines the homomorphisms

$$p^*\colon \mathcal{C}_d(X/S)^{\geq 0}_{\mathbb{Q}}\to\mathcal{C}_d(X\times_S T/T)^{\geq 0}_{\mathbb{Q}},$$
$$p^*\colon \mathcal{C}_d(X/S)_{\mathbb{Q}}\to\mathcal{C}_d(X\times_S T/T)_{\mathbb{Q}}.$$

1.4.3. REMARKS. (i) If W is a d-dimensional cycle over S, and if each irreducible component of W has finite Tor-dimension over S, then Proposition 1.4.1(ii) shows that, for each point s of S, $W(s)$ is the cycle $W\cdot_S s$. In particular, $W(s)$ has $\mathbb{Z}$-coefficients for all $s\in S$. Similarly, for a morphism of schemes $f\colon T\to S$, Proposition 1.4.1(ii) and the associativity (1.1.4.1) gives the identity

$$f^*(W) = W\cdot_S T;$$

this shows that $f^*(W)$ has $\mathbb{Z}$-coefficients under the assumption of finite Tor-dimension.
(ii) Let $Z\to S$ be irreducible and of dimension d over a normal scheme S, s a generic point of S. Taking $\mathcal{O}$ to be the localization of $k(s)[t]$ at the ideal (t), we see that $Z(s)$ is $1\cdot Z_s$. This shows that, for a dominant morphism of normal schemes, the pull-back defined in §1.1.5 agrees with the pull-back defined in Definition 1.4.2.

1.4.4. PROPOSITION. *Let $f: T \to S$, $g: U \to T$ be maps of normal schemes, $X \to S$ an S-scheme, essentially of finite type over S. Then the diagram*

$$\begin{array}{ccc} \mathcal{C}_d(X/S)_{\mathbb{Q}} & \xrightarrow{f^*} & \mathcal{C}_d(X \times_S T/T)_{\mathbb{Q}} \\ & \searrow^{(g\circ f)^*} & \downarrow{\scriptstyle g^*} \\ & & \mathcal{C}_d(X \times_S U/U)_{\mathbb{Q}} \end{array}$$

commutes.

PROOF. If T, S and U are all regular, the result is a consequence of the associativity of intersection multiplicities and Remark 1.4.3.

Take W in $\mathcal{C}_d(X/S)_{\mathbb{Q}}$. Since the pull-back of cycles is determined by the pull-back at the generic point, we may assume that $U = \operatorname{Spec} L$ for some field L. We may then reduce to proving two cases:

(a) U is a point t of T.
(b) T is a point s of S, and $U \to T$ is a map of reduced one-point schemes $u \to s$.

In the case (b), we have $f^*(W) = W(s)$, and the identity

$$g^*(W(s)) = (f \circ g)^*(W)$$

follows immediately from the definition. We now prove case (a).

We may assume that T is local, with closed point t, and that S is local, with closed point $s = f(t)$. The assertion is local on X, so we may assume that X is affine. We may therefore assume that S and T are essentially of finite type over $\mathbb{Z}$.

We have the following commutative diagram

$$\begin{array}{ccccc} t_2 & \hookrightarrow & T_2 & & \\ {\scriptstyle g_2^t}\downarrow & & \downarrow{\scriptstyle g_2} & & \\ t & \hookrightarrow & T & \xrightarrow{f} & S \\ \downarrow & & \downarrow{\scriptstyle g_1} & \nearrow & \\ t_1 & \hookrightarrow & T_1, & & \end{array}$$

where T_1 is the normalization of the closure of $f(T)$ in S, g_1 the induced morphism, and $t_1 := g_1(t)$. T_2 is the normalization of an irreducible component of $g_1^{-1}(t_1)$ containing t, g_2 is the induced map, t_2 is a point of $g_2^{-1}(t)$, and g_2^t is the restriction of g_2. From this diagram, the case (b), and the fact that the map g_2^{t*} is injective, we reduce to proving (a) in two cases:

(a1) f is dominant, and t is a generic point of $f^{-1}(s)$.
(a2) T is the normalization of a closed subset T' of S, and f the induced map.

For case (a1), we may assume that s is not a generic point of S; otherwise we are in case (b). Let $p: Y \to S$ be the normalization of the blow-up of S at s; since $Y \to S$ is projective and birational, so is $p_1: T \times_S Y \to T$. Thus, $p_1^{-1}(t)$ is non-empty; letting $q: Y_T \to T$ be the normalization of the blow-up of $T \times_S Y$ along

$p_1^{-1}(t)$, we have the commutative square

$$\begin{array}{ccc} Y_T & \xrightarrow{q} & T \\ \downarrow & & \downarrow \\ Y & \xrightarrow[p]{} & S. \end{array}$$

Let E be an irreducible component of $p^{-1}(s)$, let E' be an irreducible component of $q^{-1}(t)$ lying over E, and let $\mathcal{O}$ and $\mathcal{O}'$ be the discrete valuation rings $\mathcal{O} = \mathcal{O}_{S,E}$, $\mathcal{O}' = \mathcal{O}_{T,E'}$. This gives us the commutative square

$$\begin{array}{ccc} \operatorname{Spec} \mathcal{O}' & \xrightarrow{q} & T \\ \downarrow & & \downarrow \\ \operatorname{Spec} \mathcal{O} & \xrightarrow[p]{} & S; \end{array}$$

the result in this case follows from the definition of pull-back, and the functoriality in the regular case.

We now prove case (a2). Let $p_{T'}: Y_T \to T'$ be the normalization of the blow-up of T' at s. Then $p_{T'}$ is projective and birational, hence is the blow-up of T' along some proper subscheme Z. Let $p_S: Y_S \to S$ be be blow-up of S along Z. We arrive at the commutative diagram

$$\text{(1.4.4.1)} \qquad \begin{array}{ccc} Y_T & \xrightarrow{i_Y} & Y_S \\ {\scriptstyle p_{T'}}\downarrow & & \downarrow{\scriptstyle p_S} \\ T' & \xrightarrow[i_{T'}]{} & S \end{array}$$

with $i_{T'}$ the inclusion, and i_Y a closed embedding. This diagram identifies Y_T with the proper transform $p_S^{-1}[T']$ of T'. Blowing up Y_S along $p_S^{-1}(s)$, normalizing and changing notation, we may assume that $p_S^{-1}(s)$ is the support of a Cartier divisor on Y_S. It may of course happen that the proper transform $p_S^{-1}[T']$ is now no longer normal, but as the map $p_S^{-1}[T']^N \to p_S^{-1}[T']$ is also projective and birational, we may further blow-up Y_S along the appropriate subscheme of $p_S^{-1}[T']$, and then normalize, to normalize $p_S^{-1}[T']$. Putting this all together, we have the commutative diagram (1.4.4.1) such that Y_T and Y_S are normal, $p_{T'}$ and p_S are birational, $p_{T'}^{-1}(s)$ is the support of a Cartier divisor on Y_T, $p_S^{-1}(s)$ is the support of a Cartier divisor on Y_S, and i_Y identifies Y_T with the proper transform $p_Y^{-1}[T']$. Since Y_T is normal, the map $p_{T'}$ factors through the map $T \to T'$, giving the commutative diagram

$$\begin{array}{ccc} Y_T & \xrightarrow{i_Y} & Y_S \\ {\scriptstyle p_T}\downarrow & & \downarrow{\scriptstyle p_S} \\ T & \xrightarrow[f]{} & S. \end{array}$$

We now let $q: Y \to Y_S$ be the normalization of the blow-up of Y_S along Y_T, and let $p: Y \to S$ be the composition $p_S \circ q$. Since $p_S^{-1}(s)$ is the support of a Cartier divisor on Y_S, the same is true of $p^{-1}(s)$. As the map p_S is an isomorphism over the generic point of T', there is an irreducible component D of the pure codimension one closed subset $q^{-1}(Y_T)$ such that D dominates T. As q is proper, this implies that $q(D) = Y_T$, thus, if F' is an irreducible component of $p_{T'}^{-1}(s) = p_S^{-1}(s) \cap Y_T$, there is an irreducible component F of $D \cap p^{-1}(s)$ with $q(F) = F'$. As $p_T^{-1}(t)$ is a connected component of $p_{T'}^{-1}(s)$, the same holds for each irreducible component F' of $p_T^{-1}(t)$. Since $p^{-1}(s)$ is the support of a Cartier divisor on Y, F is necessarily a pure codimension one subset of D. Let E be a component of $p^{-1}(s)$ such that E contains F, where we now take F' to be an irreducible component of $p_T^{-1}(t)$.

Let X be the localization, $X = \operatorname{Spec} \mathcal{O}_{Y,F}$, and let D', E' be the respective intersections of D and E with X. Since F has codimension two on Y, $\mathcal{O}_{Y,F}$ is a local normal noetherian domain of Krull dimension two; as we have assumed that S is essentially of finite type over $\mathbb{Z}$, so is X. Thus, we may resolve the singularities of X by the blow-up of a sheaf of ideals; extending this sheaf of ideals to Y, blowing up, normalizing and changing notation, we may assume that X is regular to begin with. Similarly, we may assume that D' and E' are regular.

Denote the closed point of X by a, and the birational map $X \to S$ induced by p by $p_X: X \to S$. Let $W_X \to X$ be reduced X-scheme $(W \times_S X)_{\mathrm{red}}$. The map $q: D \to T$ induces the dominant map $q': D' \to T$ with $q'(a) = t$ and the map $p_X: X \to S$ induces the maps $p_{E'}: E' \to s$, $p_{a,s}: a \to s$, and $p_a: a \to S$. The restriction of f to t gives the map $f_t: t \to S$.

Let

$$i_{D'}: D' \to X, \quad i_{E'}: E' \to X, \quad i_a: a \to X,$$
$$i_{D',a}: a \to D' \quad , i_{E',a}: a \to E'$$

be the inclusions, and let $q'_a: a \to t$ be the restriction of q' to a. By definition of the pull-backs f^* and q'^*, we have

$$q'^*(f^*(W)) = i_{D'}^*\big((W \times_S X)_{\mathrm{red}}\big) = (W \times_S X)_{\mathrm{red}} \cdot_X D'. \tag{1.4.4.2}$$

Similarly, by definition of $W(s)$, we have

$$p_{E'}^*(W(s)) = i_{E'}^*\big((W \times_S X)_{\mathrm{red}}\big) = (W \times_S X)_{\mathrm{red}} \cdot_X E'. \tag{1.4.4.3}$$

Since q' is dominant, we have by (a2)

$$i_{D',a}^*(q'^*(f^*(W))) = q_a'^*(f^*(W)(t)). \tag{1.4.4.4}$$

Since D' is regular and a has codimension one on D', we have (see Proposition 1.4.1(ii)) $i_{D',a}^*(q'^*(f^*(W))) = q'^*(f^*(W)) \cdot_{D'} a$; using (1.4.4.2) and the associativity of intersections (1.1.4.1), we have $i_{D',a}^*(q'^*(f^*(W))) = (W \times_S X)_{\mathrm{red}} \cdot_X \cdot a$. From (1.4.4.4), this gives

$$q_a'^*(f^*(W)(t)) = (W \times_S X)_{\mathrm{red}} \cdot_X \cdot a. \tag{1.4.4.5}$$

Since the map $E' \to s$ is dominant, we have $i_{E',a}^*(p_{E'}^*(W(s))) = p_{a,s}^*(W(s))$. Arguing as above, starting with (1.4.4.3) instead of (1.4.4.2), this gives $p_{a,s}^*(W(s)) = (W \times_S X)_{\mathrm{red}} \cdot_X \cdot a$, hence from (1.4.4.5) we have $q_a'^*(f^*(W)(t)) = p_{a,s}^*(W(s))$. From the case (b), we have $p_{a,s}^*(W(s)) = p_a^*(W) = q_a'^*(f_t^*(W))$, giving $q_a'^*(f_t^*(W)) = q_a'^*(f^*(W)(t))$. As the map $q_a'^*$ is clearly injective, this gives the desired relation $f_t^*(W) = f^*(W)(t)$, completing the proof. □

1.4.5. Let S be a reduced scheme, $X \to S$ an S-scheme, essentially of finite type over S. By Proposition 1.4.4, sending an S-morphism $p: T' \to T$ of normal S-schemes to the homomorphism $p^*: \mathcal{C}_d(X \times_S T/T)_{\mathbb{Q}} \to \mathcal{C}_d(X \times_S T'/T')_{\mathbb{Q}}$ defines the contravariant functor

$$\mathcal{C}_d(X \times_S (-)/(-))_{\mathbb{Q}}: \mathbf{Nor}_S^{\mathrm{op}} \to \mathbf{Mod}_{\mathbb{Q}} \tag{1.4.5.1}$$

from the category $\mathbf{Nor}_S$ of normal S-schemes to $\mathbb{Q}$-vector spaces. This is in fact a Zariski sheaf on $\mathbf{Nor}_S$.

More generally, let $\mathbf{Nor}_{X\to S}$ be the category of morphisms $U \to T$, with T in $\mathbf{Nor}_S$, and $U \to T$ a localization of $X \times_S T$. Let $\mathbf{Nor}_{*\to*}$ be the category of morphisms $X \to S$, essentially of finite type, with S normal, where a map $(U \to T) \to (X \to S)$ is a morphism in $\mathbf{Nor}_{X\to S}$. The restriction map

$$\mathcal{C}_d(X \times_S T/T)_{\mathbb{Q}} \to \mathcal{C}_d(U/T)_{\mathbb{Q}}$$

defined in Remark 1.1.3 extends (1.4.5.1) to the functor

$$\mathcal{C}_d(-/-)_{\mathbb{Q}}: \mathbf{Nor}_{*\to*}^{\mathrm{op}} \to \mathbf{Mod}_{\mathbb{Q}}. \tag{1.4.5.2}$$

1.4.6. *Pull-back over S.* Now let S be a normal scheme, $f: X \to Y$ be a map of smooth S-schemes, essentially of finite type over S, and let Z be an irreducible relative codimension d cycle on Y. If each irreducible component of $f^{-1}(Z)$ has codimension d over S, we say that $f^*(Z)$ *is defined.* Let s be the generic point of S and assume $f^*(Z)$ is defined; the pull-back $f_s^*(Z_s)$ is then defined, with multiplicities being the usual alternating sums of Tor's, since the map $f_s: X_s \to Y_s$ is a map of regular schemes. Define the pull-back $f^*(Z)$ by

$$f^*(Z) = \sum_{i=1}^{r} m_i W_i$$

where $W_1, \ldots, W_r$ are the irreducible components of $f^{-1}(Z)$, and the integers m_i are determined by $f_s^*(Z_s) = \sum_{i=1}^{r} m_i W_{is}$. We extend the definition of f^* to arbitrary cycles, satisfying the condition on the codimension over S of the inverse images, by linearity. We extend the definition of f^* to $\mathbb{Q}$-cycles similarly.

1.4.7. PROPOSITION. *Let S be a normal scheme, $X \to S$, $Y \to S$ be smooth S-schemes, essentially of finite type over S, $f: X \to Y$ a map over S, and Z an element of $\mathcal{C}^d(Y/S)_{\mathbb{Q}}$ such that $f^*(Z)$ is defined.*
(i) Let s be a point of S. Then $f_s^*(Z(s))$ is defined and

$$f_s^*(Z(s)) = f^*(Z)(s).$$

(ii) Let $p: T \to S$ be a map of normal irreducible schemes, $f_T: X \times_S T \to Y \times_S T$ the map induced by f. Then $f_T^*(p^*(Z))$ is defined, and

$$p^*(f^*(Z)) = f_T^*(p^*(Z)).$$

PROOF. The fact that $f_T^*(p^*(Z))$ is defined if $f^*(Z)$ is defined is obvious. Clearly (ii) implies (i). On the other hand, (ii) follows from (i), together with the case of a dominant map $f: T \to S$. In this case, the computation of $p^*(f^*(Z))$ and $f_T^*(p^*(Z))$ can be made at the generic point of T, where the result follows from the compatibility of intersection multiplicities with flat pull-back. We now prove (i).

Using the definition of $W(s)$ for a cycle W, we reduce immediately to the case $S = \operatorname{Spec} \mathcal{O}$, where $\mathcal{O}$ is a discrete valuation ring, and s is the closed point of S.

As in this case X, Y, X_s and Y_s are regular, and we have $W(s) = W \cdot_S s$ for all equi-dimensional cycles W over S. The result is then a straightforward application of the associativity of intersection multiplicities. $\square$

2. Cycles over a reduced scheme

2.1. Gluing cycles

We extend the cycle theory of §1.4 to arbitrary reduced schemes by a gluing construction. In this section, we take all schemes to be essentially of finite type over $\mathbb{Z}$.

2.1.1. Let S be a reduced scheme, $p\colon S^N \to S$ the normalization of S, $X \to S$ an S-scheme essentially of finite type. If Z is a reduced, irreducible closed subscheme of X, of dimension d over S, there is a unique irreducible component Z' of $Z \times_S S^N$ which dominates Z; Z' then has dimension d over S^N. Sending Z to Z' and extending by linearity gives the maps

$$p^*\colon \mathcal{C}_d(X/S)_{\mathbb{Z}} \to \mathcal{C}_d(X \times_S S^N/S^N)_{\mathbb{Z}},$$
$$p^*\colon \mathcal{C}_d(X/S)_{\mathbb{Q}} \to \mathcal{C}_d(X \times_S S^N/S^N)_{\mathbb{Q}}.$$

2.1.2. Definition. Let S be a reduced scheme, $p\colon S^N \to S$ the normalization of S, $X \to S$ an S-scheme, essentially of finite type over S.

(i) Define the submonoid $\mathcal{Z}_d(X/S)^{\geq 0}_{\mathbb{Z}}$ of $\mathcal{C}_d(X/S)^{\geq 0}_{\mathbb{Z}}$ by the following condition: Let Z be in $\mathcal{C}_d(X/S)^{\geq 0}_{\mathbb{Z}}$. Then Z is in $\mathcal{Z}_d(X/S)^{\geq 0}_{\mathbb{Z}}$ if, for each field K, and for each pair of morphisms $f_1, f_2\colon \operatorname{Spec} K \to S^N$ satisfying $p \circ f_1 = p \circ f_2$, we have $f_1^*(p^*(Z)) = f_2^*(p^*(Z))$. We define $\mathcal{Z}_d(X/S)_{\mathbb{Z}}$ to be the subgroup of $\mathcal{C}_d(X/S)$ generated by $\mathcal{Z}_d(X/S)^{\geq 0}_{\mathbb{Z}}$.

The $\mathbb{Q}$-sub-cone $\mathcal{Z}_d(X/S)^{\geq 0}_{\mathbb{Q}}$ of $\mathcal{C}_d(X/S)^{\geq 0}_{\mathbb{Q}}$ is defined by the same condition:

$$\mathcal{Z}_d(X/S)^{\geq 0}_{\mathbb{Q}} = \mathbb{Q}^{\geq 0} \cdot \mathcal{Z}_d(X/S)^{\geq 0}_{\mathbb{Z}}.$$

We define $\mathcal{Z}_d(X/S)_{\mathbb{Q}}$ to be the $\mathbb{Q}$-vector subspace of $\mathcal{C}_d(X/S)_{\mathbb{Q}}$ generated by the cone $\mathcal{Z}_d(X/S)^{\geq 0}_{\mathbb{Q}}$.

(ii) If $X \to S$ is an S-scheme with connected components X_i, such that each X_i is of dimension n_i over S, we let

$$\mathcal{Z}^d(X/S)_{\mathbb{Q}} = \oplus_i \mathcal{Z}_{n_i-d}(X_i/S)_{\mathbb{Q}},$$
$$\mathcal{Z}^d(X/S)_{\mathbb{Z}} = \oplus_i \mathcal{Z}_{n_i-d}(X_i/S)_{\mathbb{Z}}.$$

2.1.3. The inclusion $\mathcal{Z}_d(X/S)^{\geq 0}_{\mathbb{Z}} \to \mathcal{Z}_d(X/S)^{\geq 0}_{\mathbb{Q}}$ induces the inclusion $\mathcal{Z}_d(X/S)_{\mathbb{Z}} \to \mathcal{Z}_d(X/S)_{\mathbb{Q}}$, and we have

$$\mathcal{Z}_d(X/S)_{\mathbb{Q}} = \mathbb{Q} \cdot \mathcal{Z}_d(X/S)_{\mathbb{Z}} \cong \mathcal{Z}_d(X/S)_{\mathbb{Z}} \otimes_{\mathbb{Z}} \mathbb{Q}.$$

2.1.4. Lemma. *Let S be a reduced scheme, $X \to S$ an S-scheme, essentially of finite type over S, $p\colon S^N \to S$ the normalization, W an element of $\mathcal{Z}_d(X/S)^{\geq 0}_{\mathbb{Q}}$, and s a point of S. Let $|W|$ be the support of W, and $|p^*(W)|$ the support of $p^*(W)$. Then*

(i) *$|p^*(W)| = (|W| \times_S S^N)_{\mathrm{red}}$; in particular, if s_1, s_2 are points of S^N with $p(s_1) = p(s_2) = s$, and if $\bar{W}$ is an irreducible component of $(|W| \times_S s)_{\mathrm{red}}$, then there are irreducible components Z_i of $(|p^*(W)|_{s_i})_{\mathrm{red}}$, $i = 1, 2$, with $p_1(Z_i) = \bar{W}$, $i = 1, 2$.*

(ii) *given* Z_1, Z_2 *as in (i), we have*

$$\frac{m(Z_1; p^*(W), s_1)}{m_{k(s_1):k(s)}(Z_1/\bar{W})} = \frac{m(Z_2; p^*(W), s_2)}{m_{k(s_2):k(s)}(Z_2/\bar{W})}.$$

PROOF. For (i), write W as a sum, $W = \sum_i r_i W_i$, $r_i \in \mathbb{Q}$, $r_i > 0$, with the W_i irreducible. For each i, let W_i' be the unique irreducible component of $(W_i \times_S S^N)_{\text{red}}$ dominating W_i; hence $p^*(W) = \sum_i r_i W_i'$. Since $S^N \to S$ is finite, so is $p_1: W_i' \to W_i$; in particular, each of the maps $W_i' \to W_i$ is surjective. Take a geometric point x of $(|W| \times_S S^N)_{\text{red}}$, and let $z := p_1(x)$. Then x is a geometric point of $(W_i \times_S S^N)_{\text{red}}$ for some i, hence there is a geometric point y of W_i' with $p_1(y) = p_1(x) = z$. If $p_2(y) = p_2(x)$, then $x = y$, and we are done; if not, let $s_1 = p_2(x)$, $s_2 = p_2(y)$. As $p_1(x) = p_1(y)$, we have $p(s_1) = p(s_2)$. Thus there is a field K, and morphisms

$$f_i: t \to S^N; \quad i = 1, 2; \quad t := \operatorname{Spec} K,$$

such that $f_i(t) = s_i$, and $p \circ f_1 = p \circ f_2$.

Since W is in $\mathcal{Z}_d(X/S)_{\mathbb{Q}}$, we have $f_1^*(p^*(W)) = f_2^*(p^*(W))$. Since $y \times_{f_2} t = z \times_S t$ is in the support of $f_2^*(p^*(W))$, this implies that $z \times_S t = x \times_{f_1} t$ is in the support of $f_1^*(p^*(W))$ as well. Thus, as the support of $f_1^*(p^*(W))$ is $(|p^*(W)| \times_{f_1} t)_{\text{red}}$, x is in $|p^*(W)|_{\text{geom}}$, showing that $|p^*(W)| \supset (|W| \times_S S^N)_{\text{red}}$. The reverse inclusion is true for trivial reasons, completing the proof of (i).

For (ii), let F be the algebraic closure of $k(s)$ in the field of rational functions on $\bar{W}$. Then, for each field extension L of $k(s)$, there is a canonical bijection between the irreducible components of $(\bar{W} \times_{k(s)} L)_{\text{red}}$ and the simple factors of the semi-simple L-algebra $F \otimes_{k(s)} L/\text{rad}$. Let K_i be the simple factor of $F \otimes_{k(s)} k(s_i)/\text{rad}$ corresponding to Z_i, and let K be a simple factor of $K_1 \otimes_{k(s)} K_2/\text{rad}$. Let $t = \operatorname{Spec} K$. As $k(s_i)$ is embedded in $F \otimes_{k(s)} k(s_i)$ as $1 \otimes k(s_i)$, we have the canonical homomorphism $k(s_i) \to K_i$, $i = 1, 2$, giving the diagram of fields

$$\begin{array}{ccc} k(s_1) & \longrightarrow & K \\ \uparrow & & \uparrow \\ k(s) & \longrightarrow & k(s_2). \end{array}$$

This in turn defines morphisms $f_i: t \to S^N$ with $p \circ f_1 = p \circ f_2$, and with $f_i(t) = s_i$. Let $F_i: \bar{W} \times_s t \to \bar{W} \times_s s_i$ and $f_i^s: t \to s_i$ be the maps induced by f_i, $i = 1, 2$.

By construction, there is a simple factor L of $F \otimes_{k(s)} K/\text{rad}$ such that K_1 and K_2 map into L under the canonical homomorphisms $F \otimes_{k(s)} k(s_i) \to F \otimes_{k(s)} K$. From this it follows that there is an irreducible component Y of $\bar{W} \times_s t$ such that $F_i(Y) = Z_i$.

By the functoriality of pull-back, we have $f_i^*(p^*(W)) = f_i^{s*}(p^*(W)(s_i))$. The multiplicity of Y in $f_i^{s*}(p^*(W)(s_i))$ is given by

$$m(Y; f_i^{s*}(p^*(W)(s_i))) = m(Z_i; p^*(W)(s_i)) \cdot m_{K:k(s_i)}(Y/Z_i). \tag{2.1.4.1}$$

By assumption, we have $f_1^*(p^*(W)) = f_2^*(p^*(W))$; with (2.1.4.1) this implies

(2.1.4.2)

$$m(Z_1; p^*(W)(s_1)) \cdot m_{K:k(s_1)}(Y/Z_1) = m(Z_2; p^*(W)(s_2)) \cdot m_{K:k(s_2)}(Y/Z_2).$$

By the multiplicativity relation (1.3.1.1), we have

$$m_{K:k(s_1)}(Y/Z_1) \cdot m_{k(s_1):k(s)}(Z_1/\bar{W}) = m_{K:k(s_2)}(Y/Z_2) \cdot m_{k(s_2):k(s)}(Z_2/\bar{W}).$$

This together with (2.1.4.2) proves (ii). □

2.2. Multiplicities over a reduced base

2.2.1. DEFINITION. Let S be a reduced scheme, essentially of finite type over $\mathbb{Z}$, and let $X \to S$ be an S-scheme, essentially of finite type.
(i) Let W be an element of $\mathcal{Z}_d(X/S)_{\mathbb{Q}}^{\geq 0}$, and s a point of S. Let $\bar{W}$ be an irreducible component of $(W \times_S s)_{\text{red}}$. Define the positive rational number $m(\bar{W}; W, s)$ by

$$m(\bar{W}; W, s) = \frac{m(Z'; p^*(W), s')}{m_{k(s'):k(s)}(Z'/\bar{W})},$$

where $s' \in S^N$ is a point lying over s, and Z' is an irreducible component of $(p^*(W)_{s'})_{\text{red}}$ lying over $\bar{W}$; by Lemma 2.1.4, $m(\bar{W}; W, s)$ is well-defined, and depends only on W, s and $\bar{W}$. We let $W(s)$ be the element of $\mathcal{C}_d(X_s/s)_{\mathbb{Q}}^{\geq 0}$ defined by

$$W(s) := \sum_{\bar{W}} m(\bar{W}; W, s) \cdot \bar{W},$$

where the sum is over all irreducible components $\bar{W}$ of $(W \times_S s)_{\text{red}}$. We extend the definition of $W(s)$ to $\mathcal{Z}_d(X/S)_{\mathbb{Q}}$ by linearity.
(ii) Let $\mathcal{Z}_d(X/S)$ be the subgroup of $\mathcal{Z}_d(X/S)_{\mathbb{Z}}$ generated by the elements W of $\mathcal{Z}_d(X/S)_{\mathbb{Z}}^{\geq 0}$ such that $W(s)$ is in $\mathcal{C}_d(X_s/s)_{\mathbb{Z}}$ for all $s \in S$.
(iii) If $X \to S$ is an S-scheme with connected components X_i, such that each X_i is of dimension n_i over S, we let

$$\mathcal{Z}^d(X/S) = \oplus_i \mathcal{Z}_{n_i - d}(X_i/S).$$

2.2.2. We note that, in case S is normal, we have $\mathcal{Z}_d(X/S)_{\mathbb{Q}} = \mathcal{C}_d(X/S)_{\mathbb{Q}}$, and the old definition of $W(s)$ agrees with the new one. In addition, the cycle $W(s)$ is characterized by the following identity in $\mathcal{C}_d(X \times_S s'/s')_{\mathbb{Q}}$: Let $p\colon S^N \to S$ be the normalization. Pick a point $s' \in S^N$ with $p(s') = s$. Then

$$p^*(W)(s') = p_{s'}^*(W(s)). \tag{2.2.2.1}$$

The identity (2.2.2.1) follows easily from the definition of $W(s)$ and the pull-back $p_{s'}^*$.

2.2.3. LEMMA. *Let S be a reduced scheme, essentially of finite type over $\mathbb{Z}$, and let $X \to S$ be an S-scheme, essentially of finite type. Let $p\colon U \to T$ be a map over S of reduced S-schemes, essentially of finite type. Then*
(i) There is a homomorphism $p^*\colon \mathcal{Z}_d(X \times_S T/T)_{\mathbb{Q}} \to \mathcal{Z}_d(X \times_S U/U)_{\mathbb{Q}}$ such that, for each point u of U, and each $W \in \mathcal{Z}_d(X \times_S T/T)_{\mathbb{Q}}$, we have

$$p^*(W)(u) = p_u^*(W(t)), \tag{2.2.3.1}$$

where $t = p(u)$, $p_u\colon u \to t$ is the restriction of p to u, and $p_u^*\colon \mathcal{C}_d(X \times_S t/t)_{\mathbb{Q}} \to \mathcal{C}_d(X \times_S u/u)_{\mathbb{Q}}$ is the pull-back homomorphism (see Definition 1.4.2).
(ii) In case U and T are normal, the homomorphism p^* agree with the pull-back homomorphism $p^*\colon \mathcal{C}_d(X \times_S T/T)_{\mathbb{Q}} \to \mathcal{C}_d(X \times_S U/U)_{\mathbb{Q}}$ of Definition 1.4.2.
(iii) p^* restricts to a homomorphism $p^*\colon \mathcal{Z}_d(X \times_S T/T) \to \mathcal{Z}_d(X \times_S U/U)$.

PROOF. The assertion (iii) follows directly from the formula (2.2.3.1). Indeed, for W in $\mathcal{Z}_d(X \times_S T/T)$ and u a point of U, the restriction $p_u : u \to p(u) = t$ is a map of regular schemes. By Remark 1.4.3, the map $p_u^*\colon \mathcal{C}_d(X \times_S t/t)_{\mathbb{Q}} \to \mathcal{C}_d(X \times_S u/u)_{\mathbb{Q}}$

restricts to $p_u^*:\mathcal{C}_d(X\times_S t/t)_{\mathbb{Z}}\to \mathcal{C}_d(X\times_S u/u)_{\mathbb{Z}}$. Thus, as $p^*(W)(u)=p_u^*(W(t))$ in $\mathcal{C}_d(X\times_S t/t)_{\mathbb{Z}}$, $p^*(W)$ is in $\mathcal{Z}_d(X\times_S T/T)$, as claimed. We now prove (i) and (ii).

Let U_{T^N} be the normalization of $(U\times_T T^N)_{\rm red}$, and let

$$q: U_{T^N}\to U,\ p^N: U_{T^N}\to T^N$$

be the maps induced by the projections $U\times_T T^N\to U$, $U\times_T T^N\to T^N$. Since U_{T^N} is normal, the map q factors through the normalization $p_U: U^N\to U$. This gives us the commutative diagram

$$\begin{array}{ccccc} U_{T^N} & \xrightarrow{q'} & U^N & \xrightarrow{p_U} & U \\ {\scriptstyle p^N}\downarrow & & {\scriptstyle q} & & \downarrow{\scriptstyle p} \\ T^N & & \xrightarrow{p_T} & & T. \end{array} \tag{2.2.3.2}$$

As p_T is a finite morphism, so is q; thus q' is also a finite morphism. The diagram (2.2.3.2) gives rise to the diagram

$$\begin{array}{ccccc} X\times_S U_{T^N} & \xrightarrow{Q'} & X\times_S U^N & \xrightarrow{P_U} & X\times_S U \\ {\scriptstyle P^N}\downarrow & & {\scriptstyle Q} & & \downarrow{\scriptstyle P} \\ X\times_S T^N & & \xrightarrow{P_T} & & X\times_S T \end{array}$$

with all horizontal morphisms being finite morphisms.

Let W be an element of $\mathcal{Z}_d(X\times_S T/T)_{\mathbb{Q}}^{\geq 0}$; we may write W as $\sum_{i=1}^n r_iW_i$, $r_i\in\mathbb{Q}, r_i>0$, where each W_i is a reduced, irreducible closed subscheme of $X\times_S T/T$, dominating an irreducible component of T. We have the cycle $p_T^*(W):=\sum_{i=1}^n r_iW_i'$, where W_i' is the unique component of $P_T^{-1}(W_i)$ dominating W_i.

Since $p^N: U_{T^N}\to T^N$ is an S-morphism of normal S-schemes, the pull-back $p^{N*}(p_T^*(W))$ is defined. By definition of this pull-back, we have $|p^{N*}(p_T^*(W))|=(P^N)^{-1}(|p_T^*(W)|)$, where $|-|$ denotes support. By Lemma 2.1.4, we have $|p_T^*(W)|=P_T^{-1}(|W|)$, hence $|p^{N*}(p_T^*(W))|=(P\circ Q)^{-1}(|W|)$. As the map Q is finite, this implies that, for each irreducible component Z of $P^{-1}(|W|)$, there is an irreducible component Z_T of $|p^{N*}(p_T^*(W))|$ with $Z=Q(Z_T)$. As $p^{N*}(p_T^*(W))$ is an element of $\mathcal{C}_d(X\times_S U_{T^N}/U_{T^N})_{\mathbb{Q}}$, and $U_{T^N}\to U$ is finite, this implies that Z dominates an irreducible component of U. The component Z_T may not be unique, but $Q'(Z_T)$ is the unique component Z' of $P_U^{-1}(Z)$ which dominates Z.

For each irreducible component Z of $P^{-1}(|W|)$, let u_Z be the generic point of the irreducible component of U which is dominated by Z, and let u_{Z_T} be the generic point of the irreducible component of U_{T^N} dominated by Z_T. We define the multiplicity $m(Z;p^*(W))$ as the rational number

$$m(Z;p^*(W)):=\frac{m(Z_T;p^{N*}(p_T^*(W)))}{m_{k(u_{Z_T}):k(u_Z)}(Z_T/Z)}.$$

We claim that $m(Z;p^*(W))$ is independent of the choice of component Z_T. Indeed, suppose we have two choices, Z_T^i, $i=1,2$, with corresponding points $u_{Z_T^i}$. Let $t_i=p^N(u_{Z_T^i})$. Since $Q'(Z_T^i)=Z'$ for $i=1,2$, we have $q'(u_{Z_T^1})=u^N=q'(u_{Z_T^2})$, where u^N is the generic point of the component of U^N dominated by Z'. If we let $t=p(u)$, we therefore have $p_U(u^N)=u_Z$ and $p_T(t_1)=t=p_T(t_2)$.

Arguing as in the proof of Lemma 2.1.4, we may find a field K and two morphisms $g_1, g_2 : \operatorname{Spec} K \to U_{T^N}$ such that, letting $v = \operatorname{Spec} K$, we have

$$g_i(v) = u_{Z_T^i},$$
$$q' \circ g_1 = q' \circ g_2,$$

and in addition, if we let $G_i : X \times_S v \to X \times_S u_{Z_T^i}$, $i = 1, 2$, be the maps induced by g_i, there is an irreducible component Y of $|W| \times_S v$ with $G_i(Y) = Z_T^i \times_{U_{T^N}} u_{Z_T^i}$, $i = 1, 2$.

Let $f_i : v \to T^N$ be the map $p^N \circ g_i$. Then

$$\begin{aligned} p_T \circ f_i &= p_T \circ p^N \circ g_i \\ &= p \circ p_U \circ q' \circ g_i, \end{aligned}$$

hence $p_T \circ f_1 = p_T \circ f_2$. Since W is in $\mathcal{Z}_d(X \times_S T/T)_{\mathbb{Q}}$, we have $f_1^*(p_T^*(W)) = f_2^*(p_T^*(W))$. By the functoriality of pull-back for normal schemes (Proposition 1.4.4), we have

$$g_1^*(p^{N*}(p_T^*(W))) = g_2^*(p^{N*}(p_T^*(W))). \tag{2.2.3.3}$$

As the multiplicity of Y in $g_i^*(p^{N*}(p_T^*(W)))$ is given by

$$m(Y; g_i^*(p^{N*}(p_T^*(W)))) = m(Z_T^i; p^{N*}(p_T^*(W))) \cdot m_{k(v):k(u_{Z_T^i})}(Y/Z_T^i)$$

(Definition 1.4.2(ii)), the identity (2.2.3.3), together with the multiplicativity relation (1.3.1.1), gives the desired identity

$$\frac{m(Z_T^1; p^{N*}(p_T^*(W)))}{m_{k(u_{Z_T^1}):k(u_Z)}(Z_T^1/Z)} = \frac{m(Z_T^2; p^{N*}(p_T^*(W)))}{m_{k(u_{Z_T^2}):k(u_Z)}(Z_T^2/Z)}.$$

We may then define the cycle $p^*(W) \in \mathcal{C}_d(X \times_S U/U)_{\mathbb{Q}}^{\geq 0}$ as

$$p^*(W) = \sum_Z m(Z; p^*(W)) \cdot Z.$$

The formula

$$q'^*(p_U^*(p^*(W))) = p^{N*}(p_T^*(W)) \tag{2.2.3.4}$$

follows immediately from the definition. From this it easily follows that, in case U and T are already normal, the new definition of p^* agrees with the old one, proving (ii).

It follows easily from (2.2.3.4) that $p^*(W)$ is in $\mathcal{Z}_d(X \times_S U/U)_{\mathbb{Q}}^{\geq 0}$. Indeed, let $f_i : v \to U^N$, $i = 1, 2$ be morphisms of a one-point scheme $v = \operatorname{Spec} K$ such that $p_U \circ f_1 = p_U \circ f_2$. We need to verify that

$$f_1^*(p_U^*(p^*(W))) = f_2^*(p_U^*(p_U^*(W))).$$

As pull-back is functorial and injective for morphisms of reduced one-point schemes, we may replace the pair f_1, f_2 with $f_1 \circ g, f_2 \circ g$ for any morphism $g : v' \to v$ of reduced one-point schemes. In particular, we may assume that we have morphisms $h_1, h_2 : v \to U_{T^N}$ with $f_i = q' \circ h_i$. Let $g_i : v \to T^N$ be the map $p^N \circ h_i$. Then $p_T \circ g_1 = p_T \circ g_2$. Since W is in $\mathcal{Z}_d(X \times_S T/T)_{\mathbb{Q}}^{\geq 0}$, we have $g_1^*(p_T^*(W)) = g_2^*(p_T^*(W))$, hence $h_1^*(p^{N*}(p_T^*(W))) = h_2^*(p^{N*}(p_T^*(W)))$. By (2.2.3.4), this implies that $h_1^*(q'^*(p_U^*(p^*(W)))) = h_2^*(q'^*(p_U^*(p^*(W))))$, hence $f_1^*(p_U^*(p^*(W))) = f_2^*(p_U^*(p^*(W)))$, as desired.

Let $p' : U^N \to T$ be the composition $p \circ p_U$, let u^* be a point of U^N, $t = p'(u^*)$, and $p'_{u^*} : u^* \to t$ the restriction of p'. We now verify the formula

$$p_U^*(p^*(W))(u^*) = p'^*_{u^*}(W(t)). \tag{2.2.3.5}$$

Pick a point u' in U_{T^N} lying over u^*, and let $t' = p^N(u')$. Let $p_{t'} : t' \to t$ be the restriction of p_T to t', and let $q'_{u'} : u' \to u^*$ be the restriction of q'. Let $p^N_{u'} : u' \to t'$ be the restriction of p^N. By (2.2.2.1), we have $p^*_{t'}(W(t)) = p^*_T(W)(t')$, hence, by functoriality of pull-back for normal schemes, we have

$$\begin{aligned} p^{N*}(p^*_T(W))(u') &= p^{N*}_{u'}(p^*_T(W)(t')) \\ &= p^{N*}_{u'}(p^*_{t'}(W(t))) \\ &= q'^*_{u'}(p'^*_{u^*}(W(t))). \end{aligned}$$

Similarly, we have $q'^*(p_U^*(p^*(W)))(u') = q'^*_{u'}(p^*_U(p^*(W))(u^*))$. By (2.2.3.4), this gives $q'^*_{u'}(p'^*_{u^*}(W(t))) = q'^*_{u'}(p^*_U(p^*(W))(u^*))$; as the map $q'^*_{u'}$ is injective, we have $p'^*_{u^*}(W(t)) = p^*_U(p^*(W))(u^*)$, as desired.

Now let $u = p_U(u^*)$, $p_{U,u^*} : u^* \to u$ the restriction of p_U. By (2.2.2.1), we have $p^*_U(p^*(W))(u^*) = p^*_{U,u^*}(p^*(W)(u))$. Combining this with (2.2.3.5), and using the identity $p'_{u^*} = p_u \circ p_{U,u^*}$ gives $p^*_{U,u^*}(p^*_u(W(t))) = p^*_{U,u^*}(p^*(W)(u))$. As p^*_{U,u^*} is injective, we have $p^*_u(W(t)) = p^*(W)(u)$, completing the proof of (2.2.3.1) and the lemma. □

2.3. The main results

Let S be a reduced scheme, $X \to S$ an S-scheme, essentially of finite type over S. Let $\mathbf{Red}_S$ denote the category of reduced S-schemes, and let $\mathbf{Red}_{X\to S}$ denote the category of maps $U \to T$, with T in $\mathbf{Red}_S$, and $U \to T$ a localization of $X \times_S T \to T$. We let $\mathbf{Red}_{*\to*}$ be the category of morphisms essentially of finite type $X \to S$, with S reduced, where a map $(U \to T) \to (X \to S)$ is a morphism in $\mathbf{Red}_{X\to S}$.

2.3.1. THEOREM. *(i) The functor (1.4.5.2) has a unique extension to a functor*

$$\mathcal{Z}_d(-/-)_{\mathbb{Q}} : \mathbf{Red}^{\mathrm{op}}_{*\to*} \to \mathbf{Mod}_{\mathbb{Q}}.$$

such that

(a) $\mathcal{Z}_d(-/-)_{\mathbb{Q}}(U \to T) = \mathcal{Z}_d(U/T)_{\mathbb{Q}}$ *if T is essentially of finite type over $\mathbb{Z}$.*

(b) *If T is a reduced S-scheme, $i : t \to T$ the inclusion of a point t of T, then, for $W \in \mathcal{Z}_d(U/T)_{\mathbb{Q}}$, we have*

$$i^*(W) = W(t) \in \mathcal{Z}_d(U_t/t)_{\mathbb{Q}}$$

(c) *For $U \to T$ in $\mathbf{Red}_{*\to*}$, the Zariski presheaf $V \mapsto \mathcal{Z}_d(V/T)_{\mathbb{Q}}$ on U is a subsheaf of the sheaf $V \mapsto \mathcal{C}_d(V/T)_{\mathbb{Q}}$ (see Remark 1.1.3).*

(ii) Let $\mathcal{Z}_d(X/S)$ be the subgroup of $\mathcal{Z}_d(X/S)_{\mathbb{Q}}$ generated by the effective W in $\mathcal{Z}_d(X/S)_{\mathbb{Q}}$ such that, for each point s of S, the $\mathbb{Q}$-cycle i_s^*W is in $\mathcal{C}_d(X_s/s)$, where $i_s : s \to S$ is the inclusion. Then, for a map of S-schemes $p : T' \to T$, the homomorphism $p^* : \mathcal{Z}_d(X \times_S T/T)_{\mathbb{Q}} \to \mathcal{Z}_d(X \times_S T'/T')_{\mathbb{Q}}$ restricts to a homomorphism $p^* : \mathcal{Z}_d(X \times_S T/T) \to \mathcal{Z}_d(X \times_S T'/T')$, giving the functor

$$\begin{aligned} \mathcal{Z}_d(-/-) : \mathbf{Red}^{\mathrm{op}}_{*\to*} &\to \mathbf{Ab} \\ (U \to T) &\mapsto \mathcal{Z}_d(U/T). \end{aligned}$$

(iii) Suppose T is irreducible, W is in $\mathcal{Z}_d(X\times_S T/T)_{\mathbb{Q}}$, and each irreducible component of W has finite Tor-dimension over T. Then, for a map of S-schemes $p: U \to T$, we have

$$p^*(W) = W \cdot_T U.$$

(iv) Let $f: X \to Y$ be a morphism of smooth S-schemes, essentially of finite type over S, and let $W \in \mathcal{Z}^d(Y/S)_{\mathbb{Q}}$ be a cycle such that $f^*(W)$ is defined and is in $\mathcal{C}^d(X/S)_{\mathbb{Q}}$. Then $f^*(W)$ is in $\mathcal{Z}^d(X/S)_{\mathbb{Q}}$; if in addition, W is in $\mathcal{Z}^d(Y/S)$ then $f^*(W)$ is in $\mathcal{Z}^d(X/S)$.

Let $p: T \to S$ be a morphism of reduced schemes, and let $f_T: X\times_S T \to Y\times_S T$ be the map induced by f. Then $p^*(W)$ is in $\mathcal{Z}^d(Y\times_S T/T)_{\mathbb{Q}}$, $p^*(f^*(W))$ is in $\mathcal{Z}^d(X\times_S T/T)_{\mathbb{Q}}$, $f_T^*(p^*(W))$ is defined, $f_T^*(p^*(W))$ is in $\mathcal{Z}^d(X\times_S T/T)_{\mathbb{Q}}$, and we have

$$f_T^*(p^*(W)) = p^*(f^*(W)).$$

In particular, if W is in $\mathcal{Z}^d(Y/S)$, and $f^*(W)$ is in $\mathcal{C}^d(X/S)_{\mathbb{Q}}$, then $f_T^*(p^*(W))$ is in $\mathcal{Z}^d(X\times_S T/T)$.

(v) Suppose we have a sequence of S-morphisms of smooth S-schemes, essentially of finite type over S,

$$X \xrightarrow{f} Y \xrightarrow{g} Z.$$

Let W be an effective cycle in $\mathcal{Z}^d(Z/S)_{\mathbb{Q}}$ such that $g^*(W)$ is defined and in $\mathcal{C}^d(Y/S)_{\mathbb{Q}}$, and $f^*(g^*(W))$ is defined and is in $\mathcal{C}^d(X/S)_{\mathbb{Q}}$. Then then $(g\circ f)^*(W)$ is also defined, $(g\circ f)^*(W)$ is in $\mathcal{C}^d(X/S)_{\mathbb{Q}}$, and

$$f^*(g^*(W)) = (g\circ f)^*(W).$$

PROOF. Let $\mathbf{Red}^{\mathrm{fin}}_{*\to *}$ be the full subcategory of $\mathbf{Red}_{*\to *}$ with objects $U \to T$, where T is essentially of finite type over $\mathbb{Z}$. Suppose we have the extension of (1.4.5.2) to a contravariant functor

$$\mathcal{Z}_d(-/-)_{\mathbb{Q}}: \mathbf{Red}^{\mathrm{fin}}_{*\to *} \to \mathbf{Mod}_{\mathbb{Q}},$$

satisfying (i)-(v). We may then define the Zariski sheaf $U \mapsto \mathcal{Z}_d(U/S)_{\mathbb{Q}}$ on X (for $f: X \to S$ in $\mathbf{Red}_{*\to *}$) to be the subsheaf of $U \mapsto \mathcal{C}_d(U/S)_{\mathbb{Q}}$ with stalk at $p \in X$ given by $\lim_{\to} \mathcal{Z}_d(U_\alpha/S_\alpha)_{\mathbb{Q}}$, where the limit is over the category of maps $(\operatorname{Spec}\mathcal{O}_{X,p} \to \operatorname{Spec}\mathcal{O}_{S,f(p)}) \to (U_\alpha \to S_\alpha)$ in $\mathbf{Red}_{*\to *}$, with $U_\alpha \to S_\alpha$ in $\mathbf{Red}^{\mathrm{fin}}_{*\to *}$. It follows from Remark 1.1.6 that this extends the definition of $\mathcal{Z}_d(-/-)_{\mathbb{Q}}$ for a normal base. It is easy to show that the properties (i)-(v) also extend; the uniqueness of the extension follows by an argument similar to that used in Remark 1.1.6. Indeed, for uniqueness, we may assume that $X \to S$ is a morphism of local schemes. If W is in $\mathcal{C}_d(X/S)$, but not in $\lim_{\to} \mathcal{Z}_d(U_\alpha/S_\alpha)_{\mathbb{Q}}$, then there is a $U_\alpha \to S_\alpha$ with $f: S \to S_\alpha$ dominant, and a $W_\alpha \in \mathcal{C}_d(U_\alpha/S_\alpha)_{\mathbb{Q}}$ such that $W = f^*W_\alpha$. In addition, we may assume that all the schemes and morphisms involved are local, and that W_α fails to be in $\mathcal{Z}_d(U_\alpha/S_\alpha)_{\mathbb{Q}}$ due to the existence of morphisms $f_1, f_2: t \to S_\alpha^N$ such that $f_1^*p^*W_\alpha \neq f_2^*p^*W_\alpha$, but $pf_1 = pf_2$, where $p: S_\alpha^N \to S_\alpha$ is the canonical morphism. If we had an extension $\mathcal{Z}'_d(-/-)_{\mathbb{Q}}$ with $W \in \mathcal{Z}'_d(X/S)_{\mathbb{Q}}$, the morphisms f_i would induce morphisms $F_1, F_2: t' \to X\times S_\alpha^N$ with $p_1F_1 = p_1F_2$, contradicting $F_1^*p_1^*W \neq F_2^*p_1^*W$.

This reduces us to the case of schemes essentially of finite type over $\mathbb{Z}$.

As an element $W \in \mathcal{Z}_d(U/T)^{\geq 0}_{\mathbb{Q}}$ is determined by the values $W(t) \in \mathcal{Z}_d(U_t/t)_{\mathbb{Q}}$ as t runs over the generic points of T, the extension of $\mathcal{C}_d(-/-)_{\mathbb{Q}}$ to $\mathcal{Z}_d(-/-)_{\mathbb{Q}}$

satisfying (a), (b) and (c) is unique. To show the existence of the extension, it suffices to show that the homomorphisms p^* defined in Lemma 2.2.3 satisfy (b), and are functorial. Indeed, the sheaf condition (c) follows from the fact that $V \mapsto \mathcal{C}_d(V/S)_{\mathbb{Q}}$ is a sheaf, and that the condition that a cycle in $\mathcal{C}_d(V/S)_{\mathbb{Q}}$ be in $\mathcal{Z}_d(V/S)_{\mathbb{Q}}$ is point-wise on V.

The formula (b) follows immediately from (2.2.3.1), and the fact that $\mathrm{id}_t^* = \mathrm{id}$. We now prove functoriality.

As the maps p^* defined in Lemma 2.2.3 clearly respect the restriction to open subschemes, it suffices to prove functoriality for $U \to T$ of the form $X \times_S T$.

Let $V \xrightarrow{q} U \xrightarrow{p} T$ be a sequence of S-morphisms of reduced S-schemes. As a cycle $W \in \mathcal{Z}_d(X \times_S V/V)_{\mathbb{Q}}^{\geq 0}$ is determined by the values $W(v)$ among generic points v of V, we may assume that V is a one-point scheme v. Let $u = q(v)$, $t = p(u)$, and let

$$p_u : u \to t,$$
$$q_v : v \to u$$

be the respective restrictions of p and q. Take a W in $\mathcal{Z}_d(X \times_S T/T)_{\mathbb{Q}}^{\geq 0}$. By (2.2.3.1), we have

$$p^*(W)(u) = p_u^*(W(t)),$$
$$q^*(p^*(W))(v) = q_v^*(p^*(W)(u)),$$
$$(p \circ q)^*(W)(v) = (p_u \circ q_v)^*(W(t)).$$

Thus, since we have functoriality for pull-back among normal S-schemes (Proposition 1.4.4), we have

$$\begin{aligned} q^*(p^*(W))(v) &= q_v^*(p_u^*(W(t))) \\ &= (p_u \circ q_v)^*(W(t)) \\ &= (p \circ q)^*(W)(v), \end{aligned}$$

hence $(p \circ q)^* = q^* \circ p^*$, as desired.

The assertion (ii) follows directly from Lemma 2.2.3(iii).

The proof of (iii) is exactly the same as the proof of Proposition 1.4.1(ii), and is left to the reader. The assertion (iv) follows directly from Proposition 1.4.7, formula (2.2.2.1) and the functoriality in the normal case, Proposition 1.4.4.

Finally, for (v), as W is effective, the pull-backs $f^*(g^*(W))$ and $(g \circ f)^*(W)$ are determined by their values at each generic point of S. This reduces us to the case of regular schemes. Since W is effective, the cycles $g^*(W)$, $f^*(g^*(W))$ and $(g \circ f)^*(W)$ are defined and have relative codimension d over S if and only if each component of $g^{-1}(|W|)$, $f^{-1}(g^{-1}(|W|)$ and $(g \circ f)^{-1}(|W|)$ has relative codimension d over S, where $|W|$ is the support of W. Since $f^{-1}(g^{-1}(|W|) = (g \circ f)^{-1}(|W|)$, $f^*(g^*(W))$ is defined and has relative codimension d if and only if the same is true of $(g \circ f)^*(W)$. If this is the case, the identity $f^*(g^*(W)) = (g \circ f)^*(W)$ is an instance of the associativity formula (1.1.4.1). □

2.3.2. THEOREM. *Let S be a reduced scheme, $X \to S$ an S-scheme, essentially of finite type over S, and W an element of $\mathcal{Z}_d(X/S)_{\mathbb{Z}}$.*
(i) There is an integer $N_W > 0$ such that $N_W \cdot W$ is in $\mathcal{Z}_d(X/S)$, i.e., the quotient group $\mathcal{Z}_d(X/S)_{\mathbb{Z}}/\mathcal{Z}_d(X/S)$ is torsion.

(ii) If all but finitely many primes $p_1, \ldots, p_n$ are invertible on S, then we may take N_W to be a power of $\prod_{i=1}^n p_i$. In particular, if S is a scheme over $\mathbb{Q}$, we have $\mathcal{Z}_d(X/S)_{\mathbb{Z}} = \mathcal{Z}_d(X/S)$.
(iii) If S is essentially of finite type over $\mathbb{Z}$, and if $S \times_{\mathbb{Z}} \mathbb{Z}[1/N]$ is regular for some $N > 0$, we may take N_W to be a power of N.

PROOF. For (i), we may suppose that S is essentially of finite type over $\mathbb{Z}$, and W effective. As S is reduced and essentially of finite type over $\mathbb{Z}$, the non-regular locus of S is a closed subscheme $i: S_1 \to S$ of S, containing no generic point of S. If s is a regular point of S, then $W(s)$ is integral by Remark 1.4.3. In addition, as S_1 has only finitely many generic points, there is an integer $N_1 > 0$ such that $N_1 \cdot W(s')$ is integral for all generic points s' of S_1. By the definition of i^*, this implies that $i^*(N_1 \cdot W)$ is in $\mathcal{Z}_d(X \times_S S_1)_{\mathbb{Z}}$. By noetherian induction, there is an integer N_2 such that $N_2 \cdot i^*(N_1 \cdot W)(s)$ is integral for all $s \in S_1$. By the functoriality of pull-back, this implies that $N_W \cdot W(s)$ is integral for all $s \in S$, where we take $N_W = N_1 N_2$.

For (ii) and (iii), if $k(s')$ has characteristic $p > 0$ for a generic point s' of S_1, then $p^\alpha \cdot W(s')$ is integral for some $\alpha \geq 0$; if $k(s')$ has characteristic 0, then $W(s')$ is already integral. Thus, if $N = \prod_{s'} \mathrm{char}(k(s'))$, where the product is over the generic points s' of S_1 with positive characteristic, we may take N_1 to be a power of N. This, together with noetherian induction, proves (ii) and (iii). □

2.3.3. REMARKS. Let S be a reduced S-scheme, $X_1 \to S$, $X_2 \to S$ smooth S-schemes, essentially of finite type.
(i) Take $Z_i \in \mathcal{C}^{d_i}(X_i/S)_{\mathbb{Q}}^{\geq 0}$, $i = 1, 2$. Define $Z_1 \times_{/S} Z_2$,

$$Z_1 \times_{/S} Z_2 \in \mathcal{C}^{d_1+d_2}(X_1 \times_S X_2/S)_{\mathbb{Q}}^{\geq 0},$$

as follows: We may assume Z_1 and Z_2 are irreducible, and then define $Z_1 \times_{/S} Z_2$ for irreducible Z_i, and extend by linearity.

Let $Z_{12} = (Z_1 \times_S Z_2)_{\mathrm{red}} \subset X_1 \times_S X_2$, and let $q_i: Z_i \to S$ be the structure morphisms. We have the cycles $q_1^*(Z_2)$ and $q_2^*(Z_1)$ on $X_1 \times_S X_2$ with support Z_{12}; it is easily checked that these agree, and define an element $Z_1 \times_{/S} Z_2$ of $\mathcal{C}^{d_1+d_2}(X_1 \times_S X_2/S)_{\mathbb{Q}}^{\geq 0}$. By the functoriality of pull-back, we have $(Z_1 \times_{/S} Z_2)(s) = Z_1(s) \times_{/s} Z_2(s)$; thus, if Z_1 and Z_2 are in $\mathcal{Z}^{d_i}(X_i/S)$, it follows from this that the cycle $Z_1 \times_{/S} Z_2$ is in $\mathcal{Z}^{d_1+d_2}(X_1 \times_S X_2/S)$.
(ii) If S is regular, and Z_1 and Z_2 are irreducible, $Z_1 \times_{/S} Z_2$ is the cycle associated to the subscheme $Z_1 \times_S Z_2$ of $X_1 \times_S X_2$. In addition, $Z_1 \times_{/S} Z_2$ satisfies, and is characterized by, the formula

$$(Z_1 \times_{/S} Z_2)(s) = Z_1(s) \times_{/s} Z_2(s) \tag{2.3.3.1}$$

for all $s \in S$. We will often omit write $Z_1 \times Z_2$ for $Z_1 \times_{/S} Z_2$ if the base scheme S is understood.
(iii) It follows from the functoriality of pull-back for equi-dimensional cycles that the product defined in (i) is functorial: If $p: T \to S$ is a map of schemes, if X_i are in $\mathbf{Sm}_S$ and we have $Z_i \in \mathcal{Z}^{d_i}(X_i/S)$, $i = 1, 2$, then

$$p^*(Z_1 \times_{/S} Z_2) = p^*(Z_1) \times_{/T} p^*(Z_2).$$

2.3.4. REMARK. Let S be a reduced scheme, $X \to S$ a smooth S-scheme, essentially of finite type, and W a reduced codimension q subscheme of X. Write W as a union of irreducible components, $W = \cup_{i=1}^{M} W_i$. Suppose that W is flat over S, and that each geometric fiber W_t is reduced (e.g. W is smooth over S). In particular, for each point s of S, and each generic point η of W_s, W is smooth over S at η. Let $S^N \to S$ be the normalization of S, and set $X^N := S^N \times_S X$, $W^N := S^N \times_S W$. Then W^N is a closed subscheme of X^N, flat over S^N, and with reduced geometric fibers. It follows from Proposition 1.4.1 that the cycle $\sum_{i=1}^{M} W_i$ is in $\mathcal{Z}^q(X/S)$.

APPENDIX B

K-Theory

In this appendix, we give a brief review of some basic facts and constructions of algebraic K-theory. We will freely use notions and constructions from the theory of simplicial sets; we will often identify a simplicial set with its geometric realization, referring to e.g. the homotopy groups of a simplicial set. For more details, we refer the reader to [**102**], [**95**], [**107**], [**116**]. See also Part II, Chapter III.

1. K-theory of rings and schemes

1.1. Preliminaries

1.1.1. *The classifying space and group homology.* For a set X and a finite set T, let X^T denote the set of maps $T \to X$. X^T is isomorphic to the $|T|$-fold product of X. We form the simplicial set $\mathrm{E}X$ by

$$\mathrm{E}X(-) := X^{(-)}.$$

A choice of a point $* \in X$ gives a contraction of $\mathrm{E}X$, the contraction being given by the sequence of maps

$$h_j^n : EX([n]) \to EX([n+1]); \quad j = 0, \dots, n$$
$$h_j^n(x_0, \dots, x_n) \mapsto (*, \dots, *, x_j, \dots, x_n)$$

(see [**95**, Chapter I, Definition 5.1]).

If G is a group, we may take the quotient of $\mathrm{E}G$ by the right diagonal action, giving the simplicial principal G-bundle $\pi_G : \mathrm{E}G \to \mathrm{B}G$; $\mathrm{B}G$ is the *classifying space* for G. Since $\mathrm{E}G$ is contractible, $\mathrm{B}G$ is connected, and

$$\pi_n(\mathrm{B}G) = \begin{cases} G & \text{for } n = 1, \\ 0 & \text{for } n > 1. \end{cases}$$

For a commutative ring A, we let $C^*(G; A)$ denote the complex with $C^n(G; A)$ the free A-module on $\mathrm{B}G([-n])$, and with differential the usual alternating sum. The *homology* of G with A-coefficients is then defined as

$$H_p(G; A) := H_p(\mathrm{B}G; A) = H^{-p}(C^*(G; A)).$$

1.1.2. *Classifying schemes.* For an S-scheme X, and a finite set T, we have the representable functor

$$\mathrm{Hom}_{\mathbf{Sch}_S}(-, X)^T : \mathbf{Sch}_S \to \mathbf{Sets},$$

represented by the self-product $X \times_S \dots \times_S X$ with $|T|$ factors. We denote this representing object by $X^{T/S}$, giving the contravariant functor $X^{(-)/S}$ from category of finite sets to $\mathbf{Sch}_S$. We let $\mathrm{E}X$ denote the simplicial scheme $X^{(-)/S} : \Delta^{\mathrm{op}} \to$

$\mathbf{Sch}_S$. If Z is an S-scheme, we have the canonical isomorphism of simplicial sets $\mathrm{Hom}_S(Z, \mathrm{E}X) \cong \mathrm{E}(\mathrm{Hom}_S(Z, X))$.

If G is a group-scheme over S, we may take the quotient of $\mathrm{E}G$ via the right diagonal action, forming the map of simplicial schemes $\pi_G : \mathrm{E}G \to \mathrm{B}G$. The quotient map on the level of n-simplices,

$$\mathrm{E}G([n]) \cong G^{n+1} \to \mathrm{B}G([n]) \cong G^n$$

is split, hence, for each S-scheme Z, we have the canonical isomorphism of simplicial sets

$$\mathrm{Hom}_S(Z, \mathrm{B}G) \cong \mathrm{B}(\mathrm{Hom}_S(Z, G)).$$

If F is an S-scheme with a left G-action, we may form the simplicial scheme $\mathrm{E}G \times_G F$, giving the bundle $\pi_{G,F} : \mathrm{E}G \times_G F \to \mathrm{B}G$.

1.1.3. *The general linear group.* For a ring A, we have the general linear group $\mathrm{GL}_N(A)$; if A is commutative we may identify $\mathrm{GL}_N(A)$ with the group of A-valued points of the group scheme (over $\mathbb{Z}$) GL_N. We let GL_N/S denote the base extension of GL_N to a group-scheme over S. Sending a matrix $g \in \mathrm{GL}_N(A)$ to the matrix

$$i_N(g) := \begin{pmatrix} g & 0 \\ 0 & 1 \end{pmatrix}$$

defines the *stabilization maps*

$$i_N : \mathrm{GL}_N/S \to \mathrm{GL}_{N+1}/S; \quad i_N : \mathrm{GL}_N(A) \to \mathrm{GL}_{N+1}(A).$$

We let $\mathrm{GL}(A)$ denote the inductive limit

$$\mathrm{GL}(A) := \varinjlim_N \mathrm{GL}_N(A).$$

Applying the constructions of §1.1.1 gives us the classifying spaces $\mathrm{BGL}_N(A)$, and the stabilization maps $\mathrm{B}i_N : \mathrm{BGL}_N(A) \to \mathrm{BGL}_{N+1}(A)$, as well as the classifying space $\mathrm{BGL}(A)$, which is homeomorphic to the inductive limit of the spaces $\mathrm{BGL}_N(A)$. We sometimes find it notationally convenient to write $\mathrm{GL}_\infty(A)$ for $\mathrm{GL}(A)$, $\mathrm{BGL}_\infty(A)$ for $\mathrm{BGL}(A)$, etc.

Applying the construction of §1.1.2, we have the classifying simplicial scheme BGL_N/S. Taking the standard action of GL_N/S on the affine space $\mathbb{A}^N_S$ gives the *universal rank N bundle* $E_N := \mathrm{EGL}_N/S \times_{\mathrm{GL}_N/S} \mathbb{A}^N_S \to \mathrm{BGL}_N/S$.

1.2. Higher algebraic K-theory

1.2.1. *The Q-construction.* Quillen [**102**] has defined higher algebraic K-theory of an *exact category M* by

$$K_p(M) := \pi_{p+1}(\mathrm{B}QM).$$

For our purposes, it will not be necessary to give a precise explanation of this formula, but we will say a few words here for the readers convenience. An exact category is an additive category, together with a collection of sequences of morphisms of the form

$$0 \to E' \to E \to E'' \to 0 \tag{1.2.1.1}$$

called *exact sequences*, which satisfy the axioms listed in [**102**, §2]. For an exact category M, QM is the category defined in [**102**, §2]. For a category $\mathcal{C}$, we have the nerve $\mathcal{N}(\mathcal{C})$ (see Part II, Chapter III, §3.2.1), and the geometric realization $\mathrm{B}(\mathcal{C})$ of $\mathcal{N}(\mathcal{C})$ [**102**, §1].

An exact category M has its *Grothendieck group*, $K_0(M)$, defined as the free abelian group on the set of isomorphism classes of objects of M, modulo the subgroup generated by differences $E - E' - E''$ for each exact sequence (1.2.1.1) in M. There is a canonical isomorphism between the Grothendieck group $K_0(M)$ and $\pi_1(\mathrm{B}QM)$ [**102**, §2, Theorem 1], so the two definitions of $K_0(M)$ are the same.

Let X be a scheme. Two primary examples of an exact category are

1. The category $\mathcal{M}_X$ of coherent sheaves on X, with the usual notion of exact sequence.
2. The full subcategory $\mathcal{P}_X$ of $\mathcal{M}_X$ of locally free coherent sheaves, where a sequence in $\mathcal{P}_X$ is exact if it is exact in $\mathcal{M}_X$.

Replacing the scheme X with a ring A, we have the exact categories $\mathcal{M}_A$ of finitely generated (left) A-modules, and $\mathcal{P}_A$ of finitely generated (left) projective A-modules.

The *algebraic K-groups of X* are defined as

$$K_p(X) := K_p(\mathcal{P}_X).$$

For a ring A, we have the algebraic K-groups of A:

$$K_p(A) := K_p(\mathcal{P}_A).$$

For details, we refer the reader to [**102**].

1.2.2. *The plus construction.* For a ring A, and integer $N \geq 3$, we have the map of simplicial sets $\iota_{A,N}: \mathrm{BGL}_N(A) \to \mathrm{BGL}_N(A)^+$, where $\mathrm{BGL}_N(A)^+$ is Quillen's *plus construction* (see e.g. [**64**]). The map $\iota_{A,N}$ is characterized, up to weak equivalence, by the following properties:

(1.2.2.1)

1. $\iota_{A,N*}$ is a surjection on π_1, with kernel the subgroup of $\pi_1(\mathrm{BGL}_N(A)) = \mathrm{GL}_N(A)$ generated by the elementary matrices.
2. For each local system L on $\mathrm{BGL}_N(A)^+$, the map

$$\iota_{A,N*}: H_*(\mathrm{BGL}_N(A); \iota_{A,N}^* L) \to H_*(\mathrm{BGL}_N(A)^+; L)$$

is an isomorphism.

It is possible to make the simplicial set $\mathrm{BGL}_N(A)^+$ and the map $\iota_{A,N}$ functorial in A (at least for $N \geq 5$, see e.g. [**114**] or [**64**]), and to extend the stabilization maps $\mathrm{B}i_N: \mathrm{BGL}_N(A) \to \mathrm{BGL}_{N+1}(A)$ to functorial stabilization maps $\mathrm{B}i_N^+: \mathrm{BGL}_N(A)^+ \to \mathrm{BGL}_{N+1}(A)^+$, compatible with the maps $\iota_{A,N}$ and $\iota_{A,N+1}$. We may also form the plus construction on $\mathrm{BGL}(A)$, $\iota_A: \mathrm{BGL}(A) \to \mathrm{BGL}(A)^+$; then $\mathrm{BGL}(A)^+$ is weakly equivalent to the inductive limit of the spaces $\mathrm{BGL}_N(A)^+$.

Grayson [**53**] has given an exposition of Quillen's construction of a natural homotopy equivalence

$$\mathrm{BGL}(A)^+ \sim (\Omega \mathrm{B}Q\mathcal{P}_A)^0, \tag{1.2.2.2}$$

where ^{0} denotes the connected component containing $0 \in K_0(A)$, giving the natural isomorphism

$$\varinjlim_N \pi_n(\mathrm{BGL}_N(A)) \cong \pi_n(\mathrm{BGL}(A)^+) \cong K_n(A); \quad p \geq 1.$$

2. K-theory and homology

We give a discussion of the Hurewicz map and how it is used in various settings to map algebraic K-theory to the homology of GL.

2.1. K-theory via diagrams

We begin by describing a version of the K-theory of a diagram of schemes. The main tool is the homotopy limit construction of [**25**]; for the properties of homotopy limits we will be using, see Part II, Chapter III, Section 3.

2.1.1. *The plus construction for schemes.* Let $\underline{[n]}$ denote the category of non-empty subsets of the set $\{0,\dots,n\}$, with morphisms the inclusions. If X is a noetherian separated scheme, and $\mathcal{U} := \{U_0,\dots,U_n\}$ is a finite cover of X, then one has the functor (see Part II, Chapter III, Example 3.2.3)

$$\underline{\mathcal{U}}\colon \underline{[n]}^{\mathrm{op}} \to \mathbf{Sch}_X$$

defined by setting $\underline{\mathcal{U}}(I) = \cap_{i\in I} U_i$, with maps being the inclusions. Applying the functor $\mathbb{Z}\times \mathrm{BGL}_N(\Gamma(-,\mathcal{O}_X))^+$ gives the functor

$$\mathbb{Z}\times \mathrm{BGL}_N(\Gamma(\underline{\mathcal{U}},\mathcal{O}_X))^+ \colon \underline{[n]} \to \mathbf{s.Sets}.$$

For a commutative ring A, we let $\mathcal{F}_A$ denote the full subcategory of $\mathcal{P}_A$ with objects the free finitely generated A-modules. Composing $\underline{\mathcal{U}}$ with the functor $U\mapsto \mathrm{B}Q\mathcal{F}_{\Gamma(U,\mathcal{O}_X)}$ gives us the functor

$$\mathrm{B}Q\mathcal{F}(\underline{\mathcal{U}})\colon \underline{[n]} \to \mathbf{s.Sets}.$$

We now replace the various functors $\mathbb{Z}\times \mathrm{BGL}_N(\Gamma(\underline{\mathcal{U}},\mathcal{O}_X))^+$, $\mathrm{B}Q\mathcal{F}(\underline{\mathcal{U}})$, etc., with fibrant models (see Part II, Chapter III, §3.4.2), and change notation. For example, one could use the singular complex of the geometric realization as a natural fibrant model; in what follows, we will replace each functor to simplicial sets with this fibrant model, without further comment.

Let $K(X,\mathcal{U})_N$ denote the homotopy limit

$$K(X,\mathcal{U})_N := \operatorname*{holim}_{\underline{[n]}} \mathbb{Z}\times \mathrm{BGL}_N(\Gamma(\underline{\mathcal{U}},\mathcal{O}_X))^+,$$

and let $K(X,\mathcal{U})$ denote the homotopy limit

$$K(X,\mathcal{U}) := \operatorname*{holim}_{\underline{[n]}} \mathbb{Z}\times \mathrm{BGL}(\Gamma(\underline{\mathcal{U}},\mathcal{O}_X))^+.$$

Similarly, let $\mathrm{B}Q\mathcal{F}(X,\mathcal{U})$ denote the homotopy limit

$$\mathrm{B}Q\mathcal{F}(X,\mathcal{U}) := \operatorname*{holim}_{\underline{[n]}} \mathrm{B}Q\mathcal{F}(\underline{\mathcal{U}}).$$

As holim sends point-wise weak equivalences to weak equivalences (for fibrant objects) [**25**, V, 5.6], the natural homotopy equivalence (1.2.2.2) gives us the weak equivalence

$$K(X,\mathcal{U}) \sim \Omega \mathrm{B}Q\mathcal{F}(X,\mathcal{U}). \tag{2.1.1.1}$$

Let $\mathcal{P}_{X,\mathcal{U}}$ be the full subcategory of $\mathcal{P}_X$ with objects the locally free coherent $\mathcal{O}_X$-modules M such that M restricts to a free $\mathcal{O}_{U_i}$-module on each U_i in $\mathcal{U}$. Suppose that each U_i is affine, $U_i = \operatorname{Spec} A_i$. The restriction maps

$$j_{U_i}^*\colon \mathrm{B}Q\mathcal{P}_{X,\mathcal{U}} \to \mathrm{B}Q\mathcal{F}_{A_i}$$

give rise to the natural map

$$\mathrm{B}Q\mathcal{P}_{X,\mathcal{U}} \to \operatorname*{holim}_{\underline{[n]}} \mathrm{B}Q\mathcal{F}(\underline{\mathcal{U}}).$$

Combining this with (2.1.1.1) gives the natural maps

$$\sigma_{X,\mathcal{U}}: \Omega\mathrm{B}Q\mathcal{P}_{X,\mathcal{U}} \to K(X,\mathcal{U}),$$
$$\sigma_{X,\mathcal{U}*}: \pi_n(\Omega\mathrm{B}Q\mathcal{P}_{X,\mathcal{U}}) \to \pi_n(K(X,\mathcal{U})).$$

As each locally free sheaf on X can be trivialized on some finite affine open cover, the category $\mathcal{P}_X$ is equivalent to the inductive limit of the categories $\mathcal{P}_{X,\mathcal{U}}$, giving via [**102**, §2] the identity

$$K_n(X) = \lim_{\overrightarrow{\mathcal{U}}} \pi_n(\Omega\mathrm{B}Q\mathcal{P}_{X,\mathcal{U}}),$$

and hence the natural map

$$\sigma_{X*}: K_n(X) \to \lim_{\overrightarrow{\mathcal{U}}} \pi_n(K(X,\mathcal{U})). \tag{2.1.1.2}$$

Here both limits are over the category of finite affine covers $\mathcal{U}$ of X, with maps being refinements.

2.1.2. Remark. For regular X, Quillen's localization theorem [**102**, §7, Proposition 3.1 and Remark 3.5] implies that σ_{X*} is an isomorphism; the extension of the localization theorem by Thomason-Trobaugh [**121**] implies that σ_{X*} is an isomorphism for X having an ample family of line bundles. We will not require this result, so we omit the details.

2.1.3. *Diagrams of schemes.* One can extend the construction of the map (2.1.1.2) to cover certain finite diagrams of schemes as follows: Let I be the category associated to a finite partially ordered set, i.e., I has finitely many objects, and there is at most one morphism between any two objects of I (in either direction).

Let $\mathbf{Sch}^+$ be the category gotten from $\mathbf{Sch}$ by "adjoining a disjoint basepoint" to each object. Precisely, we adjoin a new final object $*$, and make the canonical map $\emptyset \to *$ an isomorphism. If we have a functor $F: \mathbf{Sch} \to \mathcal{C}$ to a category $\mathcal{C}$ having an initial and final object $*$, such that $F(\emptyset)$ is isomorphic to $*$, we have the canonical extension of F to $F: \mathbf{Sch}^+ \to \mathcal{C}$, by sending $*$ to $*$.

Suppose we have a functor $X: I \to \mathbf{Sch}^+$. An *open cover* $\mathcal{U}$ of X is a collection $\{\mathcal{U}(i) \mid i \in I\}$, where $\mathcal{U}(i)$ is an open cover of $X(i)$ (for $X(i) \neq *$) $\mathcal{U}(i) := \{U_0(i), \dots, U_n(i)\}$ such that $X(i \leq j)(U_k(i)) \subset U_k(j)$. For $X(i) = *$, we take $U_k(i) = *$, $k = 0, \dots, n$.

By our condition on the category I, the open covers on X in the sense defined above are cofinal in the category of covering families of X, defined in terms of the Grothendieck site determined by taking the Zariski topology on each $X(i) \in \mathbf{Sch}_S$, and the one-point category for each $X(i) = *$.

Each open cover $\mathcal{U}$ of X gives the functor

$$\underline{\mathcal{U}}: I \times \underline{[n]}^{\mathrm{op}} \to \mathbf{Sch}^+$$

over $X \circ p_1$, defined by $\underline{\mathcal{U}}(i, J) = \cap_{j \in J} U_j(i) \subset X(i)$. We have the functor

$$\mathcal{P}_X: I^{\mathrm{op}} \to \mathbf{cat}$$

defined by $\mathcal{P}_X(i) = \mathcal{P}_{X(i)}$, and the subfunctor

$$\mathcal{P}_{X,\mathcal{U}}: I^{\mathrm{op}} \to \mathbf{cat}$$

defined by $\mathcal{P}_{X,\mathcal{U}}(i) = \mathcal{P}_{X(i),\mathcal{U}(i)}$. We have as well the functors

$$\mathbb{Z} \times \mathrm{BGL}_N(\Gamma(\underline{\mathcal{U}}, \mathcal{O}_X))^+ : I^{\mathrm{op}} \times \underline{[n]} \to \mathbf{s.Sets}$$
$$(i, J) \mapsto \mathbb{Z} \times \mathrm{BGL}_N(\Gamma(\underline{\mathcal{U}}(i, J), \mathcal{O}_{X(i)}))^+; \quad N = 5, 6, \ldots, \infty,$$

and

$$\mathrm{B}Q\mathcal{F}(\underline{\mathcal{U}}) : I^{\mathrm{op}} \times \underline{[n]} \to \mathbf{s.Sets}$$
$$(i, J) \mapsto \mathrm{B}Q\mathcal{F}(\underline{\mathcal{U}}(i, J)).$$

Define

$$K_p(X) := \pi_{p+1}\big(\operatorname*{holim}_{I^{\mathrm{op}}}(\mathrm{B}Q\mathcal{P}_X)\big);$$
$$K_p(X,\mathcal{U}) := \pi_p\big(\operatorname*{holim}_{I^{\mathrm{op}} \times \underline{[n]}}(\mathbb{Z} \times \mathrm{BGL}_\infty(\Gamma(\underline{\mathcal{U}}, \mathcal{O}_X))^+)\big).$$

The same construction as in §2.1.1 gives the isomorphism

$$K_p(X,\mathcal{U}) \cong \pi_{p+1}\big(\operatorname*{holim}_{I^{\mathrm{op}} \times \underline{[n]}}(\mathrm{B}Q\mathcal{F}(\underline{\mathcal{U}}))\big),$$

and the natural map

$$\sigma_{X*} : K_n(X) \to \varinjlim_{\mathcal{U}} K_n(X,\mathcal{U}),$$

where the limit is over covers $\mathcal{U}$ of X such that $U_j(i)$ is affine for all i and j.

2.2. Homology

2.2.1. *The Hurewicz map.* Let A be a simplicial abelian group, giving us the (cohomological) complex $C^*(A)$. The *Dold-Kan equivalence* (see [**39**], [**74**], or [**95**, Chapter V]) states that the homotopy groups of A agree with the homology groups of $C^*(A)$, i.e., $\pi_n(A) \cong H^{-n}(C^*(A))$. In particular, for a simplicial set S, we have the canonical isomorphism

$$\pi_n(\mathbb{Z}S) \cong H^{-n}(C^*(S;\mathbb{Z})) = H_n(S;\mathbb{Z}).$$

The evident map $h_S : S \to \mathbb{Z}S$ induces the map $h_{S*} : \pi_n(S) \to H_n(S;\mathbb{Z})$ on π_n; h_{S*} is none other than the classical Hurewicz map.

If now S is a presheaf of simplicial sets on a noetherian scheme X, we may form the presheaf of simplicial abelian groups $\mathbb{Z}S$ on X, and the associated complex of presheaves $C^*(S;\mathbb{Z})$. For an open cover $\mathcal{U} = \{U_0, \ldots, U_n\}$ of X, define the simplicial sets $S(\mathcal{U})$ and $(\mathbb{Z}S)(\mathcal{U})$ by

$$S(\mathcal{U}) := \operatorname*{holim}_{\underline{[n]}^{\mathrm{op}}} S \circ \underline{\mathcal{U}}; \quad (\mathbb{Z}S)(\mathcal{U}) := \operatorname*{holim}_{\underline{[n]}^{\mathrm{op}}} \mathbb{Z}S \circ \underline{\mathcal{U}},$$

and let $C^*(S;\mathbb{Z})(\mathcal{U})$ denote the total complex of the double complex of Čech cochains for $C^*(S;\mathbb{Z})$ with respect to the cover $\mathcal{U}$.

The presheaf Hurewicz map $h_S : S \to \mathbb{Z}S$ induces the map $h_S(\mathcal{U}) : S(\mathcal{U}) \to (\mathbb{Z}S)(\mathcal{U})$, factoring through the Hurewicz map for $S(\mathcal{U})$, $h_{S(\mathcal{U})} : S(\mathcal{U}) \to \mathbb{Z}(S(\mathcal{U}))$. Note that the map $\mathbb{Z}(S(\mathcal{U})) \to (\mathbb{Z}S)(\mathcal{U})$ is not in general a weak equivalence.

From the remarks of (Part II, Chapter III, Example 3.2.3) and the quasi-isomorphism (Part II, (III.3.4.3.2)), one has the canonical quasi-isomorphism

$$C^*\big((\mathbb{Z}S)(\mathcal{U})\big) \to \tau^{\leq 0} C^*(S;\mathbb{Z})(\mathcal{U}),$$

hence we have the isomorphism

$$\pi_n(\mathbb{Z}S(\mathcal{U})) \cong H^{-n}(C^*(S;\mathbb{Z})(\mathcal{U})); \quad n = 0, 1, \dots .$$

Combining this isomorphism with the map $h_S(\mathcal{U})$, we have the natural map

$$h_S(\mathcal{U})_* : \pi_n(S(\mathcal{U})) \to H^{-n}(C^*(S;\mathbb{Z})(\mathcal{U})). \tag{2.2.1.1}$$

Let $\tilde{C}^*(S;\mathbb{Z})$ denote the sheafification of the complex of presheaves $C^*(S,\mathbb{Z})$. For each open cover $\mathcal{U}$, we have the canonical map

$$H^{-n}(C^*(S;\mathbb{Z})(\mathcal{U})) \to \mathbb{H}^{-n}_{\mathrm{Zar}}(X, \tilde{C}^*(S;\mathbb{Z})),$$

which is compatible with refinements. The maps (2.2.1.1) are also compatible with refinements, giving us the natural map

$$h_{S*} : \lim_{\overrightarrow{\mathcal{U}}} \pi_n(S(\mathcal{U})) \to \mathbb{H}^{-n}_{\mathrm{Zar}}(X, \tilde{C}^*(S;\mathbb{Z})). \tag{2.2.1.2}$$

Let $\mathcal{BGL}$ and $\mathcal{BGL}^+$ be the sheafification of the respective presheaves on X

$$U \mapsto \mathrm{BGL}(\Gamma(U, \mathcal{O}_X)); \quad U \mapsto \mathrm{BGL}(\Gamma(U, \mathcal{O}_X))^+.$$

The universal property of the plus construction (1.2.2.1) implies that the map $\mathbb{H}^{-n}_{\mathrm{Zar}}(X, \tilde{C}^*(\mathcal{BGL};\mathbb{Z})) \to \mathbb{H}^{-n}_{\mathrm{Zar}}(X, \tilde{C}^*(\mathcal{BGL}^+;\mathbb{Z}))$ is an isomorphism. Composing the map (2.2.1.2) with the map (2.1.1.2) thus gives the *Hurewicz map*

$$h_X : K_n(X) \to \mathbb{H}^{-n}_{\mathrm{Zar}}(X, \tilde{\mathbb{Z}} \oplus \tilde{C}^*(\mathcal{BGL};\mathbb{Z})), \tag{2.2.1.3}$$

where the map to the summand $\tilde{\mathbb{Z}}$ is given by projection $\mathbb{Z} \times \mathrm{BGL}^+ \to \mathbb{Z}$.

2.2.2. *Stability.* Wagoner [**125**] has proven the stability result that, for a local ring $\mathcal{O}$, the map $H_p(\mathrm{GL}_N(\mathcal{O}),\mathbb{Z}) \to H_p(\mathrm{GL}(\mathcal{O}),\mathbb{Z})$ is an isomorphism for all $N \geq N_p$, where N_p is an integer depending only on p.

We have the local to global spectral sequence for Zariski hypercohomology

$$E_2^{p,q} = H^p(X, \mathcal{H}_{-q}(\mathcal{GL}_N;\mathbb{Z})) \Longrightarrow \mathbb{H}^{p+q}_{\mathrm{Zar}}(X, \tilde{C}^*(\mathcal{BGL};\mathbb{Z})); \quad N = 1, 2, \dots, \infty,$$

where $\mathcal{H}_n(\mathcal{GL}_N;\mathbb{Z})$ is the sheaf associated to the presheaf

$$U \mapsto H_n(\mathrm{GL}_N(\Gamma(U, \mathcal{O}_X));\mathbb{Z}) = H^{-n}(C^*(\mathrm{BGL}_N(\Gamma(U, \mathcal{O}_X));\mathbb{Z})).$$

If X has finite Zariski cohomological dimension d_X (e.g., if X has finite Krull dimension, see [**58**]), then this spectral sequence, together with Wagoner's stability result, shows that the map $\mathbb{H}^{-n}_{\mathrm{Zar}}(X, \tilde{C}^*(\mathcal{BGL}_N;\mathbb{Z})) \to \mathbb{H}^{-n}_{\mathrm{Zar}}(X, \tilde{C}^*(\mathcal{BGL};\mathbb{Z}))$ is an isomorphism for all $N \geq d_X + N_n + 1$. Thus, the map (2.2.1.3) gives us the maps

$$h_{X,N} : K_n(X) \to \mathbb{H}^{-n}_{\mathrm{Zar}}(X, \tilde{\mathbb{Z}} \oplus \tilde{C}^*(\mathcal{BGL}_N;\mathbb{Z})) \tag{2.2.2.1}$$

for all $N \geq d_X + N_n + 1$, stable in N.

2.2.3. REMARK. Let I be the category associated to a finite partially ordered set. We may extend the construction of §2.2.1 and §2.2.2 to functors $X : I \to \mathbf{Sch}^+$ as in §2.1.3, as follows: Let $X : I \to \mathbf{Sch}^+$ be a functor, giving the functor $X_{\mathrm{Zar}} : I \to \mathbf{Top}$, by taking the topological space $X(i)$, with the Zariski topology, for each i.

From (Part II, Chapter III, §3.3.1), we have the notion of sheaves on X_{Zar}, and hypercohomology of complexes of sheaves on X_{Zar}. We write $\mathbb{H}^0_{\mathrm{Zar}}(X, -)$ for $\mathbb{H}^0(X_{\mathrm{Zar}}, -)$.

Suppose each $X(i)$ has finite Zariski cohomological dimension $d(i)$. Since I has finite cohomological dimension, we have the functor

$$\mathbb{H}^0_{\mathrm{Zar}}(X,-)\colon \mathbf{D}(\mathrm{Sh}_X^{\mathbf{Ab}}) \to \mathbf{Ab}.$$

Given an open cover $\mathcal{U}$ of X as in §2.1.3, and an presheaf of simplicial sets S on X, we have the Čech complex $C^*(\mathcal{U}, S; \mathbb{Z})$ defined as

$$C^*(\mathcal{U}, S; \mathbb{Z}) := C^*(\operatorname*{holim}_{I^{\mathrm{op}}\times \underline{[n]}} \mathbb{Z}S(\underline{\mathcal{U}})). \tag{2.2.3.1}$$

We may also sheafify the presheaf of chain complexes $C^*(S;\mathbb{Z}) := C^*(\mathbb{Z}S)$, forming the complex of sheaves $\tilde{C}^*(S,\mathbb{Z})$.

The identity (2.2.3.1), together with the remarks of (Part II, Chapter III, Example 3.2.3), gives the canonical map

$$H^p(C^*(\mathcal{U}, S;\mathbb{Z})) \to \mathbb{H}^p_{\mathrm{Zar}}(X, \tilde{C}^*(\mathbb{Z}S)).$$

The construction of the map (2.2.1.3) then extends to give the Hurewicz map

$$h_X : K_n(X) \to \mathbb{H}^{-n}_{\mathrm{Zar}}(X, \tilde{\mathbb{Z}} \oplus \tilde{C}^*(\mathcal{BGL};\mathbb{Z})). \tag{2.2.3.2}$$

Let $d_X := \max_{i\in I} d_{X(i)} + |I| - 1$. The construction of the map (2.2.2.1) gives the map

$$h_{X,N} : K_n(X) \to \mathbb{H}^{-n}_{\mathrm{Zar}}(X, \tilde{\mathbb{Z}} \oplus \tilde{C}^*(\mathcal{BGL}_N;\mathbb{Z})).$$

for all $N \geq d_X + N_n + 1$, stable in N.

2.2.4. *Products.* Consider the tensor product representation

$$\otimes_{N,M} : \mathrm{GL}_N/S \times_S \mathrm{GL}_M/S \to \mathrm{GL}_{NM}/S.$$

This induces the map of simplicial schemes

$$\tau_{N,M} : \mathrm{BGL}_N/S \times_S \mathrm{BGL}_M/S \to \mathrm{BGL}_{NM}/S.$$

We have as well the representations

$$\otimes_{N,*} : \mathrm{GL}_N/S \times_S \mathrm{GL}_M/S \to \mathrm{GL}_{NM}/S$$

and

$$\otimes_{*,M} : \mathrm{GL}_N/S \times_S \mathrm{GL}_M/S \to \mathrm{GL}_{NM}/S$$

given by

$$\otimes_{N,*}(g,h) = g \otimes \mathrm{id}_M; \quad \otimes_{*,M}(g,h) = \mathrm{id}_N \otimes h,$$

which induce the maps of simplicial schemes

$$\tau_{N,*} : \mathrm{BGL}_N/S \times_S \mathrm{BGL}_M/S \to \mathrm{BGL}_{NM}/S;$$
$$\tau_{*,M} : \mathrm{BGL}_N/S \times_S \mathrm{BGL}_M/S \to \mathrm{BGL}_{NM}/S.$$

For a commutative ring A, we have the map

$$\hat{\mu}^A_{M,N} : C^*(\mathrm{BGL}_N(A);\mathbb{Z}) \otimes C^*(\mathrm{BGL}_M(A);\mathbb{Z}) \to C^*(\mathrm{BGL}(A)^+;\mathbb{Z})$$

induced by $\tau_{N,M} - \tau_{N,*} - \tau_{*,M}$, where we use the H-group structure on $\mathrm{BGL}(A\otimes B)^+$ to define the differences in the above formula. More precisely, $\hat{\mu}^{A,B}_{M,N}$ is induced by the composition

(2.2.4.1)

$$\mathrm{BGL}_N(A)\times\mathrm{BGL}_M(A)\xrightarrow{(\tau_{N,M},\tau_{N,*},\tau_{*,M})}\mathrm{BGL}_{NM}(A)\times\mathrm{BGL}_{NM}(A)\times\mathrm{BGL}_{NM}(A)$$
$$\to\mathrm{BGL}(A)^+\times\mathrm{BGL}(A)^+\times\mathrm{BGL}(A)^+\to\mathrm{BGL}(A)^+,$$

where the last map is $(x,y,z)\mapsto x-y-z$. We then take the integral chain complexes, and compose with the Eilenberg-MacLane map

$$C^*(\mathrm{BGL}_N(A);\mathbb{Z})\otimes C^*(\mathrm{BGL}_M(A);\mathbb{Z})\to C^*(\mathrm{BGL}_N(A)\times\mathrm{BGL}_M(A);\mathbb{Z}).$$

The subtraction in $\mathrm{BGL}(A)^+$ is only defined up to homotopy, which causes a technical problem if we want to extend this map to schemes, and to diagrams of schemes. In order to circumvent this, we must appeal to the theory of simplicial sheaves on a Grothendieck site, as given in [**73**]. We sketch the construction, using the notions from the theory of simplicial closed model categories given in [**104**].

Let $X:I\to\mathbf{Sch}^+$ be a diagram of schemes, with I a small category. We may pull back the big Zariski site on $\mathbf{Sch}^+$ to X, giving the site X_{Zar}, the category of sheaves of abelian groups, $\mathrm{Sh}^{\mathbf{Ab}}_{\mathrm{Zar}}(X)$, and the derived category $\mathbf{D}^-(\mathrm{Sh}^{\mathbf{Ab}}_{\mathrm{Zar}}(X))$. We have as well the closed simplicial model category $\mathrm{Sh}^{\Delta}_{\mathrm{Zar}}(X)$ of sheaves of simplicial sets on X_{Zar}, and the homotopy category $\mathrm{Ho}(\mathrm{Sh}^{\Delta}_{\mathrm{Zar}}(X))$, gotten by inverting the weak equivalences in $\mathrm{Sh}^{\Delta}_{\mathrm{Zar}}(X)$. Taking the integral chain complex of a sheaf of simplicial sets gives the functor

$$C^*(-;\mathbb{Z}):\mathrm{Ho}(\mathrm{Sh}^{\Delta}_{\mathrm{Zar}}(X))\to\mathbf{D}^-(\mathrm{Sh}^{\mathbf{Ab}}_{\mathrm{Zar}}(X)).$$

For elements a,b of $\mathrm{GL}(A)$, we have the *shuffle sum* $a\oplus b$, with

$$(a\oplus b)_{ij}:=\begin{cases}a_{i',j'}; & \text{if } i=2i',\ j=2j',\\ b_{i',j'}; & \text{if } i=2i'+1,\ j=2j'+1,\\ 0; & \text{otherwise.}\end{cases}$$

This induces the map of complexes of sheaves of simplicial sets on X_{Zar}:

$$\oplus:\mathcal{BGL}^+\times\mathcal{BGL}^+\to\mathcal{BGL}^+;$$

we first show that $\oplus$ are associative and commutative in $\mathrm{Ho}(\mathrm{Sh}^{\Delta}_{\mathrm{Zar}}(X))$.

Let $\alpha:\mathbb{N}\to\mathbb{N}$ be an injection; α induces the map of sheaves of simplicial sets $\alpha_*:\mathcal{BGL}\to\mathcal{BGL}$ on X by sending the matrix coefficient a_{ij} to the $(\alpha(i),\alpha(j))$ position, and extends to $\alpha_*^+:\mathcal{BGL}^+\to\mathcal{BGL}^+$. The points on the individual schemes $X(i)$ give a conservative family of points for the site X_{Zar} (see Part II, Chapter IV, §1.3.5); if p is a point of $X(i)$, the stalk of α_*^+ at p is the map

(2.2.4.2) $$\alpha_*^+(\mathcal{O}_{X(i),p}):\mathrm{BGL}(\mathcal{O}_{X(i),p})^+\to\mathrm{BGL}(\mathcal{O}_{X(i),p})^+.$$

In the course of showing that $\mathrm{BGL}(A)^+$ is an H-space, Quillen [**103**] (see also [**107**]) shows that the map

$$\alpha_*^+(A):\mathrm{BGL}(A)^+\to\mathrm{BGL}(A)^+$$

is a weak equivalence for all rings A, hence the map (2.2.4.2) is an weak equivalence. By the definition of weak equivalence in $\mathrm{Sh}^{\Delta}_{\mathrm{Zar}}(X)$, the map α_*^+ is thus a weak equivalence, i.e., an isomorphism in $\mathrm{Ho}(\mathrm{Sh}^{\Delta}_{\mathrm{Zar}}(X))$.

Thus sending α to α_*^+ extends to a map from the group completion of the monoid of injective maps $\mathbb{N}\to\mathbb{N}$; as this group completion is the trivial group, each α_*^+ is the identity in $\mathrm{Ho}(\mathrm{Sh}^{\Delta}_{\mathrm{Zar}}(X))$. In particular, reordering of the matrix coefficients induces the identity map on $\mathcal{BGL}^+$, which thus shows that $\oplus$ is associative and commutative in $\mathrm{Ho}(\mathrm{Sh}^{\Delta}_{\mathrm{Zar}}(X))$, and additionally that the identity matrix acts as the identity for $\oplus$.

Thus $\oplus$ gives the sheaf $\mathcal{BGL}^+$ the structure of a commutative monoid in $\mathrm{Ho}(\mathrm{Sh}^{\Delta}_{\mathrm{Zar}}(X))$. Consider the map

$$(\oplus, p_2): \mathcal{BGL}^+ \times \mathcal{BGL}^+ \to \mathcal{BGL}^+ \times \mathcal{BGL}^+.$$

If we evaluate at a point p, and use the fact that the sum in a connected H-space induces the sum in the homotopy groups, we see that $(\oplus, p_2)_p$ is a weak equivalence of simplicial sets. Thus $(\oplus, p_2)$ is an isomorphism in $\mathrm{Ho}(\mathrm{Sh}^{\Delta}_{\mathrm{Zar}}(X))$; if we let $\iota: \mathcal{BGL}^+ \to \mathcal{BGL}^+$ be the composition

$$\mathcal{BGL}^+ = \mathcal{BGL}^+ \times * \hookrightarrow \mathcal{BGL}^+ \times \mathcal{BGL}^+ \xrightarrow{(\oplus,p_2)^{-1}} \mathcal{BGL}^+ \times \mathcal{BGL}^+ \xrightarrow{p_1} \mathcal{BGL}^+,$$

we have the inverse map for the operation $\oplus$.

The inverse ι is functorial in the functor X, in the homotopy category.

Having cleared up this technical point, we have the functorial map in the derived category $\mathbf{D}^-(\mathrm{Sh}^{\mathbf{Ab}}_{\mathrm{Zar}}(X))$,

$$\hat{\mu}^X_{M,N}: \tilde{C}^*(\mathcal{BGL}_N; \mathbb{Z}) \otimes \tilde{C}^*(\mathcal{BGL}_M; \mathbb{Z}) \to \tilde{C}^*(\mathcal{BGL}; \mathbb{Z}),$$

defined as a composition as in (2.2.4.1); we use the inverse of the quasi-isomorphism

$$\tilde{C}^*(\mathcal{BGL}; \mathbb{Z}) \to \tilde{C}^*(\mathcal{BGL}^+; \mathbb{Z})$$

to replace $\tilde{C}^*(\mathcal{BGL}^+; \mathbb{Z})$ with $\tilde{C}^*(\mathcal{BGL}; \mathbb{Z})$.

Let

$$(2.2.4.3)\quad \mu^X_{N,M}: [\tilde{\mathbb{Z}} \oplus \tilde{C}^*(\mathcal{BGL}_N; \mathbb{Z})] \otimes [\tilde{\mathbb{Z}} \oplus \tilde{C}^*(\mathcal{BGL}_M; \mathbb{Z})] \to \tilde{\mathbb{Z}} \oplus \tilde{C}^*(\mathcal{BGL}_{NM}; \mathbb{Z})$$

be the map

$$\mu^X_{N,M}(n, a) \otimes (m, b) = (nm, nb + ma + \hat{\mu}^X_{N,M}(a \otimes b)).$$

Taking the induced map on hypercohomology gives

$$\mathbb{H}^p(X, \tilde{\mathbb{Z}} \oplus \tilde{C}^*(\mathcal{BGL}_N; \mathbb{Z})) \otimes \mathbb{H}^q(X, \tilde{\mathbb{Z}} \oplus \tilde{C}^*(\mathcal{BGL}_M; \mathbb{Z})) \xrightarrow{\mu^X_{N,M}} \mathbb{H}^{p+q}(X, \tilde{\mathbb{Z}} \oplus \tilde{C}^*(\mathcal{BGL}_{MN}; \mathbb{Z}));$$

if we fix a bijection $\mathbb{N} \times \mathbb{N} \to \mathbb{N}$, we may take the limit of the maps $\mu^X_{N,M}$, giving the map

$$\mathbb{H}^p(X, \tilde{\mathbb{Z}} \oplus \tilde{C}^*(\mathcal{BGL}; \mathbb{Z})) \otimes \mathbb{H}^q(X, \tilde{\mathbb{Z}} \oplus \tilde{C}^*(\mathcal{BGL}; \mathbb{Z})) \xrightarrow{\mu^X_H} \mathbb{H}^{p+q}(X, \tilde{\mathbb{Z}} \oplus \tilde{C}^*(\mathcal{BGL}; \mathbb{Z})).$$

These maps are compatible with the product map on the Čech complexes $C^*(\mathcal{U}, \mathbb{Z} \times \mathrm{BGL}; \mathbb{Z})$ given via (Part II, Chapter III, §3.4.4) using the collection of products $\mu^{U_I}_{N,M}$ on the global sections.

If X is a scheme, we have functorial products in K-theory,

$$(2.2.4.4)\quad \mu^X_K: \Omega\mathrm{B}Q\mathcal{P}_X \wedge \Omega\mathrm{B}Q\mathcal{P}_X \to \Omega\mathrm{B}Q\mathcal{P}_X$$

induced by the tensor product on $\mathcal{P}_X$ (see [**126**, p. 235], [**128**, §3]). This gives the product on the K-groups

$$\mu^X_K: K_p(X) \otimes K_q(X) \to K_{p+q}(X).$$

If X is a diagram of S-schemes $X: I \to \mathbf{Sch}_S^+$, as in §2.1.3, we have the extension to products

$$\Omega \mathrm{B}Q\mathcal{P}_X \wedge \Omega \mathrm{B}Q\mathcal{P}_X \to \Omega \mathrm{B}Q\mathcal{P}_X,$$

since the products (2.2.4.4) are functorial. Via the method of (Part II, Chapter III, §3.4.4), this gives the natural product

$$\mu_K^X\colon \operatorname*{holim}_I \Omega \mathrm{B}Q\mathcal{P}_X \wedge \operatorname*{holim}_I \Omega \mathrm{B}Q\mathcal{P}_X \to \operatorname*{holim}_I \Omega \mathrm{B}Q\mathcal{P}_X,$$

inducing product maps $\mu_K^X: K_p(X) \otimes K_q(X) \to K_{p+q}(X)$.

2.2.5. PROPOSITION. *Let I be the category associated to a finite partially ordered set, and let $X: I \to \mathbf{Sch}^+$ be a functor. Then the products μ_H^X and μ_K^X are compatible via the Hurewicz map* (2.2.3.2).

PROOF. Suppose first that $X = \operatorname{Spec} A$. We recall Loday's construction of products [**92**]. We have the map

$$\mu_{N,M+}^A\colon \mathrm{BGL}_N(A)^+ \times \mathrm{BGL}_M(A)^+ \to \mathrm{BGL}(A)^+$$

induced by the maps $\tau_{N,M} - \tau_{N,*} - \tau_{*,M} + *$, where $*$ is the map sending everything to the basepoint, and $-$ is with respect to the H-space structure on $\mathrm{BGL}(A)^+$. Using the given bijection $\mathbb{N} \times \mathbb{N} \to \mathbb{N}$, we may take the limit, giving the map

$$\mu_+^A\colon \mathrm{BGL}(A)^+ \times \mathrm{BGL}(A)^+ \to \mathrm{BGL}(A)^+.$$

The restriction of μ_+^A to $\mathrm{BGL}(A)^+ \times * \cup * \times \mathrm{BGL}(A)^+$ is contractible, giving the map

$$\hat{\mu}_+^A\colon \mathrm{BGL}(A)^+ \wedge \mathrm{BGL}(A)^+ \to \mathrm{BGL}(A)^+.$$

We extend $\hat{\mu}_+^A$ to

$$\mu_+^A\colon [\mathbb{Z} \times \mathrm{BGL}(A)^+] \wedge [\mathbb{Z} \times \mathrm{BGL}(A)^+] \to \mathbb{Z} \times \mathrm{BGL}(A)^+.$$

by

$$\mu_+^A(n,x) \wedge (m,y) := (nm, ny + mx + \hat{\mu}_+^A(x,y)).$$

This induces products $K_p(A) \otimes K_q(A) \to K_{p+q}(A)$ for $p,q \geq 1$; it is immediate from the definitions that these products are compatible via the Hurewicz map with the products we have defined on $\mathbb{Z} \times C_*(\mathrm{GL}(A);\mathbb{Z})$.

Now let I be a category associated to a finite partially ordered set, and let $X: I \to \mathbf{Sch}^+$ be a functor.

We have the functor

$$\underline{\mathcal{U}}\colon I \times \underline{[n]}^{\mathrm{op}} \to \mathbf{Sch}^+$$
$$\underline{\mathcal{U}}(i,J) = \cap_{j \in J} U_j(i) \subset X(i),$$

giving the functor

$$\Gamma(\underline{\mathcal{U}}, \mathcal{O}_X)\colon I^{\mathrm{op}} \times \underline{[n]} \to \mathbf{Rings}$$
$$\Gamma(\underline{\mathcal{U}}, \mathcal{O}_X)(i,J) := \Gamma(\underline{\mathcal{U}}(i,J), \mathcal{O}_{X(i)}),$$

and the commutative diagram of natural transformations

$$
\begin{array}{ccc}
(\mathbb{Z}\times \mathrm{BGL}_N)\circ\Gamma(\underline{\mathcal{U}},\mathcal{O}_X) & \xrightarrow{h} & (\mathbb{Z}\times \mathbb{Z}\mathrm{BGL}_N)\circ\Gamma(\underline{\mathcal{U}},\mathcal{O}_X) \\
\downarrow & & \downarrow \\
(\mathbb{Z}\times \mathrm{BGL}^+_N)\circ\Gamma(\underline{\mathcal{U}},\mathcal{O}_X) & \xrightarrow{h} & (\mathbb{Z}\times \mathbb{Z}\mathrm{BGL}^+_N)\circ\Gamma(\underline{\mathcal{U}},\mathcal{O}_X),
\end{array}
\tag{2.2.5.1}
$$

where h is the Hurewicz map.

The above construction of products for $\mathrm{BGL}(A)^+$ is made functorial by May [**96**, Theorem 1.6 and Theorem 2.1] (see also [**128**, §3 and §4]), the same construction gives functorial products for $\mathbb{Z}\mathrm{BGL}(A)$ and $\mathbb{Z}\mathrm{BGL}(A)^+$, compatible with the Hurewicz map, and the natural map $\mathbb{Z}\mathrm{BGL}(A)\to\mathbb{Z}\mathrm{BGL}(A)^+$. We may need to take a different model for the functor $\mathrm{BGL}(-)^+$, but the universal mapping property of $\mathrm{BGL}(-)\to\mathrm{BGL}(-)^+$ over finite diagrams of rings [**83**, Lemma 3.1 and Theorem 4.3] implies that the two homotopy limits coming from two different functorial models of $\mathrm{BGL}(-)\to\mathrm{BGL}(-)^+$ are canonically weakly equivalent. The same holds for the diagram (2.2.5.1).

Taking the homotopy limit of (2.2.5.1) over $I^{\mathrm{op}}\times\underline{[n]}$ gives the commutative diagram

$$
\begin{array}{ccc}
\operatorname*{holim}_{I^{\mathrm{op}}\times\underline{[n]}}(\mathbb{Z}\times \mathrm{BGL}_N)\circ\Gamma(\underline{\mathcal{U}},\mathcal{O}_X) & \xrightarrow{h} & \operatorname*{holim}_{I^{\mathrm{op}}\times\underline{[n]}}(\mathbb{Z}\times \mathbb{Z}\mathrm{BGL}_N)\circ\Gamma(\underline{\mathcal{U}},\mathcal{O}_X) \\
\downarrow & & \downarrow \\
\operatorname*{holim}_{I^{\mathrm{op}}\times\underline{[n]}}(\mathbb{Z}\times \mathrm{BGL}^+_N)\circ\Gamma(\underline{\mathcal{U}},\mathcal{O}_X) & \xrightarrow{h} & \operatorname*{holim}_{I^{\mathrm{op}}\times\underline{[n]}}(\mathbb{Z}\times \mathbb{Z}\mathrm{BGL}^+_N)\circ\Gamma(\underline{\mathcal{U}},\mathcal{O}_X).
\end{array}
\tag{2.2.5.2}
$$

The products constructed by May thus give products via the method of §3.4.4,

$$\operatorname*{holim}_{I^{\mathrm{op}}\times\underline{[n]}} F\wedge \operatorname*{holim}_{I^{\mathrm{op}}\times\underline{[n]}} F\to \operatorname*{holim}_{I^{\mathrm{op}}\times\underline{[n]}} F,$$

where F is any one of the functors

$$(\mathbb{Z}\times\mathbb{Z}\mathrm{BGL})\circ\Gamma(\underline{\mathcal{U}},\mathcal{O}_X),\ (\mathbb{Z}\times\mathbb{Z}\mathrm{BGL}^+)\circ\Gamma(\underline{\mathcal{U}},\mathcal{O}_X),\ (\mathbb{Z}\times\mathrm{BGL}^+)\circ\Gamma(\underline{\mathcal{U}},\mathcal{O}_X).$$

The universal mapping property of $\mathrm{BGL}_N(-)\to\mathrm{BGL}_N(-)^+$ given in [**83**, Lemma 3.1 and Theorem 4.3], together with the construction of products described in (Part II, Chapter III, §3.4.4), and the compatibility given in (*loc. cit.*, §3.4.5), implies that the May product on

$$\operatorname*{holim}_{I^{\mathrm{op}}\times\underline{[n]}}(\mathbb{Z}\times\mathbb{Z}\mathrm{BGL})\circ\Gamma(\underline{\mathcal{U}},\mathcal{O}_X)$$

is compatible, via the Dold-Kan equivalence, with the product μ^X_H we have defined on

$$\operatorname*{holim}_{I^{\mathrm{op}}\times\underline{[n]}}\mathbb{Z}\times C^*(\mathrm{BGL}\circ\Gamma(\underline{\mathcal{U}},\mathcal{O}_X);\mathbb{Z})$$

in the derived category $\mathbf{D}^-(\mathbf{Ab})$. As the map $\mathbb{Z}\mathrm{BGL}_N(A)\to\mathbb{Z}\mathrm{BGL}_N(A)^+$ is a weak equivalence for all commutative rings A, it follows from [**25**, V, 5.6] that the map

$$\operatorname*{holim}_{I^{\mathrm{op}}\times\underline{[n]}}(\mathbb{Z}\times\mathbb{Z}\mathrm{BGL})\circ\Gamma(\underline{\mathcal{U}},\mathcal{O}_X)\to \operatorname*{holim}_{I^{\mathrm{op}}\times\underline{[n]}}(\mathbb{Z}\times\mathbb{Z}\mathrm{BGL}^+)\circ\Gamma(\underline{\mathcal{U}},\mathcal{O}_X)$$

is a weak equivalence. Thus, the products on $K_*(X,\mathcal{U})$ induced by the May products on

$$\operatorname*{holim}_{(I\times\underline{[n]})^{\mathrm{op}}}(\mathbb{Z}\times \mathrm{BGL}^+)\circ\Gamma(\underline{\mathcal{U}},\mathcal{O}_X)$$

are compatible with the products μ_H^X we have defined on

$$H^{-*}(\operatorname*{holim}_{(I\times\underline{[n]})^{\mathrm{op}}}\mathbb{Z}\times C^*(\mathrm{BGL}\circ\Gamma(\underline{\mathcal{U}},\mathcal{O}_X);\mathbb{Z})),$$

and hence with the products μ_H^X on the hypercohomology $\mathbb{H}^{-*}(X,\mathcal{BGL};\mathbb{Z})$.

Waldhausen [**126**, pg. 235] has given a functorial comparison of the products for $K_*(X,\mathcal{U})$ and for $K_*(X)$ (see also [**128**, §4, §5]), which completes the discussion. □

2.2.6. REMARK. Suppose we have functors $X,Y:I\to\mathbf{Sch}_S^+$. We form the *smash product* $X\times_S Y:I\to\mathbf{Sch}_S^+$ by

$$X\times_S Y(i):=\begin{cases}X(i)\times_S Y(i); & \text{if } X(i)\neq * \text{ and } Y(i)\neq *,\\ *; & \text{otherwise.}\end{cases}$$

We then have the natural external products

$$\mathrm{B}Q\mathcal{P}_{X(i)}\wedge\mathrm{B}Q\mathcal{P}_{Y(i)}\to\mathrm{B}Q\mathcal{P}_{X(i)\times_S Y(i)},$$

giving us the product

$$\boxtimes_{X,Y}:\operatorname*{holim}_I\mathrm{B}Q\mathcal{P}_X\wedge\operatorname*{holim}_I\mathrm{B}Q\mathcal{P}_Y\to\operatorname*{holim}_I\mathrm{B}Q\mathcal{P}_{X\times_S Y},$$

which in turn induces the external product

$$\boxtimes_{X,Y}^K:K_p(X)\otimes K_{p'}(Y)\to K_{p+p'}(X\times_S Y).$$

Similarly, the product (2.2.4.3) induces the external product

$$\mathbb{H}^p(X,\tilde{\mathbb{Z}}\oplus\tilde{C}^*(\mathcal{BGL};\mathbb{Z}))\otimes\mathbb{H}^q(Y,\tilde{\mathbb{Z}}\oplus\tilde{C}^*(\mathcal{BGL};\mathbb{Z}))$$
$$\xrightarrow{\boxtimes_{X,Y}^H}\mathbb{H}^{p+q}(X\times_S Y,\tilde{\mathbb{Z}}\oplus\tilde{C}^*(\mathcal{BGL};\mathbb{Z})).$$

The same proof as Proposition 2.2.5 then shows that the external products are compatible via the Hurewicz maps for X and Y.

Part II

Categorical Algebra

Introduction: Part II

In this second part, we collect the categorical notions and constructions necessary for Part I. The first chapter is a review of the foundational notions of symmetric monoidal categories, and related constructions, followed by applications to certain constructions of graded tensor categories, including the "universal commutative external product". We include this well-known material here as it gives a unified description of a wide range of categories we will be working with: graded tensor categories, differential graded tensor categories, triangulated tensor categories, etc. This unified treatment clarifies, for example, the usual problem of sign conventions in graded categories. As general reference for Chapter I, see [**77**] and [**93**].

In Chapter II, we look at Kapranov's generalization to DG categories of the standard construction of the category of complexes for an additive category (see [**75**]). We show that the homotopy category of this category of generalized complexes is a triangulated category. We also show how a tensor structure on the underlying DG category extends to the category of complexes, and makes the homotopy category a triangulated tensor category. We consider the operation of taking the pseudo-abelian hull, as applied to triangulated categories, and show that, for a localization of the homotopy category of complexes, the resulting category is naturally a triangulated category. We conclude with the construction of some useful DG tensor categories.

In Chapter III, we collect some results on simplicial and cosimplicial objects in an additive category, including various associated (co)chain complexes, as well as some details on products for cosimplicial objects in a tensor category. We then turn to a discussion of the problem of homotopy commutativity and related questions involving cochain operations. We give a multi-simplicial construction of "categorical cochain operations" which we use to transform a cosimplicial tensor functor into a DG tensor functor. We conclude with a discussion of homotopy limits.

In Chapter IV, we recall the construction of the cosimplicial Godement resolution of sheaf on a Grothendieck site with enough points. Combining this with the constructions of Chapter III allows us to define a canonical acyclic resolution of a sheaf on a Grothendieck site so that a commutative associative multiplication on the sheaf induces a multiplicative structure on the resolution which is homotopy commutative and has the required higher homotopies. These canonical cochain complexes allow us to define in Part I, Chapter V the realization functor associated to a reasonable family of sheaves.

CHAPTER I

Symmetric Monoidal Structures

1. Foundational material

1.1. Symmetric monoidal categories

We review some foundational material on symmetric monoidal categories, and related constructions.

1.1.1. DEFINITION. (i) A *semi-monoidal category* is a triple $(\mathcal{C}, \bullet, \alpha)$, where $\mathcal{C}$ is a category, $\bullet : \mathcal{C} \times \mathcal{C} \to \mathcal{C}$ is a functor, and $\alpha : \bullet \circ (\mathrm{id}_{\mathcal{C}} \times \bullet) \to \bullet \circ (\bullet \times \mathrm{id}_{\mathcal{C}})$ is a natural isomorphism such that

$$\begin{aligned} &[\bullet \circ (\mathrm{id}_{\mathcal{C}} \times \alpha)] \circ [\alpha \circ (\mathrm{id}_{\mathcal{C}} \times \bullet \times \mathrm{id}_{\mathcal{C}})] \circ [\bullet \circ (\alpha \times \mathrm{id}_{\mathcal{C}})] \\ &\qquad = [\alpha \circ (\mathrm{id}_{\mathcal{C}} \times \mathrm{id}_{\mathcal{C}} \times \bullet)] \circ [\alpha \circ (\bullet \times \mathrm{id}_{\mathcal{C}} \times \mathrm{id}_{\mathcal{C}})] \end{aligned} \tag{1.1.1.1}$$

(the *pentagonal identity*). If

$$\bullet \circ (\mathrm{id}_{\mathcal{C}} \times \bullet) = \bullet \circ (\bullet \times \mathrm{id}_{\mathcal{C}})$$

and α is the identity, call $(\mathcal{C}, \bullet, \alpha)$ *strictly associative.*

If there is an object $1 \in \mathcal{C}$ and natural isomorphisms $\mu_l : 1 \bullet (-) \to \mathrm{id}_{\mathcal{C}}$ and $\mu_r : (-) \bullet 1 \to \mathrm{id}_{\mathcal{C}}$, such that

$$\begin{aligned} \bullet \circ (\mu_l \times \mathrm{id}_{\mathcal{C}}) &= (\mu_l \circ \bullet) \circ \alpha(1, -, -), \\ [\bullet \circ (\mathrm{id}_{\mathcal{C}} \times \mu_r)] \circ \alpha(-, -, 1) &= \mu_r \circ \bullet, \\ (\mu_r \bullet \mathrm{id}_{\mathcal{C}}) \circ \alpha(-, 1, -) &= \mathrm{id}_{\mathcal{C}} \bullet \mu_l, \end{aligned} \tag{1.1.1.2}$$

call the tuple $(\mathcal{C}, \bullet, \alpha, 1, \mu_l, \mu_r)$ a *monoidal category.* A *symmetric semi-monoidal category* is a semi-monoidal category $(\mathcal{C}, \bullet, \alpha)$, together with a natural isomorphism $\tau : \bullet \circ \tau_{\mathcal{C}} \to \bullet$, where $\tau_{\mathcal{C}} : \mathcal{C}^2 \to \mathcal{C}^2$ is the exchange of factors, such that τ^2 is the identity, and

$$[\bullet \circ (\tau \times \mathrm{id}_{\mathcal{C}})] \circ \alpha \circ [\bullet \circ (\mathrm{id}_{\mathcal{C}} \times \tau)] = \alpha \circ [\tau \circ (\bullet \times \mathrm{id})] \circ \alpha; \tag{1.1.1.3}$$

a monoidal category together with natural isomorphism τ satisfying (1.1.1.3) and with

$$\mu_l(-) = \tau(1, -) \circ \mu_r(-) \tag{1.1.1.4}$$

is a *symmetric monoidal category.* The (symmetric) (semi-)monoidal category is said to be strictly associative if the underlying semi-monoidal category is strictly associative.

(ii) A (symmetric) (semi-)monoidal functor of (symmetric) (semi-)monoidal categories is a functor which strictly intertwines the various data of the (symmetric) (semi-)monoidal categories: e.g., for a semi-monoidal functor

$$F : (\mathcal{C}_1, \bullet_1, \alpha_1) \to (\mathcal{C}_2, \bullet_2, \alpha_2),$$

we require

$$\bullet_2 \circ (F \times F) = F \circ \bullet_1$$
$$\alpha_2 \circ [F \times (F \times F)] = [(F \times F) \times F] \circ \alpha_1;$$

the strictly associative case is defined similarly.

1.1.2. EXAMPLES. The primary example of a symmetric monoidal category is the category of sets, **Sets**, with product the cartesian product $\times$, symmetry the exchange of factors, and unit the one-point set. The category of small categories, **cat**, with product the cartesian product $\times$, symmetry the exchange of factors, and unit the one-point category with only the identity morphism is a symmetric monoidal category as well.

Let A be a commutative ring. We have the category of A-modules, $\mathbf{Mod}_A$, the category of graded A-modules, $\mathbf{GrMod}_A$, with graded, degree 0 maps, and the category of differential graded A-modules $\mathbf{DG\text{-}Mod}_A$ (with differential of degree $+1$), with maps being degree 0 maps of complexes.
(i) $\mathbf{Mod}_A$ has the product tensor product over A:

$$\otimes_A : \mathbf{Mod}_A \otimes \mathbf{Mod}_A \to \mathbf{Mod}_A,$$

symmetry τ

$$\tau_{X,Y} : X \otimes_A Y \to Y \otimes_A X$$
$$\tau_{X,Y}(x \otimes y) = y \otimes x,$$

associativity isomorphism

$$\alpha_{X,Y,Z} : (X \otimes_A Y) \otimes_A Z \to X \otimes_A (Y \otimes_A Z)$$
$$\alpha_{X,Y,Z}((x \otimes y) \otimes z) = x \otimes (y \otimes z),$$

and unit A, with natural isomorphisms $\mu_{l,X} : A \otimes_A X \to X$ and $\mu_{r,X} : X \otimes_A A \to X$. This makes $\mathbf{Mod}_A$ into a symmetric monoidal category.
(ii) $\mathbf{GrMod}_A$ has the graded tensor product: For $X = \oplus_p X^p$, $Y = \oplus_q Y^q$, set

$$X \otimes_A Y = \oplus_n (X \otimes_A Y)^n; \quad (X \otimes_A Y)^n = \oplus_{p+q=n} X^p \otimes_A Y^q.$$

The graded symmetry is given by

$$\tau_{X,Y}(x_p \otimes y_q) = (-1)^{pq} y_q \otimes x_p; \quad x_p \in X^p, y_q \in Y^q.$$

The associativity isomorphisms are defined as for $\mathbf{Mod}_A$, and the unit is the module A, concentrated in degree 0, with isomorphisms μ_l, μ_r defined as for $\mathbf{Mod}_A$. This makes $\mathbf{GrMod}_A$ into a symmetric monoidal category.
(iii) $\mathbf{DG\text{-}Mod}_A$ has product $(X, d_X) \otimes (Y, d_Y) = (X \otimes_A Y, d_{X \otimes Y})$, where $X \otimes_A Y$ is the graded tensor product, and

$$d_{X \otimes Y}(x_p \otimes y_q) = d_X(x_p) \otimes y_q + (-1)^p x_p \otimes d_Y(y_q); \quad x_p \in X_p, y_q \in Y_q.$$

The symmetry is the graded symmetry, and the unit is the graded unit, with 0 differential. The associativity isomorphisms and unit isomorphisms are defined as for $\mathbf{Mod}_A$.

1.2. $\mathcal{V}$-categories

Following [**77**, Chapter 1], we have the following

1.2.1. DEFINITION. (1) Let $(\mathcal{V}, \otimes, \alpha, \tau, \mu_l, 1)$ be a symmetric monoidal category. A *$\mathcal{V}$-category* $\mathcal{C}$ consists of the following data:

(i) A collection of objects of $\mathcal{C}$, $\mathrm{Obj}(\mathcal{C})$.
(ii) For each pair A, B of objects of $\mathcal{C}$, an object $\mathrm{Hom}_{\mathcal{C}}(A, B)$ of $\mathcal{V}$.
(iii) For each triple A, B, C of objects of $\mathcal{C}$, a morphism in $\mathcal{V}$

$$\circ_{A,B,C}\colon \mathrm{Hom}_{\mathcal{C}}(B, C) \otimes \mathrm{Hom}_{\mathcal{C}}(A, B) \longrightarrow \mathrm{Hom}_{\mathcal{C}}(A, C).$$

(iv) For each object A of $\mathcal{C}$, a morphism in $\mathcal{V}$

$$\mathrm{id}_A\colon 1 \longrightarrow \mathrm{Hom}_{\mathcal{C}}(A, A).$$

These satisfy

(a) (associativity) The diagram

$$\begin{array}{ccc}
\mathrm{Hom}_{\mathcal{C}}(C, D) \otimes [\mathrm{Hom}_{\mathcal{C}}(B, C) \otimes \mathrm{Hom}_{\mathcal{C}}(A, B)] & \xrightarrow{\mathrm{id}\otimes\circ_{A,B,C}} & \mathrm{Hom}_{\mathcal{C}}(C, D) \otimes \mathrm{Hom}_{\mathcal{C}}(A, C) \\
\downarrow{\scriptstyle \alpha} & & \\
[\mathrm{Hom}_{\mathcal{C}}(C, D) \otimes \mathrm{Hom}_{\mathcal{C}}(B, C)] \otimes \mathrm{Hom}_{\mathcal{C}}(A, B) & & \downarrow{\scriptstyle \circ_{A,C,D}} \\
\downarrow{\scriptstyle \circ_{B,C,D}\otimes\mathrm{id}} & & \\
\mathrm{Hom}_{\mathcal{C}}(B, D) \otimes \mathrm{Hom}_{\mathcal{C}}(A, B) & \xrightarrow{\circ_{A,B,D}} & \mathrm{Hom}_{\mathcal{C}}(A, D)
\end{array}$$

commutes, where α is the associativity isomorphism.

(b) (unit) The diagrams

$$\begin{array}{ccc}
1 \otimes \mathrm{Hom}_{\mathcal{C}}(A, B) & \xrightarrow{\mathrm{id}_A\otimes\mathrm{id}} & \mathrm{Hom}_{\mathcal{C}}(A, A) \otimes \mathrm{Hom}_{\mathcal{C}}(A, B) \\
& \searrow{\scriptstyle \mu} & \downarrow{\scriptstyle \circ_{A,A,B}} \\
& & \mathrm{Hom}_{\mathcal{C}}(A, B)
\end{array}$$

$$\begin{array}{ccc}
\mathrm{Hom}_{\mathcal{C}}(A, B) \otimes 1 & \xrightarrow{\mathrm{id}\otimes\mathrm{id}_B} & \mathrm{Hom}_{\mathcal{C}}(A, B) \otimes \mathrm{Hom}_{\mathcal{C}}(B, B) \\
& \searrow{\scriptstyle \mu} & \downarrow{\scriptstyle \circ_{B,B,A}} \\
& & \mathrm{Hom}_{\mathcal{C}}(A, B)
\end{array}$$

commute.

(2) A *$\mathcal{V}$-functor* $F\colon \mathcal{A} \to \mathcal{B}$ of $\mathcal{V}$-categories is given as for a functor of categories by a map on objects $A \mapsto F(A)$, together with maps in $\mathcal{V}$

$$F(A, B)\colon \mathrm{Hom}_{\mathcal{A}}(A, B) \longrightarrow \mathrm{Hom}_{\mathcal{B}}(F(A), F(B))$$

satisfying

(a) (functoriality) The diagram

$$\begin{array}{ccc}
\mathrm{Hom}_{\mathcal{A}}(B,C)\otimes\mathrm{Hom}_{\mathcal{A}}(A,B) & \xrightarrow{\circ_{A,B,C}} & \mathrm{Hom}_{\mathcal{A}}(A,C) \\
\Big\downarrow{\scriptstyle F(B,C)\otimes F(A,B)} & & \Big\downarrow{\scriptstyle F(A,C)} \\
\mathrm{Hom}_{\mathcal{B}}(F(B),F(C))\otimes\mathrm{Hom}_{\mathcal{B}}(F(A),F(B)) & \xrightarrow{\circ_{F(A),F(B),F(C)}} & \mathrm{Hom}_{\mathcal{B}}(F(A),F(C))
\end{array}$$

commutes.

(b) (unit) $F(A,A)\circ\mathrm{id}_A=\mathrm{id}_{F(A)}$.

The composition of functors is given by the composition in $\mathcal{V}$.

(3) A *$\mathcal{V}$-natural transformation $\theta:F\to G$ of $\mathcal{V}$-functors $F,G:\mathcal{A}\to\mathcal{B}$* consists of giving a map

$$\theta(A):1\to\mathrm{Hom}_{\mathcal{B}}(F(A),G(A))$$

in $\mathcal{V}$ for each object A of $\mathcal{A}$ such that, for objects A, A' of $\mathcal{A}$, the diagram

$$\begin{array}{ccc}
\mathrm{Hom}_{\mathcal{A}}(A,A')\otimes 1 & \xrightarrow{G(A,A')\otimes\theta(A)} & \mathrm{Hom}_{\mathcal{B}}(G(A),G(A'))\otimes\mathrm{Hom}_{\mathcal{B}}(F(A),G(A)) \\
 & & \Big\downarrow{\scriptstyle \circ_{F(A),G(A),G(A')}} \\
\Big\downarrow{\scriptstyle \tau} & & \mathrm{Hom}_{\mathcal{B}}(F(A),G(A')) \\
 & & \Big\uparrow{\scriptstyle \circ_{F(A),F(A'),G(A')}} \\
1\otimes\mathrm{Hom}_{\mathcal{A}}(A,A') & \xrightarrow[\theta(A')\otimes F(A,A')]{} & \mathrm{Hom}_{\mathcal{B}}(F(A'),G(A'))\otimes\mathrm{Hom}_{\mathcal{B}}(F(A),F(A'))
\end{array}$$

commutes. The composition of $\mathcal{V}$-natural transformations $\theta:F\to G$, $\rho:G\to H$ is given as the collection of morphisms

$$1\xrightarrow{\mu^{-1}}1\otimes 1\xrightarrow{\rho\otimes\theta}\mathrm{Hom}_{\mathcal{B}}(G(A),H(A))\otimes\mathrm{Hom}_{\mathcal{B}}(F(A),G(A))$$
$$\xrightarrow{\circ_{F(A),G(A),H(A)}}\mathrm{Hom}_{\mathcal{B}}(F(A),H(A)).$$

This defines the "category" $\mathbf{Cat}_{\mathcal{V}}$ of $\mathcal{V}$-categories and the category $\mathbf{cat}_{\mathcal{V}}$ of small $\mathcal{V}$-categories; in fact, if we define a morphism of $\mathcal{V}$-functors $F,G:\mathcal{A}\to\mathcal{B}$ to be a $\mathcal{V}$-natural transformation, we have the structure of a **cat**-category on $\mathbf{cat}_{\mathcal{V}}$.

1.2.2. REMARK. For the symmetric monoidal categories of Example 1.1.2, each object M of $\mathcal{V}$ has an underlying set, and each $\mathcal{V}$-morphism $f:1\to M$ determines a unique element $f_1\in M$. We may thus refer to the elements of the Hom-objects as morphisms, the element corresponding to the morphism $\mathrm{id}_A:1\to\mathrm{Hom}_{\mathcal{C}}(A,A)$ as the identity morphism $\mathrm{id}_A:A\to A$, and the element corresponding to the morphism $\theta(A):1\to\mathrm{Hom}_{\mathcal{C}}(F(A),G(A))$ as the morphism $\theta(A):F(A)\to G(A)$.

For $\mathcal{V}=\mathbf{Mod}_A$ (resp., $\mathbf{GrMod}_A$, $\mathbf{DG\text{-}Mod}_A$), a $\mathcal{V}$-category is called a *pre-A-additive category* (resp., *pre-A-graded category*, resp. *pre-A-differential graded category*). A **cat**-category is called a *2-category*; for a 2-category $\mathcal{A}$, we call the objects in the category $\mathrm{Hom}_{\mathcal{A}}(X,Y)$ *morphisms*, and the morphisms in $\mathrm{Hom}_{\mathcal{A}}(X,Y)$ *2-morphisms*.

1.2.3. *The symmetric monoidal structure of* $\mathbf{cat}_{\mathcal{V}}$. Let $\mathcal{V}$ be a symmetric monoidal category, with product $\otimes$, symmetry τ and unit 1. Let $\mathcal{A}$ and $\mathcal{B}$ be $\mathcal{V}$-categories. Define $\mathcal{A}\otimes_{\mathcal{V}}\mathcal{B}$ to be the $\mathcal{V}$-category with objects being pairs (A,B), with A an object of $\mathcal{A}$ and B an object of $\mathcal{B}$. The Hom-objects are defined by

$$\mathrm{Hom}_{\mathcal{A}\otimes_{\mathcal{V}}\mathcal{B}}\big((A_1,B_1),(A_2,B_2)\big) = \mathrm{Hom}_{\mathcal{A}}(A_1,A_2)\otimes \mathrm{Hom}_{\mathcal{A}}(B_1,B_2).$$

The composition map is given by the composition (we ignore the associativity isomorphisms)

$$\begin{aligned}&[\mathrm{Hom}_{\mathcal{A}}(A_2,A_3)\otimes \mathrm{Hom}_{\mathcal{B}}(B_2,B_3)]\otimes[\mathrm{Hom}_{\mathcal{A}}(A_1,A_2)\otimes \mathrm{Hom}_{\mathcal{B}}(B_1,B_2)]\\ &\xrightarrow{\tau_{A_*,B_*}} [\mathrm{Hom}_{\mathcal{A}}(A_2,A_3)\otimes \mathrm{Hom}_{\mathcal{A}}(A_1,A_2)]\otimes[\mathrm{Hom}_{\mathcal{B}}(B_2,B_3)\otimes \mathrm{Hom}_{\mathcal{B}}(B_1,B_2)]\\ &\xrightarrow{\circ_{A_1,A_2,A_3}\otimes\circ_{B_1,B_2,B_3}} \mathrm{Hom}_{\mathcal{A}}(A_1,A_3)\otimes \mathrm{Hom}_{\mathcal{B}}(B_1,B_3)\end{aligned}$$

where τ_{A_*,B_*} is the symmetry in $\mathcal{V}$. The identity in $\mathcal{A}\otimes_{\mathcal{V}}\mathcal{B}$ is given by the composition

$$1\xrightarrow{\mu^{-1}}1\otimes 1\xrightarrow{\mathrm{id}_{\mathcal{A}}\otimes\mathrm{id}_{\mathcal{B}}}\mathrm{Hom}_{\mathcal{A}}(A,A)\otimes\mathrm{Hom}_{\mathcal{B}}(B,B).$$

We have the symmetry functor $\tau^{\mathcal{V}}_{\mathcal{A},\mathcal{B}}\colon \mathcal{A}\otimes_{\mathcal{V}}\mathcal{B}\to\mathcal{B}\otimes_{\mathcal{V}}\mathcal{A}$, given on objects by $\tau^{\mathcal{V}}_{\mathcal{A},\mathcal{B}}(A,B)=(B,A)$, and on morphisms by letting

$$\tau^{\mathcal{V}}_{\mathcal{A},\mathcal{B}}\colon \mathrm{Hom}_{\mathcal{A}\otimes_{\mathcal{V}}\mathcal{B}}\big((A_1,B_1),(A_2,B_2)\big)\to \mathrm{Hom}_{\mathcal{B}\otimes_{\mathcal{V}}\mathcal{A}}\big((B_1,A_1),(B_2,A_2)\big)$$

be the symmetry in $\mathcal{V}$

$$\tau_{**}\colon \mathrm{Hom}_{\mathcal{A}}(A_1,A_2)\otimes\mathrm{Hom}_{\mathcal{B}}(B_1,B_2)\to \mathrm{Hom}_{\mathcal{B}}(B_1,B_2)\otimes\mathrm{Hom}_{\mathcal{A}}(A_1,A_2).$$

We have associativity isomorphisms $\alpha^{\mathcal{V}}_{\mathcal{A},\mathcal{B},\mathcal{C}}$ induced by those from $\mathcal{V}$ in a similar fashion.

We let $1_{\mathcal{V}}$ be the $\mathcal{V}$-category with a single object $*$, and with Hom-module $\mathrm{Hom}_{1_{\mathcal{V}}}(*,*)$ the unit object in $\mathcal{V}$. For a $\mathcal{V}$-category $\mathcal{A}$, we have the natural isomorphisms of $\mathcal{V}$-categories

$$\begin{aligned}\mu_l^{\mathcal{V},\mathcal{A}}&\colon 1_{\mathcal{V}}\otimes_{\mathcal{V}}\mathcal{A}\to\mathcal{A}\\ \mu_r^{\mathcal{V},\mathcal{A}}&\colon \mathcal{A}\otimes_{\mathcal{V}}1_{\mathcal{V}}\to\mathcal{A}\end{aligned}$$

induced by the isomorphisms

$$\begin{aligned}\mu_{l,(A,A')}&\colon 1\otimes\mathrm{Hom}_{\mathcal{A}}(A,A')\to\mathrm{Hom}_{\mathcal{A}}(A,A')\\ \mu_{r,(A,A')}&\colon \mathrm{Hom}_{\mathcal{A}}(A,A')\otimes 1\to\mathrm{Hom}_{\mathcal{A}}(A,A').\end{aligned}$$

The category $\mathbf{cat}_{\mathcal{V}}$ is then a symmetric monoidal category with the product $\otimes_{\mathcal{V}}$, associativity isomorphism $\alpha^{\mathcal{V}}$, symmetry $\tau^{\mathcal{V}}$, unit $1_{\mathcal{V}}$ and multiplications $\mu_l^{\mathcal{V}}$ and $\mu_r^{\mathcal{V}}$.

1.3. Symmetric monoidal $\mathcal{V}$-categories

We have phased Definition 1.1.1 in terms of natural transformations, and identities among natural transformations, to make subsequent refinements easier to state. Indeed, the above notions make sense for $\mathcal{V}$-categories as long as the category $\mathbf{cat}_{\mathcal{V}}$ of small categories with structure $\mathcal{V}$ has itself the structure of a symmetric semi-monoidal category. One simply replaces the cartesian product $\mathcal{C} \times \mathcal{C}$ with the product $\mathcal{C} \otimes \mathcal{C}$ in $\mathbf{cat}_{\mathcal{V}}$, and the exchange of factors isomorphism $\tau_{\mathcal{C}}$ with the symmetry isomorphism in $\mathbf{cat}_{\mathcal{V}}$ in the above definition, and one has the definition of a (symmetric) (semi-)monoidal object in $\mathbf{cat}_{\mathcal{V}}$.

One then extends these notions to an object $\mathcal{C}$ of $\mathbf{Cat}_{\mathcal{V}}$ by requiring that each small $\mathcal{V}$-subcategory of $\mathcal{C}$ which is closed under the relevant operations is a (symmetric) (semi-)monoidal object in $\mathbf{cat}_{\mathcal{V}}$.

1.3.1. Definition. Let $\mathcal{V}$ be a symmetric monoidal category. A *symmetric monoidal $\mathcal{V}$-category* is a tuple $(\mathcal{C}, \bullet, \alpha, \tau, 1_{\mathcal{C}}, \mu_l, \mu_r)$, with $\mathcal{C}$ a $\mathcal{V}$-category, $\bullet : \mathcal{C} \otimes_{\mathcal{V}} \mathcal{C} \to \mathcal{C}$ a $\mathcal{V}$-functor, $1_{\mathcal{C}}$ an object of $\mathcal{C}$, $\alpha\colon \bullet \circ [\mathrm{id}_{\mathcal{C}} \otimes_{\mathcal{V}} \bullet] \to \bullet \circ [\bullet \otimes_{\mathcal{V}} \mathrm{id}_{\mathcal{C}}]$, $\tau\colon \bullet \circ \tau^{\mathcal{V}}_{\mathcal{C}} \to \bullet$, and $\mu_l : 1_{\mathcal{C}} \otimes \mathrm{id}_{\mathcal{C}} \to \mathrm{id}_{\mathcal{C}}$, $\mu_r : \mathrm{id}_{\mathcal{C}} \otimes 1_{\mathcal{C}} \to \mathrm{id}_{\mathcal{C}}$ natural isomorphisms, such that the relations (1.1.1.1), (1.1.1.2), (1.1.1.3) and (1.1.1.4) are satisfied as identities among $\mathcal{V}$-natural transformations, after replacing $\times$ with $\otimes_{\mathcal{V}}$.

The notions of a monoidal $\mathcal{V}$-category, a semi-monoidal $\mathcal{V}$-category and a symmetric semi-monoidal $\mathcal{V}$-category are defined analogously. We sometimes refer to a (symmetric) (semi)-monoidal $\mathcal{V}$-category as a (symmetric) (semi)-monoidal object of $\mathbf{Cat}_{\mathcal{V}}$.

1.3.2. *Bi-products.* Recall that the bi-product of objects X and Y in an pre-A-additive category $\mathcal{C}$ is a tuple

$$
\begin{gathered}
(X \oplus Y, i_X, i_Y, p_X, p_Y) \\
i_X : X \to X \oplus Y,\ i_Y : Y \to X \oplus Y \\
p_X : X \oplus Y \to X,\ p_Y : X \oplus Y \to Y,
\end{gathered}
$$

such that

$$
\begin{gathered}
p_X \circ i_X = \mathrm{id}_X,\ p_Y \circ i_Y = \mathrm{id}_Y \\
p_X \circ i_Y = 0,\ p_Y \circ i_X = 0 \\
i_X \circ p_X + i_Y \circ p_Y = \mathrm{id}_{X \oplus Y}.
\end{gathered}
$$

We call $\mathcal{C}$ an *A-additive category* if $\mathcal{C}$ is an pre-A-additive category and, in addition, finite bi-products exist in $\mathcal{C}$. In particular, there is an initial and final object 0, given as the empty bi-product.

Let $(X \oplus Y, i_X : X \to X \oplus Y, i_Y : Y \to X \oplus Y)$ be a direct sum of objects X and Y in a pre-additive category $\mathcal{C}$. The universal property of the direct sum, i.e., that the map

$$
(i_X^*, i_Y^*) : \mathrm{Hom}_{\mathcal{C}}(X \oplus Y, -) \to \mathrm{Hom}_{\mathcal{C}}(X, -) \times \mathrm{Hom}_{\mathcal{C}}(Y, -)
$$

is a natural isomorphism, gives maps $p_X : X \oplus Y \to X$ and $p_Y : X \oplus Y \to Y$, with

$$
(p_X \circ i_X, p_X \circ i_Y) = (\mathrm{id}_X, 0),\ (p_Y \circ i_X, p_Y \circ i_Y) = (0, \mathrm{id}_Y).
$$

Then $(X \oplus Y, i_X, i_Y, p_X, p_Y)$ is a bi-product of X and Y. For this reason, we often refer to a bi-product as a direct sum, suppressing the explicit mention of the projections p_X and p_Y.

We have the functors

$$
(1.3.2.1) \qquad \begin{aligned} {}^0 &: \mathbf{cat}_{\mathbf{GrMod}_A} \to \mathbf{cat}_{\mathbf{Mod}_A} \\ Z^* &: \mathbf{cat}_{\mathbf{DG\text{-}Mod}_A} \to \mathbf{cat}_{\mathbf{GrMod}_A}; \end{aligned}
$$

for a pre-graded category $\mathcal{C}$, the pre-additive category $\mathcal{C}^0$ has the same objects as $\mathcal{C}$, with $\mathrm{Hom}_{\mathcal{C}^0}(A,B) = \mathrm{Hom}_{\mathcal{C}}(A,B)^0$. For a pre-DG category $\mathcal{C}$, the pre-graded category $Z^*\mathcal{C}$ has the same objects as $\mathcal{C}$, with

$$
\mathrm{Hom}_{Z^*\mathcal{C}}(A,B)^n = \{f \in \mathrm{Hom}_{\mathcal{C}}(A,B)^n \mid df = 0\}.
$$

The category $\mathcal{C}^0$ is naturally a subcategory of $\mathcal{C}$, and similarly for $Z^*\mathcal{C}$. We let $Z^0\mathcal{C} := (Z^*\mathcal{C})^0$.

A bi-product of objects X, Y in a pre-graded category $\mathcal{C}$ is defined to be a bi-product in the pre-additive sub-category $\mathcal{C}^0$; similarly, a bi-product of objects X, Y in a pre-DG category $\mathcal{C}$ is defined to be a bi-product in the pre-additive sub-category $Z^0\mathcal{C}$

1.3.3. DEFINITION. For $\mathcal{V} = \mathbf{Mod}_A$, $\mathbf{GrMod}_A$, or $\mathbf{DG\text{-}Mod}_A$, we let $\mathbf{cat}^{\oplus}_{\mathcal{V}}$ denote the full subcategory of $\mathbf{cat}_{\mathcal{V}}$ consisting of categories $\mathcal{A}$ which admit finite bi-products. We call $\mathbf{cat}^{\oplus}_{\mathbf{Mod}_A}$ the category of *small A-additive categories*, $\mathbf{Add}_A$, $\mathbf{cat}^{\oplus}_{\mathbf{GrMod}_A}$ the category of small A-graded categories, $\mathbf{GrMod}_A$ and $\mathbf{cat}^{\oplus}_{\mathbf{DG\text{-}Mod}_A}$ the category of small A-differential-graded categories, $\mathbf{DG}_A$.

1.3.4. One can form an additive category $\mathcal{C}^{\oplus}$ from a pre-additive category by adjoining finite bi-products; as two bi-products of objects X and Y are canonically isomorphic, this operation applied to an additive category yields an equivalent category. In addition, the functor ${}^{\oplus}: \mathbf{cat}_{\mathbf{Mod}_A} \to \mathbf{Add}_A$ is left adjoint to the inclusion $\mathbf{Add}_A \to \mathbf{cat}_{\mathbf{Mod}_A}$. Similar remarks hold for $\mathbf{cat}_{\mathbf{GrMod}_A}$ and $\mathbf{cat}_{\mathbf{DG\text{-}Mod}_A}$.

1.3.5. REMARKS. (i) Let $(\mathcal{C}, \otimes, \alpha)$ be a semi-monoidal $\mathcal{V}$-category. The pentagonal relation (1.1.1.1) implies (see MacLane [**93**]) that, given n objects $X_1, \ldots, X_n$ of $\mathcal{C}$, and two orders of associativity for the product $X_1 \otimes \ldots \otimes X_n$, there is a canonical isomorphism between the two resulting objects of $\mathcal{C}$. This shows that $\mathcal{C}$ is equivalent to a canonically defined strictly associative symmetric monoidal $\mathcal{V}$-category. The same remark holds for symmetric monoidal or symmetric semi-monoidal $\mathcal{V}$-categories. In the sequel, we will systematically work with the equivalent strictly associative objects, unless otherwise noted. In particular, we will replace $\mathbf{cat}_{\mathcal{V}}$ with its equivalent strictly associative version without further comment or additional notation.
(ii) Let $(\mathcal{C}, \otimes, \tau)$ be a strictly associative symmetric semi-monoidal $\mathcal{V}$-category. The relation (1.1.1.3) implies that we can unambiguously assign an isomorphism

$$
\tau_{\sigma, X_1, \ldots, X_n} : X_1 \otimes \ldots \otimes X_n \to X_{\sigma^{-1}(1)} \otimes \ldots \otimes X_{\sigma^{-1}(n)}
$$

for each σ in the symmetric group S_n by writing σ as a composition of adjacent permutations $\sigma_{i,i+1}$, and defining $\tau_{\sigma, X_1, \ldots, X_n}$ as the corresponding composition of symmetry isomorphisms τ_{**}. In addition, the automorphisms $\tau_{\sigma, X_1, \ldots, X_n}$ are functorial in the X_i and satisfy

$$
\tau_{\rho, X_{\sigma^{-1}(1)}, \ldots, X_{\sigma^{-1}(n)}} \circ \tau_{\sigma, X_1, \ldots, X_n} = \tau_{\rho\sigma, X_1, \ldots, X_n}.
$$

1.3.6. EXAMPLES. An *A-tensor category* is an A-additive category which is a symmetric monoidal object in the category of pre-A-additive categories. An additive category which is a symmetric semi-monoidal object in $\mathbf{Mod}_A$-categories is called an *A-tensor category without unit*. Similarly, a *graded A-tensor category* is a graded

A-additive category which is a symmetric monoidal object in the category of graded pre-A-additive categories, and a *DG A-tensor category* is an differential graded category which is a symmetric monoidal object in the category of $\mathbf{DG\text{-}Mod}_A$-categories. The respective symmetric semi-monoidal objects $\mathbf{GrMod}_A$-categories, resp. $\mathbf{DG\text{-}Mod}_A$-categories are called a *graded A-tensor category without unit*, resp. *DG A-tensor category without unit.*

We often refer to a symmetric monoidal object in the category of pre-A-additive categories as an *pre-A-tensor category.* The notions of a *pre-A-tensor category without unit*, a *pre-A-DG tensor category*, etc., are defined similarly.

1.3.7. *Pseudo-tensor functors.* It is sometimes necessary to relax the condition that a functor of tensor categories be a tensor functor; we give here one version of such a construction. To fix ideas, we work in the setting of a tensor category; the same notions make sense for a symmetric monoidal $\mathcal{V}$-category, for example, a graded tensor category.

Let $\mathcal{A}$ and $\mathcal{B}$ be tensor categories. A *pseudo-tensor functor* from $\mathcal{A}$ to $\mathcal{B}$ is a pair (F,θ), with $F\colon\mathcal{A}\to\mathcal{B}$ an additive functor of the underlying additive categories, and θ a natural isomorphism of functors

$$\theta\colon F(-)\otimes_{\mathcal{B}} F(-)\to F(-\times_{\mathcal{A}}-)$$
$$F(-)\otimes_{\mathcal{B}} F(-), F(-\times_{\mathcal{A}}-)\colon \mathcal{A}\otimes\mathcal{A}\to\mathcal{B},$$

such that

1. θ is associative: The diagram

$$\begin{array}{ccc} F(X)\otimes F(Y)\otimes F(Z) & \xrightarrow{\mathrm{id}_X\otimes\theta(Y,Z)} & F(X)\otimes F(Y\otimes Z) \\ \downarrow{\scriptstyle \theta(X,Y)\otimes\mathrm{id}_Z} & & \downarrow{\scriptstyle \theta(X,Y\otimes Z)} \\ F(X\otimes Y)\otimes F(Z) & \xrightarrow[\theta(X\otimes Y,Z)]{} & F(X\otimes Y\otimes Z) \end{array}$$

commutes,

2. θ is commutative: The diagram

$$\begin{array}{ccc} F(X)\otimes F(Y) & \xrightarrow{\tau_{F(X),F(Y)}} & F(Y)\otimes F(X) \\ \downarrow{\scriptstyle \theta(X,Y)} & & \downarrow{\scriptstyle \theta(Y,X)} \\ F(X\otimes Y) & \xrightarrow[F(\tau_{X,Y})]{} & F(Y\otimes X) \end{array}$$

commutes,

3. F and θ are unital:

$$F(1_{\mathcal{A}}^{\otimes n})=1_{\mathcal{B}}^{\otimes n};\quad \theta(1_{\mathcal{A}}^{\otimes a},1_{\mathcal{A}}^{\otimes b})=\mathrm{id}_{1_{\mathcal{B}}^{\otimes a+b}},$$

and the diagram

$$\begin{array}{ccc} 1_{\mathcal{B}}\otimes F(X) & \xrightarrow{\mu_{l,F(X)}} & F(X) \\ \downarrow{\scriptstyle \theta(1_{\mathcal{A}},X)} & \nearrow{\scriptstyle F(\mu_{l,X})} & \\ F(1_{\mathcal{A}}\otimes X) & & \end{array}$$

commutes.

Composition of pseudo-tensor functors is the obvious notion:

$$(G,\theta_G)\circ(F,\theta_F) := (G\circ F, G(\theta(-,-))\circ\theta_G(F(-),F(-));$$

the composition of two pseudo-tensor functors is again a pseudo-tensor functor.

If (F,θ_F) and (G,θ_G) are pseudo-tensor functors from $\mathcal{A}$ to $\mathcal{B}$, a *natural transformation* from (F,θ_F) to (G,θ_G) is a natural transformation of the additive functors $\rho\colon F\to G$, such that $\rho(1_{\mathcal{A}}^{\otimes n})$ is the identity on $1_{\mathcal{B}}^{\otimes n}$, and such that the diagram

$$\begin{array}{ccc} F(X)\otimes F(Y) & \xrightarrow{\rho(X)\otimes\rho(Y)} & G(X)\otimes G(Y) \\ {\scriptstyle\theta_F}\downarrow & & \downarrow{\scriptstyle\theta_G(X,Y)} \\ F(X\otimes Y) & \xrightarrow[\rho(X\otimes Y)]{} & G(X\otimes Y) \end{array}$$

commutes.

Pseudo-tensor functors $(F,\theta_F)\colon\mathcal{A}\to\mathcal{B}$ and $(G,\theta_G)\colon\mathcal{B}\to\mathcal{A}$ define a *pseudo-tensor equivalence* if there are natural isomorphisms of pseudo-tensor functors

$$\rho_1\colon \mathrm{id}_{\mathcal{A}}\to(G,\theta_G)\circ(F,\theta_F);\quad \rho_2\colon\mathrm{id}_{\mathcal{B}}\to(F,\theta_F)\circ(G,\theta_G).$$

All the above notions makes sense for tensor categories without unit by ignoring the conditions on the unit.

1.3.8. Let $M=\oplus_i M^i$, $N=\oplus_j N^j$ be graded A-modules. An A-module homomorphism $f\colon M\to N$ is *graded, of degree* s, if $f=\prod_i f^i\colon M^i\to N^{i+s}$. Let $\mathrm{Hom}(M,N)^s$ be the A-module of graded, degree s maps from M to N; this makes $\mathbf{GrMod}_A$ into an A-graded category with graded Hom-module $\mathrm{Hom}(M,N):=\oplus_s\mathrm{Hom}(M,N)^s$.

If M and N are DG A-modules, we give $\mathrm{Hom}(M,N)$ the differential

$$df=(-1)^{\deg f} f\circ d_M - d_N\circ f.$$

This makes $\mathbf{DGMod}_A$ into an A-DG category, which is the usual DG category of complexes of A-modules, $\mathbf{C}(\mathbf{Mod}_A)$. The full DG subcategories $\mathbf{C}^*(\mathbf{Mod}_A)$, $*=b,+,-$, of bounded complexes, bounded below complexes and bounded above complexes are defined as usual.

For graded A-modules X, Y, Z and W, define the tensor product of graded maps $f=\oplus_i f^i\colon X^i\to Z^{i+s}$ and $g=\oplus_j\colon Y^j\to W^{j+t}$ by

$$f\otimes g=\oplus_n\oplus_{i+j=n}(-1)^{it}f^i\otimes_A g^j\colon (X\otimes_A Y)^n\to(Z\otimes W)^{n+s+t}.$$

This makes $\mathbf{GrMod}_A$ into an A-graded tensor category, and makes $\mathbf{C}(\mathbf{Mod}_A)$ into an A-DG tensor category. The full subcategories $\mathbf{C}^*(\mathbf{Mod}_A)$ of $\mathbf{C}(\mathbf{Mod}_A)$ $(*=+,-,b)$ are A-DG tensor subcategories.

2. Constructions and computations

2.1. Elementary constructions

We give some of the elementary methods for constructing symmetric monoidal $\mathcal{V}$-categories.

2.1.1. *Free objects, and similar constructions.* For $\mathcal{V}$ as in Example 1.1.2, the inclusion of the category of symmetric monoidal $\mathcal{V}$-categories into the category of $\mathcal{V}$-categories, $\mathbf{SM}_{\mathcal{V}} \to \mathbf{cat}_{\mathcal{V}}$, has a left adjoint, the functor forming the *free symmetric monoidal $\mathcal{V}$-category* on a given $\mathcal{V}$-category. We have the similar free objects for (semi-)monoidal $\mathcal{V}$-categories, symmetric semi-monoidal $\mathcal{V}$-categories, and the strictly associative version. For $\mathcal{V}$ one of the A-additive categories $\mathbf{Mod}_A$, $\mathbf{GrMod}_A$ or $\mathbf{DG\text{-}Mod}_A$, the inclusions of the additive versions $\mathbf{SM}^{\oplus}_{\mathcal{V}}$, etc., into the additive versions $\mathbf{Cat}^{\oplus}_{\mathcal{V}}$ have the left adjoint given by taking the free object as above, then applying the functor $^{\oplus}$.

We have various forgetful functors among the categories $\mathcal{V}$ considered in Example 1.1.2:

$$FD\colon \mathbf{DG\text{-}Mod}_A \to \mathbf{GrMod}_A : \text{ "forget the differential".}$$
$$FG\colon \mathbf{GrMod}_A \to \mathbf{Mod}_A : \text{ "forget the grading".}$$
$$FA\colon \mathbf{Mod}_A \to \mathbf{Sets} : \text{ "forget the } A\text{-module structure".}$$
$$FM\colon \mathbf{cat} \to \mathbf{Sets} : \text{ "forget the morphisms".}$$

The functors FA and FM have the left adjoints $\mathrm{Free}_{\mathcal{V}_2,\mathcal{V}_1}\colon \mathcal{V}_2 \to \mathcal{V}_1$, producing the free $\mathcal{V}_1$-object on the given $\mathcal{V}_2$-object.

The forgetful functors induce forgetful functors $F_{\mathcal{V}_1,\mathcal{V}_2}\colon \mathbf{cat}_{\mathcal{V}_1} \to \mathbf{cat}_{\mathcal{V}_2}$, and for $\mathcal{V}_1 = \mathbf{Mod}_A, \mathbf{cat}$, the forgetful functor has the left adjoint $\mathrm{Free}_{\mathcal{V}_2,\mathcal{V}_1}\colon \mathbf{cat}_{\mathcal{V}_2} \to \mathbf{cat}_{\mathcal{V}_1}$. Both of these are the identity on objects and operate on the Hom-objects by forgetting the $\mathcal{V}_1$ structure, resp. producing the free $\mathcal{V}_2$-object. The formation of the free object is a symmetric monoidal functor.

We denote the free pre-A-additive category on a category $\mathcal{C}$ by $A\mathcal{C}$. We sometimes write $A\mathcal{C}$ for the free A-additive category on a category $\mathcal{C}$; we will explicitly make the distinction clear.

We have in addition the functors (1.3.2.1)

$$Z^*\colon \mathbf{DG\text{-}Mod}_A \to \mathbf{GrMod}_A : \text{ "take cycles",}$$
$$^0\colon \mathbf{GrMod}_A \to \mathbf{Mod}_A : \text{ "take 0th graded piece",}$$

which have left adjoints

$$d_0\colon \mathbf{GrMod}_A \to \mathbf{DG\text{-}Mod}_A,$$
$$i_0\colon \mathbf{Mod}_A \to \mathbf{GrMod}_A,$$

where $d_0(M)$ is the complex M with 0 differential, and $i_0(M)$ is the graded A-module $i_0(M)^0 = M$, $i_0(M)^d = 0$ for $d \neq 0$. As above, these have their counterparts on the level of $\mathcal{V}$-categories, for which we use the same notation,

$$Z^*\colon \mathbf{cat}_{\mathbf{DG\text{-}Mod}_A} \to \mathbf{cat}_{\mathbf{GrMod}_A},$$

etc.

2.1.2. *Adjoining morphisms and objects.* If $(\mathcal{C}, \otimes, \alpha, \tau, \mu, 1)$ is a symmetric monoidal object in $\mathbf{cat}_{\mathcal{V}}$, we can adjoin morphisms $\{f_\alpha\colon X_\alpha \to Y_\alpha \mid \alpha \in A\}$ satisfying relations $\{Z_\beta \mid \beta \in B\}$ to form the $\mathcal{V}$-symmetric monoidal object $\mathcal{C}[\{f_\alpha\}]/\{Z_\beta\}$, with functor

$$\iota_{\{f_\alpha\},\{Z_\beta\}}\colon \mathcal{C} \to \mathcal{C}[\{f_\alpha\}]/\{Z_\beta\}.$$

The functor $\iota_{\{f_\alpha\},\emptyset}$ is universal for $\mathcal{V}$-symmetric monoidal functors $F:\mathcal{C}\to\mathcal{D}$ together with a choice of morphisms $g_\alpha: F(X_\alpha)\to F(Y_\alpha)$. Similarly, $\iota_{\{f_\alpha\},\{Z_\beta\}}$ is universal for $\mathcal{V}$-symmetric monoidal functors $F:\mathcal{C}\to\mathcal{D}$ together with a choice of morphisms $g_\alpha: F(X_\alpha)\to F(Y_\alpha)$ which satisfy relations $\tilde{F}(Z_\beta)$, where $\tilde{F}:\mathcal{C}[\{f_\alpha\}]\to\mathcal{D}$ is the extension of F determined by the g_α. The symmetric semi-monoidal, semi-monoidal, monoidal and strictly associative versions are constructed similarly, and have analogous properties.

If we have an $\mathcal{V}$-symmetric monoidal category $(\mathcal{C},\otimes,\alpha,\tau,\mu,1)$, we may adjoin a set of objects $\{X_\alpha\mid\alpha\in A\}$ to $\mathcal{C}$, with only identity morphisms, and form the functor of $\mathcal{V}$-symmetric monoidal categories $i:\mathcal{C}\to\mathcal{C}[\{X_\alpha\}]$ as the universal object for $\mathcal{V}$-symmetric monoidal functors $F:\mathcal{C}\to\mathcal{D}$, together with a choice of objects $\{Y_\alpha\}$ of $\mathcal{D}$. The analogous remarks are also valid in the setting of $\mathcal{V}$-(symmetric)(semi-)monoidal categories.

2.2. The category of pairs

One can use a 2-functor as a means of encoding generators and relations for a category. Describing this method of constructing categories is the object of this section.

2.2.1. Let $\mathcal{A}$ be a 2-category with underlying category $\mathcal{A}_0$, and let $\Pi:\mathcal{A}\to\mathbf{cat}_{\mathcal{V}}$ be a 2-functor. We assume the structure category $\mathcal{V}$ admits coproducts of a given set of objects, and that co-equalizers (quotient objects) for a pair of $\mathcal{V}$-morphisms exist; this is the case for $\mathcal{V}=\mathbf{Sets},\mathbf{Mod}_A$, $\mathbf{GrMod}_A$ and $\mathbf{DG\text{-}Mod}_A$.

Define the $\mathcal{V}$-category $(\Pi,\mathcal{A})$ as follows: Objects consist of pairs (c,a), with a an object of $\mathcal{A}$, and c an object of $\Pi(a)$. To define the morphisms, consider the $\mathcal{V}$-object

$$((c,a),(c',a')) := \coprod_{f\in\mathrm{Hom}_{\mathcal{A}}(a,a')} \mathrm{Hom}_{\Pi(a')}(\Pi(f)(c),c').$$

We let the pair (g,f) stand for the map $g:\Pi(f)(c)\to c'$ in the component f. For a 2-morphism $h:f\to f'$ in $\mathcal{A}$, we define the $\mathcal{V}$-morphism

$$\Pi(h)^*:\mathrm{Hom}_{\Pi(a')}(\Pi(f')(c),c')\to\mathrm{Hom}_{\Pi(a')}(\Pi(f)(c),c')$$

by $\Pi(h)^*(g') = g'\circ\Pi(h)(c)$. We define $\mathrm{Hom}_{(\Pi,\mathcal{A})}((c,a),(c',a'))$ to be the co-equalizer:

$$\coprod_{(f',h:f\to f')}\mathrm{Hom}_{\Pi(a')}(\Pi(f')(c),c') \rightrightarrows ((c,a),(c',a')) \to \mathrm{Hom}_{(\Pi,\mathcal{A})}((c,a),(c',a')), \tag{2.2.1.1}$$

where the two maps on the component $(f',h:f\to f')$ are

$$g'\mapsto(g',f');\quad g'\mapsto(\Pi(h)^*(g'),f).$$

Composition is given by

$$(g',f')\circ(g,f)=(g'\circ\Pi(f')(g),f'\circ f).$$

We may, of course, consider a category as a 2-category with only identity 2-morphisms; in this case we have $\mathrm{Hom}_{(\Pi,\mathcal{A})}((c,a),(c',a'))=((c,a),(c',a'))$. In general, for each morphism $f:a\to a'$ in $\mathcal{A}$, we have the natural map

$$\Psi_f:\mathrm{Hom}_{\Pi(a')}(\Pi(f)(c),c')\to\mathrm{Hom}_{(\Pi,\mathcal{A})}((c,a),(c',a')). \tag{2.2.1.2}$$

2.2.2. REMARK. Let $\mathcal{A}$ be a 2-category with underlying category $\mathcal{A}_0$. If all 2-morphisms in $\mathcal{A}$ are isomorphisms, we have the equivalence relation on the set of morphisms $\mathrm{Hom}_{\mathcal{A}_0}(a, a')$ given by

$$F \sim G \iff \text{there is a 2-morphism } \eta: F \to G.$$

Let $\bar{\mathcal{A}}$ be the category with the same objects as $\mathcal{A}_0$, and with $\mathrm{Hom}_{\bar{\mathcal{A}}}(a, a')$ the set of equivalence classes of morphisms $F: a \to a'$ in $\mathcal{A}_0$. For each $f: a \to a'$ in $\bar{\mathcal{A}}$, choose a representative morphism $F_f: a \to a'$ in $\mathcal{A}_0$.

Suppose we have a 2-functor $\Pi: \mathcal{A} \to \mathbf{cat}_{\mathcal{V}}$. Let $F: a \to a'$ be a morphism in $\mathcal{A}_0$, let c be an object of $\Pi(a)$, c' an object of $\Pi(a')$, and let

$$\Psi_F: \mathrm{Hom}_{\Pi(c')}(\Pi(F)(c), c') \to \mathrm{Hom}_{(\Pi,\mathcal{A})}((c, a), (c', a'))$$

the canonical map (2.2.1.2). Suppose that, for each morphism F in $\mathcal{A}_0$, the only 2-morphism $F \to F$ is the identity. Then the map

$$\coprod_{f \in \mathrm{Hom}_{\bar{\mathcal{A}}}(a,a')} \Psi_{F_f}: \coprod_{f \in \mathrm{Hom}_{\bar{\mathcal{A}}}(a,a')} \mathrm{Hom}_{\Pi(a')}(\Pi(F)(c), c') \\ \to \mathrm{Hom}_{(\Pi,\mathcal{A})}((c, a), (c', a'))$$

is an isomorphism. Indeed, the equivalence defining $\mathrm{Hom}_{(\Pi,\mathcal{A})}((c, a), (c', a'))$ results from identifying $\mathrm{Hom}_{\Pi(a')}(\Pi(F)(c), c')$ with $\mathrm{Hom}_{\Pi(a')}(\Pi(F')(c), c')$ by the isomorphism $\Pi(\eta)^*$ for each 2-morphism $\eta: F \to F'$. Our assumptions on the 2-morphisms in $\mathcal{A}$ yield our assertion immediately.

2.2.3. REMARK. Many structures on $\mathcal{A}$ give rise to similar structures on $(\Pi, \mathcal{A})$. For example, if $\mathcal{A}$ is a symmetric semi-monoidal 2-category, and Π is a symmetric semi-monoidal 2-functor, then $(\Pi, \mathcal{A})$ has a natural structure of a symmetric (semi)-monoidal $\mathcal{V}$-category: The product $\bullet$ on objects is given by $(c, a) \bullet (c', a') = ((c, c'), a \otimes_{\mathcal{A}} a')$, and on morphisms by $(g, f) \bullet (g', f') = (g \otimes_{\mathcal{V}} g', f \otimes_{\mathcal{A}} f')$. The symmetry $\tau_{(\Pi,\mathcal{A})}$ is given by $\tau_{(\Pi,\mathcal{A})}((c, a), (c', a')) = (\mathrm{id}_{(c',c)}, \tau_{\mathcal{A}}(a, a'))$.

2.2.4. REMARK. One can define the notion of an *inductive limit* of a 2-functor of 2-categories $F: \mathcal{A} \to \mathcal{B}$ as universal object for the data consisting of an object Z of $\mathcal{B}$, morphisms $i_a: F(a) \to Z$ for each object a of $\mathcal{A}$, and 2-morphisms $\theta_f: i_a \to i_b \circ F(f)$ for each morphism $f: a \to b$ in $\mathcal{A}$ such that

1. For each pair of composable morphisms $a \xrightarrow{f} b \xrightarrow{g} c$ in $\mathcal{A}$, we have
$$\theta_f \circ (\theta_g \circ F(f)) = \theta_{g \circ f}.$$
2. For each pair of maps $f, f': a \to b$, and each 2-morphism $\eta: f \to f'$ in $\mathcal{A}$, we have
$$\theta_f = \theta_{f'} \circ (i_b \circ F(h)).$$

Then the category of pairs $(\Pi, \mathcal{A})$ is just the inductive limit of $\Pi: \mathcal{A} \to \mathbf{cat}_{\mathcal{V}}$; as we won't be using this fact, we omit a detailed description of the inductive limit of a 2-functor, and the verification that $(\Pi, \mathcal{A})$ is the inductive limit of Π.

2.3. Some categories and a 2-category

In this section, we construct some fundamental symmetric semi-monoidal categories and an important symmetric semi-monoidal 2-category.

2.3.1. *The categories* $\mathfrak{N}$ *and* Σ. The most basic (strictly associative) (symmetric) (semi-)monoidal object is the free object on the one-point category $*$. We denote the free strictly associative semi-monoidal category on $*$ by $\mathfrak{N}$, and let Σ denote the free strictly associative symmetric semi-monoidal category on $*$. $\mathfrak{N}$ and Σ have the same set of objects, namely, the set of positive integers $\mathbb{N}$. $\mathfrak{N}$ is the semi-monoidal category associated to the semi-group $(\mathbb{N}, +)$: the only morphisms are the identities, and the monoidal product is given by the sum in $\mathbb{N}$. For Σ, we have

$$\mathrm{Hom}_\Sigma(n, m) = \begin{cases} S_n & \text{for } n = m, \\ \emptyset & \text{for } n \neq m, \end{cases}$$

with composition given by the group law in S_n (the symmetric group on n letters). The monoidal product in Σ is given by the homomorphism $+ : S_n \times S_m \to S_{n+m}$, with

$$(\sigma_1 + \sigma_2)(i) = \begin{cases} \sigma_1(i) & \text{for } 1 \leq i \leq n, \\ n + \sigma_2(i - n) & \text{for } n + 1 \leq i \leq n + m. \end{cases}$$

The symmetry isomorphism $t_{a,b} : a + b \to b + a$ is given by the permutation

$$\sigma_{a,b}(i) = \begin{cases} i + b & \text{for } 1 \leq i \leq a, \\ i - a & \text{for } a + 1 \leq i \leq a + b. \end{cases} \tag{2.3.1.1}$$

We define some other important semi-monoidal categories by allowing surjective maps of finite sets rather than just bijections.

2.3.2. *The categories* ω *and* ω_0. For a positive integer n, we denote the set $\{1, \ldots, n\}$ by $\underline{n}$. We consider $\underline{n}$ as an ordered set with the standard ordering. If A and B are ordered sets, we give the disjoint union $A \coprod B$ the order which restricts to the given order on A and on B, and with $a < b$ for $a \in A$ and $b \in B$.

We now define the category ω. The objects of ω are $1, 2, \ldots$, with $\mathrm{Hom}_\omega(n, m)$ the set $S_{n\to m}$ of surjections $f : \underline{n} \to \underline{m}$; composition is by composition of morphisms.

Let $+ : \omega \times \omega \to \omega$ be the functor $+(n, m) = n + m$; $+$ is defined on morphisms by the map

$$\begin{gathered} + : S_{n_1\to m_1} \times S_{n_2\to m_2} \to S_{n_1+m_1\to n_2+m_2}; \\ (\sigma + \rho)(i) = \begin{cases} \sigma(i) & \text{for } 1 \leq i \leq n_1, \\ m_1 + \sigma_2(i - n_1) & \text{for } n_1 + 1 \leq i \leq n_1 + n_2. \end{cases} \end{gathered} \tag{2.3.2.1}$$

The permutation isomorphisms $t_{a,b} : a + b \to b + a$ are given by (2.3.1.1).

The category ω with operation $+$ and symmetry isomorphisms t_{**} is then a strictly associative symmetric semi-monoidal category. We let ω_0 be the subcategory of ω with the same objects as ω, and with morphisms $\mathrm{Hom}_{\omega_0}(n, m)$ the subset $S^<_{n\to m}$ of *ordered* surjections. The category ω_0 with operation $+$ is a strictly associative semi-monoidal category; we have the canonical functors

$$\begin{aligned} i_0 &: \mathfrak{N} \to \omega_0, \\ i &: \Sigma \to \omega. \end{aligned}$$

2.3.3. *The category* Ω_0. Let $p : \underline{n} \to \underline{m}$ be an ordered surjection. For each $j \in \underline{m}$ we give the subset $p^{-1}(j)$ the order induced by the standard order on $\underline{n}$. Let

$$\pi_p : \underline{n} \to p^{-1}(1) \coprod \cdots \coprod p^{-1}(m)$$

be the order-preserving bijection. If σ is in S_m, we have the bijection

$$\tilde{\sigma}:p^{-1}(1)\coprod\ldots\coprod p^{-1}(m)\to(\sigma\circ p)^{-1}(1)\coprod\ldots\coprod(\sigma\circ p)^{-1}(m)$$

which sends the component $p^{-1}(j)$ to the component $(\sigma\circ p)^{-1}(\sigma(j))=p^{-1}(j)$ via the identity. We let $p^*(\sigma)$ be the bijection of $\underline{n}$:

$$p^*(\sigma)=\pi_{\sigma\circ p}^{-1}\circ(\tilde{\sigma})\circ\pi_p.$$

We have the following facts about the map $p^*:S_m\to S_n$:

(2.3.3.1)

(i) For $p\in S^<_{n\to m}$, $q\in S^<_{m\to k}$, we have $(q\circ p)^*=p^*\circ q^*$.

(ii) For $\sigma\in S_m$, $p\in S^<_{n\to m}$, the surjection

$$\sigma\cdot p:=\sigma\circ p\circ p^*(\sigma)^{-1}$$

is in $S^<_{n\to m}$.

(iii) For $p\in S_{m\to n}$, $p^*:S_m\to S_n$ is a *twisted* homomorphism, i.e.,

$$(\sigma\cdot p)^*(\tau)p^*(\sigma)=p^*(\tau\sigma)$$

for $\tau,\sigma\in S_m$.

(iv) For $f\in S_{n\to m}$, $g\in S_{m\to k}$, $\sigma\in S_k$, we have

$$\sigma\cdot(g\circ f)=(\sigma\cdot g)\circ(g^*(\sigma)\cdot f).$$

(v) For $f\in S_{n\to m}$, $g\in S_{m\to k}$ and $\sigma\in S_m$ with $g\circ\sigma=g$, we have

$$g\circ(\sigma\cdot f)=g\circ f.$$

We now define the symmetric semi-monoidal category Ω_0 with the same objects as ω, and with morphisms $\mathrm{Hom}_{\Omega_0}(a,b)$ the set of pairs (f,σ), where σ is in S_a, and $f:a\to b$ is a map in ω_0. The composition

$$\mathrm{Hom}_{\Omega_0}(b,c)\times\mathrm{Hom}_{\Omega_0}(a,b)\to\mathrm{Hom}_{\Omega_0}(a,c)$$

is given by

$$(f',\sigma')\circ(f,\sigma)=(f'\circ(\sigma'\cdot f),f^*(\sigma')\sigma).\tag{2.3.3.2}$$

The product

$$+:\mathrm{Hom}_{\Omega_0}(a,b)\times\mathrm{Hom}_{\Omega_0}(c,d)\to\mathrm{Hom}_{\Omega_0}(a+c,b+d)$$

is induced by the operation (2.3.2.1):

$$(f,\sigma)+(f',\sigma')=(f+f',\sigma+\sigma').\tag{2.3.3.3}$$

For $a,b\in\mathbb{N}$, the symmetry isomorphism $\tau_{a,b}:a+b\to b+a$ is given by

$$\tau_{a,b}=(\mathrm{id},\sigma_{a,b}),\tag{2.3.3.4}$$

where $\sigma_{a,b}$ is the permutation (2.3.1.1).

The relations of (2.3.3.1) show that the operations (2.3.3.2)-(2.3.3.4) do indeed define a symmetric semi-monoidal category. We define the functor $\mathfrak{c}:\Omega_0\to\omega$ by sending $(f,\sigma):a\to b$ to the composition $f\circ\sigma$; one easily checks that this is a symmetric semi-monoidal functor.

Let $f_{n1}:\underline{n}\to\underline{1}$ be the unique surjection; we let $f_{n1}:n\to 1$ denote the corresponding map in ω_0 and ω. We let

$$F_{n1}:n\to 1,\quad \tau_\sigma:n\to n\tag{2.3.3.5}$$

be the maps in Ω_0 defined by

$$F_{n1} = (f_{n1}, \mathrm{id}),$$
$$\tau_\sigma = (\mathrm{id}_{\underline{n}}, \sigma); \quad \sigma \in S_n.$$

We have the identities

$$\text{(2.3.3.6)} \qquad \begin{aligned} f_{21} \circ (\mathrm{id}_1 + f_{21}) &= f_{21} \circ (f_{21} + \mathrm{id}_1), \\ F_{21} \circ (\mathrm{id}_1 + F_{21}) &= F_{21} \circ (F_{21} + \mathrm{id}_1), \end{aligned}$$

the first being an identity of maps in ω_0 (or ω) and the second an identity of maps in Ω_0.

2.3.4. LEMMA. *(i) The semi-monoidal category ω_0 is isomorphic to the semi-monoidal category gotten from $\mathfrak{N}$ by adjoining a morphism $\boxtimes: 2 \to 1$, and imposing the relation*

$$\text{(2.3.4.1)} \qquad \boxtimes \circ (\mathrm{id}_1 + \boxtimes) = \boxtimes \circ (\boxtimes + \mathrm{id}_1).$$

(ii) The symmetric semi-monoidal category Ω_0 is isomorphic to the symmetric semi-monoidal category gotten from Σ by adjoining a morphism $\boxtimes: 2 \to 1$, and imposing the relation (2.3.4.1).

PROOF. Let ω_0' be the semi-monoidal category gotten from $\mathfrak{N}$ by adjoining a morphism $\boxtimes: 2 \to 1$, and imposing the relation (2.3.4.1); the relation (2.3.3.6) shows that the functor $i_1: \mathfrak{N} \to \omega_0$ extends to the functor of semi-monoidal categories $i: \omega_0' \to \omega_0$, with $i(\boxtimes) = f_{21}$.

From the relation (2.3.4.1) and an elementary induction, one sees that there is a unique morphism $\boxtimes_{n1}: n \to 1$ in ω_0'; from this it easily follows that each morphism $f: n \to m$ in ω_0' is of the form

$$f = \boxtimes_{n_1 1} + \ldots + \boxtimes_{n_m 1}; \quad \sum_j n_j = n.$$

This then implies that i is an isomorphism.

The proof of (ii) is similar, and is left to the reader. □

2.3.5. *The symmetric semi-monoidal 2-category Ω.* We now define a 2-category Ω with underlying category Ω_0

Let $f: \underline{n} \to \underline{m}$ be a surjection. We let $S(f)$ be the subgroup of S_n consisting of η such that $f \circ \eta = f$. For a morphism $F = (f, \sigma)$ in Ω_0, define $S(F) = S(f)$.

Let $(f, \sigma): a \to b$ be a morphism in Ω_0, and take η in $S(f)$. Define $\eta \cdot (f, \sigma)$ by

$$\text{(2.3.5.1)} \qquad \eta \cdot (f, \sigma) = (f, \eta\sigma).$$

For morphisms $F, F': a \to b$ in Ω_0, let $\mathrm{Hom}(F, F')$ be the set of $\eta \in S(F)$ such that $F' = \eta \cdot F$ (i.e., either the empty set or a singleton set). Composition is given by the group law in $S(F) = S(F')$:

$$\text{(2.3.5.2)} \qquad \eta_1 \cdot \eta_2 = \eta_1 \eta_2.$$

This defines the category $\mathrm{Hom}_\Omega(a, b)$ with objects $\mathrm{Hom}_{\Omega_0}(a, b)$.

We extend the composition on $\mathrm{Hom}_{\Omega_0}(-,-)$ to a composition functor

$$\mathrm{Hom}_\Omega(b, c) \times \mathrm{Hom}_\Omega(a, b) \to \mathrm{Hom}_\Omega(a, c)$$

as follows: Let $\eta\colon (f,\sigma)\to\eta\cdot(f,\sigma)$ and $\eta'\colon(f',\sigma')\to\eta'\cdot(f',\sigma')$ be morphisms in $\mathrm{Hom}_\Omega(a,b)$, $\mathrm{Hom}_\Omega(b,c)$ respectively. Define $\eta'\circ\eta$ by

$$\eta'\circ\eta = f^*(\eta'\sigma')\eta f^*(\sigma')^{-1}. \tag{2.3.5.3}$$

One easily checks, with the aid of (2.3.3.1), that this gives a well-defined composition functor, and gives Ω the structure of a 2-category.

The monoidal product $+$ on $\mathrm{Hom}_{\Omega_0}(-,-)$ extend to the product functor

$$\bullet\colon \mathrm{Hom}_\Omega(a,b)\times\mathrm{Hom}_\Omega(c,d)\to\mathrm{Hom}_\Omega(a+c,b+d) \tag{2.3.5.4}$$

by setting $\eta\bullet\eta'=\eta+\eta'$. The reader will easily verify that the operations (2.3.5.1)-(2.3.5.4) extends the symmetric semi-monoidal category Ω_0 to give Ω the structure of a symmetric semi-monoidal 2-category.

2.3.6. *The 2-category Ω and symmetric semi-monoidal $\mathcal{V}$-categories.* The 2-category Ω encodes the structure of a strictly associative symmetric semi-monoidal $\mathcal{V}$-category in the following sense: Suppose we have a symmetric semi-monoidal 2-functor $F\colon\Omega\to\mathbf{cat}_\mathcal{V}$. In particular, we have the identity $F(n)=F(1)^{\otimes_\mathcal{V} n}$ for each $n=1,2,\ldots$ in $\mathbb{N}$. Let $\mathcal{C}=F(1)$; we have the morphism $F_{21}\colon 2\to 1$ (2.3.3.5). This gives us the functor

$$\bullet := F(F_{2,1})\colon\mathcal{C}\otimes_\mathcal{V}\mathcal{C}\to\mathcal{C}.$$

The identity (2.3.3.6) in Ω_0 gives the associativity (1.1.1.1) of $\bullet$. Let $\sigma\in S_2$ be the non-trivial permutation, and let $\tau_{1,1}=\tau_\sigma\colon 2\to 2$ be the symmetry (2.3.3.4), (2.3.3.5) in Ω_0. We have the 2-morphism

$$\sigma\colon F_{21}=(f_{2,1},\mathrm{id})\to(f_{2,1},\sigma)=F_{2,1}\circ\tau_{1,1}$$

in Ω, giving the natural transformation

$$\tau := F(\sigma)\colon\bullet\to\bullet\circ F(\tau_{1,1});$$

since F is a symmetric monoidal functor, $F(\tau_{1,1})\colon\mathcal{C}\otimes_\mathcal{V}\mathcal{C}\to\mathcal{C}\otimes_\mathcal{V}\mathcal{C}$ is the symmetry $\tau_{\mathcal{C},\mathcal{C}}$ in $\mathbf{cat}_\mathcal{V}$. Since $\sigma^2=\mathrm{id}$, we have $\tau^2=\mathrm{id}$. The identity in S_3,

$$(\sigma+\mathrm{id}_1)\circ(\mathrm{id}_1+\sigma)=[(\mathrm{id}_1+\sigma)\circ(\sigma+\mathrm{id}_1)]^{-1},$$

gives the hexagonal identity (1.1.1.3) for $(\bullet,\tau)$, hence $(\mathcal{C},\bullet,\tau)$ is a strictly associative symmetric semi-monoidal category.

Conversely, it follows from Lemma 2.3.4 and Remark 1.3.5(ii) that, if we have a strictly associative semi-monoidal category $(\mathcal{C},\bullet,\tau)$, sending n to $\mathcal{C}^{\otimes_\mathcal{V} n}$, $F_{2,1}$ to $\bullet$ and σ to τ extends uniquely to a symmetric semi-monoidal 2-functor $F\colon\Omega\to\mathbf{cat}_\mathcal{V}$; 2-natural transformations correspond likewise to symmetric semi-monoidal functors. This gives a bijection between the category of strictly associative symmetric semi-monoidal $\mathcal{V}$-categories and the category of symmetric semi-monoidal 2-functors from Ω to $\mathbf{cat}_\mathcal{V}$.

We let

$$\Pi_\mathcal{C}\colon\Omega\to\mathbf{cat}_\mathcal{V} \tag{2.3.6.1}$$

denote the symmetric semi-monoidal 2-functor corresponding to a strictly associative symmetric semi-monoidal $\mathcal{V}$-category $\mathcal{C}$.

2.4. External products

External products arise naturally as the products on cochain complexes

$$\boxtimes_{X,Y}:\Gamma_X\otimes\Gamma_Y\to\Gamma_{X\times Y}$$

coming from a cohomology theory Γ; we give an abstraction of this notion to the setting of symmetric semi-monoidal categories.

2.4.1. DEFINITION. Let $(\mathcal{C},\times,t)$ and $(\mathcal{D},\otimes,\tau)$ be strictly associative symmetric semi-monoidal $\mathcal{V}$-categories. A *commutative external product on* $\mathcal{C}$ is a pair (f,θ), where $f:\mathcal{C}\to\mathcal{D}$ is a $\mathcal{V}$-functor (not necessarily symmetric monoidal), and θ is a $\mathcal{V}$-natural transformation $\theta\colon\otimes\circ(f\otimes_{\mathcal{V}}f)\to f\circ\times$ of the functors $\otimes\circ(f\otimes_{\mathcal{V}}f)$ and $f\circ\times$, such that θ is associative and commutative, i.e.,

(2.4.1.1)
$$\theta(A\times B,C)\circ(\theta(A,B)\otimes\mathrm{id}_{f(C)})=\theta(A,B\times C)\circ(\mathrm{id}_{f(A)}\otimes\theta(B,C)),$$
$$f(t_{A,B})\circ\theta(A,B)=\theta(B,A)\circ\tau_{f(A),f(B)}.$$

2.4.2. We consider the category of pairs

$$(f\colon(\mathcal{C},\times)\to(\mathcal{D},\otimes),\theta\colon\otimes\circ(f\otimes_{\mathcal{V}}f)\to f\circ\times)$$

with (f,θ) a commutative external product on $\mathcal{C}$, where a morphism $(f,\theta)\to(f',\theta')$ is a symmetric semi-monoidal $\mathcal{V}$-functor $p:\mathcal{D}\to\mathcal{D}'$, with $f'=p\circ f$ and with $\theta'=p\circ\theta$.

2.4.3. *The category* $\mathcal{C}^{\otimes,c}$. Let $(\mathcal{C},\times,t)$ be a strictly associative symmetric semi-monoidal $\mathcal{V}$-category. We have the symmetric semi-monoidal 2-functor $\Pi_{\mathcal{C}}:\Omega\to\mathbf{cat}_{\mathcal{V}}$ defined by $\mathcal{C}$ (2.3.6.1), forming the symmetric semi-monoidal $\mathcal{V}$-category $(\Pi_{\mathcal{C}},\Omega)$ (see §2.2.1). We denote this category by $(\mathcal{C}^{\otimes,c},\otimes,\tau)$.

We have the $\mathcal{V}$-functor

(2.4.3.1)
$$i_{\mathcal{C}}:\mathcal{C}\to\mathcal{C}^{\otimes,c}$$

defined on objects by $i_{\mathcal{C}}(X)=(X,1)$ and on morphisms by $i_{\mathcal{C}}(g)=(g,\mathrm{id}_1)$.

Let X and Y be objects in $\mathcal{C}$, giving the object $X\times Y$ of $\mathcal{C}$ and the object $X\otimes Y=((X,Y),2)$ of $\mathcal{C}^{\otimes,c}$. We let

$$\boxtimes_{X,Y}:X\otimes Y\to X\times Y$$

denote the morphism defined by the pair $(\mathrm{id}_{X\times Y},F_{21})$, where $F_{21}:2\to 1$ is the morphism (2.3.3.5) in Ω_0. The symmetry $\tau_{X,Y}:X\otimes Y\to Y\otimes X$ is given by the pair $(\mathrm{id}_Y\otimes_{\mathcal{V}}\mathrm{id}_X,\tau_{1,1})$, where $\tau_{1,1}:2\to 2$ is the symmetry (2.3.3.4) in Ω_0. It follows immediately from the relations (2.2.1.1) defining the category $(\Pi_{\mathcal{C}},\Omega)$ that

(2.4.3.2)

1. Sending (X,Y) to $\boxtimes_{X,Y}$ defines a $\mathcal{V}$-natural transformation

$$\boxtimes\colon\otimes\circ(i_{\mathcal{C}}\otimes_{\mathcal{V}}i_{\mathcal{C}})\to i_{\mathcal{C}}\circ\times.$$

2. We have the relation of associativity:

$$\boxtimes\circ(\mathrm{id}\otimes\boxtimes)=\boxtimes\circ(\boxtimes\otimes\mathrm{id}).$$

3. We have the relation of commutativity:

$$\boxtimes\circ\tau=t\circ\boxtimes.$$

Thus the pair

$$(i_{\mathcal{C}}:\mathcal{C}\to\mathcal{C}^{\otimes,c},\boxtimes) \tag{2.4.3.3}$$

defines a commutative external product on $\mathcal{C}$.

2.4.4. PROPOSITION. *(i) The pair (2.4.3.3) is the universal commutative external product on $\mathcal{C}$.*
(ii) Let $X_1,\dots,X_n$, $Y_1,\dots,Y_m$ be objects of $\mathcal{C}$. For each morphism $f:n\to m$ in ω, choose a morphism $F_f:n\to m$ in Ω_0 lifting f. There is a natural isomorphism

$$\Psi:\coprod_{f\in\mathrm{Hom}_\omega(n,m)}\mathrm{Hom}_{\mathcal{C}^m}(\Pi_{\mathcal{C}}(F_f)(X_1,\dots,X_n),(Y_1,\dots,Y_m))$$
$$\to\mathrm{Hom}_{\mathcal{C}^{\otimes,c}}(X_1\otimes\dots\otimes X_n,Y_1\otimes\dots\otimes Y_m).$$

PROOF. We have the free strictly associative symmetric semi-monoidal category on the one-point category, Σ. The universality of Σ gives the functor

$$i:\Sigma\to\Omega_0$$
$$i(n)=n,\ i(\sigma)=(\mathrm{id}_n,\sigma),\quad \sigma\in S_n.$$

We have the inclusion $j:\Omega_0\to\Omega$. We let

$$\pi:\Sigma\to\mathbf{cat}_{\mathcal{V}}$$
$$\Pi_0:\Omega_0\to\mathbf{cat}_{\mathcal{V}}$$

be the respective compositions $\Pi_{\mathcal{C}}\circ j\circ i$ and $\Pi_{\mathcal{C}}\circ j$.

The functors i and j give the symmetric semi-monoidal functors $i^\Pi:(\pi,\Sigma)\to(\Pi_0,\Omega_0)$ and $j^\Pi:(\Pi_0,\Omega_0)\to(\Pi,\Omega)$, and the functor $i_{\mathcal{C}}:\mathcal{C}\to(\Pi,\Omega)$ factors through the similarly defined $\mathcal{V}$-functor $i^\Sigma_{\mathcal{C}}:\mathcal{C}\to(\pi,\Sigma)$.

Let $(f:(\mathcal{C},\times)\to(\mathcal{D},\otimes),\theta)$ be a commutative external product on $\mathcal{C}$. The category (π,Σ) together with the $\mathcal{V}$-functor $i^\Sigma_{\mathcal{C}}$ is easily seen to be isomorphic to the free symmetric semi-monoidal $\mathcal{V}$-category on the $\mathcal{V}$-category $\mathcal{C}$; we therefore have the canonical symmetric semi-monoidal $\mathcal{V}$-functor $p^\Sigma:(\pi,\Sigma)\to\mathcal{D}$, with $f=p^\Sigma\circ i^\Sigma_{\mathcal{C}}$.

By Lemma 2.3.4, Ω_0 is the symmetric semi-monoidal category gotten from Σ by adjoining the morphism $F_{21}:2\to 1$, and imposing the associativity relation (2.3.3.6). This then implies that the functor p^Σ extends uniquely to the symmetric semi-monoidal $\mathcal{V}$-functor $p^{\Omega_0}:(\Pi_0,\Omega_0)\to\mathcal{D}$, with $p^{\Omega_0}(\mathrm{id}_{X\times Y},F_{21})=\theta(X,Y)$. Similarly, the 2-category Ω is generated over Ω_0 as symmetric semi-monoidal 2-category by the 2-morphism $\sigma:F_{21}\to F_{21}\circ\tau_\sigma$, where $\sigma\in S_2$ is the non-trivial permutation. Thus, the symmetric semi-monoidal $\mathcal{V}$-category $(\Pi_{\mathcal{C}},\Omega)$ is gotten from (Π_0,Ω_0) by imposing the relations

$$(t_{X,Y}\circ\mathrm{id}_{X\times Y},F_{21})=(\mathrm{id}_{Y\times X},F_{21}\circ\tau_\sigma)$$

for all objects X and Y of $\mathcal{C}$. From the commutativity relation in (2.4.1.1), we see that p^{Ω_0} extends canonically to the symmetric semi-monoidal $\mathcal{V}$-functor $p:(\Pi_{\mathcal{C}},\Omega)\to\mathcal{D}$ with $p(\boxtimes_{X,Y})=\theta(X,Y)$. This completes the proof of (i).

The statement (ii) follows directly from Remark 2.2.2. □

2.4.5. PROPOSITION. *Let $(\mathcal{C},\times,t)$ be a strictly associative symmetric semi-monoidal $\mathcal{V}$-category. Then*

(i) *The natural map (2.4.3.1) of $\mathcal{V}$-categories*

$$i_{\mathcal{C}}:\mathcal{C}\to\mathcal{C}^{\otimes,c}$$

is fully faithful.

(ii) *Let*

$$\otimes^n : (\mathcal{C}^{\otimes,c})^{\otimes_{\mathcal{V}} n+1} \to \mathcal{C}^{\otimes,c}$$
$$\times^n : \mathcal{C}^{\otimes_{\mathcal{V}} n+1} \to \mathcal{C}$$

be the n-fold products. There is a functor of symmetric semi-monoidal $\mathcal{V}$-categories

$$\rho_{\mathcal{C}} : \mathcal{C}^{\otimes,c} \to \mathcal{C}$$

satisfying

(a) $\rho_{\mathcal{C}} \circ \otimes^n i_{\mathcal{C}}^{\otimes_{\mathcal{V}} n+1} = \times^n$,

(b) *for X and Y in $\mathcal{C}$, we have*

$$\rho_{\mathcal{C}}(\boxtimes_{X,Y}) = \mathrm{id}_{X\times Y},$$
$$\rho_{\mathcal{C}}(\tau_{X,Y}) = t_{X,Y}.$$

(iii) *There is a $\mathcal{V}$-natural transformation*

$$\boxtimes : \mathrm{id}_{\mathcal{C}^{\otimes,c}} \to i_{\mathcal{C}} \circ \rho_{\mathcal{C}}.$$

PROOF. We let $*: \times \to \times$ be the identity natural transformation; thus $(\mathrm{id}_{\mathcal{C}}, *)$ is a commutative external product on $\mathcal{C}$. By the universality of the pair $(i_{\mathcal{C}}, \boxtimes)$ (Proposition 2.4.4) there is a symmetric semi-monoidal $\mathcal{V}$-functor $\rho_{\mathcal{C}} : \mathcal{C}^{\otimes,c} \to \mathcal{C}$ such that $\rho_{\mathcal{C}} \circ i_{\mathcal{C}} = \mathrm{id}_{\mathcal{C}}$, and $\rho_{\mathcal{C}} \circ \boxtimes = *$. The identities (ii)(a) and (ii)(b) follow directly. By (ii), the iterated external product

$$\boxtimes_{X_1,\dots,X_n} : X_1 \otimes \dots \otimes X_n \to X_1 \times \dots \times X_n$$

defines the natural transformation (iii). □

2.4.6. REMARK. If we work in the additive version $\mathbf{cat}_{\mathcal{V}}^{\oplus}$, we denote $(\Pi_{\mathcal{C}}, \Omega)^{\oplus}$ by $\mathcal{C}^{\otimes,c}$, unless there is cause for confusion with the notation of §2.4.3.

2.5. Adjoining morphisms to tensor categories

We now take $\mathcal{V} = \mathbf{GrMod}_A$; the case $\mathcal{V} = \mathbf{Mod}_A$ follows from the graded case by taking everything in degree 0.

2.5.1. We fix a graded A-tensor category without unit $\mathbb{E}$ with the following properties:

(2.5.1.1)

(i) There is an object $\mathfrak{e}$ of $\mathbb{E}$ which generates the objects of $\mathbb{E}$, i.e., each object of $\mathbb{E}$ is a finite direct sum of the objects $\mathfrak{e}^{\otimes a}$, $a = 1, 2 \dots$.

(ii) We have $\mathrm{Hom}_{\mathbb{E}}(\mathfrak{e}^{\otimes m}, \mathfrak{e}^{\otimes n})^q = 0$ if $m \neq n$, or if $n = m$ and $q > 0$.

Take a set I with map $\epsilon : I \to \mathbb{N}$. Let $\mathcal{B}$ be a strictly associative graded A-tensor category without unit. We let $\mathcal{B}[\mathbb{E}]$ denote the coproduct (in the category of graded tensor categories without unit) of $\mathbb{E}$ and $\mathcal{B}$.

Let $\mathcal{B}[\mathbb{E}, \{s_i | i \in I\}]$ be the (strictly associative) graded A-tensor category without unit gotten from $\mathcal{B}[\mathbb{E}]$ by adjoining morphisms $s_i : \mathfrak{e}^{\otimes\epsilon(i)} \to X_i$ of degree d_i, with $X_i \in \mathcal{B}$.

Let $\mathcal{A} = \mathcal{B}[\mathbb{E}, \{s_i\}]$, and let Σ_A be the strictly associative graded A-tensor category without unit on the one-point category $*$. There is the canonical functor $p_{\mathfrak{e}} : \Sigma_A \to \mathcal{A}$ with $p_{\mathfrak{e}}(n) = \mathfrak{e}^{\otimes n}$; this gives the canonical morphism

$$p(n) : \mathrm{Hom}_{\Sigma_A}(n, n) \to \mathrm{Hom}_{\mathcal{A}}(\mathfrak{e}^{\otimes n}, \mathfrak{e}^{\otimes n}).$$

In addition, we have the natural isomorphism of graded A-modules,

$$\mathrm{Hom}_{\Sigma_A}(n, n) \cong A[S_n],$$

with $A[S_n]$ being concentrated in degree 0.

For $i_* = (i_1, \ldots, i_k)$ in I^k, let

$$\sum i_* := \sum_j \epsilon(i_j),\ d(i_*) := \sum_j d_{i_j},$$
$$X^{\otimes i_*} := X_{i_1} \otimes \ldots \otimes X_{i_k}.$$

For a graded A-module $M := \oplus_n M^n$, we let $M[a]$ be the graded A-module with $M[a]^n := M^{n+a}$.

Let Y and Z be objects of $\mathcal{B}$. For $a = b + \sum i_*$, define the degree zero map

$$\begin{aligned}\tilde{\Psi}_{i_*}(\mathfrak{e}^{\otimes a} \otimes Y, \mathfrak{e}^{\otimes b} \otimes Z) \ :\ \mathrm{Hom}_{\mathcal{B}}(X^{\otimes i_*} \otimes Y, Z)[-d(i_*)] \otimes_A \mathrm{Hom}_{\mathbb{E}}(\mathfrak{e}^{\otimes a}, \mathfrak{e}^{\otimes a})& \\ \to\ \mathrm{Hom}_{\mathcal{A}}(\mathfrak{e}^{\otimes a} \otimes Y, \mathfrak{e}^{\otimes b} \otimes Z)&\end{aligned}$$

by letting $\tilde{\Psi}_{i_*}(\mathfrak{e}^{\otimes a} \otimes Y, \mathfrak{e}^{\otimes b} \otimes Z)(g \otimes \tau)$ be the composition

$$\begin{aligned}\mathfrak{e}^{\otimes a} \otimes Y \xrightarrow{\tau \otimes \mathrm{id}_Y} \mathfrak{e}^{\otimes a} \otimes Y = \mathfrak{e}^{\otimes b} \otimes \mathfrak{e}^{\otimes \epsilon(i_1)} \otimes \ldots \otimes \mathfrak{e}^{\otimes \epsilon(i_s)} \otimes Y& \\ \xrightarrow{\mathrm{id} \otimes (s_{i_1} \otimes \ldots \otimes s_{i_s}) \otimes \mathrm{id}_Y} \mathfrak{e}^{\otimes b} \otimes X^{\otimes i_*} \otimes Y& \\ \xrightarrow{\mathrm{id}_{\mathfrak{e}^{\otimes b}} \otimes g} \mathfrak{e}^{\otimes b} \otimes Z.&\end{aligned}$$

Define

$$\begin{aligned}\tilde{\Psi}_{i_*}(\mathfrak{e}^{\otimes a}, \mathfrak{e}^{\otimes b} \otimes Z) \ :\ \mathrm{Hom}_{\mathcal{B}}(X^{\otimes i_*}, Z)[-d(i_*)] \otimes_A \mathrm{Hom}_{\mathbb{E}}(\mathfrak{e}^{\otimes a}, \mathfrak{e}^{\otimes a})& \\ \to\ \mathrm{Hom}_{\mathcal{A}}(\mathfrak{e}^{\otimes a}, \mathfrak{e}^{\otimes b} \otimes Z)&\end{aligned}$$

by the similar formula; we let

$$\Psi(\mathfrak{e}^{\otimes a}, \mathfrak{e}^{\otimes a}) : \mathrm{Hom}_{\mathbb{E}}(\mathfrak{e}^{\otimes a}, \mathfrak{e}^{\otimes a}) \to \mathrm{Hom}_{\mathcal{A}}(\mathfrak{e}^{\otimes a}, \mathfrak{e}^{\otimes a}) \tag{2.5.1.2}$$

be the canonical map, and let

$$\begin{aligned}&\Psi(\mathfrak{e}^{\otimes a} \otimes Y, \mathfrak{e}^{\otimes b}) : 0 \to \mathrm{Hom}_{\mathcal{A}}(\mathfrak{e}^{\otimes a} \otimes Y, \mathfrak{e}^{\otimes b}), \\ &\Psi(\mathfrak{e}^{\otimes a}, \mathfrak{e}^{\otimes b}) : 0 \to \mathrm{Hom}_{\mathcal{A}}(\mathfrak{e}^{\otimes a}, \mathfrak{e}^{\otimes b});\quad a \neq b\end{aligned} \tag{2.5.1.3}$$

be the zero maps.

Given a sequence of positive integers $n_* = (n_1, \ldots, n_k)$ with $b + \sum_i n_i = m$, we have the map $\rho_{n_*} : S_k \to S_m$ defined by having a permutation $\sigma \in S_k$ act on the set with m elements by permuting the blocks of size $b, n_1, n_2, \ldots, n_k$ as σ permutes $1, \ldots, k$ (and fixing the first b elements). We have the action of S_k on I^k, and, given $i_* = (i_1, \ldots, i_k)$ in I^k, we have the weighted sign map

$$\mathrm{sgn}_{i_*} : S_k \to \{\pm 1\}$$

determined by giving the permutation exchanging adjacent elements i_s and i_t the sign $(-1)^{d_{i_s} d_{i_t}}$.

For $\eta \in S_k$, we have the identity

$$\tilde{\Psi}_{i_{\eta(1)},\dots,i_{\eta(k)}}(g \circ (\tau_{\rho_{\epsilon(i_*)}(\eta)} \otimes \mathrm{id}_Y) \otimes \tau) = \mathrm{sgn}_{i_*}(\eta)\tilde{\Psi}_{i_*}(g \otimes \rho_{\epsilon(i_*)}(\eta) \circ \tau). \tag{2.5.1.4}$$

Now suppose the set I is ordered. Let $I^k_\leq$ denote the set of ordered k-tuples $i_* = (i_1 \leq \dots \leq i_k)$ in I^k. For each $i_* \in I^k_\leq$, we let $S(i_*)$ be the subgroup of S_k which act as order-preserving permutations of the sequence i_*. Note that the restriction of $\rho_{\epsilon(i_*)}$ and sgn_{i_*} to $S(i_*)$ are homomorphisms.

We let $S(i_*)$ act on $\mathrm{Hom}_{\mathbb{E}}(\mathfrak{e}^{\otimes a}, \mathfrak{e}^{\otimes a})$ via left composition by $\rho_{\epsilon(i_*)}$ and on $\mathrm{Hom}_{\mathcal{B}}(X_{i_1} \otimes \dots \otimes X_{i_s} \otimes Y, Z)$ by

$$g \cdot \eta = \mathrm{sgn}_{i_*}(\eta) \cdot g \circ (\tau_{\rho_{\epsilon(i_*)}(\eta)} \otimes \mathrm{id}_Y).$$

By the relation (2.5.1.4), the map $\tilde{\Psi}_{i_*}(\mathfrak{e}^{\otimes a} \otimes Y, \mathfrak{e}^{\otimes b} \otimes Z)$ descends to a degree zero map

$$\Psi_{i_*}(\mathfrak{e}^{\otimes a} \otimes Y, \mathfrak{e}^{\otimes b} \otimes Z) : \mathrm{Hom}_{\mathcal{B}}(X^{\otimes i_*} \otimes Y, Z)[-d(i_*)] \otimes_{A[S(i_*)]} \mathrm{Hom}_{\mathbb{E}}(\mathfrak{e}^{\otimes a}, \mathfrak{e}^{\otimes a}) \\ \to \mathrm{Hom}_{\mathcal{A}}(\mathfrak{e}^{\otimes a} \otimes Y, \mathfrak{e}^{\otimes b} \otimes Z).$$

We have the map $\Psi_{i_*}(\mathfrak{e}^{\otimes a}, \mathfrak{e}^{\otimes b} \otimes Z)$ defined similarly.

Let

$$\bigoplus_{s=0}^{\infty} \bigoplus_{\substack{i_* \in I^s_\leq \\ \sum i_* = a-b}} \mathrm{Hom}_{\mathcal{B}}(X^{\otimes i_*} \otimes Y, Z)[-d(i_*)] \otimes_{A[S(i_*)]} \mathrm{Hom}_{\mathbb{E}}(\mathfrak{e}^{\otimes a}, \mathfrak{e}^{\otimes a}) \\ \xrightarrow{\Psi(\mathfrak{e}^{\otimes a} \otimes Y, \mathfrak{e}^{\otimes b} \otimes Z)} \mathrm{Hom}_{\mathcal{A}}(\mathfrak{e}^{\otimes a} \otimes Y, \mathfrak{e}^{\otimes b} \otimes Z) \tag{2.5.1.5}$$

be the map defined by

$$\Psi(\mathfrak{e}^{\otimes a} \otimes Y, \mathfrak{e}^{\otimes b} \otimes Z) = \sum_{s=0}^{\infty} \sum_{\substack{i_* \in I^s_\leq \\ \sum i_* = a-b}} \Psi_{i_*}(\mathfrak{e}^{\otimes a} \otimes Y, \mathfrak{e}^{\otimes b} \otimes Z).$$

Define the map

$$\bigoplus_{s=0}^{\infty} \bigoplus_{\substack{i_* \in I^s_\leq \\ \sum i_* = a-b}} \mathrm{Hom}_{\mathcal{B}}(X^{\otimes i_*}, Z)[-d(i_*)] \otimes_{A[S(i_*)]} \mathrm{Hom}_{\mathbb{E}}(\mathfrak{e}^{\otimes a}, \mathfrak{e}^{\otimes a}) \\ \xrightarrow{\Psi(\mathfrak{e}^{\otimes a}, \mathfrak{e}^{\otimes b} \otimes Z)} \mathrm{Hom}_{\mathcal{A}}(\mathfrak{e}^{\otimes a}, \mathfrak{e}^{\otimes b} \otimes Z) \tag{2.5.1.6}$$

similarly; we have already defined $\Psi(\mathfrak{e}^{\otimes a} \otimes Y, \mathfrak{e}^{\otimes b})$ and $\Psi(\mathfrak{e}^{\otimes a}, \mathfrak{e}^{\otimes b})$.

2.5.2. PROPOSITION. *The maps* (2.5.1.2), (2.5.1.3), (2.5.1.5) *and* (2.5.1.6) *are isomorphisms.*

PROOF. The relations defining a graded A-tensor category imply that the various maps $\Psi(-,-)$ are surjective.

To prove the maps $\Psi(-,-)$ are injective, let $H(\mathfrak{e}^{\otimes a} \otimes Y, \mathfrak{e}^{\otimes b} \otimes Z)$ denote direct sum

$$\bigoplus_{s=0}^{\infty} \bigoplus_{\substack{i_* \in I^s_\leq \\ \sum i_* = a-b}} \mathrm{Hom}_{\mathcal{B}}(X^{\otimes i_*} \otimes Y, Z)[-d(i_*)] \otimes_{A[S(i_*)]} \mathrm{Hom}_{\mathbb{E}}(\mathfrak{e}^{\otimes a}, \mathfrak{e}^{\otimes a}).$$

We denote the element $g\otimes\tau$ in the summand i_* by $(g\otimes\tau)_{i_*}$. We define $H(\mathfrak{e}^{\otimes a}, \mathfrak{e}^{\otimes b}\otimes Z)$ similarly as

$$\bigoplus_{s=0}^{\infty} \bigoplus_{\substack{i_*\in I^s_{\leq} \\ \sum i_*=a-b}} \mathrm{Hom}_{\mathcal{B}}(X^{\otimes i_*}, Z)[-d(i_*)] \otimes_{A[S(i_*)]} \mathrm{Hom}_{\mathbb{E}}(\mathfrak{e}^{\otimes a}, \mathfrak{e}^{\otimes a}).$$

We set

$$H(Y, Z) := \mathrm{Hom}_{\mathcal{B}}(Y, Z),$$
$$H(Y, \mathfrak{e}^{\otimes b}\otimes Z) := 0 \text{ for } b > 0,$$

and

$$H(\mathfrak{e}^{\otimes a}, \mathfrak{e}^{\otimes b}) := \mathrm{Hom}_{\mathbb{E}}(\mathfrak{e}^{\otimes a}, \mathfrak{e}^{\otimes b}).$$

For $i_*\in I^k_{\leq}$, $j_*\in I^l_{\leq}$, we let ji_* denote the sequence of indices j_*, i_*, reordered to be in increasing order. Let $\sigma(j_*, i_*)\in S_{k+l}$ be the shuffle permutation transforming the sequence j_*, i_* to the sequence ji_*; let $\mathrm{sgn}(j_*, i_*)$ be the weighted sign of $\sigma(j_*, i_*)$, where we give the permutation exchanging adjacent elements s and t the sign $(-1)^{d_s d_t}$. We let

$$\tau_{j_*,i_*}\colon X^{\otimes j_*}\otimes X^{\otimes i_*} \longrightarrow X^{\otimes ji_*}$$

denote the symmetry isomorphism associated to $\sigma(j_*, i_*)$. Let

$$\tau^{\mathfrak{e}}_{j_*,i_*}\colon \mathfrak{e}^{\otimes a-c} \longrightarrow \mathfrak{e}^{\otimes a-c}$$

be defined similarly via the identities

$$\mathfrak{e}^{\otimes a-c} = \mathfrak{e}^{\epsilon(j_*)}\otimes\mathfrak{e}^{\epsilon(i_*)} = \mathfrak{e}^{\epsilon(ji_*)},$$

where we permute the terms $\mathfrak{e}^{\epsilon(j_k)}$ and $\mathfrak{e}^{\epsilon(i_k)}$ as $\sigma(j_*, i_*)$ permutes the j_k and i_k.

If $a\geq b$ are positive integers, and τ is in $\mathrm{Hom}_{\mathbb{E}}(\mathfrak{e}^{\otimes b}, \mathfrak{e}^{\otimes b})$, we have the element $\tau\otimes\mathrm{id}_{\mathfrak{e}^{\otimes a-b}}\in\mathrm{Hom}_{\mathbb{E}}(\mathfrak{e}^{\otimes a}, \mathfrak{e}^{\otimes a})$, defined via the identity $\mathfrak{e}^{\otimes a} = \mathfrak{e}^{\otimes b}\otimes\mathfrak{e}^{\otimes a-b}$; $\mathrm{id}_{\mathfrak{e}^{\otimes c}}\otimes\tau'$ is defined similarly, for τ' in $\mathrm{Hom}_{\mathbb{E}}(\mathfrak{e}^{\otimes a-c}, \mathfrak{e}^{\otimes a-c})$.

Define the composition

$$\circ\colon H(\mathfrak{e}^{\otimes b}\otimes Z, \mathfrak{e}^{\otimes c}\otimes W)\otimes_A H(\mathfrak{e}^{\otimes a}\otimes Y, \mathfrak{e}^{\otimes b}\otimes Z) \to H(\mathfrak{e}^{\otimes a}\otimes Y, \mathfrak{e}^{\otimes c}\otimes W)$$

on homogeneous elements $(g_2\otimes\tau_2)_{j_*}\otimes(g_1\otimes\tau_1)_{i_*}$ by

$$(g_2\otimes\tau_2)_{j_*}\circ(g_1\otimes\tau_1)_{i_*} = (-1)^{d_1 d(j_*)+d_1\delta_2+d(i_*)\delta_2}\mathrm{sgn}(j_*, i_*)(g\otimes\tau)_{ji_*},$$

where

$$\tau = (\mathrm{id}_{\mathfrak{e}^{\otimes c}}\otimes\tau^{\mathfrak{e}}_{j_*,i_*}\circ(\tau_2\otimes\mathrm{id}_{\mathfrak{e}^{\otimes a-b}})\circ\tau_1,$$

d_1 is the degree of g_1, δ_i is the degree of τ_i, and g is the composition

$$X^{\otimes ji_*}\otimes Y \xrightarrow{\tau^{-1}_{ji_*}\otimes\mathrm{id}_Y} X^{\otimes j_*}\otimes X^{\otimes i_*}\otimes Y \xrightarrow{\mathrm{id}_{X^{\otimes j_*}}\otimes g_1} X^{\otimes j_*}\otimes Z \xrightarrow{g_2} W.$$

One defines the composition for the special cases $H(\mathfrak{e}^{\otimes a}, \mathfrak{e}^{\otimes b}\otimes Z)$, etc., similarly. One checks that the operation $\circ$ is associative.

Let Y_1, Y_2, Z_1, Z_2 be objects of $\mathcal{C}$. Define the product

$$\begin{aligned}\bullet: H(\mathfrak{e}^{\otimes a_1}\otimes Y_1, \mathfrak{e}^{\otimes b_1}\otimes Z_1)\otimes_A H(\mathfrak{e}^{\otimes a_2}\otimes Y_2, \mathfrak{e}^{\otimes b_2}\otimes Z_2)&\\ \to H(\mathfrak{e}^{\otimes a_1+a_2}\otimes Y_1\otimes Y_2, \mathfrak{e}^{\otimes b_1+b_2}\otimes Z_1\otimes Z_2)&\end{aligned}$$

on homogeneous elements $(g_1\otimes\tau_1)_{i_*}\otimes(g_2\otimes\tau_2)_{j_*}$ by

$$(g_1\otimes\tau_1)_{i_*}\bullet(g_2\otimes\tau_2)_{j_*} = (-1)^{\delta_1 d_2+\delta_1 d(j_*)+d(i_*)d_2}\mathrm{sgn}(i_*, j_*)(g\otimes\tau)_{ij_*},$$

where

$$\tau = (\mathrm{id}_{\mathfrak{e}^{b_1+b_2}} \otimes \tau^{\mathfrak{e}}_{i_*,j_*}) \circ (\tau_1 \otimes \tau_2),$$

d_i is the degree of g_i, δ_i is the degree of τ_i, and g is the composition

$$X^{\otimes ij_*} \otimes Y_1 \otimes Y_2 \xrightarrow{\tau^{-1}_{i_*,j_*} \otimes \mathrm{id}_{Y_1 \otimes Y_2}} X^{\otimes i_*} \otimes X^{\otimes j_*} \otimes Y_1 \otimes Y_2$$
$$\xrightarrow{\mathrm{id}_{X^{\otimes i_*}} \otimes \tau_{X^{\otimes j_*}, Y_1} \otimes \mathrm{id}_{Y_2}} X^{\otimes i_*} \otimes Y_1 \otimes X^{\otimes j_*} \otimes Y_2 \xrightarrow{g_1 \otimes g_2} Z_1 \otimes Z_2.$$

One defines the operation $\bullet$ for the special cases $H(\mathfrak{e}^{\otimes a}, \mathfrak{e}^{\otimes b} \otimes Z)$, etc., similarly. One checks that the operation $\bullet$ is associative and graded-commutative.

Thus, we may define a graded A-tensor category without unit, $\mathcal{D}$, with the same objects as $\mathcal{A}$, and with

$$\mathrm{Hom}_{\mathcal{D}}(\mathfrak{e}^{\otimes a} \otimes Y, \mathfrak{e}^{\otimes b} \otimes Z) = H(\mathfrak{e}^{\otimes a} \otimes Y, \mathfrak{e}^{\otimes b} \otimes Z),$$
$$\mathrm{Hom}_{\mathcal{D}}(\mathfrak{e}^{\otimes a}, \mathfrak{e}^{\otimes b} \otimes Z) = H(\mathfrak{e}^{\otimes a}, \mathfrak{e}^{\otimes b} \otimes Z),$$
$$\mathrm{Hom}_{\mathcal{D}}(\mathfrak{e}^{\otimes a}, \mathfrak{e}^{\otimes b}) = H(\mathfrak{e}^{\otimes a}, \mathfrak{e}^{\otimes b}),$$
$$\mathrm{Hom}_{\mathcal{D}}(\mathfrak{e}^{\otimes a} \otimes Y, \mathfrak{e}^{\otimes b}) = 0;$$

the Hom-modules for all pairs of objects are determined, up to canonical isomorphism, by this.

The identities $\mathrm{Hom}_{\mathcal{B}}(Y, Z) = H(Y, Z)$ and $\mathrm{Hom}_{\mathbb{E}}(\mathfrak{e}^{\otimes a}, \mathfrak{e}^{\otimes a}) = H(\mathfrak{e}^{\otimes a}, \mathfrak{e}^{\otimes a})$ determines the graded A-tensor functor $i\colon \mathcal{B}[\mathbb{E}] \to \mathcal{D}$ which is the identity on objects; we extend i to $i^s\colon \mathcal{A} \to \mathcal{D}$ by sending the morphism $s_i\colon \mathfrak{e} \to X_i$ to the element $\mathrm{id}_{X_i} \otimes \mathrm{id}$ of $H(\mathfrak{e}, X_i)$. Clearly, the map

$$i^s(\mathfrak{e}^{\otimes a} \otimes Y, \mathfrak{e}^{\otimes b} \otimes Z)\colon \mathrm{Hom}_{\mathcal{A}}(\mathfrak{e}^{\otimes a} \otimes Y, \mathfrak{e}^{\otimes b} \otimes Z) \to H(\mathfrak{e}^{\otimes a} \otimes Y, \mathfrak{e}^{\otimes b} \otimes Z)$$

is a left inverse to $\Psi(\mathfrak{e}^{\otimes a} \otimes Y, \mathfrak{e}^{\otimes b} \otimes Z)$, hence $\Psi(\mathfrak{e}^{\otimes a} \otimes Y, \mathfrak{e}^{\otimes b} \otimes Z)$ is injective. The other cases are proved similarly. □

2.5.3. Proposition. *(i) Let $(\mathcal{C}, \times, t)$ be a strictly associative graded A-tensor category without unit, $((\mathcal{B}^c, \otimes, \tau), \boxtimes)$ the universal graded A-tensor category with commutative external product $\mathcal{C}^{\otimes, c}$ and $\mathcal{A}$ the category $\mathcal{B}^c[\mathbb{E}, \{s_i \mid i \in I\}]$, with I an ordered set, and $s_i\colon \mathfrak{e}^{\otimes \epsilon(i)} \to X_i$ a degree d_i morphism. Let $\kappa\colon \mathcal{C} \to \mathcal{A}$ be the canonical tensor functor, and let $i\colon \mathcal{C}^* \to \mathcal{A}$ be the full additive subcategory of $\mathcal{A}$ generated by the objects $\mathfrak{e}^{\otimes k}$ and $\mathfrak{e}^{\otimes k} \otimes X$, for X in $\mathcal{C}$. Then $\mathcal{C}^*$ has the structure of a strictly associative graded A-tensor category without unit such that*

- *(a) The functor $\kappa\colon \mathcal{C} \to \mathcal{C}^*$ is a graded tensor functor.*
- *(b) There is a graded tensor functor $r\colon \mathcal{A} \to \mathcal{C}^*$ with $r \circ i = \mathrm{id}_{\mathcal{C}^*}$.*
- *(c) There is a natural transformation $\boxtimes\colon \mathrm{id}_{\mathcal{A}} \to i \circ r$.*

In addition, the functor r and natural transformation $\boxtimes$ are extensions of the functor ρ and natural transformation $\boxtimes$ of Proposition 2.4.5.

(ii) Let $\mathcal{C}^{**}$ be the full graded A-tensor subcategory of $\mathcal{C}^*$ generated by the objects of the form $\mathfrak{e}^{\otimes a} \otimes X$, where X is an object of $\mathcal{C}$, and $a \geq 0$ (we set $\mathfrak{e}^{\otimes 0} \otimes X := X$). If $\mathcal{C}$ is a graded A-tensor category with unit 1, then the structure of a graded A-tensor category without unit on $\mathcal{C}^{**}$ has a unique extension to the structure of a graded A-tensor category with unit 1, such that the functor $\mathcal{C} \to \mathcal{C}^{**}$ is a functor of graded A-tensor categories.

PROOF. We first define the structure of a graded tensor category without unit on $\mathcal{C}^*$. We retain the notation of the proof of Proposition 2.5.2; in addition, for $i_* = (i_1, \dots, i_k) \in I^k_{\leq}$, set $X^{\times i_*} := X_{i_1} \times \dots \times X_{i_k}$.

Let Y_1, Y_2, Z_1, Z_2 be objects of $\mathcal{C}$. Let

$$\boxtimes_{i_*,1} : X^{\otimes i_*} \otimes Y_1 \to X^{\times i_*} \times Y_1$$

$$\boxtimes_{j_*,2} : X^{\otimes j_*} \otimes Y_2 \to X^{\times j_*} \times Y_2$$

be the order-preserving external products. Let

$$\tau_{i_*,j_*} : X^{\times i_*} \times Y_1 \times X^{\times j_*} \times Y_2 \to X^{\times ij_*} \times Y_1 \times Y_2$$

be the shuffle permutation isomorphism.

By Proposition 2.4.4(ii), each morphism $g: X^{\otimes i_*} \otimes Y_1 \to Z_1$ in $\mathcal{B}^c$ can be written as $g = f \circ \boxtimes_{i_*,1}$, for a uniquely determined $f: X^{\times i_*} \times Y_1 \to Z_1$ in $\mathcal{C}$. Denote this f by $\rho_1(g)$; for $g: X^{\otimes j_*} \otimes Y_2 \to Z_2$, define $\rho_2(g): X^{\times j_*} \times Y_2 \to Z_2$ similarly.

We extend the tensor product $\times$ on $\mathcal{C}$ to the product $\times$ on $\mathcal{C}^*$ by setting

$$(\mathfrak{e}^{\otimes a} \otimes X) \times (\mathfrak{e}^{\otimes b} \otimes Y) = \mathfrak{e}^{\otimes a+b} \otimes (X \times Y)$$

and

$$\begin{aligned}\Psi_{i_*}(g_1 \otimes \tau_1) &\times \Psi_{j_*}(g_2 \otimes \tau_2) \\ &= (-1)^A \operatorname{sgn}(i_*, j_*) \Psi_{ij_*}([(\rho(g_1) \times \rho(g_2)) \circ \tau^{-1}_{i_*,j_*}], (\mathrm{id} \otimes \tau^{\mathfrak{e}}_{i_*,j_*}) \circ (\tau_1 \otimes \tau_2)),\end{aligned}$$

with $A := \delta_1 d_2 + \delta_1 d(j_*) + d(i_*) d_2$, where d_i is the degree of g_i, δ_i is the degree of τ_i, and the symmetry $\tau^{\mathfrak{e}}_{i_*,j_*}$ is defined as in the proof of Proposition 2.5.2. Define the symmetry isomorphism

$$(\mathfrak{e}^{\otimes a} \otimes X) \times (\mathfrak{e}^{\otimes b} \otimes Y) \to (\mathfrak{e}^{\otimes b} \otimes Y) \times (\mathfrak{e}^{\otimes a} \otimes X)$$

to be the map

$$\tau_{\mathfrak{e}^{\otimes a}, \mathfrak{e}^{\otimes b}} \otimes t_{X,Y} : \mathfrak{e}^{\otimes a+b} \otimes (X \times Y) \to \mathfrak{e}^{\otimes a+b} \otimes (Y \times X).$$

This makes $\mathcal{C}^*$ into a graded tensor category without unit.

Let $j: \mathcal{C} \to \mathcal{C}^*$ be the natural map. Then j is a tensor functor; in addition, the identity map gives the natural transformation $*$: $\otimes_{\mathcal{C}^*} \circ (j \times j) \to j \circ \times$, which is associative and commutative. Thus, the universal property of the graded tensor category $\mathcal{B}^c$ gives the graded tensor functor $\bar{r}: \mathcal{B}^c \to \mathcal{C}^*$. Since the functor $\pi: \mathcal{B}^c \to \mathcal{A}$ is universal for triples consisting of a graded tensor functor $F: \mathcal{B}^c \to \mathcal{D}$, an object d of $\mathcal{D}$, and a collection of morphisms $h_i: d^{\epsilon(i)} \to F(X_i)$, $i \in I$; the functor $\bar{r}$ extends canonically to the desired functor $r: \mathcal{A} \to \mathcal{C}^*$.

The existence of a natural transformation $\boxtimes_1: \pi \to i \circ \bar{r}$ follows from the universal property of the category $\mathcal{B}^c$; the canonical extension of $\boxtimes_1$ to a collection of maps $\boxtimes_X: X \to r(X)$, $X \in \mathcal{A}$, defines a natural transformation by the universal property of the functor π.

Finally, suppose $\mathcal{C}$ is a graded A-tensor category with unit 1 and natural isomorphisms

$$m^l: 1 \times (-) \to \mathrm{id}_{\mathcal{C}}; \quad m^r: (-) \times 1 \to \mathrm{id}_{\mathcal{C}}.$$

For an object X of $\mathcal{C}$ and integer $a \geq 0$, define

$$\mu^r_{\mathfrak{e}^{\otimes a} \otimes X} : (\mathfrak{e}^{\otimes a} \otimes X) \times 1 \to \mathfrak{e}^{\otimes a} \otimes X$$

to be the composition

$$(\mathfrak{e}^{\otimes a} \otimes X) \times 1 = \mathfrak{e}^{\otimes a} \otimes (X \times 1) \xrightarrow{\mathrm{id}_{\mathfrak{e}^{\otimes a}} \otimes m^r_X} \mathfrak{e}^{\otimes a} \otimes X.$$

We define $\mu^l_{\mathfrak{e}^{\otimes a}\otimes X}$ by

$$\mu^l_{\mathfrak{e}^{\otimes a}\otimes X} = \mu^r_{\mathfrak{e}^{\otimes a}\otimes X} \circ \tau_{1,\mathfrak{e}^{\otimes a}\otimes X}.$$

It follows immediately from definitions that, sending $\mathfrak{e}^{\otimes a} \otimes X$ to $\mu^l_{\mathfrak{e}^{\otimes a}\otimes X}$, resp. $\mu^r_{\mathfrak{e}^{\otimes a}\otimes X}$ defines natural isomorphisms

$$\mu^r : 1 \times (-) \to \mathrm{id}_{\mathcal{C}^{**}}; \quad \mu^r : (-) \times 1 \to \mathrm{id}_{\mathcal{C}^{**}}$$

which give $\mathcal{C}^{**}$ the desired structure of a graded A-tensor category with unit 1. □

2.5.4. Remark. As in Proposition 2.4.5, one can easily write down the functor r and natural transformation $\boxtimes$ explicitly. The functor $r: \mathcal{A} \to \mathcal{C}^*$ is defined on objects by $r(\mathfrak{e}^{\otimes a}\otimes X) = \mathfrak{e}^{\otimes a}\otimes\rho(X)$, and on morphisms $\Psi_{i_*}(g\otimes\tau)$, for $g: X^{\otimes i_*}\otimes Y \to Z$ and $\tau \in \mathrm{Hom}(\mathfrak{e}^{\otimes a}, \mathfrak{e}^{\otimes a})$, by

$$r(\Psi_{i_*}(g \otimes \tau)) = \Psi_{i_*}([r_{\mathcal{B}^c}(g) \circ \boxtimes_{\mathcal{B}^c}(X^{\otimes i_*} \otimes \rho(Y))] \otimes \tau).$$

Similarly, the natural transformation $\boxtimes: \mathrm{id} \to i \circ r$ is given by the external product

$$\mathrm{id}_{\mathfrak{e}^{\otimes a}} \otimes \boxtimes_Y : \mathfrak{e}^{\otimes a} \otimes Y \to \mathfrak{e}^{\otimes a} \otimes \rho(Y) = r(\mathfrak{e}^{\otimes a} \otimes Y)$$

for Y in $\mathcal{B}^c$.

If we let $\mathcal{A}'$ be the full A-tensor subcategory of $\mathcal{A}$ generated by objects of the form $\mathfrak{e}^{\otimes a}\otimes X$, with $a \geq 0$ and X in $\mathcal{C}$, then the functor r and natural transformation $\boxtimes$ restrict to the functor $r': \mathcal{A}' \to \mathcal{C}^{**}$ and natural transformation $\boxtimes': \mathrm{id}_{\mathcal{A}'} \to i' \circ r'$, where $i': \mathcal{C}^{**} \to \mathcal{A}'$ is the inclusion.

CHAPTER II

DG Categories and Triangulated Categories

In this chapter, we extend the well-known constructions of the category of bounded complexes of an abelian category and the associated homotopy category of bounded complexes to the case of a DG category. We begin with an extension of the notion of a translation in a graded category to the case of a DG tensor category. We then recall the construction of Kapranov [**75**] of the category of complexes over a DG category, and show how this extends to the case of DG tensor categories. We recall the notion of a triangulated category and show that the homotopy category of the category of complexes in a DG category is a triangulated category, and that the homotopy category of the category of complexes in a DG tensor category is a triangulated tensor category.

As in topology, a *homotopy equivalence* of DG categories induces an equivalence on the homotopy category of complexes. We also consider the operation of adjoining morphisms to a DG tensor category, and show that, in some cases, this operation preserves homotopy equivalences

We consider the pseudo-abelian hull of a triangulated category, and show that pseudo-abelian hull of a localization of the homotopy category of complexes forms a triangulated category in a natural way. We conclude with some constructions of special DG tensor categories.

We take the coefficient ring to be $\mathbb{Z}$ for simplicity of notation; all the constructions go through without change for a general commutative coefficient ring.

1. Differential graded categories

1.1. Translation structures

1.1.1. For a DG module $M := (M, d) = (\oplus_a M^a, \oplus_a d^a : M^a \to M^{a+1})$, and integer n, we have the shifted DG module $M[n]$, with $M[n]^a := M^{n+a}$ and with differential $d^a_{M[n]} := (-1)^n d^{n+a}$. We also have the shifted DG module $M^{[n]}$, with $M^{[n]a} := M^{n+a}$ and with differential $d^a_{M^{[n]}} := d^{n+a}$. Note that

$$M[a]^{[b]} = M^{[b]}[a].$$

We let $-M$ be the DG module with the same grading as M, but with $d_{-M} := -d_M$. We note that M and $-M$ are isomorphic, by the map $\psi: M \to M$ which sends $x \in M^l$ to $(-1)^l x$. In addition, we have $M^{[1]} = -M[1]$, so ψ gives isomorphisms

$$\psi: M^{[1]} \to M[1];\ \psi: M[1] \to M^{[1]}. \tag{1.1.1.1}$$

If $f: M \to M'$ is a graded map of DG modules of degree l, define $f[n]: M[n] \to M'[n]$ by $f[n]^a = (-1)^{nl} f^{n+a}$, and $f^{[n]}: M^{[n]} \to M'^{[n]}$ by $f^{[n]a} = f^{n+a}$.

The tensor product $M \otimes_{\mathbb{Z}} N$ of DG modules M and N has the differential

$$d(x \otimes y) := d_M x \otimes y + (-1)^{\deg x} x \otimes d_N y.$$

Recall from Chapter I, §1.3.8, that the graded Hom-complex $\mathrm{Hom}(M,N)$ has differential $df = (-1)^{\deg f} f \circ d_M - d_N \circ f$. This gives the identities

(1.1.1.2)

$$(M[m]) \otimes (N^{[n]}) = (M \otimes N)^{[n]}[m];\ (M^{[m]}) \otimes N = M \otimes (N[m]).$$

$$\mathrm{Hom}(M[m], N[n]) = \mathrm{Hom}(M,N)^{[-m]}[n];\ \mathrm{Hom}(M^{[m]}, N) = \mathrm{Hom}(M, N^{[-m]}).$$

We reinterpret these identities by introducing the notion of a *permutative bi-module* over a DG tensor category.

1.1.2. *Permutative bi-modules.* Let $\mathcal{T}$ be a strictly associative DG tensor category, with strict unit 1. A *permutative $\mathcal{T}$-bi-module* is an DG category $\mathcal{A}$, with functors

$$\rho_l : \mathcal{T} \otimes_{\mathbb{Z}} \mathcal{A} \to \mathcal{A}, \quad \rho_r : \mathcal{A} \otimes \mathcal{T} \to \mathcal{A},$$

and a natural isomorphism

$$\tau^\rho : \rho_l \longrightarrow \rho_r \circ \tau_{\mathcal{T},\mathcal{A}},$$

where $\tau_{\mathcal{T},\mathcal{A}} : \mathcal{T} \otimes_{\mathbb{Z}} \mathcal{A} \to \mathcal{A} \otimes_{\mathbb{Z}} \mathcal{T}$ is the symmetry in the category of DG categories, such that

(1.1.2.1)

1. The diagrams

$$\begin{array}{ccc}
\mathcal{T} \otimes_{\mathbb{Z}} \mathcal{T} \otimes_{\mathbb{Z}} \mathcal{A} & \xrightarrow{\otimes_{\mathcal{T}} \otimes_{\mathbb{Z}} \mathrm{id}_{\mathcal{A}}} & \mathcal{T} \otimes_{\mathbb{Z}} \mathcal{A} \\
{\scriptstyle \mathrm{id}_{\mathcal{T}} \otimes_{\mathbb{Z}} \rho_l} \downarrow & & \downarrow {\scriptstyle \rho_l} \\
\mathcal{T} \otimes_{\mathbb{Z}} \mathcal{A} & \xrightarrow[\rho_l]{} & \mathcal{A}
\end{array}$$

$$\begin{array}{ccc}
\mathcal{A} \otimes_{\mathbb{Z}} \mathcal{T} \otimes_{\mathbb{Z}} \mathcal{T} & \xrightarrow{\mathrm{id}_{\mathcal{A}} \otimes_{\mathbb{Z}} \otimes_{\mathcal{T}}} & \mathcal{A} \otimes_{\mathbb{Z}} \mathcal{T} \\
{\scriptstyle \rho_r \otimes_{\mathbb{Z}} \mathrm{id}_{\mathcal{T}}} \downarrow & & \downarrow {\scriptstyle \rho_r} \\
\mathcal{A} \otimes_{\mathbb{Z}} \mathcal{T} & \xrightarrow[\rho_r]{} & \mathcal{A}
\end{array}$$

and

$$\begin{array}{ccc}
\mathcal{T} \otimes_{\mathbb{Z}} \mathcal{A} \otimes_{\mathbb{Z}} \mathcal{T} & \xrightarrow{\rho_l \otimes_{\mathbb{Z}} \mathrm{id}_{\mathcal{T}}} & \mathcal{A} \otimes_{\mathbb{Z}} \mathcal{T} \\
{\scriptstyle \mathrm{id}_{\mathcal{T}} \otimes_{\mathbb{Z}} \rho_r} \downarrow & & \downarrow {\scriptstyle \rho_r} \\
\mathcal{A} \otimes_{\mathbb{Z}} \mathcal{T} & \xrightarrow[\rho_l]{} & \mathcal{A}
\end{array}$$

commute.

Write TM for $\rho_l(T, M)$ and MT for $\rho_r(M,T)$; write gf for $\rho_l(g \otimes_{\mathbb{Z}} f)$ and fg for $\rho_r(f \otimes_{\mathbb{Z}} g)$.

(2) Let S and T be in $\mathcal{T}$ and M in $\mathcal{A}$. Then the diagrams

$$\begin{array}{ccc}
TMS & \xrightarrow{\tau^\rho_{T,M} \mathrm{id}_S} & MTS \\
& {\scriptstyle \tau^\rho_{T,MS}} \searrow & \downarrow {\scriptstyle \mathrm{id}_M \tau^{\mathcal{T}}_{T,S}} \\
& & MST,
\end{array}$$

$$\begin{array}{ccc} STM & \xrightarrow{\tau^{T}_{S,T}\mathrm{id}_M} & TSM \\ & \underset{\tau^{\rho}_{S,TM}}{\searrow} & \Big\downarrow{\scriptstyle \mathrm{id}_T\tau^{\rho}_{S,M}} \\ & & TMS \end{array}$$

and

$$\begin{array}{ccc} STM & \xrightarrow{\tau^{\rho}_{S,TM}} & TMS \\ & \underset{\tau^{\rho}_{ST,M}}{\searrow} & \Big\downarrow{\scriptstyle \tau^{\rho}_{T,MS}} \\ & & MST \end{array}$$

commute.

The notion of a functor of permutative $\mathcal{T}$ bi-modules being the obvious one, we have the category of permutative $\mathcal{T}$ bi-modules, with the forgetful functor to the category of pre-DG categories.

1.1.3. *The translation category.* Let T be a symbol, and let $T^{\mathbb{Z}}$ be the free pre-additive category on the set $\{T^n \mid n \in \mathbb{Z}\}$. Let $\tau_{T^n,T^m} = (-1)^{nm}\mathrm{id}_{T^{n+m}}$. Then defining $T^n \otimes T^m := T^{n+m}$, and $a\mathrm{id}_{T^n} \otimes b\mathrm{id}_{T^m} := ab\mathrm{id}_{T^{n+m}}$, the data $(T^{\mathbb{Z}}, \otimes, \tau, T^0)$ defines a strictly associative tensor category, with strict unit T^0. We consider $T^{\mathbb{Z}}$ as a DG tensor category with all morphisms in Z^0.

The assignments

$$\begin{aligned} &(T^a, M, T^b) \to M[a]^{[b]} \\ &\mathrm{id}_{T^a} \otimes_{\mathbb{Z}} f \otimes_{\mathbb{Z}} \mathrm{id}_{T^b} \to f[a]^{[b]} \\ &\tau_{T^a,M} := \psi^a : M[a] \to M^{[a]} \end{aligned}$$

(see (1.1.1.1)) extend uniquely to give the DG category $\mathbf{C}(\mathbf{Ab})$ the structure of a permutative $T^{\mathbb{Z}}$ bi-module. In addition, the identities (1.1.1.2) become

$$\begin{aligned} T^aMT^n \otimes_{\mathbb{Z}} NT^b &= T^a(M \otimes_{\mathbb{Z}} T^nN)T^b \\ \mathrm{Hom}(T^aAT^n, T^mM) &= T^m\mathrm{Hom}(A, MT^{-n})T^{-a}. \end{aligned} \tag{1.1.3.1}$$

The map $f \mapsto f^{[m]}[n]$ is given by

$$\begin{aligned} \mathrm{Hom}(A, M) &\xrightarrow{\psi^n} (-1)^n\mathrm{Hom}(A, M) \\ &= T^n\mathrm{Hom}(A, M)T^{-n} = \mathrm{Hom}(T^nAT^m, T^nMT^m). \end{aligned} \tag{1.1.3.2}$$

Finally, we have the identity

$$T^2M = MT^2, \tag{1.1.3.3}$$

and, with respect to this identification,

$$\tau_{T^2,M} = \mathrm{id}_{T^2M}. \tag{1.1.3.4}$$

We abstract the identities (1.1.3.1)-(1.1.3.4) to the following definition.

1.1.4. DEFINITION. Let $\mathcal{C}$ be a pre-DG category.
(i) A *translation structure* on $\mathcal{C}$ is given by making $\mathcal{C}$ into a permutative $T^{\mathbb{Z}}$ bi-module such that

1. For X and Y in $\mathcal{C}$, we have $\mathrm{Hom}_{\mathcal{C}}(T^aXT^n, T^mY) = T^m\mathrm{Hom}(X, YT^{-n})T^{-a}$.

2. The map $\mathrm{id}_{T^n}(-)\mathrm{id}_{T^m} : \mathrm{Hom}_{\mathcal{C}}(X,Y) \to \mathrm{Hom}_{\mathcal{C}}(T^nXT^m, T^nYT^m)$ is given by

$$\begin{aligned}\mathrm{Hom}_{\mathcal{C}}(X,Y) &\xrightarrow{\psi^n} (-1)^n\mathrm{Hom}_{\mathcal{C}}(X,Y)\\ &= T^n\mathrm{Hom}_{\mathcal{C}}(X,Y)T^{-n} = \mathrm{Hom}_{\mathcal{C}}(T^nXT^m, T^nYT^m).\end{aligned}$$

3. For X, Y and Z in $\mathcal{C}$, the composition law

$$\mathrm{Hom}_{\mathcal{C}}(T^bY, T^cZ) \otimes_{\mathbb{Z}} \mathrm{Hom}_{\mathcal{C}}(T^aX, T^bY) \to \mathrm{Hom}_{\mathcal{C}}(T^aX, T^cZ)$$

is the composition

$$\begin{aligned}&\mathrm{Hom}_{\mathcal{C}}(T^bY, T^cZ) \otimes_{\mathbb{Z}} \mathrm{Hom}_{\mathcal{C}}(T^aX, T^bY)\\ &\qquad = T^c\mathrm{Hom}_{\mathcal{C}}(Y,Z)T^{-b} \otimes_{\mathbb{Z}} T^b\mathrm{Hom}_{\mathcal{C}}(X,Y)T^{-a}\\ &\qquad = T^c\mathrm{Hom}_{\mathcal{C}}(Y,Z) \otimes_{\mathbb{Z}} \mathrm{Hom}_{\mathcal{C}}(X,Y)T^{-a}\\ &\qquad \xrightarrow{\mathrm{id}_{T^c}(\circ_{X,Y,Z})\mathrm{id}_{T^{-a}}} T^c\mathrm{Hom}_{\mathcal{C}}(X,Z)T^{-a}\\ &\qquad = \mathrm{Hom}_{\mathcal{C}}(T^aX, T^cZ).\end{aligned}$$

4. For X in $\mathcal{C}$, we have $T^2X = XT^2$ and $\tau_{T^2,X} = \mathrm{id}_{T^2X}$.

(ii) Suppose $\mathcal{C}$ has a translation structure (ρ_l, ρ_r), and that $\mathcal{C}$ is a DG tensor category. We say that the translation structure is *compatible with the tensor structure on* $\mathcal{C}$ if

1. For X and Y in $\mathcal{C}$, $(T^aXT^c) \otimes (YT^b) = T^a(X \otimes T^cY)T^b$.
2. For morphisms $f: X \to X'$, $g: Y \to Y'$ in $\mathcal{C}$,

$$(\mathrm{id}_{T^a}f\mathrm{id}_{T^c}) \otimes (g\mathrm{id}_{T^b}) = \mathrm{id}_{T^a}(f \otimes \mathrm{id}_{T^c}g)\mathrm{id}_{T^b}.$$

3. For X and Y in $\mathcal{C}$, the diagrams

$$\begin{array}{ccc} T^aX\otimes Y & \xrightarrow{\tau^{\rho}_{T^a,X}\otimes\mathrm{id}_Y} & XT^a\otimes Y \\ & {\scriptstyle \tau^{\rho}_{T^a,X\otimes Y}}\searrow & \downarrow{\scriptstyle \mathrm{id}_X\otimes\tau^{\rho}_{T^a,Y}} \\ & & X\otimes YT^a \end{array}$$

$$\begin{array}{ccc} T^aX\otimes Y & \xrightarrow{\mathrm{id}_{T^a}\tau^{\mathcal{C}}_{X,Y}} & T^aY\otimes X \\ & {\scriptstyle \tau^{\mathcal{C}}_{T^aX,Y}}\searrow & \downarrow{\scriptstyle \tau^{\rho}_{T^a,Y}\otimes\mathrm{id}_X} \\ & & YT^a\otimes X \end{array}$$

and

$$\begin{array}{ccc} X\otimes YT^a & \xrightarrow{\tau^{\mathcal{C}}_{X,Y}\mathrm{id}_{T^a}} & Y\otimes XT^a \\ & {\scriptstyle \tau^{\mathcal{C}}_{X,YT^a}}\searrow & \uparrow{\scriptstyle \mathrm{id}_Y\otimes\tau^{\rho}_{T^a,X}} \\ & & Y\otimes T^aX \end{array}$$

commute.

4. For A, B, X and Y in $\mathcal{C}$, the tensor operation

$$\mathrm{Hom}_{\mathcal{C}}(T^aA, T^nX) \otimes_{\mathbb{Z}} \mathrm{Hom}_{\mathcal{C}}(T^bB, T^mY) \xrightarrow{\otimes} \mathrm{Hom}_{\mathcal{C}}(T^aA \otimes T^bB, T^nX \otimes T^mY)$$

is the composition

$$\begin{aligned}
&\mathrm{Hom}_{\mathcal{C}}(T^aA, T^nX) \otimes_{\mathbb{Z}} \mathrm{Hom}_{\mathcal{C}}(T^bB, T^mY)\\
&= T^n\mathrm{Hom}_{\mathcal{C}}(A, X)T^{-a} \otimes_{\mathbb{Z}} T^m\mathrm{Hom}_{\mathcal{C}}(B, Y)T^{-b}\\
&= T^n\mathrm{Hom}_{\mathcal{C}}(A, X) \otimes_{\mathbb{Z}} T^{-a}T^m\mathrm{Hom}_{\mathcal{C}}(B, Y)T^{-b}\\
&\xrightarrow{\mathrm{id}\otimes\tau_{T^{-a}, T^m\mathrm{Hom}_{\mathcal{C}}(B,Y)T^{-b}}} T^n\mathrm{Hom}_{\mathcal{C}}(A, X)T^m \otimes_{\mathbb{Z}} \mathrm{Hom}_{\mathcal{C}}(B, Y)T^{-b}T^{-a}\\
&\xrightarrow{\mathrm{id}_{T^n}\tau^{-1}_{T^m, \mathrm{Hom}_{\mathcal{C}}(A,X)}\otimes\mathrm{id}} T^{n+m}\mathrm{Hom}_{\mathcal{C}}(A, X) \otimes_{\mathbb{Z}} \mathrm{Hom}_{\mathcal{C}}(B, Y)T^{-a-b}\\
&\xrightarrow{\mathrm{id}_{T^{n+m}}(\otimes)\mathrm{id}_{T^{-a-b}}} T^{n+m}\mathrm{Hom}_{\mathcal{C}}(A \otimes B, X \otimes Y)T^{-a-b}\\
&= \mathrm{Hom}_{\mathcal{C}}(T^aT^bA \otimes B, T^nT^mX \otimes Y)\\
&\xrightarrow{(\mathrm{id}_{T^a}\tau^{-1}_{T^b, A}\otimes\mathrm{id}_B)^*\circ(\mathrm{id}_{T^n}\tau_{T^m, X}\otimes\mathrm{id}_Y)_*} \mathrm{Hom}_{\mathcal{C}}(T^aA \otimes T^bB, T^nX \otimes T^mY).
\end{aligned}$$

We write $A[n]^{[m]} = A^{[m]}[n]$ for T^nAT^m, and $f^{[m]}[n] = f[n]^{[m]}$ for $\mathrm{id}_{T^n} f \mathrm{id}_{T^m}$.

The category of pre-DG categories with translation structure is the full subcategory of the category of permutative $T^{\mathbb{Z}}$ bi-modules with the obvious objects. The category of pre-DG tensor categories with compatible translation structure is defined similarly. Ignoring the existence of the unit gives the category of pre-DG tensor categories without unit, with compatible translation structure.

1.1.5. REMARK. We have the following identities in a pre-DG category with translation structure:

1. $f[n]^{[m]} \circ g[n]^{[m]} = (f \circ g)[n]^{[m]}$.
2. $d(f[n]) = (df)[n]$; $d(f^{[n]}) = (df)^{[n]}$.

1.1.6. EXAMPLE. Recall from Chapter I, §1.3.8, that the category $\mathbf{C}(\mathbf{Ab})$ is a DG tensor category with $X \otimes Y$ the tensor product of complexes $X \otimes_{\mathbb{Z}} Y$ and with $f \otimes g = \oplus_{i,j}(-1)^{it} f^i \otimes_{\mathbb{Z}} g^j$ for graded maps $f = \oplus f^i : X^i \to Z^{i+s}$ and $g = \oplus_j g^j : Y^j \to W^{j+t}$. The translation structure $T^nMT^m = M^{[m]}[n]$ for $\mathbf{C}(\mathbf{Ab})$ described above is then compatible with the tensor structure.

1.1.7. LEMMA. *Suppose we have two translation structures on a DG category $\mathcal{C}$:*

$$(A, n, m) \mapsto A[n]_1^{[m]_1}; \quad (A, n, m) \mapsto A[n]_2^{[m]_2}.$$

Then the two structures are canonically isomorphic.

PROOF. We have the canonical isomorphism

$$\begin{aligned}
\mathrm{Hom}_{\mathcal{C}}(A[1]_1, A[1]_2) &= \mathrm{Hom}_{\mathcal{C}}(A, A)^{[-1]}[1]\\
&\xrightarrow{\psi} \mathrm{Hom}_{\mathcal{C}}(A, A).
\end{aligned}$$

This gives the isomorphism $\phi_l(A)\colon A[1]_1 \to A[1]_2$ corresponding to id_A. Similarly, we have the canonical isomorphism

$$\begin{aligned}\mathrm{Hom}_{\mathcal{C}}(A^{[1]_1}, A^{[1]_2}) &\xrightarrow{\tau^{\rho_1*}_{T,A}\circ(\tau^{\rho_2*}_{T,A})^{-1}} \mathrm{Hom}_{\mathcal{C}}(A[1]_1, A[1]_2)\\ &= \mathrm{Hom}_{\mathcal{C}}(A,A)^{[-1]}[1]\\ &\xrightarrow{\psi} \mathrm{Hom}_{\mathcal{C}}(A,A),\end{aligned}$$

giving the isomorphism $\phi_r(A)\colon A^{[1]_1} \to A^{[1]_2}$ corresponding to id_A. One easily checks that (ϕ_l, ϕ_r) gives an isomorphism of permutative $T^{\mathbb{Z}}$ bi-modules. □

1.1.8. *Free translation structures.* Let $\mathcal{A}$ be an pre-DG category. Form the pre-DG category $\mathcal{A}[*]$ with objects T^aAT^b, where A is an object of $\mathcal{A}$, and a and b are integers. Define

$$\mathrm{Hom}_{\mathcal{A}[*]}(T^aAT^b, T^nXT^m) = \mathrm{Hom}_{\mathcal{A}}(A,X)^{[-a]}[n+m-b].$$

In particular, we have

$$\begin{aligned}\mathrm{Hom}_{\mathcal{A}[*]}(T^aXT^b, T^aXT^b) = \mathrm{Hom}_{\mathcal{A}[*]}(X,X)^{[-a]}[a] = (-1)^a\mathrm{Hom}_{\mathcal{A}[*]}(X,X),\\ \mathrm{Hom}_{\mathcal{A}[*]}(T^aX, XT^a) = \mathrm{Hom}_{\mathcal{A}[*]}(X,X)^{[-a]}[a] = (-1)^a\mathrm{Hom}_{\mathcal{A}[*]}(X,X).\end{aligned}$$

Let

$$\begin{aligned}\mathrm{id}_{T^aXT^b} &:= \psi^a(\mathrm{id}_X) \in \mathrm{Hom}_{\mathcal{A}[*]}(T^aXT^b, T^aXT^b)\\ \tau_{T^a,X} &:= \psi^a(\mathrm{id}_X) \in \mathrm{Hom}_{\mathcal{A}[*]}(T^aX, XT^a).\end{aligned}$$

Define the composition

$$\begin{aligned}\mathrm{Hom}_{\mathcal{A}[*]}(T^bBT^{b'}, T^cCT^{c'}) \otimes_{\mathbb{Z}} \mathrm{Hom}_{\mathcal{A}[*]}(T^aAT^{a'}, T^bBT^{b'})\\ \to \mathrm{Hom}_{\mathcal{A}[*]}(T^aAT^{a'}, T^cCT^{c'})\end{aligned}$$

by

(1.1.8.1)

$$\begin{aligned}&\mathrm{Hom}_{\mathcal{A}[*]}(T^bBT^{b'}, T^cCT^{c'}) \otimes_{\mathbb{Z}} \mathrm{Hom}_{\mathcal{A}[*]}(T^aAT^{a'}, T^bBT^{b'})\\ &\quad = T^cT^{c'}T^{-b'}\mathrm{Hom}_{\mathcal{A}}(B,C)T^{-b} \otimes_{\mathbb{Z}} T^bT^{b'}T^{-a'}\mathrm{Hom}_{\mathcal{A}}(A,B)T^{-a}\\ &\quad = T^cT^{c'}T^{-b'}\mathrm{Hom}_{\mathcal{A}}(B,C)T^{b'-a'} \otimes_{\mathbb{Z}} \mathrm{Hom}_{\mathcal{A}[*]}(A,B)T^{-a}\\ &\quad \xrightarrow{\mathrm{id}\tau^{-1}_{T^{b'-a'},\mathrm{Hom}_{\mathcal{A}}(B,C)}\otimes\mathrm{id}} T^{c+c'-a'}\mathrm{Hom}_{\mathcal{A}}(B,C) \otimes_{\mathbb{Z}} \mathrm{Hom}_{\mathcal{A}}(A,B)T^{-a}\\ &\quad \xrightarrow{\mathrm{id}(\circ_{A,B,C})\mathrm{id}} T^{c+c'-a'}\mathrm{Hom}_{\mathcal{A}}(A,C)T^{-a} = \mathrm{Hom}_{\mathcal{A}[*]}(T^aAT^{a'}, T^cCT^{c'}).\end{aligned}$$

One easily checks that this defines a pre-DG category $\mathcal{A}[*]$ with translation structure $T^n(T^aAT^b)T^m = T^{n+a}AT^{b+m}$.

The functor $A \mapsto T^0AT^0$, $f \mapsto f$, defines a fully faithful embedding $i_{\mathcal{A}}\colon \mathcal{A} \to \mathcal{A}[*]$, which is universal for DG functors $\mathcal{A} \to \mathcal{B}$ to pre-DG categories with translation structure. Thus $\mathcal{A} \mapsto \mathcal{A}[*]$ defines a left adjoint to the functor "forget the translation structure". It follows from Lemma 1.1.7 that, if $\mathcal{A}$ already has a translation structure, then $i_{\mathcal{A}}$ is an equivalence of pre-DG categories with translation structure. We call $\mathcal{A}[*]$ the *free translation structure on $\mathcal{A}$*.

If $\mathcal{A}$ is a pre-DG tensor category, form the category $\mathcal{A}[*]^{\otimes}$ with objects the formal products $T^{a_0}A_1T^{a_1}A_2\dots A_nT^{a_n}$ where the A_i are objects of $\mathcal{A}$, and the a_i integers. Define

$$\begin{aligned}&\mathrm{Hom}_{\mathcal{A}[*]^{\otimes}}(T^{a_0}A_1\dots A_nT^{a_n}, T^{b_0}B_1\dots B_mT^{b_m})\\ &:= \mathrm{Hom}_{\mathcal{A}[*]}(T^{a_0+\dots+a_{n-1}}(A_1\otimes\dots\otimes A_n)T^{a_n}, T^{b_0+\dots+b_{m-1}}(B_1\otimes\dots\otimes B_m)T^{b_m})\end{aligned}$$

with composition law being induced by that of $\mathcal{A}[*]$.

Give $\mathcal{A}[*]^{\otimes}$ the translation structure so that the assignment

$$T^{a_0}A_1\dots A_nT^{a_n}\mapsto T^{a_0+\dots+a_{n-1}}A_1\otimes\dots\otimes A_nT^{a_n}$$

together with the identity map on morphisms defines a functor of pre-DG categories with translation structure $p_{\mathcal{A}}:\mathcal{A}[*]^{\otimes}\to\mathcal{A}[*]$. The identity map on $T^{a_0+\dots+a_{n-1}}A_1\otimes\dots\otimes A_nT^{a_n}$ defines the canonical isomorphism

$$\xi: T^{a_0+\dots+a_{n-1}}A_1\otimes\dots\otimes A_nT^{a_n}\to T^{a_0}A_1\dots A_nT^{a_n}$$

in $\mathcal{A}[*]^{\otimes}$; the isomorphisms ξ show that $p_{\mathcal{A}}$ is an equivalence of DG categories with translation structure, with inverse $j_{\mathcal{A}}:\mathcal{A}[*]\to\mathcal{A}[*]^{\otimes}$ being the identification of $\mathcal{A}[*]$ with the full subcategory of $\mathcal{A}[*]^{\otimes}$ with objects T^nXT^m.

We now describe a tensor structure for the category $\mathcal{A}[*]^{\otimes}$. Define

$$\begin{aligned}&(T^{a_0}A_1T^{a_1}A_2\dots A_nT^{a_n})\otimes(T^{b_0}B_1\dots B_mT^{b_m})\\ &\qquad := T^{a_0}A_1T^{a_1}A_2\dots A_nT^{a_n+b_0}B_1\dots B_mT^{b_m}.\end{aligned}$$

For

$$\begin{gathered}X=T^{a_0}A_1\dots A_nT^{a_n},\ Y=T^{b_0}B_1\dots B_mT^{b_m}\\ A=A_1\otimes\dots\otimes A_n,\ B=B_1\otimes\dots\otimes B_m\\ a_*=a_0+\dots+a_n,\\ b_*=b_0+\dots+b_m,\end{gathered}$$

define the symmetry isomorphism $\tau_{X,Y}:X\otimes Y\to Y\otimes X$ to be the element of

$$\mathrm{Hom}_{\mathcal{A}[*]}(T^{a_*+b_*-b_m}(A\otimes B)T^{b_m}, T^{b_*+a_*-a_n}(B\otimes A)T^{a_n})$$

given by

$$\begin{aligned}T^{a_*}T^{b_*-b_m}(A\otimes B)T^{b_m}&\xrightarrow{\mathrm{id}\tau^{-1}_{T^{b_m},A\otimes B}}T^{a_*}T^{b_*}(A\otimes B)\\ &\xrightarrow{\tau_{T^{a_*},T^{b_*}}\mathrm{id}}T^{b_*}T^{a_*}(A\otimes B)=T^{b_*}T^{a_*-a_n}T^{a_n}(A\otimes B)\\ &\xrightarrow{\tau_{T^{a_n},A\otimes B}}T^{b_*}T^{a_*-a_n}(A\otimes B)T^{a_n}\\ &\xrightarrow{\mathrm{id}\tau_{A,B}\mathrm{id}}T^{b_*}T^{a_*-a_n}(B\otimes A)T^{a_n}.\end{aligned}$$

The formula for the tensor product of morphisms is given by using shuffle permutations to rearrange the translation operators. Precisely, for

$$\begin{gathered}X=T^{a_0}A_1\dots A_nT^{a_n},\ Y=T^{b_0}B_1\dots B_mT^{b_m},\\ Z=T^{c_0}C_1\dots C_kT^{c_k},\ W=T^{d_0}D_1\dots D_lT^{d_l}\end{gathered}$$

$$\begin{aligned} A &= A_1 \otimes \ldots \otimes A_n, B = B_1 \otimes \ldots \otimes B_m, \\ C &= C_1 \otimes \ldots \otimes C_k, D = D_1 \otimes \ldots \otimes D_l, \end{aligned}$$

$$\begin{aligned} a_* &= a_0 + \ldots + a_n, b_* = b_0 + \ldots + b_m, \\ c_* &= c_0 + \ldots + c_k, d_* = d_0 + \ldots + d_l, \end{aligned}$$

the tensor product of morphisms is given by the composition

$$\begin{aligned}
&\mathrm{Hom}_{\mathcal{A}[*]^\otimes}(X, Y) \otimes_{\mathbb{Z}} \mathrm{Hom}_{\mathcal{A}[*]^\otimes}(Z, W) \\
&= T^{b_*}T^{-a_n}\mathrm{Hom}_{\mathcal{A}}(A, B)T^{a_n}T^{-a_*} \otimes_{\mathbb{Z}} T^{d_*-c_k}\mathrm{Hom}_{\mathcal{A}}(C, D)T^{-c_*+c_k} \\
&\xrightarrow{\tau^{-1}_{T^{d_*-c_k}, T^{-a_n}\mathrm{Hom}_{\mathcal{A}}(A,B)T^{a_n}T^{-a_*}} \otimes \mathrm{id}} \\
&\quad T^{b_*+d_*-c_k}T^{-a_n}\mathrm{Hom}_{\mathcal{A}}(A, B) \otimes_{\mathbb{Z}} T^{a_n}T^{-a_*}\mathrm{Hom}_{\mathcal{A}}(C, D)T^{-c_*+c_k} \\
&\xrightarrow{\mathrm{id}\tau_{T^{-a_n}, \mathrm{Hom}_{\mathcal{A}}(A,B)} \otimes \mathrm{id}} T^{b_*+d_*-c_k}\mathrm{Hom}_{\mathcal{A}}(A, B) \otimes_{\mathbb{Z}} T^{-a_*}\mathrm{Hom}_{\mathcal{A}}(C, D)T^{-c_*+c_k} \\
&\xrightarrow{\mathrm{id} \otimes \tau_{T^{-a_*}, \mathrm{Hom}_{\mathcal{A}}(C,D)T^{-c_*+c_k}}} T^{b_*+d_*-c_k}\mathrm{Hom}_{\mathcal{A}}(A, B) \otimes_{\mathbb{Z}} \mathrm{Hom}_{\mathcal{A}}(C, D)T^{-a_*-c_*+c_k} \\
&\xrightarrow{\mathrm{id}_{T^{b_*+d_*-c_k}}(\otimes)\mathrm{id}_{T^{-a_*-c_*+c_k}}} T^{b_*+d_*-c_k}\mathrm{Hom}_{\mathcal{A}}(A \otimes C, B \otimes D)T^{-a_*-c_*+c_k} \\
&\qquad\qquad\qquad\qquad\qquad\qquad\qquad = \mathrm{Hom}_{\mathcal{A}[*]^\otimes}(X \otimes Z, Y \otimes W).
\end{aligned}$$

One checks that this data defines a pre-DG tensor category with compatible translation structure.

The functor

$$\begin{aligned} i^\otimes_{\mathcal{A}} : \mathcal{A} &\to \mathcal{A}[*]^\otimes \\ A &\mapsto T^0 A T^0 \end{aligned}$$

is universal for DG tensor functors to DG tensor categories with compatible translation structure, and thus $\mathcal{A} \mapsto \mathcal{A}[*]^\otimes$ is left adjoint to the forgetful functor. As above, if $\mathcal{A}$ already has a compatible translation structure, it follows from Lemma 1.1.7 that $i^\otimes_{\mathcal{A}}$ is an equivalence of pre-DG tensor categories with compatible translation structure. We call $\mathcal{A}[*]^\otimes$ the *free compatible translation structure on* $\mathcal{A}$.

If $\mathcal{A}$ is a pre-DG tensor category without unit, the analogous construction has the analogous properties in the setting of DG tensor categories without unit. We make all the constructions of this section on the level of DG categories, DG tensor categories and DG tensor categories without unit by adjoining finite direct sums, as explained in §1.3.2.

1.1.9. REMARK. Classically (see e.g. [**123**]) a translation on an additive category $\mathcal{A}$ is given by an isomorphism $T: \mathcal{A} \to \mathcal{A}$. One can then define a graded category $(\mathcal{A}, T)$ with $\mathrm{Hom}_{(\mathcal{A},T)}(X, Y)^s = \mathrm{Hom}_{(\mathcal{A},T)}(X, T^s Y)$. We may then extend the "left" action $X \mapsto TX$ to a translation structure by adjoining new objects XT^a, and isomorphisms $\tau_{T^a,X} : T^a X \to XT^a$. We set $\mathrm{Hom}(AT^b, XT^m) := \mathrm{Hom}_{\mathcal{A}}(T^b A, T^m X)$, with the composition $f \mapsto \tau^{-1}_{T^m,X} \circ f \circ \tau_{T^a,A}$ giving the identity map. This defines a graded category $(\mathcal{A}, T)'$ with translation structure, which is equivalent, as a graded category with translation isomorphism T, to $(\mathcal{A}, T)$.

In the setting of a graded category, we will usually only use the left T action, so there is no essential difference between our notion of a translation structure, and the notion of an additive category with translation isomorphism. When the additional structure of a tensor operation, or a differential structure comes into play, then the two-sided action of T becomes useful by simplifying the sign conventions.

1.2. Complexes over a differential graded category

We give a description of Kapranov's construction [**75**] of the category of complexes over a DG category, which gives a generalization of the usual construction of the category of complexes on an additive category.

1.2.1. If $\mathcal{C}$ is a DG category, we have the functors

$$Z^p(\mathrm{Hom}_{\mathcal{C}}(-,-)^*),\quad B^p(\mathrm{Hom}_{\mathcal{C}}(-,-)^*)\quad\text{and}\quad H^p(\mathrm{Hom}_{\mathcal{C}}(-,-)^*)$$

defined by taking respectively the cocycles, coboundaries and cohomology in the complex $\mathrm{Hom}_{\mathcal{C}}(-,-)^*$. We have the graded category $Z^*\mathcal{C}$ (I.1.3.2.1). We have as well the *homotopy category* $\mathcal{C}/\mathrm{Htp}$, defined as the graded additive category with the same objects as $\mathcal{C}$ and with

$$\mathrm{Hom}_{\mathcal{C}/\mathrm{Htp}}(-,-)^n = H^n(\mathrm{Hom}_{\mathcal{C}}(-,-)^*).$$

We have natural maps of graded categories, $Z^*\mathcal{C}\to\mathcal{C}$, and $Z^*\mathcal{C}\to\mathcal{C}/\mathrm{Htp}$.

If $\mathcal{C}$ is a DG tensor category, then the categories $Z^*\mathcal{C}$ and $\mathcal{C}/\mathrm{Htp}$ are graded tensor categories, and the functors $Z^*\mathcal{C}\to\mathcal{C}$ and $Z^*\mathcal{C}\to\mathcal{C}/\mathrm{Htp}$ are functors of graded tensor categories. If $\mathcal{C}$ is a DG (tensor) category with (compatible) translation structure, then $Z^*\mathcal{C}$ and $\mathcal{C}/\mathrm{Htp}$ have a canonically induced (compatible) translation structure, and the functors $Z^*\mathcal{C}\to\mathcal{C}$ and $Z^*\mathcal{C}\to\mathcal{C}/\mathrm{Htp}$ preserve the translation structures. We have as well the additive subcategory $Z^0\mathcal{C}$ of $Z^*\mathcal{C}$ with $\mathrm{Hom}_{Z^0\mathcal{C}}(X,Y)=Z^0\mathrm{Hom}_{\mathcal{C}}(X,Y)$.

1.2.2. *Complexes for an additive category.* Let $\mathcal{A}_0$ be an additive category. We have the DG category of complexes in $\mathcal{A}_0$, $\mathbf{C}(\mathcal{A}_0)$, with objects

$$M := \{M^i,\ d^i\colon M^i\to M^{i+1}\mid i\in\mathbb{Z}\} := (M,d_M),$$

where the M^i are objects of $\mathcal{A}_0$, the d^i are morphisms in $\mathcal{A}_0$ and $d^{i+1}\circ d^i=0$. A *graded morphism* $f\colon M\to N$ of degree s is a collection

$$f^i\colon M^i\to N^{i+s};\quad i\in\mathbb{Z}.$$

This gives the Hom-complex $\mathrm{Hom}((M,d_M),(N,d_N))$ which in degree s is the group of graded, degree s morphisms from M to N, and with differential

$$df=(-1)^{\deg f}f\circ d_M-d_N\circ f.$$

Composition is induced by composition in $\mathcal{A}_0$. We have the full subcategories $\mathbf{C}^*(\mathcal{A}_0)$, $*=+,-,b$ of bounded below, bounded above and bounded complexes, defined in the usual way: (M,d_M) is in $\mathbf{C}^+(\mathcal{A}_0)$ if there is an i_0 such that $M^i=0$ for all $i<i_0$, etc.

The categories $\mathbf{C}^*(\mathcal{A}_0)$, $(*=\emptyset,+,-,b)$ have the translation structure defined as for $\mathbf{C}(\mathbf{Ab})$: $T^aMT^b:=M[a]^{[b]}$.

If $\mathcal{A}_0$ is a tensor category, make $\mathbf{C}^*(\mathcal{A}_0)$ $(* = +, -, b)$ into a DG tensor category by defining

$$(M, d_M) \otimes (N, d_N) = (M \otimes N, d_{M\otimes N}),$$
$$d_{M\otimes N}(m \otimes n) := d_M(m) \otimes n + (-1)^{\deg m} m \otimes d_N(n),$$

with $(M \otimes N)^n := \oplus_{i+j=n} M^i \otimes N^j$, and

$$(f \otimes g)^n = \oplus_{i+j=n} (-1)^{it} f^i \otimes g^j$$

if $f = \oplus_i f^i$, $g = \oplus_j g^j$ and g has degree t. The symmetry $\tau_{M,N}: M \otimes N \to N \otimes M$ is given by the sum of maps $(-1)^{ij}\tau_{M^i,N^j}: M^i \otimes N^j \to N^j \otimes M^i$. The translation structure is then compatible with the tensor structure.

Suppose we have a morphism $f:(M, d_M) \to (N, d_N)$ in $Z^0\mathbf{C}(\mathcal{A}_0)$. Form the *cone of* f by $\text{cone}(f)^n := M^{n+1} \oplus N^n$, with differential given by the matrix

$$\begin{pmatrix} -d_M^{n+1} & 0 \\ f^{n+1} & d_N^n \end{pmatrix} : \text{cone}(f)^n \to \text{cone}(f)^{n+1}.$$

The inclusions $N^n \to \text{cone}(f)^n$ and the projections $\text{cone}(f)^n \to M^{n+1}$ define the *cone sequence*

$$M \xrightarrow{f} N \xrightarrow{i_N} \text{cone}(f) \xrightarrow{j_M} M[1].$$

If f is a map in $Z^0\mathbf{C}^*(\mathcal{A}_0)$ $(* = +, -, b)$, then $\text{cone}(f)$ is also in $\mathbf{C}^*(\mathcal{A}_0)$.

1.2.3. Let $\mathcal{A}$ be a DG category, and let $\mathcal{F}(\mathcal{A})$ be the category of additive functors $F: \mathcal{A}^{\text{op}} \to \mathbf{C}(\mathbf{Ab})$. $\mathcal{F}(\mathcal{A})$ is a DG category with grading, translation structure and differential induced by that in $\mathbf{C}(\mathbf{Ab})$; sending an object E of $\mathcal{A}$ to the functor h_E, $h_E(F) := \text{Hom}_{\mathcal{A}}(F, E)$, defines a full embedding $h: \mathcal{A} \to \mathcal{F}(\mathcal{A})$. Let $\mathcal{A}_0$ be the full subcategory of $\mathcal{F}(\mathcal{A})$ generated by the translates of $h(\mathcal{A})$.

Let $\text{cone}: \text{Maps}(Z^0\mathbf{C}(\mathbf{Ab})) \to \mathbf{C}(\mathbf{Ab})$ be the functor $f \mapsto \text{cone}(f)$. We define a sequence of full subcategories

$$h(\mathcal{A}) \subset \mathcal{A}_0 \subset \ldots \subset \mathcal{A}_n$$

of $\mathcal{F}(\mathcal{A})$ by letting $\mathcal{A}_n$ be the full subcategory generated by the objects of $\mathcal{A}_{n-1}$ and $\text{cone}(\text{Maps}(Z^0\mathcal{A}_{n-1}))$.

1.2.4. DEFINITION. If $\mathcal{A}$ is a DG category, define the DG category $\overline{\mathbf{C}}^b(\mathcal{A})$ as the full subcategory of $\mathcal{F}(\mathcal{A})$

$$\overline{\mathbf{C}}^b(\mathcal{A}) := \cup_{n\to\infty} \mathcal{A}_n.$$

The homotopy category $\overline{\mathbf{C}}^b(\mathcal{A})/\text{Htp}$ is denoted $\overline{\mathbf{K}}^b(\mathcal{A})$.

1.2.5. REMARK. In case $\mathcal{A}$ is just an additive category, we have the DG category of complexes $\mathbf{C}^b(\mathcal{A})$. Sending $(\oplus_i X^i, d) \in \mathbf{C}^b(\mathcal{A})$ to the functor $(\oplus_i \text{Hom}_{\mathcal{A}}(-, X^i), d_*)$ gives an equivalence of $\mathbf{C}^b(\mathcal{A})$ with $\overline{\mathbf{C}}^b(\mathcal{A})$.

1.2.6. *The category Pre-Tr.* Let $\mathcal{A}$ be a DG category with translation structure. Kapranov [**75**] has given an explicit description of a full subcategory of $\mathcal{F}(\mathcal{A})$ containing $\overline{\mathbf{C}}^b(\mathcal{A})$. He defines Pre-Tr$(\mathcal{A})$ to be the following DG category: An object in Pre-Tr$(\mathcal{A})$ is a finite collection $\{E^i;\ i \in I \subset \mathbb{Z}\}$ of objects of $\mathcal{A}$, together with morphisms of degree $+1$

$$q^{ji}: E^i[-i] \to E^j[-j]$$

satisfying the condition

$$d_{\mathcal{A}}(q^{ji}) = \sum_m q^{jm} \circ q^{mi} \tag{1.2.6.1}$$

for each i, j. If $E := \{E^i, q^{ji}\}$, $F := \{F^i, s^{ji}\}$ are objects of Pre-Tr$(\mathcal{A})$, the graded Hom group is given by

$$\mathrm{Hom}(E, F)^l = \oplus_{i,j}\mathrm{Hom}_{\mathcal{A}}(E^i[-i], F^j[-j])^l.$$

We write an element of $\mathrm{Hom}(E, F)^l$ as

$$f = \sum_{i,j} f^{ji}; \quad f^{ji} : E^i[-i] \to F^j[-j].$$

The differential on $\mathrm{Hom}(E, F)^l$ is defined by

$$(df)^{ji} = d_{\mathcal{A}} f^{ji} + (-1)^l \sum_m s^{jm} \circ f^{mi} - \sum_n f^{jn} \circ q^{ni}$$

for $f = \sum_{i,j} f^{ji}$.

Pre-Tr$(\mathcal{A})$ has the translation structure

$$(\{E^i, q^{ji}\})^{[b]}[a] = \{(E^i)^{[b]}[a], (q^{ji})^{[b]}[a]\}$$
$$(\sum_{i,j} f^{ji})^{[b]}[a] = \sum_{i,j} (f^{ji})^{[b]}[a].$$

If we have a map $f : E \to F$ in Z^0Pre-Tr$(\mathcal{A})$, $f = \sum_{ij} f^{ji}$, let f_+^{ji} be the element of $\mathrm{Hom}(E^i[1-i], F^j[-j])^1$ corresponding to f^{ji} via the identity

$$\mathrm{Hom}(E^i[1-i], F^j[-j])^1 = \mathrm{Hom}(E^i[-i], F^j[-j])[-1]^1 = \mathrm{Hom}(E^i[-i], F^j[-j])^0.$$

We define $\mathrm{cone}(f) = \{\mathrm{cone}(f)^i\}$ by $\mathrm{cone}(f)^i = E^i[1] \oplus F^i$, with maps

$$\begin{pmatrix} q^{ji}[1] & 0 \\ f_+^{ji} & s^{ji} \end{pmatrix} : E^i[1-i] \oplus F^i[-i] \to E^j[2-j] \oplus F^j[1-j];$$

one checks directly that the condition $f \in Z^0$Pre-Tr$(\mathcal{A})$ ensures that $\mathrm{cone}(f)$ is indeed an object of Pre-Tr$(\mathcal{A})$.

We have the full embedding $i_{\mathcal{A}} : \mathcal{A} \to$ Pre-Tr$(\mathcal{A})$, defined by sending E to the collection $\{E^0 = E;\ q^{00} = 0\}$. The functor $h : \mathcal{A} \to \mathcal{F}(\mathcal{A})$ extends to the functor $h :$ Pre-Tr$(\mathcal{A}) \to \mathcal{F}(\mathcal{A})$; by $h(E)(A) = \mathrm{Hom}_{\text{Pre-Tr}(\mathcal{A})}(A, E)$; this latter is also a fully faithful embedding, and is compatible with the two cone functors, up to natural isomorphism.

1.2.7. Definition. Let $\mathbf{C}^b(\mathcal{A})$ be the strictly full DG subcategory of Pre-Tr$(\mathcal{A})$ generated from $\mathcal{A}$ by repeatedly taking cones. We denote the homotopy category $\mathbf{C}^b(\mathcal{A})/\mathrm{Htp}$ by $\mathbf{K}^b(\mathcal{A})$

1.2.8. Remarks. (i) It is immediate that $\mathbf{C}^b(\mathcal{A}) \subset$ Pre-Tr$(\mathcal{A})$ is a DG subcategory of Pre-Tr$(\mathcal{A})$, isomorphic to $\overline{\mathbf{C}}^b(\mathcal{A})$ via h. In addition, $\mathbf{C}^b(\mathcal{A})$ may be also defined as the smallest strictly full DG subcategory of Pre-Tr$(\mathcal{A})$ containing $\mathcal{A}$, and closed under taking cones of degree zero morphisms f with $df = 0$. In particular, the cone functor for Pre-Tr restricts to give the cone functor for $\mathbf{C}^b(\mathcal{A})$

(ii) Let $\mathcal{A}_0$ be an additive category, which we consider as a graded category with all morphisms in degree zero, and let $\mathcal{A} = \mathcal{A}_0[*]$ (see §1.1.8). It is easy to see that $\mathbf{C}^b(\mathcal{A}) =$ Pre-Tr$(\mathcal{A})$ and that Pre-Tr$(\mathcal{A})$ is equivalent to the category of bounded complexes $\mathbf{C}^b(\mathcal{A}_0)$ described in §1.2.2. This equivalence is defined by sending the object $\{E^i; q^{ji}\}$ of Pre-Tr$(\mathcal{A})$ to the complex E with differential $d^i_E: E^i \to E^{i+1}$ being the map $q^{i,i+1}_-[i]$, where $q^{i,i+1}_-[i]$ is the degree zero map from $E^i[-i]$ to $E^{i+1}[-i]$ corresponding to $q^{i,i+1}[i]$ via the identity

$$\mathrm{Hom}(E^i[-i], E^{i+1}[-i-1])^1 = \mathrm{Hom}(E^i[-i], E^{i+1}[-i])^0.$$

Note that $\mathrm{Hom}(E^i[-i], E^j[-j])^1 = 0$ for $j \neq i+1$.
(iii) We will use the notation $\mathbf{C}^b(-)$ to mean the usual category of bounded complexes when applied to an additive category, and the above construction when applied to a DG category with translation structure. If $\mathcal{A}$ is a DG category, we write $\mathbf{C}^b(\mathcal{A})$ for $\mathbf{C}^b(\mathcal{A}[*])$ (see §1.1.8). This notation is ambiguous, but, by the results of §1.1.8, only up to a canonical equivalence of DG categories with translation structure.

1.2.9. *The functor* Tot. Let $\mathcal{A}_0$ be a DG category with translation structure, and let $\mathcal{A}$ be the DG category with translation structure $\mathcal{A} := \mathbf{C}^b(\mathcal{A}_0)$. We define a DG functor

$$\mathrm{Tot}_{\mathcal{A}}: \mathbf{C}^b(\mathcal{A}) \to \mathcal{A} \tag{1.2.9.1}$$

which sends cones to cones as follows: Let

$$E := \{E^i, q^{ji}: E^i[-i] \to E^j[-j]\}$$

be an object of Pre-Tr$(\mathcal{A})$. Write each object E^i of $\mathbf{C}^b(\mathcal{A}_0)$ as

$$E^i = \{E^i_k, r^i_{lk}: E^i_k[-k] \to E^i_l[-l]\}.$$

By the definition of the Hom-groups in $\mathbf{C}^b(\mathcal{A})$, we may write each q^{ji} as

$$q^{ji} = \sum_{kl} q^{ji}_{lk}$$

$$q^{ji}_{lk}: E^i_k[-i-k] \to E^j_l[-j-l]; \quad \deg q^{ji}_{lk} = 1.$$

Let $\mathrm{Tot}(E)^a = \oplus_{i+k=a} E^i_k$, and let $s^{ba}: \mathrm{Tot}(E)^a[-a] \to \mathrm{Tot}(E)^b[-b]$ be the sum

$$s^{ba} = \sum_{\substack{i+k=a\\ j+l=b}} q^{ji}_{lk} + \sum_{\substack{i+k=a\\ i+l=b}} r^i_{lk}[-i].$$

Define the object $\mathrm{Tot}(E)$ of Pre-Tr$(\mathcal{A}_0)$ by

$$\mathrm{Tot}(E) = \{\mathrm{Tot}(E)^a, s^{ba}: \mathrm{Tot}(E)^a \to \mathrm{Tot}(E)^b\}.$$

The relation (1.2.6.1) for E and for the E^i implies the relation (1.2.6.1) for $\mathrm{Tot}(E)$, hence $\mathrm{Tot}(E)$ is a well-defined object of Pre-Tr$(\mathcal{A}_0)$. If we have a morphism $f: E \to E'$ in Pre-Tr$(\mathcal{A}_0)$, write f as

$$f = \sum_{ij} f^{ji}: E^i[-i] \to E'^j[-j]$$

$$f^{ji} = \sum_{k} f^{ji}_{lk}: E^i_k[-i-k] \to E'^j_l[-j-l].$$

Set

$$\mathrm{Tot}(f)^{ba} = \sum_{\substack{i+k=a\\ j+l=b}} f^{ji}_{lk}: \mathrm{Tot}(E)^a[-a] \to \mathrm{Tot}(E')^b[-b],$$

and let $\mathrm{Tot}(f)\colon \mathrm{Tot}(E) \to \mathrm{Tot}(E')$ be the map in $\text{Pre-Tr}(\mathcal{A}_0)$ given by

$$\mathrm{Tot}(f) := \sum_{ab} \mathrm{Tot}(f)^{ba}.$$

One easily checks that this defines a translation preserving DG functor

$$\mathrm{Tot}\colon \text{Pre-Tr}(\mathcal{A}) \to \text{Pre-Tr}(\mathcal{A}_0),$$

and that the functor Tot sends a cone sequence in $\text{Pre-Tr}(\mathcal{A})$ to a cone sequence in $\text{Pre-Tr}(\mathcal{A}_0)$. As the composition $\mathrm{Tot} \circ i_{\mathcal{A}}\colon \mathcal{A} \to \text{Pre-Tr}(\mathcal{A}_0)$ is the canonical inclusion of $\mathcal{A}$ into $\text{Pre-Tr}(\mathcal{A}_0)$, we see that Tot restricts to a translation preserving DG functor

$$\mathrm{Tot}\colon \mathbf{C}^b(\mathcal{A}) \to \mathbf{C}^b(\mathcal{A}_0) = \mathcal{A},$$

as desired.

1.2.10. LEMMA. *Let $\mathcal{A}_0$ be an additive category, and let $\mathcal{A} = \mathbf{C}^*(\mathcal{A}_0)$ be the category of complexes, where $* = +, -, b, \emptyset$ is a boundedness condition. Then the functor $\mathrm{Tot}\colon \mathbf{C}^b(\mathcal{A}) \to \mathcal{A}$ (1.2.9.1) is an equivalence of DG categories (with translation structure). Similarly, if $\mathcal{A}_0$ is a DG category with translation structure, and $\mathcal{A} = \mathbf{C}^b(\mathcal{A}_0)$ then the functor (1.2.9.1) is an equivalence of DG categories with translation structure.*

PROOF. We prove the second assertion; the first assertion has essentially the same proof. We consider $\mathcal{A}$ as a sub-DG category of $\mathbf{C}^b(\mathcal{A})$ via the embedding $i_{\mathcal{A}}$. For an object

$$\begin{aligned} E &:= \{E^i, q^{ji}\colon E^i[-i] \to E^j[-j]\} \\ E^i &= \{E^i_k, r^i_{lk}\colon E^i_k[-k] \to E^i_l[-l]\} \\ q^{ji} &= \sum_{k,l} q^{ji}_{lk}\colon E^i_k[-i-k] \to E^j_l[-j-l] \end{aligned}$$

as in §1.2.9, we let

$$\begin{aligned} f^i_E &\colon \mathrm{Tot}(E) \to E^i[-i], \\ g^i_E &\colon E^i[-i] \to \mathrm{Tot}(E) \end{aligned}$$

be the maps in $\mathcal{A}$ defined as the sums

$$f^i_E = \sum_a f^i_{Ea}\colon \mathrm{Tot}(E)^a[-a] \to E^i_{a-i}[-a],$$

$$g^i_E = \sum_a g^i_{Ea}\colon E^i_{a-i}[-a] \to \mathrm{Tot}(E)^a[-a],$$

with f^i_{Ea} the projection of $\mathrm{Tot}(E)^a[-a]$ on the summand $E^i_{a-i}[-a]$, and g^i_{Ea} the inclusion of the summand $E^i_{a-i}[-a]$ into $\mathrm{Tot}(E)^a[-a]$. This gives the maps

$$f_E\colon \mathrm{Tot}(E) \to E,\ g_E\colon E \to \mathrm{Tot}(E),$$

in $\mathbf{C}^b(\mathcal{A})$ defined by

$$f_E = \sum_i f^i_E,\ g_E = \sum_i g^i_E.$$

The maps f_E and g_E give an isomorphism of E with $\mathrm{Tot}(E)$ in $\mathbf{C}^b(\mathcal{A})$; clearly $\mathrm{Tot}(E) = E$ if E is in $\mathcal{A}$. This gives the desired equivalence. □

1.2.11. *Tensor structure.* Suppose $\mathcal{A}$ is a DG tensor category with compatible translation structure (Definition 1.1.4). We define the functor

$$\otimes\colon \text{Pre-Tr}(\mathcal{A}) \otimes_{\mathbb{Z}} \text{Pre-Tr}(\mathcal{A}) \to \text{Pre-Tr}(\mathcal{A})$$

by

$$(\{A^i;\ q^{ji}\} \otimes \{B^i;\ s^{ji}\})^n = \oplus_{i+j=n}(A^i[-i] \otimes B^j[-j])[n],$$

with maps

$$r^{mn} = \bigoplus_{\substack{i+j=n\\ k+j=m}} q^{ki} \otimes \mathrm{id}_{B^j[-j]} \oplus \bigoplus_{\substack{i+j=n\\ i+l=m}} \mathrm{id}_{A^i[-i]} \otimes s_{lj}.$$

This works because, as q^{ki} and s_{lj} have degree 1,

$$(q^{ki} \otimes \mathrm{id}_{B^l[-l]}) \circ (\mathrm{id}_{A^i[-i]} \otimes s_{lj}) = -(\mathrm{id}_{A^k[-k]} \otimes s_{lj}) \circ (q^{ki} \otimes \mathrm{id}_{B^j[-j]}).$$

The definition of $f \otimes g$ for morphisms f and g is induced by that of $\otimes$ on $\mathcal{A}$ by bi-linearity:

$$(\sum_{i,k} f^{ki}) \otimes (\sum_{j,l} g^{lj})^{mn} = \sum_{\substack{i+j=n\\ i+l=m}} f^{ki} \otimes g^{lj}.$$

If $A = \{A^i;\ q^{ji}\}$, $B = \{B^i;\ s^{ji}\}$, then set $\tau_{A,B} := \oplus_{i,j}\tau_{A^i[-i],B^j[-j]}$; the associativity morphisms are defined similarly. This gives Pre-Tr($\mathcal{A}$) the structure of a DG tensor category. In addition, the functor $\otimes$ restricts to the functor

$$\otimes\colon \mathbf{C}^b(\mathcal{A}) \otimes_{\mathbb{Z}} \mathbf{C}^b(\mathcal{A}) \to \mathbf{C}^b(\mathcal{A}),$$

giving $\mathbf{C}^b(\mathcal{A})$ and $\mathbf{K}^b(\mathcal{A})$ the structure of DG tensor categories without unit. If $\mathcal{A}$ is a category of complexes $\mathbf{C}^*(\mathcal{A}_0)$, with $\mathcal{A}_0$ a tensor category, the functor Tot (*cf.* (1.2.9.1) and Lemma 1.2.10) is an equivalence of DG tensor categories. Note that the grading and translation structure absorbs all the explicit signs which occur in the various structures on $\mathbf{C}^*(\mathcal{A}_0)$.

If $\mathcal{A}$ is a DG tensor category, we write $\mathbf{C}^b(\mathcal{A})$ for $\mathbf{C}^b(\mathcal{A}[*]^{\otimes})$ (see §1.1.8). As above, this notation is ambiguous if $\mathcal{A}$ is a DG tensor category with compatible translation structure, but only up to canonical equivalence.

2. Complexes and triangulated categories

2.1. Triangulated categories

As in the case of an abelian category, the homotopy category of the category of complexes is in a natural way a triangulated category.

2.1.1. *Verdier's axioms.* Let $\mathcal{A}$ be an additive category with a translation isomorphism. A *triangle* (X, Y, Z, a, b, c) in $\mathcal{A}$ is the sequence of maps

$$X \xrightarrow{a} Y \xrightarrow{b} Z \xrightarrow{c} X[1].$$

A morphism of triangles

$$(f, g, h)\colon (X, Y, Z, a, b, c) \to (X', Y', Z', a', b', c')$$

is a commutative diagram

$$\begin{array}{ccccccc} X & \xrightarrow{a} & Y & \xrightarrow{b} & Z & \xrightarrow{c} & X[1] \\ \downarrow{\scriptstyle f} & & \downarrow{\scriptstyle g} & & \downarrow{\scriptstyle h} & & \downarrow{\scriptstyle f[1]} \\ X' & \xrightarrow[a']{} & Y' & \xrightarrow[b']{} & Z' & \xrightarrow[c']{} & X'[1]. \end{array}$$

We have the notion of triangles and morphisms of triangles in a graded category with translation structure by forgetting the right T-action and restricting to degree zero morphisms.

Verdier [**123**] has defined a *triangulated category* as an additive category $\mathcal{A}$ with translation isomorphism, together with a collection $\mathcal{E}$ of triangles, called the *distinguished triangles* of $\mathcal{A}$, which satisfy

(TR1) Each triangle isomorphic to a distinguished triangle is distinguished. The triangle $(A, A, 0, \mathrm{id}_A, 0, 0)$ is distinguished. Each morphism $u: X \to Y$ is contained in a distinguished triangle.

(TR2) (X, Y, Z, u, v, w) is distinguished if and only if $(Y, Z, X[1], v, w, -u[1])$ is distinguished.

(TR3) If we have distinguished triangles (X, Y, Z, u, v, w), (X', Y', Z', u', v', w'), and a morphism $(f, g): u \to u'$, then there exists a morphism $h: Z \to Z'$ such that (f, g, h) is a morphism of triangles.

(TR4) If we have three distinguished triangles $(X, Y, Z', u, i, *)$, $(Y, Z, X', v, *, j)$, and $(X, Z, Y', w, *, *)$, with $w = v \circ u$, then there are morphisms $f: Z' \to Y'$, $g: Y' \to X'$ such that
   (a) (id_X, v, f) is a morphism of triangles
   (b) (u, id_Z, g) is a morphism of triangles
   (c) $(Z', Y', X', f, g, i[1] \circ j)$ is a distinguished triangle.

A graded functor $F: \mathcal{A} \to \mathcal{B}$ of triangulated categories is called *exact* if F takes distinguished triangles in $\mathcal{A}$ to distinguished triangles in $\mathcal{B}$. This defines the category of triangulated categories.

We get an equivalent definition if we replace the additive category with translation isomorphism with a graded category with translation structure (see Remark 1.1.9).

2.1.2. REMARK. Let $(\mathcal{A}, T, \mathcal{E})$ be an additive category with translation isomorphism T, and collection of triangles $\mathcal{E}$ which satisfies (TR1), (TR2) and (TR3). If (X, Y, Z, a, b, c) is in $\mathcal{E}$, and A is an object of $\mathcal{A}$, then the sequences

$$\ldots \xrightarrow{c[-1]_*} \mathrm{Hom}_{\mathcal{A}}(A, X) \xrightarrow{a_*} \mathrm{Hom}_{\mathcal{A}}(A, Y) \xrightarrow{b_*} \mathrm{Hom}_{\mathcal{A}}(A, Z) \xrightarrow{c_*} \mathrm{Hom}_{\mathcal{A}}(A, X[1]) \xrightarrow{a[1]_*} \ldots$$

and

$$\ldots \xrightarrow{a[1]^*} \mathrm{Hom}_{\mathcal{A}}(X[1], A) \xrightarrow{c^*} \mathrm{Hom}_{\mathcal{A}}(Z, A) \xrightarrow{b^*} \mathrm{Hom}_{\mathcal{A}}(Y, A) \xrightarrow{a^*} \mathrm{Hom}_{\mathcal{A}}(X, A) \xrightarrow{c[-1]^*} \ldots$$

are exact; the proof is an easy exercise. From this, one has the following additional properties:

1. (five-lemma): If (f, g, h) is a morphism of triangles in $\mathcal{E}$, and if two of f, g, h are isomorphisms, then so is the third.
2. If (X, Y, Z, a, b, c) and (X, Y, Z', a, b', c') are two triangles in $\mathcal{E}$, there is an isomorphism $h: Z \to Z'$ such that
$$(\mathrm{id}_X, \mathrm{id}_Y, h): (X, Y, Z, a, b, c) \to (X, Y, Z', a, b', c')$$
is an isomorphism of triangles.
3. Suppose we have three triangles
$$(X, Y, Z', u, i, *),\ (Y, Z, X', v, *, j),\ \text{ and }\ (X, Z, Y', v \circ u, *, *)$$
in $\mathcal{E}$, and morphisms $f: Z' \to Y'$, $g: Y' \to X'$ satisfying the conditions (a) and (b) of (TR4), and with $(Z', Y', X', f, g, i[1] \circ j)$ in $\mathcal{E}$. Then, for each choice of triangles
$$(X, Y, Z'', u, i', *),\ (Y, Z, X'', v, *, j'),\ \text{ and }\ (X, Z, Y'', v \circ u, *, *)$$
in $\mathcal{E}$, there are morphisms $f': Z'' \to Y''$, $g': Y'' \to X''$ satisfying (a) and (b) of (TR4), and with $(Z'', Y'', X'', f', g', i'[1] \circ j')$ in $\mathcal{E}$.

If (TR4) holds as well, then $\mathcal{E}$ is closed under taking finite direct sums.

2.1.3. DEFINITION. Let $(\mathcal{A}, \otimes)$ be a graded tensor category with translation structure, such that the underlying graded category with translation structure is a triangulated category. Suppose that, for each distinguished triangle (X, Y, Z, a, b, c) in $\mathcal{A}$, and each object W of $\mathcal{A}$, the sequence
$$X \otimes W \xrightarrow{a \otimes \mathrm{id}_W} Y \otimes W \xrightarrow{b \otimes \mathrm{id}_W} Z \otimes W \xrightarrow{c \otimes \mathrm{id}_W} X[1] \otimes W = (X \otimes W)[1]$$
is a distinguished triangle in $\mathcal{A}$. Then we call $\mathcal{A}$ a *triangulated tensor category.* The notion of a triangulated tensor category without unit is defined similarly.

2.1.4. REMARK. Let (X, Y, Z, a, b, c) be a distinguished triangle in a triangulated tensor category $\mathcal{A}$, and let W be an object of $\mathcal{A}$. Then the sequence
$$W \otimes X \xrightarrow{\mathrm{id}_W \otimes a} W \otimes Y \xrightarrow{\mathrm{id}_W \otimes b} W \otimes Z \xrightarrow{(\tau_{T,W}^{-1} \otimes \mathrm{id}_X) \circ (\mathrm{id}_W \otimes c)} (W \otimes X)[1]$$
is a distinguished triangle. Indeed, $(\tau_{X,W}, \tau_{Y,W}, \tau_{Z,W})$ gives an isomorphism of the distinguished triangle of Definition 2.1.3 with this triangle.

2.1.5. DEFINITION. Let $\mathcal{A}$ be a DG category, and let $f: E \to F$ be a map in $Z^0\mathbf{C}^b(\mathcal{A})$. The inclusion and projection $i_F: F \to E[1] \oplus F$ and $j_E: E[1] \oplus F \to E[1]$ induce the maps $i_F: F \to \mathrm{cone}(f)$ and $j_E: \mathrm{cone}(f) \to E[1]$ in $Z^0\mathbf{C}^b(\mathcal{A})$. We call a triangle in $\mathbf{C}^b(\mathcal{A})$ of the form
$$E \xrightarrow{f} F \xrightarrow{i_F} \mathrm{cone}(f) \xrightarrow{j_E} E[1]$$
a *cone sequence*; a triangle in $\mathbf{K}^b(\mathcal{A})$ is a distinguished triangle if it is isomorphic to the image of a cone sequence from $\mathbf{C}^b(\mathcal{A})$.

2.1.6. *Triangulated structure.* We now proceed to show that the distinguished triangles of Definition 2.1.5 make $\mathbf{K}^b(\mathcal{A})$ into a triangulated category; we begin with some preliminary results.

Let $f: A \to B$ be in $Z^0\mathrm{Hom}_{\mathbf{C}^b(\mathcal{A})}(A, B)$. We then have the elements
$$\mathrm{id}_B + f \in Z^0\mathrm{Hom}_{\mathbf{C}^b(\mathcal{A})}(B \oplus A, B); \quad (f, \mathrm{id}_A) \in Z^0\mathrm{Hom}_{\mathbf{C}^b(\mathcal{A})}(A, B \oplus A).$$

In addition, the natural maps

$$p_{A[1]}: B[1] \oplus A[1] \to A[1]; \quad (-f[1], \mathrm{id}_{A[1]}): A[1] \to B[1] \oplus A[1]$$
$$i_B: B \to B \oplus A; \quad \mathrm{id}_B - f: B \oplus A \to B$$

define maps

$$p_f: \mathrm{cone}(\mathrm{id}_B + f) \to A[1]; \quad i^f: A[1] \to \mathrm{cone}(\mathrm{id}_B + f)$$
$$i_f: B \to \mathrm{cone}(f, \mathrm{id}_A); \quad p^f: \mathrm{cone}(f, \mathrm{id}_A) \to B \tag{2.1.6.1}$$

in $Z^0\mathbf{C}^b(\mathcal{A})$.

2.1.6.2. LEMMA. *In $\mathbf{K}^b(\mathcal{A})$, the maps (2.1.6.1) are isomorphisms, and $i^f = p_f^{-1}$, $p^f = i_f^{-1}$.*

PROOF. We have $p_f \circ i^f = \mathrm{id}_{A[1]}$ and $p^f \circ i_f = \mathrm{id}_B$ in $\mathbf{C}^b(\mathcal{A})$. The composition

$$B[1] \oplus A[1] \oplus B \xrightarrow{p_B} B \xrightarrow{i_B} B \oplus A \oplus B[-1]$$

gives the element h of $\mathrm{Hom}_{\mathbf{C}^b(\mathcal{A})}(\mathrm{cone}(\mathrm{id}_B + f), \mathrm{cone}(\mathrm{id}_B + f))^{-1}$, with

$$dh = \mathrm{id}_{\mathrm{cone}(\mathrm{id}_B+f)} - i^f \circ p_f.$$

Similarly, the composition

$$A[1] \oplus B \oplus A \xrightarrow{p_A} A \xrightarrow{i_A} A \oplus B[-1] \oplus A[-1]$$

defines an element h' of $\mathrm{Hom}_{\mathbf{C}^b(\mathcal{A})}(\mathrm{cone}(f, \mathrm{id}_A), \mathrm{cone}(f, \mathrm{id}_A))^{-1}$, with

$$dh' = \mathrm{id}_{\mathrm{cone}(f,\mathrm{id}_A)} - i_f \circ p^f.$$

□

2.1.6.3. LEMMA. *Let $\mathcal{A}$ be a DG tensor category, let G be in $\mathbf{C}^b(\mathcal{A})$, and let*

$$E \xrightarrow{f} F \xrightarrow{i_F} \mathrm{cone}(f) \xrightarrow{j_E} E[1]$$

be a cone sequence in $\mathbf{C}^b(\mathcal{A})$. Then

$$E \otimes G \xrightarrow{f \otimes \mathrm{id}_G} F \otimes G \xrightarrow{i_F \otimes \mathrm{id}_G} \mathrm{cone}(f \otimes \mathrm{id}_G) \xrightarrow{j_E \otimes \mathrm{id}_G} E[1] \otimes G = (E \otimes G)[1]$$

is a distinguished triangle in $\mathbf{K}^b(\mathcal{A})$.

PROOF. This follows by a direct computation, using the isomorphisms $(E_i[a] \oplus F_i[b]) \otimes G_j \to (E_i \otimes G_j)[a] \oplus (F_i \otimes G_j)[b]$. □

2.1.6.4. PROPOSITION. *(i) The category $\mathbf{K}^b(\mathcal{A})$ with its set of distinguished triangles is a triangulated category.*
(ii) If $\mathcal{A}$ is a DG tensor category (without unit), then $\mathbf{K}^b(\mathcal{A})$ with its set of distinguished triangles is an triangulated tensor category (without unit).
(iii) If $\mathcal{A}$ is a category of complexes $\mathbf{C}^*(\mathcal{A}_0)$, then the functor Tot *(1.2.9.1) extends to give an equivalence of $\mathbf{K}^b(\mathcal{A})$ with the usual homotopy category $\mathbf{K}^*(\mathcal{A}_0)$; this is an equivalence of triangulated tensor categories without unit if $\mathcal{A}_0$ is a tensor category without unit.*

PROOF. We proceed to verify the axioms (TR1)-(TR4).
(TR1): The first and third conditions are satisfied directly from our construction. For the second, take $B = 0$ in Lemma 2.1.6.2; by that lemma, the map $i_0: 0 \to \mathrm{cone}(\mathrm{id}_A)$ is an isomorphism in $\mathbf{K}^b(\mathcal{A})$. The commutative diagram in $\mathbf{K}^b(\mathcal{A})$

$$\begin{array}{ccccccc}
A & = & A & \xrightarrow{i_A} & \mathrm{cone}(\mathrm{id}_A) & \xrightarrow{j_A} & A[1] \\
\| & & \| & & \uparrow{\scriptstyle i_0} & & \| \\
A & = & A & \longrightarrow & 0 & \longrightarrow & A[1]
\end{array}$$

shows that the lower row is a distinguished triangle.
(TR2): Let $f: A \to B$ be a morphism in $Z^0\mathbf{C}^b(\mathcal{A})$. Let $C = \mathrm{cone}(f)$, giving the cone sequence $A \xrightarrow{f} B \xrightarrow{i_B} C \xrightarrow{j_A} A[1]$. The identity map

$$B[1] \oplus A[1] \oplus B \to B[1] \oplus A[1] \oplus B$$

defines the isomorphism $\mathrm{cone}(i_B) \cong \mathrm{cone}(\mathrm{id}_B + f)$ in $\mathbf{C}^b(\mathcal{A})$. By Lemma 2.1.6.2, we have the isomorphism of triangles in $\mathbf{K}^b(\mathcal{A})$

$$\begin{array}{ccccccc}
B & \xrightarrow{i_B} & C & \xrightarrow{i_C} & \mathrm{cone}(i_B) & \xrightarrow{j_B} & B[1] \\
\| & & \| & & \uparrow{\scriptstyle i^f} & & \| \\
B & \xrightarrow[i_B]{} & C & \xrightarrow[j_A]{} & A[1] & \xrightarrow[-f[1]]{} & B[1].
\end{array}$$

Thus the bottom row is a distinguished triangle. Similarly, the identity map

$$A[1] \oplus B \oplus A \to A[1] \oplus B \oplus A$$

defines the isomorphism $\mathrm{cone}(j_A[-1]) \cong \mathrm{cone}\big((f, \mathrm{id}_A)\big)$ in $\mathbf{C}^b(\mathcal{A})$. The diagram

$$\begin{array}{ccccccc}
C[-1] & \xrightarrow{j_A[-1]} & A & \xrightarrow{i_A} & \mathrm{cone}(j_A[-1]) & \xrightarrow{j_{C[-1]}} & C \\
\| & & \downarrow{\scriptstyle -\mathrm{id}_A} & & \downarrow{\scriptstyle p^f} & & \| \\
C[-1] & \xrightarrow[-j_A[-1]]{} & A & \xrightarrow[f]{} & B & \xrightarrow[i_B]{} & C
\end{array}$$

gives by Lemma 2.1.6.2 an isomorphism of triangles in $\mathbf{K}^b(\mathcal{A})$, completing the verification of (TR2).
(TR3): We may replace the two triangles with isomorphic triangles; thus we may assume that the two triangles arise from cone sequences

$$(A, B, \mathrm{cone}(a), a, i_B, j_A), \quad (A', B', \mathrm{cone}(a'), a', i_{B'}, j_{A'})$$

in $\mathbf{C}^b(\mathcal{A})$. The morphism $(f, g): u \to u'$ arises then from a diagram in $Z^0\mathbf{C}^b(\mathcal{A})$

$$\begin{array}{ccc}
A & \xrightarrow{a} & B \\
\downarrow{\scriptstyle f} & & \downarrow{\scriptstyle g} \\
A' & \xrightarrow[a']{} & B',
\end{array}$$

and there is a map $H: A \to B'$ in $\mathbf{C}^b(\mathcal{A})$, of degree -1, with $dH = ga - a'f$. Let $H_+: A[1] \to B$ be the degree 0 map corresponding to H via the identity

$$\mathrm{Hom}_{\mathbf{C}^b(\mathcal{A})}(A[1], B)^0 = \mathrm{Hom}_{\mathbf{C}^b(\mathcal{A})}(A, B)[-1]^0 = \mathrm{Hom}_{\mathbf{C}^b(\mathcal{A})}(A, B)^{-1}.$$

The degree 0 map from $A[1]\oplus B$ to $A'[1]\oplus B'$ with matrix representation

$$\begin{pmatrix} f[1] & 0 \\ H_+ & g \end{pmatrix}$$

gives rise to the map $h\colon \mathrm{cone}(a)\to\mathrm{cone}(a')$ in $Z^0\mathbf{C}^b(\mathcal{A})$, giving the desired map of triangles (f,g,h) in $\mathbf{K}^b(\mathcal{A})$.
(TR4): Let

$$(X,Y,Z',u,i,j'),\ (Y,Z,X',v,i',j),\ \text{ and }\ (X,Z,Y',w,k,k')$$

be distinguished triangles with $w=v\circ u$. Lift u and v to maps $a\colon X\to Y$ and $b\colon Y\to Z$ in $Z^0\mathbf{C}^b(\mathcal{A})$, giving the lifting $c:=b\circ a$ for w. By Remark 2.1.2(2), we have an isomorphism of triangles in $\mathbf{K}^b(\mathcal{A})$

$$\begin{array}{ccccccc} X & \xrightarrow{u} & Y & \xrightarrow{i} & Z' & \xrightarrow{i'} & X[1] \\ \| & & \| & & \downarrow{\scriptstyle h} & & \| \\ X & \xrightarrow[a]{} & Y & \xrightarrow[i_Y]{} & \mathrm{cone}(a) & \xrightarrow[j_X]{} & X[1]. \end{array}$$

Doing the same for v and w, and applying Remark 2.1.2(3), we may assume that our three distinguished triangles are equal to the images in $\mathbf{K}^b(\mathcal{A})$ of the cone sequences for the maps a, b and c:

$$X\xrightarrow{a}Y\xrightarrow{i_Y}\mathrm{cone}(a)\xrightarrow{j'_X}X[1],$$
$$Y\xrightarrow{b}Z\xrightarrow{i'_Z}\mathrm{cone}(b)\xrightarrow{j_Y}Y[1],$$
$$X\xrightarrow{c}Z\xrightarrow{i_Z}\mathrm{cone}(c)\xrightarrow{j_X}X[1].$$

We have the degree 0 maps

$$f:=\mathrm{id}_{X[1]}\oplus b\colon \mathrm{cone}(a)\to\mathrm{cone}(c),$$
$$g:=a[1]\oplus\mathrm{id}_Z:\mathrm{cone}(c)\to\mathrm{cone}(b).$$

One easily computes that $df=0$, $dg=0$ and that (id_X,b,f) and (a,id_Z,g) are morphisms of triangles.

The map $a[1]\oplus c\colon X[1]\oplus X\to Y[1]\oplus Z$ determines a morphism $u\colon\mathrm{cone}(\mathrm{id}_X)\to\mathrm{cone}(b)$ in $Z^0\mathbf{C}^b(\mathcal{A})$. We have the commutative diagram in $Z^0\mathbf{C}^b(\mathcal{A})$

$$\begin{array}{ccccccc} \mathrm{cone}(\mathrm{id}_X) & \xrightarrow{u} & \mathrm{cone}(b) & \xrightarrow{i_{\mathrm{cone}(b)}} & \mathrm{cone}(u) & \xrightarrow{j_{\mathrm{cone}(\mathrm{id}_X)}} & \mathrm{cone}(\mathrm{id}_X)[1] \\ \uparrow & & \| & & \uparrow{\scriptstyle i_{\mathrm{cone}(b)}} & & \uparrow \\ 0 & \longrightarrow & \mathrm{cone}(b) & = & \mathrm{cone}(b) & \longrightarrow & 0. \end{array}$$

We have already seen that $\mathrm{cone}(\mathrm{id}_X)$ is isomorphic to 0 in $\mathbf{K}^b(\mathcal{A})$; it therefore follows from Remark 2.1.2(1) that the map $i_{\mathrm{cone}(b)}:\mathrm{cone}(b)\to\mathrm{cone}(u)$ is an isomorphism in $\mathbf{K}^b(\mathcal{A})$. The evident isomorphism

$$X[2]\oplus X[1]\oplus Y[1]\oplus Z\to X[2]\oplus Y[1]\oplus X[1]\oplus Z$$

defines the isomorphism $\mathrm{cone}(u)\cong\mathrm{cone}(f)$ in $\mathbf{C}^b(\mathcal{A})$; putting these two isomorphisms together shows that

$$\mathrm{cone}(a)\xrightarrow{f}\mathrm{cone}(c)\xrightarrow{g}\mathrm{cone}(b)\xrightarrow{i_Y[1]\circ j_Y}\mathrm{cone}(a)[1]$$

is isomorphic in $\mathbf{K}^b(\mathcal{A})$ to the cone sequence for the map f, completing the proof of (TR4).

The compatibility with the tensor structure follows from Lemma 2.1.6.3.

If $\mathcal{A}$ is a category of complexes $\mathbf{C}^*(\mathcal{A}_0)$, the functor Tot is compatible with the cone sequences and the tensor structures. Together with Lemma 1.2.10, this completes the proof. □

2.1.7. PROPOSITION. *(i) The assignments*

$$\mathcal{A} \mapsto \mathbf{C}^b(\mathcal{A}) := \mathbf{C}^b(\mathcal{A}[*]);$$
$$\mathcal{A} \mapsto \mathbf{K}^b(\mathcal{A}) := \mathbf{K}^b(\mathcal{A}[*])$$

extend to functors $\mathbf{C}^b(-)$ and $\mathbf{K}^b(-)$ from the category of DG categories to the category of DG categories with translation structure, resp. the category of triangulated categories.
(ii) The functors $\mathbf{C}^b(-)$ and $\mathbf{K}^b(-)$ extend to tensor functors on the above categories with tensor structure.
(iii) Let $\mathcal{B}$ be a DG (tensor) category Given a DG (tensor) functor $F: \mathcal{A} \to \mathbf{C}^b(\mathcal{B})$, there is a canonical extension of F to an exact (tensor) functor $\mathbf{K}^b(F): \mathbf{K}^b(\mathcal{A}) \to \mathbf{K}^b(\mathcal{B})$.

PROOF. From Kapranov's explicit construction of the functor Pre-Tr, it is clear that the assignment $\mathcal{A} \mapsto \text{Pre-Tr}(\mathcal{A}[*])$ extends to a functor from DG categories to DG categories with translation structure, compatible with the embedding $i_{\mathcal{A}}: \mathcal{A} \to \text{Pre-Tr}(\mathcal{A}[*])$. Since $\mathbf{C}^b(\mathcal{A})$ is the strictly full subcategory of Pre-Tr$(\mathcal{A}[*])$ generated by $i(\mathcal{A})$ and taking cones and translations, we have the desired functoriality of $\mathbf{C}^b(\mathcal{A})$, together with its cone functor. Since the triangulated structure on the homotopy category $\mathbf{K}^b(\mathcal{A})$ is determined by the cone functor on $\mathbf{C}^b(\mathcal{A})$, this gives the functoriality of $\mathbf{K}^b(\mathcal{A})$. The compatibility of these functors with the tensor structure follows directly from the functoriality of the tensor structure on Pre-Tr$(\mathcal{A})$, and the compatibility of the cone functor with the tensor product.

For (iii), we use the universality of the operation of taking the free translation structure on a DG category, together with the canonical equivalence $\mathbf{C}^b(\mathcal{B}) \to \mathbf{C}^b(\mathcal{B})[*]$ (see §1.1.8), and the equivalence

$$\text{Tot}: \mathbf{C}^b(\mathbf{C}^b(\mathcal{B})) \to \mathbf{C}^b(\mathcal{B})$$

(see Lemma 1.2.10), to give the canonical functor of DG categories with translation structure $\mathbf{C}^b(F): \mathbf{C}^b(\mathcal{A}) \to \mathbf{C}^b(\mathcal{B})$, compatible with the respective cone functors. Taking the associated map on the homotopy categories gives the exact functor $\mathbf{K}^b(F): \mathbf{K}^b(\mathcal{A}) \to \mathbf{K}^b(\mathcal{B})$. The same argument shows that $\mathbf{K}^b(F)$ is an exact tensor functor in case F is a DG tensor functor. □

2.2. Homotopy equivalence of DG categories

We define the notion of a homotopy equivalence of differential graded categories, and show that a homotopy equivalence of DG categories induces an equivalence on the homotopy categories of complexes.

2.2.1. DEFINITION. A functor of DG categories $F: \mathcal{A} \to \mathcal{B}$ is called a *homotopy equivalence* if

(i) F gives an isomorphism $\text{Obj}(\mathcal{A})/\text{Iso} \to \text{Obj}(\mathcal{B})/\text{Iso}$.

(ii) The map

$$F(X,Y)\colon \mathrm{Hom}_{\mathcal{A}}(X,Y) \to \mathrm{Hom}_{\mathcal{B}}(F(X),F(Y))$$

is an quasi-isomorphism for all X, Y in $\mathcal{A}$.

2.2.2. THEOREM. *Let $F\colon \mathcal{A} \to \mathcal{B}$ be a functor of DG categories.*
(i) If F is a homotopy equivalence, then the induced map $\mathbf{K}^b(F)\colon \mathbf{K}^b(\mathcal{A}) \to \mathbf{K}^b(\mathcal{B})$ is an equivalence of triangulated categories.
(ii) Suppose that $\mathcal{A}$ and $\mathcal{B}$ are DG tensor categories, and F is a DG tensor functor and a homotopy equivalence. Let $\mathbf{K}^b_F(\mathcal{B})$ be the full subcategory of $\mathbf{K}^b(\mathcal{B})$ with objects of the form $\mathbf{K}^b(F)(X)$, with X an object of $\mathbf{K}^b(\mathcal{A})$. Then $\mathbf{K}^b_F(\mathcal{B})$ is a full triangulated tensor subcategory of $\mathbf{K}^b(\mathcal{B})$, the inclusion $\mathbf{K}^b_F(\mathcal{B}) \to \mathbf{K}^b(\mathcal{B})$ is an equivalence of triangulated categories, and the functor $\mathbf{K}^b(F)\colon \mathbf{K}^b(\mathcal{A}) \to \mathbf{K}^b_F(\mathcal{B})$ induced by $\mathbf{K}^b(F)$ is a pseudo-tensor equivalence of triangulated tensor categories (see Part II Chapter I, §1.3.7).

PROOF. To prove (i), we may replace $\mathcal{A}$ and $\mathcal{B}$ with equivalent categories, so we may assume that $\mathrm{Obj}(\mathcal{A}) = \mathrm{Obj}(\mathcal{B})$ and F is the identity on objects.

Let $G = \mathbf{K}^b(F)$. We first show that the map

$$G\colon \mathrm{Hom}_{\mathbf{K}^b(\mathcal{A})}(E,F) \to \mathrm{Hom}_{\mathbf{K}^b(\mathcal{B})}(G(E),G(F)) \tag{2.2.2.1}$$

is an isomorphism for each E, F in $\mathbf{K}^b(\mathcal{A})$. If E and F are translations of objects of $\mathcal{A}$, the Hom groups are isomorphic by the definition of a homotopy equivalence of DG categories. Since $\mathbf{K}^b(\mathcal{A})$ is generated by taking repeated cones and translations in $\mathbf{C}^b(\mathcal{A})$, starting with objects of $\mathcal{A}$, the long exact sequence of Hom's coming from a distinguished triangle shows that (2.2.2.1) is an isomorphism for all E and F, as desired.

Since the map F is an isomorphism on objects, the isomorphisms (2.2.2.1) and the fact that $\mathbf{K}^b(\mathcal{B})$ is generated from $\mathcal{B}$ by taking repeated cones and translations in $\mathbf{C}^b(\mathcal{B})$ imply that each object of $\mathbf{K}^b(\mathcal{B})$ is isomorphic to an object in $G(\mathbf{K}^b(\mathcal{A}))$. Choose for each E' in $\mathbf{K}^b(\mathcal{B})$ an object $H(E')$ of $\mathbf{K}^b(\mathcal{A})$, together with an isomorphism $\Phi(E')$ of $G(H(E'))$ with E'. Using the isomorphism (2.2.2.1), H has a unique extension to a functor such that Φ defines a natural isomorphism of $G \circ H$ with the identity on $\mathbf{K}^b(\mathcal{B})$. In addition, for each E in $\mathbf{K}^b(\mathcal{A})$, the isomorphism $\Phi(G(E))\colon G(H(G(E))) \to G(E)$ determines by (2.2.2.1) an isomorphism $\Psi(E)\colon H(G(E)) \to E$; clearly Ψ determines a natural isomorphism of $H \circ G$ with the identity on $\mathbf{K}^b(\mathcal{A})$, completing the proof of (i).

For (ii), it follows from the isomorphism (2.2.2.1) as above that the full image $\mathbf{K}^b_F(\mathcal{B})$ is a triangulated tensor subcategory of $\mathbf{K}^b(\mathcal{B})$, equivalent to $\mathbf{K}^b(\mathcal{A})$; it thus follows from (i) that the inclusion $\mathbf{K}^b_F(\mathcal{B}) \to \mathbf{K}^b(\mathcal{B})$ is an equivalence of triangulated categories. For each E' in $\mathbf{K}^b_F(\mathcal{B})$, we may choose an $H(E')$ in $\mathbf{K}^b(\mathcal{A})$ lifting E', and we may take $H(1_{\mathcal{B}}^{\otimes n}) := 1_{\mathcal{A}}^{\otimes n}$.

As in the proof of (i), the assignment $E' \mapsto H(E')$ extends uniquely to an exact functor

$$H\colon \mathbf{K}^b_F(\mathcal{B}) \to \mathbf{K}^b(\mathcal{A}),$$

with $G \circ H = \mathrm{id}_{\mathbf{K}^b_F(\mathcal{B})}$. Thus $G \circ H \circ G = G$. By the isomorphism (2.2.2.1), it follows that, for each X in $\mathbf{K}^b(\mathcal{A})$ there is a *unique* morphism $\rho(X)\colon X \to H(G(X))$ with $G(\rho(X)) = \mathrm{id}_{G(X)}$; $\rho(X)$ is automatically a natural isomorphism. Similarly, for X

and Y in $\mathbf{K}^b(\mathcal{A})$, there is a unique morphism

$$\theta(X,Y): H(X) \otimes H(Y) \to H(X \otimes Y)$$

with $G(\theta(X,Y)) = \mathrm{id}_{X\otimes Y}$, and $\theta(X,Y)$ is an isomorphism The uniqueness of $\theta(X,Y)$ implies that (H,θ) defines a pseudo-tensor functor from $\mathbf{K}^b_F(\mathcal{B})$ to $\mathbf{K}^b(\mathcal{A})$, and the uniqueness of $\rho(X)$ implies that ρ defines a natural isomorphism of pseudo-tensor functors

$$\rho: \mathrm{id}_{\mathbf{K}^b(\mathcal{A})} \to H \circ G$$

(*cf.* Part II Chapter I, §1.3.7). Thus (H,θ,ρ) defines an exact pseudo-tensor inverse equivalence to G, proving (ii). □

2.2.3. *Adjoining morphisms to a DG tensor category.* Let

$$F: \mathcal{B} \to \mathcal{B}_0$$

be a DG tensor functor of DG tensor categories without unit. Let $\mathbb{E}$ be a DG tensor category without unit such that the underlying graded tensor category without unit satisfies the conditions of (I.2.5.1.1).

Let $\mathcal{B}[\mathbb{E}]$ be the coproduct of $\mathcal{B}$ and $\mathbb{E}$ as DG tensor categories without unit. As the tensor product of two DG modules M and N has underlying graded $\mathbb{Z}$-module equal to the tensor product of the underlying graded $\mathbb{Z}$-modules of M and N, it follows that the underlying graded tensor category without unit of $\mathcal{B}[\mathbb{E}]$ is the coproduct of $\mathcal{B}$ and $\mathbb{E}$ as graded tensor categories without unit. The functor F induces the functor

$$F[\mathrm{id}_{\mathbb{E}}]: \mathcal{B}[\mathbb{E}] \to \mathcal{B}_0[\mathbb{E}]$$

by taking the coproduct of F with the identity on $\mathbb{E}$.

Let I be an ordered set, with function $\epsilon: I \to \mathrm{N}$, and let $\mathcal{A}$ be the graded tensor category without unit $\mathcal{B}[\mathbb{E}, \{s_i\}]$ gotten from $\mathcal{B}[\mathbb{E}]$ by adjoining morphisms

$$s_i: \mathfrak{e}^{\otimes\epsilon(i)} \to X_i$$

of degree d_i, as in Chapter I, §2.5.1. For each $i \in I$, let $\mathcal{A}_{<i}$ be the tensor category without unit gotten by adjoining the morphisms s_j with $j < i$.

We make $\mathcal{A}$ into a DG tensor category without unit inductively, by choosing morphisms $f_i: \mathfrak{e}^{\otimes\epsilon(i)} \to X_i$ of degree $d_i + 1$ in the category $\mathcal{A}_{<i}$. Assume we have defined the structure of a DG tensor category on the graded tensor category $\mathcal{A}_{<i}$, and that $df_i = 0$. We then define $ds_i = f_i$, and continue.

We let $\mathcal{A}_0$ be the graded tensor category without unit gotten from $\mathcal{B}_0[\mathbb{E}]$ by adjoining morphisms $t_i: \mathfrak{e}^{\otimes\epsilon(i)} \to F(X_i)$ of degree d_i. We extend $F[\mathrm{id}_{\mathbb{E}}]$ to a graded tensor functor by setting $F(s_i) = t_i$. We make $\mathcal{A}_0$ into a DG tensor category without unit by setting $dt_i = F(f_i)$; this makes the extension of F into a DG tensor functor

$$F^s: \mathcal{A} \to \mathcal{A}_0.$$

For each a, the Hom-groups $\mathrm{Hom}_{\mathbb{E}}(\mathfrak{e}^{\otimes a}, \mathfrak{e}^{\otimes a})^q$ are left $\mathbb{Z}[S_a]$-modules, with $\sigma \in S_a$ acting via left composition with the symmetry isomorphism τ_σ.

2.2.4. PROPOSITION. *Suppose F is a homotopy equivalence, and that the Hom group $\mathrm{Hom}_{\mathbb{E}}(\mathfrak{e}^{\otimes a}, \mathfrak{e}^{\otimes a})^q$ is a free $\mathbb{Z}[S_a]$-module, or is zero, for each a and q. Then F^s is a homotopy equivalence.*

PROOF. Clearly F^s is an isomorphism on isomorphism classes of objects; we may assume that $\mathcal{A}$ and $\mathcal{A}_0$ have the same objects, and that F^s is the identity on objects. Let Y and Z be objects of $\mathcal{B}$.

Denote the complex $\mathrm{Hom}_{\mathbb{E}}(\mathfrak{e}^{\otimes a}, \mathfrak{e}^{\otimes a})$ by $\mathbb{E}_a$. We use the notation from Chapter I, §2.5. From Chapter I, Proposition 2.5.2, we have the isomorphism (as graded groups)

$$\bigoplus_{s=0}^{\infty} \bigoplus_{\substack{i_* \in I^s_{\leq} \\ \sum i_* = a-b}} \mathrm{Hom}_{\mathcal{B}}(X^{\otimes i_*} \otimes Y, Z)[-d(i_*)] \otimes_{\mathbb{Z}[S(i_*)]} \mathbb{E}_a \xrightarrow{\Psi(\mathfrak{e}^{\otimes a} \otimes Y, \mathfrak{e}^{\otimes b} \otimes Z)} \mathrm{Hom}_{\mathcal{A}}(\mathfrak{e}^{\otimes a} \otimes Y, \mathfrak{e}^{\otimes b} \otimes Z),$$

and the isomorphism (as graded groups)

$$\bigoplus_{s=0}^{\infty} \bigoplus_{\substack{i_* \in I^s_{\leq} \\ \sum i_* = a-b}} \mathrm{Hom}_{\mathcal{B}_0}(X^{\otimes i_*} \otimes Y, Z)[-d(i_*)] \otimes_{\mathbb{Z}[S(i_*)]} \mathbb{E}_a \xrightarrow{\Psi_0(\mathfrak{e}^{\otimes a} \otimes Y, \mathfrak{e}^{\otimes b} \otimes Z)} \mathrm{Hom}_{\mathcal{A}_0}(\mathfrak{e}^{\otimes a} \otimes Y, \mathfrak{e}^{\otimes b} \otimes Z).$$

Here $S(i_*)$ is the group of order-preserving bijections of i_*.

Now define an increasing filtration on the complexes $\mathrm{Hom}_{\mathcal{A}}(\mathfrak{e}^{\otimes a} \otimes Y, \mathfrak{e}^{\otimes b} \otimes Z)$ by ordering the summands $\mathrm{Hom}_{\mathcal{B}}(X^{\otimes i_*} \otimes Y, Z)[-d(i_*)] \otimes_{\mathbb{Z}[S(i_*)]} \mathbb{E}_a$ lexicographically with respect to the indices i_*. Then the E_1-complexes in the resulting spectral sequence are given by

$$E^{i_*}_{1,\mathcal{A}} = \mathrm{Hom}_{\mathcal{B}}(X^{\otimes i_*} \otimes Y, Z)^{[-d(i_*)]} \otimes_{\mathbb{Z}[S(i_*)]} \mathbb{E}_a.$$

We have the similarly defined filtration on $\mathrm{Hom}_{\mathcal{A}_0}(\mathfrak{e}^{\otimes a} \otimes Y, \mathfrak{e}^{\otimes b} \otimes Z)$; clearly the functor F^s respects these filtrations, and defines a map of the corresponding spectral sequences:

$$E^{i_*}_{*}(F^s) \colon E^{i_*}_{*,\mathcal{A}} \to E^{i_*}_{*,\mathcal{A}_0}.$$

The map $E^{i_*}_1(F^s)$ on the E_1 complexes $E_1(F^s)$ is given by

$$\mathrm{Hom}_{\mathcal{B}}(X^{\otimes i_*} \otimes Y, Z)^{[-d(i_*)]} \otimes_{\mathbb{Z}[S(i_*)]} \mathbb{E}_a \xrightarrow{F(X^{\otimes i_*} \otimes Y, Z) \otimes \mathrm{id}_{\mathbb{E}_a}} \mathrm{Hom}_{\mathcal{B}_0}(X^{\otimes i_*} \otimes Y, Z)^{[-d(i_*)]} \otimes_{\mathbb{Z}[S(i_*)]} \mathbb{E}_a.$$

As $\mathbb{E}_a$ is a complex of free $\mathbb{Z}[S(i_*)]$-modules, and F is a homotopy equivalence, each map $F(X^{\otimes i_*} \otimes Y, Z) \otimes \mathrm{id}_{\mathbb{E}_a}$ is a quasi-isomorphism. Thus, F^s induces a quasi-isomorphism on the E_1-complexes. Since each morphism in $\mathcal{A}$ (resp. $\mathcal{A}_0$) involves only finitely many of the morphisms s_i (resp.t_i), it follows that F^s gives a quasi-isomorphism

$$F^s(\mathfrak{e}^{\otimes a} \otimes Y, \mathfrak{e}^{\otimes b} \otimes Z) \colon \mathrm{Hom}_{\mathcal{A}}(\mathfrak{e}^{\otimes a} \otimes Y, \mathfrak{e}^{\otimes b} \otimes Z) \to \mathrm{Hom}_{\mathcal{A}_0}(\mathfrak{e}^{\otimes a} \otimes Y, \mathfrak{e}^{\otimes b} \otimes Z).$$

The other cases $F^s(1^{\otimes a} \otimes Y, 1^{\otimes b})$, $F^s(1^{\otimes a}, 1^{\otimes b} \otimes Z)$, etc., are handled similarly. □

2.2.5. COROLLARY. *Let $F \colon \mathcal{B} \to \mathcal{B}_0$ be a DG tensor functor of DG tensor categories without unit. Let $F^s \colon \mathcal{A} \to \mathcal{A}_0$ be the extension of $F[\mathrm{id}_{\mathbb{E}}]$ to the DG tensor functor* *of DG tensor categories without unit:*

$$\mathcal{A} = \mathcal{B}[\mathbb{E}, \{s_i : \mathbf{e}^{\epsilon(i)} \to X_i \mid i \in I, \deg(s_i) = d_i, ds_i = f_i\}],$$
$$\mathcal{A}_0 = \mathcal{B}_0[\mathbb{E}, \{t_i : \mathbf{e}^{\epsilon(i)} \to X_i \mid i \in I, \deg(t_i) = d_i, dt_i = g_i\}],$$
$$F^s(s_i) = t_i, \ F^s(f_i) = g_i.$$

Suppose F is a homotopy equivalence and a surjection on objects. Then the induced functor $\mathbf{K}^b(F^s) : \mathbf{K}^b(\mathcal{A}) \to \mathbf{K}^b(\mathcal{A}_0)$ is an equivalence of triangulated tensor categories.

PROOF. This follows from Theorem 2.2.2 and Proposition 2.2.4. □

2.3. Localization

In this section, we recall the construction of Verdier [**123**] of the localization of a triangulated category, and we show how to extend localization to triangulated tensor categories.

2.3.1. *Thick subcategories and multiplicative systems.*

2.3.1.1. DEFINITION. Let $\mathcal{B}$ be a full triangulated subcategory of a triangulated category $\mathcal{A}$. $\mathcal{B}$ is called *thick* (épaisse) if the following condition is satisfied:

> Let $X \xrightarrow{f} Y \to Z \to X[1]$ be a distinguished triangle in $\mathcal{A}$, with Z in $\mathcal{B}$. If f factors as $X \xrightarrow{f_1} B' \xrightarrow{f_2} Y$ with B' in $\mathcal{B}$, then X and Y are in $\mathcal{B}$.

2.3.1.2. DEFINITION. Let $\mathcal{A}$ be a triangulated category. A set of morphisms $\mathcal{S}$ in $\mathcal{A}$ is called a *multiplicative system of morphisms* if the following properties hold:

(FR1) If $f, g \in \mathcal{S}$ and if f and g are composable, then $f \circ g \in \mathcal{S}$. For all X in $\mathcal{A}$, id_X is in $\mathcal{S}$.

(FR2) In $\mathcal{A}$, each diagram

$$\begin{array}{ccc} & & Y \\ & & \downarrow{\scriptstyle s\in\mathcal{S}} \\ Z & \xrightarrow[f]{} & X \end{array}$$

can be extended to a commutative diagram

$$\begin{array}{ccc} P & \xrightarrow{g} & Y \\ {\scriptstyle t\in\mathcal{S}}\downarrow & & \downarrow{\scriptstyle s\in\mathcal{S}} \\ Z & \xrightarrow[f]{} & X. \end{array}$$

The symmetrically defined property holds as well.

(FR3) For morphisms f and g in $\mathcal{A}$, the following conditions are equivalent:
   (a) There is an $s \in \mathcal{S}$ with $s \circ f = s \circ g$.
   (b) There is a $t \in \mathcal{S}$ with $f \circ t = g \circ t$.

(FR4) If s is in $\mathcal{S}$, then $s[1]$ is in $\mathcal{S}$.

(FR5) Let (X, Y, Z, u, v, w) and (X', Y', Z', u', v', w') be distinguished triangles in $\mathcal{A}$, and let

$$\begin{array}{ccc} X & \xrightarrow{u} & Y \\ {\scriptstyle f}\downarrow & & \downarrow{\scriptstyle g} \\ X' & \xrightarrow[u']{} & Y' \end{array}$$

be a commutative diagram, with f and g in $\mathcal{S}$. Then there is an $h \in \mathcal{S}$ such that $(f, g, h): (X, Y, Z, u, v, w) \to (X', Y', Z', u', v', w')$ is a morphism of triangles.

A multiplicative system of morphisms is called *saturated* if

> A morphism f is in $\mathcal{S}$ if and only if there are morphisms g and g' such that $g \circ f$ and $f \circ g'$ are in $\mathcal{S}$.

2.3.2. If $\mathcal{B}$ is a thick subcategory of $\mathcal{A}$, the set of morphisms $s: X \to Y$ in $\mathcal{A}$ which fit into a distinguished triangle $X \xrightarrow{s} Y \to Z \to X[1]$ with Z in $\mathcal{B}$ forms a saturated multiplicative system of morphisms. Conversely, if $\mathcal{S}$ is a saturated multiplicative system of morphisms in $\mathcal{A}$, the full subcategory $\mathcal{B}$ of $\mathcal{A}$ consisting of objects Z which fit into into a distinguished triangle $X \xrightarrow{s} Y \to Z \to X[1]$ with s in $\mathcal{S}$ forms a thick subcategory of $\mathcal{A}$. This gives a 1-1 correspondence between the collection of thick subcategories of $\mathcal{A}$ and the collection of saturated multiplicative systems of morphisms in $\mathcal{A}$.

The intersection of thick subcategories of $\mathcal{A}$ is a thick subcategory of $\mathcal{A}$, and similarly for the intersection of saturated multiplicative systems of morphisms. Thus, for each set $\mathcal{T}$ of objects of $\mathcal{A}$, there is a smallest thick subcategory $\mathcal{B}$ containing $\mathcal{T}$, called the thick subcategory *generated* by $\mathcal{T}$.

2.3.3. *Localization of triangulated categories.* Let $\mathcal{S}$ be a saturated multiplicative system in a triangulated category $\mathcal{A}$. For each X in $\mathcal{A}$, we have the category $\mathcal{S}_X$ of morphisms s in $\mathcal{S}$ with range $\operatorname{rng}(s)$ equal to X, and the category $\mathcal{S}^X$ of morphisms s in $\mathcal{S}$ with domain $\operatorname{dom}(s)$ equal to X. Form the category $\mathcal{A}[\mathcal{S}^{-1}]$ with the same objects as $\mathcal{A}$, with

$$\operatorname{Hom}_{\mathcal{A}[\mathcal{S}^{-1}]}(X, Y) = \varinjlim_{s \in \mathcal{S}_X^{\mathrm{op}}} \operatorname{Hom}_{\mathcal{A}}(\operatorname{dom}(s), Y).$$

Composition of diagrams

$$\begin{array}{ccccc} & & Y' & \xrightarrow{g} & Z \\ & & \downarrow{\scriptstyle t} & & \\ X' & \xrightarrow[f]{} & Y & & \\ {\scriptstyle s}\downarrow & & & & \\ X & & & & \end{array}$$

is defined by filling in the middle via (FR2):

$$\begin{array}{ccccc} X'' & \xrightarrow{f'} & Y' & \xrightarrow{g} & Z \\ {\scriptstyle s'}\downarrow & & \downarrow{\scriptstyle t} & & \\ X' & \xrightarrow[f]{} & Y & & \\ {\scriptstyle s}\downarrow & & & & \\ X. & & & & \end{array}$$

One can describe $\mathrm{Hom}_{\mathcal{A}[\mathcal{S}^{-1}]}(X, Y)$ by a calculus of left fractions as well, i.e.,

$$\mathrm{Hom}_{\mathcal{A}[\mathcal{S}^{-1}]}(X, Y) = \varinjlim_{s \in \mathcal{S}^Y} \mathrm{Hom}_{\mathcal{A}}(X, \mathrm{rng}(s)).$$

If $\mathcal{B}$ is a thick subcategory of $\mathcal{A}$, define $\mathcal{A}/\mathcal{B}$ to be $\mathcal{A}[\mathcal{S}^{-1}]$, where $\mathcal{S}$ is the saturated multiplicative system of morphisms corresponding to $\mathcal{B}$. It is easy to see that the translation structure on $\mathcal{A}$ induces one on $\mathcal{A}[\mathcal{S}^{-1}]$. Let $Q_{\mathcal{S}} : \mathcal{A} \to \mathcal{A}[\mathcal{S}^{-1}]$ and $Q_{\mathcal{B}} : \mathcal{A} \to \mathcal{A}/\mathcal{B}$ be the canonical functors. The main theorem of this paragraph is

THEOREM [Verdier [**123**]]. *(i) $\mathcal{A}[\mathcal{S}^{-1}]$ is a triangulated category, where a triangle T in $\mathcal{A}[\mathcal{S}^{-1}]$ is distinguished if T is isomorphic to the image under $Q_{\mathcal{S}}$ of a distinguished triangle in $\mathcal{A}$.*
(ii) The functor $Q_{\mathcal{S}}$ is universal for exact functors $F : \mathcal{A} \to \mathcal{C}$ such that $F(s)$ is an isomorphism for all $s \in \mathcal{S}$, and the functor $Q_{\mathcal{B}}$ is universal for exact functors $F : \mathcal{A} \to \mathcal{C}$ such that $F(B)$ is isomorphic to 0 for all B in $\mathcal{B}$.
(iii) $\mathcal{S}$ is equal to the collection of maps in $\mathcal{A}$ which become isomorphisms in $\mathcal{A}[\mathcal{S}^{-1}]$ and $\mathcal{B}$ is the subcategory of objects of $\mathcal{A}$ which becomes isomorphic to zero in $\mathcal{A}/\mathcal{B}$.

2.3.4. *Localization of triangulated tensor categories.* If $\mathcal{A}$ is a triangulated tensor category, and $\mathcal{B}$ a thick subcategory, call $\mathcal{B}$ a *thick tensor subcategory* if A in $\mathcal{A}$ and B in $\mathcal{B}$ implies that $A \otimes B$ and $B \otimes A$ are in $\mathcal{B}$. If $\mathcal{S}$ is a saturated multiplicative system, call $\mathcal{S}$ a saturated *tensor* multiplicative system if s in $\mathcal{S}$ and A in $\mathcal{A}$ implies that $\mathrm{id}_A \otimes s$ and $s \otimes \mathrm{id}_A$ are in $\mathcal{S}$. One easily sees that the correspondence of §2.3.3 between saturated multiplicative systems and thick subcategories restricts to a correspondence between saturated tensor multiplicative systems and thick tensor subcategories.

Let $\mathcal{S}$ be a saturated tensor multiplicative system in a tensor category $\mathcal{A}$. It follows immediately that the tensor operation

$$\otimes_{\mathcal{A}} : \mathrm{Hom}_{\mathcal{A}}(X, Y) \otimes_{\mathbb{Z}} \mathrm{Hom}_{\mathcal{A}}(Z, W) \to \mathrm{Hom}_{\mathcal{A}}(X \otimes Y, Z \otimes W)$$

passes to the inductive limit defining the Hom-groups in $\mathcal{A}[\mathcal{S}^{-1}]$, giving $\mathcal{A}[\mathcal{S}^{-1}]$ the structure of a tensor category. Similarly, the condition that the collection of distinguished triangles in $\mathcal{A}$ is closed under right tensor product with objects of $\mathcal{A}$ passes to $\mathcal{A}[\mathcal{S}^{-1}]$, making $\mathcal{A}[\mathcal{S}^{-1}]$ a triangulated tensor category, with exact tensor functor $Q_{\mathcal{S}} : \mathcal{A} \to \mathcal{A}[\mathcal{S}^{-1}]$ which is universal for exact tensor functors $\mathcal{A} \to \mathcal{C}$ which invert the morphisms in $\mathcal{S}$. The quotient $Q_{\mathcal{B}} : \mathcal{A} \to \mathcal{A}/\mathcal{B}$ of $\mathcal{A}$ by a thick tensor subcategory thus has the analogous properties.

2.4. The pseudo-abelian hull of a triangulated category

One step in the construction of the category of (pure) motives involves adjoining objects to an additive category corresponding to idempotent endomorphisms. In this section, we show how this procedure functions in the setting of a triangulated category.

2.4.1. DEFINITION. Let $\mathcal{C}$ be an additive category. Form the category $\mathcal{C}_\#$ with objects pairs (X,p), with $p \in \mathrm{Hom}_\mathcal{C}(X,X)$ satisfying $p^2 = p$. A morphism $f\colon (X,p) \to (Y,q)$ is a morphism $f\colon X \to Y$ in $\mathcal{C}$ with $f = qfp$, i.e., $\mathrm{Hom}_{\mathcal{C}_\#}((X,p),(Y,q))$ is the summand $q\mathrm{Hom}_\mathcal{C}(X,Y)p$ of $\mathrm{Hom}_\mathcal{C}(X,Y)$; composition is induced by the composition in $\mathcal{C}$. The category $\mathcal{C}$ is embedded as a full subcategory of $\mathcal{C}_\#$ by sending X to (X,id_X). $\mathcal{C}_\#$ is called the *pseudo-abelian hull* of $\mathcal{C}$.

If $\mathcal{C}$ is a graded category, we make the same definition, requiring in addition that p have degree zero.

2.4.2. LEMMA. *$\mathcal{C}_\#$ is an additive category. If $\mathcal{C}$ is a graded, resp. tensor category, then so is $\mathcal{C}_\#$; if $\mathcal{C}$ is a graded tensor category, then so is $\mathcal{C}_\#$. A translation structure on $\mathcal{C}$ extends canonically to a translation structure on $\mathcal{C}_\#$, and similarly for a translation structure compatible with a tensor structure. The embedding $\mathcal{C} \to \mathcal{C}_\#$ is a functor of graded, resp. tensor resp. graded tensor categories.*

PROOF. The direct sum of (X,p) and (Y,q) is given as $(X \oplus Y, p \oplus q)$. One easily checks that this gives $\mathcal{C}_\#$ the structure of an additive category. If $\mathcal{C}$ is graded, the summand $p\mathrm{Hom}_\mathcal{C}(X,Y)q$ is a graded subgroup of $\mathrm{Hom}_\mathcal{C}(X,Y)$, giving $\mathcal{C}_\#$ a canonical graded structure. If $\mathcal{C}$ has a translation structure, defining $(X,p)[a]^{[b]}$ to be $(X[1]^{[b]}, p[1]^{[b]})$ gives $\mathcal{C}_\#$ a translation structure. If $\mathcal{C}$ is a tensor category, setting $(X,p)\otimes(Y,q) = (X\otimes Y, p\otimes q)$ makes $\mathcal{C}_\#$ into a tensor category. One checks directly that, if $\mathcal{C}$ is a graded tensor category, the graded and tensor structures on $\mathcal{C}_\#$ give $\mathcal{C}_\#$ the structure of a graded tensor category, and if $\mathcal{C}$ has a compatible translation structure, the translation structure on $\mathcal{C}_\#$ is compatible with the tensor structure. □

2.4.3. DEFINITION. Let $\mathcal{C}$ be a triangulated category. We define a triangle T in $\mathcal{C}_\#$ to be distinguished if there is a map of distinguished triangles

$$(p,q,r)\colon (X,Y,Z,f,g,h) \to (X,Y,Z,f,g,h)$$

in $\mathcal{C}$, such that T is isomorphic to

$$(X,p) \xrightarrow{qfp} (Y,q) \xrightarrow{rgq} (Z,r) \xrightarrow{p[1]hr} (X[1],p[1])$$

in $\mathcal{C}_\#$. We call the map (p,q,r) a *lifting* of the distinguished triangle

$$((X,p),(Y,q),(Z,r),qfp,rgq,p[1]hr)$$

of $\mathcal{C}_\#$.

2.4.4. PROPOSITION. *If $\mathcal{C}$ is a triangulated category, then $\mathcal{C}_\#$ satisfies (TR1), (TR2) and (TR3). If $\mathcal{C}$ is a triangulated tensor category, then the distinguished triangles in $\mathcal{C}_\#$ are closed under right tensor product with arbitrary objects of $\mathcal{C}_\#$.*

PROOF. We already have the translation structure on $\mathcal{C}_\#$. The axiom (TR2) for $\mathcal{C}_\#$ follows directly from (TR2) for $\mathcal{C}$. The distinguished triangles in $\mathcal{C}_\#$ are closed under isomorphism by definition, and $((X,p),(X,p),0,\mathrm{id}_{(X,p)},0,0)$ is a distinguished triangle for all (X,p) in $\mathcal{C}_\#$, with lifting $(p,p,0)\colon (X,X,0,\mathrm{id}_X,0,0) \to (X,X,0,\mathrm{id}_X,0,0)$.

Let

$$((X,p),(Y,q),(Z,r),qfp,rgq,p[1]hr),$$
$$((\bar{X},\bar{p}),(\bar{Y},\bar{q}),(\bar{Z},\bar{r}),\bar{q}\bar{f}\bar{p},\bar{r}\bar{g}\bar{q},\bar{p}[1]\bar{h}\bar{r})$$

be triangles in $\mathcal{C}_\#$, with liftings

$$(p,q,r)\colon (X,Y,Z,f,g,h) \to (X,Y,Z,f,g,h),$$
$$(\bar{p},\bar{q},\bar{r})\colon (\bar{X},\bar{Y},\bar{Z},\bar{f},\bar{g},\bar{h}) \to (\bar{X},\bar{Y},\bar{Z},\bar{f},\bar{g},\bar{h}).$$

Take a map of maps in $\mathcal{C}_\#$, $(\bar{p}up,\bar{q}vq)\colon qfp \to \bar{q}\bar{f}\bar{p}$. This gives the map of maps $(\bar{p}up,\bar{q}vq)\colon f \to \bar{f}$ in $\mathcal{C}$. Let $w\colon Z \to Z$ be a map in $\mathcal{C}$ such that

$$(\bar{p}up,\bar{q}vq,w)\colon (X,Y,Z,f,g,h) \to (\bar{X},\bar{Y},\bar{Z},\bar{f},\bar{g},\bar{h})$$

is a map of triangles. Then

$$(\bar{p}up,\bar{q}vq,\bar{r}wr)\colon (X,Y,Z,f,g,h) \to (\bar{X},\bar{Y},\bar{Z},\bar{f},\bar{g},\bar{h})$$

is also a map of triangles, giving the map of triangles

$$((X,p),(Y,q),(Z,r),qfp,rgq,p[1]hr) \xrightarrow{(\bar{p}up,\bar{q}vq,\bar{r}wr)} ((\bar{X},\bar{p}),(\bar{Y},\bar{q}),(\bar{Z},\bar{r}),\bar{q}\bar{f}\bar{p},\bar{r}\bar{g}\bar{q},\bar{p}[1]\bar{h}\bar{r})$$

in $\mathcal{C}_\#$. This verifies the axiom (TR3) for $\mathcal{C}_\#$.

Let $qfp\colon (X,p) \to (Y,q)$ be a morphism in $\mathcal{C}_\#$, and let $X \xrightarrow{f} Y \xrightarrow{g} Z \xrightarrow{h} X[1]$ be a distinguished triangle in $\mathcal{C}$. We may take f satisfying $qf = fp$ (e.g. use qfp for f), hence there is a map $r\colon Z \to Z$ giving the map of triangles

$$\begin{array}{ccccccc} X & \xrightarrow{f} & Y & \xrightarrow{g} & Z & \xrightarrow{h} & X[1] \\ \downarrow{\scriptstyle p} & & \downarrow{\scriptstyle q} & & \downarrow{\scriptstyle r} & & \downarrow{\scriptstyle p[1]} \\ X & \xrightarrow[f]{} & Y & \xrightarrow[g]{} & Z & \xrightarrow[h]{} & X[1]. \end{array} \tag{2.4.4.1}$$

We will show that we may take r to be an idempotent; assuming this, (TR1) for $\mathcal{C}_\#$ follows.

To show that we may take r to be an idempotent, let r' be another choice of map making (2.4.4.1) commute. Then $h \circ (r' - r) = p[1] \circ h - p[1] \circ h = 0$, hence $r' - r = gs$ for some map $s\colon Z \to Y$. Also

$$gsg = (r' - r) \circ g = g \circ q - g \circ q = 0.$$

Conversely, if $s\colon Z \to Y$ is a map with $gsg = 0$, then $(p,q,r+gs)$ also gives a map of triangles.

Since $p^2 = p$ and $q^2 = q$, the map r^2 is another morphism $Z \to Z$ making (2.4.4.1) commute, hence we have $r^2 = r + gs$, with s as above. Then $rgs = gsr$ and $(gs)^2 = 0$, hence

$$r^n gs = rgs; \quad r^n = r^{n-1} + r^{n-2}gs,$$

for $n \geq 2$.

Let $\rho = r^2 - 2rgs$. Then we may use ρ instead of r in the diagram (2.4.4.1), and

$$\begin{aligned}\rho^2 &= r^4 - 4r^3gs \\ &= r^3 + r^2gs - 4rgs \\ &= r^2 + rgs + rgs - 4rgs \\ &= \rho.\end{aligned}$$

If $\mathcal{C}$ is a triangulated tensor category, it follows directly from the definition of the tensor structure on $\mathcal{C}_\#$ that the collection of distinguished triangles in $\mathcal{C}_\#$ is closed under right tensor product by arbitrary objects of $\mathcal{C}_\#$. □

2.4.5. *Distinguished octahedra.* Let

$$(A, B, C', a, u, v),\ (B, C, A', b, u', v'),$$
$$(A, C, B', ba, u'', v''),\ (C', B', A', f, g, u[1] \circ v')$$

be triangles in a category $\mathcal{C}$, such that

1. $(\mathrm{id}_A, b, f) \colon (A, B, C') \to (A, C, B')$ is a morphism of triangles
2. $(a, \mathrm{id}_C, g) \colon (A, C, B') \to (B, C, A')$ is a morphism of triangles.

We call the tuple (or equivalently, the resulting diagram)

$$(A, B, C, A', B', C'; a, b, u, v, u', v', u'', v'', f, g)$$

an *octahedron.* A map of octahedra consists of maps

$$(p, q, r, p', q', r') \colon (A, B, C, A', B', C') \to (\bar{A}, \bar{B}, \bar{C}, \bar{A}', \bar{B}', \bar{C}')$$

such that the four triples of maps

$$(p, q, r'),\ (q, r, p'),\ (p, r, q'),\ (r', q', p')$$

are maps of triangles. If all four triangles in the octahedron are distinguished, we say the octahedron is *distinguished.*

2.4.6. LEMMA. *Let $\mathcal{A}$ be a DG category, and let $\mathcal{C}$ be the localization of $\mathbf{K}^b(\mathcal{A})$ with respect to a thick subcategory. Let*

(2.4.6.1)
$$\begin{array}{ccccc} A & \xrightarrow{a} & B & \xrightarrow{b} & C \\ {\scriptstyle p}\downarrow & & {\scriptstyle q}\downarrow & & \downarrow{\scriptstyle r} \\ A & \xrightarrow[a]{} & B & \xrightarrow[b]{} & C \end{array}$$

be a commutative diagram in $\mathcal{C}$, and let

$$(A, B, C', a, u, v),\ (B, C, A', b, u', v'),\ (A, C, B', ba, u'', v'')$$

be distinguished triangles. Then there is a distinguished octahedron

$$(A, B, C, A', B', C'; a, b, u, v, u', v', u'', v'', f, g)$$

and maps $p' \colon A' \to A'$, $q' \colon B' \to B'$, and $r' \colon C' \to C'$ such that

$$(p, q, r, p', q', r') \colon (A, B, C, A', B', C') \to (A, B, C, A', B', C')$$

forms a map of octahedra.

PROOF. If we have two completions of a map $f: X \to Y$ to distinguished triangles

$$X \xrightarrow{f} Y \xrightarrow{u} Z \xrightarrow{v} X[1]$$
$$X \xrightarrow{f} Y \xrightarrow{u'} Z' \xrightarrow{v'} X[1]$$

then these triangles are isomorphic, by a map of triangles of the form

$$(\mathrm{id}_X, \mathrm{id}_Y, w): (X, Y, Z) \to (X, Y, Z').$$

Thus, if we are able to prove the result for one choice of distinguished triangles (A, B, C', a, u, v), etc., it is true for all choices.

Similarly, if we have a commutative diagram in $\mathcal{C}$

$$\begin{array}{ccccc} A & \xrightarrow{a} & B & \xrightarrow{b} & C \\ \downarrow{s_a} & & \downarrow{s_b} & & \downarrow{s_c} \\ \bar{A} & \xrightarrow[\bar{a}]{} & \bar{B} & \xrightarrow[\bar{b}]{} & \bar{C}, \end{array} \tag{2.4.6.2}$$

with the maps s_a, s_b, s_c in $\mathcal{S}$, then, letting $\bar{p} = s_a p s_a^{-1}$, $\bar{q} = s_b q s_b^{-1}$, and $\bar{r} = s_c r s_c^{-1}$, a morphism of distinguished octahedra

$$(\bar{A}, \bar{B}, \bar{C}, \bar{A}', \bar{B}', \bar{C}'; \bar{a}, \bar{b}, \bar{u}, \bar{v}, \bar{u}', \bar{v}', \bar{u}'', \bar{v}'', \bar{f}, \bar{g})$$
$$\xrightarrow{(\bar{p}, \bar{q}, \bar{r}, \bar{p}', \bar{q}', \bar{r}')} (\bar{A}, \bar{B}, \bar{C}, \bar{A}', \bar{B}', \bar{C}'; \bar{a}, \bar{b}, \bar{u}, \bar{v}, \bar{u}', \bar{v}', \bar{u}'', \bar{v}'', \bar{f}, \bar{g})$$

gives the morphism of distinguished octahedra

$$(A, B, C, A', B', C'; a, b, u, v, u', v', u'', v'', f, g)$$
$$\xrightarrow{(p, q, r, p', q', r')} (A, B, C, A', B', C'; a, b, u, v, u', v', u'', v'', f, g)$$

with

$$A' = \bar{A}',\ B' = \bar{B}',\ C' = \bar{C}';\ p' = \bar{p}',\ q' = \bar{q}', r' = \bar{r}', f = \bar{f},\ g = \bar{g};$$
$$u = \bar{u} s_b,\ v = s_a[1]^{-1}\bar{v},\ u' = \bar{u}' s_c,\ v' = s_b[1]^{-1}\bar{v}',\ u'' = \bar{u}'' s_c,\ v'' = s_a[1]^{-1}\bar{v}''.$$

Using the properties of a saturated multiplicative system of morphisms listed in Definition 2.3.1.2, we may find a commutative diagram (2.4.6.2) with $\bar{a}$, $\bar{b}$ and $\bar{c}$ maps in $\mathbf{K}^b(\mathcal{A})$. Thus, we may assume a, b and c are maps in $\mathbf{K}^b(\mathcal{A})$. Using these properties again, we may find a commutative diagram in $\mathbf{K}^b(\mathcal{A})$:

$$\begin{array}{ccccc} A & \xrightarrow{a} & B & \xrightarrow{b} & C \\ \uparrow{s_a} & & \uparrow{s_b} & & \uparrow{s_c} \\ \bar{A} & \xrightarrow{\bar{a}} & \bar{B} & \xrightarrow{\bar{b}} & \bar{C} \\ \downarrow{\bar{p}} & & \downarrow{\bar{q}} & & \downarrow{\bar{r}} \\ A & \xrightarrow{a} & B & \xrightarrow{b} & C \end{array} \tag{2.4.6.3}$$

with the maps s_a, s_b and s_c in $\mathcal{S}$, and with $p = \bar{p} \circ s_a^{-1}$, $q = \bar{q} \circ s_b^{-1}$, and $r = \bar{r} \circ s_c^{-1}$.

For a sequence of maps $X \xrightarrow{\alpha} Y \xrightarrow{\beta} Z$ in $\mathbf{K}^b(\mathcal{A})$, form the distinguished octahedron

$$O(X,Y,Z;\alpha,\beta) := \\ (X,Y,Z,\mathrm{cone}(\tilde{\beta}),\mathrm{cone}(\tilde{\beta}\tilde{\alpha}),\mathrm{cone}(\tilde{\alpha}),u,v,u',v',u'',v'',f,g),$$

where $\tilde{\alpha}$ and $\tilde{\beta}$ are liftings of α and β to $\mathbf{C}^b(\mathcal{A})$, with maps u, v, u', v', u'' and v'' being those coming from the cone sequences for $\tilde{\alpha}$, $\tilde{\beta}\tilde{\alpha}$ and $\tilde{\beta}$, and with the maps f and g being given by $f := \mathrm{id}_{A[1]} \oplus \tilde{\beta}$, $g := \tilde{\alpha}[1] \oplus \mathrm{id}_C$. We have shown in the proof of Proposition 2.1.6.4 that $O(X,Y,Z;\alpha,\beta)$ is indeed a distinguished octahedron.

Let

$$\begin{array}{ccccc} X & \xrightarrow{\alpha} & Y & \xrightarrow{\beta} & Z \\ {\scriptstyle p}\downarrow & & {\scriptstyle q}\downarrow & & \downarrow{\scriptstyle r} \\ X' & \xrightarrow{\alpha'} & Y' & \xrightarrow{\beta'} & Z' \end{array} \tag{2.4.6.4}$$

be a commutative diagram in $\mathbf{K}^b(\mathcal{A})$. Lift (2.4.6.4) to the diagram

$$\begin{array}{ccccc} X & \xrightarrow{\tilde{\alpha}} & Y & \xrightarrow{\tilde{\beta}} & Z \\ {\scriptstyle \tilde{p}}\downarrow & & {\scriptstyle \tilde{q}}\downarrow & & \downarrow{\scriptstyle \tilde{r}} \\ X' & \xrightarrow{\tilde{\alpha}'} & Y' & \xrightarrow{\tilde{\beta}'} & Z' \end{array}$$

in $\mathbf{C}^b(\mathcal{A})$; there are then degree -1 maps $h_1 : Y \to Z'$ and $h_2 : X \to Y'$ with

$$dh_1 = \tilde{\beta}'\tilde{q} - \tilde{r}\tilde{\beta},\ dh_2 = \tilde{\alpha}'\tilde{p} - \tilde{q}\tilde{\alpha}.$$

Let $h_3 = h_1\tilde{\alpha} + \tilde{\beta}'h_2$, so

$$dh_3 = \tilde{\beta}'\tilde{\alpha}'\tilde{p} - \tilde{r}\tilde{\beta}\tilde{\alpha}.$$

Let $h_1^+ : Y[1] \to Z'$ be the degree zero map determined by h_1, and define h_2^+ and h_3^+ similarly. We map $\mathrm{cone}(\tilde{\alpha})$, $\mathrm{cone}(\tilde{\beta})$ and $\mathrm{cone}(\tilde{\beta}\tilde{\alpha})$ to $\mathrm{cone}(\tilde{\alpha}')$, $\mathrm{cone}(\tilde{\beta}')$ and $\mathrm{cone}(\tilde{\beta}'\tilde{\alpha}')$ by the matrices

$$p' := \begin{pmatrix} \tilde{q}[1] & 0 \\ h_1^+ & \tilde{r} \end{pmatrix} : \mathrm{cone}(\tilde{\beta}) \to \mathrm{cone}(\tilde{\beta}')$$

$$q' := \begin{pmatrix} \tilde{p}[1] & 0 \\ h_3^+ & \tilde{r} \end{pmatrix} : \mathrm{cone}(\tilde{\beta}\tilde{\alpha}) \to \mathrm{cone}(\tilde{\beta}'\tilde{\alpha}')$$

$$r' := \begin{pmatrix} \tilde{p}[1] & 0 \\ h_2^+ & \tilde{q} \end{pmatrix} : \mathrm{cone}(\tilde{\alpha}) \to \mathrm{cone}(\tilde{\alpha}')$$

We have the degree -1 maps

$$h_4 := \begin{pmatrix} h_2[1] & 0 \\ 0 & 0 \end{pmatrix} : \mathrm{cone}(\tilde{\beta}\tilde{\alpha}) \to \mathrm{cone}(\tilde{\beta}')$$

$$h_5 := \begin{pmatrix} 0 & 0 \\ 0 & h_1 \end{pmatrix} : \mathrm{cone}(\tilde{\alpha}) \to \mathrm{cone}(\tilde{\beta}'\tilde{\alpha}')$$

One checks that the maps $h_1, \ldots, h_5$ define the homotopies required to show that

$$(p,q,r,p',q',r') : O(X,Y,Z;\alpha,\beta) \to O(X',Y',Z';\alpha',\beta')$$

defines a map of distinguished octahedra. If p, q and r are in $\mathcal{S}$, then so are p', q' and r'.

We apply this construction to the diagram (2.4.6.3), giving the maps of distinguished octahedra in $\mathbf{K}^b(\mathcal{A})$

$$O(A,B,C;a,b) \xleftarrow{(s_a,s_b,s_c,s'_a,s'_b,s'_c)} O(\bar{A},\bar{B},\bar{C};\bar{a},\bar{b}) \xrightarrow{(\bar{p},\bar{q},\bar{r},\bar{p}',\bar{q}',\bar{r}')} O(A,B,C;a,b).$$

Setting $p' = \bar{p}\circ s_a'^{-1}$, $q' = \bar{q}\circ s_b'^{-1}$, and $r' = \bar{r}\circ s_c'^{-1}$ gives the map of distinguished octahedra in $\mathcal{C}$,

$$(p,q,r,p',q',r')\colon O(A,B,C;a,b) \to O(A,B,C;a,b).$$

□

2.4.7. THEOREM. *Let $\mathcal{A}$ be a DG category, and let $\mathcal{C}$ be the localization of $\mathbf{K}^b(\mathcal{A})$ with respect to a thick subcategory. Then $\mathcal{C}_\#$ is a triangulated category. If $\mathcal{A}$ is a DG tensor category (with or without unit), and $\mathcal{C}$ is the localization of $\mathbf{K}^b(\mathcal{A})$ with respect to a thick tensor subcategory, then $\mathcal{C}_\#$ is a triangulated tensor category (with or without unit).*

PROOF. The assertion on the tensor structures follows from the first part of the theorem, together with Proposition 2.4.4, so we need only prove the first assertion. By Proposition 2.4.4, we need only verify the axiom (TR4).

Let then

$$(A,p) \xrightarrow{a} (B,q) \xrightarrow{b} (C,r) \tag{2.4.7.1}$$

be a sequence of maps in $\mathcal{C}_\#$ (we may assume $a = qap$, $b = rbq$ as maps in $\mathcal{C}$), giving the commutative diagram

$$\begin{array}{ccccc} A & \xrightarrow{a} & B & \xrightarrow{b} & C \\ {\scriptstyle p}\downarrow & & {\scriptstyle q}\downarrow & & \downarrow{\scriptstyle r} \\ A & \xrightarrow{a} & B & \xrightarrow{b} & C \end{array}$$

in $\mathcal{C}$. By Proposition 2.4.4, we may complete the maps a, b and ba to distinguished triangles

$$((A,p),(B,q),(C',r''),a,r''uq,p[1]vr''),$$
$$((B,q),(C,r),(A',p''),b,p''u'r,q[1]v'p''),$$
$$((A,p),(C,r),(B',q''),ba,q''u''r,p[1]v''q''),$$

with liftings

$$(p,q,r'')\colon (A,B,C',a,u,v) \to (A,B,C',a,u,v),$$
$$(q,r,p'')\colon (B,C,A',b,u',v') \to (B,C,A',b,u',v'),$$
$$(p,r,q'')\colon (A,C,B',ba,u'',v'') \to (A,C,B',ba,u'',v'')$$

to $\mathcal{C}$.

By Lemma 2.4.6, there is a map of distinguished octahedra

$$(A,B,C,A',B',C';a,b,u,v,u',v',u'',v'',f,g)$$
$$\xrightarrow{(p,q,r,p',q',r')} (A,B,C,A',B',C';a,b,u,v,u',v',u'',v'',f,g)$$

in $\mathcal{C}$. We now argue as in the proof of (TR2) in Proposition 2.4.4 to change p', q' and r' to idempotents. Indeed, since p, q and r are idempotent,

$$(A, B, C, A', B', C', a, b, u, v, u', v', u'', v'', f, g) \xrightarrow{(p,q,r,p'^2,q'^2,r'^2)} (A, B, C, A', B', C', a, b, u, v, u', v', u'', v'', f, g)$$

is also a map of distinguished octahedra. Thus, there are maps $s: A' \to C$, $t: B' \to C$, and $w: C' \to B$ such that

$$p'^2 = p' + u's,\ q'^2 = q' + u''t,\ r'^2 = r' + uw$$
$$u'su' = 0,\ u''tu'' = 0,\ uwu = 0$$
$$fuw = u''tf,\ gu''t = u'sg.$$

Replacing (p', q', r') with

$$(p'^2 - 2p'u's, q'^2 - 2q'u''t, r'^2 - 2r'uw),$$

and changing notation, we may assume that p', q' and r' are idempotent (see the proof of Proposition 2.4.4). The octahedron in $\mathcal{C}_\#$

$$((A,p),(B,q),(C,r),(A',p'),(B',q'),(C',r'), a, b, r'uq, p[1]vr', p'u'r, q[1]v'p', q'u''r, p[1]v''q', q'fr', p'gq')$$

is then distinguished. As it suffices by Remark 2.1.2(3) to find one distinguished octahedron containing the diagram (2.4.7.1), the proof is complete. □

2.4.8. *Splitting idempotents in triangulated categories.* Bökstedt and Neeman [**21**] have given an elegant construction for splitting idempotents in certain triangulated categories; we give a modification of their method in this section, with some examples and applications.

Let $\mathcal{A}$ be a pre-additive category, S a set, A an object of $\mathcal{A}$. We let A^S denote the direct sum (coproduct over 0) $\oplus_{i\in S} A$, assuming the direct sum exists, i.e., that there are maps $i_s: A \to A^S$, $s \in S$, such that

$$\mathrm{Hom}_{\mathcal{A}}(A^S, Z) \xrightarrow{\prod_{s\in S} i_s^*} \prod_{s\in S} \mathrm{Hom}_{\mathcal{A}}(A, Z)$$

is an isomorphism for all Z in $\mathcal{A}$.

2.4.8.1. LEMMA. *Let $\mathcal{C}$ be a triangulated category, X an object of $\mathcal{C}$, $p: X \to X$ an idempotent endomorphism. Suppose that $X^{\mathbb{N}}$ exists in $\mathcal{C}$. Then there is a direct sum decomposition of X as $X \cong X(p) \oplus X(1-p)$ such that p is identified with the projection on $X(p)$:*

$$X(p) \oplus X(1-p) \xrightarrow{j_{X(p)}} X(p) \xrightarrow{i_{X(p)}} X(p) \oplus X(1-p).$$

PROOF. For $j \in \mathbb{N}$, let $i_j: X \to X^{\mathbb{N}}$ be the corresponding inclusion. For each $j \geq 2 \in \mathbb{N}$, let $s_j: X \to X^{\mathbb{N}}$ be the map $s_j = i_j - i_{j-1} \circ p$; let $s_1 = i_1$. By the universal property of $X^{\mathbb{N}}$, there is a unique map $1 - tp: X^{\mathbb{N}} \to X^{\mathbb{N}}$ with $(1 - tp) \circ i_j = s_j$ for all $j = 1, 2, \ldots$.

Complete $1 - tp$ to a distinguished triangle

$$X^{\mathbb{N}} \xrightarrow{1-tp} X^{\mathbb{N}} \xrightarrow{q} X(p) \xrightarrow{r} X^{\mathbb{N}}[1].$$

For each Z in $\mathcal{C}$, we have the exact sequence

$$\prod_{i\in\mathbb{N}} \mathrm{Hom}_{\mathcal{C}}(X[1], Z) \xrightarrow{(1-tp)[1]^*} \prod_{i\in\mathbb{N}} \mathrm{Hom}_{\mathcal{C}}(X[1], Z) \to \mathrm{Hom}_{\mathcal{C}}(X(p), Z)$$
$$\to \prod_{i\in\mathbb{N}} \mathrm{Hom}_{\mathcal{C}}(X, Z) \xrightarrow{(1-tp)^*} \prod_{i\in\mathbb{N}} \mathrm{Hom}_{\mathcal{C}}(X, Z)$$

with

$$(1-tp)^*(f_1, f_2, \dots, f_n, \dots) = (f_1 - f_2\circ p, f_2 - f_3\circ p, \dots, f_n - f_{n+1}\circ p, \dots),$$

and similarly for $(1-tp)[1]^*$. Thus $(1-tp)[1]^*$ is surjective; the projection on the factor $i = 1$ identifies the kernel of $(1-tp)^*$ with the summand $\mathrm{Hom}_{\mathcal{C}}(X, Z)p$ of $\mathrm{Hom}_{\mathcal{C}}(X, Z)$, giving the natural isomorphism $\mathrm{Hom}_{\mathcal{C}}(X, Z)p \cong \mathrm{Hom}_{\mathcal{C}}(X(p), Z)$.

Taking $Z = X$, the endomorphism p gives the map $i_{X(p)}\colon X(p) \to X$ with $i^*_{X(p)}\colon \mathrm{Hom}_{\mathcal{C}}(X, Z) \to \mathrm{Hom}_{\mathcal{C}}(X(p), Z)$ identifying $\mathrm{Hom}_{\mathcal{C}}(X(p), Z)$ with the summand $\mathrm{Hom}_{\mathcal{C}}(X, Z)p$ of $\mathrm{Hom}_{\mathcal{C}}(X, Z)$. Replacing p with $1-p$ gives the morphism $i_{X(1-p)}\colon X(1-p) \to X$; the morphism

$$i_{X(p)} + i_{X(1-p)}\colon X(p) \oplus X(1-p) \to X$$

thus gives the natural isomorphism

$$(i_{X(p)} + i_{X(1-p)})^*\colon \mathrm{Hom}_{\mathcal{C}}(X, Z) \to \mathrm{Hom}_{\mathcal{C}}(X(p) \oplus X(1-p), Z).$$

By the Yoneda lemma, the map $i_{X(p)} + i_{X(1-p)}$ is an isomorphism; it is easy to check that this isomorphism identifies p with the projection onto $X(p)$. □

As an immediate consequence, we have

2.4.8.2. THEOREM. *Let $\mathcal{C}$ be a triangulated category such that $X^{\mathbb{N}}$ exists for all X in $\mathcal{C}$. Then the embedding of additive categories $\mathcal{C} \to \mathcal{C}_\#$ is an equivalence of additive categories. In particular, the category $\mathcal{C}_\#$ is a triangulated category, equivalent to $\mathcal{C}$. If $\mathcal{C}$ is in addition a triangulated tensor category, then the embedding $\mathcal{C} \to \mathcal{C}_\#$ is an equivalence of triangulated tensor categories.*

2.4.9. EXAMPLES. We give a few examples of triangulated categories satisfying the hypothesis of Theorem 2.4.8.2.
(i) Let $\mathcal{A}$ be a DG category such that $X^{\mathbb{N}}$ exists for all X in $\mathcal{A}$. Then $X^{\mathbb{N}}$ exists in $\mathbf{K}^b(\mathcal{A})$ for all X in $\mathbf{K}^b(\mathcal{A})$. Similarly, if $\mathcal{A}$ is an additive category such that $X^{\mathbb{N}}$ exists for all X in $\mathcal{A}$, then $X^{\mathbb{N}}$ exists in $\mathbf{K}^*(\mathcal{A})$ for all X, for $* = \emptyset, +, -, b$.
(ii) Let $\mathcal{C}$ be a triangulated category such that direct sums of arbitrary sets of objects of $\mathcal{C}$ exist. Let $\mathcal{B}$ be a thick subcategory which is closed under taking arbitrary direct sums. Then [**21**, Lemma 1.5] arbitrary direct sums exist in the localization $\mathcal{C}/\mathcal{B}$; in particular, $X^{\mathbb{N}}$ exists for all X in $\mathcal{C}/\mathcal{B}$.
(iii) Let $\mathcal{A}$ be an abelian category such that arbitrary direct sums exist. Then $X^{\mathbb{N}}$ exists for all X in the unbounded derived category $\mathbf{D}(\mathcal{A})$. This follows directly from (ii), noting that arbitrary direct sums exist in the unbounded homotopy category $\mathbf{K}(\mathcal{A})$, and that an arbitrary direct sum of acyclic complexes is acyclic.
(iv) Let $\mathcal{A}$ be an abelian category having enough injectives, such that $X^{\mathbb{N}}$ exists for all X in $\mathcal{A}$. Then $X^{\mathbb{N}}$ exists for all X in the derived category $\mathbf{D}^*(\mathcal{A})$, for $* = +, b$. Indeed, if we let $\mathcal{I}$ be the full subcategory of $\mathcal{A}$ consisting of injective objects, then the natural functor $\mathbf{K}^*(\mathcal{I}) \to \mathbf{D}^*(\mathcal{A})$ is an equivalence of triangulated categories; more precisely, if $f\colon A \to B$ is a quasi-isomorphism in $\mathbf{K}^*(\mathcal{A})$, and Y is in $\mathbf{C}^*(\mathcal{I})$,

then $f^*: \mathrm{Hom}_{\mathbf{K}^b(\mathcal{A})}(B,Y) \to \mathrm{Hom}_{\mathbf{K}^b(\mathcal{A})}(A,Y)$ is an isomorphism. By (i), $X^{\mathbb{N}}$ exists in $\mathbf{K}^*(\mathcal{A})$ for all X in $\mathbf{K}^*(\mathcal{I})$; taking a quasi-isomorphism $X^{\mathbb{N}} \to \tilde{X}^{\mathbb{N}}$ with $\tilde{X}^{\mathbb{N}}$ in $\mathbf{K}^*(\mathcal{I})$, we see that $\tilde{X}^{\mathbb{N}}$ represents the direct sum $\oplus_{i\in\mathbb{N}}X$ in $\mathbf{K}^*(\mathcal{I})$.

2.4.10. COROLLARY. *Let $\mathcal{B}$ be an abelian category, let $\mathcal{A}$ be a DG category and let $\mathcal{C}$ be a localization of the triangulated category $\mathbf{K}^b(\mathcal{A})$. Let $F:\mathcal{C} \to \mathbf{D}^*(\mathcal{B})$ be an exact functor, where $* = \emptyset, +, b$ is a boundedness condition. Suppose in addition*

(In case $* = +, b$) $X^{\mathbb{N}}$ exists in $\mathcal{B}$ for all X in $\mathcal{B}$, and $\mathcal{B}$ has enough injectives.
(In case $* = \emptyset$) Arbitrary direct sums exist in $\mathcal{B}$.

Then there is an extension of F to an exact functor $F_\#:\mathcal{C}_\# \to \mathbf{D}^*(\mathcal{B})$, which is unique up to natural isomorphism. If $\mathcal{A}$ is a DG tensor category, $\mathcal{B}$ an abelian tensor category, and F is a functor of triangulated tensor categories, then so is $F_\#$.

PROOF. The category $\mathcal{C}_\#$ is a triangulated category (resp. triangulated tensor category) by Theorem 2.4.7. The existence of $F_\#$, either as a functor of triangulated categories, or of triangulated tensor categories, follows directly from Theorem 2.4.8.2 and Example 2.4.9(iii) for $* = \emptyset$, or Example 2.4.9(iv) for $* = +, b$. The uniqueness up to natural isomorphism follows from the uniqueness, up to canonical isomorphism, of finite direct sums in an additive category. □

3. Constructions

In this final section of the chapter, we give two constructions of DG tensor categories. The first is a type of "fat point"; in topological terms, this would be a contractible space with a free action of the infinite symmetric group. This category with be of use in our construction of the motivic DG tensor category, as it has the effect of absorbing the non-trivial cohomology that usually arise from cohomology operations.

The second construction is the first of two methods for producing a categorical external product which is homotopy commutative, and admits all higher homotopies. The second method, which is multi-simplicial, will be taken up in the next chapter.

3.1. The homotopy one point DG tensor category

We construct a DG tensor category $\mathbb{E}$ which plays the role of a "homotopy point".

3.1.1. *The category of $\mathbb{Z}/2$-Sets.* Let $\mathbb{Z}/2$-**Sets** be the category with objects being pointed sets $(S, *)$ with a pointed $\mathbb{Z}/2$-action, such that the action on $S\backslash\{*\}$ is free; morphisms are pointed maps of sets respecting the $\mathbb{Z}/2$-action. We write $\mathbb{Z}/2 = \{\pm 1\}$, and write $-x$ for $(-1)\cdot x$. A *graded* $\mathbb{Z}/2$-set is a $\mathbb{Z}/2$-set $(S^*, *)$ together with a decomposition as $\mathbb{Z}/2$-sets

$$S^* = \bigvee_{n\in\mathbb{Z}} (S^n, *);$$

here $\vee$ is the pointed coproduct. The notion of a map of graded $\mathbb{Z}/2$-sets being the obvious one, we have the category **Gr-**$\mathbb{Z}/2$**-Sets** of graded $\mathbb{Z}/2$-sets.

The category **Gr-**$\mathbb{Z}/2$**-Sets** is a symmetric monoidal category. The product $((S \times T)^*, *)$ of graded $\mathbb{Z}/2$-sets $(S^*, *)$ and $(T^*, *)$ is given by

$$(S \times T)^n = \bigvee_{p+q=n} S^p \wedge T^q / \equiv,$$

where $\wedge$ is the pointed product (smash product) and the equivalence relation $\equiv$ is given by $(s,t) \equiv (-s,-t)$. We give $S^p \wedge T^q/\equiv$ the action

$$-(s,t) = (-s,t) = (s,-t);$$

one easily checks that this is free. The symmetry

$$t_{S,T}\colon ((S \times T)^*, *) \to ((T \times S)^*, *)$$

is given by $t_{S,T}(s,t) = (-1)^{pq}(t,s)$, for $s \in S^p$, $t \in T^q$.

3.1.2. DEFINITION. A *graded symmetric monoidal category* is a symmetric monoidal object in the category $\mathbf{cat}_{\mathbf{Gr}-\mathbb{Z}/2-\mathbf{Sets}}$. Similarly, a *graded symmetric semi-monoidal category* is a symmetric semi-monoidal object in $\mathbf{cat}_{\mathbf{Gr}-\mathbb{Z}/2-\mathbf{Sets}}$.

3.1.3. *Equivalence relations.* If we have a (graded) $\mathbb{Z}/2$-set $(S,*)$ and a $\mathbb{Z}/2$-equivariant equivalence relation $\equiv$ on S, we have the quotient $\mathbb{Z}/2$-set $(S,*)/\equiv$, which is gotten from the pointed set with $\mathbb{Z}/2$-action $(S/\equiv, *)$ by identifying $\bar{x} \in S/\equiv$ with the base point $*$ if $-\bar{x} = \bar{x}$. This gives the categorical quotient.

We have the functor

$$(-)_{\mathbb{Z}}\colon \mathbb{Z}/2\text{-}\mathbf{Sets} \to \mathbf{Ab} \tag{3.1.3.1}$$

which sends $(S,*)$ to the quotient of the free abelian group on S by the relations:

$$n \cdot * = 0$$
$$n \cdot (-x) = -n \cdot x.$$

The functor $(-)_{\mathbb{Z}}$ extends to the graded setting, and sends (graded) product to (graded) tensor product. As consequence, the functor $(-)_{\mathbb{Z}}$ gives the functor $(-)_{\mathbb{Z}}$ from the category of graded symmetric monoidal categories to graded tensor categories, and from graded symmetric semi-monoidal categories to graded tensor categories without unit.

3.1.4. REMARK. The functor (3.1.3.1) is *not* in general compatible with the operation of taking the quotient by a $\mathbb{Z}/2$-equivariant equivalence relation; one does however have the canonical isomorphism

$$((S,*)/\equiv)_{\mathbb{Z}} \cong ((S,*)_{\mathbb{Z}})/\equiv_{\mathbb{Z}})/2 - \text{torsion}.$$

Another way to say the same thing is to redefine the functor $(-)_{\mathbb{Z}}$ as a functor to the category of abelian groups which are 2-torsion free; this functor is then compatible with quotients.

3.1.5. We note that taking the degree zero component of the Hom-sets in a graded symmetric (semi-)monoidal category, and taking the quotient by the $\mathbb{Z}/2$-action defines a functor $\mathbf{gr}^0$ from graded symmetric (semi-)monoidal categories to symmetric (semi-)monoidal categories. Similarly, we may generate a graded symmetric (semi-)monoidal category $\pm\mathcal{C}$ from a symmetric (semi-)monoidal category $\mathcal{C}$ by taking the free $\mathbb{Z}/2$-set on the Hom-sets of $\mathcal{C}$, concentrated in degree zero, and adding a base-point. In particular, if x is an object in a graded symmetric semi-monoidal category $(\mathcal{C}, \bullet)$, we have the natural map

$$\iota_n(x)\colon \{\pm 1\} \times S_n \to \mathrm{Hom}_{\mathcal{C}}(x^{\bullet n}, x^{\bullet n})^0$$

sending $\sigma \in S_n$ to the permutation isomorphism $t_\sigma\colon x^{\bullet n} \to x^{\bullet n}$.

3.1.6. DEFINITION. Let $(\mathcal{C}, \bullet)$ be a graded symmetric semi-monoidal category. Call $(\mathcal{C}, \bullet)$ *punctual* if

(i) There is an object $\mathfrak{e}$ of $\mathcal{C}$ such that $\mathrm{Obj}(\mathcal{C}) = \{\mathfrak{e}^{\bullet n} \mid n = 1, 2, \ldots\}$ and the objects $\mathfrak{e}^{\bullet n}$ and $\mathfrak{e}^{\bullet m}$ are distinct if $m \neq n$. We call $\mathfrak{e}$ the *generator* of $\mathcal{C}$.
(ii) $\mathrm{Hom}_{\mathcal{C}}(\mathfrak{e}^{\bullet n}, \mathfrak{e}^{\bullet m})^* = *$ if $n \neq m$.
(iii) $\mathrm{Hom}_{\mathcal{C}}(\mathfrak{e}^{\bullet n}, \mathfrak{e}^{\bullet n})^p = *$ if $p > 0$ and the map

$$\iota_n(\mathfrak{e})\colon \{\pm 1\} \times S_n \to \mathrm{Hom}_{\mathcal{C}}(\mathfrak{e}^{\bullet n}, \mathfrak{e}^{\bullet n})^0 \setminus \{*\}$$

is an isomorphism.
(iv) For each n, the action of $S_n \cong 1 \times S_n$ on $\mathrm{Hom}_{\mathcal{C}}(\mathfrak{e}^{\bullet n}, \mathfrak{e}^{\bullet n})^p \setminus \{*\}$ by both left and right composition via the map $\iota_n(\mathfrak{e})$ is a free action for all p such that $\mathrm{Hom}_{\mathcal{C}}(\mathfrak{e}^{\bullet n}, \mathfrak{e}^{\bullet n})^p \neq \{*\}$.

As an example, let $\pm\mathfrak{N}$ be the graded symmetric semi-monoidal category generated by the symmetric semi-monoidal category $\mathfrak{N}$. Then $\pm\mathfrak{N}$ is punctual with generator 1.

3.1.7. We may adjoin objects, morphisms and relations to a $\mathbb{Z}/2$-**Sets** category. In particular, if $\mathcal{C}$ is a graded symmetric (semi)-monoidal category, and a and b are objects of $\mathcal{C}$, we may form the graded symmetric (semi)-monoidal category $\mathcal{C}[h]$ gotten from $\mathcal{C}$ by adjoining a morphism $h\colon a \to b$ of some degree p. We have the canonical functor $i_h\colon \mathcal{C} \to \mathcal{C}[h]$, and i_h satisfies the usual universal mapping property to graded symmetric (semi)-monoidal categories. Explicitly, $\mathcal{C}[h]$ is the $\mathbb{Z}/2$-**Sets** category gotten from $\mathcal{C}$ by adjoining morphisms $\mathrm{id}_c \bullet h \bullet \mathrm{id}_d$, $\mathrm{id}_c \bullet h$ and $h \bullet \mathrm{id}_d$, where c and d run over objects of $\mathcal{C}$, and imposing the relations of graded commutativity

$$\begin{aligned}(\mathrm{id}_c \bullet h \bullet \mathrm{id}_b \bullet \mathrm{id}_d) &\circ (\mathrm{id}_c \bullet \mathrm{id}_a \bullet h \bullet \mathrm{id}_d),\\ &= (-1)^p(\mathrm{id}_c \bullet \mathrm{id}_b \bullet h \bullet \mathrm{id}_e) \circ (\mathrm{id}_c \bullet h \bullet \mathrm{id}_a \bullet \mathrm{id}_e)\end{aligned}$$

$$\begin{aligned}(\mathrm{id}_c \bullet h \bullet \mathrm{id}_d \bullet \mathrm{id}_e) &\circ (\mathrm{id}_c \bullet \mathrm{id}_a \bullet f \bullet \mathrm{id}_e)\\ &= (-1)^{pq}(\mathrm{id}_c \bullet \mathrm{id}_{d'} \bullet h \bullet \mathrm{id}_e) \circ (\mathrm{id}_c \bullet f \bullet \mathrm{id}_a \bullet \mathrm{id}_e);\\ &\qquad \text{for } f\colon d \to d',\ \deg(f) = q,\end{aligned}$$

$$\begin{aligned}(\mathrm{id}_c \bullet h \bullet \mathrm{id}_d \bullet \mathrm{id}_e) &\circ (\mathrm{id}_c \bullet \tau_{d,a} \bullet \mathrm{id}_e)\\ &= (\mathrm{id}_c \bullet \tau_{b,d} \bullet \mathrm{id}_e) \circ (\mathrm{id}_c \bullet \mathrm{id}_d \bullet h \bullet \mathrm{id}_e),\end{aligned}$$

together with similar relations for $h \bullet \mathrm{id}_d \bullet \mathrm{id}_e$ and $\mathrm{id}_c \bullet \mathrm{id}_d \bullet h$; one must then identify with $*$ all expressions E with $E = -E$, modulo the above relations. This suffices to give the symmetric semi-monoidal case, for the symmetric monoidal case, one must adjoin additional relations related to the action of the unit.

3.1.8. REMARK. The operation of adjoining a morphism h to a graded symmetric (semi-)monoidal category may not in general be compatible with the functor $(-)_{\mathbb{Z}}$ due to the 2-torsion issue discussed in Remark 3.1.4. It is the case, however, that the graded tensor category without unit $(\mathcal{C}[h])_{\mathbb{Z}}$ is canonically isomorphic to the graded tensor category without unit formed from $(\mathcal{C})_{\mathbb{Z}}[h_{\mathbb{Z}}]$ by taking the quotient of the Hom groups by their 2-torsion subgroup:

$$(\mathcal{C}[h])_{\mathbb{Z}} \cong (\mathcal{C})_{\mathbb{Z}}[h_{\mathbb{Z}}]/2\text{-torsion}.$$

Consequently, the functor $(i_h)_{\mathbb{Z}}:\mathcal{C}_{\mathbb{Z}} \to (\mathcal{C}[h])_{\mathbb{Z}}$ has the universal mapping property of $i_{h_{\mathbb{Z}}}:\mathcal{C}_{\mathbb{Z}} \to (\mathcal{C})_{\mathbb{Z}}[h_{\mathbb{Z}}]$ if we restrict to graded tensor categories without unit which have no 2-torsion in the Hom-groups.

As an example, we have the one-point category 1, which we give a unique structure of a symmetric monoidal category; the extension ± 1 of 1 to a $\mathbb{Z}/2$-**Sets** category has the structure of a graded symmetric monoidal category. Now suppose we adjoin morphisms

$$h_i:1 \to 1; \qquad i = 1, \dots, n,$$
$$x_j:1 \to 1; \qquad j = 1, \dots, m,$$

with the h_i of degree 1 and the x_j of degree 0, forming the graded symmetric monoidal category $\pm 1[h_1, \dots, h_n; x_1, \dots, x_m]$. The graded tensor category

$$(\pm 1[h_1, \dots, h_n; x_1, \dots, x_m])_{\mathbb{Z}}$$

is then isomorphic to an exterior algebra on the h_i, tensored with a polynomial algebra on the x_i. If however, we first form the graded tensor category $(\pm 1)_{\mathbb{Z}}$, and then adjoin the h_i and the x_j, forming the graded tensor category

$$(\pm 1)_{\mathbb{Z}}[h_1, \dots, h_n; x_1, \dots, x_m],$$

we have the relations

$$h_i \otimes h_i \neq 0,$$
$$2(h_i \otimes h_i) = 0.$$

This seems to indicate that the "correct" answer is to first adjoin morphisms as a graded symmetric (semi-)monoidal category, and then form the tensor category, rather than the other way around.

3.1.9. LEMMA. *Let $(\mathcal{C}, \bullet)$ be the graded symmetric semi-monoidal category gotten from $\pm\mathfrak{N}$ by adjoining morphisms $h_i:n_i \to n_i$ of degree $p_i < 0$, $i \in \mathcal{I}$. Then*
(i) $(\mathcal{C}, \bullet)$ is punctual.
(ii) If $\mathcal{C}[h]$ is gotten from $\mathcal{C}$ by adjoining a morphism $h:n \to n$ of degree $p < 0$, then the natural map

$$i_{m,q}:\mathrm{Hom}_{\mathcal{C}}(m, m)^q \to \mathrm{Hom}_{\mathcal{C}[h]}(m, m)^q$$

is injective for all m and q, an isomorphism for all q if $m < n$, and an isomorphism for all $q > p$ if $m = n$. In addition,

$$\mathrm{Hom}_{\mathcal{C}[h]}(n, n)^p = i_{n,p}(\mathrm{Hom}_{\mathcal{C}}(n, n)^p) \coprod \coprod_{\sigma,\rho \in S_n} \pm t_\sigma \circ h \circ t_\rho.$$

PROOF. For $a \in \mathbb{N}$, let 1_a denote the identity morphism on a; for $a, b \geq 0$ integers and $i \in \mathcal{I}$, we let $1_a \bullet h_i \bullet 1_b$ denote the corresponding morphism $a + n_i + b \to a + n_i + b$ in $\mathcal{C}$, where $1_a \bullet h_i$ is the map h_i if $a = 0$, and similarly for $h_i \bullet 1_b$. Then every morphism $f:m \to m$ of degree q in $\mathcal{C}$ can be written as a composition

$$f = t_{\sigma_0} \circ (1_{a_1} \bullet h_{i_1} \bullet 1_{b_1}) \circ t_{\sigma_1} \circ \dots \circ t_{\sigma_{s-1}} \circ (1_{a_s} \bullet h_{i_s} \bullet 1_{b_s}) \circ t_{\sigma_s} \tag{3.1.9.1}$$

with the $\sigma_i \in S_m$, and

$$a_j + n_{i_j} + b_j = m,$$
$$\sum_j p_{i_j} = q.$$

The representation of f as such a composition is of course not unique, however, if we define $\sigma(f) \in S_m \cup \{*\}$ by

$$\sigma(f) = \sigma_0 \cdot \sigma_1 \cdot \ldots \cdot \sigma_s$$

if $f \neq *$, and $\sigma(*) = *$, then we claim that $\sigma(f)$ depends only on f. Indeed, by §3.1.7, the relations among the t_σ and the maps $1_a \bullet h_i \bullet 1_b$ are generated by relations of the form

(a) $(1_a \bullet h_i \bullet 1_b) \circ (1_c \bullet h_j \bullet 1_d) = \pm(1_c \bullet h_j \bullet 1_d) \circ (1_a \bullet h_i \bullet 1_b)$ if $a + n_i \leq c$, $a + n_i + b = c + n_j + d$,
(b) $t_\sigma \circ (1_a \bullet h_i \bullet 1_b) = (1_c \bullet h_i \bullet 1_d) \circ t_\sigma$ if $\sigma(a+j) = c+j$ for $j = 1, \ldots, n_i$, and $d = m - c - n_i$.
(c) $t_\sigma \circ t_\rho = t_{\sigma\rho}$.

As these relations leave the product defining $\sigma(f)$ unchanged, our claim is verified. As

$$\sigma(t_\rho \circ f) = \rho \cdot \sigma(f), \ \sigma(f \circ t_\rho) = \sigma(f) \cdot \rho$$

for $f \neq *$, it follows that the action of S_m on $\mathrm{Hom}_{\mathcal{C}}(m, m) \setminus \{*\}$ is free. The remaining identities required to show that $\mathcal{C}$ is punctual follow from the representation (3.1.9.1) of an arbitrary morphism in $\mathcal{C}$. This completes the proof of (i).

For (ii), we note that the functor $i: \mathcal{C} \to \mathcal{C}[h]$ is split by the functor $\mathcal{C}[h] \to \mathcal{C}$ sending h to $*$, hence i is injective on the Hom sets. The assertions of (ii) then follows from the representation (3.1.9.1), the corresponding representation of a morphism in $\mathcal{C}[h]$, and noting that there are no relations of type (a) or (b) among the morphisms $\pm t_\sigma \circ h \circ t_\rho$. □

We note that the category $\pm\mathfrak{N}_{\mathbb{Z}}$ is just the free graded tensor category without unit generated by one object; we denote this category by $\mathcal{E}$, and the generating object by $\mathfrak{e}$. Applying Remark 3.1.8 and Lemma 3.1.9 we find

3.1.10. PROPOSITION. *Let $(\mathcal{A}, \otimes, \tau)$ be the graded tensor category without unit gotten from $\mathcal{E}$ by adjoining morphisms $h_i: \mathfrak{e}^{\otimes n_i} \to \mathfrak{e}^{\otimes n_i}$ of degree $p_i < 0$ and taking the quotient of the Hom-groups by their 2-torsion subgroup. Let $\mathcal{A}[h]$ be the graded tensor category without unit gotten from $\mathcal{A}$ by adjoining a morphism $h: \mathfrak{e}^{\otimes n} \to \mathfrak{e}^{\otimes n}$ of degree $p < 0$ and taking the quotient of the Hom-groups by their 2-torsion. Then*
(i) $\mathcal{A} = \mathcal{C}_{\mathbb{Z}}$ for a punctual graded symmetric semi-monoidal category $\mathcal{C}$; in particular, sending $\sigma \in S_n$ to the symmetry isomorphism τ_σ gives an isomorphism

$$\mathbb{Z}[S_n] \cong \mathrm{Hom}_{\mathcal{A}}(\mathfrak{e}^{\otimes n}, \mathfrak{e}^{\otimes n})^0.$$

(ii) The abelian groups $\mathrm{Hom}_{\mathcal{A}}(\mathfrak{e}^{\otimes m}, \mathfrak{e}^{\otimes n})^q$ are zero if $n \neq m$ or if $q > 0$.
(iii) If $\mathrm{Hom}_{\mathcal{A}}(\mathfrak{e}^{\otimes n}, \mathfrak{e}^{\otimes n})^q \neq 0$, then $\mathrm{Hom}_{\mathcal{A}}(\mathfrak{e}^{\otimes n}, \mathfrak{e}^{\otimes n})^q$ is a free $\mathbb{Z}[S_n]$-module, for the left, resp. right, action of $\mathrm{Hom}_{\mathcal{A}}(\mathfrak{e}^{\otimes n}, \mathfrak{e}^{\otimes n})^0 = \mathbb{Z}[S_n]$, with basis a set of representatives in $\mathrm{Hom}_{\mathcal{C}}(\mathfrak{e}^{\otimes n}, \mathfrak{e}^{\otimes n})^q$ for the action of $\mathrm{Hom}_{\mathcal{C}}(\mathfrak{e}^{\otimes n}, \mathfrak{e}^{\otimes n})^0 = \{\pm 1\} \times S_n$ by left, resp. right, composition.
(iv) The natural map

$$i_{m,q}: \mathrm{Hom}_{\mathcal{A}}(\mathfrak{e}^{\otimes m}, \mathfrak{e}^{\otimes m})^q \to \mathrm{Hom}_{\mathcal{A}[h]}(\mathfrak{e}^{\otimes m}, \mathfrak{e}^{\otimes m})^q$$

is a split injection for all m and q, an isomorphism for all q if $m < n$, and an isomorphism for all $q > p$ if $m = n$.

(v) The morphism h generates a free $\mathbb{Z}[S_n] \otimes \mathbb{Z}[S_n]^{\mathrm{op}}$ summand of the group $\mathrm{Hom}_{\mathcal{A}[h]}(\mathfrak{e}^{\otimes n}, \mathfrak{e}^{\otimes n})^p$, and we have the direct sum decomposition

$$\mathrm{Hom}_{\mathcal{A}[h]}(\mathfrak{e}^{\otimes n}, \mathfrak{e}^{\otimes n})^p = i_{n,p}(\mathrm{Hom}_{\mathcal{A}}(\mathfrak{e}^{\otimes n}, \mathfrak{e}^{\otimes n})^p) \oplus \oplus_{\sigma,\rho \in S_n} \mathbb{Z} \cdot \tau_\sigma \circ h \circ \tau_\rho.$$

3.1.11. *The construction.* We now define a sequence of DG tensor categories without unit

$$\mathcal{E} = \mathcal{E}_0 \to \mathcal{E}_1 \to \mathcal{E}_2 \to \ldots \to \mathcal{E}_n \to \ldots .$$

We form $\mathcal{E}_{i+1}$ from $\mathcal{E}_i$ as a graded tensor category without unit by adjoining morphisms h of the type considered in Proposition 3.1.10, taking the quotient by the 2-torsion, and then setting $dh = f$ for certain morphisms f in $\mathcal{E}_i$ with $df = 0$. Each $\mathcal{E}_{i+1}$ will be given as a inductive limit of a sequence

$$\mathcal{E}_i = \mathcal{E}_{i+1,1} \to \mathcal{E}_{i+1,2} \to \ldots$$

formed in this way. Throughout the construction, we will take the quotient by the 2-torsion subgroups without further comment.

Form $\mathcal{E}_1$ from $\mathcal{E}$ by adjoining a morphism $h_\tau : \mathfrak{e}^{\otimes 2} \to \mathfrak{e}^{\otimes 2}$ of degree -1, with $dh_\tau = \tau_{\mathfrak{e},\mathfrak{e}} - \mathrm{id}_{\mathfrak{e}^{\otimes 2}}$. Let $\mathcal{E}_{2,1} = \mathcal{E}_1$. Let $H_{2,1}$ be the set of non-zero morphisms $f : \mathfrak{e}^{\otimes 2} \to \mathfrak{e}^{\otimes 2}$ of degree -1 with $df = 0$. For each $f \in H_{2,1}$, adjoin a morphism $h_f : \mathfrak{e}^{\otimes 2} \to \mathfrak{e}^{\otimes 2}$ of degree -2, with $dh_f = f$. This forms the DG tensor category without unit $\mathcal{E}_{2,2}$.

Suppose we have formed the category $\mathcal{E}_{2,r}$ for some $r \geq 2$. Let $H_{2,r}$ be the set of non-zero morphisms $f : \mathfrak{e}^{\otimes 2} \to \mathfrak{e}^{\otimes 2}$ of degree $-r$ with $df = 0$. For each $f \in H_{2,r}$, adjoin a morphism $h_f : \mathfrak{e}^{\otimes 2} \to \mathfrak{e}^{\otimes 2}$ of degree $-r-1$, with $dh_f = f$. This forms the DG tensor category without unit $\mathcal{E}_{2,r+1}$. We let $\mathcal{E}_2$ be the inductive limit

$$\mathcal{E}_2 = \varinjlim_r \mathcal{E}_{2,r}.$$

Now suppose we have formed the category $\mathcal{E}_{k-1}$ for some $k \geq 3$. Let $\mathcal{E}_{k,1} = \mathcal{E}_k$. Suppose we have formed the category $\mathcal{E}_{k,r}$ for some $r \geq 1$. Let $H_{k,r}$ be the set of non-zero morphisms $f : \mathfrak{e}^{\otimes k} \to \mathfrak{e}^{\otimes k}$ of degree $-r$ such that $df = 0$. For each $f \in H_{k,r}$, adjoin a morphism $h_f : \mathfrak{e}^{\otimes k} \to \mathfrak{e}^{\otimes k}$ of degree $-r-1$, with $dh_f = f$. This forms the DG tensor category without unit $\mathcal{E}_{k,r+1}$. We let $\mathcal{E}_k$ be the inductive limit

$$\mathcal{E}_k = \varinjlim_r \mathcal{E}_{k,r}.$$

We let $\mathbb{E}$ be the inductive limit

$$\mathbb{E} = \varinjlim_k \mathcal{E}_k.$$

We call the category $\mathbb{E}$ the *homotopy one-point DG tensor category.*

3.1.12. PROPOSITION. *(i) $\mathbb{E}$ is a DG tensor category without unit, with objects generated by the object $\mathfrak{e}$. As a graded tensor category, $\mathbb{E} = \mathcal{C}_{\mathbb{Z}}^{\oplus}$ for a punctual graded symmetric semi-monoidal category $\mathcal{C}$.*
(ii) For $m \neq n$, we have $\mathrm{Hom}_{\mathbb{E}}(\mathfrak{e}^{\otimes m}, \mathfrak{e}^{\otimes n})^q = 0$ for all q.
(iii) For all $q > 0$, $\mathrm{Hom}_{\mathbb{E}}(\mathfrak{e}^{\otimes n}, \mathfrak{e}^{\otimes n})^q = 0$. The map sending $\sigma \in S_n$ to the symmetry isomorphism τ_σ gives an isomorphism

$$\mathrm{Hom}_{\mathbb{E}}(\mathfrak{e}^{\otimes n}, \mathfrak{e}^{\otimes n})^0 \cong \mathbb{Z}[S_n].$$

If $\mathrm{Hom}_{\mathbb{E}}(\mathfrak{e}^{\otimes n}, \mathfrak{e}^{\otimes n})^q \neq 0$, then $\mathrm{Hom}_{\mathbb{E}}(\mathfrak{e}^{\otimes n}, \mathfrak{e}^{\otimes n})^q$ is a free $\mathbb{Z}[S_n]$-module, for the action of $\mathrm{Hom}_{\mathbb{E}}(\mathfrak{e}^{\otimes n}, \mathfrak{e}^{\otimes n})^0$ by left, resp. right, composition, with basis being a set

of representatives in $\mathrm{Hom}_{\mathcal{C}}(\mathfrak{e}^{\otimes n}, \mathfrak{e}^{\otimes n})^q \setminus \{*\}$ *for the action of* $\mathrm{Hom}_{\mathcal{C}}(\mathfrak{e}^{\otimes n}, \mathfrak{e}^{\otimes n})^0 = \{\pm 1\} \times S_n$ *by left, resp. right, composition.*
(iv) We have $H^0(\mathrm{Hom}_{\mathbb{E}}(\mathfrak{e}^{\otimes n}, \mathfrak{e}^{\otimes n})^*) = \mathbb{Z}$*, with generator the class of the identity map, and* $H^q(\mathrm{Hom}_{\mathbb{E}}(\mathfrak{e}^{\otimes n}, \mathfrak{e}^{\otimes n})^*) = 0$ *for* $q \neq 0$*.*

PROOF. The assertions (i)-(iii) follows by construction and Proposition 3.1.10. For (iv), the assertion about the cohomology H^q for $q \neq 0$ follows from the construction, together with Proposition 3.1.10(iii).

To compute the H^0, take an integer $n \geq 2$. Let σ_n be the element of S_n which exchanges 1 and 2. For $\sigma \in S_n$, we have

$$d(t_\sigma \circ (h_\tau \otimes \mathrm{id}_{\mathfrak{e}^{\otimes n-2}}) \circ t_\sigma^{-1}) = \mathrm{id}_{\mathfrak{e}^{\otimes n}} - t_{\sigma\sigma_n\sigma^{-1}}. \tag{3.1.12.1}$$

As the normal subgroup of S_n generated by σ_n is all of S_n, this shows that the class of the identity map generates the H^0. On the other hand, the only morphism of degree -1 adjoined to $\mathcal{E}$ to form $\mathbb{E}$ is the morphism h_τ. By degree considerations, the group of boundaries

$$d(\mathrm{Hom}_{\mathbb{E}}(\mathfrak{e}^{\otimes n}, \mathfrak{e}^{\otimes n})^{-1}) \subset \mathrm{Hom}_{\mathbb{E}}(\mathfrak{e}^{\otimes n}, \mathfrak{e}^{\otimes n})^0 = \mathbb{Z}[S_n]$$

is the $\mathbb{Z}[S_n] \otimes \mathbb{Z}[S_n]^{\mathrm{op}}$-submodule of $\mathbb{Z}[S_n]$ generated by the elements in (3.1.12.1). In particular, the boundaries are contained in the augmentation ideal of $\mathbb{Z}[S_n]$, hence $H^0 = \mathbb{Z} \cdot \mathrm{id}$, as claimed. □

The category $\mathbb{E}$ has the following universal mapping property:

3.1.13. PROPOSITION. *Let* $\mathcal{A}$ *be a DG tensor category without unit, and let* P *be an object of* $\mathcal{A}$ *such that*

(a) *For each* $n > 0$ *and each* $\sigma \in S_n$, *we have*

$$\tau_\sigma = \mathrm{id}_{P^{\otimes n}} \in H^0(\mathrm{Hom}_{\mathcal{A}}(P^{\otimes n}, P^{\otimes n}))$$

where $\tau_\sigma : P^{\otimes n} \longrightarrow P^{\otimes n}$ *is the symmetry isomorphism.*

(b) *For* $q < 0$ *and* $n > 0$, $H^q(\mathrm{Hom}_{\mathcal{A}}(P^{\otimes n}, P^{\otimes n})) = 0$.

(c) *For* $q < 0$ *and* $n > 0$, $\mathrm{Hom}_{\mathcal{A}}(P^{\otimes n}, P^{\otimes n})^q$ *is 2-torsion free.*

Then there is a DG tensor functor $\rho_P : \mathbb{E} \longrightarrow \mathcal{A}$ *with*

$$\rho_P(\mathfrak{e}^{\otimes n}) = P^{\otimes n}; \qquad n = 1, 2, \ldots. \tag{3.1.13.1}$$

In addition, the condition (3.1.13.1) *determines* ρ_P *uniquely up to homotopy.*

PROOF. This follows from the construction (see §3.1.11) of $\mathbb{E}$ as a freely generated graded tensor category, modulo 2-torsion, together with Proposition 3.1.12 and Remark 3.1.8. □

3.2. Homotopy commutativity

For a tensor category $\mathcal{C}$, we give a construction of a DG tensor category which is homotopy equivalent to the universal commutative external product $\mathcal{C}^{\otimes,c}$ of Chapter I, §2.4.2. This DG tensor category is freely generated over the free tensor category on $\mathcal{C}$ (as a tensor category) which makes it appropriate for a type of "acyclic models" argument.

3.2.1. *The homotopy commutative* Ω_0. Recall the symmetric semi-monoidal categories ω and Ω_0 of Chapter I, §2.3, and the functor $p: \Omega_0 \to \omega$ which is the identity on objects, and sends a morphism (f, σ) to $f \circ \sigma$.

Let $\mathbb{Z}\Omega_0$ be the pre-additive category generated by Ω_0, i.e., objects are the objects of Ω_0, and the morphisms $\mathrm{Hom}_{\mathbb{Z}\Omega_0}(a,b)$ are $\mathbb{Z}[\mathrm{Hom}_{\Omega_0}(a,b)]$ The symmetric semi-monoidal structure on Ω_0 gives $\mathbb{Z}\Omega_0$ the structure of a pre-tensor category without unit; we consider $\mathbb{Z}\Omega_0$ as a pre-DG tensor category with all morphisms in degree zero, and with zero differential.

We form the pre-DG tensor category without unit $\mathbb{Z}\Omega^\natural$ as follows: For all $a \geq 1$, we set

$$\mathrm{Hom}_{\mathbb{Z}\Omega^\natural}(a,a)^p = \begin{cases} 0; & \text{for } p \neq 0, \\ \mathbb{Z}[\mathrm{Hom}_{\Omega_0}(a,a)] = \mathbb{Z}[S_a]; & \text{for } p = 0, \end{cases}$$

with zero differential.

Recall from (I.2.3.3.5) the morphism $F_{21}: 2 \to 1$ and the symmetry $\tau_\sigma: 2 \to 2$, where $\sigma \in S_2$ is the non-trivial permutation.

Form the complex $\mathrm{Hom}_{\mathbb{Z}\Omega^\natural}(2,1)^*$ by first adjoining a free right $\mathrm{Hom}_{\mathbb{Z}\Omega^\natural}(2,2)$-module with generator $h: 2 \to 1$ of degree -1, with differential $dh = F_{21} - F_{21} \circ \tau_\sigma$. We write $h \circ \tau_\sigma$ for $h \cdot \sigma$ and define $d(h \circ \tau_\sigma) = dh \circ \tau_\sigma$.

We now define the complex $\mathrm{Hom}_{\mathbb{Z}\Omega^\natural}(2,1)^*$ inductively. Suppose we have defined the complex $\mathrm{Hom}_{\mathbb{Z}\Omega^\natural}(2,1)^*$ in degrees $-r \leq * \leq 0$, with $\mathrm{Hom}_{\mathbb{Z}\Omega^\natural}(2,1)^* = 0$ for $* > 0$. Define $\mathrm{Hom}_{\mathbb{Z}\Omega^\natural}(2,1)^{-r-1}$ to be the free right $\mathrm{Hom}_{\mathbb{Z}\Omega^\natural}(2,2)$-module on $Z^{-r}(\mathrm{Hom}_{\mathbb{Z}\Omega^\natural}(2,1)^*) \setminus \{0\}$. Denote the element of $\mathrm{Hom}_{\mathbb{Z}\Omega^\natural}(2,1)^{-r-1}$ corresponding to $g \in Z^{-r}(\mathrm{Hom}_{\mathbb{Z}\Omega^\natural}(2,1)^*)$ by h_g, and define the differential as above by $d(h_g \circ \tau_\sigma) = g \circ \tau_\sigma$.

Now suppose we have defined the complexes $\mathrm{Hom}_{\mathbb{Z}\Omega^\natural}(a,1)^*$ for $n > a \geq 2$, together with a right action (right composition) by $\mathrm{Hom}_{\mathbb{Z}\Omega^\natural}(a,a)^0 = \mathbb{Z}[S_a]$. In addition, we assume

(3.2.1.1)

1.

$$\mathrm{Hom}_{\mathbb{Z}\Omega^\natural}(a,1)^r = \begin{cases} 0 & \text{for } r > 0, \\ \mathbb{Z}\mathrm{Hom}_{\Omega_0}(a,1) & \text{for } r = 0. \end{cases}$$

2. $H^r(\mathrm{Hom}_{\mathbb{Z}\Omega^\natural}(a,1)^*) = 0$ for $r < 0$, and the surjection $\mathbb{Z}\mathrm{Hom}_{\Omega_0}(a,1) \to H^0(\mathrm{Hom}_{\mathbb{Z}\Omega^\natural}(a,1)^*)$ identifies $H^0(\mathrm{Hom}_{\mathbb{Z}\Omega^\natural}(a,1)^*)$ with $\mathbb{Z}[\mathrm{Hom}_\omega(a,1)]$.
3. $\mathrm{Hom}_{\mathbb{Z}\Omega^\natural}(a,1)^r$ is free as a right $\mathrm{Hom}_{\mathbb{Z}\Omega^\natural}(a,a)^0$-module.

Let $f: \underline{n} \to \underline{b}$ be a order-preserving surjection, i.e, a morphism $f: n \to b$ in ω_0. For $2 \leq b \leq n-1$, we define $\mathrm{Hom}_{\mathbb{Z}\Omega^\natural}(n,b)^*_f$ by

$$\mathrm{Hom}_{\mathbb{Z}\Omega^\natural}(n,b)^*_f := \otimes_{i=1}^b \mathrm{Hom}_{\mathbb{Z}\Omega^\natural}(|f^{-1}(i)|, 1)^*. \tag{3.2.1.2}$$

For $f: n \to b$ in ω_0, we have the subgroup $S(f)$ of S_n consisting of those permutations η with $f\eta = f$; explicitly, $S(f)$ is the product $S_{f^{-1}(1)} \times \ldots \times S_{f^{-1}(b)}$ embedded in S_n in the obvious way. $S(f)$ acts on $\mathrm{Hom}_{\mathbb{Z}\Omega^\natural}(n,b)^*_f$ on the right via the right action of $S_{f^{-1}(1)} \times \ldots \times S_{f^{-1}(b)}$ as right composition on each factor.

Define the complex $\mathrm{Hom}_{\mathbb{Z}\Omega^\natural}(n,b)^*$ by

$$\text{(3.2.1.3)} \qquad \mathrm{Hom}_{\mathbb{Z}\Omega^\natural}(n,b)^* = \bigoplus_{f\in\mathrm{Hom}_{\omega_0}(n,b)} \mathrm{Hom}_{\mathbb{Z}\Omega^\natural}(n,b)^*_f \otimes_{\mathbb{Z}[S(f)]} \mathbb{Z}[S_n].$$

It follows from (3.2.1.1) that $\mathrm{Hom}_{\mathbb{Z}\Omega^\natural}(n,b)^*_f$ is a free $\mathbb{Z}[S(f)]^{\mathrm{op}}$-module, hence $\mathrm{Hom}_{\mathbb{Z}\Omega^\natural}(n,b)^*$ is a free right $\mathbb{Z}[S_n] = \mathrm{Hom}_{\mathbb{Z}\Omega^\natural}(n,n)^0$-module, and the cohomology of $\mathrm{Hom}_{\mathbb{Z}\Omega^\natural}(n,b)^*$ is given as

$$\text{(3.2.1.4)} \qquad \begin{aligned} H^0(\mathrm{Hom}_{\mathbb{Z}\Omega^\natural}(n,b)^*) &= \bigoplus_{f\in\mathrm{Hom}_{\omega_0}(n,b)} \mathbb{Z}[\mathrm{Hom}_{\Omega^0}(n,b)_f] \otimes_{\mathbb{Z}[S(f)]} \mathbb{Z}[S_n] \\ &= \mathbb{Z}[\mathrm{Hom}_\omega(n,b)]. \\ H^p(\mathrm{Hom}_{\mathbb{Z}\Omega^\natural}(n,b)^*) &= 0 \text{ for } p\neq 0. \end{aligned}$$

For $\rho\in S_b$, $f\in S_{n\to b}$, recall from (2.3.3.1)(ii) the construction of the element $\rho\cdot f$ of $S_{n\to b}$, and the map $f^*: S_b\to S_n$. For the sequence of non-negative integers $d := (d_1,\dots,d_b)$, let $\mathrm{sgn}^d(\rho)$ be the weighted sign of the permutation ρ, where we give j weight d_j. We make $\mathrm{Hom}_{\mathbb{Z}\Omega^\natural}(n,b)^*$ a left $\mathbb{Z}[S_b]$-module by the action

$$\rho\cdot : \mathrm{Hom}_{\mathbb{Z}\Omega^\natural}(n,b)^*_f \otimes_{\mathbb{Z}[S(f)]} \mathbb{Z}[S_n] \to \mathrm{Hom}_{\mathbb{Z}\Omega^\natural}(n,b)^*_{\rho\cdot f} \otimes_{\mathbb{Z}[S(f)]} \mathbb{Z}[S_n]$$

defined by

$$\rho\cdot[(x_1\otimes\ldots\otimes x_b)\otimes\sigma] := \mathrm{sgn}^{d(x)}(\rho)(x_{\rho^{-1}(1)}\otimes\ldots\otimes x_{\rho^{-1}(b)})\otimes f^*(\rho)\sigma,$$

where $d(x)$ is the sequence $(\deg(x_1),\dots,\deg(x_b))$. We define

$$\mathrm{Hom}_{\mathbb{Z}\Omega^\natural}(n,1)^*_0 := \bigoplus_{b=2}^{n-1} \mathrm{Hom}_{\mathbb{Z}\Omega^\natural}(b,1)^* \otimes_{\mathbb{Z}[S_b]} \mathrm{Hom}_{\mathbb{Z}\Omega^\natural}(n,b)^*.$$

Then $\mathrm{Hom}_{\mathbb{Z}\Omega^\natural}(n,1)^p_0 = 0$ for $p>0$, $\mathrm{Hom}_{\mathbb{Z}\Omega^\natural}(n,1)^p_0$ is a free right $\mathrm{Hom}_{\mathbb{Z}\Omega^\natural}(n,n)^*$-module for each $p\leq 0$, and

$$\text{(3.2.1.5)} \qquad H^0(\mathrm{Hom}_{\mathbb{Z}\Omega^\natural}(n,1)^*_0) \cong \bigoplus_{b=1}^{n-1} \mathbb{Z}[\mathrm{Hom}_\omega(b,1)] \otimes_{\mathbb{Z}[S_b]} \mathbb{Z}[\mathrm{Hom}_\omega(n,b)].$$

Using the isomorphism (3.2.1.5) and the composition in ω, we have the map

$$\mathrm{Hom}_{\mathbb{Z}\Omega^\natural}(n,1)^0_0 \to \mathbb{Z}[\mathrm{Hom}_\omega(n,1)],$$

which is evidently surjective. As generators for the kernel of

$$\circ : \bigoplus_{b=1}^{n-1} \mathbb{Z}[\mathrm{Hom}_\omega(b,1)] \otimes_{\mathbb{Z}[S_b]} \mathbb{Z}[\mathrm{Hom}_\omega(n,b)] \to \mathbb{Z}[\mathrm{Hom}_\omega(n,1)],$$

we have the elements

$$s_{g,g'} := f_{b1}\otimes g - f_{b'1}\otimes g',$$

where $g: n\to b$ and $g': n\to b'$ are maps in ω, and $f_{b1}: b\to 1$ is the unique morphism in ω. Choose $\tilde{s}_{g,g'}$ in $\mathrm{Hom}_{\mathbb{Z}\Omega^\natural}(n,1)^0_0$ lifting $s_{g,g'}$, and adjoin to $\mathrm{Hom}_{\mathbb{Z}\Omega^\natural}(n,1)^{-1}_0$ a free right $\mathbb{Z}[S_n]$-module with basis $h_{g,g'}$, and with

$$d(h_{g,g'}\circ\tau_\sigma) = \tilde{s}_{g,g'}\circ\tau_\sigma.$$

We then adjoin a free right $\mathbb{Z}[S_n]$-module to $\mathrm{Hom}_{\mathbb{Z}\Omega^\natural}(n,1)^*_0$ in each degree $r<-1$, to kill the cohomology in degrees ≤ -1, as in the case $n=2$. This forms the complex $\mathrm{Hom}_{\mathbb{Z}\Omega^\natural}(n,1)^*$.

Define the composition

$$\mathrm{Hom}_{\mathbb{Z}\Omega^\flat}(b,1)^* \otimes \mathrm{Hom}_{\mathbb{Z}\Omega^\flat}(n,b)^* \to \mathrm{Hom}_{\mathbb{Z}\Omega^\flat}(n,1)^*$$

for $n > b \geq 1$ by mapping to the corresponding summand in $\mathrm{Hom}_{\mathbb{Z}\Omega^\flat}(n,1)^*_0$, and then including in the complex $\mathrm{Hom}_{\mathbb{Z}\Omega^\flat}(n,1)^*$ by the canonical map. This gives us the composition

$$\circ\colon \mathrm{Hom}_{\mathbb{Z}\Omega^\flat}(a,b)^*_f \otimes \mathrm{Hom}_{\mathbb{Z}\Omega^\flat}(n,a)^*_g \to \mathrm{Hom}_{\mathbb{Z}\Omega^\flat}(n,b)^*_{f\circ g}$$

by

$$(x_1 \otimes \ldots x_b) \circ (y_1 \otimes \ldots \otimes y_a) = \\ [x_1 \circ (y_1 \otimes \ldots \otimes y_{a_1})] \otimes \ldots \otimes [x_b \circ (y_{a_{b-1}+1} \otimes \ldots \otimes y_b)],$$

where the order-preserving surjection f is given by

$$f^{-1}(i) = \{a_{i-1}+1 < \ldots < a_i\}; \quad i = 1, \ldots, b, \; a_0 = 0, a_b = b.$$

Now take $n > a > b \geq 2$, and define the composition

$$\left(\mathrm{Hom}_{\mathbb{Z}\Omega^\flat}(a,b)^*_f \otimes \rho\right) \otimes \left(\mathrm{Hom}_{\mathbb{Z}\Omega^\flat}(n,a)^*_g \otimes \mathbb{Z}[S_n]\right) \to \mathrm{Hom}_{\mathbb{Z}\Omega^\flat}(n,b)^*_{f\circ(\rho\cdot g)} \otimes \mathbb{Z}[S_n]$$

$(\rho \in S_a)$ by

$$(x \otimes \rho) \otimes (y_1 \otimes \ldots \otimes y_a \otimes \sigma) \mapsto \mathrm{sgn}^{d(y)}(\rho)(x \circ (y_{\rho^{-1}(1)} \otimes \ldots \otimes y_{\rho^{-1}(a)})) \otimes f^*(\rho)\sigma.$$

Taking the tensor product of $\mathrm{Hom}_{\mathbb{Z}\Omega^\flat}(a_1,b_1)^*_{f_1}$ and $\mathrm{Hom}_{\mathbb{Z}\Omega^\flat}(a_2,b_2)^*_{f_2}$, and mapping to the obvious summand in $\mathrm{Hom}_{\mathbb{Z}\Omega^\flat}(a_1+a_2, b_1+b_2)^*_{f_1+f_2}$ gives a tensor operation on morphisms. It is tedious, but easy, to check that this data defines a pre-DG tensor category $\mathbb{Z}\Omega^\flat$.

By construction, we have a functor of pre-DG tensor categories $j\colon \mathbb{Z}\Omega_0 \to \mathbb{Z}\Omega^\flat$, which gives an identification

$$\mathrm{Hom}_{\mathbb{Z}\Omega_0}(a,b) = \mathbb{Z}[\mathrm{Hom}_{\Omega_0}(a,b)] \cong \mathrm{Hom}_{\mathbb{Z}\Omega^\flat}(a,b)^0 = Z^0(\mathrm{Hom}_{\mathbb{Z}\Omega^\flat}(a,b)^*).$$

Also by construction (see (3.2.1.4)), the complex $\mathrm{Hom}_{\mathbb{Z}\Omega^\flat}(a,b)^*$ satisfies

(3.2.1.6)

1. $\mathrm{Hom}_{\mathbb{Z}\Omega^\flat}(a,b)^n = 0$ for $n > 0$ or $a < b$.
2. $H^n(\mathrm{Hom}_{\mathbb{Z}\Omega^\flat}(a,b)^*) = 0$ for $n < 0$.
3. The natural map $\mathbb{Z}[\mathrm{Hom}_{\Omega_0}(a,b)] \to H^0(\mathrm{Hom}_{\mathbb{Z}\Omega^\flat}(a,b)^*)$ gives an identification of $H^0(\mathrm{Hom}_{\mathbb{Z}\Omega^\flat}(a,b)^*)$ with $\mathbb{Z}[\mathrm{Hom}_\omega(a,b)]$.

3.2.2. REMARK. (see e.g. [**65**]) Letting $\mathcal{O}(n) = \mathrm{Hom}_{\mathbb{Z}\Omega^\flat}(n,1)^*$, the collection

$$\mathcal{O}(1), \mathcal{O}(2), \ldots$$

forms an *operad* in the category of cochain complexes, i.e., each $\mathcal{O}(n)$ is a right $\mathbb{Z}[S_n]$-module, and there are "substitution laws":

$$\circ_{i,n,m}\colon \mathcal{O}(n) \otimes \mathcal{O}(m) \to \mathcal{O}(n+m-1); \quad i = 1, \ldots, m,$$

satisfying

1. $\circ_{i,n,m+l-1} \circ (\mathrm{id} \otimes \circ_{j,m,l}) = \circ_{i+j-1,n+m-1,l} \circ (\circ_{i,n,m} \otimes \mathrm{id})$.

2. If we let $\rho_{i,n,m}: S_n \times S_m \to S_{n+m-1}$ be the homomorphism gotten by identifying the ordered set
$$1_n < 2_n < \ldots < (i-1)_n < 1_m < \ldots < m_m < (i+1)_n < \ldots < n_n$$
with $\underline{n+m-1}$, we have
$$\sigma(x) \circ_{i,n,m} \tau(y) = \rho_{i,n,m}(\sigma, \tau)(x \circ_{i,n,m} y).$$

It is not difficult to show that each operad $\mathcal{O}(*)$ in cochain complexes gives rise to a pre-DG tensor category $\mathcal{O}$ with objects $1, 2, \ldots$, and Hom-complexes
$$\mathrm{Hom}_{\mathcal{O}}(n, 1)^* = \mathcal{O}(n),$$
where the general Hom-complex $\mathrm{Hom}_{\mathcal{O}}(a, b)^*$ is given by a formula as in (3.2.1.3):
$$\mathrm{Hom}_{\mathcal{O}}(a, b)^* = \oplus_{f \in \mathrm{Hom}_{\omega_0}(a,b)} [\otimes_{i=1}^b \mathrm{Hom}_{\mathcal{O}}(|f^{-1}(i)|, 1)^*] \otimes_{\mathbb{Z}[S(f)]} \mathbb{Z}[S_a].$$

3.2.3. *Decomposition into type.* Let $\mathcal{A}$ be a pre-DG category, with product $\otimes$. We say that $\mathcal{A}$ has a *decomposition into type* if

1. $\mathcal{A}$ has the same objects as ω, i.e., $1, 2, \ldots$.
2. For each a and b in $\mathcal{A}$, the Hom-complex $\mathrm{Hom}_{\mathcal{A}}(a, b)^*$ has a direct sum decomposition
$$\mathrm{Hom}_{\mathcal{A}}(a, b) = \oplus_{f \in \mathrm{Hom}_{\omega}(a,b)} \mathrm{Hom}_{\mathcal{A}}(a, b)^*_f;$$
in particular, $\mathrm{Hom}_{\mathcal{A}}(a, b)^* = \{0\}$ if $a < b$.

The direct sum decomposition in (2) satisfies

(i) $h \in \mathrm{Hom}_{\mathcal{A}}(a, b)^*_f, g \in \mathrm{Hom}_{\mathcal{A}}(b, c)^*_{f'} \Longrightarrow g \circ h \in \mathrm{Hom}_{\mathcal{A}}(a, c)^*_{f' \circ f}$.

(ii) $h \in \mathrm{Hom}_{\mathcal{A}}(a, b)^*_f, g \in \mathrm{Hom}_{\mathcal{A}}(a', b')^*_{f'}$
$$\Longrightarrow g \otimes h \in \mathrm{Hom}_{\mathcal{A}}(a \otimes a', b \otimes b')^*_{f'+f}.$$

(iii) $\mathrm{Hom}_{\mathcal{A}}(a, a)^*_f$ is concentrated in degree zero and is isomorphic to $\mathbb{Z}$, with generator $\tilde{f}$, such that $\tilde{f} \circ \tilde{f}' = \widetilde{ff'}$ for all f, f' in $\mathrm{Hom}_{\omega}(a, a)$.

Note that the condition (iii) is satisfied if there is a functor $\mathbb{Z}\Omega_0 \to \mathcal{A}$ which induces an isomorphism $\mathrm{Hom}_{\mathbb{Z}\Omega_0}(a, a) \to \mathrm{Hom}_{\mathcal{A}}(a, a)$ for each a.

Let $f: a \to b$ be a morphism in ω_0, and $\sigma \in S_a$. Giving the summand $\mathrm{Hom}_{\mathbb{Z}\Omega^{\mathfrak{h}}}(a, b)^*_f \otimes \mathbb{Z}[\sigma]$ of $\mathrm{Hom}_{\mathbb{Z}\Omega^{\mathfrak{h}}}(a, b)^*$ the type $f \circ \sigma$ defines a decomposition into type for $\mathbb{Z}\Omega^{\mathfrak{h}}$ (*cf.* (3.2.1.3)). This follows easily from the definition of composition and tensor product in $\mathbb{Z}\Omega^{\mathfrak{h}}$, together with the computation of $\mathrm{Hom}_{\mathbb{Z}\Omega^{\mathfrak{h}}}(a, a)^*$ in §3.2.1.

3.2.4. *The homotopy commutative $\mathcal{C}^{\otimes}$.* We refer to Chapter I, §2.4.3 for the notation. Let $(\mathcal{C}, \times, t)$ be a tensor category without unit. We have the 2-functor $\Pi_{\mathcal{C}}: \Omega \to \mathbf{cat_{Ab}}$ (I.2.3.6.1), and the restriction $\Pi^0_{\mathcal{C}}$ to Ω_0.

We let $(\Pi_{\mathcal{C}}, \Omega_0)^{\oplus}$ and $(\Pi_{\mathcal{C}}, \Omega)^{\oplus}$ denote the additive categories generated from the pre-additive categories $(\Pi_{\mathcal{C}}, \Omega_0)$ and $(\Pi_{\mathcal{C}}, \Omega)$. We may consider $(\Pi_{\mathcal{C}}, \Omega_0)^{\oplus}$ and $(\Pi_{\mathcal{C}}, \Omega)^{\oplus}$ as graded categories, or as DG categories with trivial graded and differential structure; the tensor structure on induces the structure of a tensor category (resp. graded tensor category, resp. DG tensor category) without unit on $(\Pi_{\mathcal{C}}, \Omega_0)^{\oplus}$ and $(\Pi_{\mathcal{C}}, \Omega)^{\oplus}$. The tensor category without unit $(\Pi_{\mathcal{C}}, \Omega)^{\oplus}$ is the category $\mathcal{C}^{\otimes, c}$ constructed in Chapter I, §2.4.3 and Remark 2.4.6.

The 2-functor $\Omega_0 \to \Omega$ gives rise to the functor $c_0: (\Pi_{\mathcal{C}}, \Omega_0) \to (\Pi_{\mathcal{C}}, \Omega)$; We let
$$\mathfrak{c}_{0\mathcal{C}}: (\Pi_{\mathcal{C}}, \Omega_0)^{\oplus} \to (\Pi_{\mathcal{C}}, \Omega)^{\oplus} = \mathcal{C}^{\otimes, c}$$

denote the tensor functor induced by c_0.

By (3.2.1.6), we have the canonical isomorphism of rings

$$\mathbb{Z}[S_a] \to Z^0(\mathrm{Hom}_{\mathbb{Z}\Omega^{\mathfrak{h}}}(a,a)^*);$$

we henceforth identify $\mathrm{Hom}_{\mathbb{Z}\Omega^{\mathfrak{h}}}(a,a)^0$ with $\mathbb{Z}[S_a]$ via this isomorphism.

Let $f_0: a \to b$ be a morphism in ω_0, take $\sigma \in S_a$, giving the morphism $F := (f_0,\sigma): a \to b$ in Ω_0, and let $p(F) := f_0\sigma$ be the resulting morphism in ω. Note that f_0 is uniquely determined by F. For objects $x_1,\dots,x_n$, $y_1,\dots,y_m$ of $\mathcal{C}$, and morphism $f: n \to m$ in ω, write $x := (x_1,\dots,x_n)$ and $y := (y_1,\dots,y_m)$, and set

$$\mathrm{Hom}(x,y)_f := \oplus_{\substack{F\in\mathrm{Hom}_{\Omega_0}(n,m)\\ p(F)=f}} \mathrm{Hom}_{\mathcal{C}^{\otimes n}}(\Pi(F)(x),y).$$

Denote a morphism g in the summand of $\mathrm{Hom}((x_1,\dots,x_n),(y_1,\dots,y_m))_f$ indexed by F by g_F.

If η is in $S(f)$, and $F = (f_0,\sigma)$ is a morphism from a to b in Ω_0, the natural isomorphism $\Pi(\eta): \Pi(F) \to \Pi(\eta\cdot F)$ gives the symmetry isomorphism

$$\Pi(\eta)(x): \Pi(F)(x) \to \Pi(\eta\cdot F)(x)$$

in $\mathcal{C}^{\otimes n}$. We define a left action of $S(f_0)$ on $\mathrm{Hom}(x,y)_f$ by

$$\eta\cdot g_F = (g\circ\Pi(\eta)(x)^{-1})_{\eta\cdot F}.$$

Since $S(f_0)$ acts freely on the set $p^{-1}(f)$, $S(f_0)$ acts freely on the abelian group $\mathrm{Hom}(x,y)_f$. The identification of $\mathrm{Hom}_{\mathbb{Z}\Omega^{\mathfrak{h}}}(a,a)^0$ with $\mathbb{Z}[S_a]$ gives the right action of $S(f_0)\subset S_a$ on $\mathrm{Hom}_{\mathbb{Z}\Omega^{\mathfrak{h}}}(a,b)^*_f$ by composition on the right.

Define the DG tensor category $\mathcal{C}^{\otimes,\mathfrak{h}}$ with the same objects as $(\Pi_{\mathcal{C}},\Omega_0)^{\oplus}$, and where the Hom-complexes are given as follows: For $x = (x_1,\dots,x_n)$ and $y = (y_1,\dots,y_m)$ set

$$\mathrm{Hom}(x,y)^* := \oplus_{f\in\mathrm{Hom}_\omega(n,m)}\mathrm{Hom}(x,y)^*_f, \tag{3.2.4.1}$$

with

$$\mathrm{Hom}(x,y)^*_f := \mathrm{Hom}_{\mathbb{Z}\Omega^{\mathfrak{h}}}(n,m)^*_f \otimes_{\mathbb{Z}[S(f_0)]} \mathrm{Hom}(x,y)_f.$$

Since $d\eta = 0$ for $\eta\in S(f_0)\subset S_a\subset Z^0(\mathrm{Hom}_{\mathbb{Z}\Omega^{\mathfrak{h}}}(a,a)^*)$, the differential on $\mathrm{Hom}_{\mathbb{Z}\Omega^{\mathfrak{h}}}(n,m)^*_f$ gives a well-defined differential on $\mathrm{Hom}(x,y)^*$.

Composition is given by

$$(h\otimes g_F)\circ(h'\otimes g'_{F'}) = (h\circ h')\otimes(g\circ\Pi(F)(g'))_{F'\circ F},$$

and the tensor operation is given by

$$(h\otimes g_F)\otimes(h'\otimes g'_{F'}) = (h\otimes h')\otimes(g\otimes g')_{F+F'};$$

one easily checks that these operations respect the $S(f_0)$-action, and that the result satisfies the axioms of a DG tensor category.

The inclusion $\Omega_0 \to \mathbb{Z}\Omega^{\mathfrak{h}}$ induces the functor $j_{\mathcal{C}}: (\Pi_{\mathcal{C}},\Omega_0)^{\oplus} \to \mathcal{C}^{\otimes,\mathfrak{h}}$, giving the diagram

$$\begin{array}{ccc} (\Pi_{\mathcal{C}},\Omega_0)^{\oplus} & \xrightarrow{j_{\mathcal{C}}} & \mathcal{C}^{\otimes,\mathfrak{h}} \\ {\scriptstyle c_0}\downarrow & & \\ (\Pi_{\mathcal{C}},\Omega)^{\oplus} & = & \mathcal{C}^{\otimes,c}. \end{array} \tag{3.2.4.2}$$

3.2.5. PROPOSITION. *There is a unique DG tensor functor*

$$\mathfrak{c}^{\mathfrak{h}} : \mathcal{C}^{\otimes,\mathfrak{h}} \to \mathcal{C}^{\otimes,c}$$

which fills in (3.2.4.2) *to form a commutative diagram. In addition,* $\mathfrak{c}^{\mathfrak{h}}$ *is a homotopy equivalence.*

PROOF. Write x for $(x_1, \dots, x_n)$, y for $(y_1, \dots, y_m)$. Since the action of $S(f_0)$ on $\mathrm{Hom}(x,y)_f$ is free, it follows from (3.2.1.6) that

$$H^p(\mathrm{Hom}_{\mathcal{C}^{\otimes,\mathfrak{h}}}(x,y)^*) = \begin{cases} 0 & \text{for } p \neq 0, \\ \oplus_{f \in \mathrm{Hom}_\omega(n,m)} \mathbb{Z} \otimes_{\mathbb{Z}[S(f_0)]} \mathrm{Hom}(x,y)_f & \text{for } p = 0. \end{cases}$$

The relation imposed on $\mathrm{Hom}(x,y)_f$ by tensoring with $\mathbb{Z}$ over $\mathbb{Z}[S(f_0)]$ is just the relation

$$g_F \sim (g \circ \Pi(\eta)^{-1})_{\eta \cdot F}$$

for $\eta \in S(f_0)$, which is the same as the relation imposed on the Hom-groups in $(\Pi_{\mathcal{C}}, \Omega_0)$ to form $(\Pi_{\mathcal{C}}, \Omega)$. As a DG tensor functor from $\mathcal{C}^{\otimes,\mathfrak{h}}$ to $\mathcal{C}^{\otimes,c}$ must necessarily factor through H^0, the proposition is proved. □

CHAPTER III

Simplicial and Cosimplicial Constructions

In this chapter, we collect a number of useful results on simplicial and cosimplicial objects in a category. We give some constructions, via multi-simplicial objects, of DG tensor categories which have a homotopy commutative external product, and which give rise to categorical cochain operations. We conclude with a discussion of homotopy limits, both for DG categories and for simplicial sets.

1. Complexes arising from simplicial and cosimplicial objects

1.1. Simplicial and cosimplicial objects

1.1.1. The fundamental object is Δ, the category with objects the ordered sets $[n] := \{0 < 1 < \ldots < n\}$ with morphisms order-preserving maps. The maps in Δ are generated by the *coface maps*

(1.1.1.1)
$$\delta_i^m : [m] \to [m+1],$$
$$\delta_i^m(j) = \begin{cases} j & \text{if } 0 \le j < i, \\ j+1 & \text{if } i \le j \le m. \end{cases}$$

and the *codegeneracy maps*

(1.1.1.2)
$$\sigma_i^m : [m] \to [m-1]; \qquad 0 \le i \le m-1,$$
$$\sigma_i^m(j) = \begin{cases} j & \text{for } 0 \le j \le i, \\ j-1 & \text{for } i < j \le m. \end{cases}$$

Let $\mathcal{C}$ be a category, c.s.$\mathcal{C}$, s.$\mathcal{C}$ the categories of cosimplicial, resp. simplicial objects in $\mathcal{C}$; i.e., functors $F^* : \Delta \to \mathcal{C}$, resp. $F_* : \Delta^{\mathrm{op}} \to \mathcal{C}$. We have as well the full subcategory $\Delta^{\le n}$ with objects $[0], \ldots, [n]$, and the functor categories c.s.$^{\le n}\mathcal{C}$, s.$^{\le n}\mathcal{C}$ of truncated (co)simplicial objects in $\mathcal{C}$. The inclusions

$$j_n : \Delta^{\le n} \to \Delta,$$
$$j_{n,m} : \Delta^{\le n} \to \Delta^{\le m}; \qquad n \le m,$$

induce the restriction functors

(1.1.1.3)
$$j_n^* : \text{c.s.}\mathcal{C} \to \text{c.s.}^{\le n}\mathcal{C},$$
$$j_{n,m}^* : \text{c.s.}^{\le m}\mathcal{C} \to \text{c.s.}^{\le n}\mathcal{C}; \qquad n \le m,$$

and similarly for the simplicial versions.

We let $\Delta_{\text{n.d.}}$ be the subcategory of Δ with the same objects, and with

$$\mathrm{Hom}_{\Delta_{\text{n.d.}}}([m],[n]) \subset \mathrm{Hom}_{\Delta}([m],[n])$$

consisting of the *injective* maps. For an integer $n \ge 0$, we let $\Delta_{\text{n.d.}}/[n]$ be the category of injective maps $f : [m] \to [n]$ in Δ.

1.1.2. We have the free additive category $\mathbb{Z}\mathcal{C}$ generated by $\mathcal{C}$; objects are finite direct sums of objects of $\mathcal{C}$, with $\mathrm{Hom}_{\mathbb{Z}\mathcal{C}}(X,Y) = \mathbb{Z}[\mathrm{Hom}_{\mathcal{C}}(X,Y)]$ for objects X, Y of $\mathcal{C}$, where $\mathbb{Z}[\mathcal{S}]$ denotes the free abelian group on a set $\mathcal{S}$. We may then form the category of bounded complexes $\mathbf{C}^b(\mathbb{Z}\mathcal{C})$ and the homotopy category $\mathbf{K}^b(\mathbb{Z}\mathcal{C})$.

1.1.3. *Complexes associated to truncated simplicial objects.* Let $F_*: \Delta^{\leq n\mathrm{op}} \to \mathcal{C}$ be a functor. Form the object $\mathbb{Z}_n^{\oplus}(F_*)$ of $\mathbf{C}^b(\mathbb{Z}\mathcal{C})$ by setting

$$\mathbb{Z}_n^{\oplus}(F_*)^{-m} = \bigoplus_{\substack{f:[m]\to[n]\\ f\in\Delta_{\mathrm{n.d.}}/[n]}} F_m.$$

The differential $d^{-m}: \mathbb{Z}_n^{\oplus}(F_*)^{-m} \to \mathbb{Z}_n^{\oplus}(F_*)^{-m+1}$ is given by

$$d^{-m} = \bigoplus_{\substack{f:[m]\to[n]\\ i=0,\dots,m}} d_{f,i}^{-m},$$

where $d_{f,i}^{-m}$ maps the summand F_m corresponding to f to the summand F_{m-1} corresponding to $f\circ\delta_i^{m-1}$, via the map $(-1)^iF(\delta_i^{m-1}): F_m \to F_{m-1}$. It follows from the identities

$$\delta_i^m \circ \delta_j^{m-1} = \delta_{j+1}^m \circ \delta_i^{m-1}; \quad 0 \leq i \leq j \leq m,$$

that $\mathbb{Z}_n^{\oplus}(F_*)$ is indeed a complex.

We have as well the object $\mathbb{Z}_n(F_*)$ of $\mathbf{C}^b(\mathbb{Z}\mathcal{C})$ defined by setting $\mathbb{Z}_n(F_*)^{-m} = F_m$, with differential $d^{-m}: \mathbb{Z}_n(F_*)^{-m} \to \mathbb{Z}_n(F_*)^{-m+1}$ given by the usual alternating sum

$$d^{-m} = \sum_{i=0}^{m}(-1)^iF(\delta_i^{m-1}).$$

Sending $\mathbb{Z}_n^{\oplus}(F_*)^{-m}$ to F_m by the sum of the projections

$$\pi^{-m} = \sum_{f:[m]\to[n]} \pi_f: \mathbb{Z}_n^{\oplus}(F_*)^{-m} \to F_m$$

defines the map in $\mathbf{C}^b(\mathbb{Z}\mathcal{C})$

$$\pi_n: \mathbb{Z}_n^{\oplus}(F_*) \to \mathbb{Z}_n(F_*). \tag{1.1.3.1}$$

1.1.4. For $N > n$, we let $\delta_0^{N,n}: [n] \to [N]$ be the composition $\delta_0^{N-1} \circ \ldots \circ \delta_0^n$.

Let $F_*: \Delta^{\leq N\mathrm{op}} \to \mathcal{C}$ be a functor and take $n < N$; one easily checks that sending F_m in the summand $f:[m] \to [n]$ to F_m in the summand $\delta_0^{N,n} \circ f: [m] \to [N]$ via the identity gives a map of complexes

$$\chi^{N,n}: \mathbb{Z}_n^{\oplus}(j_{n,N}^*F_*) \to \mathbb{Z}_N^{\oplus}(F_*). \tag{1.1.4.1}$$

We have the canonical identification of $\mathbb{Z}_n(j_{n,N}^*F_*)$ with the "stupid truncation" $\sigma^{\geq -n}\mathbb{Z}_N(F_*)$, giving the canonical map of complexes $j_{N,n*}: \mathbb{Z}_n(j_{n,N}^*F_*) \to \mathbb{Z}_N(F_*)$; one sees immediately that the diagram

$$\begin{array}{ccc}
\mathbb{Z}_n^{\oplus}(j_{n,N}^*F_*) & \xrightarrow{\chi^{N,n}} & \mathbb{Z}_N^{\oplus}(F_*) \\
\downarrow{\scriptstyle \pi_n} & & \downarrow{\scriptstyle \pi_N} \\
\mathbb{Z}_n(j_{n,N}^*F_*) & \xrightarrow[j_{N,n*}]{} & \mathbb{Z}_N(F_*)
\end{array}$$

commutes. Similarly, if $F_*: \Delta^{\mathrm{op}} \to \mathcal{C}$ is a functor, we have the truncations

$$j_N^* F_*: \Delta^{\leq N \mathrm{op}} \to \mathcal{C}.$$

The map (1.1.3.1) gives the natural map

$$\pi_N: \mathbb{Z}_N^{\oplus}(j_N^* F_*) \to \mathbb{Z}_N(j_N^* F_*) \tag{1.1.4.2}$$

and the commutative diagram

$$\begin{array}{ccc} \mathbb{Z}_n^{\oplus}(j_n^* F_*) & \xrightarrow{\chi^{N,n}} & \mathbb{Z}_N^{\oplus}(j_N^* F_*) \\ {\scriptstyle \pi_n}\downarrow & & \downarrow{\scriptstyle \pi_N} \\ \mathbb{Z}_n(j_n^* F_*) & \xrightarrow[j_{N,n*}]{} & \mathbb{Z}_N(j_N^* F_*). \end{array} \tag{1.1.4.3}$$

We often omit the truncation j_N^* from the notation, if the meaning is clear from the context.

The following is a reformulation of a well-known result of Dold [**39**], which we include for the convenience of the reader.

1.1.5. LEMMA. *(i) Let $F_*: \Delta^{\leq N \mathrm{op}} \to \mathbf{Ab}$ be a functor. Then for all $0 \leq n < N$, the map (1.1.4.1) induces an isomorphism on H^{-m} for $m < n$ and a surjection for $m = n$.*
(ii) Let $F_*: \Delta^{\leq n \mathrm{op}} \to \mathbf{Ab}$ be a functor. Then the map (1.1.3.1) induces an isomorphism on H^p for $-n < p \leq 0$. For $p = n$, the map $H^{-n}(\pi_n): H^{-n}(\mathbb{Z}_n^{\oplus}(F_*)) \to H^{-n}(\mathbb{Z}_n(F_*))$ is injective, and identifies $H^{-n}(\mathbb{Z}_n^{\oplus}(F_*))$ with $\cap_{i=0}^n \ker[F(\delta_i^{n-1}): F_n \to F_{n-1}]$.

PROOF. It suffices to prove (i) for $n = N-1$. We first construct a left splitting $\sigma: \mathbb{Z}_N^{\oplus}(F_*) \to \mathbb{Z}_{N-1}^{\oplus}(j_{N-1,N}^* F_*)$ to $\chi^{N,N-1}$.

We have the codegeneracy map $\sigma_{N-1}^N: [N] \to [N-1]$ (1.1.1.2), which induces the map $\sigma: \mathbb{Z}_N^{\oplus}(F_*) \to \mathbb{Z}_{N-1}^{\oplus}(j_{N-1,N}^* F_*)$ by sending F_m in the summand $f: [m] \to [N]$ to F_m in the summand $\sigma_{N-1}^N \circ f: [m] \to [N-1]$ via the identity map, if $\sigma_{N-1}^N \circ f$ is injective, and to zero otherwise. One verifies without difficulty that σ defines a map of complexes, with

$$\sigma \circ \chi^{N,N-1} = \mathrm{id}_{\mathbb{Z}_{N-1}^{\oplus}(j_{N-1,N}^* F_*)}.$$

For a map $g: [m] \to [N-1]$ in Δ, let $g + 1: [m+1] \to [N]$ be the map

$$(g+1)(i) := \begin{cases} g(i) & 0 \leq i \leq m \\ N & i = m+1. \end{cases}$$

Clearly $g + 1$ is injective if g is.

Define the map $h_m: \mathbb{Z}_N^{\oplus}(F_*)^{-m} \to \mathbb{Z}_N^{\oplus}(F_*)^{-m-1}$ by sending F_m in the summand $f: [m] \to [N]$ to F_{m+1} in the summand $(\sigma_{N-1}^N \circ f) + 1: [m+1] \to [N]$ via the map $(-1)^{m+1} F(\sigma_m^{m+1}): F_m \to F_{m+1}$ if $\sigma_{N-1}^N \circ f$ is injective and $f(m) = N$, and to zero otherwise. It follows by an elementary computation that

$$d^{-m-1} \circ h_m + h_{m-1} \circ d^{-m} = \chi^{N,N-1} \circ \sigma - \mathrm{id}$$

on $\mathbb{Z}_N^{\oplus}(F_*)^{-m}$, for $m = 0, \ldots, N-1$, which proves (i).

For (ii), we may use (i); thus it suffices to show that the map (1.1.3.1) induces an isomorphism on H^{-p} for $p = n-1$, and gives the desired injection for $p = n$.

Let $\mathbb{Z}_n(F_*)_0$ be the normalized cochain complex,

$$\mathbb{Z}_n(F_*)_0^{-p} := \cap_{i=1}^p \ker[F(\delta_i^{p-1}) : \mathbb{Z}_n(F_*)^{-p} \to \mathbb{Z}_n(F_*)^{-p+1}],$$

with differential $\partial^{-p} := F(\delta_0^{p-1})$. The inclusion $\mathbb{Z}_n(F_*)_0 \to \mathbb{Z}_n(F_*)$ induces an isomorphism in cohomology H^{-p} for $p \le n-1$ (see e.g., [**95**, Chapter V]). By (i), we have the exact sequence

$$\text{(1.1.5.1)} \quad \mathbb{Z}_n(F_*)_0^{-n} \xrightarrow{F(\delta_0^{n-1})} H^{-n+1}(\mathbb{Z}_{n-1}^{\oplus}(j_{n-1,n}^* F_*)) \\ \xrightarrow{\chi^{n,n-1}} H^{-n+1}(\mathbb{Z}_n^{\oplus}(F_*)) \to 0.$$

On the other hand, a direct computation shows that π_{n-1} gives an identification

$$H^{-n+1}(\mathbb{Z}_{n-1}(j_{n-1,n}^* F_*)) \\ \cong \ker[F(\delta_0^{n-2}) : \mathbb{Z}_n(F_*)_0^{-n+1} \to \mathbb{Z}_n(F_*)_0^{-n+2}].$$

This in turn gives, via π_n, the identification of the sequence (1.1.5.1) with the canonical sequence defining $H^{-n+1}(\mathbb{Z}_n(F_*))$,

$$\mathbb{Z}_n(F_*)_0^{-n} \xrightarrow{F(\delta_0^{n-1})} \ker[F(\delta_0^{n-2}) : \mathbb{Z}_n(F_*)_0^{-n+1} \to \mathbb{Z}_n(F_*)_0^{-n+2}] \\ \to H^{-n+1}(\mathbb{Z}_n(F_*)_0) \to 0.$$

This proves that π_n gives an isomorphism on H^{-n+1}, and the desired injection on H^{-n}, completing the proof. □

1.2. Multiplication of cosimplicial objects

We describe how one gives a multiplicative structure to cosimplicial objects in certain symmetric monoidal categories.

1.2.1. *External products.* We recall the standard construction of Alexander-Whitney products for cosimplicial objects in a tensor category.

Let $\mathcal{A}$ be an additive category, and let $X : \Delta \to \mathcal{A}$ be a cosimplicial object of $\mathcal{A}$. We may form the object X^* of $\mathbf{C}^+(\mathcal{A})$:

$$X^* := X^0 \xrightarrow{d_0} \ldots \xrightarrow{d_{n-1}} X^n \xrightarrow{d_n} \ldots; \quad X^n = X([n]),$$

where $d_n : X^n \to X^{n+1}$ is the usual alternating sum $\sum_{i=0}^{n+1}(-1)^i X(\delta_i^n)$. We may also form the various truncations of X^*:

$$X^{m \le * \le n} := X^m \xrightarrow{d_m} \ldots \xrightarrow{d_{n-1}} X^n.$$

If p and q are positive integers, we have the maps

$$\text{(1.2.1.1)} \quad \begin{aligned} f_p^{p,q} &: [p] \to [p+q], \\ f_q^{p,q} &: [q] \to [p+q], \end{aligned}$$

given by $f_p^{p,q}(i) = i$ and $f_q^{p,q}(j) = j + p$.

Suppose that $\mathcal{A}$ is a tensor category with tensor product $\otimes$. Let $X : \Delta \to \mathcal{A}$ and $Y : \Delta \to \mathcal{A}$ be cosimplicial objects of $\mathcal{A}$, giving the diagonal cosimplicial object

$$X \otimes Y : \Delta \to \mathcal{A}, \\ (X \otimes Y)(x) = X(x) \otimes Y(x),$$

where x is an object, or a morphism, in Δ. We have as well the tensor product double complex $X^* \otimes Y^*$, and the complex $(X \otimes Y)^*$.

Let

$$\cup^n_{p,q} : X^p \otimes Y^q \to X^n \otimes Y^n$$

be the map $\cup^n_{p,q} = X(f^{p,q}_p) \otimes Y(f^{p,q}_q)$, and define

$$\cup^n_{X,Y} : \oplus_{p+q=n} X^p \otimes Y^q \to X^n \otimes Y^n \tag{1.2.1.2}$$

by $\cup^n_{X,Y} = \sum_{p+q=n} \cup^n_{p,q}$. The relation (of linear combinations of maps from $[p] \coprod [q]$ to $[p+q+1]$)

$$\begin{aligned}\sum_{i=0}^{p+1}(-1)^i[(f^{p+1,q}_{p+1} \circ \delta^p_i) \coprod f^{p+1,q}_q] + \sum_{i=0}^{q+1}(-1)^{i+p}[f^{p,q+1}_p \coprod (f^{p,q+1}_{q+1} \circ \delta^q_i)]\\ = \sum_{i=0}^{p+q+1} (-1)^i \delta^{p+q}_i \circ (f^{p,q}_p \coprod f^{p,q}_q)\end{aligned} \tag{1.2.1.3}$$

implies that the maps (1.2.1.2) define the map of complexes

$$\cup_{X,Y} : \mathrm{Tot}(X^* \otimes Y^*) \to (X \otimes Y)^*. \tag{1.2.1.4}$$

One easily verifies that the maps $\cup_{X,Y}$ are associative, in the obvious sense; it follows from the Eilenberg-Zilber theorem [**42**] that the maps $\cup_{X,Y}$ are graded-commutative, up to functorial homotopy.

1.2.2. *Multiplication in a symmetric monoidal category.* Let $(\mathcal{A}, \otimes, \tau, \mu, 1)$ be a symmetric monoidal category. We have the diagonal functor

$$\Delta_{\mathcal{A}} : \mathcal{A} \to \mathcal{A} \times \mathcal{A}$$

$$\Delta_{\mathcal{A}}(X) = (X, X)$$

$$\Delta_{\mathcal{A}}(f : X \to Y) = (f, f).$$

A *commutative multiplication* in $\mathcal{A}$ is a natural transformation $m : \otimes \circ \Delta_{\mathcal{A}} \to \mathrm{id}_{\mathcal{A}}$ such that

$$m \circ (m \times \mathrm{id}_{\mathcal{A}}) = m \circ (\mathrm{id}_{\mathcal{A}} \times m)$$

$$m \circ (\tau \circ \Delta_{\mathcal{A}}) = m$$

$$m(1) = \mu_1 : 1 \otimes 1 \to 1.$$

1.2.3. *Cup products.* Suppose that $\mathcal{A}$ is a symmetric monoidal category with multiplication m, and X is a cosimplicial object in $\mathcal{A}$.

Define the map of cosimplicial objects $m_X : X \otimes X \to X$ by

$$m_X([n]) = m_{X([n])} : X([n]) \otimes X([n]) \to X([n]);$$

we let $m^*_X : (X \otimes X)^* \to X^*$ be the map induced by m_X.

The symmetric monoidal structure on $\mathcal{A}$ induces the structure of a tensor category on the free additive category $\mathbb{Z}\mathcal{A}$ generated by $\mathcal{A}$. We may then define the map in $\mathbf{C}^+(\mathbb{Z}\mathcal{A})$,

$$m(X^*) : \mathrm{Tot}(X^* \otimes X^*) \to X^*, \tag{1.2.3.1}$$

by $m(X^*) = m^*_X \circ \cup_{X,X}$.

For a complex X^*, we $X^{*\le n}$ denote the truncation of X to degrees d, $d \le n$, and $X^{m\le *\le n}$ the truncation to degrees d, $m \le d \le n$. Taking truncations gives the maps

$$m^{n'}(X^{m\le*\le n})\colon \mathrm{Tot}(X^{m\le*\le n} \otimes X^{*\le n'}) \to X^{m\le*\le n} \tag{1.2.3.2}$$

for all $n' \ge n$. For $m' \le m$, and $n \le n' \le n''$, the diagrams

$$\begin{array}{ccc}
\mathrm{Tot}(X^{m\le*\le n} \otimes X^{*\le n'}) & \xrightarrow{m^{n'}(X^{m\le*\le n})} & X^{m\le*\le n} \\
\downarrow & & \downarrow \\
\mathrm{Tot}(X^{m'\le*\le n} \otimes X^{*\le n'}) & \xrightarrow[m^{n'}(X^{m'\le*\le n})]{} & X^{m'\le*\le n}
\end{array} \tag{1.2.3.3}$$

and

$$\begin{array}{ccc}
\mathrm{Tot}(X^{m\le*\le n'} \otimes X^{*\le n''}) & \xrightarrow{m^{n''}(X^{m\le*\le n'})} & X^{m\le*\le n'} \\
\downarrow & & \downarrow \\
\mathrm{Tot}(X^{m\le*\le n} \otimes X^{*\le n''}) & \xrightarrow[m^{n''}(X^{m'\le*\le n})]{} & X^{m\le*\le n}
\end{array} \tag{1.2.3.4}$$

commute, and for $n \le n' \le n''$, the diagram

$$\begin{array}{ccc}
\mathrm{Tot}(X^{m\le*\le n} \otimes X^{*}) & \xrightarrow{m(X^{m\le*\le n})} & X^{m\le*\le n} \\
\downarrow & & \| \\
\mathrm{Tot}(X^{m\le*\le n} \otimes X^{*\le n''}) & \xrightarrow{m^{n''}(X^{m\le*\le n})} & X^{m\le*\le n} \\
\downarrow & & \| \\
\mathrm{Tot}(X^{m\le*\le n} \otimes X^{*\le n'}) & \xrightarrow{m^{n'}(X^{m\le*\le n})} & X^{m\le*\le n}
\end{array} \tag{1.2.3.5}$$

commutes.

When the indices are obvious, we write

$$\cup_X\colon \mathrm{Tot}(X^{m\le*\le n} \otimes X^{*\le n'}) \to X^{m\le*\le n} \tag{1.2.3.6}$$

for the map (1.2.3.2).

1.2.4. REMARK. The maps (1.2.3.1) are associative, which one checks by a direct computation. The maps $m(X^*)$ are not in general commutative, but are commutative up to homotopy; this follows from the Eilenberg-Zilber theorem [**42**].

Thus, suppose we have a graded tensor functor $F\colon \mathbf{K}^b(\mathbb{Z}\mathcal{A}) \to \mathcal{B}$. Then the maps $F(m^n(X^{*\le n}))$ and $F(m^n(X^{m\le*\le n}))$ give $\mathrm{Hom}_{\mathcal{B}}(1_{\mathcal{B}}, F(X^{*\le n}))$ the structure of a (possibly non-unital) graded-commutative ring, and make $\mathrm{Hom}_{\mathcal{B}}(1_{\mathcal{B}}, F(X^{m\le*\le n}))$ a graded $\mathrm{Hom}_{\mathcal{B}}(1_{\mathcal{B}}, F(X^{*\le n}))$-module.

2. Categorical cochain operations

In this section, we use simplicial methods to construct external products in certain tensor categories, which are associative, and are graded-commutative up to homotopy and "all higher homotopies" (compare with the construction of Chapter II, §3.2). This may be viewed as a categorical version of the construction of the "Eilenberg-MacLane operad" of Hinich and Schechtman [**65**]. We will apply these results in §2.2 to show how an associative, commutative product on cosimplicial objects gives rise to products on the associated cochain complexes which are graded-commutative up to homotopy and "all higher homotopies". Combining this result with the multiplicative structure on the cosimplicial Godement resolution of a sheaf, discussed in Chapter IV, allows us to solve the fundamental coherence problem in constructing realization functors for the motivic triangulated category.

2.1. A homotopy commutative DG tensor category

2.1.1. *The extended total complex.* If $\mathcal{F}: \Delta^{n\mathrm{op}} \times \Delta^m \to \mathbf{Ab}$ is a functor, we may form the *extended total complex* of $\mathcal{F}$, $(\mathcal{F}^*, d)$, with $\mathcal{F}^s$ the abelian group:

$$\mathcal{F}^s = \prod_{\Sigma_j q_j - \Sigma_i p_i = s} F([p_1], \dots, [p_n], [q_1], \dots, [q_m]).$$

Denote an element g of $\mathcal{F}^s$ as a tuple

$$g = (\dots, g^{p_1, \dots, p_n}_{q_1, \dots, q_m}, \dots),$$
$$g^{p_1, \dots, p_n}_{q_1, \dots, q_m} \in \mathcal{F}([p_1], \dots, [p_n], [q_1], \dots, [q_m]).$$

Write

$$\mathcal{F}(\delta^{p_j}_i; \mathrm{id}) := \mathcal{F}(\mathrm{id}_{[p_1]}, \dots, \delta^{p_j}_i, \dots, \mathrm{id}_{[p_n]}; \mathrm{id}_{[q_1]}, \dots, \mathrm{id}_{[q_m]})$$
$$\mathcal{F}(\mathrm{id}; \delta^{q_j}_i) := \mathcal{F}(\mathrm{id}_{[p_1]}, \dots, \mathrm{id}_{[p_n]}; \mathrm{id}_{[q_1]}, \dots, \delta^{q_j}_i, \dots, \mathrm{id}_{[q_m]}).$$

The differential $d^s: \mathcal{F}^s \to \mathcal{F}^{s+1}$ is given by

$$d^s g = (\dots, d^s g^{p_1, \dots, p_n}_{q_1, \dots, q_m}, \dots),$$

with

$$d^s g^{p_1, \dots, p_n}_{q_1, \dots, q_m} := (-1)^s \sum_{j=1}^n (-1)^{p_1 + \dots + p_{j-1}} \sum_{i=0}^{p_j} (-1)^i g^{p_1, \dots, p_n}_{q_1, \dots, q_m} \circ \mathcal{F}(\delta^{p_i}_i; \mathrm{id}) - \sum_{j=1}^m (-1)^{q_1 + \dots + q_{j-1}} \sum_{i=0}^{q_j} (-1)^i \mathcal{F}(\mathrm{id}; \delta^{q_j}_i) \circ g^{p_1, \dots, p_n}_{q_1, \dots, q_m}.$$

For $n = 0$, this is the usual complex associated to a functor $F: \Delta^m \to \mathbf{Ab}$, except with minus the usual differential; if $m = 0$, this complex has the same underlying graded group as the complex associated to a functor $F: \Delta^{n\mathrm{op}} \to \mathbf{Ab}$, but with differential differing by the sign $(-1)^s$ in degree s. The first two complexes are isomorphic, by sending x in degree s to $(-1)^s x$, and the second two complexes are isomorphic, by sending x in degree s to x if $s \equiv 0, 1 \bmod 4$ and to $-x$ if $s \equiv 2, 3 \bmod 4$.

2.1.2. *The complex of multi-simplicial maps.* Let Δ_{un} be the full subcategory of **Sets** with the same objects as Δ, i.e., $\mathrm{Hom}_{\Delta_{\mathrm{un}}}([n], [m]) = \mathrm{Hom}_{\mathbf{Sets}}([n], [m])$. The operation of ordered disjoint union, where we identify two finite ordered sets of the same cardinality by the unique ordered bijection, makes Δ_{un} into a strictly associative symmetric semi-monoidal category; this gives us the symmetric semi-monoidal 2-functor (see Chapter I, §2.3.5, §2.3.6, and (I.2.3.6.1))

$$\pi_{\Delta_{\mathrm{un}}}:\Omega\to\mathbf{cat}. \tag{2.1.2.1}$$

We let $\pi^0_{\Delta_{\mathrm{un}}}$ denote the restriction of $\pi_{\Delta_{\mathrm{un}}}$ to the underlying category Ω_0.

Following (Chapter I, §2.2), we form the category of pairs $(\pi^0_{\Delta_{\mathrm{un}}},\Omega_0)$, which, by Remark 2.2.3 of Chapter I, has the natural structure of a symmetric semi-monoidal category.

The projection on the second factor gives the symmetric semi-monoidal functor $p_{\Omega_0}:(\pi^0_{\Delta_{\mathrm{un}}},\Omega_0)\to\Omega_0$; this in turn gives the natural decomposition of the Hom-sets as

$$\begin{aligned}(2.1.2.2)\quad &\mathrm{Hom}_{(\pi^0_{\Delta_{\mathrm{un}}},\Omega)}(([p_1],\dots,[p_n],n),(([q_1],\dots,[q_m]),m))\\ &=\coprod_{F\in\mathrm{Hom}_{\Omega_0}(n,m)}\mathrm{Hom}_{(\pi_{\Delta_{\mathrm{un}}},\Omega_0)}(([p_1],\dots,[p_n],n),(([q_1],\dots,[q_m]),m))_F,\end{aligned}$$

where

$$\mathrm{Hom}_{(\pi^0_{\Delta_{\mathrm{un}}},\Omega_0)}(([p_1],\dots,[p_n],n),(([q_1],\dots,[q_m]),m))_F$$

is the set of pairs (g,F), with

$$g:\pi^0_{\Delta_{\mathrm{un}}}(F)([p_1],\dots,[p_n])\to([q_1],\dots,[q_m]) \tag{2.1.2.3}$$

a map in $\mathbf{Sets}^m$.

Recall from (Chapter I, §2.3.3) that a morphism $F:n\to m$ in Ω_0 is a pair (f,σ) with $f:\underline{n}\to\underline{m}$ an ordered surjection, and $\sigma\in S_n$. Given $F:n\to m$ in Ω_0, we may write $\pi^0_{\Delta_{\mathrm{un}}}(F)([p_1],\dots,[p_n])$ as an m-tuple:

$$\begin{aligned}&\pi^0_{\Delta_{\mathrm{un}}}(F)([p_1],\dots,[p_n])\\ &\qquad=(\pi^0_{\Delta_{\mathrm{un}}}(F)([p_1],\dots,[p_n])_1,\dots,\pi^0_{\Delta_{\mathrm{un}}}(F)([p_1],\dots,[p_n])_m),\end{aligned}$$

and each $\pi^0_{\Delta_{\mathrm{un}}}(F)([p_1],\dots,[p_n])_j$ is a disjoint union

$$\pi^0_{\Delta_{\mathrm{un}}}(F)([p_1],\dots,[p_n])_j=[p_{i^j_1}]\coprod\cdots\coprod[p_{i^j_{s_j}}],$$

with $\{i^j_1,\dots,i^j_{s_j}\}=F^{-1}(j)$. The map (2.1.2.3) may then be written as

$$\begin{aligned}g&=(g_1,\dots,g_n)\\ g_i&:[p_i]\to[q_{F(i)}].\end{aligned}$$

The composition in $(\pi^0_{\Delta_{\mathrm{un}}},\Omega_0)$ may then be described as follows: For

$$\begin{aligned}g^1&:\pi^0_{\Delta_{\mathrm{un}}}(F^1)([p_1],\dots,[p_n])\to([q_1],\dots,[q_m]),\\ g^2&:\pi^0_{\Delta_{\mathrm{un}}}(F^2)([q_1],\dots,[q_m])\to([r_1],\dots,[r_k]),\end{aligned}$$

write

$$\begin{aligned}g^1&=(g^1_1,\dots,g^1_n)\\ g^1_i&:[p_i]\to[q_{F^1(i)}],\end{aligned}$$

$$\begin{aligned}g^2&=(g^2_1,\dots,g^2_m)\\ g^2_j&:[q_j]\to[r_{F^2(j)}].\end{aligned}$$

Then

$$(g^2, F^2) \circ (g^1, F^1) = (g^2 \circ g^1, F^2 \circ F^1),$$

where $F^2 \circ F^1$ is the composition in Ω_0 and

$$(g^2 \circ g^1) = (g_1, \dots, g_n),\ g_i\colon [p_i] \to [r_{F^2(F^1(i))}],$$
$$g_i = g^2_{F^1(i)} \circ g^1_i.$$

We let

$$\operatorname{Hom}_{(\pi^0_{\Delta_{\mathrm{un}}}, \Omega_0)^{\mathrm{ord}}}\big(([p_1], \dots, [p_n], n), ([q_1], \dots, [q_m], m)\big)_F$$
$$\subset \operatorname{Hom}_{(\pi^0_{\Delta_{\mathrm{un}}}, \Omega_0)}\big(([p_1], \dots, [p_n], n), ([q_1], \dots, [q_m], m)\big)_F$$

denote the subset consisting of pairs (g, F), with $g = (g_1, \dots, g_n)$, $g_i\colon [p_i] \to [q_{F(i)}]$, such that, for each $i = 1, \dots, n$, the map $g_i\colon [p_i] \to [q_{F(i)}]$ is order-preserving. We set

$$\operatorname{Hom}_{(\pi^0_{\Delta_{\mathrm{un}}}, \Omega_0)^{\mathrm{ord}}}\big(([p_1], \dots, [p_n], n), ([q_1], \dots, [q_m], m)\big)$$
$$= \coprod_F \operatorname{Hom}_{(\pi^0_{\Delta_{\mathrm{un}}}, \Omega_0)^{\mathrm{ord}}}\big(([p_1], \dots, [p_n], n), ([q_1], \dots, [q_m], m)\big)_F.$$

For $n, m \in \mathbb{N}$, let

(2.1.2.4) $$\operatorname{Hom}_{(\pi^0_{\Delta_{\mathrm{un}}}, \Omega_0)^{\mathrm{ord}}}(n, m)\colon \Delta^{n\mathrm{op}} \times \Delta^m \to \mathbf{Sets}$$

be the functor $\operatorname{Hom}_{(\pi^0_{\Delta_{\mathrm{un}}}, \Omega_0)^{\mathrm{ord}}}((-, n), (-, m))$, i.e.,

$$\operatorname{Hom}_{(\pi^0_{\Delta_{\mathrm{un}}}, \Omega_0)^{\mathrm{ord}}}(n, m)([p_1], \dots, [p_n]; [q_1], \dots, [q_m])$$
$$= \operatorname{Hom}_{(\pi^0_{\Delta_{\mathrm{un}}}, \Omega_0)^{\mathrm{ord}}}\big(([p_1], \dots, [p_n], n), ([q_1], \dots, [q_m], m)\big),$$
$$\operatorname{Hom}_{(\pi^0_{\Delta_{\mathrm{un}}}, \Omega_0)^{\mathrm{ord}}}(n, m)(f, g) = \operatorname{Hom}_{(\pi^0_{\Delta_{\mathrm{un}}}, \Omega_0)^{\mathrm{ord}}}((f, \mathrm{id}_n), (g, \mathrm{id}_m)).$$

The decomposition (2.1.2.2) gives the decomposition

$$\operatorname{Hom}_{(\pi^0_{\Delta_{\mathrm{un}}}, \Omega_0)^{\mathrm{ord}}}(n, m) = \coprod_{F \in \operatorname{Hom}_{\Omega_0}(n,m)} \operatorname{Hom}_{(\pi^0_{\Delta_{\mathrm{un}}}, \Omega_0)^{\mathrm{ord}}}(n, m)_F.$$

Let

(2.1.2.5) $$\mathbb{Z}[\operatorname{Hom}_{(\pi_{\Delta_{\mathrm{un}}}, \Omega_0)^{\mathrm{ord}}}(n, m)_F] : \Delta^{n\mathrm{op}} \times \Delta^m \to \mathbf{Ab},$$

be the free abelian group on the functor (2.1.2.4), and let $\operatorname{Hom}_{\Delta_{\Omega_0}}(n, m)^*_F$ be the extended total complex of the functor (2.1.2.5). Define the complex

$$\operatorname{Hom}_{\Delta_{\Omega_0}}(n, m)^* := \oplus_{F \in \operatorname{Hom}_{\Omega_0}(n,m)} \operatorname{Hom}_{\Delta_{\Omega_0}}(n, m)^*_F.$$

2.1.3. *Cohomological triviality.* We now proceed to compute the cohomology of the complexes $\operatorname{Hom}_{\Delta_{\Omega_0}}(n, m)^*$.

2.1.3.1. LEMMA. *Let $k > 0$ and $q \geq 0$ be integers, and let*

$$\operatorname{Hom}_{\Delta_{\mathrm{un}}}([p_1] \coprod \cdots \coprod [p_k], [q])^{\mathrm{ord}}$$

denote the subset of $\mathrm{Hom}_{\Delta_{\mathrm{un}}}([p_1]\coprod\cdots\coprod[p_k],[q])$ *consisting of maps whose restriction to each component* $[p_i]$ *is order-preserving. Let* $C^*_{q,k}$ *denote the complex associated to the functor*

$$C_{q,k}\colon \Delta^{k\mathrm{op}} \to \mathbf{Ab},$$
$$C_{q,k}([p_1],\dots,[p_k]) = \mathbb{Z}[\mathrm{Hom}_{\Delta_{\mathrm{un}}}([p_1]\coprod\cdots\coprod[p_k],[q])^{\mathrm{ord}}].$$

Then

$$H^a(C^*_{q,k}) = \begin{cases} 0 & \text{for } a\neq 0,\\ \mathbb{Z} & \text{for } a=0,\end{cases}$$

and $H^0(C^*_{q,k})$ *is generated by the class of the map* $[0]\coprod\cdots\coprod[0]\to[q]$ *which has image* $0\in[q]$.

PROOF. The complex $C^*_{q,1}$ is the chain complex of ordered affine simplicial chains for the standard q-simplex Δ_q, whence the result for $k=1$. We proceed by induction on k.

For a non-negative integer b, let $C_{q,k-1,b}$ be the functor

$$C_{q,k-1,b}\colon \Delta^{k-1\mathrm{op}} \to \mathbf{Ab}$$
$$C_{q,k-1,b}([p_1],\dots,[p_{k-1}]) = \mathbb{Z}[\mathrm{Hom}_{\Delta_{\mathrm{un}}}([p_1]\coprod\cdots\coprod[p_{k-1}]\coprod[b],[q])^{\mathrm{ord}}],$$

and let $C^*_{q,k-1,b}$ be the resulting total complex.

We may form the double complex $C^{**}_{q,k}$ from $C_{q,k}$ by forming the total complex with respect to the first $k-1$ variables, i.e.,

$$C^{-a,-p}_{q,k} = \bigoplus_{p_1+\dots+p_{k-1}=a} C_{q,k}([p_1],\dots,[p_{k-1}],[p]).$$

The total complex associated to $C^{**}_{q,k}$ is just $C^*_{q,k}$. The subcomplexes $F^bC^*_{q,k}$ of $C^*_{q,k}$ given by taking the total complex of $C^{*,*\geq b}_{q,k}$ give a filtration of $C^*_{q,k}$; the resulting E_1-spectral sequence is

$$E_1^{a,b} = H^a(C^*_{q,k-1,-b}) \Longrightarrow H^{a+b}(C^*_{q,k}).$$

On the other hand, we have the isomorphism of complexes:

$$C^*_{q,k-1,b} \cong \bigoplus_{f\in\mathrm{Hom}_\Delta([b],[q])} C^*_{q,k-1};$$

this, together with our induction hypothesis, gives the computation of the E_1-terms as

$$E_1^{a,-b} = \begin{cases} 0 & \text{for } a\neq 0,\\ \mathbb{Z}[\mathrm{Hom}_\Delta([b],[q])] & \text{for } a=0.\end{cases}$$

Additionally, the complex $E_1^{0,*}$ is isomorphic to the complex $C^*_{q,1}$. Thus the spectral sequence degenerates at E_2 and gives the desired result. □

2.1.3.2. LEMMA. *The complexes* $\mathrm{Hom}_{\Delta_{\Omega_0}}(n,m)^*$ *satisfy*

$$H^a(\mathrm{Hom}_{\Delta_{\Omega_0}}(n,m)^*) = \begin{cases} 0 & \text{if } q\neq 0,\\ \mathbb{Z}[\mathrm{Hom}_{\Omega_0}(n,m)] & \text{if } q=0.\end{cases}$$

Furthermore, for each $F \in \mathrm{Hom}_{\Omega_0}(n,m)$, projection on the factor $p_1 = \ldots = p_n = q_1 = \ldots = q_m = 0$ gives an isomorphism

$$H^0(\mathrm{Hom}_{\Delta_{\Omega_0}}(n,m)^*_F) \\ \to \mathbb{Z}[\mathrm{Hom}_{\mathbf{Sets}}([0]\coprod\cdots\coprod[0],[0]) \times \ldots \times \mathrm{Hom}_{\mathbf{Sets}}([0]\coprod\cdots\coprod[0],[0])] \\ \cong \mathbb{Z}.$$

PROOF. It suffices to show that, for each morphism $F: n \to m$ in Ω_0, the complex $\mathrm{Hom}_{\Delta_{\Omega_0}}(n,m)^*_F$ is acyclic in non-zero degrees, and that the above projection gives an isomorphism on H^0. Using the action of the symmetric group S_n on $\mathrm{Hom}_{\Omega_0}(n,m)$, we may assume that $F = (f, \mathrm{id})$, where $f: \underline{n} \to \underline{m}$ is an order-preserving surjection. We may thus write f as a product $f = f_1 + \ldots + f_m$, with each $f_i: \underline{n}_i \to \underline{1}$ being the unique surjection.

For each $j = 1, \ldots, m$, and each collection of non-negative integers $p_1, \ldots, p_n$, let $[p^j_*]$ denote the disjoint union

$$[p_{n_1+\ldots+n_{j-1}+1}]\coprod[p_{n_1+\ldots+n_{j-1}+2}]\coprod\cdots\coprod[p_{n_1+\ldots+n_j}]$$

(we take $n_0 = 0$). Using the definition of the complexes $\mathrm{Hom}_{\Delta_{\Omega_0}}(a,b)^*_G$, we have the isomorphism

$$\mathrm{Hom}_{\Delta_{\Omega_0}}(n,m)^s_F \\ = \prod_{\Sigma_{i=1}^m q_i - \Sigma_{i=1}^n p_i = s} \mathbb{Z}[\mathrm{Hom}_{\Delta_{\mathrm{un}}}([p^1_*],[q_1])^{\mathrm{ord}} \times \ldots \times \mathrm{Hom}_{\Delta_{\mathrm{un}}}([p^m_*],[q_m])^{\mathrm{ord}}] \\ \cong \prod_{\Sigma_{i=1}^m q_i - \Sigma_{i=1}^n p_i = s} \mathbb{Z}[\mathrm{Hom}_{\Delta_{\mathrm{un}}}([p^1_*],[q_1])^{\mathrm{ord}}] \otimes_{\mathbb{Z}} \ldots \otimes_{\mathbb{Z}} \mathbb{Z}[\mathrm{Hom}_{\Delta_{\mathrm{un}}}([p^m_*],[q_m])^{\mathrm{ord}}].$$

Denote the complex $\mathrm{Hom}_{\Delta_{\Omega_0}}(n,m)^*_F$ by T^*. Fix integers $a_1, \ldots, a_k \geq 0$, and let $T^*_{a_1,\ldots,a_k}$ be the complex with

$$T^s_{a_1,\ldots,a_k} := \\ \prod_{\substack{\Sigma_{i=k+1}^m q_i - \Sigma_{i=1}^n p_i = s \\ q_1 = a_1, \ldots, q_k = a_k}} \mathbb{Z}[\mathrm{Hom}_{\Delta_{\mathrm{un}}}([p^1_*],[q_1])^{\mathrm{ord}}] \otimes_{\mathbb{Z}} \ldots \otimes_{\mathbb{Z}} \mathbb{Z}[\mathrm{Hom}_{\Delta_{\mathrm{un}}}([p^m_*],[q_m])^{\mathrm{ord}}],$$

and differential defined as in T^*. We now show that $T^*_{a_1,\ldots,a_k}$ is acyclic in non-zero degrees and has $H^0 = \mathbb{Z}$; we proceed by descending induction on k.

For $k = m$, the complex $T^*_{a_1,\ldots,a_m}$ has term in degree s the finite sum

$$T^s_{a_1,\ldots,a_m} = \\ \bigoplus_{\Sigma_{i=1}^n p_i = -s} \mathbb{Z}[\mathrm{Hom}_{\Delta_{\mathrm{un}}}([p^1_*],[a_1])^{\mathrm{ord}}] \otimes_{\mathbb{Z}} \ldots \otimes_{\mathbb{Z}} \mathbb{Z}[\mathrm{Hom}_{\Delta_{\mathrm{un}}}([p^m_*],[a_m])^{\mathrm{ord}}];$$

thus $T^*_{a_1,\ldots,a_m}$ is isomorphic to the tensor product complex

$$C^*_{a_1,n_1} \otimes_{\mathbb{Z}} \ldots \otimes_{\mathbb{Z}} C^*_{a_m,n_m},$$

where $C^*_{q,k}$ is the complex considered in Lemma 2.1.3.1. The desired computation of the cohomology of $T^*_{a_1,\ldots,a_m}$ follows from Lemma 2.1.3.1.

Now suppose $k < m$ and that $T^*_{a_1,\ldots,a_k,b}$ has the desired cohomology for all indices $(a_1,\ldots,a_k,b)$. Filter the complex $T^*_{a_1,\ldots,a_k}$ by the subcomplexes

$$T^s_{a_1,\ldots,a_k,q_{k+1}\geq b} := \prod_{\substack{\Sigma_{i=k+1}^m q_i - \Sigma_{i=1}^n p_i = s \\ q_1=a_1,\ldots,q_k=a_k,\ q_{k+1}\geq b}} \mathbb{Z}[\mathrm{Hom}_{\Delta_{\mathrm{un}}}([p^1_*],[q_1])^{\mathrm{ord}}] \otimes_{\mathbb{Z}} \ldots \otimes_{\mathbb{Z}} \mathbb{Z}[\mathrm{Hom}_{\Delta_{\mathrm{un}}}([p^m_*],[q_m])^{\mathrm{ord}}].$$

The quotient complex $T^*_{a_1,\ldots,a_k,q_{k+1}\geq b}/T^*_{a_1,\ldots,a_k,q_{k+1}\geq b+1}$ is then isomorphic to the complex $T^*_{a_1,\ldots,a_k,b}[-b]$. This gives us the E_1-spectral sequence

$$E_1^{a,b} = H^a(T^*_{a_1,\ldots,a_k,b}) \Longrightarrow H^{a+b}(T^*_{a_1,\ldots,a_k});$$

since the complexes $T^*_{a_1,\ldots,a_k,b}$ are zero for $b < 0$, and since the degree s terms $T^s_{a_1,\ldots,a_k}$ are a *product* (rather than a direct sum), this spectral sequence is (weakly) convergent.

By our induction hypothesis, we have $E_1^{a,b} = 0$ for $a \neq 0$ and $E_1^{0,b} \cong \mathbb{Z}$. One easily checks that the differential $E_1^{0,b} \to E_1^{0,b+1}$ is zero for b even and the identity for b odd, hence the spectral sequence degenerates at E_2, and gives the desired result for the cohomology of $T^*_{a_1,\ldots,a_k}$.

Similarly, if we take $a_1 = \ldots = a_{k+1} = 0$, and if we assume that the projection on the factor $p_1 = \ldots = p_n = 0$, $q_{k+1} = \ldots = q_m = 0$ gives an isomorphism

$$H^0(T^*_{a_1,\ldots,a_k,a_{k+1}}) \to \mathbb{Z}[\mathrm{Hom}_{\mathbf{Sets}}([0]\coprod\cdots\coprod[0],[a_1]) \times \ldots \times \mathrm{Hom}_{\mathbf{Sets}}([0]\coprod\cdots\coprod[0],[a_{k+1}])],$$

it follows that the projection

$$H^0(T^*_{a_1,\ldots,a_k}) \to \mathbb{Z}[\mathrm{Hom}_{\mathbf{Sets}}([0]\coprod\cdots\coprod[0],[a_1]) \times \ldots \times \mathrm{Hom}_{\mathbf{Sets}}([0]\coprod\cdots\coprod[0],[a_k])]$$

is an isomorphism as well. The lemma then follows by taking $k = 0$. □

2.1.4. Let $(\mathcal{C},\times,t)$ be a tensor category without unit.

Recall from (Chapter I, §2.3.6, §2.4.3, and Remark 2.4.6) the 2-functor

$$\Pi_{\mathcal{C}}:\Omega \to \mathbf{cat}_{\mathbf{Ab}},$$

the restriction $\Pi^0_{\mathcal{C}}$ to Ω_0, the pre-tensor categories $(\Pi_{\mathcal{C}},\Omega_0)$ and $(\Pi_{\mathcal{C}},\Omega)$, and the tensor categories $(\Pi_{\mathcal{C}},\Omega_0)^{\oplus}$ and $(\Pi_{\mathcal{C}},\Omega)^{\oplus} = \mathcal{C}^{\otimes,c}$.

The inclusion $\Omega_0 \to \Omega$ induces the pre-tensor functor $c_0:(\Pi_{\mathcal{C}},\Omega_0) \to (\Pi_{\mathcal{C}},\Omega)$, which in turn induces the tensor functor $\mathfrak{c}_0(\mathcal{C}):(\Pi_{\mathcal{C}},\Omega_0)^{\oplus} \to (\Pi_{\mathcal{C}},\Omega)^{\oplus} = \mathcal{C}^{\otimes,c}$. We have as well the inclusion functor $i_{\mathcal{C}}:\mathcal{C} \to \mathcal{C}^{\otimes,c}$, which is the universal commutative external product on $\mathcal{C}$ (see Chapter I, Proposition 2.4.4 and Proposition 2.4.5).

We have the additive category $\mathbb{Z}\Delta_{\mathrm{un}}$ generated by Δ_{un}; the symmetric semi-monoidal structure on Δ_{un} gives $\mathbb{Z}\Delta_{\mathrm{un}}$ the structure of a tensor category without unit.

We make $\mathbb{Z}\Delta_{\mathrm{un}}$ a *graded* category by giving a map $f:[p] \to [q]$ degree $q-p$. If we have maps $f_1:[p_1] \to [q_1]$ and $f_2:[p_2] \to [q_2]$, define $f_1 \otimes f_2:[p_1]\coprod[p_2] \to [q_1]\coprod[q_2]$ by

$$f_1 \otimes f_2 := (-1)^{p_1(q_2-p_2)}(f_1 \coprod f_2).$$

Define $\tau_{[p],[q]} : [p] \coprod [q] \to [q] \coprod [p]$ by

$$\tau_{[p],[q]} = (-1)^{pq} t_{[p],[q]}$$

where $t_{[p],[q]}$ is the symmetry in Δ_{un}. This makes $\mathbb{Z}\Delta_{\mathrm{un}}$ into a graded tensor category without unit.

The 2-functor (2.1.2.1) extends to the 2-functor $\Pi_{\mathbb{Z}\Delta_{\mathrm{un}}} : \Omega \to \mathbf{cat}_{\mathrm{Gr}\mathbf{Ab}}$ and gives the restriction $\Pi^0_{\mathbb{Z}\Delta_{\mathrm{un}}} : \Omega^0 \to \mathbf{cat}_{\mathrm{Gr}\mathbf{Ab}}$. The graded tensor structure on $\mathbb{Z}\Delta_{\mathrm{un}}$ makes the category of pairs $(\Pi^0_{\mathbb{Z}\Delta_{\mathrm{un}}}, \Omega_0)$ into a graded pre-tensor category without unit.

2.1.5. *The category of multi-simplices.* We proceed to construct the DG tensor category $\mathcal{C}^{\otimes,\mathfrak{sh}}$. The objects are finite direct sums of pairs $((x_1, \ldots, x_n), n)$, with $n \in \mathbb{N}$ and $x_1, \ldots, x_n$ objects of $\mathcal{C}$.

For pairs

$$(x, n) := ((x_1, \ldots, x_n), n),$$
$$(y, m) := ((y_1, \ldots, y_m), m),$$

and a morphism $F : n \to m$ in Ω_0, define the complex

$$((x,n),(y,m))^{\Delta}_F := \mathrm{Hom}_{\mathcal{C}^{\otimes m}}(\Pi_{\mathcal{C}}(F)(x), y) \otimes_{\mathbb{Z}} \tau^{\le 0}\mathrm{Hom}_{\Delta_{\Omega_0}}(n,m)_F, \tag{2.1.5.1}$$

where, for a complex C, $\tau^{\le 0}C$ is the *canonical truncation*

$$\tau^{\le 0}C^p := \begin{cases} C^p; & \text{for } p < 0 \\ \ker(C^0 \xrightarrow{d^0} C^1); & \text{for } p = 0 \\ 0; & \text{for } p > 0. \end{cases}$$

We have the map

$$\circ : ((y,m),(z,k))^{\Delta}_G \otimes_{\mathbb{Z}} ((x,n),(y,m))^{\Delta}_F \to ((x,n),(z,k))^{\Delta}_{G\circ F} \tag{2.1.5.2}$$

induced by the composition in the categories $(\Pi^0_{\mathcal{C}}, \Omega_0)$ and $(\Pi^0_{\mathbb{Z}\Delta_{\mathrm{un}}}, \Omega_0)$:

$$\begin{aligned}(h_2 \otimes (\ldots g_{2,r_1\ldots r_k}^{q_1,\ldots,q_m} \ldots)) \circ (h_1 \otimes (\ldots g_{1,q_1\ldots q_m}^{p_1,\ldots,p_n} \ldots))\\ = h_2 \circ \Pi_{\mathcal{C}}(G)(h_1) \otimes (\ldots \Sigma_{(q_1,\ldots,q_m)} g_{2,r_1\ldots r_k}^{q_1,\ldots,q_m} \circ \Pi_{\mathbb{Z}\Delta_{\mathrm{un}}}(G)(g_{1,q_1\ldots q_m}^{p_1,\ldots,p_n}) \ldots)\end{aligned}$$

(note that the sum is finite).

Similarly, the tensor products in $(\Pi^0_{\mathcal{C}}, \Omega_0)$ and $(\Pi^0_{\mathbb{Z}\Delta_{\mathrm{un}}}, \Omega_0)$ induces the tensor product

$$\begin{aligned}\bullet : ((x_1,n_1),(y_1,m_1))^{\Delta}_{F_1} \otimes_{\mathbb{Z}} ((x_2,n_2),(y_2,m_2))^{\Delta}_{F_2}\\ \to ((x_1,x_2),n_1+n_2),((y_1,y_2),m_1+m_2))^{\Delta}_{F_1+F_2}\end{aligned} \tag{2.1.5.3}$$

by

$$\begin{aligned}(h_1 \otimes (\ldots g_{1,q_1\ldots q_{m_1}}^{p_1,\ldots,p_{n_1}} \ldots)) \bullet (h_2 \otimes (\ldots g_{2,q'_1\ldots q'_{m_2}}^{p'_1,\ldots,p'_{n_2}} \ldots))\\ = (h_1 \otimes h_2) \otimes (\ldots g_{1,q_1\ldots q_{m_1}}^{p_1,\ldots,p_{n_1}} \otimes g_{2,q'_1\ldots q'_{m_2}}^{p'_1,\ldots,p'_{n_2}} \ldots).\end{aligned}$$

The differential structure on the complex $((x,n),(y,m))^{\Delta}_F$ is given by the differential in the factor $\tau^{\le 0}\mathrm{Hom}_{\Delta_{\Omega_0}}(n,m)_F$.

Let $\tau_{a,b} : a+b \to b+a$ be the symmetry (2.3.3.4) in Ω_0. The symmetry

$$\tau_{(x,a),(y,b)} \in ((x,y,a+b),(y,x,b+a))^{\Delta}_{\tau_{a,b}} \tag{2.1.5.4}$$

is given by

$$\tau_{(x,a),(y,b)} = \mathrm{id}_{y,x} \otimes \prod_{[p_1],\dots,[p_a];[q_1],\dots,[q_b]} (-1)^{\Sigma_{i,j} p_i q_j} \mathrm{id}_{([q_1],\dots,[q_b])} \times \mathrm{id}_{([p_1],\dots,[p_a])}.$$

Let $F: n \to m$ a morphism in Ω_0, and let $\eta: F \to \eta \cdot F$ be a 2-morphism in Ω (see §2.3.5 of Chapter I). For $g \in \mathrm{Hom}_{\Delta_{\Omega_0}}(n,m)^d_F$,

$$g = \prod_{p_1,\dots,p_n;q_1,\dots,q_m} g^{p_1,\dots,p_n}_{q_1,\dots,q_m},$$

$$g^{p_1,\dots,p_n}_{q_1,\dots,q_m} : \pi_{\Delta_{\mathrm{un}}}(F)([p_1],\dots,[p_n]) \to ([q_1],\dots,[q_m]),$$

define

$$\eta \cdot g \in \mathrm{Hom}_{\Delta_{\Omega_0}}(n,m)^d_{\eta\cdot F}$$

by

$$\eta \cdot g = \prod_{p_1,\dots,p_n;q_1,\dots,q_m} (\eta \cdot g)^{p_1,\dots,p_n}_{q_1,\dots,q_m},$$

where

$$(\eta \cdot g)^{q_1,\dots,q_m}_{p_1,\dots,p_n} : \pi_{\Delta_{\mathrm{un}}}(\eta \cdot F)([p_1],\dots,[p_n]) \to ([q_1],\dots,[q_m])$$

is the map

$$(\eta \cdot g)^{p_1,\dots,p_n}_{q_1,\dots,q_m} = g^{p_1,\dots,p_n}_{q_1,\dots,q_m} \circ \Pi_{\mathbb{Z}\Delta_{\mathrm{un}}}(\eta)([p_1],\dots,[p_n])^{-1}.$$

Similarly, for $h \in \mathrm{Hom}_{\mathcal{C}^{\otimes m}}(\Pi_{\mathcal{C}}(F)(x), y)$, define

$$\eta \cdot h := h \circ \Pi_{\mathcal{C}}(h)(x)^{-1}.$$

For a 2-morphism $\eta: F \to \eta \cdot F$, and for $g \otimes h \in ((x,n),(y,m))^{\Delta}_F$, we define

$$\eta \cdot (g \otimes h) := (\eta \cdot g) \otimes (\eta \cdot h) \in ((x,n),(y,m))^{\Delta}_{\eta\cdot F}. \tag{2.1.5.5}$$

We then define $\mathrm{Hom}_{\mathcal{C}^{\otimes,\mathfrak{sh}}}((x,n),(y,m))$ as the quotient complex of the sum of the complexes (2.1.5.1)

$$\mathrm{Hom}_{\mathcal{C}^{\otimes,\mathfrak{sh}}}((x,n),(y,m)) := \oplus_{F \in \mathrm{Hom}_{\Omega_0}(n,m)} ((x,n),(y,m))^{\Delta}_F / \sim, \tag{2.1.5.6}$$

where $\sim$ is the equivalence relation $(g \otimes h) \sim \eta \cdot (g \otimes h)$.

One checks by direct computation that the Hom-complexes (2.1.5.6), with composition (2.1.5.2), tensor product (2.1.5.3) and symmetry (2.1.5.4) defines a DG tensor category $\mathcal{C}^{\otimes,\mathfrak{sh}}$.

We have the inclusion functor $i^{\mathfrak{sh}}_{\mathcal{C}}: \mathcal{C} \to \mathcal{C}^{\otimes,\mathfrak{sh}}$ defined by

$$i^{\mathfrak{sh}}_{\mathcal{C}}(x) = (x,1)$$
$$i^{\mathfrak{sh}}_{\mathcal{C}}(f: x \to y) = f \otimes (\dots \mathrm{id}_{[p]} \dots).$$

2.1.6. *The homotopy commutative external product.* For $[p], [q] \in \Delta$, have the Alexander-Whitney map $f_{[p],[q]} := f_p^{p,q} \cup f_q^{p,q} : [p] \coprod [q] \to [p+q]$ (1.2.1.1),

$$f_{[p],[q]}(i) = \begin{cases} i; & \text{if } i \text{ is in } [p], \\ i+p; & \text{if } i \text{ is in } [q]. \end{cases}$$

Let $F_{21}: 2 \to 1$ be the map (I.2.3.3.5) in Ω_0, and let $\boxtimes^\delta \in \mathrm{Hom}_{\Delta_{\Omega_0}}(2,1)^0_{F_{21}}$ be the morphism defined by the product $\prod_{p,q} f_{[p],[q]}$. One easily checks the associativity relation:

$$\boxtimes^\delta \circ (\mathrm{id}_1 \otimes \boxtimes^\delta) = \boxtimes^\delta \circ (\boxtimes^\delta \otimes \mathrm{id}_1);$$

the relation $d\boxtimes^\delta = 0$ follows from the identity (1.2.1.3). For x, y in $\mathcal{C}$, we let

$$\boxtimes^{\mathfrak{sh}}_{x,y} : (x, y, 2) \to (x \times y, 1) \tag{2.1.6.1}$$

be the map given by

$$\mathrm{id}_{x \times y} \otimes \boxtimes^\delta \in ((x, y, 2), (x \times y, 1))_{F_{21}}.$$

If $q: \mathcal{C} \to \mathcal{D}$ is a symmetric semi-monoidal functor, we have the induced functor

$$q^{\otimes,\mathfrak{sh}} : \mathcal{C}^{\otimes,\mathfrak{sh}} \to \mathcal{D}^{\otimes,\mathfrak{sh}} \tag{2.1.6.2}$$

defined by

$$q^{\otimes,\mathfrak{sh}}((x_1, \dots, x_n), n) = ((q(x_1), \dots, q(x_n)), n);$$
$$q^{\otimes,\mathfrak{sh}}(g \otimes h) = q^{\otimes}(g) \otimes h.$$

2.1.7. THEOREM. *(i) Sending $\mathcal{C}$ to $\mathcal{C}^{\otimes,\mathfrak{sh}}$ and $q: \mathcal{C} \to \mathcal{D}$ to $q^{\otimes,\mathfrak{sh}} : \mathcal{C}^{\otimes,\mathfrak{sh}} \to \mathcal{D}^{\otimes,\mathfrak{sh}}$ gives a functor from tensor categories without unit to DG tensor categories without unit.*
(ii) There is a natural (in $\mathcal{C}$) DG tensor functor $\mathfrak{c}_{\mathcal{C}} : \mathcal{C}^{\otimes,\mathfrak{sh}} \to \mathcal{C}^{\otimes,c}$ with

$$\mathfrak{c}_{\mathcal{C}}(\boxtimes^{\mathfrak{sh}}_{x,y}) = \boxtimes_{x,y}$$

for all x, y in $\mathcal{C}$, and with

$$\mathfrak{c}_{\mathcal{C}} \circ i^{\mathfrak{sh}}_{\mathcal{C}} = i^{c}_{\mathcal{C}}.$$

(iii) The functor $\mathfrak{c}_{\mathcal{C}}$ is a homotopy equivalence.

PROOF. We leave the elementary verification of (i) to the reader. Let $F: n \to m$ be a map in Ω_0. The canonical map of complexes

$$H^0 : \tau^{\leq 0}\mathrm{Hom}_{\Delta_{\Omega_0}}(n, m)_F \to H^0(\mathrm{Hom}_{\Delta_{\Omega_0}}(n, m)_F) \tag{2.1.7.1}$$

is by Lemma 2.1.3.2 a quasi-isomorphism; in addition, we have the isomorphism

$$H^0(\mathrm{Hom}_{\Delta_{\Omega_0}}(n, m)_F) \to \mathbb{Z}. \tag{2.1.7.2}$$

Let

$$\psi_F : \tau^{\leq 0}\mathrm{Hom}_{\Delta_{\Omega_0}}(n, m)_F \to \mathbb{Z}$$

be the composition of the maps (2.1.7.1) and (2.1.7.2), and let

$$\bigoplus_{F \in \mathrm{Hom}_{\Omega_0}(n,m)} ((x, n), (y, m))^{\Delta}_F \xrightarrow{\tilde{\Psi}} \bigoplus_{F \in \mathrm{Hom}_{\Omega_0}(n,m)} \mathrm{Hom}_{\mathcal{C}^{\otimes m}}(\Pi_{\mathcal{C}}(F)(x), y)$$
$$= \mathrm{Hom}_{(\Pi^0_{\mathcal{C}}, \Omega_0)}((x, n), (y, m))$$

be the map $\tilde{\Psi}((h\otimes g)_F) = (\psi_F(g)\cdot h)_F$, where $(x)_F$ denotes the element x in the summand indexed by F. One easily checks that, for a 2-morphism $\eta\colon F\to\eta\cdot F$ in Ω, the map $\tilde{\Psi}$ satisfies

$$\tilde{\Psi}(\eta\cdot h\otimes\eta\cdot g)_{\eta\cdot F} = \eta\cdot\tilde{\Psi}((h\otimes g)_F),$$

hence $\tilde{\Psi}$ descends to the map

$$\Psi\colon \mathrm{Hom}_{\mathcal{C}^{\otimes,\mathfrak{sh}}}((x,n),(y,m)) \to \mathrm{Hom}_{\mathcal{C}^{\otimes,c}}((x,n),(y,m)).$$

One checks directly that Ψ is compatible with the tensor structure in $\mathcal{C}^{\otimes,\mathfrak{sh}}$ and $\mathcal{C}^{\otimes,c}$, giving us the functor $\mathfrak{c}_{\mathcal{C}}$. The relations in (ii) follows directly from the definitions.

For (iii), we have the commutative diagram

$$\begin{array}{ccc}
\oplus_F((x,n),(y,m))_F & \longrightarrow & \mathrm{Hom}_{(\Pi^0_{\mathcal{C}},\Omega_0)}((x,n),(y,m)) \\
\downarrow & & \downarrow \\
\mathrm{Hom}_{\mathcal{C}^{\otimes,\mathfrak{sh}}}((x,n),(y,m)) & \longrightarrow & \mathrm{Hom}_{\mathcal{C}^{\otimes,c}}((x,n),(y,m)),
\end{array}$$

with the top map a quasi-isomorphism. The vertical arrows are the quotient maps induced by the equivalence relation defined by the action of the 2-morphisms in Ω. As this action is given via the action of a finite group, which acts freely on the unique non-zero cohomology H^0, the bottom map is a quasi-isomorphism as well, completing the proof. □

2.2. Categorical cochain operations

We now show how a functor from $\mathcal{C}$ to cosimplicial objects in a tensor category $\mathcal{B}$ gives rise to a functor from $\mathcal{C}^{\otimes,\mathfrak{sh}}$ to complexes in $\mathcal{B}$ (see also [**65**]).

2.2.1. Let $\mathcal{B}$ be a tensor category with operation $\otimes$. We have the category c.s.$\mathcal{B}$ of cosimplicial objects of $\mathcal{B}$, i.e, functors from Δ to $\mathcal{B}$. Let $\mathfrak{F}\colon\mathcal{C}\to$ c.s.$\mathcal{B}$ be a functor. We define $\mathrm{cc}\mathfrak{F}\colon\mathcal{C}\to\mathbf{C}^+(\mathcal{B})$ to be the cochain complex associated to $\mathfrak{F}$, i.e, $\mathrm{cc}\mathfrak{F}^n = \mathfrak{F}([n])$ and $\delta^n\colon\mathfrak{F}^n\to\mathfrak{F}^{n+1}$ is the usual alternating sum $\sum_{i=0}^{n+1}(-1)^i\mathfrak{F}(\delta^i_n)$.

2.2.2. *Multiplicative structure.* Let $\mathcal{C}$ be an additive category. We may view a functor $\mathfrak{F}\colon\mathcal{C}\to$ c.s.$\mathcal{B}$ as a functor from $\mathcal{C}\times\Delta$ to $\mathcal{B}$. If we have two functors $\mathfrak{F}_1,\mathfrak{F}_2\colon\mathcal{C}\to$ c.s.$\mathcal{B}$, define $\mathfrak{F}_1\otimes\mathfrak{F}_2\colon\mathcal{C}^{\otimes 2}\to$ c.s.$\mathcal{B}$ by taking the diagonal cosimplicial object associated to the functor

$$\begin{aligned}
&\mathfrak{F}_1\boxtimes\mathfrak{F}_2\colon\mathcal{C}^{\otimes 2}\times\Delta^2\to\mathcal{B}\\
&\mathfrak{F}_1\boxtimes\mathfrak{F}_2(X_1,X_2;[m_1],[m_2]) = \mathfrak{F}_1(X_1)([m_1])\otimes\mathfrak{F}_2(X_2)([m_2])
\end{aligned}$$

In particular, given $\mathfrak{F}\colon\mathcal{C}\to$ c.s.$\mathcal{B}$, we have the functor $\mathfrak{F}^{\otimes n}\colon\mathcal{C}^{\otimes n}\to$ c.s.$\mathcal{B}$.

Let $(\mathcal{C},\times,a,t)$ be a tensor category without unit. A *multiplication* on $\mathfrak{F}\colon\mathcal{C}\to$ c.s.$\mathcal{B}$ is a natural transformation

$$\mu\colon\mathfrak{F}^{\otimes 2}\to\mathfrak{F}\circ\times$$

which is commutative and associative, in the obvious sense. Concretely, μ is given by maps $\mu_{X,Y}\colon\mathfrak{F}(X)\otimes\mathfrak{F}(Y)\to\mathfrak{F}(X\times Y)$ in c.s.$\mathcal{B}$, which are natural in X and Y, and which have the evident associativity and commutativity properties.

The semi-monoidal structure on $\mathcal{C}$ gives via (I.2.3.6.1) the functor $\Pi_{\mathcal{C}}\colon\Omega\to\mathbf{cat}$ and the category of pairs $(\Pi_{\mathcal{C}},\mathcal{C})$. The associativity and commutativity of the multiplication μ implies that the association

$$\begin{gathered}(X_1,\dots,X_n;n)\mapsto \mathfrak{F}^{\otimes n}(X_1,\dots,X_n)\\ (f_1\colon X_1\to Y_1,\dots,f_n\colon X_n\to Y_n;n)\mapsto \mathfrak{F}^{\otimes n}(f_1,\dots,f_n)\\ (F_{21},\mathrm{id}_{X\times Y})\mapsto \mu(X,Y)\\ (\tau_{1,1},(\mathrm{id}_Y,\mathrm{id}_X))\mapsto \tau_{\mathfrak{F}(X),\mathfrak{F}(Y)}\colon \mathfrak{F}(X)\otimes\mathfrak{F}(Y)\to\mathfrak{F}(Y)\otimes\mathfrak{F}(X)\end{gathered}$$

extends uniquely to a functor

$$\mathfrak{F}^{\otimes}\colon(\amalg_{\mathcal{C}},\mathcal{C})\to \mathrm{c.s.}\mathcal{B} \tag{2.2.2.1}$$

2.2.3. *The functor* $\mathrm{cc}\mathfrak{F}^{\otimes,\mathfrak{sh}}$. Suppose that our functor $\mathfrak{F}$ of §2.2.1 has a multiplication μ, as in §2.2.2. We proceed to define a functor of DG tensor categories (without unit)

$$\mathrm{cc}\mathfrak{F}^{\otimes,\mathfrak{sh}}\colon\mathcal{C}^{\otimes,\mathfrak{sh}}\to \mathbf{C}^+(\mathcal{B}).$$

For an object $(X_1,\dots,X_n)$ of $\mathcal{C}^{\otimes,\mathfrak{sh}}$, set

$$\mathrm{cc}\mathfrak{F}^{\otimes,\mathfrak{sh}}((X_1,\dots,X_n))=\mathrm{cc}\mathfrak{F}(X_1)\otimes\ldots\otimes\mathrm{cc}\mathfrak{F}(X_n).$$

Let $F\colon n\to m$ be a morphism in Ω_0, let $g=(\dots g^{p_1,\dots,p_n}_{q_1,\dots,q_m}\dots)$ be in $\mathrm{Hom}_{\Delta_{\Omega_0}}(n,m)^s_F$, and let $X_1,\dots,X_n$ be objects of $\mathcal{C}$. Write $\amalg_{\mathcal{C}}(F)(X_1,\dots,X_n)=(Y_1,\dots,Y_m)$, and let $h_i\colon Y_i\to Z_i$, $i=1,\dots,m$, be morphisms in $\mathcal{C}$. This gives us the morphism

$$H:=(h_1\otimes\ldots\otimes h_m)\otimes g\colon(X_1,\dots,X_n;n)\to(Z_1,\dots,Z_m;m)$$

in $\mathcal{C}^{\otimes,\mathfrak{sh}}$. Write F as $F=(f,\sigma)$, with $\sigma\in S_n$ and $f\in S^<_{n\to m}$. We write the set $f^{-1}(j)$ as $\{a_j,a_j+1,\dots,a_{j+1}-1\}$ with $1=a_1<a_2<\ldots<a_m<a_{m+1}=n+1$ (recall that f is surjective and order-preserving).

For positive integers $c_1,\dots,c_m$ we have the *weighted sign map*

$$\mathrm{sgn}^{c_1,\dots,c_m}\colon S_m\to\{\pm1\}$$

gotten by having $\rho\in S_m$ act on $[\Sigma_i c_i]\cong[c_1]\amalg\ldots\amalg[c_m]$ by permuting the blocks $[c_i]$ and taking the sign.

We define $\mathrm{cc}\mathfrak{F}^{\otimes,\mathfrak{sh}}(H)$ as follows: We have

$$\mathrm{cc}\mathfrak{F}^{\otimes,\mathfrak{sh}}((X_1,\dots,X_n))^t=\bigoplus_{\substack{p_1,\dots,p_n\\ \Sigma_i p_i=t}}\mathfrak{F}(X_1)([p_1])\otimes\ldots\otimes\mathfrak{F}(X_n)([p_n]).$$

For each pair of tuples $(p_1,\dots,p_n)$, $(q_1,\dots,q_m)$, the map $g^{p_1,\dots,p_n}_{q_1,\dots,q_m}$ is a collection of ordered maps $g_i\colon[p_i]\to[q_{F(i)}]=[q_{f(\sigma(i))}]$, $i=1,\dots,n$. We may then form the composition

$$\begin{aligned}&\mathfrak{F}(X_1)([p_1])\otimes\ldots\otimes\mathfrak{F}(X_n)([p_n])\\ &\xrightarrow{\mathrm{sgn}^{p_1,\dots,p_n}(\sigma)\tau_\sigma}\mathfrak{F}(X_{\sigma^{-1}(1)})([p_{\sigma^{-1}(1)}])\otimes\ldots\otimes\mathfrak{F}(X_{\sigma^{-1}(n)})([p_{\sigma^{-1}(n)}])\\ &\xrightarrow{\mathfrak{F}(X_{\sigma^{-1}(1)})(g_{\sigma^{-1}(1)})\otimes\ldots\otimes\mathfrak{F}(X_{\sigma^{-1}(n)})(g_{\sigma^{-1}(n)})}\\ &\qquad\mathfrak{F}(X_{\sigma^{-1}(1)})([q_1])\otimes\ldots\otimes\mathfrak{F}(X_{\sigma^{-1}(n)})([q_m]).\end{aligned}\tag{2.2.3.1}$$

Writing out the last line, the image of the map (2.2.3.1) is the tensor product

$$\begin{aligned}&\mathfrak{F}(X_{\sigma^{-1}(1)})([q_1]) \otimes \ldots \otimes \mathfrak{F}(X_{\sigma^{-1}(a_2-1)})([q_1]) \otimes \\ &\mathfrak{F}(X_{\sigma^{-1}(a_2)})([q_2]) \otimes \ldots \otimes \mathfrak{F}(X_{\sigma^{-1}(a_3-1)})([q_2]) \otimes \\ &\vdots \\ &\mathfrak{F}(X_{\sigma^{-1}(a_m)})([q_m]) \otimes \ldots \otimes \mathfrak{F}(X_{\sigma^{-1}(n)})([q_m]).\end{aligned}$$

For each $j = 1, \ldots, m$, the identity $Y_j = X_{\sigma^{-1}(a_j)} \times \ldots \times X_{\sigma^{-1}(a_{j+1}-1)}$ determines the map

$$\phi_j = (\mathrm{id}_{Y_j}, F_{a_{j+1}-a_j,1}) \colon (X_{\sigma^{-1}(a_j)}, \ldots, X_{\sigma^{-1}(a_{j+1}-1)}) \to Y_j$$

in $(\Pi^0_{\mathcal{C}}, \mathcal{C})$. We may therefore compose (2.2.3.1) with the composition

$$\begin{aligned}&\mathfrak{F}(X_{\sigma^{-1}(1)})([q_1]) \otimes \ldots \otimes \mathfrak{F}(X_{\sigma^{-1}(n)})([q_m]) \\ &\xrightarrow{\mathfrak{F}^{\otimes}(\phi_1)\otimes\ldots\otimes\mathfrak{F}^{\otimes}(\phi_m)} \mathfrak{F}(Y_1)([q_1]) \otimes \ldots \otimes \mathfrak{F}(Y_m)([q_m]) \\ &\xrightarrow{\mathfrak{F}(h_1)([q_1])\otimes\ldots\otimes\mathfrak{F}(h_m))([q_m])} \mathfrak{F}(Z_1)([q_1]) \otimes \ldots \otimes \mathfrak{F}(Z_m)([q_m])\end{aligned}$$

to give the map $\mathrm{cc}\mathfrak{F}^{\otimes,\mathfrak{sh}}(H)^{p_1,\ldots,p_n}_{q_1,\ldots,q_m}$. Taking the sum over all indices $(p_1, \ldots, p_n)$, $(q_1, \ldots, q_m)$ gives the map

$$\mathrm{cc}\mathfrak{F}^{\otimes,\mathfrak{sh}}(H) \colon \mathrm{cc}\mathfrak{F}^{\otimes,\mathfrak{sh}}\big((X_1, \ldots, X_n)\big) \to \mathrm{cc}\mathfrak{F}^{\otimes,\mathfrak{sh}}\big((Z_1, \ldots, Z_n)\big).$$

We extend the definition of $\mathrm{cc}\mathfrak{F}^{\otimes,\mathfrak{sh}}(H)$ to arbitrary morphisms H by linearity.

2.2.4. THEOREM. *The association*

$$\begin{aligned}(X_1, \ldots, X_n) &\mapsto \mathrm{cc}\mathfrak{F}^{\otimes,\mathfrak{sh}}\big((X_1, \ldots, X_n)\big) \\ H &\mapsto \mathrm{cc}\mathfrak{F}^{\otimes,\mathfrak{sh}}(H)\end{aligned}$$

for $X_1, \ldots, X_n$ objects of $\mathcal{C}$, and H a morphism in $\mathcal{C}^{\otimes,\mathfrak{sh}}$, defines a functor of DG tensor categories without unit

$$\mathrm{cc}\mathfrak{F}^{\otimes,\mathfrak{sh}} \colon \mathcal{C}^{\otimes,\mathfrak{sh}} \to \mathbf{C}^+(\mathcal{B}).$$

PROOF. The proof is a straightforward verification, which we leave to the reader; in fact the data for the DG category $\mathcal{C}^{\otimes,\mathfrak{sh}}$ was chosen precisely with this result in mind. □

3. Homotopy limits

We conclude this chapter with a discussion of homotopy limits. We give a description of the homotopy limit of a functor with values in a DG category, for lack of a suitable reference. We then recall the Bousfield-Kan construction of homotopy limits for simplicial sets [**25**], and relate the two constructions.

3.1. Cohomology for diagrams

We give a review of some notions, constructions and results from [**3**, exposé V].

3.1.1. *Cohomology over a category.* Let I be a small category, giving us the category $\mathbf{Ab}^I$ of functors $F: I \to \mathbf{Ab}$. The category $\mathbf{Ab}^I$ is an abelian category, with kernels and cokernels given pointwise, i.e., $F' \to F \to F''$ is exact if and only if $F'(i) \to F(i) \to F''(i)$ is exact for all $i \in I$.

We have the exact forgetful functor $\mathbf{Ab}^I \to \prod_I \mathbf{Ab}$; applying [**3**, V, Proposition 0.2], the category $\mathbf{Ab}^I$ has enough injectives. We have the DG category $\mathbf{C}(\mathbf{Ab}^I)$, which we identify with the category of functors to $Z^0\mathbf{C}(\mathbf{Ab})$, and the homotopy category $\mathbf{K}(\mathbf{Ab}^I)$, as well as the derived category $\mathbf{D}(\mathbf{Ab}^I)$.

Given A, B in $\mathbf{C}(\mathbf{Ab}^I) = [Z^0\mathbf{C}(\mathbf{Ab})]^I$, we have the complex $\mathcal{H}om_I(A, B)$ of natural transformations $f: A \to B$.

Define the functor $H^0(I, -): \mathbf{Ab}^I \to \mathbf{Ab}$ to be the projective limit functor

$$H^0(I, F) := \varprojlim_I F.$$

Letting $\mathbb{Z}_I$ be the constant functor with value $\mathbb{Z}$, we have the identity

$$H^0(I, F) = \mathrm{Hom}_{\mathbf{Ab}^I}(\mathbb{Z}_I, F);$$

in particular, $H^0(I, -)$ is left exact. We define the *cohomology over I*,

$$H^p(I, -): \mathbf{Ab}^I \to \mathbf{Ab},$$

to be the pth right-derived functor of $H^0(I, -)$. We then have the identity

$$H^p(I, F) = \mathrm{Ext}^p_{\mathbf{Ab}^I}(\mathbb{Z}_I, F)$$

and $H^0(I, -)$ extends to a cohomological functor

$$H^0(I, -): \mathbf{K}(\mathbf{Ab}^I) \to \mathbf{Ab}.$$

3.1.2. Let F be in $\mathbf{C}(\mathbf{Ab}^I)$. Define the functor $H^p(F): I \to \mathbf{Ab}$ by $H^p(F)(i) := H^p(F(i))$.

We have the spectral sequence

$$E_2^{p,q} := H^p(I, H^q(F)) \Longrightarrow H^{p+q}(I, F) \tag{3.1.2.1}$$

which is convergent if F is in $C^+(\mathbf{Ab}^I)$, or if I has finite cohomological dimension. Thus, the cohomological functor $H^0(I, -)$ on $\mathbf{K}^+(\mathbf{Ab}^I)$ defines the cohomological functor

$$H^0(I, -): \mathbf{D}^+(\mathbf{Ab}^I) \to \mathbf{Ab},$$

and extends to the cohomological functor

$$H^0(I, -): \mathbf{D}(\mathbf{Ab}^I) \to \mathbf{Ab},$$

if I has finite cohomological dimension.

3.2. Homotopy limits in a DG category

3.2.1. *Nerves.* If $(S, <)$ is a partially ordered set, we may consider S as a category with a unique morphism from i to j exactly when $i \le j$, and no morphisms otherwise; we write the morphism from i to j as $i \le j$. Each order-preserving map of partially ordered sets is thus a functor on the corresponding categories, and conversely. For example, we may consider the category Δ as having objects the categories $[n]$, with maps being the functors from $[n]$ to $[m]$.

Let I be a small category. Define the *nerve* of I, $\mathcal{N}(I)$, as the simplicial set

$$\mathcal{N}(I)(-) := \mathrm{Hom}_{\mathbf{cat}}(-, I) \colon \Delta^{\mathrm{op}} \to \mathbf{Sets}.$$

Explicitly, $\mathcal{N}(I)([n])$ is the set of composable sequences of n morphisms in I:

$$i_0 \xrightarrow{f_1} i_1 \to \ldots \xrightarrow{f_n} i_n.$$

A simplex $i_0 \xrightarrow{f_1} \ldots \xrightarrow{f_n} i_n$ in $\mathcal{N}(I)$ is *degenerate* if some f_i is an identity morphism; non-degenerate otherwise; we let $\mathcal{N}_{\mathrm{n.d.}}(I)([n])$ be the set of non-degenerate simplices.

We call a category I *finite* if $\mathcal{N}(I)$ has only finitely many non-degenerate simplices, i.e., the geometric realization of $\mathcal{N}(I)$ is a finite CW complex.

The formation of the nerve is functorial in the category I. A natural transformation $\omega\colon f \to g$ of functors $f, g\colon J \to I$ is the same as a functor

$$(f, g, \omega)\colon J \times \{0 < 1\} \to I,$$

where $\{0 < 1\}$ is the category associated to the ordered set $\{0 < 1\}$, and

$$(f, g, \omega)_{J\times 0} := f; \quad (f, g, \omega)_{J\times 1} := g; \quad (f, g, \omega)(\mathrm{id}_X \times (0 < 1)) := \omega(X).$$

Taking nerves gives the map of simplicial sets $\mathcal{N}(f, g, \omega)\colon \mathcal{N}(J \times \{0 < 1\}) \to \mathcal{N}(I)$; since

$$\mathcal{N}(J \times \{0 < 1\}) = \mathcal{N}(J) \times \mathcal{N}(\{0 < 1\}) = \mathcal{N}(J) \times [0, 1],$$

the map $\mathcal{N}(f, g, \omega)$ gives a homotopy between $\mathcal{N}(f)$ and $\mathcal{N}(g)$. In particular, a choice of an initial object or a final object of I gives a contraction of $\mathcal{N}(I)$.

3.2.2. *Additive homotopy limits.* Let C be in $\mathbf{C}(\mathbf{Ab}^I) = [Z^0\mathbf{C}(\mathbf{Ab})]^I$. For $s\colon [n] \to I$ in $\mathcal{N}(I)([n])$, define $C(s) := C(s(n))$. Let $g\colon [n] \to [k]$ be a map in Δ, and let $t\colon [k] \to I$ be a functor (i.e., t is an element of $\mathcal{N}(I)([k])$). Then, as $g(n) \leq k$, we have the morphism $t(g(n) \leq k)\colon t(g(n)) \to t(k)$ in I. We let

$$C(g)\colon C(t \circ g) \to C(t)$$

denote the map $C(t(g(n) \leq k))\colon C(t(g(n))) \to C(t(k))$.

Let C^δ be the following cosimplicial object of $\mathbf{C}(\mathbf{Ab})$:

$$C^\delta([n])^m := \prod_{s \in \mathcal{N}(I)([n])} C(s(n))^m$$

with differential the product of the differentials $d^m(C(s(n)))$. For $g\colon [n] \to [k]$ in Δ, let $C^\delta(g)\colon C^\delta([n]) \to C^\delta([k])$ be the map defined by

$$\pi_t(C^\delta(g)) = C(g) \circ \pi_{t\circ g},$$

where $\pi_t\colon C^\delta([k]) \to C(t)$ is the projection on the factor $C(t)$, and similarly for $\pi_{t\circ g}$.

We may then form the double complex $C^{\delta *}$ associated to cosimplicial complex C^δ, and define $\mathrm{holim}_I(C)$ to be the extended total complex of $C^{\delta *}$.

3.2.3. EXAMPLE. Let X be a topological space, $\mathcal{U} := \{U_0, \ldots, U_n\}$ an open cover of X. Let $O(X)$ be the category of open subsets of X.

Let $\underline{[n]}$ denote the category of non-empty subsets of the set $\{0, \ldots, n\}$, with maps the inclusions of subsets. We have the functor $\underline{\mathcal{U}}\colon \underline{[n]}^{\mathrm{op}} \to O(X)$ defined by $\underline{\mathcal{U}}(I) := \cap_{i\in I} U_i$.

An abelian presheaf P on X gives the functor $P \circ \underline{\mathcal{U}}\colon \underline{[n]} \to \mathbf{Ab}$. Then $\operatorname{holim}_{\underline{[n]}} P \circ \underline{\mathcal{U}}$ is the standard (unordered) Čech complex for P with respect to the cover $\mathcal{U}$.

3.2.4. Let S be a simplicial set, $\mathbb{Z}S$ the simplicial abelian group with n-simplices the free abelian group on the set $S([n])$. The complex $C_*(S;\mathbb{Z})$ associated to $\mathbb{Z}S$ is the complex of integral simplicial chains of S; we consider $C_*(S;\mathbb{Z})$ as a complex $C^*(S;\mathbb{Z})$ with differential of degree $+1$, and with $C^k(S,\mathbb{Z}) := C_{-k}(S;\mathbb{Z})$. We may form the complex, $C^*(S;\mathbb{Z})_{\text{n.d.}}$, of *non-degenerate simplices*, gotten by taking the quotient of $C^n(S;\mathbb{Z})$ by the images $S(\sigma)(C^m(S;\mathbb{Z}))$ where $\sigma\colon[n]\to[m]$ is a non-injective map in Δ.

More generally, if we have a simplicial abelian group A, we let $C^*(A)$ denote the chain complex $C^m(A) := A([-m])$, with usual alternating sum as differential; we have the quotient complex of non-degenerate simplices $C^*(A)_{\text{n.d.}}$ defined as the quotient of $C^*(A)$ as above.

3.2.5. For $i\in I$, let I/i be the category of morphisms $j\to i$ in I. For a morphism $s\colon i\to i'$ in I, we have the functor

$$\begin{gathered} s_*\colon I/i \to I/i' \\ s_*(t\colon j\to i) = s\circ t\colon j\to i' \end{gathered}$$

This gives us the functor

$$\begin{gathered} \mathcal{N}(I/-)\colon I\to \mathbf{s.Sets} \\ i\mapsto \mathcal{N}(I/i) \end{gathered}$$

Taking the complex $C^*(\mathcal{N}(I/i);\mathbb{Z})$ of integral simplices gives us the functor $C^*(\mathcal{N}(I/-);\mathbb{Z})\colon I\to \mathbf{C}^-(\mathbf{Ab})$.

3.2.6. LEMMA. *Let $\mathcal{F}$ be in $C(\mathbf{Ab}^I)$.*
(i) There is a natural isomorphism

$$\mathcal{H}om_I(C^*(\mathcal{N}(I/-);\mathbb{Z}),\mathcal{F}) \to \operatorname*{holim}_I \mathcal{F}.$$

(ii) The natural map

$$\mathcal{H}om_I(C^*(\mathcal{N}(I/-);\mathbb{Z})_{\text{n.d.}},\mathcal{F}) \to \mathcal{H}om_I(C^*(\mathcal{N}(I/-);\mathbb{Z}),\mathcal{F})$$

is a quasi-isomorphism.
(iii) Suppose that $\mathcal{F}$ is in $\mathbf{C}^+(\mathbf{Ab}^I)$ or that I is finite. There is a natural isomorphism

$$H^p(I,\mathcal{F}) \to H^p(\operatorname*{holim}_I \mathcal{F}).$$

PROOF. For (i), we may assume that $\mathcal{F}$ is in $\mathbf{Ab}^I$. Define the map $\mathcal{N}(I)([n])\to \coprod_{i\in I}\mathcal{N}(I/i)([n])$ by sending $i_0 \xrightarrow{f_1} \dots \xrightarrow{f_n} i_n$ to

$$\begin{array}{ccc} i_0 \xrightarrow{f_1} \cdots & \xrightarrow{f_n} & i_n \\ {\scriptstyle f_n\circ\ldots\circ f_1}\searrow & & \swarrow{\scriptstyle \mathrm{id}_{i_n}} \\ & i_n & \end{array}$$

in $\mathcal{N}(I/i_n)$. This induces the injective map

$$\iota\colon \operatorname*{holim}_I \mathcal{F} \to \prod_{i\in I} \mathcal{H}om(C^*(\mathcal{N}(I/i);\mathbb{Z}), \mathcal{F}(i)),$$

where $\mathcal{H}om$ is the complex of maps in $\mathbf{C}(\mathbf{Ab})$; one easily sees that ι has image in the subcomplex $\mathcal{H}om_I(C^*(\mathcal{N}(I/-);\mathbb{Z}), \mathcal{F})$.

Suppose we have an n-simplex

$$\sigma := \begin{array}{ccc} i_0 & \xrightarrow{f_1} \cdots \xrightarrow{f_n} & i_n \\ & {}_{s_0}\searrow \quad \swarrow_{s_n} & \\ & i & \end{array}$$

in $\mathcal{N}(I/i)$. We have the element

$$\sigma' := \begin{array}{ccc} i_0 & \xrightarrow{f_1} \cdots \xrightarrow{f_n} & i_n \\ & {}_{f_n\circ\ldots\circ f_1}\searrow \quad \swarrow_{\mathrm{id}_{i_n}} & \\ & i_n & \end{array}$$

of $\mathcal{N}(I/i_n)$, with $s_{n*}(\sigma') = \sigma$. Thus, if $f\colon C^*(\mathcal{N}(I/-);\mathbb{Z}) \to \mathcal{F}$ is a map, we have $f(i)(\sigma) = \mathcal{F}(s)(f(i_n)(\sigma'))$. This implies that ι is surjective, proving (i).

For (iii), first assume that $\mathcal{F}$ is in $\mathbf{Ab}^I$. As I/i has the final object $i \xrightarrow{\mathrm{id}} i$, $\mathcal{N}(I/i)$ is contractible, hence the augmentation $C^*(\mathcal{N}(I/i);\mathbb{Z}) \to \mathbb{Z}$ is a quasi-isomorphism, i.e., the augmentation $C^*(\mathcal{N}(I/-);\mathbb{Z}) \to \mathbb{Z}_I$ forms a resolution of $\mathbb{Z}_I$. It thus suffices to show that the individual terms $C^m(\mathcal{N}(I/-);\mathbb{Z})$ are projective objects of $\mathbf{Ab}^I$.

Suppose then we have a map $p\colon F \to G$ in $\mathbf{Ab}^I$ such that $p(i)\colon F(i) \to G(i)$ is surjective for all $i \in I$, and a map $g\colon C^m(\mathcal{N}(I/-);\mathbb{Z}) \to G$. Take $i \in I$, and

$$\sigma := \begin{array}{ccc} i_0 & \xrightarrow{f_1} \cdots \xrightarrow{f_n} & i_n \\ & {}_{s_0}\searrow \quad \swarrow_{s_n} & \\ & i & \end{array}$$

in $\mathcal{N}(I/i)([n])$. As above, we have the simplex

$$\sigma' := \begin{array}{ccc} i_0 & \xrightarrow{f_1} \cdots \xrightarrow{f_n} & i_n \\ & {}_{f_n\circ\ldots\circ f_1}\searrow \quad \swarrow_{\mathrm{id}_{i_n}} & \\ & i_n & \end{array}$$

in $\mathcal{N}(I/i_n)$, with $s_{n*}(\sigma') = \sigma$ in $\mathcal{N}(I/i)$. Thus

$$g(i)(\sigma) = G(s_n)(g(i_n)(\sigma')).$$

Choose $f(\sigma') \in F(i_n)$ lifting $g(i_n)(\sigma')$, and define $f(i)(\sigma) := F(s_n)(f(\sigma'))$. One easily checks that this gives a map $f\colon C^m(\mathcal{N}(I/-);\mathbb{Z}) \to F$ with $p \circ f = g$, as desired. The result (iii) for $\mathcal{F}$ in $\mathbf{C}^+(\mathbf{Ab}^I)$ follows from the case of $\mathbf{Ab}^I$ by devissage.

It is well-known (see e.g. [**95**, Chapter V]) that, for a simplicial group A, the map $C^*(A) \to C^*(A)_{\text{n.d.}}$ is a homotopy equivalence. In particular, the augmentation $C^*(\mathcal{N}(I/-);\mathbb{Z})_{\text{n.d.}} \to \mathbb{Z}_I$ gives a resolution of $\mathbb{Z}_I$. Arguing as above, we see that $C^n(\mathcal{N}(I/-);\mathbb{Z})_{\text{n.d.}}$ is projective for each n, proving (ii).

If I is finite, $\mathcal{N}(I)$ is a finite dimensional complex, hence $C^*(\mathcal{N}(I/-);\mathbb{Z})_{\text{n.d.}}$ gives a finite projective resolution of $\mathbb{Z}_I$. We thus have the natural isomorphism

$$\begin{aligned} &H^p(\mathcal{H}om_I(C^*(\mathcal{N}(I/-);\mathbb{Z}),\mathcal{F})) \\ &\qquad \cong H^p(\mathcal{H}om_I(C^*(\mathcal{N}(I/-);\mathbb{Z})_{\text{n.d.}},\mathcal{F})) \cong H^p(I,\mathcal{F}) \end{aligned}$$

for all $\mathcal{F}$ in $\mathbf{C}(\mathbf{Ab}^I)$. □

3.2.7. *Non-degenerate holim.* If $\mathcal{F}: I \to Z^0\mathbf{C}(\mathbf{Ab})$ is a functor, we have the subcomplex $\operatorname{holim}_{I,\ \text{n.d.}} \mathcal{F}$ of $\operatorname{holim}_I \mathcal{F}$ gotten by taking the product

$$\prod_{\sigma\in\mathcal{N}(I)([n])_{\text{n.d.}}} \mathcal{F}(\sigma)$$

and including in $\operatorname{holim}_I \mathcal{F}^n$ by filling in the remaining factors with 0's. The proof of Lemma 3.2.6 identifies $\operatorname{holim}_{I,\ \text{n.d.}} \mathcal{F}$ with $\mathcal{H}om_I(C^*(\mathcal{N}(I/-);\mathbb{Z})_{\text{n.d.}},\mathcal{F})$, hence the inclusion

$$\operatorname*{holim}_{I,\ \text{n.d.}} \mathcal{F} \to \operatorname*{holim}_{I} \mathcal{F}$$

is a quasi-isomorphism.

Suppose I is a finite category. Then $\operatorname{holim}_{I,\ \text{n.d.}} \mathcal{F}^n$ is a *finite* product of terms of the form $\mathcal{F}(i)$, $i \in I$. Thus, suppose we have a functor $\mathcal{F}: I \to Z^0\mathbf{C}^b(\mathcal{A})$, where $\mathcal{A}$ is a DG category. Using the same formula as for the case $\mathcal{A} = \mathbf{Ab}$, we have the complex

$$\operatorname*{holim}_{I,\ \text{n.d.}} \mathcal{F} \in \mathbf{C}^b(\mathcal{A}),$$

functorial in $\mathcal{F}$ and in $\mathcal{A}$. Explicitly, for $\sigma \in \mathcal{N}(I)([n])$, set

$$\mathcal{F}(\sigma) := \begin{cases} \mathcal{F}(\sigma(n)); & \text{for } \sigma \in \mathcal{N}(I)([n])_{\text{n.d.}}, \\ 0; & \text{otherwise,} \end{cases}$$

and define

$$\mathrm{ho}^n_{\text{n.d.}} \mathcal{F} := \bigoplus_{\sigma\in\mathcal{N}(I)([n])_{\text{n.d.}}} \mathcal{F}(\sigma).$$

For $g: [m] \to [n]$ in Δ, we have the (degree 0) map

$$\mathrm{ho}_{\text{n.d.}}\mathcal{F}(g): \mathrm{ho}^m_{\text{n.d.}}\mathcal{F}[-m] \to \mathrm{ho}^n_{\text{n.d.}}\mathcal{F}[-m]$$

induced by the maps

$$\mathcal{F}(\sigma(g(m) \le n)): \mathcal{F}(\sigma \circ g) \to \mathcal{F}(\sigma); \quad \sigma \in \mathcal{N}(I)([n]).$$

We have the degree zero map

$$\sum_{i=0}^{m} (-1)^i \mathrm{ho}_{\text{n.d.}}\mathcal{F}(\delta^m_i): \mathrm{ho}_{\text{n.d.}}\mathcal{F}^m[-m] \to \mathrm{ho}_{\text{n.d.}}\mathcal{F}^{m+1}[-m]; \tag{3.2.7.1}$$

let $d^m: \mathrm{ho}_{\text{n.d.}}\mathcal{F}^m[-m] \to \mathrm{ho}_{\text{n.d.}}\mathcal{F}^{m+1}[-m-1]$ be the be the degree 1 map induced by (3.2.7.1). We then have the object $\mathrm{ho}_{\text{n.d.}}\mathcal{F} := \{\mathrm{ho}^m_{\text{n.d.}}\mathcal{F}; d^m\}$ of $\mathbf{C}^b(\mathbf{C}^b(\mathcal{A}))$; we let $\operatorname{holim}_{I,\text{n.d.}} \mathcal{F}$ in $\mathbf{C}^b(\mathcal{A})$ be the total complex $\mathrm{Tot}(\mathrm{ho}_{\text{n.d.}}\mathcal{F})$ (Chapter II, §1.2.9).

3.2.8. *Functoriality.* Let I and J be small categories, and let $\iota: J \to I$, $X: I \to Z^0\mathbf{C}(\mathbf{Ab})$, and $Y: J \to Z^0\mathbf{C}(\mathbf{Ab})$ be functors. Let $f: X \circ \iota \to Y$ be a natural transformation.

The collection of maps $f(\sigma): X(\iota \circ \sigma) \to Y(\sigma)$, $\sigma \in \mathcal{N}(J)([n])$, defines the maps

$$(f,\iota)^n: \prod_{\tau \in \mathcal{N}(I)([n])} X(\tau) \to \prod_{\sigma \in \mathcal{N}(J)([n])} Y(\sigma); \quad n = 0, 1, \dots,$$

which in turn define the natural map

$$\operatorname*{holim}_J f \circ \iota^*: \operatorname*{holim}_I X \to \operatorname*{holim}_J Y.$$

Restricting to the subcomplex $\operatorname{holim}_{I,\text{ n.d.}} X$ gives the natural map

$$\operatorname*{holim}_{J,\text{ n.d.}} f \circ \iota^*_{\text{n.d.}}: \operatorname*{holim}_{I,\text{ n.d.}} X \to \operatorname*{holim}_{J,\text{ n.d.}} Y.$$

If I and J are finite categories, $\mathcal{A}$ a DG category, $\iota: J \to I$, $X: I \to Z^0\mathbf{C}^b(\mathcal{A})$, and $Y: J \to Z^0\mathbf{C}^b(\mathcal{A})$ functors, and $f: X \circ \iota \to Y$ a natural transformation, we have the map

$$\bigoplus_{\tau \in \mathcal{N}(I)([n])_{\text{n.d.}}} X(\tau) \to \bigoplus_{\sigma \in \mathcal{N}(J)([n])_{\text{n.d.}}} Y(\sigma)$$

defined by the collection of maps $f(\sigma): X(\iota \circ \sigma) \to Y(\sigma)$, $\sigma \in \mathcal{N}(J)([n])_{\text{n.d.}}$, as above. This defines the natural map

$$\operatorname*{holim}_{J,\text{ n.d.}} f \circ \iota^*_{\text{n.d.}}: \operatorname*{holim}_{I,\text{ n.d.}} X \to \operatorname*{holim}_{J,\text{ n.d.}} Y.$$

These maps make holim_I and $\operatorname{holim}_{I,\text{ n.d.}}$ into functors.

3.2.9. *The homotopy limit distinguished triangle.* As in §3.2.7, let $\mathcal{A}$ be a DG category, let I be a finite category, and let $\mathcal{F}: I \to Z^0\mathbf{C}^b(\mathcal{A})$ be a functor, giving us the non-degenerate homotopy limit $\operatorname{holim}_{I,\text{ n.d.}} \mathcal{F}$ in $\mathbf{C}^b(\mathcal{A})$.

Since I is finite, there exist minimal objects of I, i.e., an object i such that $\operatorname{Hom}_I(j,i) = \emptyset$ for all $j \neq i$. Since I is finite, we have $\operatorname{Hom}_I(i,i) = \{\text{id}\}$.

Take a minimal $i \in I$, giving us the full subcategory $I \setminus \{i\}$ of I, with inclusion functor $j_{I\setminus\{i\}}$. We have as well the inclusion functor $j_i: \{i\} \to I$. The identity natural transformations on $\mathcal{F} \circ j_{I\setminus\{i\}}$ and $\mathcal{F} \circ j_i$ give the maps

$$j^*_{I\setminus\{i\}}: \operatorname*{holim}_{I,\text{ n.d.}} \mathcal{F} \to \operatorname*{holim}_{I\setminus\{i\},\text{ n.d.}} \mathcal{F}_{|I\setminus\{i\}},$$

$$j^*_i: \operatorname*{holim}_{I,\text{ n.d.}} \mathcal{F} \to \operatorname*{holim}_{\{i\},\text{ n.d.}} \mathcal{F}(i) = \mathcal{F}(i).$$

Let $I^{i/}$ be the category of morphisms $s: i \to j$ in I, with $j \neq i$, where a morphism $(s: i \to j) \to (s': i \to j')$ is a map $t: j \to j'$ with $s' = t \circ s$. Let $\mathcal{F}^{i/}: I^{i/} \to Z^0\mathbf{C}^b(\mathcal{A})$ be the functor $\mathcal{F}^{i/}(s: i \to j) = \mathcal{F}(j)$.

Mapping $\mathcal{F}(i)$ to $\mathcal{F}^{i/}(s: i \to j) = \mathcal{F}(j)$ by the map $\mathcal{F}(s)$, and summing over s gives the map

$$j^*_{I^{i/},i}: \mathcal{F}(i) \to \operatorname*{holim}_{I^{i/},\text{ n.d.}} \mathcal{F}^{i/}.$$

For an n-simplex $i_0 \xrightarrow{s_1} i_1 \xrightarrow{s_2} \dots \xrightarrow{s_n} i_n$ in $\mathcal{N}(I \setminus \{i\})$, and for $s: i \to i_0$, let σs be the element

$$\begin{array}{ccc} i_0 \xrightarrow{s_1} & \cdots & \xrightarrow{s_n} i_n \\ {\scriptstyle s}\nwarrow & & \nearrow{\scriptstyle s_n \dots s_1 s} \\ & i & \end{array}$$

of $\mathcal{N}(I^{i/})([n])$. Sending $\mathcal{F}(\sigma)$ to $\mathcal{F}^{i/}(\sigma s)$ by the identity on $\mathcal{F}(i_n)$ and summing over all $s: i \to j$ in $I^{i/}$ defines the map

$$j^*_{I^{i/}, I\setminus\{i\}}: \operatorname*{holim}_{I\setminus\{i\},\text{ n.d.}} \mathcal{F}_{|I\setminus\{i\}} \to \operatorname*{holim}_{I^{i/},\text{ n.d.}} \mathcal{F}^{i/}.$$

We have the identity $\operatorname{holim}_I \mathcal{F} = \operatorname{cone}(j^*_{I^{i/},i} - j^*_{I^{i/},I\setminus\{i\}})[-1]$, with the natural map $\operatorname{holim}_I \mathcal{F} \to \mathcal{F}(i) \oplus \operatorname{holim}_{I\setminus\{i\},\text{ n.d.}} \mathcal{F}_{|I\setminus\{i\}}$ being $(j_i^*, j^*_{I\setminus\{i\}})$. This gives us the *homotopy limit distinguished triangle* in $\mathbf{K}^b(\mathcal{A})$

$$(3.2.9.1)\quad \operatorname*{holim}_{I,\text{ n.d.}} \mathcal{F} \xrightarrow{(j_i^*, j^*_{I\setminus\{i\}})} \mathcal{F}(i) \oplus \operatorname*{holim}_{I\setminus\{i\},\text{ n.d.}} \mathcal{F}_{|I\setminus\{i\}} \xrightarrow{j^*_{I^{i/},i} - j^*_{I^{i/},I\setminus\{i\}}} \operatorname*{holim}_{I^{i/},\text{ n.d.}} \mathcal{F}^{i/} \to \operatorname*{holim}_{I,\text{ n.d.}} \mathcal{F}[1],$$

natural in $\mathcal{F}$. One immediate application of (3.2.9.1) is

3.2.10. PROPOSITION. *Let I be a finite category, $\mathcal{A}$ a DG category, and $\mathcal{D}$ a localization of $\mathbf{K}^b(\mathcal{A})$. Let $f: \mathcal{F} \to \mathcal{G}$ be a natural transformation of functors $\mathcal{F}, \mathcal{G}: I \to Z^0\mathbf{C}^b(\mathcal{A})$ such that $f(i): \mathcal{F}(i) \to \mathcal{G}(i)$ is an isomorphism in $\mathcal{D}$ for all $i \in I$. Then*

$$\operatorname*{holim}_{I,\text{ n.d.}} f: \operatorname*{holim}_{I,\text{ n.d.}} \mathcal{F} \to \operatorname*{holim}_{I,\text{ n.d.}} \mathcal{G}$$

is an isomorphism in $\mathcal{D}$.

PROOF. We define the *dimension* of a finite category J to be the maximal n for which there is a sequence of non-identity morphisms $j_0 \xrightarrow{s_1} \dots \xrightarrow{s_n} j_n$ in J; this is the same as the maximal n for which $\mathcal{N}(J)_{\text{n.d.}}([n])$ is non-empty. Let $N := \dim I$; we may assume the result for all finite J with $\dim J \leq N$, and with $|\mathcal{N}(J)_{\text{n.d.}}([N])| < |\mathcal{N}(I)_{\text{n.d.}}([N])|$.

Since I is finite, there is a minimal element i with $|\mathcal{N}(I \setminus \{i\})_{\text{n.d.}}([N])| < |\mathcal{N}(I)_{\text{n.d.}}([N])|$; we obviously have $\dim I^{i/} < \dim I$.

The distinguished triangle (3.2.9.1) and the similar triangle with $\mathcal{F}$ replaced by $\mathcal{G}$ give distinguished triangles in $\mathcal{D}$; the map f induces a map of distinguished triangles. Thus, our induction assumption together with the five lemma in the triangulated category $\mathcal{D}$ shows that $\operatorname{holim}_{I,\text{ n.d.}} f$ is an isomorphism in $\mathcal{D}$. □

3.3. Cohomology and homotopy limits

3.3.1. *Hypercohomology.* Let I be a small category, and $X: I \to \mathbf{Top}$ a functor. Pulling back the site $\mathbf{Top}$ via X gives the site X, the category $\tilde{X}$ of sheaves on X and the category $\hat{X}$ of presheaves on X. In particular, the sheaf category is a full subcategory of the presheaf category. We let $\mathrm{Sh}_X^{\mathbf{Ab}}$ denote the category of sheaves of abelian groups on X; we have the constant sheaf $\mathbb{Z}_X$ on X. The functor

$H^0(X,-)$ from $\mathrm{Sh}_X^{\mathbf{Ab}}$ to $\mathbf{Ab}$ is defined as $H^0(X,-) := \mathrm{Hom}_{\mathrm{Sh}_X^{\mathbf{Ab}}}(\mathbb{Z}_X, -)$. We have as well the functor $H^0\colon \mathbf{Top}^I \to \mathbf{Ab}^I$ defined by

$$H^0(S)(i) := H^0(X(i), S(i)),$$

and the identity

$$H^0(I, H^0(S)) = H^0(X, S). \tag{3.3.1.1}$$

We let

$$\mathbb{H}^0(X,-)\colon \mathbf{D}^+(\mathrm{Sh}_X^{\mathbf{Ab}}) \to \mathbf{Ab}$$

denote the extension of $H^0(X,-)$ to complexes. If I has finite cohomological dimension, and each $X(i)$ has finite cohomological dimension, then $\mathbb{H}^0(X,-)$ extends to

$$\mathbb{H}^0(X,-)\colon \mathbf{D}(\mathrm{Sh}_X^{\mathbf{Ab}}) \to \mathbf{Ab}.$$

The identity (3.3.1.1) gives us the following explicit complex computing the hypercohomology $\mathbb{H}^p(X,\mathcal{F})$ for $\mathcal{F}$ a complex of abelian sheaves on X. We may form the Godement resolution $\mathcal{F}(i) \to G\mathcal{F}(i)$ of the complex $\mathcal{F}(i)$ for each $i \in I$. As the Godement resolution is functorial, taking pointwise global sections gives the functor $G\mathcal{F}\colon I \to \mathbf{C}(\mathbf{Ab})$. We may then form the homotopy limit $\mathrm{holim}_I\, G\mathcal{F}$.

In case $\mathcal{F}$ is bounded below, $\mathrm{holim}_I\, G\mathcal{F}$ is a representative in $\mathbf{C}^+(\mathbf{Ab})$ of the object $RH^0(X,\mathcal{F})$ of $\mathbf{D}^+(\mathbf{Ab})$. Similarly, if each $X(i)$ has finite cohomological dimension and I has finite cohomological dimension, then $\mathrm{holim}_I\, G\mathcal{F}$ is a representative in $\mathbf{C}(\mathbf{Ab})$ of the object $RH^0(X,\mathcal{F})$ of $\mathbf{D}(\mathbf{Ab})$ for general $\mathcal{F}$. In particular, we have the natural isomorphism

$$H^p(\operatorname*{holim}_I G\mathcal{F}) \cong \mathbb{H}^p(X,\mathcal{F}).$$

Using the representative $G\mathcal{F}$, the spectral sequence (3.1.2.1) gives us the spectral sequence

$$E_2^{p,q} = H^p(I, [i \mapsto \mathbb{H}^q(X(i), S(i))]) \Longrightarrow \mathbb{H}^{p+q}(X,S) \tag{3.3.1.2}$$

for S in $\mathbf{D}^+(\mathrm{Sh}_X^{\mathbf{Ab}})$, or in $\mathbf{D}(\mathrm{Sh}_X^{\mathbf{Ab}})$ if the above hypotheses are satisfied.

If I is finite, one can also apply the distinguished triangle (3.2.9.1) to the representative $G\mathcal{F}$, giving the distinguished triangle in $\mathbf{D}(\mathbf{Ab})$

$$\begin{aligned} RH^0(X,\mathcal{F}) &\to RH^0(X(i),\mathcal{F}(i)) \oplus RH^0(X_{|I\setminus\{i\}}, \mathcal{F}_{|I\setminus\{i\}}) \\ &\to RH^0(X^{i/}, \mathcal{F}^{i/}) \to RH^0(X,\mathcal{F})[1], \end{aligned} \tag{3.3.1.3}$$

for $i \in I$ a minimal element.

3.4. Homotopy limits of simplicial sets

3.4.1. *Homotopy limits.* We recall some of the basic constructions and results of [**25**]. For each $n = 0, 1, 2, \ldots$, we have the simplicial set

$$\Delta_n\colon \Delta^{\mathrm{op}} \to \mathbf{Sets}$$
$$\Delta_n(-) := \mathrm{Hom}_\Delta(-, [n]).$$

Sending n to Δ_n thus gives the functor $\Delta_*\colon \Delta \to \mathbf{s.Sets}$. The category of simplicial sets has the internal Hom defined by

$$\mathcal{H}om(X,Y) := \mathrm{Hom}_{\mathbf{s.Sets}}(X \times \Delta_*, Y)\colon \Delta^{\mathrm{op}} \to \mathbf{Sets},$$

that is

$$\mathcal{H}om(X,Y)([n]) = \mathrm{Hom}_{\mathbf{s.Sets}}(X \times \Delta_n, Y),$$

and similarly for morphisms. If we have two functors $X, Y: I \to \mathbf{s.Sets}$ we may form the simplicial set $\mathcal{H}om_I(X,Y)$ similarly by

$$\mathcal{H}om_I(X,Y) := \mathrm{Hom}_{\mathbf{s.Sets}^I}(X \times (-), Y): \Delta^{\mathrm{op}} \to \mathbf{Sets}.$$

If X is a functor $X: I \to \mathbf{s.Sets}$, the *homotopy limit* of X over I is the simplicial set $\mathrm{holim}_I X := \mathcal{H}om_I(\mathcal{N}(I/-), X)$. This gives the functor

$$\operatorname*{holim}_I : \mathbf{s.Sets}^I \to \mathbf{s.Sets}.$$

The construction of §3.2.8 gives the similarly defined map

$$\operatorname*{holim}_J f \circ \iota^* : \operatorname*{holim}_I X \to \operatorname*{holim}_J Y \tag{3.4.1.1}$$

given functors $\iota: I \to J$, $X: I \to \mathbf{s.Sets}$, and $Y: J \to \mathbf{s.Sets}$, and natural transformation $f: X \circ \iota \to Y$.

3.4.2. *Closed simplicial model categories.* We refer the reader to [**104**] for the basic notions of closed model categories and closed simplicial model categories. We will not attempt to discuss these notions here; we only list the few basic concepts we will have occasion to use.

There is the notion of a *fibrant* simplicial set (see e.g., [**25**, VIII, §3]). A map $X \to Y$ of simplicial sets is a *weak equivalence* if the map on the geometric realizations $|X| \to |Y|$ induces an isomorphism on all homotopy groups. An object X of $\mathbf{s.Sets}^I$ is defined to be fibrant if $X(i)$ is fibrant for all $i \in I$, and a map $X \to Y$ in $\mathbf{s.Sets}^I$ is a weak equivalence if $X(i) \to Y(i)$ is a weak equivalence for each $i \in I$ (see [**25**, XI, §8, proof of Proposition 8.1]).

A simplicial abelian group is fibrant. If S is a simplicial set, then the singular complex of the geometric realization of S, $Sin(|S|)$, is fibrant and the canonical map $S \to Sin(|S|)$ is a weak equivalence (see e.g., [**25**, VIII, §3]). As this construction is functorial, the canonical map $S \to Sin(|S|)$ gives a canonical fibrant model for S in $\mathbf{s.Sets}^I$ as well.

If $f: X \to Y$ is a weak equivalence of fibrant objects in $\mathbf{s.Sets}^I$, then

$$\operatorname*{holim}_I f: \operatorname*{holim}_I X \to \operatorname*{holim}_I Y$$

is a weak equivalence of fibrant objects in $\mathbf{s.Sets}$ (see [**25**, V, 5.6]).

3.4.3. *Homotopy and homology.* We now relate the operation holim_I for simplicial sets to holim_I for complexes of abelian groups. To distinguish these two, we sometimes denote the holim for complexes of abelian groups by $\mathrm{holim}^{\mathbf{Ab}}$. If I is finite and $\mathcal{A}$ is a DG category, we sometimes denote the holim for a functor $X: I \to Z^0\mathbf{C}^b(\mathcal{A})$ by $\mathrm{holim}^{\mathcal{A}}_{I,\ \mathrm{n.d.}} X$.

Let S be a simplicial set and T a simplicial abelian group. The Dold-Kan equivalence [**39**], [**74**], see also [**95**, Chapter V] of the homotopy category of simplicial abelian groups, and the homotopy category of (cohomological) complexes of abelian groups which are supported in degrees ≤ 0, gives the natural homotopy equivalence of complexes

$$C^*(\mathcal{H}om(S,T)) \sim \tau^{\leq 0}\mathcal{H}om(C^*(S;\mathbb{Z}), C^*(T))$$

where $\mathcal{H}om(C^*(S;\mathbb{Z}), C^*(T))$ is the internal Hom in the category $\mathbf{C}(\mathbf{Ab})$, and $\tau^{\leq 0}$ is the canonical truncation

$$[\tau^{\leq 0}C]^p := \begin{cases} C^p; & \text{if } p < 0, \\ \ker[C^0 \xrightarrow{d^0} C^1]; & \text{if } p = 0, \\ 0; & \text{if } p > 0. \end{cases}$$

Since the homotopy equivalence is natural, the analogous result extends to functors $S: I \to \mathrm{s}.\mathbf{Sets}$, $T: I \to \mathrm{s}.\mathbf{Ab}$, i.e., there is a natural homotopy equivalence of complexes

$$(3.4.3.1) \qquad C^*(\mathcal{H}om_I(S,T)) \sim \tau^{\leq 0}\mathcal{H}om_I(C^*(S;\mathbb{Z}), C^*(T)).$$

Thus, it follows from Lemma 3.2.6 and (3.4.3.1) that we have the homotopy equivalence

$$(3.4.3.2) \qquad C^*(\operatorname{holim}_I A) \sim \tau^{\leq 0} \operatorname{holim}_I^{\mathbf{Ab}} C^*(A),$$

for simplicial abelian groups A, natural in A.

3.4.4. *Products.* Let $\mathbf{s.Sets}^*$ be the category of pointed simplicial sets, and let $X, Y: I \to \mathbf{s.Sets}^*$ be functors. This gives us the functor $X \wedge Y: I \to \mathbf{s.Sets}^*$ with $(X \wedge Y)(i) := X(i) \wedge Y(i)$.

Suppose we have a "multiplication", i.e., a natural transformation $\mu: X \wedge Y \to Z$. We give $\operatorname{holim}_I X$ a base-point by taking the base-point in each $X(i)$, and similarly for $\operatorname{holim}_I Y$ and $\operatorname{holim}_I Z$. Define the pointed map

$$\operatorname{holim}_I \mu: \operatorname{holim}_I X \wedge \operatorname{holim}_I Y \to \operatorname{holim}_I Z$$

by sending

$$f: \mathcal{N}(I/-) \times \Delta^n \to X; \quad g: \mathcal{N}(I/-) \times \Delta^n \to Y$$

to

$$\mu \circ (f \wedge g) \circ \iota_{\mathcal{N}(I/-)\times\Delta^n}: \mathcal{N}(I/-) \times \Delta^n \to Z,$$

where

$$\iota_{\mathcal{N}(I/-)\times\Delta^n}: \mathcal{N}(I/-) \times \Delta^n \to (\mathcal{N}(I/-) \times \Delta^n) \wedge (\mathcal{N}(I/-) \times \Delta^n)$$

is the diagonal embedding.

If μ is associative (resp. commutative), so is $\operatorname{holim}_I \mu$.

Now suppose we have functors $A, B, C: I \to Z^0\mathbf{C}^*(\mathbf{Ab})$ (where $* = +$ or $* = -$, and I has finite cohomological dimension if $* = -$), and a multiplication $\mu: A \otimes B \to C$, where $(A \otimes B)(i) := A(i) \otimes B(i)$. Define the *Alexander-Whitney product*

$$(3.4.4.1) \qquad \operatorname{holim}_I^{\mathbf{Ab}} \mu: \operatorname{holim}_I^{\mathbf{Ab}}(A) \otimes \operatorname{holim}_I^{\mathbf{Ab}}(B) \to \operatorname{holim}_I^{\mathbf{Ab}}(C)$$

as follows: We have the maps $f_n^{n,m}: [n] \to [n+m]$ and $f_m^{n,m}: [m] \to [n+m]$ given by $f_n^{n,m}(i) = i$ and $f_m^{n,m}(j) = n + j$ (see §1.2.1). Let $\sigma := (i_0 \xrightarrow{f_1} \dots \xrightarrow{f_{n+m}} i_{n+m})$ be an $n+m$-simplex in $\mathcal{N}(I)$, giving the n-simplex

$$\sigma \circ f_n^{n,m} := (i_0 \xrightarrow{f_1} \dots \xrightarrow{f_n} i_n)$$

and the m-simplex

$$\sigma \circ f_m^{n,m} := (i_n \xrightarrow{f_{n+1}} \dots \xrightarrow{f_{n+m}} i_{n+m}).$$

Given

$$f := \prod_{\tau \in \mathcal{N}([n])} f(\tau) \in A(\tau); \quad g := \prod_{\rho \in \mathcal{N}([m])} g(\rho) \in B(\rho),$$

let

$$\operatorname{holim}_I^{\mathbf{Ab}} \mu(f \otimes g) := \prod_{\sigma \in \mathcal{N}([n+m])} \mu(f(\sigma \circ f_n^{n,m}) \otimes g(\sigma \circ f_m^{n,m})).$$

If we have a collection of multiplications μ which are associative, the relations described in §1.2.1 show that $\operatorname{holim}_I^{\mathbf{Ab}} \mu$ is associative; if μ is commutative (with respect to an involution on C), then $\operatorname{holim}_I^{\mathbf{Ab}} \mu$ is commutative up to functorial homotopy. Taking $\mu: A \otimes B \to A \otimes B$ to be the identity, we have the map

$$\operatorname{holim}_I^{\mathbf{Ab}} \mu: \operatorname{holim}_I^{\mathbf{Ab}} A \otimes \operatorname{holim}_I^{\mathbf{Ab}} B \to \operatorname{holim}_I^{\mathbf{Ab}} (A \otimes B),$$

which is commutative (up to homotopy) with respect to the symmetry isomorphism on $A \otimes B$, and satisfies the obvious associativity condition.

Restricting to simplices in $\mathcal{N}_{\text{n.d.}}(I)$ gives a multiplication for $\operatorname{holim}_{I,\text{ n.d.}}^{\mathbf{Ab}}$, compatible with the multiplication for $\operatorname{holim}_I^{\mathbf{Ab}}$ via the canonical inclusion

$$\operatorname{holim}_{I,\text{ n.d.}}^{\mathbf{Ab}}(-) \to \operatorname{holim}_I^{\mathbf{Ab}}(-).$$

If I is finite, $\mathcal{A}$ a DG tensor category, $A, B, C: I \to Z^0\mathbf{C}^b(\mathcal{A})$ functors and $\mu: A \otimes B \to C$ a multiplication, the same formula gives products

$$\operatorname{holim}_I^{\mathcal{A}} \mu: \operatorname{holim}_I^{\mathcal{A}} A \otimes \operatorname{holim}_I^{\mathcal{A}} B \to \operatorname{holim}_I^{\mathcal{A}}(C).$$

3.4.5. *Comparison of products.* Let $g: [m+n] \to [m] \times [n]$ be an injective order-preserving map. The map g determines an isomorphism

$$\tilde{g}: \{1, \dots, m+n\} \to \{1, \dots, m\} \coprod \{1, \dots, n\}$$

by sending i to $g_1(i)$ if $g_1(i) > g_1(i-1)$ and to $g_2(i)$ if $g_2(i) > g_2(i-1)$; define $\operatorname{sgn}(g)$ to be the sign of the shuffle permutation determined by $\tilde{g}$. We then have the triangulation of $\Delta_m \times \Delta_n$, $\sum_g \operatorname{sgn}(g) g$.

Suppose we have functors $A, B, C: I \to \mathrm{s.}\mathbf{Ab}$ and a bilinear map $\mu: A \times B \to C$. This then induces maps $\mu^\wedge: A \wedge B \to C$ and $\mu^\otimes: A \otimes B \to C$.

We have the *Eilenberg-MacLane map*

$$\theta_{A,B}: C^*(A) \otimes C^*(B) \to C^*(A \otimes B),$$

defined by sending $\sigma_p \otimes \tau_q$ ($\sigma_p \in A([p])(i)$, $\tau_q \in B([q])(i)$) to

$$\sum_{g=(g_1,g_2)} \operatorname{sgn}(g) A(g_1)(\sigma_p) \otimes B(g_2)(\tau_q),$$

where $g = (g_1, g_2): [p+q] \to [p] \times [q]$ runs over the injective, order-preserving maps. This gives the map

$$C^*(\mu) := C^*(\mu^\otimes) \circ \theta_{A,B}: C^*(A) \otimes C^*(B) \to C^*(C).$$

The map

$$\operatorname*{holim}_I(\mu^\wedge)\colon \operatorname*{holim}_I(A)\wedge\operatorname*{holim}_I(B)\to\operatorname*{holim}_I(C)$$

descends to the product

$$\operatorname*{holim}_I(\mu^\wedge)\colon \operatorname*{holim}_I(A)\otimes\operatorname*{holim}_I(B)\to\operatorname*{holim}_I(C)$$

and the map $C^*(\mu)$ gives the product

$$\operatorname*{holim}_I^{\mathbf{Ab}}(C^*(\mu))\colon \operatorname*{holim}_I^{\mathbf{Ab}}(C^*(A))\otimes\operatorname*{holim}_I^{\mathbf{Ab}}(C^*(B))\to\operatorname*{holim}_I^{\mathbf{Ab}}(C^*(C)).$$

We may take complexes and apply the Eilenberg-MacLane map for $\operatorname{holim}_I(\mu^\wedge)$, giving the product ($hA := \operatorname{holim}_I A$, $hB := \operatorname{holim}_I B$)

$$C^*(\operatorname*{holim}_I(\mu^\wedge))\circ\theta_{hA,hB}\colon C^*(\operatorname*{holim}_I(A))\otimes C^*(\operatorname*{holim}_I(B))\to C^*(\operatorname*{holim}_I(C)).$$

These two products are compatible, up to homotopy, via the homotopy equivalence (3.4.3.2); this follows from the fact that the Eilenberg-MacLane map and the Alexander-Whitney product (3.4.4.1) are homotopy inverses via a functorial homotopy (see [**95**, Chapter VI]).

More precisely, let S and T be simplicial sets, A and B simplicial abelian groups. The tensor structure on the category of complexes gives the natural map

$$\begin{aligned}&\mathcal{H}om(C^*(S;\mathbb{Z}),C^*(A))\otimes\mathcal{H}om(C^*(T;\mathbb{Z}),C^*(B))\\&\qquad\to\mathcal{H}om(C^*(S;\mathbb{Z})\otimes C^*(T;\mathbb{Z}),C^*(A)\otimes C^*(B)).\end{aligned}$$

We have the *Alexander-Whitney map*

$$\begin{gathered}C^*(S\times T;\mathbb{Z})\to C^*(S;\mathbb{Z})\otimes C^*(T;\mathbb{Z})\\ \sigma_n\times\tau_n\mapsto\sum_{p+q=n}S(f_p^{p,q})(\sigma)\otimes T(f_q^{p,q})(\tau);\\ \sigma_n\in S([n]),\ \tau_n\in T([n]).\end{gathered}$$

This gives the product

$$\begin{aligned}&\mathcal{H}om(C^*(S;\mathbb{Z}),C^*(A))\otimes\mathcal{H}om(C^*(T;\mathbb{Z}),C^*(B))\\&\qquad\to\mathcal{H}om(C^*(S\times T;\mathbb{Z}),C^*(A)\otimes C^*(B)).\end{aligned}$$

If $S=T$, we may pull-back by the diagonal, giving the product map

$$\begin{aligned}&\mathcal{H}om(C^*(S;\mathbb{Z}),C^*(A))\otimes\mathcal{H}om(C^*(S;\mathbb{Z}),C^*(B))\\&\qquad\to\mathcal{H}om(C^*(S;\mathbb{Z}),C^*(A)\otimes C^*(B));\end{aligned}\tag{3.4.5.1}$$

as all the maps are natural, we have the similarly defined products for functors from I to simplicial sets, resp. simplicial abelian groups. Via Lemma 3.2.6, this gives us a product for the functor $\operatorname{holim}_I^{\mathbf{Ab}}$, which one checks is the product (3.4.4.1).

We have as well the simplicially defined product

$$\mathcal{H}om(S,A)\times\mathcal{H}om(T,B)\to\mathcal{H}om(S\times T,A\times B),$$

which descends to the product

$$\mathcal{H}om(S,A)\otimes\mathcal{H}om(T,B)\to\mathcal{H}om(S\times T,A\otimes B).$$

Taking $S=T$ and pulling back by the diagonal gives the product

$$\mathcal{H}om(S,A)\otimes\mathcal{H}om(S,B)\to\mathcal{H}om(S,A\otimes B).$$

Using the Eilenberg-MacLane map, the Alexander-Whitney map, and the homotopy equivalence (3.4.3.1), we have the product map

$$\tau^{\leq 0}\mathcal{H}om(C^*(S;\mathbb{Z}), C^*(A)) \otimes \tau^{\leq 0}\mathcal{H}om(C^*(S;\mathbb{Z}), C^*(B)) \\ \to \tau^{\leq 0}\mathcal{H}om(C^*(S;\mathbb{Z}), C^*(A) \otimes C^*(B)).$$

One can then use an acyclic models argument to give a natural homotopy between this latter product and the product map (3.4.5.1). Applying this to the definitions of holim_I and $\mathrm{holim}_I^{\mathbf{Ab}}$ as Hom objects gives the desired compatibility.

CHAPTER IV

Canonical Models for Cohomology

We review some basic material about Grothendieck sites and topoi, with the construction of the cosimplicial Godement resolution and a description of its properties being the main goal.

1. Sheaves, sites, and topoi

1.1. Grothendieck topologies

We begin by recalling the notions of a Grothendieck pre-topology, a Grothendieck site, and a topos.

1.1.1. Let $\mathcal{C}$ be a category. A *Grothendieck pre-topology on* $\mathcal{C}$ consists of giving, for each object X of $\mathcal{C}$, a collection of families of morphisms

$$\mathrm{Cov}(X) := \{\{f_\alpha : U_\alpha \to X \mid \alpha \in A\}\}$$

satisfying the following axioms:

1. If $\{f_\alpha : U_\alpha \to X \mid \alpha \in A\}$ is in $\mathrm{Cov}(X)$, and if $Y \to X$ is in $\mathcal{C}$, then the fiber product $U_\alpha \times_X Y$ exists for each $\alpha \in A$, and the family $\{p_2 : U_\alpha \times_X Y \to Y \mid \alpha \in A\}$ is in $\mathrm{Cov}(Y)$
2. If $\{f_\alpha : U_\alpha \to X \mid \alpha \in A\}$ is in $\mathrm{Cov}(X)$, and if $\{g_\beta : V_{\alpha\beta} \to U_\alpha \mid \beta \in B_\alpha\}$ is in $\mathrm{Cov}(U_\alpha)$ for each $\alpha \in A$ then $\{f_\alpha \circ g_\beta : V_{\alpha\beta} \to X \mid \alpha \in A, \beta \in B_\alpha\}$ is in $\mathrm{Cov}(X)$
3. The identity map $\mathrm{id}_X : X \to X$ is in $\mathrm{Cov}(X)$.

The elements of $\mathrm{Cov}(X)$ are called the *covering families of* X.

A Grothendieck pre-topology generates a *Grothendieck topology* (see [**4**, II, Chapter 1]); as we will not need the notion of a Grothendieck topology, we omit its description, and by abuse of notation, refer to a category with a Grothendieck pre-topology as a *Grothendieck site.*

1.1.1.1. EXAMPLES. (i) Let X be a topological space, and let $\mathcal{C}$ be the category with objects the open subsets of X, and with maps $U \to V$ the inclusions $U \subset V$. For U in $\mathcal{C}$, a family $\{i_\alpha : U_\alpha \to U\}$ is in $\mathrm{Cov}(U)$ if and only if the U_α cover U, i.e., $U = \cup_\alpha U_\alpha$. This forms the site X_{top}. Let **Top** denote the category of topological spaces, and define $\mathrm{Cov}(X)$ to be the set of families of maps $\{f_\alpha : U_\alpha \to X\}$ which are isomorphic over X to covering families of open subsets. This forms the site **Top**.

(ii) Let X be a scheme, and let $\mathcal{C}$ be the category of étale maps of finite type $U \to X$ (where a morphism is a commutative triangle). Define $\mathrm{Cov}(U)$ to be the collection of families $\{f_\alpha : U_\alpha \to U\}$ such that the map of underlying topological spaces $\coprod_\alpha f_\alpha : \coprod_\alpha U_\alpha \to U$ is surjective. This defines the site $X_{\mathrm{\acute{e}t}}$. Using the same definition of $\mathrm{Cov}(X)$ for X a scheme gives the site $\mathbf{Sch}_{\text{ét}}$ with underlying category the category of noetherian schemes.

1.1.2. *Presheaves and sheaves.* Let $(\mathcal{C}, \mathfrak{T})$ be a Grothendieck site. A *presheaf* on $\mathcal{C}$, with values in a category $\mathcal{A}$, is simply a functor $P: \mathcal{C} \to \mathcal{A}$; with morphisms of presheaves being natural transformations, this forms the category of presheaves on $\mathcal{C}$ with values in $\mathcal{A}$.

A presheaf P with values in **Sets** is a *sheaf for the topology* $\mathfrak{T}$ if, for each object U of $\mathcal{C}$, and each covering family $\{f_\alpha: U_\alpha \to U \mid \alpha \in A\}$ in $\mathrm{Cov}(U)$, the sequence of sets

$$\emptyset \to P(U) \xrightarrow{\prod_\alpha P(f_\alpha)} \prod_\alpha P(U_\alpha) \underset{P(p_2)}{\overset{P(p_1)}{\rightrightarrows}} \prod_{\alpha,\beta} P(U_\alpha \times_U U_\beta)$$

is exact. More generally, a presheaf P with values in a category $\mathcal{A}$ is a sheaf if, for each object A of $\mathcal{A}$, the presheaf of sets P_A, $P_A(X) := \mathrm{Hom}_{\mathcal{A}}(A, P(X))$, is a sheaf for the topology $\mathfrak{T}$. This forms the category $\mathrm{Sh}^{\mathcal{A}}_{(\mathcal{C},\mathfrak{T})}$ of sheaves on $\mathcal{C}$ with values in $\mathcal{A}$ as the full subcategory of the presheaf category.

We denote the category of presheaves of sets on $\mathcal{C}$ by $\hat{\mathcal{C}}$, the category of sheaves of sets by $\tilde{\mathcal{C}}$, and $\iota: \tilde{\mathcal{C}} \to \hat{\mathcal{C}}$ the canonical inclusion. By [**4**, II, Théorème 3.4], ι has the left adjoint $\eta: \hat{\mathcal{C}} \to \tilde{\mathcal{C}}$, the *sheafification* of a presheaf.

For an object X of $\mathcal{C}$, we let $\hat{X}$ denote the representing presheaf $\hat{X}(Y) = \mathrm{Hom}_{\mathcal{C}}(Y, X)$ and let $\tilde{X}$ be the sheafification $\tilde{X} := \eta\hat{X}$. Sending X to $\tilde{X}$ defines the functor

$$\sigma: \mathcal{C} \to \tilde{\mathcal{C}}. \tag{1.1.2.1}$$

1.1.3. EXAMPLE. If $\mathcal{C}$ is a category, one can form the finest topology on $\mathcal{C}$ for which all the representable functors $Y \mapsto \mathrm{Hom}_{\mathcal{C}}(Y, X)$ are sheaves. This is called the *canonical* topology on the category $\mathcal{C}$.

1.2. Hypercovers

1.2.1. *The coskeleton.* Let $\mathcal{C}_{\mathfrak{T}}$ be a Grothendieck site, $\mathcal{U} := \{f_\alpha: U_\alpha \to U \mid \alpha \in A\}$ in $\mathrm{Cov}(U)$. One can then form the augmented simplicial object in $\tilde{\mathcal{C}}$, $\tilde{\mathcal{U}}_* \to \tilde{U}$, with

$$\tilde{\mathcal{U}}_n = \coprod_{(\alpha_0,\dots,\alpha_n)\in A^{n+1}} \widetilde{U_{\alpha_0} \times_U \dots \times_U U_{\alpha_n}},$$

and with the usual face and degeneracy maps. This generalizes to the notion of a *hypercover* of a simplicial object of $\tilde{\mathcal{C}}$. We will give here a brief sketch of this notion; for more details, we refer the reader to [**3**, Chapter V].

For a category $\mathcal{B}$, we have the category $\mathrm{s}.\mathcal{B}$ of simplicial objects of $\mathcal{B}$. We have the full subcategory $\Delta^{\leq n}$ of Δ with objects $[0], \dots, [n]$, and the category $\mathrm{s}.^{\leq n}\mathcal{B}$ of *n-truncated* simplicial objects in $\mathcal{B}$, i.e., the category of functors $\Delta^{\leq n\mathrm{op}} \to \mathcal{B}$. The restriction to $\Delta^{\leq n\mathrm{op}}$ defines the functor $i_n^*: \mathrm{s}.\mathcal{B} \to \mathrm{s}.^{\leq n}\mathcal{B}$. Assuming the existence of finite projective limits in $\mathcal{B}$, the functor i_n^* admits the right adjoint $i_{n*}: \mathrm{s}.^{\leq n}\mathcal{B} \to \mathrm{s}.\mathcal{B}$; let $\mathrm{cosk}^n: \mathrm{s}.\mathcal{B} \to \mathrm{s}.\mathcal{B}$ be the composition $i_{n*} \circ i_n^*$. The identity map on $i_n^*(X)$ defines by adjunction the natural map $\mathfrak{c}^n: X \to \mathrm{cosk}^n X$.

Now let F be an object in the sheaf category $\tilde{\mathcal{C}}$, and take $\mathcal{B}$ to be the category $\tilde{\mathcal{C}}/F$ of maps $F' \to F$ in $\tilde{\mathcal{C}}$. A simplicial object in $\tilde{\mathcal{C}}/F$ is then just a simplicial object of $\tilde{\mathcal{C}}$, with augmentation to F.

We have the canonical topology on $\tilde{\mathcal{C}}$ (Example 1.1.3), with covering families $\mathrm{Cov}_{\mathrm{can}}$.

1.2.2. DEFINITION. An object $F_* \to F$ of $\mathrm{s}.\tilde{\mathcal{C}}/F$ is a *hypercover* of F if for each $n = 0, 1, \ldots$, the natural map $c^n_{n+1}: F_{n+1} \to (\mathrm{cosk}^n F_*)_{n+1}$ is in $\mathrm{Cov}_{\mathrm{can}}((\mathrm{cosk}^n F_*)_{n+1})$.

It follows from the compatibility of the functor cosk^n with projective limits that, if $F_* \to F$ is a hypercover of F, and if $F' \to F$ is a map in $\tilde{\mathcal{C}}$, then the fiber product $F' \times_F F_* \to F'$ is a hypercover of F'.

1.2.3. DEFINITION. Let $f: F' \to F$ be a morphism in $\tilde{\mathcal{C}}$, $F'_* \to F'$ and $F_* \to F$ hypercovers. A morphism $f_*: F'_* \to F_*$ of hypercovers over f is a morphism in $\mathrm{s}.\tilde{\mathcal{C}}/F'$,

$$f_*: F'_* \to F' \times_F F_*.$$

This defines the category of hypercovers in $\tilde{\mathcal{C}}$.

Using the functor (1.1.2.1), we may define a hypercover of an object X of $\mathcal{C}$ as an augmented simplicial object $X_* \to X$ such that $\tilde{X}_* \to \tilde{X}$ is a hypercover of $\tilde{X}$ in $\tilde{\mathcal{C}}$; morphisms of hypercovers of objects of $\mathcal{C}$ are defined similarly, giving the category of hypercovers in $\mathcal{C}$.

1.3. Topoi and points

1.3.1. A *topos* is a category which is equivalent to the category $\tilde{\mathcal{C}}$ for some Grothendieck site $(\mathcal{C}, \mathfrak{T})$. A *morphism of topoi* $u: T_1 \to T_2$ is a triple consisting of functors $u_*: T_1 \to T_2$, $u^*: T_2 \to T_1$, and a natural isomorphism

$$\phi: \mathrm{Hom}_{T_1}(u^*(-), -) \to \mathrm{Hom}_{T_2}(-, u_*(-)),$$

with the additional requirement that the functor u^* is *left-exact*, i.e., preserves finite projective limits. With the obvious notion of composition, topoi form a category.

If T is a topos, one can form the site T_{can} by using the canonical topology (Example 1.1.3). It is a theorem of Giraud that *all* sheaves of sets on T_{can} are representable (see [**4**, IV, Theorem 1.2]). From this, one shows that a morphism of topoi $u: T_1 \to T_2$ is determined by the functor $u^*: T_2 \to T_1$, and that conversely, each functor $u^*: T_2 \to T_1$ which preserves finite projective limits and arbitrary inductive limits comes from a morphism of topoi [**4**, IV, Corollary 1.7]. Also, finite projective limits in a topos are representable.

With its unique topology, the one-point category forms a site, and the category of sheaves on this site is the same as the category of presheaves, which in turn is equivalent to the category of sets; thus the category **Sets** is a topos.

1.3.2. DEFINITION. Let T be a topos. A *point* of a topos T is a morphism of topoi $p: \mathbf{Sets} \to T$ We denote the category of points of T by $\mathrm{Point}(T)$.

1.3.3. A *fiber functor on* T is a functor

$$F: T \to \mathbf{Sets}$$

which preserves finite projective limits and arbitrary inductive limits.

By §1.3.1, sending a point p of a topos T to the fiber functor p^* on T defines an equivalence of the category of points of T with the opposite of the category of fiber functors on T.

If p is a point of T, and X is an object (or morphism) of T, we often write X_p for $p^*(X)$, and ϕ_p for the fiber functor p^*; X_p is called the *stalk* of X at p.

1.3.4. EXAMPLES. (i) Suppose we have a topology (in the usual sense) on a set X, giving the site X_{top} as in Example 1.1.1.1(i), and the category of sheaves on X, $\tilde{X}_{\text{top}}$. Let p be a point of X. Sending a sheaf P to the stalk P_p,

$$P_p := \varinjlim_{p \in U} P(U),$$

defines the functor $p^*: \tilde{X}_{\text{top}} \to \mathbf{Sets}$. Sending a set S to the skyscraper sheaf at p with stalk S defines the functor $p_*: \mathbf{Sets} \to \tilde{X}_{\text{top}}$; we have the obvious adjunction isomorphism

$$\phi: \mathrm{Hom}_{\mathbf{Sets}}(p^*(-), -) \to \mathrm{Hom}_{\tilde{X}_{\text{top}}}(-, p_*(-)).$$

The triple (p_*, p^*, ϕ) then defines the morphism of topoi

$$p: \mathbf{Sets} \to \tilde{X}_{\text{top}}.$$

(ii) Let X be a finite type k-scheme (k a field), and let $p: \operatorname{Spec} \bar{k} \to X$ be a *geometric point* of X. Sending a sheaf P on X for the étale topology to the stalk

$$P_p := \varinjlim_{(U,u) \to (X,p)} P(U),$$

where $(U, u) \to (X, p)$ is an étale pointed map of finite type k-schemes, and sending a set S to the skyscraper sheaf at p with value S defines the point of $\tilde{X}_{\text{ét}}$, $p: \mathbf{Sets} \to \tilde{X}_{\text{ét}}$

1.3.5. DEFINITION. Let T be a topos. We say that T *has enough points* if there is a set $\mathcal{P}$ of points of T such that a map $f: P \to P'$ is an isomorphism (resp. monomorphism, resp. epimorphism) if and only if the maps $f_p: P_p \to P'_p$ are isomorphisms (resp. monomorphisms, resp. epimorphisms) for all $p \in \mathcal{P}$. A set $\mathcal{P}$ of points of T which satisfies the above condition is called a *conservative family of points of* T. If $\mathcal{C}_{\mathfrak{T}}$ is a Grothendieck site, we call a conservative family of points of $\tilde{\mathcal{C}}_{\mathfrak{T}}$ a conservative family of points of $\mathcal{C}_{\mathfrak{T}}$.

1.3.6. REMARKS. (i) Let $\mathcal{P}$ be a set of points of a topos T. The collection of morphisms $p: \mathbf{Sets} \to T$ for $p \in \mathcal{P}$ defines the morphism of topoi

$$i: \coprod_{p \in \mathcal{P}} \mathbf{Sets} \to T.$$

Then $\mathcal{P}$ forms a conservative family of points of T if and only if the functor

$$\prod_{p \in \mathcal{P}} p^*: T \to \prod_{p \in \mathcal{P}} \mathbf{Sets}$$

is a conservative functor [**4**, I, 6.1.1].
(ii) Let X be a topological space. The set of points of X forms via Example 1.3.4(i) a conservative family of points of $\tilde{X}_{\text{top}}$.
(iii) Let X be a finite type k-scheme. The set of geometric points of X (maps $\operatorname{Spec} \bar{k} \to X$ up to k-isomorphism $\sigma: \operatorname{Spec} \bar{k} \to \operatorname{Spec} \bar{k}$) forms via Example 1.3.4(ii) a conservative family of points of $\tilde{X}_{\text{ét}}$.

1.3.7. *Points, neighborhoods and pro-objects.* We recall some basic facts and notions on points of a topos, their interpretation in terms of neighborhoods and pro-objects in the underlying category of a site. For reference, see [**4**, IV, §6.8].

Let T be a topos, $p\colon \mathbf{Sets} \to T$ a point of T, $\phi_p\colon T \to \mathbf{Sets}$ the corresponding fiber functor. A *neighborhood of* p is a pair (X, u), with X an object of T, and $u \in X_p$. A morphism $(X, u) \to (Y, v)$ of neighborhoods of p is defined to be a morphism $f\colon X \to Y$ in T such that $f_p(u) = v$. This defines the category $V(p)$ of neighborhoods of p, projection on the first factor determining a functor $V(p) \to T$.

Since finite projective limits in T are representable, and the fiber functor ϕ preserves finite projective limits, it follows that finite projective limits in $V(p)$ are representable, and the functor $V(p) \to T$ commutes with such limits. From this, it follows that the opposite category $V(p)^{\mathrm{op}}$ is filtering. In addition, for each object F of T, there is a canonical isomorphism

$$\phi_p(F) = F_p \cong \varinjlim_{(X,u)\in V(p)^{\mathrm{op}}} F(X). \tag{1.3.7.1}$$

Now suppose that $T = \tilde{\mathcal{C}}$ for a Grothendieck site $\mathcal{C}$, let p be a point of $\tilde{\mathcal{C}}$, and X an object of $\mathcal{C}$. We write X_p for $\tilde{X}_p$, and similarly for morphisms. We let $V_{\mathcal{C}}(p)$ be the category of pairs (X, u) with X in $\mathcal{C}$, and $u \in X_p$; morphisms are defined as in $V(p)$. The category $V_{\mathcal{C}}(p)^{\mathrm{op}}$ is again filtering, and one has the isomorphism analogous to (1.3.7.1)

$$\phi_p(F) = F_p \cong \varinjlim_{(X,u)\in V_{\mathcal{C}}(p)^{\mathrm{op}}} F(X). \tag{1.3.7.2}$$

The projection on the first factor defines the functor $V_{\mathcal{C}}(p) \to \mathcal{C}$, and thus defines a pro-object of $\mathcal{C}$ (a functor $I \to \mathcal{C}$, where I^{op} is filtering, and has a small, cofinal subcategory). Conversely, if $f\colon I \to \mathcal{C}$ is a pro-object of $\mathcal{C}$, then f defines a fiber functor via the formula (1.3.7.2) if and only if the following condition is satisfied:

(1.3.7.3)

Let Y be in $\mathcal{C}$, $\{g_\alpha\colon Y_\alpha \to Y \mid \alpha \in A\}$ in $\mathrm{Cov}(Y)$. Given an $i_0 \in I$, and a morphism $f(i_0) \to Y$ in $\mathcal{C}$, there is an $i \in I$, a morphism $s\colon i \to i_0$ in I, an $\alpha \in A$, and a commutative diagram

$$\begin{array}{ccc} f(i) & \longrightarrow & Y_\alpha \\ {\scriptstyle f(s)}\downarrow & & \downarrow{\scriptstyle g_\alpha} \\ f(i_0) & \longrightarrow & Y \end{array}$$

in $\mathcal{C}$.

Sending a point p of $\tilde{\mathcal{C}}$ to the corresponding pro-object of $\mathcal{C}$ defines a fully faithful embedding

$$i_{\mathrm{Point}}\colon \mathrm{Point}(\mathcal{C}) \to \text{Pro-}\mathcal{C}$$

with essential image the full subcategory of pro-objects which satisfy the condition (1.3.7.3).

1.3.8. If F is a presheaf (of sets) on $\mathcal{C}$, define

$$F_p := \varinjlim_I F(f(-)),$$

where $f: I \to \mathcal{C}$ is the pro-object associated to p. We have as well the associated sheaf $\tilde{F}$; the canonical map of presheaves gives the canonical map

$$F_p \to \tilde{F}_p. \tag{1.3.8.1}$$

1.3.9. LEMMA. *The map* (1.3.8.1) *is an isomorphism.*

PROOF. Let X be in $\mathcal{C}$, and let $\mathcal{U} := \{f_\alpha : U_\alpha \to X\}$ be in $\mathrm{Cov}(X)$. Define $LF(\mathcal{U})$ by the exactness of

$$\emptyset \to LF(\mathcal{U}) \longrightarrow \prod_\alpha F(U_\alpha) \rightrightarrows \prod_{\alpha,\beta} F(U_\alpha \times_X U_\beta).$$

If $X = f(i)$ for some $i \in I$, then, by the condition (1.3.7.3), there is a map $t: j \to i$ in I, an α, and a map $g: f(j) \to U_\alpha$ with $f(t) = f_\alpha \circ g$.

Let $LF(X)$ be the inductive limit of the $LF(\mathcal{U})$, over the category $\mathrm{Cov}(X)$ (with maps being refinements). From [**4**, II, Remarque 3.3], sending X to $LF(X)$ defines a presheaf on $\mathcal{C}$; from the remark above, it follows that the natural map $F_p \to LF_p$ is an isomorphism. In addition, from [**4**, II, Théorème 3.4], the presheaf LLF is the sheafification $\tilde{F}$ of F, whence the lemma. □

1.3.10. *Sheaves with additional structure.* One can consider a somewhat more general situation, replacing sheaves of sets with sheaves in a suitable category $\mathcal{A}$. Suppose for example $\mathcal{A}$ is defined as a category of sets "with structure of type Σ", given as the existence of certain operations satisfying certain axioms which are described as various commutative diagrams, and suppose we have a morphism of topoi $u: \tilde{C}_1 \to \tilde{C}_2$, where C_1 and C_2 are sites, and $\tilde{C}_i$ are the categories of sheaves of sets. As the functors u^* and u_* both preserve finite projective limits (in particular, products), they will preserve the structure Σ defining $\mathcal{A}$, and thus give functors $u^*_{\mathcal{A}}: \mathrm{Sh}^{\mathcal{A}}_{C_2} \to \mathrm{Sh}^{\mathcal{A}}_{C_1}$ and $u^{\mathcal{A}}_*: \mathrm{Sh}^{\mathcal{A}}_{C_1} \to \mathrm{Sh}^{\mathcal{A}}_{C_2}$, which will still be adjoint functors. For example, the functors u^* and u_* extend to functors on sheaves of abelian groups, modules over a fixed ring R, and sets with G-action, for a fixed group G.

For a ring R, we write $\tilde{C}_R$ for the category of sheaves of R-modules.

2. Canonical resolutions

2.1. The cosimplicial Godement resolution

2.1.1. Let $u = (u_*, u^*, \phi): T_1 \to T_2$ be a morphism of topoi. The adjunction properties of u_* and u^* give rise to natural transformations

$$\alpha: \mathrm{id}_{T_1} \to u_* u^*; \quad \beta: u^* u_* \to \mathrm{id}_{T_2}. \tag{2.1.1.1}$$

2.1.2. LEMMA. *The natural transformations* (2.1.1.1) *satisfy*

1. $(u_* \circ \beta) \circ (\alpha \circ u_*) = \mathrm{id}$,
2. $(\beta \circ u^*) \circ (u^* \circ \alpha) = \mathrm{id}$.

PROOF. Let X be an object of T_1, and let $A = u_* X$ and $B = u^* u_* X$. We have

$$\alpha(u_* X) = \phi_{A,B}(\mathrm{id}_B), \ \beta(X) = \phi^{-1}_{A,X}(\mathrm{id}_A).$$

By the naturality of ϕ, we have

$$\begin{aligned} u_*(\beta(X)) \circ \alpha(u_*(X)) &= u_*(\phi_{A,X}^{-1}(\mathrm{id}_A)) \circ \phi_{A,B}(\mathrm{id}_B) \\ &= \phi_{A,X}(\phi_{A,X}^{-1}(\mathrm{id}_A)_*(\mathrm{id}_B)) \\ &= \phi_{A,X}(\phi_{A,X}^{-1}(\mathrm{id}_A)) \\ &= \mathrm{id}_A, \end{aligned}$$

which proves (1). The identity (2) is similar, and is left to the reader.

□

Let $G^n: T_1 \to T_1$ be the functor $(u_*u^*)^{n+1}$. For each codegeneracy map $\sigma_i^n: [n] \to [n-1]$, $i = 0, \ldots, n-1$, let $G(\sigma_i^n): G^n \to G^{n-1}$ be the natural transformation

$$u_* \circ (u^*u_*)^i \circ \beta \circ (u^*u_*)^{n-i-1} \circ u^*.$$

For each coface map $\delta_i^{n-1}: [n-1] \to [n]$, $i = 0, \ldots, n$, let $G(\delta_i^{n-1}): G^{n-1} \to G^n$ be the natural transformation

$$(u_*u^*)^i \circ \alpha \circ (u_*u^*)^{n-i}.$$

The identities of Lemma 2.1.2 (together with the well-known presentation of Δ, see e.g. [**95**, Chapter 1]) imply that functors G^n and the natural transformations $G(\sigma_i^n)$ and $G(\delta_i^{n-1})$ extend uniquely to the cosimplicial object

$$G_u: \Delta \to \mathrm{Funct}(T_1, T_1)$$

in the category of functors from T_1 to itself. Similarly, the natural transformation α gives the augmentation

$$\epsilon_1: \mathrm{id}_{T_1} \to G_u. \tag{2.1.2.1}$$

Let X be a cosimplicial object in a category $\mathcal{C}$, and Y a simplicial set. If $\mathcal{C}$ has a final object $*$, and if finite products over $*$ exist, we may form the cosimplicial object X^Y of $\mathcal{C}$ defined by $x \mapsto X(x)^{Y(x)}$ for x a morphism or an object of Δ. We have the simplicial set $[0,1] := \mathrm{Hom}_\Delta(-, [1])$, with maps $i_0, i_1: * \to [0,1]$ induced by the two inclusions $\delta_0^0, \delta_1^0: [0] \to [1]$. Recall that a homotopy of maps of cosimplicial objects of $\mathcal{C}$, $f, g: X \to Y$, is given by a map $h: X \to Y^{[0,1]}$, with $i_0^*h = f$ and $i_1^*h = g$. For example, we may take $\mathcal{C}$ to be the functor category $\mathrm{Funct}(T_1, T_2)$, so we may speak of a *homotopy equivalence* of cosimplicial objects of $\mathrm{Funct}(T_1, T_2)$.

2.1.3. PROPOSITION. *Applying u^* to the map* (2.1.2.1) *induces a homotopy equivalence of cosimplicial objects in the functor category $\mathrm{Funct}(T_1, T_2)$*

$$\epsilon_2: u^* \to u^*(G_u). \tag{2.1.3.1}$$

PROOF. Let $\Delta*$ be the category of order-preserving, pointed maps of the pointed (with base-point $*$) ordered sets $[n]* := \{* < 0 < \ldots < n\}$, $n = -1, 0, 1, \ldots$. The morphisms in $\Delta*$ are generated from Δ by the addition of the codegeneracy maps

$$\sigma_{-1}^n: [n]* \to [n-1]*,$$

$$\sigma_{-1}^n(i) := \begin{cases} *; & \text{if } i = *, 0, \\ i-1; & \text{if } i > 0. \end{cases}$$

Sending $[n]$ to $[n]*$ embeds Δ as a subcategory of $\Delta*$.

Suppose we have a category $\mathcal{C}$ with a final object, and finite products over the final object. Let $X_0 \xrightarrow{\epsilon} X$ be an augmented cosimplicial object of $\mathcal{C}$, and suppose we have an extension of X to a functor $X*:\Delta* \to \mathcal{C}$ with $X*([-1]*) = X_0$, and $\epsilon = X*([-1]* \hookrightarrow [0]*)$. Then the map ϵ is a homotopy equivalence (where we consider X_0 as the constant cosimplicial object). Indeed, let $\pi_n:[n]* \to [-1]*$ be the unique map; then the maps $X*(\pi_n):X([n]) \to X_0$ give a splitting π to ϵ. In addition, let

$$\sigma_{n,j}:\{* < 0 < \ldots < n\} \to \{* < 0 < \ldots < n\}; \quad j = 0, \ldots, n+1,$$

be the map

$$\sigma_{n,j}(i) := \begin{cases} *; & \text{for } i < j, \\ i - j; & \text{for } i \geq j. \end{cases}$$

Letting $p_{n,j}:[n] \to [1]$, $j = 0, \ldots, n+1$ be the map

$$p_{n,j}(i) := \begin{cases} 0; & \text{for } i < j, \\ 1; & \text{for } i \geq j, \end{cases}$$

we have

$$\mathrm{Hom}_\Delta([n],[1]) = \{p_{n,0}, \ldots, p_{n,n+1}\}.$$

Then sending $X([n])$ to $X([n])^{[0,1]([n])}$ by the map $X(\sigma_{n,j})$ in the factor indexed by $p_{n,j}$ gives a homotopy of $\epsilon \circ \pi$ with id_X, completing the verification of our claim.

We have the natural transformations

$$\beta \circ (u^* u_*)^n \circ u^*: u^* G^n \to u^* G^{n-1}; \quad n > 0$$
$$\beta \circ u^*: u^* G^0 \to u^*.$$

The identities of Lemma 2.1.2 imply these maps give an extension of the augmented cosimplicial object (2.1.3.1) to a functor $u^*G*:\Delta* \to \mathrm{Funct}(T_1, T_2)$. By the above discussion, this shows that ϵ_2 is a homotopy equivalence. □

2.1.4. Let $\mathcal{C}_\mathfrak{T}$ be a Grothendieck site, forming the topos of sheaves $\tilde{\mathcal{C}}$. Suppose that $\tilde{\mathcal{C}}$ has a conservative family of points $\mathcal{P}$. Let $\tilde{\mathcal{C}}^\delta$ be the *discrete* topos associated to $\mathcal{P}$:

$$\tilde{\mathcal{C}}^\delta := \coprod_{p \in \mathcal{P}} \mathbf{Sets},$$

and let

$$i:\tilde{\mathcal{C}}^\delta \to \tilde{\mathcal{C}} \tag{2.1.4.1}$$

the morphism of topoi as in Remark 1.3.6(i).

For a commutative ring R, we have the category $\tilde{\mathcal{C}}_R$ of sheaves of R-modules on $\mathcal{C}$, and the category $\tilde{\mathcal{C}}_R^\delta$ of sheaves (or presheaves) of R-modules on $\mathcal{P}$. The functors i_* and i^* induce adjoint functors

$$i_*:\tilde{\mathcal{C}}_R^\delta \to \tilde{\mathcal{C}}_R; \quad i^*:\tilde{\mathcal{C}}_R \to \tilde{\mathcal{C}}_R^\delta.$$

We let $\hat{\mathcal{C}}_R$ denote the category of presheaves of R-modules on $\mathcal{C}$.

We recall that a map $f:S \to T$ of simplicial sets is a *weak equivalence* if the map $|f|:|S| \to |T|$ on the geometric realizations induced by f gives an isomorphism on the homotopy groups (see [**25**] for details). We call a map $f:S \to T$ of cosimplicial

R-modules a weak equivalence if f induces a quasi-isomorphism on the associated complexes of R-modules.

2.1.5. LEMMA. *(i) The functor i_* (both for sheaves of sets, and for sheaves of R-modules) is left exact and preserves epimorphisms; the same holds as well for the compositions*

$$\tilde{\mathcal{C}}^\delta \xrightarrow{i_*} \tilde{\mathcal{C}} \xrightarrow{\iota} \hat{\mathcal{C}},$$

$$\tilde{\mathcal{C}}^\delta_R \xrightarrow{i_*} \tilde{\mathcal{C}}_R \xrightarrow{\iota} \hat{\mathcal{C}}_R.$$

(ii) Let $X_* \to X$ be an augmented simplicial object of $\tilde{\mathcal{C}}$ such that $(X_*)_p \to X_p$ is a weak equivalence of simplicial sets for each $p \in \mathcal{P}$. Then, for each object S of $\tilde{\mathcal{C}}^\delta_R$, the natural map of cosimplicial R-modules

$$\mathrm{Hom}_{\tilde{\mathcal{C}}}(X, i_*S) \to \mathrm{Hom}_{\tilde{\mathcal{C}}}(X_*, i_*S)$$

is a weak equivalence, where we make $\mathrm{Hom}_{\tilde{\mathcal{C}}}(-, i_*S)$ a functor to $\mathbf{Mod}_R$ via the R-module structure on i_*S.
(iii) Let $F_1 \to F_2$ be a monomorphism in $\tilde{\mathcal{C}}$. Then, for each S in $\tilde{\mathcal{C}}^\delta$, the induced map of sets $\mathrm{Hom}_{\tilde{\mathcal{C}}}(F_2, i_*S) \to \mathrm{Hom}_{\tilde{\mathcal{C}}}(F_1, i_*S)$ is a surjection.

PROOF. (i) The same proof works for sheaves of sets, and sheaves of R-modules; to fix ideas, we work with sheaves of sets.

Both i_* and ι, being right adjoints, preserve projective limits. As ι has the left adjoint η (the sheafification functor), and the natural map $E \to \eta\iota E$ is an isomorphism for all sheaves E, it suffices to show that ιi_* preserves epimorphisms.

For an object X of $\mathcal{C}$, we have the sheafification $\tilde{X}$ of the representing presheaf $\hat{X}$, $\hat{X}(Y) = \mathrm{Hom}_{\mathcal{C}}(Y, X)$. By the Yoneda lemma, for each presheaf S, we have the natural isomorphism $\mathrm{Hom}_{\hat{\mathcal{C}}}(\hat{X}, S) \cong S(X)$.

Now suppose we have a surjection $S_1 \to S_2$ in $\tilde{\mathcal{C}}^\delta$. Then we have isomorphisms

$$\begin{aligned}(i_*S_i)(X) &\cong \mathrm{Hom}_{\hat{\mathcal{C}}}(\hat{X}, \iota i_*S_i)\\ &\cong \mathrm{Hom}_{\tilde{\mathcal{C}}^\delta}(i^*\tilde{X}, S_i)\end{aligned}$$

On the other hand, if S is a set, the functor $\mathrm{Hom}_{\mathbf{Sets}}(S, -)$ sends surjections to surjections, hence the map $(i_*S_1)(X) \to (i_*S_2)(X)$ is surjective for all X, proving (i).

For (ii), we have

$$\begin{aligned}\mathrm{Hom}_{\tilde{\mathcal{C}}}(X_*, i_*S) &\cong \mathrm{Hom}_{\tilde{\mathcal{C}}^\delta}(i^*X_*, S)\\ &= \prod_{p\in\mathcal{P}} \mathrm{Hom}_{\mathbf{Sets}}((X_*)_p, S(p)),\end{aligned} \tag{2.1.5.1}$$

and similarly for $\mathrm{Hom}_{\tilde{\mathcal{C}}}(X, i_*S)$. Now, if $f: A \to B$ is a weak equivalence of simplicial sets, the induced map on the simplicial R-modules freely generated by A and B, $Rf: RA \to RB$, is a weak equivalence of simplicial R-modules, hence, as the homotopy groups of a simplicial R-module are the same as the homology of the associated complex [**95**, Chapter V], the map of associated complexes of free R-modules, $Rf^*: RA^* \to RB^*$, is a quasi-isomorphism. For an R-module M, we have the isomorphisms of cosimplicial R-modules

$$\mathrm{Hom}_{\mathbf{Sets}}(A, M) \cong \mathrm{Hom}_{\mathbf{Mod}_R}(RA, M); \quad \mathrm{Hom}_{\mathbf{Sets}}(B, M) \cong \mathrm{Hom}_{\mathbf{Mod}_R}(RB, M),$$

hence, by the universal coefficient theorem, the map of cosimplicial R-modules $\mathrm{Hom}_{\mathbf{Sets}}(B, M) \to \mathrm{Hom}_{\mathbf{Sets}}(A, M)$ is a weak equivalence. Applying these remarks to the augmentation $(X_*)_p \to X_p$, we see that

$$\mathrm{Hom}_{\mathbf{Sets}}((X_*)_p, S(p)) \to \mathrm{Hom}_{\mathbf{Sets}}(X_p, S(p))$$

is a weak equivalence for all $p \in \mathcal{P}$. Since taking products is an exact functor in $\mathbf{Mod}_R$, the identity (2.1.5.1) implies that

$$\mathrm{Hom}_{\tilde{\mathcal{C}}}(X, i_*S) \to \mathrm{Hom}_{\tilde{\mathcal{C}}}(X_*, i_*S)$$

is a weak equivalence, proving (ii).

For (iii), we have the natural isomorphisms

$$\mathrm{Hom}_{\tilde{\mathcal{C}}}(F_j, i_*S) \cong \mathrm{Hom}_{\tilde{\mathcal{C}}^\delta}(i^*F_j, S); \quad j = 1, 2.$$

As the functor i^* is exact, the map $i^*F_1 \to i^*F_2$ is a monomorphism in $\tilde{\mathcal{C}}^\delta$. As monomorphisms in **Sets** are split, the map

$$\mathrm{Hom}_{\tilde{\mathcal{C}}^\delta}(i^*F_2, S) \to \mathrm{Hom}_{\tilde{\mathcal{C}}^\delta}(i^*F_1, S)$$

is surjective. □

2.2. Cohomology and cohomology with support

We describe how the Godement resolution gives a computation of sheaf cohomology.

2.2.1. We suppose that $\tilde{\mathcal{C}}$ has a conservative family of points $\mathcal{P}$, as in §2.1.4, giving the morphism of topoi (2.1.4.1) $i: \tilde{\mathcal{C}}^\delta \to \tilde{\mathcal{C}}$, and the *Godement resolution* $G: \tilde{\mathcal{C}} \to$ c.s.$\tilde{\mathcal{C}}/\mathcal{C}$.

Let R be a commutative ring. The functor G then extends to the functor

$$G_R: \tilde{\mathcal{C}}_R \to \text{c.s.}\tilde{\mathcal{C}}_R/\tilde{\mathcal{C}}_R. \tag{2.2.1.1}$$

For F in $\tilde{\mathcal{C}}_R$, we let $F \to G_R^*F$ be the augmented cochain complex associated to the augmented cosimplicial object $F \to G_RF$. This defines the functor

$$G_R^*: \tilde{\mathcal{C}}_R \to \mathbf{C}^+(\tilde{\mathcal{C}}_R). \tag{2.2.1.2}$$

For F^* in $\mathbf{C}(\tilde{\mathcal{C}}_R)$, we have the presheaf $X \mapsto H^p(F^*(X))$ on $\mathcal{C}$; taking the associated sheaf defines the *cohomology sheaves* $\mathcal{H}^p(F^*)$. A map $f: F_1^* \to F_2^*$ is a *quasi-isomorphism* if f induces an isomorphism on the cohomology sheaves $\mathcal{H}^p(F_j^*)$ for all p. Form the *derived category* $\mathbf{D}^*(\tilde{\mathcal{C}}_R)$ ($*$ a boundedness condition) by localizing the homotopy category $\mathbf{K}^*(\tilde{\mathcal{C}}_R)$ with respect to quasi-isomorphisms.

For an object X of $\mathcal{C}$, we have the functor

$$\Gamma(X, -): \tilde{\mathcal{C}}_R \to \mathbf{Mod}_R$$
$$\Gamma(X, F) := F(X).$$

This extends to the derived functor

$$R\Gamma(X, -): \mathbf{D}^+(\tilde{\mathcal{C}}_R) \to \mathbf{D}^+(\mathbf{Mod}_R).$$

Since the restriction functor $i_X^*: \tilde{\mathcal{C}}_R \to (\tilde{\mathcal{C}}/X)_R$ is exact, the cohomology $H^*(X, F_{|X})$ is given by the cohomology of $R\Gamma(X, F)$.

2.2.2. LEMMA. *(i) The augmentation* id $\to G_R^*$ *induces a natural isomorphism* id $\cong \mathcal{H}^0(G_R^*)$.
(ii) For all $p > 0$, $\mathcal{H}^p(G_R^*(-)) = 0$.
(iii) Let $f: X \to Y$ *be a map in* $\mathcal{C}$. *Then, for all* $p > 0$, $n \geq 0$, $H^p(Y, f_* G_R^n(-)_{|X}) = 0$.
(iv) Let $f: X \to Y$ *be a morphism in* $\mathcal{C}$. *Let* P *be a projective* R*-module,* $\tilde{P}$ *the associated constant sheaf on* $\mathcal{C}$, *giving us the Hom-sheaf* $\mathcal{H}om(\tilde{P}_{|Y}, f_* G_R^n(-)_{|X})$ *on* Y. *Then* $H^p(Y, \mathcal{H}om(\tilde{P}_{|Y}, f_* G_R^n(-)_{|X})) = 0$ *for all* $p > 0$, *and the natural map*

$$\mathrm{Hom}(\tilde{P}_{|Y}, f_* G_R^n(-)_{|X}) \to H^0(Y, \mathcal{H}om(\tilde{P}_{|Y}, f_* G_R^n(-)_{|X}))$$

is an isomorphism.

PROOF. To prove (i) and (ii), it follows from the exactness of the functor $i^*: \tilde{\mathcal{C}}_R \to \tilde{\mathcal{C}}_R^\delta$, and the fact that $\mathcal{P}$ is a conservative family, that $\mathcal{H}^0(G_R^*(F)) \cong F$ (resp. $\mathcal{H}^p(G_R^* F) = 0$) if and only if $\mathcal{H}^0(i^* G_R^*(F)) \cong i^* F$ (resp. $\mathcal{H}^p(i^* G_R^* F) = 0$). These latter two properties follow from Proposition 2.1.3.

For (iii), it follows from [**3**, V, Théorème 7.4.1] that, for a sheaf F on X with values in $\mathbf{Mod}_R$, the cohomology $H^p(X, F)$ can be computed as the inductive limit of the cohomologies $H^p(F(X_*)^*)$, where $X_* \to X$ runs over hypercovers of X, where $F(X_*)^*$ is the cohomological complex associated to the cosimplicial object in $\mathcal{A}$:

$$n \mapsto F(X_n),$$

and the inductive limit is taken over the category of hypercovers of X. By [**3**, V, Théorème 7.3.2(3)], a hypercover $S_* \to S$ in **Sets** is a weak equivalence. In addition, if $u: T_1 \to T_2$ is a map of topoi, and $F_* \to F$ is a hypercover in T_2, then $u^* F_* \to u^* F$ is a hypercover in T_1. Thus, from Lemma 2.1.5, if $X_* \to X$ is a hypercover of an object X in $\mathcal{C}$, and F is a sheaf on $\mathcal{C}$ with values in $\mathbf{Mod}_R$, the map

$$G_R^n(F)(X) = \mathrm{Hom}_{\tilde{\mathcal{C}}}(\tilde{X}, G_F^n(F)) \to \mathrm{Hom}_{\tilde{\mathcal{C}}}(\tilde{X}_*, G_R^n(F)) = G_R^n(F)(X_*)$$

is a weak equivalence. By the Dold-Kan equivalence of the homotopy category of simplicial abelian groups with the homotopy category of complexes of abelian groups [**39**], [**74**] this implies that $G_R^n(X_*)^*$ has no higher cohomology and the map $G_R^n(F)(X) \to H^0(G_R^n(F)(X_*)^*)$ is an isomorphism. Taking the limit over hypercovers of X, we see that $H^p(X, G_R^n(F)_{|X}) = 0$ for all $p > 0$.

If now $f: X \to Y$ is a map in $\mathcal{C}$, and $U \to Y$ is an open for the topology on $\mathcal{C}$, it follows from the previous paragraph that $H^p(X \times_Y U, G_R^n(F)) = 0$ for all $p > 0$. In particular, the sheafification $R^q f_* G_R^n(F)_{|X}$ of the presheaf $(U \to Y) \mapsto H^q(X \times_Y U, G_R^n(F))$ is zero for all $q > 0$. Thus, the local to global spectral sequence

$$E_2^{p,q} := H^p(Y, R^q f_* G_R^n(F)_{|X}) \Longrightarrow H^{p+q}(X, G_R^n(F)_{|X})$$

degenerates at E_2, and gives the isomorphism

$$H^p(Y, f_* G_R^n(F)_{|X}) \cong H^p(X, G_R^n(F)_{|X}).$$

As we have already seen that $H^p(X, G_R^n(F)_{|X}) = 0$ for all $p > 0$, this proves (iii).

For (iv), let $\tilde{P}_Y$ denote the sheafification (on Y) of the constant presheaf P. We have the local to global spectral sequence

$$E_2^{p,q} := H^p(Y, Ext^q(\tilde{P}_Y, f_* G_R^n(F))) \Longrightarrow \mathrm{Ext}^{p+q}(\tilde{P}, f_* G_R^n(-)),$$

where $\mathcal{E}xt^q(\tilde{P}_Y, f_*G^n_R(F))$ is the sheafification of the presheaf

$$U \mapsto \mathrm{Ext}^q_R(P, f_*G^n_R(F)(U)).$$

Since P is projective, this spectral sequence degenerates at E_2, and thus gives the isomorphism

$$\mathrm{Hom}(\tilde{P}_Y, f_*G^n_R(-)) \to H^0(Y, \mathcal{H}om(\tilde{P}_Y, f_*G^n_R(F))).$$

We extend $\tilde{P}_Y$ to the constant sheaf $\tilde{P}$ on $\mathcal{C}$. Since we have $\tilde{P}_p = P$ for all $p \in \mathcal{P}$ (Lemma 1.3.9), we have the identity $\mathcal{H}om(\tilde{P}, F)_p = \mathrm{Hom}_R(P, F_p)$ for all sheaves of R-modules F on $\mathcal{C}$. Thus

$$\mathcal{H}om(\tilde{P}_Y, f_*G^n_R(F)) \cong f_*G^n_R\mathcal{H}om(\tilde{P}, F),$$

hence the cohomology vanishing follows from (iii). □

Let X be in $\mathcal{C}$, and $j: U \to X$ a monomorphism in $\mathcal{C}$. We write $W := X \setminus U$ as a formal symbol, in analogy with the case in which $j: U \to X$ is the inclusion of a subset U of a topological space X. If F is a sheaf of R-modules on X, one defines the *cohomology of F with support in W* as the cohomology of the cone

$$H^p_W(X, F) := H^p(\mathrm{cone}\big(R\Gamma(X, F) \xrightarrow{j^*} R\Gamma(U, F)\big)[-1]).$$

Suppose we have, for each X in $\mathcal{C}$, a monomorphism $j_X: U_X \to X$, such that, for each morphism $f: X \to Y$ in $\mathcal{C}$, the composition $f \circ j_X$ factors through $j_Y: U_Y \to Y$. Since j_Y is a monomorphism, there is a unique morphism $f_U: U_X \to U_Y$ making the diagram

$$\begin{array}{ccc} U_X & \xrightarrow{j_X} & X \\ {\scriptstyle f_U}\downarrow & & \downarrow{\scriptstyle f} \\ U_Y & \xrightarrow[j_Y]{} & Y \end{array}$$

commute. This gives us functor $j_*j^*: \tilde{\mathcal{C}}_R \to \tilde{\mathcal{C}}_R$ defined by

$$j_*j^*F_{|X} = j_{X*}j_X^*(F_{|X}).$$

We have as well the natural map $\rho(F): F \to j_*j^*F$. For a sheaf of R-modules F on $\mathcal{C}$, let $G^W_R(F)^*$ be defined as the kernel of $\rho(G_R(F)^*)$.

2.2.3. LEMMA. *Let X be in $\mathcal{C}$, F in $\tilde{\mathcal{C}}_R$.*
(i) The map $j_X^*: G_R(F)^*(X) \to G_R(F)^*(U_X)$ is degree-wise surjective.
(ii) The sequence

$$G^W_R(F)^*(X) \to G_R(F)^*(X) \xrightarrow{j_X^*} G_R(F)^*(U_X)$$

canonically extends to a distinguished triangle in $\mathbf{D}^+(\mathbf{Mod}_R)$, isomorphic to the sequence

$$\mathrm{cone}(j_X^*)[-1] \to R\Gamma(X, F) \xrightarrow{j_X^*} R\Gamma(U_X, F) \to \mathrm{cone}(j_X^*).$$

PROOF. By Lemma 2.2.2, there is a natural isomorphism in $\mathbf{D}^+(\mathbf{Mod}_R)$,

$$G_R(F)^*(X) \to R\Gamma(X, F),$$

so (ii) follows from (i). The assertion (i) follows from Lemma 2.1.5. □

2.2.4. REMARK. The analogous result holds for a collection of projective systems $i \mapsto j_X(i): U_X(i) \to X$ of monomorphisms to X (functorial in X), by defining

$$G_R^W(F)^*(X) \to G_R(F)^*(X) \xrightarrow{j_X^*} G_R(F)^*(U_X)$$

to be the inductive limit of the sequences

$$G_R^{W(i)}(F)^*(X) \to G_R(F)^*(X) \xrightarrow{j_X(i)^*} G_R(F)^*(U_X(i)).$$

2.3. Multiplicative structure

2.3.1. We now assume that the category $\mathcal{C}$ has a final object $*$, and that products over $*$ exist, giving the functor $\times: \mathcal{C} \times \mathcal{C} \to \mathcal{C}$. This gives $\mathcal{C}$ the structure of a symmetric monoidal category, with unit $*$. It follows from the axioms for covers that the topology on $X \times Y$ is finer than the product topology, for each pair of objects X, Y of $\mathcal{C}$.

The category $\tilde{\mathcal{C}}$ has the final object e, the sheafification of the constant presheaf with value the one-point set. As projective limits exist in $\tilde{\mathcal{C}}$, the product over e defines the operation $\times: \tilde{\mathcal{C}} \times \tilde{\mathcal{C}} \to \tilde{\mathcal{C}}$. Explicitly, $(F \times F')(X) = F(X) \times F'(X)$. We have the similarly defined product in the presheaf category.

2.3.2. LEMMA. *Let X and Y be objects of $\mathcal{C}$. There is a natural isomorphism*

$$\tilde{X} \times \tilde{Y} \cong \widetilde{X \times Y}.$$

PROOF. By the universal property of the product over $*$, we have the natural isomorphism $\hat{X} \times \hat{Y} \cong \widehat{X \times Y}$. As sheafification is compatible with products, this gives the isomorphism $\tilde{X} \times \tilde{Y} \cong \widetilde{X \times Y}$. □

2.3.3. *Multiplication of sheaves.* We let $\mathcal{C} \times \mathcal{C}$ be the site with underlying category $\mathcal{C} \times \mathcal{C}$, and with the product pre-topology:

$$\mathrm{Cov}((X,Y)) = \{\{(U_\alpha, V_\beta) \xrightarrow{(f_\alpha, g_\beta)} (X,Y)\} \mid \{U_\alpha \to X\} \in \mathrm{Cov}(X), \{V_\beta \to Y\} \in \mathrm{Cov}(Y)\}$$

If we have sheaves F and F' on $\mathcal{C}$, we form the sheaf $p_1^*F \times p_2^*F'$ on $\mathcal{C} \times \mathcal{C}$ with $p_1^*F \times p_2^*F'((X,Y)) = F(X) \times F'(Y)$. If F'' is a third sheaf, a *multiplication* is a natural transformation

$$\mu: p_1^*F \times p_2^*F' \to F'' \circ \times,$$

i.e., a collection of maps $\mu_{X,Y}: F(X) \times F'(Y) \to F''(X \times Y)$ which is natural with respect to pairs of maps $(f,g): (X,Y) \to (X',Y')$. If $F = F' = F''$, we have the notion of an associative, or commutative multiplication (see §1.2.2).

2.3.4. LEMMA. *Let*

$$\mu: p_1^*F \times p_2^*F' \to F'' \circ \times$$

be a multiplication. Then there is a multiplication

$$G^0\mu: p_1^*i_*i^*F \times p_2^*i_*i^*F' \to i_*i^*F'' \circ \times$$

which is compatible with μ *via the natural transformation* α, *i.e., the diagram*

$$\begin{array}{ccc} p_1^*i_*i^*F \times p_2^*i_*i^*F' & \xrightarrow{G^0\mu} & i_*i^*F'' \circ \times \\ {\scriptstyle p_1^*\alpha\times p_2^*\alpha}\uparrow & & \uparrow{\scriptstyle \alpha\circ\times} \\ p_1^*F \times p_2^*F' & \xrightarrow[\mu]{} & F'' \circ \times \end{array} \tag{2.3.4.1}$$

commutes. In addition, $G^1\mu := G^0G^0\mu$ *is compatible with* $\sigma_0 := i_*\beta i^*$, i.e., *the diagram*

$$\begin{array}{ccc} p_1^*G^1F \times p_2^*G^1F' & \xrightarrow{G^1\mu} & G^1F'' \circ \times \\ {\scriptstyle p_1^*\sigma_0\times p_2^*\sigma_0}\downarrow & & \downarrow{\scriptstyle \sigma_0\circ\times} \\ p_1^*F \times p_2^*F' & \xrightarrow[\mu]{} & F'' \circ \times \end{array} \tag{2.3.4.2}$$

commutes.

PROOF. Let X and Y be in $\mathcal{C}$, and take F and F' in $\tilde{\mathcal{C}}$. Using the fact that i^* preserves finite projective limits, together with Lemma 2.3.2, we have the natural isomorphisms

$$\begin{aligned} i_*i^*F(X) \times i_*i^*F'(Y) &\cong \mathrm{Hom}_{\tilde{\mathcal{C}}}(\tilde{X}, i_*i^*F) \times \mathrm{Hom}_{\tilde{\mathcal{C}}}(\tilde{Y}, i_*i^*F') \\ &\cong \mathrm{Hom}_{\tilde{\mathcal{C}}^\delta}(i^*\tilde{X}, i^*F) \times \mathrm{Hom}_{\tilde{\mathcal{C}}^\delta}(i^*\tilde{Y}, i^*F') \\ &= \prod_{q\in\mathcal{P}} \mathrm{Hom}_{\mathbf{Sets}}(\tilde{X}_q, F_q) \times \prod_{q\in\mathcal{P}} \mathrm{Hom}_{\mathbf{Sets}}(\tilde{Y}_q, F'_q), \end{aligned} \tag{2.3.4.3}$$

$$\begin{aligned} i_*i^*(F'')(X \times Y) &\cong \mathrm{Hom}_{\tilde{\mathcal{C}}}(\widetilde{X \times Y}, i_*i^*F'') \\ &\cong \mathrm{Hom}_{\tilde{\mathcal{C}}^\delta}(i^*\widetilde{X \times Y}, i^*F'') \\ &\cong \mathrm{Hom}_{\tilde{\mathcal{C}}^\delta}(i^*\tilde{X} \times i^*\tilde{Y}, i^*F'')) \\ &= \prod_{p\in\mathcal{P}} \mathrm{Hom}_{\mathbf{Sets}}(\tilde{X}_p \times \tilde{Y}_p, F''_p). \end{aligned} \tag{2.3.4.4}$$

Let p be in $\mathcal{P}$, and let $U_p : I_p \to \mathcal{C}$ be the corresponding pro-object of $\mathcal{C}$ as described in §1.3.7. Take

$$(\bar{s}, \bar{t}) \in \tilde{X}_p \times \tilde{Y}_p,$$

and

$$(f, g) \in \prod_{q\in\mathcal{P}} \mathrm{Hom}_{\mathbf{Sets}}(\tilde{X}_q, F_q) \times \prod_{q\in\mathcal{P}} \mathrm{Hom}_{\mathbf{Sets}}(\tilde{Y}_q, F'_q).$$

Taking the p-component (f_p, g_p) of f and g, we may evaluate at $(\bar{s}, \bar{t})$, giving $(f_p(\bar{s}), g_p(\bar{t})) \in F_p \times F'_p$. Since $\tilde{X}_p$ and $\tilde{Y}_p$ are given as the inductive limits

$$\tilde{X}_p = \varinjlim_{j\in I_p} \tilde{X}(U_p(j)); \quad \tilde{Y}_p = \varinjlim_{j\in I_p} \tilde{Y}(U_p(j)),$$

we may lift $(\bar{s}, \bar{t})$ to

$$(s, t) \in \tilde{X}(U_p(j)) \times \tilde{Y}(U_p(j)) = \widetilde{X \times Y}(U_p(j)).$$

Increasing j if necessary, this gives the lifting of $(f_p(\bar{s}), g_p(\bar{t}))$ to

$$\widetilde{(f_p(\bar{s}), g_p(\bar{t}))} \in F(U_p(j)) \times F'(U_p(j)) = (p_1^* F \times p_2^* F')(U_p(j) \times U_p(j)).$$

Composing with $\mu_{U_p(j),U_p(j)}$, and pulling back by the diagonal

$$\Delta_{U_p(j)}: U_p(j) \to U_p(j) \times U_p(j)$$

gives

$$(F''(\Delta_{U_p(j)}) \circ \mu_{U_p(j),U_p(j)})(\widetilde{(f_p(\bar{s}), g_p(\bar{t}))}) \in F''(U_p(j)). \tag{2.3.4.5}$$

We then define $\mu_p(f_p(\bar{s}), g_p(\bar{t}))$ in F_p'' to be the image of (2.3.4.5) in F_p''.

It is easy to check that $\mu_p(f_p(\bar{s}), g_p(\bar{t}))$ is independent of the choices made; taking the product over all $p \in \mathcal{P}$ gives the map

$$\prod_p \mu_p: \prod_{q\in\mathcal{P}} \mathrm{Hom}_{\mathbf{Sets}}(\tilde{X}_q, F_q) \times \prod_{q\in\mathcal{P}} \mathrm{Hom}_{\mathbf{Sets}}(\tilde{Y}_q, F_q') \to \prod_{p\in\mathcal{P}} \mathrm{Hom}_{\mathbf{Sets}}(\tilde{X}_p \times \tilde{Y}_p, F_p'').$$

It is easy to check that $\prod_p \mu_p$ is natural in X and Y, giving via the isomorphisms (2.3.4.3) and (2.3.4.4) the desired multiplication

$$G^0\mu: p_1^* i_* i^* F \times p_2^* i_* i^* F' \to i_* i^* F'' \circ \times.$$

To check the commutativity of the diagram (2.3.4.1), take

$$s \in F(X) = \mathrm{Hom}_{\tilde{\mathcal{C}}}(\tilde{X}, F); \quad t \in F'(Y) = \mathrm{Hom}_{\tilde{\mathcal{C}}}(\tilde{Y}, F').$$

For $f \in \widetilde{X \times Y}(U)$, let $f_X \in \tilde{X}(U)$, $f_Y \in \tilde{Y}(U)$ be the images of f under the projections $\widetilde{X \times Y} \to \tilde{X}$ and $\widetilde{X \times Y} \to \tilde{Y}$. This gives $f_X \times f_Y \in \widetilde{X \times Y}(U \times U)$; then $\Delta_U^*(f_X \times f_Y) = f \in \widetilde{X \times Y}(U)$. From this, it follows that

$$\mu_{X,Y}(s,t)(f) = \Delta_U^*(\mu_{U,U}(s(U)(f_X), t(U)(f_Y))).$$

In addition, if S is a sheaf, and $v \in S(V)$ for some object V of $\mathcal{C}$, then the section $\alpha(v) \in i_* i^* S(V)$ is given, via the isomorphism

$$i_* i^* S(V) \cong \prod_{p\in\mathcal{P}} \mathrm{Hom}_{\mathbf{Sets}}(\tilde{V}_p, S_p),$$

as the product over p of the system of induced maps

$$i \mapsto v_p(i): \tilde{V}(U_p(i)) \to S(U_p(i)).$$

Putting these two identifications together gives the commutativity of the diagram (2.3.4.1).

To check the commutativity of the diagram (2.3.4.2), first let S be a sheaf on $\mathcal{C}^\delta$, $S = \coprod_{p\in\mathcal{P}} S_p$. Then

$$i^* i_* S = \coprod_{q\in\mathcal{P}} (i^* i_* S)_q.$$

The natural transformation β is given by the sequence of identifications and natural maps

$$\begin{aligned}
(i^*i_*S)_q &= \varinjlim_{j\in I_q} i_*S(U_q(j)) \\
&= \varinjlim_{j\in I_q} \prod_{p\in\mathcal{P}} \mathrm{Hom}_{\mathbf{Sets}}(\widetilde{U_q(j)}_p, S_p) \\
&\to \varinjlim_{j\in I_q} \mathrm{Hom}_{\mathbf{Sets}}(\widetilde{U_q(j)}_q, S_q) \\
&= \varinjlim_{j\in I_q} \mathrm{Hom}_{\mathbf{Sets}}(\varinjlim_{i\in I_q} \widetilde{U_q(j)}(U_q(i)), S_q) \\
&\to \varinjlim_{j\in I_q} \mathrm{Hom}_{\mathbf{Sets}}(\varinjlim_{i\in I_q} \mathrm{Hom}_{\mathcal{C}}(U_q(i), U_q(j)), S_q) \\
&= \varinjlim_{j\in I_q} \mathrm{Hom}_{\mathbf{Sets}}(\varinjlim_{i\to j\in I_q/j} \mathrm{Hom}_{\mathcal{C}}(U_q(i), U_q(j)), S_q) \\
&\xrightarrow{ev_{i\to j}} S_q,
\end{aligned}$$

where the map $ev_{i\to j}$ is the map which evaluates an element in the $\mathrm{Hom}_{\mathbf{Sets}}$ at the map $U_q(i) \to U_q(j)$ induced by the structure map $i \to j$. Using this identification of β, the commutativity of (2.3.4.2) follows by a sequence of identifications similar to the proof of the commutativity of (2.3.4.1). □

2.3.5. PROPOSITION. *Let F, F' and F'' be sheaves on $\mathcal{C}$, with a multiplication $\mu: p_1^*F \times p_2^*F' \to F'' \circ \times$. Then there is a multiplication of cosimplicial sheaves*

$$G\mu: p_1^*G(F) \times p_2^*G(F') \to G(F'') \circ \times$$

which is natural in μ, and is compatible with μ via the augmentations

$$\begin{aligned}
&p_1^*\alpha \times p_2^*\alpha: p_1^*F \times p_2^*F' \to p_1^*G(F) \times p_2^*G(F'), \\
&\alpha: F'' \to GF''.
\end{aligned}$$

If $F = F' = F''$, and μ is associative and commutative, then $G\mu$ is also associative and commutative.

PROOF. The multiplication $G\mu$ on the cosimplices of degree n is gotten by iterating the transformation $\mu \mapsto G^0\mu$ of Lemma 2.3.4 n times. That this defines a map of cosimplicial sets follows directly from the commutative diagrams in Lemma 2.3.4. The commutativity and associativity of $G\mu$ follow easily from the explicit description of $G^0\mu$. □

2.3.6. If F and G are in $\tilde{\mathcal{C}}_R$, we form the sheaf $p_1^*F \otimes_R p_2^*G$ on $\mathcal{C} \times \mathcal{C}$ by taking the sheafification of the presheaf $(X, Y) \mapsto F(X) \otimes_R G(Y)$. This extends to the operation

$$p_1^*(-) \otimes_R p_2^*(-): \mathbf{C}^+(\tilde{\mathcal{C}}_R) \otimes \mathbf{C}^+(\tilde{\mathcal{C}}_R) \to \mathbf{C}^+(\widetilde{\mathcal{C} \times \mathcal{C}_R}).$$

Given sheaves of R-modules F, F' and F'', a *multiplication* is a map $\mu: p_1^*F \otimes_R p_2^*F' \to F'' \circ \times$; we have the similar notion for complexes of sheaves. We have the following analog of Proposition 2.3.5.

2.3.7. PROPOSITION. *Let F, F' and F'' be sheaves of R-modules on $\mathcal{C}$, (resp., complexes of sheaves of R-modules) with a multiplication $\mu: p_1^*F \otimes_R p_2^*F' \to F'' \circ \times$.*

There is a multiplication of cosimplicial sheaves of R-modules, (resp. cosimplicial $\mathbf{C}(\mathbf{Mod}_R)$*-valued sheaves)*

$$G_{\otimes}\mu: p_1^*G(F) \otimes_R p_2^*G(F') \to G(F'') \circ \times$$

which is natural in μ*, and is compatible with* μ *via the augmentations*

$$p_1^*\alpha \otimes_R p_2^*\alpha: p_1^*F \otimes_R p_2^*F' \to p_1^*G(F) \otimes_R p_2^*G(F'),$$
$$\alpha \circ \times : F'' \circ \times \to GF'' \circ \times.$$

If $F = F' = F''$*, and* μ *is associative and commutative, then* $G_{\otimes}\mu$ *is also associative and commutative.*

PROOF. The case of complexes of sheaves of R-modules follows directly from case of sheaves of R-modules. In this case, the multiplication of sheaves of R-modules determines the multiplication $\mu_0: p_1^*F \times p_2^*F' \to F'' \circ \times$ of sheaves of sets by composing μ with the natural transformation $\otimes_{F,F'}$. One sees directly from the definition of the multiplication $G\mu$ that, if F F' and F'' are sheaves of R-modules, then the multiplications $G^n\mu_0$ are R-bilinear, giving the multiplications

$$G^n_{\otimes}\mu: p_1^*G^n(F) \otimes_R p_2^*G^n(F') \to G^n(F'') \circ \times$$

by the universal mapping property of $\otimes_R$. That the maps $G^n_{\otimes}\mu$ define a map of augmented cosimplicial objects follows from Proposition 2.3.5 and the uniqueness in the universal mapping property of $\otimes_R$. The remainder of the assertions follow from Proposition 2.3.5 and the commutativity and associativity properties of the natural transformation $\otimes_{**}$. □

2.4. Flatness

Fix a commutative ring R, and a Grothendieck site $\mathcal{C}$. We call a sheaf of R-modules F *flat* if the functor $(-) \otimes_R F: \tilde{\mathcal{C}}_R \to \tilde{\mathcal{C}}_R$ is exact; we call a complex C^* of sheaves of R-modules flat if C^n is flat for each n.

We proceed to give a criterion for the Godement resolution of a flat sheaf to be flat.

2.4.1. LEMMA. *Let* p *be a point of* $\tilde{\mathcal{C}}$*,* F *and* F' *in* $\tilde{\mathcal{C}}_R$*. Then* $(F \otimes_R F')_p$ *is canonically isomorphic to the R-module* $F_p \otimes_R F'_p$*.*

PROOF. The sheaf $F \otimes_R F'$ is the sheaf associated to the presheaf $X \mapsto F(X) \otimes_R F'(X)$. Let $f: I \to \mathcal{C}$ be the pro-object corresponding to the point p (see §1.3.7). Then, by Lemma 1.3.9, we have the canonical isomorphism

$$\varinjlim_{i \in I} F(f(i)) \otimes_R F'(f(i)) \cong (F \otimes_R F')_p.$$

As

$$F_p = \varinjlim_{i \in I} F(f(i)); \quad F'_p = \varinjlim_{i \in I} F'(f(i)),$$

and as tensor products commute with filtered inductive limits, we have the canonical isomorphism $F_p \otimes_R F'_p \cong (F \otimes_R F')_p$. □

2.4.2. LEMMA. *Suppose that* $\tilde{C}$ *has a conservative family of points* $\mathcal{P}$*. Then a sheaf* F *of R-modules is flat if and only if* F_p *is a flat R-module for all* $p \in \mathcal{P}$*.*

PROOF. Let

$$i\colon \tilde{C}^\delta := \coprod_{p\in\mathcal{P}} \mathbf{Sets} \to \tilde{C}$$

be the morphism of topoi associated to the family of points $\mathcal{P}$ (§2.1.4), giving us the functor

$$i^*\colon \tilde{C}_R \to \prod_{p\in\mathcal{P}} \mathbf{Mod}_R,$$

and the right adjoint to i^*,

$$i_*\colon \prod_{p\in\mathcal{P}} \mathbf{Mod}_R \to \tilde{C}_R.$$

If F_p is flat for all $p \in \mathcal{P}$, then, as i^* is conservative, it follows from Lemma 2.4.1 that F is flat. For the converse, let X be in $\mathcal{C}$, and $S = \coprod_p S_p$ an object of $\tilde{C}^\delta$. For X in $\mathcal{C}$, we have

$$\begin{aligned} i_*S(X) &\cong \mathrm{Hom}_{\tilde{C}}(\tilde{X}, i_*S) \\ &\cong \mathrm{Hom}_{\tilde{C}^\delta}(i^*\tilde{X}, S) \\ &= \prod_p \mathrm{Hom}_{\mathbf{Sets}}(\tilde{X}_p, S_p). \end{aligned}$$

Thus, if we take S_q to be the one-point set for all $q \neq p$, we have

$$i_*S(X) = \mathrm{Hom}_{\mathbf{Sets}}(\tilde{X}_p, S_p). \tag{2.4.2.1}$$

If now F is flat, take an injective map of R-modules $0 \to N \to M$, and let

$$0 \to N^\delta \to M^\delta$$

be the sequence of objects of $\tilde{C}^\delta$ which is the sequence $0 \to N \to M$ at p, and zero at all $q \neq p$. Then, as i_* is left-exact, we have the exact sequence in $\tilde{C}_R$,

$$0 \to i_*N^\delta \to i_*M^\delta,$$

from which we have the exact sequence of R-modules

$$0 \to (i_*N^\delta)_p \otimes_R F_p \to (i_*M^\delta)_p \otimes_R F_p,$$

using the flatness of F and Lemma 2.4.1.

Let $f\colon I \to \mathcal{C}$ be the pro-object associated to the point p. This gives us the pro-object $f_p\colon I \to \mathbf{Sets}$ by $f_p(i) := \widetilde{f(i)}_p$. Let

$$T := \varprojlim_I f_p.$$

Since I is filtering, T is non-empty, indeed, for each $i \in I$, the category of objects over i is non-empty. The collection of maps $f(j) \to f(i)$ for $j \to i$ a map in I thus gives a canonical element in $\widetilde{f(i)}_p$, natural in i.

Pick an element $t \in T$. From (2.4.2.1), the projection of $\mathrm{Hom}_{\mathbf{Sets}}(\widetilde{f(i)}_p, N)$ onto the factor N corresponding to the image of t in $\widetilde{f(i)}_p$ defines a splitting to the natural map $N \to (i_*N^\delta)_p$, and similarly for M. Thus, the sequence

$$0 \to N \to M$$

is a direct summand of the sequence

$$0 \to (i_*N^\delta)_p \to (i_*M^\delta)_p,$$

from which it follows that

$$0 \to N \otimes_R F_p \to M \otimes_R F_p$$

is injective, i.e., that F_p is a flat R-module. □

2.4.3. PROPOSITION. *Let $\mathcal{P}$ be a conservative family of points for $\tilde{C}$, and let F be a flat sheaf of R-modules on $\mathcal{C}$. Suppose that R satisfies the following property: A product of flat R module is flat. Then for each n, and each X in $\mathcal{C}$, $G^nF(X)$ is a flat R-module. In addition the sheaf G^nF is a flat sheaf on $\mathcal{C}$.*

PROOF. As a filtered inductive limit of flat R-modules is flat, the second assertion follows from the first, Lemma 2.4.2, and the description of the stalks of a sheaf as an inductive limit. For the first assertion, by the inductive nature of the definition of the sheaves G^nF, it suffices to prove the case $n = 0$.

We have the canonical isomorphism, as in the proof of Lemma 2.4.2,

$$\begin{aligned} G^0F(X) := i_*i^*F(X) &\cong \prod_{p\in\mathcal{P}} \mathrm{Hom}_{\mathbf{Sets}}(\tilde{X}_p, F_p) \\ &\cong \prod_{p\in\mathcal{P}} \prod_{x\in X_p} F_p. \end{aligned}$$

By Lemma 2.4.2, each F_p is a flat R-module, so the flatness of $G^0F(X)$ results from our hypothesis on R. □

2.4.4. REMARK. If R is noetherian, then the hypothesis on R in Proposition 2.4.3 is satisfied. Indeed, let $\{M_\alpha \mid \alpha \in A\}$ be a set of flat R-modules, and let $M := \prod_{\alpha\in A} M_\alpha$. Then M is flat.

PROOF. It suffices to check that the functor $- \otimes_R M$ is exact on the full subcategory of finitely generated R modules. Let

$$R^a \to R^b \to N \to 0$$

be a presentation of a finitely generated R-module N, with a and b finite. We have the canonical isomorphism

$$\begin{aligned} R^s \otimes_R M &\cong \prod_{i=1}^{s} \prod_{\alpha} M_\alpha \\ &\cong \prod_{\alpha} \prod_{i=1}^{s} M_\alpha \\ &\cong \prod_{\alpha} R^s \otimes_R M_\alpha \end{aligned}$$

for all finite s, giving the presentation

$$\prod_\alpha R^a \otimes_R M_\alpha \to \prod_\alpha R^b \otimes_R M_\alpha \to N \otimes_R M \to 0$$

of $N \otimes_R M$. Thus we have the natural isomorphism

$$N \otimes_R M \cong \prod_\alpha N \otimes M_\alpha.$$

The result follows from the fact that the operation of taking products is exact in $\mathbf{Mod}_R$. □

Bibliography

[1] **Motives**. Proceedings of the AMS-IMS-SIAM Joint Summer Research Conference held at the University of Washington, Seattle, Washington, July 20–August 2, 1991. Edited by Uwe Jannsen, Steven Kleiman and Jean-Pierre Serre. Proceedings of Symposia in Pure Mathematics, **55**, Part 1. American Mathematical Society, Providence, RI, 1994. xiv+747 pp. MR **94i:**11003

[2] **Théorie des intersections et théorème de Riemann-Roch**. Séminaire de Géométrie Algébrique du Bois-Marie 1966–1967 (SGA 6). Dirigé par P. Berthelot, A. Grothendieck et L. Illusie. Avec la collaboration de D. Ferrand, J. P. Jouanolou, O. Jussila, S. Kleiman, M. Raynaud et J. P. Serre. Lecture Notes in Mathematics, Vol. 225. Springer-Verlag, Berlin-New York, 1971. xii+700 pp. MR **50:**7133

[3] **Théorie des topos et cohomologie étale des schémas**. Tome 2. Séminaire de Géométrie Algébrique du Bois-Marie 1963–1964 (SGA 4). Dirigé par M. Artin, A. Grothendieck et J. L. Verdier. Avec la collaboration de N. Bourbaki, P. Deligne et B. Saint-Donat. Lecture Notes in Mathematics, Vol. 270. Springer-Verlag, Berlin-New York, 1972. iv+418 pp. MR **50:**7131

[4] **Théorie des topos et cohomologie étale des schémas**. Tome 1. Séminaire de Géométrie Algébrique du Bois-Marie 1963–1964 (SGA 4). Dirigé par M. Artin, A. Grothendieck et J. L. Verdier. Avec la collaboration de N. Bourbaki, P. Deligne et B. Saint-Donat. Lecture Notes in Mathematics, Vol. 269. Springer-Verlag, Berlin-New York, 1972. xix+525 pp. MR **50:**7130

[5] **Revêtements étales et groupe fondamental**. Séminaire de Géométrie Algébrique du Bois Marie 1960–1961 (SGA 1). Dirigé par Alexandre Grothendieck. Augmenté de deux exposés de M. Raynaud. Lecture Notes in Mathematics, Vol. 224. Springer-Verlag, Berlin-New York, 1971. xxii+447 pp. MR **50:**7129

[6] Atiyah, M. F.; Tall, D. O. *Group representations, λ-rings and the J-homomorphism.* Topology **8** (1969) 253–297. MR **39:**5702

[7] Bass, H.; Tate, J. *The Milnor ring of a global field.* Algebraic K-theory, II: "Classical" algebraic K-theory and connections with arithmetic (Proc. Conf., Seattle, Wash., Battelle Memorial Inst., 1972), pp. 349–446. Lecture Notes in Math., Vol. 342, Springer, Berlin, 1973. MR **56:**449

[8] Baum, Paul; Fulton, William; MacPherson, Robert. *Riemann-Roch for singular varieties.* Inst. Hautes Études Sci. Publ. Math. No. 45 (1975), 101–145. MR **54:**317

[9] Beilinson, A.A. *Polylogarithm and cyclotomic elements.* Preprint (1989).

[10] Beilinson, A. A. *Height pairing between algebraic cycles.* K-theory, arithmetic and geometry (Moscow, 1984–1986), 1–25, Lecture Notes in Math., Vol. 1289, Springer, Berlin-New York, 1987. MR **89h:**11027

[11] Beilinson, A. A. *Notes on absolute Hodge cohomology.* Applications of algebraic K-theory to algebraic geometry and number theory, Part I (Boulder, Colo., 1983), 35–68, Contemp. Math., **55**, Amer. Math. Soc., Providence, R.I., 1986. MR **87m:**14019

[12] Beilinson, A. A. *Higher regulators and values of L-functions.* Current problems in mathematics, Vol. 24, 181–238, Itogi Nauki i Tekhniki, Akad. Nauk SSSR, Vsesoyuz. Inst. Nauchn. i Tekhn. Inform., Moscow, 1984. MR 86h:11103

[13] Beilinson, A.A.; Deligne, Pierre. *Interprétation motivique de la conjecture de Zagier reliant polylogarithmes et régulateurs.* Motives (Seattle, WA, 1991), 97–121, Proc. Sympos. Pure Math., **55**, Part 2, Amer. Math. Soc., Providence, RI, 1994. MR **95a:**19008

[14] Beilinson, A. A.; Goncharov, A. B.; Schechtman, V. V.; Varchenko, A. N. *Aomoto dilogarithms, mixed Hodge structures and motivic cohomology of pairs of triangles on the plane.* The Grothendieck Festschrift, Vol. I, 135–172, Progr. Math., **86**, Birkhäuser Boston, Boston, MA, 1990. MR **92h:**19007

[15] Beilinson, A.A.; MacPherson, R.; Schechtman, V. *Notes on motivic cohomology*. Duke Math. J. **54** (1987), no. 2, 679–710. MR **88f:**14021

[16] Bloch, Spencer. *Private communication.* (1994).

[17] Bloch, Spencer; Kriz, Igor. *Mixed Tate motives*. Ann. of Math. (2) **140** (1994), no. 3, 557–605. MR **96c:**11062

[18] Bloch, Spencer. *The moving lemma for higher Chow groups*. J. Algebraic Geom. **3** (1994), no. 3, 537–568. MR **96c:**14007

[19] Bloch, Spencer. *Algebraic cycles and higher K-theory*. Adv. in Math. **61** (1986), no. 3, 267–304. MR **88f:**18010

[20] Bloch, Spencer; Ogus, Arthur. *Gersten's conjecture and the homology of schemes*. Ann. Sci. École Norm. Sup. (4) **7** (1974), 181–201. MR **54:**318

[21] Bökstedt, Marcel; Neeman, Amnon. *Homotopy limits in triangulated categories*. Compositio Math. **86** (1993), no. 2, 209–234. MR **94f:**18008

[22] Borel, Armand. *Stable real cohomology of arithmetic groups*. Ann. Sci. École Norm. Sup. (4) **7** (1974), 235–272. MR **52:**8338

[23] Borel, Armand; Yang, Jun. *The rank conjecture for number fields*. Math. Res. Lett. **1** (1994), no. 6, 689–699. MR **95k:**11077

[24] Bousfield, A. K.; Gugenheim, V. K. A. M. *On PL de Rham theory and rational homotopy type*. Mem. Amer. Math. Soc. **8** (1976), no. 179, ix+94 pp. MR **54:**13906

[25] Bousfield, A. K.; Kan, D. M. **Homotopy limits, completions and localizations**. Lecture Notes in Mathematics, Vol. 304. Springer-Verlag, Berlin-New York, 1972. v+348 pp. MR **51:**1825

[26] Browder, William. *Algebraic K-theory with coefficients $\mathbb{Z}/p$*. Geometric applications of homotopy theory (Proc. Conf., Evanston, Ill., 1977), I, pp. 40–84, Lecture Notes in Math., Vol. 657, Springer, Berlin, 1978. MR **80b:**18011

[27] Cartan, Henri. *Théories cohomologiques*. Invent. Math. **35** (1976), 261–271. MR **55:**4139

[28] Cenkl, Bohumil; Porter, Richard. *de Rham theorem with cubical forms*. Pacific J. Math. **112** (1984), no. 1, 35–48. MR **85j:**55009

[29] Chow, Wei-Liang. *On equivalence classes of cycles in an algebraic variety*. Ann. of Math. (2) **64** (1956), 450–479. MR **18:**509a

[30] Deligne, Pierre. *À quoi servent les motifs?* Motives (Seattle, WA, 1991), 143–161, Proc. Sympos. Pure Math., **55**, Part 1, Amer. Math. Soc., Providence, RI, 1994. MR **95c:**14013

[31] Deligne, Pierre. *Catégories tannakiennes*. The Grothendieck Festschrift, Vol. II, 111–195, Progr. Math., **87**, Birkhäuser Boston, Boston, MA, 1990. MR **92d:**14002

[32] Deligne, Pierre. *Le groupe fondamental de la droite projective moins trois points*. Galois groups over $\mathbb{Q}$ (Berkeley, CA, 1987), 79–297, Math. Sci. Res. Inst. Publ., **16**, Springer, New York-Berlin, 1989. MR **90m:**14016

[33] Deligne, Pierre. *Letter to C. Soulé*. (1985).

[34] Deligne, Pierre. *Valeurs de fonctions L et périodes d'intégrales*. With an appendix by N. Koblitz and A. Ogus. Proc. Sympos. Pure Math., **XXXIII**, Automorphic forms, representations and L-functions (Proc. Sympos. Pure Math., Oregon State Univ., Corvallis, Ore., 1977), Part 2, pp. 313–346, Amer. Math. Soc., Providence, R.I., 1979. MR **81d:**12009

[35] Deligne, Pierre. **Cohomologie étale**. Séminaire de Géométrie Algébrique du Bois-Marie SGA $4\frac{1}{2}$. Avec la collaboration de J. F. Boutot, A. Grothendieck, L. Illusie et J. L. Verdier. Lecture Notes in Mathematics, Vol. 569. Springer-Verlag, Berlin-New York, 1977. iv+312pp. MR **57:**3132

[36] Deligne, Pierre. *Les constantes des équations fonctionnelles des fonctions L*. Modular functions of one variable, II (Proc. Internat. Summer School, Univ. Antwerp, Antwerp, 1972), pp. 501–597. Lecture Notes in Math., Vol. 349, Springer, Berlin, 1973. MR **50:**2128

[37] Deligne, Pierre; Milne, James S.; Ogus, Arthur; Shih, Kuang-yen. **Hodge cycles, motives, and Shimura varieties**. Lecture Notes in Mathematics, Vol. 900. Philosophical Studies Series in Philosophy, **20**. Springer-Verlag, Berlin-New York, 1982. ii+414 pp. MR **84m:**14046

[38] Deligne, Pierre. *Théorie de Hodge. II*. Inst. Hautes Études Sci. Publ. Math. No. 40 (1971), 5–57. MR **58:**16653a

[39] Dold, Albrecht. *Homology of symmetric products and other functors of complexes*. Ann. of Math. (2) **68** (1958) 54–80. MR **20:**3537

[40] Dupont, Johan L.; Sah, Chih Han. *Scissors congruences. II*. J. Pure Appl. Algebra **25** (1982), no. 2, 159–195. MR **84b:**53062b

[41] Ekedahl, Torsten. *On the adic formalism.* The Grothendieck Festschrift, Vol. II, 197–218, Progr. Math., **87**, Birkhäuser Boston, Boston, MA, 1990. MR **92b:**14010

[42] Eilenberg, Samuel; Zilber, J. A. *On products of complexes.* Amer. J. Math. **75** (1953). 200–204. MR **14:**670c

[43] Friedlander, Eric; Voevodsky, Vladimir. *Bivariant cycle theories.* Cycles, Transfers and Motivic Homology Theories, Annals of Math. Studies, Princeton Univ. Press. To appear.

[44] Fulton, William. **Intersection theory**. Ergebnisse der Mathematik und ihrer Grenzgebiete (3), **2**. Springer-Verlag, Berlin-New York, 1984. xi+470 pp. MR **85k:**14004

[45] Fulton, William; Lang, Serge. **Riemann-Roch algebra**. Grundlehren der Mathematischen Wissenschaften, **277**. Springer-Verlag, New York-Berlin, 1985. x+203 pp. MR **88h:**14011

[46] Gillet, Henri. *Riemann-Roch theorems for higher algebraic K-theory.* Adv. in Math. **40** (1981), no. 3, 203–289. MR **83m:**14013

[47] Gillet, Henri; Soulé, Christophe. *Descent, motives and K-theory.* J. Reine Angew. Math. **478** (1996), 127–176.

[48] Gillet, Henri; Soulé, Christophe. *Filtrations on higher K-theory.* preprint (1983).

[49] Godement, Roger. **Topologie algébrique et théorie des faisceaux**. Troisième édition revue et corrigée. Publications de l'Institut de Mathématique de l'Université de Strasbourg, XIII. Actualités Scientifiques et Industrielles, No. 1252. Hermann, Paris, 1973. viii+283 pp. MR **49:**9831

[50] Goncharov, A. B. *Chow polylogarithms and regulators.* Math. Res. Lett. **2** (1995), no. 1, 95–112. MR **96b:**19007

[51] Goncharov, A. B. *Geometry of configurations, polylogarithms, and motivic cohomology.* Adv. Math. **114** (1995), no. 2, 197–318. MR **94m:**19003

[52] Goncharov, A. B. *Polylogarithms and motivic Galois groups.* Motives (Seattle, WA, 1991), 43–96, Proc. Sympos. Pure Math., **55**, Part 2, Amer. Math. Soc., Providence, RI, 1994. MR **96g:**19005

[53] Grayson, Daniel. *Higher algebraic K-theory. II (after Daniel Quillen).* Algebraic K-theory (Proc. Conf., Northwestern Univ., Evanston, Ill., 1976), pp. 217–240. Lecture Notes in Math., Vol. 551, Springer, Berlin, 1976. MR **58:**28137

[54] Grothendieck, Alexander. *Standard conjectures on algebraic cycles.* 1969 Algebraic Geometry (Internat. Colloq., Tata Inst. Fund. Res., Bombay, 1968) pp. 193–199 Oxford Univ. Press, London. MR **42:**3088

[55] Grothendieck, Alexandre. **Éléments de géométrie algébrique. IV**. Étude locale des schémas et des morphismes de schémas IV. Inst. Hautes Études Sci. Publ. Math. No. 32 1967 361 pp. MR **39:**220

[56] Grothendieck, Alexandre. **Éléments de géométrie algébrique. III**. Étude cohomologique des faisceaux cohérents. I. Inst. Hautes Études Sci. Publ. Math. No. 11 1961 167 pp. MR **29:**1209

[57] Grothendieck, Alexandre. *La théorie des classes de Chern.* Bull. Soc. Math. France **86** (1958) 137–154. MR **22:**6818

[58] Grothendieck, Alexandre. *Sur quelques points d'algébre homologique.* Tôhoku Math. J. (2) **9** (1957) 119–221. MR **21:**1328

[59] Guillén, F.; Navarro Aznar, V. *Une critère d'extension d'un foncteur défini sur les schémas lisses.* preprint (1995).

[60] Guillén, F.; Navarro Aznar, V.; Pascual Gainza, P.; Puerta, F. **Hyperrésolutions cubiques et descente cohomologique**. Papers from the Seminar on Hodge-Deligne Theory held in Barcelona, 1982. Lecture Notes in Mathematics, Vol. 1335. Springer-Verlag, Berlin-New York, 1988. xii+192 pp. MR **90a:**14024

[61] Hanamura, Masaki. *Cohomological and homological motives.* preprint (1996).

[62] Hanamura, Masaki. *Mixed motives and algebraic cycles, II.* preprint (1996).

[63] Hanamura, Masaki. *Mixed motives and algebraic cycles, I.* Math. Res. Lett. **2** (1995), no. 6, 811–821. MR **97d:**14014

[64] Hiller, Howard L. *λ-rings and algebraic K-theory.* J. Pure Appl. Algebra **20** (1981), no. 3, 241–266. MR **82e:**18016

[65] Hinich, V. A.; Schechtman, V. V. *On homotopy limit of homotopy algebras.* K-theory, arithmetic and geometry (Moscow, 1984–1986), 240–264, Lecture Notes in Math., Vol. 1289, Springer, Berlin-New York, 1987. MR **89d:**55052

[66] Hironaka, Heisuke. *Resolution of singularities of an algebraic variety over a field of characteristic zero. I, II.* Ann. of Math. (2) **79** (1964), 109–203; ibid. (2) 79 1964 205–326. MR **33:**7333

[67] Huber, Annette. **Mixed motives and their realization in derived categories**. Lecture Notes in Mathematics, Vol. 1604. Springer-Verlag, Berlin, 1995. xvi+207 pp.

[68] Huber, Annette; Wildeshaus, Jörg. *Classical motivic polylogarithm according to Beilinson and Deligne.* preprint (1997).

[69] Jannsen, Uwe. *Mixed motives, motivic cohomology, and Ext-groups.* Proceedings of the International Congress of Mathematicians, Vol. 1, 2 (Zürich, 1994), 667–679, Birkhäuser, Basel, 1995. MR **97j:**14022

[70] Jannsen, Uwe. *Motivic sheaves and filtrations on Chow groups.* Motives (Seattle, WA, 1991), 245–302, Proc. Sympos. Pure Math., **55**, Part 1, Amer. Math. Soc., Providence, RI, 1994. MR **95c:**14006

[71] Jannsen, Uwe. **Mixed motives and algebraic K-theory**. With appendices by S. Bloch and C. Schoen. Lecture Notes in Mathematics, Vol. 1400. Springer-Verlag, Berlin, 1990. xiv+246 pp. MR **91g:**14008

[72] Jannsen, Uwe. *Continuous étale cohomology.* Math. Ann. **280** (1988), no. 2, 207–245. MR **89a:**14022

[73] Jardine, J. F. **Simplicial objects in a Grothendieck topos**. Applications of algebraic K-theory to algebraic geometry and number theory, Part I, II (Boulder, Colo., 1983), 193–239, Contemp. Math., **55**, Amer. Math. Soc., Providence, R.I., 1986. MR **88g:**18008

[74] Kan, Daniel M. *Functors involving c.s.s. complexes.* Trans. Amer. Math. Soc. **87** (1958) 330–346. MR **24:**A1720

[75] Kapranov, M. M. *On the derived categories of coherent sheaves on some homogeneous spaces.* Invent. Math. **92** (1988), no. 3, 479–508. MR **89g:**18018

[76] Kato, Kazuya. *Milnor K-theory and the Chow group of zero cycles.* Applications of algebraic K-theory to algebraic geometry and number theory, Part I, II (Boulder, Colo., 1983), 241–253, Contemp. Math., **55**, Amer. Math. Soc., Providence, R.I., 1986. MR **88c:**14012

[77] Kelly, Gregory Maxwell. **Basic concepts of enriched category theory**. London Mathematical Society Lecture Note Series, **64**. Cambridge University Press, Cambridge-New York, 1982. 245 pp. MR **84e:**18001

[78] Kleiman, Steven L. *The transversality of a general translate.* Compositio Math. **28** (1974), 287–297. MR **50:**13063

[79] Kleiman, Steven L. *Motives.* Algebraic geometry, Oslo 1970 (Proc. Fifth Nordic Summer-School in Math., Oslo, 1970), pp. 53–82. Wolters-Noordhoff, Groningen, 1972. MR **52:**3152

[80] Kratzer, Ch. *λ-structure en K-théorie algébrique.* Comment. Math. Helv. **55** (1980), no. 2, 233–254. MR **81m:**18011

[81] Kriz, I.; May, J. Peter. *Derived categories and motives.* Math. Res. Lett. **1** (1994), no. 1, 87–94. MR **95a:**14009

[82] Lang, Serge. **Algebra**. Second edition. Addison-Wesley Publishing Co., Reading, Mass., 1984. xv+714 pp. MR **86j:**00003

[83] Levine, Marc. *Lambda-operations, K-theory and motivic cohomology.* Algebraic K-theory (Toronto, ON, 1996), 131–184, Fields Inst. Commun., **16**, Amer. Math. Soc., Providence, RI, 1997.

[84] Levine, Marc. *Bloch's higher Chow groups revisited.* K-theory (Strasbourg, 1992). Astérisque No. 226 (1994), 10, 235–320. MR **96c:**14008

[85] Levine, Marc. *Tate motives and the vanishing conjectures for algebraic K-theory.* Algebraic K-theory and algebraic topology (Lake Louise, AB, 1991), 167–188, NATO Adv. Sci. Inst. Ser. C Math. Phys. Sci., **407**, Kluwer Acad. Publ., Dordrecht, 1993. MR **97e:**19008

[86] Lichtenbaum, Stephen. *Quasi-motives of curves.* Algebraic K-theory (Toronto, ON, 1996), 185–197, Fields Inst. Commun., **16**, Amer. Math. Soc., Providence, RI, 1997.

[87] Lichtenbaum, Stephen. *Motivic complexes.* Motives (Seattle, WA, 1991), 303–313, Proc. Sympos. Pure Math., **55**, Part 1, Amer. Math. Soc., Providence, RI, 1994. MR **95a:**19007

[88] Lichtenbaum, Stephen. *Suslin homology and Deligne 1-motives.* Algebraic K-theory and algebraic topology (Lake Louise, AB, 1991), 189–196, NATO Adv. Sci. Inst. Ser. C Math. Phys. Sci., **407**, Kluwer Acad. Publ.. Dordrecht, 1993. MR **96m:**14026

[89] Lichtenbaum, Stephen. *The construction of weight-two arithmetic cohomology.* Invent. Math. **88** (1987), no. 1, 183–215. MR **88d:**14011

[90] Lichtenbaum, Stephen. *Values of zeta-functions at nonnegative integers.* Number theory, Noordwijkerhout 1983 (Noordwijkerhout, 1983), 127–138, Lecture Notes in Math., Vol. 1068, Springer, Berlin, 1984.

[91] Lipman, Joseph. *Desingularization of two-dimensional schemes.* Ann. Math. (2) **107** (1978), no. 1, 151–207. MR **58:**10924

[92] Loday, Jean-Louis *K-théorie algébrique et représentations de groupes.* Ann. Sci. École Norm. Sup. (4) **9** (1976), no. 3, 309–377. MR **56:**5686

[93] MacLane, Saunders. **Categories for the working mathematician**. Graduate Texts in Mathematics, Vol. 5. Springer-Verlag, New York-Berlin, 1971. ix+262 pp. MR **50:**7275

[94] Manin, Ju. I. *Correspondences, motifs and monoidal transformations.* Mat. Sb. (N.S.) **77** (119) 1968 475–507. MR **41:**3482

[95] May, J. Peter. **Simplicial objects in algebraic topology**. Reprint of the 1967 original. Chicago Lectures in Mathematics. University of Chicago Press, Chicago, IL, 1992. viii+161 pp. MR **93m:**55025

[96] May, J. Peter. *Pairings of categories and spectra.* J. Pure Appl. Algebra **19** (1980), 299–346. MR **82c:**55010

[97] Miller, Edward Y. *de Rham cohomology with arbitrary coefficients.* Topology **17** (1978), no. 2, 193–203. MR **57:**7575

[98] Milne, James S. **Étale cohomology**. Princeton Mathematical Series, **33**. Princeton University Press, Princeton, N.J., 1980. xiii+323 pp. **81j:14002**

[99] Milnor, John. *Algebraic K-theory and quadratic forms.* Invent. Math. **9** (1969/1970) 318–344. MR **41:**5465

[100] Nagata, Masayoshi. *A general theory of algebraic geometry over Dedekind domains. I. The notion of models.* Amer. J. Math. **78** (1956), 78–116. MR **18:**600e

[101] Nesterenko, Yu. P.; Suslin, A. A. *Homology of the general linear group over a local ring, and Milnor's K-theory.*Izv. Akad. Nauk SSSR Ser. Mat. **53** (1989), no. 1, 121–146. MR **90a:**20092

[102] Quillen, Daniel. *Higher algebraic K-theory. I.* Algebraic K-theory, I: Higher K-theories (Proc. Conf., Battelle Memorial Inst., Seattle, Wash., 1972), pp. 85–147. Lecture Notes in Math., Vol. 341, Springer, Berlin 1973. MR **49:**2895

[103] Quillen, Daniel. *Cohomology of groups.* Actes du Congrès International des Mathématiciens (Nice, 1970), Tome 2, pp. 47–51. Gauthier-Villars, Paris, 1971. MR **58:**7627a

[104] Quillen, Daniel. **Homotopical algebra**. Lecture Notes in Mathematics, No. 43 Springer-Verlag, Berlin-New York 1967 iv+156 pp. MR **36:**6480

[105] Raynaud, Michel; Gruson, Laurent. *Critères de platitude et de projectivité. Techniques de "platification" d'un module.* Invent. Math. **13** (1971), 1–89. MR **46:**7219

[106] Roberts, Joel. *Chow's moving lemma.* Appendix 2 to: "Motives" (*Algebraic geometry, Oslo 1970* (Proc. Fifth Nordic Summer School in Math.), pp. 53–82, Wolters-Noordhoff, Groningen, 1972) by Steven L. Kleiman. Algebraic geometry, Oslo 1970 (Proc. Fifth Nordic Summer School in Math.), pp. 89–96. Wolters-Noordhoff, Groningen, 1972. MR **52:**3154

[107] Rosenberg, Jonathan. **Algebraic K-theory and its applications**. Graduate Texts in Mathematics, **147**. Springer-Verlag, New York, 1994. x+392 pp. **95e:19001**

[108] Rost, Marcus. *Chow groups with coefficients.* Documenta Math. **1** (1996), 319–393 (electronic).

[109] Saavedra Rivano, Neantro. **Catégories Tannakiennes**. Lecture Notes in Mathematics, Vol. 265. Springer-Verlag, Berlin-New York, 1972. ii+418 pp. MR **49:**2769

[110] Saito, Morihiko. *Mixed Hodge modules.* Publ. Res. Inst. Math. Sci. **26** (1990), no. 2, 221–333. MR **91m:**14014

[111] Saito, Morihiko. *Modules de Hodge polarisables.* Publ. Res. Inst. Math. Sci. **24** (1988), no. 6, 849–995 (1989). MR **90k:**32038

[112] Serre, Jean-Pierre. **Algèbre locale. Multiplicités**. Cours au Collège de France, 1957–1958, rédigé par Pierre Gabriel. Seconde édition, 1965. Lecture Notes in Mathematics, **11** Springer-Verlag, Berlin-New York 1965 vii+188 pp. MR **34:**1352

[113] Serre, Jean-Pierre. *Facteurs locaux des fonctions zêta des variétés algébriques (définitions et conjectures).* Sém. Delange-Pisot-Poitou 1969/70, exposé 19.

[114] Soulé, Christophe. *Opérations en K-théorie algébrique.* Canad. J. Math. **37** (1985), no. 3, 488–550. MR **87b:**18013

[115] Spanier, Edwin H. **Algebraic topology**. McGraw-Hill Book Co., New York-Toronto, Ont.-London 1966 xiv+528 pp. MR **35:**1007

[116] Srinivas, V. **Algebraic K-theory**. Second edition. Progress in Mathematics, **90**. Birkhäuser Boston, Inc., Boston, MA, 1996. xviii+341 pp. MR **97c:**19001

[117] Suslin, A.A; Voevodsky, Vladimir. *Relative cycles and Chow sheaves*. Cycles, Transfers and Motivic Homology Theories, Annals of Math. Studies, Princeton Univ. Press. To appear.

[118] Suslin, A.A. *Higher Chow groups of affine varieties and étale cohomology* preprint(1994).

[119] Suslin, A. A. *K_3 of a field, and the Bloch group*. Galois theory, rings, algebraic groups and their applications (Russian). Trudy Mat. Inst. Steklov. **183** (1990), 180–199, 229. MR **91k:**19003

[120] Thomason, R. W. *Absolute cohomological purity*. Bull. Soc. Math. France **112** (1984), no. 3, 397–406. MR **87e:**14018

[121] Thomason, R. W.; Trobaugh, Thomas. *Higher algebraic K-theory of schemes and of derived categories*. The Grothendieck Festschrift, Vol. III, 247–435, Progr. Math., **88**, Birkhäuser Boston, Boston, MA, 1990. MR **92f:**19001

[122] Totaro, Burt. *Milnor K-theory is the simplest part of algebraic K-theory*. K-Theory **6** (1992), no. 2, 177–189. MR **94d:**19009

[123] Verdier, J.L. *Catégories triangulées, état 0*. Cohomologie étale. Séminaire de Géométrie Algébrique du Bois-Marie SGA $4\frac{1}{2}$. P. Deligne, avec la collaboration de J. F. Boutot, A. Grothendieck, L. Illusie et J. L. Verdier. pp. 262–311. Lecture Notes in Mathematics, Vol. 569. Springer-Verlag, Berlin-New York, 1977. iv+312pp. MR **57:**3132

[124] Voevodsky, Vladimir *The triangulated category of motives over a field*. Cycles, Transfers and Motivic Homology Theories, Annals of Math. Studies, Princeton Univ. Press. To appear.

[125] Wagoner, J. B. *Stability for homology of the general linear group of a local ring*. Topology **15** (1976), no. 4, 417–423. MR **54:**5320

[126] Waldhausen, Friedhelm. *Algebraic K-theory of generalized free products. III, IV*. Ann. of Math. (2) **108** (1978), no. 2, 205–256. MR **58:**16845b

[127] Weibel, Charles. *Étale Chern classes at the prime* 2. Algebraic K-theory and algebraic topology (Lake Louise, AB, 1991), 249–286, NATO Adv. Sci. Inst. Ser. C Math. Phys. Sci., **407**, Kluwer Acad. Publ., Dordrecht, 1993. MR **96m:**19013

[128] Weibel, Charles. *A survey of products in algebraic K-theory*. Algebraic K-theory, Evanston 1980 (Proc. Conf., Northwestern Univ., Evanston, Ill., 1980), pp. 494–517, Lecture Notes in Math., **854**, Springer, Berlin, 1981. MR **82k:**18011

[129] Zagier, Don. *Polylogarithms, Dedekind zeta functions and the algebraic K-theory of fields*. Arithmetic algebraic geometry (Texel, 1989), 391–430, Progr. Math., **89**, Birkhäuser Boston, Boston, MA, 1991. MR **92f:**11161

[130] Zariski, Oscar. *The connectedness theorem for birational transformations*. Algebraic geometry and topology. A symposium in honor of S. Lefschetz, pp. 182–188. Princeton University Press, Princeton, N. J., 1957. MR **19:**767f

Subject Index

Index of Notation

Selected Titles in This Series

(Continued from the front of this publication)

25 **Jack K. Hale,** Asymptotic behavior of dissipative systems, 1988
24 **Lance W. Small, Editor,** Noetherian rings and their applications, 1987
23 **E. H. Rothe,** Introduction to various aspects of degree theory in Banach spaces, 1986
22 **Michael E. Taylor,** Noncommutative harmonic analysis, 1986
21 **Albert Baernstein, David Drasin, Peter Duren, and Albert Marden, Editors,** The Bieberbach conjecture: Proceedings of the symposium on the occasion of the proof, 1986
20 **Kenneth R. Goodearl,** Partially ordered abelian groups with interpolation, 1986
19 **Gregory V. Chudnovsky,** Contributions to the theory of transcendental numbers, 1984
18 **Frank B. Knight,** Essentials of Brownian motion and diffusion, 1981
17 **Le Baron O. Ferguson,** Approximation by polynomials with integral coefficients, 1980
16 **O. Timothy O'Meara,** Symplectic groups, 1978
15 **J. Diestel and J. J. Uhl, Jr.,** Vector measures, 1977
14 **V. Guillemin and S. Sternberg,** Geometric asymptotics, 1977
13 **C. Pearcy, Editor,** Topics in operator theory, 1974
12 **J. R. Isbell,** Uniform spaces, 1964
11 **J. Cronin,** Fixed points and topological degree in nonlinear analysis, 1964
10 **R. Ayoub,** An introduction to the analytic theory of numbers, 1963
9 **Arthur Sard,** Linear approximation, 1963
8 **J. Lehner,** Discontinuous groups and automorphic functions, 1964
7.2 **A. H. Clifford and G. B. Preston,** The algebraic theory of semigroups, Volume II, 1961
7.1 **A. H. Clifford and G. B. Preston,** The algebraic theory of semigroups, Volume I, 1961
6 **C. C. Chevalley,** Introduction to the theory of algebraic functions of one variable, 1951
5 **S. Bergman,** The kernel function and conformal mapping, 1950
4 **O. F. G. Schilling,** The theory of valuations, 1950
3 **M. Marden,** Geometry of polynomials, 1949
2 **N. Jacobson,** The theory of rings, 1943
1 **J. A. Shohat and J. D. Tamarkin,** The problem of moments, 1943